The Insects

Structure and Function

FIFTH EDITION

The Insects has been the standard textbook in the field since the first edition was published over 40 years ago. Building on the strengths of Chapman's original text, this long-awaited new edition has been revised and expanded by a team of eminent insect physiologists, bringing it fully up to date for the molecular era.

The chapters retain the successful structure of the earlier editions, focusing on particular functional systems rather than on taxonomic groups and making it easy for students to delve into topics without extensive knowledge of taxonomy. The focus is on form and function, bringing together basic anatomy and physiology and examining how these relate to behavior. This, combined with nearly 600 clear illustrations, provides a comprehensive understanding of how insects work.

Now also featuring a richly illustrated prologue by George McGavin, this is an essential text for students, researchers and applied entomologists alike.

R. F. Chapman (1930–2003) was an eminent insect physiologist and Professor in the Division of Neurobiology at the University of Arizona. His first four editions of *The Insects* have formed the standard text in the field for more than 40 years.

Stephen J. Simpson is ARC Laureate Fellow in the School of Biological Sciences and Academic Director of the Perkins Centre for the study of obesity, diabetes and cardiovascular disease at the University of Sydney. His core research aims are to understand swarming in locusts and to develop and implement an integrative framework for studying nutrition. In 2012 he was awarded the Wigglesworth Medal from the Royal Entomological Society of London.

Angela E. Douglas is Daljit S. and Elaine Sarkaria Professor of Insect Physiology and Toxicology at Cornell University, New York. Her research and teaching is motivated by the mechanisms underlying insect function, and her core research interests are the overlapping topics of insect nutrition and interactions between insects and beneficial microorganisms. She is a Fellow of The Royal Entomological Society and The Entomological Society of America.

The Insects

Structure and Function

FIFTH EDITION

R. F. CHAPMAN

Formerly of the University of Arizona, USA

Edited by

STEPHEN J. SIMPSON

The University of Sydney, Australia

ANGELA E. DOUGLAS

Cornell University, New York, USA

CAMBRIDGE
UNIVERSITY PRESS

CAMBRIDGE
UNIVERSITY PRESS

University Printing House, Cambridge CB2 8BS, United Kingdom

One Liberty Plaza, 20th Floor, New York, NY 10006, USA

477 Williamstown Road, Port Melbourne, VIC 3207, Australia

314-321, 3rd Floor, Plot 3, Splendor Forum, Jasola District Centre, New Delhi - 110025, India

79 Anson Road, #06-04/06, Singapore 079906

Cambridge University Press is part of the University of Cambridge.

It furthers the University's mission by disseminating knowledge in the pursuit of education, learning and research at the highest international levels of excellence.

www.cambridge.org
Information on this title: www.cambridge.org/9780521113892

© Cambridge University Press 1998, 2013

First published by Edward Arnold 1969
Second edition 1971, 6th printing 1980
Third edition 1982, 5th printing 1991
Fourth edition published by Cambridge University Press 1998, 7th printing 2011
Fifth edition 2013

A catalogue record for this publication is available from the British Library

Library of Congress Cataloging in Publication data
Chapman, R. F. (Reginald Frederick)
   The insects : structure and function / R. F. Chapman. – 5th edition / edited by
Stephen J. Simpson, Angela E. Douglas.
      pages   cm
   Includes bibliographical references and indexes.
   ISBN 978-0-521-11389-2
   1. Insects.   I. Simpson, Stephen J.   II. Douglas, A. E. (Angela Elizabeth), 1956–   III. Title.
   QL463.C48 2013
   595.7–dc23   2012018826

ISBN 978-0-521-11389-2 Paperback

CONTENTS

CONTRIBUTORS

Nichole D. Bond
Buck Institute for Research on Aging
University of California, San Fransisco
USA

Lars Chittka
School of Biological and Chemical Sciences
Queen Mary, University of London
UK

Bronwen W. Cribb
Centre for Microscopy & Microanalysis and
School of Biological Sciences
The University of Queensland, Brisbane
Australia

Angela E. T. Douglas
Department of Entomology
Cornell University
Ithaca, NY
USA

Julian A. T. Dow
Institute of Molecular Cell and Systems Biology
College of Medical, Veterinary & Life Sciences
University of Glasgow
UK

Allen G. Gibbs
School of Life Sciences
University of Nevada, Las Vegas
USA

Jon F. Harrison
School of Life Sciences
Arizona State University, AZ
USA

Ralf Heinrich
Abtl. Zelluläre Neurobiologie
Schwann-Schleiden-Forschungszentrum, Göttingen
Germany

Deborah K. Hoshizaki
Division Kidney, Urologic & Hematologic Diseases
NIDDK, National Institutes of Health
Bethesda, MD
USA

Michael F. Land
School of Life Sciences
University of Sussex, Brighton
UK

Tom Matheson
Department of Biology
University of Leicester
UK

George C. McGavin
Oxford University Museum of Natural History
Oxford
UK

Jeremy McNeil
Department of Biology
University of Western Ontario, London
Canada

David J. Merritt
School of Biological Sciences
The University of Queensland, Brisbane
Australia

Hans Merzendorfer
Fachbereich Biologie/Chemie, Osnabrück
Germany

Jocelyn G. Millar
Department of Entomology
University of California, Riverside
USA

Stuart Reynolds
Department of Biology & Biochemistry
University of Bath
UK

Stephen Rogers
Department of Zoology
University of Cambridge
UK

Leigh W. Simmons
Centre for Evolutionary Biology
School of Animal Biology
The University of Western Australia, Crawley
Australia

Stephen J. Simpson
School of Biological Sciences
The University of Sydney
Australia

Michael T. Siva-Jothy
Department of Animal and Plant Sciences
University of Sheffield
UK

John C. Sparrow
Department of Biology
University of York
UK

Michael R. Strand
Department of Entomology
Center for Tropical and Emerging Global Diseases
University of Georgia, GA
USA

Graham K. Taylor
Department of Zoology
Oxford University
UK

John S. Terblanche
Department of Conservation Ecology & Entomology
Faculty of AgriSciences
Stellenbosch University
South Africa

Peter Vukusic
School of Physics
University of Exeter
UK

Lutz T. Wasserthal
Institut für Zoologie I
Universität Erlangen-Nürnberg
Germany

PREFACE

Reginald Chapman's *The Insects: Structure and Function* has been the preeminent textbook for insect physiologists for the past 43 years (since the moon landing, in fact). For generations of students, teachers and researchers *The Insects* has provided the conceptual framework explaining how insects work. Without this book, the lives of entomologists worldwide would have been substantially more difficult. Nevertheless, the most recent (fourth) edition of this remarkable book was published in 1998, and a great deal has happened since then. Sadly, Reg died in 2003 and there was no reasonable prospect of any other person taking on the next revision single-handed. We have decided to take a different approach: to invite a team of eminent insect physiologists to bring their expertise to the collective enterprise of writing the fifth edition of *The Insects*.

Our aim has been to protect the identity of *The Insects* by working with Reg's original text. Certain areas have needed more revision than others, and some sections have been shrunk to accommodate advances in others. Our sole major deviation from the style of previous editions has been to remove all citations to primary literature from the main text. These in-text citations had accreted across successive revisions, and were somewhat patchy in coverage throughout the book. With the availability of online literature search engines today, students and researchers alike are better served by a short list of key references at the end of each chapter to provide a lead-in to the literature.

It has been the greatest pleasure for us to work with 23 colleagues from seven countries over the last four years, as the fifth edition of *The Insects* has taken shape. This project brings into sharp relief the intellectual strength and vigor of our discipline – the new discoveries over the last 14 years since the fourth edition are nothing short of breathtaking. We have also come to admire, more than ever, the breadth of Reg's knowledge and understanding of insects. He was a remarkable man.

STEVE SIMPSON and ANGELA DOUGLAS

ACKNOWLEDGMENTS

We wish to express our considerable gratitude to all our authors for their insight, expertise and commitment to this venture. We also thank Pedro Telleria-Teixeira for his tireless efforts in helping prepare the manuscript for submission, and to Cambridge University Press for taking it from there. Finally, we thank Elizabeth Bernays for her encouragement to take on the task. We hope that Reg would have been pleased with the result.

PROLOGUE

GEORGE C. McGAVIN

The ancestor of the Arthropoda was in all probability a segmented worm-like marine creature that lived in oceans during the late Precambrian. By the early to mid-Cambrian (540–520 million years ago) the early arthropods had already evolved into a range of clearly recognizable groups with distinct body plans. Arthropods are characterized by a number of features: the possession of a periodically molted, chitinous cuticle that acts as a rigid exoskeleton for the internal attachment of striated muscles; segmental paired legs; and the aggregation and/or fusing of body segments into discrete functional units, of which the most universal is the head. Besides the head there may be a trunk, as in the Myriapoda, or a separate thorax and abdomen as in the Crustacea and Hexapoda.

Based on the ubiquity of α-chitin in arthropod cuticles, similarities in musculature and tendon systems and recent molecular data, the overwhelming consensus of opinion is that this very large taxon is monophyletic. However, the relationships within the Arthropoda have been the subject of much controversy for more than 100 years. Recent molecular and genetic data confirm that the Hexapoda (comprising the Insecta and three other non-insect hexapod classes) are monophyletic, but that Crustacea are not. The monophyletic Hexapoda and paraphyletic Crustacea are now thought to form a single superclade called the Pancrustacea (Fig. 1). The mandibles of these two groups have similar origins, and the development of the nervous system is similar, as is the structure and wiring of the compound eyes.

A little over 1.5 million species of living organism have been scientifically described to date. The vast majority (66%) are arthropods such as crustaceans, arachnids, myriapods and insects. Insects represent 75% of all animals, and one insect order – the beetles (Coleoptera) – is famously species-rich, but another comprising the wasps, bees and ants (Hymenoptera) may rival the beetles if taxonomists ever complete their studies. One thing is clear, however – the full extent of Earth's biodiversity remains a mystery. From attempts over 30 years ago to estimate the number of extant species to the present day we still only have a rough idea of how many species live alongside us. Estimates range from as few as five million to perhaps as many as 10–12 million species. The task of enumerating them may become substantially easier as the loss and degradation of natural habitats, especially the forests of the humid tropics, continues unabated. It is certain that the majority of insect species will become extinct before they are known to science.

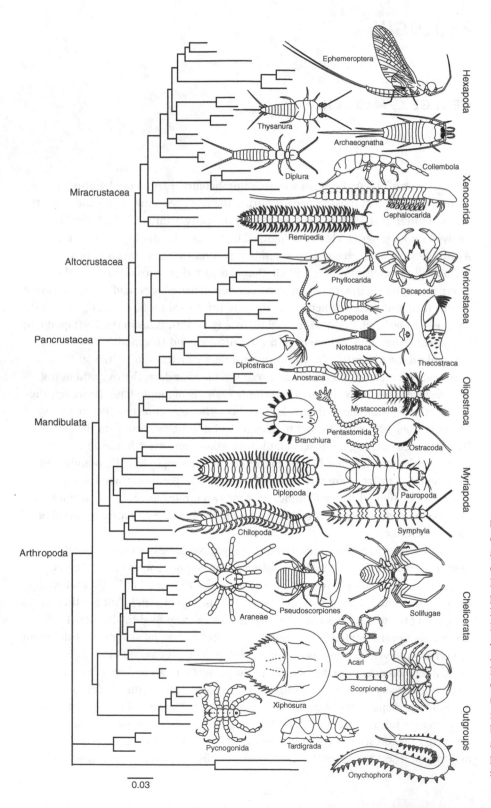

Figure 1 Phylogram of relationships for 75 arthropod and five outgroup species. Reprinted by permission from Macmillan Publishers Ltd: Regier, J. C., Shultz, J. W., Zwick, A., Hussey, A., Ball, B., Wetzer, R., Martin, J. W. and Cunningham, C. W. (2010) Arthropod relationships revealed by phylogenomic analysis of nuclear protein-coding sequences. *Nature* 463, 1079–1083.

Insects are the dominant multicellular life form on the planet, ranging in size from minute parasitic wasps at around 0.2 mm to stick insects measuring 35 cm in length. Insects have evolved diverse lifestyles and although they are mainly terrestrial, there are a significant number of aquatic species. Insects have a versatile, lightweight and waterproof cuticle, are generally small in size and have a complex nervous system surrounded by an effective blood–brain barrier. Insects were the first creatures to take to the air and have prodigious reproductive rates. These factors, together with the complex interactions they have with other organisms, have led to their great success both in terms of species richness and abundance. The very high diversity of insects today is the result of a combination of high rates of speciation and the fact that many insect taxa are persistent – that is, they show relatively low rates of extinction.

In comparison to insects, vertebrate species make up less than 3% of all species. As herbivores they are altogether out-munched by the myriad herbivorous insects. In tropical forests, for example, 12–15% of the total leaf area is eaten by insects as compared with only 2–3% lost to vertebrate herbivores. Termites remove more plant material from the African savannahs than all the teeming herds of wildebeest and other ungulates put together. Vertebrates also fail to impress as predators. Ants are the major carnivores on the planet, devouring more animal tissue per annum than all the other carnivores. In many habitats ants make up one-quarter of the total animal biomass present.

Insects pollinate the vast majority of the world's 250 000 or so species of flowering plant. The origin of bees coincides with the main radiation of the angiosperms approximately 100 million years ago, and without them there would be no flowers, fruit or vegetables. At least 25% of all insect species are parasites or predators of other insect species. Insects are also important in nutrient recycling by disposing of carcasses and dung.

Insects are the principal food source for many other animals. Virtually all birds and a large number of other vertebrates feed on them. An average brood of great tit chicks will consume around 120 000 caterpillars while they are in the nest and a single swallow chick may consume upwards of 200 000 bugs, flies and beetles before it fledges.

Insects can also have a huge negative impact on humans. One-sixth of all crops grown worldwide are lost to herbivorous insects and the plant diseases they transmit. About one in six human beings alive today is affected by an insect-borne illness such as plague, sleeping sickness, river blindness, yellow fever, filariasis and leishmaniasis. About 40% of the world's population are at risk of malaria. More than 500 million people become severely ill and more than one million die from this disease every year. To complete the destructive side of their activities, insects can cause great damage to wooden structures and a wide range of natural materials and fabrics.

But without insects performing essential ecosystems services, the Earth would be a very different place and most terrestrial vertebrates that depend on them directly as food would become extinct. The loss of bees alone might cause the extinction of one-quarter of all life on Earth. A total loss of insects would see the human population plummet to perhaps a few hundred thousand individuals subsisting mainly on cereals.

Their small size and high reproductive rates make insects ideal model systems in molecular, cellular, organismal, ecological and evolutionary studies. Indeed, many of the most important discoveries in genetics, physiology, behavior, ecology and evolutionary biology have relied on insects.

There may come a day when humans venture far enough into space to visit other planets on which life has developed. If we do and there are multicellular organisms present, it is likely they will look a lot like insects.

Mini-biographies of the insect orders

The Insecta and three other classes, the Protura, Diplura and Collembola, together comprise the arthropod superclass, Hexapoda. The Class Insecta is divided into 30 orders, which are outlined below.

THE PRIMITIVE WINGLESS INSECTS (INFRACLASS APTERYGOTA)

ARCHAEOGNATHA

- Bristletails
- ~500 species
- Body length: 7–15 mm

Bristletails are the most primitive living insects, having persisted for more than 400 million years. They are mainly nocturnal, living in leaf litter and under stones in a wide range of habitats from coastal to mountainous regions. The body, which is elongate with a cylindrical cross-section, is covered in tiny scales and has a characteristically humped thorax.

The head has a pair of long antennae, large contiguous compound eyes and three well-developed ocelli (single-faceted, simple eyes). The mouthparts are simple, with long maxillary palps. The mandibles have a single point of articulation (termed monocondylar) and are used to pick at lichens and algae. This jaw articulation is a very primitive feature separating the bristletails from all other insects, including the Thysanura, which have two points of articulation (dicondylar).

The abdomen has accessory walking appendages called styles (present on abdominal segments 2–9), which support the abdomen when bristletails run over uneven or steep surfaces. Surface water can be absorbed through one or two pairs of eversible vesicles located on the underside of abdominal segments 1–7. The abdomen has a pair of multi-segmented cerci and a much longer central filament. Bristletails can jump by rapid flexion of the abdomen.

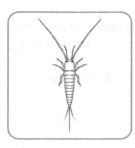

THYSANURA (ZYGENTOMA)

- Silverfish
- <400 species
- Body length: 2–22 mm

Although very similar to bristletails, silverfish are actually more closely related to the winged insects. The body, which may have a covering of scales, is rather more flattened and the thorax is not humped. Silverfish are scavengers in soil, leaf litter, on trees and sometimes in buildings, where they can be minor pests.

The head has a pair of long antennae, small compound eyes and may have ocelli. The maxillary palps are shorter than those of the Archaeognatha and the jaws, although still of a primitive design, have two points of articulation and act in the transverse plane.

Styles may be present on abdominal segments 2–9, but usually on fewer segments (7–9). Pairs of water-absorbing, eversible vesicles usually occur on the abdominal segments (2–7), although in some species they are absent. The end of the abdomen has a pair of cerci and a central filament. Silverfish are fast running but do not jump.

THE WINGED INSECTS

The infraclass Pterygota is made up of three very unequal divisions. The mayflies (Ephemeroptera), comprising <0.3% of all insects species, and the dragonflies and damselflies (Odonata), comprising ~0.5% of all insect species, are each a division. Species in these two divisions are unable to fold their wings back along the body. Together they are sometimes termed the Paleoptera, although this is not a natural (monophyletic) grouping. The third, and by far the largest division, comprising all other insect species, is the Neoptera, which are monophyletic.

DIVISION I

EPHEMEROPTERA

- Mayflies
- ~2500 species
- Body length: 5–34 mm
- Wingspan: up to 50 mm

The Ephemeroptera are the oldest (basal) group of winged insects on Earth today and are unique in having a pre-adult winged stage called the subimago – they are the only insects that molt after they have developed functional wings. This habit was probably much more common in extinct Carboniferous and Permian taxa, where immature stages had wing-like structures and molted them throughout their lives.

The order is divided into two suborders, the Schistonota (split-back mayflies) and the Pannota (fused-back mayflies). Schistonotan nymphs have their wing pads free along the midline, whereas in pannotan nymphs, the wing pads are fused along the midline of the body.

Mayflies are soft-bodied with nearly cylindrical bodies, longish legs and typically two pairs of wings, which, when at rest, are held over the body. The head has a pair of short bristle-like antennae, a pair of large compound eyes and three ocelli. Adults have reduced non-functional mouthparts. The end of the abdomen bears a pair of elongate cerci and, usually, a single, long central filament.

The lifecycle is dominated by the aquatic, nymphal stages and adults live for a very short time, often less than a day.

DIVISION II

ODONATA

- Damselflies and dragonflies
- <6000 species
- Body length: up to 150 mm
- Wingspan: 18–200 mm

These fast-flying insects, often seen near water, are instantly recognizable. Odonates have a distinctive elongate body and are often brightly colored or metallic. They have a large, mobile head with very large compound eyes, three ocelli, short, hair-like antennae and biting mouthparts. They have two pairs of similarly sized wings, which can be used out of phase with each other, allowing great maneuverability.

The nymphal stages (called naiads) are aquatic and actively hunt or ambush prey. The mouthparts are unique in that there is a prehensile labial mask, which can be rapidly extended. Spine-like palps on the labium impale prey items and the mask is then folded back toward the mouth.

The order is split into two major suborders, the dragonflies (Anisoptera) and the damselflies (Zygoptera). A third suborder (Ansiozygoptera) comprises only two Oriental species. Dragonflies have round heads and very large eyes, while damselflies have broader heads with widely separated eyes. The large eyes give odonates near all-round vision and, as would be expected of aerial hunters, they are able to resolve distant objects better than any other insect.

DIVISION III: NEOPTERA

In all neopterans, flexor muscles attached to a third axillary sclerite at the base of the wings allow the wings to be folded back along the body. The evolution of a wing-folding mechanism allowed much better exploitation of the terrestrial environment without the risk of wing damage.

Subdivison: Hemimetabola

PLECOPTERA

- Stoneflies
- ~2000 species
- Body length: 3–48 mm
- Maximum wingspan: about 100 mm

Stoneflies are slender insects with soft, slightly flattened bodies. The head has bulging eyes, two or three ocelli and thread-like antennae. The mouthparts are weakly developed or non-functional. They have two pairs of membranous wings, which are held flat or folded around the body at rest. They are not strong fliers and seldom travel far from water. The elongate abdomen has a pair of single- or multi-segmented cerci.

The order is divided into two suborders, the Arctoperlaria and the Antarctoperlaria. With the exception of one family, all Arctoperlaria are found in the Northern Hemisphere. All families in the Antarctoperlaria are found in the Southern Hemisphere.

Stonefly nymphs are aquatic and can swim using lateral body movements. Many graze algae from rocks.

BLATTODEA (BLATTARIA)

- Cockroaches
- ~4000 species
- Body length: 3–100 mm

Cockroaches are fast-running, flattened, broadly oval and leathery-bodied insects. The head, which is directed downwards and largely concealed by the pronotum, has biting mouthparts, well-developed compound eyes, two ocelli-like spots and long antennae. The front pairs of wings are toughened as protective "tegmina" to cover the larger, membranous hindwings. The abdomen carries a pair of one- or multi-segmented cerci. Eggs are typically laid in a toughened case or ootheca, a feature shared with the closely related, but entirely predatory Mantodea.

The vast majority of cockroaches are nocturnal, omnivorous or saprophagous species living in soil and leaf litter communities. Only about 40 species are considered pests because of their close association with humans, and only half of these have a significant impact. The main problem is that they can carry a huge diversity of pathogenic organisms on their tarsi and other body parts. When they feed they regurgitate partly digested food and leave behind their feces and a characteristic offensive odor. Exposure to high levels of cockroach allergens in house dust can produce serious health problems such as allergies, dermatitis, eczema and asthma.

MANTODEA

- Mantids
- ~2300 species
- Body length: 8–150 mm

These distinctive predatory insects have a triangular, highly mobile head with large compound eyes, thread-like antennae and usually three ocelli. The prothorax is typically elongate and carries the specialized, raptorial front legs. The front wings are narrow and toughened, protecting the much larger membranous hindwings. Eggs are laid in a papery, foam- or cellophane-like ootheca.

True binocular vision allows mantids to calculate the distance of their prey using triangulation. The coxa of the front legs is very elongate and the femur is enlarged and equipped with rows of sharp spines and teeth. The tibia, which is also spined or toothed, folds back on the inner face of the femur like a jack knife. The strike, which takes place in two phases, lasts less than 100 milliseconds. In the initial phase, the tibiae are fully extended in readiness for the second phase, which takes the form of a rapid sweeping action. The femora are quickly extended and, at the same time, the tibiae are flexed around the prey.

Mantids, which are mainly diurnal, predate a wide range of insects, spiders and other arthropods, which they ambush or stalk. Larger species have even been recorded catching and eating vertebrates such as frogs, mice and even small birds.

ISOPTERA

- Termites
- <3000 species
- Body length: 3–20 mm, mostly under 15 mm; queens can be up to 100 mm

Generally pale and soft-bodied, termites are social insects living in permanent colonies with different castes of both sexes. Workers and soldiers are wingless, while the reproductives (kings and queens) have two pairs of equal-sized wings, which are shed after a nuptial flight.

The foodstuff of termites, cellulose, is an abundant biomolecule but is difficult to break down. Termites have evolved symbiotic relationships with cellulase-producing microorganisms to make use of this resource. The gut of lower termites harbors protists, while those of the higher termites (Termitidae) contain bacterial symbionts.

Termites can build impressively large nest structures, including the large multi-vented chimneys that ventilate the subterranean nests of African *Macrotermes* species and the wedge-shaped nests in northern Australia made by the magnetic termite, *Amitermes meridionalis*.

Confined to regions between 45–50° north and south of the Equator, termites have an immense impact on soil enrichment and carbon cycling. They may consume up to one-third of the annual production of dead wood, leaves and grass and be present in huge numbers, comprising 10% of all animal biomass present.

GRYLLOBLATTODEA (NOTOPTERA)

- Rock crawlers or ice crawlers
- 26 species (1 family: Grylloblattidae)
- Body length: 12–30 mm

These slender, wingless, slightly hairy insects were first discovered in the Canadian Rockies in 1913 and are a relict group confined to certain high-altitude regions across the Northern Hemisphere. The head has small compound eyes, although these are sometimes absent, no ocelli, slender, thread-like antennae and simple, chewing mouthparts. The abdomen is cylindrical, with a pair of slender, multi-segmented cerci.

Grylloblattids live under stones, decaying wood and leaf litter in cold temperate forests and sometimes in caves. There may be eight nymphal instars and complete nymphal development may take up to 5–6 years. As nymphs get older they become darker colored and add segments to their antennae at each molt. The adults typically live for less than two years.

MANTOPHASMATODEA

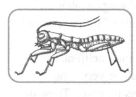

- Gladiators, African rock crawlers or heel-walkers
- 15 species (1 family: Mantophasmatidae)
- Body length: 12–35 mm

Discovered in 2002, the species that make up this small order live in dry, rocky habitats in southern Africa and may be related to the Grylloblattodea. The head has well-developed compound eyes and long, slender antennae and biting mouthparts, but lacks ocelli. The name "heel-walkers" refers to the way the claws are held clear of the ground when walking.

These elongate, wingless insects can be found under stone and among tufts of grasses and other plants. At night they emerge to catch other insects, holding small prey using their spiny front and middle legs.

DERMAPTERA

- Earwigs
- ~1900 species
- Body length: 5–54 mm

Mostly drab, nocturnal and generally reluctant to fly, the majority of these elongate and slightly flattened insects are immediately recognizable on account of their distinctive abdominal forcep-like cerci. The head, which may have a pair of compound eyes but no ocelli, has biting mouthparts and long antennae. The front wings are short, leathery and veinless, covering the large, semicircular hindwings.

The order is divided into three very unequal suborders. The largest – which accounts for 99% of all known species – is the Forficulina, which prefer confined, humid microhabitats such as soil, leaf litter or beneath bark. The Hemimerina is made up of 11 species of African cockroach-like earwigs, which are ectoparasites in the fur of giant rats. The Arixenina comprises five blind, wingless South Asian species that feed on skin fragments and excreta in the fur or roosts of two species of molossid bat.

The terminal forceps, which are usually straight in females and curved in males, are used in a variety of ways but mainly as weapons for defense and prey handling, but also for courtship displays. The flexible and telescopic abdominal segments allow earwigs to use their forceps in all directions and they often use this ability to assist in folding the large hindwings.

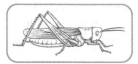

ORTHOPTERA

- Crickets, grasshoppers and relatives
- ~22 500 species
- Body length: 5–155 mm

These distinctive, elongate insects typically have enlarged hindlegs used for jumping. The head has well-developed compound eyes and may have ocelli. They have biting mouthparts and an enlarged, saddle- or shield-shaped pronotum. The front wings are toughened and typically narrower than hindwings, which are folded in longitudinal pleats beneath. The abdomen has a pair of short, terminal cerci.

The order is divided into two suborders, the Ensifera and the Caelifera. The Ensifera, comprising the crickets and katydids, have long or very long antennae and sing by rubbing structures on their front wings together. They are mainly nocturnal and solitary and most species mimic dead or living leaves. Many species are herbivorous, but some are partly or wholly predaceous. The ovipositor is always prominent and sword-, sickle- or stiletto-shaped.

The Caelifera, comprising grasshoppers and locusts (which show density-dependent polyphenism), have short antennae and the females never have prominent ovipositors. Songs are produced by a row of pegs on the hind femora rubbing against the edge of the front wings. They are generally ground-living, diurnal, grass- and/or forb-feeders and can be cryptically colored or brightly colored to advertise their unpalatability. Several, such as the desert locust, *Schistocerca gregaria*, are serious crop pests.

PHASMATODEA

- Stick and leaf insects
- >3000 species
- Body length: up to 566 mm, mostly 10–100 mm

The elongate body of stick insects can be short and smooth or large and very spiny or leaf-like. The head is characteristically domed and carries relatively long, thread-like antennae, chewing mouthparts, a pair of small compound eyes and, in winged species, ocelli. The front wings are short and toughened while the fan-shaped membranous hindwings are large. Many species are short-winged or wingless, and in others wing length varies between the sexes.

Stick insects are slow-moving, herbivorous and mostly nocturnal. Their shape and cryptic coloring make them very difficult to see among foliage and affords them protection from predators. Some species freeze motionless when disturbed, holding the middle and hindlegs along the body and stretching out the front legs, while others sway to imitate the movement of the vegetation. Leaf insects, which are broad and flattened with fantastic leaf-like expansions, are contained in one family, the Phylliidae, comprising about 50 species confined to Southeast Asia, New Guinea and Australia.

EMBIOPTERA (EMBIIDINA, EMBIODEA)

- Webspinners
- ~350 species
- Body length: 3–20 mm, mostly under 12 mm

Webspinners are narrow-bodied, cylindrical or slightly flattened gregarious insects living in warm temperate and tropical regions. The head has small, kidney-shaped compound eyes, thread-like antennae and biting mouthparts. The front legs of all life-stages and both sexes have swollen basal tarsal segments containing glands, which produce silk to make communal galleries in soil, litter and under bark.

As colonies grow, galleries and tunnels are extended to take in new food sources such as dead plant material, litter, lichens and mosses. Only adult females and nymphs feed. Males do not feed as adults and only use their jaws to grasp the female during copulation. Females are wingless but the males usually have two equal-sized pairs of long, narrow wings. The wings have hollow veins that can be inflated with hemolymph to make them stiff for flight. When the veins are not inflated, the wings can fold forwards without damage when the male has to run backwards through the galleries.

ZORAPTERA

- Angel insects
- 32 species
- Body length: 2–3 mm

Mostly associated with rotting wood, these small, delicate-bodied insects are termite-like. The adults are dimorphic, being either blind, pale and wingless (resembling the nymphs) or darkly pigmented with eyes and two pairs of pale, sparsely veined wings. The head carries a pair of short, thread-like antennae and may have ocelli.

Zorapterans are gregarious under bark or in piles of wood dust, leaf litter or in termite nests, where they eat fungal threads, spores, mites and other small arthropods. As populations grow, winged morphs disperse to new locations and the wings are then shed. All the known species are currently assigned to a single genus, *Zorotypus*.

PSOCOPTERA

- Barklice and booklice
- <4500 species
- Body length: 1–10 mm, mostly under 6 mm

Barklice and booklice are very common insects, which on account of their small size and cryptic coloration, are often overlooked. The head is relatively large, with bulging compound eyes, long, thread-like antennae, biting mouthparts and, in winged species, three ocelli. The thorax is slightly humped and the wings, when present, are held roof-like over the body at rest.

Psocoptera can be found in a very wide range of terrestrial habitats, including caves and the nests of birds, bees and wasps, but are particularly abundant in litter and soil and on the bark and foliage of trees and shrubs. Most species graze on algae, lichens and molds and fungal spores, but some can be pests of stored products.

Three suborders are recognized – the Trogiomorpha, considered the most primitive, the Troctomorpha and the Psocomorpha, the most advanced suborder, containing more than 80% of the known species.

PHTHIRAPTERA

- Parasitic lice
- ~5000 species
- Body length: 1–10 mm, mostly under 6 mm

These small, wingless, dorso-ventrally flattened ectoparasites live permanently on bird or mammal hosts, where they feed on skin debris, secretions, feathers or blood.

The eyes are very small or absent, there are no ocelli and the antennae are short, with a maximum of five segments. The legs are short and robust, with the tarsi and claws typically modified for grasping hair or feathers. Several species are significant vectors of human and animal diseases.

The nymphs pass through three instars or nymphal stages, taking anything from two weeks to a few months to reach adulthood. Many lice have symbiotic relationships with bacteria which live in special mycetocytes associated with the digestive system. These bacteria allow the lice to digest feather protein (keratin) and blood.

There are four suborders within the Phthiraptera. The Amblycera are a primitive group of chewing lice living on birds and mammals. The Rhyncophthirina are ectoparasites of elephants and warthogs. The largest suborder, the Ischnocera, are chewing lice mainly found on birds, while the Anoplura are sucking lice which include the human head and body louse and the pubic louse.

HEMIPTERA

- True bugs
- >82 000 species
- Body length: 1–100 mm, mostly under 50 mm

True bugs range from minute, wingless scale insects to giant water bugs with raptorial front legs capable of catching fish and frogs. Compound eyes are often prominent and ocelli may be present. Bugs lack maxillary and labial palps and the mandibles and maxillae, which are enclosed by the labium, take the form of elongate, grooved stylets through which saliva can be injected and liquids sucked up. Two pairs of wings are usually present.

There are four distinct suborders. The Auchenorrhyncha, comprising planthoppers, leafhoppers, froghoppers, treehoppers, lantern bugs and cicadas, and the Sternorrhyncha, including jumping plant lice, whiteflies, phylloxerans, aphids and scale insects, are herbivorous. The Coleorrhyncha is represented by a single family of cryptic bugs found in the Southern Hemisphere. The majority of species belonging to the fourth suborder, the Heteroptera, are herbivorous but the suborder contains a significant number of predatory taxa and even some blood-sucking species. A characteristic feature of heteropterans is the possession of defensive stink glands.

Many bug species are significant plant pests and some transmit human and animal diseases.

THYSANOPTERA

- Thrips
- ~5500 species
- Body length: 0.5–12 mm, mostly under 3 mm

Thrips are small or very small, slender-bodied insects with prominent, large-faceted eyes, short antennae and asymmetrical piercing and sucking mouthparts. One mandible is very small and non-functional while the other is sharp and stylet-like and used to penetrate plant tissue or sometimes the bodies of minute insects. The other mouthparts form hemipteran-like stylets and are used to suck up liquid food. They usually have two pairs of very narrow, hair-fringed wings, but wings can be reduced, vestigial or absent. Three ocelli are present in winged individuals. The tarsi have an eversible bladder-like structure between the claws. Many species are serious plant pests.

Although these insects are most closely related to the Hemiptera, they are unusual in that there are one or more pupa-like resting stages between the two, true nymphal stages and the adult. In some cases there are three pre-adult stages of which the first may still be capable of feeding. The next two pre-adult stages become more pupa-like with a degree of tissue reorganization; a cocoon may even be formed.

Subdivison: Holometabola

The following neopteran orders comprise the most advanced and successful of all insects. The immature stages are called larvae and look very different and have different lifestyles to the adults. The wings develop internally and metamorphosis from larva to adult takes place during a pupal stage.

MEGALOPTERA

- Alderflies and dobsonflies
- ~300 species
- Body length: 10–150 mm
- Wingspan: 18–170 mm

The two families that comprise this small order (alderflies [Sialidae] and dobsonflies [Corydalidae]) are the most primitive insects with complete metamorphosis. The head has conspicuous compound eyes and long, thread-like antennae. Ocelli are present in corydalids but absent in sialids. Despite having well-developed jaws, adult megalopterans do not feed. In some male dobsonflies the jaws may be several times the length of the head and used in male-to-male combat or for grasping the female. Megalopterans have two pairs of similarly sized wings, which are held roof-like over the body.

Megaloptera are found near cool, clean streams in temperate regions. Dobsonflies prefer running water while alderflies can be found in ponds and canals as well as streams. The predaceous larvae are aquatic with simple or branched, abdominal gills.

Larval development can take anything from 12 months in alderflies but sometimes more than 48 months in dobsonflies. Pupation takes place on land within a simple chamber made in moist soil, sand or mossy vegetation or under rotting wood. The pupae have functional jaws, can move around freely and even protect themselves.

RAPHIDIOPTERA

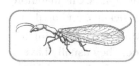

- Snakeflies
- ~220 species
- Body length: 6–28 mm

Confined to cool, temperate woodlands, this order comprises just two families, the Raphidiidae and the Inocellidae. The large head, which is supported by an elongate prothorax, is slightly flattened, broad in the middle and tapers to the rear. The antennae are slender and the compound eyes are conspicuous. Ocelli are present in the Raphidiidae but absent in the Inocellidae. The biting mouthparts are forward-facing to seize prey. The two pairs of wings are similarly sized, clear and have similar venation to that of megalopterans, but the veins are forked close to the wing margins as in neuropterans. Both pairs of wings have a small, dark or pale mark called the pterostigma on the front edge, toward the wing tips. The females, which are a little larger than males, have a long, slender and conspicuous ovipositor.

Snakeflies are closely related to alderflies but differ in that the larvae are completely terrestrial and the adult stage feeds.

NEUROPTERA

- Antlions, lacewings and relatives
- ~5000 species
- Body length: 2–90 mm
- Wingspan: 5–150 mm

Adult neuropterans have biting mouthparts, a pair of conspicuous, laterally placed compound eyes and may have ocelli. The antennae are generally long and thread-like, and in some owlflies and antlions the end of the antennae may be swollen to form a club. The adults of some families have prothoracic glands capable of producing substances that repel some predators. There are usually two pairs of similarly sized wings held roof-like over the body at rest. The venation in most neuropterans is net-like, with the main veins forking at the wing margins.

The majority of species are predatory and mainly active in the evening or after dark. The larvae of all species, which have their mandibles and maxillae united to form a pair of sharp, sucking tubes, are highly predaceous and can be found in a wide range of habitat types.

The species of some families are very similar to other insects. Adult Mantispidae have enlarged, raptorial front legs like those of praying mantids. Owlflies (Ascalaphidae) are aerial predators and look very like dragonflies.

COLEOPTERA

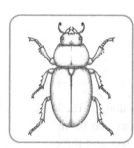

- Beetles
- ~370 000 species
- Body length: 0.1–180 mm, mostly under 25 mm

This very large order makes up at least 40% of all insect species.

The head has conspicuous compound eyes, antennae usually with less than 11 segments and biting mouthparts. Ocelli are typically absent. The prothorax is usually large and freely articulated with the rest of the thorax. The toughened front wings, or elytra, meet in the body midline and cover the larger membranous hindwings, which are folded lengthwise and crosswise underneath.

Beetles can be found in every conceivable terrestrial and freshwater habitat. The possession of protective elytra has allowed beetles to burrow, dig and squeeze into places that other insects cannot reach without compromising their ability to fly.

The order contains many scavengers, predators and a few specialized parasites, but the vast majority of beetle species are herbivorous and here lies the second major reason for their great success. The rise to dominance of the flowering plants (Angiospermae) in the Cretaceous provided herbivorous beetles with multiple opportunities to radiate.

The order is divided into four very unequal suborders. The Archostemata comprises a very small group of specialist wood-borers. The Mxyophaga is made up of around 60 species of small aquatic species. The Adephaga, comprising about 10% of all beetles, is made up of 12 families of ground-living and aquatic species where the larvae and adults are predaceous. The species that make up the largest suborder, the Polyphaga, have very diverse lifestyles and eating habits.

STREPSIPTERA

- Strepsipterans
- ~600 species
- Body length: 0.4–35 mm, mostly under 6 mm

Strepsipterans are highly specialized endoparasites of other insects in more than 30 insect families belonging to the Orders Thysanura, Blattodea, Mantodea, Orthoptera, Hemiptera, Diptera and Hymenoptera. The adults are dimorphic. Females are typically endoparasitic without eyes, antennae, mouthparts, legs or wings. Males are free-living, with raspberry-like eyes, branched antennae and wings. The front wings are small and strap-like while the hindwings are fan-shaped.

The order is divided into two suborders, the Mengenillidia (one small family – the Mengenillidae) and the Stylopidia (seven families). In the Mengenillidae, full-grown male and female larvae leave the hosts, which are species of silverfish, and pupation takes places outside. In this family, unusually, both adult males and females are free-living, but the females are not grub-like and have normal features of adult insects, such as legs and antennae. Females of all other families are totally endoparasitic. They never leave the confines of their host's body and are surrounded by the cuticle of their own pupal stage.

MECOPTERA

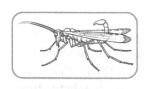

- Scorpionflies
- ~600 species
- Body length: 3–28 mm

Scorpionflies are elongate insects found mostly in damp woodlands. The head, which is characteristically extended downwards to form a beak, has biting mouthparts, slender, thread-like antennae, large compound eyes and three ocelli. They usually have two pairs of large, narrow wings, but some species are short-winged or wingless.

The legs are long and slender in most species, but in the hangingflies (Bittacidae) they are very long and used for prey capture. The fifth tarsal segment of the hindlegs is enlarged and raptorial to seize small insects. The majority of other scorpionflies feed on dead or dying insects and will also feed on carrion, nectar, sap and fruit juices.

SIPHONAPTERA

- Fleas
- ~2500 species
- Body length: 1–8 mm, mostly under 5 mm

Found wherever there are suitable hosts, fleas are a distinctive and readily recognizable group. Well over 90% of flea species feed on the blood of land mammals – the remainder are bird ectoparasites. Fleas are small, wingless, tough-bodied and laterally flattened. The head typically has very short, three-segmented antennae, which fit into grooves and short mouthparts for piercing skin and sucking blood. Fleas may have a pair of simple, lateral eyes similar to ocelli, and are negatively phototactic (avoid light). The enlarged hindlegs are part of the flea's unique and powerful jumping mechanism, which incorporates an energy store made of a rubber-like protein called resilin.

Comb-like structures on the cheeks and the posterior edge of the pronotum of many species, together with numerous backward-pointing spines and bristles on the body, help the fleas to remain in the host's fur. As holometabolous insects, the larvae are very different from the adults and feed on detritus and dried blood in the host's lair or den. Pupation takes place inside a loose cocoon. Many flea species are disease vectors.

DIPTERA

- True flies
- ~122 000 species
- Body length: 0.5–60 mm
- Wingspan: up to 75 mm

Most of the species that make up this huge and diverse order are beneficial to ecosystem function as pollinators, parasites and predators, and are vital to the processes of decomposition and nutrient recycling. However, the activities of relatively few species have a greater impact on man and other animals than any other insect group. Perhaps as many as one person in six is affected by a fly-borne disease such as malaria, yellow fever, dengue fever and leishmaniasis.

Flies have a mobile head with large compound eyes and three ocelli. The mouthparts, which vary according to diet, are adapted for lapping and sponging liquids or piercing and sucking.

A characteristic feature of the order is the possession of a single pair of membranous front wings, although some ectoparasitic species are wingless. The hindwings in all species are reduced to form a pair of balancing organs called halteres.

The order is divided into two suborders, the Nematocera and the Brachycera. Nematocera is the more primitive suborder, and includes crane flies, mosquitoes, black flies, midges and fungus gnats, with delicate thread-like antennae. The Brachycera are more robust, with short, stout antennae of less than six segments, and include the orthorrhaphan groups, typified by horse flies and robber flies, and the cyclorrhaphan species such as fruit flies, hover flies, blowflies and flesh flies. Larval habits vary from fully aquatic to terrestrial and many larvae are serious plant pests.

TRICHOPTERA

- Caddis flies
- >11 000 species
- Body length: 2–38 mm

Caddis flies are mainly nocturnal and can be found almost everywhere there is freshwater. The elongate adults are rather moth-like in appearance with long, slender legs. The body and wings, particularly the front wings, are covered with hairs. The head has a pair of large compound eyes, long, thin antennae and two or three ocelli may be present. The weakly developed mouthparts allow adults to lick up water and nectar, but many do not feed as adults. The front and hind pairs of wings are held over the body in a characteristically tent-like manner.

The caterpillar-like larvae, which are aquatic, show a range of feeding habits. Some species may be free-living or spin food-catching nets, but most live inside portable tube-like cases made from sand grains, small stones or bits of vegetation held together with silk secreted from glands in the head.

The order, which is most closely related to the Lepidoptera, is divided into two suborders, the Annulipalpia, mostly with net-spinning larvae, and the Integripalpia, comprised of species with mostly tube-case-building larvae.

LEPIDOPTERA

- Butterflies and moths
- ~200 000 species
- Wingspan: 3–300 mm, mostly under 75 mm

Members of this readily recognizable order occur everywhere there is vegetation. The body and wings of these familiar insects are covered with minute scales, which may be colored or iridescent. The compound eyes are large and the mouthparts typically take the form of a coiled proboscis through which liquids such as nectar can be sucked. The larvae, known as caterpillars, are typically herbivorous and have a number of abdominal prolegs in addition to the three pairs of thoracic legs. When fully grown they spin a silk cocoon in which they pupate. Some species are significant plant pests.

The order is divided into four suborders, three of them – the Zeugloptera (one family – Micropterigidae), the Aglossata (one family – Agathiphagidae) and Heterobathmiina (one family – Heterobathmiidae) – with only a handful of species. The fourth – Glossata – contains the vast majority of the species. Within the Glossata, the superfamily Papilionoidea comprises the four true butterfly families, the Papilionidae, Pieridae, Lycaenidae and Nymphalidae.

The first Lepidoptera appeared in the Jurassic and then radiated greatly with the rise of flowering plants in the Cretaceous.

HYMENOPTERA

- Sawflies, wasps, bees and ants
- >150 000 species
- Body length: 0.25–70 mm

Abundant and ubiquitous, it is almost certain that the true number of living species of Hymenoptera may exceed 500 000. Species within the order exhibit an incredible diversity of lifestyles: solitary or social, herbivorous, carnivorous or parasitic. The Hymenoptera must be regarded as the most beneficial of all insects for the control of natural insect populations exerted by parasitic and predatory wasp species and the pollination services of bees.

The head carries a pair of thread-like antennae, a pair of well-developed compound eyes and, usually, three ocelli. The mouthparts are adapted for chewing and biting but, in many species, liquids are ingested. In the bees (superfamily

Apoidea) the maxillae and the labium are extended and modified to form a tongue through which nectar is sucked. Most species are strong fliers with two pairs of membranous wings which are joined in flight by a row of small hooks called hamuli.

The order is divided into two suborders, the Symphyta (sawflies) and the Apocrita (wasps, ants and bees). Sawflies have herbivorous larvae and the adults do not have a constricted waist. The females have a saw-like ovipositor for laying eggs into plant tissue. In the Apocrita, the first segment of the abdomen is fused to the thorax and the second, and sometimes the third, abdominal segments are very narrow, which gives the distinctive wasp-waisted appearance. Parasitic apocritans have a slender and sometimes very elongate ovipositor for penetrating and laying eggs in other insects. The aculeate apocritans (stinging wasps, ants and bees) have a modified ovipositor in the form of a sting with an associated venom gland.

Illustrations by Karen Hiscock-Lawrence, KHL Creative.

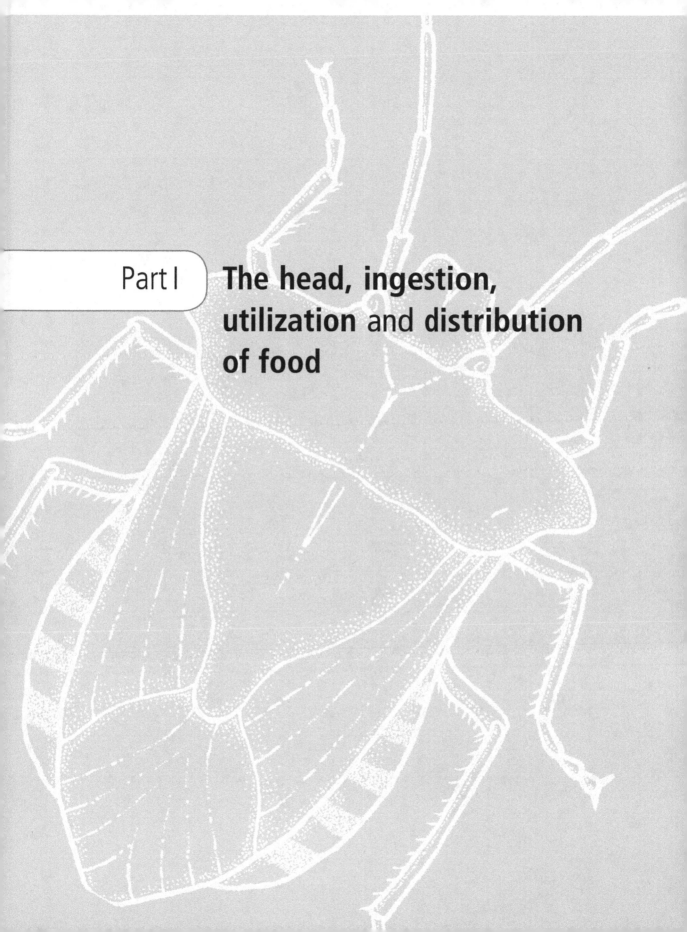

Part I The head, ingestion, utilization and distribution of food

1 Head

REVISED AND UPDATED BY **STEPHEN J. SIMPSON**

INTRODUCTION

Insects and other arthropods are built up on a segmental plan, and their characteristic feature is a hard, jointed exoskeleton. The cuticle, which forms the exoskeleton, is continuous over the whole of the outside of the body and consists of a series of hard plates, the sclerites, joined to each other by flexible membranes, which are also cuticular. Sometimes the sclerites are articulated together so as to give precise movement of one on the next. Each segment of the body primitively has a dorsal sclerite, the tergum, joined to a ventral sclerite, the sternum, by lateral membranous areas, the pleura. Arising from the sternopleural region on each side is a jointed appendage.

In insects, the segments are grouped into three units, the head, thorax and abdomen, in which the various basic parts of the segments may be lost or greatly modified. Typical walking legs are only retained on the three thoracic segments. In the head, the appendages are modified for sensory and feeding purposes and in the abdomen they are lost, except that some may be modified as the genitalia and in Apterygota some pregenital appendages are retained. This chapter introduces the structures of the head (Section 1.1), neck (Section 1.2) and antennae (Section 1.3). Chapter 2 concerns the mouthparts and feeding.

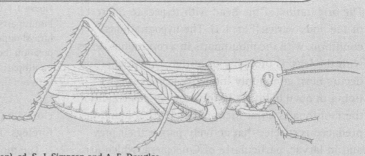

The Insects: Structure and Function (5th edition), ed. S. J. Simpson and A. E. Douglas.
Published by Cambridge University Press. © Cambridge University Press 2013.

1.1 Head

The insect head is a strongly sclerotized capsule joined to the thorax by a flexible, membranous neck. It bears the mouthparts, comprising the labrum, mandibles, maxillae and labium, and also the antennae, compound eyes and ocelli. On the outside it is marked by grooves, most of which indicate ridges on the inside; some of these inflexions extend deep into the head, fusing with each other to form an internal skeleton. These structures serve to strengthen the head and provide attachments for muscles, as well as supporting and protecting the brain and foregut.

The head is derived from the primitive pre-oral and post-oral segments. Molecular studies of *Drosophila* suggest that there are seven head segments: labral, ocular, antennal, intercalary, mandibular, maxillary and labial. The last three segments are post-oral and are innervated by the three neuromeres of the subesophageal ganglion (see Chapter 20). They are often called the gnathal segments because their appendages form the mouthparts of the insect. The pre-oral segments are innervated by the brain, but their nature and number remains contentious. The protocerebrum (forebrain) innervates the compound eyes, the deutocerebrum (midbrain) innervates the antennae, and the labrum receives its innervation from the tritocerebrum (hindbrain) (see Chapter 20).

1.1.1 Orientation

The orientation of the head with respect to the rest of the body varies (Fig. 1.1). The hypognathous condition, with the mouthparts in a continuous series with the legs, is probably primitive. This orientation occurs most commonly in phytophagous species living in open habitats. In the prognathous condition the mouthparts point forwards and this is found in predaceous species that actively pursue their prey, and in larvae, particularly of Coleoptera, which use their mandibles in burrowing. In Hemiptera, the

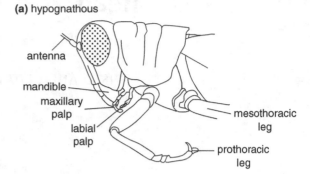

(a) hypognathous

antenna
mandible
maxillary palp
labial palp
mesothoracic leg
prothoracic leg

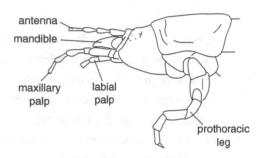

(b) prognathous

antenna
mandible
maxillary palp
labial palp
prothoracic leg

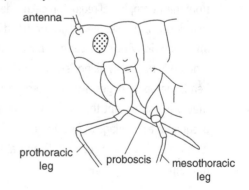

(c) opisthorhynchous

antenna
prothoracic leg
proboscis
mesothoracic leg

Figure 1.1 Orientation of the head. (a) Hypognathous – mouthparts ventral, in a continuous series with the legs (grasshopper). (b) Prognathous – mouthparts in an anterior position (beetle larva). (c) Opisthorhynchous – sucking mouthparts with the proboscis extending back between the front legs (aphid).

elongate proboscis slopes backwards between the forelegs. This is the opisthorhynchous condition.

The mouthparts (labrum with a basal segment called the clypeus, mandibles, labium and maxillae)

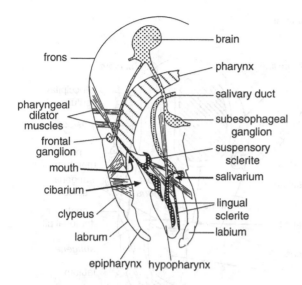

Figure 1.2 Pre-oral cavity and some musculature. Diagrammatic vertical section through the head of an insect with biting and chewing mouthparts. Sclerites associated with the hypopharynx are black with white spots. Muscles attached to these sclerites move the hypopharynx (after Snodgrass, 1947).

enclose a cavity, the pre-oral cavity, which is divided into two sections by the hypopharynx (Fig. 1.2). The larger of these sections, the cibarium, is enclosed between the hypopharynx and the labrum/clypeus, with the true mouth (the opening to the foregut) at its inner end. Between the hypopharynx and the labium is a smaller cavity known as the salivarium, into which the salivary duct opens.

1.1.2 Rigidity

The head is a continuously sclerotized capsule with no outward appearance of segmentation, but it is marked by a number of grooves. Most of these grooves are sulci (singular: sulcus), marking lines along which the cuticle is inflected to give increased rigidity. The term "suture" should be retained for grooves marking the line of fusion of two formerly distinct plates. The groove which ends between the points of attachment of maxillae and labium at the back of the head is generally believed to represent the

line of fusion of the maxillary and labial segments, and is therefore known as the postoccipital suture.

Since the sulci are functional mechanical developments to resist the various strains imposed on the head capsule, they are variable in position in different species and any one of them may be completely absent. However, the needs for strengthening the head wall are similar in the majority of insects, so some of the sulci are fairly constant in occurrence and position (Fig. 1.3). The most constant is the epistomal (frontoclypeal) sulcus, which acts as a brace between the anterior mandibular articulations. At each end of this sulcus is a pit, the anterior tentorial pit, which marks the position of a deep invagination to form the anterior arm of the tentorium. The lateral margins of the head above the mandibular articulations are strengthened by a horizontal inflexion indicated externally by the subgenal sulcus. This sulcus is generally a continuation of the epistomal sulcus to the postoccipital suture. The part of the subgenal sulcus above the mandible is called the pleurostomal sulcus; the part behind the mandible is the hypostomal sulcus. Another commonly occurring groove is the circumocular sulcus, which strengthens the rim of the eye and may develop into a deep flange protecting the inner side of the eye. Sometimes this sulcus is connected to the subgenal sulcus by a vertical subocular sulcus; the inflexions associated with these sulci act as a brace against the pull of the muscles associated with feeding. The circumantennal ridge, marked by a sulcus externally, strengthens the head at the point of insertion of the antenna, while running across the back of the head, behind the compound eyes, is the occipital sulcus.

The areas of the head defined by the sulci are given names for descriptive purposes, but they do not represent primitive sclerites. Since the sulci are variable in position, so too are the areas which they delimit. The front of the head, the frontoclypeal area, is divided by the epistomal sulcus into the frons above and the clypeus below (Fig. 1.3). It is common to regard the arms of the ecdysial cleavage line as delimiting

(a) anterior

(b) lateral

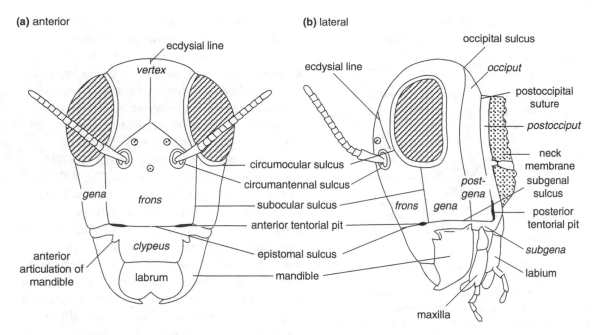

Figure 1.3 Common lines or grooves on the insect head and the areas which they define (italicized) (modified after Snodgrass, 1960).

the frons dorsally, but this is not necessarily so. From the frons, muscles run to the pharynx, the labrum and the hypopharynx; from the clypeus arise the dilators of the cibarium. The two groups of muscles are always separated by the frontal ganglion and its connectives to the brain (Fig. 1.2). Dorsally the frons continues into the vertex and posteriorly this is separated from the occiput by the occipital sulcus. The occiput is divided from the postocciput behind it by the postoccipital suture, while at the back of the head, where it joins the neck, is an opening, the occipital foramen, through which the alimentary canal, nerve cord and some muscles pass into the thorax.

The lateral area of the head beneath the eyes is called the gena, from which the subgena is cut off below by the subgenal sulcus, and the postgena behind by the occipital sulcus. The region of the subgena above the mandible is called the pleurostoma, and that part behind the mandible is the hypostoma.

In hypognathous insects with a thick neck, the posterior ventral part of the head capsule is membranous. The postmentum of the labium is contiguous with this membrane, articulating with the subgena on either side. The hypostomal sulci bend upwards posteriorly and are continuous with the postoccipital suture (Fig. 1.4a). In insects with a narrow neck, permitting greater mobility of the head, and in prognathous insects, the cuticle of the head below the occipital foramen is sclerotized. This region has different origins. In Diptera, the hypostomata of the two sides meet in the midline below the occipital foramen to form a hypostomal bridge that is continuous with the postocciput (Fig. 1.4b). In other cases, such as Hymenoptera and the water bugs *Notonecta* and *Naucoris* (Hemiptera), a similar bridge is formed by the postgenae, but the bridge is separated from the postocciput by the postoccipital suture (Fig. 1.4c). Where the head is held in the prognathous position, the lower ends of the postocciput fuse and extend forwards to form a median ventral plate, the gula (Fig. 1.4d), which may be a continuous sclerotization with the labium. Often the gula is

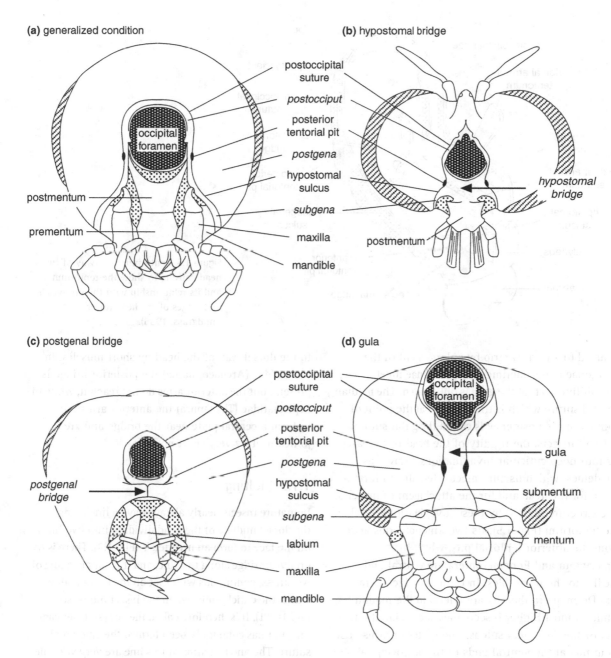

(a) generalized condition

postoccipital suture
postocciput
posterior tentorial pit
postgena
hypostomal sulcus
subgena
maxilla
mandible

occipital foramen

postmentum
prementum

(b) hypostomal bridge

hypostomal bridge

postmentum

(c) postgenal bridge

postoccipital suture
postocciput
posterior tentorial pit
postgena
hypostomal sulcus
subgena
labium
maxilla
mandible

postgenal bridge

(d) gula

occipital foramen

gula
submentum
mentum

Figure 1.4 Sclerotization at the back of the head. Notice the position of the bridge below the occipital foramen with reference to the posterior tentorial pit. Membranous areas stippled, compound eyes cross-hatched. The names of areas defined by sulci are italicized (after Snodgrass, 1960). (a) Generalized condition, no ventral sclerotization; (b) hypostomal bridge (*Deromyia*, Diptera); (c) postgenal bridge (*Vespula*, Hymenoptera); (d) gular bridge formed from the postoccipital sclerites (*Epicauta*, Coleoptera).

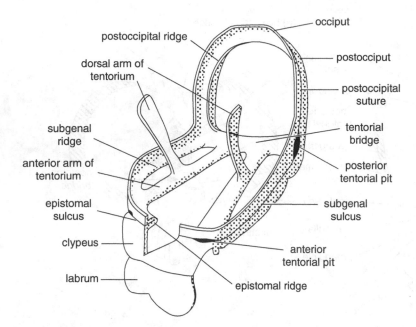

Figure 1.5 Tentorium. Cutaway of the head capsule to show the tentorium and its relationship with the grooves and ridges of the head (after Snodgrass, 1935).

reduced to a narrow strip by enlargement of the postgenae and sometimes the postgenae meet in the midline, so that the gula is obliterated. The median ventral suture which is thus formed at the point of contact of the postgenae is called the gular suture.

In all insects, the rigidity of the head is increased by four deep cuticular invaginations, known as apodemes, which usually meet internally to form a brace for the head and for the attachment of muscles. The structure formed by these invaginations is called the tentorium (Fig. 1.5). Its two anterior arms arise from the anterior tentorial pits, which in Apterygota and Ephemeroptera are ventral and medial to the mandibles. In Odonata, Plecoptera and Dermaptera the pits are lateral to the mandibles, while in most higher insects they are facial at either end of the epistomal sulcus. The posterior arms arise from pits at the ventral ends of the postoccipital suture and they unite to form a bridge running across the head from one side to the other. In Pterygota the anterior arms also join up with the bridge, but the development of the tentorium as a whole is very variable. Sometimes a pair of dorsal arms arise from the anterior arms and they may be attached

to the dorsal wall of the head by short muscles. In Machilidae (Archaeognatha) the posterior bridge is present, but the anterior arms do not reach it, while in Lepismatidae (Thysanura) the anterior arms unite to form a central plate near the bridge and are joined to it by very short muscles.

1.1.3 Molting

Immature insects nearly always have a line along the dorsal midline of the head dividing into two lines on the face to form an inverted Y (Fig. 1.3). There is no groove or ridge along this line, and it is simply a line of weakness, continuous with that on the thorax, along which the cuticle splits when the insect molts (see Fig. 16.11). It is therefore called the ecdysial cleavage line, but has commonly been termed the epicranial suture. The anterior arms of this line are very variable in their development and position and, in Apterygota, they are reduced or absent. The ecdysial cleavage line may persist in the adult insect, and sometimes the cranium is inflected along this line to form a true sulcus. Other ecdysial lines may be present on the ventral surface of the head of larval insects.

1.2 Neck

The neck or cervix is a membranous region which gives freedom of movement to the head. It extends from the postocciput at the back of the head to the prothorax, and possibly represents the posterior part of the labial segment together with the anterior part of the prothoracic segment. Laterally in the neck membrane are the cervical sclerites. Sometimes there is only one, as in Ephemeroptera, but there may be two or three. In *Schistocerca* (Orthoptera) the first lateral cervical sclerite, which articulates with the occipital condyle at the back of the head, is very small. The second sclerite articulates with it by a ball and socket joint, allowing movement in all planes. Posteriorly it meets the third (posterior) cervical sclerite; movement at this joint is restricted to the vertical plane. The third cervical sclerite connects with the prothoracic episternum, relative to which it can move in all planes.

Muscles arising from the postocciput and the pronotum are inserted on the cervical sclerites (Fig. 1.6a) and their contraction increases the angle between the sclerites so that the head is pushed forward (Fig. 1.6b). A muscle arising ventrally and inserted onto the second cervical sclerite may aid in retraction or lateral movements of the head. Running through the neck are longitudinal muscles, dorsal muscles from the antecostal ridge of the mesothorax to the postoccipital ridge, and ventral muscles from the sternal apophyses of the prothorax to the postoccipital ridge or the tentorium. These muscles serve to retract the head onto the prothorax, while their differential contraction will cause lateral movements of the head. *Schistocerca* has 16 muscles on each side of the neck, each of which is innervated by several axons, often including an inhibitory fiber. This polyneuronal innervation, together with the versatility of the cervical articulations and the complexity of the musculature, permit movement of the head in a highly versatile and accurately controlled manner.

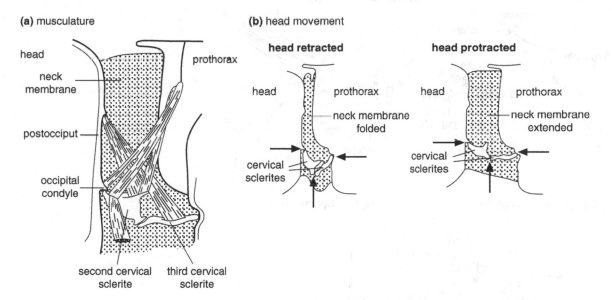

Figure 1.6 Neck and cervical sclerites of a grasshopper. (a) Seen from the inside to show the muscles (after Imms, 1957). (b) Diagrams showing how a change in the angle between the second and third cervical sclerites retracts or protracts the head (the first cervical sclerite is small and is not shown). Arrows indicate points of articulation.

1.3 Antennae

All insects possess a pair of antennae, but they may be greatly reduced, especially in larval forms. Among the non-insect Hexapoda, Collembola and Diplura have antennae, but Protura do not.

1.3.1 Antennal structure

The antenna consists of a basal scape, a pedicel and a flagellum. The scape is inserted into a membranous region of the head wall and pivoted on a single marginal point, the antennifer (Fig. 1.7a), so it is free to move in all directions. Frequently the flagellum is divided into a number of similar annuli joined to each other by membranes so that the flagellum as a whole is flexible. The term "segmented" should be avoided with reference to the flagellum of insects since the annuli are not regarded as equivalent to leg segments.

The antennae of insects are moved by levator and depressor muscles arising on the anterior tentorial arms and inserted into the scape, and by flexor and extensor muscles arising in the scape and inserted into the pedicel (Fig. 1.8a). There are no muscles in the flagellum, and the nerve which traverses the flagellum is purely sensory. This is the annulated type of antenna. In Collembola and Diplura the musculature at the base of the antenna is similar to that in insects, but, in addition, there is an intrinsic musculature in each unit of the flagellum (Fig. 1.8b), and, consequently, these units are regarded as true segments.

The number of annuli is highly variable between species. Adult Odonata, for example, have five or fewer annuli, while adult *Periplaneta* (Blattodea) have over 150, increasing from about 48 in the first-stage larva.

The form of the antenna varies considerably depending on its precise function (Fig. 1.7). Sometimes the modification produces an increase

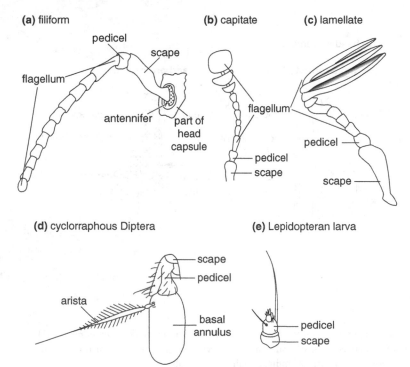

(a) filiform
pedicel
scape
flagellum
antennifer part of head capsule

(b) capitate
flagellum
pedicel
scape

(c) lamellate
pedicel
scape

(d) cyclorraphous Diptera
scape
pedicel
arista
basal annulus

(e) Lepidopteran larva
pedicel
scape

Figure 1.7 Antennae. Different forms occurring in different insects. Not all to the same scale.

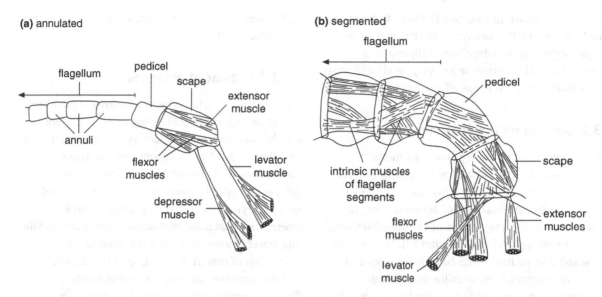

(a) annulated

flagellum pedicel scape extensor muscle

annuli

flexor muscles levator muscle

depressor muscle

(b) segmented

flagellum pedicel

intrinsic muscles of flagellar segments scape

flexor muscles extensor muscles

levator muscle

Figure 1.8 Antenna. Proximal region showing the musculature. (a) Typical insect annulated antenna. There are no muscles in the flagellum (*Locusta*, Orthoptera). (b) Segmented antenna of a non-insect hexapod (*Japyx*, Diplura) (after Imms, 1940).

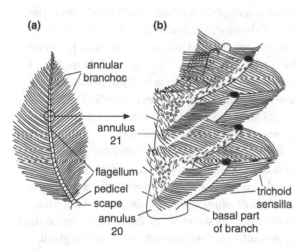

(a)

annular branches

annulus 21

flagellum
pedicel
scape
annulus 20

(b)

trichoid sensilla
basal part of branch

Figure 1.9 Antenna. Plumose form providing space for large numbers of sensilla (male of the moth *Telea polyphemus*) (after Boeckh *et al.*, 1960). (a) The whole antenna seen from above. Two slender branches arise on opposite sides of each annulus. (b) Detail of two annuli from the side showing the bases of the branches and arrangement of long trichoid olfactory sensilla along the branches.

in surface area, allowing a large number of sensilla to be accommodated on the antenna (Fig. 1.9) and, in the case of the plumose antennae of some male moths, enabling them to sample a large volume of air. Sexual dimorphism in the antennae is common, those of the male often being more complex than those of the female. This often occurs where the male is attracted to or recognizes the female by her scent. Conversely, in chalcid parasitoid wasps (Hymenoptera) scent plays an important part in host-finding by the female, and in this case the female's antennae are more specialized than the male's.

The antennae of larval hemimetabolous insects are similar to those of the adult, but with fewer annuli. The number increases at each molt (see Fig. 15.10). In *Periplaneta*, for example, there are only 48 annuli in the first-stage larva compared with over 150 in the adult. The antennae of larval holometabolous insects are usually considerably different from those of the adult. The larval antennae of Neuroptera and Megaloptera have a number of annuli, but in larval Coleoptera and Lepidoptera the antennae are reduced to three simple segments. In some larval Diptera and Hymenoptera the antennae are very small and may be no more than swellings of the head wall.

1.3.2 Sensilla on the antennae

The antennae are primarily sensory structures and they are richly endowed with sensilla in most insects. It is characteristic of insects that the pedicel contains a chordotonal organ, Johnston's organ, which responds to movement of the flagellum with respect to the pedicel (see Chapter 23). In addition, the scape and pedicel often have hair plates and groups of campaniform sensilla that provide information on the positions of the basal segments with respect to the head and to each other. Scattered mechanosensory hairs are also often present on these segments.

The principal sensilla on the flagellum of most insects are olfactory, and these have a variety of forms (see Chapter 24). It is common for contact chemoreceptors, mechanoreceptors and thermohygroreceptors also to be present. Where the flagellum is made up of a series of similar annuli, successive annuli often have a similar arrangement of sensilla, but the sensilla are often concentrated in particular regions. In *Melanoplus* (Orthoptera), for instance, there are no basiconic or coeloconic pegs on the proximal annuli; most of these sensilla are found on the annuli in the middle of the flagellum. In Pieridae (Lepidoptera), most of the antennal sensilla are aggregated on the terminal club. The terminal annulus often has a group of contact chemoreceptors at its tip. The total numbers of sensilla on an antenna are often very large. Adult male *Periplaneta*, for instance, have about 250 000 sensilla on each antenna and male corn-borer moths, *Ostrinia*, about 8000. When the antennae are sexually dimorphic, as in many Lepidoptera, the more complex antenna bears a much larger number of sensilla. For example, male *Telea* have over

65 000 sensilla on one antenna, while the female has only about 13 000.

1.3.3 Functions of antennae

The antennae function primarily as sense organs. They are active tactile sensors in some insects ("feelers") and they are the primary olfactory organs of all insects (see Chapter 24). Very long antennae, such as occur in the cockroach, are associated with their use as feelers. Johnston's organ is important in the regulation of air speed in flying insects (see Fig. 9.30) and in some insects – such as male mosquitoes, female *Drosophila* and worker honey bees – it is involved in the perception of near-field sounds (see Fig. 23.10).

Sometimes the antennae have other functions. The adult water beetle, *Hydrophilus*, submerges with a film of air over its ventral surface which it renews at intervals when it comes to the surface. At the water surface the body is inclined to one side and a funnel of air, connecting the ventral air bubble to the outside air, appears between the head, the prothorax and the distal annuli of the antenna, which is held along the side of the head. The four terminal annuli of the antenna are enlarged and are clothed with hydrofuge hairs which facilitate the formation of the air funnel. In the newly hatched larvae of *Hydrophilus* the antennae assist the mandibles in masticating the prey. This is facilitated by a number of sharp spines on the inside of the antennae.

In fleas and Collembola the antennae are used in mating. Male fleas use the antennae to clasp the female from below and the inner surfaces bear large numbers of adhesive discs. These discs, about 5 μm in diameter, are set on stalks above the general surface of the cuticle and within each one there is a gland, presumably secreting an adhesive material. Species with sessile or semi-sessile females lack these organs. In many Collembola the males have prehensile antennae with which they hold onto the antennae of the female and, in *Sminthurides aquaticus*, the male may be carried about by the female, holding onto her antennae, for several days.

Summary

- The insect head is a rigid capsule that is joined to the thorax by a flexible neck.

- The head bears the mouthparts, the antennae, the compound eyes and the ocelli, and contains the dorsally located brain and the ventral subesphageal ganglion.

- The head has an internal skeleton that provides strength and attachment sites for feeding, antennal and neck muscles.

- The antennae are the primary site of olfactory reception and also serve as active sensors in many insects.

Recommended reading

Angelini, D. R. and Kauffman, T. C. (2005). Comparative developmental genetics and the evolution of arthropod body plans. *Annual Review of Genetics* 39, 95–119.

Schmidt-Ott, U., González-Gaitán, M., Jäckle, H. and Technau, G. M. (1994). Number, identity, and sequence of the *Drosophila* head segments as revealed by neural elements and their deletion patterns in mutants. *Proceedings of the National Academy of Sciences USA* 91, 8363–8367.

Scholtz, G. (2001). Evolution of developmental patterns in arthropods: the analysis of gene expression and its bearing on morphology and phylogenetics. *Zoology* 103, 99–111.

Snodgrass, R. E. (1947). *The Insect Cranium and the "Epicranial Suture"*. Washington, DC: Smithsonian Institution.

Snodgrass, R. E. (1960). *Facts and Theories Concerning the Insect Head*. Washington, DC: Smithsonian Institution.

Staudacher, E. M., Gebhardt, M. and Dürr, V. (2005). Antennal movements and mechanoreception: neurobiology of active tactile sensors. *Advances in Insect Physiology* 32, 49–205.

Zacharuk, R. Y. (1985). Antennae and sensilla. In *Comprehensive Insect Physiology, Biochemistry and Pharmacology*, vol. 6, ed. G. A. Kerkut and L. I. Gilbert, pp. 1–69. Oxford: Pergamon Press.

References in figure captions

Boeckh, J., Kaissling, K.-E. and Schneider, D. (1960). Sensillen und Bau der Antennengeissel von *Telea polyphemus*. *Zoologische Jahrbücher (Anatomie)* **78**, 559–584.

Imms, A. D. (1940). On the antennal structure in insects and other arthropods. *Quarterly Journal of Microscopical Science* **81**, 273–320.

Imms, A. D. (1957). *A General Textbook of Entomology*. London: Methuen.

Snodgrass, R. E. (1935). *Principles of Insect Morphology*. New York, NY: McGraw-Hill.

Snodgrass, R. E. (1947). *The Insect Cranium and the "Epicranial Suture."* Washington, DC: Smithsonian Institution.

Snodgrass, R. E. (1960). *Facts and Theories Concerning the Insect Head*. Washington, DC: Smithsonian Institution.

2 Mouthparts and feeding

REVISED AND UPDATED BY **STEPHEN J. SIMPSON**

INTRODUCTION

The mouthparts of insects comprise an intricate "toolkit" for feeding. The basic elements of the toolkit comprise the unpaired labrum in front, a median hypopharynx behind the mouth, a pair of mandibles and maxillae laterally and a labium forming the lower lip. These component parts have been modified into a remarkable diversity of forms that allow the insects as a group to exploit an extraordinarily wide range of food sources. This chapter begins by describing how the mouthparts have become modified to suit different feeding niches (Section 2.1). Next (Section 2.2) the mechanics of feeding are described, and in Sections 2.3 and 2.4 aspects of the regulation and consequences of feeding behavior are explained. The chapter ends with an extended consideration of the head glands and their various secretions, introducing, among other things, the world of saliva – the "salioverse" (Section 2.5).

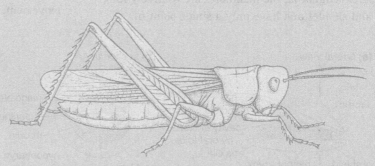

The Insects: Structure and Function (5th edition), ed. S. J. Simpson and A. E. Douglas.
Published by Cambridge University Press. © Cambridge University Press 2013.

2.1 Ectognathous mouthparts

In the non-insect hexapods, Collembola, Diplura and Protura, the mouthparts lie in a cavity of the head produced by the genae, which extend ventrally as oral folds and meet in the ventral midline below the mouthparts (Fig. 2.1). This is the entognathous condition. In the Insecta the mouthparts are not enclosed in this way, but are external to the head; the ectognathous condition. The form of the mouthparts is related to diet, but two basic types can be recognized: one adapted for biting and chewing solid food, and the other adapted for sucking up fluids.

2.1.1 Biting mouthparts

Labrum. The labrum is a broad lobe suspended from the clypeus in front of the mouth and forming the upper lip (Figs. 1.2, 2.2a). On its inner side it is membranous and may be produced into a median lobe, the epipharynx, bearing some sensilla. The labrum is raised away from the mandibles by two muscles arising in the head and inserted medially into the anterior margin of the labrum. It is closed against the mandibles in part by two muscles arising in the head and inserted on the posterior lateral margins on two small sclerites, the tormae, and, at least in some insects, by a resilin spring in the cuticle at the junction of the labrum with the clypeus. Differential use of the muscles can produce a lateral rocking movement of the labrum.

Mandibles. In the entognathous groups and the Archaeognatha, the mandibles are relatively long and slender and have only a single point of

articulation with the head capsule (Fig. 2.3). The mandible is rotated about its articulation by anterior and posterior muscles arising on the head capsule and on the anterior tentorial arms. The principal adductor muscles are transverse and ventral, those of the two sides uniting in a median tendon.

In Thysanura and the Pterygota, the mandibles are articulated with the cranium at two points, having a second, more anterior articulation with the subgena in addition to the original posterior one (Figs. 1.3, 2.2b). These mandibles are usually short and strongly sclerotized and the cuticle of the cusps is often hardened by the presence of zinc or manganese (see Fig. 16.9). These cusps may become worn down during feeding, but the distribution of the harder areas of cuticle promotes self-sharpening.

The original anterior and posterior rotator muscles of Apterygota have become abductors and adductors in the Pterygota, the adductor often becoming very powerful. The apterygote ventral adductor is retained in most orthopteroids and arises from the hypopharyngeal apophysis, but in Acrididae and the higher insects this muscle is absent; in insects with sucking mouthparts it may be modified as a protractor muscle of the mandible.

Maxillae. The maxillae occupy a lateral position, one on each side of the head behind the mandibles. The proximal part of the maxilla consists of a basal cardo, which has a single articulation with the head, and a flat plate, the stipes, hinged to the cardo (Fig. 2.2c). Both cardo and stipes are loosely joined to the head by a membrane so they are capable of movement. Distally on the stipes are two lobes, an

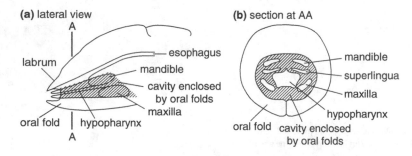

(a) lateral view

labrum — esophagus — mandible — cavity enclosed by oral folds — maxilla

oral fold | hypopharynx

(b) section at AA

oral fold | cavity enclosed by oral folds — mandible — superlingua — maxilla — hypopharynx

Figure 2.1 Entognathous mouthparts (modified after Denis, 1949). (a) Lateral view showing the mouthparts within the cavity formed by the oral folds. The extent of the cavity is indicated by hatching. (b) Transverse section at AA in (a).

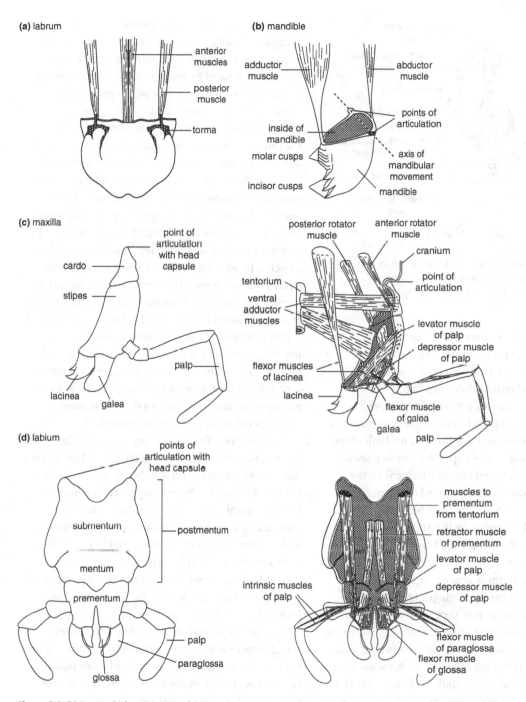

Figure 2.2 Biting and chewing mouthparts of a pterygote insect. Surfaces normally in contact with the hemocoel, the inside of the cuticle, are shaded (after Snodgrass, 1935, 1944). (a) Labrum seen from the posterior, epipharyngeal surface. (b) Mandible – notice the dicondylic articulation. (c) Maxilla from the outside (left) and inside (right). (d) Labium from the outside (left) and inside (right).

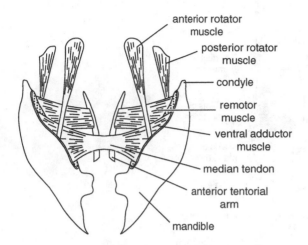

anterior rotator muscle

posterior rotator muscle

condyle

remotor muscle

ventral adductor muscle

median tendon

anterior tentorial arm

mandible

Figure 2.3 Monocondylic mandibles as found in Archaeognatha and non-insect hexapods. Not all muscles are shown (after Snodgrass, 1935).

inner lacinea and an outer galea, one or both of which may be absent. More laterally on the stipes is a jointed, leg-like palp made up of a number of segments; in Orthoptera there are five.

Anterior and posterior rotator muscles are inserted on the cardo, and ventral adductor muscles arising on the tentorium are inserted on both cardo and stipes. Arising in the stipes are flexor muscles of lacinea and galea and another lacineal flexor arises in the cranium, but neither lacinea nor galea has an extensor muscle. The palp has levator and depressor muscles arising in the stipes and each segment of the palp has a single muscle causing flexing of the next segment.

Labium. The labium is similar in structure to the maxillae, but with the appendages of the two sides fused in the midline so that they form a median plate (Fig. 2.2d). The basal part of the labium, equivalent to the maxillary cardines and possibly including a part of the sternum of the labial segment, is called the postmentum. This may be subdivided into a proximal submentum and a distal mentum. Distal to the postmentum, and equivalent to the fused maxillary stipites, is the prementum. The prementum closes the pre-oral cavity from behind. Terminally it

bears four lobes, two inner glossae and two outer paraglossae, which are collectively known as the ligula. One or both pairs of lobes may be absent or they may be fused to form a single median process. A palp arises from each side of the prementum, often being three-segmented.

The musculature corresponds with that of the maxillae, but there are no muscles to the postmentum. Muscles corresponding with the ventral adductors run from the tentorium to the front and back of the prementum; glossae and paraglossae have flexor muscles, but no extensors, and the palp has levator and depressor muscles arising in the prementum. The segments of the palp each have flexor and extensor muscles. In addition, there are other muscles with no equivalent in the maxillae. Two pairs arising in the prementum converge onto the wall of the salivarium at the junction of labium with hypopharynx (Fig. 1.2). A pair of muscles opposing these arises in the hypopharynx and the combined effect of them all may be to regulate the flow of saliva or to move the prementum. Finally, a pair of muscles arising in the postmentum and inserted into the prementum serves to retract or flex the prementum.

Hypopharynx. The hypopharynx is a median lobe immediately behind the mouth (Fig. 1.2). The salivary duct usually opens behind it, between it and the labium. Most of the hypopharynx is membranous, but the adoral face is sclerotized distally, and proximally contains a pair of suspensory sclerites which extend upwards to end in the lateral wall of the stomodeum. Muscles arising on the frons are inserted into these sclerites, which distally are hinged to a pair of lingual sclerites. These, in turn, have inserted into them antagonistic pairs of muscles arising on the tentorium and labium. The various muscles serve to swing the hypopharynx forwards and back; in the cockroach there are two more muscles running across the hypopharynx, which dilate the salivary orifice and expand the salivarium.

In Apterygota, larval Ephemeroptera and Dermaptera there are two lateral lobes of the hypopharynx called the superlinguae.

2.1.2 Variation in form

The form of the mouthparts varies greatly between species. The biting surface of the mandible is often differentiated into a more distal incisor region and a proximal molar region whose development varies with diet. The mandibles of carnivorous insects are armed with strong shearing cusps; in grasshoppers feeding on vegetation other than grasses, there is a series of sharp, pointed cusps, while in grass-feeding species the incisor cusps are chisel-edged and the molar area has flattened ridges for grinding. In species that do not feed as adults, the mouthparts may be greatly reduced. The mandibles of adult Ephemeroptera, for example, are vestigial or absent altogether, and the maxillae and labium are also greatly reduced, being represented mainly by the palps.

The greatest divergence from the basic form occurs in the larvae of holometabolous insects. While larval Lepidoptera and Coleoptera usually have well-developed biting and chewing mouthparts, larval Diptera and Hymenoptera show some extreme modifications and reductions. Mosquito larvae, for example, have brushes of hairs on the mandibles and maxillae which are especially long in some species and serve to filter particulate material, including food, from the water. The larvae of cyclorrhaphous flies exhibit extreme reduction of the head. The principal structures are a pair of heavily sclerotized mouthhooks with which the larva rasps at its food; sensory papillae probably represent the palps. Among Hymenoptera, larval Symphyta have well-developed mouthparts, similar to caterpillars, but in some parasitic species the mandibles are represented only by simple spines, and other mouthparts are not differentiated into separate sclerites.

2.1.3 Sucking mouthparts

The mouthparts of insects which feed on fluids are modified in various ways to form a tube through which liquid can be drawn into the mouth, and usually another through which saliva passes. The muscles of the cibarium or pharynx are strongly developed to form a pump. In Hemiptera and many Diptera, which feed on fluids within plants or animals, some components of the mouthparts are modified for piercing, and the elongate structures are called stylets. The combined tubular structures are referred to as the proboscis, although specialized terminology is used in some groups.

In Hemiptera, mandibles, maxillae and labium are all elongate structures, while the labrum is relatively short. The food canal is formed by the opposed maxillae, which are held together by a system of tongues and grooves (Figs. 2.4a, 2.8a). These allow the stylets to slide freely on each other, while maintaining the integrity of the food canal. The maxillae also contain the salivary canal. On either side of the maxillae are the mandibular stylets. These are the principal piercing structures and they are often barbed at the tip. When the insect is not feeding, the slender maxillary and mandibular stylets are held within a groove down the anterior side of the labium. The hemipteran labium is known as the rostrum. It is usually segmented, allowing it to fold as the stylets penetrate the host. There are no palps. Since the Hemiptera are hemimetabolous, the larvae and adults have similar feeding habits and both have sucking mouthparts.

Thysanoptera are also fluid-feeders as larvae and adults. Their stylets are normally held in the cone-shaped lower part of the head formed by the clypeus, labrum and labium. Only the left mandible is present. It is used to penetrate plant cells. The maxillary stylets are held together to form the food canal. There is no salivary canal; the salivary duct opens into the front of the esophagus.

The adult Diptera exhibit a great variety of modifications of the mouthparts, but in all of them the food canal is formed between the apposed labrum and labium and the salivary canal runs through the hypopharynx (Fig. 2.4e,f). The mandibles and maxillae are styliform in species that suck the blood of vertebrates, but are generally lacking in other species, including the blood-sucking Cyclorrhapha. Where they are present, they are the piercing organs; in blood-sucking

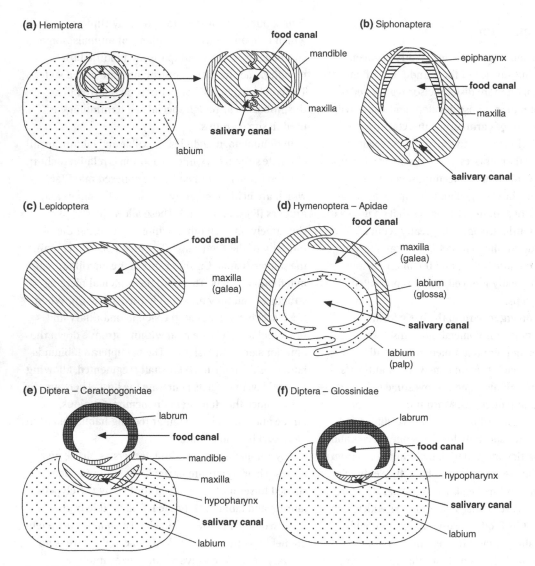

Figure 2.4 Sucking mouthparts. Diagrammatic cross-sections of the proboscis showing the principal structures used to form tubes for delivery of saliva and intake of food. Homologous structures are indicated with the same shading in all the diagrams. In some cases the structures contain an extension of the hemocoel – this is not shown. (a) Hemiptera (bugs) (compare Fig. 2.8a); (b) Siphonaptera (fleas); (c) Lepidoptera (butterflies and moths); (d) Hymenoptera, Apidae (bees); (e) Diptera, Ceratopogonidae (biting midges); (f) Diptera, Glossinidae (tsetse flies).

Cyclorrhapha tooth-like structures at the tip of the labium penetrate the host tissues by a rasping action. Similar prestomal teeth occur in other Cyclorrhapha, including the housefly, *Musca*. In many species, the tip of the labium is expanded to form a lobe, the labellum which, in Brachycera, is traversed by a series of grooves known as pseudotracheae because they are held open by cuticular ribs, giving them a superficial similarity to tracheae. The pseudotracheae converge centrally on the distal end of the food canal. Diptera have maxillary palps, but no labial palps.

The food canal of fleas is formed between an extension of the epipharynx and the maxillary stylets (Fig. 2.4b). A salivary canal extends along the inside of each maxilla, which also form the piercing organs. Both maxillary and labial palps are present.

The proboscis of adult Lepidoptera is formed from the galeae held together by a system of cuticular hooks ventrally and a series of plates dorsally (Fig. 2.4c). Since most Lepidoptera are nectar-feeders, they do not require piercing mechanisms, and the rest of the mouthparts, apart from the labial palps, are reduced or absent. There is no salivary canal, although adult Lepidoptera do have salivary glands.

Adults of most Hymenoptera have biting and chewing mouthparts, but the bees are nectar-feeders and are described as lapping the nectar. This is achieved by an elongation and flattening of the galeae and labial palps which surround the fused glossae (Fig. 2.4d). The space outside the glossal tongue forms the food canal. The salivary canal is in the posterior folds of the tongue.

Larval Neuroptera and some predaceous larval Coleoptera that digest their prey extra-orally have a food canal in each of the mandibles. These function in a similar way to those of biting and chewing insects, but they are sickle-shaped. In larval Neuroptera, a groove on the inside of each mandible is converted to a tube by the juxtaposition of a slender lacinea (Fig. 2.5a). A similar groove is present in the mandibles of some larval Dytiscidae (Coleoptera), but instead of being closed by the lacinea the lips of the groove almost join to form a tube (Fig. 2.5b). Larval Lampyridae (Coleoptera) have a tube running through each mandible and opening by a hole near the tip and another near the base within the cibarial cavity (Fig. 2.5c).

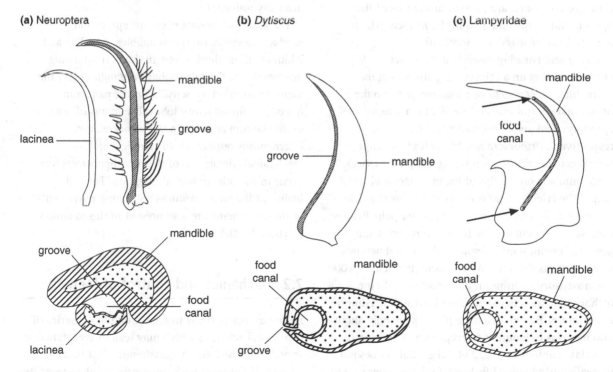

Figure 2.5 Sucking mouthparts of larval holometabolous insects. (a) An antlion (Neuroptera); (b) *Dytiscus* larva (Coleoptera, Dytiscidae); (c) a firefly larva (Coleoptera, Lampyridae). Arrows show the positions at which the food canal opens to the outside (based on Cicero, 1994).

Associated with the production of a tube for feeding is the development of a pump for drawing up the fluids and a salivary pump for injecting saliva (Fig. 3.12). Often the feeding pump is developed from the cibarium, which by extension of the lateral lips of the mouth becomes a closed chamber connecting with the food canal. The cibarial muscles from the clypeus enlarge so a powerful pump is produced. In Lepidoptera and Hymenoptera the cibarial pump is combined with a pharyngeal pump which has dilators arising on the frons.

2.1.4 Sensilla on the mouthparts

Most of the sensilla on the mouthparts are contact chemoreceptors (see Chapter 24), but mechanoreceptors are also common and olfactory sensilla are often present on the palps. Chordotonal organs (see Chapter 23), which probably function as pressure receptors, are present at the tips of the mandibular cusps and also in the lacinea, which is heavily sclerotized and tooth-like.

Biting and chewing insects have contact chemoreceptors on all the mouthparts except the mandibles. They also have chemoreceptors on the dorsal and ventral walls of the cibarium, often called epipharyngeal and hypopharyngeal sensilla, respectively. Orthoptera and Blattodea have large numbers of sensilla in groups (Fig. 2.6a), with especially large numbers on the tips of the maxillary and labial palps. The cricket, *Gryllus bimaculatus*, for example, has over 3000 sensilla on each maxillary palp. Because each sensillum contains at least four neurons, the potential chemosensory input to the central nervous system is considerable; an adult locust has about 16 000 chemosensory neurons on the mouthparts. In the orthopteroid insects, the numbers increase each time the insect molts. By contrast, caterpillars have only about 100 neurons in mouthpart receptors and the closely associated antennae (Fig. 2.6b); the number does not increase during larval life. Fluid-feeding insects usually have chemoreceptors at the tip of the labium, on the palps when these are present, and in the walls of the

cibarium (Fig. 2.6c). In addition, at least some planthoppers have an olfactory sensillum toward the tip of the rostrum. No chemoreceptors are present on the mandibular and maxillary stylets. Aphids have only mechanoreceptors at the tip of the labium; their only chemoreceptors on the mouthparts are in the cibarium.

In piercing and sucking insects (Hemiptera and Culicidae, for example) only the cibarial sensilla come directly into contact with the food as it is ingested; the labium does not enter the tissues of the host so its sensilla can only monitor the outer surface of the food, either plant or animal. In blood-sucking species that use labial teeth to rasp through the tissues, such as *Glossina* and *Stomoxys*, however, the labial sensilla come into direct contact with the blood. This is also true of nectar-feeding insects, Lepidoptera, Apoidea and many Diptera, including female mosquitoes. Cibarial sensilla are known to be present in some of these species and are probably universal.

The axons of contact chemoreceptors and mechanoreceptors on the mandibles, maxillae and labium end in arborizations in the corresponding neuromeres of the subesophageal ganglion, but the axons from olfactory sensilla on the palps run directly to the olfactory lobes. The axons of sensilla on the labrum arborize in the tritocerebrum. Interneurons responding to chemical and mechanical stimulation of the mouthpart receptors occur in the subesophageal ganglion. The cell bodies of the motor neurons regulating movements of the mouthparts are also present in this ganglion (Section 2.2.1).

2.2 Mechanics and control of feeding

Before an insect starts to feed it exhibits a series of behavioral activities which may lead to acceptance or rejection of the food. A grasshopper first touches the surface of the plant with the sensilla at the tips of its palps. This behavior enables the insect to monitor the chemicals on the surface of the plant and perhaps

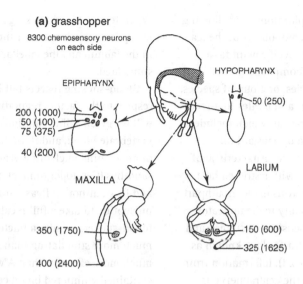

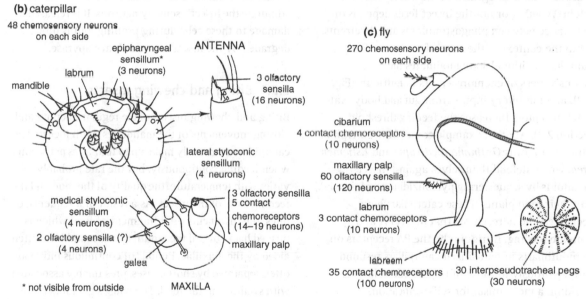

Figure 2.6 Chemoreceptors on the mouthparts of various insects. Numbers show the number of sensilla and, in brackets, the number of chemosensitive neurons in each group of receptors. Sensilla are contact chemoreceptors unless otherwise stated. The numbers do not include sensilla scattered over the mouthparts, which may be present in addition to those in groups. (a) An adult locust, *Locusta*; (b) a caterpillar, *Pieris*; (c) an adult fly, *Drosophila*.

also the odor of the plant. This may lead the insect to reject the plant without further investigation or to make an exploratory bite, releasing chemicals from within the plant, which flow across the hydrophilic inner surface of the cibarium. This, in turn, may result in rejection, or the insect may start to feed. Essentially similar behaviors are seen in caterpillars and leaf-eating beetles.

The principal chemicals inducing feeding (phagostimulants) of leaf-eating insects are sucrose and hexose sugars, but the exploratory bite following palpation may be induced by components of the leaf wax. Insects that feed only on specific plant taxa may require the presence of a compound that is characteristic of the plant species, or group of species. For example, many caterpillars and beetles that feed on plants in the family Brassicaceae, which includes cabbage, are stimulated to feed by mustard oil glucosides (glucosinolates) that are characteristic of this plant family. Some species with restricted host ranges have chemosensory neurons in the mouthpart sensilla which respond specifically to the indicator chemicals. Other neurons are sensitive to sugars, and others to compounds that inhibit feeding, known as feeding deterrents (see Chapter 24). Information from all the sensilla is integrated in the central nervous system. Whether or not the insect feeds depends on the balance between phagostimulants and deterrents, within the context of the state of the insect. For example, nutritional state modulates the responsiveness of chemoreceptors to nutrients (Fig. 4.12), and inhibitory inputs from gut and body-wall stretch receptors increases the feeding threshold (Section 2.3). Another example comes from caterpillars of the tiger moths *Grammia incorrupta* and *Estigmene acrea*, which defend themselves against insect parasitoids by sequestering pyrrolizidine alkaloids (PA) from their food plants. These caterpillars have specialized chemoreceptors that detect PA and stimulate feeding. Intriguingly, the PA receptors on the mouthparts of parasitized caterpillars respond more vigorously to those of unparasitized larvae, providing a mechanism for self-medication.

Among nectar-feeding insects, sugars are phagostimulants. Before starting to feed, the insects exhibit a sequence of behaviors comparable with that of the leaf-chewing insects. If the tarsi contact sugar above a certain threshold concentration, the proboscis is extended. This is true of flies, bees and butterflies. Stimulation of the trichoid sensilla on the outside of the labellum of a fly causes the insect to spread the labellar lobes so that the interpseudotracheal papillae (Fig. 2.6c) contact the food. Their stimulation initiates ingestion.

When female blood-sucking insects, such as mosquitoes, feed on nectar the stylets remain enclosed in the labium and the labellar chemoreceptors are stimulated.

Blood-sucking insects fall into two classes with respect to the factors regulating ingestion: those that will gorge on saline solutions that are isotonic with vertebrate blood, and those that require the presence of an adenine nucleotide. Among the former are sandflies (Ceratopogonidae), anopheline mosquitoes and fleas, although fleas require the presence of a nucleotide to take a full meal. Most of the other blood-suckers require a nucleotide. ATP is generally much more stimulating than ADP, and this in turn is much more effective than AMP. ATP is normally contained within red blood cells, where it would not stimulate the insect's sensory neurons. It is released by damage to these cells during probing, but is quickly degraded by the insect's own salivary apyrase.

2.2.1 Biting and chewing insects

Biting and chewing insects make regular opening and closing movements of the mandibles; both locusts and caterpillars commonly make up to four bites per second when feeding continuously, but the rate probably varies with temperature, the quality of the food and the feeding state of the insect. As a result of a sequence of bites, the insect cuts off a fragment of food which is pushed back toward the mouth by the mandibles, often aided by the maxillae. Periods of continuous biting are often separated by short pauses, presumably associated with swallowing the food. The pauses get longer as a meal progresses, as negative feedbacks from stretch receptors in the gut build up throughout the course of the meal and initial levels of arousal decline.

The motor pattern controlling movement of the mouthparts is generated in the subesophageal ganglion. Premotor central pattern-generating circuits produce the basic mandibular chewing rhythm, which is modulated by feedback from mechano- and chemoreceptors on the mouthparts responding to physical and chemical properties of the food. In caterpillars of *Manduca*, rhythmical chewing activity occurs spontaneously in the absence of inhibitory input from receptors in the wall of the thoracic segments, but in grasshoppers such spontaneous activity does not occur. Mechanosensory modulation of the basic motor patterns includes stimulation of contraction in the mandibular abductor muscles by mechanical stimulation of the labrum. Sensilla that detect the positions of the open mandibles then stimulate the mandibular adductor muscle motor neurons via the ganglion so the mandibles close. This stimulates other mechanoreceptors, probably those at the tips of the cusps, which presumably starts another cycle of opening. In locusts the motor neurons and muscles controlling the other mouthparts – except for the muscles to the labial palps – and the salivary glands appear to be coupled to the mandibular motor pattern circuitry. Chemosensory inputs modify the rate of chewing.

A modulatory effect on the activity of the muscles of the mouthparts is exerted by serotonin. In the cockroach, *Periplaneta*, and some other insects all the nerves to these muscles have a branching network of fine serotonergic axons over their surfaces. These neurons, of which the axons are a part, are only active during feeding and it is presumed that the electrical activity leads to the release of serotonin from the fine branches where they locally affect the activity of the muscles moving the mouthparts. The precise effect of this modulation is not known.

2.2.2 Fluid-feeding insects

Proboscis extension by nectar-feeding flies and bees depends on the activity of muscles associated with the mouthparts. In Lepidoptera, the proboscis is caused to unroll by increased pressure of the hemolymph. This pressure is generated in the stipes associated with each galea. A valve isolates the hemocoel of the stipes when the latter contracts. Coiling results from the elasticity of the cuticle of the galeae together with the activity of intrinsic muscles.

Insects feeding on the internal fluids of other organisms must first penetrate the host tissues. Homoptera, which feed on plants, first secrete a blob of viscous saliva which solidifies around the labellar lobes, forming a salivary flange which remains even after the insect has departed (Fig. 2.7). This probably

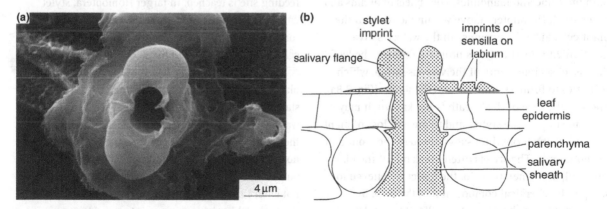

Figure 2.7 Salivary flange produced from saliva by a planthopper. (a) Scanning electron micrograph showing the flange remaining on a leaf surface when the insect stops feeding. The saliva hardens around the mouthparts. The hole in the center was occupied by the stylets; the small holes to the right were produced by sensilla at the tip of the labium. The structures running obliquely from the top left are leaf veins (after Ribeiro, 1995). (b) Diagrammatic section through a flange showing its relationship to the leaf surface and the salivary sheath.

(a)

(b)

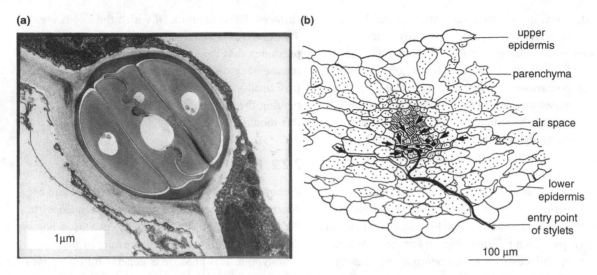

Figure 2.8 Feeding by an aphid (after Tjallingii and Esch, 1993). (a) Transverse section through the stylets and salivary sheath in a leaf. The maxillary stylets interlock to form the food canal (center) and the salivary canal (above) (compare Fig. 2.4a). Each mandibular stylet has a narrow lumen, an extension of the hemocoel, containing mechanoreceptor neurons. The dark ring surrounding the stylets is the salivary sheath. Outside it, the pale fibrous material is plant cell wall. Notice that the stylets are contained within the cell wall; they do not enter the surrounding cytoplasm. (b) Pathways taken by the stylets of an aphid at the start of feeding. Abortive pathways are shown in white with the ends of the paths indicated by arrows. The final pathway, reaching the phloem, is shown in black. Phloem sieve tubes, black; xylem, cross-hatched; parenchyma, stippled.

serves to prevent the stylets from slipping over the plant surface as pressure is applied. The mandibles penetrate the leaf cuticle and epidermis; subsequent deeper penetration may involve the maxillae alone, or both maxillae and mandibles. The stylets of aphids and other small Homoptera move within the walls of the plant cells aided by enzymes in the watery saliva (Section 2.5). As the stylets progress through the leaf, the insect secretes more of the viscous saliva, which solidifies to form a sheath around the stylets (Fig. 2.8a). The significance of this sheath is not known. It may serve to support the stylets and to prevent loss of plant sap and of the more fluid saliva through the wound in the epidermis. The stylets often break out of the plant cell wall, and when they do, the insect is believed to sample the chemical composition of the fluid by drawing it up to the cibarial sensilla. If the stylets are not in a sieve element of the phloem, the insect withdraws its stylets for a short distance and moves them in another direction. This process is repeated, and the path of the stylets may become very irregular

(Fig. 2.8b), until the phloem is entered, at which point the stylets are pushed into it and ingestion begins (Fig. 2.8b). It may take anything from five minutes to three hours from the beginning of probing until a feeding site is reached. In larger Homoptera, stylet entry occurs in a similar way and a salivary sheath is formed, but in most cases the stylet pathway is through cells rather then between them.

Blood-sucking insects, like the Homoptera among plant-feeders, must penetrate the host tissues before starting to feed. In many species that feed on warm-blooded vertebrates, the proboscis is moved into the feeding position in response to the warmth of the host. In mosquitoes and triatomine bugs, the stylets are pushed into the tissues and the labium folds up, but does not enter the wound. This separation of the stylets from the ensheathing labium appears to provide some additional stimulus necessary for the insects to take a full blood meal. In tsetse and stable flies, rasping movements of the prestomal (labial) teeth tear into the tissues. Blood vessels comprise less than 5% of the volume of mammalian skin, and blood is not available until a vessel is ruptured.

Salivary secretions play a critical role in blocking the hemostatic responses of the host, which usually prevent blood loss. Three different processes contribute to hemostasis: platelet aggregation, blood coagulation and vasoconstriction. The saliva of blood-sucking insects is a hugely complex mixture of chemicals that inhibit all three as well as anesthetize the host (Table 2.1). ADP released from injured blood cells is an important signal for platelet aggregation and blood-sucking insects generally have a salivary apyrase that degrades ADP to orthophosphate and AMP. The saliva also contains a vasodilator to counteract the vasoconstriction induced by the host, but these vasodilators differ from one insect species to another. Factor X and thrombin are chemicals that regulate coagulation; peptides that counteract the effects of both chemicals are present in the saliva.

The intake of fluids depends on their viscosity and whether or not they are under pressure. Viscosity increases with the concentration of dissolved solutes, so nectar containing 40% sucrose is six times more viscous than water at the same temperature. The phloem and xylem of plants are very dilute so their viscosities are not markedly different from water. The viscosity of vertebrate blood, however, varies with the diameter of the tube through which it is being drawn. For tubes less than 100 μm in diameter, the viscosity falls as tube diameter is reduced down to about 6 μm. At smaller diameters it increases sharply because flow depends on the distortion of the red blood cells. In addition, the fluid may be under positive or negative pressure. Blood pressure in human capillaries is about 3 kPa, whereas the phloem in plants is under much higher pressure, 0.2–1 MPa. Xylem, on the other hand, is under strong negative pressure, as much as −2 MPa and higher. The cibarial and pharyngeal pumps which draw fluid through the proboscis are consequently different in insects with different feeding habits. Capillarity may be important in bees, helping to move nectar up the food canal.

Once a phloem-feeding insect, such as an aphid or planthopper, has reached the phloem, the pressure of the phloem is sufficient to push fluid into the insect, which appears then to play no active role in ingestion. These insects can, however, pump fluid into the gut when the food is not under pressure and do, periodically, feed from parenchyma or xylem. Consequently, they can feed through membranes from artificial food media in laboratory experiments.

The negative pressure of xylem tends to draw fluid out of the insect, and the massive cibarial pumps that are characteristic of habitual xylem-feeding insects, such as cicadas and cercopids, are necessary to overcome these pressures (Fig. 3.12). Even so, the leafhopper *Homalodisca* exhibits markedly reduced feeding rates at negative xylem pressures in excess of −1.5 MPa.

In nectar-feeding bees, the glossa is repeatedly extended into the nectar while the galeae and labial palps which surround it (Fig. 2.4d) remain motionless. During extension of the glossa, nectar moves onto it by capillarity. The fluid moves the hairs on the glossa to a position at a right angle to the surface of the glossa so that the volume of fluid that is held is increased. When the glossa is retracted into the tube formed by the galeae and labial palps the fluid is drawn into the cibarium by the pump. In *Bombus* these licking movements occur at a frequency of about 5 Hz. The intake rate of sucrose solutions at concentrations up to about 40% is around 1.75 μl^{-1}. It declines at higher concentrations due to the increasing viscosity.

Blood-sucking insects produce pressure differentials well in excess of the capillary blood pressure of the host, so the latter is unlikely to play a significant role in feeding. Calculated pressure differences, which are produced by the cibarial and pharyngeal pumps, are approximately 8 kPa in the mosquito (*Aedes*), 20 kPa in the louse (*Pediculus*), 80 kPa in the bed bug (*Cimex*) and 100–200 kPa in *Rhodnius*. In the latter, the pumping rate is about 7 Hz, producing an ingestion rate of about 450 nl^{-1}. In the mosquito, *Aedes*, the intake rate is about 16 nl^{-1}.

Table 2.1 **Molecularly characterized salivary proteins of blood-feeding insects (excluding immunity related)**

Name	Biological action	Activity	Molecular family	Organism	Family or order
Apyrase	Degrades ADP and ATP	Anti-platelet	Cimex-type	*Cimex lectularius*	Cimicidae
Cimex nitrophorin	Carriers of NO	Vasodilator, anti-platelet	Inositol-phosphate phosphatase, truncated	*Cimex lectularius*	Cimicidae
Sialokinin	Endothelium-dependent vasodilator	Vasodilator	Tachykinins	*Aedes aegypti*	Culicidae
Apyrase	Degrades ADP and ATP	Anti-platelet	5′-nucleotidase	*Aedes aegypti*	Culicidae
Purine nucleosidase	Degrades adenine to hypoxanthine	Destroys mast-cell degranulating adenine and inosine	Purine nucleosidase	*Aedes aegypti*	Culicidae
Anti Xa serpin	Factor Xa inhibitor	Anti-clotting	Serpin	*Aedes aegypti*	Culicidae
Aegyptin	Collagen inhibitor	Anti-platelet	Unique	*Aedes aegypti*, *Anopheles stephensi*	Culicidae
Peroxidase	Degrades catecholamines	Vasodilatory	Peroxidase	*Anopheles albimanus*	Culicidae
Anophelin	Thrombin inhibitor	Anti-clotting	Unique	*Anopheles albimanus*	Culicidae
D7 large	Sequesters leukotrienes	Anti-inflammatory	D7 – OBP superfamily	*Anopheles gambiae*	Culicidae
D7 small	Sequesters biogenic amines	Vasodilatory, anti-inflammatory	D7 – OBP superfamily	*Anopheles gambiae*	Culicidae
Hamadarin	Factor XII inhibitor	Anti-bradykinin	D7 – OBP superfamily	*Anopheles stephensi*	Culicidae
Anophensin	Factor XII inhibitor	Anti-bradykinin	gSG7	*Anopheles stephensi*	Culicidae
Endonuclease	Degrades double-stranded DNA	Spreading factor	Endonuclease	*Culex quinquefasciatus*	Culicidae
Hyaluronidase	Degrades skin matrix components	Spreading factor	Hyaluronidase	*Tabanus yao*	Diptera
Adenosine deaminase	Converts adenosine to inosine	Destroys mast-cell degranulating adenosine	Adenosine deaminase	*Glossina morsitans*	Glossinidae

Name	Target	Action	Type	Species	Family
Tsetse thrombin inhibitor	Thrombin inhibitor	Anti-clotting	Unique	Glossina morsitans	Glossinidae
Thrombostasin	Anti-thrombin	Anti-clotting	Thrombostasin	Haematobia irritans	Muscidae
Stomoxys Ag5	Unknown target (complement activation?)	Binds Ig	Antigen 5	Stomoxys calcitrans	Muscidae
Adenosine deaminase	Converts adenosine to inosine	Destroys mast-cell degranulating adenosine	Adenosine deaminase	Aedes aegypti, Phlebotomus dubosqi	Nematocera
Maxadilan	PACAP receptor agonist	Vasodilator	Unique	Lutzomyia longipalpis	Psychodidae
5'-nucleotidase	Degrades AMP to adenosine	Produces vasodilatory and anti-platelet adenosine	5'-nucleotidase	Lutzomyia longipalpis	Psychodidae
Apyrase	Degrades ADP and ATP	Anti-platelet	Cimex-type	Phlebotomus papatasi	Psychodidae
Adenosine, AMP	Agonist of purinergic receptors	Vasodilator, anti-platelet	Nucleotides	Phlebotomus papatasi, P. argentipes	Psychodidae
Lysophosphatidylcholine	Inhibits platelet aggregation	Anti-platelet	Lipid	Rhodnius prolixus	Reduviidae
Rhodnius nitrophorins	Carriers of NO	Vasodilator, anti-platelet	Lipocalin of the nitrophorin family	Rhodnius prolixus	Reduviidae
Inositol phosphatase	Degrades inositol phosphates	Unknown	SHIP	Rhodnius prolixus	Reduviidae
RPAI	Sequesters ADP	Anti-platelet	Lipocalin of the triabin superfamily	Rhodnius prolixus	Reduviidae
BABP	Anti-serotonin, anti-adrenergic	Vasodilatory	Lipocalin	Rhodnius prolixus	Reduviidae
Prolixin-S	Inhibits factor VIII	Anti-clotting	Lipocalin of the nitrophorin family	Rhodnius prolixus	Reduviidae
Triatoma anesthetic	Reduces nerve sodium current	Anesthetic	Unknown	Triatoma infestans	Reduviidae

Table 2.1 (*cont.*)

Name	Biological action	Activity	Molecular family	Organism	Family or order
Sialidase	Degrades sialic acid	Anti-neutrophil?	Unknown	*Triatoma infestans*	Reduviidae
Apyrase	Degrades ADP and ATP	Anti-platelet	5'-nucleotidase	*Triatoma infestans*	Reduviidae
Triapsin	Serine protease	Unknown – matrix degradation?	Trypsin	*Triatoma infestans*	Reduviidae
Triplatin	Possible GPVI antagonist or ADP binder	Anti-platelet	Lipocalin of the triabin superfamily	*Triatoma infestans*	Reduviidae
Pallidipin	Possible ADP binding	Anti-platelet	Lipocalin of the triabin superfamily	*Triatoma pallidipennis*	Reduviidae
Triabin	Thrombin inhibitor	Anti-clotting	Lipocalin of the triabin superfamily	*Triatoma pallidipennis*	Reduviidae
SVEP	Possible PACAP receptor agonist	Vasodilator	Ricin lectin family	*Simulium vittatum*	Simuliidae
Simullidin	Anti-thrombin	Anti-clotting	Possible D7 distant member	*Simulium vittatum*	Simuliidae
Apyrase	Degrades ADP and ATP	Anti-platelet	CD-39	*Xenopsylla cheopis*	Siphonaptera
Chrysoptin	Degrades ADP and ATP	Anti-platelet	5'-nucleotidase	*Chrysops* spp.	Tabanidae
Vasotab, Vasotab TY	Possible ion channel effect	Vasodilator	Kazal	*Hybomitra bimaculata, Tabanus yao*	Tabanidae
Tabkunin	Anti-thrombin	Anti-clotting	Kunitz	*Tabanus yao*	Tabanidae
Tabinhibitin	RGD disintegrin	Anti-platelet	Antigen 5	*Tabanus yao*	Tabanidae
Tabserin	Serine protease	Anti-clotting, target unknown	Trypsin	*Tabanus yao*	Tabanide
Tabfiglysin	Serine protease	Fibrinogenlytic	Trypsin	*Tabanus yao*	Tabanide

Source: After Ribeiro and Arcà, 2009.

The louse, *Pediculus*, has a food canal that is smaller than the diameter of a human red blood cell (about 7.5 µm); as a result, the feeding rate of an adult female louse is about five times lower than that of a mosquito.

Little is known about the control of feeding in fluid-feeding insects, but serotonin is released into the hemolymph during feeding by *Rhodnius*, probably from neurohemal organs on the abdominal nerves. It triggers plasticization of the abdominal cuticle (Section 16.4.3).

2.2.3 Prey capture by predaceous insects

Predators catch their prey either by sitting and waiting for it to come their way, or by actively pursuing it. The mantis is an example of an insect that sits and waits. Like many predators, mantids have wide heads and the eyes are relatively far apart. This facilitates the accurate determination of the distance of the prey from the insect (see Fig. 22.11). In addition, the head is very mobile, meaning movements of prey can be followed without the whole mantis moving. The mantis strikes when the prey is of a suitable size, is at a suitable distance in front of the head and is moving. *Sphodromantis*, a relatively large insect, strikes when the prey subtends an angle of 20–40° at the eyes and is 30–40 mm in front of the head. In *Tenodera* the attack zone is in an area defined by the length of the foreleg and below the body axis (Fig. 2.9). The forelegs of mantids are raptorial and armed with spines. The mantis follows the prey with its eyes (see Fig. 22.12) and moves its prothorax to orient directly toward the prey. The effect of this is to make the area of attack two-dimensional; the mantis has to judge the position of the prey directly in front and not to a position on either side of the head. The attack involves a strike by the forelegs and a forwards lunge of the whole body that reduces the distance between the head of the mantid and its prey. The strike involves first the extension of the coxa and tibia and then a rapid extension of the femur and flexure of the tibia to grasp the prey. This last movement occurs in about 30 ms.

Dragonfly larvae also sit and wait for their prey and then capture it with the modified labium. The postmentum and prementum are both elongate and the labial palps are claw-like structures set distally on the prementum (Fig. 2.10a). At rest, the labium, often referred to as a labial mask, is folded beneath the head. If a potential prey item comes within range, the mask is extended and the prey caught by the labial palps (Fig. 2.10c). As in the mantis, the strike is very rapid, being completed in 25 ms or less, but in this case the speed of the strike depends on prior energy storage. Before the strike, the anal sphincter is closed and the dorso-ventral abdominal muscles contract (Fig. 2.10b). This results in an increase in hemolymph pressure. At the same time the flexor and extensor muscles of the prementum, which arise in the head, contract together (co-contract) so that the labial mask is held against the head and, presumably, tension builds up at the postmentum/prementum joint. Relaxation of the flexor muscle results in the sudden extension of the prementum due to the continued activity of the premental extensor muscles and release of the cuticular tension at the postmentum/prementum joint. There are no extensor muscles of the postmentum and its extension is produced by the high hemolymph pressure.

Some insects that sit and wait for prey construct traps. Antlions (larval Myrmeleontidae, Neuroptera), for instance, dig pits 2–5 cm in diameter with sloping sides in dry sand. The insect then buries itself in the sand at the bottom of the pit with only the head exposed. If an ant walks over the edge of such a pit it has difficulty regaining the top because of the instability of the sides. In addition, the larva, by sharp movements of the head, flicks sand at the ant, causing it to fall to the bottom of the pit and be captured.

Larvae of the fly *Arachnocampa* (Mycetophilidae) are luminescent and use the light to lure prey into a sticky trap consisting of vertical threads of silk studded with sticky drops.

Adult dragonflies are active aerial hunters, pursuing other insects in flight. Their thoracic

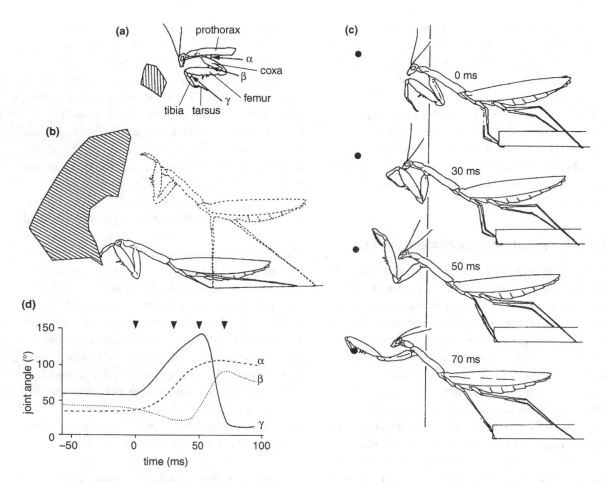

Figure 2.9 Prey capture by a mantis, *Tenodera* (after Corrette, 1990). (a) Area of capture relative to the head. The vertically hatched area shows the region in which prey can be captured by striking with the forelegs without any other movement of the body. (b) Area of capture made possible by changes in orientation coupled with the lunge, shown by oblique hatching. Before the lunge, the insect may occupy any position between the two extremes shown in the diagram. (c) Diagrams of the changes in position associated with capturing prey: 0 ms, start of strike; 30 ms, start of lunge; 70 ms, capture. The vertical line is a reference showing the forward movement of the body during the lunge. The black spot represents the position of the target. (d) Changes in the angles between leg segments (shown in (a)) at times corresponding with the positions shown in (c). Notice the rapid increase in β (extension of the femur) coincident with the decrease in γ (flexion of the tibia to grasp the prey).

segments are rotated forwards ventrally, bringing the legs into an anterior position which facilitates grasping (Fig. 2.11). Tiger beetles (Coleoptera, Cicindelidae) hunt on the ground and have long legs, which enable them to move quickly, and prognathous mouthparts with large mandibles. In the trap-jaw ant, *Odontomachus*, mandible closure is triggered by mechanical stimulation of a hair on the inside of the mandible. This ant hunts with its jaws wide open, and when it encounters prey they snap shut in less than 1 ms. Such a rapid closure could not be achieved by direct muscular action, and it is probable that there is some method of developing tension at the mandibular articulation with the head, analogous to that found in jumping mechanisms.

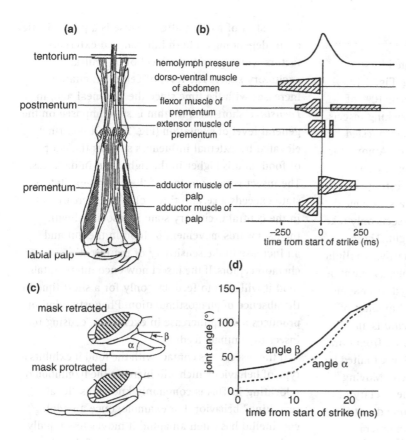

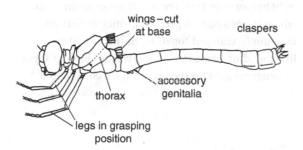

Figure 2.10 Strike of a dragonfly larva (partly after Tanaka and Hisada, 1980). (a) Dorsal view of an extended labial mask with its principal muscles. (b) Timing of activity of the muscles relative to the strike. (c) Lateral view showing the labial mask retracted and extended. (d) Changes in the angles between the prementum and postmentum (angle α) and the postmentum and prothorax (angle β) during the strike. The positions of the angles are shown in (c).

Figure 2.11 Diagram of a male dragonfly to show the oblique development of the thorax bringing the legs into an anterior position, which facilitates grasping the prey.

These active hunters have well-developed eyes, since only vision can give a sufficiently rapid directed response to moving prey. The visual response is usually not specific and the predator will pursue any moving object of suitable size. Thus a dragonfly will turn toward a small stone thrown into the air, and the wasp, *Philanthus*, orients to a variety of moving objects, although it only catches objects having the smell of bees.

Mechanical stimulation is sometimes important in finding prey. *Notonecta* is able to locate prey trapped in the air/water interface as a result of the ripples which radiate from the struggling object. The vibrations are perceived by mechanoreceptors on the swimming legs. Coccinellid larvae preying on aphids only respond to prey on contact.

Many Hymenoptera use a sting to inject venom into their prey. Among predaceous Heteroptera (Hemiptera), the prey is rapidly subdued, apparently by fast-acting lytic processes caused by enzymes from the salivary glands or midgut rather than by specific toxins.

2.3 Regulation of feeding

Most insects eat discrete meals separated by relatively long periods of non-feeding (Fig. 2.12, Table 2.2). A meal may weigh more than 10% of the body weight, and in some blood-sucking insects, such as *Rhodnius*, the quantity of blood ingested greatly exceeds the weight of the insect. Among the chewing insects, cannibalistic Mormon crickets (*Anabrus simplex*) hold the record, ingesting an entire conspecific of the same size in a single meal lasting five minutes. Nectar-feeding insects only take very small meals relative to body weight. Those shown in Table 2.2 use the nectar primarily as a flight fuel. When an ample supply is available, as might be the case with insects feeding on honeydew or some extrafloral nectaries, only a very small amount of time is spent feeding. The blowfly *Lucilia* is an example of this. However, insects feeding from floral nectaries, such as *Vanessa* in Table 2.2, are limited by the small amounts of nectar found there. Moving from flower to flower and reaching the nectary of each one occupies about half of the time devoted to foraging by the butterfly *Vanessa*. Homoptera (Hemiptera) such as aphids and planthoppers that feed on plant phloem or xylem either do not have discrete meals or have exceedingly prolonged meals, feeding almost continuously. Presumably they need to do so because of the very low concentrations of nutrients in their food.

The start of feeding after a pause is a probabilistic event depending on both internal and external factors which govern the level of a "central excitatory state." The level of central excitation increases with the time since the last meal and, in *Locusta*, a short-term rhythm is superimposed on the general level of excitation (Fig. 2.13). It is further elevated by external influences such as the odor of food, and is higher in the light than in darkness. The insect is ready to feed if the central excitatory state exceeds a certain threshold. The increase in the central excitatory state is associated with the backwards movement of food in the gut and an increase in the sensitivity of the insect's chemoreceptors. If the insect now encounters suitable food it will start to feed, but only for a short time in the absence of phagostimulation. Phagostimulation produces a sharp increase in excitation, causing the insect to continue feeding.

If the insect loses contact with the food it exhibits a type of behavior which will increase the likelihood of relocating it. This is commonly known as "local searching" behavior. For example, once an adult coccinellid has eaten an aphid it moves less rapidly and turns more frequently. The result of this change in behavior is to keep the insect close to the point at which it encountered prey and, since aphids are commonly clumped together, to increase the likelihood of encountering more food. Similar types of behavior are observed when a fly consumes a

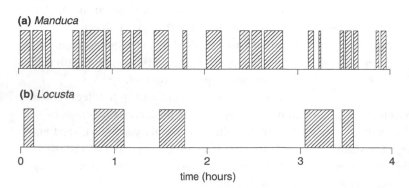

(a) *Manduca*

(b) *Locusta*

0   1   2   3   4

time (hours)

Figure 2.12 Examples of the pattern of feeding of two phytophagous insects feeding on acceptable plants under constant conditions in the laboratory. (a) A caterpillar, *Manduca*, feeding on tobacco (after Reynolds *et al.*, 1986). (b) A locust, *Locusta*, feeding on wheat (after Blaney *et al.*, 1973).

Table 2.2 **Feeding patterns of some insects with different feeding habits. Based on laboratory observations under optimal light and temperature conditions**

Insect	Order	Food	Insect weight (mg)	Meal size (mg)	% body weight	Meal duration (min)	Time between feeding meals	Time %
Locusta (larva)	Orthoptera	Leaves	700	118	17	6–8	70–80 min	7–11
Nilaparvata (female)	Hemiptera	Plant fluids	22.5	–	–	Continuous	–	85
Zelus (female)	Hemiptera	Insect contents	105	~20	~20	130	>24 h	<9
Rhodnius (larva)	Hemiptera	Vertebrate blood	40	300	750	15	days	<1
Manduca (larva)	Lepidoptera	Leaves	3500	80	2	10–25	15–25 min	25–60
Vanessa (adult)	Lepidoptera	Nectar	600	30	5	21	130 min	14[a]
Lucilia (adult)	Diptera	Nectar	25	2	8	0.5–1	25–40 min	1–3.5[b]
Apis (adult, worker)	Hymenoptera	Nectar	90	30	33	30–80[c]	–	48[c]
Apis (adult, worker)	Hymenoptera	Pollen	90	20	22	210[c]	–	12[c]

Notes:
[a] Feeding on flowers in the laboratory – used for energy supply; in other insects listed food is for growth and/or egg development, as well as energy.
[b] Feeding on glucose solution – used for energy supply; in other insects listed food is for growth and/or egg development, as well as energy.
[c] Foraging on flowers in the field. Includes foraging time.

small sugar drop or when a grasshopper loses contact with its food.

Negative feedbacks accrue throughout a meal, leading to meal termination when negative and positive feedbacks balance. A major source of negative feedback is volumetric, coming from stretch receptors on the alimentary canal or in the body wall, which measure the degree of distension of the gut and body. Nectar-feeding insects fill the crop and blood-sucking insects, when feeding on blood, fill the midgut. In grasshoppers, stretch receptors at the anterior end of the gut and on the ileum affect this regulation; the crop of flies is covered by a network of nerves associated with stretch receptors; in *Rhodnius* chordotonal organs in the body wall inhibit further feeding as blood in the midgut causes the abdomen to expand. The axons from these receptors pass to the central nervous system, either directly or via the stomatogastric nervous system. The inputs from these receptors are inhibitory and are believed to function by reducing the level of the central excitatory state below the threshold for feeding.

Other sources of inhibition as a meal continues include declining inputs from mouthpart chemoreceptors responding to phagostimulants (although in grasshoppers the rapid tapping of

(a)

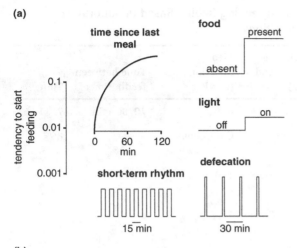

(b)

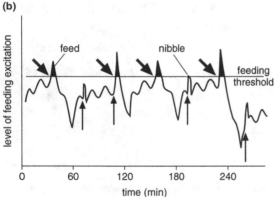

Figure 2.13 Control of the pattern of feeding of a locust. (a) Effects of various factors on the probability that the insect will start to feed. The scale to the left applies to each of the diagrams (after Simpson and Ludlow, 1986). (b) Model, based on real data, showing how the factors in (a) interact to produce the feeding pattern. Feeding (shown in solid black) starts when the level of excitation exceeds the threshold. Small vertical arrows show increases in excitation following defecation; large, oblique arrows above the threshold show the point at which excitation is increased when the insect bites the food (after Simpson, 1995).

mouthpart palps on the food during feeding may help counteract such sensory adaptation), summation of input from chemoreceptors on the mouthparts and perhaps within the alimentary canal responding to deterrent compounds in the food, rapid changes in blood composition throughout the meal and release of neurohormones.

The intervals between meals are very variable. In caterpillars they are commonly of the order of 15–30 minutes; in grasshoppers 1–2 hours (Fig. 2.12). In both caterpillars and grasshoppers, feeding on a protein-rich diet results in extended intermeal intervals, in part as a result of elevated levels of free amino acids in the hemolymph. Adult female mosquitoes and tsetse flies take a blood meal once every few days in relation to the cycle of oogenesis; larval *Rhodnius* only feed once in each larval stage. Consequently, the overall percentage of time spent feeding is usually very low (Table 2.2).

Variation in intermeal duration coupled with changes in meal length result in variation in the total amount of food consumed over a period. In grasshopper nymphs, for example, differences may be considered on three time scales. More food is eaten during the light period than the dark, even when the temperature is constant; food intake reaches a peak in mid-stadium and ceases altogether some time before the molt; and over 50% of the food consumed over the entire developmental period of the insect is eaten during the final stadium (Fig. 2.14). Comparable changes occur in other insects. Further changes may occur during the adult period in relation to oogenesis (see Chapter 3).

Molecular dissection of the pathways regulating the initiation of feeding behavior in *Drosophila* has revealed a role for circuits expressing the gene *hugin*, which encodes prepropeptides that cleave to produce a series of neuropeptides that have sequence homology and functional similarities to some mammalian appetite regulatory peptides. The *hugin* neurons arborize in the subesophageal ganglion, where they intermingle closely with projections from chemosensory neurons on the mouthparts and appear to play a role in the initiation of feeding. It is evident that numerous neuropeptides are involved in the regulation of feeding in *Drosophila* and other insect species, but their mechanisms of action and interactions remain to be discovered. These peptides include FMRFamide-like peptides, FXPLRamide-related peptides, neuropeptide F, short neuropeptide

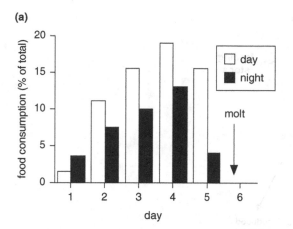

(a)

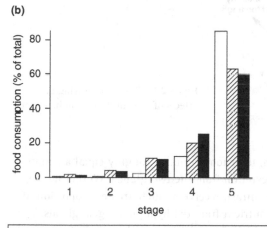

(b)

- □ *Pieris* – holometabolous, phytophagous
- ▨ *Schistocerca* – hemimetabolous, phytophagous
- ■ *Rhodnius* – hemimetabolous, blood-sucking

Figure 2.14 Long-term variation in food intake. (a) Daily variation in the amounts of food consumed by a first-stage larva of a grasshopper, *Schistocerca*. Expressed as percentages of the total amount eaten during the whole of the first stage (data from Chapman and Beerling, 1990). (b) The amount of food consumed by each larval stage of insects with different types of development and feeding habits. Expressed as percentages of the total amounts of food consumed during the whole of larval development (after Waldbauer, 1968).

F, *Drosphila* sex peptide, tachykinin-related peptides and sulfakinins. Additionally, biogenic amines such as serotonin, octopamine and dopamine modulate aspects of feeding behavior.

2.4 Other consequences of feeding

In addition to its primary function of providing nutrients, feeding has other effects on the physiology and behavior of insects. Typically, after a full meal the insect becomes quiescent; it will not feed even in the presence of feeding stimuli, diuresis may occur and food may begin to pass backwards through the gut. In grasshoppers, all these activities are regulated, at least in part, by one or more hormones released from the corpora cardiaca as a consequence of distending the foregut and stimulating the stretch receptors (Fig. 2.15). These are the same receptors that lead to the cessation of feeding.

The reduction in readiness to feed is associated with a reduction in the sensitivity of contact chemoreceptors concerned with feeding. This results partly – in grasshoppers and caterpillars – from high levels of sugars or amino acids in the hemolymph following a meal. In grasshoppers there is also closure of the terminal pores of the contact chemoreceptors. These effects are probably not the immediate cause of the failure to feed, but have the effect of reducing sensory input during a period in which further feeding might reduce the effectiveness with which food from the previous meal is digested and absorbed. The effect of feeding on diuresis is best known in *Rhodnius*, where stimulation of the abdominal stretch receptors leads to the release of diuretic hormone into the hemolymph (Section 18.4).

2.5 Head glands

Associated with each of the gnathal segments (mandibular, maxillary and labial) may be a pair of glands, although they are not usually all present together.

2.5.1 Mandibular, hypopharyngeal and maxillary glands

Mandibular glands are found in Apterygota, Blattodea, Mantodea, Isoptera, Coleoptera and

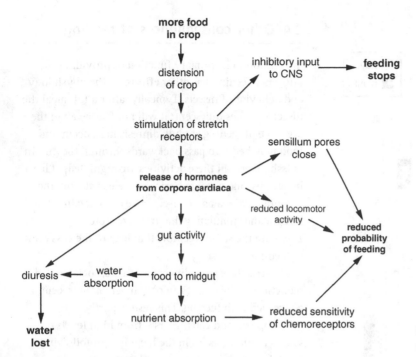

Figure 2.15 Diagram showing the effects of foregut distension in a locust.

Hymenoptera and are usually sac-like structures in the head opening near the bases of the mandibles. They are large in larval Lepidoptera, where they are the functional salivary glands; they are absent from adult Lepidoptera. They are especially important in social Hymenoptera, where they are sources of pheromones (see Section 27.2.7).

Hypopharyngeal glands occur in Hymenoptera and are particularly well developed in worker honey bees. They are vestigial in queens and absent from males. There is one gland on each side of the head consisting of a long, coiled duct with numerous small glandular lobes attached to it. The ducts open at the base of the hypopharynx. These glands produce an invertase as well as components of brood food.

Secretions from the hypopharyngeal and mandibular glands of worker honey bees are fed to larvae and regulate their development into queen or worker bees. Queen larvae receive mainly the secretion of the mandibular glands during their first three days of development, then, for the last two

days, their food contains roughly equal amounts of secretion from the two glands. Worker larvae, by contrast, receive a much greater proportion of the nutrient from the hypopharyngeal glands. The "royal jelly" on which queen larvae are fed contains many different compounds, including large amounts of 10-hydroxy-*trans*-2-decanoic acid, nucleic acids and all the common amino acids. It also contains about ten times more pantothenic acid (a B vitamin) and biopterin than food fed to worker larvae. Sugars, which act as phagostimulants, comprise over 30% of royal jelly; the food of worker larvae has, at first, only about 12% sugar. Consequently, queen larvae eat much more food and grow bigger. Full queen determination, however, relies on elevated levels of a protein, royalactin, in royal jelly. Worker bees feed either type of larva, apparently regulating the food they give according to the type of larva they visit. The consequences of the differences in food on larval development are discussed in Section 15.5.

Maxillary glands are found in Protura, Collembola, Heteroptera (Hemiptera) and some larval Neuroptera and Hymenoptera. They are usually small, opening near the bases of the maxillae, and may be concerned with lubrication of the mouthparts.

2.5.2 Labial glands

The most commonly occurring head glands are the labial glands, which are present in all the major orders of insects except the Coleoptera. In most insects they function as salivary glands.

Structure. The labial glands of most insects are acinous glands (Fig. 2.16), but in Lepidoptera, Diptera and Siphonaptera (fleas) they are tubular, with the ducts swelling to form the terminal glandular parts (Fig. 2.17). Sometimes, as in cockroaches, there is also a salivary reservoir, while in Heteroptera the gland consists of a number of separate lobes (Fig. 2.18).

The cells of the gland and duct are differentiated to perform three functions: the production and secretion of enzymes and other chemicals present in the saliva; movement of water from the hemolymph into the lumen by the active secretion of sodium or potassium; and modification of the fluid as it passes down the salivary duct by resorption of ions and, perhaps, some water. In acinous glands, the central, or zymogen, cells have extensive endoplasmic reticulum and Golgi bodies and probably produce the enzymes. The peripheral, or parietal, cells have an extensive microvillar border to the channel leading to the lumen of the duct. These cells are responsible for the movement of water into the lumen of the gland. In tubular glands both functions appear to be performed by the same cells. The movement of water results from the creation of an osmotic gradient across the cells by H^+–ATPase pumps in the microvillar plasma membrane. The pumps move protons into the lumen, and the protons are then exchanged for sodium or potassium at cation/H^+ antiporters, resulting in a high ionic concentration in the lumen. The duct cells of acinous glands, or the cuboid cells of tubular glands, remove cations from the salivary secretion; those of the cockroach have sodium/potassium pumps in the basal plasma membrane.

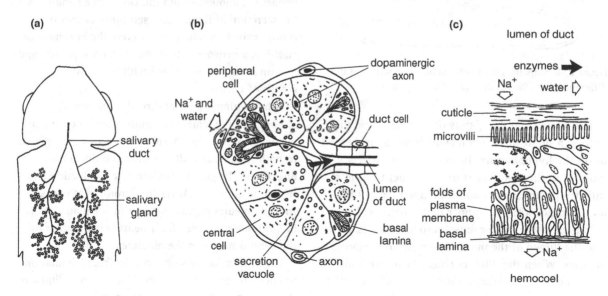

Figure 2.16 Acinous salivary gland. Open arrows show the movement of sodium and water; black arrows show the movements of enzymes. (a) General arrangement in *Locusta*. (b) Section through an acinus of *Nauphoeta* (after Maxwell, 1978). (c) Section through a duct cell of *Schistocerca* (after Kendall, 1969).

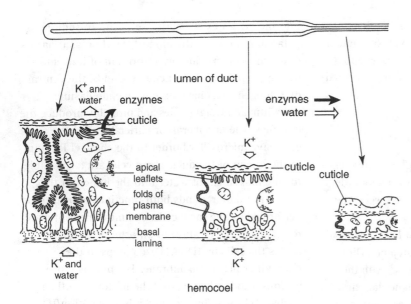

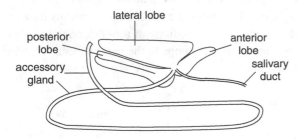

Figure 2.17 Tubular salivary gland of *Calliphora*. At the top is a representation of the gland showing the positions of the cells depicted below. Left: secretory cell; center: cuboid cell; right: duct cell. Open arrows show the movement of potassium and water; black arrows show the movements of enzymes (after Oschman and Berridge, 1970).

Figure 2.18 Salivary glands of a hemipteran showing the different lobes (*Oncopeltus*) (after Miles, 1960).

The ducts from the glands run forwards and unite to form a single median duct which opens just behind or on the hypopharynx. In fluid-feeding insects, muscles in the head insert onto the duct to form a salivary pump. At this point the lower wall of the duct is rigid, while the upper part is flexible. It is pulled up by the dilator muscles so that fluid is drawn into the lumen of the pump. There are no compressor muscles. When the dilators relax, the upper wall springs down by virtue of the elasticity of the cuticle lining the pump and forces saliva out. In some insects, at least, valves ensure the forward flow of saliva.

Control of secretion. Acinous glands are innervated by axons from the subesophageal ganglion and from the stomatogastric system. In both cockroach and locust, one of the innervating neurons produces dopamine, another serotonin. Dopamine stimulates the secretion of fluid, while serotonin causes the central cells to produce and secrete the enzymes. In addition, a network of branches of an octopaminergic neuron is closely associated with the salivary glands of *Locusta*.

Tubular glands are not directly innervated, although in the female mosquito, *Aedes aegypti*, a plexus of nerves closely surrounds part of the gland. In both *Aedes* and *Calliphora*, serotonin, acting directly on the gland, regulates the production and release of saliva. In *Aedes* the serotonin is released from the neural plexus adjacent to the gland, while in *Calliphora* it comes from neurohemal organs on the ventral nerves in the abdomen.

The production of saliva results from stimulation of chemoreceptors on the mouthparts; in *Calliphora* stimulation of the interpseudotracheal pegs is necessary. In most insects production and release occur together, but in insects with a salivary reservoir the two processes may be separately controlled. Production stops when sensory input is removed. In *Calliphora* it takes less than two minutes to clear the serotonin from the hemolymph when feeding stops and the interpseudotracheal pegs are no longer stimulated.

Functions of saliva. Saliva serves to lubricate the mouthparts. More is produced if the food is dry. It also contains enzymes which start digestion of the food. The presence of particular enzymes is related to diet, but an amylase, converting starch to sugar, and an invertase, converting sucrose to glucose and fructose, are commonly present (Table 2.3). In blood-sucking insects, the saliva contains no digestive enzymes, but it does have a remarkable array of components – collectively termed the "sialome" – to aid feeding and overcome the hemostatic responses of the host (Table 2.1).

In a number of insects, specific enzymes are present in the saliva, which facilitate penetration and digestion of the food (Table 2.3). For example, leaf-cutting ants, *Acromyrmex*, have a salivary chitinase that attacks chitin in the fungus on which the insects feed; larval warble flies, *Hypoderma*, which bore into the subcutaneous tissues of cattle, secrete a collagenase which facilitates movement of the larva through the tissues of the host.

Plant-sucking Hemiptera produce two types of saliva, a typical watery saliva carrying enzymes, and another which hardens to form the salivary sheath (Fig. 2.8). The enzymes in watery saliva are produced in the posterior lobe of the glands, and water probably comes from the accessory gland. The watery saliva is produced during penetration of the plant tissues; some of its enzymes facilitate penetration through the middle lamellae of plant cell walls.

Table 2.3 Digestive enzymes in the saliva of insects with different feeding habits

| Insect | Order | Food | Enzymes | | | | |
			Amylase	Invertase	Proteinase	Lipase	Diet-related
Periplaneta	Blattodea	Detritus	+	+	+	+	
Locusta	Orthoptera	Leaves	+	+	−	+	
Oncopeltus	Hemiptera	Seeds	+	+	+	+	
Aphids	Hemiptera	Plant fluids	±	+	±	?	Pectinesterase[a] Polygalacturonidase[a]
Rhodnius	Hemiptera	Vertebrate blood	−	−	−	−	See Table 2.1
Zelus	Hemiptera	Insects	+	?	+	+	
Platymeris	Hemiptera	Insects	?	?	+	−	Hyaluronidase[b]
Hypoderma (larva)	Diptera	Vertebrate tissues	?	?	+	?	Collagenase[c]
Acromyrmex (adult)	Hymenoptera	Fungus	+	+	−	+	Chitinase[d]

Notes:
+: present; −: absent; ±: present in some species, not others; ?: not known.
[a] Digests components of plant cell walls.
[b] Digests insect connective tissue.
[c] Digests vertebrate collagen.
[d] Digests chitin in fungi.

Aphids, for example, have both a pectinesterase and a galacturonidase. In addition, the saliva in some species contains an amylase and a proteinase which contribute to extra-oral digestion of the plant tissue. Amino acids may be present in relatively large amounts in the saliva and it is possible that the aphids excrete unutilized dietary components in this way. The sheath material comes from the anterior and lateral lobes of the gland in the milkweed bug, *Oncopeltus*. It contains a catechol oxidase and a peroxidase which perhaps counter the effects of products produced by plants to inhibit insect feeding.

Male scorpion flies (Mecoptera) have enlarged salivary glands and produce large quantities of saliva, which are eaten by the female during copulation.

Trophallaxis. The mutual or unilateral exchange of alimentary fluid, including saliva, is called trophallaxis. Oral trophallaxis is most common, but anal trophallaxis also occurs. Oral transfer of regurgitated liquid from one adult to another is a common feature of social Hymenoptera, and may have been of importance in the evolution of social behavior. Trophallaxis can result in the rapid transfer of a chemical through a colony. In *Formica fusca*, for example, traces of honey eaten by one worker can be found in every member of the colony 24 hours later. This rapid transfer is especially important in the distribution of pheromones regulating colony structure (see Section 27.2.7). Transfer between adult ants often involves nectar, storage of which in the crop is made possible by a valve-like proventriculus, which regulates the backwards movement of fluid into the midgut. This storage is at its most extreme in honeypot ants, such as *Myrmecocystus*, where some workers are fed until their crops, and therefore their gasters where the crop resides, become enormously distended and their movement is greatly restricted. These "repletes" are fed by other workers when nectar is plentiful, but serve as the source of food during dry periods.

The larvae of wasps, ants and some bees have enlarged salivary glands (Fig. 2.19) which appear to be the source of the fluids they regurgitate. Larvae of vespid wasps produce saliva containing sugars at

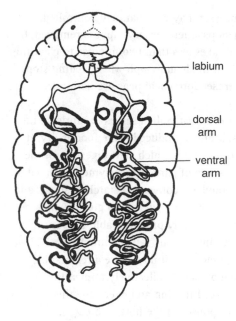

Figure 2.19 Salivary glands of a larval social wasp, *Vespula*. The dorsal arm of each gland is shown black for clarity (after Maschwitz, 1966).

concentrations of 10 mg ml^{-1} or more, principally glucose and trehalose, about 1% proteins and 18–24 amino acids at total concentrations ranging over 25–95 μmol L^{-1}. Larval ants produce salivary secretions from which carbohydrates are absent, but which contain high concentrations of amino acids and proteins, including a number of enzymes. These enzymes may add to the amounts present in the midguts of adult ants, although the significance of the transfer is unclear. The nutrient components of the larval saliva of wasps and ants have nutritional value for the adult insect obtaining them, and may be critically important.

Anal trophallaxis, or proctodeal feeding, is especially important in termites with symbiotic organisms in the rectum. In Kalotermitidae, which have symbiotic flagellates in the rectum, the flagellates are lost when the insect molts and they are regained from the anal fluid of other individuals in the colony. Production of a fluid containing the flagellates, as well as wood fragments, is stimulated when a termite touches the perianal region of another with its antennae. The fluid may also have direct nutritional value.

Silk production. In larval Lepidoptera and Trichoptera the labial glands produce silk. Silk production may begin immediately after hatching in species in which the larvae disperse on silken threads, a process known as ballooning. Other species use silk for larval shelters, and many moths spin extensive silken cocoons in which they pupate. In *Bombyx mori*, the silk moth, the glands hypertrophy at the end of the last larval stage, the increase in size resulting from 18–20 endomitotic divisions of the cells during which the nuclei become complexly branched. The posterior part of the gland, which consists only of some 500 large cells, produces the main silk protein, fibroin, as well as a polypeptide whose function is unknown. The single amino acid, glycine, comprises more than 40% of the fibroin. The central part of the gland, with about 300 cells, produces the proteins that cement the silk threads together. These are called sericins and contain large amounts of serine. The fibroin is molded to form a thread as it passes through the silk press, formed by fusion of the hypopharynx and labium around the salivarium. In Lepidoptera, the ducts from Lyonnet's gland join the ducts of the silk glands. This small gland possibly has a lubricating function as the silk passes through the press.

In Hymenoptera that spin a silken cocoon or, if they are social species, cap their cells with silk, the labial glands function first as salivary glands and change to silk production just before pupation. Weaver ants, *Oecophylla*, use silk from larvae approaching metamorphosis to bind the leaves of their nest together. Psocoptera have two pairs of labial glands: one pair produces saliva and the other produces silk.

A few insects produce silk from other glands. In Neuroptera the Malpighian tubules produce silk in the final larval stage for production of the cocoon. Embiids, which live in tunnels of silk, have silk-producing glands on the fore tarsi.

 ## Summary

- The extraordinary diversity of feeding niches occupied by insects has involved an equivalent diversification of mouthpart structures. The resulting mouthpart designs are all modifications of a common ground plan, involving the same basic "toolkit" of components – labrum, hypopharynx, mandibles, maxillae and labium.

- The mechanics of feeding involves a complex interplay between the physical and chemical properties of the food and the sensory, neural and motor system responsible for coordinating the movements of the mouthparts.

- The timing and size of bouts of feeding (meals) is determined by the physiological state of the insect, chemical and mechanosensory cues from the food and negative feedbacks that accrue during and after a meal.

- The secretions of head glands play many and varied roles in the biology of insects, from determination of caste status in social Hymenoptera to silk production.

- Salivary secretions attain their greatest complexity in blood-feeding insects, where they contain an array of compounds that orchestrate suppression of platelet aggregation, blood coagulation and vasoconstriction, as well as anesthetize the host.

Recommended reading

FEEDING MECHANISMS

Chapman, R. F. and de Boer, G. (eds.) (1995). *Regulatory Mechanisms in Insect Feeding.* New York, NY: Chapman & Hall.

Clissold, F. J. (2007). The biomechanics of chewing and plant fracture: mechanisms and implications. *Advances in Insect Physiology* **34**, 317–372.

REGULATION OF FEEDING

Audsley, N. and Weaver, R. J. (2007). Neuropeptides associated with the regulation of feeding in insects. *General and Comparative Endocrinology* **162**, 93–104.

Bernays, E. A. and Simpson, S. J. (1982). Control of food intake. *Advances in Insect Physiology* **16**, 59–118.

Chapman, R. F. and de Boer, G. (ed.) (1995). *Regulatory Mechanisms in Insect Feeding.* New York, NY: Chapman & Hall.

Melcher, C., Bader, R. and Pankratz, M. J. (2007). Amino acids, taste circuits, and feeding behavior in *Drosophila*: towards understanding the psychology of feeding in flies and man. *Journal of Endocrinology* **192**, 467–472.

Nässel, D. R. and Winther, A. M. E. (2010). *Drosophila* neuropeptides in regulation of physiology and behavior. *Progress in Neurobiology* **92**, 42–104.

Rogers, S. M. and Newland, P. L. (2003). The neurobiology of taste in insects. *Advances in Insect Physiology* **31**, 141–204.

Simpson, S. J. and Raubenheimer, D. (2012). *The Nature of Nutrition: A Unifying Framework from Animal Adaptation to Human Obesity.* Princeton, NJ: Princeton University Press.

HEAD GLANDS AND SALIVA

Kamakura, M. (2011). Royalactin induces queen differentiation in honeybees. *Nature* **473**, 478–483.

Hunt, J. A. and Nalepa, C. A. (eds.) (1994). *Nourishment and Evolution in Insect Societies.* Boulder, CO: Westview Press.

Prudhomme, J.-C., Couble, P., Garel, J.-P. and Daillie, J. (1985). Silk synthesis. In *Comprehensive Insect Physiology, Biochemistry and Pharmacology*, vol. **10**, ed. G. A. Kerkut and L. I. Gilbert, pp. 571–594. Oxford: Pergamon Press.

Ribeiro, J. M. C. and Arcà, B. (2009). From sialomes to the sialoverse: an insight into salivary potion of blood-feeding insects. *Advances in Insect Physiology* **37**, 59–118.

References in figure captions and tables

Blaney, W. M., Chapman, R. F. and Wilson, A. (1973). The pattern of feeding of *Locusta migratoria* (L.) (Orthoptera, Acrididae). *Acrida* **2**, 119–137.

Chapman, R. F. and Beerling, E. A. M. (1990). The pattern of feeding of first instar nymphs of *Schistocerca americana. Physiological Entomology* **15**, 1–12.

Cicero, J. M. (1994). Composite, haustellate mouthparts in netwinged beetle and firefly larvae (Coleoptera, Cantharoidea: Lycidae, Lampyridae). *Journal of Morphology* **219**, 183–192.

Corrette, B. J. (1990). Prey capture in the praying mantis *Tenodera aridifolia sinensis*: coordination of the capture sequence and strike movements. *Journal of Experimental Biology* **148**, 147–180.

Denis, R. (1949). Sous-classe des Aptérygotes. In *Traité de Zoologie*, vol. 9, ed. P.-P. Grassé, pp. 111–275. Paris: Masson et Cie.

Kendall, M. D. (1969). The fine structure of the salivary glands of the desert locust *Schistocerca gregaria* Forskål. *Zeitschrift für Zellforschung und Mikroskopische Anatomie* **98**, 399–420.

Maschwitz, U. (1966). Das Speichelsekret der Wespenlarven und seine biologische Bedeutung. *Zeitschrift für Vergleichende Physiologie* **53**, 228–252.

Maxwell, D. J. (1978). Fine structure of axons associated with the salivary apparatus of the cockroach, *Nauphoeta cinerea*. *Tissue & Cell* **10**, 699–706.

Miles, P. W. (1960). The salivary secretions of a plant-sucking bug, *Oncopeltus fasciatus* (Dall.) (Heteroptera: Lygaeidae): III. Origins in the salivary glands. *Journal of Insect Physiology* **4**, 271–282.

Oschman, J. J. and Berridge, M. J. (1970). Structural and functional aspects of salivary fluid secretion in *Calliphora*. *Tissue & Cell* **2**, 281–310.

Reynolds, S. E., Yeomans, M. R. and Timmins, W. A. (1986). The feeding behaviour of caterpillars (*Manduca sexta*) on tobacco and on artificial diet. *Physiological Entomology* **11**, 39–51.

Ribeiro, J. M. C. (1995). Insect saliva: function, biochemistry and physiology. In *Regulatory Mechanisms in Insect Feeding*, ed. R. F. Chapman and G. de Boer, pp. 74–97. New York, NY: Chapman & Hall.

Ribeiro, J. M. C. and Arcà, B. (2009). From sialomes to the sialoverse: an insight into salivary potion of blood-feeding insects. *Advances in Insect Physiology* **37**, 59–118.

Simpson, S. J. (1995). Regulation of a meal: chewing insects. In *Regulatory Mechanisms in Insect Feeding*, ed. R. F. Chapman and G. de Boer, pp. 137–156. New York, NY: Chapman & Hall.

Simpson, S. J. and Ludlow, A. R. (1986). Why locusts start to feed: a comparison of causal factors. *Animal Behaviour* **34**, 480–496.

Snodgrass, R. E. (1935). *Principles of Insect Morphology*. New York, NY: McGraw-Hill.

Snodgrass, R. E. (1944). *The Feeding Apparatus of Biting and Sucking Insects Affecting Man and Animals*. Washington, DC: Smithsonian Institution.

Tanaka, Y. and Hisada, M. (1980). The hydraulic mechanism of the predatory strike in dragonfly larvae. *Journal of Experimental Biology* **88**, 1–19.

Tjallingii, W. F. and Esch, T. H. (1993). Fine structure of aphid stylet routes in plant tissues in correlation with EPG signals. *Physiological Entomology* **18**, 317–328.

Waldbauer, G. P. (1968). The consumption and utilisation of food by insects. *Advances in Insect Physiology* **5**, 229–288.

3 Alimentary canal, digestion and absorption

REVISED AND UPDATED BY **ANGELA E. DOUGLAS**

INTRODUCTION

The lumen of the alimentary canal is a portion of the external environment over which the insect has great control, and this control is exerted predominantly by the gut epithelium. The principal functions of the alimentary canal are to process ingested food, mostly by chemical modification but also, in some insects, by mechanical disruption, and then to assimilate the products of digestion. The alimentary canal also comes into contact with toxins and microorganisms, including pathogens, associated with the food. Some of these microorganisms become resident in the gut and are beneficial to the insect. It therefore has to combine three capabilities: to mediate chemical transformations involving enzymes capable of degrading the insect's tissues; to provide an absorptive surface for the assimilation of nutrients and water; and to protect against pathogens and noxious compounds. These diverse capabilities are achieved through regional differentiation of the alimentary tract and integration of structure and function at both the molecular and whole-organ levels. Furthermore, the insect alimentary tracts include spectacular examples of adaptation in structural organization and function to different insect diets and habits.

This chapter is divided into four sections. Section 3.1 describes the structural organization of the insect alimentary tract. It is followed by Section 3.2 on digestion and Section 3.3 on absorption of nutrients, ions and water, and concludes with Section 3.4 on the gut as an immunological organ. Throughout the chapter the variation in structure and function with insect phylogeny and diet is addressed.

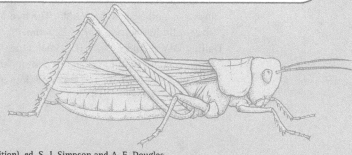

The Insects: Structure and Function (5th edition), ed. S. J. Simpson and A. E. Douglas.
Published by Cambridge University Press. © Cambridge University Press 2013.

3.1 The alimentary canal

3.1.1 General organization

The alimentary canal of insects is divided into
three main regions: the foregut (or stomodeum),
which is ectodermal in origin; the midgut (or
mesenteron), which is endodermal; and the hindgut
(or proctodeum), which again is ectodermal (Fig. 3.1).
Unlike the multilayered gut wall of vertebrates, all
parts of the insect gut consist of a single layer of
epithelial cells, bounded by a basement
membrane and striated muscle.

Since the foregut and hindgut are ectodermal
in origin, the cells secrete cuticle that is continuous
with that covering the outside of the body. The lining
cuticle is known as the intima. It is shed and renewed
at each molt. The midgut is not lined by cuticle, but
most insects secrete a non-cellular layer called the
peritrophic envelope which separates the
epithelial cells from the food.

Usually the gut is a continuous tube
running from the mouth to the anus, with
regional differentiation and various evaginations,
including diverticula in the foregut of some groups,
and ceca in the midgut and the Malpighian tubules
at the junction between midgut and hindgut in most
insects. The connection between the midgut and the
hindgut is occluded in some insects that feed on a
fluid diet containing little or no solid waste
material. This is the case for: some plant-sucking
Heteroptera, where the occlusion is between
different parts of the midgut; larval Neuroptera
which digest their prey extra-orally; and larvae of
social Hymenoptera, which never foul the nest, and
deposit a pellet of fecal matter at the larva-pupa molt.

3.1.2 The foregut

The cells of the foregut are usually flattened
and undifferentiated since they are not generally
involved in absorption or secretion. The cuticular
lining tends to be unsclerotized, consisting only
of endocuticle and epicuticle, but in many insects
sclerotized spines or teeth project from its surface.
The spines commonly point backwards, although
the exact arrangement varies from species to
species and in different parts of the foregut
(Fig. 3.2a), and they are believed to assist with the
backward movement of food toward the midgut.

The foregut is commonly differentiated into
four regions: the pharynx, esophagus, crop and
proventriculus.

The pharynx is concerned with the ingestion and
backwards passage of food and has a well-developed
musculature, which is described further in Section 3.1.6.

The esophagus is usually a simple tube
connecting the pharynx and crop. It is often poorly
defined in hemimetabolous insects, but in many adult
holometabolous insects it is a long, slender tube
running between the flight muscles and back to
the abdomen. Exceptionally, the esophagus of

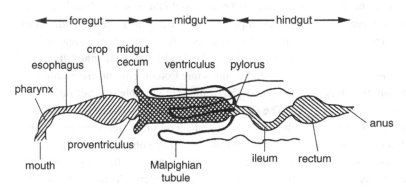

Figure 3.1 Alimentary canal. Diagram
showing the major subdivisions in a
generalized insect.

(a)

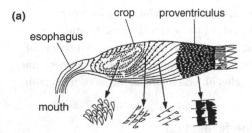

(b)

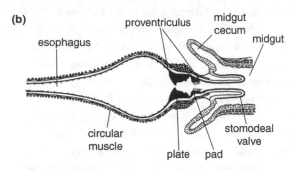

Figure 3.2 Adaptations of the foregut for mechanical disruption of food. (a) Sagittal section through the foregut of a locust showing the pattern of cuticular spines on the intima. Enlargements show details of the spines. In the proventriculus, the spines are replaced by larger sclerotized plates with backward-directed teeth at the posterior edges (after Williams, 1954). (b) Longitudinal section of the foregut of *Periplaneta* showing the development of the proventriculus to form a grinding apparatus (after Snodgrass, 1935).

some insects feeding on highly resinous plants bears single or paired diverticula, in which the resin is stored. Examples include caterpillars of *Myrascia*, which feed on Myrtaceae (Fig. 3.3d) and sawfly larvae in the genus *Neodiprion*, which feed on pines. In *Neodiprion* the diverticula have powerful circular muscles that enable the larva to eject the contents through the mouth in response to attack by natural enemies.

The crop is a storage organ which, in most insects, is an extensible part of the gut immediately following the esophagus (Fig. 3.3a); in adult Diptera and Lepidoptera it is a lateral diverticulum of the esophagus (Fig. 3.3b,c). The walls of the crop are folded longitudinally and transversely. The folds become flattened as the crop is filled, usually

permitting a very large increase in volume. In *Periplaneta*, however, there is little change in volume because when the crop does not contain food it is filled with air. The effectiveness of the crop as a store, especially in fluid-feeding insects, depends on the impermeability of its cuticular lining to hydrophilic molecules.

The proventriculus is very variable in form. It often comprises a simple valve at the proximal end of the midgut, projecting a short distance into the midgut lumen. In grasshoppers (Acrididae) it forms a constriction just proximal to the midgut and can limit the movement of solid food from midgut to foregut, while permitting the movement of liquids in both directions. Ants have a specialized proventriculus that separates the partly digested food in the midgut from food in the crop, which includes material used in trophallaxis. The proventriculus is also specialized in honey bees (Fig. 3.4), allowing them to retain nectar in the crop while passing pollen grains to the midgut, and modified in the Orthoptera, Blattodea and some Coleoptera to form a grinding apparatus comprising strong cuticular plates or teeth (Fig. 3.2b).

3.1.3 The midgut

The tubular part of the midgut is known as the ventriculus, which often exhibits marked regional differentiation at the anatomical level (e.g., in many Heteroptera) and at the cellular level (e.g., in Diptera, including *Drosophila*). Many insects bear ceca, usually at the proximal end of the midgut (Fig. 3.1). For example: crickets (Gryllidae) and some dipteran larvae have two ceca; Acrididae (grasshoppers) have six; the cockroaches and larval Culicidae (Diptera) have eight; and some Coleoptera and Heteroptera have many ceca that are variable in position.

The cells of the midgut are actively involved in the production and secretion of digestive enzymes and the absorption of nutrients. The majority of the cells are tall and columnar, with the membrane on the luminal side bearing microvilli that can increase

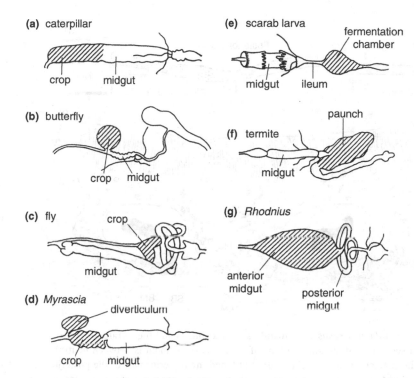

(a) caterpillar

crop midgut

(b) butterfly

crop midgut

(c) fly

crop

midgut

(d) *Myrascia*

diverticulum

crop midgut

(e) scarab larva

fermentation chamber

midgut ileum

(f) termite

paunch

midgut

(g) *Rhodnius*

anterior midgut

posterior midgut

Figure 3.3 Storage in the gut. The different regions in which food or food components are temporarily stored are shown by hatching. (a)–(d) Storage in the foregut. (a) A typical caterpillar; (d) a caterpillar that feeds on resinous plants and stores the resin in the esophageal diverticulum. (e)–(f) Storage in the hindgut. (g) Storage in the midgut.

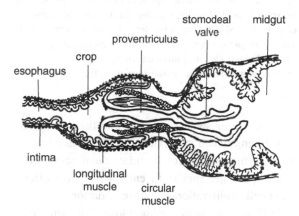

stomodeal valve midgut

proventriculus

crop

esophagus

intima

longitudinal muscle circular muscle

Figure 3.4 Foregut of a worker honey bee in longitudinal section showing the development of the proventriculus. The anteriorly directed part enables the insect to extract pollen grains from nectar in the crop; the posterior part, projecting into the midgut, forms a valve (after Snodgrass, 1956).

the surface area of the cell by up to two orders of magnitude (Fig. 3.5). These cells are called enterocytes (also sometimes termed columnar cells or principal cells), and they mediate the secretion of

digestive enzymes and absorption of nutrients. Other prevalent cell types in the midgut epithelium are enteroendocrine cells, which secrete hormones that regulate the function of the midgut, and intestinal stem cells (Fig. 3.6a), from which the differentiated cells are derived. The stem cells are often located in groups, known as nidi. In some beetles the nidi are restricted to crypts, which are visible as small papillae on the outside of the midgut (Fig. 3.6b).

The midgut of lepidopteran larvae has a further distinctive cell type, called the goblet cell, in which the luminal cell membrane is invaginated to form a flask-shaped cavity (Fig. 3.7a) bearing microvilli. The goblet cells mediate the alkalinization of the midgut via a $K^+/2H^+$ antiporter that is driven by an electrogenic V-ATPase (Fig. 3.7b). Midgut cells with a large central cavity also occur in Ephemeroptera, Plecoptera and Trichoptera, but their function is not fully established. The role of proliferation and

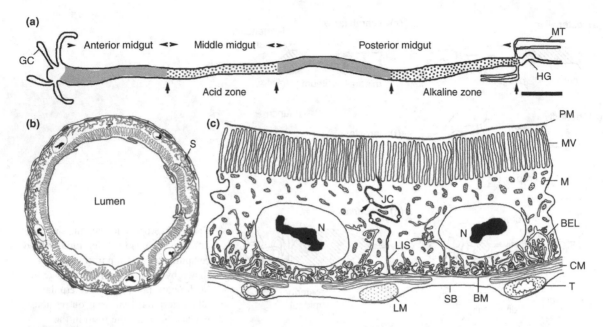

Figure 3.5 Larval midgut epithelium of *Drosophila melanogaster* showing various segments and the intraluminal pH zones. The luminal content of the anterior segment and the anterior part of the posterior segment is between neutral to mild alkalinity (pH >7 and <8); the middle segment is highly acidic (pH <3.0); and the posterior part of the posterior segment is highly alkaline (pH >10). HG, hindgut; MT, Malpighian tubules. (b) Schematic representation of a cross-section of the posterior region of the larval posterior midgut showing the arrangement of peritrophic membrane, epithelial cells, basement membrane, muscle fibers, regenerative cells or stem cells (marked S) and serosal barrier. (c) General organization of two adjacent epithelial cells, lateral dimensions compressed. The arrow denotes very small lateral intercellular spaces. BEL, basal extracellular labyrinth; BM, basement membrane; CM, circular muscle; JC, junctional complex; LM, longitudinal muscle; M, mitochondria; MV, microvilli; PM, peritrophic membrane; SB, serosal barrier; T, tracheole (reproduced from Shanbhag and Tripathi, 2009).

differentiation of the intestinal stem cells in shaping the midgut has been studied in various insects, especially Lepidoptera and *Drosophila melanogaster*. The midgut of adult *Drosophila* is produced at metamorphosis from stem cells present in the larval midgut, by differentiation regulated by EGFR (epidermal growth factor receptor) signaling, and the stem cells continue to divide throughout adult life, generating the enterocytes and enteroendocrine cells, as specified by Notch/Delta and Wingless signaling (Notch promotes differentiation, while expression of *wg* gene in the circular muscles underlying the midgut epithelium sustains the stem cell phenotype). The adult midgut epithelium has a turnover time of 1–2 weeks, with

cells continuously sloughed off into the gut lumen. This phenomenon is promoted by the presence of resident bacteria, which stimulate the production of reactive oxygen species by enterocytes, promoting stem cell proliferation and differentiation.

In *Drosophila* larvae, midgut growth is mediated mostly by increase in cell size. Lepidopteran larvae, in contrast, display bursts of stem cell division just prior to each molt, and the new cells intercalate into pre-existing midgut epithelium. The larval epithelial layer is replaced entirely at metamorphosis by a pupal epithelium (derived, as in *Drosophila*, from larval stem cells) and the pupal epithelium persists through adult life. The midgut epithelium of the larval Lepidoptera undergoes apoptosis and

(a)

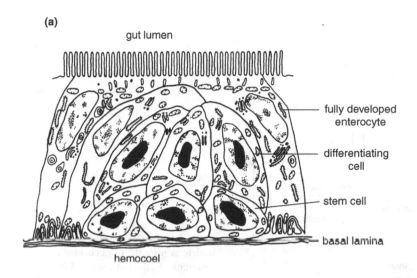

gut lumen

fully developed
enterocyte

differentiating
cell

stem cell

basal lamina

hemocoel

(b)

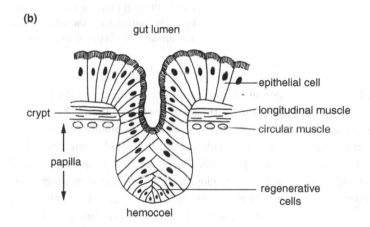

gut lumen

epithelial cell

crypt

longitudinal muscle

circular muscle

papilla

regenerative
cells

hemocoel

Figure 3.6 Stem cells of the midgut.
(a) Group of stem cells (often known
as a nidus) at the base of the midgut
epithelium, showing the differentiation
of enterocytes (after Fain-Maurel *et al.*,
1973). (b) Midgut crypt of a beetle
extending through the muscle layer to
form a papilla (after Snodgrass, 1935).

autophagy, so recycling nutrients, and the remnant
is discarded as meconium at adult eclosion.

The role of enterocytes in enzyme production and
release is evident at the subcellular level. When they
are synthesizing enzymes, the enterocytes have
conspicuous stacks of rough endoplasmic reticulum
and Golgi bodies. In most insects, synthesis appears
to occur at the time of secretion into the gut lumen
so that stores of enzymes do not accumulate in the
cells (see Section 3.2.7). In some insects, including
the blood-feeding dipteran *Stomoxys calcitrans*
(stable fly), enzymes are stored, probably as inactive

precursors (zymogens), in membrane-bound vesicles
in the distal parts of the cells, and then secreted
within a few minutes of feeding.

Different routes of enzyme secretion have been
reported. Membrane-bound vesicles may move to the
periphery of the cell, fuse with the cell membrane and
release their contents into the gut lumen by
exocytosis. The gut principal cells have also been
interpreted to display apocrine secretion, i.e., the
pinching-off of cytoplasmic extensions of the cell,
carrying their contents into the gut lumen and
ultimately releasing them, but this putative process is

(a)

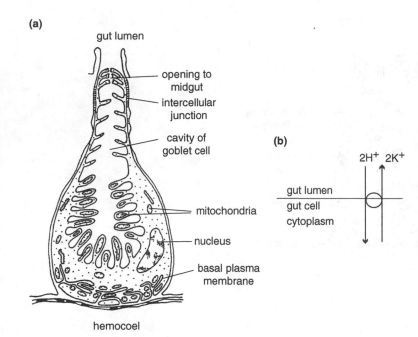

gut lumen

opening to midgut

intercellular junction

cavity of goblet cell

(b)

2H+ 2K+

gut lumen
gut cell
cytoplasm

mitochondria

nucleus

basal plasma membrane

hemocoel

Figure 3.7 The goblet cell and its role in determining the physicochemical conditions in the gut lumen of the caterpillar midgut. (a) A goblet cell (after Cioffi, 1979); (b) ion exchange mediated by K^+/H^+ antiporter at the apical plasma membrane (after Lepier *et al.*, 1994).

contentious because the principal evidence is microscopical and could be an artifact of fixation methods. Intracellular vesicles containing enzymes may also be transported to the microvilli, from which they bud off before releasing their contents. Different enzymes may leave a single cell by different routes, and different methods of secretion may occur in one insect.

3.1.4 The peritrophic envelope

The midgut of most insects is separated from the main volume of the gut lumen by a non-cellular structure called the peritrophic envelope (also known as the peritrophic membrane or peritrophic matrix). The composition and mode of formation of the peritrophic envelope varies among insects, but it generally comprises a proteoglycan gel on a framework of chitin microfibrils.

The chitin chains in the peritrophic envelope are mostly α-chitin, which comprises antiparallel microfibrils and is aggregated into larger bundles that confer considerable tensile strength. The

dominant chitin synthetase in midgut cells of *Manduca sexta* is the product of the *chs2* gene, and is localized to the apex of midgut cell microvilli, consistent with its role in the production of the chitin of the peritrophic envelope. The proteins associated with the envelope vary in the ease with which they can be extracted: some are released by simple buffers or detergent, while others require strong denaturants (e.g., urea) to be isolated or are apparently covalently linked to the chitin microfibrils. Most research has been conducted on one group of proteins, known as the peritrophins, which can be extracted in urea. The peritrophins have chitin-binding domains and six, eight or ten cysteine residues that form disulfide bridges, creating a binding pocket in which hydrophobic residues form hydrogen bonds with the chitin microfibrils. (The peritrophins lack the RR motif characteristic of many chitin-binding proteins of the exoskeleton – see Chapter 16.) Some peritrophins are extensively glycosylated, with the carbohydrate accounting for up to 50% of the total molecular mass. These proteins are called invertebrate intestinal mucins (IIMs), and they include proteins structurally similar to a group of proteins, the mucins, in vertebrate mucus.

The peritrophic envelope is akin to a molecular sieve, in which the pores are aqueous channels crossing the proteoglycan gel. The dimensions of these pores have been investigated by electron microscopy and permeability studies with compounds of known size. For example, studies with fluorescently labeled dextrans of different dimensions suggest that the peritrophic envelope of several larval Lepidoptera is permeable to 21–29 nm diameter molecules, and the envelope of Orthoptera to 24–36 nm molecules. The peritrophic envelope is permeable to inorganic ions, small organic molecules (sugars, amino acids, etc.), peptides and small proteins, but is an effective barrier to the passage of lipids, large proteins and polysaccharides.

The peritrophic envelope of many insects can be categorized as either Type I, produced by the entire length of the midgut, or Type II, produced by specialized tissues called cardia in the anterior midgut, but there is much variation within each type.

The Type I peritrophic envelope is found in Coleoptera, Blattodea, Ephemeroptera, Hymenoptera, Odonata, Orthoptera, Phasmida, larval Lepidoptera and adult Diptera. Production of the peritrophic envelope is not synchronized over the whole midgut but is made up of patches that join together. In some species the microfibrils forming the laminae are assembled at the bases of the microvilli, which form a template for the developing grid of chitin microfibrils (Fig. 3.8a,b), a process known as delamination. In an actively feeding locust a new lamina is produced about every 15 minutes, so the peritrophic envelope becomes multilaminar. The inner layers

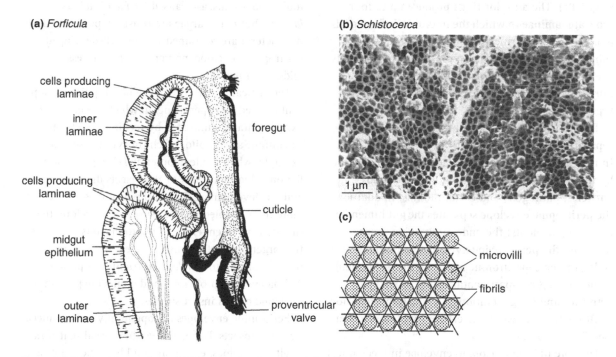

Figure 3.8 Peritrophic envelope. (a) Diagram showing the origin of the peritrophic envelope in cells at the anterior end of the midgut of *Forficula* (after Peters *et al.*, 1979). (b) Scanning electron micrograph of the peritrophic envelope of *Schistocerca*. The envelope has been washed so that only the chitinous lattice remains (after Chapman, 1985). (c) Diagram showing how the fibers which will form the lattice of the peritrophic envelope are laid down round the microvilli.

move back with the food as more laminae are produced, with the result that the envelope has more laminae in the distal midgut than in the proximal midgut. The robustness of the envelope varies from species to species. For example, among the Acrididae (grasshoppers), *Locusta*, which is graminivorous, has a thin, delicate envelope, whereas *Schistocerca*, a polyphagous species, has a much thicker and tougher envelope.

The Type II peritrophic envelope is characteristic of Dermaptera, Isoptera, some Lepidoptera and larval Diptera. In larvae of the mosquito, *Aedes*, the most anterior 8–10 rings of cells in the midgut epithelium (just 300–400 cells) produce the entire peritrophic envelope, which comprises a single lamina about 0.7 μm thick with outer and inner granular layers bounding the chitin microfibrils. The earwig, *Forficula*, in contrast, has several separate rings of cells in the most anterior part of the midgut (Fig. 3.8c). The anterior rings produce up to four separate laminae in which the microfibrils form grids, and the posterior rings produce several more very thin laminae in which the microfibrils are randomly oriented. The complete peritrophic envelope of *Forficula* thus consists of several layers of the two types of lamina.

The peritrophic envelope functions as a barrier, separating the midgut cells from the ingested food. In this way it can protect the gut wall from damage by ingested material, including abrasive food particles, pathogens and certain toxins. In addition, the peritrophic envelope separates the gut lumen into two compartments: the endo-peritrophic space (between the peritrophic envelope and the midgut cells) and ecto-peritrophic space (in the gut lumen), so permitting compartmentalization of enzyme activities and the generation of a countercurrent flow of fluids that increases the efficiency of absorption (see Section 3.3.5).

The role of the peritrophic envelope in mechanical protection appears self-evident, and is illustrated by research on a mutant of *Bombyx mori* that cannot produce a larval peritrophic membrane: the midgut

epithelium of these insects was severely damaged. Chemical protection is illustrated by various phytophagous insects, including *Schistocerca* locusts and some larval Lepidoptera, in which the peritrophic envelope forms an effective barrier against tannins, a major constituent of many plants on which these insects feed. Tannin molecules are much smaller than the pores in the envelopes of these species, and how these molecules are retained by the envelope is not fully resolved. The peritrophic envelope of some insects, however, confers no protection against tannins. For example, the envelope of *Locusta* is permeable to tannic acid; when these insects ingest food containing tannic acid they incur major lesions in the midgut epithelium, resulting in high mortality. The peritrophic envelope also protects against many potential microbial pathogens. For example, young honey bee larvae, in which the envelope is not yet developed, are highly susceptible to American foulbrood – a disease caused by *Paenibacillus larvae* – but older larvae are resistant, partly because the bacteria are restrained by the well-developed peritrophic envelope, preventing their access to the midgut epithelium.

The critical importance of the peritrophic envelope is illustrated by experiments in which the envelope is experimentally eliminated by feeding insects on calcofluor, an inhibitor of chitin synthetase. For example, when the lepidopteran *Trichoplusia ni* was fed on calcofluor, most of the insects died as larvae, and the development time of the surviving larvae was more than doubled. It is no surprise, therefore, that the peritrophic envelope of phytophagous insects is the target of plant defenses. For example, maize plants resistant to *Spodoptera frugiperda* produce a 22 kDa cysteine protease that disrupts the peritrophic envelope of this insect species.

Peritrophic envelopes are apparently not produced by some insects. For example, they are absent from adult mosquitoes, except after a blood meal ingested by females, and unknown among nectar-feeding adult Lepidoptera. Exceptionally among insects, the Hemiptera and Thysanoptera lack a peritrophic envelope but possess, instead, a lipoprotein layer called the perimicrovillar membrane, which lies external to the midgut cells. In most hemipteran groups, the perimicrovillar membrane forms complex folds, so appearing as multiple layers in cross-section, and extends both into the gut lumen and between individual microvilli of the midgut cells. In aphids it is attached to the apex of midgut cells by structures similar to septate junctions. The origin of the perimicrovillar membrane remains to be established definitively, but is proposed to be derived from membrane synthesized in the Golgi vesicles of midgut cells. In all species tested, the perimicrovillar membrane bears α-glucosidase activity.

3.1.5 The hindgut

The hindgut is usually differentiated into three regions: the pylorus, ileum and rectum (Fig. 3.1). As with the foregut, the hindgut is bounded by cuticular intima, but the hindgut intima is generally very thin (~10 µm) and much more permeable than that of the foregut.

The pylorus is the site of entry of the Malpighian tubules to the gut, and the pylorus is, therefore, the zone in which food residue from the midgut and secretions from the Malpighian tubules are mixed. In many insects, the pylorus is bounded by circular muscle and also acts as a sphincter, controlling the passage of food from midgut to hindgut and preventing reflux of the hindgut contents back into the midgut.

In most insects the ileum (also sometimes referred to as the intestinum) is a narrow tube between the pylorus and the rectum. Sometimes the distal part of the ileum is recognizably different and is called the colon. In many insects, only a single cell type is present in the ileum. The cells have extensive folding of the apical plasma membrane with abundant closely associated mitochondria. The basal plasma membrane may also be folded, although the folds are less extensive than those apically (Fig. 3.9). Among insects that have symbiotic microorganisms in their

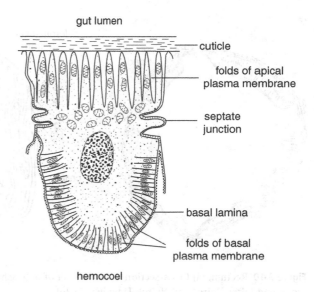

gut lumen

cuticle

folds of apical plasma membrane

septate junction

basal lamina

folds of basal plasma membrane

hemocoel

Figure 3.9 Cell of the ileum (part of the hindgut) of *Schistocerca* (based on Irvine *et al.*, 1988).

hindgut, the ileum is expanded to house the microbes, forming the "paunch" in termites and "fermentation chamber" in larval Scarabaeidae (chafers and dung beetles) (Fig. 3.3e,f). The cuticle lining these chambers is produced into elongate spines or bulbous lobes which probably serve for the attachment of microorganisms. Similar structures are present on the ileal cuticle of cockroaches and crickets that have a bacterial flora.

The rectum is usually an enlarged sac with a thin epithelium except for certain regions, the rectal pads, in which the epithelial cells are columnar. The rectal pads are the primary site of water absorption from the hindgut, and the cuticle in these structures is unsclerotized and much thinner than at other regions of the rectum, e.g., just 1 µm thick in cockroaches. There are usually six rectal pads arranged radially around the rectum (Fig. 3.10a), and they may extend longitudinally along the rectum or they may be papilliform, as in the Diptera. The epithelial cells in each rectal pad are isolated from the hemocoel by sheath cells and a basal cellular layer (Fig. 3.10b) or, less commonly, by a pad of connective tissue, as for example in the blowfly (Calliphoridae). Consistent

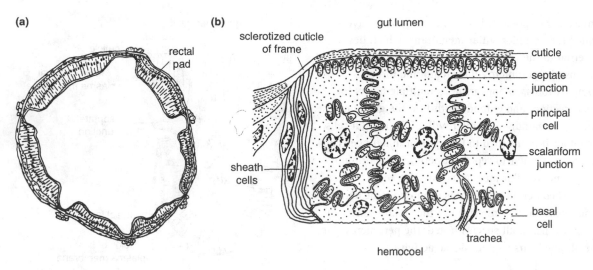

Figure 3.10 Rectum. (a) Cross-section of the rectum of a grasshopper showing the six rectal pads. (b) Section through a rectal pad (after Noirot and Noirot-Timothée, 1976).

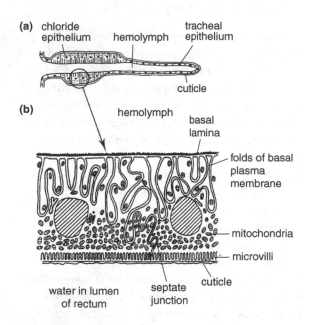

Figure 3.11 Chloride cells in the rectum of a dragonfly larva (Fig. 17.29 shows the position of the gills within the rectum). (a) Section of a rectal gill showing the position of the chloride epithelium. Note that tracheal epithelium refers to the epithelium of the gill through which gas exchange with the tracheae (not shown) occurs. (b) Details of the chloride cells. Notice that the luminal (water) side of the gill is at the bottom (after Schmitz and Komnick, 1976).

with the role of the rectal pads in water absorption, the cell membrane of the epithelial cells is thrown out into folds enclosing mitochondria, and the cells are extensively supplied with tracheoles.

The rectum is the site of respiratory gas exchange in some aquatic insects, notably larvae of the Anisoptera (dragonflies, Odonata), in which the proximal rectum is modified to form a branchial chamber. The gills are repeatedly branching filaments, with the terminal branches just 5–10 μm in diameter, and heavily invested with tracheoles. At the base of each gill are chloride cells, which actively take up Cl⁻ and other inorganic ions from the water as it is pumped in and out of the rectum during respiration. The chloride cells have microvilli at the apical margins beneath a very thin cuticle, and large numbers of mitochondria in the adjacent cytoplasm (Fig. 3.11).

3.1.6 Muscles of the gut

The muscles of the alimentary canal fall into two categories: extrinsic visceral muscles that arise on the body wall and are inserted into the gut, and intrinsic visceral muscles that are associated only with the gut. The intrinsic visceral muscles comprise circular and longitudinal muscles. The circular muscles are not usually inserted into the gut epithelium, but are continuous all around the gut so that their contraction tends to produce longitudinal folding of the epithelium. The longitudinal muscles extend proximally–distally along parts of the alimentary canal. Ultrastructurally, the insect visceral muscles are striated (unlike vertebrate visceral muscles) because their myofilaments are regularly arranged. The extrinsic visceral muscles resemble typical skeletal muscles (see Chapter 10), but the intrinsic muscles have reduced sarcoplasmic reticulum and T-system and have 12 actin filaments around each myosin.

Extrinsic visceral muscles are localized in the foregut and hindgut and generally function as dilators of the gut. Those in the head form pumps which suck fluids into the gut and push food back to the esophagus. The muscles inserting at the pharynx arise on the frons and the tentorium, and those inserting at the cibarium arise on the clypeus. The pharyngeal muscles are antagonized by intrinsic circular muscles. The cibarium has no compressor muscles and compression results from the elasticity of the cuticle lining of the cibarium. Insects with biting and chewing mouthparts generally force food toward the midgut by relatively weakly developed dilator muscles associated with the pharynx, while fluid-feeders utilize one or both of the cibarial and pharyngeal pumps. In most Hemiptera, Thysanoptera and Diptera, the cibarial pump is well developed, while the muscles of the pharyngeal pump are often relatively weakly developed (Fig. 3.12a); in adult mosquitoes, larval *Dytiscus* (Coleoptera) and probably also adult Lepidoptera and bees, both pumps are well developed (Fig. 3.12b,c,d). Exceptionally, in Hemiptera feeding on fluid foods under high hydrostatic pressure (e.g., plant phloem sap), the function of the extrinsic musculature and cibarium is modulated to act not as a pump but as a valve that controls and limits food uptake.

The extrinsic visceral muscles of the hindgut are usually present as dilators of the rectum. They are especially well developed in larval dragonflies that pump water over the gills in the branchial chamber of the rectum (see Fig. 17.29a).

The intrinsic muscles tend to be well developed in the foregut of most insects. The circular muscles are

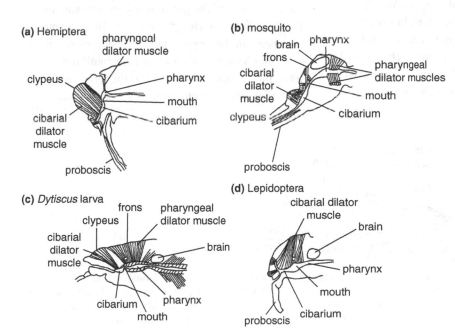

(a) Hemiptera — pharyngeal dilator muscle, clypeus, pharynx, mouth, cibarium, cibarial dilator muscle, proboscis

(b) mosquito — brain, pharynx, frons, cibarial dilator muscle, pharyngeal dilator muscles, mouth, clypeus, cibarium, proboscis

(c) *Dytiscus* larva — frons, clypeus, pharyngeal dilator muscle, cibarial dilator muscle, brain, cibarium, pharynx, mouth

(d) Lepidoptera — cibarial dilator muscle, brain, pharynx, mouth, cibarium, proboscis

Figure 3.12 Foregut muscles in fluid-feeding insects. The cibarial muscles are well developed in all these insects, but the pharyngeal muscles are also important in larval *Dytiscus* and mosquito (after Chapman, 1985).

external to the longitudinal muscles, and they are well developed around the pharynx (acting as antagonist to the extrinsic muscles). They are also significant around the proventriculus where this forms a valve or has a grinding function. Generally, intrinsic muscles are poorly developed in the midgut. The main longitudinal muscles are outside the circular muscles, and appear to extend for the full length of the midgut without any insertions into it; they are inserted anteriorly into the posterior end of the foregut and posteriorly into the anterior end of the hindgut. The intrinsic musculature of the hindgut is complex in caterpillars, and probably in other insects. There are commonly strong muscles associated with the pyloric valve, where the Malpighian tubules join the gut. Other specific arrangements enable the insect to produce discrete fecal pellets from the cylinder of food passing along the gut. Usually the circular muscles are outside the longitudinal muscles.

3.1.7 Innervation of the gut

The muscles of the foregut and anterior midgut are innervated by the stomodeal (or stomatogastric) nervous system. The principal ganglion of this system is the frontal ganglion, which lies on the dorsal wall of the pharynx anterior to the brain (Fig. 3.13). It connects with the tritocerebrum on either side by nerves called the frontal connectives. In Orthoptera and related groups, a median nerve,

the recurrent nerve, extends back from the frontal ganglion beneath the brain to join with the hypocerebral ganglion. This ganglion is closely associated with the corpora cardiaca, and from it a nerve on each side passes to the ingluvial ganglion on the side of the crop. From this ganglion, nerves extend back to the midgut. The cell bodies of most of the motor neurons controlling the muscles are in the ganglia of the stomodeal system, but a few are in the tritocerebrum. Central pattern-generator circuits in these ganglia coordinate the pattern of muscular contraction, which helps move food through the gut and plays a role in molting. The muscles and other tissues of the hindgut are innervated from the terminal abdominal ganglion. Various biogenic amines and neuropeptides are released from neuroendocrine cells innervating the gut, as well as from enteroendocrine cells, and serve to regulate gut motility and digestive enzyme secretion. It has been proposed that some of these neuroendocrine cells are chemoreceptive, detecting nutrients within the gut lumen and triggering release of digestive enzymes via neuropeptides. Multipolar cells, which function as stretch receptors, are present on the outside of the gut in many insects and are involved in monitoring gut fullness. They are most abundant on the foregut, but are also present on the mid- and hindgut. Trichoid and campaniform mechanoreceptors are present in the terminal region of the fermentation chamber of scarab larvae.

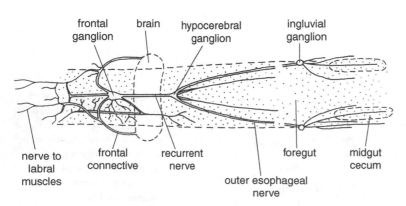

frontal ganglion   brain   hypocerebral ganglion   ingluvial ganglion

nerve to labral muscles   frontal connective   recurrent nerve   outer esophageal nerve   foregut   midgut cecum

Figure 3.13 Stomodeal nervous system of a grasshopper (dorsal view) (based on Anderson and Cochrane, 1978, and R. Allum, unpublished).

3.1.8 The passage of food through the gut

Food is pushed back from the pharynx by the muscles of the pharyngeal or cibarial pump, and subsequently passed along the gut by peristalsis. These movements are controlled from the stomodeal nervous system and are modulated by various neurohormones. For example, when *Locusta* feeds after an interval of several hours without food, the solid food remains in the foregut while the insect is feeding. Backward movement of the food bolus begins shortly after the foregut becomes fully distended. This is at least partly controlled by a hormone from the corpora cardiaca.

In the females of blood-sucking flies that feed on both blood and nectar, the destination of the food is determined by its chemical composition, detected by the receptors on the mouthparts. Stimulation with sugars at high concentrations causes the meal to be directed to the crop, which in flies is a lateral diverticulum of the esophagus (Fig. 3.3c). Stimulation with ATP, ADP or dilute sugar solutions results in the meal going to the midgut. In the blowfly, *Phormia*, and probably also the cockroach, *Periplaneta*, crop emptying is regulated by the osmotic pressure of the hemolymph. At high hemolymph osmotic pressures, the food is passed more slowly to the midgut than at lower pressures. Hemolymph osmotic pressure is affected by that of the food because the sugars are rapidly absorbed. Consequently, meals of concentrated sugars are retained in the crop for longer periods than meals of dilute sugars.

In the midgut of most insects, the passage of food is aided by the peritrophic envelope which, as it moves down the gut, carries the enclosed food with it. Spines on the intima of the hindgut aid the backwards movement of the envelope in insects which possess them. Blood-sucking bugs, such as *Rhodnius* and *Cimex*, and flies, such as mosquitoes and stable flies, which take large, infrequent meals (Table 2.2) store food in the anterior midgut (Fig. 3.3g), where water is absorbed but no digestion occurs. The anterior midgut also acts as a temporary food store in plant-sucking bugs.

The movements of the hindgut are important in the elimination of undigested material. In *Schistocerca* the ileum usually forms an S-bend, and at the point of inflexion the muscles constrict the gut contents. This probably breaks the peritrophic envelope and the food column so that a fecal pellet is formed. Contractions of the distal part of the ileum and the rectum force the pellet out of the anus. In grasshoppers the feces are enclosed in the peritrophic envelope, but in caterpillars the envelope is broken up in the hindgut.

The time taken for food to pass through the gut is very variable. In a grasshopper with continuous access to food, liquids from the food reach the midgut and are absorbed within five minutes of the start of a meal, but solid particles may take 15 minutes or more. The food from one meal has usually left the foregut in about 90 minutes, being pushed back by newly eaten food, and by this time some of the meal is present in the hindgut and may appear in the feces (Fig. 3.14). In the absence of more food, the foregut is completely empty in about five hours, and the midgut is empty after about eight hours. In *Periplaneta* solid food passes through the gut in about 20 hours, but some can still be found in the crop after some days of starvation.

Termites and larval scarab beetles feeding on wood and other intractable plant material retain food fragments for long periods in the hindgut, where their microorganisms digest the cellulose (see Section 3.2.4).

3.2 Digestion

3.2.1 General patterns in insect digestion

Digestion refers to the chemical transformation of large and complex molecules in food to smaller molecules that can be absorbed across the gut wall for insect nutrition. Digestion is mediated principally by enzymes, the profile of which has coevolved with the diet of the insect and the

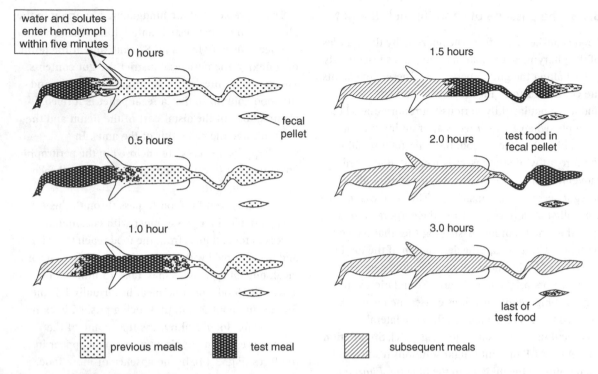

water and solutes enter hemolymph within five minutes

0 hours

0.5 hours

1.0 hour

1.5 hours

2.0 hours

3.0 hours

fecal pellet

test food in fecal pellet

last of test food

previous meals   test meal   subsequent meals

Figure 3.14 Movement of food through the gut of a grasshopper with continuous access to food (after Baines *et al.*, 1973).

physicochemical conditions (especially pH) in the insect gut. All insects have intrinsic digestive enzymes – i.e., enzymes produced by the insect. These may include proteinases, carbohydrases and lipases. In some insects, however, refractory materials, especially plant cell wall material, are not digested by intrinsic enzymes, but broken down by microbial fermentation. The incidence of digestion by intrinsic enzymes versus microbial fermentation cannot be inferred exclusively from the insect diet, beyond the generality that insects on digestively tractable diets, e.g., nectar or animal flesh, use intrinsic enzymes exclusively. Among species using plant material, some digest and assimilate the easily utilized compounds (sugars, starch, protein, etc.) using intrinsic enzymes and void the remainder, including the cellulose. Others utilize the plant cell wall material by one or both of intrinsic degradative enzymes in the midgut and microbial fermentation in the hindgut. The capacity to degrade refractory

plant material is more widespread among insects feeding on wood or detritus (e.g., termites, silverfish and woodroaches) than among insects, such as caterpillars and grasshoppers, feeding on living plant tissue (leaves, grass, etc.).

Digestion by intrinsic enzymes occurs principally in the midgut of most insects (Section 3.2.3). Nevertheless, digestion prior to ingestion by salivary enzymes is important in some groups (Section 3.2.3), and some insects (notably Orthoptera and carabid beetles) have appreciable enzymatic activity in the foregut, presumably mediated by ingested salivary enzymes and the forward movement of enzymes from the midgut. Microbial digestion is almost invariably restricted to the hindgut, distal to insect digestion and absorption. In this way, the insect host can prevent microbial consumption of food items utilizable by the insect host.

The key conditions influencing digestive function are temperature, pH and redox potential.

Enzymatic reactions increase with increasing temperature, from undetectable activity in cold conditions to a maximum, above which the enzyme is progressively denatured and activity declines rapidly to zero (Fig. 3.15). The maximum activity of most insect digestive enzymes in vitro is in the 35–45°C range, even in insects that are intolerant to such high temperatures. In other words, the thermal limit of many insects is determined by factors other than the thermal profile of their digestive enzymes. The activity of digestive enzymes in insects is very low or undetectable at temperatures lower than 10°C.

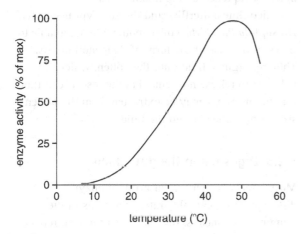

Figure 3.15 Enzyme activity in relation to temperature. The activity of the α-glucosidase of a locust (after Evans and Payne, 1964).

The pH of the gut varies along the length of the insect gut. In many insects the pH range is relatively small – e.g., between 5.8 and 7.3 in the midgut of *Locusta* and other Orthoptera. Insects generally do not have a strongly acidic gut region equivalent to the vertebrate stomach. Exceptionally, the middle midgut region of the cyclorraphous Diptera (including *Drosophila*) is pH 3–4 (Fig. 3.5), and this has been interpreted as an adaptation to lyse bacteria in the diet of these insects. At the opposite extreme, the midgut pH of larval Lepidoptera is always above 8 and can exceed 12 units, at least in part through the uptake of H^+ by goblet cells (see Section 3.1.3 and Fig. 3.7). The hindgut is usually slightly more acid than the midgut, partly due to the secretions of the Malpighian tubules. The digestive enzymes of insects are most active only within a limited range of pH, and many have their optimum at pH 6–7, but in lepidopteran larvae optimal activity occurs at about pH 10 (Fig. 3.16), corresponding with the midgut pH. As well as affecting the activity of the insect's own enzymes, the pH of the gut influences the potentially harmful effects of some ingested compounds. For example, the capacity of the fall webworm, *Hyphantria cunea*, to feed on the cyanogenic leaves of black cherry, *Prunus serotina*, without apparent harm has been attributed to the high pH of the gut contents, inhibiting the conversion of the plant cyanogen to cyanide.

(a) proteolytic enzymes

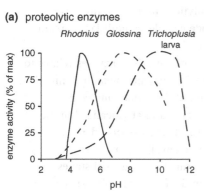

(b) amylase

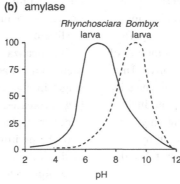

Figure 3.16 Enzyme activity in relation to pH. (a) Proteolytic activity of *Rhodnius*, a blood-sucking bug; *Glossina*, a blood-sucking fly; larva of *Trichoplusia*, a leaf-eating caterpillar. The endopeptidases of *Rhodnius* are cathepsins, those of *Glossina* and *Trichoplusia* are trypsins. (b) Amylase activity of the larva of *Rhynchosciara*, a fly larva feeding on decaying plant material; larva of *Bombyx*, a leaf-eating caterpillar.

Redox potential is a measure of the oxidizing or reducing conditions in a system. A high positive potential indicates strongly oxidizing conditions; a high negative potential indicates strongly reducing conditions. Correlated with the small size of the alimentary tract and aerobic condition of the gut epithelium, the gut lumen of most insects is believed to have a positive redox potential. Some lepidopterans, including *Manduca* larvae and the clothes moth, *Tineola bisselliella*, sustain reducing conditions in the midgut, probably by the secretion of antioxidants, especially ascorbate and reduced glutathione; and the redox potential of the hindgut fermentation chambers are invariably negative, for example ranging from −50 to −230 mV in the paunch of termites, and *ca.* −100 mV in the scarabeid beetle, *Pachnoda ephippiata*.

3.2.2 Extra-oral digestion

For many insects, not all the digestive activity occurs within the gut lumen. The food can be partially or completely digested even before it is ingested, by digestive enzymes secreted in saliva or regurgitated from the midgut onto the food. Digestion before ingestion is termed extra-oral digestion or extra-intestinal digestion.

All heteropteran bugs feeding on solid food display extra-oral digestion. Because these insects are only capable of ingesting liquid food, they inject saliva containing potent digestive enzymes into their food, converting it to liquid, which is then sucked up. For example, when the reduviid bug *Platymeris* injects saliva into a cockroach prey, the cockroach stops struggling at 3–5 seconds and is motionless by 15 seconds as a result of the rapid digestive function of the digestive enzymes in the *Platymeris* saliva. The bug then ingests the liquefied prey. The enzymes in the saliva of heteropterans vary with diet, but include proteases, especially serine proteases, and amylases. Other insects for which extra-oral digestion by salivary enzymes is important include adult Asilidae and Empididae among the Diptera.

The enzymes responsible for extra-oral digestion by predaceous beetles, such as *Dytiscus* and the Lampyridae, and also by larval Neuroptera, are injected into the prey from the biting mandibles with which these insects capture their prey (Fig. 2.5). Beetles have no salivary glands – the secreted enzymes are presumably derived from the midgut. Adult carabids in the tribes Carabini and Cychrini release enzymes onto their prey as they masticate, such that the food is well digested before the fluid contents are ingested.

Proteolytic enzymes persist in the excreta of larval blowflies and so the meat in which they live is partially liquefied before it is ingested.

Heliconiine butterflies and the eucalypt nectar fly, *Drosophila flavohirta*, collect pollen but are unable to ingest it because of the form of their mouthparts. They regurgitate fluid onto the pollen, which is believed to release nutrients that the insects can then imbibe; nutrients may be extracted from the pollen simply by diffusion into the fluid.

3.2.3 Digestion in the gut lumen

Most insects have a similar array of enzymes that digest proteins, carbohydrates and lipids in the midgut. Nevertheless, the enzymes produced reflect the type of food eaten by each species and stage. For example, insects feeding on plant seeds and stored products rich in carbohydrates generally have high amylase activities; insects feeding on protein-rich animal products, such as animal flesh, blood or wool, have high proteolytic activity; and digestive proteases are undetectable in adult Lepidoptera feeding on nectar.

The localization of digestive enzymes relates to their mode of action. Enzymes mediating initial stages in digestion of food macromolecules (e.g., proteases, α-amylase) are localized in the endoperitrophic space and associated with the peritrophic envelope; enzymes mediating the intermediate and final stages of digestion occur in the ectoperitrophic space

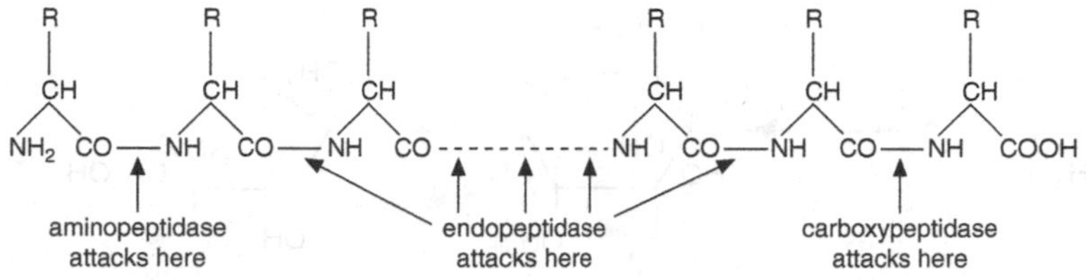

Figure 3.17 Sites of hydrolysis of proteins by different classes of proteases.

(e.g., acetylglucosaminidase) and are associated with the surface of the gut epithelial cells (e.g., maltase, dipeptidases). This pattern is consistent with the permeability of the peritrophic envelope to peptides, amino acids, oligosaccharides, etc., but not to macromolecules such as proteins and polysaccharides. The role of the perimicrovillar membrane of the Hemiptera (which do not produce a peritrophic envelope [Section 3.1.4]) in digestive function is less well understood, beyond the observation that the two chief digestive enzymes, sucrase and aminopeptidase, in the distal midgut of aphids are borne on this membrane.

3.2.4 Protein digestion

The digestion of proteins involves endopeptidases, which attack peptide bonds within the protein molecule, and exopeptidases, which remove the terminal amino acids from the molecule (Fig. 3.17). Within these general categories, the enzymes are classified according to the nature of their active sites and the sites at which they cleave protein molecules.

Three types of endoproteases have been demonstrated in insects: serine proteases (e.g., trypsin and chymotrypsin), which have serine at the active site; cysteine (thiol) proteases, including the B and L families of cathepsins, with cysteine residues at the active site; and aspartic (carboxyl) proteases with aspartate residues at the active sites. The exopeptidases are metalloproteases and fall into two categories: carboxypeptidases that attack peptides from the –COOH end, and aminopeptidases that attack the chain from the –NH$_2$ end.

Insect groups vary in their complement of digestive proteases. Among the Coleoptera, cysteine proteases are dominant. The pH optimum of these enzymes, at 5–7 units, matches the conditions in the midgut of these insects. Larval Lepidoptera and many Diptera tend to have serine proteases, especially trypsins and chymotrypsins, which have a high pH optimum. Trypsin cleaves peptide linkages involving the carboxyl groups of arginine and lysine residues, and chymotrypsin, which is less specific, preferentially cleaving bonds involving the carboxyl groups of tyrosine, phenylalanine and tryptophan. The only insects that definitively have digestive aspartic proteases are the cyclorrhaphous Diptera, including the housefly and drosopholid flies. Their digestive proteases are members of family A1 of aspartic proteases, and characterized by the presence of two catalytic aspartic residues, activity at low pH and inhibition by pepstatin. Sequence analysis demonstrates that they are related to cathepsin D, a major intracellular aspartic protease (also in family A1) localized to the lysosome and widely distributed across the animal kingdom. The dominant digestive cathepsin D of the housefly, *Musca domestica*, (CAD3) lacks two features of the lysosomal cathepsins: an eight-amino acid signature sequence, known as the proline loop, and glycosylation sites. Apart from the cyclorrhaphous Diptera, the only other cathepsin D with a known digestive function is the vertebrate stomach pepsin. Cathepsin Ds have been described in the midgut of various other insects, but they are almost certainly lysosomal, playing a role, for example, in gut remodeling, especially at metamorphosis, and do not contribute to digestion.

Many insects have multiple proteases. For example, 11 proteases, mostly serine proteases, mediate the digestion of the blood meal by the mosquito *Anopheles gambiae*, presumably reflecting

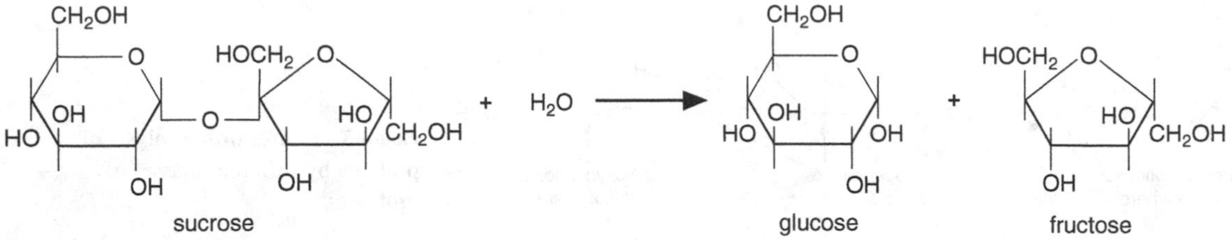

Figure 3.18 Hydrolysis of sucrose by the α-glucosidase, sucrase, to the constituent monosaccharides, glucose and fructose.

the complexity of the ingested proteins, and larvae of *Helicoverpa armigera* (Lepidoptera) have at least nine digestive serine proteases, all detectable in the proteome of the endoperitrophic space – i.e., associated with the food bolus enclosed within the peritrophic envelope. One factor proposed to contribute to the diversity and complexity of digestive proteases in herbivorous insects is the high incidence of proteinase inhibitors (PIs) in plants. PIs against serine, cysteine and aspartic proteases are known; and they all lead to high mortality of susceptible insects. However, many insects adapt readily to chronic ingestion of plant PIs by increased feeding, increased production of susceptible enzymes (so saturating the inhibitory effect) and synthesis of alternative PI-resistant proteinases. In this way, multiple proteinases with different susceptibilities to PIs can confer resistance to PI-based plant defenses. It has also been proposed that cysteine proteases have been recruited for digestive function in insects feeding on legume seeds and other plant tissues that are defended by serine proteinase inhibitors.

Wool, hair and feathers contain large amounts of a single protein, keratin, which has multiple disulfide linkages. To break down the keratin, various insects, including the clothes moth (*Tineola*), carpet beetles (Dermestidae) and numerous biting lice (Ischnocera), exploit a complex mixture of digestive proteases, including a highly active cysteine desulfydrase released into the gut lumen. This enzyme generates hydrogen sulfide from cysteine, contributing to the strong reducing conditions in the gut which, in turn, promote the breaking of disulfide bonds in the keratin.

3.2.5 Carbohydrate digestion

Carbohydrates are generally absorbed as monosaccharides (glucose, fructose, etc.), so disaccharides and polysaccharides in food require digestion. Starch and glycogen, the main storage polysaccharides of plants and animals, respectively, are digested by amylases that hydrolyze α-1,4-glucosidic linkages. There may be separate endo- and exo-amylases, acting on starch internally or terminally. Other common carbohydrases are the α-glucosidases and β-glucosidases, which hydrolyze specific disaccharides and oligosaccharides. Examples include a sucrase (α-glucosidase) (Fig. 3.18) that breaks down sucrose in the midgut of phloem sap-feeding aphids, a β-galactosidase in the digestive juice of the palm weevil, *Rhynchophorus palmarum*, and a maltase (α-glucosidase) in the gut lumen of the larval housefly, *Musca domestica*. These enzymes generally mediate hydrolysis because water is the typical acceptor for the sugar residues, but sugars at high concentrations can act as acceptors, resulting in the formation of oligosaccharides. Thus, the aphid sucrase can mediate the formation of trisaccharide and longer-chain oligosaccharides, which are commonly voided in the honeydew, and the palm weevil galactosidase can, similarly, catalyze transgalactosylation reactions.

A very abundant potential source of carbohydrate for plant-feeding insects is cellulose (Fig. 3.19), which is a β-1,4-glucose polymer. Cellulose is the dominant component of plant cell walls, accounting for 20–30% (by weight) of primary cell walls, and

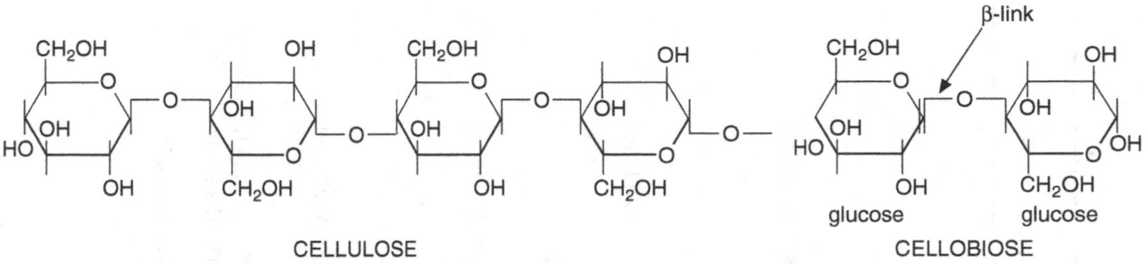

Figure 3.19 Cellulose (β-1,4-glucose polymer) and its disaccharide hydrolysis product, cellobiose.

more than 50% of secondary cell walls, in which the cellulose microfibrils are bounded by hemicelluloses and lignin to form lignocellulose. Cellulose is degraded by the combined action of three sets of enzymes: endoglucanases (EC 3.2.1.4), which attack the bonds between glucose residues within the chain; exoglucanases (EC 3.2.1.74 and 3.2.1.91), which attack bonds near the ends of the cellulose molecule; and β-glucosidases (EC 3.2.1.21), which hydrolyze cellobiose (Fig. 3.19).

Three potential sources of cellulase activity have been proposed for insects: intrinsic; microbial symbionts located in a fermentation chamber; and enzymes ingested with the food. Until comparatively recently, it was believed that animals lack the genetic capacity to produce cellulases, and that all cellulose degradation in insects is mediated by microorganisms. It is now realized, from compelling genomic and enzymological evidence, that various insects and other animals produce digestive cellulases. Even so, the importance of cellulolytic microbial symbionts for lower termites is well established, and the contribution of microbes to cellulose digestion in other insects is increasingly recognized. Ingested cellulases are currently considered not to be generally important, although this source of digestive function may be significant for certain species, including the longicorn beetle, *Monochamus marmorator*.

Unambiguous evidence for intrinsic cellulase activity has been obtained for Blattodea (cockroaches), Isoptera (termites), Coleoptera (longicorn and mustard beetles) and Orthoptera (crickets). In addition, genome sequencing projects have identified candidate genes annotated as cellulases in a total of 27 species from six orders (Table 3.1). Cellulases of glycoside hydrolase family 9 (GH9) have been reported in all insect orders represented, and Coleoptera additionally have GH5 and GH49. The genetic capacity to produce cellulases is not, however, universal among insects. The genomes of all *Drosophila* and mosquito species that have been sequenced to date lack an identifiable cellulase gene.

In the lower termite, *Mastotermes darwiniensis*, intrinsic cellulase activity accounts for approximately 25% of the total activity in the gut (partitioned equally between the midgut and the foregut/salivary glands). The remainder is mediated by protists, specifically trichomonads and hypermastigotes, in the paunch of the hindgut (Fig. 3.3f). The protists engulf fragments of plant material and ferment the cellulose, producing short-chain fatty acids (SCFAs), such as acetate, carbon dioxide and hydrogen. The SCFAs are absorbed in the hindgut and provide a large proportion of the respiratory substrate used by the insect. These organisms may constitute more than 25% of the wet weight of the insect. Because the cellulose degradation takes place within the protist cells, free cellulase activity is negligible in the gut lumen. Woodroaches (family Cryptocercidae), which comprise the cockroach sister group of the termites, also bear cellulolytic hypermastigote protists.

Table 3.1 **List of insects reported with cellulase genes**

Insect species	Common name	Family	GH family	Function	GenBank accession no.
Orthoptera					
Teleogryllus emma	Emma field cricket	Gryllidae	GH9	Endo-β-1,4-glucanase	EU126927
Blattaria					
Polyphaga aegyptiaca	Egyptian desert cockroach	Polyphagidae	GH9	Endo-β-1,4-glucanase	AF220583, AF220584, AF220585
Blattella germanica	German cockroach	Blattellidae	GH9	Endo-β-1,4-glucanase	AF220595
Panesthia angustipennis spadica	Cockroach (no common name)	Blaberidae	GH9	Endo-β-1,4-glucanase	AB438950, AB438951, AB438952
Panesthia cribrata	Australian cockroach	Blaberidae	GH9	Endo-β-1,4-glucanase	AF220596, AF220597
Salganea esakii	Cockroach (no common name)	Blaberidae	GH9	Endo-β-1,4-glucanase	AB438946, AB438947, AB438948
Periplaneta americana	American cockroach	Blattidae	GH9	Endo-β-1,4-glucanase	AF220586, AF220587
Cryptocercus clevelandi	Cockroach	Cryptocercidae	GH9	Endo-β-1,4-glucanase	AF220590, AF220589, AF220588
Mastotermes darwiniensis	Giant northern termite	Mastotermitidae	GH9	Endo-β-1,4-glucanase	AJ511339, AJ511340, AJ511341, AJ511342, AJ511343, AF220593, AF220594
Hodotermopsis sjoestedti (synonym *H. japonica*)	Damp wood termite	Termopsidae	GH9	Endo-β-1,4-glucanase	AF220592, AB118662, AB118794, AB118795, AB118796

Species	Common name	Family		Enzyme	Accession
Neotermes koshunensis	Termite (no common name)	Kalotermitidae	GH9	Endo-β-1,4-glucanase	AB118797, AB118798, AB118799, AF220591
Reticulitermes speratus	Japanese termite	Rhinotermitidae	GH9	Endo-β-1,4-glucanase	AB008778, AB019095
Reticulitermes flavipes	Eastern subterranean termite	Rhinotermitidae	GH9	Endo-β-1,4-glucanase	AY572862
Coptotermes formosanus	Formosan subterranean termite	Rhinotermitidae	GH9	Endo-β-1,4-glucanase	AB058667, AB058669, AB058670, AB058671
Coptotermes acinaciformis	Australian subterranean termite	Rhinotermitidae	GH9	Endo-β-1,4-glucanase	AF336120
Odontotermes formosanus	Black-winged subterranean termite	Termitidae	GH9	Endo-β-1,4-glucanase	AB118800, AB118801, AB118802
Nasutitermes takasagoensis	Termite (no common name)	Termitidae	GH9	Endo-β-1,4-glucanase	AB013272, AB019585, AB118803, AB013273
Nasutitermes walkeri	Termite (no common name)	Termitidae	GH9	Endo-β-1,4-glucanase	AB013273
Sinocapritermes mushae	Termite (no common name)	Termitidae	GH9	Endo-β-1,4-glucanase	AB118804, AB118805, AB118806
Phthiraptera					
Pediculus humanus humanus	Body louse	Pediculidae	GH9	(GH9 homolog)[a]	XM_002426420

Table 3.1 (cont.)

Insect species	Common name	Family	GH family	Function	GenBank accession no.
Hemiptera					
Acyrthosiphon pisum	Pea aphid	Aphididae	GH9	(GH9 homolog)[a]	XM_001944739
Coleoptera					
Tribolium castaneum	Red flour beetle	Tenebrionidae	GH9	(GH9 homolog)[a]	XM_001810641
Apriona germari	Mulberry longhorn beetle	Cerambycidae	GH5	Endo-β-1,4-glucanase	AY771358
			GH45	Endo-β-1,4-glucanase	AY162317, AY162317, AY451326
Psacothea hilaris	Yellow-spotted longicorn beetle	Cerambycidae	GH5	Endo-β-1,4-glucanase	AB080266
Phaedon cochleariae	Mustard beetle	Chrysomelidae	GH45	Endo-β-1,4-glucanase	Y17907
Hymenoptera					
Apis mellifera	Honey bee	Apoidae	GH9	(GH9 homolog)[a]	XM_396791
Nasonia vitripennis	Jewel wasp	Pteromalidae	GH9	(GH9 homolog)[a]	XM_001606404

[a] Predicted from genome sequences of respective insects. Because P-glucosidases are commonly present in insects, information of the relevant sequences is not shown in the present table.

Abbreviations: GH, glycoside hydrolase.

Source: Reproduced from Table 1 of Watanabe and Tokuda (2010).

The microbiota of the higher termites (family Termitidae, accounting for 75% of all termites) comprises entirely bacteria, which have traditionally been considered to lack any cellulolytic capability (with the implication that all cellulose digestion in higher termites is mediated by intrinsic cellulases in the midgut). However, recent data suggest that hindgut bacteria in *Nasutitermes takasagoensis* may have cellulases. Cellulolytic bacteria have also been isolated from the gut of larval bark beetles, *Ips pini* and *Dendroctonus frontalis*, and the longicorn beetle, *Saperda vestita*.

Some insects gain access to cellulose digestion by associating with cellulolytic fungi, which they maintain in their immediate environs (e.g., their nest) and carry with them when dispersing, in pockets known as mycangia, in their cuticle. Termites of the subfamily Macrotermitinae cultivate fungi of the genus *Termitomyces* in fungus gardens, formed from feces containing chewed but only partially digested plant fragments. The fungus grows on this comb, producing cellulolytic enzymes, and the termites then feed on the fungus and the comb. Similarly, the leaf-cutter ants and other Attini maintain the fungus *Attomyces*, which composts plant fragments provided by the ants; and the ants, especially the larvae, feed on the fungal mycelium. In this way, these ants are proxy herbivores, as elegantly revealed by stable isotope analysis (Fig. 3.20). The ambrosia beetles (weevils of the subfamilies Platypodinae and Scolytinae) and larval Cerambycidae and wood wasps (Siricidae) excavate tunnels in trees, in which they cultivate fungi. The fungi digest the lignocellulose, especially of the xylem vessels, and concentrate nitrogen; and, as with the termites and ants, the insects feed on the fungus. All these insects exploit the degradative capacity of a fungal biomass that is considerably greater than could be accommodated within their guts.

The β-1,3-glucanases comprise a further group of carbohydrases abundant in the digestive tract of various insect herbivores and detritivores, including Collembola, Trichoptera, Blattodea, Orthoptera,

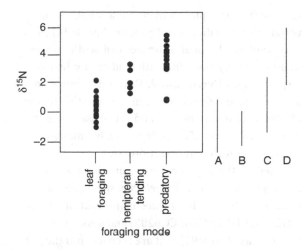

Figure 3.20 Mean $\delta^{15}N$ values (‰) of ant species in a Borneo rainforest, classified according to predominant feeding mode and calibrated against the range of $\delta^{15}N$ of plants (A), hemipterans (B), chewing herbivorous insects (C) and arthropod predators (D). (Redrawn from Fig. 2 of Davidson *et al.*, 2003, omitting species with mixed or uncertain foraging strategies.)

Coleoptera and Diptera. As with cellulases, these enzymes are apparently absent from vertebrates. The β-1,3-glucanases contribute to the degradation of hemicelluloses in plant cell walls and also to β-1,3 and β-1,6-glucans in fungal cell walls. They are, therefore, potentially important in releasing the contents of fungal cells for access by other digestive enzymes, as well as in the degradation of carbohydrate polymers.

3.2.6 Digestion of lipids

The enzymes mediating lipid digestion of insects are lipases (EC 3.1.1.3). These enzymes cleave the carboxylester linkages in triacylglycerols (TAGs), diacylglycerols (DAGs), galactolipids and phospholipids. Lipolysis occurs predominantly in the proximal midgut of most insects, yielding free fatty acids, glycerol, partial acylglycerols and lysophospholipids, which are absorbed into the midgut cells. (Lysophospholipids are phospholipids in which one of the two O-acyl groups at positions sn-1 or sn-2 is replaced by a free alcohol.) As in vertebrates, the digestive lipases of insects belong to two families: the neutral lipases and acidic lipases (represented by the pancreatic and gastric lipases, respectively, of vertebrates). However, there is no direct correspondence between the digestive lipases of insects and vertebrates, and the lipase genes in insects have diversified independently among and even within different insect orders.

Based on the sequence and expression pattern of lipase genes, *Drosophila melanogaster* has been deduced to have at least three major neutral lipases (CG6271, CG6277 and CG6283) and possibly one acidic lipase (CG8093) that are secreted into the gut lumen, where they mediate lipid digestion. Multiple lipases have also been implicated in lipid digestion by other Diptera, including the mosquito *Anopheles gambiae*, as well as Lepidoptera, Coleoptera and Hemiptera. For example, lipid digestion in the midgut of larval *Epiphyas postvittana* (an apple moth, Lepidoptera) is mediated by six functional neutral lipases and three acidic lipases. The sequence of the digestive neutral lipases suggests that they can hydrolyze phospholipids and galactolipids, but not TAGs; and TAG hydrolysis may be mediated exclusively by the acidic lipases. For the phytophagous Lepidoptera, at least, this profile of lipase specificity correlates well with the dominant lipids in leaf tissue, which are the phospholipids and galactolipids of chloroplast thylakoid membranes. Although TAGs are minor lipids in green plant tissues, they are the major lipids in plant seeds and other storage organs and also in animal tissues. The lipases responsible for TAG digestion in seed-feeding and carnivorous insects have yet to be studied in detail.

3.2.7 Variation in enzyme activity

The enzymatic activity in the gut is a crucial factor determining the nutrient availability for the insect, and it is therefore tightly regulated according to the diet composition and nutritional requirements of the insect. The regulation of enzyme activity is particularly apparent in female mosquitoes that feed on both sugar-rich nectar and protein-rich vertebrate blood. Within hours of taking a blood meal, the digestive protease activity in the mosquito midgut increases rapidly, reaching its maximum after about two days and then declining; protease activity is undetectable after a sugar meal.

In many insects, digestive enzyme activity does not increase consistently with increasing substrate availability in the diet. Several factors can contribute to this effect. The first is technical: enzyme activity assays do not generally discriminate between activity available for digestion and stored enzyme in cells. For example, the midguts of starved insects often have very high enzyme activity, even though the digestive activity is negligible. A second important issue is that the digestive function is not invariably regulated to maximize digestion of all ingested foods. When an insect is feeding on a nutritionally unbalanced diet, such that one dietary component is in excess, the enzymes mediating the degradation of that dietary component can be down-regulated. For example, locusts *Locusta migratoria* feeding on diets with excess protein or carbohydrate display reduced activity of digestive α-chymotrypsin and α-amylase, respectively (Fig. 3.21).

Some digestive enzymes are apparently synthesized constitutively, while others are produced in response to the presence of substrate in the midgut (a process known as secretogogue or prandial regulation), or under the control of hormones. The pattern of enzyme production varies among both insect species and enzymes within one insect. For example, *Aedes aegypti* mosquitoes have three trypsin genes expressed in the midgut. The synthesis of two trypsins, known as the late trypsins, is under secretogogue regulation, although the detail is complex. Initial production (within three hours of feeding) is from a preformed mRNA in response to protein in the blood; subsequent production (8–10 hours after feeding) comes from *de novo* trypsin gene expression induced by amino acid products of trypsin-mediated digestion of blood proteins. The

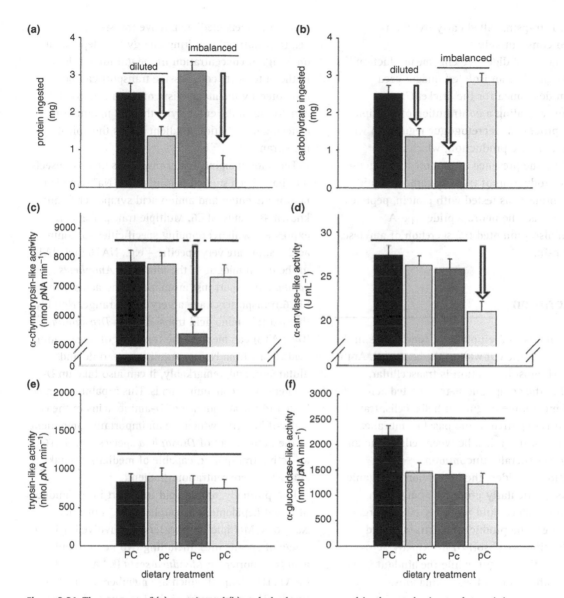

Figure 3.21 The amount of (a) protein and (b) carbohydrate consumed in the meal prior to determining enzyme activity. The four treatment diets contained combinations of high or low concentrations of protein and carbohydrate (both high, PC; both low, pc; high protein, low carbohydrate, Pc; low protein, high carbohydrate, pC). (c)–(e) Total enzyme-like activity of four digestive enzymes (α-chymotrypsin, trypsin, α-amylase, α-glucosidase) is shown in the foregut contents, where activity levels for all enzymes were greatest and where the majority of digestion occurs in locusts. (c) α-chymotrypsin-like activity and (d) α-amylase-like activity were strongly down-regulated only when their nutrient substrate was present in the diet in excess relative to the other macronutrient (i.e., on Pc and pC for α-chymotrypsin and α-amylase, respectively). Treatment diet did not affect the activity of (e) trypsin-like activity or (f) α-glucosidase-like activity. The down arrows indicate a significant difference ($p < 0.05$) and bars with a line across the top are not statistically different. (c–f) Values are ANCOVA-adjusted (GIT weight) means ± s.e. (c) $p = 0.010$; (d) $p = 0.039$; (e) $p = 0.989$; (f) $p = 0.308$. Reproduced from figure 1 of Clissold *et al.* (2010).

other midgut trypsin, called early trypsin, is synthesized constitutively.

The regulation of digestive enzyme production has been investigated in an isolated midgut ceca preparation developed for the cricket *Gryllus bimaculatus*, revealing a contribution of multiple regulatory processes. Secretogogue regulation was evident for amylase production, which was stimulated by the presence of maltose and glucose, but not for production of trypsin, carboxypeptidase or aminopeptidase, as tested with protein, peptides and amino acids. The neuropeptide type-A allatostatin also promoted the secretion of amylase, but not trypsin.

3.3 Absorption

Absorption refers to the uptake of compounds in the gut lumen across the gut wall to the hemocoel. At the cellular level, most absorption is transcellular, meaning that the compound is transported across the apical cell membrane of gut epithelial cells, transits the cell and is exported at the basal membrane. Paracellular transport, i.e., between cells of the gut epithelium, is generally uncommon.

This section considers the absorption of organic compounds, particularly products of digestion (sugars, amino acids, lipid products, etc.), inorganic ions and water. The products of extra-oral and midgut digestion are absorbed principally across enterocytes of the midgut, while the hindgut is the main site of absorption of water and ions.

3.3.1 Transport of amino acids and protein

Most research on the absorption of sugars and amino acids has focused on their passage across the apical membrane of enterocytes. Multiple sugar and amino acid transporters have been identified in various insects. Some transporters mediate active transport, i.e., transport dependent on energy to move a substance against its concentration gradient; others mediate passive transport, i.e., transport not requiring energy and dependent on a higher concentration in the gut lumen than in the enterocyte cell. Passive transport can be promoted by the metabolism of the transported compound in the enterocyte, thereby maintaining a steep concentration gradient across the apical membrane.

The midgut amino acid transporters in most insects that have been studied belong to the Na^+-coupled neurotransmitter and amino acid symporter family, known as family SLC6. Multiple transporters are expressed, with overlapping specificities for amino acids. Some are very specific – e.g., NAT6 and NAT8 in the distal midgut of the mosquito *Anopheles gambiae* transport just aromatic amino acids. Other SLC6 transporters have a very broad range. Notably, the neutral amino acid transporter in *Drosophila* (DmNAT6) can mediate the transport of most amino acids apart from lysine, arginine, aspartate and glutamate; and, remarkably, it can also take up D-isomers of several amino acids. This capability can be linked to the abundance of D-amino acids in the cell walls of bacteria, which are an important component of the natural diet of *Drosophila* species. DmNAT6 is an active transporter, capable of mediating uptake against the concentration gradient.

Exceptionally, amino acid transport in the midgut of larval Lepidoptera is coupled to K^+ ions, not Na^+ ions. Multiple transporters are involved, with a range of specificities, including two neutral amino acid transporters in *Manduca sexta* (KAAT1 and CAATCH1), both of which are members of the SL6 family. The cotransport of the K^+ ions and amino acid into enterocyte cells is coupled to the ATPase-dependent extrusion of K^+ ions from adjacent goblet cells, which play a crucial role in alkalinization of the gut lumen (Fig. 3.7). The coupled functions of electrogenic K^+ transport and K^+/amino acid uptake are mediated by different cells, presumably because the high EMF generated by the goblet cells could compromise the function of the SL6 and other transporters.

Amino acid transporters are also expressed in the apical membrane of the insect hindgut, where they mediate the uptake of amino acids in the primary urine produced in the Malpighian tubules (see Chapter 18). For example, glycine, serine, alanine and threonine are actively reabsorbed into the cells of the rectal pads of the locust by an Na^+-cotransporter of the SLC6 family. Proline is also taken up, and is a major respiratory substrate of rectal cells.

Amino acid transport across the basolateral membrane of gut epithelial cells is not well understood, but active transport is not generally involved. Both amino acid uniporters and amino acid exchangers have been described.

There is also evidence for the movement of proteins across the gut wall to the hemolymph. For example, the hemolymph of blood-feeding insects, including tsetse fly, mosquitoes and fleas, routinely contain antibodies derived from the vertebrate blood on which they feed; and the hemolymph of the silkworm, *Bombyx mori*, has active urease proteins derived from their diet of mulberry leaves. It is unclear whether proteins are transported by the transcellular or paracellular route; it may vary with protein and insect species. There is unambiguous evidence that absorbed proteins are cleared from the hemocoel over a period of hours, mediated by degradation and transfer to other organs – e.g., fat body, ovaries. It is uncertain whether these proteins have any nutritional value to the insect.

3.3.2 Transport of sugars

Carbohydrates are absorbed mainly as monosaccharide sugars across the midgut epithelium, but the molecular basis of sugar transport has been little-studied. There is a general expectation of parallels with glucose absorption in the mammalian intestine, in which transport is transcellular and mediated by a Na^+/glucose cotransporter SGLT1 (a member of the Na^+/solute symporter family) and two facilitative glucose transporters, GLUT2 and GLUT5 (both members of the major facilitator superfamily,

MFS). Consistent with this expectation, glucose transport across the midgut of the hymenopteran parasite *Aphidius ervi* is mediated by a SGLT1-like transporter on the apical membrane, together with a GLUT2-like transporter on both the apical and basolateral membranes of the enterocytes; and a second passive transporter similar to GLUT5 is implicated in fructose uptake. This condition is not, however, universal among insects. For example, genome annotation of the pea aphid, *Acyrthosiphon pisum*, revealed no Na^+/solute symporter with plausible specificity for sugars, but 29 candidate sugar transporters in the MFS family, equivalent to GLUT. These included an abundantly expressed gene *ApSt3*, a hexose uniporter with specificity for glucose and fructose in the distal midgut. Aphids presumably do not need to take up sugars against the concentration gradient because their diet of plant phloem sap is sugar-rich, and a concentration gradient from gut lumen to epithelial cell and hemocoel is maintained by the excess sugar in the gut lumen. The metabolism of the absorbed sugars, both as a respiratory fuel and in trehalose and glycogen synthesis, may also promote sugar absorption.

3.3.3 Transport of lipids and related compounds

The products of lipid digestion include free fatty acids, glycerol, mono- and diacylglycerols and lysophospholipids. These compounds can move relatively freely across membranes, and they are absorbed principally across the midgut epithelium, although absorption in the foregut, e.g., the crop of the cockroach, *Periplaneta americana*, can also occur.

Two factors promote lipid absorption. The first is emulsification, which assists their diffusion across the aqueous boundary layer to contact the apical membrane of the gut epithelial cells. Insects lack the bile salts that are crucial for emulsification in the vertebrate gut (this, incidentally, is why the insect-neutral lipases do not have a colipase cofactor, which is required by the pancreatic lipase of vertebrates to hydrolyze the bile salt-covered lipid in the vertebrate gut). Instead, emulsification in the insect gut is promoted by the formation of fatty acid–amino acid complexes and glycolipid complexes, and the formation of micelles of fatty acids and lysophospholipids.

The second factor promoting absorption is the resynthesis of lipids from absorbed compounds in the enterocytes. This metabolic activity maintains a steep concentration gradient, promoting uptake. In all insects studied to date the enterocytes synthesize 1,2-diacylglycerols (DAGs), triacylglycerols (TAGs) and phospholipids. Lipids can also be generated in the enterocytes from absorbed glucose, galactose or other sugars. Lipids are exported from enterocytes to the hemolymph predominantly as DAGs, and are transported to other tissues bound to the protein lipophorin. Lipophorin has also been implicated in the transport of hydrocarbons, carotenoids, sterols and phospholipids. Sterols appear to be absorbed unchanged, but, in some caterpillars, sterols are esterified in the gut cells.

3.3.4 Absorption of inorganic ions

For sodium to move from the gut lumen into the hemolymph of a locust, it must move against the electrochemical gradient (Fig. 3.22). Energy for this active movement is provided by V-ATPase pumps in the apical plasma membranes of the anterior ceca, and by major sodium–potassium exchange pumps in the basal plasma membranes of the rectal cells. Potassium, on the other hand, probably moves passively down its electrochemical gradient into the hemocoel. The electrochemical gradients for chloride and calcium ions are less extreme. In the rectum, chloride ions are actively removed from the lumen by pumps in the apical membranes of the cells and pass passively from the cells to the hemolymph. Calcium is actively moved from the gut lumen to the hemocoel by cells in the midgut.

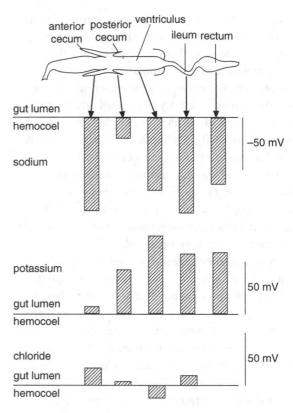

Figure 3.22 Electrochemical gradients across the gut epithelium of a locust. The insects were recently fed. Bars extending into the lumen (upwards) indicate a positive gradient from gut to hemolymph – diffusion into the hemocoel will occur. Bars extending into the hemocoel indicate a negative gradient from gut to hemocoel – passive diffusion into the hemocoel will not occur (after Dow, 1981).

3.3.5 Absorption of water

There are two major water-absorbing zones in the insect gut, one in the midgut, where water is absorbed from the food, and the other in the rectum, where water is absorbed from the feces before they exit the body. Absorption in the rectum is a key component of the regulation of body water. Water absorption in the midgut serves to concentrate the food, enhancing both the efficiency of digestion and the maintenance of concentration gradients favorable to the absorption of nutrients across the wall of the gut. It also creates water flows which are important in nutrient absorption and perhaps enzyme conservation. Specifically, absorption in the proximal midgut, especially the midgut ceca (where these structures occur) drives the forward flow of water in the endoperitrophic space, carrying forward the products of digestion. The water may be derived directly from the food, as in well-fed grasshoppers, or drawn forwards as it leaves the Malpighian tubules, as in cockroaches. In larval Diptera, and perhaps in some other insects, the forward movement occurs in the ectoperitrophic space, while fluid in the endoperitrophic space moves backwards. This countercurrent probably improves the efficiency of utilization of the digested materials in the gut.

Water absorption depends on the establishment of an osmotic gradient across the epithelium, such that water passes from a zone of low solute concentration to high solute concentration. In many insects the flux of water from gut lumen to hemolymph is based on the accumulation of Na^+ or K^+ ions in the extracellular space between the epithelial cells, which has only a few openings to the hemolymph. Water is thus drawn through, or between the cells from the gut lumen into the extracellular space. This creates a hydrostatic pressure which forces water out through the openings into the hemolymph (Fig. 3.23a). This process is energetically very demanding, and is supported by large numbers of mitochondria close to the cell membrane.

Water absorption from the midgut often occurs in localized areas. In cockroaches, grasshoppers and the larvae of some flies (mosquitoes and Sciaridae), it occurs in the midgut ceca, while in blood-sucking insects water is removed from the stored meal in the anterior midgut. Deep invaginations of the basal plasma membrane, which may extend more than halfway toward the apex of the cell, are closely associated with large numbers of mitochondria. These provide the energy by which K^+ (or Na^+ in the case of *Rhodnius*) is actively pumped into the

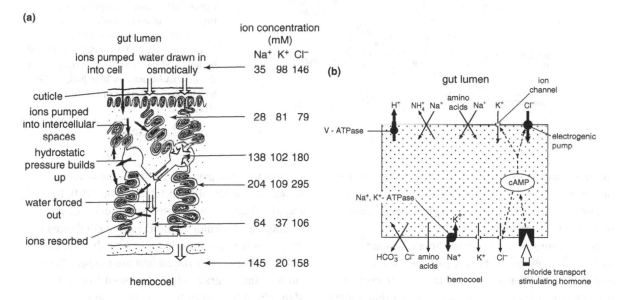

Figure 3.23 Absorption of inorganic ions from the rectum of a grasshopper. (a) Diagram of the principal rectal cells showing the ionic concentrations measured in one experiment (after Phillips *et al.*, 1986). (b) Diagram showing the different pumps and channels involved in the movement of ions from the rectal lumen to the hemocoel (after Phillips and Audsley, 1995).

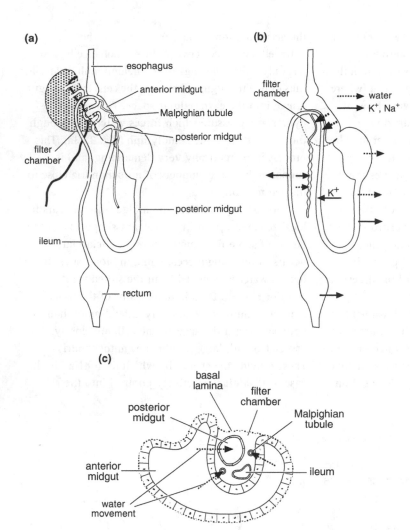

(a)

esophagus
anterior midgut
Malpighian tubule
posterior midgut
filter chamber
posterior midgut
ileum
rectum

(b)

filter chamber
······▶ water
——▶ K⁺, Na⁺
K⁺

(c)

basal lamina
filter chamber
posterior midgut
Malpighian tubule
anterior midgut
ileum
water movement

Figure 3.24 Water absorption from the midgut of a cercopid. (a) General arrangement of the gut showing the filter chamber. (b) Postulated movement of salts and water across the wall of the gut (after Cheung and Marshall, 1973). (c) Transverse section of the filter chamber.

intercellular spaces to create a high osmotic gradient between the gut lumen and the intercellular spaces.

The Hemiptera that feed on xylem sap of plants must ingest large volumes because the concentrations of nutrients in these fluids are very low. Modifications of the gut provide for the rapid elimination of the excess water taken in. In cercopids, for example, the anterior midgut forms a thin-walled bladder that wraps around the posterior midgut and the proximal ends of the Malpighian tubules. This arrangement, which is called a filter chamber,

enables water to pass directly from the anterior midgut to the Malpighian tubules (Fig. 3.24). In this way the food is concentrated and dilution of the hemolymph is avoided; and the urine eliminated by the insect is hypo-osmotic to the hemolymph. In this instance, the cells of the midgut are unspecialized, but an osmotic gradient is established from the Malpighian tubules to the midgut. In some bryocorine Miridae the anterior midgut makes contact with a large accessory salivary gland. After feeding, a clear fluid is exuded from the mouthparts, suggesting that water is withdrawn from the midgut directly to the salivary glands and then eliminated via the mouthparts.

Water absorption from blood stored in the anterior midgut of *Rhodnius* is dependent on the active movement of sodium into the hemocoel. It is believed that, in this case, Na^+ ions are transported out of the cell by Na^+/K^+ exchange pumps in the basal plasma membrane.

In terrestrial and saltwater insects water is absorbed from the rectum. Here, as in the midgut, V-ATPase in the apical plasma membrane probably provides the energy that drives the inward movement of ions and amino acids, principally chloride and proline in the locust (Fig. 3.23b). Water moves across the epithelium due to the increased osmotic pressure. Chloride moves passively into the lateral intercellular spaces and sodium is pumped out by sodium/potassium ATPase so that water is drawn osmotically into the spaces. Hydrostatic pressure forces the solution out to the hemolymph, but ions are reabsorbed as the fluid passes out from between the principal cells. This permits the recycling of the ions and the maintenance of absorption.

3.4 The alimentary tract as an immunological organ

The alimentary tract is one of the most immunologically active organs of the insect, along with the fat body. In most insects, the alimentary tract plays a central role in the response to the many potential pathogens that gain entry to insects via the oral route, and it is also the habitat for a community of resident microorganisms. The immunological response of the alimentary tract is important for both protection against pathogens and management of the resident microbiota.

Midgut immunity has been investigated in detail in *Drosophila melanogaster*. Two complementary inducible immunological responses are important:

the production of reactive oxygen species (ROS) and anti-microbial peptides (AMPs).

ROS are produced by a dual oxidase (DUOX) with two enzymatic activities: NADPH oxidase, which mediates the NADPH-dependent reduction of oxygen to generate superoxide ions, which can dismutate spontaneously to form hydrogen peroxide (H_2O_2); and the peroxidase activity which combines hydrogen peroxide with chloride ions to form hypochlorous acid (HClO). HClO is a very potent microbicide (it is the active component of household bleach). There is genetic evidence that the peroxidase activity of DUOX is absolutely required to protect the insect gut against microbial invasion. ROS and HClO are cytotoxic, damaging DNA, proteins and lipids, and the DUOX system is tightly regulated by complex signaling cascades to minimize autotoxicity. In addition, the ROS are cleared by a catalase (immune-related catalase, IRC). Flies with genetically reduced IRC activity ("knockdown flies") suffer high mortality when infected by pathogens that are not lethal to wild-type flies. It is believed that the high gut ROS levels in IRC-knockdown flies cause redox imbalance and cell damage.

AMPs are amphipathic peptides, 12–50 amino acids in length, that partition readily into membranes and kill microorganisms by membrane permeabilization or via cytoplasmic targets, e.g., inhibition of DNA, RNA or protein synthesis. The *Drosophila* gut cells secrete various AMPs, including drosomycin and diptericin, that are active principally against Gram-negative bacteria. This branch of the local immune response is mediated by the recognition of a major constituent of the bacterial cell wall, peptidoglycan, with signaling via the IMD (immune deficiency) pathway. Expression of AMP genes in the healthy insect is dampened through negative regulation by the transcription factor Caudal, so protecting the resident microbiota from AMP attack. Knockdown of *caudal* gene expression results in altered composition of the microbiota, which becomes dominated by species that are deleterious to the insect. These data illustrate how the amount and composition of AMPs can function to regulate the composition of the resident microbiota. It appears that the insect gut epithelium does not produce AMPs regulated by the Toll signaling pathway that are active principally against Gram-positive bacteria or fungi.

Summary

- The alimentary tract comprises three regions: the foregut, midgut and hindgut. The foregut and hindgut are bounded by a chitinous extension of the exoskeleton, and the midgut of most insects bears a peritrophic envelope, comprising proteoglycans on a chitinous framework.

- The foregut of many insects includes a crop, where food is stored. In some insects, the proventriculus of the foregut is modified for mechanical grinding of food particles.

- The midgut is the principal site of digestion and absorption, mediated by the enterocytes, which are the most abundant cells of the midgut epithelium. Digestive enzymes include proteases, carbohydrases (including amylases, and in some insects cellulases) and lipases. Transporters mediating active transport (against the concentration gradient) and passive transport (down the concentration gradient) of sugars and amino acids are known. The uptake of fatty acids, acylglycerols and other products of lipid digestion is aided by emulsification. The midgut is immunologically active, displaying the inducible production of reactive oxygen species and anti-microbial peptides in response to potential pathogens.

- The hindgut includes the ileum (intestinum) and rectum, and mediates the uptake of inorganic ions, water and other solutes derived from the Malpighian tubules. The ileum is also the site of a fermentation chamber bearing cellulolytic microorganisms in some insects, such as termites.

Recommended reading

Clissold, F. J., Tedder, B. J., Conigrave, A. D. and Simpson, S. J. (2010). The gastrointestinal tract as a nutrient-balancing organ. *Proceedings of the Royal Society of London B* **277**, 1751–1759.

Hegedus, D., Erlandson, M., Gillott, C. and Toprak, U. (2009). New insights into peritrophic matrix synthesis, architecture, and function. *Annual Review of Entomology* **54**, 285–302.

Horne, I., Haritos, V. S. and Oakeshott, J. G. (2009). Comparative and functional genomics of lipases in holometabolous insects. *Insect Biochemistry and Molecular Biology* **39**, 547–567.

Jeffers, L. A. and Roe, R. M. (2008). The movement of proteins across the insect and tick digestive system. *Journal of Insect Physiology* 54, 319–332.

Ohlstein, B. and Spradling, A. (2006). The adult *Drosophila* posterior midgut is maintained by pluripotent stem cells. *Nature* 439, 470–474.

Tokuda, G. and Watanabe, H. (2007). Hidden cellulases in termites: revision of an old hypothesis. *Biology Letters* 3, 336–339.

Turunen, S. and Crailsheim, K. (1996). Lipid and sugar absorption. In *Biology of the Midgut*, ed. M. J. Lehane and P. F. Billingsley, pp. 293–320. London: Chapman & Hall.

Watanabe, H. and Tokuda, G. (2010). Cellulolytic systems in insects. *Annual Review of Entomology* 55, 609–632.

References in figure captions and table

Anderson, M. and Cochrane, D. G. (1978). Studies on the mid-gut of the desert locust *Schistocerca gregaria*: II. Ultrastructure of the muscle coat and its innervation. *Journal of Morphology* 156, 257–278.

Baines, D. M., Bernays, E. A. and Leather, E. M. (1973). Movement of food through the gut of fifth-instar males of *Locusta migratoria migratorioides* (R. & F.). *Acrida* 2, 319–332.

Chapman, R. F. (1985). Structure of the digestive system. In *Comprehensive Insect Physiology, Biochemistry and Pharmacology*, vol. 4, ed. G. A. Kerkut and L. I. Gilbert, pp. 165–211. Oxford: Pergamon Press.

Cheung, W. W. K. and Marshall, A. T. (1973). Studies on water and ion transport in homopteran insects: ultrastructure and cytochemistry of the cicadoid and cercopoid midgut. *Tissue & Cell* 5, 651–669.

Cioffi, M. (1979). The morphology and fine structure of the larval midgut of a moth (*Manduca sexta*) in relation to active ion transport. *Tissue & Cell* 11, 467–479.

Clissold, F. J., Tedder, B. J., Conigrave, A. D. and Simpson, S. J. (2010). The gastrointestinal tract as a nutrient-balancing organ. *Proceedings of the Royal Society of London B* 277, 1751–1759.

Davidson, D. W., Cook, S. C., Snelling, R. R. and Chua, T. H. (2003). Explaining the abundance of ants in lowland tropical rainforest canopies. *Science* 300, 969–972.

Dow, J. A. T. (1981). Ion and water transport in locust alimentary canal: evidence from *in vivo* electrochemical gradients. *Journal of Experimental Biology* 93, 167–179.

Evans, W. A. L. and Payne, D. W. (1964). Carbohydrases of the alimentary tract of the desert locust, *Schistocerca gregaria*. *Journal of Insect Physiology* 10, 657–674.

Fain-Maurel, M. A., Cassier, P. and Alibert, J. (1973). Étude infrastructurale et cytochimique de l'intestin moyen de *Petrobius maritimus* Leach en rapport avec ses fonctions excrétrice et digestives. *Tissue & Cell* 5, 603–631.

Irvine, B., Audsley, N., Lechleitner, R., Meredith, J., Thomson, B. and Phillips, J. (1988). Transport properties of locust ileum *in vitro*: effects of cyclic AMP. *Journal of Experimental Biology* 137, 361–385.

Lepier, A., Azuma, M., Harvey, W. R. and Wieczorek, H. (1994). K^+/H^+ antiport in the tobacco hornworm midgut: the K^+-transporting component of the K^+ pump. *Journal of Experimental Biology* **196**, 361–373.

Noirot, C. and Noirot-Timothée, C. (1976). Fine structure of the rectum in cockroaches (Dictyoptera): general organization and intercellular junctions. *Tissue & Cell* **8**, 345–368.

Peters, W., Heitmann, S. and D'Haese, J. (1979). Formation and fine structure of peritrophic membranes in the earwig, *Forficula auricularia* (Dermaptera: Forficulidae). *Entomologia Generalis* **5**, 241–254.

Phillips, J. E. and Audsley, N. (1995). Neuropeptide control of ion and fluid transport across locust hindgut. *American Zoologist* **35**, 503–514.

Phillips, J. E., Hanrahan, J., Chamberlin, A. and Thompson, B. (1986). Mechanisms and control of reabsorption in insect hindgut. *Advances in Insect Physiology* **19**, 329–422.

Schmitz, M. and Komnick, H. (1976). Rectal Chloridepithelien und osmoregulatorische Salzaufnahme durch den Enddarm von Zygopteren und Anisopteren Libellenlarven. *Journal of Insect Physiology* **22**, 875–883.

Shanbhag, S. and Tripathi, S. (2009). Epithelial ultrastructure and cellular mechanisms of acid and base transport in the *Drosophila* midgut. *Journal of Experimental Biology* **212**, 1731–1744.

Snodgrass, R. E. (1935). *Principles of Insect Morphology*. New York, NY: McGraw-Hill.

Snodgrass, R. E. (1956). *Anatomy of the Honey Bee*. London: Constable.

Watanabe, H. and Tokuda, G. (2010). Cellulolytic systems in insects. *Annual Review of Entomology* **55**, 609–632.

Williams, L. H. (1954). The feeding habits and food preferences of Acrididae and the factors which determine them. *Transactions of the Royal Entomological Society of London* **105**, 423–454.

4

Nutrition

REVISED AND UPDATED BY **ANGELA E. DOUGLAS**
AND STEPHEN J. SIMPSON

INTRODUCTION

Nutrition concerns the chemicals required by an organism for growth, tissue
maintenance, reproduction and the energy necessary to maintain these functions.
Many of these chemicals are ingested with the food, but others are synthesized by the
insect itself. In some insects, microorganisms contribute to the insect's nutritional
requirements. Achieving optimal nutrition involves a complex interplay between feeding
behavior and post-ingestive processing of food. Insects must eat appropriate amounts of
suitable foods, but avoid ingesting harmful excesses of toxins and nutrients. In this chapter
we begin by considering the nature of nutritional requirements (Section 4.1). In Section 4.2
we set out the ways in which insects maintain nutritional balance, both by adjusting
feeding behavior and post-ingestive processing. Next, we discuss the consequences for
growth, development, reproduction and lifespan of failing to maintain nutrient
balance (Section 4.3). Finally, we detail the nutritional contributions made
by symbiotic microorganisms (Section 4.4).

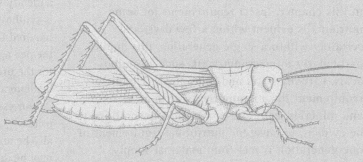

4.1 Required nutrients

4.1.1 How nutritional requirements are identified

Most insects have qualitatively similar nutritional requirements because their chemical compositions and metabolic capabilities are broadly uniform. Variation among insects arises from adaptations to particular diets or associations with microorganisms that provide specific nutrients.

The best route to establish the nutritional requirements of an insect is to quantify its performance (survival, growth, reproduction) on chemically defined artificial diets containing and lacking the nutrient of interest. Because the composition of chemically defined diets is known and can be manipulated, it is possible to determine whether a nutrient is required in the diet and across what range of levels. Such a "one nutrient at a time" approach does not deal adequately with interactions among nutrients, however, where the level of one nutrient influences the requirement for others (Section 4.2). Suitable chemically defined diets have not been developed for many insects. Some species can be reared on meridic (semi-defined) diets that include complex constituents, such as wheat germ, protein hydrolysate or plant extracts. With careful experimental design and interpretation, nutritional information can be gleaned from insects on meridic diets, as is illustrated by various examples in this chapter. Insect requirements for some nutrients is evident within a few days, and certainly within a single generation; this applies particularly to macronutrients, such as amino acids, carbohydrates and some lipids. The dietary requirement for micronutrients, such as vitamins and minerals, can take two or more generations to be evident, because the nutritional requirement of one generation is met from maternal supply, generally via the egg.

4.1.2 Amino acids

Amino acids are required for the production of proteins which are used for structural purposes, as enzymes, for transport and storage, and as receptor molecules. In addition, some amino acids are involved in morphogenesis. Tyrosine is essential for cuticular sclerotization (Section 16.6.2) and tryptophan for the synthesis of visual screening pigments. Others, γ-aminobutyric acid and glutamate, are neurotransmitters (Section 20.2.3). Various amino acids are used as a respiratory fuel, with proline as a particularly important energy source for flight in various insects, including the tsetse fly, *Glossina*, Colorado potato beetle, *Leptinotarsa decemlineata*, and various scarabeid beetles.

Most insects derive their amino acid requirement from dietary protein. The nutritional value of protein depends on its digestibility and amino acid composition. The key nutritional requirement is for the 9–10 amino acids that contribute to protein but cannot be synthesized by insects. These are called essential amino acids (the amino acids that insects can synthesize are termed non-essential amino acids); and, if the supply of just one essential amino acid is insufficient, the insect cannot grow, whatever the total dietary supply of amino acids. Nine of the essential amino acids are common to all animals and are often termed "the rat essentials" because the landmark nutritional studies were conducted on laboratory rats. Many insects (and juvenile rats) need a dietary supply of arginine because this amino acid is synthesized, via the urea cycle, at lower rates than required for rapid growth. Exceptionally, however, the pea aphid, *Acyrthosiphon pisum*, lacks the genes for the urea cycle; arginine synthesis in other hemipteran insects remains to be investigated.

Non-essential amino acids are synthesized primarily in the fat body, although other tissues can also be important (see below). The molecular skeleton may be derived from glucose or acetate being incorporated into compounds that are intermediates in glycolysis or the tricarboxylic acid cycle. From these compounds, amino acids are formed by the addition of ammonia, as for example in the synthesis of glutamine from glutamate via glutamine synthetase, or by transamination, i.e., the transfer of an amino group from a pre-existing amino acid. Glutamate often plays a central role in transamination reactions. The pattern of non-essential amino acid synthesis can vary widely across different organs of a single insect. For example, ammonia derived from the degradation of protein in a blood meal consumed by the female mosquito, *Aedes aegypti*, is incorporated principally into alanine in the fat body, and into proline and glutamine in the midgut.

Although insects can synthesize all non-essential amino acids, their sustained growth is generally impaired by gross dietary imbalance. For example, development of some Diptera, including *Culex* mosquitoes, is arrested on diets lacking proline; aspartic acid and glutamic acid are required for growth of *Phormia* blowflies; and alanine and either glycine or serine are necessary, in addition to the essential amino acids, for optimal growth of the silkworm, *Bombyx mori*. Generally, proteins contain approximately essential and non-essential amino acids at a 1:1 ratio, although there is much variation, and insects tend to perform well on a balanced mix of dietary amino acids.

A further constraint on non-essential amino acid nutrition of insects is that the synthesis of some non-essential amino acids also depends on the availability of precursors. This applies particularly to tyrosine and cysteine. Cysteine is a sulfur amino acid, produced in animals, including insects, from the essential sulfur amino acid, methionine, by the transulfuration pathway. Consequently, cystine and cysteine are not necessary in the diet if there is ample methionine. Tyrosine is an aromatic amino acid and a key substrate for cuticle sclerotization, as well as a constituent of protein. Insects cannot synthesize the aromatic ring of tyrosine, and many produce tyrosine by hydroxylation of the essential amino acid

phenylalanine – i.e., they are dependent on sufficient dietary supply of phenylalanine to meet their needs for both phenylalanine and tyrosine. Some polyphagous grasshoppers are able to use some phenolic compounds, such as protocatechuic acid and gallic acid, in cuticle sclerotization (Section 16.5.3), so conserving phenylalanine and tyrosine for protein synthesis. For many insects, however, these compounds are non-nutrient and potentially harmful.

4.1.3 Carbohydrates

Carbohydrates, including simple sugars, starch and other polysaccharides, are important components of the diet for most insects. They are the usual respiratory fuel, can be converted to lipid and provide the carbon skeleton for the synthesis of various amino acids. In addition, insect cuticle characteristically contains the polysaccharide chitin (Section 16.6.2). All insects can synthesize glucose sugar by gluconeogenesis, utilizing metabolic intermediates derived from lipid or amino acid breakdown, and some insects can grow readily on artificial diets containing no carbohydrates. This is true of the larva of the screw-worm fly, which feeds on live animal tissues; and carbohydrate can be replaced by wax in the diet of the wax moth, *Galleria*. Nevertheless, most insects so far examined require some carbohydrate in the diet, and perform best at a particular dietary proportion of carbohydrate relative to other nutrients that differs according to species (Section 4.2).

The utilization of different carbohydrates depends on the insect's ability to hydrolyze polysaccharides (Section 3.2.5), the readiness with which different compounds are absorbed and the possession of enzyme systems capable of introducing these substances into the metabolic processes. Some insects have a very broad digestive capability, enabling them to use a very wide range of carbohydrates. *Tribolium* (Coleoptera), for example, can utilize starch, the alcohol mannitol, the trisaccharide raffinose, the disaccharides sucrose, maltose and cellobiose, as well as various monosaccharides. Other insects feeding on stored products, and some phytophagous insects, e.g., *Schistocerca* and *Locusta* (Orthoptera), can also use a wide range of carbohydrates, but other phytophagous insects are more restricted. The grasshopper *Melanoplus* is unable to utilize polysaccharides, and stem-boring larvae of the moth *Chilo* can only use sucrose, maltose, fructose and glucose. Pentose sugars do not require digestion since they are monosaccharides, but nevertheless do not generally support growth and may be actively toxic, perhaps because they interfere with the absorption or oxidation of other sugars. There may be differences in the ability of larvae and adults to utilize carbohydrates. For instance, the larva of *Aedes* can use starch and glycogen, but the adult cannot.

4.1.4 Lipids

Fatty acids, phospholipids and sterols are components of cell membranes, as well as having other specific functions. Insects are able to synthesize many fatty acids and phospholipids, so they are not usually essential dietary constituents, but some insects do require a dietary source of polyunsaturated fatty acids (PUFAs), and all insects require sterols.

Fatty acids form a homologous series with the general formula $C_nH_{2n+1}COOH$. In insects they are present mainly as diacylglycerides and triacylglycerides, and are usually dominated by skeletons of 16 and 18 carbon atoms: palmitic (16:0), palmitoleic (16:1), stearic (18:0), oleic (18:1) and the polyunsaturated fatty acids linoleic (18:2) and linolenic (18:3) (Fig. 4.1). (16:0 = 16 carbon atoms, no double bonds; 18:1 = 18 carbon atoms, one

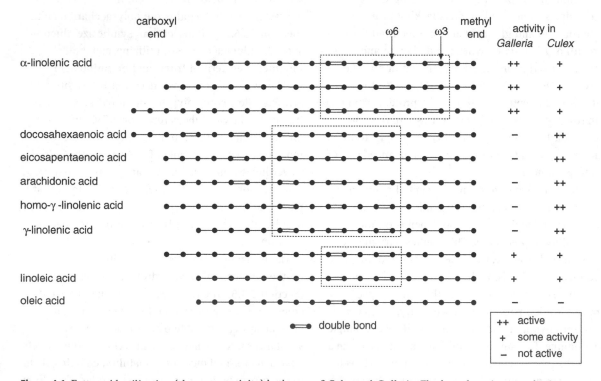

Figure 4.1 Fatty acid utilization (shown as activity) by larvae of *Culex* and *Galleria*. The boxed sections emphasize arrangements of double bonds that are important to the two insects. *Culex* larvae utilize acids with double bonds in different positions from those utilized by *Galleria* larvae (after Dadd, 1985).

double bond.) There is, however, great variation. For example, triglycerides of *Anthonomus* weevils include 23 fatty acids ranging in chain length from 6 to 20 carbon atoms, and the dominant fatty acids in aphid triglycerides are myristic (14:0) and hexanoic (6:0) acids.

A dietary requirement for PUFAs has been demonstrated in various Lepidoptera, Coleoptera and Orthoptera. When fed on diets lacking PUFAs, these insects variously develop poorly, die at ecdysis or display depressed fecundity. The deleterious effects of diets lacking PUFAs are commonly evident in the first generation, but it can be important to test for a dietary requirement through at least two generations because lipids supplied in the egg of some species can sustain the insect's requirements through one full generation. Some insects are independent of a dietary supply of these nutrients, as revealed by multi-generation diet experiments and parallel demonstrations of lipid biosynthesis. These insects include *Zootermopsis* termite species; general scavengers such as the house cricket, *Acheta domestica*, and cockroach, *Periplaneta americana*; and plant-sap feeders such as the pea aphid, *Acyrthosiphon pisum*, and whitefly, *Bemisia tabaci*.

One size class (C20) of PUFAs is important as precursors of eicosanoids, which are oxygenated metabolites, especially of arachidonic acid (20:4). The eicosanoids include the prostaglandins, the epoxyeicosatrienoic acids and lipoxygenase metabolites. They have been implicated in various aspects of insect physiology, including induction of egg-laying in crickets, regulation of cellular immunity in Lepidoptera, Diptera and Homoptera and control of fluid secretion rates in tick salivary glands. They can be synthesized readily by insects from C20 PUFAs, but the dietary requirement for the C20 PUFAs varies widely among insects. Some species, including larval mosquitoes, require a dietary supply; some (e.g., the lepidopteran *Ephestia*) can synthesize them from dietary supply of C18 PUFAs, while others (see above) can synthesize C20 PUFAs from acetate.

Insects are unable to synthesize sterols, which are essential components of animal cell membranes and precursors for ecdysteroid hormones. As a consequence, most insects require a dietary supply of sterol, although a minority obtain their sterols from symbiotic microorganisms (see Section 4.4). The key sterol for many species is cholesterol, which acts as a membrane constituent and precursor in the synthesis of 20-hydroxyecdysone and other C27 ecdysteroid hormones. The availability of dietary cholesterol varies with the type of food: Most animal products contain ample cholesterol, while plant and fungal food stuffs have little or no cholesterol, but contain various other sterols. As a consequence, many phytophagous and fungivorous insects metabolize the ingested sterols to cholesterol via poorly understood dealkylation reactions believed to be mediated in the gut (Fig. 4.2). Some phytophagous and fungivorous insects, however, have limited capacity for sterol dealkylation. They may incorporate dietary sterols other than cholesterol into membranes with little or no modification, so sparing their cholesterol requirement as ecdysteroid precursor. In addition, certain species (including some Diptera, heteropteran bugs and the honey bee) synthesize ecdysteroids from phytosterols without removing the side-chain alkyl group; their C29 ecdysteroid Makisterone A has the same structure as 20-hydroxyecdysone but with an extra 24(R)-methyl group. Various insects additionally utilize sterol derivatives as pheromones and defensive compounds, e.g., cholestanone is the trail pheromone of tent caterpillars (*Malacosoma*) and the highly toxic steroid cardenolides are synthesized and stored in the defensive glands of chrysomelid beetles.

A few insects with specialized feeding habits have unusual and very specific sterol requirements. Most notably, the cactophilic *Drosophila pachea* feeds exclusively on necrotic senita cactus of the Senoran Desert (western United States), and has an absolute requirement for Δ^7-sterols, specifically 7-cholestenol and 7-campestenol.

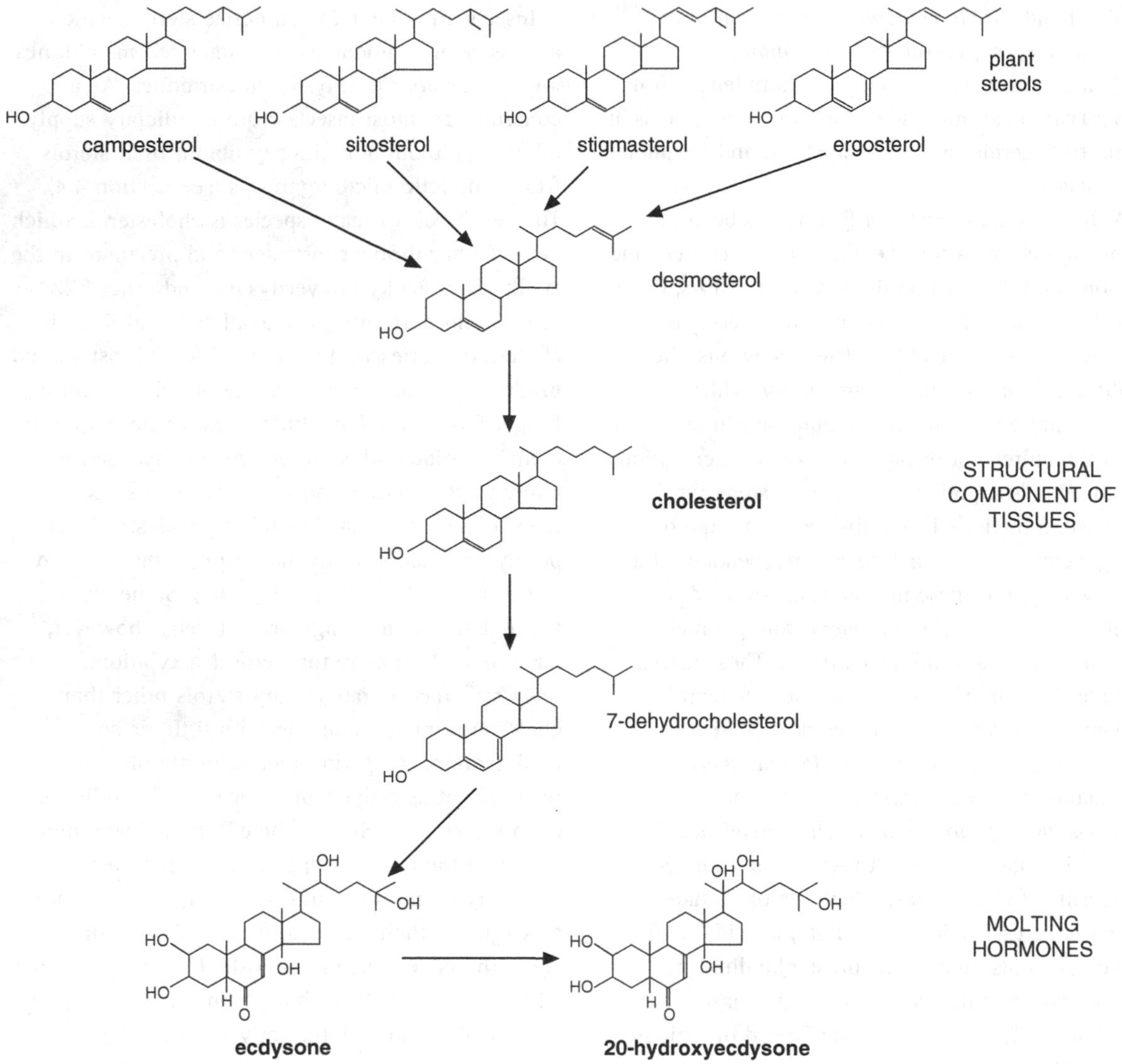

Figure 4.2 Metabolic pathways by which plant sterols are changed to cholesterol and the molting hormones in phytophagous insects. Many intermediate steps are omitted.

4.1.5 Vitamins

Vitamins are organic compounds required in trace amounts for sustained growth. Vitamins are classified as either water soluble or lipid soluble. Water-soluble vitamins have a far shorter half-life in insect tissues than lipid-soluble vitamins, which tend to accumulate in lipid stores.

The chief water-soluble vitamins required by insects are vitamin C (ascorbic acid) and the B vitamins. The response to dietary deficiency of these vitamins appears to be much less specific in insects than mammals. For example, mammals exhibit well-defined symptoms, such as beriberi (vitamin B_1 deficiency) and scurvy (vitamin C deficiency), while insects generally display non-specific growth defects in response to a shortfall in the dietary supply of any water-soluble vitamins.

Our understanding of vitamin function in insects is largely extrapolated from vertebrate research. Vitamin C acts as an antioxidant and, presumably as in vertebrates, promotes the synthesis of collagen and the extracellular matrix in insects. The B vitamins function as cofactors in various metabolic pathways, including: decarboxylations (vitamin B_1, thiamine); flavoproteins (vitamin B_2, riboflavin) and cytochromes (vitamin B_3, niacin) in ATP production; acyl group transfer reactions (vitamin B_5, pantothenate); amino acid metabolism (vitamin B_6, biotin); and one-carbon transfer reactions (vitamin B_9, folic acid). Additional water-soluble vitamins required in very small amounts by some insects include: choline, carnitine, cyanocobalamin (also known as vitamin B_{12}) and lipoic acid.

Among the four lipid-soluble vitamins of mammals, insects have a requirement for the vitamin A complex (β-carotene and related carotenoids) and vitamin E (tocopherols), but apparently not for vitamin D (calciferols) or vitamin K (phylloquinone). Vitamin A is required as a functional component of visual pigments (Section 22.3.1). The amounts required by some insects are tiny: for example, in one classic study, the dietary requirement of the housefly, *Musca domestica*, was revealed only in the 15th generation on vitamin A-free diet. By contrast, *Schistocerca* reared on a carotene-free diet display reduced growth and delayed development in a single generation. Vitamin E is important for reproduction of insects, including spermatogenesis in the house cricket, *Acheta domesticus*, and egg maturation in the beetle *Cryptolaemus montrousieri*.

4.1.6 Minerals

Various metal ions are required as coenzymes and in metalloenzymes. Examples include copper in cytochrome oxidase and phenoloxidase, iron in catalase, molybdenum in xanthine dehydrogenase, magnesium in glucose-6-phosphatase and zinc in alcohol dehydrogenase and metalloproteases. These elements are nearly always present as impurities in artificial diets. Insects additionally require sodium, potassium, phosphate and chloride for cellular ionic balance. Importantly, the requirement of most insects for calcium and iron is considerably lower than in vertebrates, which utilize these elements in bone and hemoglobin synthesis, respectively. (Insects that utilize hemoglobin as a respiratory pigment presumably have a proportionately higher iron requirement than those that do not.) For this reason, commercial salts mixtures, e.g., Wesson's mix, designed for vertebrates, are not suitable for many insects.

4.2 Balance of nutrients

Although some growth occurs on foods containing widely differing levels of nutrients, optimal performance requires the nutrient levels to be appropriately balanced. There are two main reasons for this. First, an imbalance may require that an insect ingest and process excessive quantities of food in order to obtain enough of a particular component that is present only in low concentrations in the diet. Excesses of other food components that are ingested as a consequence, including nutrients and non-nutrient compounds, can prove costly in various ways. Some nutrient excesses can be toxic, whereas excess ingested carbohydrates can result in deleterious effects of obesity. Second, interconversions from one compound to another, where possible (e.g., protein cannot be made from carbohydrate, but the reverse conversion is possible), can be metabolically costly and the rates at which they can occur are limited.

A balanced diet allows the insect to acquire its optimal intake of all nutrients simultaneously. The nutritional requirement for multiple nutrients can be represented as a point in a multi-dimensional nutrient space. This "intake target" moves over time as the quantity and mix of nutrients changes with activity, growth, development, reproduction and senescence. It also shifts over evolutionary time as insects adapt to different diets. The intake target is generally adapted to the natural foods of the species (Fig. 4.3). Insects that feed on other animals have high amino acid and fat requirements relative to carbohydrates, reflecting the relatively high protein and low digestible carbohydrate content of animal tissues. Plant-feeding species generally require approximately equal amounts of amino acids and carbohydrates, although species that consume the majority of their adult protein needs as larvae have a higher protein target during larval stages (Fig. 4.3). Insects feeding on high carbohydrate diets, such as phloem-feeders and the grain beetles, have high dietary requirements for carbohydrate relative to protein and amino acids, and have developed associations with symbiotic bacteria that help provision the host with nitrogenous nutrients that are at low levels in the food (Fig. 4.3, Section 4.4).

Apart from these gross needs, an appropriate balance between specific components is also necessary. *Schistocerca gregaria* develops well on lettuce, averaging an 82% increase in mass during the final larval stadium. Supplementing the lettuce with 1 mg day^{-1} phenylalanine resulted in the insects increasing their mass by 130% relative to the controls. These insects consumed similar amounts of lettuce to those eating lettuce without any additions, but they utilized the food more efficiently. Most of the phenylalanine was incorporated into the adult cuticle, presumably having been converted to tyrosine for sclerotizing the new cuticle.

Another example of a single amino acid affecting the overall utilization of protein occurs in oogenesis of blood-feeding mosquitoes. Here, isoleucine is the critical amino acid. Female *Aedes* feeding on the blood of guinea pigs, which has high isoleucine content, produce about 35 eggs per milligram of blood ingested. When feeding on human blood, containing similar amounts of amino acids except for a low concentration of isoleucine, they only produce about 24 eggs per milligram (and see Fig. 4.4).

The former use about 34% of the ingested amino acids, while those feeding on human blood utilize less than 20%; the rest is excreted.

Amounts of minor dietary components also need to be in balance. The concentration of RNA needed for optimal development of *Drosophila* is doubled if folic acid is not present, and an increase in dietary casein from 4% to 7% necessitates a doubling of the concentrations of nicotinic acid and pantothenic acid and a six-fold increase in folic acid for optimal growth.

Nutrients also interact with non-nutrient chemicals in the diet. For example, phenolic compounds, which are common components of leaves, may reduce the digestibility of proteins in caterpillars. However, the effects may vary in different insects, depending on their feeding habits. Whereas tannic acid is detrimental to grass-feeding grasshoppers, it may serve as a nutrient for others. Both tannic acid and gallic acid can be used by the grasshopper *Anacridium* as sparing agents for phenylalanine in cuticular sclerotization. The deleterious impact of tannic acid on survival of nymphal *Locusta migratoria* depends upon the ratio of protein to carbohydrate in the diet. When the ratio is near balanced (approximately 1:1 protein to carbohydrate), locusts tolerate up to 10% dry weight of tannic acid in the diet. However, mortality increases with the degree of imbalance in dietary protein to carbohydrate ratio. When the diet is too high in protein relative to carbohydrate, survival is compromised by tannic acid causing post-ingestive damage such as lesions in the gut lining; when the dietary protein to carbohydrate ratio becomes too low, tannic acid acts as a powerful antifeedant, leading to starvation (Fig. 4.5).

4.2.1 Changes in nutrient requirements

The nutritional requirements of an insect change with time because of the varying demands of growth, reproduction, diapause or migration. In larval insects it is generally true that the nitrogen content of the

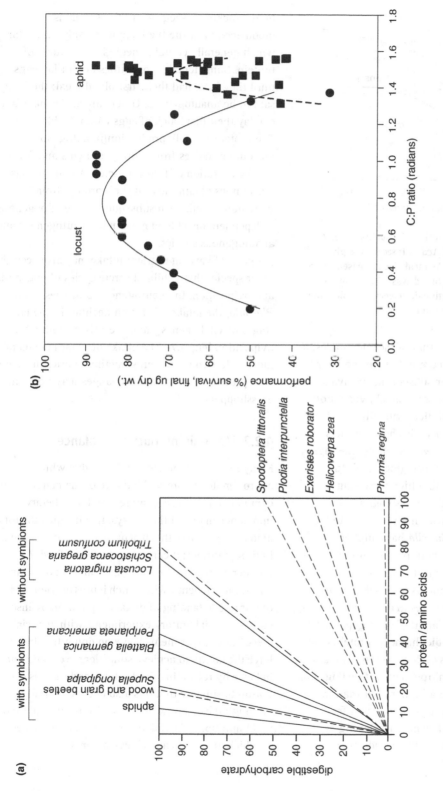

Figure 4.3 Performance in relation to the ratio of protein (or amino acids) to carbohydrate in the diets of juvenile migratory locusts and pea aphids. The optimal dietary composition and the range of diet compositions tolerated reflect the natural foods of the two species (after Raubenheimer and Simpson, 1993; Abisgold *et al.*, 1994). Also shown, a selection of data from a comparative analysis of intake target ratios for protein and carbohydrate in 117 species of larval insect. Dashed lines (diet rails) indicate endopterygote species, while unbroken rails are species with exopterygote development. Irrespective of developmental strategy or taxon, species with endosymbiotic bacteria have a lower P:C than those without. Species which feed as larvae on animal-based food (e.g., larvae of the parasitoid wasp *Exeristes roborator* and the blowfly *Phormia regina*) have high P:C, as do species which obtain most of their adult protein needs as larvae (e.g., caterpillars of the moths *Plodia interpunctella*, *Spodoptera littoralis* and *Helicoverpa zea*) (from Raubenheimer and Simpson, 1999).

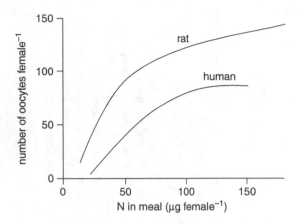

Figure 4.4 Effects of food quantity and quality on egg production by the mosquito, *Aedes*. Insects were given different-sized blood meals. As meal size increased, more eggs were produced. Human blood was utilized much less efficiently because of its relative deficiency in isoleucine (see text) (data from Briegel, 1985).

early stages is greater than that of the later stages, at least in part due to the accumulation, in the later stages, of lipid reserves for subsequent survival, development and reproduction. Larval gypsy moths, given a choice of artificial diets with different levels of proteins and lipids, alter their choice of diets as they get older in a manner which reflects the higher lipid levels of later stage insects (Fig. 4.6a). In addition, the efficiency with which these stages utilize ingested nitrogen decreases (Fig. 4.6b). Changes may also occur within a stadium. For example, the cockroach *Supella* has a higher carbohydrate intake relative to protein in the first half of a larval stadium than in the second.

Among those holometabolous insects that do not feed as adults, sexual differences in diet selection may already be apparent in the larvae. Female gypsy moth larvae continue to select a diet higher in protein than males do in the later stages of development, and also maintain a higher level of nitrogen utilization (Fig. 4.6).

As adults, females have a higher need than males for dietary protein for egg production, and feed on more protein-rich foods when developing eggs. This is most obvious in mosquitoes and other blood-sucking insects where the female is blood-feeding while the male feeds only on nectar, which generally contains negligible amounts of protein. Some female mosquitoes do not lay eggs until they have had their first blood meal; they are said to be anautogenous. Others are autogenous and can lay their first batch of eggs without a blood meal. The protein for yolk production in autogenous mosquitoes comes from storage proteins and from the degeneration of flight muscles. Anautogenous mosquitoes obtain most of their protein from vertebrate blood. Each subsequent cycle of oogenesis is dependent on a blood meal in both autogenous and anautogenous species.

Cycles of varying nutrient intake may also occur in other species that exhibit discrete cycles of oogenesis and oviposition. In anautogenous blowflies, such as *Phormia*, the intake of protein declines in the later stages of vitellogenesis and then rises again after oviposition (Fig. 4.7). The intake of sugar may remain more or less constant, but sometimes varies inversely with protein intake. Similar changes may occur in grasshoppers.

4.2.2 Maintaining nutrient balance

Many insects are known to select a diet which approximates an optimal balance of the major components. An insect can respond to a dietary imbalance in one of three ways. It can adjust the total amount ingested so that it acquires enough of the most limiting nutrient; it can move from one food to another with a different nutrient balance; or it can adjust the efficiency with which it uses nutrients. Most of our understanding of dietary regulation by insects comes from laboratory experiments with artificial diets. These experiments leave no doubt that insects have the ability to achieve some degree of nutritional balance by regulating food intake, and there is every reason to suppose that this also occurs naturally, although the complex makeup of most natural foods and limitations of availability may restrict the degree to which an insect can achieve balance.

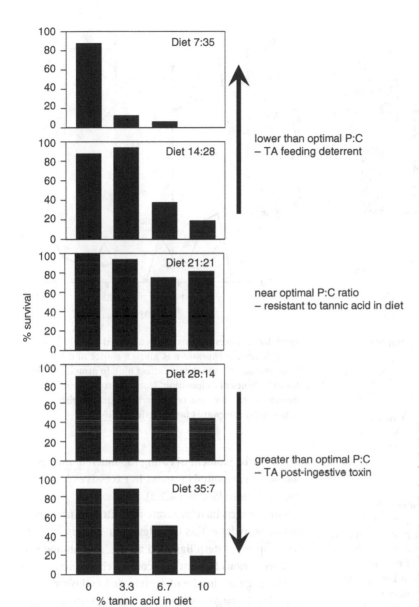

Figure 4.5 The interaction between nutritional balance and tannic acid in the diet of *Locusta migratoria* nymphs. The intake target for protein and carbohydrate is near to 21:21 (21% protein, 21% carbohydrate). Note how locusts were resistant to tannic acid in a balanced diet, but that survival was reduced by addition of tannic acid to the food as the ratio of protein to carbohydrate shifted away from the intake target (from Simpson and Raubenheimer, 2001).

Species including grasshoppers, caterpillars, cockroaches, aphids and ants have been shown to increase the amount eaten if the entire nutrient composition of a diet is diluted with some inert non-nutritional substance. *Melanoplus sanguinipes*, feeding on dried wheat sprouts, was able to maintain its intake of wheat close to optimal even when the food was diluted 7:1 with cellulose. This necessitated an almost seven-fold increase in the total amount of food consumed (Fig. 4.8a). At the greatest dilution, the insects were unable to compensate. Caterpillars of *Spodoptera* compensated for dilution of the nutrients

(a)

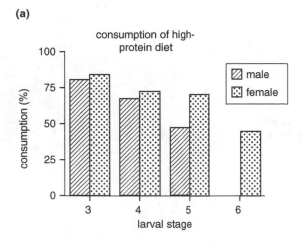

(b)

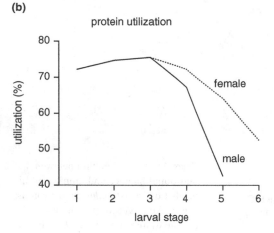

Figure 4.6 Changes in consumption and utilization of protein in successive larval stages of the gypsy moth, *Lymantria*. (a) The percentage of the high-protein diet eaten by each stage. Relatively less of this diet was eaten by the later stages. Insects were given a choice of two artificial diets. One contained high concentrations of both protein and lipid, the other contained a low protein concentration with high lipid. Females have six larval stages, males only five (data from Stockhoff, 1993). (b) The efficiency of utilization of ingested protein declines in the later stages (after Montgomery, 1982).

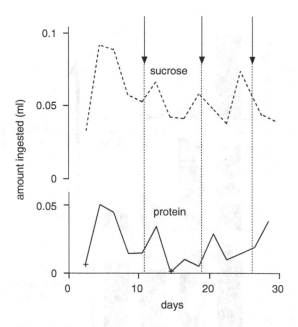

Figure 4.7 Changes in consumption related to reproduction. An adult female *Phormia* was given a choice of 0.1 M sucrose and a brain–heart extract high in protein. Arrows show the times of oviposition. Notice that "protein" consumption is low just before oviposition and rises immediately afterwards (after Dethier, 1961).

they face the problem of eating too much of nutrients and other compounds present in excessive concentrations (Section 4.2.3).

Some insects have separate appetites for protein and carbohydrate; this includes plant-feeders such as grasshoppers, caterpillars and aphids, and omnivores such as various Diptera and cockroaches. These species regulate intake of protein and carbohydrate to an intake target by selecting among nutritionally complementary foods if these are available (Fig. 4.9). Predatory ground beetles (Carabidae) have separate protein and lipid appetites, and regulate intake of both when allowed to self-select their diet. In this way, grasshoppers, caterpillars, flies and cockroaches can correct for a previous imbalance of carbohydrate or protein (Figs. 4.9b, 4.10), and carabid beetles are able to adjust the amounts of lipids and protein consumed.

in their diet from 30% to 10% by eating three times as much (Fig. 4.8b). Insects can also compensate for deficiencies in a class of major nutrients by increasing the total amount eaten, but in so doing

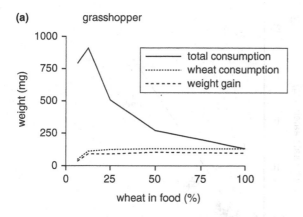

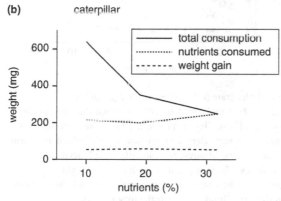

Figure 4.8 Dietary compensation by insects with no choice of food. (a) A grasshopper, *Melanoplus*. Amount of food eaten in five days by final-stage larvae feeding on dried, ground wheat. The wheat was mixed with indigestible cellulose in different proportions. On all but the lowest concentration, wheat consumption and weight gain were almost constant because the insect ate more (data from McGinnis and Kasting, 1967). (b) A caterpillar, *Spodoptera*. Amount of food eaten in the final larval stage. Nutrients were diluted with indigestible cellulose. Weight gain was similar irrespective of the percentage of nutrients in the diet (data from Wheeler and Slansky, 1991).

4.2.3 Nutritional compromises when balance cannot be achieved

When confined to an imbalanced diet that does not allow the intake target to be reached, insects have to compromise between over-consuming excess nutrients and eating too little of the nutrients in deficit. The form of this compromise varies according

to the nutrients involved and the ecology and life-stage of the species. For example, given a choice of foods containing different amounts of inorganic salts, *Locusta migratoria* modifies its choice so that it maintains a target intake of both salts and the major nutrients. In the absence of a choice, however, different amounts of salts are ingested because the insect maintains the optimal intake of major nutrients. In other words, regulation of macronutrient intake takes priority over salt regulation. When forced to ingest diets containing an unbalanced ratio of protein and carbohydrate, however, *Locusta* does not substantially over-consume the excess nutrient to decrease its deficit of the other nutrient (Fig. 4.11a). In contrast, another, more polyphagous species, the desert locust, *Schistocerca gregaria*, will ingest a greater amount of either protein or carbohydrate than will *Locusta* to gain more of the more limiting nutrient (Fig. 4.11b). This pattern of host-plant generalists being more willing to tolerate nutrient excesses than host-plant specialists is widespread across species of grasshoppers and caterpillars and is accompanied by a greater tendency by generalists to store excesses as body reserves rather than void them.

4.2.4 Pre-ingestive mechanisms of nutritional regulation

The ability to adjust food intake to nutritional requirements implies some feedback of nutritional status on food selection and feeding behavior. In the locust, blood-borne nutrient feedback from eating a diet rich in amino acids depresses the sensitivity of the peripheral contact chemoreceptors to amino acids in the diet (Fig. 4.12). The sensitivity of chemoreceptors to sucrose is, however, unaffected. Conversely, if the insect feeds on a diet with high levels of carbohydrate, the sensitivity of its receptors to sucrose is depressed. Hence, the insect tastes what it needs and eats what it tastes. The increase in blood osmolality that follows feeding also reduces further feeding, either by extending the interval between meals or by reducing the amount eaten within a meal, depending on the state of the insect.

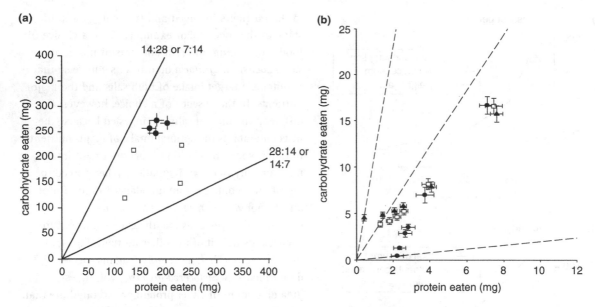

Figure 4.9 Two experiments demonstrating regulation of a protein:carbohydrate (P:C) intake target. (a) Locusts (*Locusta migratoria*) given one of four food combinations (P:C 14:28 or 7:14 and 28:14 or 14:7) reached the same point in nutrient space by altering the relative amounts of the two foods eaten (from Chambers *et al.*, 1995). The open squares indicate the expected outcome if there was no homeostatic regulation. (b) Female German cockroaches (*Blattella germanica*) were subjected to a 48-hour pretreatment in which they were confined to one of three nutritionally imbalanced foods. The dotted lines represent nutritional rails of the three pretreatment foods. The first point in each series (i.e., that closest to the origin) shows the amounts of protein and carbohydrate ingested by the end of the pretreatment period. Subsequent points in each series show how cumulative intake converged upon the same intake target trajectory over the following 4, 10, 24, 48 and 120 hours in which the insects had access to all three foods, and so could move freely within the nutrient space (after Raubenheimer and Jones, 2006).

Learning has an important role in regulating nutrient intake. Sugars, amino acids and salts are tasted by many insects (see Chapter 24), but proteins, sterols and vitamins probably are not. The insect regulates the intake of these compounds by learning to associate some quality of the food with feedback on its own nutrient status. Visual characteristics, odor and taste due to chemicals other than the nutrients may provide the stimuli and the association may be positive, resulting in feeding, or negative, leading to rejection of the food. Most work on this aspect of compensatory feeding has been on locusts and grasshoppers, but it is likely that other insects exhibit similar behavior. Locusts have been shown to associate different odors or color cues with high protein and high carbohydrate foods and to prefer the cues that were previously associated with a food that meets current nutritional needs. Grasshoppers have also been shown to develop an aversion to food containing non-utilizable sterols.

Longer-term changes during larval development probably also reflect nutritional feedback. On the other hand, regulation of changes related to ovarian cycles in adult females is at least partly controlled neurally. Abdominal distension caused by the developing oocytes is important in reducing the rate of protein intake in *Phormia*, while in *Musca* distension of the oviducts may be important. Studies on *Drosophila* have provided molecular insights into the mechanisms of protein appetite in female reproduction. The rapidly induced preference for protein-rich yeast in female flies after mating results

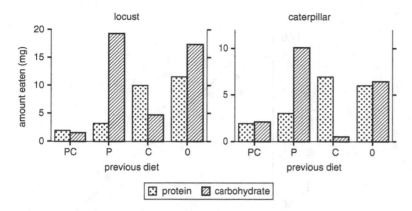

Figure 4.10 Dietary compensation by insects with a choice of foods. Insects were given a choice of artificial diet containing either protein or carbohydrate after a period of four hours during which only one of four different diets was available. The previous diets contained protein and carbohydrate (PC), protein with no carbohydrate (P), carbohydrate with no protein (C) or neither of these components (0). After feeding on PC, little feeding occurred; after 0, carbohydrate and protein were almost equally acceptable. After feeding on P, the insects selected carbohydrate and after feeding on C, they selected protein. The histogram for the locust shows the amounts eaten in one hour, for the caterpillar the amounts eaten in eight hours (after Simpson *et al.*, 1988).

not from depletion of protein reserves as eggs are developed, but as a direct result of a sex peptide that is introduced with the male's seminal fluid during mating, which stimulates special sensory neurons in the female's reproductive tract. An additional mechanism responding to the protein demands of egg development then controls how much yeast is eaten, involving TOR/S6 kinase and serotonin signaling pathways in the central nervous system.

4.2.5 Post-ingestive regulation

The importance of post-ingestive regulation of nutrient balance is demonstrated by experiments with locusts. These insects maintain a relatively constant increase in body nitrogen despite a more than three-fold increase in the amount of nitrogen ingested. Very little of the protein is present unchanged in the feces. Most of the excess is excreted as uric acid or some other, unknown, nitrogenous end product of catabolism (Fig. 4.13). Other means of post-ingestive regulation reported in locusts include differential secretion of digestive enzymes to lower the efficiency of digestion of excess

carbohydrate or protein in the diet (Fig. 3.21); adjusting the timing of gut emptying to alter the ratio of protein and carbohydrate absorbed from the gut; increasing metabolic rate to "burn off" excess ingested carbohydrate; and selecting environmental temperatures that favor utilization of either protein or carbohydrate. In *Pieris* there is an inverse relationship between the levels of particular amino acids in the diet and the metabolic utilization of each amino acid, suggesting the presence of a mechanism regulating oxidation of specific amino acids according to needs.

4.3 Nutritional effects on growth, development, reproduction and lifespan

Variations in the quantity or quality of an acceptable diet can have profound effects on an insect's development and lifespan. Eating too little is obviously detrimental to performance. Commonly, as food intake decreases or nutrient balance becomes sub-optimal, the duration of development is extended, sometimes with insertion of extra larval stadia, and the insect becomes smaller and lighter

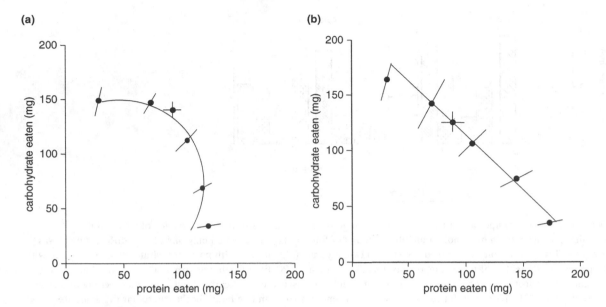

Figure 4.11 Nutrient intake arrays for the first three days of the fifth larval stadium of (a) *Locusta migratoria* and (b) *Schistocerca gregaria*. Points with bidirectional error bars represent the intake target selected by insects given a simultaneous choice of nutritionally complementary foods (data from Raubenheimer and Simpson, 2003).

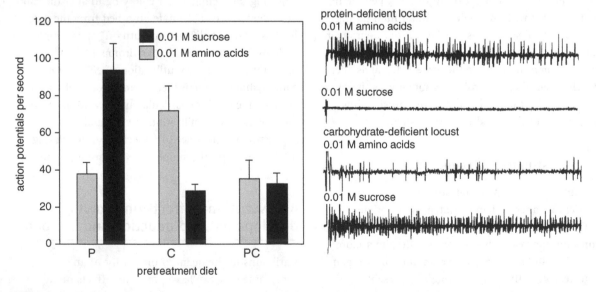

Figure 4.12 Nutritional feedbacks onto food selection behavior and gustation in locusts. Locusts were pretreated for four hours on one of three synthetic foods, P (protein-rich), C (carbohydrate-rich) or PC (containing both protein and carbohydrate). Subsequent changes in food selection behavior were accompanied by changes in the responsiveness of gustatory receptors on the maxillary palps. P-pretreated locusts (carbohydrate-deprived) selected C diet in a choice over P, and had selectively increased gustatory responsiveness to stimulation by 0.01 M sucrose. In contrast, C-pretreated insects (protein-deprived) selectively fed on P over C and had elevated gustatory responsiveness to a 0.01 M mixture of amino acids (after Simpson *et al.*, 1991).

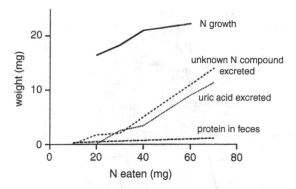

Figure 4.13 Dietary regulation by post-ingestive processes. Final-stage larvae of *Locusta* were given diets containing different amounts of protein. The insects maintained the amount of growth within narrow limits as the amount of protein eaten increased by excreting increasing amounts of nitrogenous materials, mostly as uric acid and other, unknown, nitrogen-containing compounds. Most of the protein was digested and absorbed, as indicated by the small amounts of unchanged protein in the feces. Only very small amounts of free amino acids were excreted (not shown) (after Zanotto *et al.*, 1993).

in weight. The quantities of protein and amino acids ingested are important for optimal growth and reproduction. There are many examples of insects feeding on natural foods in which growth and reproduction are positively correlated with the nitrogen content of the food. However, it is not correct to assume that the more nitrogen the better; rather, there are performance costs to ingesting excesses as well as deficits of protein and other nutrients.

Insects develop the equivalent of obesity on a diet with an excess of carbohydrate. That there are evolutionary costs to excess lipid storage is evident in laboratory selection experiments. Hence, the propensity to store excess ingested carbohydrate as fat in caterpillars of the diamondback moth, *Plutella xylostella*, changed with consecutive generations of rearing on diets of differing protein:carbohydrate ratios. High-carbohydrate–low-protein (obesogenic) diets resulted in caterpillars becoming less prone to depositing fat, whereas caterpillars maintained on a high-protein–low-carbohydrate regime developed greater lipid stores when offered excess carbohydrate food (Fig. 4.14).

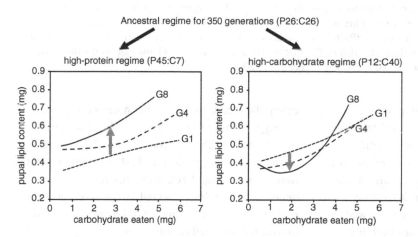

Figure 4.14 Changes in the efficiency with which ingested carbohydrate was converted to body fat in diamondback moth caterpillars, *Plutella xylostella*, reared for eight generations under either a high-protein or a high-carbohydrate regime. At generations 1, 4 and 8, caterpillars were exposed for the final larval stadium to one of five diets containing different proportions of protein and carbohydrate and their food intake and body fat growth were measured. Over the eight generations, caterpillars from the high-protein, low-carbohydrate environment progressively laid down more body fat per intake of carbohydrate (upwards gray arrow), while the opposite occurred under high-carbohydrate, low-protein environments (downward gray arrow) (after Warbrick-Smith *et al.*, 2006).

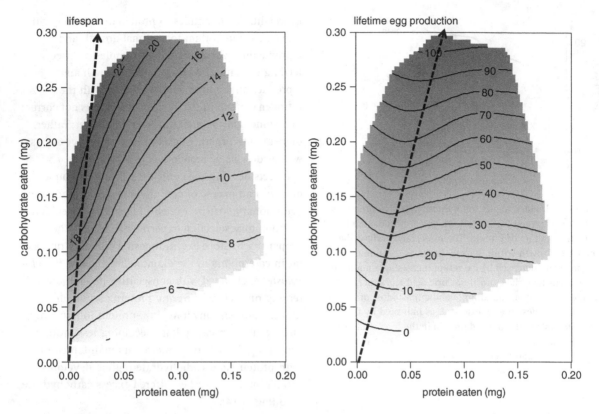

Figure 4.15 Response surfaces showing how the intake of protein and carbohydrate affects lifespan and reproductive output in mated female *Drosophila melanogaster*. Flies were given *ad libitum* access to one of 28 diets varying in the ratio and total concentration of protein to carbohydrate (P:C). Fitted surfaces for longevity and reproductive output are plotted onto nutrient intake arrays. Dashed lines show the dietary P:C that maximized lifespan (1:16) and egg production (1:4). Flies allowed to select their diet chose a P:C ratio of 1:4 (after Lee *et al.*, 2008).

Nutritional balance affects reproductive performance and lifespan. When confined to one of 28 diets varying in protein and carbohydrate content, adult female *Drosophila* lived longest on a diet containing a 1:16 ratio of protein to carbohydrate, but produced most eggs across their lifetime on a 1:4 protein to carbohydrate ratio. When allowed to select their own diet composition, they mixed a diet comprising a 1:4 protein to carbohydrate ratio (Fig. 4.15). Similar differences between lifespan and reproduction are found in crickets, tephritid fruit flies, ants and bees.

The immune response of insects is affected by dietary nutrient balance. *Spodoptera littoralis* caterpillars infected with either a bacterial or viral pathogen survived better as the ratio of protein to carbohydrate in the diet was increased. Uninfected larvae, in contrast, performed best on an intermediate nutrient ratio (Fig. 4.16). When offered the opportunity to select, infected larvae mixed a higher protein diet than did uninfected larvae – a case of nutritional self-medication.

Differences in nutrition may also produce profound differences in coloration, morphology and other features of the phenotype. This phenomenon is termed "diet-induced polyphenism." For example, caterpillars of the spring brood of *Nemoria arizonaria* resemble the oak catkins on which they feed. They

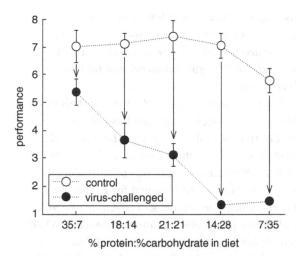

Figure 4.16 Performance of *Spodoptera littoralis* caterpillars challenged with nucleopolyhedrovirus and control caterpillars when confined to one of five diets differing in protein to carbohydrate ratio. Performance was calculated as the product of the proportion of insects surviving by average growth (mg) per day. Each descending arrow indicates pathogen-induced performance loss of insects feeding on each diet (from Lee *et al.*, 2006).

are yellow with a rough cuticle, and have two rows of reddish spots along the midline. Caterpillars of the summer brood resemble stems. They are green-gray and are without the rows of spots, feed on leaves and have larger head capsules and mandibles than the spring brood. These differences result entirely from differences in the quality of food eaten by the insects, although the chemical cues are not yet known. In the honey bee, the quality of food given to larvae by the workers determines whether the larvae will become queens or workers, with the effect being mediated by a protein, royalactin (Section 15.5).

4.4 Contribution of symbiotic microorganisms to insect nutrition

An estimated 10% of all insects utilize diets that are nutritionally so poor or unbalanced that they depend on resident, beneficial microorganisms

for sustained growth and reproduction. In some insects, the microorganisms degrade complex dietary components to a form that can be assimilated by the insect. Specific examples, notably the cellulose-degrading microbiota that supplement intrinsic cellulase enzymes in lower termites and some other insects, are discussed in Section 3.2.5. Other microorganisms have a biosynthetic function, and they are considered here.

The chief nutrients that insects derive from microorganisms are essential amino acids (especially in insects feeding on plant sap), vitamins (insects feeding on vertebrate blood through the lifecycle) and sterols (for example in various insects utilizing wood). The role of microorganisms in essential amino acid synthesis is one aspect of various ways in which microorganisms provide solutions to a widespread nutritional problem for insects – nitrogen hunger, as is considered first.

4.4.1 Symbiotic microorganisms and provision of nitrogen

Symbiotic microorganisms promote insect utilization of low-nitrogen foods in various ways. One way is to concentrate the nitrogen. For example, the fungi cultivated by Macrotermitinae produce nitrogen-rich nodules on which the termites feed. Other ways involve metabolic capabilities of the microorganisms that are absent from the insect, including the utilization of insect nitrogenous waste compounds (e.g., uric acid), synthesis of "high-value" nitrogenous compounds (e.g., essential amino acids) and nitrogen fixation.

Microbial utilization of insect nitrogenous waste products conserves nitrogen, in that the microorganisms utilize nitrogenous compounds that would otherwise be excreted, so reducing their use of nitrogenous compounds valuable to the insect host. This interaction has been demonstrated for bacteria, including *Bacteroides* and *Citrobacter* species in the hindgut of termites such as *Reticulitermes flavipes*,

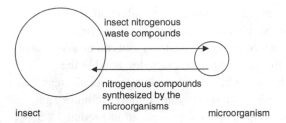

insect nitrogenous
waste compounds

nitrogenous compounds
synthesized by the
microorganisms

insect   microorganism

Figure 4.17 Nitrogen recycling in insect-microbial symbioses. The microorganisms transform nitrogenous waste products of the insect (uric acid, ammonia, urea, etc.) into nitrogenous compounds valuable to animal metabolism (e.g., essential amino acids), and these compounds are translocated back to the animal tissues.

and clavicipitacean fungi associated with the hemolymph and fat body of some planthoppers, notably *Nilaparvata lugens*. In termites, the nitrogen is recycled, i.e., the bacteria metabolize nitrogen from the insect-derived uric acid to amino acids, which are transferred back to the termite and incorporated into protein. In this way, microbial nitrogen recycling transforms waste nitrogen to high-value nitrogenous compounds, including essential amino acids of high nutritional value to the insect host (Fig. 4.17).

Essential amino acid synthesis is also important for insects feeding on plant sap through the lifecycle. Plant-sap feeding has evolved multiple times in the Hemiptera, but apparently in no other insects, and it is invariably associated with the possession of microorganisms. In many hemipterans, the microorganisms are restricted to a particular cell type called the mycetocyte (also known as bacteriocyte). Nevertheless, some plant-sap feeding hemipterans bear microorganisms in different locations (e.g., in the posterior midgut of pentatomid stink bugs), and the mycetocyte symbioses are not restricted to hemipterans, but occur in several different insect orders (Table 4.1). In all mycetocyte symbioses of insects, the microorganisms are transmitted vertically from mother to offspring,

usually via the egg, and both insect and microbial partners require the association. The location of the mycetocytes is, however, very variable and includes the gut, hemocoel and fat body (Fig. 4.18).

The possession of symbiotic microorganisms by plant-sap feeding hemipterans has been correlated and attributed to the low essential amino acid content of phloem and xylem sap. Direct evidence that symbiotic microorganisms provide essential amino acids has been obtained for aphids, most of which bear the γ-proteobacterium *Buchnera aphidicola* in mycetocytes in the hemocoel. When aphids are reared on diets from which individual essential amino acids are omitted, the growth of aphids containing their normal complement of *Buchnera* is not affected, but aphids experimentally deprived of *Buchnera* by antibiotic treatment grow very poorly. In support of these dietary data, radiotracer experiments have confirmed that *Buchnera* synthesize essential amino acids from radioactively labeled precursors, and release these labeled essential amino acids to the aphid tissues. These physiological data have been confirmed spectacularly by genome sequencing analyses showing that *Buchnera* has the genetic capacity to synthesize essential amino acids, despite its small genome size (0.6 Mb, less than 20% of the size of the genome of the related bacterium *Escherichia coli*). *Buchnera* may also provide the insect with riboflavin.

The taxonomic identity of symbiotic microorganisms varies among different groups of plant-sap feeding hemipteran insects (Table 4.1), and all are believed to provide essential amino acids, although this has not been demonstrated by physiological methods. The genomic approach has been applied to investigate the nutritional role of symbiotic bacteria in a xylem-feeding hemipteran, the auchenorhynchan *Homalodisca coagulata* (the glassy-winged sharpshooter). As with various other auchenorhynchan groups, this insect bears two bacteria: *Baumannia cicadellinicola*, a

Table 4.1 **Survey of mycetocyte symbioses in insects**

Insect	Microorganisms
Homoptera	
Auchenorrhyncha (including leafhoppers, planthoppers, cicadas)	*Sulcia muelleri* (Bacteroidetes); and *Baumannia cicadellinicola* (γ-proteobacteria) in Cicadellinae, *Zinderia insecticola* (β-proteobacteria) in Cercopidae, *Hodgkinia cicadicola* (α-proteobacteria) in Cicadae and Pyrenomycete fungi in some Fulgoridae
Aphids	*Buchnera aphidicola* (γ-proteobacteria) or pyrenomycete fungi
Whitefly	*Portiera aleyrodidarum* (γ-proteobacteria)
Psyllids (jumping lice)	*Carsonella ruddii* (γ-proteobacteria)
Scale insects & mealybugs	*Tremblaya principes* (β-proteobacteria), *Moranella endobia* (γ-proteobacteria)
Heteroptera	
Cimicids (bed bugs)	*Wolbachia* (α-proteobacteria) in *Cimer lectularius*
Triatomine bugs	*Arsenophonus triatominarum* (γ-proteobacteria)
Anoplura (sucking lice)	*Riesia pediculicola* (γ-proteobacteria) in human head louse & body louse
Diptera Pupiparia	*Wigglesworthia* (γ-proteobacteria) in tsetse flies
Blattidae (cockroaches)	*Blattabacterium* (flavobacteria)
Mallophaga (biting lice)	Not known
Psocoptera (booklice)	*Rickettsia* (α-proteobacteria)
Beetles	
Weevils	Various γ-proteobacteria (including *Nardonella*) *Symbiotaphrina* (yeasts)
Anobiid timber beetles	
Hymenoptera	
Camponoti (carpenter ants)	*Blochmannia* (γ-proteobacteria)

γ-proteobacterium, and *Sulcia muelleri*, a member of the Bacteroidetes. These bacteria have very small genomes and restricted but complementary predicted metabolic capabilities. *Sulcia* can synthesize essential amino acids and *Baumannia* can synthesize various vitamins and cofactors. Furthermore, the putative capabilities of the two bacteria are interdependent. For example, *Sulcia* synthesizes homoserine, the substrate for synthesis of the essential amino acid methionine by *Baumannia*, and *Baumannia* provides the polyisoprenoids required for menaquinone synthesis by *Sulcia* (Fig. 4.19). In other words, this insect is

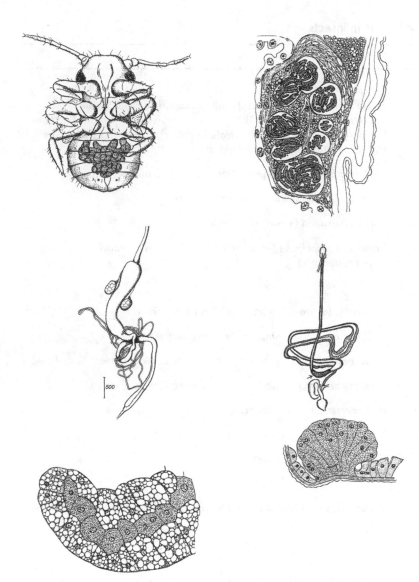

Figure 4.18 Location of mycetocyte symbioses in various insects. Clockwise from top-left: The psyllid *Psylla buxi* with mycetocytes in abdominal hemocoel. Elongate bacteria in the mycetocytes comprising the stomach disc of the louse *Pediculus capitis*. Gut of adult tsetse fly *Glossina* sp. with thickened section of midgut (arrow) comprising mycetocytes. *Glossina* mycetocytes, enlarged cells of midgut epithelium. Lobe of fat body of the cockroach *Blatta orientalis* showing row of mycetocytes bounded by vesicular fat body cells. Paired mycetomes suspended from anterior midgut of the bostrichid beetle *Scobicia chevieri*. Reproduced from Buchner (1965) *Endosymbioses of Animals with Plant Microorganisms*.

dependent on cross-feeding of metabolites between the two bacteria.

A final, potentially important way for insects to gain nitrogen is by association with bacteria that can fix atmospheric nitrogen. Nitrogen-fixing bacteria have been demonstrated in a few insects, including a minority of termites, the wood-feeding larvae of the stag beetle, *Dorcus rectus*, and in fruit flies, particularly the medfly, *Ceratitis capitata*, which contains large populations of nitrogen-fixing enterobacteria in its guts. Rather little research, however, has been conducted on the nutritional significance of nitrogen-fixing bacteria to insects.

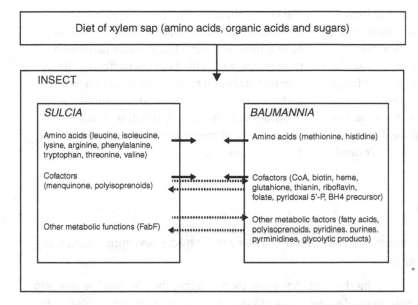

Figure 4.19 The chief metabolic capabilities of the symbiotic bacteria *Sulcia* and *Baumannia* in the insect host *Homalodisca coagulata*, as predicted from genomic sequencing. Solid arrows indicate nutrients transferred to insect, and dashed arrows indicate metabolites shared between the symbiotic bacteria. (Redrawn from Fig. 3 of McCutcheon and Moran, 2007.)

4.4.2 Microbial production of vitamins

Some of the best evidence for microbial provisioning of vitamins comes from research on insects feeding on vertebrate blood through the lifecycle, including the tsetse flies, *Glossina*, other Diptera Pupiparia, the anopluran "sucking" lice and the cimicids (bed bugs). Vertebrate blood is notoriously deficient in B vitamins, and the microbial symbionts in these insects, all of which are localized to mycetocytes (Section 4.4.1), have been implicated in B vitamin provisioning. Insects that utilize blood solely as adults (e.g., fleas, female mosquitoes) obtain sufficient B vitamins from other dietary sources and are not dependent on symbiotic microorganisms.

The chief evidence for microbial production of B vitamins comes from dietary studies, demonstrating that insects bearing the bacteria are independent of a dietary supply of B vitamins, while insects experimentally deprived of their bacteria require these nutrients. Supporting genomic evidence is available for the tsetse fly symbiosis. Specifically, the genome of *Wigglesworthia*, the mycetocyte symbiont of *Glossina brevipalpis*, has genes coding

for the synthesis of pantothenate (vitamin B_5), biotin (vitamin B_7), thiamin (vitamin B_1), riboflavin and FAD (vitamin B_2), pyridoxine (vitamin B_6), nicotinamide (vitamin B_3) and folate (vitamin B_9). B vitamin provisioning has also been proposed for various wood-feeding insects associated with mycetocyte symbionts.

4.4.3 Microorganisms and the sterol nutrition of insects

Some insects obtain sterols from fungal symbionts. Bacteria generally have no capacity for sterol synthesis, and symbiotic bacteria do not contribute directly to the sterol requirements of insects.

The principal evidence for the role of fungal symbionts in insect sterol nutrition comes from the composition of insect sterols: sterols of fungi are generally dominated by $\Delta^{5,7}$-sterols such as ergosterol, and the presence of these sterols in the insect sterol profile is a firm indication of fungal source. The leaf-cutting ant, *Acromyrmex octospinosus*, and the ambrosia beetle, *Xyleborus ferrugineus*, possess these distinctive fungal sterols, suggesting that they derive their sterol requirements from the fungi that they cultivate. Anobiid beetles *Lasioderma serricorne* and *Stegobium paniceum* also derive ergosterol and related sterols from their fungal partners (ascomycetes assigned to the genus *Symbiotaphrina*), which are located in gut ceca, and they metabolize the fungal sterols to the dominant insect sterols, 7-dehydrocholesterol and cholesterol.

The clavicipitacaean fungi associated with the fat body of the planthopper *Nilaparvata lugens* have a truncated sterol biosynthetic pathway due to nonsense mutations in genes mediating the terminal steps of the ergosterol biosynthetic pathway, resulting in the accumulation of ergosta-5,7,24(28)-trienol, which is metabolized by the insect to 24-methylenecholesterol, cholesterol and other steroids.

Summary

- Insects have qualitatively similar requirements for macro- and micronutrients as other animals.

- Insects possess mechanisms that allow them to adjust their feeding behavior and food choices to regulate the intake of macronutrients and some micronutrients to an intake target.

- Once ingested, insects are able to adjust nutrient balance further by differentially digesting, retaining and metabolizing nutrients.

- Diet balance affects growth, development, reproduction, immune responses and lifespan, with some of these performance variables having different nutritional optima.

- Symbiotic microorganisms provide novel metabolic capabilities to insects and have allowed insects to adapt to otherwise unpromising nutritional environments.

Recommended reading

NUTRIENT REQUIREMENTS
Behmer, S. T. and Nes, W. D. (2003). Insect sterol nutrition and physiology: a global overview. *Advances in Insect Physiology* 31, 1–72.

Cohen, A. C. (2004). *Insect Diets: Science and Technology*. Boca Raton, FL: CRC Press.

Dadd, R. H. (1985). Nutrition. In *Comprehensive Insect Physiology, Biochemistry and Pharmacology*, vol. 4, ed. G. A. Kerkut and L. I. Gilbert, pp. 313–391. Oxford: Pergamon Press.

Stanley, D. (2006). Prostaglandins and other eicosanoids in insects: biological significance. *Annual Review of Entomology* 51, 25–44.

NUTRIENT BALANCE

Raubenheimer, D. and Simpson, S.J. (1999). Integrating nutrition: a geometrical approach. *Entomologia Experimentalis et Applicata* 91, 67–82.

Simpson, S.J. and Raubenheimer, D. (2012). *The Nature of Nutrition: A Unifying Framework from Animal Adaptation to Human Obesity.* Princeton, NJ: Princeton University Press.

DIET-INDUCED POLYPHENISM

Simpson, S.J., Sword, G.A. and Lo, N. (2011). Polyphenism in insects. *Current Biology* 21, R738–R749.

SYMBIOSIS

Buchner, P. (1965). *Endosymbioses of Animals with Plant Micro-organisms.* Chichester: John Wiley & Sons.

Douglas, A.E. (2006). Phloem sap feeding by animals: problems and solutions. *Journal of Experimental Botany* 57, 747–754.

Nardi, J.B., Mackie, R.I. and Dawson, J.O. (2002). Could microbial symbionts in arthropod guts contribute significantly to nitrogen fixation in terrestrial ecosystems? *Journal of Insect Physiology* 48, 751–763.

Nasir, H. and Noda, H. (2003). Yeast-like symbiotes as a sterol source in anobiid beetles (Coleoptera, Anobiidae): possible metabolic pathways from fungal sterols to 7-dehydrocholesterol. *Archives of Insect Biochemistry and Physiology* 52, 175–182.

References in figure captions

Abisgold, J.D., Simpson, S.J. and Douglas, A.E. (1994). Nutrient regulation in the pea aphid *Acyrthosiphon pisum*: application of a novel geometric framework to sugar and amino acid consumption. *Physiological Entomology* 19, 95–102.

Briegel, H. (1985). Mosquito reproduction: incomplete utilization of the blood meal protein for oogenesis. *Journal of Insect Physiology* 31, 15–21.

Buchner, P. (1965) *Endosymbioses of Animals with Plant Microorganisms.* Chichester: John Wiley & Sons.

Chambers, P.G., Simpson, S.J. and Raubenheimer, D. (1995). Behavioural mechanisms of nutrient balancing in *Locusta migratoria*. *Animal Behaviour* 50, 1513–1523.

Dadd, R.H. (1985). Nutrition: organisms. In *Comprehensive Insect Physiology, Biochemistry and Pharmacology*, vol. 4, ed. G.A. Kerkut and L.I. Gilbert, pp. 313–390. Oxford: Pergamon Press.

Dethier, V.G. (1961). Behavioral aspects of protein ingestion by the blowfly *Phormia regina*. *Biological Bulletin of the Marine Biological Laboratory, Woods Hole* 121, 456–470.

Lee, K. P., Cory, J. S., Wilson, K., Raubenheimer, D. and Simpson, S. J. (2006). Flexible diet choice offsets protein costs of pathogen resistance in a caterpillar. *Proceedings of the Royal Society of London B* 273, 823–829.

Lee, K. P., Simpson, S. J., Clissold, F. J., *et al.* (2008). Lifespan and reproduction in *Drosophila*: new insights from nutritional geometry. *Proceedings of the National Academy of Sciences USA* 105, 2498–2503.

McGinnis, A. J. and Kasting, R. (1967). Dietary cellulose: effect on food consumption and growth of a grasshopper. *Canadian Journal of Zoology*, 45, 365–367.

Montgomery, M. E. (1982). Life-cycle nitrogen budget for the gypsy moth, *Lymantria dispar*, reared on artificial diet. *Journal of Insect Physiology* 28, 437–442.

Raubenheimer, D. and Jones, S. A. (2006). Nutritional imbalance in an extreme generalist omnivore: tolerance and recovery through complementary food selection. *Animal Behaviour* 71, 1253–1262.

Raubenheimer, D. and Simpson, S. J. (1993). The geometry of compensatory feeding in the locust. *Animal Behaviour* 45, 953–964.

Raubenheimer, D. and Simpson, S. J. (1999). Integrating nutrition: a geometrical approach. *Entomologia Experimentalis et Applicata* 91, 67–82.

Raubenheimer, D. and Simpson, S. J. (2003). Nutrient balancing in grasshoppers: behavioural and physiological correlates of dietary breadth. *Journal of Experimental Biology* 206, 1669–1681.

Simpson, S. J. and Raubenheimer, D. (2001). The geometric analysis of nutrient-allelochemical interactions: a case study using locusts. *Ecology* 82, 422–439.

Simpson, S. J., Simmonds, M. S. J. and Blaney, W. M. (1988). A comparison of dietary selection behaviour in larval *Locusta migratoria* and *Spodoptera littoralis*. *Physiological Entomology* 13, 225–238.

Simpson, S. J., James, S., Simmonds, M. S. J. and Blaney, W. M. (1991). Variation in chemosensitivity and the control of dietary selection behaviour in the locust. *Appetite* 17, 141–154.

Stockhoff, B. A. (1993). Ontogenetic change in dietary selection for protein and lipid by gypsy moth larvae. *Journal of Insect Physiology* 39, 677–686.

Warbrick-Smith, J., Behmer, S. T., Lee, K. P., Raubenheimer, D. and Simpson, S. J. (2006). Evolving resistance to obesity in an insect. *Proceedings of the National Academy of Sciences USA* 103, 14045–14049.

Wheeler, G. S. and Slansky, F. (1991). Compensatory responses of the fall armyworm (*Spodoptera frugiperda*) when fed water- and cellulose-diluted diets. *Physiological Entomology* 16, 361–374.

Zanotto, F. P., Simpson, S. J. and Raubenheimer, D. (1993). The regulation of growth by locusts through post-ingestive compensation for variation in the levels of dietary protein and carbohydrate. *Physiological Entomology* 18, 425–434.

5

Circulatory system, blood and the immune system

REVISED AND UPDATED BY **ANGELA E. DOUGLAS**
AND **MICHAEL T. SIVA-JOTHY**

INTRODUCTION

The cells of all animals, including insects, are bathed in an extracellular fluid (ECF); and most cells of the animal exchange solutes with the ECF, not the external environment. In most animals, including insects, solute exchange between cells and ECF is facilitated by the bulk flow of ECF, powered by one-to-many pumps (including hearts). The physiological system mediating bulk flow is known as the circulatory system, the topic of this chapter. Most insects have one major pump, the dorsal vessel, and multiple accessory pumps, with the circulating portion of the ECF known as the blood or hemolymph. The insect circulatory system is open, meaning that the hemolymph flows freely around insect organs, in contrast to the closed circulatory system of vertebrates in which blood is retained within vessels. The structural organization of the insect circulatory system and the determinants of blood flow are described in Sections 5.1 and 5.2, respectively, and the composition of hemolymph is considered in Section 5.3.

Hemolymph has four key functions. It is the vehicle for transport of hormones and nutrients between tissues, and a site for storage of some nutrients and water; these two functions are considered in Section 5.3. It is also a crucially important component of the insect immune system. The humoral (non-cellular) immune function is addressed in Section 5.3 and the cellular function in Section 5.4. Other insect organs also contribute to insect immunity, notably the gut (Chapter 3), fat body (Chapter 6) and Malpighian tubules (Chapter 18).

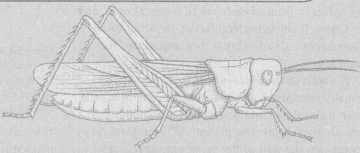

The Insects: Structure and Function (5th edition), ed. S. J. Simpson and A. E. Douglas.
Published by Cambridge University Press. © Cambridge University Press 2013.

5.1 The circulatory system

5.1.1 The dorsal vessel

The dorsal vessel has the primary role in the circulation of hemolymph around the body. In most insects it runs along the dorsal midline, just below the terga, for almost the whole length of the body (Fig. 5.1). Exceptionally, in the thorax of adult Lepidoptera and some Hymenoptera, it loops down between the longitudinal flight muscles (see Fig. 19.3). It may be bound to the dorsal body wall or suspended from it by elastic filaments. Anteriorly, it leaves the dorsal wall and is more closely associated with the alimentary canal, passing under the brain and just above the esophagus.

The dorsal vessel is open anteriorly, ending abruptly in most insects, but as an open gutter in orthopteroids. Posteriorly, it is closed, except in larval mayflies (Ephemeroptera), where three vessels diverge to the caudal filaments from the end of the heart.

The dorsal vessel is divided into two regions: a posterior heart in which the wall of the vessel is perforated by incurrent and sometimes also by excurrent openings (ostia), and an anterior aorta which is a simple, unperforated tube (Fig. 5.2). The heart is usually restricted to the abdomen, but may extend as far forwards as the prothorax in cockroaches (Blattodea). In orthopteroids it has a chambered appearance because it is slightly enlarged into ampullae at the points where the ostia pierce the wall. In the larvae of dragonflies (Odonata) and the cranefly, *Tipula*, the heart is divided into chambers by valves in front of each pair of incurrent ostia, and in *Cloeon* (Ephemeroptera) larvae, the ostial valves themselves are so long that they meet across the lumen.

The wall of the dorsal vessel is contractile and usually consists of one or two layers of muscle cells with a circular or spiral arrangement. Longitudinal muscle strands are also present; in Heteroptera they insert into the wall of the vessel anteriorly and posteriorly and do not connect with other tissues.

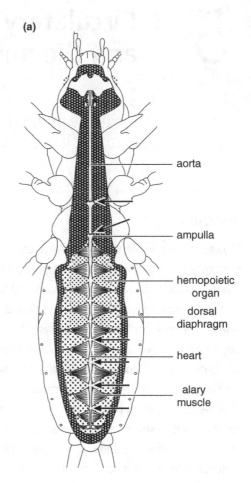

(a)

aorta

ampulla

hemopoietic organ

dorsal diaphragm

heart

alary muscle

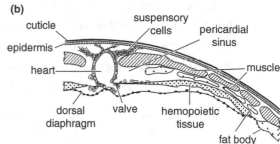

(b)

cuticle

suspensory cells

pericardial sinus

epidermis

heart

muscle

dorsal diaphragm

valve

hemopoietic tissue

fat body

Figure 5.1 Dorsal vessel and hemopoietic organs of the mole cricket, *Gryllotalpa*, in ventral dissection. The dorsal diaphragm is continuous over the ventral wall of the heart, but is omitted from the drawing for clarity. Arrows show positions of incurrent ostia.

The muscles of the heart are sometimes oriented in many different directions, but this appearance may arise through the insertion of the alary muscles into

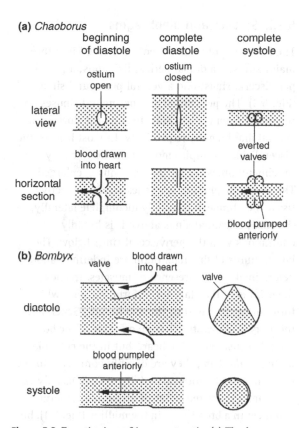

Figure 5.2 Functioning of incurrent ostia. (a) The larva of *Chaoborus*; the valves are prevented from opening outwards at systole by a unicellular thread (not shown) attached to the inside of the heart (after Wigglesworth, 1972). (b) The larva of *Bombyx*; the heart is shown in horizontal (left) and transverse (right) sections.

the heart (see below). Muscles of the dorsal vessel have short sarcomere lengths, with A-bands about 2 μm long. Each thick filament is surrounded by 9–12 thin filaments, as commonly found in visceral muscles of insects. The muscles are ensheathed inside and out by a basal lamina which also covers the nephrocytes (Section 18.8).

The incurrent ostia are vertical, slit-like openings in the lateral wall of the heart. The maximum number found in any insect is 12 pairs – nine abdominal and three thoracic. All 12 pairs are present in Blattodea and Orthoptera; many Lepidoptera have seven or eight. In aculeate Hymenoptera (bees, wasps and ants) there are only five pairs, while the housefly,

Musca, has only four. Lice (Phthiraptera) and many Heteroptera have only two or three pairs and the heart is restricted to the posterior abdominal segments.

The anterior and posterior lips of each ostium are reflexed into the heart to form a valve which permits the flow of hemolymph into the heart at diastole (expansion of the heart), but prevents its outward passage at systole (contraction of the heart) (Fig. 5.2a). During diastole the lips are forced apart by the inflowing hemolymph. When diastole is complete the lips are forced together by the pressure of hemolymph in the heart, and they remain closed throughout systole. In larvae of the phantom midge, *Chaoborus*, the valves tend to be evaginated by the pressure exerted during systole, but they are prevented from completely everting by a unicellular thread attached to the inside of the heart. In the silkworm, *Bombyx*, only the hind lip of each ostium is extended as a flap within the heart (Fig. 5.2b). During systole this is pressed against the wall of the heart and prevents the escape of hemolymph; and, when the heartbeat is reversed (see below) hemolymph flows out of the "incurrent" ostia. Excurrent ostia are restricted to the hearts of a few insect groups, including the Thysanura (silverfish), Orthoptera, Plecoptera (stoneflies) and Embioptera. In the last two orders they are unpaired, but in Orthoptera they are paired ventro-lateral openings in the heart wall and lack internal valves. Their number varies, but grasshoppers have five abdominal and two thoracic pairs. Most Blattodea and Mantodea have no excurrent ostia, but the hemolymph leaves the heart via segmental vessels that extend laterally. For example, the cockroach, *Periplaneta*, has five pairs of abdominal segmental vessels in late-stage nymphs; adults additionally have two pairs of thoracic segmental vessels. Mantids have only abdominal vessels, each of which has a muscular valve that permits only the outward flow of hemolymph. The walls of the vessels are non-muscular.

In most insects, the heart is innervated by nerves running around the body wall from the ventral segmental ganglia. In Odonata, Blattodea, Phasmatodea, Orthoptera, larval Lepidoptera and some adult Coleoptera, branches of the segmental nerves combine to form a lateral cardiac nerve running along each side of the heart. In *Locusta*, for example, groups of neurons with cell bodies in the midline in each abdominal ganglion send axons to the heart. In addition, a pair of neurosecretory cells in the subesophageal ganglion each sends one axon forwards into the circumesophageal connective and another back along the length of the ventral nerve cord. This axon branches in each of the abdominal ganglia, sending a branch into a lateral nerve which extends dorsally to the heart. These branches contribute to the lateral cardiac nerve and have varicose terminals, typical of a neurosecretory cell, along the heart. The cells produce a FMRFamide-like peptide (Fig. 20.6). The lateral cardiac nerves also receive innervation from neurons whose somata are not in the central nervous system, but lie adjacent to the heart itself. They are called cardiac neurons. *Periplaneta* has about 32 such neurons, some of which are neurosecretory.

In the majority of holometabolous insects, the segmental nerves extend to the heart, but do not form lateral cardiac nerves. Exceptionally, the heart of some insects (e.g., *Anopheles* mosquitoes) receive no direct innervation at all, although there are segmental nerves to the alary muscles.

5.1.2 Sinuses and diaphragms

The hemocoel of many insects is divided into three major sinuses: a dorsal pericardial sinus; a perivisceral sinus; and a ventral perineural sinus (Fig. 5.3). The pericardial and perineural sinuses are separated from the visceral sinus by the dorsal and ventral diaphragms, respectively. In most insects, the visceral sinus occupies most of the body cavity, but in ichneumonids the perineural sinus is enlarged. The dorsal diaphragm is a fenestrated connective tissue membrane. It is usually incomplete laterally, so that the pericardial sinus above it is broadly continuous with the perivisceral sinus below. The lateral limits of the diaphragm are indefinite and are determined by the presence of muscles, tracheae or the origins of the alary (or aliform) muscles which form an integral part of the diaphragm. Generally, the alary muscles stretch from one side of the body to the other just below the heart, but in the cecropia moth, *Hyalophora*, they are directly connected to the heart muscles by intercalated discs. They usually fan out from a restricted point of origin on the tergum and meet in a broad zone in the midline (Fig. 5.1), but sometimes, as in grasshoppers, the origin of the muscles is also broad. In most orthopteroids, only the portion of the alary muscle near the point of origin is contractile, the rest, and greater part, being made up of bundles of connective tissue which branch and

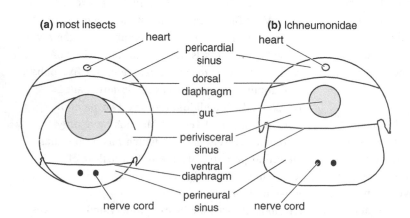

Figure 5.3 Main sinuses of the hemocoel shown in diagrammatic cross-section (after Richards, 1963).

anastomose. Some of the connective tissue fibers form a plexus which extends to the heart wall. Orthopteroids may have as many as ten abdominal and two thoracic pairs of alary muscles, but in other insects the number is reduced. Most terrestrial Heteroptera, for instance, have 4–7 pairs. The alary muscles are visceral muscles with 10–12 thin filaments to every thick filament.

The ventral diaphragm is a horizontal septum just above the nerve cord separating the perineural sinus from the main perivisceral sinus (Fig. 5.3). It is present in both larvae and adults of Odonata, Blattodea, Orthoptera, Neuroptera (lacewings) and Hymenoptera, but is only found in adults of Mecoptera (scorpion flies), Lepidoptera and nematocerous Diptera. In the Lepidoptera the nerve cord is attached to the ventral surface of the ventral diaphragm by connective tissue. When present, the diaphragm is usually restricted to the abdomen, but it extends into the thorax in some grasshoppers and Hymenoptera. The structure of the ventral diaphragm varies. For instance, in the thorax of grasshoppers it is a delicate membrane with little or no muscle, but in the abdomen it is a solid muscular sheet. Laterally it is attached to the sternites, usually at one point in each segment, resulting in broad gaps along the margins where the perivisceral and perineural sinuses are continuous. Its structure may vary with developmental stage, e.g., in *Corydalis* (Neuroptera) it forms a solid sheet in the larva, but a fenestrated membrane in the adult.

5.1.3 Accessory pulsatile organs

Insects have multiple pulsatile structures that assist the bulk flow of hemolymph, especially through the appendages, including wings, legs, antennae and, in adult females, the ovipositor.

A pulsatile organ drawing hemolymph from the wings is present in both wing-bearing segments of most adult insects, but only in the mesothorax of Diptera and Coleoptera Polyphaga. A hemolymph space, or reservoir, beneath the posterior part of the tergum (scutellum) which is largely or completely isolated from the remaining hemocoel of the thorax connects with the posterior veins of the wing via the axillary cord of the wing. The muscular pump is formed from the ventral wall of the reservoir, which may be derived from the heart or a separate structure (Fig. 5.4). The exact form of the wing pulsatile organs varies among the different insect groups.

Most holometabolous insects, as well as Hemiptera, have wing pulsatile organs in which a muscular diaphragm, independent of the actual heart, bounds the subscutellar reservoir on the ventral side. It is suspended from the scutellum by a number of filamentous strands (Fig. 5.4a–c). Contraction of the muscles, which are innervated from the ventral nerve cord, causes the diaphragm to flatten, drawing hemolymph from the wings. Relaxation of the muscles is associated with an upward movement of the diaphragm, presumably due to the elasticity of the suspensory strands, and hemolymph is forced into the body cavity (Fig. 5.4b,c). In hemimetabolous groups, other than Hemiptera, and in Coleoptera and Hymenoptera Symphyta, the wing pulsatile organ is an expansion or diverticulum of the dorsal vessel with a pair of incurrent ostia opening from the subscutellar reservoir. Odonata, for example, have, in each pterothoracic segment, an ampulla which connects with the aorta by a narrow vessel (Fig. 5.4d). It is suspended from the tergum by elastic ligaments, and its dorsal wall is muscular. When the muscles contract, the ampulla is compressed and hemolymph is driven into the aorta. At the same time, the volume of the subscutellar reservoir is increased and hemolymph is sucked from the wings: The posterior veins therefore deal with the afferent flow while the anterior veins carry the efferent flow. When the muscles relax, the elastic ligaments restore the shape of the ampulla so that hemolymph is sucked into it from the reservoir.

A pulsatile organ is also found at the base of each antenna. It consists of an ampulla from which a fine tube extends almost to the tip of the antenna. The ampullae of Thysanura (silverfish), Archaeognatha

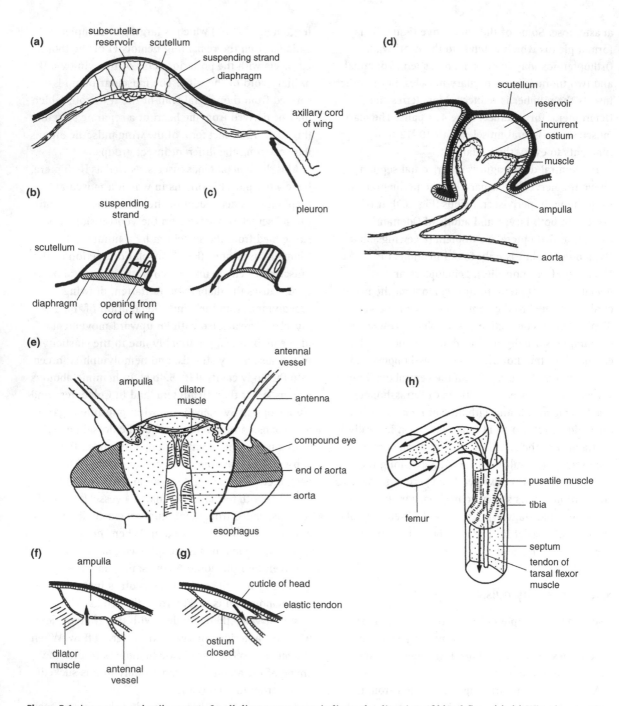

Figure 5.4 Accessory pulsatile organs. In all diagrams arrows indicate the direction of blood flow. (a)–(c) Wing heart not connected to the aorta, such as occurs in most holometabolous insects. (a) Transverse section through the thorax showing the connection of the subscutellar reservoir to the axillary cord of the wing. (b), (c) Diagrammatic longitudinal section, anterior to left. When muscles of the diaphragm contract (b), the diaphragm is flattened and blood is drawn in from the wing.

(bristletails) and some Plecoptera have no muscles, and it is presumed that in these insects the ampulla serves simply to direct the flow of hemolymph from the opening of the aorta into the antenna. In most insects, however, the ampullae have dilator muscles. Compression, which drives hemolymph into the antenna, results from the activity of elastic filaments on both sides of the wall of the ampulla. Only Dermaptera (earwigs) have compressor muscles. *Periplaneta* has a single muscle connecting the two ampullae so that when it contracts both ampullae are dilated and hemolymph flows into each one through an ostium (Fig. 5.4e–g). The contractions are myogenic in origin, but may be modulated by neural input since nerve endings do occur in the muscle. However, most of the nerve endings are concentrated in the ampulla, which appears to function as a neurohemal organ. The axons originate in the subesophageal ganglion from a dorsal unpaired median neuron and from a pair of somata placed laterally. In Lepidoptera, the aorta ends anteriorly in a sac from which the antennal vessels arise.

Most insects have a longitudinal septum in the legs which divides the lumen into two sinuses and permits a bidirectional flow of hemolymph within the leg (see below). In Hemiptera, the septum twists through 90° at the proximal end of the tibia, and there is a muscle at this point (Fig. 5.4h). When the muscle contracts, it compresses one sinus, forcing hemolymph into the thorax, and at the same time enlarges the other so that hemolymph is drawn into the leg. Its activity is probably myogenic, although it may also be subject to neural modulation.

Some insects have very long cerci in which the maintenance of hemolymph flow might require some special feature. Ephemeroptera have small, non-contractile vessels extending from the posterior end of the heart into the cerci. Plecoptera also have cercal hemolymph vessels, but these do not connect with the heart. They open directly into the perivisceral cavity, but the remainder of the cercal lumen connects with the perivisceral cavity via the lumen of the paraproct. Changes in the volume of the paraproct produced by a small muscle draw hemolymph from the outer lumen of the cercus and pump it into the perivisceral cavity. This flow draws hemolymph into the cercus through the central vessel.

5.2 Circulation

5.2.1 Movement of the hemolymph

The movement of hemolymph around the insect body has been studied by analysis of the distribution of markers injected into the body cavity or ingested by the insect. Generally, injected substances become uniformly distributed through the hemolymph within 5–20 minutes, varying with environmental conditions, especially temperature, and the physiological status of the insect.

Hemolymph circulation is maintained primarily by the dorsal vessel. In hemimetabolous and larval holometabolous insects, the hemolymph is pumped forwards through the heart at systole, entering the perivisceral sinus through the anterior opening of the aorta in the head (and through the excurrent

---

Caption for Figure 5.4 (*cont.*) When the muscles relax (c), elastic suspensory elements draw the diaphragm up and force blood out anteriorly (after Krenn and Pass, 1993). (d) Wing heart connected to the aorta as in most hemimetabolous insects. Longitudinal section of the thorax. The subscutellar reservoir connects with the axillary cord of the wing on either side (modified after Whedon, 1938). (e)–(g) Antennal pulsatile organ of cockroach. (e) General arrangement as seen from above. Top of the head cut away and brain removed. (f) Dilator muscle contracts, enlarging lumen of ampulla so that blood is drawn in from the hemocoel in the head. Lowered pressure causes constriction at the origin of the antennal vessel so that the backflow of blood from the antenna is restricted. (g) The muscle relaxes and the ampulla is flattened by the elasticity of its inner wall and the pull of the tendon. The ostium is closed by a valve and blood forced into the antennal vessel (after Pass, 1985). (h) Leg pulsatile organ of *Triatoma*. Contraction of the muscle compresses the blood sinus on one side of the septum and enlarges that on the other so that blood flows down the leg on one side of the septum and up on the other (after Kaufman and Davey, 1971).

ostia where these exist). The valves on the incurrent ostia prevent the escape of hemolymph through these openings. The force of hemolymph leaving the aorta anteriorly tends to push hemolymph backwards in the perivisceral sinus. The backwards flow is aided by the movements of the dorsal diaphragm and by the inflow of hemolymph into the heart, through the incurrent ostia, at diastole (Fig. 5.5a–c). Movements of the ventral diaphragm presumably help to maintain the supply of hemolymph to the ventral nerve cord.

In adult Lepidoptera, Coleoptera and Diptera, and perhaps in some other insects, the hemolymph is shunted backwards and forwards between the thorax and abdomen, rather than circulated. This is possible because the hemocoel in the two parts is separated – by a movable flap of fatty tissue in Lepidoptera or large air sacs in cyclorrhaphous Diptera – and because the heartbeat exhibits periodic reversals. As the abdomen contracts, the heart pumps hemolymph forwards into the head (Fig. 5.5d). When the heart reverses, the abdomen actively expands, drawing hemolymph past the barrier of fat (Fig. 5.5e). Movements of the hemolymph are coordinated with ventilatory movements. Any activity which tends to induce pressure differences between different parts of the body must affect the circulation, and, especially in adult Lepidoptera, Coleoptera and Diptera, there is a close functional link between ventilatory and hemolymph movements.

Many insects appear to have a well-defined, but variable circulation through the wings, although in some, apparently, circulation only occurs in the young adult. In the absence of this wing circulation, the tracheae in the wings collapse, and the wing structure becomes brittle and dry. In most cases, hemolymph is drawn out of the wings from the axillary cords by the thoracic pulsatile organs. In the German cockroach, *Blattella*, the anal veins connect with the axillary cord through channels between the upper and lower wing membranes. As hemolymph is drawn out posteriorly, it is drawn in anteriorly from a hemolymph sinus from which the major anterior veins arise (Fig. 5.6).

In insects lacking leg pulsatile organs, the flow of hemolymph through the legs is thought to be maintained by pressure differences at the base. Hemolymph passes into the posterior compartment of the legs from the perineural sinus and out into the perivisceral sinus from the anterior compartment.

5.2.2 Heartbeat

Systole, the contraction phase of the heartbeat, results from the contractions of the intrinsic muscles of the heart wall. In most insects this activity begins posteriorly and spreads forwards as a wave.

In *Periplaneta* and Orthoptera, however, the entire length of the heart contracts, usually synchronously although peristaltic contractions may sometimes occur. Diastole, the expansion phase when hemolymph enters the heart, results from relaxation of the heart muscles assisted by the elastic filaments that support the heart. In general, the alary muscles are not responsible for diastole and often contract at a lower frequency than the heart. In *Rhodnius*, however, complete diastole is dependent on the alary muscles. After diastole there is a third phase in the heart cycle known as diastasis, in which the heart remains in the expanded state (Fig. 5.7). Increases in the frequency of the heartbeat result from reductions in the period of diastasis. There is considerable within- and between-species variation in the frequency of the heartbeat. In general, the heartbeat frequency declines with increasing age and size of larval insects, and also varies with age within a larval stadium. For example, in silkworms (*Bombyx*), the rate declines from 80 beats min^{-1} in second-stage larvae to about 50 in the fifth stage, and it drops sharply just before each molt except the last (Fig. 5.8a). In the pupa, the heartbeat falls to 10–20 beats min^{-1} and remains at this low level until shortly before adult eclosion. A similar, but slightly less marked, change occurs during the development of the migratory locust, *Locusta*, but the rate of beating tends to rise before a molt (Fig. 5.8b). In Lepidoptera, a very sharp increase in the heart

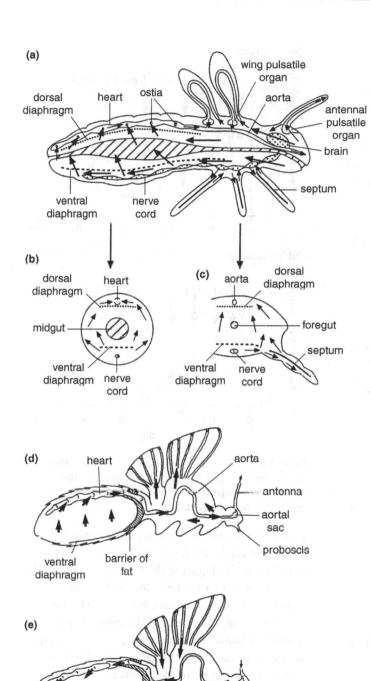

Figure 5.5 Blood circulation. (a)–(c) An insect with a fully developed circulatory system. Arrows indicate the course of the circulation. (a) Longitudinal section; (b) transverse section of abdomen; (c) transverse section of thorax. (d),(e) An insect in which the blood oscillates between the thorax and abdomen. (d) Abdominal contraction with the heart beating forwards pushes blood into the anterior regions of the insect; (e) abdominal expansion with the heart beating backwards draws blood into the abdomen (after Wasserthal, 1980).

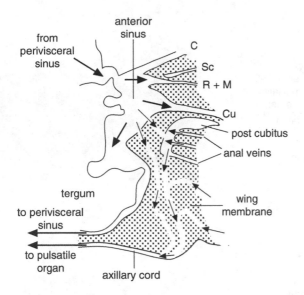

from perivisceral sinus

anterior sinus

C

Sc

R + M

Cu

post cubitus

anal veins

tergum

to perivisceral sinus

to pulsatile organ

wing membrane

axillary cord

Figure 5.6 Blood circulation in the base of the forewing of *Blattella*. Areas in which the two membranes of the wing are fused together are shaded. Well-defined channels, such as veins, are outlined by a solid line, less definite channels have no lines. Axillary sclerites are omitted (after Clare and Tauber, 1942).

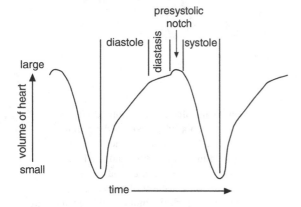

Figure 5.7 Changes in the volume of a section of the heart during beating. The presystolic notch results from an increase in hydrostatic pressure, producing a slight increase in volume within this part of the heart due to the start of systole in the more posterior segments.

rate occurs at eclosion, falling to a sustained but moderate rate when eclosion is complete.

In addition to these intrinsic changes, the rate of heartbeat is affected by environmental factors.

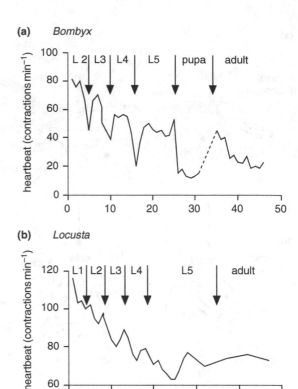

(a) *Bombyx*

(b) *Locusta*

Figure 5.8 The frequency of heartbeat at different stages of development. L1–L5 indicate larval stages; arrows indicate ecdysis. (a) *Bombyx* (after Masera, 1933); (b) *Locusta* (after Roussel, 1972).

Heartbeat increases with temperature within the thermal limits for the insect. In *Locusta* the rate of heartbeat is also higher in the light than in the dark.

It is common for the heart to stop beating, sometimes for a few seconds, but sometimes for 30 seconds or more. The heart of the young pupae of *Anopheles* sometimes stops beating altogether, and in old pupae no activity of the heart is observed.

It is also common for the heartbeat to undergo periodic reversals, with waves of contraction starting at the anterior end. When this occurs, hemolymph is forced out of the "incurrent" ostia; in the mole cricket, *Gryllotalpa*, powerful currents pass out of the subterminal incurrent ostia. In female *Anopheles*, 31% of heartbeats start at the anterior end of the heart. Reversal of heartbeat is rare in holometabolous larvae, but begin in the pupal stage or even at the larva–pupa ecdysis. Usually the rate of heartbeat is lower when the heart is pumping backwards. In the adult blowfly, *Calliphora*, the rates are about 175 beats min^{-1} backwards compared with about 375 beats min^{-1} forwards. In pharate adult Lepidoptera, periods of fast, forward beating lasting a few minutes alternate with periods when the heart reverses and beats more slowly, but during the period of wing expansion following eclosion no reversals occur (Fig. 5.9). Subsequently, periodic reversals are an essential feature of hemolymph circulation, at least in the Lepidoptera (see above). The distribution of hemolymph pressures is an important determinant of the direction of the heartbeat (if pressure at the front of the heart is so high that back pressure is set up, the heartbeat is reversed), but other factors, including oxygen supply, may also contribute to the incidence of beat reversal.

The activity of the pulsatile organs may be different from that of the heart. The antennal ampullae of *Periplaneta* pulse at about 28 beats min^{-1}, considerably slower than the heart rate. The wing pulsatile organs of Lepidoptera are only active during heart reversal (Fig. 5.5). At eclosion the pulsatile organs pulsate more rapidly and without interruption.

The activity of the heart is basically myogenic, although the myogenic pattern may be modulated neurally or hormonally. As the segmental nerves leading to the heart contain the ramifying terminals of neurosecretory axons, it might be expected that their secretions exert modulatory effects. In addition, hormones released into the hemolymph at points remote from the heart are known to affect it. For example, the increase in the rate of beating at the time of eclosion in Lepidoptera is at least partly due to peptides released in the hemolymph at this time. These cardioacceleratory peptides are produced in neurosecretory cells in the ganglia of the ventral nerve cord and released into the hemolymph at the perivisceral neurohemal organs. The same peptides also increase the heartbeat during flight. Their effects are synergized by very low levels of octopamine, which is also present in the hemolymph during wing inflation and flight. A number of other neurohormones affect the heart rate in vitro, but they are not known to be involved in regulation of heartbeat in the intact insect.

5.3 Hemolymph

The hemolymph fills the hemocoel and circulates around the body, bathing the tissues directly. It consists of fluid plasma in which cells, known as hemocytes, are suspended. The properties and composition of the plasma, together with its function in transport, storage and immunity, are considered in this section. The diversity and functions of hemocytes are addressed in Section 5.4.

5.3.1 Hemolymph volume

The hemolymph volume, expressed as a percentage of the total body weight of the insect, varies widely among insects. In the heavily sclerotized tenebrionid beetle *Onymachus*, hemolymph constitutes about

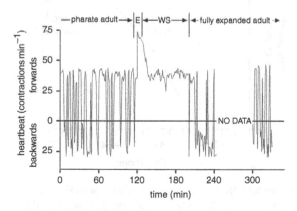

Figure 5.9 Changes in the heartbeat of an individual *Manduca* at the time of eclosion. E = eclosion, WS = wing spreading (after Tublitz and Truman, 1985).

11% of the beetle's total mass; in mid-stadium larvae of *Locusta*, the figure is about 18%, while in mature adults it is about 12%; in cockroaches it is about 17%; and in caterpillars 35–40%.

In most insects hemolymph water accounts for 20–25% of the total body water (the remainder being principally within cells). Generally, the water content and hydrostatic pressure of hemolymph increase just prior to ecdysis. For example, the hemolymph volume in *Schistocerca* locusts almost doubles in advance of each molt, and is then reduced again after the molt (Fig. 5.10). In *Periplaneta* the pre-molt increase in volume is associated with an increase in activity of an antidiuretic hormone and the post-ecdysial fall in volume is produced by a transient rise in the hemolymph titer of diuretic hormone (Fig. 5.11). Similarly, diuresis reduces the hemolymph volume of the cabbage butterfly *Pieris* by about 70% in the hours following eclosion. In caterpillars, however, the hemolymph water content is high, accounting for up to 50% of the total water content, and the associated high pressure provides for a hydrostatic skeleton that maintains body shape and contributes to locomotion. Caterpillar locomotion is considered further in Chapter 8.

An important function of the hemolymph is to provide a reservoir of water to sustain the levels of water in the tissues. This is illustrated vividly by experiments conducted on the desert tenebriod *Onymachus*. When maintained in desiccating conditions, hemolymph volume is reduced by about 60%, but tissue water content does not change (Fig. 5.12). Some of the tissue water is drawn from the hemolymph; the remainder comes from fat metabolism. Similar changes can occur over shorter periods in some insects. For example, it is almost

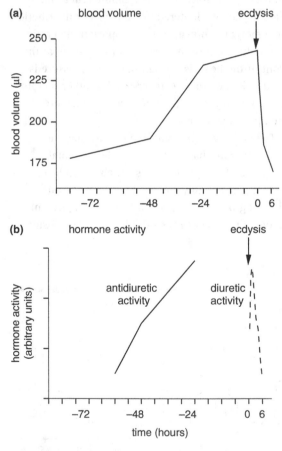

Figure 5.11 Regulation of blood volume in relation to ecdysis in *Periplaneta* (after Mills and Whitehead, 1970). (a) Changes in blood volume; (b) antidiuretic and diuretic activity of the hemolymph, suggesting that changes in hormonal activity regulate the changes in volume.

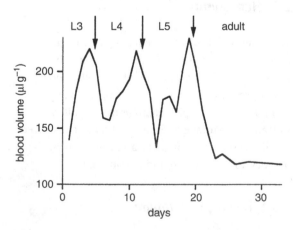

Figure 5.10 Changes in the blood volume (expressed as volume per unit weight) during the development of *Schistocerca*. L3, L4 and L5 refer to successive larval stages. Arrows indicate the time of ecdysis (after Lee, 1961).

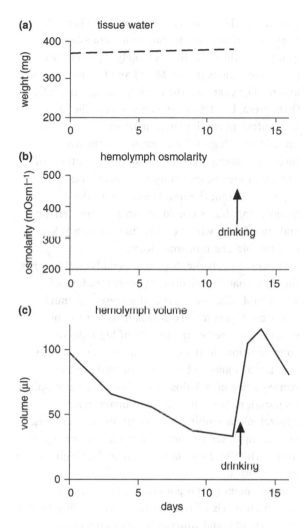

Figure 5.12 Regulation of tissue water and hemolymph osmotic pressure in *Onymacris*. The insects were without food or water at 26°C and 10–15% relative humidity for the first 12 days. On day 12 they were given distilled water to drink and then maintained at 50–60% relative humidity (after Nicolson, 1980). (a) Tissue water remained almost constant over the 12 days without food or water. (b) Hemolymph osmolarity rose very slightly during the first 12 days of concentration, despite the marked reduction in hemolymph volume (c).

impossible to obtain hemolymph samples from red desert locusts, *Nomadacris*, collected in the heat of the afternoon, but the insects have ample hemolymph in the morning and late evening.

5.3.2 pH of hemolymph

The pH of insect hemolymph is usually between 6.4 and 6.8, although values >7 pH units have been recorded in a dragonfly larva and in the larvae of the midge *Chironomus*. During activity the hemolymph tends to become more acidic due to the liberation of acid metabolites, including carbon dioxide. The buffering capacity of the hemolymph is mediated by various compounds, including bicarbonate and inorganic phosphate, the carboxylic and amino groups on various organic compounds, especially amino acids. The extrusion of protons from the hemolymph can also contribute to maintaining the hemolymph pH. In experiments on *Schistocerca*, the hemolymph recovered its normal pH within 24 hours after injection of acid ($25\,\mu l\,0.4\,M$ HCl); and this was accompanied by a significant reduction in pH of the gut contents, suggesting that the acid load in the hemolymph was relieved, at least partly, by protons extruded from the hemolymph.

5.3.3 Osmotic pressure and osmotically active solutes

The osmotic pressure of insect hemolymph is generally 0.7–1.2 MPa, with some variation with species, physiological condition and developmental stage of the insect. (In particular, overwintering insects in diapause sustain high hemolymph osmotic pressures, often greater than 1.2 MPa.) For a given insect, the osmotic pressure is regulated within narrowly defined limits.

The most important osmotic effectors are inorganic ions and small organic molecules, including sugars and amino acids. These compounds are also essential substrates for the maintenance and growth of insect tissues and organs. Consequently, the concentration of these solutes in hemolymph is controlled, despite variation in hemolymph volume (Section 5.3.1), by exchange with tissues (uptake by tissues when hemolymph volume is declining, and release from tissues into hemolymph when it is increasing).

Importantly, the osmotic pressure of hemolymph cannot be determined accurately from its solute content because some solutes are bound to macromolecules, reducing their osmotic effect. For example, in the larvae of the mosquito *Aedes*, the osmotic activity of sodium is lower than expected from its concentration, possibly because some is bound to large anionic molecules.

The relative importance of the different compounds varies among insect groups (Fig. 5.13). In many insects the most abundant inorganic ions in the hemolymph are Na^+ and Cl^-. Anions additional to Cl^- include phosphate and bicarbonate, usually at low concentrations; and the cations K^+, Mg^{2+} and Ca^{2+} are also present. There is, however, considerable variation among insect groups, following both phylogenetic and dietary factors (Fig. 5.13).

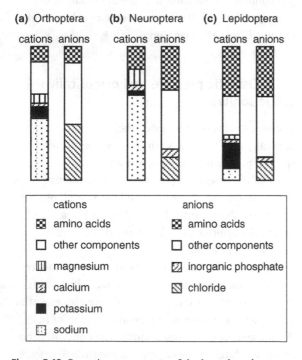

(a) Orthoptera   **(b)** Neuroptera   **(c)** Lepidoptera

cations anions   cations anions   cations anions

cations	anions
amino acids	amino acids
other components	other components
magnesium	inorganic phosphate
calcium	chloride
potassium	
sodium	

Figure 5.13 Osmotic components of the hemolymph in different groups of insects expressed as percentages of the total osmolar concentration. Each vertical column represents 50% of the total concentration (after Sutcliffe, 1963).

The Na^+ and Cl^- contents tend to be higher in hemimetabolous insects than holometabolous groups, in which the Na^+ is largely replaced by other inorganic cations (K^+ or Mg^{2+}) and Cl^- by organic anions. For example, the ratio of Na^+:K^+ is 10:1 in Orthoptera, but 1:5–10 in Lepidoptera. The high K^+ content of lepidopteran hemolymph has been linked to the high K^+:Na^+ ratio of plant tissues. Insect hemolymph contains low concentrations of transition metals, including zinc, manganese, copper and iron; the iron is not free in the hemolymph, but is bound to two proteins, ferritin and transferrin, which ensure that the iron is kept in a soluble and non-toxic form.

In most insects the dominant hemolymph sugar is the disaccharide trehalose, at concentrations of 10–50 mM. Glucose is generally present at much lower concentrations (<5 mM). It remains to be established whether the reports of high glucose concentrations in the hemolymph of some species, e.g., larval *Phormia* blowflies, are valid or the consequence of trehalose hydrolysis during sample preparation. Some insects additionally have appreciable hemolymph concentrations of polyols. For example, sorbitol accumulates in the hemolymph of the whitefly, *Bemisia tabaci*, reared at high temperatures.

Insect hemolymph plasma is characterized by very high levels of free amino acids, ranging from 25 to 75 mM, and functioning as buffers in osmoregulation and as substrates for protein synthesis. Most or all of the 20 amino acids contributing to insect protein are detectable, although there is considerable variation in the relative abundance of the various amino acids among insect groups and with developmental stage. Of particular interest is tyrosine, which is poorly soluble and occurs predominantly as the more soluble glucoside in hemolymph. However, just before the insect molts, free tyrosine is released and accumulates and subsequently decreases sharply as it is used in tanning and melanization of the new cuticle (see Fig. 16.18).

Organic acids are present in some quantity in the plasma of insect hemolymph. The major components are acids associated with the citric acid cycle, including citrate, α-ketoglutarate, succinate and malate.

5.3.4 Hemolymph proteins

The plasma of insect hemolymph is protein-rich. The protein content varies among insect groups and with developmental age over an order of magnitude between 20 and 200 mg ml^{-1}, and proteome analyses reveal up to several hundred different proteins. Protein composition varies with developmental age and physiological condition. Many of the most abundant proteins play key roles in the transport and storage of nutrients between organs and in systemic immunity. The principal source of hemolymph proteins is the fat body, with some contribution from hemocytes (especially for immune-related proteins) and other organs.

One very important group of storage proteins are the hexamerins, which are very large proteins comprising (as the name suggests) six subunits (each 80 kDa). The hexamerins with a very wide phylogenetic distribution are the arylphorins, in which the aromatic amino acids phenylalanine and tyrosine comprise 18–26% of the total, but which contain little methionine. Lepidoptera also have methionine-rich hexamerins that contain 4–8% methionine. The hexamerins are synthesized in the fat body and released into the hemolymph, where they can accumulate to 50% of the total hemolymph protein. They are subsequently used as a source of amino acids, such as during pupation of holometabolous insects. Hexamerins are related to hemocyanins (respiratory pigments of many Crustacea and some insects, e.g., Plecoptera), and are believed to have evolved by the loss of the Cu^{2+}-prosthetic group and oxygen-binding site of the hemocyanin molecule, and the shift in amino acid composition to favor aromatic amino acids or methionine.

Because lipids are not soluble in water, their transport in hemolymph involves combination with specific proteins of the large lipid transfer protein (LLTP) superfamily to form lipoproteins. Insects have a single class of lipoprotein called high-density lipophorin (HDLp), which transports diacylglycerol (DAG), together with small amounts of phospholipids and sterols. The HDLp particle is a re-usable shuttle: It circulates in the hemolymph between different tissues, alternately taking up lipids (e.g., from gut epithelium, fat body) and delivering lipids (e.g., to ovaries, flight muscle) without being internalized or degraded. The protein core of the HDLp particle is one copy each of apolipophorin I and apolipophorin II, derived by a single cleavage of the precursor protein apolipophorin II/I. Under conditions of high lipid demand (e.g., during sustained flight), the HDLp particle is transformed to a low-density lipophorin (LDLp) by association with several copies of a third apolipoprotein called apolipophorin III, together with substantially increased loading of DAG. The diameter of the LDLp is more than double that of the HDLp (Fig. 5.14).

The hemolymph of reproductive female insects contains appreciable amounts of a further protein of the LLTP superfamily: the glycolipophosphoprotein vitellogenin, which is usually transported as a homodimer of 200–660 kDa. Vitellogenin is a nutrient reserve for embryos. It is synthesized in the fat body and transported via the hemolymph to the ovaries, where it is taken up by developing oocytes (at a discrete developmental stage known as vitellogenesis). The vitellogenins of the honey bee, *Apis mellifera*, have been studied extensively, revealing that the hemolymph of adult workers (which are reproductively sterile) also contains appreciable amounts of vitellogenin. Overall, vitellogenin accounts for 70% of the protein in the hemolymph of adult queens and 50% of the hemolymph protein in young workers that reside in the nest ("nurse bees"). When workers are about two weeks old they switch to foraging, and this is accompanied by a rapid decline in hemolymph

(a)

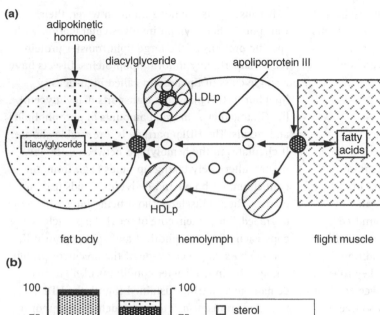

(b)

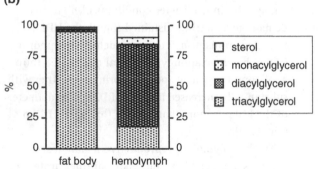

Figure 5.14 Lipid in the hemolymph. (a) Diagrammatic representation of the movement of lipids from the fat body to flight muscles in *Manduca* by the formation of low-density lipophorin (LDLp) (based on Shapiro *et al.*, 1988). (b) Proportions of major lipid components in the fat body and hemolymph of the pupa of *Hyalophora*. Most lipid is stored as triacylglycerol, but it is transported as diacylglycerol (data from Gilbert, 1967).

vitellogenin titer, to less than 20% of the total protein. Studies using RNAi-mediated knockdown of vitellogenin gene expression demonstrate that declining vitellogenin level is the signal responsible for the transition from nurse bee to foraging (Fig. 5.15).

The hemolymph additionally bears various immunity-related products that function in protecting the insect against systemic infection. These products are components of the humoral (i.e., non-cellular) part of the insect immune system. To a large extent, the humoral effectors are inducible, i.e., they are undetectable (or nearly so) in uninfected insects, and increase to high titers in response to mechanical wounding and infection by bacteria, fungal pathogens, nematodes, etc., or by the oviposition of parasitoid eggs. In this section,

two vitally important inducible humoral effectors are considered: the phenoloxidase system, which is induced very rapidly by a proteolytic cascade (see below) and anti-microbial peptides, whose production requires *de novo* gene expression and is somewhat slower.

Phenoloxidase is an enzyme that catalyzes the oxidation of phenols to quinones, and cytotoxic molecules (such as 5,6-dihyroxyindole) in reactions that also generate reactive oxygen and nitrogen intermediates. Subsequent reactions with phenoloxidase-derived quinones in the hemolymph lead to the synthesis of a brown-black insoluble pigment called melanin, which is also cytotoxic and contributes to the physical isolation of invading organisms. For example, various insects, including mosquitoes and aphids, enclose and inactivate

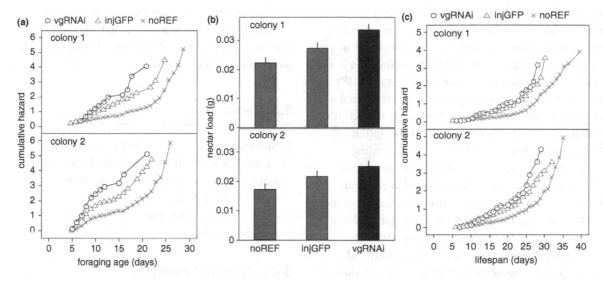

Figure 5.15 Impact of RNAi knockdown of expression of the vitellogenin gene in workers of the honey bee *Apis mellifera*. (a) Frequency of transition from nurse bee (in hive) to foraging bee activity; (b) nectar collection; (c) lifespan. vgRNAi: RNAi-mediated knockdown of vitellogenin gene expression; injGFP: injected negative control for RNAi against non-target gene (green fluorescent protein, GFP); noREF: uninjected control. Reproduced from Nelson *et al.* (2007), figures 2, 3 and 4.

bacterial clumps and parasitoid eggs in melanotic capsules, comprising non-cellular material dominated by melanin.

The phenoloxidase protein is stored as an inactive precursor, prophenoloxidase, in circulating hemocytes. In response to microbial products (e.g., peptidoglycan) or wounding, the prophenoloxidase is released from hemocytes and activated by limited proteolysis through the action of a specific serine protease. Prophenoloxidase activation and melanin formation also play a crucial role in the insect response to wounding by contributing to coagulation of blood and clot formation, thereby promoting bleeding control and wound healing. Studies with mutant *Drosophila* indicate that the primary clot requires the protein hemolectin (which has some sequence homology to the von Willebrand factor important in mammalian blood coagulation) but not phenoloxidase. In this insect the phenoloxidase-derived melanin is required for strengthening and stabilizing the initial clot, to form the mature, hardened clot that limits the risk

of infection after wounding. In some other insects, notably *Anopheles* mosquitoes, phenoloxidase is important for the formation of the initial clot.

Many of the humoral effectors in the hemolymph are derived from the fat body. They include anti-microbial peptides (AMPs), opsonins, which are essential components of the melanization clotting system (see above), catalase, which limits the reactive oxygen species produced by immune-related respiratory burst of immune cells, and transferrin, which reduces the availability of the free iron required for rapid microbial proliferation. By far the best characterized of these humoral immune effectors are the AMPs. Most AMPs are 15–50 amino acid residues in length and have an amphipathic structure, i.e., they include both charged and hydrophobic sequences, and this property enables them to partition into biological membranes. (The composition and net charge on microbial and insect membranes are different, and insect AMPs preferentially partition into bacterial membranes.) Some AMPs insert into the microbial membrane, generating a pore that depolarizes the cell membrane; others penetrate through the microbial membrane to bind and disrupt an intracellular target. In response to infection, AMPs are synthesized in the fat body and released into the hemolymph. AMPs are gene products, and their production is regulated at the transcriptional level, specifically by transcription factors of the NF-κB family. The transcription factor Relish is activated by the Imd signaling cascade in response to peptidoglycan fragments of Gram-negative bacteria that bind to receptors (peptidoglycan recognition proteins, PGRPs) on the surface of insect cells; and the transcription factor Dif is activated by the Toll signaling pathway, ultimately in response to molecular patterns of Gram-positive bacteria and fungi. Although there is cross-talk between the Toll and Imd pathways and other components of the intracellular signaling network, the AMPs regulated by the Imd pathway are primarily active against Gram-negative bacteria; and those regulated by the Toll pathway are active against Gram-positive bacteria and fungi.

The Imd and Toll signaling pathways are broadly conserved among the insects that have been studied (apart from the pea aphid, which lacks an intact Imd pathway), but the number and identity of AMPs regulated by these pathways vary widely. Some AMPs are taxonomically restricted. For example, only drosophilid flies are known to produce the anti-fungal AMPs drosomycin and metchnikowin, and gloverins are specifically lepidopteran AMPs. Other AMPs (e.g., cecropins, attacins) are known in various insects, and defensins occur widely across the animal kingdom.

The hemolymph of uninfected insects additionally contains lysozyme. This enzyme hydrolyzes β-1,4 linkages between N-acetylmuramic acid and N-acetyl-D-glucosamine residues in peptidoglycan of bacterial cell walls. In the uninfected insect, the amount and activity of lysozyme in the hemolymph is low but generally detectable. It cleaves the peptidoglycan of invading bacteria, generating peptidoglycan fragments which act as elicitors of the humoral response. Specifically, the fat body is triggered to synthesize and release high titers of lysozyme (an example of a positive feedback loop), phenoloxidase and AMPs.

5.4 Hemocytes

Hemocytes are the cells residing in hemolymph. They mediate the cellular arm of the insect immune response, especially phagocytosis and, in many insects, encapsulation. In this section the types of hemocytes, their origin and their function in immunity are addressed.

5.4.1 Types of hemocytes

The hemocytes have variously been classified by the criteria of morphology, function and molecular markers; and there is no universally recognized scheme for the classification of these cells (Fig. 5.16). Even so, the following types are recognized in many insects:

(1) The prohemocyte: the stem cell, from which other hemocytes are derived. The prohemocyte is a small cell with very little cytoplasm. In most insects the prohemocytes are retained predominantly in hematopoietic organs (see below) and make a small contribution to the hemocyte population circulating in the hemolymph.

(2) The granulocyte: a phagocytically active, adherent cell type that contains large numbers of intracellular granules. On activation, the granulocyte releases the granule contents. Granulocytes represent the most abundant hemocytes in many Lepidoptera. In mosquitoes, granulocytes inducibly express phenoloxidase activity following immune challenge.

(3) The plasmatocyte: a large cell type that is strongly adherent and spreads across foreign surfaces (for example, to glass slides). Plasmatocytes have been described in various

(a)

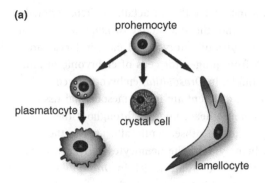

(b)

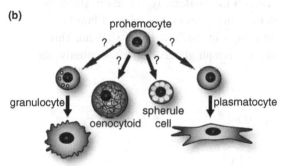

(c)

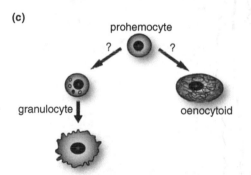

Lepidoptera, but are not evident in some other insect groups, including *Anopheles* or *Aedes* mosquitoes or in aphids.

(4) The oenocytoid: a non-adherent cell that is defined functionally as constitutively containing phenoloxidase activity.

(5) The spherule: cells with large, refractile spherules in the cytoplasm and that contain cuticular components.

A different nomenclature is used to describe the hemocytes of *Drosophila* (Fig. 5.16a) in which three differentiated cell types are recognized: the phagocytically active plasmatocyte, which are broadly equivalent to the mammalian macrophages (and functionally very different from the plasmatocyte of other insect groups, see above) and account for >90% of circulating hemocytes of larval *Drosophila* under most conditions; the crystal cells (equivalent to oenocytoids of other insects), which contain prophenoloxidase and other components of the phenoloxidase cascade (see Section 5.3.4); and the lamellocytes, which are large, flat cells that mediate the cellular encapsulation of large foreign entities, such as parasitoid eggs.

5.4.2 Origin of hemocytes

Hemocytes are produced in structures known as hematopoietic organs. (Since, in insects, only the blood cells, not the plasma, are produced in these

Figure 5.16 Types of hemocytes and their proposed lineage relationships in selected insects. (a) *Drosophila* larvae contain three differentiated hemocyte types in circulation: phagocytic plasmatocytes; phenoloxidase (PO)-containing crystal cells; and capsule-forming lamellocytes. Plasmatocytes are rounded cells in circulation but readily bind and spread symmetrically on foreign surfaces. Each cell type derives from progenitor prohemocytes and proliferation of each cell type occurs from hemocytes already in circulation or in hematopoietic organs. (b) Hemocytes in larval lepidopterans include four differentiated hemocyte types in circulation: phagocytic granulocytes; capsule-forming plasmatocytes; spherule cells; and PO-containing oenocytoids. Granulocytes and plasmatocytes are rounded cells in circulation but both bind to foreign surfaces with granulocytes spreading symmetrically and plasmatocytes spreading asymmetrically. Hematopoietic organs contain putative progenitor prohemocytes and plasmatocytes, whereas proliferation of other hemocyte types occurs primarily in circulation. (c) Hemocytes of mosquitoes (*Aedes aegypti* and *Anopheles gambiae*) include three circulating types: prohemocytes; phagocytic granulocytes; and PO-containing oenocytoids. Prohemocytes are putative progenitor cells but lineage relationships have not been experimentally defined. Reproduced from Strand (2008).

structures, they should strictly be called hematocytopoietic organs, but the more general term is usually used.) The position of hematopoietic organs varies among insects. For example, in the cricket, *Gryllus*, and the mole cricket, *Gryllotalpa*, they are paired, segmental structures that open directly into the heart (Fig. 5.1), and in *Locusta*, *Periplaneta* and larvae of cyclorrhaphous flies and of the beetle *Melolontha* they consist of irregular accumulations of cells close to the heart, but not connected with it (Fig. 5.17a,b). By contrast, the lepidopteran hematopoietic organs comprise four structures, one in close proximity to each imaginal wing disc (Fig. 5.17c,d).

In all insects studied in detail, the hematopoietic organs are not the sole source of hemocytes. All of the hemocytes of embryos and most in larvae are derived from progenitor cells of embryonic origin. For example, in *Drosophila*, embryo-derived hemocytes differentiate in the mesodermal tissue of the embryo head, migrate throughout the embryo and are subsequently allocated to the larval hemolymph. The hemocytes derived from the hemopoietic organ (which in *Drosophila* comprises the bilateral lymph gland along the anterior dorsal vessel) are released into the hemolymph in the final larval stadium. This is when the lymph glands degrade. Similarly, the

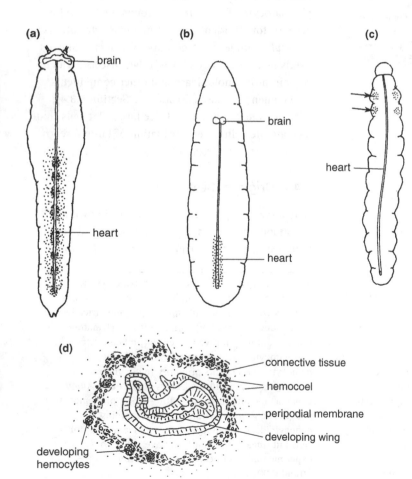

Figure 5.17 Hemopoietic organs in different insects. Stippling indicates hemopoietic tissue (see also Fig. 5.2). (a) *Locusta*; (b) *Calliphora* larva; (c), (d) caterpillar. (c) Showing the positions of the organs (arrows); (d) section through one wing disc (after Monpeyssin and Beaulaton, 1978).

hemocytes from the hematopoietic organ of Lepidoptera contribute to the hemocyte pool late in larval development. The genes that specify hemocytes include homologs of the GATA (Serpent) and Runx (Lozenge) families of transcription factors, which are involved in vertebrate hematopoiesis.

Hemocytes are usually quantified as the number of cells per unit hemolymph volume, which is known as the "total hemocyte count." Values vary widely among insects, and with developmental age and physiological condition. Aphids are reported to have just 1800 hemocytes per microliter, and there are 15 000 hemocytes per microliter in the bark bug, *Halys dentate*, 80 000 cells per microliter in the cockroach, *Periplaneta americana*, and 300 000 cells per microliter in *Spodoptera* just prior to the onset of metamorphosis. In hemimetabolous insects, the numbers are generally similar in larval and adult insects, but in holometabolous species it is usual for larvae to have more cells per unit volume of hemolymph than adults. In general, adult females have a higher number of hemocytes than males.

These generalities about the variation in hemocyte counts notwithstanding, values of hemocyte counts can be misleading for two reasons. First, hemocyte counts are normalized to hemolymph volume, which can vary widely because of the role of the hemolymph in water balance (see Section 5.3.1). Second, some hemocytes are not in circulation, but attached to epithelial tissues, forming a distinct pool of sessile cells. The sessile hemocytes can be released into circulation following an immune challenge, a process that may be accompanied by one or both of proliferation and differentiation. For example, the encapsulation of parasitoid eggs in larval *Drosophila* is mediated principally by lamellocytes that are derived from sessile hemocytes associated with the epithelial cells and imaginal discs. It appears that the cells undergo the final stages of differentiation into lamellocytes as they are released into circulation.

5.4.3 Functions of hemocytes

Hemocytes play a crucial role in both development and immunity. Studies on *Drosophila* in which hemocytes are ablated have revealed that these cells are absolutely required for normal development. In the embryo, the hemocytes phagocytose cells that have undergone programmed cell death as a normal part of development, and secrete extracellular matrix and other proteins that promote organogenesis. In particular, hemocyte-derived material is required for normal development of the ventral nerve cord and Malpighian tubules. In holometabolous insects, hemocytes are also important in elimination of larval tissues that have undergone programmed cell death at metamorphosis.

The hemocytes play a central role in wound healing and protection against infection in both the insect embryo and throughout the life of the insect. This is partly through the secretion of compounds, such as phenoloxidase that contribute to humoral immunity (Section 5.3.4), and also through direct involvement, i.e., cellular immunity. Hemocytes remove a variety of small particles, including isolated bacterial cells and yeasts, as well as synthetic beads, by phagocytosis. After internalization, microbes that lack appropriate defenses are delivered to lysosomes and destroyed.

Where the hemolymph is infected by large numbers of bacteria or fungal spores, the hemocytes do not phagocytose the microbes, but produce nodules (Fig. 5.18). Within a few minutes of entry, bacteria are localized into aggregations by humoral compounds, sometimes accompanied by melanization (see Section 5.3.4). Subsequently, plasmatocytes become associated with the aggregation, forming a layer around the outside of the clump of melanized and dying cells. Eventually the clusters of coagulated microbes are enclosed within a hardened melanized nodule often containing hemocyte debris and necrotic cells, as well as dead microbes.

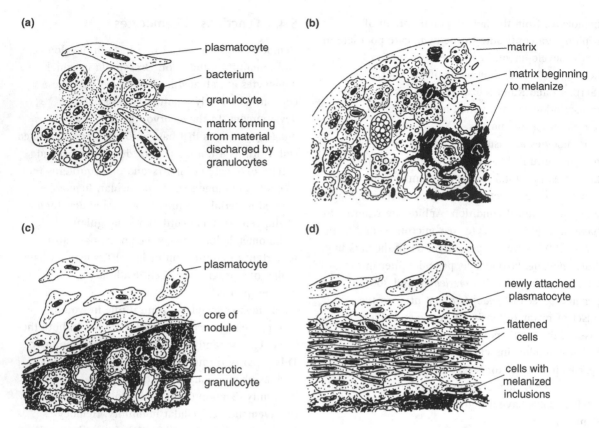

Figure 5.18 Nodule formation (after Ratcliffe and Gagen, 1977). (a) One minute after injection of bacteria. Granulocytes have degranulated and bacteria are trapped in the flocculent material produced by the cells. (b) 30 minutes after injection. The clumps of granulocytes and bacteria have compacted, and melanization of the matrix is beginning. (c) Plasmatocytes arrive at the nodule and melanization of the matrix is advanced. (d) 24 hours after injection. Nodulation is complete. Three regions are recognizable in the layers of plasmatocytes.

In many insects, large invaders, such as parasitoid eggs or larvae and nematodes, evoke a cellular response known as encapsulation, which is essentially the same as nodule formation but produces a larger melanized structure. Encapsulation involves large numbers of hemocytes, generally the plasmatocytes, together with some granulocytes, in Lepidoptera and lamellocytes in *Drosophila*. For example, in the lepidopteran *Pseudoplusia includens*, granulocytes first adhere to the foreign structure. The granulocytes are believed to produce factors that activate and recruit plasmatocytes, which generate a multilayered sheath that completely surrounds the foreign structure. The process is usually complete

within 24 hours. Melanin is often produced in this layer as a consequence of the activation of the prophenoloxidase system.

The microbes in nodules and parasites in capsules are almost certainly killed by a combination of being entombed in a hard capsule and thus isolated from the nutrients in the hemolymph, as well as the effects of the cytotoxins, including the products of the prophenoloxidase cascade. The nodules and capsules remain inside the insect's hemocoel until it dies.

If the epidermis is damaged, a clot forms beneath the wound. Formation of the clot involves components from both the hemocytes and the plasma. Granulocytes release material which forms a gel. This gel is stabilized by plasma lipophorins; phenoloxidases from the hemocytes may also be important.

Some time after clotting has occurred, plasmatocytes migrate to the site. In *Rhodnius* the cells become linked to each other by zonulae adherens within 24 hours of the wound occurring and subsequently tight junctions and septate desmosomes are formed. In this way, hemocytes become bound together to form a continuous tissue. The epidermal cells migrate over the clot to repair the wound.

Summary

- The circulatory system of insects comprises the dorsal vessel, which mediates the circulation of the hemolymph (blood) around the insect body, and multiple accessory pulsatile organs that promote blood flow through appendages, including legs, wings and antennae. The circulatory system is open, meaning that hemolymph is in direct contact with the insect organs.

- Hemolymph is generally pumped forwards from the abdomen to head by peristalsis of the dorsal vessel. The heartbeat is myogenic in origin, with modulation by neuronal input in some insect groups. Blood flow is sometimes reversed by waves of contraction starting at the anterior end of the dorsal vessel.

- The hemolymph accounts for 10–40% of the volume of the insect. It contains inorganic ions, sugars (especially trehalose), amino acids and other organic compounds that collectively maintain the osmotic pressure of the hemolymph. Hemolymph proteins at 20–200 mg ml^{-1} include proteins that function in storage (hexamerins, vitellogenin), transport (lipophorins that mediate lipid transport) and immunity, notably phenoloxidase and anti-microbial peptides that are induced by immunological challenge.

- The hemolymph contains hemocytes (blood cells) measured in cells per microliter. Some hemocytes are phagocytic and function to eliminate microbial pathogens and tissue debris generated during embryonic development, metamorphosis and by wounding. Other hemocytes protect against large foreign bodies (e.g., parasitoid eggs, nematodes) by encapsulation.

Recommended reading

Eleftherianos, I. and Revenis, C. (2011). Role and importance of phenoloxidase in insect hemostasis. *Journal of Innate Immunity* 3, 28–33.

Glenn, J. D., King, J. G. and Hillyer, J. F. (2010). Structural mechanics of the mosquito heart and its function in bidirectional hemolymph transport. *Journal of Experimental Biology* 213, 541–550.

Krzemien, J., Crozatier, M. and Vincent, A. (2010). Ontogeny of the *Drosophila* larval hematopoietic organ, hemocyte homeostasis and the dedicated cellular immune response to parasitism. *International Journal of Developmental Biology* **54**, 1117–1125.

Lemaitre, B. and Hoffmann, J. (2007). The host defense of *Drosophila melanogaster*. *Annual Review of Immunology* **25**, 697–743.

Pass, G. (2000). Accessory pulsatile organs: evolutionary innovations in insects. *Annual Review of Entomology* **45**, 495–518.

Strand, M. R. (2008). The insect cellular immune response. *Insect Science* **15**, 1–14.

References in figure captions

Clare, S. and Tauber, O. E. (1942). Circulation of haemolymph in the wings of the cockroach *Blattella germanica* L: III. Circulation in the articular membrane, the pteralia, and wing folds as directive and speed controlling mechanisms in wing circulation. *Iowa State College Journal of Science* **16**, 349–356.

Gilbert, L. I. (1967). Lipid metabolism and function in insects. *Advances in Insect Physiology* **4**, 70–211.

Kaufman, W. R. and Davey, K. G. (1971). The pulsatile organ in the tibia of *Triatoma phyllosoma pallidipennis*. *Canadian Entomologist* **103**, 487–496.

Krenn, H. W. and Pass, G. (1993). Winghearts in Mecoptera. *International Journal of Insect Morphology and Embriology* **22**, 63–76.

Lee, R. M. (1961). The variation of blood volume with age in the desert locust (*Schistocerca gregaria* Forsk.). *Journal of Insect Physiology* **6**, 36–51.

Masera, E. (1933). Il ritmo del vaso pulsante nel *Bombyx mori*. *Rivista di Biologia* **15**, 225–234.

Mills, R. R. and Whitehead, D. L. (1970). Hormonal control of tanning in the American cockroach: changes in blood cell permeability during ecdysis. *Journal of Insect Physiology* **16**, 331–340.

Monpeyssin, M. and Beaulaton, J. (1978). Hemocytopoiesis in the oak silkworm *Antheraea pernyi* and some other Lepidoptera: I. Ultrastructural study of normal processes. *Journal of Ultrastructure Research* **64**, 35–45.

Nelson, C. M., Ihle, K. E., Fondrk, M. K., Page Jr., R. E. and Amdam, G. V. (2007). The gene *vitellogenin* has multiple coordinating effects on social organization. *PLoS Biology* **5**, e62.

Nicolson, S. W. (1980). Water balance and osmoregulation in *Onymacris plana*, a tenebrionid beetle from the Namib Desert. *Journal of Insect Physiology* **26**, 315–320.

Pass, G. (1985). Gross and fine structure of the antennal circulatory organ in cockroaches (Blattodea, Insecta). *Journal of Morphology* **185**, 255–268.

Ratcliffe, N. A. and Gagen, S. J. (1977). Studies on the *in vivo* cellular reactions of insects: an ultrastructural analysis of nodule formation in *Galleria mellonella*. *Tissue & Cell* **9**, 73–85.

Richards, A. G. (1963). The ventral diaphragm of insects. *Journal of Morphology* 113, 17–47.

Roussel, J.-P. (1972). Rythme et régulation du coeur chez *Locusta migratoria migratorioides* L. *Acrida* 1, 17–39.

Shapiro, J. P., Law, J. H. and Wells, M. A. (1988). Lipid transport in insects. *Annual Review of Entomology* 33, 297–318.

Strand, M. R. (2008). The insect cellular immune response. *Insect Science* 15, 1–14.

Sutcliffe, D. W. (1963). The chemical composition of haemolymph in insects and some other arthropods, in relation to their phylogeny. *Comparative Biochemistry and Physiology* 9, 121–135.

Tublitz, N. J. and Truman, J. W. (1985). Insect cardioactive peptides II: neurohormonal control of heart activity by two cardioacceleratory peptides in the tobacco hawkmoth, *Manduca sexta. Journal of Experimental Biology* 114, 381–395.

Wasserthal, L. T. (1980). Oscillating haemolymph "circulation" in the butterfly, *Papilio machaon* L. revealed by contact thermography and photocell measurements. *Journal of Comparative Physiology B* 139, 145–163.

Whedon, A. D. (1938). The aortic diverticula of the Odonata. *Journal of Morphology* 63, 229–261.

Wigglesworth, V. B. (1972). *The Principles of Insect Physiology*. London: Chapman & Hall.

6 | Fat body

REVISED AND UPDATED BY **DEBORAH K. HOSHIZAKI, ALLEN G. GIBBS AND NICHOLE D. BOND**

INTRODUCTION

The fat body is a dynamic organ that plays a central role in the metabolic function of the insect. It is of mesodermal origin and is located in the hemocoel, with all the cells in close contact with the insect hemolymph, facilitating exchange of metabolites. The fat body has sometimes been described as equivalent to a combination of the adipose tissue (storage function) and liver (major metabolic functions) of vertebrates, but this comparison does not do full justice to the fat body, which is also a major endocrine organ, and central to systemic immunity of insects. In addition, the fat body monitors and responds to the physiological needs of the insect during different developmental stages and under different environmental conditions, thereby coordinating insect growth with metamorphosis and reproduction.

This chapter is divided into three sections. Section 6.1 describes the structure and development of the fat body. It is followed by Section 6.2 on the storage and utilization of energy and nutrients, and, finally, Section 6.3 on the role of the fat body as an endocrine organ and nutrient sensor. The effectors of the humoral immune system derived from the fat body are considered in Chapter 5 (Section 5.3.4).

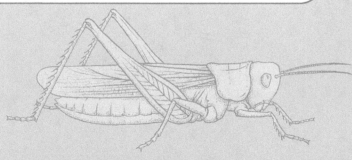

The Insects: Structure and Function (5th edition), ed. S. J. Simpson and A. E. Douglas.
Published by Cambridge University Press. © Cambridge University Press 2013.

6.1 Fat body structure and development

The fat body consists of thin sheets or ribbons, usually only one or two cells thick, or of small nodules suspended in the hemocoel by connective tissue and tracheae. All of its cells are consequently in immediate contact with the hemolymph, facilitating the exchange of metabolites. There is generally a peripheral, or parietal, fat body layer immediately beneath the body wall, and often a perivisceral layer surrounding the alimentary canal can also be distinguished (Fig. 6.1). The fat body is most conspicuous in the abdomen, but components extend into the thorax and head. The structure of the fat body is generally uniform among individuals of individual species, but there is considerable variation among species, especially across different insect orders.

In hemimetabolous insects, the larval fat body persists in the adult without major changes. In holometabolous insects, the fat body undergoes a striking transformation during metamorphosis in which the tissue dissociates into individual cells. In the majority of the holometabolous insects, the adult fat cells are rebuilt from the larval fat cells, but in Hymenoptera and the higher Diptera, the adult fat cells develop *de novo* (Section 6.2.5).

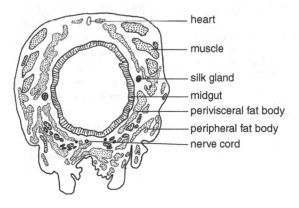

Figure 6.1 Distribution of fat body in a caterpillar in transverse section of the abdomen.

6.1.1 Trophocytes

The principal cells of the fat body are trophocytes (adipocytes), which store energy. In many insect orders these are the only cells present within the fat body. The trophocytes are held together by desmosomes to form sheets of tissues. The cytoplasm of adjacent trophocytes is connected through gap junctions, and the whole tissue is clothed in a basal lamina that is attached to the cells by hemidesmosomes (Fig. 6.2). The fat body of some insects additionally contains one or more of urate cells, mycetocytes and oenocytes. In a number of Lepidoptera and Diptera, the larval fat body is regionally differentiated to perform different functions. For example, in the larva of *Helicoverpa* (Lepidoptera) during the period just before pupation, protein synthesis occurs only in the peripheral fat body, while the storage of arylphorin and a very high-density lipoprotein, colored blue by non-covalently bound biliverdin, is restricted to the cells of the perivisceral fat body. In *Drosophila melanogaster* (Diptera), different pigments for the adult eye are synthesized and sequestered in different parts of the larval fat body, and in the larva of *Chironomus* (Diptera), hemoglobin synthesis appears to occur in the peripheral fat body, while storage might occur in the perivisceral fat body cells.

The form of the trophocyte varies according to developmental stage and nutritional status of the insect. In a larva soon after ecdysis, the trophocytes are generally small, with relatively little cytoplasm and little development of organelles. There are few mitochondria following the cell division preceding ecdysis (Fig. 6.3a,d), but, in a well-fed insect, there follows a preparative phase during which the trophocytes develop their capacity for synthesis. In the final larval stage of the moth *Calpodes* (Lepidoptera) the preparative period lasts about 66 hours. During this period there is extensive replication of DNA, but no nuclear division (Fig. 6.3a, b). Most of the cells become octaploid, although some cells exhibiting 16- and 32-ploidy also occur. A similar development occurs in *Rhodnius* (Hemiptera), while in *Calliphora* (Diptera), and probably in other Diptera, polyteny occurs (the

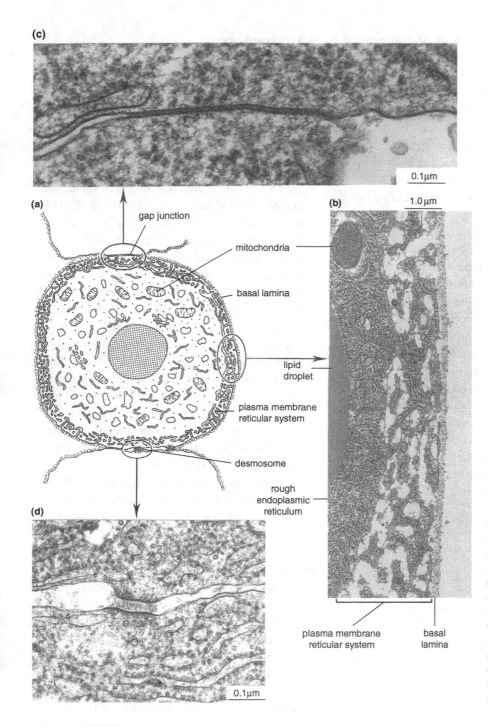

(c)

0.1µm

(a)

gap junction

mitochondria

basal lamina

lipid droplet

plasma membrane reticular system

desmosome

rough endoplasmic reticulum

(b)

1.0 µm

(d)

0.1µm

plasma membrane reticular system

basal lamina

Figure 6.2 Structure of a mature trophocyte. (a) Diagram of a trophocyte. (b) Transmission electron micrograph of the plasma membrane reticular system of a trophocyte from the larva of *Calpodes*. (c) Gap junction between two trophocytes in the fat body of *Calpodes*. (d) Desmosome joining two trophocytes in the fat body of *Calpodes* (b, c and d after Dean *et al.*, 1985).

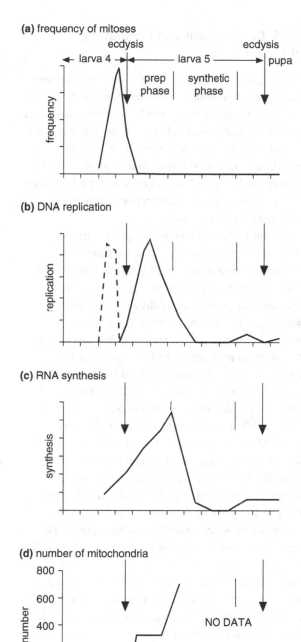

Figure 6.3 Changes occurring in the cells of the fat body of the caterpillar of *Calpodes* during a molt/intermolt cycle

chromosomes divide, but do not separate). At the same time RNA synthesis occurs (Fig. 6.3c), ribosomes increase in number, rough endoplasmic reticulum proliferates and the numbers of mitochondria increase by division (Fig. 6.3d). The trophocytes now have the apparatus necessary to begin synthesis. During the preparative period, the cell membrane invaginates in a series of folds which interconnect to form the plasma membrane reticular system (Fig. 6.2). In *Calpodes* the membranes in this system are separated from each other by 100–150 nm. The reticular system occupies the peripheral 1–1.5 μm of the cells and it presents an exceptionally large surface area to the hemolymph. However, this surface is negatively charged and the effect of this is to limit the access of some large charged molecules to the interior of the reticulum. It is possible that the reticulum is concerned with the docking and unloading of lipophorins. When the larva is approaching the molt to pupa, the components of the cell that have been involved in protein synthesis regress.

Immediately after eclosion, the trophocytes of adult insects commonly contain extensive lipid droplets, accumulations of glycogen and protein granules. The trophocytes in males do not show further development and they probably play no further major role in protein synthesis. In females, however, changes occur which are comparable with those occurring in larval stages. This reflects the need, in many species, for the synthesis of vitellogenins.

6.1.2 Urate cells

Urate cells, or urocytes, are present in Collembola (springtails), Thysanura (silverfish), Blattodea

(prep phase is preparatory phase) (mainly after Locke, 1970). (a) The frequency of mitoses. Mitosis is limited to the period immediately before ecdysis. (b) DNA replication. The broken line is the presumed phase of replication associated with mitosis. The solid line shows measured replication which was not associated with cell division. (c) RNA synthesis in the cytoplasm. This occurs primarily in the preparatory phase. (d) Numbers of mitochondria in one cell (from Dean *et al.*, 1985).

(cockroaches) and larval Apocrita (Hymenoptera) (bees and wasps). These cells characteristically contain large crystalloid spherules of uric acid. Uric acid also accumulates as small granules in all fat body cells of larval and pupal Lepidoptera and in larval mosquitoes. For Collembola, which lack Malpighian tubules, and for Apocrita larvae, which are confined within their nest cells, the accumulation of uric acid might provide a mechanism to sequester nitrogenous waste products. This might also be true in Lepidoptera, where uric acid accumulates during the larval wandering phase and continues to accumulate during the first part of the pupal period, but then is transferred to the rectum to be excreted in the meconium. In the cockroaches, however, uric acid provides a store of nitrogen that can be recycled (Section 18.5.2).

6.1.3 Hemoglobin cells

Respiratory proteins such as hemoglobin have been thought to be unnecessary in insects because the tracheal system efficiently supplies oxygen to the respiring tissues. However, genes for hemoglobin have been detected in every insect genome sequenced to date, and intracellular hemoglobins have been identified in the fat body and tracheal system of various insects. Specialized hemoglobin cells have been described for botfly larvae (Oestridae, Diptera) and backswimmers (Notonectidae, Hemiptera) that occur in potentially hypoxic habitats (Section 17.4). Hemoglobin cells are large, measuring 20–80 µm in *Anisops* (Notonectidae) and up to 400 µm in diameter in *Gasterophilus* (Oestridae, Diptera). The hemoglobin cells are closely associated with tracheae.

6.1.4 Other cells

Mycetocytes are cells containing microorganisms that are localized in the fat body of cockroaches and some Hemiptera (Section 4.4). Oenocytes, derived from the epidermis, are also associated with the fat body in some groups, such as Hemiptera (Section 16.1.2).

6.1.5 Development and maturation of the fat body

The origin of the insect fat body has primarily been studied in *Drosophila melanogaster*, where technical advances in cell biology tools (e.g., green fluorescent protein tags), the availability of the complete genome sequence and extensive genetic tools have allowed cell lineage tracing studies. These studies reveal the mesodermal origin of the fat cells and the formation of the fat body by the coalescence of individual embryonic fat cell clusters.

The fat body of the *D. melanogaster* larva is derived from cell clusters that arise from the embryonic mesoderm. The organization of the cell clusters depends upon patterning genes, including the pair-rule genes that subdivide the mesoderm and establish segment identity. These, in turn, serve to establish the expression of a transcription factor, Serpent (a member of the family of GATA transcription factors characterized by their ability to bind to the DNA sequence GATA), which specifies the fat cell fate. The progenitor fat cells proliferate and coalesce to form the three morphological domains of the fat body: the dorsal fat cell projections, which extend in the anterior direction from the posterior–dorsal region of the lateral fat body; the lateral fat body, which spans the lateral region of the embryo; and the ventral commissure, which extends from the anterior fat body and spans the ventral midline. The embryonic fat body persists into the larva and the domain structure identified in the embryo is maintained in the larva. This organization is likely to characterize all Diptera, although a detailed comparative study has not been carried out.

Metamorphosis in holometabolic insects is characterized by a complete change in body plan controlled in part by pulses of the steroid hormone 20-hydroxyecdysone (20E) (Section 15.3) that induce programmed cell death of most larval tissues, except for the fat body. For example, during metamorphosis of the skipper butterfly, *Calpodes ethlius*

(Lepidoptera), the individual fat cells reorganize into nodular clumps surrounding the tracheoles, and then undergo classical autophagy, where the intracellular remodeling takes place. The mitochondria, microbodies and rough endoplasmic reticulum become sequestered in organelle-specific autophagic vacuoles and are destroyed by hydrolytic enzymes, but the cell retains its integrity. After intracellular remodeling, a new round of organelle biogenesis takes place to generate the adult fat body.

The larval fat body of *D. melanogaster* and probably other Diptera is also refractive to 20E-triggered programmed cell death. In the pupal *D. melanogaster*, the larval fat body persists but undergoes tissue remodeling. 20E and the competence factor BFTZ-F1 act in concert to induce within the fat body the Matrix Metalloproteinase 2 (MMP2), which is likely necessary for the breakdown of the fat body extracellular matrix. The resultant cells become dispersed throughout the pupa and are carried forward into the adult. At 3–4 days post-eclosion, the larval fat cells are replaced by adult fat cells that arise *de novo* from cells of the dorsal thoracic and eye-antennal imaginal discs and larval histoblasts.

6.2 Storage and utilization of energy and nutrients

The fat body functions in many aspects of energy storage and synthesis of proteins, lipids and carbohydrates. Lipids are stored as triglycerides, and carbohydrates are stored as glycogen, i.e., polymers of glucose. The mobilization of these macronutrients is controlled by the adipokinetic hormone (AKH) family of peptides and the family of insulin-like peptides (ILPs). AKH is produced by the corpus cardiaca, whereas ILPs are produced in the median neurosecretory cells of the brain, the corpus allatum, the corpus cardiaca and peripheral tissues including the fat body. The ratio of triglyceride to glycogen in the fat body varies among insect species, and with lifecycle stage and environmental stress, with triglycerides being the primary component.

In holometabolous insects, animals do not feed during metamorphosis, and in some cases the adult itself may not feed, e.g., the silkworm moth, *Bombyx mori* (Lepidoptera). Therefore, adequate nutrients must be stored in the fat body during the larval stage to support development of the adult tissues and to support the adult until feeding. This life history trait places the larval fat body in a unique position where energy storage and utilization is monitored for both immediate use and later use by the pupa and adult. Thus, the cells of the larval fat body serve as a physical link to carry larval nutrients into the adult. These larval fat cells provide nutrient reserves important for ovary maturation in the adult and can be mobilized in the event of adult starvation stress.

6.2.1 Lipids

The fat body is the principal storage site of lipids in insects. Most of the lipid is present as triacylglycerol (triglyceride), which commonly constitutes more than half of the dry weight of the fat body. The amount stored varies with the stage of development and state of feeding of the insect. Lipid stores normally increase during periods of active feeding and decline when feeding stops (Fig. 6.4) or when large quantities of lipid are used during oogenesis or prolonged flight.

The lipids are stored in lipid droplets within the trophocytes. Many triglycerides are synthesized from diacylglycerides derived from fatty acids or proteins. Fatty acids can be rapidly taken up by the fat body and converted to triglycerides. Dietary carbohydrates can also be converted to triglycerides in the fat body. Lipids are important because they are a more efficient form of energy storage than carbohydrates, for two reasons. First, lipids contain approximately twice the amount of energy per gram than carbohydrates, and second, glycogen contains substantial water of hydration, while lipid droplets contain little or no

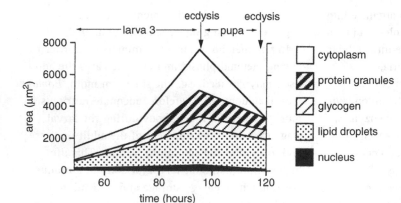

Figure 6.4 Changes in the amounts of the major components of a trophocyte during the final larval and pupal stages of *Drosophila*. The major increases occur during the period of feeding. Amounts are expressed as the areas occupied by the components in cross-sections of the tissue. Data from Butterworth *et al.* (1965).

water. Although water of hydration may be released when glycogen is metabolized under dry conditions, its additional weight is disadvantageous for activities such as flight.

Factors stimulating lipid synthesis in insects have not been defined, but the Target of Rapamycin (TOR) signaling pathway has been implicated in the regulation of nutrient uptake, storage and metabolism. There is considerable evidence that juvenile hormone inhibits lipid synthesis, but its mechanism of action is not known.

6.2.2 Proteins and amino acids

The fat body is the principal site of synthesis of hemolymph proteins, described in Section 5.3.4. In the larva of *Calpodes*, the fat body synthesizes 14 out of 26 hemolymph polypeptides, amounting to about 90% of the total hemolymph protein. In adult females the fat body produces vitellogenin, the protein that will form most of the yolk protein in the eggs.

Diapause proteins are also produced by the fat body. For example, adult Colorado potato beetles, *Leptinotarsa* (Coleoptera), enter diapause under short-day conditions. The adult beetles synthesize vitellogenins and diapause proteins under all conditions. If newly eclosed beetles experience long days, vitellogenins are synthesized at a high rate, and

production of diapause proteins is low. In short days (less than ten hours of light), however, relatively little vitellogenin, but more of the diapause proteins are produced.

As the insect prepares to pupate, protein synthesis in the fat body stops. Proteins, originally synthesized in and secreted by the fat body, are now removed from the hemolymph and stored as granules in the fat body (Fig. 6.5). Some protein uptake does occur during the phase of protein synthesis, but the uptake is non-selective and proteins are hydrolyzed within the cells. Protein breakdown ceases at the end of the period of synthesis, and the uptake of proteins is selective; different proteins are taken up to different extents. In both *Helicoverpa* (Lepidoptera) and *Sarcophaga* (Diptera), this selective uptake of specific proteins is dependent on the appearance of specific receptor proteins in the plasma membranes of the fat body. A precursor of the receptor protein is already present in the larval fat body of *Sarcophaga*, and its conversion to the receptor for the uptake of storage protein is activated by molting hormone in the hemolymph before pupation. In *Helicoverpa*, the receptor protein for the blue-colored protein is formed *de novo* at the time of pupation. This receptor is only present in the membranes of cells in the perivisceral fat body, not in the peripheral fat body.

The factors regulating protein synthesis and storage are not known with certainty, although both juvenile hormone and ecdysteroids are involved. For example, the synthesis of arylphorin in the larval silkworm *Bombyx* is suppressed by juvenile hormone. Synthesis is initiated when juvenile hormone is no longer detectable in the hemolymph of the final larval stadium hemolymph. In adult insects of most orders, synthesis of vitellogenin is stimulated by juvenile hormone, although in Diptera this function is performed by ecdysteroids (see Fig. 13.10).

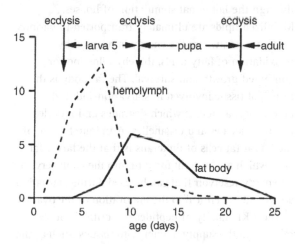

Figure 6.5 Amounts of blue-colored very high-density lipoprotein present in the fat body and hemolymph in various stages of development of *Helicoverpa* (after Haunerland *et al.*, 1990).

The fat body plays an important role in amino acid metabolism, principally because it is a major site of transamination between amino acids (Section 4.1.2). It also contributes to the regulation of the amino acid tyrosine content of the hemolymph. Tyrosine functions in cuticle sclerotization and accumulates in the hemolymph just before a molt (Section 16.6.2). In the intermolt period of some insects, tyrosine is taken up from the hemolymph and stored in large vacuoles in the trophocytes. This has been most comprehensively studied in the fourth larval stadium of *Calpodes*, where the uptake of tyrosine begins about one day after ecdysis. Shortly before the next ecdysis, tyrosine is released into the hemolymph (Fig. 6.6). Additionally or alternatively, some insects store conjugated tyrosine (e.g., as glucoside) in the hemolymph (see Fig. 16.17).

6.2.3 Carbohydrates

Carbohydrate is stored in the fat body as glycogen and circulates in the hemolymph in the form of trehalose. The amount of stored glycogen in the fat body is determined in part by the levels of trehalose in the hemolymph. Both glycogen and trehalose are synthesized in the fat body from UDP-glucose, which is derived from dietary carbohydrates or amino acids. As the level of trehalose rises, its synthesis in the fat body is inhibited and UDP-glucose is diverted to glycogen synthesis. Glycogen also serves as a source of cryoprotectants in overwintering insects.

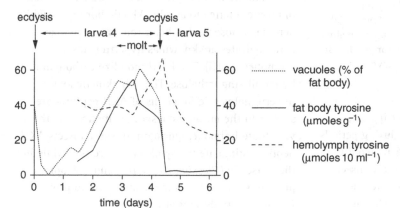

Figure 6.6 Changes in the tyrosine content of the fat body and hemolymph of *Calpodes* larva. Tyrosine in the fat body is sequestered in vacuoles and the proportion of the fat body occupied by vacuoles is paralleled by changes in the tyrosine content. At ecdysis the vacuoles disappear as they release tyrosine into the hemolymph (after McDermid and Locke, 1983).

(a) total sugar

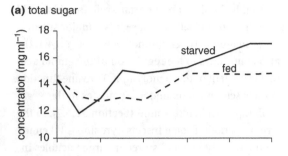

(b) glucose

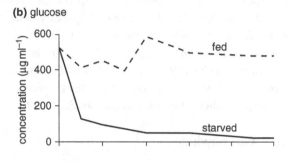

(c) active phosphorylase

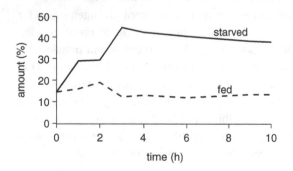

Figure 6.7 The effects of starvation on carbohydrate metabolism in a well-fed final-stage larva of *Manduca* (after Gies *et al.*, 1988). (a) Concentration of total sugars, mainly trehalose, in the hemolymph. (b) Concentration of glucose in the hemolymph. (c) Percentage of active glycogen phosphorylase in the fat body.

In caterpillars, and probably in other insects, glycogen accumulates in the fat body during periods of active feeding. This store becomes depleted during sustained activity or over a molt, when the insect is not feeding, or if it is starved. For example, the glycogen content of the fat body of a well-fed

migratory locust, *Locusta* (Orthoptera), is about 20 mg g^{-1} fresh weight. Of this, 75% is consumed after two hours of flight. In *Manduca* (Lepidoptera) larvae, the hemolymph concentration of trehalose is maintained by conversion of glycogen to trehalose in the fat body (Fig. 6.7).

6.2.4 Mobilization of energy stores

Starvation and high activity levels stimulate the release of energy reserves from the fat body. During starvation, stored triglycerides can be metabolized through the hormonal stimulation of lipases. Mobilized lipids are ultimately transported to various target tissues, where the energy is released by β-oxidation of fatty acids, thereby allowing for continued growth and survival. The fat body is the principal tissue involved in starvation-induced autophagy, a process which degrades and recycles macromolecules and organelles. A unique feature of the larval fat cells of dipterans is that the larval fat cells which are brought forward into the adult serve as an energy reservoir that can be used during starvation stress and are critical to the maturation of the ovaries.

The AKH family of peptides are critical for regulating the supply of energy to tissues, such as the flight muscles to maintain long-distance flight. Depending on the insect species, lipids, carbohydrates, proline or a combination of these substrates are released from the fat body during times of metabolic need. Peptides that primarily mobilize lipids are referred to as adipokinetic hormones, while peptides whose predominant role is to mobilize carbohydrates are known as hypertrehalosemic hormones (HrTH). The HrTH mobilize carbohydrates by mobilizing trehalose. The adipokinetic and hypertrehalosemic functions of AKH peptides are similar to the metabolic responses induced by the vertebrate hormone, glucagon. In some insects AKH peptides stimulate the synthesis of proline, while in the tsetse fly (*Glossina morsitans*) and Coleoptera, proline is released into the hemolymph to provide fuel for the flight muscles.

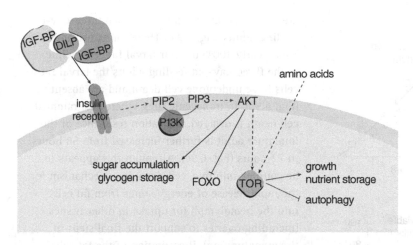

Figure 6.8 Insulin/IGF signaling (IIS) and the Target of Rapamycin (TOR) signaling in the fat cell serve as a nutrient-sensing system to maintain energy homeostasis. The release of *Drosophila* insulin-like polypeptide (DILP) from the insulin-like growth factor binding proteins (IGF-BPs) allows the DILPs to be accessible to the insulin receptor and activates the IIS pathway. The IIS pathway promotes growth through the control of protein synthesis and autophagy via TOR, contributes to energy homeostasis by regulating carbohydrate storage and enhances translation by repressing FOXO. TOR signaling can also be activated by free amino acids present in the cell. FOXO can act as a translational repressor.

In addition to AKH signaling, the insulin signaling pathway regulates nutrient uptake, storage and metabolism. In insects, the insulin and insulin-like growth factor (IGF) signaling pathways function as a single pathway, the insulin/IGF pathway (IIS). In the IIS pathway both IGFs and ILPs bind to a single receptor, the insulin receptor (InR) (Fig. 6.8). In *D. melanogaster* ILP accessibility is regulated by IGF-binding proteins (IGF-BPs), which sequester the ILPs and protect them from degradation. Release of ILP from the complex allows binding to InR and activation of the IIS pathway. The IIS pathway promotes nutrient storage by inserting glucose transporters into the cell membrane to increase accumulation of sugars, phosphorylation of glycogen synthase to increase glycogen storage and by inactivation of the translational repressor FOXO (Fig. 6.8).

6.2.5 Larval energy stores in adults

In holometabolous insects distinct developmental stages are tightly linked to feeding (larvae and most adults) and non-feeding periods (pupae and some adults). In *D. melanogaster*, and probably other Diptera, the larval fat cells generated by the remodeling of the fat body during metamorphosis (Section 6.1.5) are present in the newly eclosed adult. Within 24 hours, 85% of the larval fat cells are destroyed within the adult by the process of programmed cell death. Presumably the role of the fat cells is to provide an energy reserve to support the adult prior to feeding. In *D. melanogaster*, for example, the adults do not feed for the first eight hours.

For Lepidoptera, stable carbon isotope studies indicate that all the essential amino acids in the eggs deposited by adult females are derived from larval feeding, whereas non-essential amino acids are produced from the adult diet. Presumably, some of the essential amino acids are derived from storage proteins, but this has not been demonstrated directly.

Direct demonstration of a role for larval fat cells as a nutrient reserve in the adult comes from starvation experiments. Immature adult *D. melanogaster* (within ten minutes of emergence)

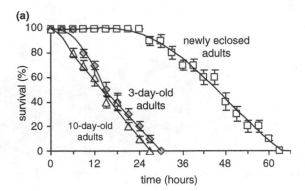

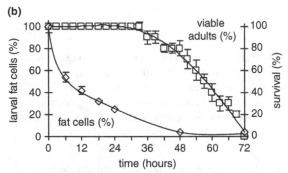

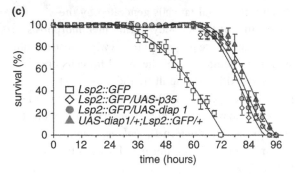

Figure 6.9 The role of larval fat cells in adult *Drosophila melanogaster*. (a) Survival of starved adults of different ages. (b) Number of larval fat cells in fed adults, and survival of starved adults over six days after eclosion. (c) Effect of retention of larval fat cells on survival of starved flies. Three genetic routes to prevent programmed cell death of larval fat cells were applied (diamond, circle, triangle), with wild-type controls (square). The LD50 for control flies was <60 hours and for genetically manipulated flies was 84 hours (reproduced from Aguila *et al.*, 2007).

are more resistant to starvation than three-day-old feeding adults (Fig. 6.9a). These immature adults have nearly 100% of their larval fat cells, while in the three-day-old feeding adults the larval fat cells have undergone cell death and are absent (Fig. 6.9b). In transgenic flies, where programmed cell death is delayed, starvation resistance of the immature adult is further increased from 58 hours to 72 hours (Fig. 6.9c). Although this appears to be counter-intuitive, cell death is a mechanism for the rapid release of energy stores from fat cells into the hemolymph for uptake in other tissues (including ovaries to support the final steps in tissue maturation). Preservation of the fat cells maintains the energy reservoir for a measured mobilization in response to starvation. Thus, the larval fat cells serve as a nutrient reservoir that can be mobilized in response to stress to support the survival of the adult.

6.3 Function as an endocrine organ and nutritional sensor

The insect fat body has emerged as a dynamic tissue that functions as an endocrine organ to ensure proper energy homeostasis and as a sensor to integrate larval nutritional status with maturation signals for metamorphosis. The fat body coordinates the growth of multiple tissues with the energy demands of the organism and is involved in determining body size. For example, a reduction in the ability of the fat body to sense amino acids leads to a reduction in both cell growth and cell proliferation of other tissues, such as the larval salivary glands and imaginal discs.

6.3.1 The fat body as an endocrine organ

In holometabolous insects the larval stage is characterized by rapid growth of larval tissues (e.g., fat body, salivary glands, Malpighian tubules, trachea, muscle and epidermis) by increasing cell size via endomitosis, and the cellular proliferation of the precursors to the adult tissues (imaginal discs) using the mitotic cell cycle. The insect fat body has a central role in fueling this growth through its participation in intermediary metabolism (i.e., generation of ATP) and in coordinating growth with nutritional status through the production of growth factors.

Since the initial efforts to culture insect cells and tissues, the addition of fat body or fat body conditioned medium was found to be an important requirement for in vitro cell proliferation and differentiation. For example, the maintenance of *D. melanogaster* imaginal discs in culture required fat body conditioned medium in addition to insulin and a juvenile hormone analog. These supplements are also needed for optimal growth and normal cell division in cultured lepidopteran imaginal discs and *Manduca sexta* midgut stem cells.

The fat body of *D. melanogaster* produces at least two classes of growth factors: the imaginal disc growth factor family (IDGFs), and the adenosine deaminase-related growth factors (ADGFs). The ADGFs stimulate imaginal disc proliferation, and one of them, ADGF-D, is produced primarily by the fat body and brain. The IDGF family is produced by embryonic yolk cells and the fat body of the embryo and larva and might correspond to the mitogenic factors responsible for growth in fat body conditioned media. IDGF1 and IDGF2 promote imaginal disc cell proliferation in vitro and are likely to act through the *Drosophila* insulin receptor to promote cell growth. It appears that control of larval cell growth, as well as the imaginal cell proliferation, is regulated by fat body mitogenic growth factors.

In addition to the secretion of growth factors, the fat body is a key player in the synthesis of the insect molting hormone, 20-hydroxyecdysone (20E). Although the precursors of 20E are synthesized in the prothoracic gland (Section 15.4.2), the final step in 20E synthesis occurs in the peripheral tissues, e.g.,

the fat body. Specifically, P450 monooxygenase CYP314A1 (Shade protein in *D. melanogaster*) hydroxylates alpha-ecdysone to form the active form of the hormone, 20E. Orthologs of the CYP314A1 gene have been found in Hymenoptera, Coleoptera, Lepidoptera and Diptera, but functional studies of these monooxygenases have only been carried out in dipterans and lepidopterans. Studies in *D. melanogaster* have shown that conversion of alpha-ecdysone to 20E during the larval and pupal stages occurs predominantly in the fat body, with other organs such as the Malpighian tubules and midgut contributing small amounts of monooxygenase activity.

6.3.2 Monitoring nutritional status

Two major signaling pathways, the insulin/IGF signaling (IIS) and the Target of Rapamycin (TOR) pathway are active in the fat body of *D. melanogaster* and are involved in maintaining a stable equilibrium between energy storage and utilization (Fig. 6.8). In insects, the IIS and TOR signaling pathways are integrated to regulate growth, nutrient storage and autophagy, a catabolic process used to degrade the cell's own components through the lysosomal machinery. Autophagy is a major mechanism by which a starving cell reallocates nutrients from unnecessary processes to more essential processes. In addition to activation by IIS, the TOR signaling pathway can be activated by amino acids.

Several putative amino acid transporters have been identified in the fat body. One of these, Minidisks, is predominately expressed in the fat body and is necessary for imaginal disc proliferation in *D. melanogaster*. Another amino acid transporter, Slimfast, is required for proper growth of the endoreplicating cells, which make up most of the larval tissues. Fat body lacking Slimfast also has down-regulated TOR activity, thus connecting this amino acid transporter to the nutrient-sensing mechanism in the fat body.

Summary

- The fat body is in the hemocoel, and the proximity of the cells to the hemolymph facilitates the exchange of metabolites. The dominant cell type of the fat body is the trophocyte, but the fat body of some insects additionally have urate cells, mycetocytes or oenocytes.

- The fat body is of mesodermal origin. In hemimetabolous insects the fat body persists into adulthood. In holometabolous insects the fat body cells disaggregate at metamorphosis, and either the organ reforms from these cells in the adult, or the larval fat body cells are eliminated with the adult fat body cells generated *de novo*.

- The fat body is rich in lipids, especially triacylglycerols, and glycogen. It is also responsible for the synthesis of many hemolymph proteins and major storage proteins, including vitellogenin and arylphorins.

- Fat body cells have active insulin-like peptide and TOR signaling, which play central roles in mediating the regulated growth and nutrition of the insect.

Recommended reading

Arrese, E. L. and Soulages, J. L. (2010). Insect fat body: energy, metabolism, and regulation. *Annual Review of Entomology* 55, 207–225.

Boggs, C. L. (2009). Understanding insect life histories and senescence through a resource allocation lens. *Functional Ecology* 23, 27–37.

Hoshizaki, D. K. (2005). Fat-cell development. In *Comprehensive Molecular Insect Science*, ed. L. I. Gilbert, K. Iatrou and S. Gill, pp. 315–345. Oxford: Elsevier.

Mirth, C. and Riddiford, L. (2007). Size assessment and growth control: how adult size is determined in insects. *Bioessays* 29, 344–355.

References in figure captions

Aguila, J. R., Suszko, J., Gibbs, A. G. and Hoshizaki, D. K. (2007). The role of larval fat cells in adult *Drosophila melanogaster*. *Journal of Experimental Biology* 210, 956–963.

Butterworth, F. M., Bodenstein, D. and King, R. C. (1965). Adipose tissue of *Drosophila melanogaster* I: an experimental study of larval fat body. *Journal of Experimental Zoology* 158, 141–154.

Dean, R. L., Locke, M. and Collins, J. V. (1985). Structure of the fat body. In *Comprehensive Insect Physiology, Biochemistry and Pharmacology*, vol. 3, ed. G. A. Kerkut and L. I. Gilbert, pp. 155–210. Oxford: Pergamon Press.

Gies, A., Fromm, T. and Ziegler, R. (1988). Energy metabolism in starving larvae of *Manduca sexta. Comparative Biochemistry and Physiology* 91A, 549–555.

Haunerland, N. H., Nair, K. N. and Bowers, W. S. (1990). Fat body heterogeneity during development of *Heliothis zea. Insect Biochemistry* 20, 829–837.

Hoshizaki, D. K. (2005). Fat-cell development. In *Comprehensive Molecular Insect Science*, ed. L. I. Gilbert, K. Iatrou and S. Gill, pp. 315–345. Oxford: Elsevier.

Locke, M. (1970). The molt/intermolt cycle in the epidermis and other tissues of an insect *Calpodes ethlius* (Lepidoptera, Hesperiidae). *Tissue & Cell* 2, 197–223.

McDermid, H. and Locke, M. (1983). Tyrosine storage vacuoles in insect fat body. *Tissue & Cell* 15, 137–158.

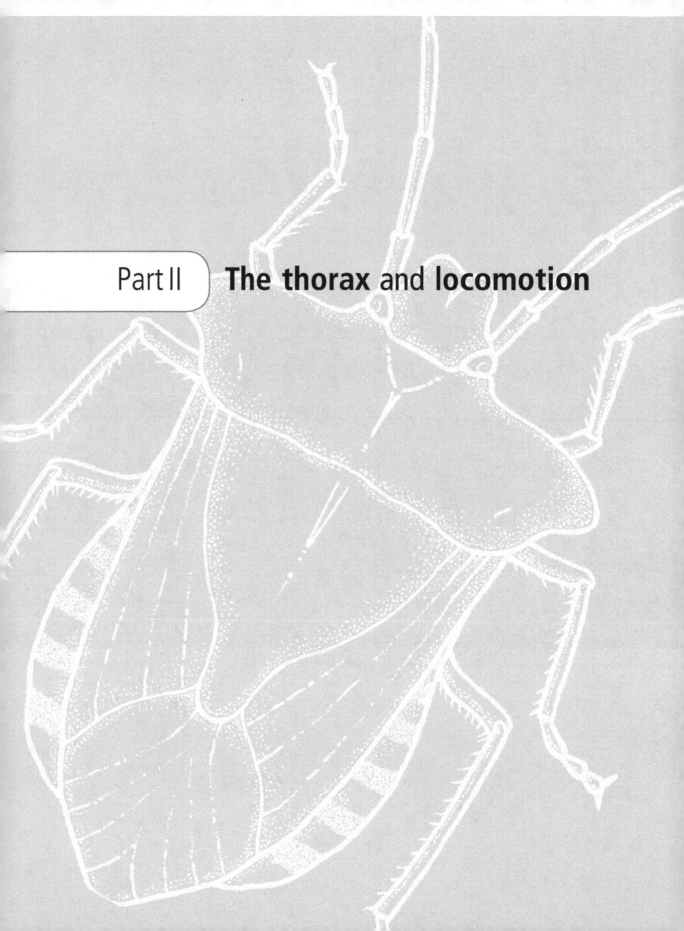

Part II The thorax and locomotion

7 | Thorax

REVISED AND UPDATED BY **GRAHAM K. TAYLOR**

INTRODUCTION

The skeleton of the thoracic segments is modified to give efficient support for the legs and wings, and the musculature is adapted to produce the movements of these appendages. This chapter provides an overview of the segmentation (Section 7.1), morphology (Section 7.2) and musculature (Section 7.3) of the thorax. This is important to understanding the functioning of the legs (Chapter 8) and the wings (Chapter 9) in locomotion and other activities.

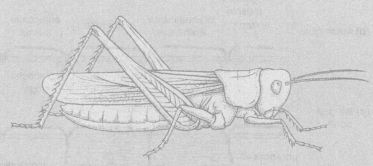

7.1 Segmentation of the thorax

The thorax consists of three segments known as the pro-, meso- and metathoracic segments. In most insects all three segments bear a pair of legs, but this is not the case in larval Diptera, larval Apocrita (Hymenoptera), some larval Coleoptera and a small number of adult insects which are apodous. In addition, winged insects have a pair of wings on the meso- and metathoracic segments and these two segments are then collectively known as the pterothorax.

In larval holometabolous insects the cuticle is soft and flexible, or only partially sclerotized, and the longitudinal muscles are attached to the intersegmental folds (Fig. 7.1a). Insects with this arrangement move as a result of successive changes in the shapes of the thoracic and abdominal segments (see Section 8.3), these changes of shape being permitted by the flexible cuticle.

When the cuticle is sclerotized, sclerites in the intersegmental folds that have the longitudinal muscles attached to them are usually fused with the segmental sclerites behind. This intersegmental region is incorporated anteriorly into the dorsal

sclerite of each adult segment, the original fold being marked by the antecostal sulcus, where the cuticle is inflected. The narrow rim in front of the sulcus is called the acrotergite (Figs. 7.1b, 7.2). An acrotergite never occurs at the front of the prothorax because the anterior part of this segment forms part of the neck and the muscles from the head pass directly to the acrotergite of the mesothorax.

An area at the back of each segment remains membranous, forming a new intersegmental membrane. This does not correspond with the original intersegmental groove and so a secondary

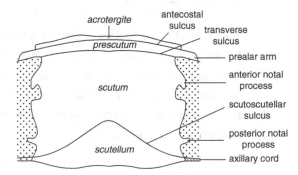

Figure 7.2 Notum of a wing-bearing segment. Stippled areas are membrane at the base of the wing (axillary sclerites not shown). Names of sclerites in italics (after Snodgrass, 1935).

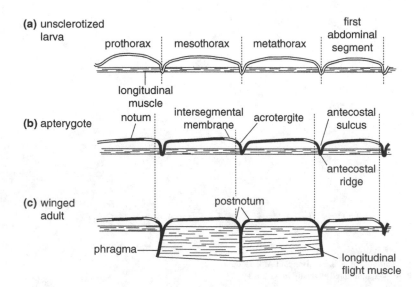

Figure 7.1 Segmentation and the derivation of the postnotum and phragmata in pterygote insects. Sclerotized areas are indicated by a solid line, membranous areas by a double line.

segmentation is superimposed on the first (Fig. 7.1b,c). Neither anatomical boundary corresponds with the boundaries of the embryonic parasegments defined by developmental studies, as the boundaries of the anatomical segments develop in the middle of the parasegments (see Section 14.2.8).

This basic condition with a membranous area at the posterior end of each segment occurs in the abdomen where this is sclerotized, in the meso- and metathoracic segments of larval insects with a sclerotized thorax, in the Apterygota and in adult Blattodea and Isoptera where the wings are not moved by indirect muscles. With this arrangement, contraction of the longitudinal muscles produces telescoping of the segments.

7.2 Morphology of the thorax

The dorsal sclerite of a segment is called the tergum, or, in the thorax, the notum. This is where the wings articulate with the body. The opposite, ventral sclerite is known as the sternum. The pleuron is the lateral surface between the tergum and the sternum and is where the legs articulate with the body.

7.2.1 Tergum

The tergum of the prothoracic segment is known as the pronotum. It is often small, serving primarily for attachment of the muscles of the first pair of legs, but in Orthoptera, Blattodea and Coleoptera it forms a large plate affording some protection to the pterothoracic segments. The meso- and metanota are relatively small in wingless insects and larvae, but in winged insects they become modified for the attachment of the wings. In the majority of winged insects the downward movement of the wings depends on an upwards distortion of the dorsal wall of the thorax (Section 9.3.3). This is made possible by a modification of the basic segmental arrangement. The acrotergites of the metathorax and the first

abdominal segment extend forwards to join the tergum of the segment in front and in many cases become secondarily separated from their original segment by a narrow membranous region. Each acrotergite and antecostal sulcus is now known as a postnotum (Fig. 7.1c). There may thus be a mesopostnotum and a metapostnotum if both pairs of wings are more or less equally important in flight, but where the hindwings provide most power, as in Orthoptera and Coleoptera, only the metapostnotum is developed. The Diptera, on the other hand, using only the forewings for flight, have a well-developed mesopostnotum, but no metapostnotum. To provide attachment for the large longitudinal muscles moving the wings, the antecostal ridges at the front and back of the mesothorax and the back of the metathorax usually develop into extensive internal plates, the phragmata (Figs. 7.1c, 7.3). Which of the phragmata are developed depends again on which wings are most important in flight.

Various strengthening ridges develop on the tergum of a wing-bearing segment. These are local adaptations to the mechanical stresses imposed by the wings and their muscles. The ridges appear externally as sulci which divide the notum into areas. Often, a transverse sulcus divides the notum into an anterior prescutum and a scutum, while a V-shaped sulcus posteriorly cuts off the scutellum (Fig. 7.2). These areas are commonly demarcated, but, because of their origins as functional units, plates of the same name in different insects are not necessarily homologous. In addition, the lateral regions of the scutum may be cut off by a sulcus or there may be a median longitudinal sulcus. Commonly, the prescutum connects with the pleuron by an extension, the prealar arm, in front of the wing, while behind the wing a postalar arm connects the postnotum to the epimeron (Fig. 7.4b). Laterally the scutum is produced into two processes, the anterior and posterior notal processes, which articulate with the axillary sclerites in the wing base (see Fig. 9.10). The posterior fold of the scutellum continues as the axillary cord along the trailing edge of the wing.

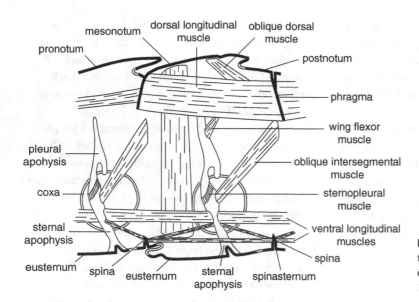

Figure 7.3 The main muscles, other than the leg muscles, in the mesothorax of a winged insect (after Snodgrass, 1935).

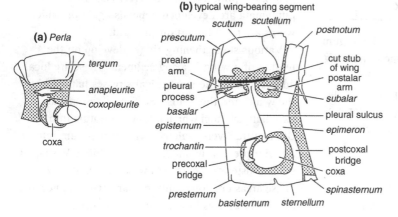

Figure 7.4 Lateral view of thoracic segments. Anterior to left, membranous regions stippled. Names of sclerites in italics (after Snodgrass, 1935). (a) The prothorax of *Perla* (Plecoptera). (b) A typical wing-bearing segment.

7.2.2 Sternum

The primary sclerotizations on the ventral side are segmental and intersegmental plates, which often remain separate in the thorax. The intersegmental sclerite is produced internally into a spine and is called the spinasternum, while the segmental sclerite is called the eusternum (Fig. 7.3). Various degrees of fusion occur so that four basic arrangements may be found:

(1) all elements separate – eusternum of prothorax; first spina; eusternum of mesothorax; second spina; eusternum of metathorax (Fig. 7.5a – notice that in the diagram eusternum is divided into basisternum and sternellum);

(2) eusternum of mesothorax and second spina fuse, the rest remaining separate;

(3) eusternum of prothorax and first spina also fuse so that there are now three main elements: compound prosternum, compound mesosternum, eusternum of metathorax;

(4) complete fusion of meso- and metathoracic elements to form a pterothoracic plate (Fig. 7.5b).

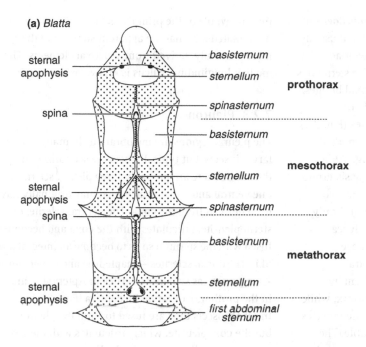

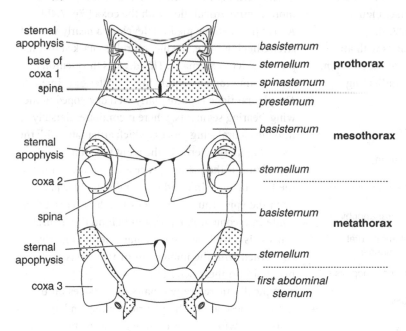

Figure 7.5 Ventral view of the thorax. Names of sclerites in italics. The points at which the sternal apophyses and spina invaginate are slightly exaggerated in size for clarity. Membranous regions stippled. (a) All elements separate (*Blatta* [Blattodea]) (after Snodgrass, 1935). (b) Complete fusion of meso- and metathoracic elements (*Nomadacris* [Orthoptera]) (after Albrecht, 1956).

The sternum of the pterothoracic segments does not differ markedly from that of the prothorax, but usually the basisternum is bigger, providing for the attachment of the large dorso-ventral flight muscles. The sternum is attached to the pleuron by pre- and postcoxal bridges.

Arising from the eusternum are a pair of apophyses, the so-called sternal apophyses (Fig. 7.3). The origins of these on the sternum are marked externally by pits joined by a sulcus (Fig. 7.5b) so that the eusternum is divided into a basisternum and sternellum, while in higher insects the two apophyses arise together in the midline and only separate internally, forming a Y-shaped furca (Fig. 7.6). Distally the sternal apophyses are associated with the inner ends of the pleural apophyses, usually being connected to them by short muscles. This adds rigidity to the thorax, while variation in the degree of contraction of the muscles makes this rigidity variable and controllable. The sternal apophyses also serve for the attachment of the bulk of the ventral longitudinal muscles, although a few fibers retain their primitive intersegmental connections with the spinasterna (Fig. 7.3).

Some insects have a longitudinal sulcus with an internal ridge running along the middle of the sternum. This is regarded by some authorities as indicating

that the whole of the primitive sternum has become invaginated and that the apparent sternum in these insects is really derived from subcoxal elements. The median longitudinal sulcus is known as the discrimen.

7.2.3 Pleuron

The pleural regions are membranous in many larval insects, but typically become sclerotized in the adult. There are probably three pleural sclerites, one ventral and two dorsal, which may originally have been derived from the coxa. The ventral sclerite, or sternopleurite, articulates with the coxa and becomes fused with the sternum so as to become an integral part of it. The dorsal sclerites – anapleurite and coxopleurite – are present as separate sclerites in Apterygota and in the prothorax of larval Plecoptera (Fig. 7.4a). In other insects they are fused to form the pleuron, but the coxopleurite, which articulates with the coxa, remains partially separate in the lower pterygote orders, forming the trochantin and making a second, more ventral articulation with the coxa (Fig. 7.4b). Above the coxa the pleuron develops a nearly vertical strengthening ridge, the pleural ridge, marked by the pleural sulcus externally. This divides the pleuron into an anterior episternum and a posterior epimeron. The pleural ridge is particularly well developed in the wing-bearing segments, where it continues dorsally into the pleural wing process which articulates with the second axillary sclerite in the wing base (Fig. 7.4b).

In front of the pleural process in the membrane at the base of the wing and only indistinctly separated from the episternum are one or two basalar sclerites, while in a comparable position behind the pleural process is a well-defined subalar sclerite. Muscles concerned with the movement of the wings are inserted into these sclerites.

Typically there are two pairs of spiracles on the thorax. These are in the pleural regions and are associated with the mesothoracic and metathoracic segments. The mesothoracic spiracle often occupies a position on the posterior edge of the propleuron, while the smaller metathoracic spiracle may similarly

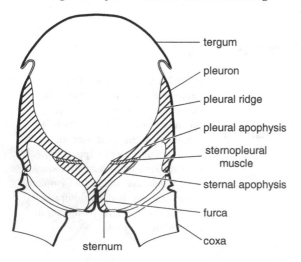

Figure 7.6 Cross-section of a thoracic segment showing the pleural ridges and sternal apophyses (after Snodgrass, 1935).

move onto the mesothorax. Diplura have three or four pairs of thoracic spiracles. *Heterojapyx*, for instance, has two pairs of mesothoracic and two pairs of metathoracic spiracles.

7.3 Muscles of the thorax

The longitudinal muscles of the thorax, as in the abdomen, run from one antecostal ridge to the next. They are relatively poorly developed in sclerotized larvae, in adult Odonata, Blattodea and Isoptera, which have direct wing depressor muscles, and also in secondarily wingless groups such as Siphonaptera. In these cases they tend to telescope one segment into the next, while the more lateral muscles rotate the segments relative to each other. In unsclerotized insects, contraction of the longitudinal muscles shortens the segment.

In most winged insects, however, the dorsal longitudinal muscles are the main wing depressors and they are well developed (Section 9.3.3; Fig. 7.4), running from phragma to phragma so that their contraction distorts the segments. The ventral longitudinal muscles run mainly from one sternal apophysis to the next in adult insects, producing some ventral telescoping of the thoracic segments.

Dorso-ventral muscles run from the tergum to the pleuron or sternum. They are primitively concerned with rotation or compression of the segment, but in winged insects they are important flight muscles (Section 9.3.3). In larval insects an oblique intersegmental muscle runs from the sternal apophysis to the anterior edge of the following tergum or pleuron, but in adults it is usually only present between prothorax and mesothorax.

The other important muscles of the thorax are concerned with movement of the legs and wings. They are dealt with separately (see Chapters 8 and 9).

Summary

- The thorax consists of a prothoracic segment, a mesothoracic segment and a metathoracic segment, each of which bears a pair of legs in most insects. In winged insects, the two posterior segments each bear a pair of wings.

- The wings articulate with the body at the dorsal sclerite, called the tergum or notum. The opposite, ventral sclerite is known as the sternum. The legs articulate with the body at the lateral surface between the notum and the sternum, called the pleuron.

- The musculature of the thorax is primarily adapted to produce and control movements of the wings and legs.

Recommended reading

Matsuda, R. (1970). *Morphology and Evolution of the Insect Thorax*. Ottawa: Entomological Society of Canada.

Minelli, A. and Fusco, G. (2004). Evo-devo perspectives on segmentation: model organisms, and beyond. *Trends in Ecology and Evolution* **19**, 423–429.

Peel, A. D., Chipman, A. D. and Akam, M. (2005). Arthropod segmentation: beyond the *Drosophila* paradigm. *Nature Reviews Genetics* **6**, 905–916.

Snodgrass, R. E. (1935). *Principles of Insect Morphology.* New York, NY: McGraw-Hill.

Snodgrass, R. E. (1958). *Evolution of Arthropod Mechanisms.* Washington, DC: Smithsonian Institution.

References in figure captions

Albrecht, F. O. (1956). *The Anatomy of the Red Locust,* Nomadacris septemfasciata *Serville.* London: Anti-locust Research Centre.

Snodgrass, R. E. (1935). *Principles of Insect Morphology.* New York, NY: McGraw-Hill.

8 Legs and locomotion

REVISED AND UPDATED BY **GRAHAM K. TAYLOR**

INTRODUCTION

Insects typically have three pairs of jointed legs, one pair on each of the thoracic segments. The legs are moved by sets of extrinsic and intrinsic muscles, and are furnished with a range of sensory receptors, as well as with other mechanical structures such as hairs, spines and adhesive pads. In most insects, the primary function of the legs is to enable walking on land, but modifications of their structure allow them to be used in other kinds of locomotion, including jumping, swimming and walking on water. The legs are also specialized for a range of other tasks in different insects, including grasping, grooming, pollen collection and silk production.

This chapter is divided into five sections. Section 8.1 describes the structure of the legs, considering their segments, musculature, sensory apparatus and attachment devices. Section 8.2 discusses maintenance of stance and the movements and coordination of the legs in walking and running. Section 8.3 considers other mechanisms of terrestrial locomotion, including jumping and crawling. Section 8.4 describes the mechanism and function of the legs in aquatic locomotion, both in walking on the surface or the bottom, and in swimming. Section 8.5 discusses modifications of the legs for functions other than locomotion.

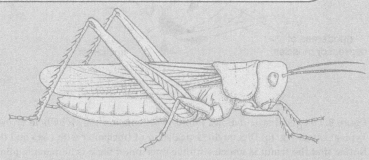

The Insects: Structure and Function (5th edition), ed. S. J. Simpson and A. E. Douglas.
Published by Cambridge University Press. © Cambridge University Press 2013.

8.1 Structure of the legs

Each leg consists typically of six segments, articulating with each other by mono- or dicondylic articulations set in a membrane, the corium. The six basic segments listed from the most proximal to the most distal are the coxa, trochanter, femur, tibia, tarsus and pretarsus (Fig. 8.1a). The legs of insects have an extensive sensory system. Some of the sensory elements are proprioceptors, monitoring

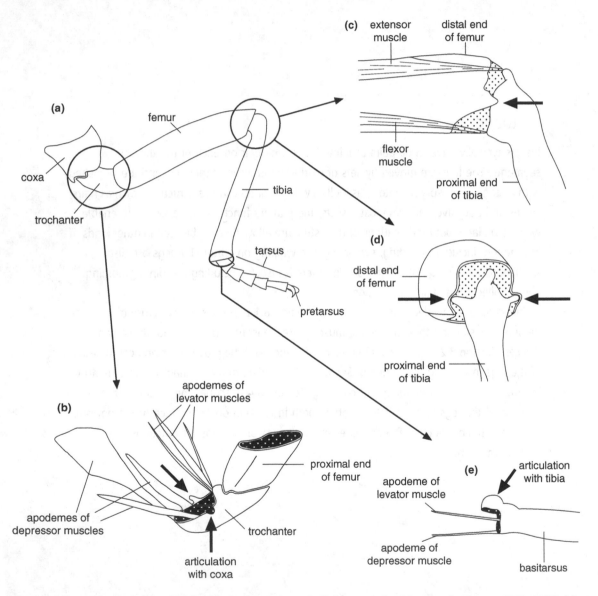

Figure 8.1 Leg and articulations. Points of articulation shown by bold arrows (mainly after Snodgrass, 1935, 1952). (a) Typical insect leg. (b) Dicondylic articulation of trochanter with coxa and the apodemes of muscles moving the trochanter. Notice that the femur is united with the trochanter; there is no moving joint. (c), (d) Dicondylic articulation of tibia and femur: (c) side view; (d) end view. (e) Monocondylic ball articulation of tarsus with tibia.

the positions of the leg segments and the stance of the insect. Other mechanoreceptors and chemoreceptors are involved in the perception of environmental stimuli. The legs of many insects are also furnished with smooth or hairy pads that function as attachment devices and allow insects to accomplish such feats as walking on a windowpane.

8.1.1 Segmentation of the legs

The coxa is often in the form of a truncated cone and articulates basally with the wall of the thorax. There may be only a single articulation with the pleuron (Fig. 8.2a), in which case movement of the coxa is very free, but frequently there is a second articulation with the trochantin (Fig. 8.2b). This restricts movement to some extent, but because the trochantin is flexibly joined to the episternum, the coxa is still relatively mobile. In some higher forms there are rigid pleural and sternal articulations limiting movement of the coxa to rotation about these two points (Fig. 8.2c). In the Lepidoptera, the coxae of the middle and hindlegs are fused with the thorax and this is also true of the hind coxae in Adephaga (Coleoptera). The part of the coxa bearing

the articulations is often strengthened by a ridge indicated externally by the basicostal sulcus, which marks off the basal part of the coxa as the basicoxite (Fig. 8.3a). The basicoxite is divided into anterior and posterior parts by a ridge strengthening the articulation, the posterior part being called the meron. This is very large in Neuroptera, Mecoptera, Trichoptera and Lepidoptera (Fig. 8.3b), while in the higher Diptera it becomes separated from the coxa altogether and forms a part of the wall of the thorax.

The trochanter is a small segment with a dicondylic articulation with the coxa such that it can only move in the vertical plane (Fig. 8.1b). In Odonata there are two trochanters and this also appears to be the case in Hymenoptera, but here the apparent second trochanter is, in fact, a part of the femur.

The femur is often small in larval insects, but in most adults it is the largest and stoutest part of the leg. It is often more or less fixed to the trochanter and moves with it. In this case, there are no muscles moving the femur with respect to the trochanter, but sometimes a single muscle arising in the trochanter is able to produce a slight backward movement, or reduction, of the femur (Fig. 8.5b).

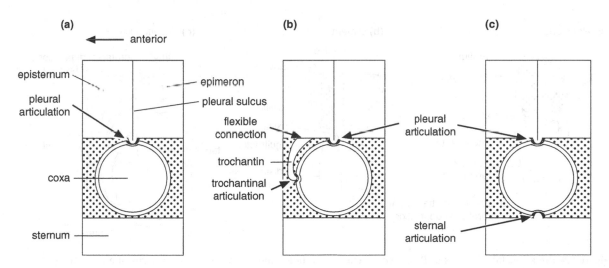

Figure 8.2 Three types of coxal articulation with the thorax. Points of articulation shown by arrows. Membranous regions stippled (after Snodgrass, 1935).

The tibia is the long shank of the leg, articulating with the femur by a dicondylic joint so that it moves in a vertical plane (Fig. 8.1c,d). In most insects the head of the tibia is bent so that the shank can flex right back against the femur (Fig. 8.1a).

The tarsus in most insects is subdivided into 2–5 tarsomeres. These are differentiated from true segments by the absence of muscles. The basal tarsomere, or basitarsus, articulates with the distal end of the tibia by a single condyle (Fig. 8.1e), but between the tarsomeres there is no articulation; they are connected by flexible membrane so that they are freely movable. In Protura, some Collembola and the larvae of most holometabolous insects, the tarsus is unsegmented (Fig. 8.4c) or, in the latter, may be fused with the tibia.

The pretarsus, in the majority of insects, consists of a membranous base supporting a median lobe, the arolium, which may be membranous or partly sclerotized, and a pair of claws which articulate with a median process of the last tarsomere known as the unguifer. Ventrally there is a basal sclerotized

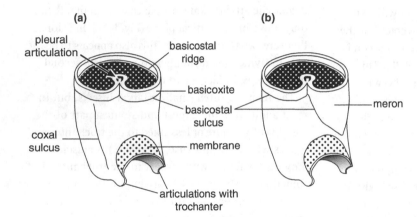

Figure 8.3 Coxa, oblique lateral view (after Snodgrass, 1935). (a) Typical insect; (b) coxa with a large meron.

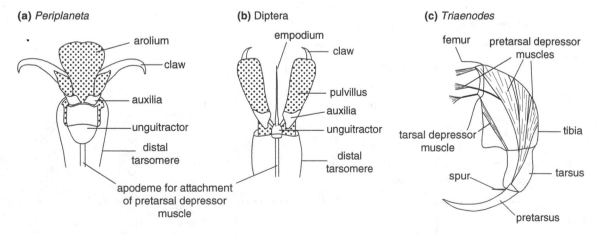

Figure 8.4 Pretarsus. (a) Pretarsus of *Periplaneta*, ventral view (after Snodgrass, 1935). (b) Pretarsus of a dipteran, ventral view (after Snodgrass, 1935). (c) Distal part of prothoracic leg of larval *Triaenodes* (Trichoptera) showing a simple pretarsal segment (after Tindall, 1964).

plate, the unguitractor, and between this and the claws are small plates called auxiliae (Fig. 8.4a). In Diptera a membranous pulvillus arises from the base of each auxilia, while a median empodium, which may be spine- or lobe-like, arises from the unguitractor (Fig. 8.4b). There is no arolium in Diptera other than Tipulidae. The development of the claws varies in different insect groups. Commonly they are more or less equally well developed, but in Thysanoptera they are minute and the pretarsus consists largely of the bladder-like arolium (Fig. 8.8e). In other groups, the claws develop unequally; one may fail to develop altogether, so that in Ischnocera (Phthiraptera), for instance, there is only a single claw. In many holometabolous larvae, the entire pretarsus consists of a single claw-like segment (Fig. 8.4c).

8.1.2 Muscles of the legs

The muscles that move the legs fall into two categories: extrinsic, arising outside the leg, and intrinsic, located wholly within the leg and running from one segment to another. The coxa is moved by extrinsic muscles arising in the thorax. A fairly typical arrangement is shown in Fig. 8.5a, with promotor and remotor muscles arising from the tergum, abductor and adductor muscles from the pleuron and sternum, and rotator muscles from the sternum. The roles of the muscles vary, depending on the activities of other muscles and also on the type of articulation. In *Apis* (Hymenoptera), which has rigid pleural and sternal articulations, promotor and remotor muscles from the tergum are absent. In the pterothoracic segments, muscles (insertions marked with oblique hatching in Fig. 8.5a) run from the coxae to the basalar and subalar sclerites. They are concerned with movements of the wings as well as the legs.

The intrinsic musculature is much simpler than the coxal musculature, consisting typically only of pairs of antagonistic muscles in each segment (Fig. 8.5b). In *Periplaneta* (Blattodea), there are three levator muscles of the trochanter arising in the coxa and three depressor muscles, two again with origins in the

coxa and a third arising on the pleural ridge and the tergum.

The femur is usually immovably attached to the trochanter, but the tibia is moved by extensor and flexor muscles arising in the femur and inserted into apodemes from the membrane at the base of the tibia (see Fig. 8.17a). Levator and depressor muscles of the tarsus arise in the tibia and are inserted into the proximal end of the basitarsus, but there are no muscles within the tarsus moving the tarsomeres.

It is characteristic of insects that the pretarsus has a depressor muscle, but no levator muscle. The fibers of the depressor occur in small groups in the femur and (Phthraptera) tibia and are inserted into a long apodeme which arises on the unguitractor (Figs. 8.5b, 8.17a). Levation of the pretarsus results from the elasticity of its basal parts.

The innervation of the leg muscles is complex. Most muscles are innervated by both fast and slow axons and by inhibitory axons, but not all of the fibers within a muscle are innervated by all three types of motor neuron. For example, in *Periplaneta*, each of the meso- and metathoracic coxae has four muscles which depress the trochanter (Fig. 8.6). Two of these muscles (numbered 136 and 137 in Fig. 8.6) are innervated only by a fast axon, which also supplies parts of the other two muscles (135d' and 135e' in Fig. 8.6). These parts are also innervated by a slow axon, which additionally supplies other parts of these muscles with no fast nerve supply (135d and 135e in Fig. 8.6). Three inhibitory fibers innervate those parts of the muscles that receive input from the slow axon but not the fast axon.

The extensor tibiae muscle in the hindleg of a grasshopper has an even more complex supply. In addition to fast, slow and inhibitory axons, it receives input from an axon which releases the neuromodulator, octopamine. This axon is called the dorsal unpaired median axon of the extensor tibiae muscle. A majority of fibers are innervated only by the fast axon, but this is probably not a general feature of leg muscles, as the extensor tibiae muscle of grasshoppers is specialized for jumping. Nearly all the muscles moving the coxa,

(a) extrinsic muscles

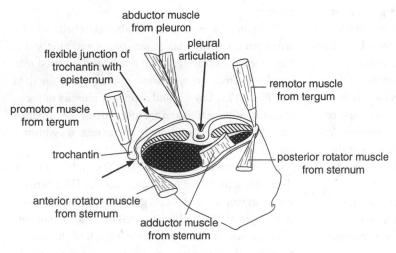

(b) intrinsic muscles

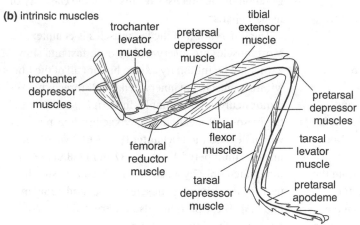

Figure 8.5 Leg muscles. (a) Extrinsic muscles of coxa as seen from the midline of the insect. Muscles arising from the areas marked with diagonal hatching are omitted (from Snodgrass, 1935). (b) Intrinsic muscles; note that one trochanter depressor muscle is extrinsic (after Snodgrass, 1927).

trochanter and tibia in each leg of a locust are innervated by the same inhibitory neuron, which is consequently known as the common inhibitor. Two other inhibitory neurons innervate the muscles of the distal parts of the legs, both running to the same muscles.

8.1.3 Sensory system of the legs

The legs of insects have an extensive sensory system. Some of the sensory elements are proprioceptors, monitoring the positions of the leg segments and the stance of the insect. Other mechanoreceptors and chemoreceptors are involved in the perception of environmental stimuli.

The leg proprioceptors include hair plates, campaniform sensilla and chordotonal organs. The hair plates on the legs contribute to the insect's gravitational sense. *Periplaneta* has hair plates at the proximal end of the coxa and also at the coxa–trochanter joint (Fig. 8.7a). There are groups of campaniform sensilla on the trochanter, a group proximally on the femur and another on the tibia, and a small number on the dorsal

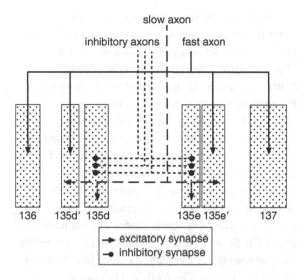

Figure 8.6 Muscle innervation. Diagrammatic representation of the motor neurons to the depressor muscles of the trochanter in the mesothoracic coxa of *Periplaneta*. Muscle units are numbered as in the text (after Pearson and Iles, 1971).

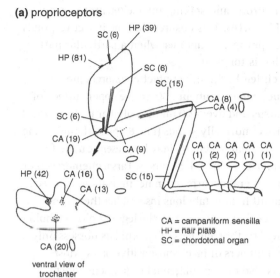

Figure 8.7 Sensory system of the leg. (a) Proprioceptors on the foreleg of *Periplaneta*. Numbers in brackets show the number of sensilla in each group. Ellipses show orientations of campaniform sensilla. (b) Exteroceptors on the fore tarsus of the grasshopper, *Schistocerca americana*. Posterior view: there are 140 long mechanosensitive sensilla and 200 small chemosensitive sensilla on the upper surface and sides of the tarsus. Ventral view: sensilla on the pulvilli and arolium. There are 180 sensilla. About 60% are chemoreceptors.

surface at the distal end of each tarsomere. In total, there are about 140 sensilla in hair plates and 80 campaniform sensilla on each front leg. The other legs have similar numbers. In addition, *Periplaneta* has a chordotonal organ associated with each joint of the leg, and multipolar neurons at the trochanter–femur and femur–tibia joints. There may also be strand receptors. Similar arrangements of proprioceptors occur on the legs of the migratory locust, *Locusta* (Orthoptera), the stick insect, *Carausius* (Phasmatodea), and the adult tobacco hornworm, *Manduca* (Lepidoptera).

Many exteroceptive sensilla are also present on the legs. Mechanosensitive trichoid sensilla are distributed all over the legs, and the final larval stage of *Schistocerca americana* (Orthoptera) has about 140 of these sensilla on each tarsus (Fig. 8.7b); others are present on the femur and tibia. The axons from the sensilla in different areas of the leg converge to separate interneurons so that spatial information is maintained within the central nervous system. This is true not only of the first-order spiking local interneurons, but also of non-spiking interneurons and spiking intersegmental neurons (see Fig. 8.16). This ensures that the insect responds in an appropriate manner when a particular part of a leg is touched.

Each leg has many contact chemoreceptors. *S. americana* has about 200 on the upper surface of the tarsus and over 100 on the pulvillar pads on which the insect normally stands (Fig. 8.7b). Although some contact chemoreceptors may be present on the femur and tibia, most are on the tarsus. Tarsal chemoreceptors are of general occurrence in hemimetabolous insects and adult holometabolous insects, but there is no evidence that they occur on the legs of holometabolous larvae. The thoracic legs of caterpillars possess only small numbers of mechanosensitive hairs. Most insects also have a subgenual organ, sensitive to substrate vibration.

8.1.4 Attachment devices

For the legs to be effective in terrestrial locomotion, they must be able to gain purchase on the substrate, but an attachment device that is used for locomotion must be able to detach quickly from the substrate as well as to attach quickly to it. Presumably for this reason, insects have not evolved to use fixed hooks, suckers or glue for the attachment of their legs to the substrate during locomotion. Instead, the leg attachment devices that insects use for locomotion fall into one of two categories: smooth pads and hairy pads. Such structures occur widely across the phylogeny of insects, although they are absent from Odonata. They may be found on different parts of the pretarsus, tarsus or tibia, and therefore appear to have evolved convergently on multiple occasions. Smooth pads are well suited to adhering to smooth surfaces, but hairy pads are better able to accommodate rough surfaces. Another possible advantage of having hairy pads is that they may be easier to keep clean. In the leaf beetle, *Gastrophysa viridula* (Coleoptera), foot grooming is stimulated by a reduction in the friction force between the hairy pads and the substrate.

Smooth pads occur in a variety of different forms (Fig. 8.8). Sometimes, the arolium between the pretarsal claws serves as a smooth pad (Fig. 8.8a). This is the case, for example, in Hymenoptera, Blattodea, Tipulidae (Diptera) and some Orthoptera. Paired membranous lobes ventral to the pretarsal claws are known as pulvilli, and are adapted to function as smooth pads in several groups, including Siphonaptera, Trichoptera and some Heteroptera (Fig. 8.8b). Smooth pad-like structures called euplantulae are found on the ventral side of one or more tarsomeres in various groups, including Orthoptera, Blattodea, Phasmatodea and Mantodea (Fig. 8.8h). Some aphids (Aphididae) have an eversible soft pad resembling a euplantula at the joint between the tibia and tarsus (Fig. 8.8f), and in sawflies (Hymenoptera) the tarsal thorns and claw pads are both transformed into smooth structures (Fig. 8.8i). Most soft pads share in common a specific internal structure, being composed of a fibrous material with the fibers oriented roughly perpendicular to the substrate. Foam-like arolia are found in froghoppers (Hemiptera: Cercopoidea), and a more complex hierarchical structure is found in some grasshoppers (Orthoptera). In each case, this results in a structure that is compliant under compression but relatively non-compliant under tension. These properties are important in allowing the pad to conform closely to the shape of the surface that it contacts, while providing high-tensile strength when the insect is hanging from or pulling on the substrate.

The tarsomeres are densely covered in hairs in Strepsiptera, Raphidioptera, Megaloptera and some Coleoptera (Fig. 8.8d). Hairy pulvilli are found in some Lepidoptera and in most Diptera except members of the Tipulidae (Fig. 8.8b,c). Specialized hairy structures called fossulae spongiosae are found at the distal ends of the tibiae of the front and middle legs in *Rhodnius* (Heteroptera) and some other Reduviidae (Fig. 8.8g). In each of these cases the adhesive structures are areas of membranous cuticle covered by large numbers of small setae. For example, in the

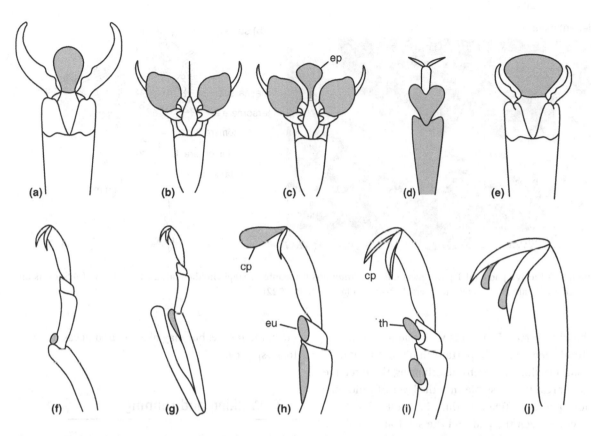

Figure 8.8 Diversity of leg attachment devices (gray-shaded areas) in insects. (a) Arolium (smooth); (b) pulvilli (smooth or hairy); (c) empodial pulvillus (ep) (hairy); (d) hairy adhesive soles of tarsomeres; (e) eversible pretarsal bladder (smooth); (f) eversible structure between tibia and tarsus (smooth); (g) fossula spongiosa (hairy); (h) euplantulae (eu) and claw pad (cp) (both smooth); (i) tarsal thorns transformed into adhesive structures (th), claw pad (cp) (both smooth); (j) adhesive claw setae (Beutel and Gorb, 2001).

lady beetle *Epilachna* (Coleoptera) there are two pads on the underside of each tarsus. Each pad carries about 800 setae which are 70–120 µm long. Many of the setae are expanded at the tip to form flattened, foot-like structures 5–10 µm in diameter (Fig. 8.9a). In the fly *Calliphora* (Diptera) the adhesive setae are also on the tarsi. They are much smaller than those in the beetle, only 9–15 µm high with a "foot" about 1 µm in diameter. The fly has about 42 000 adhesive hairs altogether. The flexibility of the setae enables them to make contact with irregular surfaces much more efficiently than would be true of a single, larger structure, which greatly increases the power of adhesion. The males of many species of beetle have

more adhesive setae than the females. These additional setae are used by the male to grasp the female during mating.

A fluid is secreted at the tips of the hairs or on the surface of the pads during walking. This secretion is an emulsion that contains both a water-soluble fraction and a lipid-soluble fraction, incorporating hydrocarbons, fatty acids and alcohols. Secretion and reabsorption of pad fluid seem to be driven mechanically by capillary forces. This secretion seems to play an integral part in adhesion, but its function is still far from fully understood. In the stick insect *Carausius* the biphasic nature of the secretion has been shown to enhance resistance

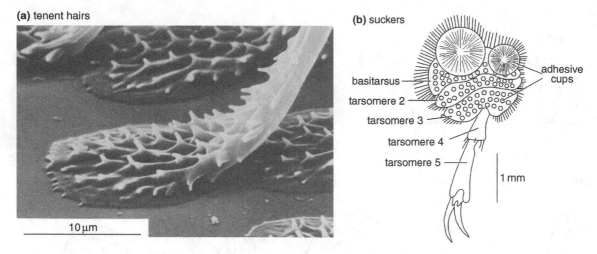

(a) tenent hairs

10 µm

(b) suckers

basitarsus
tarsomere 2
tarsomere 3
tarsomere 4
tarsomere 5

adhesive cups

1 mm

Figure 8.9 Adhesive pads. (a) Tip of tenent hair of *Philonthus* (Coleoptera, Staphylinidae) (after Stork, 1983). (b) Suckers on the foreleg of male *Dytiscus* (Coleoptera, Dytiscidae) (after Miall, 1922).

to shearing forces. In insects with smooth pads, the fluid seems to be important in filling tiny gaps on rough surfaces, thereby maximizing the effective contact area. It is possible that the layer of fluid on a smooth pad is sufficiently thin that molecular contact between the pad and the substrate is still possible. However, the measured adhesion of isolated pad secretions deposited by the hairy pulvilli of *Calliphora* (Diptera) on glass is similar to the adhesion measured in the center of a single attachment pad. This suggests that secretion-mediated capillary forces dominate in the adhesion of the hairy pads of flies, although van der Waals forces may also play some part in the adhesion process.

In male dytiscid beetles a different mechanism of adhesion occurs. The first three tarsomeres of the foreleg of male *Dytiscus* (Coleoptera) are enlarged to form a circular disc. On the inside, this disc is set with stalked cuticular cups, most of which are only about 0.1 mm in diameter, but two of which are much larger than the rest, one being about 1 mm across (Fig. 8.9b). It seems that these cups act as true suckers, although it is not certain how the suction is created. The suckers are used by the male to grasp the female, but may also be used occasionally to grasp prey.

8.2 Walking and running

Besides allowing the insect to stand with its body raised off the substrate, the primary function of the legs of most insects is in terrestrial locomotion, and more particularly in walking and running. The movements of the legs that are involved in these activities are well known, but the details of how these movements are coordinated and controlled are still only partially understood, and such understanding as we do have is derived from only a few model species. Although centrally generated motor patterns appear to be important, and play a dominant role in faster movements, it is obvious that sensory feedback from the periphery must also play an important role in walking, given the unpredictable nature of the terrain that insects cross. Increasingly, however, it is recognized that passive viscoelastic damping complements sensory feedback in stabilizing movements of the legs and body, especially during fast movements when there is little time for sensory feedback to operate.

8.2.1 Maintenance of stance

Even standing still requires muscular activity and the continual adjustment of this activity to compensate for small shifts in position. Some leg muscles, like the extensor tibiae muscles of the front and middle legs of *Schistocerca*, are continuously active in a stationary insect. Slow axons innervating these muscles fire at low rates (5–30 Hz), their activity varying with leg position. The fast axons are not active in a stationary insect, and the inhibitory neurons are only sporadically active, but do fire in response to contact or vibration.

When an insect is standing still, any force tending to change the angles between the segments of the legs, such as a sudden gust of wind, is opposed by a muscular reflex called the postural position reflex, and in most cases this is mediated by the chordotonal organs. Most studies have examined the femur–tibia joint. In this case, any tendency to reduce the angle between the femur and tibia is opposed by the tibial extensor muscle, while an increase in the angle is opposed by the tibial flexor muscle.

Similar reflexes have been described at other joints of the leg. These reflexes are produced in response to passive changes in the angles between segments, whatever the initial degree of flexion or extension, but the input from the chordotonal organ, and the activity of the motor neurons to the muscles, varies according to the position. When *Locusta* (Orthoptera) is standing on a horizontal surface, the angles between the femora and tibiae of the front and middle legs are usually close to 90°. In the hindlegs, the angle is more variable, but is usually less than 45°. The postural resistance reflexes contribute to the maintenance of these positions in the middle and hindlegs, but not in the forelegs.

Campaniform sensilla on the legs monitor strains in the cuticle and produce reflex responses in muscles tending to alleviate those strains. Some campaniform sensilla at the proximal end of the tibia of *Periplaneta* are oriented parallel with the long axis of the leg; others are at right angles to it (Fig. 8.7a). Axial forces, such as would be produced by the weight of the insect when it is standing still are sensed by the most proximal, transversely oriented sensilla, although their responses to such axial forces are relatively weak. Bending the tibia, however, produces strong responses. When the insect is standing on a horizontal surface with the tibia inclined away from the body, the mass of the body will tend to cause an upward bending of the tibia, compressing the proximal sensilla (Fig. 8.10b). Twisting the tibia causes both groups of sensilla to respond. Stimulation of the proximal sensilla excites slow motor neurons to the extensor tibiae and extensor trochanteris muscles, and inhibits the slow motor neurons to the flexor tibiae and flexor trochanteris muscles (Fig. 8.10c). Stimulation of the distal sensilla has the opposite effect.

A comparable reflex system compensates for stress in the trochanter of the stick insect *Carausius* when the femur is bent anteriorly or posteriorly. Here there are two groups of campaniform sensilla on the trochanter, which reflexively activate the retractor or protractor muscles of the coxa. The pathways between proprioceptive sense cells and motor neurons are, in some cases at least, monosynaptic. Similar systems for the control of stance are almost certainly present in all insects.

8.2.2 Patterns of leg movement

It is convenient to divide each step during walking or running into two phases. During the stance phase, the leg is retracted so that it moves backwards relative to the body, with the foot on the ground, thereby propelling the body forwards. During the swing phase, the leg is protracted so that it swings forwards with the foot off the ground.

The pattern of stepping often varies with the speed of movement. At the very lowest speeds, an insect may have most of its feet on the ground for

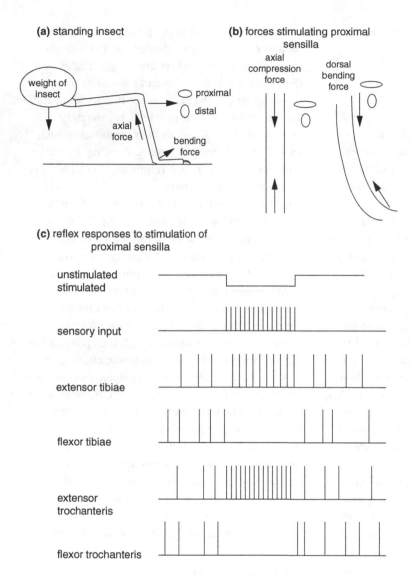

(a) standing insect

weight of insect

proximal

distal

axial force

bending force

(b) forces stimulating proximal sensilla

axial compression force

dorsal bending force

(c) reflex responses to stimulation of proximal sensilla

unstimulated
stimulated

sensory input

extensor tibiae

flexor tibiae

extensor trochanteris

flexor trochanteris

Figure 8.10 Functioning of the proximal campaniform sensilla of a leg in a standing insect (based on Zill and Moran, 1981; Zill *et al.*, 1981). (a) Cross-section of an insect. The two orientations of the sensilla in the proximal group are shown by the ellipses. (b) Diagram showing the effects of an axial force on dorsal bending. The transversely oriented sensilla, compressed along their short axes, are stimulated. (c) Activity in the motor neurons of tibial and trochanteral extensor muscles is enhanced when the sensilla are stimulated. The activity of the flexor muscles is inhibited. Vertical lines represent action potentials.

most of the time. In this case, the legs may be protracted singly in a metachronal wave running first down one side of the body and then down the other, in an antero-posterior direction. At higher stepping rates, insects usually adopt an alternating tripod gait, in which the body is always supported on a tripod formed by the fore- and hindlegs of one side and the middle leg of the other. As one set of three legs is protracted, so the other set is retracted (Fig. 8.11b). Both sets of legs may be planted on the ground together at lower speeds, and at very high speeds there may be periods when all of the legs are off the ground at the same time.

At the very highest speeds, some insects, such as the cockroach *Periplaneta* switch to quadrupedal and even bipedal gaits, running on the hindlegs and raising the front of the body at an increasing angle to

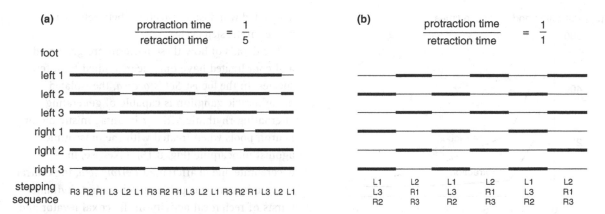

Figure 8.11 Stepping patterns. Diagram showing the disposition of the feet with different protraction time:retraction time ratios. Thick lines indicate retraction with the foot on the ground, thin lines protraction with the foot in the air (based on Hughes, 1952).

the ground. Some insects are effectively quadrupedal due to modification of their fore- or hindlegs for other purposes. Mantids (Mantodea) often walk with the forelegs off the ground, and it is common for grasshoppers (Orthoptera), in which the hindleg is modified for jumping, to use only the anterior two pairs of legs in walking. At high speed, *Tropidopola* (Orthoptera) moves the legs of each segment together and maintains stability by using the tip of the abdomen as an additional point of support.

Higher walking and running speeds can be produced by increasing the frequency with which the legs are moved, or by increasing the length of each stride. Except at very low stepping rates, stride frequency is increased primarily by shortening the duration of the stance phase (Fig. 8.12). In the slow-moving cockroach *Blaberus discoidalis* (Blattodea), stride frequency increases to a maximum of about 13 Hz as the insect's speed increases, but then reaches a plateau (Fig. 8.13a). Further increases in speed result from increases in stride length. The faster-moving *Periplaneta* moves its legs much more quickly, but shows relatively little increase in stride frequency as it runs faster. In this species, increases in stride length account for most of the increase in speed.

8.2.3 Coordination of leg movement

Alternating movements of protraction and retraction are the result of regular patterns of activity of antagonistic muscles in the different segments of the leg. In *Periplaneta*, for example, the levator muscle of the trochanter is active during protraction, lifting the foot off the ground (Fig. 8.14b); the trochanter depressor muscle is inactive, but it starts to contract before the foot is placed on the ground while the levator muscle is still active. At the same time, the contralateral depressor muscle is continuously active as this foot is on the ground. Similar patterns of activity have been recorded for other pairs of antagonistic muscles and for other insects (Fig. 8.14a).

Recruitment of the fast and slow motor neurons that innervate each muscle varies according to the speed of movement. During slow running in the cockroach *Blaberus*, the angular velocity of the joints in the stance phase varies in proportion to the average firing rates of the slow motor neurons that innervate the coxal depressor and extensor tibiae muscles. Modulation of the average firing rate of these slow motor neurons appears to be used to control running speed and direction, but variation in their firing

(a) protraction and retraction times

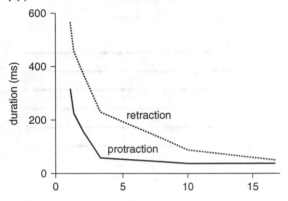

(b) protraction/retraction

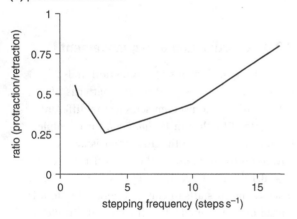

Figure 8.12 Mechanics of locomotion. Changes in duration of retraction and protraction times of a leg in *Periplaneta* in relation to the stepping frequency (after Delcomyn and Usherwood, 1973). (a) Protraction and retraction times; (b) ratio of protraction/retraction.

rate within a burst does not appear to impact upon the movement of the joints. Fast motor neurons are recruited in addition to the slow motor neurons during fast running in *Blaberus*, but this does not result in any sudden acceleration of the leg, and the angular velocities of the joints continue to vary with the average firing velocity of the slow motor neurons during the stance phase. Instead, firing of the fast coxal depressor motor neuron is timed to coincide with the transition from flexion to extension, and is

associated with faster transitions between the two phases of the step cycle.

The details of how these patterns are generated and coordinated have only been studied for a few species. In the locust *Schistocerca*, the isolated metathoracic ganglion is capable of generating alternating rhythmic activity in antagonistic motor neuron pools when treated with the muscarinic agonist pilocarpine (Fig. 8.15). Likewise, in decerebrate and deafferented *Periplaneta*, the central nervous system is capable of generating rhythmic bursts of reciprocal activity in the coxal levator and depressor motor neurons, even in the absence of sensory input from the legs. It is unclear whether these centrally generated motor patterns really correspond to walking, but it makes sense that central motor commands should play a dominant role in coordinating the leg movements of the fast-moving cockroach *Periplaneta*, in which there is comparatively little time for sensory feedback to influence pattern generation on the time scale of a single step cycle. In contrast, the slow-moving stick insect *Carausius* is known to rely much more heavily upon sensory feedback in its pattern generation.

In *Carausius*, each intact leg is capable of generating its own oscillation rhythm independently. The isolated thoracic central nervous system of *Carausius* can be induced to generate alternating rhythms in the antagonistic motor neuron pools of each leg joint when treated with pilocarpine, but the activity is not coordinated between the joints and does not therefore resemble walking. Hence, it is usually inferred that each joint has its own central pattern generator in *Carausius*, but that sensory input plays a key role in intrasegmental coordination, because without it the rhythmic motor activity of the different joints is uncoordinated. Coordination of the different legs is apparently achieved through intersegmental connections, the influences of which vary from segment to segment.

The detailed pattern of leg movement is not fixed. It varies with the load on each leg, and according to the nature of the terrain. This flexibility involves

(a) stepping frequency

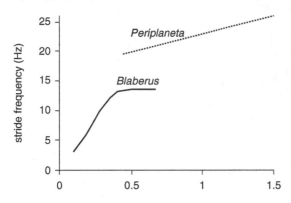

(b) drag

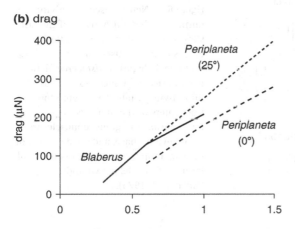

(c) power output

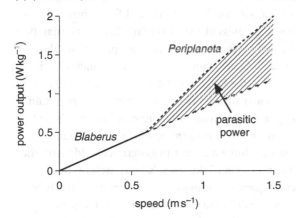

pattern modulation by input from the proprioceptors, and to some extent also from other mechanoreceptors on the leg. As in standing, different groups of proprioceptors may be involved in different processes. For example, as a slowly walking cockroach puts its foot down at the end of protraction, the proximal campaniform sensilla on the tibia are stimulated by the dorsal bending of the leg (Fig. 8.10). The slow motor neuron to the extensor tibiae muscle starts to fire soon afterwards and it is probable that the input from the campaniform sensilla influences both the timing and the rate of firing of the motor neuron. This reduces the degree of dorsal bending and the proximal sensilla stop firing. If the rate of firing of the slow neuron to the extensor tibiae muscle exceeds 300 Hz, the distal campaniform sensilla fire. Their activity contributes to the inhibition of the extensor tibiae motor neuron and prevents excessive muscle contraction that could damage the tibia. If the leg encounters obstacles during its movement, the input from these sensilla is affected. The timing of activity of the campaniform sensilla is altered during fast walking. The chordotonal organs in the legs are also important in regulating the stepping movements.

The proprioceptors often connect directly, via monosynaptic pathways, with the motor neurons (Fig. 8.16). In contrast, the input from exteroceptive mechanoreceptors on the leg is integrated, primarily, by spiking local interneurons. The axon from one hair synapses with several interneurons, and each interneuron receives input from many sensilla in its receptive field. The spiking interneurons connect, in turn, with a network of non-spiking interneurons. In the locust, spiking local interneurons in the midline of the ganglion make inhibitory synapses with non-spiking interneurons, whereas the interneurons in another group (antero-medial) generally make

Figure 8.13 Mechanics of locomotion in *Blaberus*, slow moving, and *Periplaneta*, fast moving (after Full and Tu, 1990; Full and Koehl, 1993). (a) Stepping (stride) frequency at different speeds. (b) Drag at different speeds and, for *Periplaneta*, different angles of attack. (c) Power output at different speeds. The hatched area shows the parasitic power exerted by *Periplaneta* in order to overcome drag.

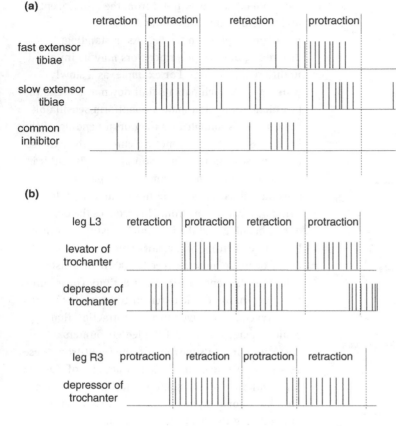

Figure 8.14 Neuromuscular activity during walking. (a) Activity of the fast, slow and inhibitory axons to the extensor tibiae muscle of the prothoracic leg of *Schistocerca* (data from Burns and Usherwood, 1979). (b) Activity of muscles moving the trochanters of the hindlegs of *Periplaneta*. Antagonistic muscles in L3 are in antiphase, and the depressor of L3 is in antiphase with the depressor of R3 (data from Delcomyn and Usherwood, 1973).

excitatory connections. Each spiking local interneuron synapses with more than one non-spiking interneuron, and each of the latter receives input from several spiking interneurons. Finally, each non-spiking interneuron connects with several motor neurons and each motor neuron receives input from several non-spiking interneurons. Some of these inputs are excitatory, but others are inhibitory. As a result of all these interconnections, each non-spiking interneuron responds to a stimulus in the context of the activity generated by many other interneurons and the output to the motor neurons is varied accordingly.

8.2.4 Stability

A system is said to be statically stable if its initial tendency is to return in the direction of static equilibrium when subjected to a small perturbation from it. The alternating tripod gait is statically stable if, when viewed from above, the center of mass of the insect falls inside the triangle formed by the three legs that are planted on the ground. Static stability is further enhanced by the fact that the body is slung between the legs, so that the center of mass is low. Static stability is certainly important in avoiding toppling at low speeds, but when an insect is running quickly, the ground reaction force does not in general coincide with the line of action of the insect's body weight. In consequence, a running insect is never actually at static equilibrium, which means that the static stability of the tripod gait is irrelevant. Instead, what matters is whether the moments about the center of mass balance out over the course of each

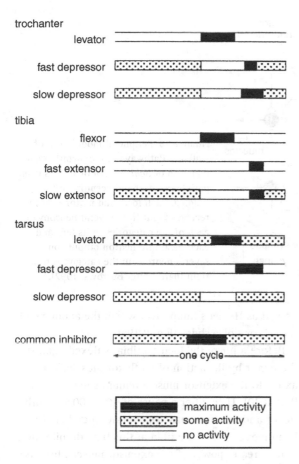

Figure 8.15 Central nervous control of walking. Rhythmic activity in the motor neurons in an isolated metathoracic ganglion of *Schistocerca* during one complete cycle equivalent to protraction and retraction of the leg. Dark bars show periods of maximal activity; dotted bars show periods when some activity may occur; open bars, no activity. Notice that the neurons to antagonistic muscles are in antiphase, periods of maximal activity in one coinciding with minimal activity in the other. As this output is to a hindleg, flexion of the tibia occurs when the foot is off the ground due to the levation of the trochanter and tarsus (after Ryckebusch and Laurent, 1993; data for the fast extensor tibiae neuron is less well documented than for other neurons).

stride. For this reason, stability in running is better thought of as requiring dynamic stabilization of a limit cycle motion. This dynamic stabilization is achieved, in part, by the viscoelastic properties of the legs. For example, the cockroach *Blaberus* is able to recover lateral perturbations in velocity within a single step cycle. The rapidity of this response suggests that this is the result of mechanical damping rather than reflex-based control. This kind of passive dynamic response is sometimes referred to as a preflex, and acts as an important supplement to sensory feedback in stabilizing body and leg movements, especially when very fast responses are required.

8.3 Other mechanisms of terrestrial locomotion

Although the local movements of most insects are achieved by walking or running, insects from several different orders have developed specializations for jumping. Typically, these jumping movements are propelled by the hindlegs, and are usually powered using energy released from elastic storage rather than by direct muscle contractions, but some jumping insects flip their whole body to lever themselves into the air. The soft-bodied larvae of many holometabolous insects move by changing the shape of their body, which results in various types of crawling motion.

8.3.1 Jumping with the legs

Most jumping insects use their hindlegs, and in some cases their middle legs, to propel themselves into the air. Catapult mechanisms allow slow muscle contractions to build up and store elastic energy, and are important to most accomplished jumping insects. An exception is the bushcricket, *Pholidoptera griseoaptera* (Orthoptera), which uses direct contractions of its large extensor tibiae muscles to power jumps of up to 0.7 m. Its hindlegs are 1.5 times longer than its body, which allows the insect to reach a high take-off velocity without imposing unduly high strain rates on the muscles. This increases the force that the muscles apply, and thereby increases the total amount of work they do in contraction. Leg

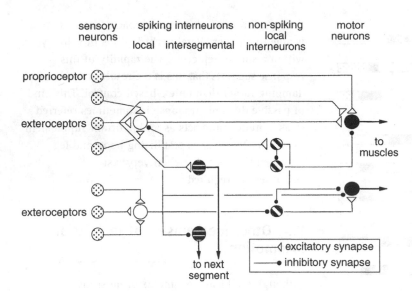

sensory neurons   spiking interneurons   non-spiking local interneurons   motor neurons

local   intersegmental

proprioceptor

exteroceptors

exteroceptors

to muscles

to next segment

◁ excitatory synapse
● inhibitory synapse

Figure 8.16 Peripheral modulation of walking. Pathways in the metathoracic ganglion of *Schistocerca* of signals from peripheral mechanoreceptors of the hindleg. Each local and motor neuron receives input from several neurons in each of the categories to its left, and each local interneuron outputs onto several neurons in the categories to its right (partly after Burrows, 1992).

length is not expected to affect take-off velocity in animals that rely upon elastic energy storage for jumping, because the work done is nearly independent of the strain rate in such systems. Consequently, although very long legs are characteristic of most jumping vertebrates, this is by no means the case for most jumping insects.

Some insects with long hindlegs, such as the locust *Schistocerca*, do nevertheless rely upon a catapult mechanism for jumping. The structure of the hindleg of *Schistocerca*, and especially of the femoro-tibial joint, is adapted to permit the development of maximum force by the extensor tibiae muscle, the storage of the energy they produce and its rapid release, resulting in the sudden extension of the tibia. The extensor tibiae muscle consists of a series of short fibers inserted obliquely into a long, flat apodeme (Fig. 8.17). Collectively, because of their oblique arrangement, they have a large cross-sectional area and can develop a force up to 16 N, compared with only 0.7 N by the flexor tibiae muscle. Just above the articulation with the tibia, the cuticle of the femur is heavily sclerotized, forming a dark area known as the semilunar process (Fig. 8.18). Beneath the articulation, the cuticle of the femur is thickened internally to form a process,

known as Heitler's lump, over which the apodeme of the flexor tibiae slides (Fig. 8.19a).

Before a jump, the locust's tibia is flexed against the femur by the action of its flexor muscle. The axons to the extensor muscle remain silent (Fig. 8.20). Then, after a pause of 100–200 ms, both flexor and extensor muscles contract together. There is no movement of the tibia at this time despite the much greater power of the extensor muscle, because the lever ratio between the flexor and extensor muscles (f/e in Fig. 8.19) greatly favors the flexor muscle. When the tibia is closed up against the femur, this ratio is about 21:1 (Fig. 8.19d). However, as both muscles exert parallel forces on the distal end of the femur, this becomes distorted and energy is stored in the semilunar processes (Fig. 8.18). Rapid extension of the tibia occurs when the flexor muscles suddenly relax due to the cessation of their motor input and to the activity of their inhibitory nerve supply (Fig. 8.20). The semilunar process springs back to its relaxed shape so that the femur–tibia articulation moves distally at the same time as the extensor muscle pulls the head of the tibia proximally.

Because of the length of the tibia, a small movement of the head of the tibia produces a large movement of the distal end. This mechanical

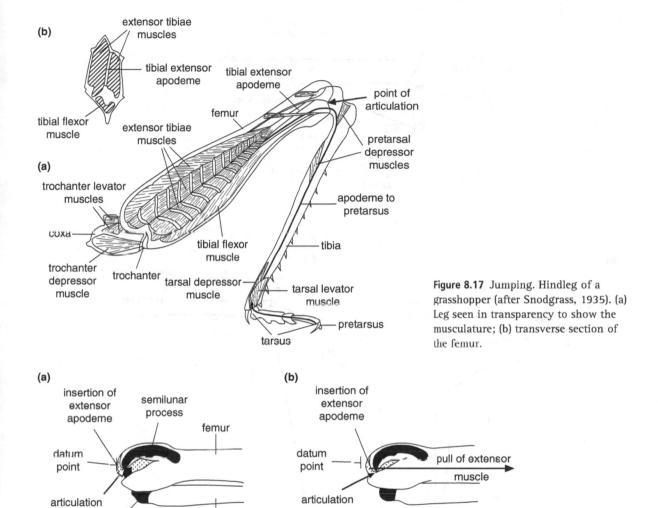

Figure 8.17 Jumping. Hindleg of a grasshopper (after Snodgrass, 1935). (a) Leg seen in transparency to show the musculature; (b) transverse section of the femur.

Figure 8.18 Jumping. Specializations of the hind femoro-tibial joint of a locust. Membranous areas stippled, heavily sclerotized areas black (after Bennet-Clark, 1975). (a) The tibia is flexed, but the extensor muscle remains relaxed. (b) The extensor muscle contracts at the same time as the flexor muscle. Because the insertion of the extensor apodeme is almost in line with the articulation, contraction of the extensor muscle causes distortion of the head of the femur, straining the semilunar process. Notice that the femur has shortened slightly with reference to the datum point which shows the position of the origin of the extensor apodeme in (a), and the semilunar process is bowed.

advantage varies with the position of the tibia, but is greatest – about 150:1 – at the start of the movement. As a result of the force exerted by the distal ends of the two tibiae, the insect is hurled into the air. The insect's center of gravity is close to the line joining the insertions of the metathoracic coxae, so little torque is produced when the legs extend, and the insect moves through the air without rolling. The initial thrust exerted by the tibiae is directly downwards because just before the jump, the flexed femur–tibia is moved so that the tibia is parallel with the ground. In the course of extension, however, a

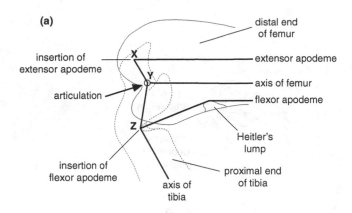

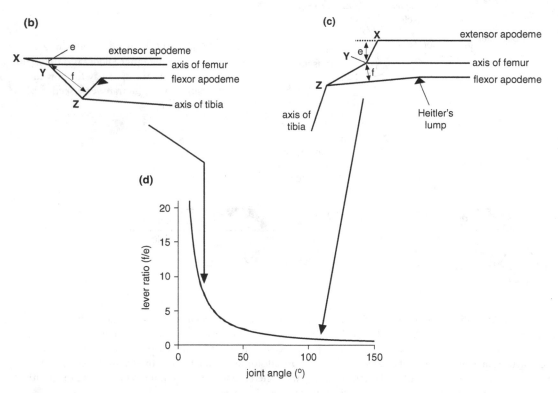

Figure 8.19 Jumping. Mechanical relationship between the tibial extensor and flexor muscles of a locust. X shows the position at which the apodeme of the extensor muscle is attached to the head of the tibia, and Z the position at which the apodeme of the flexor muscle is attached. Y is the point of articulation of the tibia with the femur (after Heitler, 1974, 1977). (a) Diagram showing the position of Heitler's lump and arrangement of the apodemes. The distal end of the femur is shown in outline (solid line) and the proximal end of the tibia (dotted). (b), (c) Changes in positions of apodemes and their insertions with the tibia flexed (b) and extended (c). (d) Changes in the lever ratio of the tibial flexor and extensor muscles as the tibia swings away from the femur. The very high lever ratio (f/e) when the tibia is fully flexed enables the flexor muscle to hold the tibia in position despite the much larger size of the extensor muscle which is contracting at the same time.

backward component develops, pushing the insect forwards.

In many other insects jumping is produced by torques applied about the coxo-trochanteral joint, rather than the femoro-tibial joint. This is true of the several families of Hemiptera that use their hindlegs for jumping, which include some of the most accomplished of all jumping insects. For example, the froghopper *Philaenus spumarius* (Cercopoidea) is capable of reaching take-off velocities of $4.7 \, \mathrm{m \, s^{-1}}$ within 0.9 ms, while the planthopper *Issus coleoptratus* (Issidae) can reach a take-off velocity of $5.5 \, \mathrm{m \, s^{-1}}$ within 0.8 ms. Leafhoppers (Cicadellidae) that have been examined to date achieve somewhat lower take-off velocities, and in long-legged leafhoppers the acceleration phase can last up to 5 ms. One possible advantage to having this longer acceleration phase is that the ground reaction forces are lower, which could assist by reducing energy losses to the substrate when jumping from a compliant surface such as a leaf.

Froghoppers store elastic energy in the metathorax by bending a pair of bow-shaped pleural arches through contraction of the trochanteral depressor muscle. The pleural arches are composite structures comprising stiff chitinous cuticle and elastic resilin, although the resilin stores only a small fraction of the total elastic energy. Release of the mechanism results in rapid rotation of the leg about the coxo-trochanteral joint, and it is held in its cocked position by the interlocking of microtrichia that cover prominent protrusions on the coxa and femur. Planthoppers also have a microtrichia-covered protrusion on the coxa, but have only a patch of smooth cuticle on the dorsal surface of the femur, which might perhaps lock to the coxa through adhesion. In contrast, the hindlegs of leafhoppers appear to have no mechanical locking mechanism.

The jumping mechanism of fleas (Siphonaptera) is similar in principle to that of froghoppers, involving elastic storage by a resilin pad in the internal skeleton of the thorax, and rapid rotation of the hindlegs about the coxo-trochanteral joint. As in froghoppers, the ground reaction force appears to be transmitted via the hind tibiae and tarsi, but whereas the hindlegs of froghoppers are slung under the body for jumping, the hindlegs of fleas are displaced laterally. This difference has important consequences for the control of jumping direction. A similar jumping mechanism is used by snow fleas (Boreidae), which, however, use their middle legs as well as their hindlegs during jumping. Other less-accomplished jumpers such as the stick insect *Sipyloidea* (Phasmatodea) and the ant *Harpegnathos saltator* (Hymenoptera) also use their middle legs and hindlegs to jump short distances, apparently through direct muscle contraction. *Harpegnathos* may be unique among insects in using jumping to catch prey in flight.

8.3.2 Other mechanisms of jumping

Jumping mechanisms that do not involve the legs occur in click beetles (Elateridae) and the larvae of various Diptera. Last-stage larvae of *Piophila* (Diptera) are able to jump by bending the head beneath the abdomen, causing the mandibles to engage in a transverse fold near the posterior spiracles at the end of the abdomen. The longitudinal muscles on the outside of the resulting loop contract and build up tension until suddenly the mandibles are released and the larva jerks straight, striking the ground so that it is thrown into the air, sometimes as high as 0.2 m. A similar phenomenon occurs in the larvae of some Clusiidae and Tephritidae (Diptera). In cecidomyiid larvae (Diptera), anal hooks catch in a forked prosternal projection producing a leap by building up muscular tension and then suddenly releasing it.

Elaterids (Coleoptera) jump if they are turned on their backs, and the jump serves as a means by which they can right themselves. The insect first arches its back between the prothorax and the mesothorax so that it is supported anteriorly by

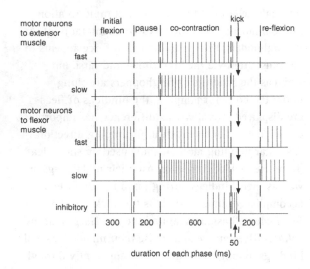

Figure 8.20 Jumping. Activity of the motor neurons controlling the extensor and flexor tibiae muscles of the hindleg of a locust in relation to a jump. Vertical lines represent action potentials (after Heitler and Burrows, 1977).

the prothorax and posteriorly by the elytra, with the middle of the body off the ground. This movement is produced by the median dorsal muscle and results in the withdrawal of a median prosternal peg from the pit in which it is normally at rest. The massive prothoracic intersegmental muscle acts antagonistically to the median dorsal muscle, but when it starts to contract, a small process on the upper side of the peg catches on a lip on the anterior edge of the mesosternum and no movement occurs. Tension builds up in the muscle because of its continued contraction and energy is stored within it and, possibly, in the associated cuticle. Finally, the prosternal peg slips off its catch and this energy is released as kinetic energy, rotating the prothorax and posterior end of the body upwards towards each other, and lifting the center of mass of both parts of the body upwards at considerable speed.

8.3.3 Crawling

The larvae of many holometabolous insects move by changes in the shape of the body rather than by movements of the legs as in walking or running by adult insects. This type of locomotion can be differentiated as crawling. In the majority of crawling forms, the cuticle is soft and flexible and does not, by itself, provide a suitable skeleton on which the muscles can act. Instead, the hemolymph within the body provides a hydrostatic skeleton. Muscles lining the body wall of caterpillars keep the body turgid and, because of the incompressibility of the body fluids, compression of one part of the body due to muscular contraction is compensated for by expansion of some other part. The place and form of these compensating changes is controlled by the differences in tension of the muscles throughout the body.

Lepidopteran caterpillars typically have, in addition to the thoracic legs, a pair of prolegs on each of abdominal segments 3–6 and another pair on segment 10. The prolegs are hollow, cylindrical outgrowths of the body wall, the lumen being continuous with the hemocoel (Fig. 8.21a). An apical area, less rigid than the sides, is known as the planta and it bears one or more rows or circles of outwardly curved hooks, or crochets, with which the proleg obtains a grip. Retractor muscles from the body wall are inserted into the center of the planta so that when they contract it is drawn inwards and the crochets are disengaged. The leg is evaginated by turgor pressure when the muscles relax. On a smooth surface the prolegs can function as suckers. The crochets are turned up and the planta surface is first pressed down onto the substratum and then the center is slightly drawn up so as to create a vacuum.

Caterpillars move by serial contractions of the longitudinal muscles coupled with leg movements starting posteriorly and continuing as a wave to the front of the body. The two legs of a segment, including those on the thoracic segments, move in synchrony. Each segment is lifted by contraction of the dorsal longitudinal muscles of the segment in front and its own dorso-ventral muscles, while at the same time the prolegs are retracted (Fig. 8.21b). Subsequently,

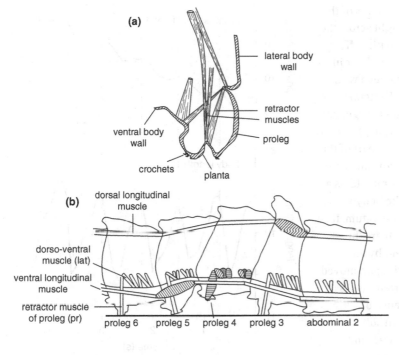

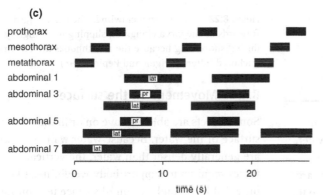

Figure 8.21 Caterpillar crawling. (a) Transverse section through part of an abdominal segment of a caterpillar showing a proleg (after Hinton, 1955). (b) Longitudinal section through the abdomen of a caterpillar showing a wave of contraction which passes along the body from behind forwards (left to right) and produces forward movement. Contracted muscles are shown hatched. There are no prolegs on abdominal segment 2 (based on Hughes, 1965). (c) Central nervous control of crawling. Rhythmic activity in nerves from an isolated nerve cord of *Manduca*. Each black bar shows the periods of activity of motor neurons in a segmental nerve. The thoracic nerves shown innervate the femoral levator muscles and extensor muscles of more distal leg segments. They produce protraction of the leg. The abdominal nerves shown innervate the lateral body wall muscles (lat in b) and the proleg retractor muscles (pr in b). There are no prolegs on abdominal segments 1 and 7 (after Johnston and Levine, 1996).

contraction of the ventral longitudinal muscles brings the segment down again and completes the forward movement as the legs are extended and obtain a fresh grip. As the wave of contraction passes forwards along the body, at least three segments are in different stages of contraction at any one time. These patterns are the product of rhythmic activity in the ventral nerve cord, which occurs even in the absence of sensory input (Fig. 8.21c), although it is probable that this central control is normally modified by local reflexes involving the stretch receptors. Many geometrid larvae have prolegs only on abdominal segments 6 and 10. These insects loop along, drawing the hind end of the body up to the thorax and then extending the head and thorax to obtain a fresh grip.

In the apodous larvae of cyclorrhaphous Diptera, movement again depends on changes in the shape of the body as a result of muscles acting against the body fluids. The posterior segments of the body have raised pads, usually running right across the ventral surface of a segment and armed with stiff, curved setae, which may be distributed evenly or in rows or patches. Each pad, or welt, is provided with retractor muscles. In the larva of *Musca* (Diptera) there are locomotory welts on the anterior edges of segments 6–12, and also on the posterior edge of segment 12 and behind the anus. In movement, the anterior part of the body is lengthened and narrowed by the contraction of oblique muscles, while the posterior part maintains a grip with the welts. Hence the front of the body is pushed forwards over, or through, the substratum. It is then anchored by the mouthhooks, which are thrust against the substratum until they are held by an irregularity of the surface and the posterior part moved forwards by a wave of longitudinal shortening. As a result, the larva exhibits a regular sequence of lengthening and shortening (Fig. 8.22). In soil-dwelling larvae of Tipulidae and Bibionidae, and probably in other burrowing forms, the anterior region is anchored by the broadening of the body, which accompanies shortening.

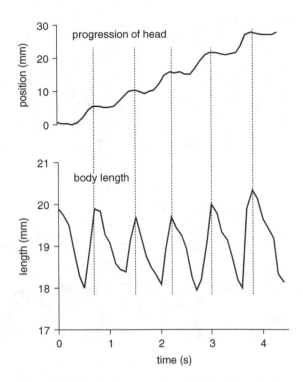

Figure 8.22 Blowfly larva crawling. The head is pushed forwards as the larva elongates, slipping back slightly during shortening because the mouthhooks are not firmly anchored (after Berrigan and Pepin, 1995).

8.4   Aquatic locomotion

Aquatic insects have evolved a number of different mechanisms for moving about on or under the water. Walking on the surface of the water is common among some Heteroptera, and a variety of insects are capable of using surface tension forces to climb up the menisci that form around emergent vegetation or the water's edge. Some insects even secrete surfactants to enable them to propel themselves across the water's surface like a soap boat. Bottom-dwelling insects use walking movements much like those of terrestrial insects, but some are capable of jetting water to propel a fast escape response. Free-swimming insects use a variety of different swimming techniques, but in the most accomplished swimmers the hindlegs, and sometimes also the middle legs, are modified structurally for swimming.

8.4.1   Movement on the surface

Some insects are able to move on or in the film at the surface of the water. Because water walking insects are generally denser than water, the vertical forces required to support body weight must be provided by a combination of surface tension forces and pressure forces. The upward component of the static pressure force is equal to the weight of water that the wetted surface of the body displaces. The upward component of the surface tension force that acts upon a partially submerged body is equal to the weight of water that the meniscus displaces. It follows that surface tension forces dominate over pressure forces in weight support if the legs are thin with respect to the capillary length of water, as is the case for most water-walking insects.

Water striders such as *Gerris* (Heteroptera) stand on the surface film and row over the surface. All of the legs possess hydrofuge properties distally, and do not break the surface film. At the start of a power stroke, the forelegs are lifted off the surface and the long middle legs sweep backwards, producing an indentation of the water surface that spreads backwards as a wave (Fig. 8.23). The transfer of momentum to the water through this surface wave was once thought to explain water strider propulsion. In fact, the amount of momentum transferred to the surface wave is small in comparison with the amount of momentum that is transferred to vortices in the wake. In principle, water striders therefore row themselves through the water in a similar way to insects that swim beneath the surface (see below), a key difference being that they use menisci, rather than flattened appendages, as paddles.

Moving from the water's surface to emergent vegetation, say, is made difficult for insects by the steep meniscus at the water's edge. Larger insects such as water striders may be able to jump over the meniscus, but many smaller insects rely upon a technique known as meniscus climbing. Positively buoyant objects, such as bubbles, are drawn up a meniscus by their buoyancy. Insects that make use of meniscus climbing are not positively buoyant, but by pulling upwards on the steepest section of a meniscus, they are able to climb it just as a buoyant object would do. In order to obtain weight support, the insect must also push downward on the water's surface with a part of its body that is on a shallower section of the meniscus. Furthermore, in order to maintain torque equilibrium the insect must pull upwards on the surface at the opposite end of its body. Hence, a meniscus-climbing insect will characteristically pull up on the water's surface at

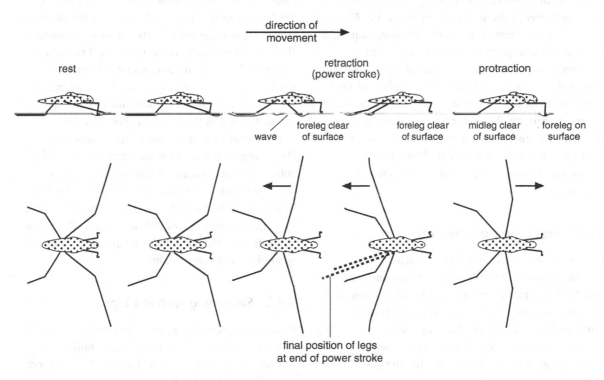

Figure 8.23 Movement on the water surface. Positions of the legs of *Gerris*. Arrows show the directions of movement of the legs relative to the body; insect moving from left to right (after Nachtigall, 1974).

its front and back, while pushing down on the water's surface at its middle. For example, the water treader *Mesovelia* (Heteroptera) pulls upward on the water's surface with its forelegs and hindlegs, but pushes downwards with its middle legs. Larvae of the waterlily leaf beetle *Pyrrhalta* (Coleoptera) achieve the same end simply by arching their body. Water-walking insects with otherwise hydrophobic legs may use retractable hydrophilic claws to pull upwards on the water's surface during meniscus climbing.

Some small insects make use of the Marangoni effect to propel themselves across the water's surface. For example, rove beetles of the genus *Stenus* (Coleoptera) live on grass stems bordering mountain streams in situations such that the beetles fall into the water quite frequently. The beetle can walk on the surface of the water, but only slowly. More rapid locomotion is produced following the secretion of chemicals from the pygidial glands opening beneath the last abdominal tergite. Five chemicals are released, of which the most important is probably a terpenoid called stenusin. This substance lowers the surface tension of the water and at the same time it makes the surface of the beetle hydrophobic, so drag is reduced as the insect is drawn through the water by the higher surface tension in front. It may move at up to $70 \, cm \, s^{-1}$ using its abdomen as a rudder. Marangoni propulsion has also been observed in some Heteroptera.

8.4.2 Movement on the bottom

Bottom-dwelling aquatic insects, such as *Aphelocheirus* (Heteroptera) and larval Odonata and Trichoptera, can walk over the substratum in the same way as terrestrial insects. The larva of *Limnephilus* (Trichoptera) basically uses an alternation of triangles of support, but because of the irregularity of the surface the stepping pattern tends to be irregular. The forelegs may step together instead of alternating and the

hindlegs may follow the same pattern. Normally the power for walking comes primarily from traction by the fore- and middle legs and pushing by the hindlegs, but under difficult conditions the hindlegs may be extended far forwards outside the middle legs so that they help the other legs to pull the larva along.

Larval Anisoptera can walk across the substratum using their legs, but they are also able to make sudden escape movements by forcing water rapidly out of the branchial chamber, causing the body to be driven forwards. The branchial chamber is compressed by longitudinal and dorso-ventral contractions of the abdomen, the contractions being strongest in segments 6–8, in which the branchial chamber lies. Before this contraction, the anal valves close and then open slightly, leaving an aperture about $0.01 \, mm^2$ in area. The contractile movement lasts about 100 ms and water is forced through the anus at a velocity of about $0.25 \, m \, s^{-1}$, propelling the larva forwards at 0.3–$0.5 \, m \, s^{-1}$. As the abdomen contracts, the legs are retracted so as to lie along the sides of the body, offering minimal resistance to forward movement. Successive contractions may occur at frequencies up to 2.2 Hz, continuing for up to 15 seconds. Coordination involves giant fibers running in the ventral nerve cord.

The larval case of the caddis fly, *Triaenodes* (Trichoptera), is built of plant material arranged in a spiral, the most anterior whorl of which extends dorsally beyond the rest of the case above the thorax. This dorsal whorl provides a certain amount of lift, helping to carry the case off the bottom. This lift is controlled by the movements of the legs, which tend to produce a downward thrust.

8.4.3 Swimming with the legs

Adult Coleoptera, larval and adult Heteroptera and larval and pupal Diptera make up the bulk of the free-swimming insects. Coleoptera and Heteroptera typically use their hindlegs for swimming, sometimes together with their middle legs. The points of attachment of the hindlegs are displaced posteriorly in comparison with terrestrial relatives, and in dytiscids and gyrinids (Coleoptera), the coxae are immovably fused to the thorax. Intrinsic muscles of the legs tend to be reduced, and the movements of the distal parts of the legs during swimming are largely passive. Hindlegs that are used for swimming are relatively shorter than in terrestrial relatives, but their tarsi are relatively longer. The two legs of a segment normally move together, contrasting with the alternating movements of contralateral legs in terrestrial insects. *Hydrophilus* (Coleoptera) is an exception. This beetle uses the middle and hindlegs in swimming, the middle leg of one side being retracted simultaneously with the hindleg of the opposite side, but out of phase with the contralateral middle leg.

Retraction of the hindlegs produces the power stroke on which the insect is driven forwards. Protraction of the legs during the recovery stroke also produces hydrodynamic forces, however, which will tend to drive the insect backwards. Hence, if the insect is to move forwards, the thrust produced on the power stroke must exceed the opposing forces produced on the recovery stroke. The thrust that a leg exerts in water is proportional to its area and the square of the velocity with which it moves. Hence, to produce the most efficient forward movement, a leg should present a large surface area and move rapidly on the backstroke, while presenting only a small surface and moving relatively slowly on the recovery stroke.

The hind tibiae and tarsi, and sometimes also those of the middle legs, are flattened antero-posteriorly to form a paddle. In *Acilius* (Coleoptera) and *Dytiscus* the area of the paddle is increased by inflexible hairs, while in *Gyrinus* (Coleoptera) cuticular blades 1 μm thick and 30–40 μm wide are used to increase paddle area (Fig. 8.24). In *Acilius* the backstroke is faster than the forward stroke, so that for a given leg area the forward thrust on the body exceeds the backward thrust. In *Gyrinus*, on the other hand, the backstroke is slower than the forward stroke, so that for a given area the backward thrust should be greater

than the forward thrust. However, the cuticular blades fringing the leg of *Gyrinus* are placed asymmetrically so that they open like a venetian blind, turning to overlap and produce a solid surface during the power stroke. During the recovery stroke the tarsomeres collapse like a fan and are concealed in a hollow of the tibia, which in turn is partly concealed in a hollow of the femur (Fig. 8.25a). This ensures that the insect produces a net forward thrust.

Such feathering mechanisms are widespread. For example, the fringing hairs on the swimming legs of *Dytiscus* are only spread to expose their maximum area on the power stroke (Fig. 8.26a–d). On the recovery stroke, the femoro-tibial joint flexes so that the tibia and tarsus trail out behind (Fig. 8.26e–h). At the same time the tibia rotates through 45° so that the previously dorsal surface becomes anterior and the fringing hairs fold back. The tarsus, which articulates with the tibia by a ball-and-socket joint, rotates through 100° in the opposite direction. These movements are passive, resulting from the form of the legs and the forces exerted by the water, and they ensure that the tibia and tarsus are presented edge-on to the movement, producing a minimum of hydrodynamic force during the recovery stroke.

8.4.4 Other mechanisms of swimming

Appendages other than the legs are sometimes used in swimming. For example, the tiny mymarid wasp *Caraphractus cinctus* (Hymenoptera) parasitizes the eggs of dytiscids, which are laid under water. It swims jerkily through the water by rowing with its wings, making about two strokes per second. Larval Ephemeroptera and Zygoptera move by vertical undulations of the caudal gills and the abdomen. Many dipteran larvae flex and straighten the body alternately to either side, often increasing the thrust by a fin-like extension of the hind end. Mosquito larvae, for instance, have a fan of dense hairs on the last abdominal segment and as a result of the lateral flexing of the body move along, tail first (Fig. 8.27). The larvae of other Diptera make similar movements to those of

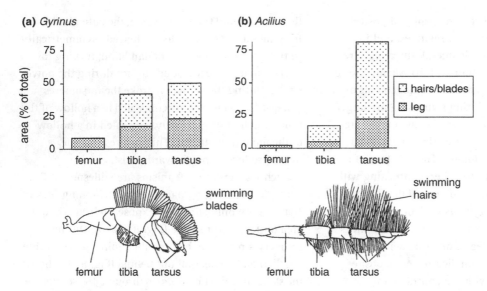

Figure 8.24 Hindleg modifications of aquatic beetles. Structural adaptations and surface area of each part of the leg (after Nachtigall, 1962). (a) *Gyrinus*, (b) *Acilius*.

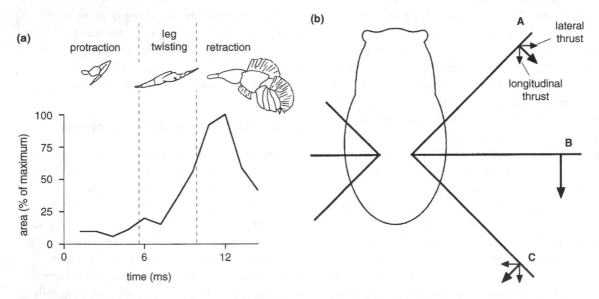

Figure 8.25 Swimming. Changes in thrust and the effective area of the hindleg. (a) Effective area of the leg of *Gyrinus* during one complete stroke (after Nachtigall, 1962). (b) Thrust exerted with the hindleg at different points of the power stroke. It is assumed that the velocity of the leg at A and C is only half its velocity at B, where, since the leg is at right angles to the body, only longitudinal thrust is produced. Equal and opposite forces act on the body (based on Nachtigall, 1965).

retraction
(power stroke)   protraction

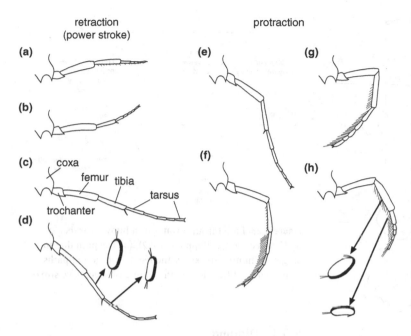

(a)

(b)

(c) coxa
        femur tibia
                        tarsus
        trochanter
(d)

(e)

(f)

(g)

(h)

Figure 8.26 Swimming. Successive positions of the right hindleg of *Dytiscus* during swimming as seen from above (assuming the body of the insect to be transparent). Insets are cross-sections of the tibia and tarsus showing their orientation during retraction (d) and protraction (h). The thick line represents the morphologically anterior side of the leg (after Hughes, 1958). (a)–(d) Stages of retraction (the power stroke) with the femur swinging back relative to the coxa. (e)–(h) Stages of protraction (the recovery stroke) with the femur moving forwards.

mosquito larvae. When suspended from the surface film or browsing on the bottom, mosquito larvae can glide slowly along as a result of the rapid vibrations of the mouth brushes in feeding. This is the normal method of progression in *Aedes communis* (Diptera).

8.4.5 Buoyancy

Many free-swimming insects are positively buoyant. When they stop swimming they come to rest at the surface of the water in a characteristic position that results from the distribution of air stores on and in the body. Most species float head down. *Notonecta* (Heteroptera), for instance, rests at an angle of 30° to the surface. As it kicks with its swimming legs this angle is increased to 55° so that the insect is driven down, but as it loses momentum during the recovery stroke of the legs it tends to rise again (Fig. 8.28). If the driving movements of the legs are repeated rapidly, before the insect rises very much, the path may be straightened out. By controlling the rate of leg movement the insect

can dive, move at a constant level or rise to the surface (Fig. 8.28b). A few insects, such as larval *Chaoborus* (Diptera) and *Anisops* (Heteroptera), can control their buoyancy so that they can remain suspended in mid-water. Early-stage mosquito larvae are usually less dense than the medium, so they rise to the surface when they stop swimming. This is also true of the pupae, but last-stage larvae may be slightly denser and sink when they stop actively moving.

8.5 Other uses of legs

The legs of many insects are modified for a variety of other functions besides locomotion, including stridulation (Chapter 26), burrowing, grasping and grooming behaviors. Modifications of the legs for grasping are common in predatory insects, and in ectoparasites that need to hold onto their host. Insects commonly use their legs for grooming,

(a) *Ceratopogon*   **(b)** *Aedes*

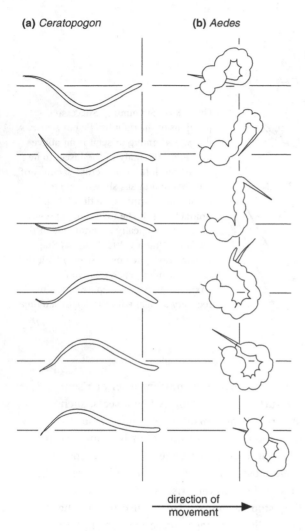

direction of movement →

Figure 8.27 Swimming by larval Diptera (after Nachtigall, 1965). (a) *Ceratopogon*, (b) *Aedes*.

and in Apoidea (Hymenoptera) the grooming mechanism has become specialized for collecting pollen from the hairs of the body. Embioptera have modified forelegs with a large number of silk glands on the basal tarsomere, which they use for web spinning. Various insect groups have reduced one or more pairs of legs, and a few groups of insect are completely apodous in one or more stages of their life history.

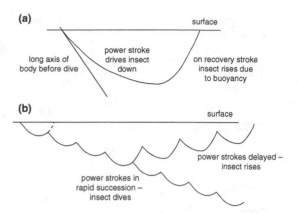

Figure 8.28 Path through water of a buoyant insect such as *Notonecta* (after Popham, 1952). (a) The path due to a single swimming stroke by the legs; (b) different paths produced by differences in the timing of successive strokes.

8.5.1 Digging

Legs modified for digging are best known in the Scarabaeoidea (Coleoptera) and the mole cricket, *Gryllotalpa* (Orthoptera). In *Gryllotalpa* the forelimb is very short and broad, the tibia and tarsomeres bearing stout lobes which are used in excavation. In the scarab beetles, the femora are short, the tibiae are again strong and toothed, but the tarsi are often weakly developed. Larval cicadas (Cicadidae) are also burrowing insects. They have large, toothed fore femora, the principal digging organs, and strong tibiae which may serve to loosen the soil (Fig. 8.29a). The tarsus is inserted dorsally on the tibia and can fold back. In the first-stage larva it is three-segmented, but it becomes reduced in later instars and may disappear completely.

8.5.2 Grasping

Modifications of the legs for grasping are frequent in predatory insects. Often, pincers are formed by the apposition of the tibia on the femur. This occurs in the forelegs of mantids (Mantodea) and mantispids (Neuroptera), in some Hemiptera such as Phymatidae

(a) digging

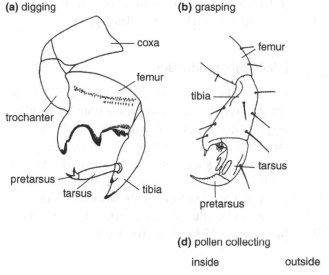

coxa
femur
trochanter
pretarsus
tarsus
tibia

(b) grasping

femur
tibia
tarsus
pretarsus

(c) grooming

spur
comb
tibia
basitarsus

(d) pollen collecting

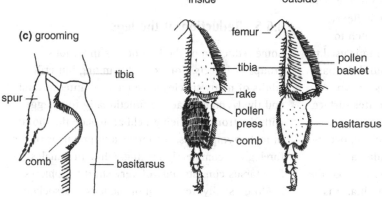

inside   outside

femur
tibia
rake
pollen press
comb
pollen basket
basitarsus

Figure 8.29 Adaptations of legs. (a) Digging; foreleg of a larval cicada (after Pesson, 1951). (b) Grasping; leg of *Haematopinus* (Phthiraptera) (after Séguy, 1951). (c) Grooming; foreleg of a mutillid (Hymenoptera) (after Schönitzer and Lawitzky, 1987). (d) Hind tibia and tarsus of a worker honey bee showing the pollen-collecting apparatus (partly after Snodgrass, 1956).

and Nepidae, and in some Empididae and Ephydridae among the Diptera. In some Empididae the middle legs are modified in this way. In *Bittacus* (Mecoptera) the fifth tarsomere on all the legs closes back against the fourth to form a grasping structure.

The ability to hold on is important in ectoparasitic insects. These usually have well-developed claws and the legs are frequently stout and short, as in Hippoboscidae (Diptera), Ischnocera and Anoplura (Phthiraptera). In the latter two groups, the tarsi are only one- or two-segmented and often there is only a single claw which folds against a projection of the tibia to form a grasping organ (Fig. 8.29b).

8.5.3 Grooming

Insects commonly use the legs or mandibles to groom parts of the body, removing particles of detritus in the process. The eyes and antennae are often groomed, and so are the wings. In *Schistocerca* stimulation of the trichoid sensilla on the wings elicits targeted hindleg scratching behavior, which might keep the wings clean, and might also help in fending off predators or conspecifics.

Cockroaches clean their antennae by passing them through the mandibles, which chew lightly at the surface. Many other insects use the forelegs for this purpose, then clean the legs with the mandibles.

Neuroptera and Diptera hold an antenna between the two forelegs, which are drawn forwards together toward its tip. Mosquitoes have a comb, consisting of several rows of setae, at the distal end of the fore tibia. The combs are scraped along the proboscis or antennae in rapid strokes. In many other insects each antenna is cleaned by the ipsilateral foreleg, which is often modified as a toilet organ.

Schistocerca has a cleaning groove between the first and second pads of the first tarsomere, which is fitted over the lowered antenna and then drawn slowly along it by an upward movement of the head and extension of the leg. In *Apis* and other Hymenoptera there is a basal notch in the basitarsus lined with spine-like hairs forming a comb. A flattened spur extends down from the tip of the tibia in such a way that when the metatarsus is flexed against the tibia the spur closes off the notch to form a complete ring (Fig. 8.29c). This ring is used to clean the antenna. First, it is closed around the base of the flagellum and then the antenna is drawn through it so that the comb cleans the outer surface and the spines on the spur scrape the inner surface. A similar, though less well-developed organ, occurs in Coleoptera of the families Staphylinidae and Carabidae. Lepidoptera have a mobile lobe called the strigil on the ventral surface of the fore tibia. It is often armed with a brush of hairs and is used to clean the antenna and possibly the proboscis.

The hindlegs of Apoidea are modified to collect pollen from the hairs of the body and accumulate it in the pollen basket. Pollen collecting is facilitated by pectinate hairs which are characteristic of the Apoidea. In the honey bee, *Apis*, pollen collected on the head region is brushed off with the forelegs and moistened with regurgitated nectar before being passed back to the hindlegs, which also collect pollen from the abdomen using the comb on the basitarsus (Fig. 8.29d). The pollen on the combs of one side is then removed by the rake of the opposite hindleg and collected in the pollen press between the tibia and basitarsus. By closure of the press, pollen is forced outwards and upwards onto the outside of the tibia and is held in place by the hairs of the pollen basket. On returning to the nest, the pollen is kicked off into a cell by the middle legs.

8.5.4　Silk production

Insects in the order Embioptera are unique in having silk glands in the basitarsus of the front legs in all stages of development of both sexes. The basal tarsomere is greatly swollen, and within it are numerous silk glands, each with a single layer of cells surrounding a reservoir (Fig. 8.30). There may be as many as 200 glands within the tarsomere, each connected by a duct to its own seta with a pore at the tip through which the silk is extruded.

8.5.5　Reduction of the legs

Some reduction of the legs occurs in various groups of insects. For example, among butterflies, adults of many species have reduced anterior tarsi, and the Nymphalidae are functionally four-legged, with the front legs being held alongside the thorax. In male nymphalids, the tarsus and pretarsus of the foreleg are completely lacking, while in females the fore tarsus consists only of very short tarsomeres.

More usually, reduction of the legs is associated with a sedentary lifestyle or some other specialized habit, such as burrowing, in which legs would be an encumbrance. Thus female coccids are sedentary and are held in position by the stylets of the proboscis. The legs are reduced, sometimes to simple spines, and in some species they are absent altogether in the later stages of development. Female Psychidae (Lepidoptera) show varying degrees of leg reduction, some species being completely apodous. These insects never leave the bags constructed by their larvae. Legs are also completely absent from female Strepsiptera, which are endoparasitic upon other insects.

Apart from the Diptera, all the larvae of which are apodous, legless larvae are usually associated with particular modes of life. There is a tendency for

(a)   **(b)**

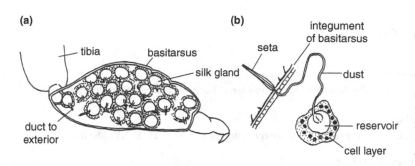

Figure 8.30 Silk production in the foreleg of an embiid. (a) Basitarsus seen in transparency to show the silk glands. (b) A single silk gland showing its connection to a seta.

larvae of leaf-mining Lepidoptera, Coleoptera and Tenthredinoidea (Hymenoptera) to be apodous. Parasitic larvae of Hymenoptera and Strepsiptera are apodous, and in larval Meloidae (Coleoptera) the legs are greatly reduced. Finally, in the social and semi-social Hymenoptera in which the larvae are provided with food by the parent, apodous forms are also the rule.

 ## Summary

- Insects typically have six legs, each containing six segments, the coxa, trochanter, femur, tibia, tarsus and pretarsus. The legs are furnished with a variety of exteroceptors, proprioceptors and chemoreceptors, and most insect orders have evolved smooth or hairy pads as attachment devices.

- Walking and running movements typically involve the coordinated movements of all six legs. Sensory feedback is important to patterning leg movements, particularly when the required movements are slow or unpredictable. The central nervous system plays an important role in pattern generation, and centrally generated commands may dominate during fast motions. Viscoelastic damping is also important in providing dynamic stabilization of fast movements.

- The hindlegs are commonly modified for other locomotor functions besides walking, including jumping and swimming. In some groups the middle legs are also modified for these functions. Most jumping insects use elastic storage of energy, either in the legs or in the thorax. Mechanical locking mechanisms are often present. In swimming insects, the legs are usually modified to increase the surface area that they present on the power stroke, and to reduce the surface area that they present on the recovery stroke.

- Crawling mechanisms are used by the soft-bodied larvae of many holometabolous insects, and involve changes in the shape of the body.

- Insects commonly use their legs for grooming, and in many predators the forelegs are modified for grasping. Modifications for grasping are also found in the legs of ectoparasites. Reduction or even loss of the legs occurs in a minority of insects, including leaf-mining and endoparasitic forms.

Recommended reading

COORDINATION
Burrows, M. (1996). *The Neurobiology of an Insect Brain.* Oxford: Oxford University Press.

MORPHOLOGY
Snodgrass, R. E. (1935). *Principles of Insect Morphology.* New York, NY: McGraw-Hill.

ATTACHMENT DEVICES
Gorb, S. N. (2007). Smooth attachment devices in insects: functional morphology and biomechanics. *Advances in Insect Physiology* **34**, 81–115.

WALKING ON LAND
Büschges, A. and Gruhn, M. (2007). Mechanosensory feedback in walking: from joint control to locomotor patterns. *Advances in Insect Physiology* **34**, 193–230.

WALKING ON WATER
Bush, J. W. M. and Hu, D. L. (2006). Walking on water: biolocomotion at the interface. *Annual Review of Fluid Mechanics* **38**, 339–369.

References in figure captions

Bennet-Clark, H. C. (1975). The energetics of the jump of the locust *Schistocerca gregaria. Journal of Experimental Biology* **63**, 53–83.

Berrigan, D. and Pepin, D. J. (1995). How maggots move: allometry and kinematics of crawling in larval Diptera. *Journal of Insect Physiology* **41**, 329–337.

Beutel, R. G. and Gorb, S. N. (2001). Ultrastructure of attachment specializations of hexapods (Arthropoda): evolutionary patterns inferred from a revised ordinal phylogeny. *Journal of Zoological Systematics and Evolutionary Research* **39**, 177–207.

Burns, M. D. and Usherwood, P. N. R. (1979). The control of walking in Orthoptera II: motor neurone activity in normal free-walking animals. *Journal of Experimental Biology* **79**, 69–98.

Burrows, M. (1992). Local circuits for the control of leg movements in an insect. *Trends in Neuroscience* **15**, 226–232.

Delcomyn, F. and Usherwood, P. N. R. (1973). Motor activity during walking in the cockroach *Periplaneta americana. Journal of Experimental Biology* **59**, 629–642.

Full, R. J. and Koehl, M. A. R. (1993). Drag and lift on running insects. *Journal of Experimental Biology* **176**, 89–101.

Full, R. J. and Tu, M. S. (1990). Mechanics of six-legged runners. *Journal of Experimental Biology* **148**, 129–146.

Heitler, W. J. (1974). The locust jump: specialization of the metathoracic femoral–tibial joint. *Journal of Comparative Physiology* **89**, 93–104.

Heitler, W. J. (1977). The locust jump III: structural specializations of the metathoracic tibiae. *Journal of Experimental Biology* **67**, 29–36.

Heitler, W. J. and Burrows, M. (1977). The locust jump I: the motor programme. *Journal of Experimental Biology* **66**, 203–219.

Hinton, H. E. (1955). On the structure, function, and distribution of the prolegs of the Panorpoidea, with a criticism of the Berlese–Imms theory. *Transactions of the Royal Entomological Society of London* **106**, 455–545.

Hughes, G. M. (1952). The co-ordination of insect movements: I. The walking movements of insects. *Journal of Experimental Biology* **29**, 267–284.

Hughes, G. M. (1958). The co-ordination of insect movements: III. Swimming in *Dytiscus, Hydrophilus,* and a dragonfly nymph. *Journal of Experimental Biology* **35**, 567–583.

Hughes, G. M. (1965). Locomotion: terrestrial. In *The Physiology of Insecta*, 1st edn., vol. 3, ed. M. Rockstein, pp. 227–254. New York, NY: Academic Press.

Johnston, R. M. and Levine, R. B. (1996). Crawling motor patterns induced by pilocarpine in isolated larval nerve cords of *Manduca sexta*. *Journal of Neurophysiology* **76**, 3178–3195.

Miall, L. C. (1922). *The Natural History of Aquatic Insects*. London: Macmillan.

Nachtigall, W. (1962). Funktionelle Morphologie, Kinematic und Hydromechanik des Ruderapparates von *Gyrinus*. *Zeitschrift für Vergleichende Physiologie* **45**, 193–226.

Nachtigall, W. (1965). Locomotion: swimming (hydrodynamics) of aquatic insects. In *The Physiology of Insecta*, 1st edn., vol. 2, ed. M. Rockstein, pp. 255–281. New York, NY: Academic Press.

Nachtigall, W. (1974). Locomotion: mechanics and hydrodynamics of swimming in aquatic insects. In *The Physiology of Insecta*, 2nd edn., vol. 3, ed. M. Rockstein, pp. 381–432. New York, NY: Academic Press.

Pearson, K. G. and Iles, J. F. (1971). Innervation of coxal depressor muscles in the cockroach, *Periplaneta americana*. *Journal of Experimental Biology* **54**, 215–232.

Pesson, P. (1951). Ordre des Homoptères. In *Traité de Zoologie*, vol. **10**, ed. P.-P. Grassé, pp. 1390–1656. Paris: Masson et Cie.

Popham, E. J. (1952). A preliminary investigation into the locomotion of aquatic Hemiptera and Coleoptera. *Proceedings of the Royal Entomological Society of London A* **27**, 117–119.

Ryckebusch, S. and Laurent, G. (1993). Rhythmic patterns evoked in locust leg motor neurons by the muscarinic agonist pilocarpine. *Journal of Neurophysiology* **69**, 1583–1595.

Schönitzer, K. and Lawitzky, G. (1987). A phylogenetic study of the antenna cleaner in Formicidae, Mutillidae, and Tipulidae (Insecta, Hymenoptera). *Zoomorphology* **107**, 273–285.

Séguy, E. (1951). Ordre des Anoploures ou Poux. In *Traité de Zoologie*, vol. **10**, ed. P.-P. Grassé, pp. 1365–1384. Paris: Masson et Cie.

Snodgrass, R. E. (1927). *Morphology and Mechanism of the Insect Thorax*. Washington, DC: Smithsonian Institution.

Snodgrass, R. E. (1935). *Principles of Insect Morphology*. New York, NY: McGraw-Hill.

Snodgrass, R. E. (1952). *A Textbook of Arthropod Anatomy*. Ithaca, NY: Cornell University Press.

Snodgrass, R. E. (1956). *Anatomy of the Honey Bee*. London: Constable.

Stork, N. E. (1983). The adherence of beetle tarsal setae to glass. *Journal of Natural History* **17**, 583–597.

Tindall, A. R. (1964). The skeleton and musculature of the larval thorax of *Triaenodes bicolor* Curtis (Trichoptera: Limnephilidae). *Transactions of the Royal Entomological Society of London* **116**, 151–210.

Zill, S. N. and Moran, D. T. (1981). The exoskeleton and insect proprioception: I. Responses of tibial campaniform sensilla to external and muscle-generated forces in the American cockroach, *Periplaneta americana*. *Journal of Experimental Biology* **91**, 1–24.

Zill, S. N., Moran, D. T. and Varela, F. G. (1981). The exoskeleton and insect proprioception: II. Reflex effects of tibial campaniform sensilla in the American cockroach, *Periplaneta americana*. *Journal of Experimental Biology*, **94**, 43–55.

9 Wings and flight

REVISED AND UPDATED BY **GRAHAM K. TAYLOR**

INTRODUCTION

Wings are the defining character of pterygotes, and were a key innovation in the evolutionary history of insects. Flight is important for a vast range of activities, and the detailed structure and form of the wings primarily reflects their adaptation for flight. The wings are connected to the thorax via the most complex joints of any animal, and the thoracic musculature provides control of the wings as well as high power output. By varying their detailed wingbeat kinematics, many insects are able to achieve excellent agility and maneuverability. The aerodynamics of insects differ greatly from the aerodynamics of fixed-wing aircraft, and it is these differences that explain the ability of insects to lift their weight with comparatively small wings. Even so, flight is the most energetically demanding of an insect's activities, and the power for flight may be provided by oxidation of several different substrates. Flight is only useful if it is stable and controlled, and a range of different sensory systems is used to provide the necessary feedback.

This chapter is divided into seven sections. Sections 9.1 and 9.2 describe the structure and form of the wings. Section 9.3 considers how the movements of the wings are generated, and is followed by Section 9.4 on wing kinematics. Section 9.5 reviews the aerodynamic mechanisms of insect flight, and is followed by Section 9.6 on power. Section 9.7 describes the sensory systems that are used by different insects in flight control.

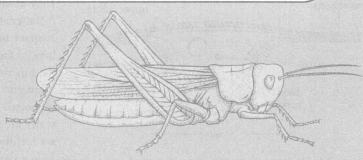

The Insects: Structure and Function (5th edition), ed. S. J. Simpson and A. E. Douglas.
Published by Cambridge University Press. © Cambridge University Press 2013.

9.1 Structure of the wings

Fully developed and functional wings occur only in adult insects, although the developing wings are present in larvae. In hemimetabolous larvae they are visible as external pads, but they develop internally in holometabolous species (Chapter 15). The Ephemeroptera are exceptional in having two full-winged stages. The final larval stage molts to a subimago, which resembles the adult except for having fringed and slightly translucent wings and rather shorter legs. It is able to make a short flight, after which it molts and the adult stage emerges. In the course of this molt the cuticle of the wings is shed with the rest of the cuticle.

The fully developed wings of all insects appear as thin, rigid flaps arising dorso-laterally from between the pleura and nota of the meso- and metathoracic segments. Each wing consists of a thin membrane supported by a system of veins. The membrane is formed by two layers of integument closely apposed, while the veins are formed where the two layers remain separate and the cuticle may be thicker and more heavily sclerotized (Fig. 9.1). Within each of the major veins is a nerve and a trachea, and, since the cavities of the veins are connected with the hemocoel, hemolymph can flow into the wings (Chapter 5).

The detailed structure of the wing is determined primarily by the need to optimize the interaction of aerodynamic, elastic and inertial forces during flight, subject to the constraint that the wings may need to be folded for stowage when not in use. As a first approximation, the wing is separated into zones that are primarily adapted for deformability and zones that are primarily adapted for rigidity, although different zones often operate together as integrated mechanisms. Rigidity is provided by the vein system, which is interrupted by deformable zones representing flexion lines or fold lines. These allow the wing to change shape appropriately in flight or to fold away at rest. Creases near the wing tips may also allow the wing to crumple without damage when it contacts an obstacle.

9.1.1 Wing membrane

The wing membrane is typically semi-transparent and often exhibits iridescence as a result of its nanoscale structure (Chapter 25). Sometimes the wing membrane is patterned by pigments contained in the epidermal cells. This is true in some Mecoptera and Tephritidae (Diptera), while in many insects which have hardened forewings, such as Orthoptera and Coleoptera, the whole forewing is pigmented.

The surface of the wing membrane is often set with small non-innervated spines called microtrichia. In Trichoptera, larger macrotrichia clothe the whole of the wing membrane, giving it a hairy appearance. In Lepidoptera, the wings are clothed in scales which vary in form from hair-like to flat plates. They usually cover the body as well as the wings. A flattened scale consists of two lamellae with an airspace between: the inferior lamella, that is, the lamella facing the wing membrane, being smooth; the superior lamella usually having longitudinal and transverse ridges (Fig. 9.2). The two lamellae are supported by internal struts called trabeculae. The scales are set in sockets in the wing membrane and are inclined to the surface, overlapping each other to form a complete covering. In primitive Lepidoptera, the scales are randomly distributed on the wings, but

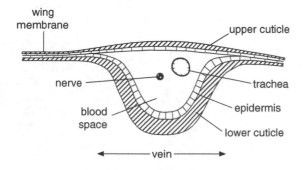

Figure 9.1 Diagrammatic section through part of a wing including a transverse section of a vein.

(a)

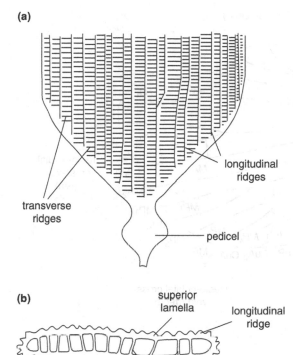

transverse
ridges

longitudinal
ridges

pedicel

(b)

superior
lamella

longitudinal
ridge

inferior
lamella

trabeculae

Figure 9.2 Lepidopteran scale. (a) Basal half of a scale
showing the pedicel that attaches to the wing membrane.
(b) Transverse section of a scale (after Bourgogne, 1951).

in butterflies (Papilionoidea) and some other groups
they are arranged in rows.

Pigments in the scales are responsible for the
colors of many Lepidoptera, the pigment being in the
wall or the cavity of the scale. In other instances,
physical colors result from the structure of the scale
(Chapter 25). Some specialized scales are associated
with glands (Chapter 27), while scales may also be
important in smoothing the airflow over the wings
and body. On the body they are also important as an
insulating layer, helping to maintain high thoracic
temperatures (Chapter 19). Scales also occur on the
wing veins and body of mosquitoes (Diptera) and
on the wings of some Psocoptera and a few
Trichoptera and Coleoptera. Scales and hairs on the

wing membrane are not innervated, but
mechano- and chemosensitive hairs are often present
on the veins.

9.1.2 Veins and venation

The principal support of the wing membrane is
provided by a number of well-marked veins running
along the length of the wing and connected to each
other by a variable number of cross-veins. There is a
tendency for the wings of lower orders of insects to
be pleated in a fan-like manner, with the longitudinal
veins alternately on the crests or in the troughs of
folds. A vein on a crest is called convex, while a vein
in a trough is called concave. The basic longitudinal
veins that can be distinguished in modern insects
are shown in Fig. 9.3a. From the leading edge of
the wing backwards they are:

costa (abbreviated to C) on or just behind the
leading edge
subcosta (Sc)
radius (R)
radial sector (Rs)
anterior media (MA) – media (M) where anterior
media and posterior media cannot be
distinguished
posterior media (MP) – media (M) where anterior
media and posterior media cannot be
distinguished
anterior cubitus (CuA)
posterior cubitus (CuP)
anals (1A, 2A, etc.)

Any of these veins may branch, the branches
then being given subscripts 1, 2, 3, etc. It is important
to recognize that these branches are not necessarily
homologous in different groups of insects.

The veins divide the area of the wing into a
series of cells that are most satisfactorily named after
the vein forming the anterior boundary of the cell
(Fig. 9.3a). A cell entirely surrounded by veins is said
to be closed, while one which extends to the wing

(a)

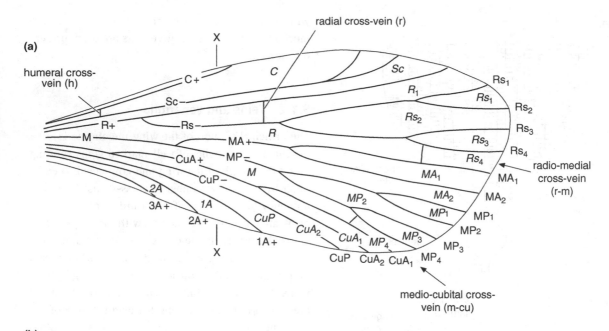

(b)

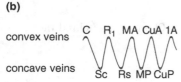

Figure 9.3 Wing venation. (a) Diagram of wing venation showing the main cross-veins and the names of the cells (italicized) enclosed by the veins. See text for abbreviations. (b) Section at X–X in (a) showing the concave and convex veins with the depth of pleating greatly exaggerated.

margins is open. On the anterior margin of the wing in some groups is a pigmented cell, the pterostigma (Fig. 9.4a). This is present on both pairs of wings of Odonata and on the forewings of many Hymenoptera, Psocoptera, Megaloptera and Mecoptera. The mass of the pterostigma is frequently greater than that of an equivalent area of adjacent wing and its inertia influences the movement of the whole wing membrane. In Odonata it is believed to reduce wing flutter during gliding, thus raising the maximum speed at which gliding can occur. In smaller insects it provides some passive control of the angle of attack of the wing during flapping flight, giving enhanced efficiency at the beginning of the wing stroke without the expenditure of additional energy.

In some very small insects, the venation may be greatly reduced. In Chalcidoidea (Hymenoptera), for instance, only the subcosta and part of the radius are present (Fig. 9.5f). Conversely, an increase in venation may occur by the branching of existing veins to produce accessory veins or by the development of additional, intercalary veins between the original ones, as in the wings of Orthoptera. Large numbers of cross-veins are present in some insects, and they may form a reticulum as in the wings of Odonata and at the base of the forewings of Tettigoniidae and Acridoidea (Orthoptera).

The form of an individual vein reflects its role in the production of useful aerodynamic forces by the wing as a whole. On the leading edge of the wing, the longitudinal veins form a rigid spar supporting the

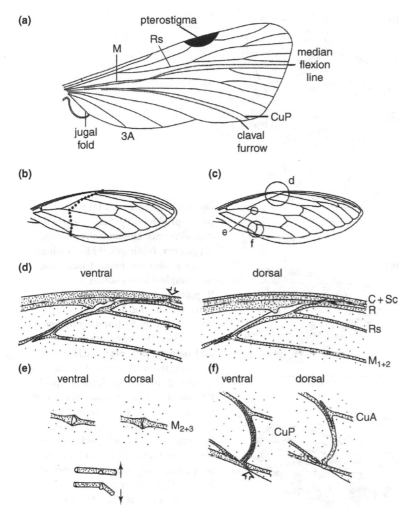

Figure 9.4 Fold lines and flexion lines (mainly after Wootton, 1981). (a) Diagram illustrating the main flexion lines. (b) A transverse flexion line, shown by dots, on the forewing of a cicada. (c) Forewing of a cicada showing the areas enlarged in (d), (e) and (f). Notice, in (d)–(f), that breaks in the veins are incomplete above, but complete below so that the wing will flex down, but not up. (d) Enlargement of part of the leading edge of the wing shown in (c). Open arrow shows the complete break on the ventral surface of C + Sc. (e) Break in M_{2+3}. (f) Break in the cubital cross-vein complete on the lower surface (open arrow).

wing as it moves through the air. In Lepidoptera, for example, the subcostal vein is circular in cross-section and so is equally resistant to bending in any direction (Fig. 9.6a), and in dragonflies the cross-veins along the leading edge of the wing form angle brackets which contribute to its rigidity (Fig. 9.6d). Behind the leading edge, the wing is often longitudinally corrugated (Fig. 9.3b). This, in itself, confers some degree of resistance to longitudinal bending, and the elliptical cross-section of some of the veins (Fig. 9.6b) confers further resistance to vertical bending. The arrangement of folds and veins in the hindwing of Orthoptera also limits vertical

flexibility while facilitating folding (Fig. 9.6c). Cross-veins are often circular in cross-section (Fig. 9.6e). Where flexibility is required the veins are annulated (Fig. 9.6f) or have short, narrow or unsclerotized regions (Fig. 9.4b–f).

9.1.3 Flexion lines

The production of useful aerodynamic forces requires that the wing remains relatively rigid on the downstroke, but some flexibility is necessary on the upstroke in order either to invert or feather the wing at stroke reversal. Flexural stiffness varies across the

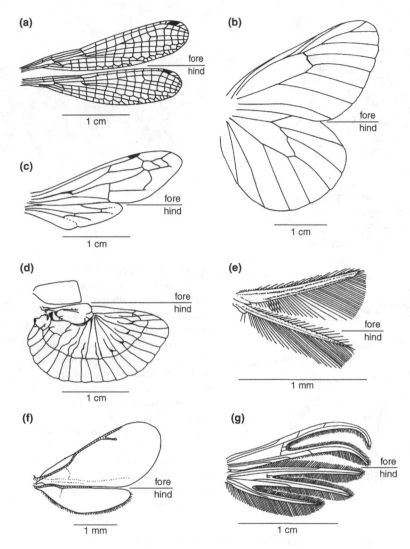

Figure 9.5 Wing forms. (a) Both wings power-producing; petiolate wings of a damselfly; not anatomically coupled (Zygoptera, *Ischnura*). (b) Both wings power-producing; broad-based wings with amplexiform coupling (Lepidoptera, *Aporia*). (c) Forewing power-producing; hindwing reduced and coupled to the forewing by hamuli in a hornet (Hymenoptera, *Vespa*). (d) Hindwing power-producing; forewing reduced to a short tegmen in an earwig; not anatomically coupled (Dermaptera, *Echinosoma*). (e) Fringed wings and reduced venation of a thrips; frenate-type wing coupling (Thysanoptera, *Thrips*); notice the small size. (f) Reduced venation of a chalcid wasp; hindwing coupled to forewing by hamuli (Hymenoptera, *Eulophus*); notice the small size. (g) Deeply divided wings of a plume moth; frenate wing coupling (Lepidoptera, *Alucita*).

wings in a highly anisotropic manner, but wing flexion mainly occurs along flexion lines defined by points at which breaks or flexibility are built into the veins.

Two longitudinal flexion lines are of widespread occurrence. These are the median flexion line, which usually arises close to the media and runs just behind the radial sector for much of its length, and the claval furrow, which runs close to the CuP (Fig. 9.4a). This furrow allows the posterior part of the wing to flap up and down with respect to the rest of the wing.

There is also commonly a transverse flexion line, such as the nodal flexion line of cicadas (Fig. 9.4b). This line permits the distal regions of the wing to bend down on the upstroke, but does not permit upward flexion during the downstroke. This is achieved in cicadas by lines of weakness on the ventral sides of some veins while sclerotization along the dorsal surface is continuous (Fig. 9.4c–f). In many other insects, the arched (cambered) cross-section of the whole wing seems adequate to prevent dorsal bending while allowing the wing to flex ventrally.

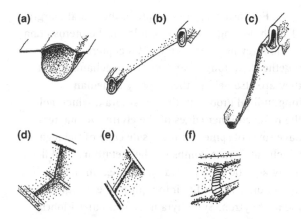

Figure 9.6 Vein morphology (after Wootton, 1992).
(a) Principal supporting vein, resistant to bending
and twisting in any direction (Lepidoptera, *Papilio*).
(b) Supporting veins near the leading edge of a dragonfly
wing, resistant to bending up or down (Odonata,
Calopteryx). (c) Veins from the pleated area of a hindwing;
ridge vein is resistant to bending up, or down, but the
trough vein is compliant in all directions (Orthoptera,
Schistocerca). (d) Cross-veins forming a rigid angle bracket
and linking the anterior longitudinal veins to form a stout
spar supporting the leading edge of the wing of a dragonfly
(Odonata, *Calopteryx*). (e) Normal cross-vein, circular in
cross-section (Odonata, *Calopteryx*). (f) Annulate cross-vein
permitting flexibility (Diptera, *Eristalis*).

9.1.4 Fold lines

When at rest, the wings are held over the back in
most insects. This may involve folding of the wing
membrane along the radial veins or longitudinal
flexion lines, but may also involve folding of the
wings along lines of high flexibility that are adapted
to this specific function and known as fold lines
(Fig. 9.4).

Where the anal area of the hindwing is large, as in
Orthoptera and Blattodea, the whole of this part may
be folded under the anterior part of the wing along a
vannal fold a little posterior to the claval furrow. In
addition, in Orthoptera and Blattodea, the anal area is
folded like a fan along the veins, the anal veins being
convex, at the crests of the folds, and the accessory
veins concave. Whereas the claval furrow and jugal

fold are probably homologous in different species,
the vannal fold varies in position in different taxa.

Most Neoptera have a jugal fold line just behind
vein 3A on the forewings (Fig. 9.4a). It is sometimes
also present on the hindwings. Paper wasps
(Vespidae) and some other Hymenoptera have a
longitudinal fold line close to the cubital vein. The
form of the cuticle where the fold line crosses the
claval furrow ensures that only two positions of the
fold are stable: with the wing fully extended or fully
flexed. The former condition occurs when the wing is
pulled into the flight position, the latter when the
wing is in the rest position.

The hindwings of Coleoptera, Dermaptera
and a few Blattodea fold transversely as well as
longitudinally, so they can be accommodated
beneath the protective forewings. Wings that fold
transversely all do so by using the same basic
mechanism, consisting of four adjoining panels
constrained to rotate hingewise about four fold lines
converging at a single point (Fig. 9.7). Depending
upon the number and disposition of concave versus
convex fold lines, the distal part of the wing may
fold up inside or outside the rest of the wing. It is
usual for several such mechanisms to be found
within a single wing.

Folding is usually produced by a muscle arising
on the pleuron and inserted into the third axillary
sclerite in such a way that, when it contracts, the
sclerite pivots about its points of articulation with
the posterior notal process and the second axillary
sclerite. As a result, the distal arm of the third axillary
sclerite rotates upwards and inwards, so that finally its
position is completely reversed. The anal veins are
articulated with this sclerite in such a way that when it
moves they are carried with it and become flexed over
the back of the insect. Activity of the same muscle in
flight affects the power output of the wing and so it is
also important in flight control. In Staphylinidae
(Coleoptera), the abdomen is used to fold the wings.

In orthopteroid insects, the elasticity of the cuticle
causes the vannal area of the wing to fold radially
along the veins. In Dermaptera, resilin also serves as

(a)

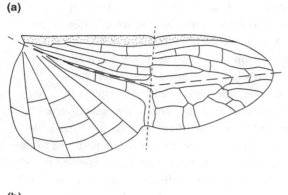

(b)

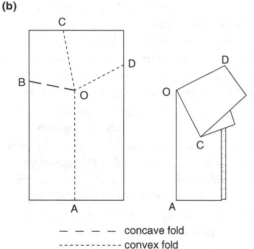

— — — — concave fold
------------ convex fold

Figure 9.7 (a) Drawing of the wing of a cockroach (Blattodea), showing the four convergent fold lines (dashed). (b) Schematic of the associated folding mechanism. Redrawn after Haas and Wootton (1996), courtesy of S. M. Walker.

the major driving mechanism for transverse wing folding, and is important in making the mechanism bistable so that the wing does not collapse in flight. Consequently, energy is expended in unfolding the wings. In general, wing extension probably results from the contraction of muscles attached to the basalar sclerite or, in some insects, to the subalar sclerite. However, in Dermaptera, unfolding of the two hindwings is performed sequentially by the cerci.

The wings are sometimes held in the folded position by being coupled together or fastened to the body. For instance, in Psocoptera, the costal margin of the hindwing is held by a fold on the pterostigma of the forewing. The elytra of Coleoptera are held together by a tongue-and-groove mechanism, but they are also held to the body by a median longitudinal groove in the metathorax, which holds the reflexed inner edges of the elytra. Dermaptera have rows of spines on the inside edge of the tegmen which catch onto combs on the metathorax, while many aquatic Heteroptera have a peg on the mesothorax which fits into a pit in the margin of the hemelytron. Symphyta have specialized lobes, the cenchri, on the metanotum, which engage with rough areas on the undersides of the forewings to hold them in place.

9.1.5 Areas of the wing

The flexion lines and fold lines divide the wing into different areas. The region containing the bulk of the veins in front of the claval furrow is called the remigium (Fig. 9.8a). The area behind the claval furrow is called the clavus, except in hindwings in which this area is greatly expanded, when it is known as the vannus. Finally, the jugum is cut off by the jugal fold where this is present.

In some Diptera there are three separate lobes in the region of the wing base, known from proximally outwards as the thoracic squama, alar squama and alula (Fig. 9.8b). There is some confusion in the terminology and homologies of these lobes, but it is probable that the thoracic squama is derived from the posterior margin of the scutellum, the alar squama represents the jugum and the alula is a part of the claval region which has become separated from the rest. Some Coleoptera have a lobe called an alula folded beneath the elytron. It appears to be equivalent to the jugum.

The wing margins and angles are also named (Fig. 9.8a). The leading edge of the wing is called the costal margin; the trailing edge is the anal margin; and the outer edge is the apical margin. The angle between the costal and apical margins is the apical

(a)

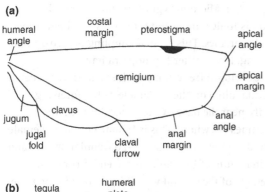

(b)

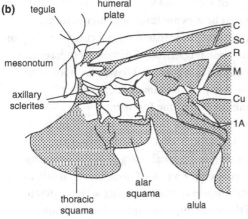

Figure 9.8 Wing areas. (a) The terminology applied to different parts of the wing. (b) Base of the right wing of a tabanid (Diptera) showing the arrangement of the lobes at the base of the wing (after Oldroyd, 1949).

angle; that between the apical and anal margins is the anal angle; and the angle at the base of the wing is called the humeral angle.

9.1.6 Sensilla on the wings

Many insects have hair sensilla along the wing veins. In general, these are probably mechanoreceptors responding to touch and possibly to the flow of air over the wings in flight. In Orthoptera, the forewing hairs respond to chemical and mechanical stimuli and are responsible for eliciting targeted hindleg scratching movements in response to disturbance when the insect is at rest. Contact chemoreceptors are also known to be present on the forewings of many species of Diptera.

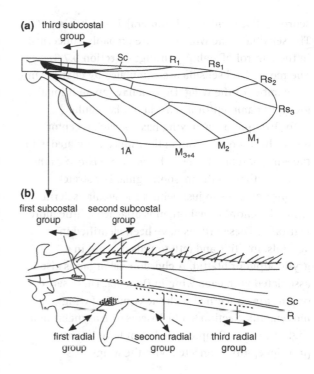

Figure 9.9 Distribution of groups of campaniform sensilla at the base of the wing of a fly (Diptera, *Empis*). Arrows indicate the orientation of the long axes of the sensilla (after Pringle, 1957). (a) Whole wing showing, enclosed within the box, the position of the area enlarged in (b). (b) Proximal parts of the anterior veins.

Campaniform sensilla are present at the base of the wing, mainly in groups on the subcostal and radial veins (Fig. 9.9). These groups are often present on both upper and lower surfaces of the wing, but in Acrididae and Blattodea they are only present ventrally. The sensilla in the groups are generally oval, all those in a group being similarly oriented, so that they are sensitive to distortions of the wing base in a particular direction. More distally on the veins are other scattered campaniform sensilla, but these are large and almost circular, so that, unlike those in the basal groups, they have no directional sensitivity. The number of sensilla in each group varies, there being more in more highly maneuverable species. Thus *Apis* (Hymenoptera) has about 700 campaniform sensilla at the base of each forewing, while the scorpion fly, *Panorpa* (Mecoptera), has only about 60. The sensilla at the wing base are probably concerned in the control of stability in flight (Section 9.7.3). The more distal sensilla are stimulated by changes in the camber of the wing. Diptera have an additional group of campaniform sensilla on the tegula.

In Orthoptera, each wing has a stretch receptor and a chordotonal organ in the thorax associated with the wing base: in the mesothorax of *Schistocerca* they arise together on the mesophragma. The stretch receptor extends to just behind the subalar sclerite, while the chordotonal organ is attached a little more ventrally. These organs have been identified in acridids, gryllids and tettigoniids, but not in a gryllotalpid or blattid. A chordotonal organ is also associated with each wing in Odonata. These sense organs are concerned with the control of wing movements (Section 9.3.6). Insects with asynchronous flight muscles (Chapter 10) do not have internal proprioceptors connected with the wings.

9.1.7　Wing coupling

In the majority of insects the fore- and hindwings are linked anatomically so that they move together as a single unit. This wing coupling may take various forms, but, in many species, it involves lobes or spines at the wing base. A primitive arrangement is found in some Mecoptera of the family Choristidae in which there is a jugal lobe at the base of the forewing and a humeral lobe at the base of the costal margin of the hindwing. Both lobes are set with setae known as the jugal and frenular bristles, respectively (Fig. 9.10a), and, although they do not firmly link the wings, they overlap sufficiently to prevent the wings moving out of phase.

In some Trichoptera, only the jugum is present; it lies on top of the hindwing and the coupling mechanism is not very efficient. However, the Hepialidae (Lepidoptera) have a strong jugal lobe which lies beneath the costal margin of the hindwing so that this is held between the jugum and the rest of the forewing (Fig. 9.10b). This is called jugate wing

coupling. In Micropterygidae (Lepidoptera), the jugum is folded under the forewings and holds the frenular bristles. This type of coupling is jugo-frenate coupling. Many other Lepidoptera have a well-developed frenulum which engages with a catch or retinaculum on the underside of the forewing, usually near the base of the subcostal vein but sometimes elsewhere. This is frenate coupling. Female noctuids, for instance, have 2–20 frenular bristles and a retinaculum of forwardly directed hairs on the underside of the cubital vein (Fig. 9.10c); in the male, the frenular bristles are fused together to form a single stout spine and the retinaculum is a cuticular clasp on the radial vein (Fig. 9.10d). Thysanoptera have the wings coupled in a comparable way by hooked spines at the base of the hindwing, which catch a membranous fold of the forewing.

Other insects have the wings coupled by more distal modifications that hold the costal margin of the hindwing to the anal margin of the forewing. Hymenoptera have a row of hooks, the hamuli, along the costal margin of the hindwing which catch into a fold of the forewing; Psocoptera have a hook at the end of the CuP of the forewing, which hooks onto the hind costa; and the forewing of Heteroptera has a short gutter edged with a brush of hairs on the underside of the clavus, which holds the costal margin of the hindwing. Homoptera exhibit a variety of modifications linking the anal margin of the forewing to the costal margin of the hindwing.

The wings of the Papilionoidea and some Bombycoidea (Lepidoptera) are coupled by virtue of an extensive area of overlap between the two. This is known as amplexiform wing coupling. A similar arrangement is present in some Trichoptera, often occurring together with some other method of coupling.

9.1.8　Articulation of the wings with the thorax

Where the wing joins the thorax, its dorsal and ventral cuticular layers are membranous and flexible. In these membranes are the axillary sclerites, which

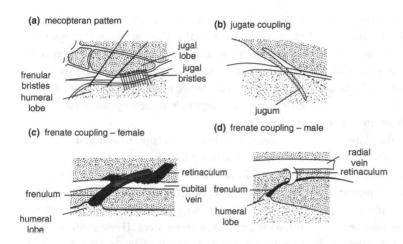

(a) mecopteran pattern

frenular bristles
humeral lobe
jugal lobe
jugal bristles

(b) jugate coupling

jugum

(c) frenate coupling – female

frenulum
humeral lobe
retinaculum
cubital vein

(d) frenate coupling – male

frenulum
humeral lobe
radial vein
retinaculum

Figure 9.10 Wing coupling mechanisms involving the jugal and humeral regions of the wings. All diagrams represent the mechanisms as seen from below, with the attachment to the thorax immediately to the left. Membrane of the forewing with dark stippling, that of the hindwing with light stippling (after Tillyard, 1918). (a) Primitive mecopteran pattern (Mecoptera, *Taeniochorista*). (b) Jugate coupling in a hepialid moth (Lepidoptera, *Charagia*). (c) Frenate coupling in a female sphingid moth (Lepidoptera, *Hippotion*). (d) Frenate coupling in a male sphingid moth (Lepidoptera, *Hippotion*).

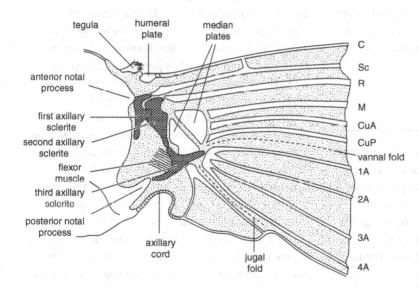

tegula
humeral plate
median plates
anterior notal process
first axillary sclerite
second axillary sclerite
flexor muscle
third axillary sclerite
posterior notal process
axillary cord
jugal fold

C
Sc
R
M
CuA
CuP
vannal fold
1A
2A
3A
4A

Figure 9.11 Wing articulation with the thorax. Axillary sclerites with dark stippling for clarity (modified after Snodgrass, 1935).

transmit movements of the thorax produced by the flight muscles to the wing. Typically there are three axillary sclerites (Fig. 9.11). The first is in the dorsal membrane and articulates proximally with the anterior notal process and distally with the subcostal vein and the second axillary sclerite. The second extends to both dorsal and ventral membranes and articulates ventrally with the pleural wing process (see Fig. 7.5b) and distally with the base of the radius. It is also connected with the third axillary sclerite, which articulates proximally with the posterior notal process and distally with the anal veins. The third axillary sclerite is Y-shaped, with a muscle inserted into the crutch of the Y. In Hymenoptera and

Orthoptera there is a fourth axillary sclerite between the posterior notal process and the third axillary sclerite. The precise arrangement of the axillary sclerites relative to each other and the flexion lines of the wing is extremely complex and plays a significant role in changes in wing form during flight.

In addition to the axillary sclerites, there are other plates in the wing base. Connected with the third axillary, and perhaps representing a part of it, may be one or two median plates from which the media and the cubitus arise. At the base of the costa is a humeral plate and often, proximal to it, is another plate derived from the edge of the articular membrane, called the tegula. In *Locusta* (Orthoptera) this has been shown to be an important sensory structure that modulates the basic pattern of wing movements. This may also be true in other insects. It is well developed in Lepidoptera, Hymenoptera and Diptera (Fig. 9.11).

Odonata have only two large plates hinged to the tergum and supported by two arms from the pleural wing process. The plates are called the humeral and axillary plates.

9.2 Form of the wings

The gross form of the wings is determined primarily by aerodynamic considerations, but other ecological factors may provide different selective pressures. Reduction or loss of one or both wing pairs is widespread, and in several orders the forewings are thickened so as to provide protection for the hindwings when the wings are folded over the body.

9.2.1 Wings adapted for flight

Wings with narrow, petiolate bases are found in relatively slow-flying insects, such as some damselflies (Zygoptera) and antlions (Neuroptera) (Fig. 9.5a). This shape probably reduces aerodynamic interference between contralateral wing pairs, which might enhance maneuverability. Wings with broad bases, on the other hand, are associated with the capacity for rapid flight. They occur in Orthoptera, many Hemiptera and Lepidoptera (Fig. 9.5b), as well as in dragonflies (Anisoptera) and ascalaphids (Neuroptera). This shape probably increases the efficiency and magnitude of aerodynamic force production.

In Odonata, Isoptera, Mecoptera and male Embioptera the two pairs of wings are roughly similar in form, but in most other groups of insects the fore- and hindwings differ from each other. Sometimes the hindwings are small, relative to the forewings, as in Ephemeroptera, Hymenoptera (Fig. 9.5c,f) and male coccids (Hemiptera: Homoptera), while in some Ephemeroptera, such as *Cloeon*, and some male coccids they are absent altogether. In Diptera the hindwings are reduced and modified to form dumbbell-shaped halteres (Section 9.2.2). In other insects, most of the power for flight is provided by the hindwings which have a much bigger area than the forewings (Fig. 9.5d). This is the case in Blattodea, Mantodea, Orthoptera, Dermaptera and most Plecoptera and Coleoptera, and also in male Strepsiptera, where the reduced forewings resemble the halteres of Diptera.

The wings of very small insects are often reduced to straps with one or two supporting veins and long fringes of hairs (Fig. 9.5e). These forms occur in Thysanoptera, in Trichogrammatidae and Mymaridae among the Hymenoptera, and in some of the small Staphylinoidea among the Coleoptera. It might be expected that such wings would function as paddles rather than as aerofoils because of their low Reynolds number and peculiar form, but there is no evidence for this. Even thrips and small Hymenoptera appear to depend on aerodynamic lift forces in the same way as larger insects. The wings of plume moths, Pterophoridae and Orneodidae, are deeply cleft and fringed with scales (Fig. 9.5g). Wing fringes are common in Lepidoptera and mosquitoes (Culicidae), and in some Tineoidea (Lepidoptera) they are so extensive as to greatly increase the effective area of the wing.

Swallow-tailed butterflies and some Lycaenidae (Lepidoptera) have a projection from the hind margin of the hindwing, while in the Nemopteridae (Neuroptera) and some Zygaenidae (Lepidoptera) the hindwings are slender ribbons trailing out behind the insect. This probably tends to divert the attention of a predator away from the head and thorax, or to deter attack by increasing apparent body size, at least in some of these insects. An irregular outline to the wings, such as occurs in some butterflies, serves to break up the outline of a resting insect and presumably has a camouflage function.

9.2.2 Halteres

The hindwings of Diptera are modified to form halteres, which are sense organs concerned with the maintenance of stability in flight. Each haltere consists of a basal lobe, a stalk and an end-knob which projects backwards from the end of the stalk so that its center of mass is also behind the stalk (Fig. 9.12). The whole structure is rigid except for some flexibility of the ventral surface near the base, which allows some freedom of movement, while the

cuticle of the end-knob is thin but kept distended by the turgidity of large vacuolated cells inside it. In crane flies (Tipulidae) and robber flies (Asilidae) the haltere is relatively long, exceeding 12% of forewing length, but in mosquitoes and Cyclorrhapha it is only about 6% of the forewing length.

On the basal lobe of the haltere are groups of campaniform sensilla that can be homologized with the groups at the base of a normal wing. Dorsally there are two large groups of sensilla: the basal and scapal plates (Fig. 9.12). In *Calliphora* (Diptera) there are about 100 sensilla in each group. The sensilla of the scapal plate are parallel with the axis which passes through the main point of articulation and the center of mass of the haltere (indicated as the long axis of the haltere in Fig. 9.12); those of the basal plate are oriented with their long axes at about 30° to the axes of the longitudinal rows in which they are arranged. Near the basal plate is a further small group of campaniform sensilla known as Hicks papillae. These are set below the general surface of the cuticle and are oriented parallel with the long axis of the haltere. There is also a single round, undifferentiated papilla near the scapal plate. On the ventral surface

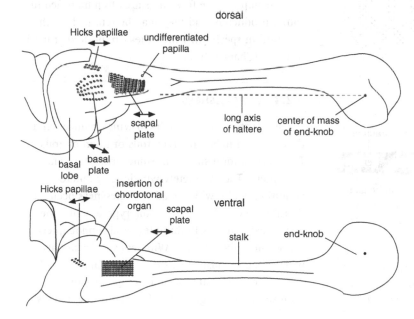

Figure 9.12 Halteres. Dorsal and ventral views of the haltere of a fly showing the groups of sensilla. The orientation of the campaniform sensilla is indicated by the arrows (Diptera, *Lucilia*) (after Pringle, 1948).

there is another scapal plate with about 100 sensilla and a group of ten Hicks papillae. These are oriented parallel with the long axis of the haltere. Also attached to the ventral surface is a large chordotonal organ oriented at about 45° to the long axis of the haltere. A smaller chordotonal organ runs vertically across the base. These sensilla react to the forces acting at the base of the haltere during flight. They sense the vertical movements of the haltere and also the torque produced by turning movements of the fly (Section 9.7.4).

9.2.3 Wings adapted to other functions

The forewings of many insects are thicker than the hindwings and serve to protect the latter when they are folded at rest. Forewings modified in this way are known as elytra or tegmina. Leathery tegmina occur in Blattodea, Mantodea, Orthoptera and Dermaptera, while in heteropteran Hemiptera only the basal part of the wing is hardened, such wings being known as hemelytra (Fig. 9.13a). The basal part of the

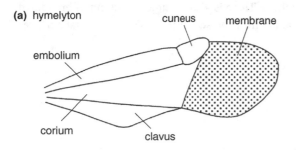

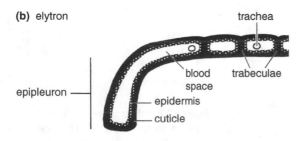

Figure 9.13 Protective forewings. (a) Hemelytron of a mirid (Heteroptera) (after Comstock, 1918). (b) Diagrammatic transverse section through part of an elytron of a beetle.

hemelytron may be subdivided into regions by well-marked veins and, in mirids, where the development is most complete, the anterior part of the wing is cut off as a proximal embolium and distal cuneus, the center of the wing is the corium, and the anal region is cut off as the clavus. In lygaeids only the corium and the clavus are differentiated.

The elytra of Coleoptera are usually very heavily sclerotized and the basic wing venation is lost, although it may be indicated internally by the arrangement of tracheae. The two surfaces of the elytron are separated by a blood space (Fig. 9.13b) across which run cuticular columns, the trabeculae, arranged in longitudinal rows and marked externally by striations. There are usually nine or ten such striae, although the number may be as high as 25 in some ground beetles (Carabidae). The elytra of beetles do not overlap in the midline, but meet and are held together by a tongue-and-groove joint, while in some Carabidae, Curculionidae and Ptinidae they are fused together so that they cannot open. In these species the hindwings are also reduced. At the sides, the elytra are often reflexed downwards, the vertical part being called the epipleuron and the horizontal part the disc.

In Orthoptera, the forewings are often modified for sound production and they may be retained for this function in species in which they are no longer used in flight (Chapter 26).

9.2.4 Winglessness

Some insects have both pairs of wings reduced and they are said to be brachypterous or micropterous. This occurs, for instance, in some Orthoptera and Hemiptera. The completely wingless, or apterous, condition is also widespread. Winglessness occurs as a primitive condition in the Apterygota, while the ectoparasitic orders Phthiraptera and Siphonaptera are secondarily wingless. Wingless species are also widespread in most other orders, but apparently do not occur in Odonata or Ephemeroptera. Sometimes both sexes are wingless, but frequently the male is winged and only the female is apterous. This is the case in coccids, Embioptera, Strepsiptera, and among Hymenoptera in Mutillidae and some Chalcididae. In the ants and termites, only the reproductive caste is winged and here the wings are shed after the nuptial flight, breaking off by a basal suture so that only a wing scale remains. The break is achieved in different ways, but termites frequently rest the wing on the ground and then break it off by twisting the wing base. After loss of the wings, the flight muscles degenerate.

Commonly, the development of the wings varies within a species either geographically or seasonally. The extent to which this is genetically determined is often unclear, but in many species environmental factors have a dominant effect. Such wing polyphenism occurs in various groups. In many other insects, however, and notably in Hemiptera, species may either be apterous or macropterous (with fully developed wings), without any intermediates (Chapter 15).

9.3 Movement of the wings

The muscles concerned with wing movement fall, functionally, into three groups: direct muscles are inserted into the wing base and their contractions have direct effects on wing movement; indirect muscles move the wings indirectly by causing distortions of the thorax; accessory muscles modulate the wing stroke by changing the shape or mechanical properties of the thorax. Movements of the wings also involve elasticity of the thorax, of the wing base and of the muscles themselves. In approximately three-quarters of known insect species, the flight muscles are asynchronous and rely upon stretch activation (Chapter 10), so the elasticity of the system is integral to their function.

9.3.1 Types of flight muscle

The direct wing muscles are inserted onto the axillary sclerites or the basalar and subalar sclerites (see Fig. 7.5), which connect to the axillary sclerites by ligaments. One muscle, arising on the pleuron and inserted into the third axillary sclerite, flexes the wing backwards. It is also active in flight, producing some remotion of the wing during both the up- and downstrokes and, in this way, effecting steering. The basalar and subalar muscles commonly consist of several units arising on different parts of the pleuron, as well as the sternum and the coxa (Fig. 9.14). Both muscles are involved in wing twisting (Section 9.4.4) as well as wing depression. In addition, the basalar muscle is involved in wing extension from the flexed position. Odonata have two muscles arising from the episternum inserted into the humeral plate and two from the edge of the epimeron inserted into the axillary plate. The relatively large number of direct muscles (Table 9.1) reflects their role in changing the form of the wing during the wingbeat cycle to provide the precision necessary for efficient aerodynamics and steering.

The indirect wing muscles are usually the main power-producing muscles for flight. In all insects, wing elevation is produced by indirect dorso-ventral muscles inserted into the tergum of the wing-bearing segment. There are usually several muscles with different points of origin (Fig. 9.14). These muscles are not always homologous. In many insects they arise on the sternum or the coxae, but in Auchenorrhyncha and Psyllidae (Hemiptera) the tergosternal muscles are small and are functionally replaced as wing elevators by the oblique dorsal muscles. These arise on the postphragma and so are normally obliquely longitudinal, but in the groups mentioned the phragma extends ventrally, carrying the origins of the muscles with it so that they come to exert their pull vertically instead of horizontally. Wing depression is produced, in most insects, by dorsal longitudinal indirect muscles usually comprising five or six muscle units. These extend from the anterior to the posterior phragma of each wing-bearing segment (Fig. 9.14). The mode of action of these muscles is illustrated in the next

Table 9.1 **Numbers of muscles associated with one mesothoracic wing and the number of neurons by which they are innervated**[a]

| | Indirect muscles | | | | | | Direct muscles | |
| | Depressor muscles | | Unifunctional levator muscles | | Bifunctional levator muscles | | | |
Order	Number (units)	Neurons/ unit	Number	Neurons/ muscle	Number	Neurons/ muscle	Number	Neurons/ muscle[b]
Odonata	0	–	2	3	2	3, 5	9	3–10
Orthoptera	1 (5)	1	2	1	4	1–3	4	1–8
Lepidoptera	2 (5)	1	2	2	3	2–3	4	1
Diptera	1 (5–6)	1	5	1	2	1	17	1–2

Notes:
[a] In some cases the number of units in a muscle is not known so that more than one neuron/muscle may indicate polyneuronal innervation of a single unit, or may indicate a number of separate units. These numbers do not include DUM neurons.
[b] In most cases the innervation is known for only some of the direct muscles.

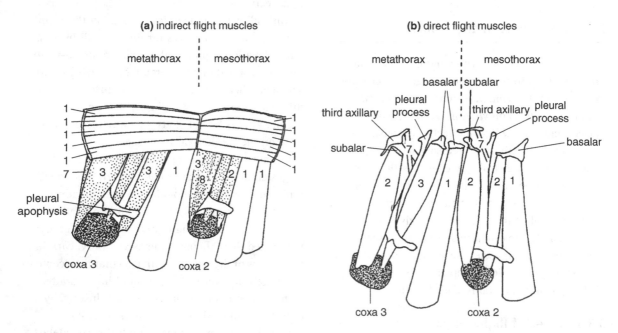

Figure 9.14 Flight muscles of a locust and their innervation. Numbers on each muscle show the numbers of motor axons, including inhibitory axons, innervating each unit. Dorso-ventral muscles not arising from the coxae have their origins on the sternum. (a) Indirect flight muscles; bifunctional muscles that move the wings and the legs are stippled. (b) Direct flight muscles.

section. The accessory muscles are pleurosternal and tergopleural muscles. The pleurosternal muscles control the lateral elasticity of the thorax. The tergopleural muscles are suitably positioned to alter thoracic stiffness across the episternal cleft, and although they are not inserted directly on any wing sclerite, they are inserted within the region of the wing hinge, the mechanics of which they directly affect.

9.3.2 Innervation of the flight muscles

The structure and innervation of the flight muscles reflects their functions. Most of the indirect flight muscles are dedicated exclusively to flight. In keeping with this, they usually comprise one or a small number of similar muscle units each innervated by a single fast motor neuron (Fig. 9.14; Table 9.1). In contrast, the direct and accessory muscles as well as bifunctional indirect muscles may have polyneuronal innervation. For example, one of the tergocoxal muscles of *Schistocerca* is innervated by six motor neurons and the common inhibitory neuron (see Fig. 10.9). In addition, these muscles may contain units that are physiologically different from each other. For example, the third axillary muscle of *Manduca* (Lepidoptera) has three units, two that are of the physiologically "intermediate" type and one that is a "tonic" muscle (Section 10.1.2). Each intermediate unit is innervated by a separate fast neuron, while the tonic unit has a slow neuron. This permits increased versatility in the activity of the muscle, reflecting the need for versatility in the control of flight.

Because the main power-producing muscles are innervated by a single motor neuron, the total number of neurons controlling power production in flight is small; the forewings of the locust are controlled by about 20, and those of a fly by about 26. A larger number of neurons is concerned with modulating the power output via the direct wing muscles.

9.3.3 Movements produced by the flight muscles

The upward movement of the wings is produced by the indirect dorso-ventral muscles. By contracting, they pull the tergum down and hence also move down the point of articulation of the wing with the tergum. The effect of this is to move the wing membrane up, with the pleural process acting as a fulcrum (Fig. 9.15a–d).

The downward movement of the wings in Odonata and Blattodea is produced by direct muscles inserted into the basalar and subalar sclerites. These muscles pull on the wings outside the fulcrum of the pleural process and so pull the wings down (Fig. 9.15b). By contrast, in Diptera and Hymenoptera, the downward movement is produced by the dorsal longitudinal indirect muscles. Because the dorsum of the pterothorax is an uninterrupted plate, without membranous junctions (see Fig. 7.1), contraction of the dorsal longitudinal muscles does not produce a telescoping of the segments as in the abdomen. Instead, the center of the tergum becomes bowed upwards (Fig. 9.15e,f). This moves the tergal articulation of the wing up and the wing membrane flaps down (Fig. 9.15d). In Coleoptera and Orthoptera the downward movement is produced by the direct and indirect longitudinal muscles acting together. The direct muscles are also concerned in twisting the wing during the course of the stroke (see Section 9.3.6).

Details of the up–down movement of the wings may be much more complex than this simple account suggests. In cyclorrhaphous Diptera contraction of the dorsal longitudinal muscles during the downstroke causes the scutellar lever to swing upwards (Fig. 9.16). The scutellar lever articulates with the first axillary sclerite, and its upward movement causes the sclerite to rotate until it hits the paranotal plate, pushing it and the scutum upwards and stretching the dorso-ventral muscles. The second axillary sclerite rotates in a socket on the inner face of the pleural wing process. As it swings, the wing is

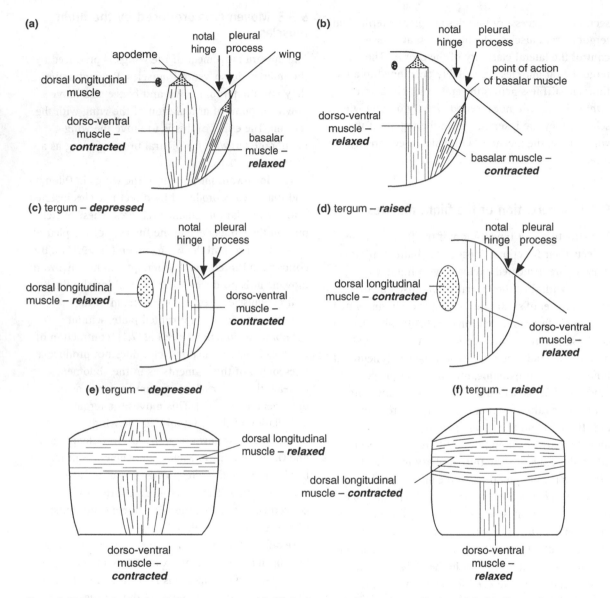

Figure 9.15 Muscular basis of wing movements. (a), (b) An insect, such as a dragonfly, in which the direct wing muscles cause depression of the wings; (a) indirect dorso-ventral muscles cause wing elevation; (b) direct dorso-ventral muscles cause depression. (c)–(f) An insect, such as a fly, in which both up and down movements of the wing are produced by indirect muscles; (c), (d) cross-sections of the thorax; (e), (f) sagittal sections of the wing-bearing segment from the inside corresponding with (c) and (d), respectively. In (f), contraction of the dorsal longitudinal muscles raises the tergum (as seen in cross-section in (d)) and the wing flaps down.

depressed until a process on the underside of the radial vein, known as the radial stop, fits into a groove on top of the pleural process (Fig. 9.16). It is

not clear whether the radial stop and pleural process make contact on every wingbeat, but when they do so, the pleural process becomes a pivot about which

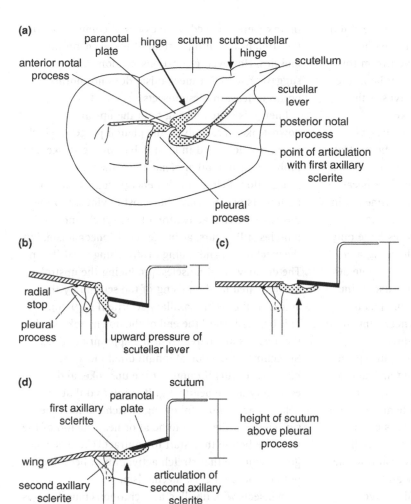

Figure 9.16 Wing articulation in a fly.
(a) Lateral view of the thorax.
Membranous parts are stippled (after
Pringle, 1957). (b)–(d) Diagrammatic
representations of the movements of the
axillary sclerites during wing
depression. The first axillary sclerite
(dotted) is pushed upwards by the
scutellar lever and (c) makes contact
with the underside of the paranotal
plate (black) pushing it upwards (d).
This raises the height of the scutum.
At the same time the radial stop hits the
top of the pleural process (c) and the
wing bends down (d) (after Miyan
and Ewing, 1985).

the wing bends. Some species are able to effect
contact at two different positions on the pleural wing
process, and this "gear changing" is thought to be
involved in producing different power output from
the wings.

Although the movement of the wings on the
thorax involves some condylic movement at the
pleural process, a great deal of movement is
permitted by the presence of resilin ligaments, such
as the wing hinge ligament of Orthoptera (Chapter
16). In this way, the problems of friction and
lubrication which would occur at a normal
articulation moving at the high frequency of the
wings are avoided.

A number of the muscles moving the wings arise in
the coxae, which are themselves movable. Whether
these muscles move the legs or the wings appears to
be determined by the activity of other muscles and
the position of the appendages: If the wings are
closed the muscles move the legs, but in flight, with
the legs in the flight position, the wings are moved.

9.3.4 Movements due to elasticity

The capacity to store elastic energy and subsequently
to release it at high rates is an essential feature of the
flight mechanism of most insects. At the beginning of
a wing stroke (up or down), energy is expended to

overcome the inertia of the wing, while at the end of a stroke the wing has momentum and must be stopped. Insects store the energy derived from this momentum in elastic systems which may be in the cuticle or in the flight muscles themselves. This energy is then used on the return stroke.

In *Schistocerca*, and probably in most other insects, much of the energy involved in the upstroke is stored as elastic forces in the pad of resilin which forms the main wing hinge. This is possible because the aerodynamic forces produced at this time act in the same direction as the wing movement, so assisting its movement. Thus the muscles have only to overcome the forces of inertia of the wing and elasticity of the wing base, and as a result some 86% of the energy they produce is stored for use in the downstroke. The elastic properties of this pad are almost perfect and all but 3% of the energy imparted to it when it is stretched during the upward movement of the wings is available for pulling the wing down. The radial stop in the wing articulation of cyclorrhaphous Diptera (see above) may also enable energy to be stored in the cuticle at the end of the downstroke. Elastic (resilin) ligaments connect the subalar sclerite to the second and third axillary sclerites of *Calliphora*, which also has resilin in the apodemes to which the pleuroaxillary and tergopleural muscles attach. Dragonflies have a similar elastic apodeme where the subalar muscle attaches to the subalar sclerite.

Elastic energy is also stored as a result of distortions of the thorax. The lateral stiffness of the thorax, produced by the sternopleural articulation and, to a lesser extent, the tergopleural articulation, was formerly thought to be of major importance. The position of the wings was thought to be unstable for much of the stroke as a result of this lateral stiffness, and the wings were thought only to be stable in the extreme up and down positions. This arrangement was called a "click" mechanism, the wings "clicking" automatically to one of the stable positions once they passed the position of maximum instability. The accompanying sudden changes in shape of the

thorax were believed to be essential components of the stretch-activation system of the asynchronous muscles. However, critical observations of flies suggest that a click mechanism does not normally occur, at least in these insects.

Energy is stored during both the up- and downstrokes, probably in the pleural process, which is slightly deformed in both halves of the stroke, but probably also in other elements of the wing articulation. Release of this energy contributes to both parts of the stroke, but most obviously to the upstroke. Stretch activation of the synchronous muscles still occurs, as in the "click" mechanism, but the mechanism and timing of stretching are different. The dorso-ventral muscles, producing the upstroke, are activated by the raising of the scutum by the action of the second axillary sclerite (Fig. 9.16). This occurs toward the end of the downstroke so that the muscles are activated at the appropriate time. Stretching of the dorsal longitudinal muscles, however, occurs throughout the upstroke, and especially at the beginning. It is inferred that the delay in activation, following stretching, must be sufficient for the wing to be at or near the top of the upstroke before they start to contract. There is also good evidence that stretch activation in bees does not depend on a click mechanism.

In insects with fibrillar, asynchronous muscles, it is probable that the muscles themselves are the principal site of energy storage. These muscles are characterized by a greater resistance to stretch compared with other muscles due to the elastic properties of the contractile system (Chapter 10).

9.3.5 Initiation and maintenance of wing movements

In most insects the wings start to beat as a result of loss of tarsal contact with the substratum, which occurs naturally when the insect jumps into the air. Movement of the wings is inhibited when the legs are touching the ground, contact being sensed directly by hair sensilla on the undersides of the tarsi, and indirectly by campaniform sensilla at the leg joints. This inhibition is overridden in insects that engage in preflight warm-up.

Loss of tarsal contact with the substratum is sufficient to maintain the movement of the wings of *Drosophila* (Diptera), but in most other insects flight soon stops unless the insect receives further stimulation. This is provided by the movement of wind against the head. In *Schistocerca*, a wind speed less than the flight speed of the insect is sufficient stimulus to maintain the wingbeat, with the air movement being sensed by fields of wind-sensitive hairs on the head. In Diptera, and in the water beetle, *Dytiscus* (Coleoptera), the wind is sensed by movements of the antennae. Wind stimuli also result in the legs being drawn up close to the body in a characteristic manner. In locusts, stimulation of the cephalic hair fields causes the forelegs to assume the flight position, while Diptera hold their legs in the flight position when their antennae are stimulated in flight.

Fibrillar, asynchronous muscles only oscillate at high frequency when stretch-activated, so the start of flight in insects with such muscles depends upon having some starter mechanism to initiate the process. In bees, the separate units in the dorso-ventral muscles are stimulated synchronously by neural inputs. This results in a muscle twitch of large amplitude, effecting the stretch of the dorsal longitudinal muscles which, at about the same time, are activated via their motor neurons.

Octopamine has an important arousal effect in insects, and its hemolymph titer rises rapidly in the first few minutes of flight. It acts directly on the wingbeat central pattern generator, enhances the response of sensory neurons and interneurons involved in flight control, alters the kinetics of muscle contraction and enhances flight muscle glycolysis. Octopamine has generally been attributed a central role in initiating flight, but this view has recently begun to be challenged. Acute application of octopamine is sufficient to initiate fictive flight in various insect preparations, but octopamine does not appear to be necessary for initiation of flight in *Drosophila*, because mutants lacking octopamine can still fly. Furthermore, none of the identified interneurons initiating flight in locusts appear to be octopaminergic. There is also evidence from *Drosophila*, *Schistocerca* and *Manduca* that the action of octopamine is modulated, perhaps at different sites, by the action of its precursor tyramine.

Results of experiments with *Schistocerca* suggest that acetylcholine acting via muscarinic receptors is the key neurotransmitter involved in flight initiation, with octopamine, tyramine and possibly certain other amines suggested to act as neuromodulators to facilitate the cholinergic flight-initiating pathway and shape the final motor pattern.

9.3.6 Control of wing movements

In locusts and other insects with synchronous flight muscles, muscle contraction is regulated directly by the motor neurons: each time a neuron fires, the muscle it innervates contracts. The basic pattern of muscular contractions involved in the flight of the locust can be produced in the complete absence of input from peripheral sensilla. This rhythm is generated by a central pattern generator, formed by a complex of interneurons which drive the motor neurons. The centrally generated rhythm is slower and differs in certain details from that normally required for efficient flight, so peripheral feedback is still required to regulate wingbeat frequency. Inputs from the tegulae and stretch receptors at the base of each wing reset the rhythm and keep the pattern generator active.

The stretch receptors fire close to the time of maximum elevation of the wings, sometimes just after depression has begun. Their input inhibits activity of the motor neurons to the levator muscles, and promotes activity of the motor neurons to the depressor muscles. The chordotonal organs associated with the tegulae, in contrast, start to fire soon after the beginning of the downstroke and may continue to be active through most of the stroke.

Their input signals completion of the downstroke, perhaps by monitoring the velocity of the wing movement, and they elicit an immediate upstroke. In this way, the stretch receptors and tegulae control the wingbeat frequency even though the basic rhythm is generated within the central nervous system.

Superimposed on this basic system are inputs concerned with the maintenance of steady flight and with steering. In the locust, about 20 descending interneurons, with cell bodies in the brain and axons extending as far as the fourth abdominal neuromere, are known to be involved in conveying information from the head to the motor neurons in the thoracic and abdominal segments. They convey information from the compound eyes, the wind-sensitive hairs, the ocelli, the antennae and proprioceptive hairs on the neck. Each descending interneuron tends to be most sensitive to one or two of the sensory inputs, although many also respond to other inputs. Each one also tends to respond best to deviations in a particular direction, for example, changes in the visual field resulting from rolling in one direction rather than the other (Fig. 9.17).

Within each of the thoracic ganglia, numerous other interneurons are involved in steering. In the mesothoracic ganglion of *Locusta*, there are at least 28 of these neurons, which are additional to those producing the basic pattern of oscillation. They receive inputs from the descending interneurons and from proprioceptors associated with the wings. They are driven and inhibited by inputs from the pattern generator, and they output, sometimes directly, to the motor neurons of the flight muscles (Fig. 9.18). Essentially similar arrays of interneurons occur in the anterior abdominal neuromeres to control the activity of the abdominal muscles involved in steering.

The force exerted by a muscle can be increased either by increasing the number of units that are active or by increasing the strength of the pull exerted by each unit. Although the units of the indirect flight muscles are only innervated by fast axons, they contract more strongly if, instead of being stimulated by a single action potential, they are stimulated by two or more spikes following close together. This provides a means by which graded information can be transmitted to muscles despite the all-or-nothing code of the nervous system.

The twisting of the wings by controller muscles in the locust is precisely timed by the pattern of motor impulses to the muscles (Fig. 9.19). Only in the case of one muscle, the mesothoracic subalar muscle, is the firing of the motor neuron very variable in its timing and this is the muscle which varies forewing pronation to control lift. The precise coordination of the other neurons does not arise from a fixed pattern of connections between them as some function during walking in sequences completely different from that involved in flight.

Control of the wingbeat is different in insects in which the wingbeat is produced by asynchronous muscles. Here, too, the muscles act in a precise sequence, but this sequence is not directly related to nervous input and the timing of firing of the motor neurons does not coincide with any particular phase of the wingbeat cycle. Instead, the nervous input to the flight muscles serves as a general stimulator, maintaining the muscle contractions. In general, the motor neurons only produce single spikes, and their rate of firing varies from 5 spikes s^{-1} to 25 spikes s^{-1} while wingbeat frequency is commonly in excess of 100 Hz. Wingbeat frequency in these insects is controlled to some extent by the muscles that control the mechanical properties of the thorax: An increase in the lateral stiffness of the thorax produces an increased wingbeat frequency, while a decrease in stiffness leads to a reduced frequency. There is evidence, however, that the rate of firing of the motor neurons is positively correlated with the wingbeat frequency (Section 9.6.3).

9.4 Wing kinematics

The wings are flexible structures that change shape while pivoting and rotating about a movable hinge, and their kinematics defy simple description.

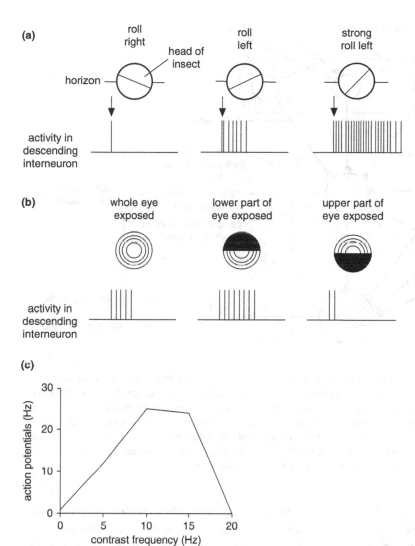

Figure 9.17 Maintenance of stability. Examples of the activity of visually sensitive descending interneurons to perturbations of the visual field comparable with those occurring in flight in a locust (*Schistocerca*). (a) A wind-sensitive interneuron that responds to roll to the left. Arrows mark the onset of the wind (after Rowell and Reichert, 1991). (b) A neuron responding to image patterns simulating forward flight. The neuron responds most strongly when only the lower part of the eye is stimulated by the moving pattern, as would be the case in normal flight. Interneurons with these characteristics are probably involved in the optomotor response (based on Baader *et al.*, 1992). (c) The sensitivity of a neuron similar to that in (b) to different rates of pattern movement (contrast frequency) over the eye (data from Baader *et al.*, 1992).

Nevertheless, the stroke plane, which is the average plane through which the wing moves relative to the body in each stroke cycle, can be used to define several gross kinematic parameters for comparison between wingbeats (Fig. 9.20). The stroke plane angle measures the inclination of the stroke plane relative either to the longitudinal axis of the body or to the horizontal. The stroke amplitude is the angle swept by the wing between the top and bottom of the stroke, measured within the stroke plane. The mid-stroke angle measures the average direction of

the wing's long axis during a stroke cycle, and is measured within the stroke plane relative to a transverse axis. The stroke plane deviation angle measures the amplitude of out-of-plane motion of the wing.

Superimposed upon these approximately planar wing movements is a sequence of rotations about the wing's long axis (Fig. 9.21). The wing rotates nose-down at the top of the stroke in a movement called pronation. It rotates nose-up at the bottom of the stroke in a movement called supination. The

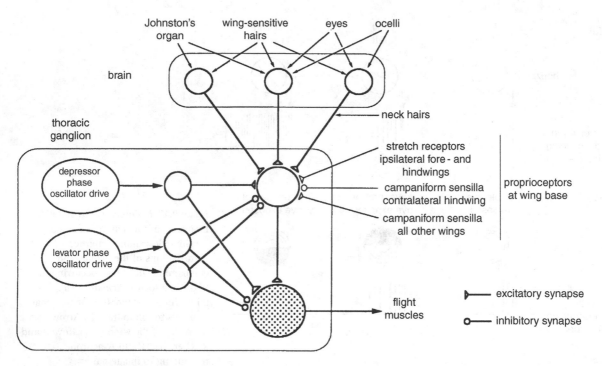

Figure 9.18 Neural control of wingbeat in an insect with synchronous flight muscles. Sensory inputs from the sense organs of the head and neck affect the activity of interneurons in the brain (open circles), which vary in the selectivity of their responses. These interneurons have axons extending (descending) as far as the anterior abdominal neuromeres. In each ganglion (only one is shown) of the flight system, these descending neurons synapse with a premotor interneuron (large open circle) which also receives input from the proprioceptors at the wing bases. These different inputs modulate the activity of the premotor interneuron, which is driven or inhibited by the outputs from the pattern-generating systems. The premotor interneuron activates the appropriate motor neuron (stippled) controlling a flight muscle (after Rowell and Reichert, 1991).

wings twist markedly during pronation and supination, as a result of their flexibility and inertia, and they may also adopt different degrees or directions of camber in response to aerodynamic, elastic and inertial loading.

Considerable variation in wing kinematics exists between species, and many of the same parameters are modified between wingbeats to effect flight stabilization and maneuver control. In addition, the relative phasing of the wings is an important variable in flight control. For example, the forewing and hindwings of Odonata normally beat out of phase with each other, but they beat in phase during take-off and other maneuvers involving high accelerations. In locusts, at least, phase differences

between the left and right wings are correlated with turning responses. Movements of the legs and abdomen are also important in flight control in many insects, serving either to shift the center of gravity and thereby change balance, or to generate aerodynamic forces and thereby act like a rudder.

9.4.1 Wingbeat frequency

Wingbeat frequency is negatively correlated with body mass. Among Odonata and orthopteroid insects, which are usually moderate to large sized, wingbeat frequency is usually within the range 15–40 Hz (Fig. 9.22). Hemipteroid insects usually have higher wingbeat frequencies than this and, in the small

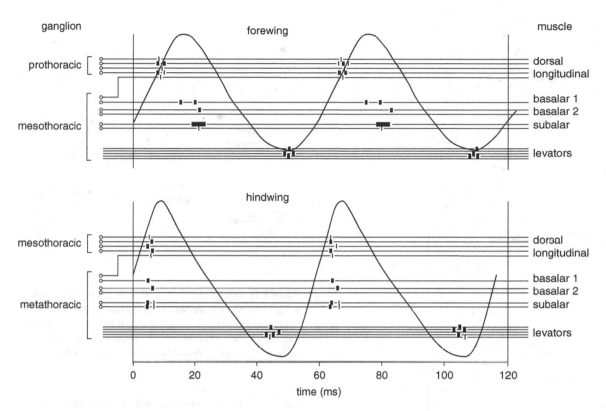

Figure 9.19 Neural control of wingbeat in an insect with synchronous flight muscles. Diagram illustrating the timing of firing of motor neurons to the flight muscles of the fore- and hindwings of a locust in relation to the wingbeat cycle. Each neuron is shown by a horizontal line with its origin in the appropriate ganglion on the left-hand side. Each dot on a line represents an action potential occurring at that time; a small dot indicates that activity in the neuron may or may not occur, a large dot that it always occurs. The heavy bar on the motor neuron to the forewing subalar muscle indicates that firing occurs within this period, but not at a precisely fixed time as with the other units. The heavy curves indicate the angular displacement of the wings (after Wilson and Weis-Fogh, 1962).

whiteflies (Aleyrodidae), the frequency may exceed 200 Hz. Hymenoptera and Diptera may beat their wings even more rapidly, with frequencies exceeding 400 Hz in some species. Lepidoptera have wingbeat frequencies ranging from about 4 Hz to 80 Hz. Insects with high wingbeat frequencies generally have asynchronous flight muscles (Fig. 9.22).

Variation of wingbeat frequency with thoracic temperature has been recorded in some small- to moderate-sized insects among the Diptera, Lepidoptera, Hymenoptera and Coleoptera, as well as in larger Odonata and Orthoptera. In general, the increase is slight with Q_{10} (the change in relative rate for a 10°C rise in temperature) varying between 1.0 and 1.4. In some larger insects, thoracic temperature is regulated during flight over a range of ambient temperatures, and large changes in wingbeat frequency would not be expected in these insects. In insects with asynchronous flight muscles, these changes are the result of changes in the resonant properties of the thorax rather than changes in the neural output to the muscles, whereas, in Orthoptera, the increase in wingbeat frequency is, at least partly, due to temperature effects on the central nervous system. Adult age also affects wingbeat frequency in some hemimetabolous insects, such as locusts.

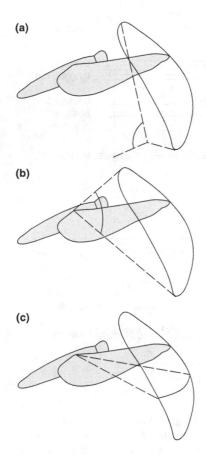

(a)

(b)

(c)

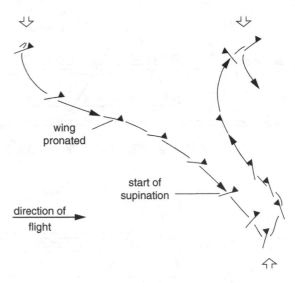

wing
pronated

start of
supination

direction of
flight

Figure 9.21 Wing twisting in a fly. The orientation of the wing is shown at intervals of approximately 0.5 ms. The upper side of the leading edge is shown by a triangle. Notice the rapid rotations at the top and bottom of the stroke (indicated by open arrows) (after Nachtigall, 1966).

Figure 9.20 Definition sketch of three angles used to describe the gross wing kinematics of insects. The closed path shows the trajectory of the wing tip relative to the body through a complete stroke cycle. (a) The stroke plane angle measures the inclination of the stroke plane, relative either to the longitudinal axis of the body or to the horizontal. (b) The stroke amplitude is the angle swept by the wing between the top and bottom of the stroke, measured within the stroke plane. (c) The stroke plane deviation angle measures the amplitude of out-of-plane motion of the wing. Redrawn after Walker *et al.* (2009).

Wingbeat frequency is involved in regulating aerodynamic power output in a number of species, but the correlation between the two is not always positive. In *Drosophila*, for example, the relationship between aerodynamic power output and wingbeat frequency is humped. Changes in wingbeat frequency are in any case tightly linked to changes in stroke amplitude in most species, and as wingbeat frequency is constrained to be the same on both sides of the insect it cannot play a role in generating lateral force asymmetries.

9.4.2　Stroke plane angle

The stroke plane is almost horizontal in hovering insects but is inclined toward the vertical in species that do not hover. The stroke plane tends to tilt forwards as an insect flies faster. In some insects, such as *Drosophila*, this is achieved principally by tilting the long axis of the body. The angle between the stroke plane and the long axis of the body also remains more or less constant in many insects with low wingbeat frequencies such as Orthoptera and Blattodea. In Odonata, Heteroptera, Lepidoptera, Diptera and Hymenoptera, however, the relationship can be more variable, and marked changes occur in some species when the insect is maneuvering.

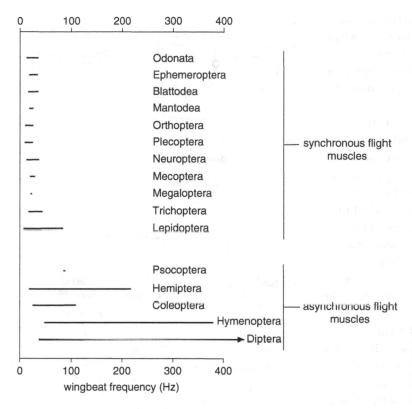

Figure 9.22 Wingbeat frequencies. Horizontal lines show the ranges of frequencies at which insects in different orders flap their wings (after Brodsky, 1994).

Differences in the stroke plane angle of the wings on the two sides of the body may produce lateral turning movements.

9.4.3 Stroke amplitude

Stroke amplitude typically falls within the range 70–130°, but wings whose primary function is protective, such as the elytra of beetles and tegmina of grasshoppers, beat with lower amplitudes because they are not the primary power producers. The forewings of *Locusta*, for instance, commonly move through 70–80° compared with 110–130° for the hindwings, and the elytra of *Oryctes* (Coleoptera) have an amplitude of only about 20°, with nearly all the power coming from the hindwings.

Stroke amplitude varies greatly in flight and is probably the single most important variable involved in regulating aerodynamic power output. Greater amplitudes are associated with greater power output and may result in the wings "clapping" together above the body (Section 9.5.4), but increases in stroke amplitude are usually effected by extending the ventral part of the stroke, so changes in stroke amplitude are commonly linked to changes in mid-stroke angle. These shift the line of action of the flight force, causing the body to pitch at the same time as aerodynamic power output is increased or decreased. Variation in stroke amplitude on the two sides of the body is used in steering by many insects, the insect turning away from the side of greatest amplitude.

9.4.4 Wing rotation

Rotation about the long axis of the wing occurs at the top and bottom of each stroke as the wing changes from upstroke to downstroke and vice versa.

The extent of wing rotation may be relatively small, as in grasshoppers, but in insects with narrow wings, and especially in insects that hover, the wings may rotate through large angles at high speeds, reaching angular velocities in excess of $10^{5\circ}\,s^{-1}$ in some insects.

Wing rotation is produced, at least in part, by differential action of the basalar and subalar muscles acting at the base of the wing. The former pulls the leading edge down, while the latter pulls the trailing edge down. In addition, the momentum of the wing assists this process and, in Diptera, is sufficient to account for much of the rotation. The axis about which the wing twists is close to the radial vein, but the center of mass of the wing lies behind the torsional axis. Consequently, as the wing decelerates and then reverses, inertia causes it to twist about the axis (Fig. 9.23).

Rapid rotation may be a source of useful aerodynamic lift (Section 9.5.2), and some insects are able to alter the forces they produce by controlling the timing of wing rotation with respect to stroke reversal. This affects the direction and magnitude of the aerodynamic force, apparently by changing the dynamics of wing–wake interaction (Section 9.5.3). Besides these dynamic effects of wing rotation, the angle of attack of the wing during the translational phase of the stroke is also an important control variable in many insects. The extent of pronation is typically increased on the inside wing during lateral turning maneuvers, which should serve to reduce the aerodynamic forces on the inside wing by reducing its angle of attack.

9.4.5 Wing deformation

Flexion, bending and torsion of the wing result from a combination of aerodynamic, inertial and elastic forces, although aerodynamic forces have been found to have only minor effects on wing bending in *Manduca*. Twisting and cambering of the wing can be important in maintaining efficient aerodynamic force production. In the hindwing fans of Orthoptera,

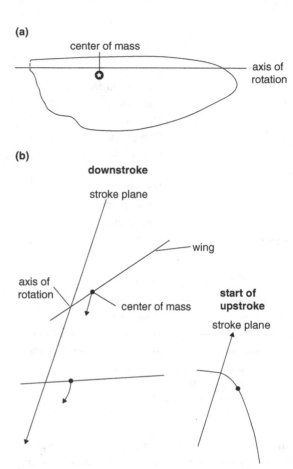

Figure 9.23 Wing rotation in a fly. (a) The center of mass of the wing lies behind the axis of rotation, which approximately coincides with the radial vein. (b) As the wing decelerates at the end of the downstroke, the inertia of the center of mass causes the wing to start rotating and, as the wing starts to move up, flex downwards about the axis of rotation (after Ennos, 1988).

Phasmatodea, Blattodea and Mantodea, twist and camber arise automatically during the downstroke through an "umbrella effect." Aerodynamic loading produces tension in the anal margin, which compresses the radial veins and causes them to undergo Euler buckling. Odonata use a different mechanism of automatic camber generation, involving the basal complex of veins. An upward force applied distal to the basal complex depresses the anal margin, and to a lesser extent the apical

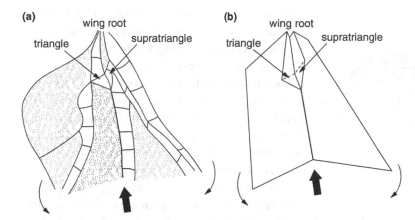

(a) wing root, triangle, supratriangle

(b) wing root, triangle, supratriangle

Figure 9.24 (a) Drawing of the basal triangle–supratriangle complex of a dragonfly. (b) Schematic of the associated mechanism of wing flexion and camber generation. Application of an upwards force outboard of the basal complex causes the apical margin and anal margin of the wing to flex downward, thereby generating positive camber. Redrawn after Wootton *et al.* (1998), courtesy of S. M. Walker.

margin, which produces camber under the aerodynamic loading experienced during the downstroke (Fig. 9.24). An analogous structure formed by a different set of veins is present in some extinct relatives of modern dragonflies. Camber tends to limit distortion of the wing during the downstroke, but may facilitate supination on the upstroke. In addition, many of the flexion lines also permit downward, but not upward, bending of the distal parts of the wing (Fig. 9.4b) so that they too contribute to supination on the upstroke.

9.5 Aerodynamic mechanisms

Flapping is the exclusive mode of flight for the great majority of insects, and besides enabling forward flight, it allows many insects to hover or to fly backwards. Hovering is commonly associated with a particular behavior such as feeding, defending territory, mating or oviposition, and is often observed before landing, enabling the insect to land precisely on a particular spot. Gliding is used by some Odonata, Orthoptera and Lepidoptera, and ranges from a pause in wing movement lasting only a fraction of a second, to prolonged glides lasting many seconds. Locusts are able to lock their forewings in an outstretched position, which may

facilitate gliding, and it is suggested that the inability of dragonflies to fold their wings is a secondary adaptation to gliding. However, even in species that glide, flapping remains the most important mode of flight.

The aerodynamics of flapping differ in two important ways from the aerodynamics of gliding. First, whereas the air tends to remain attached on a gliding wing, flowing smoothly over the aerofoil's surface, air flowing over a flapping wing tends to become entrained in a swirling vortex bound to the upper surface of the wing. This is known as a separated flow. Second, whereas the attached flow over a gliding wing looks approximately similar from one moment to the next, the separated flow over a flapping wing varies continually. This is known as an unsteady flow. Although once a point of controversy, it is now widely accepted that insects make extensive use of unsteady separated flow mechanisms to generate far greater aerodynamic forces than would be possible with steady, or quasi-steady, attached flow.

9.5.1 Leading-edge vortices

The most ubiquitous, and probably the most important, separated flow mechanism is the leading-edge vortex. Whereas the flow over a fixed

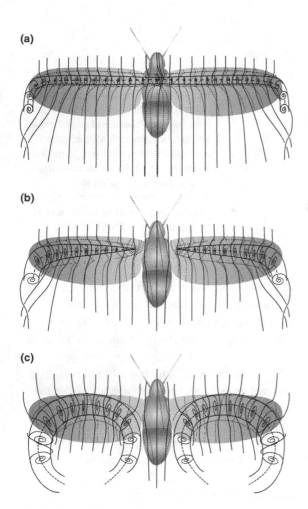

(a)

(b)

(c)

Figure 9.25 The flow over a flapping wing typically separates at the leading edge and becomes entrained within a swirling vortex sitting on top of the wing. Different leading-edge vortex topologies are possible, according to whether the vortex: (a) extends across the thorax; (b) attaches at the base of each wing; or (c) forms a separate horseshoe-shaped vortex system on both wings (reproduced from Bomphrey *et al.*, 2009).

wing remains attached to the wing's surface at low angles of attack and becomes stalled at high angles of attack, the flow over a flapping wing typically separates at the leading edge and becomes entrained within a swirling vortex sitting on top of the wing (Fig. 9.25).

The dynamics of vortex formation are governed by a dimensionless number called the Strouhal number, calculated as wingbeat frequency multiplied by peak-to-peak wing tip excursion divided by air speed. Other parameters influence the dynamics, but for wings operating at high angles of attack, leading-edge vortex formation is difficult to avoid when the Strouhal number is much above 0.2. Accordingly, leading-edge vortices have been observed in most insects for which flow visualizations are available, including Lepidoptera, Diptera and Odonata. The presence of a leading-edge vortex on top of the wing results in a local reduction in pressure, which causes an upward-acting suction force known as vortex lift.

In Orthoptera and Odonata, leading-edge vortex formation may be avoided, at least at times, by appropriately modifying the angle of attack. This presumably reduces the aerodynamic forces, but may serve to enhance aerodynamic efficiency.

9.5.2 Rotational lift

The aerodynamic forces that act upon a translating wing are modified if the wing rotates about a spanwise axis. The axis of rotation of an insect's wing is located anteriorly, so rotating the leading edge upward about this axis causes an increase in the wing's effective angle of attack at the point where this should be measured, three-quarters of the way back from the leading edge. The aerodynamic forces acting on a wing increase with increasing angle of attack, so the expected effect of rotating the wing leading-edge upward, as in supination, is to increase the aerodynamic forces. This is a quasi-steady effect of rotation, which amounts in essence to a proper accounting of the incident flow.

The unsteady effects of wing rotation are less well known for the large angular amplitudes associated with insect flight, but rotating a wing leading-edge upward delays the onset of stall and thereby extends the production of useful aerodynamic lift to higher angles of attack. This unsteady effect is known as the

Kramer effect, and may be responsible for the transient lift enhancement that is observed during wing rotation in insect flight.

9.5.3 Wing–wake interactions

Wing–wake interactions are likely to be of particular importance in hovering flight, where a wing has the possibility of re-encountering the wake that it left behind on the previous stroke. Wing–wake interactions are also important in functionally four-winged insects, where the hindwing operates in the wake of the forewing. Such interactions are probably responsible for the changes in aerodynamic forces that result from altering the timing of wing rotation with respect to stroke reversal in hovering Diptera, and from altering the phasing of the forewing and hindwing pairs in Orthoptera and Odonata. Wing–wake interactions are the least well understood of the various unsteady aerodynamic mechanisms of insect flight, because the effect of the interaction depends upon the fine details of the movement of the wing and of the wake it leaves behind. The effects of wing–wake interaction may therefore be quite specific to the system being considered.

9.5.4 Clap and fling

The first unsteady aerodynamic mechanism to be described in insects was the "clap and fling." This refers to the clapping together and subsequent flinging apart of the wings at the top of the stroke. This mechanism is best known from the chalcid wasp *Encarsia* (Hymenoptera), which has a wingspan of about 1.3 mm and a wingbeat frequency of about 400 Hz. Large Orthoptera and Lepidoptera have also been observed to clap their wings dorsally, so the mechanism is not restricted to small insects. However, it is unclear how important the clap and fling mechanism is under natural conditions. For example, *Drosophila* are commonly observed to clap their wings dorsally in tethered flight, but rarely do

so in free flight. This is presumably an artifact of the large stroke amplitudes that are associated with tethering.

The aerodynamics of the clap and fling are complicated, but are best understood as an example of wing–wing interaction. As the wings clap together, they squeeze out a jet of air between them, which the insect can use to augment thrust. Some insects may enhance their maneuverability by redirecting this jet of air. The more important part of the mechanism, however, is the fling. As the wings pronate, their leading edges are flung apart, while the trailing edges remain in contact. This allows the generation of rotational lift, the growth of which is enhanced by suction of air into the expanding gap between the wings, and by a favorable aerodynamic interaction resulting from the proximity of the wings and their associated vortex systems.

9.6 Power for flight

Flight demands a great deal of power. The activity of the flight muscles is entirely aerobic, but only a small proportion of the energy expended by the muscles is effectively available for flight. Most of the energy consumed by the muscles is lost as heat, and the measured efficiency of insect flight muscle is usually in the range 5–15%. After taking account of further mechanical, inertial and aerodynamic losses, only around 3% of the energy consumed by the flight muscles actually contributes to flight. The forces exerted by flight muscles are in no way unusual, and the high power output necessary for flight is achieved not through efficiency of the flight muscles, but through their high frequencies of contraction.

9.6.1 Fuels for flight

The fuels providing energy for flight vary in different insects. Some insects, such as bees, appear to fuel flight with carbohydrate only. Others, such as locusts, depend mainly on fats, but use carbohydrates during short flights and the early stages of sustained flight (Fig. 9.26). Still others, such as the tsetse fly, *Glossina* (Diptera), rely primarily or exclusively on amino acids, notably proline. Fat is more suitable than carbohydrate as a reserve for insects that make long flights because it produces roughly twice as much energy per unit weight. In addition, glycogen, a common carbohydrate reserve, is strongly hydrated, so that it is eight times heavier than isocaloric amounts of fat. Differences in fuel use also reflect diet and foraging ecology.

Fuel reserves within the muscles themselves are utilized initially, but in most insects these are adequate only for very short flights and further supplies of fuel are drawn from elsewhere. Beyond the muscles themselves, the fat body is the principal store of fuel. The blood-sucking bugs, *Rhodnius* and *Triatoma* (Heteroptera), are exceptional in storing relatively large amounts of triacylglycerol in the flight muscles.

Carbohydrates in the flight muscles are usually the immediate source of energy at the start of flight, and their oxidation involves the usual processes of glycolysis and the citric acid cycle. In the locust, the muscle concentration of glucose starts to decline immediately after flight begins, but, in blowflies, there is a transient increase due to the rapid production of glucose from trehalose. Subsequently, after about 30 minutes of flight, the glucose concentration stabilizes in both species. These stable levels indicate that glucose is provided from other sources, initially from glycogen in the muscle itself, and then from elsewhere.

Trehalose in the hemolymph forms an important carbohydrate reserve in many insects, and is also the form in which carbohydrates are transported from other sites. As a result, the hemolymph concentration of trehalose may fall, as it does in the locust, where it becomes stable after about 30 minutes of flight. In the fly *Calliphora*, however, the concentration first rises and then returns to its original level. The stable levels in both species result from the synthesis of trehalose from glycogen in the fat body. In some other insects, sugars in the gut, in the crop of *Tabanus* (Diptera) and the honey stomach of *Apis*, may be converted to trehalose immediately after absorption and transported directly to the flight muscles.

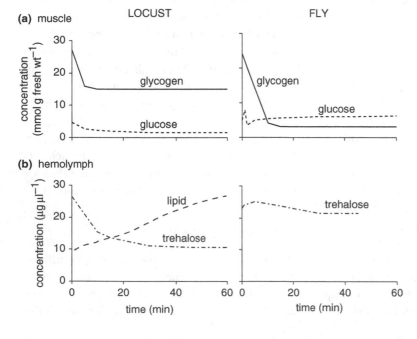

Figure 9.26 Fuel consumption in flight. Changes in the concentrations of fuels during the first hour of flight in the migratory locust and the blowfly. (a) Glycogen and glucose in flight muscle. (b) Trehalose and lipid in the hemolymph.

Insects engaging in long-range migration, such as locusts, some butterflies and planthoppers, switch from using carbohydrates as the main source of fuel to using lipids. Octopamine is known to stimulate carbohydrate catabolism in flight muscles, and the switch to lipid metabolism may be regulated by inhibition of octopaminergic cells in the thoracic ganglia. In general, the lipids are obtained from the fat body where they are stored as triacylglycerides. They are transported through the hemolymph as diacylglycerides, and their concentration increases and then stabilizes in the first 2–3 hours of flight. At the flight muscles, the glycerides are degraded in a series of steps into two-carbon units, and enter the citric acid cycle as acetyl-coenzyme A, condensing with oxaloacetate to form citrate.

Oxidation of amino acids may occur to a minor extent in the flight muscle of most insects, but proline is the major substrate utilized by the flight muscles in a few species. This occurs in the tsetse fly, *Glossina*, and some beetles, such as *Leptinotarsa* (Coleoptera). The initial reserve of proline is small, and in the tsetse fly sufficient to last only for about two minutes, so it is probable that proline is synthesized during flight. Proline is first converted to glutamate, which then undergoes transamination with pyruvate to produce alanine and oxoglutarate. The latter enters the citric acid cycle while the former is returned to the fat body for the resynthesis of proline.

9.6.2 Mobilization of fuel for flight

The energy sources used by flight muscles are usually stored in a form that is not immediately metabolizable and is often at some point remote from the muscles. Consequently, the mobilization of these reserves must be coordinated with muscle activity. The nerve impulse that initiates contraction of the flight muscles activates the fibrillar ATPase by the release of Ca^{2+} from the sarcoplasmic reticulum. This calcium also promotes the activation of phosphorylase involved in the breakdown of glycogen. As a consequence, utilization of carbohydrate proceeds at a fast rate.

In locusts and some other insects, information from the brain leads to the release of adipokinetic hormone from the corpora cardiaca, but this release is inhibited if the carbohydrate concentration in the hemolymph is high. This hormone causes the mobilization of lipids in the fat body by activating a lipase. In locusts the release of diacylglycerides into the hemolymph is apparent within five minutes of the start of flight and their concentration more than doubles. In locusts, adipokinetic hormone also mediates glycogen metabolism by activating a glycogen phosphorylase. In *Manduca*, however, activation of the glycogen phosphorylase is regulated by the titer of hemolymph carbohydrates. A fall in the level of hemolymph trehalose during flight, as it is utilized by the flight muscles, stimulates activation of the enzyme.

9.6.3 Regulation of muscle power output

The motor neurons of insects with synchronous flight muscles are able to regulate muscle power output directly, because of the direct correspondence between muscle action potentials and muscle contractions. In insects with asynchronous flight muscle, however, the pattern of muscle contraction is decoupled from the firing of the motor neurons. Such insects still need to be able to control muscle power output, and this might in principle be achieved by having a graded, rather than binary, response to Ca^{2+} in the flight muscles. In this case, varying the firing frequency of the motor neurons might regulate the power output of asynchronous muscle power output by adjusting the tonic level of Ca^{2+}.

9.7 Sensory systems for flight control

Successful flight requires stability against disturbances. Rotation about the long axis of the body is called rolling; rotation about the transverse axis is called pitching; and rotation about the

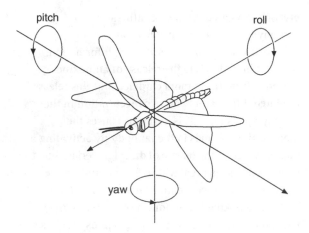

pitch

roll

yaw

Figure 9.27 Instability. Diagram showing the axes about which an insect may rotate when in flight (after Weis-Fogh, 1956).

dorso-ventral axis is called yawing (Fig. 9.27). The aerodynamic forces associated with the wingbeat provide some degree of passive damping in yaw, which will tend to slow down unintended yaw rotations, but will not in itself correct for a yaw disturbance. Some degree of passive stability in pitch may be provided through a pendulum effect, by having the wings operate above the center of gravity of the insect for most of the stroke. Drag on the elongated cerci in insects such as Ephemeroptera may also provide pitch and yaw stability by acting like the tail of a kite. Nevertheless, it is safe to assume that active control is essential for providing stability in most insects. This requires the use of sensory feedback to detect and correct for unintended deviations in body orientation, speed and trajectory. Of primary importance in this respect are the visual system, the antennae, the hair fields on the front of the head and the wing campaniform sensilla. The halteres are of fundamental importance to flight stabilization in Diptera.

9.7.1 Vision

Vision is the single most important sensory modality for flight control in most insects, and is involved in the control of landing and collision avoidance, as well as in the regulation of ground speed, orientation and trajectory. Whereas the ocelli mainly function as fast and highly sensitive horizon detectors, sensory input from the compound eyes is used in responses to optic flow and image expansion cues. The sensory processing that is associated with these functions makes the compound eyes intrinsically slower in their response, and the ocelli and compound eyes therefore complement each other in their dynamic range.

In Orthoptera, Diptera and Odonata, the ocelli contribute to a phasic dorsal light response. Changes in roll attitude are sensed by detecting changes in the direction of ambient illumination, and are compensated by appropriate movements of the head and body. Ocellar input also modulates the response of interneurons involved in processing input from the compound eyes.

Landing and collision avoidance are both mediated by interneurons sensitive to image expansion, known as looming-sensitive neurons. These are known from Diptera, Orthoptera and Lepidoptera, and take their input from the compound eyes. Different pathways may be used to mediate landing and collision avoidance, with a different response exhibited according to whether the insect perceives frontal or lateral image expansion.

Optic flow refers to the apparent movement of objects in different parts of the visual field. When integrated across the visual field, this provides an important source of information on the angular velocity of the head with respect to the visual environment, which can be used to stabilize gaze and flight. The flight control pathway is complex, involving multi modal sensory integration by the head–neck motor system, as well as multiple levels of visual processing. In all cases, however, the first stage in visual processing is the detection of local visual motion.

Local visual motion is sensed by elementary motion detectors, which in Diptera operate by performing a spatiotemporal correlation of light intensities sampled by adjacent ommatidia. The

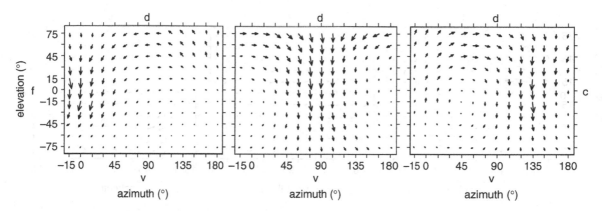

Figure 9.28 Vertical system (VS) cells of the visual system in *Calliphora* detect rotations about various near-horizontal axes, corresponding to various combinations of pitch and roll. Running from left to right, the three graphs show the pattern of local motion preference of the visual receptive fields of tangential cells VS1, VS6 and VS8, plotted in a rectangular projection of the right visual hemisphere. The organization of each receptive field shows a pattern of local motion preference reminiscent of the directional gradient of velocity vectors in a specific rotational optic flow field. The lettering surrounding the graphs denotes the dorsal (d), ventral (v), frontal (f) and caudal (c) directions. Reproduced from Taylor and Krapp (2007).

response of a correlation-type motion detector is linear in contrast frequency rather than velocity, which means that the signal is ambiguous in respect of the magnitude of local visual motion. This is not true, however, of the response of gradient-type motion detectors, which might possibly occur in some insects.

The next stage in optic flow processing involves the integration of elementary motion detector responses from across the visual field. The interneurons that perform this integration have local response sensitivities that are matched to the specific pattern of optic flow experienced during particular self-motions. For example, in *Calliphora* the ten lobula plate tangential cells of the vertical system each respond to a different rotation about some axis close to the horizontal (Fig. 9.28). They are sensitive, therefore, to different combinations of pitch and roll rotation. In contrast, the three cells of the horizontal system respond primarily to yaw rotations and horizontal translation (Fig. 9.29).

During pure rotational movements, the direction and magnitude of the local visual motion in an optic flow field is independent of distance. This is not true during motions involving an element of translation, as nearer objects appear to move faster than distant ones. Translational optic flow fields therefore contain information upon distance as well as self-motion, which is a further source of ambiguity in respect of flight control. Nevertheless, translational optic flow still provides a useful cue for regulating ground speed, altitude and distance to objects in the environment.

9.7.2 Airflow sensors

The antennae are known to be involved in regulating air speed in Odonata, Orthoptera, Lepidoptera, Hymenoptera and Diptera. In flight, the antennae are held horizontally and directed forwards. The oncoming airflow tends to push the flagellum backwards, but in most species an antennal position reaction compensates for this deflection by swinging the pedicel forwards. At higher air speeds the compensation is greater and so the antennae are pointed more directly forwards (Fig. 9.30). This reaction seems to be mediated by pedicellar campaniform sensilla, and serves to keep the highly sensitive Johnston's organ, which responds to movement of the pedicel–flagellum joint, within its

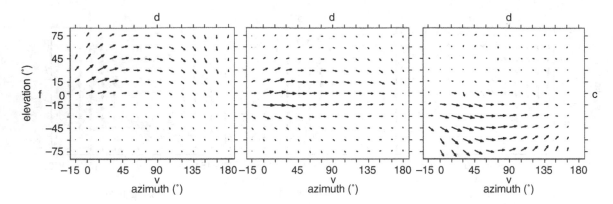

Figure 9.29 Horizontal system (HS) cells of the visual system in *Calliphora* indicate specific translation-related parameters and rotations around near-vertical axes. Running from left to right, the three graphs show the pattern of local motion preference of the visual receptive fields of tangential cells HSN, HSE and HSS, in a rectangular projection of the right visual hemisphere. On average these cells prefer horizontal front-to-back motion, and show a gradual change in their directional preferences reminiscent of translational optic flow. The lettering surrounding the graphs denotes the dorsal (d), ventral (v), frontal (f) and caudal (c) directions. Reproduced from Taylor and Krapp (2007).

working range. The Johnston's organ only responds to movement, and it is probable that, in flight, vibrations of the flagellum produced by the flapping wings stimulate the Johnston's organ. Scolopidia within the Johnston's organ are stimulated differentially depending on the degree of deflection of the flagellum and it is probably on this basis that air speed is sensed.

In locusts, the fields of wind-sensitive hairs on the head are known to be involved in flight control, as well as in maintenance of the wingbeat. The directional sensitivity of the hairs is relatively consistent within a field, but varies between fields, and they therefore respond differentially to rotation of the airstream. In locusts, presenting an asymmetric airflow to the hair fields on the head induces a stabilizing yaw reaction, which is eliminated by covering certain of the hair fields.

9.7.3 Wing load sensors

Although there are in principle many wing mechanoreceptors that might be involved in flight control, only the campaniform sensilla are known to be involved in the stabilization of flight. The

chordotonal organ, stretch receptor and tegula all play a part in regulating the wingbeat, but have not been studied from the perspective of flight stabilization. The campaniform sensilla respond to specific wing deformations associated with twisting of the wings. In Diptera, they may encode information about the travel of torsional waves along the wing. In locusts, firing of the campaniform sensilla changes as the pitch of the insect is varied, and appears to control compensatory twisting of the forewing during the downstroke.

9.7.4 Inertial sensors

There is some evidence that the antennae of Lepidoptera might act as inertial sensors, and the mass of the head apparently allows the head–neck system to sense angular accelerations in Odonata. However, the most important inertial sensors in insects are the halteres of Diptera, which play an essential role in maintaining rotational stability in this order.

The halteres beat vertically in antiphase to the forewings and function as vibratory gyroscopes, sensing the angular velocity of the thorax. Because of the mass of the end-knob, and because of the nature

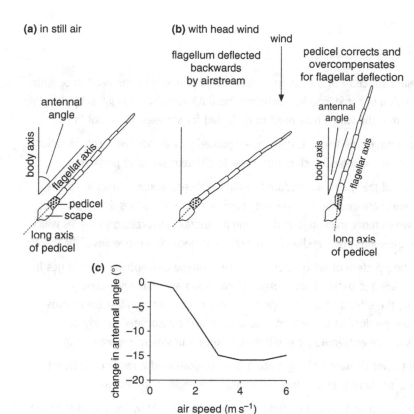

(a) in still air

(b) with head wind

wind

flagellum deflected
backwards
by airstream

pedicel corrects and
overcompensates
for flagellar deflection

antennal
angle

body axis

flagellar axis

pedicel
scape

long axis
of pedicel

antennal
angle

body axis

flagellar axis

long axis
of pedicel

(c)

change in antennal angle (°)

air speed (m s⁻¹)

Figure 9.30 Control of air speed (after Gewecke, 1974). (a) Position of the antenna in still air. (b) Changes in the position of an antenna resulting from a head wind. Long axis of pedicel shown as a dotted line. (c) Changes in the antennal angle of a tethered locust associated with increasing wind speed (equivalent to increasing air speed of a freely flying insect).

of their articulation with the thorax, the halteres vibrate in an almost planar motion. However, the mass of the end-knob means that it tends to continue moving in the same direction in inertial space when the insect rotates, and the resulting Coriolis force exerts a torque upon the haltere stalk. These torques are sensed by campaniform sensilla at the base of the haltere, and vary cyclically through the course of each haltere oscillation.

Because each haltere beats vertically, the torques induced by yawing peak twice per cycle, whereas the torques produced by pitching and rolling peak only once per cycle. The beating planes of the halteres are inclined backward by about 30°, and the left and right halteres therefore measure roll motions in different frames of reference. Consequently, the response of the halteres to rolling is expected to be asymmetric between the left and right sides, whereas their response to pitching is expected to be symmetric. Unilateral manipulation of a haltere does not result in any change to the kinematics of the contralateral wing, but the independent operation of ipsilateral control loops on both sides should enable three-axis control, through the summation of forces generated by the left and right wings.

Summary

- The detailed structure of the wings is determined primarily by the need to optimize the interaction of aerodynamic, elastic and inertial forces during flight, subject to the constraint that the wings may need to be folded for stowage when not in use.

- The gross form of the wings is determined primarily by aerodynamic considerations, although other ecological factors may provide different selective pressures.

- Movements of the wings are produced by three different groups of muscles: direct muscles which are inserted into the wing base and move the wings directly, indirect muscles which move the wings by distorting the thorax, and accessory muscles which modulate the wing stroke by changing the shape or mechanical properties of the thorax.

- The resulting pattern of wing kinematics defies simple description, but changes in kinematics are the basis of active flight stabilization and maneuver control. Ultimately, the effects of wing shape and movement must manifest themselves through the production of aerodynamic force, but the associated aerodynamic mechanisms are exceedingly complicated, involving unsteady, separated flows.

- The high power demands of flight are met by utilization of a range of different substrates, so there is biochemical adaptation for flight in the thorax.

- The stabilization and control of flight is made possible only by the use of feedback from a range of different sensory modalities. This involves multimodal integration of sensory information by sensory interneurons, descending neurons, the head–neck motor system and the flight motor itself.

Recommended reading

GENERAL

Dudley, R. (2000). *The Biomechanics of Insect Flight: Form, Function, Evolution.* Princeton, NJ: Princeton University Press.

MORPHOLOGY

Wootton, R. J. (1992). Functional morphology of insect wings. *Annual Review of Entomology* **37**, 113–140.

KINEMATICS

Taylor, G. K. (2001). Mechanics and aerodynamics of insect flight control. *Biological Reviews* **76**, 449–471.

AERODYNAMICS

Sane, S. P. (2003). The aerodynamics of insect flight. *Journal of Experimental Biology* **206**, 4191–4208.

SENSORY SYSTEMS

Taylor, G. K. and Krapp, H. G. (2007). Sensory systems and flight stability: what do insects measure and why? *Advances in Insect Physiology* **34**, 231–316.

References in figure captions

Baader, A., Schäfer, M. and Rowell, C. H. F. (1992). The perception of the visual flow field by flying locusts: a behavioural and neuronal analysis. *Journal of Experimental Biology* **165**, 137–160.

Bomphrey, R. J., Taylor, G. K. and Thomas, A. L. R. (2009). Smoke visualization of free-flying bumblebees indicates independent leading-edge vortices on each wing pair. *Experiments in Fluids* **46**, 811–821.

Bourgogne, J. (1951). Ordre des Lépidoptères. In *Traité de Zoologie*, vol. 10, ed. P.-P. Grassé, pp. 174–448. Paris: Masson et Cie.

Brodsky, A. K. (1994). *The Evolution of Insect Flight.* Oxford: Oxford University Press.

Comstock, J. H. (1918). *The Wings of Insects.* New York, NY: Comstock Publishing Co.

Ennos, A. R. (1988). The inertial cause of wing rotation in Diptera. *Journal of Experimental Biology* **140**, 161–169.

Gewecke, M. (1974). The antennae of insects as air-current sense organs and their relationship to the control of flight. In *Experimental Analysis of Insect Behaviour*, ed. L. Barton Browne, pp. 100–113. Berlin: Springer-Verlag.

Haas, F. and Wootton, R. J. (1996). Two basic mechanisms in insect wing folding. *Proceedings of the Royal Society of London B* **263**, 1651–1658.

Miyan, J. A. and Ewing, A. W. (1985). How Diptera move their wings: a re-examination of the wing base articulation and muscle systems concerned with flight. *Philosophical Transactions of the Royal Society of London* **311**, 271–302.

Nachtigall, W. (1966). Die Kinematic der Schlagflügelbewegungen von Dipteren. Methodische und analytische Grundlagen zur Biophysik des Insektenfluges. *Zeitschrift für Vergleichende Physiologie* **52**, 155–211.

Oldroyd, H. (1949). *Diptera: I. Introduction and Key to Families.* London: Royal Entomological Society of London.

Pringle, J. W. S. (1948). The gyroscopic mechanism of the halteres of Diptera. *Philosophical Transactions of the Royal Society of London B* **233**, 347–384.

Pringle, J. W. S. (1957). *Insect Flight.* Cambridge: Cambridge University Press.

Rowell, C. H. F. and Reichert, H. (1991). Mesothoracic interneurons involved in flight steering in the locust. *Tissue & Cell* **23**, 75–139.

Snodgrass, R. E. (1935). *Principles of Insect Morphology.* New York, NY: McGraw-Hill.

Taylor, G. K. and Krapp, H. G. (2007). Sensory systems and flight stability: what do insects measure and why? *Advances in Insect Physiology* 34, 231–316.

Tillyard, R. J. (1918). The panorpoid complex: I. The wing-coupling apparatus, with special reference to the Lepidoptera. *Proceedings of the Linnean Society of New South Wales* 43, 286–319.

Walker, S. M., Thomas, A. L. R. and Taylor, G. K. (2009). Deformable wing kinematics in the desert locust: how and why do camber, twist and topography vary through the stroke? *Journal of the Royal Society Interface* 6, 735–747.

Weis-Fogh, T. (1956). Biology and physics of locust flight: II. Flight performance of the desert locust (*Schistocerca gregaria*). *Philosophical Transactions of the Royal Society of London B* 239, 459–510.

Wilson, D. M. and Weis-Fogh, T. (1962). Patterned activity of co-ordinated motor units, studied in flying locusts. *Journal of Experimental Biology* 39, 643–667.

Wootton, R. J. (1981). Support and deformity in insect wings. *Journal of Zoology, London* 193, 447–468.

Wootton, R. J. (1992). Functional morphology of insect wings. *Annual Review of Entomology* 37, 113–140.

Wootoon, R. J., Kukalová-Peck, J., Newman, D. J. S. and Muzón, J. (1998). Smart engineering in the mid-Carboniferous: how well could Palaeozoic dragonflies fly? *Science* 282, 749–751.

10 | Muscles

REVISED AND UPDATED BY JOHN C. SPARROW

INTRODUCTION

Muscles power all the movements, external and internal, in insects. All insect muscles are striated, like vertebrate cardiac and skeletal muscle. In their structure, protein content, contractility and regulation, insect muscles show high levels of homology to these vertebrate muscles. However, the varieties of movements and modes of locomotion in insects, particularly larval crawling, flight, feats of jumping and stridulation (or singing) have led to a wide range of muscle specializations in structure, function and regulation. The study of specialized insect muscles has allowed us to understand how these muscles cope with the demands made on them in achieving insect movement. However, insects have also been important as model organisms for understanding muscle more generally. Hence our knowledge of muscle contraction owes much to work particularly on the waterbug, *Lethocerus*, whereas the fruit fly, *Drosophila*, is providing new insights into muscle function, development and human diseases.

This chapter covers insect muscle from its structure and molecular function to its integration into the controlled regulation of insect movements. Section 10.1 describes insect muscle ultrastructure with the location of specific proteins. Section 10.2 describes how insect muscles contract. Section 10.3 concerns muscle innervation, the activation of muscle contraction and its regulation. Section 10.4 deals with the energetics of muscle function and Section 10.5 links muscle control to locomotion in the whole organism. Muscle development is considered in Section 10.6, together with the role of the muscles in major lifecycle events.

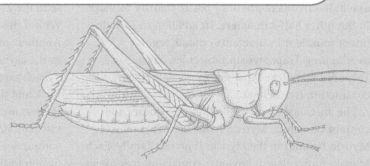

The Insects: Structure and Function (5th edition), ed. S. J. Simpson and A. E. Douglas.
Published by Cambridge University Press. © Cambridge University Press 2013.

10.1 Structure

10.1.1 Basic muscle structure

Each muscle is made up of a number of fibers, which are long, usually multinucleate cells running the length of the muscle. The characteristic feature of muscle fibers is the presence of myofibrils, which are embedded in the cytoplasm (sarcoplasm) and extend continuously from one end of the fiber to the other. The fibrils are long serial arrays of contractile units known as sarcomeres. In turn each sarcomere is composed of interdigitated molecular filaments, consisting mainly of two proteins: Myosin and Actin (Fig. 10.1a,b), which through their cyclical, ATP-dependent interactions generate the contractile forces and movements that shorten each sarcomere and the overall myofibril, producing muscle contractions.

Each sarcomere is bounded by electron-dense Z-discs (Fig. 10.1a), which connect neighboring sarcomeres. From either side of each Z-disc (also called Z-line), so-called "thin filaments" extend toward, but do not reach, the center of the sarcomere. Each thin filament has at its core a single helical (F-Actin) polymer made up of globular Actin units (G-Actin). As in all eukaryotic cells, the G-Actin monomers within an F-Actin are polymerized in the same orientation, giving a polarity to each F-Actin with a barbed (or +) end and pointed (or –) end. In all muscles the barbed ends of the F-Actin are embedded in the Z-discs, so the thin filaments in one half-sarcomere are of opposite polarity to those in the other half-sarcomere. In addition to F-Actin, insect muscle thin filaments contain regular overlapping Tropomyosin molecules, as well as Troponin complexes that act to regulate muscle contraction (see below).

The so-called "thick filaments" are stouter and contain numerous Myosin molecules. This muscle Myosin belongs to the Myosin II protein family. Each Myosin molecule is a hexamer containing two copies each of the Myosin heavy-chain polypeptide and the two different Myosin light chains. Together these form an asymmetric, elongated molecular structure with two globular "heads" at one end and a rod domain at the other. The Myosin head, or motor domain, contains the globular region of the Myosin heavy chain and its associated light chains. This domain contains the ATPase and Actin-binding activities that drive contraction. The extended Myosin rod domains are formed by association of the two Myosin heavy-chain polypeptides and allow Myosins to assemble side by side and in regular helical patterns with other Myosins on the thick filament surface. They do so such that each thick filament is bipolar, with Myosin molecules in one half all aligned in one direction, opposite to those in the other half (Fig. 10.1c). Insect thick filaments are usually much thicker than those of vertebrates due to their association with other thick filament proteins, notably Paramyosin, which often fills the thick filament core but is not universally required. In its absence some insect thick filaments may appear to have a hollow core (compare Fig. 10.1e,f), although this is usually only true for part of their length. Thick filaments usually stay in register within the sarcomere and there is evidence that they are cross-connected at the M-line. Other known thick filament-associated proteins include Myofilin, Stretchin, Projectin and Kettin/Sallimus. Connections – C-filaments – seen by electron microscopy between the ends of the thick filaments and the Z-disc in flight muscle may be present in all muscles. Kettin/Sallimus molecules appear to span the Z-disc and extend to the thick filament tips. Within this link they are associated with Projectin. Another thick filament protein, Flightin, occurs only in the asynchronous flight muscles of *Drosophila*.

Each sarcomere comprises a lattice of interdigitated thick and thin filaments. In the filament lattices of skeletal muscles other than flight muscles, and in visceral muscle, each thick (Myosin) filament is surrounded by 10–12 Actin filaments, with a ratio of thin to thick filaments of 6:1 (Fig. 10.1f), though there is considerable variation between muscles in the

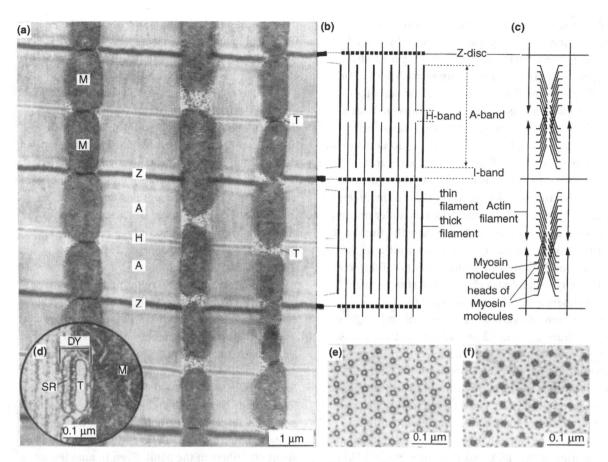

Figure 10.1 Basic structure of a muscle. (a) Electron micrograph of a longitudinal section of part of muscle (asynchronous flight muscle of the wasp, *Polistes*). Abbreviations: A = A-band, H = H-band, I = I-band, M = mitochondrion, T = transverse tubule, Z = Z-disc (after Smith, 1968). (b) Diagram showing the arrangement of the filaments that produces the banding pattern seen in (a). The filaments are aligned with the bands of two sarcomeres in (a). (c) Diagrammatic representation of the orientation of the Actin (thin) and Myosin (thick) molecules in a muscle. (d) Electron micrograph of a dyad (DY). Abbreviations: M = mitochondrion, SR = sarcoplasmic reticulum, T = transverse tubule (from the flight muscle of a dragonfly, *Celithemis*) (after Smith, 1966). (e) Electron micrograph of a transverse section of a flight muscle with a ratio of thin:thick filaments of 3:1 (after Smith, 1972). (f) Electron micrograph of a transverse section of an intersegmental muscle with a ratio of thin:thick filaments of 6:1 (after Smith, 1972).

regularity of the lattice. In flight muscles, whether these are synchronous or asynchronous (see below), there is invariably a very regular lattice of six thin filaments surrounding each thick one, with a ratio of 3:1 (Fig. 10.1e).

The Z-discs largely comprise the overlapping F-Actin filament ends of the thin filaments from neighbouring sarcomeres crosslinked by α-Actinin.

In addition, Z-discs contain a variety of proteins, many of which are yet uncharacterized, including Kettin/Sallimus, Zetalin, MSP300 and others. The major function of the Z-disc protein complex is to transmit and withstand the contractile forces generated within the neighbouring sarcomeres and myofibril.

As a result of these sarcomeric arrangements and the alignment of the components across a whole fiber, the muscles have a banded appearance when viewed under phase contrast or in stained preparations. Each sarcomere also has a banded appearance (Fig. 10.1a–c). Immediately inside the darkly staining Z-lines are the lightly staining (isotropic) I-bands containing the proximal parts of the thin filaments. Between the I-bands is the (anisotropic) A-band, which is the region containing the thick filaments. This band comprises three regions. At its center is the M-line within the rather paler H-band, which consists of the proximal/central parts of the thick filaments. The remainder of the A-band comprises the regions where the thick and thin filaments overlap. The relative widths of the I-, A- and H-bands are a characteristic of different muscle types and also change as a structural consequence of filament-sliding during contraction/relaxation. Sarcomere resting lengths range from 2 μm to 12 μm, varying between muscle types and species. It is in the thick–thin filament overlap region that the Myosin "heads" form crossbridges with the F-Actin of the thin filaments, and through the ATP-dependent crossbridge cycle generate forces and relative movement of the filaments (see Section 10.2.1). Due to the inherent polarities of the thick and thin filaments, the thin filaments move toward the center of the sarcomere, causing the sarcomere to shorten. This is the basic mechanism by which force and movement are generated by muscles.

Each fiber is bounded by the sarcolemma comprising the plasma membrane of the cell plus the basal lamina. The fiber cytoplasm is often called the sarcoplasm. The endoplasmic reticulum, which has no connection to the sarcolemma, is known as the sarcoplasmic reticulum. The plasma membrane is deeply invaginated into the fiber, often as regular radial canals between the Z-discs and the H-bands (see below). This system of invaginations is called the transverse tubular, or T-system. It is extensive: For example, in the moth *Philosamia* about 70% of the muscle plasma membrane is invaginated within the fiber. The T-system is associated with the sarcoplasmic reticulum (Fig. 10.1d). Where the two membrane systems are very close, contacts between their membranes are visible as electron-dense material; the arrangement is called a dyad. Functionally the T-tubules and dyads allow the rapid transmission of electrical signals from the sarcolemma to the center of the fiber, where they cause the release of calcium ions from the sarcoplasmic reticulum, which activate sarcomeric contraction. Muscle fiber nuclei and the bulk of the sarcoplasm occur in different positions in the cell in different types of muscle. The arrangement of the myofibrils within the fiber varies (see below), but they are always in close contact with the mitochondria.

Single insect muscles vary in the number of fibers they contain, from a single fiber (e.g., some embryonic muscles in *Drosophila*) to many, but are usually made up of a number of fibers, sometimes a very large number. For example, the metathoracic dorsal longitudinal muscle of the adult grasshopper *Schistocerca* has over 3000 fibers, and even in a fourth-stage larva there are over 500 fibers in this muscle. A tergocoxal muscle in the same insect contains about 50 fibers in the fourth-stage larva and about 400 fibers in the adult. Even in muscles not concerned with flight, fiber numbers may be large. For example, in the coxal depressor muscle of the cockroach *Periplaneta* there are about 765 fibers in the fifth (out of ten) larval stage. This number only increases to about 870 in the adult. In contrast to these numbers, the number of fibers in asynchronous flight muscles is small (see below).

The muscle fibers are collected into units separated from neighboring units by a tracheolated membrane. Each muscle consists of one or a few such units. For example, there are five units in the dorsal longitudinal flight muscles of grasshoppers, and three in the tergocoxal muscle referred to above. Each muscle unit may have its own independent nerve supply, but in other cases several muscle units have a common innervation and so function together as the motor unit.

10.1.2 Variations in structure

Three broad categories of muscle can be distinguished: skeletal muscle, visceral muscle and the heart. Skeletal muscles are attached at either end to the cuticle and move one part of the skeleton relative to another. Visceral muscles move the viscera and have only one or, commonly, no attachment to the body wall. Many form circular muscles around the gut and ducts of the reproductive system.

Skeletal muscles can be differentiated functionally into synchronous and asynchronous muscles. Most skeletal muscles are synchronous: They exhibit a direct relationship between motor neuron activity and contraction (see Section 10.5.2). Asynchronous muscles do not have this direct relationship and only occur in the flight muscles of Thysanoptera, Psocoptera, Homoptera and Heteroptera (Hemiptera), Hymenoptera, Coleoptera and Diptera and in the timbal muscles of some Cicadidae (Homoptera).

Synchronous skeletal muscles The vast majority of insect muscles are synchronous muscles, in which each contraction is driven by a single neural stimulus (see Section 10.3.2). The form and arrangement of the myofibrils in synchronous muscles is very variable. Tubular muscles have the myofibrils arranged radially around a central core of cytoplasm containing the nuclei. This arrangement is common in leg and trunk muscles of many insects, but also occurs in the flight muscles of Odonata and Blattodea (Fig. 10.2a). In close-packed muscles, on the other hand, the myofibrils are only 0.5–1.0 μm in diameter and are packed throughout the whole fiber; the nuclei are flattened and peripheral. Fibers of this type occur in some larval insects, in Apterygota and in the flight muscles of Orthoptera, Trichoptera and Lepidoptera (Fig. 10.2b).

The abundance and arrangement of mitochondria is related to the level and type of muscle activity. In the tubular muscles of Odonata and the close-packed flight muscles of Orthoptera they are large and numerous, occupying about 40% of the fiber volume (Fig. 10.3a). This is also true of the muscles which oscillate at high frequencies to produce the sounds of cicadas and bushcrickets (Fig. 10.3c). In muscles that do not oscillate rapidly the mitochondria generally occupy a much smaller proportion of the fiber volume (Fig. 10.3e–h), making it possible for a larger proportion of the fiber to be occupied by the contractile elements. The mitochondria may be in pairs on either side of a Z-line or scattered irregularly between the fibers. Fibers with abundant mitochondria may be colored pink by their high cytochrome content.

The development of the sarcoplasmic reticulum is correlated with the mechanical properties of the muscle, and in particular with the rates of fiber relaxation. In muscles which tend to maintain a sustained contraction, such as the locust extensor tibiae muscle (Fig. 10.3e) and the accessory or direct flight muscles of Diptera, it is relatively poorly developed. On the other hand, in fast-contracting synchronous muscles, such as those associated with the sound-producing apparatus of male *Neoconocephalus* (Orthoptera) and cicadas, the sarcoplasmic reticulum comprises 15% or more of the total fiber volume (Fig. 10.3c). A characteristic of synchronous flight muscles, whether they are tubular or close-packed, is that the distance from the sarcoplasmic reticulum to the myofibrils is short, generally less than 0.5 μm and even less in very fast-contracting muscle. This close proximity facilitates the rapid release and resequestration of calcium ions during cycles of contraction and relaxation (Section 10.3.2).

In synchronous muscles, the invaginations of the T-system occur midway between the Z-disc and the center of the sarcomere. Sarcomere length in many synchronous muscles ranges from about 3 μm to 9 μm, but is usually only about 3–4 μm in flight muscles. The I-band usually constitutes 30–50% of the resting length of the sarcomere, but *in situ* the extent of muscle shortening may be much less than

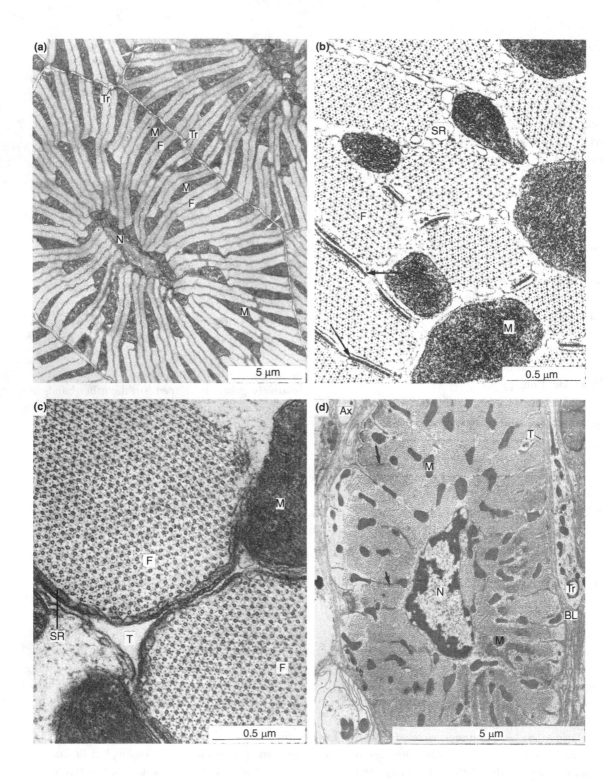

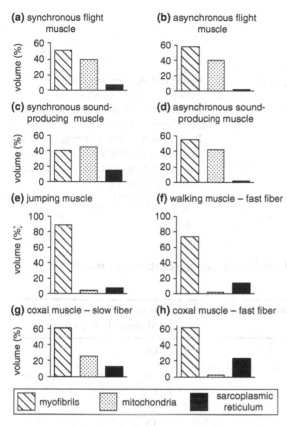

Figure 10.3 Proportions of the muscle volume occupied by the different major components, myofibrils, mitochondria and sarcoplasmic reticulum, in different types of muscle: (a) synchronous flight muscle; (b) asynchronous flight muscle; (c) synchronous sound-producing muscle; (d) asynchronous sound-producing muscle; (e) jumping muscle (extensor tibiae of locust); (f) skeletal muscle with fast fibers; (g) skeletal muscle with slow fibers; (h) skeletal muscle with fast fibers.

this. For instance, the I-bands of locust flight muscle constitute about 20% of the sarcomere length, but, during flight, the muscle shortens by

only about 5%. Fiber diameter is commonly greater in close-packed muscle than in tubular muscle; up to 100 μm in the former compared with 10–30 μm in the latter.

Structurally and functionally, synchronous skeletal muscles form a continuum with slow, or tonic, fibers at one extreme and fast fibers at the other. Slow fibers have little sarcoplasmic reticulum, the volume occupied by mitochondria is relatively large (Fig. 10.3g) and the ratio of thin to thick filaments is high (6:1). The filaments occupy the greater part of the fibers and may not be grouped into discrete myofibrils, and the sarcomeres are long. Fast fibers have the opposite characteristics: extensive sarcoplasmic reticulum; a small volume of mitochondria (Fig. 10.3f,h); low ratios (3:1) of thin to thick filaments; and short sarcomeres.

Some muscles have a uniform complement of fibers. For example, the posterior coxal depressor muscle of *Periplaneta* (muscle 136 in Fig. 8.6) consists entirely of fast fibers. Others include more than one type of fiber. Muscle 135d′ in Fig. 8.6, for example, has a bundle of about 250 fast fibers ventrally and almost 700 slow fibers dorsally. The extensor tibiae muscle of the locust hindleg contains different fiber types mixed in various proportions in different parts of the muscle (Fig. 10.4; Table 10.1). The tergal depressor of the trochanter, the "jump muscle," in *Drosophila* consists of 26–30 fibers, of which the four most anterior fibers appear different from the rest.

To a large extent, the anatomical characteristics of the fibers are reflected in the types of innervation they receive, with fast fibers being innervated by fast axons and slow fibers by slow axons (see below, Table 10.1).

Caption for Figure 10.2 Muscle types. Electron micrographs of transverse sections of different muscles. Abbreviations: Ax = axon, BL = basal lamina, F = myofibril, M = mitochondrion, N = nucleus, SR = sarcoplasmic reticulum, T = transverse tubule, Tr = trachea. (a) Part of a tubular muscle with radial arrangement of myofibrils within each fiber. Opposing white arrows indicate intercellular space between fibers (flight muscle of dragonfly, *Enallagma*) (after Smith, 1968). (b) Part of a close-packed muscle fiber. Arrows point to dyads (after Smith, 1984). (c) Fibrillar muscle showing parts of two myofibrils. Notice their large size (after Smith, 1984). (d) One fiber of visceral muscle. The myofilaments are not grouped into myofibrils. Arrows point to dyads (spermatheca of cockroach, *Periplaneta*) (after Smith, 1968).

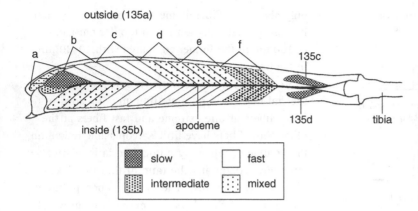

Figure 10.4 Arrangement of fibers with different properties in the extensor tibiae muscle of a locust. Shading shows the dominant type of fiber in each region. Lettering indicates the type of innervation (see Table 10.1) and numbers indicate the anatomically distinct parts of the muscle (after Hoyle, 1978).

Table 10.1 **Fiber types in the extensor tibiae muscle of the locust. The percentages of muscle fibers having different innervation and ultrastructure**

Axons innervating[a]	Structural type of muscle fibers	Region of location in extensor tibiae (see Fig. 10.4)	Percentage of total number of fibers in the muscle
F	Fast	a, b, c, d, e, f	68
F D	Fast	b, c, d	6
F S	Intermediate	a, d, e, f	3
F S I	Intermediate	a, c, d, e, f	12
F D S	Fast	a, d	0.5
F I	Fast	a, d, e, f	2
S I	Slow	a, f, I35c, I35d	8
S	Slow	a	0.5

Notes:
[a] F, fast axon.
S, slow axon.
D, octopamine axon.
I, inhibitory axon.
Source: After Hoyle (1978).

Asynchronous skeletal muscles The specialized asynchronous muscles (in which neural stimulation is asynchronous with respect to contraction; see Section 10.3.2) typically have large cylindrical myofibrils, 1–5 μm in diameter, with corresponding increases in fiber diameters, ranging from 30 μm in carabid beetles to 1.8 mm in *Rutilia* (Diptera). Because the myofibrils are so conspicuous and easily separated, muscles with this characteristic are sometimes called fibrillar, but not all fibrillar muscles are asynchronous in their activation. The myofibrils are distributed through the entire cross-section of the fiber (Fig. 10.2c). The nuclei are scattered between the myofibrils and under the sarcolemma.

Asynchronous muscles may contain only a few fibers because these fibers are so big. For example, the dorsal longitudinal flight muscles of Muscidae and Drosophilidae (Diptera) consist of only six fibers. Further, sarcomere length is short, 1–2 μm in *Tenebrio* (Coleoptera), 1.7–2.5 μm in *Lethocerus* (Heteroptera, Hemiptera) and 3.3–3.5 μm in *Drosophila*. The I-band is typically narrow, making up less than 10% of the sarcomere. The thick filaments may appear to taper toward the Z-line and are attached to it through so-called C-filaments.

The plasma membrane is invaginated in a T-system as in other muscles, but the positions of the invaginations are variable. Regular reticulate patterns of T-tubule association with the sarcomeric structure occur as in the wasp *Polistes* (Hymenoptera), where they are aligned with the H-band, and in *Drosophila* where they lie midway between the M- and Z-lines. In the fibrillar muscles of *Tenebrio* (Coleoptera) and *Megoura* (Homoptera; Hemiptera) the system is more complex and less regular. In these insects invaginations of the plasma membrane are produced by indenting tracheoles (see Section 10.4.5; and Chapter 17); from these invaginations fine tubules of plasma membrane extend to entwine each myofibril. The sarcoplasmic reticulum associated with these T-systems consists of a number of unconnected vesicles scattered without reference to sarcomere pattern. It occupies only a very small proportion of the fiber volume (Fig. 10.3b,d).

Mitochondria are large, as in all flight muscles, and occupy 30–40% of the fiber volume. Almost the whole surface of each myofibril can be in direct contact with mitochondria. These may be regularly arranged, as between the Z-discs and H-bands in *Polistes*, or without any regular arrangement, as in *Calliphora* and *Drosophila*.

Visceral muscles Visceral muscles differ in structure from skeletal muscles in several respects. Adjacent fibers are held together by desmosomes, which are absent from skeletal muscle, and in some cases the fibers may branch and anastomose, such as occurs in the *Drosophila* ovarian sheath. Further, each fiber is uninucleate and the contractile material is not grouped into fibrils but packs the whole fiber (Fig. 10.2d). As in other muscles, each fiber consists of thick and thin filaments in a rather irregular lattice, though often with rings of 10–12 Actin filaments around each Myosin filament (Fig. 10.1f). A T-system with a regular arrangement is present in *Periplaneta* (Blattodea), but in *Carausius* (Phasmatodea) and *Ephestia* (Lepidoptera) it is irregularly disposed. The sarcoplasmic reticulum is poorly developed; mitochondria are small and often few in number.

All insect muscles are striated, so visceral muscle resembles skeletal muscle in contrast to the "smooth" visceral muscle of vertebrates. The Z-discs and H-bands of insect visceral muscle are irregular and sarcomere length is very variable. While visceral muscle always appears striated by light microscopy, in electron micrographs the much less regular sarcomere and filament lattice often make the muscle appear non-striated. As in some body wall muscles, the Z-discs often appear perforated, perhaps to allow thick filaments from one sarcomere to slide into the next, as in supercontractile skeletal muscles (Section 10.2.2). In cardiac muscle the sarcomeres are short, about 3 μm, but they may be as long as 10 μm in other visceral muscles.

Visceral muscles may be innervated from the stomodeal (stomatograstric) nervous system (see Chapter 3) or from the ganglia of the ventral nerve cord, but are sometimes without innervation, as in the heart of *Anopheles* (Diptera) larvae.

Cardiac muscle The insect heart usually consists of a simple tube that contains a layer of contractile myocardial cells. These are usually mononucleate cells with striated longitudinal and circular myofibrils. Though activated largely myogenically (i.e., without neural input), the heart rate is influenced by nerves that innervate the heart in most insects.

10.1.3 Muscle insertion

Skeletal muscles are fixed at either end through myotendinous junctions to the integument, spanning a joint in the skeleton so that muscle contraction moves one part of the skeleton relative to another. In most cases, muscles are attached to invaginations of the cuticle called apodemes.

At the point of attachment to the epidermis, the plasma membranes of muscle and epidermal (tendon) cells interdigitate and are held together by desmosomes (Fig. 10.5). Within the epidermal cell, microtubules run from the desmosomes to hemidesmosomes on the outer plasma membrane, and from each hemidesmosome a dense attachment fiber passes to the epicuticle through a pore canal. In earlier studies, microtubules and attachment fibers were not recognized as separate structures and were termed tonofibrillae. In most muscles, Actin filaments reach the terminal plasma membrane of the muscle fiber, inserting into the dense material of desmosomes. In asynchronous flight muscle, the terminal region of each myofibril consists of a dense body which has the appearance and protein content of a much-enlarged Z-disc with a core of extended Actin filaments. Each of these regions is inserted onto an extension of an epidermal "tendon" cell containing microtubules as in other muscle attachments.

The myotendinous attachment fibers in the cuticle are not digested by molting fluid, so during molting they remain attached to the old cuticle across the exuvial space between the new and old cuticles. As a result, the insect is able to continue its activities after apolysis during the development of the new cuticle. The connections to the old cuticle are broken at about the time of ecdysis.

Muscle attachment fibers extending to the epicuticle can only be produced at a molt and most myotendinous attachments form at this time. Attachment can occur later, however, if cuticle production continues in the postecdysial period, but in this case the attachment fibers are only connected to the newly formed procuticle and do not reach the epicuticle.

10.2 Muscle contraction

10.2.1 Mechanics of contraction

Flight muscles from the waterbug *Lethocerus*, and more recently *Drosophila*, have played significant roles in furthering the understanding of muscle contraction generally, but the basic mechanisms are the same across all animals.

Muscle contraction results from the thin and thick filaments sliding relative to each other so that each

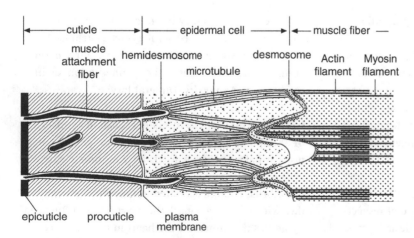

Figure 10.5 Attachment of a muscle fiber to the integument (after Caveney, 1969).

sarcomere is shortened (Fig. 10.6a). This is driven by the ATP-dependent cyclical biochemical/biomechanical activities of individual Myosin motor domains. In the absence of nucleotide the Myosin forms "rigor" crossbridges with the F-Actin of the thin filament. The binding of ATP releases the Myosin from the thin filament, whereupon the Myosin hydrolyzes the ATP (to ADP and inorganic phosphate, Pi, which remain bound to the Myosin) and changes its conformation. In this state it has a high affinity for rebinding to F-Actin. When it does so, the Myosin releases the Pi and ADP, undergoing a conformational change which imparts a force/movement to the F-Actin. As all the Actomyosin interactions in a half sarcomere have the same orientation, these movements cause the thick and thin filaments to move relative to one another and the sarcomere shortens. Through simultaneous shortening of all the sarcomeres of the myofibrils, the muscle shortens. In this way the ATP from energy metabolism, largely oxidative phosphorylation, is converted into the work done by the muscle.

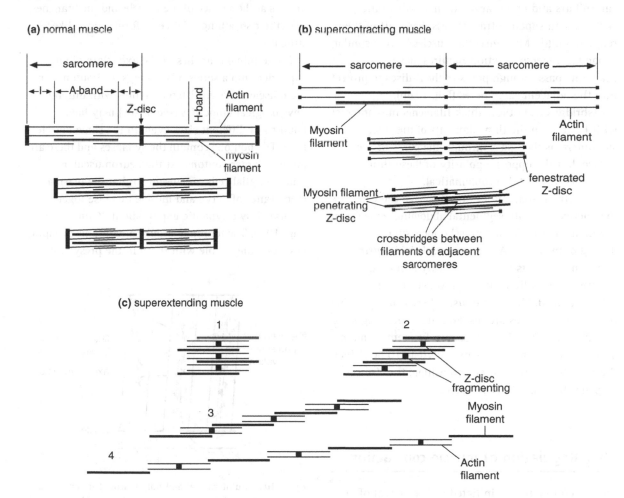

Figure 10.6 Mechanics of muscle contraction. (a) Normal muscle showing the sliding of the filaments producing shortening of the sarcomeres (I = I-band). (b) Supercontracting muscle. When the muscle is fully contracted, the Myosin filaments extend through perforations of the Z-discs (after Osborne, 1967). (c) Superextending muscle. Numbers indicate successive stages of extension. When the muscle is extended, the Z-discs become fragmented (after Jorgensen and Rice, 1983).

In most muscles the extent of contraction appears to be limited by I-band width. Shortening can continue only until the end of the thick filaments reach the Z-disc. The asynchronous flight muscles have very narrow I-bands and are essentially isometric, shortening by only 2–4%.

10.2.2 Supercontraction and superextension

However, most body wall muscles of dipteran larvae, caterpillars and other larvae with a hydrostatic skeleton can supercontract to less than half of their relaxed length. Most visceral muscles have a similar capacity. Supercontraction requires that thick filaments pass through pores in the Z-discs to project into the adjacent sarcomeres (Fig. 10.6b). Crossbridges from these thick filaments may interact with F-Actin in the thin filaments of the next sarcomere as they pass through the pores of the Z-disc, but the opposite polarity of these thin filaments makes this problematical.

In contrast, some other muscles have the capacity to superextend. Intersegmental abdominal muscles in segments 4–7 of female grasshoppers superextend during oviposition. A similar muscle in *Locusta* between segments 5 and 6 can vary in length from 1.2 mm when fully contracted to over 11 mm when fully extended. This very unusual behavior is possible because the Z-discs are discontinuous and fragment into Z-bodies, to which the Actin filaments remain attached, when the muscle extends (Fig. 10.6c). When the muscles contract after oviposition, the Z-discs may not be completely restored.

10.3 Regulation of muscle contraction

Muscle contraction is initiated by the arrival of an action potential at the nerve–muscle junction. At an excitatory synapse this leads, first, to excitation and then to activation of the muscle fiber.

10.3.1 Innervation

Characteristically in insects, each axon innervating a muscle has many nerve endings spaced at intervals of 30–80 µm along each fiber (Fig. 10.7). This is called multiterminal innervation. The form of the nerve ending varies. The fine nerve branches in flight muscles of Diptera run longitudinally over the surface of the muscle; in the flight muscles of the flour beetle (*Tenebrio*), they are completely invaginated into the muscle. In Orthoptera, the axon divides at the surface of the muscle and the branches, with their sheathing glial cells, form a claw-like structure.

The terminal branches of the axons are often expanded into a series of swellings, or boutons, on the muscle surface. Different axons, with different physiological properties (see below) may have boutons of different sizes even on the same muscle fiber. The boutons contain the synapses and there are more in larger boutons. At the neuromuscular junction, glial cells are absent so that the plasma membranes of nerve and muscle lie close together, separated by a synaptic gap of about 30 nm (Fig. 10.8). The terminal axoplasm contains synaptic vesicles, comparable with those in the presynaptic

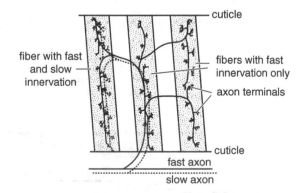

Figure 10.7 Multiterminal and polyneuronal innervation. Diagram illustrating the innervation of three fibers of a muscle unit. All three receive branches of the fast axon, while one (on the left) also has endings from the slow axon (after Hoyle, 1974).

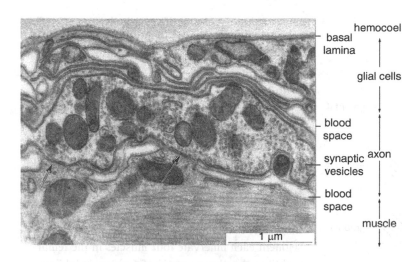

basal lamina — hemocoel

glial cells

blood space

synaptic vesicles — axon

blood space

muscle

Figure 10.8 Electron micrograph of a nerve–muscle junction. Arrows show points at which the glial sheath of the neuron is absent, permitting close contact between the axon terminal and the muscle membrane (after Smith and Treherne, 1963).

terminal of a synapse in the central nervous system. They vary in diameter from 20 nm to 60 nm.

In muscles with a single, discrete function, such as some of the indirect flight muscles, the fibers are innervated by a single axon, but it is common for a single muscle fiber to be innervated by more than one motor axon. Such multiple innervations are called polyneuronal and allow the muscle to function more variably. Some examples of flight muscles in which at least some fibers are innervated by more than one axon are given in Fig. 9.16 and Table 9.1. The locust extensor tibiae muscle is an example of a muscle made of a large number of fibers differing in their innervation. Because this muscle is of primary importance in jumping, 68% of the fibers are innervated only by a single axon, the "fast" axon (see below and Table 10.1). Most other fibers are also innervated by this axon, but also receive inputs from one or two others. A small fraction, about 8%, of the fibers are not innervated by the "fast" axon, but by a separate "slow" axon, usually accompanied by a second axon. The additional axons may have inhibitory or more subtle modulatory effects on the activity of the muscle (see below).

While it is usual for the different neurons innervating a single fiber to have qualitatively or quantitatively different effects, there are examples where this is not so. The transverse sternal muscles in

the abdomen of the bushcricket, *Decticus*, receive input from three excitatory neurons with similar properties (see below), and one of these muscles in the cricket, *Gryllus*, is innervated by four neurons with similar effects.

Different units of a muscle sometimes serve different functions; in this case they have completely separate nerve supplies. Thus, the posterior part of the basalar muscle of the beetle, *Oryctes*, is concerned only with wing depression and has only a fast innervation, but the anterior part also controls wing twisting and its innervation is complex, consisting of up to four axons, one of which is inhibitory.

The cell bodies of the motor neurons controlling skeletal muscles are usually in the ganglion of the segment in which the muscle occurs. Sometimes, however, a muscle is innervated by a motor neuron with its cell body in a different ganglion. An extreme example of this occurs in the abdominal transverse sternal muscles of crickets and bushcrickets. These muscles are continuous across the midline, but the nerve supply of each side is separate. In the bushcricket, *Decticus*, each side of the muscles in abdominal segments 4–7 is innervated by excitatory neurons from three ganglia. In addition, each side of each muscle receives input from an inhibitory cell arising in the ganglion of its own segment (Fig. 10.9).

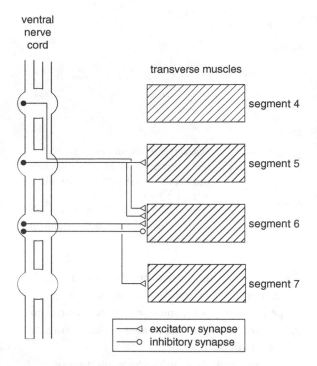

Figure 10.9 Polyneuronal innervation. Innervation of a transverse abdominal muscle of *Teleogryllus* by neurons with cell bodies in different segmental ganglia. Muscles of one side only shown; muscles on the other side of the body are similarly innervated. The complete innervation of the muscle in segment 6 is shown. Each of the other muscles receives a similar innervation. Some neurons innervate muscles in two segments (after Consoulas *et al.*, 1993).

Sometimes a single neuron innervates more than one muscle. This is most likely to occur where different muscles commonly act in unison. For example, each of the meso- and metathoracic coxae of *Periplaneta* has four depressor muscles (Fig. 8.6). A single fast axon innervates all four, although it does not innervate all units in two of the muscles (135d,e). All the units of two muscles (135d,e) are innervated by a single slow axon, and some are also innervated by inhibitory axons, one of which goes to most of the fibers, the distribution of the others being more restricted. The innervation of sternal muscles in *Decticus* and *Teleogryllus* (Fig. 10.9) is another example of one neuron innervating different muscles, in this case in different body segments. This arrangement does not mean that activity of the muscles is linked in a fixed way, because they often receive additional input from separate inhibitory neurons.

Extreme examples of a neuron innervating more than one muscle are the common inhibitors in the locust meso- and metathoracic ganglia. These neurons innervate the 12 and 13 muscles moving the middle and hindlegs, respectively (Fig. 10.10). These include muscles that normally act antagonistically to each other and which have separate excitatory innervation. Two other inhibitory neurons in each segment each innervate four muscles in the femur and tibia. Another example of a common inhibitor has its cell body in the brain of the cricket. It innervates six out of seven antennal muscles.

Many of the muscles of the foregut are innervated by neurons with cell bodies in the ganglia of the stomodeal nervous system.

10.3.2 Excitation of the muscle

Activation of the muscle fiber With the exception of some visceral muscles, muscles are stimulated to contract by the arrival of an action potential at the neuromuscular junctions (NMJs). For skeletal muscles L-glutamate is the usual chemical transmitter across the synaptic gap and this also appears to be true with visceral muscles, though L-aspartate may do so at some NMJs. As at other chemical synapses, the transmitter is present in synaptic vesicles at the nerve ending. While some spontaneous vesicle and hence transmitter discharge normally occurs into the synaptic gap, the rate of vesicle release is greatly enhanced by the arrival of the action potential.

The unstimulated muscle fiber has a difference in electrical potential across the sarcolemma. This resting potential is within the range 30–70 mV, the inside being negative with respect to the outside. It depends largely on the maintenance of an excess of

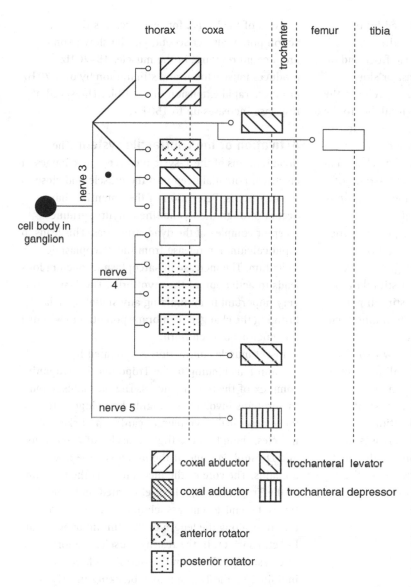

thorax · coxa · trochanter · femur · tibia

nerve 3

cell body in ganglion

nerve 4

nerve 5

▨ coxal abductor   ◨ trochanteral levator

▧ coxal adductor   ⦚ trochanteral depressor

⊠ anterior rotator

⠿ posterior rotator

Figure 10.10 Innervation of many muscles by a single neuron. Muscles innervated by the common inhibitor neuron of the metathoracic leg of a locust. Each rectangle represents a muscle with its origin in the segment to the left and insertion in the segment to the right. Nerves 3, 4 and 5 are nerves emanating from the metathoracic ganglion. The axon of the inhibitor neuron has a branch in each nerve. Note that the neuron innervates muscles that are antagonistic to each other. Unshaded muscles are those whose antagonists are not innervated by the neuron (after Hale and Burrows, 1985).

potassium ions inside the membrane associated with an inflow of chloride ions. The resting potential is generally maintained by an energy-requiring sodium/potassium pump and the passive movement of chloride ions. In larval *Spodoptera* (Lepidoptera) it appears to be maintained by an H^+/K^+-ATPase together with a K^+/Cl^- cotransporter system.

Excitatory transmitter arrival at the postsynaptic membrane of the NMJ changes its permeability, leading to a rise (that is, a depolarization) in the muscle membrane potential. The inward current producing the depolarization is carried by calcium ions. The short-lived increase in potential produced by these changes is called the postsynaptic potential. The postsynaptic potential spreads from the synapse but decreases rapidly; its effect is therefore localized and large numbers of nerve endings are necessary to stimulate the whole fiber (multiterminal innervation).

The size of the muscle twitch produced by the excitatory action potential depends on the anatomical characteristics of the muscle fiber and on whether stimulation occurs via the fast or slow axons. The terms "fast" and "slow" do not refer to the speed of conduction of the action potential, but to the size of postsynaptic potential, and hence muscle twitch, that is produced. The difference probably resides in the amount of neurotransmitter released at the nerve–muscle junction following the arrival of the action potential. Stimulation via the fast axon is presumed to release a large amount of neurotransmitter and produces a large postsynaptic potential of consistent size followed by a brief, powerful contraction of the muscle (Fig. 10.11a). Contractions tend to fuse if the rate of stimulation exceeds ten per second and at 20–25 stimuli per second the muscle undergoes a smooth, maintained contraction: it is in a state of tetanus.

A single action potential from the slow axon, on the other hand, probably releases a small amount of neurotransmitter and produces only a small postsynaptic potential followed by a very small twitch. With increasing frequency of action potentials, the postsynaptic potential increases in size (Fig. 10.11b) and the velocity and force with which the muscle contracts increases progressively; the response is said to be graded. In the extensor tibiae

muscles of the locust, for instance, less than five action potentials per second via the slow axon produce no response in the muscle, 15–20 Hz produces muscle tonus and stimulation by over 70 Hz produces rapid extension of the tibia. The speed of response increases up to 150 Hz.

Activation of the contractile system The invaginations of the T-system convey the changes in electrical potential deep into the muscle and close to the myofibrils. Perception of the potential change occurs through the Ryanodine/dihydropyridine receptor complex at the dyad structures. This leads to rapid calcium ion release from the sarcoplasmic reticulum. The increase in sarcoplasmic calcium ions leads to activation of the myofibrils. The T-system is very important in minimizing any signaling delay by bringing the changed membrane potential to within a few microns of each fibril.

Insect muscle contraction is activated by calcium ions binding to the Tropomyosin–Troponin complex of the thin filaments. The mechanism and the proteins involved are highly homologous to those described in vertebrate cardiac and skeletal muscles, though in the flight muscles of *Lethocerus* and *Drosophila* some of the proteins have novel domains. The core of the mechanism is the position of the Tropomyosin, a long elongated molecule that forms, by end-to-end association, a helical strand that runs along the length of the thin filament on the F-Actin surface. In the resting muscle, the formation of crossbridges between the Actin and Myosin is inhibited by the Tropomyosin blocking the Myosin binding sites on the Actin. Tropomyosin is held there by the Troponin complex. The configuration of the complex is altered by the binding of calcium to the Troponin C component, so that Tropomyosin can move across the Actin surface. Since Tropomyosin no longer blocks their binding sites Myosin heads can now engage in crossbridge cycling and contraction occurs.

A cycle of muscle contraction and relaxation is associated with calcium release and then its

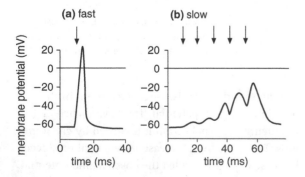

Figure 10.11 Electrical changes at the muscle membrane following stimulation by (a) a fast axon, and (b) a slow axon. Arrows show times of neural stimulation (after Hoyle, 1974).

sequestration. The rapid removal of calcium from the system is associated with the well-developed sarcoplasmic reticulum and, in general, twitch duration is inversely correlated with the quantity of sarcoplasmic reticulum present (Fig. 10.12a).

In asynchronous muscles, best known from the oscillatory indirect flight muscle systems of many insects and most studied in *Lethocerus*, the release of calcium ions is required, but not sufficient. These muscles require strain activation. That is, in addition to increased calcium ion concentrations the muscle must also be activated by an applied strain, which in vivo usually originates by the contraction of antagonistic muscles. Once activated these muscles will continue to contract, providing the oscillatory strain continues to be applied and a sufficient calcium ion concentration is maintained within the fiber by the asynchronous neural stimulation. Strain activation is detectable (Fig. 10.13) as development of increased tension following, after a characteristic delay, an applied strain (a stretch). As this response is seen in isolated demembranated fibers it is clearly an inherent property of the sarcomeres themselves. The kinetics of delayed tension development correlate with the wingbeat frequency of the insects from which the fibers originate, so the contractile system is "tuned" to the natural oscillatory frequency of the wingbeat (see Section 10.5.2).

Inhibition of muscle contraction In addition to the normal excitatory innervation, some muscle fibers of some muscles are innervated by inhibitory neurons. At an inhibitory nerve–muscle junction, a neural transmitter, probably γ-aminobutyric acid (GABA), is released. It causes a change in permeability at the postsynaptic membrane, but, unlike the process occurring at an excitatory synapse, results in an influx of chloride ions. As a result, the membrane potential becomes even more negative, the membrane is hyperpolarized and the fiber tension decreases.

Neuromodulation Many muscles, in addition to being innervated by excitatory and perhaps also inhibitory neurons, receive input from neurons that release compounds modifying the muscle's response to normal excitation. Four chemicals have been commonly identified as such neuromodulators: octopamine, 5-hydroxytryptamine (serotonin), proctolin and FMRFamide-containing peptides. These have frequently been studied by their effects on visceral muscle, especially that of male and female reproductive systems, and associated behaviors. While these chemicals are released at

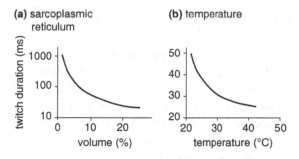

(a) sarcoplasmic reticulum   **(b)** temperature

Figure 10.12 Twitch duration. (a) Relationship with the degree of development of the sarcoplasmic reticulum in various insect muscles (after Josephson, 1975). (b) Effect of temperature in the locust dorsal longitudinal flight muscle (after Neville and Weis-Fogh, 1963).

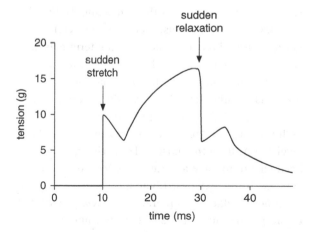

Figure 10.13 Stretch activation. The effects of sudden small changes in length on the tension developed by asynchronous muscle (after Pringle, 1965).

neuromuscular junctions, they may all also have humoral roles through release generally into the hemolymph.

The octopamine-producing cells are situated in the midline of a ganglion and are unpaired, their axons branching to innervate muscles on either side of the body. Because of their positions they are called dorsal, unpaired, median cells (abbreviated to DUM) with a suffix to indicate the muscles that they innervate. Thus: DUMeti innervates the extensor tibiae muscle in the hindleg of a grasshopper; DUMdl innervates the dorsal longitudinal muscle; and DUMovi the muscles of the oviduct. Octopaminergic cells are also known to supply the antennal muscles and, in *Periplaneta*, the muscles of the male accessory glands. Not all DUM cells are concerned with modulating muscular activity and, in the desert locust, *Schistocerca*, only eight out of a total of 90 DUM cells in the metathoracic ganglion are known to terminate at muscles.

The axons from DUM cells commonly accompany motor axons to the muscles. Their terminal branches form series of swellings, called varicosities, which contain vesicles varying in diameter from 60 nm to 230 nm and with an electron-dense core (unlike synaptic vesicles which are electron-lucent). They may or may not contain octopamine, but are clearly related to the secretion of the compound. Unlike the terminals of motor axons, there are no discrete nerve–muscle junctions and the axon terminals are separated from the muscle by thickened sarcolemma.

Octopamine may have its effect both pre- and postsynaptically (Fig. 10.14), possibly through causing more of the normal neurotransmitter, L-glutamate, to be released, and it may elevate the level of cAMP in the muscle. These changes have been shown to have a variety of effects on muscle activity. In the extensor tibiae muscle of Stenopelmatidae (Orthoptera) it produces a sustained tension. Intrinsic rhythms in the same muscle in grasshoppers are eliminated by octopamine, although its effects are dependent on the level of activity of the slow axon to the muscles. In the dorsal longitudinal

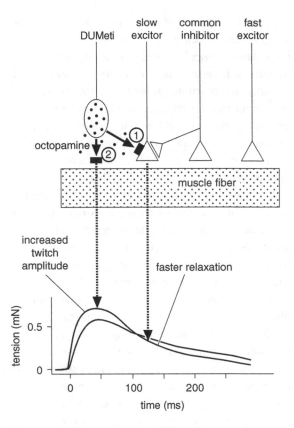

Figure 10.14 Neuromodulation. Effect of octopamine on muscle activity. The effect may be presynaptic (1) or postsynaptic (2). The latter increases the speed and amplitude of the twitch, the former produces faster relaxation (after Evans, 1985).

flight muscles, octopamine increases the force generated when the muscle contracts and it also increases the rates of contraction and relaxation. In locusts, octopamine does not induce these changes after the teneral period. A lack of octopamine (or octopamine and tyramine) in mutant *Drosophila* reduces the force produced in jumping.

Clearly, octopamine is likely to have a very important role in modulating muscular activity during walking, jumping and flying, contributing to the behavioral versatility of insects. However, there is no certain information on its real role in the whole insect.

Cells which produce serotonin and which have axons to the muscles have been described in a variety of insects including locusts, a cricket, a cockroach and *Drosophila*. Effects have been seen in the oviducts of *Locusta*. In *Periplaneta* the mouthparts have two or three pairs of serotonin-producing neurons in the subesophageal ganglion. Their axons branch to form a fine network over the surfaces of all the nerves to the mouthparts, forming a neurohemal organ. In *Locusta* the homologous fibers extend over the surface of the mandibular adductor muscles and some labral and antennal muscles. Serotinergic neurons also run to the muscles of the fore-, mid- and hindgut in the locust and cricket, and to the muscles of the reproductive system.

In *Locusta* the serotinergic neurons of the mouthpart muscles are active during feeding. The serotonin they release is believed to modulate the activity of the muscles since, in vitro, it increases the amplitude and rate of contraction and the rate of relaxation. Like octopamine, serotonin increases cAMP in the muscle. Caterpillars do not have serotinergic neurons immediately associated with the muscles, but there may be a serotinergic neurohemal organ in the head which serves a similar function.

Proctolin, a short neuropeptide (RYLPT), is present in some slow motor neurons, apparently being released at the same time as the principal neurotransmitter. It may act presynaptically, by increasing the rate at which the transmitter is released, and postsynaptically, enhancing the tension produced by neural stimulation. It is known to be present in some axons to the antennal muscles of *Gryllus*, *Locusta* and *Periplaneta* and to an opener muscle of the ovipositor of *Locusta*. Neurons innervating a variety of visceral muscles in the reproductive system and alimentary canal of *Rhodnius* and *Locusta* are proctolinergic. Proctolin also stimulates heart rates in a variety of insects.

FMRFamide-containing peptides are found widely throughout the animal kingdom and regulate a multitude of physiological activities, including visceral and cardiac muscles in insects. FMRFamides can be divided into several different groups based on their structure and function. Such complexity may require for their resolution the genetic approaches available with *Drosophila*, where they are encoded by a number of genes.

Modulation of muscle activity by these and perhaps other compounds is probably a widespread, possibly universal, phenomenon in insects. Even where muscles do not receive a direct neural supply of the compounds, it is likely that they are affected humorally by their presence in the hemolymph.

Control of visceral muscles The principles of muscle control are the same in innervated visceral muscles as in skeletal muscles. L-glutamate may be involved as a neurotransmitter, but it is conceivable that different transmitters are involved in different muscles.

In some insects, the heart is not innervated and the contractions of its muscles are purely myogenic. This does not mean that they are uncontrolled, and, for example, in *Manduca* it is known that heart muscle activity is modulated by a neurosecretion from the corpora cardiaca. Some of the neuromodulatory chemicals mentioned above have effects on heart muscle activity; some of them may well do so humorally in all non-innervated muscles, though other blood-borne factors may yet remain undiscovered.

Myogenic contractions are commonly slow and rhythmic, but fast myogenic contractions also occur in some muscles at some times. These are produced by action potentials generated spontaneously within some muscle fibers. Not all the fibers in a muscle appear able to produce these action potentials, and electrical activity spreads from cell to cell, decreasing as the distance from the active cell increases.

Some muscles exhibit both neurogenic and myogenic contractions. This is the situation in muscles associated with the oviducts of orthopterans and is likely to be widespread.

10.4 Energetics of muscle contraction

10.4.1 Definitions

Force is an influence causing a mass to change its state of motion, so that it accelerates or decelerates. A muscle generates a force as it contracts because of the resistance of the object being moved. The unit of force is a Newton (N). $N = kg\,m\,s^{-2}$.

Tension is produced by opposing forces pulling on an object. Tension in a muscle normally rises to a peak as the muscle begins to shorten and then falls again as shortening continues. If there is no measurable change in length of the muscle, it is said to be contracting *isometrically*. During isometric contraction, tension increases to a maximum. If the muscle decreases in length, while tension remains constant, the contraction is said to be *isotonic*.

Work is the application of a force over a distance, or a measure of the energy transferred by a force.

Energy is the ability to do work. The unit of energy is the same as the unit of work, the joule (J). $J = kg\,m^2\,s^{-2}$.

Power is the rate of doing work, or the rate at which energy is supplied. The unit of power is the watt (W). $W = kg\,m^2\,s^{-3}$.

10.4.2 Tension and force

The tension exerted by insect muscles is not exceptional. For instance, the mandibular muscles of various insects exert tensions of $3.6–6.9\,g\,cm^2$, and the extensor tibiae muscle of *Decticus* (Orthoptera) is $5–9\,g\,cm^2$ compared with the values of $6–10\,g\,cm^2$ in humans.

Because a muscle has intrinsic elasticity, tension does not fall to zero in the absence of stimulation. Some of this elasticity is attributed to the muscle attachments to the cuticle and some to the sarcolemma, but the greater part is due to elastic elements in the contractile system itself. Energy is stored in this elastic system when the muscle is stretched, but the energy stored depends on the stiffness of the system. Thus flight muscles, and especially fibrillar muscles, as studied in *Lethocerus*, *Drosophila* and other insects, have a much higher elastic storage because they are stiffer than other muscles.

The force exerted by a muscle is proportional to its cross-sectional area and, in general, this is not very great in insects. In some muscles, however, such as the extensor tibiae of a locust, a considerable cross-sectional area is achieved by an oblique insertion of the muscle fibers into a large apodeme (see Fig. 10.4). As a result, this muscle can exert a force of up to 15 N.

10.4.3 Twitch duration

The duration of each muscle twitch, the time for it to shorten and relax, is temperature dependent (Fig. 10.12b). This is of critical importance in flight because twitch duration limits the rate at which antagonistic muscles can operate efficiently. To produce the aerodynamic forces necessary for flight, the wings must beat at a certain minimum rate. If the muscle twitch durations of antagonistic muscles overlap to a significant extent, some of the muscle energy is wasted. Efficient flight by *Schistocerca* requires a wingbeat frequency of about 20 Hz, with a period of about 50 ms. Only above 30°C is twitch duration short enough to avoid significant overlap of antagonistic muscle twitches (Fig. 10.12b) and it is therefore not surprising that sustained flight only occurs at relatively high temperatures. Some large insects, such as *Lethocerus*, go through extensive "futile" flight muscle contractions to raise their body temperature prior to flight.

Twitch duration also varies in different fiber types, being shorter in fast than slow fibers. Fast fibers (short twitch) have myofibrils that are small in cross-sectional area with a relatively large proportion of sarcoplasmic reticulum. These factors increase the rates at which calcium reaches the most distant thin filament Troponin complexes and can be resequestered (see above). Twitch duration is not correlated with fiber diameter, sarcomere length or the volume of the fiber occupied by mitochondria.

10.4.4 Power output

A muscle's mechanical efficiency is defined as the ratio of mechanical power output to metabolic energy consumption. Direct measurements of the mechanical power output of flight muscles of *Manduca* and *Bombus* (insects with synchronous and asynchronous flight muscles, respectively) give maximum values of 130 W kg^{-1} and 110 W kg^{-1}, respectively. These values indicate mechanical efficiencies of 10% or less. Not all this power is available for mechanically useful work. In a flying insect, energy is required not only to move the wing to produce aerodynamic force, but also to start it moving from an extreme up or down position, and for braking at the end of a half stroke (see Fig. 10.20). The wing's inertia toward the end of the half stroke stretches the antagonistic muscles, and this energy will be stored in elastic elements within the muscle, including the stiff connecting filaments between the Z-discs and the ends of the thick filaments. As flight muscles only shorten by very small amounts, approximately 2% of their length, some Actomyosin crossbridges may remain attached throughout the cycle of elongation and contraction and function as part of the elastic element energy storage. Direct evidence for this has not been found. Energy may also be stored in elastic elements of the wing hinge (see Chapter 9), as this is stretched toward the end of the half stroke, and in other thoracic cuticular structures. This stored energy then contributes to the beginning of the next half stroke.

Maximum efficiency is achieved only under certain conditions. Work output per cycle of contraction and relaxation rises to a maximum and then declines as the frequency of cycling increases. For the flight muscle of *Manduca* at 35°C, maximum efficiency occurs at a cycle frequency of about 30 Hz.

Work per cycle, at the optimal cycle frequency, increases with temperature, at least up to 40°C.

The power output of a muscle may be increased by multiple stimulation via the motor axon, and such double (or multiple) firing is commonly recorded when insects appear to produce more power. This occurs, for instance, in the double firing of the axon to the second basalar muscle of *Schistocerca* (see Fig. 9.18), which may result in the muscle more than doubling the amount of work it does. The extra force exerted varies with the timing of the second impulse relative to the first. In the basalar muscle at 40°C, the force is maximal when the second stimulus follows about 8 ms after the first. In the bushcricket *Neoconocephalus* maximum power output from a tergocoxal muscle is achieved when the motor neuron fires three times with intervals of 4 ms between spikes. However, multiple firing does not increase the power output of all muscles at their normal operating frequencies.

A great deal of power is required to lift the insect off the ground during flight. Because of the low level of muscle efficiency, most insects are able to produce sufficient power only by a high wingbeat frequency, reflecting the oscillation frequency of the flight muscles. Energy consumption by flying insects is very high. The metabolic rates of flying insects are commonly 100 times higher than those of resting insects.

10.4.5 Oxygen supply

Muscular contraction requires metabolic energy (ATP), and the muscles have a good tracheal supply. This is particularly true of the flight muscles, where the respiratory system is specialized to maintain the supply of oxygen to the muscles during flight (Section 17.1.4). In most muscles the tracheoles are in close contact with the outside of the muscle fiber. This arrangement provides an adequate supply of oxygen to relatively small muscles or those whose oxygen demands are not high, but in the flight muscles of all insects, except perhaps those of Odonata and Blattodea, the tracheoles indent the muscle membrane, becoming functionally, but not anatomically, intracellular within the muscle fiber (see Fig. 17.1). The tracheoles follow the invaginations of the T-system and so penetrate to the center of the fibers. Fine tracheoles, less that 200 nm in diameter, branch off from the tracheoles forming the secondary supply to the muscles. They come very close to the mitochondria, and probably every mitochondrion in the flight muscles is supplied by one or two tracheoles of this tertiary system so that the distance oxygen has to pass through the tissue is reduced to a minimum. In Odonata and cockroaches, on the other hand, tracheoles remain superficial to the muscle fibers and these fibers, with a radius of about 10 μm, are believed to approach the limiting size for the diffusion of oxygen at sufficiently high rates. though it has been suggested that the flight muscles of these insects may also have an invaginating system of terminal tracheoles.

Locust flight muscles use some 80 liters $O_2\,kg^{-1}\,h^{-1}$ during flight and consumption by the flight muscles of some other insects may exceed 400 liters $O_2\,kg^{-1}\,h^{-1}$. The special adaptations of the thoracic tracheal system enable these demands to be met. In locusts, ventilation of the pterothorax (the part of the thorax containing flight muscles and to which the wings are attached) produces a supply of oxygen well in excess of needs, and the specialized system of tracheae and tracheoles in the muscles ensures that oxygen reaches the site of consumption. The fine branches of the tertiary tracheal supply to the muscles may be fluid-filled when the insect is not flying, but, in flight, the fluid is withdrawn so that air extends to the very tips of the branches close to the mitochondria. The cuticle lining these fine branches may be very permeable.

10.5 Muscular control in the intact insect

Insects have only small numbers of separate motor units in their muscles compared with vertebrates. Consequently, precision and flexibility of movement is achieved not by employing different numbers of units but by changes in the strengths of contraction of individual units. This fine control is effected through the polyneuronal innervation of the muscles and through neuromodulation.

10.5.1 Muscle tonus

The muscles of a stationary insect are not completely relaxed. As in any animal, they must maintain some degree of tension if the insect is to maintain its stance and be ready to make an immediate response. Maintaining this tension or tonus may involve three different mechanisms.

In many cases it is dependent on a low level of neural input to slow (tonic) muscle fibers. The tonic fibers of some muscles, when stimulated by a high-frequency burst from the slow axon, sustain a higher tension than before the burst at a low level of stimulation (Fig. 10.15). This is known as a "catch" tension. It is eliminated by the activity of a fast or inhibitory axon. Second, some muscles may exhibit a steady tension in the absence of neural input. This

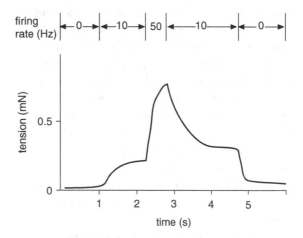

Figure 10.15 The development of "catch" tension in a muscle stimulated via a slow axon (mesothoracic extensor tibiae muscle of the desert locust, *Schistocerca*). Notice that the tension exerted when the muscle is stimulated at 10 Hz is higher after a brief burst of stimulation at 50 Hz (after Evans and Siegler, 1982).

has been demonstrated in some spiracle muscles and in the extensor tibiae muscle of the locust. Finally, some muscles are known to undergo slow rhythmic changes in tension which are myogenic in origin. These muscles also respond to neural stimulation.

10.5.2 Locomotion

Most behavioral activities result from the coordinated activity of sets of muscles. This is most obvious in locomotion, which involves the oscillation of an appendage such as a leg or a wing. For example, during slow walking by a cockroach, only the slow axon to the coxal depressor muscles is active and so only the muscles numbered 135 in Fig. 8.6 are involved; the strength and speed of contraction depends on the frequency of nerve firing. At walking speeds of more than ten cycles per second, the slow axon is reinforced by the activity of the fast axon, which also activates muscles 136 and 137. It is presumed that the inhibitory axons fire at the end of contraction and so ensure complete and rapid relaxation as the antagonistic muscle contracts. The presence of three separate inhibitors, perhaps innervating different fibers in the muscles, gives increased flexibility to the system, but the situation is complicated by the fact that they also innervate other muscles.

The control of muscle activity in jumping by locusts and the control of stridulation in grasshoppers provide other examples of the interaction between fast, slow and inhibitory axons (see Figs. 8.24, 26.16).

Skeletal muscles usually occur in antagonistic pairs, such as the extensor and flexor muscles of the tibia, and the levator and depressor muscles of the wings. Some muscles are opposed only by the elasticity of the cuticle. Thus, depression of the pretarsus is produced by a muscle, but extension results entirely from cuticle elasticity at the base of the segment. A similar mechanism operates through the mesothoracic tergal depressor of trochanter muscle in Diptera following a jump. The timbal

muscle in cicadas has no antagonist; it is stretched when the timbal buckles outwards (Section 26.3.3).

When antagonistic muscles oscillate, they are driven by motor neurons firing in antiphase (except for asynchronous flight muscles, see below). The precise timing of neural activity and the interplay of fast and slow excitatory neurons and inhibitory neurons permits a wide range of modulation. Section 8.2.3 describes neural regulation in walking insects.

The high wingbeat frequencies necessary for flight are produced by the rapid oscillation of pairs of antagonistic muscles. This is achieved despite the relatively low rate of shortening of about $40 \, \text{mm s}^{-1}$ in *Schistocerca* and only $11 \, \text{mm s}^{-1}$ in *Sarcophaga*. Three factors combine to reduce the duration of the muscle twitch and so make flight possible: (1) the loading of the muscle; (2) the temperature of the muscle; and (3) the very slight contraction necessary to move the wing.

A maximum rate of shortening is achieved if the tension and loading of the muscle are maximal at the beginning of contraction. Loading of flight muscles involves the inertia of the wings, mechanical leverage of the wings (which changes in the course of a stroke), damping of the movement of the wings by the air and elastic loading due to the straining of the thorax and stretching the antagonistic muscles. Inertia is highest at the beginning of the stroke when the wing may be momentarily stationary, or even moving in the opposite direction (Fig. 10.16). Mechanical leverage of the wings and loading due to the elasticity of the thorax are also greatest at this time and so will favor a high rate of shortening.

Finally, the wings are articulated with the thorax in such a way that only a very small contraction of the muscles is necessary to produce a large movement of the wing: The flight muscles of *Sarcophaga*, for instance, only shorten by 1–2% in the course of a wing stroke. As a result, the muscle twitch is brief despite the low rate of shortening.

In insects with synchronous flight muscles, each contraction of the flight muscles is produced by the arrival of a nerve impulse and the motor neurons to

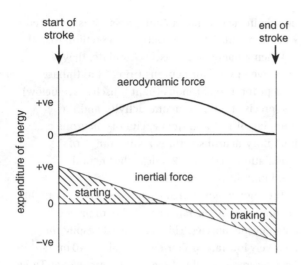

Figure 10.16 The forces involved in moving a wing. At the start of the stroke, most energy is used overcoming the inertia of the wing. In midstroke, useful aerodynamic forces are produced. At the end of the stroke, the wing has considerable inertia (after Alexander, 1995).

the antagonistic muscles fire approximately in antiphase (Fig. 10.17a). These muscles occur in Odonata and orthopteroid insects, as well as in Neuroptera, Trichoptera and Lepidoptera among the holometabolous insects. Wingbeat frequencies are generally less than 50 Hz, although in some Lepidoptera the frequency approaches 100 Hz.

These muscles are generally activated by one or two action potentials in the motor axons, but in some Lepidoptera with very low wingbeat frequencies (5–10 Hz) the motor burst is characteristically 3–7 action potentials. The occurrence of two or more action potentials is associated with high lift requirements by the insects and results in higher power output by the muscles. Where longer bursts occur, the longer period of excitation keeps the muscles contracting for relatively long periods, and burst length is inversely proportional to wingbeat frequency. Maneuvers in flight are achieved by alterations in the power output of specific muscles and by changes in the times at which different direct muscles, which affect wing twisting, are activated

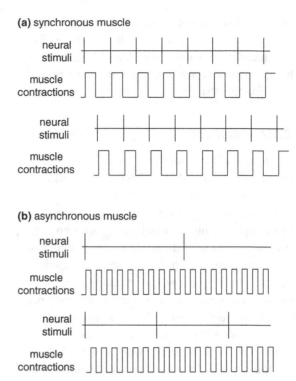

Figure 10.17 Contraction of antagonistic muscle pairs (as in flight muscles) in relation to neural stimulation. (a) Synchronous muscle. Each contraction results from the arrival of an action potential. The stimuli to antagonistic muscles are in antiphase. (b) Asynchronous muscle. Muscle oscillations occur independently of the arrival of action potentials which serve only to keep the muscle in an activated state. The rate of neural input to the antagonistic muscles may differ, as in this example.

relative to each other. Fig. 9.18 illustrates some aspects of muscle coordination in flight.

Asynchronous muscle is an insect invention not known to occur elsewhere. Wingbeat frequency is determined primarily by the mechanical properties of the thorax and the asynchronous muscles themselves. It is a characteristic of these muscles that neural stimulation is asynchronous with the contraction and that contraction frequency exceeds the frequency of nerve impulses (Fig. 10.17b). Neural inputs serve to maintain the muscles in an active state by maintaining a sufficient sarcoplasmic free calcium ion concentration (see Section 10.3.2).

Each neural signal to an asynchronous muscle normally consists of a single spike; double firing does not usually occur. The frequency of the single spikes varies from 5 Hz to about 25 Hz in different insects, while wingbeat frequency varies from about 50 Hz to over 200 Hz. The ratio of action potentials to wing cycles varies from about 1:5 to 1:40, but the rate of firing to antagonistic muscles is not necessarily the same. For example, during flight in the honey bee the dorsal longitudinal muscles are stimulated at a lower frequency (about 86%) than the dorso-ventral muscles. Nevertheless, an increase in the firing rate of the motor neurons does sometimes correlate with an increase in wingbeat frequency.

Asynchronous flight muscles move the wings indirectly through distortions of the thoracic cuticle (see Chapter 9). The contraction of one set of flight muscles distorts the cuticle. This moves the wings and applies a lengthening force to the antagonistic muscles. This increase in passive tension within the antagonistic muscle leads to its strain activation, through which it responds, after a characteristic delay, by developing its active tension. In the first muscle shortening is followed by a transient decrease in its ability to generate force (strain deactivation). Activation of the antagonistic muscle now leads to a new cycle. Thus a self-sustaining sequence of strain activation, delayed tension development and strain deactivation out of phase, communicated through the cuticle, occurs between the antagonistic muscles. It produces an oscillatory behavior that is maintained as long as adequate calcium ion concentrations are maintained in the sarcoplasm. The oscillation frequency depends on the passive stiffness of the thoracic cuticle, that of the muscles themselves, the damping effect due to wing movement and on the kinetics of muscle contraction and the delays in strain activation/deactivation.

The oscillatory behavior of indirect, asynchronous flight muscles allows very efficient mechanical behavior since oscillatory systems can store considerable energy and be maintained by the addition of minimal energy. It is also metabolically efficient since there is no requirement to pump calcium back into the sarcoplasmic reticulum after each contraction. This vastly reduces the ATP required, so that a larger fraction of it can be used to power flight. While asynchronous muscles may have enabled the evolution of high wingbeat frequencies found in small insects, they may have initially evolved in larger insects where their more efficient energetics offered advantages.

10.6 Changes during development

Muscle development, or myogenesis, as in all animals begins with the specification of myoblasts from mesodermal cells during the embryonic period. Two types of myoblasts are known, pioneer or founder myoblasts, which specify the type and position of the developing fibers, and fusion-competent myoblasts which fuse with the founder cells to form the multinucleate fibers. Once formed, the embryonic muscle pattern remains constant during later stages. Fibers grow considerably as the organism grows but by increases in fiber volume, not by nuclear divisions or by further myoblast fusion.

There are major differences in muscle development between holometabolous and hemimetabolous insects because of pupal metamorphosis in the former. Much is known about insect muscle development from studies of *Drosophila*, a holometabolous insect. During embryonic myogenesis in holometabolous insects a large fraction of the myoblasts in each segment is put aside as separate "nests" of cells or within the imaginal discs and increase in number but do not differentiate. At metamorphosis, most larval muscles are destroyed. A few remain, such as the larval muscles which act as templates for dorso-longitudinal flight muscle myogenesis or continue to function during eclosion and then degenerate, such as the "persistent" larval muscles (see below). All other muscles are formed *de novo* in the pupa, including the dorso-ventral flight muscles, from the division and fusion of the myoblasts set aside during embryogenesis. As during embryonic myogenesis, muscle type, pattern and size are determined by founder/pioneer cells. Muscle development in hemimetabolous insects is much less well understood, but following embryogenesis seems to be largely one of growth and the emergence of specialized function during later nymphal stages.

Muscles often continue development in the early days of adult life and this is reflected in the insect's inability to undertake extended flights during this period. The period after eclosion during which an adult insect's ability to fly is not fully developed is called the teneral period. Flight muscles may regress in adults that have completed their flight period. Some muscles may be associated specifically with hatching or with ecdysis, regressing or completely disappearing after a brief period of use. These changes are considered in this section, but see Section 15.3 for changes occurring at metamorphosis.

10.6.1 Growth during the development of hemimetabolous insects

In hemimetabolous insects, muscles increase in size throughout larval/nymphal development and the increase continues for the first few days after adult eclosion (Fig. 10.18). Growth during the larval/nymphal stages results primarily from an increase in the number of fibers forming the muscles, whereas growth after eclosion is produced by an increase in the size of the fibers. This difference is probably a reflection of the fact that new attachments to the cuticle are normally formed at molts (see above). The difference in the number of fibers between early-stage larvae and adults is relatively small in muscles which perform the same functions throughout life, but much more marked in the case of flight muscles.

An increase in fiber size involves all the elements of the muscle: myofibrils, mitochondria and sarcoplasmic reticulum (Fig. 10.18c,d), but the relative rates of increase of these components varies from muscle to muscle. In a tergocoxal muscle of the grasshopper *Schistocerca nitens*, all the components increase in proportion during larval development, but in the metathoracic dorsal longitudinal muscle the percentage of volume occupied by myofibrils approximately doubles during the last three larval stages. In both muscles, the proportion of the muscle volume occupied by mitochondria increases during the first two weeks of adult life. Fiber growth in young adults involves an increase in the size of the myofibrils and, in the dorsal longitudinal muscle of *Locusta*, a doubling of the number of filaments in each fibril.

These anatomical changes are sometimes accompanied by changes in enzyme activity (Fig. 10.19). The activity of enzymes involved in flight metabolism increases through the final stages of muscle development, tending to reach a plateau about seven days after the final molt. At the same time, the activity of lactate dehydrogenase declines.

10.6.2 Post-eclosion growth of the flight muscles of holometabolous insects

At eclosion, when the cuticle expands, the flight muscles of cyclorrhaphous flies rapidly increase in length by 25% or more. The change in length is correlated with a comparable length increase in the sarcomeres and their filaments. The number of filaments within a myofibril does not increase at this time. Similar changes in flight muscles length may occur in all insects that exhibit a marked increase in body size at eclosion, but is not the case in *Drosophila*, another dipteran.

Increases in flight muscle volume continue over the days following eclosion in cyclorrhaphous flies and some Hymenoptera. In the tsetse fly, *Glossina*, for example, post-eclosion changes comparable with those described in Orthoptera occur, but flight muscle maturation depends on the insects taking several blood meals. The mass of the muscles increases considerably, mostly due to an increase in the

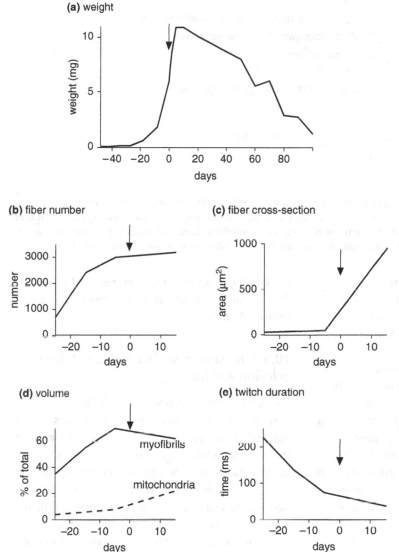

Figure 10.18 Muscle growth in hemimetabolous insects. Changes in the dorsal longitudinal flight muscle associated with the development of flight behavior. Arrow (day 0) indicates the time of adult eclosion. The period before eclosion includes several larval stadia. (a) Change in wet weight of the muscle in the cricket, *Teleogryllus* (after Ready and Josephson, 1982). (b) Increase in fiber number in the grasshopper, *Schistocerca nitens*. (c) Increase in cross-sectional area of fibers in the grasshopper, *S. nitens*. (d) Changes in the percentage of muscle occupied by mitochondria and myofibrils in the grasshopper, *S. nitens*. (e) Decrease in twitch duration in the grasshopper, *S. nitens*.

Notice that in (b)–(d) the final two points represent the mid-point of the last larval stadium and day 15 of adult life. For this reason they do not show unequivocally that growth occurred after the final molt. Data from other sources, however, show that a marked increase in fiber cross-section and changes in the volume densities of the components do occur in the early adult (after Mizisin and Ready, 1986).

absolute size and relative proportion of each fiber occupied by the myofibrils (Fig. 10.20). Within each myofibril, the number of thick filaments more than doubles. The mitochondria also increase in size. At the same time the wingbeat frequency increases, partly due to changes in the cuticle but also due to increased muscle stiffness. Comparable changes have been documented in other flies, and in the honey bee, *Apis*, the mitochondrial volume increases 12-fold during the first three weeks of adult life. It is not the case in *Drosophila* where flies can fly within a few hours after eclosion. Similarly, in Lepidoptera no post-eclosion changes occur; the flight muscles are fully functional immediately after wing expansion.

10.6.3 Regressive changes in flight muscles

The flight muscles of the reproductive castes of termites and ants regress completely after the nuptial flight, beginning when the wings are shed. The

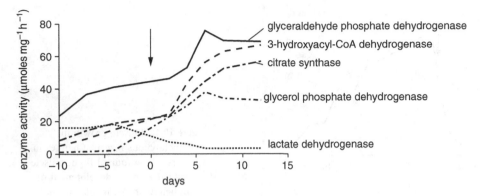

Figure 10.19 Enzyme activity in the respiratory pathways in the dorsal longitudinal flight muscle of the locust, *Locusta*, in relation to the development of flight behavior. Arrow at day 0 indicates adult eclosion. About seven days after eclosion, the insect is capable of sustained flight. Enzyme activity is expressed as μmoles substrate digested per mg muscle protein per hour. The actual values for lactate dehydrogenase are one-tenth of those shown (after Beenakkers *et al.*, 1975).

products from the muscles may contribute to oogenesis in the period before workers are available to feed the queen. Flight muscle degeneration following dealation (casting the wings) also occurs in some crickets.

In some other insects – including aphid species, other Hemiptera, some crickets and bark beetles – flight muscle histolysis follows a dispersal flight even though the wings are not shed. Among bark beetles, such degeneration may be followed, after a period of reproduction, by redevelopment of the muscles, and further flight. In the Colorado potato beetle, temporary regression of the flight muscles is associated with reproductive diapause.

In general, muscle histolysis is correlated with, and may be caused by, the increase in juvenile hormone titer that regulates reproductive development. In *Leptinotarsa*, however, the converse is true. Here, the insects enter diapause, and the flight muscles degenerate, when the juvenile hormone titer is low; they regenerate when it rises. In the aphid *Acyrthosiphon*, and probably in other insects, breakdown of the indirect flight muscles is a genetically programmed event.

Regression of the flight muscles occurs in older insects of all species. It has been studied in some

crickets (Fig. 10.18a), in mosquitoes (*Aedes*) and *Drosophila*. As the size of the mitochondria declines, so does flight activity.

10.6.4 Muscles associated with hatching, eclosion and molting

Muscles specifically associated with hatching are known to occur in grasshoppers and crickets. Newly hatched grasshopper larvae have a pair of ampullae in the neck membrane which are used when the insect escapes from the egg shell and makes its way to the surface of the soil. Once the insect reaches the surface, the ampullae are no longer used and their associated retractor muscles degenerate. The muscles associated with molting (see below) are probably also active at this time, but they do not degenerate after hatching. In *Rhodnius* the muscles that develop at each molt (see below) are also present when the insect hatches, but regress shortly afterwards.

Muscles associated with molting are known from several hemimetabolous species and may be commonly present. Their function is to facilitate ecdysis and expansion of the cuticle at molting by increasing the hydraulic pressure of the hemolymph. In grasshoppers, and probably in other insects, some of these muscles are attached to membranous regions of the cuticle and

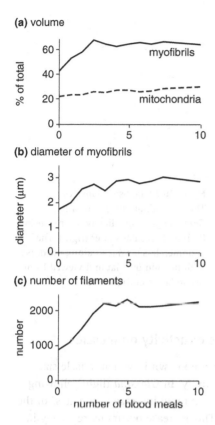

(a) volume

(b) diameter of myofibrils

(c) number of filaments

number of blood meals

Figure 10.20 Post-eclosion changes in flight muscles of the tsetse fly, *Glossina*. Muscle development is dependent on the insect having several blood meals (after Anderson and Finlayson, 1976). (a) Percentage of the muscle volume occupied by myofibrils and mitochondria. (b) Diameter of myofibrils. (c) Number of thick filaments per myofibril.

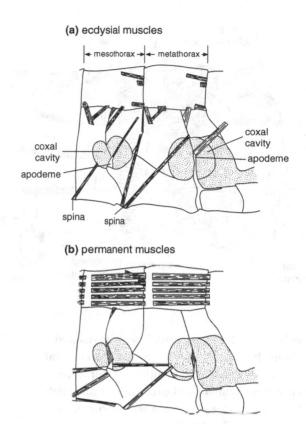

(a) ecdysial muscles

(b) permanent muscles

Figure 10.21 Molting muscles. Muscles in the pterothorax of a locust in the first larval stage. Note that the flight muscles are not well developed at this time. Membranous areas stippled (after Bernays, 1972). (a) Muscles which disappear after the final molt. (b) Muscles which remain throughout life.

they may prevent excessive ballooning of these areas so that the greatest forces are exerted on the presumptive sclerites, ensuring maximum expansion. Grasshoppers have many of these accessory muscles in the neck, thorax and abdomen, which regress after the final molt (Fig. 10.21).

In the blood-sucking bug *Rhodnius*, some ventral intersegmental muscles in the abdomen are fully differentiated only when the insect molts. Within a few days of molting they regress and lose their contractile function. They redevelop, in readiness for the next molt, when the insect has a blood meal.

Among holometabolous insects, muscles that function only at the time of eclosion are known in Diptera and Lepidoptera and they probably also occur in other groups. In the blowfly, *Sarcophaga*, and in *Drosophila* there are special muscles in the abdomen that appear to be involved in escape from the puparium and subsequent cuticle expansion (Fig. 10.22). These muscles, which are known as persistent larval muscles, are internal to the definitive muscles and extend from the front of one presumptive sclerite to the front end of the next. In newly emerged adults contraction of these muscles causes the sclerites to buckle because they are not yet

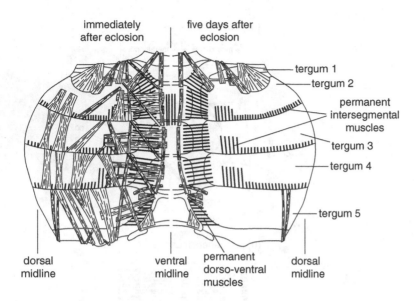

immediately
after eclosion

five days after
eclosion

tergum 1
tergum 2
permanent
intersegmental
muscles
tergum 3
tergum 4
tergum 5

dorsal
midline

ventral
midline

permanent
dorso-ventral
muscles

dorsal
midline

Figure 10.22 Eclosion muscles. The abdominal muscles of a fly, *Sarcophaga*, immediately after eclosion (left) and five days later (right). The abdomen has been split along the dorsal midline, laid out flat and viewed from inside (after Cottrell, 1962).

hardened, and the abdominal cavity is reduced in volume. It is these movements that hydrostatically cause the eversion of the ptilinum (to open the puparial operculum) and expansion of the wings. These muscles break down when the cuticle is expanded and hardened.

The first stage of breakdown of these muscles in Lepidoptera is triggered by the falling concentration of molting hormone. It may begin before eclosion, but is delayed if the motor neurons controlling the muscles are active (as they normally are at this time). Under these circumstances, breakdown does not occur until expansion of the adult wings and cuticle is complete, and the first signs of degeneration appear about five hours after eclosion. In the tobacco hornworm moth, *Manduca*, the absence of molting hormone accounts for the complete disappearance of the muscles in the days after eclosion, but in the silk moth, *Antheraea*, eclosion hormone provides the signal inducing their final degeneration. In the flesh fly *Sarcophaga*, eclosion hormone starts a slow degeneration, but some other factor triggers the rapid degeneration of thoracic eclosion muscles. The effect of the hormones is to switch on genetically programmed cell death.

10.6.5 Effects of activity on muscles

A few examples are known in which muscle size is affected by activity. In *Glossina* flight following eclosion results in a rapid increase in the mass of the flight muscles. This increase occurs more slowly in the absence of induced activity.

The hardness of food affects the size of the mandibular adductor muscles in caterpillars of the moth *Pseudaletia*. How quickly the change occurs is unknown, but insects feeding on harder food develop larger heads and larger muscles than those eating soft foods. Similar changes seem to occur in grasshoppers.

Regression of unused muscles can also occur but is not a direct consequence of lack of use. Regression of flight muscles in numerous insects follows loss of the wings and is probably controlled by an increase in the juvenile hormone titer in the hemolymph. Grasshoppers can autotomize their hindlegs, a break appearing between the trochanter and the femur. Thoracic muscles associated with this leg subsequently atrophy as the result of severing the nerve to the leg even though this nerve does not innervate the muscles that degenerate.

Summary

This chapter has described the structure of insect muscles, their innervation, contraction and functions. It is very clear that the success of insects has depended on the evolution of specialized muscle functions, some unique to the order.

- Insect muscles allow insects to walk, crawl, swim, jump (often prodigiously), pounce, "sing" and fly.

- Overall insect muscle structure, sarcomeric proteins, contraction mechanisms and regulation show very high homologies to those of all other animals.

- To produce some of their locomotory achievements insects have evolved some very specific adaptations of muscle structure, activation, energy storage and energy production.

- Perhaps the most notable development is the evolution of the asynchronous flight muscles which, through modifications of muscle structure, strain activation, energy storage (in stiff fibers) and energy production, have allowed insects of all sizes to take so successfully to the air.

- Many other important adaptations of muscle function, including supercontraction and superextension, are found widely in insects.

- Insects have widely exploited the storage of energy in the cuticle of work done by relatively slow muscles to achieve fast movements for catching prey and jumping to avoid predators.

- Much has been learned about muscle function generally from studies of insects. Much more remains to be discovered of how the specific muscle adaptations are achieved and how they provide appropriate kinetics through muscle structure, biochemistry and neural control.

Recommended reading

Kerkut, G. A. and Gilbert, L. I. (eds.) (1985). *Comprehensive Insect Physiology, Biochemistry and Pharmacology.* Oxford: Pergamon Press.

Sink, H. (ed.) (2006). *Muscle Development in Drosophila.* New York, NY: Springer/Landes Bioscience.

Usherwood, P. N. R. (ed.) (1975). *Insect Muscle.* London: Academic Press.

Vigoreaux, J. O. (ed.) (2006). *Nature's Versatile Engine: Insect Flight Muscle Inside and Out.* Austin, TX: Landes Bioscience.

References in figure captions and table

Alexander, R. M. (1995). Springs for wings. *Science* **268**, 50–51.

Anderson, M. and Finlayson, L. H. (1976). The effect of exercise on the growth of mitochondria and myofibrils in the flight muscles of the tsetse fly, *Glossina morsitans*. *Journal of Morphology* **150**, 321–326.

Beenakkers, A. M. T., van den Broek, A. T. M. and de Ronde, T. H. A. (1975). Development of catabolic pathways in insect flight muscles: a comparative study. *Journal of Insect Physiology* **21**, 849–859.

Bernays, E. A. (1972). The muscles of newly hatched *Schistocerca gregaria* larvae and their possible functions in hatching, digging and ecdysial movements (Insecta: Acrididae). *Journal of Zoology, London* **166**, 141–158.

Caveney, S. (1969). Muscle attachment related to cuticle architecture in Apterygota. *Journal of Cell Science* **4**, 541–559.

Consoulas, C., Hustert, R. and Theophilidis, G. (1993). The multisegmental motor supply to transverse muscles differs in a cricket and a bushcricket. *Journal of Experimental Biology* **185**, 335–355.

Cottrell, C. B. (1962). The imaginal ecdysis of blowflies: observations on the hydrostatic mechanisms involved in digging and expansion. *Journal of Experimental Biology* **39**, 431–448.

Evans, P. D. (1985). Octopamine. In *Comprehensive Insect Physiology, Biochemistry and Pharmacology*, vol. **11**, ed. G. A. Kerkut and L. I. Gilbert, pp. 499–530. Oxford: Pergamon Press.

Evans, P. D. and Siegler, M. V. S. (1982). Octopamine mediated relaxation of catch tension in locust skeletal muscle. *Journal of Physiology* **324**, 93–112.

Hale, J. P. and Burrows, M. (1985). Innervation patterns of inhibitory motor neurones in the thorax of the locust. *Journal of Experimental Biology* **117**, 401–413.

Hoyle, G. (1974). Neural control of skeletal muscle. In *The Physiology of Insecta*, vol. **4**, ed. M. Rockstein, pp. 176–236. New York, NY: Academic Press.

Hoyle, G. (1978). Distributions of nerve and muscle fibre types in locust jumping muscle. *Journal of Experimental Biology* **73**, 205–233.

Jorgensen, W. K. and Rice, M. J. (1983). Superextension and supercontraction in locust ovipositor muscles. *Journal of Insect Physiology* **29**, 437–448.

Josephson, R. K. (1975). Extensive and intensive factors determining the performance of striated muscle. *Journal of Experimental Zoology* **194**, 135–154.

Mizisin, A. P. and Ready, N. E. (1986). Growth and development of flight muscle in the locust (*Schistocerca nitens*, Thünberg). *Journal of Experimental Zoology* **237**, 45–55.

Neville, A. C. and Weis-Fogh, T. (1963). The effect of temperature on locust flight muscle. *Journal of Experimental Biology* **40**, 111–121.

Osborne, M. P. (1967). Supercontraction in the muscles of the blowfly larva: an ultrastructural study. *Journal of Insect Physiology* **13**, 1471–1482.

Pringle, J. W. S. (1965). Locomotion: flight. In *The Physiology of Insecta*, vol. **2**, ed. M. Rockstein, pp. 283–329. New York, NY: Academic Press.

Ready, N. E. and Josephson, R. K. (1982). Flight muscle development in a hemimetabolous insect. *Journal of Experimental Zoology* **220**, 49–56.

Smith, D. S. (1966). The organization and function of the sarcoplasmic reticulum and T-system of muscle cells. *Progress in Biophysics and Molecular Biology* **16**, 109–142.

Smith, D. S. (1968). *Insect Cells: Their Structure and Function*. Edinburgh: Oliver & Boyd.

Smith, D. S. (1972). *Muscle: A Monograph*. New York, NY: Academic Press.

Smith, D. S. (1984). The structure of insect muscles. In *Insect Ultrastructure*, vol. 2, ed. R. C. King and H. Akai, pp. 111–150. New York, NY: Plenum Press.

Smith, D. S. and Treherne, J. E. (1963). Functional aspects of the organisation of the insect nervous system. *Advances in Insect Physiology* **1**, 401–484.

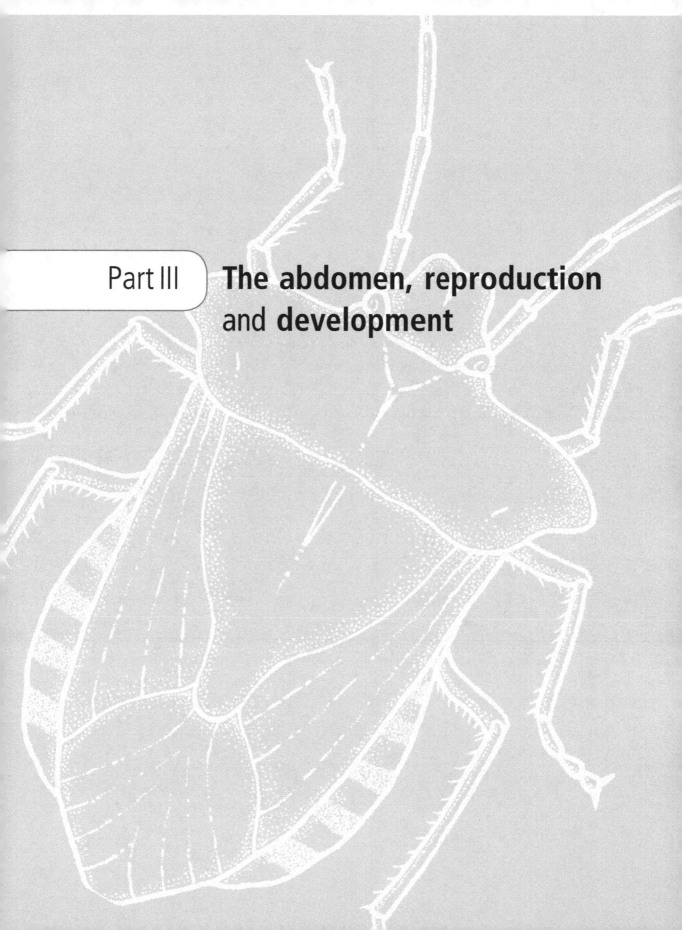

Part III   **The abdomen, reproduction**
and **development**

11 Abdomen

REVISED AND UPDATED BY **LEIGH W. SIMMONS**

INTRODUCTION

The insect abdomen is more obviously segmental in origin than either the head or the thorax, consisting of a series of similar segments, but with the posterior segments modified for mating and oviposition. In general, the abdominal segments of adult insects are without appendages except for those concerned with reproduction and a pair of terminal, usually sensory, cerci. Pregenital appendages are, however, present in Apterygota and in many larval insects, as well as in non-insectan hexapods. Aquatic larvae often have segmental gills, while many holometabolous larvae, especially among the Diptera and Lepidoptera, have lobe-like abdominal legs called prolegs. This chapter provides a general description of the insect abdomen (Section 11.1), followed by a discussion of the structure and function of abdominal appendages (Section 11.2).

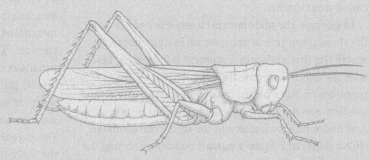

11.1 Segmentation

11.1.1 Number of segments

The basic number of segments in the abdomen is 11, plus the postsegmental telson, which bears the anus (the telson is sometimes referred to as a 12th segment). Only in adult Protura and the embryos of some hemimetabolous insects is the full complement visible. In all other instances there is some degree of reduction. The telson, if it is present at all, is generally represented only by the circumanal membrane, but in larval Odonata three small sclerites surrounding the anus may represent the telson.

In general, more segments are visible in the more generalized hemimetabolous orders than in the more specialized holometabolous insects. Thus, in Acrididae all 11 segments are visible (Fig. 11.1a), whereas in adult Muscidae only segments 2–5 are visible and segments 6–9 are normally telescoped within the others (Fig. 11.1b). Collembola are exceptional in having only six abdominal segments, even in the embryo.

The definitive number of segments is present at hatching in all hexapods except Protura. All the segments differentiate in the embryo and this type of development is called epimorphic. In Protura, on the other hand, the first-stage larva hatches with only eight abdominal segments plus the telson; the remaining three segments are added at subsequent molts, arising behind the last abdominal segment, but in front of the telson. This type of development is called anamorphic.

In general, the abdomen is clearly marked off from the thorax, but this is not the case in Hymenoptera, where the first abdominal segment is intimately fused with the thoracic segments and is known as the propodeum. The waist of Hymenoptera Apocrita is thus not between the thorax and abdomen, but between the first abdominal segment and the rest of the abdomen. Often, segment 2 forms a narrow petiole connecting the two parts. The swollen part of the abdomen behind the waist is called the gaster (Fig. 11.2).

11.1.2 Structure of abdominal segments

A typical abdominal segment, such as the third, consists of a sclerotized tergum and sternum joined by membranous pleural regions which are commonly hidden beneath the sides of the tergum, as in Figs. 11.1 and 11.2. In many holometabolous larvae, however, there is virtually no sclerotization and the abdomen consists of a series of membranous segments. This is true in many Diptera and Hymenoptera, some Coleoptera and most lepidopterous larvae. In these, the only sclerotized areas are small plates bearing trichoid sensilla. Even where well-developed terga and sterna are present these may be divided into a number of small sclerites as in the larva of the beetle *Calosoma* (Fig. 11.3). In contrast, the tergum, sternum and pleural elements sometimes fuse to form a complete sclerotized ring. This is true in the genital segments of many adult male insects, in segment 10 of Odonata, Ephemeroptera and Dermaptera and segment 11 of Machilidae.

Typically, the posterior part of each segment overlaps the anterior part of the segment behind (Fig. 11.4a), the two being joined by a membrane, but segments may fuse together, wholly or in part. For instance, in Acrididae the terga of segments 9 and 10 fuse together (Fig. 11.1a), while in some Coleoptera the second sternum fuses with the next two and the sutures between them are largely obliterated.

The more anterior segments have a spiracle on either side. This may be set in the pleural membrane (Fig. 11.3), or in a small sclerite within the membrane, or on the side of the tergum (Fig. 11.1) or sternum. The reproductive opening in male insects is usually on segment 9, while in the majority of female insects the opening of the oviduct is on or behind segment 8 or 9. The Ephemeroptera and Dermaptera are unusual in having the opening behind segment 7. These genital segments may be highly modified, in the male to produce copulatory apparatus and in the females of some orders to form an ovipositor. This

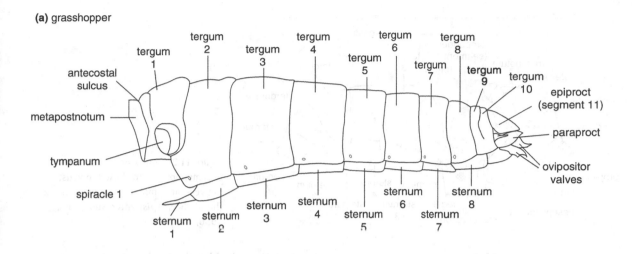

(a) grasshopper

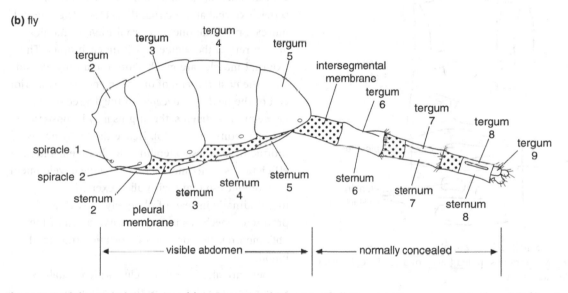

(b) fly

Figure 11.1 Abdomen in lateral view. (a) An insect in which parts of all 11 segments are present in the adult (female red locust, *Nomadacris* [Orthoptera]) (after Albrecht, 1956). (b) An insect with a reduced number of segments in the adult. Segments 6–9, which form the ovipositor, are normally retracted within the anterior segments (female housefly, *Musca* [Diptera]) (after Hewitt, 1914).

may be formed by the sclerotization and telescoping of the posterior abdominal segments, or it may involve modified abdominal appendages.

In front of these genital segments the abdominal segments are usually unmodified, although segment 1 is frequently reduced or absent. Behind the genital segments, segment 10 is usually developed, but segment 11 is often represented only by a dorsal lobe, the epiproct, and two latero-ventral lobes, the paraprocts. In Plecoptera, Blattodea and Isoptera the epiproct is reduced and fused with the tergum of segment 10, while in most holometabolous insects segment 11 is lacking altogether and segment 10 is terminal.

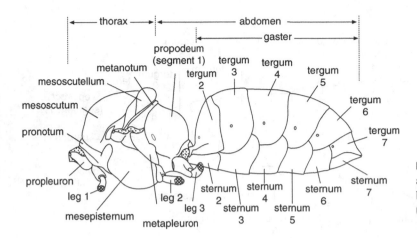

Figure 11.2 Thorax and abdomen of a hymenopteran to show the waist between abdominal segments 1 and 2 (honey bee, *Apis*) (after Snodgrass, 1956).

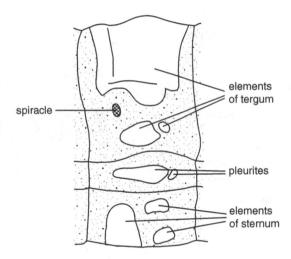

Figure 11.3 Small, sclerotized plates in an abdominal segment. Membrane stippled (larva of a beetle, *Calosoma*) (after Snodgrass, 1935).

Modifications of the terminal abdominal segments often occur in aquatic insects and are concerned with respiration (see Chapter 17).

11.1.3 Musculature

Where the cuticle of the abdomen is largely membranous, as in many holometabolous larvae, most longitudinal muscles run from one intersegmental fold to the next. In most insects with well-sclerotized abdominal segments, the dorsal and ventral abdominal longitudinal muscles are in two series, external and internal (Fig. 11.4). The internal muscles run from one antecostal ridge to the next and so retract the segments within each other. The external muscles are much shorter and only extend from the posterior end of one segment to the anterior end of the next and, because of the degree of overlap between the segments, the origins may be posterior to the insertions (Fig. 11.4b). Hence they may act as protractor muscles, extending the abdomen, and their efficiency is sometimes increased by the development of apodemes so that their pull is exerted longitudinally instead of obliquely. If such a protractor mechanism is absent, extension of the abdomen results from the hydrostatic pressure of blood.

There are also lateral muscles, which usually extend from the tergum to the sternum, but sometimes arise on or are inserted into the pleuron. They are usually intrasegmental, but sometimes cross from one segment to the next. Their effect is to compress the abdomen dorso-ventrally (Fig. 11.4c). Dilation of the abdomen often results from its elasticity and from blood pressure, but in some insects some of the lateral muscles function as dilators. This occurs when the tergal origins of the muscles are carried ventrally by extension of the terga, while the sternal insertions may also be carried dorsally on apodemes (Fig. 11.4d).

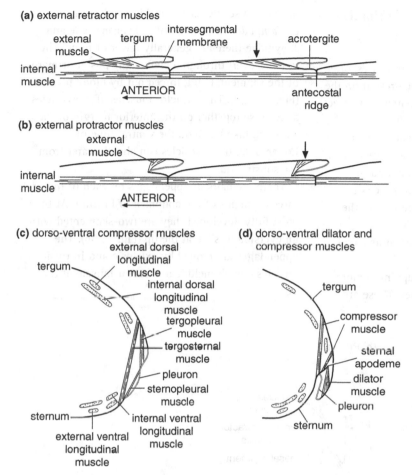

(a) external retractor muscles

external muscle · tergum · intersegmental membrane · acrotergite · internal muscle · ANTERIOR · antecostal ridge

(b) external protractor muscles

external muscle · internal muscle · ANTERIOR

(c) dorso-ventral compressor muscles

tergum · external dorsal longitudinal muscle · internal dorsal longitudinal muscle · tergopleural muscle · tergosternal muscle · pleuron · sternopleural muscle · sternum · internal ventral longitudinal muscle · external ventral longitudinal muscle

(d) dorso-ventral dilator and compressor muscles

tergum · compressor muscle · sternal apodeme · dilator muscle · pleuron · sternum

Figure 11.4 Abdominal musculature (from Snodgrass, 1935). (a) Diagram of the dorsal longitudinal musculature in an abdominal segment. Typical arrangement of external and internal muscles, both acting as retractors. (b) Origin of external muscle (arrow) shifted posteriorly so that it acts as a protractor. (c) Transverse section of the right-hand side showing a typical arrangement with the tergosternal muscles acting only as compressors. (d) Transverse section of the right-hand side showing tergosternal muscles differentiated into compressor and dilator muscles. Notice how the insertion of the dilator muscle onto the sternum is shifted dorsally by the apodeme.

In addition to the longitudinal and lateral muscles, others are present in connection with abdominal appendages, especially the genitalia, and the spiracles (Section 17.2), while transverse bands of muscle form the dorsal and ventral diaphragms (Chapter 5).

11.2 Abdominal appendages and outgrowths

Insects are generally believed to have been derived from an arthropod ancestor with a pair of appendages on each segment. Typical legs, such as are found on the thorax, never occur on the abdomen of insects, but various appendages do occur and some of these are probably derived from typical appendages. Molecular studies indicate that the prolegs of caterpillars are homologous with the thoracic legs. Some other appendages are probably secondary structures which have developed quite independently of the primitive appendages.

The structure and functioning of the male and female genitalia are considered in Chapters 12 and 13. Apart from the genitalia and the cerci, abdominal appendages or other outgrowths of the body wall tend to occur in larvae rather than in adults. The appendages of Apterygota and other primitive hexapods are present in all stages of development, however.

11.2.1 Abdominal appendages of primitive hexapods

Styliform appendages Styliform structures, often associated with eversible vesicles, are present on the abdomen of Apterygota and some related non-insect hexapods. On abdominal segments 2–9 of Machilidae, 7–9 or 8–9 of Lepismatidae, 1–7 of Japygidae and 2–7 of Campodeidae there are pairs of small, unjointed styli, each inserted on a basal sclerite which is believed to represent the coxa (Fig. 11.5a). Since similar styli are present on the coxae of the thoracic legs of *Machilis* (Archaeognatha), these styli are regarded as coxal epipodites.

Associated with the styli, but occupying a more median position, are eversible vesicles. These are present on segments 1–7 of Machilidae and 2–7 of *Campodea* (Diplura), but in Lepismatidae and Japygidae there are generally fewer or none. The vesicles evert through a cleft at the posterior margin of the segment, being forced out by blood pressure (Fig. 11.5c,d). The retractor muscles of the vesicles arise close together on the anterior margin of the sternum. Like those on the ventral tube of Collembola, these vesicles can absorb water from the substratum.

There are pairs of appendages on each of the first three segments of the abdomen of Protura. At their most fully developed they are two-segmented with an eversible vesicle at the tip (Fig. 11.5b). The appendages are moved by extrinsic and intrinsic muscles, which include a retractor muscle of the vesicle.

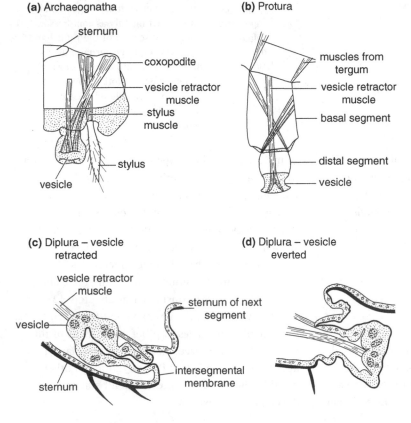

(a) Archaeognatha

sternum
coxopodite
vesicle retractor muscle
stylus muscle
stylus
vesicle

(b) Protura

muscles from tergum
vesicle retractor muscle
basal segment
distal segment
vesicle

(c) Diplura – vesicle retracted

vesicle retractor muscle
vesicle
sternum
sternum of next segment
intersegmental membrane

(d) Diplura – vesicle everted

Figure 11.5 Styliform appendages and eversible vesicles. (a) Archaeognatha; abdominal appendage of *Nesomachilis*; the sternum and coxopodite are seen from the inside (from Snodgrass, 1935). (b) Protura; abdominal appendage of *Acerentomon* (from Snodgrass, 1935). (c) Diplura; section through an eversible vesicle of *Campodea*, retracted (after Drummond, 1953). (d) Diplura; section through an eversible vesicle of *Campodea*, everted (after Drummond, 1953).

Abdominal appendages of Collembola The Collembola have pregenital appendages on three abdominal segments. From the first segment a median lobe projects forwards and down between the last pair of legs. This is known as the ventral tube, and at its tip are a pair of eversible vesicles which in many Symphypleona are long and tubular. The unpaired basal part of the ventral tube is believed to represent the fused coxae of the segmental appendages and the vesicles are thus coxal vesicles. The vesicles are everted by blood pressure from within the body and are withdrawn by retractor muscles.

The ventral tube appears to have two functions. In some circumstances it functions as an adhesive organ enabling the insect to walk over smooth or steep surfaces. To facilitate this on a dry surface the vesicles are moistened by a secretion from cephalic glands opening onto the labium and connecting with the ventral tube by a groove in the cuticle in the ventral midline of the thorax. The ventral tube also enables Collembola to adhere to the surface film on water since it is the only part of the cuticle which is wettable; all the rest is strongly hydrofuge.

The second function of the vesicles of the ventral tube is the absorption of water from the substratum (Section 18.9.2).

The appendages of the third and fourth segments of the abdomen of many Collembola form the retinaculum and the furca, which are used in locomotion.

11.2.2 Larval structures associated with locomotion and attachment

Leg-like outgrowths of the body wall, known as prolegs, are common features of the abdomen of holometabolous larvae. These appendages are expanded by blood pressure and moved mainly by the muscles of the adjacent body wall together with others inserted at the base of the proleg and a retractor muscle extending to the sole or planta surface (see Fig. 8.21).

Well-developed prolegs are a feature of lepidopterous larvae, which usually have a pair on each of abdominal segments 3–6 and 10 (see Fig. 15.5). Megalopygidae have more prolegs than other Lepidoptera, with prolegs on segments 2–7 and 10. Those on segments 2 and 7 have no crochets. More frequently the number of prolegs is reduced, and in Geometridae there are usually only two pairs, on segments 6 and 10. Prolegs are completely absent from some leaf-mining larvae and from the free-living Eucleidae, some of which, however, have weak ventral suckers on segments 1–7.

Distally, where it makes contact with the substratum, the proleg is flattened, forming the planta surface. This is usually armed with hook-shaped structures called crochets (see Fig. 8.21). The arrangement of the crochets varies. They may form a complete ring, or be arranged in transverse or longitudinal rows, reflecting the behavior of the larva and the nature of the substrate on which it lives.

Digitiform prolegs without crochets occur on the first eight abdominal segments of larval Mecoptera. They have no intrinsic musculature, but are moved by changes in blood pressure and by the action of muscles on adjacent parts of the ventral body wall. Prolegs without crochets also occur on the abdomen of larval Symphyta and particularly in the Tenthredinoidea. The number varies from six to nine pairs.

Larval Trichoptera have anal prolegs on segment 10 (see Fig. 11.8b). Their development varies, but in the Limnephilidae, where they are most fully developed, there are two basal segments and a terminal claw, having both levator and depressor muscles. These appendages, together with a dorsal and two lateral retractile papillae on the first abdominal segment, enable the larva to hold onto its case.

Sometimes, when prolegs are not developed, their position is occupied by a raised pad armed with spines. Such a pad is called a creeping welt and is functionally comparable with a proleg. Creeping welts and prolegs are present in many dipterous larvae, some of which have several prolegs on each segment (Fig. 11.6a) while others have creeping welts which extend all around the segment. The larvae of a number of families of Diptera have abdominal suckers which may be derived from prolegs. Thus the larva of the psychodid *Maruina* has a sucker on each of abdominal segments 1–8 and these suckers enable the larva to maintain its position along the sides of waterfalls. In another larva, of *Horaiella*, a single large sucker, bounded by a fringe of hairs, extends over the ventral surface of several segments.

(a) tabanid prolegs

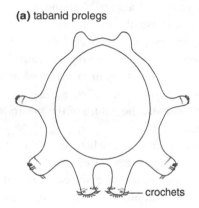

crochets

(b) blepharocerid sucker

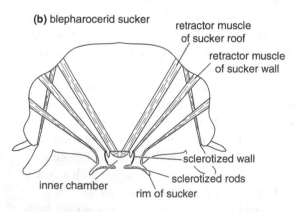

retractor muscle of sucker roof

retractor muscle of sucker wall

sclerotized wall

sclerotized rods

rim of sucker

inner chamber

Figure 11.6 Prolegs and suckers of larval Diptera (after Hinton, 1955). (a) Cross-section of an abdominal segment of a tabanid larva showing several pairs of prolegs, some with crochets. (b) Transverse section through the sixth abdominal segment of a blepharocerid larva showing the ventral sucker.

Larval Blepharoceridae, which live in fast-flowing streams and waterfalls, have a sucker on each of abdominal segments 2–7. Each sucker has an outer flaccid rim with an incomplete anterior margin. The central disc of the sucker is supported by closely packed sclerotized rods and in the middle a hole leads into an inner chamber with strongly sclerotized walls and an extensively folded roof (Fig. 11.6b). Muscles inserted into the roof and the rim of the sclerotized walls of the inner chamber increase the volume of the chamber when they contract and if at the same time the rim of the sucker is pressed down onto the substratum a partial vacuum is created so that the sucker adheres to the surface.

Even if a well-formed sucker is not present, many dipterous larvae can produce a sucker-like effect by raising the central part of the ventral surface while keeping the periphery in contact with the substratum, the sucker being sealed and made effective by a film of moisture.

11.2.3 Sensory structures

Most insects have mechanosensitive sensilla on the abdominal segments, and grasshoppers also have small contact chemoreceptors scattered among the mechanoreceptors. In addition, the appendages of segment 11 often form a pair of structures called cerci, which usually function as sense organs.

Cerci are present and well developed in the Apterygota and the hemimetabolous orders other than the hemipteroids. In holometabolous insects, cerci are present in the adults of Mecoptera and some Diptera; they are not present in holometabolous larvae. The cerci may be simple, unsegmented structures as in Orthoptera (Fig. 11.7a), or multi-segmented as in Blattodea and Mantodea (Fig. 11.7b). They may be very short and barely visible or long and filamentous – as long as or longer than the body, as in Thysanura, Ephemeroptera and Plecoptera. Even within a group, such as the Acridoidea, the range of form of the cerci is considerable. In cockroaches, where the cerci are

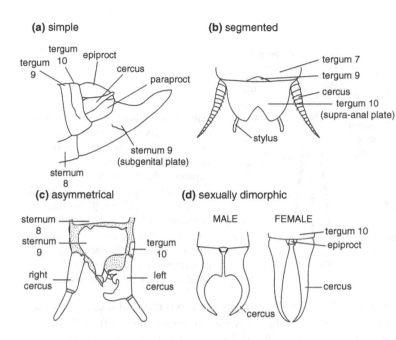

(a) simple

tergum 9
tergum 10
epiproct
cercus
paraproct
sternum 9
(subgenital plate)
sternum 8

(b) segmented

tergum 7
tergum 9
cercus
tergum 10
(supra-anal plate)
stylus

(c) asymmetrical

sternum 8
sternum 9
right cercus
tergum 10
left cercus

(d) sexually dimorphic

MALE   FEMALE
tergum 10
epiproct
cercus
cercus

Figure 11.7 Different types of cerci. (a) Simple cercus (lateral view of tip of abdomen of male red locust, *Nomadacris* [Orthoptera]). (b) Segmented (dorsal view of the tip of the abdomen of male *Periplaneta* [Blattodea]). (c) Asymmetrical (ventral view of tip of abdomen of male *Idioembia* [Embioptera]). (d) Sexually dimorphic (male and female forceps [cerci]) *Forficula* [Dermaptera].

segmented, additional segments are added at each molt. The first instar cercus of *Periplaneta* has three segments, while in the adult male it has 18 or 19 and in the adult female 13 or 14. Growth results from division of the basal segment.

The cerci are usually set with large numbers of trichoid sensilla. Sometimes these sensilla are filiform and are sensitive to air movements. This is true in cockroaches and crickets, where different filiform hairs are maximally sensitive to air movements from different directions (Chapter 23). In the female of the sheep blowfly, *Lucilia*, there are also a small number of contact chemoreceptors and olfactory receptors on the cerci. This may also be true in other insects.

Sometimes the cerci differ in the two sexes of a species, and they may play a role in copulation. Thus the cerci of female *Calliptamus* (Orthoptera) are simple cones, but in the male they are elongate, flattened structures with two or three lobes at the apex armed with strong inwardly directed points. There is similar dimorphism in Embioptera, where the male cerci are generally asymmetrical with the basal segment of the left cercus, forming a clasping organ

(Fig. 11.7c). Among the earwigs the cerci form powerful forceps which are usually straight and unarmed in the female, but incurved and toothed in the male (Fig. 11.7d). Similar forceps-like cerci in the Japygidae are used in catching prey. In some holometabolous insects they form part of the external genitalia.

The cerci of larval Zygoptera are modified to form the two lateral gills (see Fig. 17.28), while in the ephemeropteran *Prosopistoma* the long, feather-like cerci, together with the median caudal filament, can be used to drive the insect forwards by beating against the water.

11.2.4 Gills

Gills are present on the abdominal segments of the larvae of many aquatic insects. Ephemeroptera usually have six or seven pairs of plate-like or filamentous gills (see Figs. 15.4, 18.4) which are moved by muscles. They may play a direct role in gaseous exchange, but perhaps are more important in maintaining a flow of water over the body. Gill tufts may also be present on the first two or three

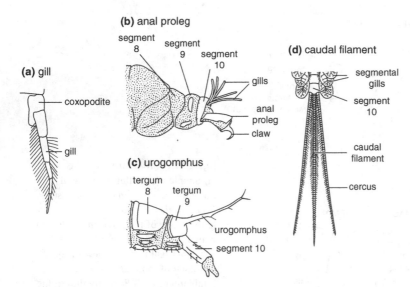

(a) gill

coxopodite

gill

(b) anal proleg

segment 8

segment 9

segment 10

gills

anal proleg

claw

(c) urogomphus

tergum 8

tergum 9

urogomphus

segment 10

(d) caudal filament

segmental gills

segment 10

caudal filament

cercus

Figure 11.8 Abdominal appendages of pterygote larvae. Membrane stippled. (a) Gill (*Sialis* [Megaloptera], dorsal view). (b) Gills and anal proleg (*Hydropsyche* [Trichoptera] lateral view of terminal abdominal segments). (c) Urogomphus (*Oodes* [Coleoptera] lateral view of terminal abdominal segments). (d) Caudal filament and segmented cerci (*Heptagenia* [Ephemeroptera] dorsal view of terminal abdominal segments).

abdominal segments, or in the anal region of larval Plecoptera. The larva of *Sialis* (Megaloptera) has seven pairs of five-segmented gills, each arising from a basal sclerite on the side of the abdomen (Fig. 11.8a), and a terminal filament of similar form arises from segment 9. Similar, but unsegmented gills are present in other larval Megaloptera and in some larval Coleoptera. Larval Trichoptera have filamentous gills in dorsal, lateral and ventral series.

Some aquatic larvae have papillae, often incorrectly called gills, surrounding the anus. They are concerned with salt regulation (Section 18.3) and are found in larval mosquitoes and chironomids, where a group of four papillae surrounds the anus (see Fig. 18.3), and in some larval Trichoptera.

11.2.5 Secretory structures

Some insects have glands opening on the abdomen which probably have a defensive function in most cases (see Chapter 27). Most aphids have a pair of tubes, known as siphunculi or cornicles, projecting from the dorsum of segment 5 or 6, or from between them. Each cornicle has a terminal opening which is normally closed by a flap of cuticle controlled by an opener muscle, and the whole structure can be moved by a muscle inserted at the base so that the cornicle

can be pointed in various directions, even forwards. Aphids release an alarm pheromone from the cornicles if they are attacked by parasites or predators (Section 27.2.6). This causes a response in other aphids of the same species, but the response differs in different species. Individuals of *Schizaphis graminum* usually drop off the plant when they perceive the pheromone; *Myzus persicae* may drop off the plant or walk away from the feeding site; other species jerk about in a manner which is presumed to discourage attack without withdrawing their stylets from the host. The effective radius of the pheromone may extend up to about 3 cm from the emitting aphid.

The first abdominal appendages are well developed in the embryos of insects belonging to a number of groups. They are known as pleuropodia (Section 14.2.12), but they do not persist after hatching. Perhaps their primary function is the secretion of enzymes that digest the serosal cuticle prior to hatching.

11.2.6 Secondary sexual structures

Secondary sexual structures are often found in Diptera. When entering mating swarms female empids *Rhamphomyia* inflate their abdomens with air, stretching the pleural membranes to exaggerate their body size and attract males. Some female chironomids evert long glandular strings from their abdomens when they enter mating swarms and male tephritid flies sometimes bear modified setae on the abdomen that are conspicously colored. These abdominal structures are thought to serve as ornaments in mate attraction. The cerci of Forficula (Dermaptera) appear to be secondary sexual traits used in aggressive interactions between males, and in the courtship of females (Fig. 11.7a). In fireflies (Lampyridae: Coleoptera) dorsal areas of the abdominal segments form lanterns. These bioluminescent organs are diverse in size and location, though they are generally located close to the body surface behind a translucent window of cuticle. They occur in larvae, adult males and/or adult females, depending on species, and function in mate attraction.

In both Panorpa (Mecoptera) and Cyphoderris (Orthoptera) the male has a clamp-like structure formed by modifications to the posterior and anterior edges of two adjacent tergites. The clamp serves to force females to copulate or prolong copulation. Such adaptations may not be in the best interests of females, leading to sexual conflict. In *Gerris* (Hemiptera) females possess posteriolateral abdominal spines that prevent males from grasping them during premating struggles, and thus protect females from enforced copulation.

11.2.7 Other abdominal structures

Apart from the segmentally arranged prolegs and gills, some insects have other abdominal appendages which often appear to have a defensive role, but whose function is sometimes unknown. Some groups of insects have a median process projecting from the last segment. In Thysanura and Ephemeroptera this is in the form of a median caudal filament which resembles the two cerci (Fig. 11.8d). Larval Zygoptera have a median

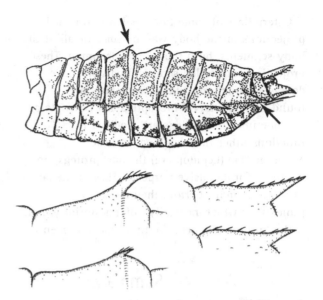

Figure 11.9 Abdomen of a last-instar *L. dubia* (Anisoptera) larva in lateral view (top). Arrows indicate position of the dorsal V and lateral IX abdominal spines. Natural shape variation in spine morphology is illustrated in representative examples shown in magnification below. The bottom left shows the dorsal V spine (lateral view), and the bottom right the lateral IX spine (dorsal view), of last-instar larvae collected from lakes with (upper) and without (lower) predatory fish (from Arnqvist and Johansson, 1998).

terminal gill on the epiproct, while in larval hawk moths (Sphingidae) a terminal spine arises from the dorsum of segment 10.

Some larval Coleoptera have a pair of processes called urogomphi, which are outgrowths of the tergum of segment 9 (Fig. 11.8c). They may be short spines or multiarticulate filaments and they may be rigid with the tergum or arise from the membrane behind it so that they are mobile.

Larval Anisoptera (Odonata) can have spines on the dorsal posterior edge of the abdominal tergae. These spines serve an anti-predator role, and exhibit considerable phenotypic plasticity. In larval *Leucorrhinia* defensive spines grow more solid and elongated when water-borne environmental cues of fish predators are present (Fig. 11.9).

Caterpillars of some families have branched projections of the body wall on some or all of the body segments, both thorax and abdomen. These projections are called scoli and possibly have defensive functions. Similar structures are found on some larval Mecoptera and Coleoptera.

Sometimes the prolegs are modified for functions other than walking. In some Notodontidae (Lepidoptera) the anal prolegs are modified for defensive purposes. Thus in the larvae of the puss moth *Cerura*, they are slender projections which normally point posteriorly, but, if the larva is touched, the tip of the abdomen is flexed forwards and a slender pink process is everted from the end of each projection. At the same time the larva raises its head and thorax from the ground and emits formic acid from a ventral gland in the prothorax. This reaction is presumed to be a defensive display.

In a few larval Diptera, prolegs may be used for holding prey. The larva of *Vermileo* lives in a pit in dry soil and feeds in the same way as an antlion. It lies ventral side up, and prey which fall into the pit are grasped against the thorax by a median proleg on the ventral surface of the first abdominal segment.

 ## Summary

- The adult abdomen generally consists of 11 segments plus the telson, which bears the anus. Abdominal segments typically consist of a sclerotized tergum and sternum, and a membranous pleural region. Intersegmental membranes connect adjacent segments, although these may also be fused wholly or in part.

- The more anterior segments have a spiracle on either side. Genital segments may be highly modified, to form copulatory structures in the male, and in some orders, the ovipositor in the female.

- There are longitudinal muscles that extend and retract the abdomen, and lateral muscles that compress the abdomen dorso-ventrally. Other muscles are also present that connect with the abdominal appendages, especially the genitalia, and the spiracles.

- Abdominal appendages associated with locomotion occur in primitive hexapods, and leg-like outgrowths of the body wall are a common feature of larval holometabolous insects. There is considerable taxonomic variation both in the location of these "prolegs" and in their degree of development.

- Mechano- and chemoreceptors are found on the abdomen of many insects. These are particularly prevalent on the cerci. Some insects have glands opening on the abdomen that secrete toxic compounds in response to predators, or release pheromones to communicate with conspecifics.

- Abdominal structures can serve anti-predator roles, function as ornaments for attracting mates, as clamps for holding females or as protective structures in females for the prevention of mating.

Recommended reading

Arnqvist, G. and Johansson, F. (1998). Ontogenetic reaction norms of predator-induced defensive morphology in dragonfly larvae. *Ecology* **79**, 1847–1858.

Day, J. C. (2011). *Fireflies and Glow-worms*. Abingdon: Brazen Head Publishing.

Drummond, F. H. (1953). The eversible vesicles of *Campodea* (Thysanura). *Proceedings of the Royal Entomological Society of London A* **28**, 115–118.

Matsuda, R. (1976). *Morphology and Evolution of the Insect Abdomen*. Oxford: Pergamon Press.

Sivinski, J. (1997). Ornaments in the Diptera. *Florida Entomologist* **80**, 142–164.

Tomkins, J. L. and Simmons, L. W. (1998). Female choice and manipulations of forceps size and symmetry in the earwig *Forficula auricularia* L. *Animal Behaviour* **56**, 347–356.

References in figure captions

Albrecht, F. O. (1956). *The Anatomy of the Red Locust,* Nomadacris septemfasciata *Serville.* London: Anti-locust Research Centre.

Arnqvist, G. and Johansson, F. (1998). Ontogenetic reaction norms of predator-induced defensive morphology in dragonfly larvae. *Ecology* **79**, 1847–1858.

Drummond, F. H. (1953). The eversible vesicles of *Campodea* (Thysanura). *Proceedings of the Royal Entomological Society of London A* **28**, 115–118.

Hewitt, C. G. (1914). *The House-fly,* Musca domestica *Linn.* Cambridge: Cambridge University Press.

Hinton, H. E. (1955). On the structure, function, and distribution of the prolegs of the Panorpoidea, with a criticism of the Berlese–Imms theory. *Transactions of the Royal Entomological Society of London* **106**, 455–545.

Snodgrass, R. E. (1935). *Principles of Insect Morphology*. New York, NY: McGraw-Hill.

Snodgrass, R. E. (1956). *Anatomy of the Honey Bee*. London: Constable.

12 | Reproductive system: male

REVISED AND UPDATED BY **LEIGH W. SIMMONS**

INTRODUCTION

Insects vary greatly in their reproductive anatomy, behavior and physiology. This chapter provides a general overview of the male reproductive system, including the mechanisms of copulation and insemination. It describes in Section 12.1 the basic elements of the internal anatomy, which include the testes that produce sperm, the vas deferens and seminal vesicles for their storage prior to ejaculation and the accessory glands that manufacture seminal fluids that mix with sperm to form the ejaculate and, in species in which insemination is indirect, the spermatophore in which the ejaculate is housed when passed to the female. Section 12.2 provides a detailed overview of the ultrastructure of insect spermatozoa and the processes of spermatogenesis. Section 12.3 describes the various means by which insemination is accomplished, covering a description of the male external genitalia, copulation and the transfer of sperm to the female reproductive tract. Finally, Section 12.4 discusses the various effects of mating on female nutrition and physiology that are brought about by accessory gland products that are transferred to females as part of the ejaculate.

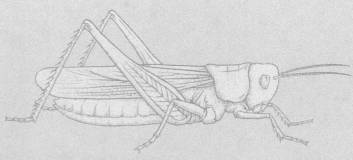

The Insects: Structure and Function (5th edition), ed. S. J. Simpson and A. E. Douglas. Published by Cambridge University Press. © Cambridge University Press 2013.

12.1 Anatomy of the internal reproductive organs

The male reproductive organs typically consist of a pair of testes connecting with paired seminal vesicles and a median ejaculatory duct (Fig. 12.1). In most insects there are also a number of accessory glands which open into the vasa deferentia or the ejaculatory duct.

Testis. The testes may lie above or below the gut in the abdomen and are often close to the midline. Usually, each testis consists of a series of testis tubes or follicles ranging in number from one in Coleoptera Adephaga to over 100 in grasshoppers (Acrididae). Sometimes, as in Lepidoptera, the follicles are incompletely separated from each other (Fig. 12.2c), and the testes of Diptera consist of simple, undivided sacs, which may be regarded as single follicles.

Sometimes the follicles are grouped together into several separate lobes (Fig. 12.1b). In the cerambycid *Prionoplus*, for example, each testis comprises 12–15 lobes each with 15 follicles. The testes of Apterygota are often undivided sacs, but it is not certain in this case that they are strictly comparable with the gonads of other insects since the germarium occupies a lateral position in the testis instead of being terminal.

The wall of a follicle is a thin epithelium, sometimes consisting of two layers of cells, standing on a basal lamina. The follicles are bound together by a peritoneal sheath, and if the two testes are close to each other they may be bound together. In some Lepidoptera the two testes fuse completely to form a single median structure.

Vas deferens and seminal vesicle. From each testis follicle a fine, usually short, vas efferens connects

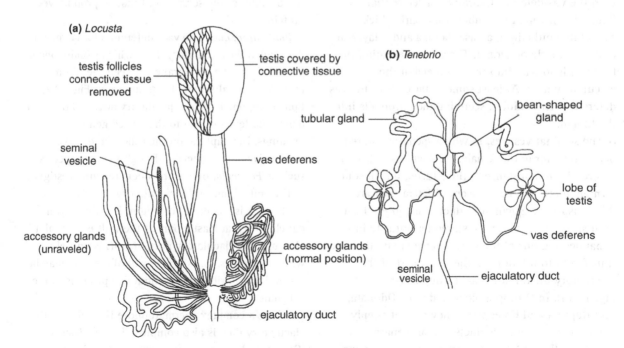

(a) *Locusta*

testis follicles connective tissue removed

testis covered by connective tissue

(b) *Tenebrio*

seminal vesicle

tubular gland

bean-shaped gland

vas deferens

accessory glands (unraveled)

accessory glands (normal position)

lobe of testis

vas deferens

seminal vesicle

ejaculatory duct

ejaculatory duct

Figure 12.1 Basic structure of the internal reproductive organs of the male. (a) An insect with a large number of testis follicles and accessory glands. The testes lie close together in the midline, but are distinct. The glands are of different types (but this is not shown in the diagram) (*Locusta* [Orthoptera]) (from Uvarov, 1966). (b) An insect with several distinct testis lobes and only two pairs of accessory glands (*Tenebrio*, Coleoptera) (from Imms, 1957).

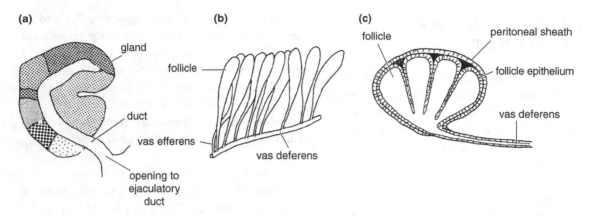

(a)

gland

duct

vas efferens

opening to
ejaculatory
duct

(b)

follicle

vas deferens

(c)

follicle

peritoneal sheath

follicle epithelium

vas deferens

Figure 12.2 Male reproductive organs. (a) Accessory gland. Diagram of the bean-shaped gland of *Tenebrio* (see Fig. 12.1b) showing that different regions contain cells producing different secretions (shown by different shading). Two additional cell types also occur, but are not in the plane of this section (after Dailey *et al.*, 1980). (b) Testis. Diagram of a series of testis follicles opening independently into the vas deferens, as in Orthoptera. (c) Testis. Diagrammatic section through a testis in which the follicles are incompletely separated from each other and have a common opening to the vas deferens, as in Lepidoptera (from Snodgrass, 1935).

with the vas deferens (plural: vasa deferentia) (Fig. 12.2b), which is a tube with a fairly thick bounding epithelium, a basal lamina and a layer of circular muscle outside it. The vasa deferentia run backwards to lead into the distal end of the ejaculatory duct. At least some of the cells of the vas deferens are glandular, secreting their products into the lumen.

The seminal vesicles, in which sperm are stored before transfer to the female, are dilations of the vasa deferentia in many insects (Fig. 12.1b), but in some Hymenoptera and nematoceran Diptera they are dilations of the ejaculatory duct. Lepidoptera have both structures: sperm are stored temporarily in expanded regions of the vasa deferentia and then are transferred to dilations in the upper part of the ejaculatory duct, known as the duplex (see Fig. 12.14). In Orthopteroidea and some Odonata, Phthiraptera and Coleoptera, they are not simply expansions of the male ducts, but are separate structures (Fig. 12.1a). In some insects, the seminal vesicles are epidermal in origin and in these cases they are lined with cuticle. The cellular lining of the seminal vesicles is glandular and in *Drosophila* is

responsible for the secretion of male reproductive proteins.

Ejaculatory duct. The vasa deferentia join a median duct called the ejaculatory duct, which usually opens posteriorly in the membrane between the ninth and tenth abdominal segments (gonopore in Fig. 12.7). Ephemeroptera have no ejaculatory duct and the vasa deferentia lead directly to the paired genital openings. Dermaptera, on the other hand, have paired ejaculatory ducts, although in some species, such as *Forficula*, one of the ducts remains vestigial.

The epithelium of the ejaculatory duct is one cell thick. As it is epidermal in origin, it is lined with cuticle. Often at least a part of the wall is muscular, although the ejaculatory duct in *Apis* is entirely without muscles. Parts of the wall of the duct may be glandular, contributing reproductive proteins to the ejaculate.

Where a complex spermatophore is produced, the ejaculatory duct is also complex. Thus in *Locusta* (Orthoptera), the ejaculatory duct consists of upper and lower ducts connected via a funnel-like constriction (see Fig. 12.13). The upper part of the duct into which the accessory glands open has a columnar epithelium and thin cuticle and the lumen is a vertical slit in cross-section. In the funnel, the cuticle is thicker and usually forms nine ridges on either side. These curve upwards posteriorly as they run back to meet in the dorsal midline and they project so that they almost completely divide the lumen. The lumen of the lower duct is circular in cross-section and leads to the ejaculatory sac and spermatophore sac. Scattered muscle fibers are present in the wall of the upper duct but are absent elsewhere. The ejaculatory duct of the milkweed bug, *Oncopeltus*, is also extremely complex, being specialized for erection of the penis (see Fig. 12.10).

The ejaculatory duct in Lepidoptera is extended inwards as an unpaired duct of mesodermal origin and so not lined with cuticle. This is called the simplex. It bifurcates distally to connect with the accessory glands and vasa deferentia – this part is known as the duplex (see Fig. 12.14). The simplex produces most of the spermatophore since Lepidoptera have only a single pair of accessory glands. In *Calpodes* it is divided into seven sections partially separated from each other by constrictions, and some of these are further differentiated into zones of cells with different structures. These different sections and zones produce a variety of secretions which contribute to the spermatophore. The constrictions make it possible for the secretions of different sections to be used separately and in sequence.

Accessory glands. The male accessory glands may be ectodermal or mesodermal in origin, in which case they are known as ectadenia or mesadenia, respectively. Ectadenia, which open into the ejaculatory duct, occur in many Coleoptera, in the Diptera Nematocera and some Homoptera. Mesadenia, which open into the vasa deferentia or the distal end of the ejaculatory duct, are found in Orthoptera and many other orders. In some species of Heteroptera and Coleoptera, both ectadenia and mesadenia are present.

The number and arrangement of accessory glands varies considerably between different groups of insects. In Lepidoptera there is a single pair of glands (see Fig. 12.14); in *Tenebrio* there are two pairs (Fig. 12.1b). In contrast, *Schistocerca* and *Locusta* have 15 pairs of accessory glands, not counting the seminal vesicles with which they are closely associated (Fig. 12.1a), and *Gryllus* has over 600. Sometimes there are no morphologically distinct accessory glands. This is the case in Apterygota, Ephemeroptera and Odonata, and muscoid Diptera.

Each accessory gland consists of a single layer of epithelial cells whose fine structure varies depending on their stage of development and also on the nature of the secretion produced. Where the glands are few in number, different regions within them may be functionally distinguishable. In *Tenebrio*, for example, the tubular glands produce three classes of compounds, and the bean-shaped glands have eight morphologically distinguishable cell types secreting a number of different products (Fig. 12.2a). In Lepidoptera the accessory glands are regionally differentiated to produce two different secretions (see Fig. 12.14). On the other hand, where many glands are present, several glands may produce the same proteins. This is the case in the Orthoptera. In *Schistocerca* 7 of the 15 pairs of glands appear to produce a single product, while most of the other pairs each produce a unique product. Where there are few or no accessory glands, their role is sometimes taken over by glandular cells in the ejaculatory duct. This is most obvious in the lepidopteran simplex, but also occurs in muscoid Diptera.

Outside the epithelium is a muscle layer which, in *Gryllus*, consists of a single layer of fibers wound around the gland in a tight spiral, but with a more complex arrangement around the openings of the glands to the ejaculatory duct. In this insect, both sets of muscles are innervated by proctolinergic DUM neurons (Section 10.3.2), with cell bodies in the terminal abdominal ganglion. At least some of the muscles are also innervated by other neurons that may be inhibitory. In *Periplaneta* the DUM cells innervating the accessory glands are octopaminergic.

The accessory glands become functional in the adult insect. Their secretions are involved in producing the spermatophore, where one is present, in providing nourishment and protection to the sperm, and also in modifying female behavior and physiology (see below).

12.2 Spermatozoa

12.2.1 Structure of mature spermatozoa

The mature sperm of most insects are flagellate, often about 300 µm long and less than a micron in diameter, with head and tail regions of approximately the same diameter (Fig. 12.3). However, insect sperm show patterns of extreme morphological diversity, from simple aflagellate disc-like structures seen in some Protura, to the giant flagellate sperm of *Drosophila bifurca*, which reach 60 mm in length. Flagellate sperm have a typical cell membrane about 10 nm thick, coated on the outside by a layer of glycoprotein known as the glycocalyx. In fleas this is about 13 nm thick, and in grasshoppers about 30 nm. The glycocalyx is made up of rods at right angles to the surface of the sperm (Fig. 12.4a). Lepidopteran sperm have a series of projections called lacinate appendages running along their length. They are made up of thin laminae stacked parallel with, but external to, the surface membrane (Fig. 12.4b).

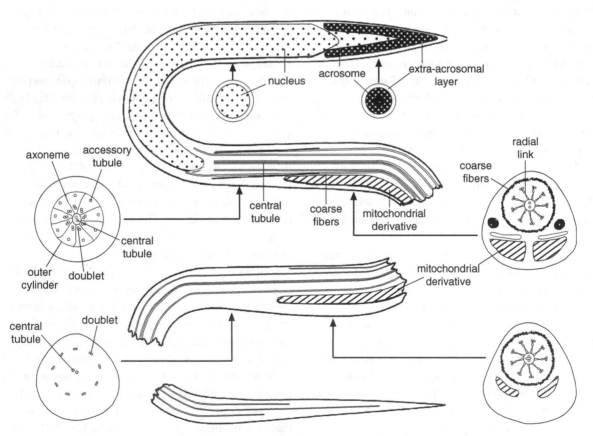

Figure 12.3 Diagram showing the structure of a sperm in longitudinal section with representative cross-sections at the points shown.

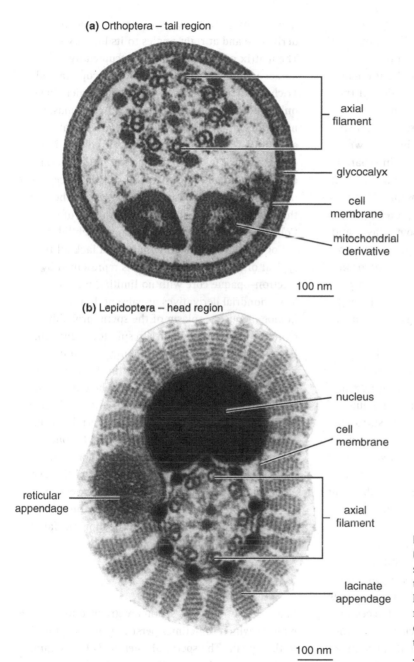

(a) Orthoptera – tail region

axial
filament

glycocalyx

cell
membrane

mitochondrial
derivative

100 nm

(b) Lepidoptera – head region

nucleus

cell
membrane

reticular
appendage

axial
filament

lacinate
appendage

100 nm

Figure 12.4 Sperm structure. (a) Electron micrograph of a transverse section through the tail region showing the glycocalyx (grasshopper, after Longo *et al.*, 1993). (b) Electron micrograph of a transverse section of a lepidopteran sperm showing the lacinate appendages (after Jamieson, 1987).

These structures are no longer present when the sperm are released from the cyst (Section 12.2.4), but the material of which they are composed may subsequently be used in binding the sperm together.

The greater part of the head region is occupied by the nucleus (Fig. 12.3). In mature sperm of most species the nucleus is homogeneous in appearance, but sometimes, as in the grasshopper *Chortophaga*, it has a honeycomb appearance. The DNA is apparently arranged in strands parallel with the long axis of the sperm. In Lepidoptera a second kind of sperm is produced in addition to the normal nucleate (eupyrene) sperm. These sperm are without nuclei (apyrene) so they cannot effect fertilization of the egg. Spermatids which give rise to them have numerous micronuclei instead of a single nucleus and these micronuclei subsequently break down completely. Apyrene sperm are formed in separate cysts from the eupyrene sperm and are highly motile.

In front of the nucleus is the acrosome. This is a membrane-bound structure of glycoprotein with, in most insects, a granular extra-acrosomal layer and an inner rod or cone. Neuropteran sperm have no acrosome and occasional species with no acrosome occur in other orders. The acrosome is probably involved with attachment of the sperm to the egg and possibly also with lysis of the egg membrane, thus permitting sperm entry.

Immediately behind the nucleus, the axial filament, or axoneme, arises. In most cases this consists of two central tubules with a ring of nine doublets and nine accessory tubules on the outside (Figs. 12.3, 12.4). The central tubules are surrounded by a sheath and are linked radially to the doublets. Additional fibers are usually present between the accessory tubules. Some unusual exceptions to this 9 + 9 + 2 arrangement occur. Accessory tubules are lacking in Collembola, Japygidae (Diplura), Mecoptera and Siphonaptera.

The sperm of some insects have two axial filaments. This occurs in Psocoptera, Phthiraptera, Thysanoptera and many bugs. In *Sciara* (Diptera) there is no well-organized axial filament, but 70–90 tubule doublets, each with an associated accessory tubule, are arranged in a spiral which encloses the mitochondrial derivative posteriorly. It is presumed that the axial filament or the equivalent structure causes the undulating movements of the tail which drive the sperm forwards.

The sperm of Pterygota have two mitochondrial derivatives which flank the axial filament. Within these the cristae become arranged as a series of lamellae projecting inwards from one side of the derivative and at right angles to its long axis. The matrix of the derivative is occupied by a paracrystalline material. Sperm of Mecoptera and Trichoptera and species of some other orders have only one mitochondrial derivative, while phasmids have none at all (but this does not mean that they are without respiratory enzymes, see below). More or less normal mitochondria persist in the sperm of Apterygota and non-insect Hexapoda, except that they fuse together and become elongated. There are three such mitochondria in the sperm tail of Collembola, and two in Diplura and Machilidae.

Coccid sperm occur in bundles and lack all the typical organelles. The nucleus is represented by an electron-opaque core with no limiting membrane. Mitochondrial derivatives are absent, but the homogeneous cytoplasm of the sperm probably contains the enzymes with a respiratory function. Each sperm has 45–50 microtubules in a spiral around a central mass of chromatin. They run the whole length of the sperm and may be concerned with its motility, replacing the typical axial filament.

Sperm of Kalotermitidae and Rhinotermitidae have no flagellum at all. The sperm of *Reticulitermes* (Rhinotermitidae) is spherical with no acrosome, but it has a few normal mitochondria and two short axial filaments, although these do not extend into a tail. It is presumed that this sperm is non-motile. Non-motile sperm also occur in the dipteran family Psychodidae and in *Eosentomon* (Protura).

12.2.2 Sperm bundles

In a number of insects, sperm are grouped together in bundles which sometimes persist even after transfer to the female. The sperm of *Thermobia* (Thysanura) normally occur in pairs, the two individuals being twisted around each other with their membranes joined at points of contact. Pairs of sperm also occur in some Coleoptera.

Coccids have much more specialized sperm bundles. In these insects, each cyst (see Section 12.2.3) commonly produces 32 sperm which may become separated into two bundles of about 16. Some species have 64 sperm in a bundle. Each bundle becomes enclosed in a membranous sheath and the cyst wall degenerates. The sheath of many species, such as *Pseudococcus*, is longer than the sperm, which occupy only the middle region, and the head-end of the sheath has a corkscrew-like form. The sperm bundles of *Parlatoria*, on the other hand, are only the same length as the sperm, which are all oriented in the same direction within the bundle. Movement of the bundles does not normally occur until they enter the female. It results from the combined activity of the sperm within. Subsequently, the sheath of the bundle ruptures and the sperm are released.

In some Orthoptera and Odonata different types of sperm bundle, known as spermatodesms, are formed. The spermatodesms of tettigoniids comprise about ten sperm anchored together by their acrosomes. These bundles are released from the testis and the sperm heads then become enclosed in a muff of mucopolysaccharide secreted by the gland cells of the vas deferens. In acridids the spermatodesms are completely formed within the testis cyst and may include all the sperm within the cyst. The spermatids come to lie with their heads oriented toward a cyst-cell and extracellular granular material around the acrosome of each coalesces to form a cap in which the heads of all the sperm are embedded. The spermatodesms of Acrididae persist until they are transferred to the female.

12.2.3 Spermatogenesis

At the distal end of each testis follicle is the germarium, in which the germ cells divide to produce spermatogonia (cells which divide mitotically to produce spermatocytes; spermatocytes divide meiotically to produce spermatids) (Fig. 12.5). In Orthoptera, Blattodea, Homoptera and Lepidoptera, the spermatogonia probably obtain nutriment from a large apical cell with which they have cytoplasmic connections, while in Diptera and Heteroptera an apical syncytium performs a similar function.

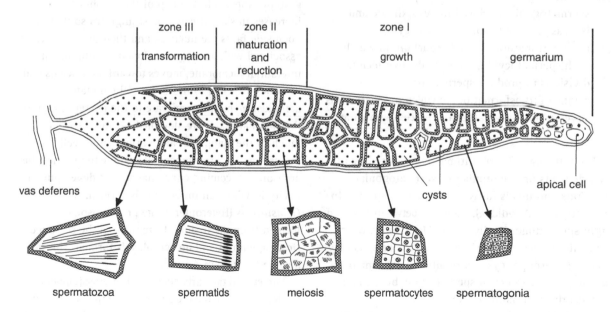

Figure 12.5 Diagram of a testis follicle showing the sequence of stages of development of the sperm (from Wigglesworth, 1965).

Transfer of mitochondria from this syncytium to the spermatogonia has been observed in Diptera.

The apical connections are soon lost and the spermatogonia associate with other cells which form a cyst around them (Fig. 12.5). One, or sometimes more, spermatogonia are enclosed in each cyst and, in *Prionoplus*, there are initially two cyst-cells around each spermatogonium. They may supply nutriment to the developing sperm and, in *Popillia* (Coleoptera), nutrient transfer may be facilitated by the sperm at one stage having their heads embedded in the cyst-cells. In Heteroptera large cells with irregular nuclei, called trophocytes, are scattered among the cysts.

As more cysts are produced at the apex of a follicle, they displace those which have developed earlier so that a range of developmental stages are present in each follicle, with the earliest stages distally in the germarium and the oldest in the proximal part of the follicle adjacent to the vas deferens. Three zones of development are commonly recognized below the germarium (Fig. 12.5):

(1) a zone of growth, in which the primary spermatogonia, enclosed in cysts, divide and increase in size to form spermatocytes;
(2) a zone of maturation and reduction, in which each spermatocyte undergoes the two meiotic divisions to produce spermatids;
(3) a zone of transformation, in which the spermatids develop into spermatozoa, a process known as spermiogenesis.

The number of sperm ultimately produced by a cyst depends on the number of spermatogonial divisions and this is fairly constant for a species. In grasshoppers (Acrididae), there are between five and eight spermatogonial divisions; *Melanoplus*, which typically has seven divisions before meiosis, usually has 512 sperm per cyst. Normally four spermatozoa are produced from each spermatocyte, but in many coccids the spermatids which possess heterochromatic chromosomes degenerate so that only two sperm are formed from each spermatocyte

and 32 are present in each cyst. In *Sciara* (Diptera) only one spermatid is formed from each spermatocyte because of an unequal distribution of chromosomes and cytoplasm at the meiotic divisions.

Spermiogenesis The spermatid produced at meiosis is typically a rounded cell containing normal cell organelles. It subsequently becomes modified to form the sperm and this process of spermiogenesis entails a complete reorganization of the cell. It is convenient to consider separately each organelle of the mature sperm.

Acrosome. The acrosome is derived, at least in part, from Golgi material, which in spermatocytes is scattered through the cytoplasm in the form of dictyosomes. There may be 30–40 of these in the cell and they consist of several pairs of parallel membranes with characteristic vacuoles and vesicles. After the second meiotic division the dictyosomes in *Acheta* fuse to a single body called the acroblast, which consists of 6–10 membranes forming a cup with vacuoles and vesicles both inside and out. In the later spermatid, a granule, called the pro-acrosomal granule, appears in the cup of the acroblast and increases in size. The acroblast migrates so that the open side faces the nucleus, and then the granule, associated with a newly developed membrane, the interstitial membrane, moves toward the nucleus and becomes attached to it. As the cell elongates, the acroblast membranes migrate to the posterior end of the spermatid and are sloughed off together with much of the cytoplasm and various other cell inclusions. The pro-acrosomal granule then forms the acrosome, becoming cone-shaped and developing a cavity in which an inner cone is formed. In *Gelastocoris* (Heteroptera), the pro-acrosome is formed from the fusion of granules in the scattered Golgi apparatus and no acroblast is formed. This may also be the case in Acrididae.

Nucleus. In the early spermatid of grasshoppers the nucleus appears to have a typical interphase structure with the fibrils, which constitute the basic morphological units of the chromosomes, unoriented. As the sperm develops, the nucleus becomes very long and narrow and the chromosome fibrils become aligned more or less parallel with its long axis. The nucleoplasm between the fibrils is progressively reduced until finally the whole of the nucleus appears to consist of a uniformly dense material. A similar linear arrangement of the chromosomes occurs in other groups.

Mitochondria. In the spermatid, the mitochondria fuse to form a single large body, the nebenkern, consisting of an outer limiting membrane and a central pool of mitochondrial components. The nebenkern separates into two mitochondrial derivatives associated with the developing axial filament immediately behind the nucleus. They elongate to form a pair of ribbon-like structures. At the same time, their internal structure is reorganized so that the cristae form a series of parallel lamellae along one side and the matrix is replaced by paracrystalline material.

Centriole and axial filament. Young spermatids contain two centrioles oriented at right angles to each other and each composed, as in most cells, of nine triplets of tubules. One gives rise to the axial filament, but ultimately both centrioles disappear. The tubules of the axial filament grow out from the centriole and finally extend the length of the sperm's tail. The accessory tubules arise from tubule doublets, appearing first as side arms, which become C-shaped and then separate off and close up to form cylinders.

Biochemical changes The repeated cell divisions during spermatogenesis entail the synthesis of large amounts of DNA and RNA, but synthesis of DNA stops before meiosis occurs, while RNA synthesis continues into the early spermatid. Subsequently, no further synthesis occurs and RNA is eliminated first from the nucleus and then from the cell as the nucleus elongates. The reduction in RNA synthesis is associated with a rise in the production of an arginine-rich histone which forms a complex with DNA, stopping it from acting as a primer for RNA synthesis, and, perhaps, insulating the genetic material from enzymic attack during transit to the egg.

Control of spermatogenesis The spermatocytes reach meiosis before the final molt in most insect species and, in species which do not feed as adults, the whole process of spermatogenesis may be complete before adult eclosion (Fig. 12.6). In many species, however, spermatogenesis continues for an extended period and new sperm continue to be produced throughout adult life. The time taken for

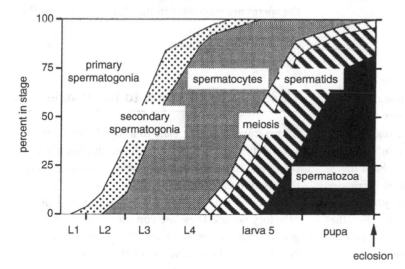

Figure 12.6 Spermatogenesis sometimes continues through the life of the insect. Notice that, in this example, the earliest developing cells undergo meiosis during the fourth larval stage, but later developing cells are in this stage in the late pupa (*Bombyx*, Lepidoptera) (after Engelmann, 1970).

the completion of spermatogenesis varies, but in *Melanoplus* the spermatogonial divisions take eight or nine days and spermiogenesis, ten. In the skipper butterfly, *Calpodes*, spermiogenesis takes about four days.

The factors regulating spermatogenesis are not well understood. At least in Lepidoptera, ecdysteroids have some role, but this differs from species to species. Spermatogenesis in the larva proceeds as far as prophase of the first meiotic division, apparently without any hormonal involvement, but is then delayed. In *Manduca* further development is initiated by a peak of ecdysteroid during the wandering period of the last larval stage and then it continues through the early pupal period. High titers of ecdysteroid late in the pupal stage may inhibit further meiosis in spermatocytes developing into eupyrene sperm. In contrast, in *Heliothis* a factor produced by the testis sheath is necessary for meiosis to occur. This factor is not an ecdysteroid, although subsequently the sheath does produce ecdysteroids that regulate both the fusion of the two testes into a single structure and the development of the internal genital tract during the pupal period. Juvenile hormone may regulate maturation of the testes and accessory glands in the newly eclosed adult insect and may also affect reproductive behavior.

12.2.4 Transfer of sperm to the seminal vesicle

In some Heteroptera, in *Chortophaga* (Orthoptera) and possibly in other insects the sperm make a complex circuit of the testis follicle before they leave the testis, moving in a spiral path to the region of the secondary spermatocytes and then turning back and passing into the vas deferens. In *Chortophaga* the movement occurs after the spermatodesm is released from the cyst, but in the heteropteran *Leptocoris* the sperm are still enclosed in the cyst. In this case, the displacement starts while the spermatids are still differentiating and is at least partly due to the elongation of the cyst that occurs during sperm development.

The fate of the cyst-cells is variable. In *Prionoplus* they break down in the testis, but in *Popillia*, although the sperm escape from the cysts as they leave the testis, the cyst-cells accompany the sperm in the seminal fluid into the bursa of the female. Here they finally break down and it has been suggested that they release glycogen used in the maintenance of the sperm.

In some Lepidoptera the release of sperm from the testis occurs in the pharate adult. At first this is inhibited by the high titer of ecdysteroid in the hemolymph, but release is permitted when this drops to a low level and then occurs with a circadian rhythmicity. Sperm start to move into the vas deferens toward the end of the light phase and remain there during the scotophase. Then, early in the next light phase, they are moved to the seminal vesicles. The cells of the upper vas deferens also show secretory activity, which is at a maximum when the sperm move out of the testes and again when the sperm are transferred to the seminal vesicles. The sperm are inactive in the vas deferens and are carried along by peristaltic movements of the wall of the tube. They remain immobile in the seminal vesicle, where they are often very tightly packed and in some cases, as in *Apis*, the heads of the sperm are embedded in the glandular wall of the vesicle.

12.3 Transfer of sperm to the female

Sperm transfer from a male to a female involves several different activities: the location of one sex by the other, courtship, pairing, copulation and, finally, the insemination of the female. Location of one sex by the other may involve the visual, auditory and olfactory senses. These are considered in Chapters 25–27. This section deals with the mechanisms of copulation and insemination of the female.

12.3.1 External reproductive organs of the male

The external reproductive organs of the male are concerned in coupling with the female genitalia and with the intromission of sperm. They are known collectively as the genitalia.

Male genital morphology typically exhibit patterns of highly divergent evolution. There is considerable variation in the terminology used in describing the genitalia in different orders, and problems in homologizing the different structures. Nevertheless, the basic elements are derived from a pair of primary phallic lobes which are present in the posterior ventral surface of segment 9 of the embryo (Fig. 12.7a). They are commonly regarded as representing limb buds and the structures arising from them as derived from typical appendages. However, they may represent ancestral penes rather than appendages of segmental origin. These phallic lobes divide to form an inner pair of mesomeres and outer parameres, collectively known as the phallomeres (Fig. 12.7b). The mesomeres unite to form the aedeagus, the intromittent organ. The inner wall of the aedeagus, which is continuous with the ejaculatory duct, is called the endophallus, and the opening of the duct at the tip of the aedeagus is the phallotreme (Fig. 12.7c). The gonopore is at the outer end of the ejaculatory duct, where it joins the endophallus and hence is internal, but in many insects the endophallic duct is eversible and so the gonopore assumes a terminal position during copulation. The parameres develop into claspers,

which are very variable in form. They may be attached with the aedeagus on a common base, the phallobase, and in many insects these basic structures are accompanied by secondary structures on segments 8, 9 or 10. The term "phallus" is sometimes used as a collective noun for the parameres and the aedeagus, but is also often used to mean the aedeagus alone; "penis" is sometimes used instead of "phallus."

No intromittent organ is present in Collembola or Diplura. Male Archaeognatha and Thysanura have terminal segments similar to those in females, but with a median phallus which is bilobed in Thysanura. In these groups, sperm are not transferred directly to the female (see Section 12.3.3). Paired penes are present in Ephemeroptera and some Dermaptera, but in the majority of pterygote insects there is a single median aedeagus. This is protected from injury in various ways. In grasshoppers and fulgorids, the sternum of the last abdominal segment extends to form a subgenital plate (Fig. 11.7a). In many Endopterygota, protection is afforded by withdrawal of the genital segments within the preceding abdominal segments.

Structures on the intromittent organ can function in coercing unreceptive females to copulate, in prolonging copulations and in defending females from take-over attempts by other males. In *Collosobruchus* genital spines anchor the male into position during copulation. These spines cause considerable damage to the internal reproductive tract of females, generating sexual conflict over copulation duration. Other components of the male external genitalia are also concerned with grasping

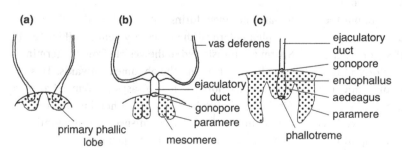

Figure 12.7 Diagrams illustrating the origin and development of the phallic organ (after Snodgrass, 1957).

the female. They are often called claspers. They may be derived from the parameres, from cerci, as in Dermaptera and many Orthoptera, or from the paraprocts, as in Zygoptera and some Tridactyloidea. In many Plecoptera, and occasionally in other orders, there are no claspers, the sexes being held together by the fit of the intromittent organ into the female bursa.

Many male Diptera have the terminal abdominal segments rotated so that the relative positions of the genitalia are altered. In Culicidae, some Tipulidae, Psychodidae, Mycetophilidae and some Brachycera, segment 8 and the segments behind it rotate through 180° soon after eclosion. As a result, the aedeagus comes to lie above the anus instead of below it and the hindgut is twisted over the reproductive duct (Fig. 12.8a). The rotation may occur in either a clockwise or an anticlockwise direction. In *Calliphora*, and probably in all Schizophora, the terminal segments are rotated through 360° in the pupa so that the genitalia are in their normal positions at eclosion, but the movement is indicated by some asymmetry of the preceding sclerites and by the ejaculatory duct looping right around the gut (Fig. 12.8b). The extent to which different segments rotate varies in different groups. Among the Syrphidae a total twist of 360° is achieved by two segments rotating through 90° and one through 180° so that there is an obvious external asymmetry. Temporary rotation of the genital segments during copulation occurs in some other insects, such as Heteroptera (see Fig. 12.10).

Male Odonata differ from all other insects in having the intromittent organs on abdominal segments 2 and 3. Appendages which are used to clasp the female are present on segment 10, but the genital apparatus on segment 9 is rudimentary. A depression on the ventral surface of segment 2, known as the genital fossa, opens posteriorly into a vesicle derived from the anterior end of segment 3. In Anisoptera, the vesicle connects with a three-segmented penis and laterally there are various accessory lobes which guide and hold the tip of the

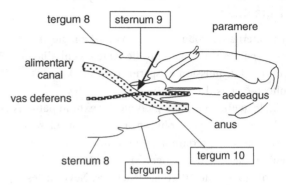

(a) 180° rotation

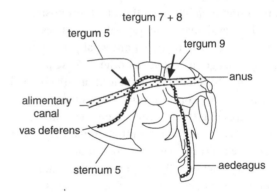

(b) 360° rotation

Figure 12.8 Diagrams illustrating rotation of the terminal segments in male Diptera. Arrows show points at which the relative positions of alimentary canal and vas deferens change (from Séguy, 1951). (a) Rotation of the ninth and following segments through 180°. Notice that segments 9 and 10 are inverted (boxed labels) (*Aedes*). (b) Rotation through 360° indicated by the vas deferens twisting over the hindgut in a muscid.

female abdomen during intromission; the whole complex is termed the accessory genitalia (Fig. 12.9). Sperm are transferred to the vesicle from the terminal gonoduct by bending the abdomen forwards. This may occur before the male grasps the female, as in *Libellula*, or after he has grasped her, but before copulation, as in *Aeschna*. In species of dragonfly that copulate when settled, the penis serves to

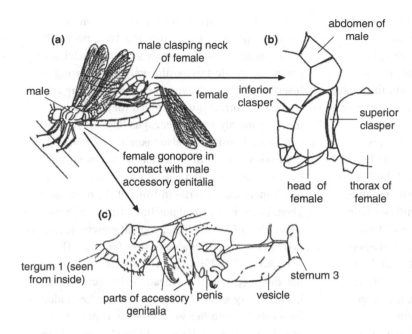

Figure 12.9 Mating in Odonata. (a) Male and female copulating (*Aeschna*) (after Longfield, 1949). (b) Position of the male claspers around the neck of the female (*Aeschna*) (after Tillyard, 1917). (c) Male accessory genitalia, terga of left side removed (based on Chao, 1953).

remove sperm of other males already in the female's spermatheca before he introduces his own sperm.

12.3.2 Pairing and copulation

When the sexes meet, the male will either engage the female in copulation directly, or in some species will deliver courtship behavior during which he will stimulate the female via acoustic, olfactory, visual and/or tactile means, in order to encourage the female to engage in copulation. One sex commonly mounts onto the back of the other. In cockroaches and some gryllids and tettigoniids the female climbs onto the back of the male. Another common position is with the male on the female; this occurs, for instance, in Tabanidae. Sometimes, as in Acrididae, although the male sits on top of the female, his abdomen is twisted underneath her during copulation. The abdomen of the male is also twisted under the female in insects, such as the scorpion fly (*Panorpa*), which lie side by side at the start of copulation. In other groups, the insects pair end to end and, in this case, the terminal segments of the

male are often twisted through 180°. The end-to-end position is achieved with the male on his back in some tettigoniids and a few Diptera, while in Culicidae male and female lie with their ventral surfaces in contact.

In pairing, the male often grasps the female with his feet. In *Aedes aegypti*, for instance, the insects lie with their ventral surfaces adjacent and the male holds the female's hindlegs in a hollow of the distal tarsomere by flexing back the pretarsus. His middle and hindlegs push up the female abdomen until genital contact is established and then his middle legs may hook onto the wings of the female, while his hindlegs hang free. Some male Hymenoptera, such as *Ammophila*, hold the female with their mandibles instead of, or, in some species, as well as, the legs.

The legs of males are sometimes modified for grasping the female. For example, the forelegs of *Dytiscus* and some other beetles bear suckers (see Fig. 8.8b); spines on the middle femora of *Hoplomerus* (Hymenoptera) fit between the veins on the female's wings; and the hind femora of male *Osphya* (Coleoptera) are modified to grip the female's abdomen and elytra. Male fleas (Siphonaptera) and some male

Collembola and Ischnocera (biting lice) have modified antennae with which they hold the female.

Male Odonata are exceptional in their manner of holding the female. At first, the male grasps the thorax of the female with his second and third pairs of legs, while the first pair touch the basal segments of her antennae. He then flexes his abdomen forwards and fits two pairs of claspers on his abdominal segment 10 into position on the head or thorax of the female. This completed, he lets go with his legs and the two insects fly off "in tandem." The claspers consist of superior and inferior pairs and, in Anisoptera, the superior claspers fit around the neck of the female while the inferior claspers press down on top of her head (Fig. 12.9b). In most Zygoptera, the claspers grip a dorsal lobe of the pronotum and, in some Coenagriidae, they appear to be cemented on by a sticky secretion.

Copulation may occur immediately after the insects have paired or there may be a considerable interval before they copulate. In the cockroach *Eurycotis*, where the female climbs on the male's back, the signal that the insects are in an appropriate position for copulation comes from mechanosensitive hairs on the male's first abdominal tergite. The hairs are stimulated when the female attempts to feed on a secretion produced by glands close to the hairs and this only occurs when she is appropriately placed for the male to copulate. Once the male and female genitalia are linked, the insects may alter their positions and it is common among Orthoptera and Diptera for an end-to-end position to be adopted at this time.

The details of copulation vary from group to group depending on the structure of the genitalia, and only a few examples are given. In Acrididae, the tip of the male's abdomen is twisted below the female and the edges of the epiphallus (a plate on top of the genital complex) grip the sides of the female's subgenital plate and draw it down into the male's anal depression. The male uses his cerci to hold the female's abdomen and the aedeagus is inserted between the ventral valves of the ovipositor.

The male of *Oncopeltus* mounts the female, the genital capsule is rotated through 180°, mainly by muscular action and the male's parameres (claspers) grasp the female's ovipositor valves. Following insertion of the aedeagus, the insects assume an end-to-end orientation in which they are held together mainly by the aedeagus (Fig. 12.10). An end-to-end position is also taken up by *Blattella* (Blattodea), but at first the female climbs on the back of the male, who engages the hook on his left phallomere on a sclerite in front of the ovipositor. Then, in the end-to-end position, the lateral hooks on either side of the anus and a small crescentic sclerite take a firm grip on the ovipositor (Fig. 12.11).

Copulation in Odonata involves the male flexing his abdomen so that the head of the female touches his accessory genitalia; she then brings her abdomen forwards beneath her so as to make contact with the accessory genitalia (Fig. 12.9a,c). Some species, such as *Crocothemis*, copulate and complete sperm transfer in flight and, in these insects, copulation is brief, lasting less than 20 seconds. Many species, however, settle before copulating and the process may last for a few minutes or an hour or more. During most of this time the male is removing sperm already present in the female's spermatheca; sperm transfer takes only a few seconds.

The duration of copulation in other insects is equally variable. In mosquitoes the process is complete within a few seconds, while in *Oncopeltus* the insects may remain coupled for five hours, in *Locusta* for 8–10 hours and in *Anacridium* (Orthoptera) for up to 60 hours. Insemination is completed much more rapidly than this; in *Locusta* sperm reach the spermatheca within two hours of the start of copulation. The duration of post-insemination coupling can depend on the density of males in the population. In *Nezara* copulation lasts for 1–2 days when the population consists mainly of females, but is extended to as long as seven days when males are in the majority. Extended copulations serve to guard against further mating by the female and thus protect the copulating male from sperm competition.

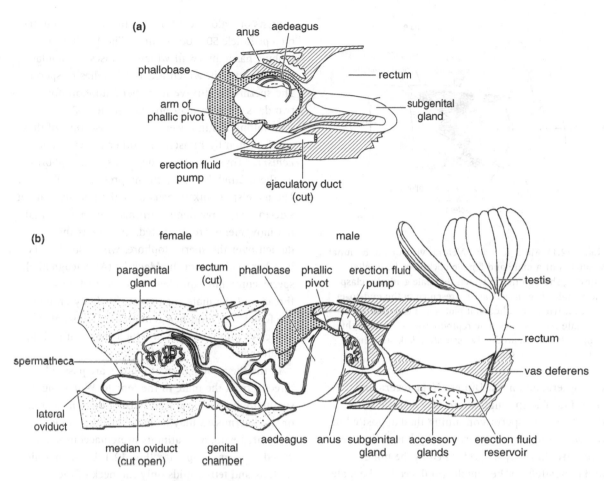

Figure 12.10 Male genitalia and direct insemination (*Oncopeltus* [Hemiptera]) (after Bonhag and Wick, 1953). (a) Sagittal section of the male genital capsule with the aedeagus retracted. (b) Sagittal section through the terminal segments of a copulating pair. Notice the inversion of the male genital capsule (with the anus now ventral to the phallic pivot) and the insertion of the aedeagus into the spermatheca.

12.3.3 Insemination

In the insects, the transfer of sperm to the female (insemination) is a quite separate process from fertilization of the eggs, which in some cases does not occur until months or even years later. During this interval the sperm are stored in the female's spermatheca. Sperm may be transferred in a spermatophore produced by the male; they may be passed directly into the spermatheca without a spermatophore being produced; or they may be passed into the bursa copulatrix of the female

from which they migrate to the spermatheca under their own motility and/or via the aid of the spermathecal ducts.

Spermatophores The primitive method of insemination in insects involves the production by the male of a spermatophore, a capsule in which the sperm are conveyed to the female. Spermatophores, of varying complexity, are produced by the Apterygota, Orthoptera, Blattodea, some Heteroptera, all the Neuroptera except Coniopterygidae, some Trichoptera, Lepidoptera, some Hymenoptera and

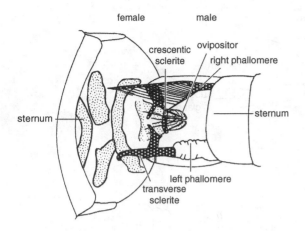

female   male

crescentic ovipositor
sclerite   right phallomere

sternum

sternum

left phallomere
transverse
sclerite

Figure 12.11 Male genitalia. Ventral view of the terminal segments of a male and female cockroach (*Blattella*) showing the manner in which the male genitalia clasp the female. The insects are represented in the end-to-end position with the subgenital plates and endophallus of the male removed. Female reproductive sclerites, light stippling; male reproductive sclerites, dark stippling (after Khalifa, 1950).

Coleoptera, and a few Diptera Nematocera and *Glossina*. The spermatophore may be little more than a drop of sperm-containing fluid deposited in the environment, but more usually it is a discrete structure that may be preformed by the male and deposited on the female, produced by the male when he encounters a female or produced by male secretions in the female ducts during copulation.

Structure and transmission. In Collembola the male deposits spermatophores on the ground quite independently of the presence of a female. The spermatophore consists of a sperm-containing droplet, without any surrounding membrane, mounted on top of a stalk which is often about 500 μm high. Sometimes spermatophores are produced in aggregations of Collembola, so there is a good chance of a female finding one and inserting it into her reproductive opening, while, in other cases, the male grasps the female by her antennae and leads her over the spermatophore. The spermatophores of *Campodea* (Diplura) are also produced in the absence of the female and, like those of Collembola, each one

consists of a globule 50–70 μm in diameter mounted on a peduncle 50–100 μm high (Fig. 12.12a). The globule has a thin wall which encloses a granular fluid, floating in which are 1–4 bundles of sperm. The sperm can survive in a spermatophore for two days. A male may produce some 200 spermatophores in a week, but at least some of these will be eaten by himself and other insects. Male *Lepisma* (Thysanura) also deposit spermatophores on the ground, but only in the presence of females. The male spins silk threads over the female, making side-to-side movements with his abdomen, so that her movements are restricted, and she is then guided over the spermatophore, which she inserts into her genital duct. In *Machilis* (Archaeognatha), sperm-containing droplets are deposited on a thread and the male then twists his body around the female, guiding her genitalia with his antennae and cerci into positions in which they can pick up the droplets.

In the Pterygota spermatophores are passed directly from the male to the female. Where the spermatophore is produced outside the female, one or two sperm sacs may be embedded in, or closely associated with, a gelatinous proteinaceous mass called the spermatophylax (Fig. 12.12b). In phasmids, gryllids and tettigoniids only the neck of the spermatophore penetrates the female ducts; the body of the structure remains outside and is liable to be eaten by the female or other insects (see below). In Blattodea the body of the spermatophore, although still outside the female ducts, is protected by the female's enlarged subgenital plate. The signal to transfer a spermatophore is provided to a male cricket by mechanical stimulation of small trichoid sensilla in a cavity enclosed by his epiphallus. These hairs are stimulated by the female's copulatory papillae when the insects copulate.

The spermatophore is specialized in some Acrididae to form a tube which is effectively a temporary elongation of the intromittent organ (Fig. 12.12c). It consists of two basal bladders in the ejaculatory and sperm sacs of the male (Fig. 12.13)

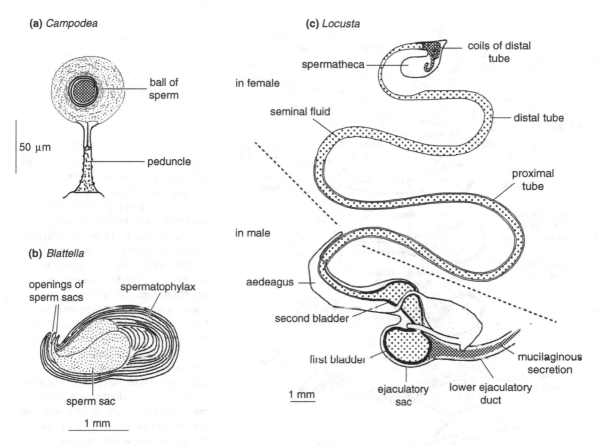

(a) *Campodea*

ball of sperm

50 μm

peduncle

(b) *Blattella*

openings of sperm sacs

spermatophylax

sperm sac

1 mm

(c) *Locusta*

in female

coils of distal tube

spermatheca

seminal fluid

distal tube

proximal tube

in male

aedeagus

second bladder

first bladder

mucilaginous secretion

ejaculatory sac

lower ejaculatory duct

1 mm

Figure 12.12 Spermatophores. (a) Produced independently of the female and left on the ground (*Campodea* [Diplura]) (after Bareth, 1964). (b) Deposited at the opening of the female duct (*Blattella* [Blattodea]) (after Khalifa, 1950). (c) Produced within the male and female ducts during copulation and connecting directly with the spermatheca. See Fig. 12.13a for details of male genitalia (*Locusta* [Orthoptera]) (after Gregory, 1965).

leading to a tube which extends into the female's spermathecal bulb.

In most other insect groups, many species practice direct insemination without producing a spermatophore. Where a spermatophore is retained, it is often formed in the female's bursa copulatrix.

Spermatophore production. The spermatophore is produced from secretions of glands of the male's reproductive system, usually the accessory glands, but where these are absent or reduced, from glands in the ejaculatory duct (or the simplex in Lepidoptera). The secretions are molded in the ducts of the male and sometimes also the female. Secretions from different glands, or different cells within a gland, are produced in sequence to form separate parts of the spermatophore.

Production of the spermatophore by *Locusta* begins within two minutes of the start of copulation. Most of it is formed in the male, although the ducts of the female serve to mold the tubular part. The first secretion builds up in the upper ejaculatory duct and is forced down through the funnel, the shape of which produces a series of folds, molding the secretion into a cylinder (Fig. 12.13). A white semi-fluid secretion is then forced into the core of the cylinder so that it becomes a tube (Fig. 12.13b).

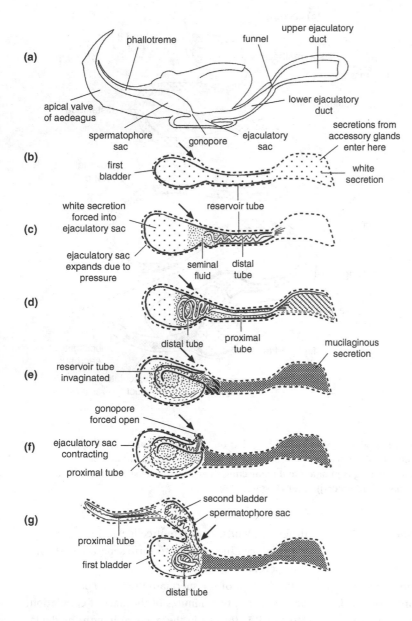

Figure 12.13 Spermatophore production in *Locusta* (after Gregory, 1965). Compare the fully developed spermatophore in Fig. 12.12c. (a) Sagittal section through the male ejaculatory duct and genitalia. Diagrams (b)–(g) are aligned with this. Arrow in (b)–(g) indicates the position of the gonopore. (b) The first secretion is molded to form the first bladder and reservoir tube by the white secretion. Ejaculatory duct shown as broken line. (c) After movement of the seminal fluid and sperm into the bladder, further secretions form the distal tube. (d) The proximal tube of the spermatophore is formed. (e) A mucilaginous secretion invaginates the reservoir tube inside the first bladder (distal tube not shown, see (d)). (f) Continued pressure and contractions of the ejaculatory sac start to force the end of the proximal tube through the gonopore (previously closed) (distal tube not shown, see (d)). (g) Continued contraction of the ejaculatory sac everts the reservoir tube into the spermatophore sac where it forms the second bladder. Similar pressure forces the proximal and then the distal tubes into the female ducts (see Fig. 12.12c).

This is enlarged in the ejaculatory sac to form the first bladder, while the part remaining in the ejaculatory duct (known at this time as the reservoir tube) ultimately forms the second bladder in the spermatophore sac. At this stage seminal fluid is passed into the rudimentary spermatophore and then a separate cylinder of material is formed and pushed into the bladder, where it becomes coiled up (Fig. 12.13c). This will form the distal tube and a further series of secretions forms the proximal tube (Fig. 12.13d). As the last part of the proximal tube enters the bladder it draws the wall of the reservoir tube with it, so that this becomes invaginated, and finally the whole of the tube except for the tip is pushed inside the bladder by a mucilaginous secretion (Fig. 12.13e).

At this time the ejaculatory sac starts to contract and squeezes the tube of the spermatophore out of the bladder, while the pressure of mucilage in the ejaculatory duct forces the tip backwards through the gonopore, which is now open, and out through the aedeagus into the duct of the female's spermatheca (Fig. 12.13f,g). This process involves the tube being turned inside out and finally the second bladder is everted and molded in the

sperm sac. In *Gomphocerus* the tube is not turned inside out, but elongates by expansion.

The spermatophore of Lepidoptera is formed wholly within the female ducts after the start of copulation. Here, the different glands function in sequence, starting with the section of the simplex nearest the aedeagus (Fig. 12.14). The aedeagus projects into the female's bursa copulatrix and the secretion of the lower region of the simplex forms a pearly body (not shown in the figure) and the wall of the spermatophore. This is followed by the contents

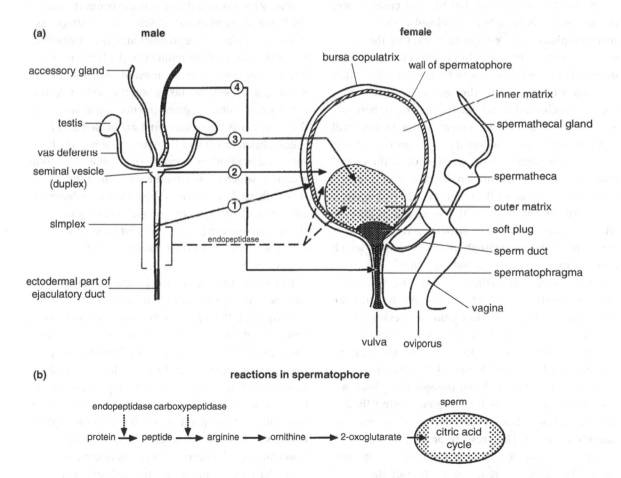

Figure 12.14 (a) Male and part of the female reproductive systems of the silk moth, *Bombyx*, showing how secretions from different parts of the male system contribute to different parts of the spermatophore (shown by corresponding shading). Note that the female system is shown at a much larger scale. Numbers show the sequence in which secretions are produced from different regions (based on Osanai *et al.*, 1987, 1988). (b) Chemical reactions occurring in the spermatophore which produce the respiratory substrate for sperm activity.

of the seminal vesicle, the sperm and seminal fluid, partly mixing with them to form the inner matrix of the spermatophore. The secretion from the lower parts of the accessory glands forms the outer matrix, and, finally, the secretions from the ends of the glands form the spermatophragma, which blocks the duct to the female's bursa copulatrix.

Transfer of sperm to the spermatheca. Immediately following the transfer of the spermatophore, the sperm migrate to the spermatheca, where they are stored. Sometimes they are able to escape from the sperm sac through a pore, but in other cases, where the sperm sac is completely enclosed within the spermatophore, they escape as a result of the spermatophore rupturing. In Lepidoptera and *Sialis*, the inside of the bursa copulatrix is lined with spines or bears a toothed plate, the signum dentatum, to which muscles are attached. The spermatophore is gradually abraded by movements of the spines until it is torn open. In *Rhodnius* the first sperm reach the spermatheca within about ten minutes of the end of mating, while in *Acheta* transfer takes about an hour, and in *Zygaena* (Lepidoptera) 12–18 hours.

The transfer of sperm to the spermatheca may be either active or passive. In *Acheta* sperm are held in the body of the spermatophore (the ampulla), which remains external to the female, and the spermatophore is specialized to force the sperm out into the female ducts. An outer reservoir of fluid, the evacuating fluid, with a low osmotic pressure is separated by an inner layer with semipermeable properties from an inner proteinaceous mass called the pressure body, which has a high osmotic pressure (Fig. 12.15a). When the spermatophore is deposited, water passes from the evacuating fluid into the pressure body because of the difference in osmotic pressure. The pressure body produces a transparent material and swells, forcing the sperm out of the ampulla and down the tube of the spermatophore into the spermatheca (Fig. 12.15b–d). In *Locusta*, also, the sperm in the spermatophore are initially outside the female (in this case, in the first bladder of the spermatophore). From here they are pumped along the spermatophore by contractions of the ejaculatory sac, first appearing in the spermatheca about 90 minutes after the start of copulation.

In many insects the spermatophore is placed in the bursa copulatrix of the female and the transfer of sperm to the spermatheca is probably brought about by the contractions of the female ducts. An opaque secretion from the male accessory glands of *Rhodnius* injected into the bursa with the spermatophore induces rhythmic contractions of the oviducts, probably by way of a direct nervous connection from the bursa to the oviducal muscles. The contractions cause shortening of the oviduct and, it is suggested, cause the origin of the oviduct in the bursa to make bite-like movements in the mass of semen in the bursa so that sperm are taken into the oviduct. As this process continues, the more anterior sperm are forced forwards along the oviduct and are passed into the spermatheca. In Diptera Nematocera, it is probable that fluid is absorbed from the spermatheca, creating a current which transports the sperm into the spermatheca. In contrast to these examples, however, the sperm of *Bombyx* are activated before they leave the spermatophore (see below), and they may contribute actively to their transfer to the spermatheca.

Fate of the spermatophore. Females of some species eject the spermatophore some time after insemination. *Blattella* and *Rhodnius*, for instance, drop the old spermatophores some 12 and 18 hours, respectively, after copulation. The female *Sialis* pulls the spermatophore out and eats it; this commonly happens in Blattodea, where specific post-copulatory behavior often keeps the female occupied for a time to ensure that the sperm have left the spermatophore before she eats it. In some grasshoppers, such as *Gomphocerus*, the spermatophore is ejected by muscular contraction of the spermathecal duct.

The spermatophore is dissolved by proteolytic enzymes in other insects, such as Lepidoptera and Trichoptera, and in many Trichoptera one or two days after copulation only the sperm sac remains.

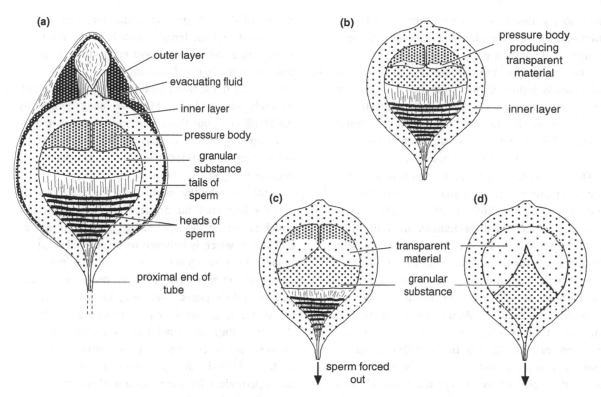

Figure 12.15 Spermatophore of the house cricket (*Acheta*, Orthoptera) (after Khalifa, 1949). (a) Horizontal section through the ampulla which remains outside the female. (b) The pressure body starts to swell and releases a transparent material due to the absorption of evacuating fluid from outside the inner layer. In (b)–(d) the outer layer and evacuating fluid are not shown. (c) Production of the transparent material continues, starting to push sperm out of the ampulla. (d) Sperm have been completely forced out of the ampulla.

In the wax moth *Galleria* digestion is complete in ten days, but the neck of the spermatophore persists. The spermatophore of *Locusta* breaks when the two sexes separate, either where the tube fits tightly in the spermathecal duct or at its origin with the bladders in the male. The part remaining in the male is ejected within about two hours by the contractions of the copulatory organ, while in the female the distal tube disappears within a day, presumably being dissolved, but the proximal tube dissolves much more slowly, and persists for several days until it is ejected, probably by contractions of the spermathecal duct. In *Chorthippus* the activity of the enzyme-secreting cells of the spermatheca is controlled by the corpora allata.

Direct insemination Various groups of insects have dispensed with a spermatophore; sperm are transferred directly to the bursa copulatrix, the female ducts, and even into the spermatheca, by the penis, which may be long and flagelliform for this purpose. Such direct insemination occurs in some members of the Orders Heteroptera, Mecoptera, Trichoptera, Hymenoptera, Coleoptera and Diptera.

Direct insemination occurs in *Aedes aegypti*; in this insect, the paraprocts expand the genital orifice of the female while the aedeagus is erected by the action of muscles attached to associated apodemes. The aedeagus only penetrates just inside the female opening, where it is held by spines engaging with a valve of the spermatheca. A stream of fluid from the accessory glands is driven along the ejaculatory duct and into the female by contractions of the glands, and sperm are injected into the stream by the contractions of the seminal vesicles. Thus a mass of semen is deposited inside the atrium of the female and from here the sperm are transferred to the spermatheca. The sperm of *Drosophila* are similarly deposited in the vagina and then pass to the seminal receptacle and spermatheca.

Oncopeltus has a long penis that reaches into the spermatheca and deposits sperm directly into it (Fig. 12.10). Erection of the phallus in this insect is a specialized mechanism involving the displacement of an erection fluid into the phallus from a reservoir of the ejaculatory duct. The fluid is forced back from the reservoir by pressure exerted by the body muscles, and this pressure is maintained throughout copulation. At the end of the ejaculatory duct, the fluid is forced into a vesicle and then pumped into the phallus. In those Coleoptera and Hymenoptera with a long penis, erection is probably produced by an increase in blood pressure resulting from the sudden contraction of the abdominal walls.

Hemocoelic insemination In some Cimicoidea and Miridae (Hemiptera) the sperm, instead of being deposited in the female reproductive tract, are injected into the hemocoel. A good deal of variation occurs between the species practicing this method and they can be arranged in a series showing progressive specialization. In *Alloeorhynchus flavipes* the penis enters the vagina, but a spine at its tip perforates the wall of the vagina so that the sperm are injected into the hemocoel. They are not phagocytosed immediately, but disperse beneath the integument and later collect under the peritoneal membrane surrounding the ovarioles, possibly being directed chemotactically. The sperm adjacent to the lowest follicle penetrate the follicular epithelium and fertilize the eggs via the micropyles.

Primicimex shows a further separation from the normal method of insemination. Here, the left

clasper of the male penetrates the dorsal surface of the abdomen of the female, usually between tergites 4 and 5 or 5 and 6. The clasper ensheathes the penis, and sperm are injected into the hemocoel. They accumulate in the heart and are distributed around the body with the blood. Many are phagocytosed by the blood cells, but those that survive are stored in two large pouches at the base of the oviducts. The holes made in the integument by the claspers become plugged with tanned cuticle.

In other species the sperm are not injected directly into the hemocoel, but are received into a special pouch called the mesospermalege or organ of Ribaga or Berlese, which is believed to be derived from blood cells. Other genera have a cuticular pouch, called the ectospermalege, for the reception of the clasper and the penis. There may be one or two ectospermalegia and their positions vary, but in *Afrocimex* they are situated in the membrane between segments 3 and 4 and segments 4 and 5 on the left-hand side. *Xylocoris galactinus* has a mesospermalege for the reception of sperm immediately beneath the ectospermalege (Fig. 12.16a). The mesospermalege is formed from vacuolated cells surrounding a central lacuna into which the sperm are injected and from here they move down a solid core of cells, forming the conducting lobe into the hemocoel. They finally arrive at the conceptacula seminis at the bases of the lateral oviducts, where they accumulate. In *Cimex* this migration takes about 12 hours and after the female takes her next blood meal the sperm are carried intracellularly in packets to the ovaries through special conduit cells. At the base of each ovariole they accumulate in a corpus seminalis derived from the follicular cells.

In Strepsiptera sperm also pass into the hemocoel to fertilize the eggs, but they do so via the genital canals of the female (see Fig. 14.30).

In *Anthocoris* and *Orius* perforation of the integument is not necessary because a copulatory tube opens on the left, between the sternites of segments 7 and 8, and passes to a median sperm

(a) *Xylocoris*

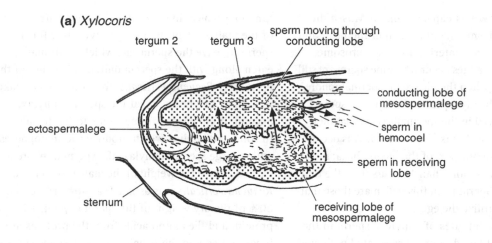

(b) *Orius*   **(c)** *Orius*

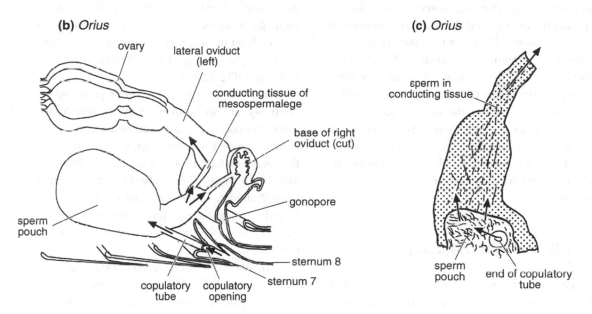

Figure 12.16 Hemocoelic insemination in Hemiptera. Arrows show direction of sperm movement. (a) Longitudinal section through the ectospermalege and mesospermalege of *Xylocoris galactinus* about one hour after copulation (after Carayon, 1953a). (b) Diagram of the internal reproductive organs of female *Orius*. (c) Longitudinal section of part of the mesospermalege of *Orius* showing the sperm pouch and conducting tissue (after Carayon, 1953b).

pouch, where the sperm accumulate (Fig. 12.16b,c). From here the mesospermalege forms a column of conducting tissue along which the sperm pass to the oviducts so that they are never free in the hemocoel.

Sperm capacitation   In a number of species sperm undergo changes after they are transferred to the female spermatheca and, in some cases at least, these changes are essential before the sperm can fertilize an egg. This process of maturation of sperm within the female is known as capacitation. In *Musca* the sperm lose the plasma membrane from the head and most of the granular material from the acrosome.

More obvious changes occur where the sperm are still grouped together when they leave the spermatophore. In grasshoppers the glycoprotein binding the sperm together is dissolved in the spermatheca so that they separate. The glycocalyx is also removed. This occurs within six hours in one part of the spermatheca; in another part it takes more than 15 hours from the time of insemination. Sperm from this region are those that will ultimately fertilize the eggs.

The breakup of bundles of eupyrene sperm in the spermatophore of Lepidoptera is associated with, and probably caused by, the activation of apyrene sperm. A specific protease from the lower simplex digests the glycoprotein covering the flagellar membrane of the apyrene sperm. The same protease liberates arginine from proteins in the seminal fluid, and the arginine is metabolized to 2-oxoglutarate, a preferred substrate for the sperm (Fig. 12.14). Cyclic AMP is also necessary to activate the sperm. Once the apyrene sperm of *Bombyx* are activated, they appear to break up the eupyrene sperm bundles by their physical activity. A protease liberating arginine and cAMP are also necessary for sperm activation in grasshoppers.

12.4 Other effects of mating

In the course of sperm transfer, secretions of the male accessory glands are transferred to the female. In *Drosophila* 133 seminal fluid proteins have been found, and these may have various effects on the behavior and physiology of females after insemination.

12.4.1 Nutritive effects of mating

In insects such as crickets and tettigoniids, where the spermatophore includes a spermatophylax, the latter is usually eaten by the female and serves two functions which, in different ways, favor the survival of the male's offspring. First, it gives time for the sperm to leave the sperm sac, which is ultimately eaten along with the spermatophylax. Females of the tettigoniid *Requena verticalis*, for example, can take as long as five hours to eat the spermatophylax. Second, it provides extra nutrients to the female. In some tettigoniids which produce spermatophores with a large spermatophylax, the spermatophore can exceed 20% of the weight of the male producing it; in *Requena verticalis* it is as much as 40%. More than 80% of the dry weight of the spermatophylax is protein, and the amino acids from the proteins are incorporated into the female tissues after she eats the spermatophore, increasing the weight of oocytes in the ovaries. More than one-third of the protein of each egg produced by a zaprochiline tettigoniid is derived from the male. This material permits the female to produce bigger and, in some species, more eggs (Fig. 12.17).

Even where the spermatophore is produced internally, and so is not eaten by the female, the male may contribute to the nutrient pool of the female. Amino acids, and even some proteins, from the spermatophore of *Melanoplus* are subsequently found in the eggs. This is also known to occur in some Blattodea, Lepidoptera and Coleoptera. Likewise, when insemination is direct, proteins from the seminal fluid can contribute to egg production.

In grasshoppers belonging to the subfamilies Cyrtacanthacridinae, Melanoplinae and Pyrgomorphidae, several spermatophores are produced during a single copulation. *Melanoplus* produces about seven and *Schistocerca* six at each normal mating. These spermatophores are simple sac-like structures quite different from those of *Locusta* and related species. A single spermatophore contains sufficient sperm to fertilize several batches of eggs and it is possible that mating after the first oviposition is more important in maintaining fecundity than in maintaining fertility. Although only one spermatophore is produced each time

(a) number of oocytes

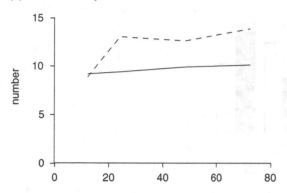

(b) egg weight

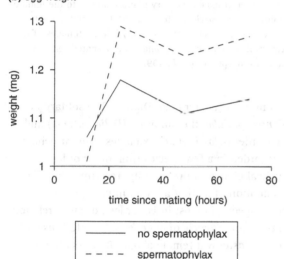

time since mating (hours)

——— no spermatophylax

– – – spermatophylax

Figure 12.17 Nutritional value of the spermatophylax in a tettigoniid. Insects were either allowed to eat the spermatophylax in the normal way, or the spermatophylax was removed so that it could not be eaten (after Simmons, 1990). (a) Total number of developing oocytes in the ovaries. (b) Weight of one chorionated egg.

Chorthippus mates, repeated mating by the female enhances both the rate of egg production and the number of eggs per pod irrespective of whether the insects have ample food or only a limited supply. There is also evidence that multiple mating by *Drosophila mojavensis* contributes material for oogenesis.

Some butterflies exhibit puddling behavior, aggregating at muddy pools and drinking. A principal function of this behavior is to accumulate sodium. In many species only males exhibit this behavior and they may be anatomically, and presumably physiologically, adapted to accumulate sodium which they then transfer to the females during copulation. Males of the skipper butterfly *Thymelicus lineola* transfer more than 10% of their body weight as sodium to females in this way. The sodium increases female fertility.

12.4.2 Enhancement of female oviposition

In addition to the possible nutrient effects of seminal fluids that accompany sperm transfer in many insects, copulation may also be a trigger for oviposition and sometimes oogenesis. Virgin females usually do not lay eggs, or lay relatively few. Mating, or the experimental injection of a component of the male accessory glands, causes them to oviposit (Fig. 12.18).

In *Drosophila* and the beetle, *Acanthoscelides*, the oviposition-enhancing compound is a peptide. In some other insects, grasshoppers, crickets and a mosquito, the active compound is known to be a protein. In addition, males of some crickets and lepidopterans transfer prostaglandin-synthesizing chemicals to the female during copulation. These have also been shown to stimulate oviposition.

Within the female, the peptides and proteins may act in one of two ways. Either they may stimulate the tissue of the reproductive tract to produce a hormone, or they may enter the hemocoel and affect a distant target site. In *Rhodnius* the oviposition-enhancing factor induces the tissues of the spermatheca and associated ducts to produce a hormone, known as the spermathecal factor. This causes some neurosecretory cells in the brain to release a peptide that produces contractions of the ovarian sheath leading to ovulation.

The sex peptide of *Drosophila* is derived from a larger peptide transferred to the female and then

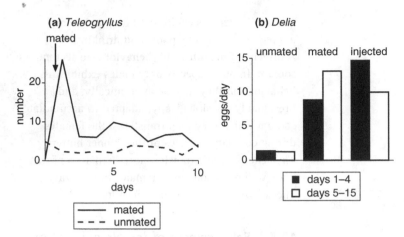

Figure 12.18 Enhancement of oviposition as a result of mating and transfer of male accessory gland material to the female. (a) Number of eggs per day laid by females of the Australian field cricket, *Teleogryllus*, after mating. Unmated females lay only a few eggs each day (after Stanley-Samuelson, 1994). (b) Average number of eggs per day laid by females of the onion fly, *Delia*, when unmated, mated or after injection of an extract of male accessory glands into unmated females. Bars show averages for the first four days and the next 11 days (data from Spencer *et al.*, 1992).

cleaved in the bursa copulatrix. After absorption into her hemolymph, it acts directly on the brain, at least with respect to its function in inhibiting female receptivity (see Section 12.4.3). The fecundity-enhancing compound produced by *Bombyx* acts directly on the terminal abdominal ganglion. In the mosquito, *Aedes aegypti*, a secretion from the male accessory glands also induces the female to respond to the stimuli of potential oviposition sites to which she was previously unresponsive.

12.4.3 Reduction of female's readiness to remate

Males of many species employ some mechanism to reduce the likelihood that a female with whom he has just mated will mate again. This can take the form of mate guarding, but often involves a male-induced effect on the female's ability or willingness to mate again. In some species the reduction in receptivity is permanent and the female never mates again; in others receptivity subsequently returns. Species in which the female mates only once are common

among Hymenoptera and Diptera. The solitary bee (*Centris pallida*), the onion fly (*Delia antiqua*) and tsetse flies (*Glossina*) are examples. Similar behavior is recorded in a few species from other orders. Cyclical changes in receptivity, resulting from the resumption of mating after an interval, occur in many species in most insect orders, often in relation to the female's cycle of oogenesis. Female *Drosophila melanogaster* will remate about 48 hours after a previous mating.

These changes in female behavior induced by the male may be associated with the physical blocking of the female ducts by the spermatophore or some other component of the male accessory gland secretion forming a mating plug. In other instances, components of the male accessory gland secretion affect female behavior without physically preventing further sperm transfer.

Mating plugs, distinct from spermatophores, are produced in a number of insects. In mosquitoes belonging to the genera *Anopheles*, *Aedes* and *Psorophora* a plug formed from the accessory gland secretions of the male is deposited in the genital chamber of the female, even though these insects do not produce a spermatophore. A similar plug is produced by some species of *Drosophila* and by the bumble bees, *Bombus*. A comparable structure is present in some Lepidoptera, where it is produced immediately after the spermatophore (Fig. 12.14). It is called the spermatophragma or sphragis and it effectively prevents further mating by the female. It may also ensure that sperm in the bursa copulatrix are transferred to the spermatheca rather than moving back toward the vulva.

Loss of receptivity by the female is sometimes regulated directly through neural pathways. For instance, females of *Nauphoeta* and *Gomphocerus* are unreceptive while there is a spermatophore in the spermatheca; cutting the ventral nerve cord results in the return of receptivity. In *Aedes* receptivity is also first switched off by mechanical stimulation resulting from filling the bursa copulatrix with seminal fluid. Subsequently, however, the female remains unreceptive for the rest of her life as a result of a substance, known as matrone, in the secretions of the male accessory glands. Matrone passes from the bursa copulatrix into the hemolymph and then acts directly on the central nervous system of the female. Receptor sites probably exist in the terminal abdominal ganglion. The male accessory gland secretions also affect host-seeking by female mosquitoes. A sex peptide also switches off female receptivity in *Drosophila*, but the sustained depression of sexual activity over a period of days is probably also dependent on the presence of sperm in the spermatheca. In the housefly, *Musca*, and the stable fly, *Stomoxys*, where accessory glands are absent, a secretion from the ejaculatory duct inhibits female receptivity (see below). Compounds that inhibit receptivity are also known to occur in Lepidoptera. Not only is female receptivity reduced, in *Helicoverpa* some chemical component of the spermatophore leads to an immediate reduction in the synthesis of sex attractant pheromone by the female (Fig. 12.19). While loss of sexual receptivity by females may serve male interests in preventing sperm competition, it may not be in the best interests of females, and can generate sexual conflict. In *Drosophila* sex peptide is toxic to females, reducing their lifespan.

12.4.4 Transfer of other ecologically relevant compounds

Some examples are known of males transferring chemicals which have ecological relevance to the species beyond the immediate context of reproduction. For example, males of the "Spanish fly" (*Lytta*, Coleoptera) synthesize cantharidin, presumably as a defensive substance. It is probably synthesized and stored in the accessory reproductive glands. Females do not synthesize the compound, but

(a) after normal mating

(b) after injection of accessory gland extract

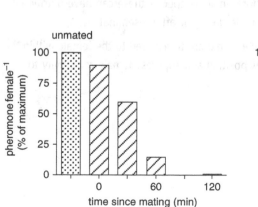
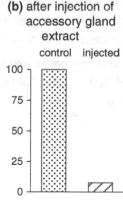

Figure 12.19 Reduction in pheromone production as a result of mating and transfer of male accessory gland material to the female of the moth, *Helicoverpa*. (a) Reduction in the amount of the sex attractant pheromone component, Z-11-hexadecanal, in the female following mating (after Raina, 1989). (b) Effect of injecting an extract of the male accessory glands on the amount of Z-11-hexadecanal in the female (data from Kingan et al., 1993).

large quantities are transferred to the female during copulation.

A similar transfer of compounds that confer protection against predators is known to occur in some Lepidoptera. Larvae of the moth *Utetheisa* feed on plants containing pyrrolizidine alkaloids, and adult males of the milkweed butterfly, *Danaus*, obtain the same chemicals by licking at the plant surface. In both species, males transfer large amounts of these compounds to the females during copulation. They are subsequently found in the eggs, where they have been shown to have a protective function. The spermatophore of *Utetheisa* is large, as much as 10% of the body weight, presumably in relation to its function in conveying nutrients and alkaloids to the female.

 ## Summary

- The internal anatomy of the male reproductive tract includes the testes, where sperm are manufactured, the vas deferens and seminal vesicles, where mature sperm are stored, and the ejaculatory duct and accessory glands that secrete proteins that contribute to the seminal fluid.

- The head of the mature spermatozoa consists of an acrosome and nucleus. The tail consists of the axoneme and mitochondrial derivative.

- Sperm cysts are produced at the apex of the testis follicle and grow to form spermatocytes as they move through the follicle. Following two meiotic divisions to produce spermatids, they develop into spermatozoa before being shed into the vas deferens and from there move to the seminal vesicle.

- Insect external genitalia exhibit considerable variability, and there are problems in homologizing structures across taxa. Some species lack intromittent structures and transfer sperm via spermatophores, while others have an aedeagus and endophalus that delivers the ejaculate directly into the female reproductive tract.

- In species where insemination is indirect, accessory glands produce structural proteins that form the spermatophore. In some species these can have nutritional value and are consumed by the female during or after insemination.

- Accessory glands also produce proteins that are transferred to the female within the seminal fluid, and can stimulate oviposition and suppress female receptivity to further mating.

Recommended reading

Avila, F. W., Sirot, L. K., LaFlamme, B. A., Rubinstein, C. D. and Wolfner, M. F. (2011). Insect seminal fluid proteins: identification and function. *Annual Review of Entomology* **56**, 21–40.

Chen, P. S. (1984). The functional morphology and biochemistry of insect male accessory glands and their secretions. *Annual Review of Entomology* **29**, 233–255.

Happ, G. M. (1992). Maturation of the male reproductive system and its endocrine regulation. *Annual Review of Entomology*, **37**, 303–320.

Jamieson, B. G. M., Dallai, R. and Afzelius, B. A. (1999). *Insects: Their Spermatozoa and Phylogeny.* Enfield: Science Publishers.

Mann, T. (1984). *Spermatophores.* Heidelberg: Springer-Verlag.

Simmons, L. W. (2001). *Sperm Competition and its Evolutionary Consequences in the Insects.* Princeton, NJ: Princeton University Press.

Thornhill, R. and Alcock, J. (1983). *The Evolution of Insect Mating Systems.* Cambridge, MA: Harvard University Press.

Vahed, K. (1998) The function of nuptial feeding in insects: a review of empirical studies. *Biological Reviews* **73**, 43–78.

Werner, M. and Simmons, L. W. (2008). Insect sperm motility. *Biological Reviews* **83**, 191–208.

References in figure captions

Bareth, C. (1964). Structure et dépôt des spermatophores chez *Campodea remyi. Compte Rendus Hebdomadaire de Séances de l'Academie des Science, Paris* **259**, 1572–1575.

Bonhag, P. F. and Wick, J. R. (1953). The functional anatomy of the male and female reproductive systems of the milkweed bug, *Oncopeltus fasciatus* (Dallas) (Heteroptera: Lygaeidae). *Journal of Morphology* **93**, 177–283.

Carayon, J. (1953a). Organe de Ribaga et fécondation hémocoelienne chez les *Xylocoris* du groupe *galactinus* (Hemipt. Anthocoridae). *Compte Rendus Hebdomadaire de Séances de l'Academie des Sciences, Paris* **236**, 1009–1101.

Carayon, J. (1953b). Existence d'un double orifice genital et d'un tissu conducteur des spermatozoides chez les Anthocorinae (Hemipt. Anthocoridae). *Compte Rendus Hebdomadaire de Séances de l'Academie des Sciences, Paris* **236**, 1206–1208.

Chao, H.-F. (1953). The external morphology of the dragonfly *Onychogomphus ardens* Needham. *Smithsonian Miscellaneous Collections* **122**, 6, 1–56.

Dailey, P. J., Gadzama, N. M. and Happ, G. M. (1980). Cytodifferentiation in the accessory glands of *Tenebrio molitor*: VI. A congruent map of the cells and their secretions in the layered elastic product of the male bean-shaped gland. *Journal of Morphology* **166**, 289–322.

Engelmann, F. (1970). *The Physiology of Insect Reproduction.* Oxford: Pergamon Press.

Gregory, G. E. (1965). The formation and fate of the spermatophore in the African migratory locust, *Locusta migratoria migratorioides* Reiche and Fairmaire. *Transactions of the Royal Entomological Society of London* 117, 33–66.

Imms, A. D. (1957). *A General Textbook of Entomology*, 9th edn., revised by O. W. Richards and R. G. Davies. London: Methuen.

Jamieson, B. G. M. (1987). *The Ultrastructure and Phylogeny of Insect Spermatozoa.* Cambridge: Cambridge University Press.

Khalifa, A. (1949). The mechanism of insemination and the mode of action of the spermatophore in *Gryllus domesticus. Quarterly Journal of Microscopical Science* 90, 281–292.

Khalifa, A. (1950). Spermatophore production in *Blattella germanica* L. (Orthoptera: Blattidae). *Proceedings of the Royal Entomological Society of London A* 25, 53–61.

Kingan, T. G., Thomas-Laemont, P. A. and Raina, A. K. (1993). Male accessory gland factors elicit change from "virgin" to "mated" behavior in the female corn earworm moth *Helicoverpa zea. Journal of Experimental Biology* 183, 61–76.

Longfield, C. (1949). *The Dragonflies of the British Isles.* London: Warne.

Longo, G., Sottile, L., Viscuso, R., Giuffrida, A. and Provitera, R. (1993). Ultrastructural changes in sperm in *Eyprepocnemis plorans* (Charpentier) (Orthoptera: Acrididae) during storage of gametes in female genital tract. *Invertebrate Reproduction and Development* 24, 1–6.

Osanai, M., Kasuga, H. and Aigaki, T. (1987). The spermatophore and its structural changes with time in the bursa copulatrix of the silkworm, *Bombyx mori. Journal of Morphology* 193, 1–11.

Osanai, M., Kasuga, H. and Aigaki, T. (1988). Functional morphology of the glandula prostatica, ejaculatory valve, and ductus ejaculatorius of the silkworm, *Bombyx mori. Journal of Morphology* 198, 231–241.

Raina, A. K. (1989). Male-induced termination of sex pheromone production and receptivity in mated females of *Heliothis zea. Journal of Insect Physiology* 35, 821–826.

Séguy, E. (1951). Ordre des Diptères. In *Traité de Zoologie*, vol. 10, ed. P.-P. Grassé, pp. 449–744. Paris: Masson et Cie.

Simmons, L. W. (1990). Nuptial feeding in tettigoniids: male costs and the rates of fecundity increase. *Behavioral Ecology and Sociobiology* 27, 43–47.

Snodgrass, R. E. (1935). *Principles of Insect Morphology.* New York, NY: McGraw-Hill.

Snodgrass, R. E. (1957). *A Revised Interpretation of the External Reproductive Organs of Male Insects.* Washington, DC: Smithsonian Institution.

Spencer, J. L., Bush, G. L., Keller, J. E. and Miller, J. R. (1992). Modification of female onion fly, *Delia antiqua* (Meigen), reproductive behavior by male paragonial gland extracts (Diptera: Anthomyiidae). *Journal of Insect Behavior* 5, 689–697.

Stanley-Samuelson, D. W. (1994). Prostaglandins and related eicosanoids in insects. *Advances in Insect Physiology* 24, 115–212.

Tillyard, R. J. (1917). *The Biology of Dragonflies.* Cambridge: Cambridge University Press.

Uvarov, B. P. (1966). *Grasshoppers and Locusts*, vol. 1. Cambridge: Cambridge University Press.

Wigglesworth, V. B. (1965). *The Principles of Insect Physiology.* London: Methuen.

13

Reproductive system: female

REVISED AND UPDATED BY **LEIGH W. SIMMONS**

INTRODUCTION

This chapter provides an overview of the female reproductive system. Section 13.1 describes the basic internal anatomy of the female reproductive tract. A detailed review of the physiological mechanisms governing the development of eggs, from their origin in the ovaries as oogonia to their final release into the oviducts at ovulation is then provided in Sections 13.2 and 13.3; in Sections 13.4 and 13.5 the fertilization and oviposition of eggs are described, including a description of the female external genitalia that are involved in oviposition, and the role of female accessory glands in reproduction.

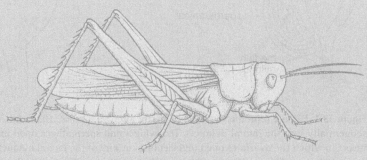

The Insects: Structure and Function (5th edition), ed. S. J. Simpson and A. E. Douglas.
Published by Cambridge University Press. © Cambridge University Press 2013.

13.1 Anatomy of the internal reproductive organs

The female reproductive system consists of a pair of ovaries, which connect with a pair of lateral oviducts. These join to form a median oviduct opening posteriorly into a genital chamber. Sometimes the genital chamber forms a tube, the vagina, and this is often developed to form a bursa copulatrix for reception of the penis and/or spermatophore. Opening from the genital chamber or the vagina is a

spermatheca for storing sperm, and, frequently, a pair of accessory glands is also present (Fig. 13.1).

Ovary. The two ovaries lie in the abdomen above or lateral to the gut. Each consists of a number of egg-tubes, or ovarioles, comparable with the testis follicles in the male. The oocytes develop in the ovarioles. The ovaries of Collembola are probably not homologous with those of insects, but are sac-like, with a lateral germarium and no ovarioles.

The number of ovarioles in an ovary varies in relation to size and lifestyle of the insect, as well as

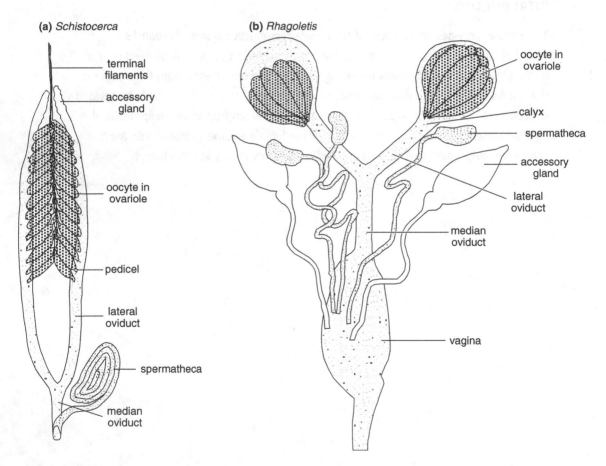

(a) *Schistocerca*

- terminal filaments
- accessory gland
- oocyte in ovariole
- pedicel
- lateral oviduct
- spermatheca
- median oviduct

(b) *Rhagoletis*

- oocyte in ovariole
- calyx
- spermatheca
- accessory gland
- lateral oviduct
- median oviduct
- vagina

Figure 13.1 Female reproductive system (partly after Snodgrass, 1935). (a) An insect with many ovarioles opening sequentially into the lateral oviducts. The oviduct and spermatheca open independently (*Schistocerca* [Orthoptera]). (b) An insect in which the ovarioles open together into the end of the lateral oviduct, which forms the calyx. The genital chamber is a continuation of the median oviduct, forming the vagina. The spermathecae arise at the junction of the vagina with the median oviduct (*Rhagoletis* [Diptera]).

its taxonomic position. In general, larger species within a group have more ovarioles than small ones; thus small grasshoppers commonly have only four ovarioles in each ovary, while larger ones may have more than 100. Similarly, in the higher Diptera, *Drosophila* has 10–30 ovarioles in each ovary, whereas *Calliphora* has about 100. By contrast, most Lepidoptera have four ovarioles on each side irrespective of their size. Variation in relation to size may also occur within a species. In the mosquito, *Aedes punctor*, the number of ovarioles in the two ovaries combined varies from about 30 to 175 as the insect's size increases, but the cockroach, *Periplaneta americana*, has eight on each side irrespective of its size.

Viviparous species exhibit extreme reduction in the numbers of ovarioles. The viviparous Diptera, *Melophagus*, *Hippobosca* and *Glossina*, have only two in each ovary and some viviparous aphids have only one functional ovary with a single ovariole. At the other extreme are the queens of some species of social insects: Each ovary of the queen termite, *Eutermes*, has over 2000 ovarioles, and that of a queen honey bee 150–180. Species which disperse as first-stage larvae may also have large numbers of ovarioles. Scale insects have several hundred ovarioles and the blister beetle, *Meloe*, 1000.

Other than in Diptera, there is no sheath enclosing the ovary as a whole, but each ovariole has a sheath which is made up of two layers: an outer ovariole sheath, or tunica externa, and an inner tunica propria (see Fig. 13.4b,c). The tunica externa is an open network of cells, sometimes including muscle cells. The cells of this net are rich in lipids and glycogen and are metabolically active, but there is no evidence that they are directly concerned with oocyte development. Tracheoles also form part of the external sheath, but they do not penetrate it and all the oxygen utilized by the developing oocytes diffuses in from these elements. In *Periplaneta* mycetocytes (Section 4.4) are present in the sheath.

The tunica propria is a basal lamina with elastic properties, possibly secreted by the cells of the terminal filament and the follicle cells. It surrounds the whole of each ovariole and the terminal filament. During the early stages of development it increases in thickness, but during the period of yolk uptake, when the oocytes enlarge rapidly, it becomes stretched and very thin. The tunica propria has a supporting function, maintaining the shape of the ovariole, and, in addition, because of its elasticity, playing a part in ovulation (Section 13.3). It also has the potential to function as a molecular sieve since, in *Phormia*, it does not permit the passage of molecules larger than 500 kDa.

Each ovariole is produced, distally, into a long terminal filament consisting of a cellular core bounded by the tunica propria (Figs. 13.1a, 13.4). Usually the individual filaments from each ovary combine to form a suspensory ligament and sometimes the ligaments of the two sides merge into a median ligament. The ligaments are inserted into the body wall or the dorsal diaphragm and so suspend the developing ovaries in the hemocoel.

Proximally, the ovariole narrows to a fine duct, the pedicel, which connects with the oviduct. In an immature insect the lumen of the ovariole is cut off from the oviduct by an epithelial plug, but this is destroyed at the time of the first ovulation and subsequently is replaced by a plug of follicular tissue.

The ovarioles may connect with the oviduct in a linear sequence and, if there are only a few, as in some Apterygota and Ephemeroptera, they may appear to be segmental. This appearance is probably coincidental and there is no suggestion of a segmental arrangement in insects with a larger number of ovarioles (Fig. 13.1a). In other groups, such as the Lepidoptera and Diptera, the ovarioles open together into an expansion of the oviduct known as the calyx (Fig. 13.1b).

Oviducts. The oviducts are tubes with walls consisting of a single layer of cuboid or columnar cells standing on a basal lamina and surrounded by a muscle layer. Usually, the two lateral oviducts join a median oviduct which is ectodermal in origin and hence lined with cuticle. However, the

Ephemeroptera are exceptional in having the lateral oviducts opening separately, each with its own gonopore. The median oviduct is usually more muscular than the lateral ducts, with circular and longitudinal muscles. It opens at the gonopore which, in Dermaptera, is ventral on the posterior end of segment 7, but in most other groups opens into a genital chamber invaginated above the sternum of segment 8 (Fig. 13.2a). Sometimes the genital chamber becomes tubular and it is then effectively a continuation of the oviduct through segment 9. Such a continuation is called the vagina and its opening to the exterior, the vulva (but notice the different terminology in ditrysian Lepidoptera, Fig. 13.2b). The vagina may not be distinguishable in structure from the median oviduct, but its anterior end, and the position of the true gonopore, is marked by the insertion of the spermatheca.

Frequently the vagina is developed to form a pouch, the bursa copulatrix, which receives the penis, while in viviparous Diptera the anterior part of the chamber is enlarged to form the uterus, in which larval development occurs.

The females of the ditrysian Lepidoptera are unusual in having two reproductive openings (Fig. 13.2b). One, on segment 9, serves for the discharge of eggs and is known as the oviporus; the other, on segment 8, is the copulatory opening, known as the vulva. The latter leads to the bursa copulatrix, which is connected with the oviduct by a sperm duct. Two openings also occur in the water beetles *Agabus*, *Ilybius* and *Hydroporus*, but here both openings are terminal, with the bursa copulatrix opening immediately above the vagina.

Spermatheca. A spermatheca, used for the storage of sperm from the time the female is inseminated until the eggs are fertilized, is present in most female insects. Sometimes there are two, as in *Blaps* (Coleoptera) and *Phlebotomus* (Diptera); most of the higher flies have three (Fig. 13.1b); and *Diplatys* (Dermaptera) have ten, which share a complex of four interconnecting spermathecal ducts. In Orthoptera and other lower insect orders the spermatheca opens into the genital chamber independently of the oviduct (Fig. 13.2a), but, where the genital chamber forms a vagina, the spermathecal opening is internal

(a) *Locusta*

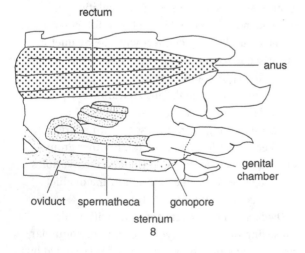

(b) ditrysian lepidopteran

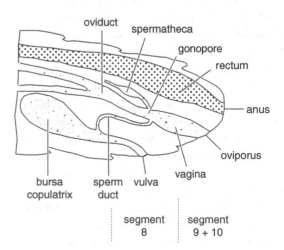

Figure 13.2 Female reproductive system. Diagrammatic sagittal sections of the end of the abdomen. (a) An insect in which the genital chamber is open, not tubular (*Locusta* [Orthoptera]) (after Uvarov, 1966). (b) A ditrysian lepidopteran in which the genital chamber forms the vagina (after Imms, 1957).

and is effectively within the oviduct (Fig. 13.2b). Some insects have a second sperm storage organ. In *Drosophila* sperm are also stored in the seminal receptacle. The spermatheca is ectodermal in origin and is lined with cuticle. Typically it consists of a storage pouch with a muscular duct leading to it. A gland is often associated with it, or the spermathecal epithelium may itself be glandular. The contents of the spermatheca, derived from the glands, are known to contain several proteins and include a carbohydrate–protein complex. The functions of these secretions probably provide nutrients for the sperm during storage, and/or they may be concerned with sperm activation.

Accessory glands. Female accessory glands often arise from the genital chamber or the vagina. Where such glands are apparently absent the walls of the oviducts may themselves be glandular. This is the case in grasshoppers, where the lateral oviducts also usually have a wholly glandular anterior extension (Fig. 13.1a).

Accessory glands often produce a substance for attaching the eggs to the substratum during oviposition and hence are often called colleterial (glue) glands. In a number of insects they produce an ootheca that protects the eggs after oviposition (Section 13.5.4). Glands associated with the genitalia, which are often modified to form a sting, perform a variety of functions in female Hymenoptera (see Chapter 27).

13.2 Oogenesis

Each ovariole consists of a distal germarium in which oocytes are produced from oogonia, and a more proximal vitellarium in which yolk is deposited in the oocytes. These two regions reflect two phases of oocyte growth: the first is regulated directly by the oocyte's genome and contains species-specific information (all the substances whose synthesis is regulated by the DNA of the germ line are known collectively as the euplasm); the second is mainly regulated by genes outside the oocyte, producing pools of molecules that will subsequently be involved in embryonic growth. The vitellarium in a mature insect forms by far the greater part of the ovariole.

The germarium contains prefollicular tissue and the stem line oogonia and their derivatives. The stem line oogonia are derived directly from the original germ cells (Section 14.2), and in *Drosophila* there are only two or three of these in each ovariole. When they divide, one of the daughter cells remains a functional stem line cell, while the other becomes a definitive oogonium (sometimes called a cystocyte) and develops into an oocyte. Oocytes pass back down the ovariole, enlarging as they do so, and as each oocyte leaves the germarium it is clothed by the prefollicular tissue which forms the follicular epithelium. At first this may be two- or three-layered, but ultimately consists of a single layer of cells. Oocyte growth continues, the follicular epithelium keeping pace by cell division so that its cells become cuboid or columnar. In *Drosophila* the number of follicle cells surrounding each oocyte increases from an initial figure of about 16 to about 1200. During yolk accumulation, growth of the oocyte is very rapid, but at this time the follicle cells no longer divide and they become stretched over the oocyte as a flattened squamous epithelium. Nuclear division may continue in follicle cells without cell division, so the cells become binucleate or endopolyploid, permitting the high levels of synthetic activity in which these cells are involved. The functions of follicle cells change during oocyte development (Fig. 13.3). At first they produce some minor yolk proteins and perhaps some of the enzymes that will later be involved in processing the yolk. The follicle cells also produce ecdysone, or a precursor of ecdysone, which, at least in some insects, accumulates in the oocyte (see Chapter 14; see also Section 13.2.4). In the later stages of oogenesis they produce the vitelline envelope and the ligands responsible for determination of the terminals of the embryo and its dorsal–ventral axis (Section 14.2). Finally, they produce the egg shell, or chorion (Section 13.2.5).

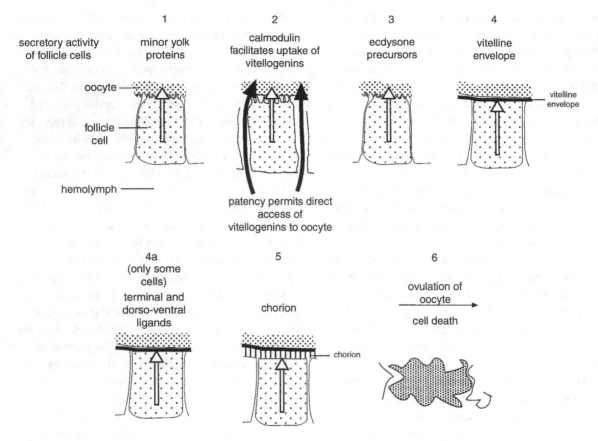

Figure 13.3 Diagram of the changes in function of cells in the follicle epithelium from just before the start of yolk accumulation in the oocyte (1) to cell death at the time of ovulation (6). Notice that in (2) the cells are widely separated from each other, allowing the hemolymph to reach the oocyte.

Each ovariole typically contains a linear series of oocytes in successive stages of development, with the most advanced in the most proximal position furthest from the germarium (Fig. 13.4). An oocyte with its surrounding follicular epithelium is termed a follicle and successive follicles are separated by interfollicular tissue derived from the prefollicular tissue. In many species, new follicles are produced as the oldest oocytes mature and are ovulated (but see below). As a result, the number of follicles in an ovariole may be approximately constant for a species, although there is variation between species. Thus *Schistocerca* commonly has about 20 follicles in each ovariole even in senile females which have

oviposited several times. In *Drosophila* there are usually six follicles per ovariole, while *Melophagus* has only one follicle in each ovariole at any one time.

13.2.1 Types of ovariole

There are two broad categories of ovarioles: panoistic, in which there are no special nurse cells; and meroistic, in which nurse cells, or trophocytes, are present. Further, there are two types of meroistic ovariole: telotrophic, in which all the trophocytes remain in the germarium; and polytrophic, in which trophocytes are closely associated with each oocyte and are enclosed within the follicle. Trophocytes are

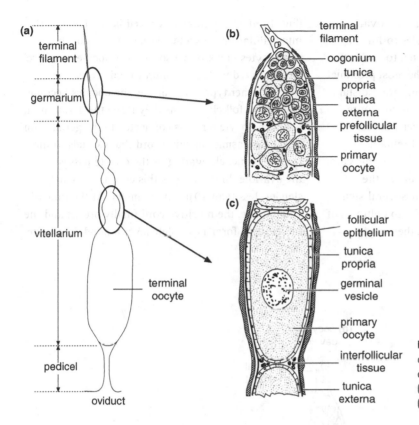

Figure 13.4 Parts of a panoistic ovariole. (a) Diagram of a whole ovariole; (b) the distal region (germarium); (c) a proximal region (part of the vitellarium).

sister-cells of the oocytes so they have the same genome, retain their connections with the oocyte because cell division is incomplete and supplement the oocyte in the synthesis of the euplasm. As a result, the oocytes in meroistic ovarioles have much larger amounts of euplasm than those in panoistic ovarioles. This has important consequences for the development of the embryo (Section 14.2).

Rapid synthesis of euplasmic constituents can also be achieved by replication of the chromosomes in the trophocytes without further cell division (endopolyploidy, see below) and by the amplification of genes extrachromosomally in DNA bodies. This is known to occur in some species with meroistic ovarioles. An increase in the DNA content of these bodies occurs up to the early stages of meiosis, but then they fragment and ultimately disappear.

As the oocyte grows, its nucleus increases proportionately in size and is now known as the germinal vesicle. Transcription by the nuclear DNA appears to be suppressed soon after the beginning of yolk uptake. During yolk deposition, as the oocyte grows much more rapidly, the germinal vesicle becomes relatively smaller and, finally, the nuclear membrane breaks down.

Panoistic ovarioles. Panoistic ovarioles (Fig. 13.4), which have no specialized nurse cells, are found in the more primitive orders of insects, the Thysanura, Odonata, Plecoptera, Orthoptera and Isoptera. Among the holometabolous insects, only Siphonaptera have ovarioles of this type. The prefollicular tissue may be cellular, but sometimes, as in *Thermobia* (Thysanura), it consists of small, scattered nuclei in a common cytoplasm.

Telotrophic ovarioles. Telotrophic ovarioles are characterized by the presence of trophic tissue as well as oogonia and prefollicular tissue in the terminal regions. This arrangement is found in Hemiptera and

Coleoptera Polyphaga. In Hemiptera each ovariole has only a single stem cell that divides to form a cluster of cells which remain connected to each other by cytoplasmic bridges (Fig. 13.5). The most proximal cells are the precursors of the oocytes; more distal cells are the trophocytes. These may divide more frequently than the oocytes so a larger number of trophocytes is produced. The cells all remain connected to a central region called the trophic core. A basically similar phenomenon occurs in the Polyphaga, except that here there are several stem cells in an ovariole and each gives rise to columns of cells, the most proximal of which is the oocyte

(Fig. 13.6). In the more advanced beetle families the intercellular bridges and cell membranes of the trophocytes break down so that the oocytes come to be connected with a common trophic core.

As in other types of ovariole, the oocytes become clothed by follicle cells as they leave the germarium, but each oocyte remains connected to the germarium by a cytoplasmic nutritive cord that extends to the trophic tissue, elongating as the oocyte passes down the ovariole. In the beetles this cord is extremely slender, less than 10 μm in diameter. At the time of yolk uptake, the nutritive cord finally breaks and the follicle cells form a complete layer around the oocyte.

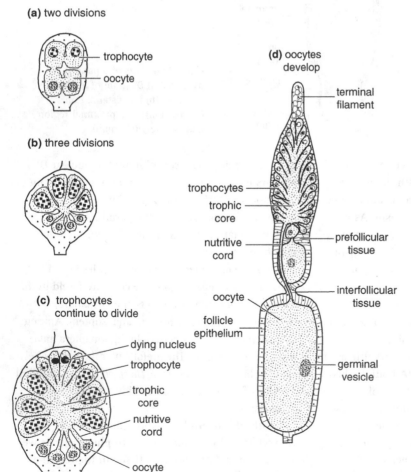

(a) two divisions

trophocyte
oocyte

(b) three divisions

trophocytes
trophic core
nutritive cord

(c) trophocytes continue to divide

dying nucleus
trophocyte
trophic core
nutritive cord
oocyte

(d) oocytes develop

terminal filament

prefollicular tissue

interfollicular tissue

oocyte
follicle epithelium

germinal vesicle

Figure 13.5 Telotrophic ovariole of a hemipteran. (a)–(c) Successive stages in the formation of a cluster of oocytes and trophocytes (after Büning, 1993). (d) Diagram of a later stage in the development of the oocytes which remain connected to the trophic core by the nutritive cords (after Huebner and Diehl-Jones, 1993).

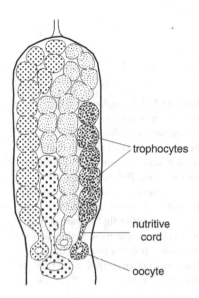

Figure 13.6 Telotrophic ovariole of a beetle. Each oocyte is shown with its associated cluster of trophocytes (Coleoptera, Polyphaga) (after Büning, 1993).

Unlike panoistic and polytrophic ovarioles, where the production of new oocytes is a continuous process, division of the germ cells does not continue after oocyte growth begins. As a result, the potential number of oocytes produced by an ovariole is fixed.

Polytrophic ovarioles. As in telotrophic ovarioles, divisions of the cells derived from stem cells in polytrophic ovarioles are incomplete, so clusters of cells are formed. However, unlike telotrophic ovarioles, the trophocytes move down the ovariole with their associated oocyte and become enclosed in the follicle (Fig. 13.7). Polytrophic ovarioles occur in Dermaptera, Psocoptera, Phthiraptera and throughout the holometabolous orders, except for most Coleoptera and the Siphonaptera. Divisions of the trophocytes within a cluster are synchronized by cues (as yet unknown) which pass through the cytoplasmic bridges. The number of divisions, and so the number of trophocytes associated with each oocyte, is characteristic for each species, although in those species with larger numbers of trophocytes some variation may occur. *Aedes* and *Melophagus*

(Diptera) have seven trophocytes associated with each oocyte as a result of three successive cell divisions; *Drosophila* and *Dytiscus* (Coleoptera) have 15 trophocytes from four cell divisions; *Habrobracon* (Hymenoptera) 31; and vespid wasps 63. No more than six successive cell divisions occur in the trophocyte clusters in any insect, so the maximum number of cells in a cluster is 64 (1 oocyte plus 63 trophocytes). Dermaptera are exceptional in having only one trophocyte associated with each oocyte, but this results from the separation of pairs of cells from larger, interconnected groups.

As each oocyte with its trophocytes leaves the germarium, the oocyte always occupies a proximal position with respect to the base of the ovariole. All the cells become enclosed within a common follicular epithelial layer which soon becomes flattened over the trophocytes, but is thicker, with cuboid cells around the oocyte. A fold of follicular epithelium pushes inward, separating the oocyte from the trophocytes except for a median pore. In Neuroptera, Coleoptera and Hymenoptera the trophocytes are pinched off in a separate, but connected, follicle from the beginning (Fig. 13.7b).

At first the trophocytes are larger than the oocyte, and the trophocyte nuclei enlarge considerably. In *Drosophila* the trophocyte nuclei increase in volume about 2000-fold and the chromosomes undergo up to ten doublings to produce polytene chromosomes. In the trophocytes of the moth *Antheraea* ploidy levels up to 2^{16} occur. Unlike earlier cell divisions, these mitoses are not synchronized in the different trophocytes of a follicle. In *Drosophila* the trophocytes adjacent to the oocyte have larger nuclei, and their chromosomes undergo one more replication than the anterior (distal) trophocytes. These cells subsequently lose their DNA, but the anterior cells do not. In most cases the whole genome is replicated to an equal extent, but there is also evidence for selective replication of those elements that are particularly important in oocyte development.

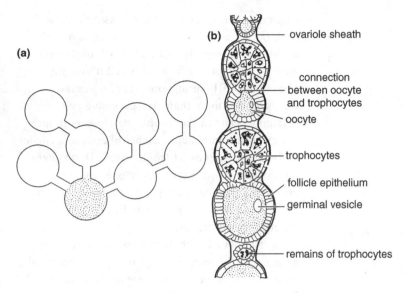

(a)

(b)
- ovariole sheath
- connection between oocyte and trophocytes
- oocyte
- trophocytes
- follicle epithelium
- germinal vesicle
- remains of trophocytes

Figure 13.7 Polytrophic ovariole. (a) Diagram showing the interconnections that remain between the oocyte (stippled) and its associated trophocytes as a result of incomplete cell divisions. The oocyte always occupies the most posterior (proximal) position as it moves down the ovariole (after King, 1964). (b) Diagram of part of an ovariole in which the trophocytes are in separate follicles from the oocytes (*Bombus* [Hymenoptera]) (after Hopkins and King, 1966).

13.2.2 Transport from trophocytes to oocyte

During the early stages of oocyte growth, mRNAs and ribonucleoproteins are transported to the oocyte. Subsequently, ribosomes and other major constituents of the euplasm are transferred. Finally, in polytrophic ovarioles, all the cytoplasmic contents are moved to the oocyte when the trophocytes collapse. In *Drosophila* this final movement causes the oocyte to almost double in volume in less than 30 minutes. The RNAs transferred include those that determine the long axis of the embryo (Section 14.2).

The movement of material into the oocyte from the trophocytes may be along an electrical potential gradient. In insects with polytrophic ovarioles, there is a difference in electrical charge between the trophocytes and the oocyte so that charged molecules are transported electrophoretically, contributing to the movement of material into the oocyte. Alternatively, or perhaps additionally, transport may involve special proteins associated with cytoskeletal elements running between the cells. It is known that in *Rhodnius* microtubules extend from the nurse cells to the oocytes and these could form the basis for such a transport mechanism. In polytrophic ovarioles the final collapse of the trophocytes, perhaps mediated by contractile units within the cells, might force the contents into the oocyte.

13.2.3 Meiosis

Meiosis begins early in the development of the oocyte in most insects, but the meiotic divisions are not completed in the ovary, and oocytes usually leave the ovarioles in the metaphase of the first maturation division. In *Locusta* meiosis does not proceed beyond the first prophase until the time of ovulation, and the resumption of meiosis coincides with sperm entry in many insects, although other factors are apparently responsible for the effect. In *Locusta* continuation of the process is stimulated by ecdysone from the follicle cells to which the oocyte becomes sensitive just before chorionogenesis, and, in *Drosophila*, ovulation triggers the resumption of meiosis through some hydration effect.

In viviparous species, such as *Hemimerus* (Dermaptera), and in Heteroptera which practice hemocoelic insemination, however, maturation of the oocytes and fertilization take place in the ovary.

13.2.4 Yolk

The euplasm in insect oocytes usually constitutes much less than 10% of the total oocyte content. The remaining 90% or more is yolk, consisting largely of lipids and proteins. Lipids often comprise about 40% of the dry weight of the terminal oocyte and most of the lipid is stored as triacylglycerol. The protein content of the yolk is usually approximately equal to the lipid content, and 60–90% of yolk proteins are derived from vitellogenins. These are proteins manufactured outside the oocyte, only by females, and with specific uptake mechanisms at the oocyte membrane. Most insects produce only one or two vitellogenins, which range in molecular weight in different species, from 210 000 to 652 000. The vitellogenins are glycolipophosphoproteins in which lipid makes up 7–15% of the total mass and carbohydrates 1–14%. The carbohydrates in the molecule are oligosaccharides of mannose. The higher Diptera differ from other insects in producing three to five smaller yolk proteins varying in molecular weight from 44 000 to 51 000. When vitellogenins enter the oocyte they often undergo some relatively minor modifications and then are known as vitellins. Lipophorin may also constitute a significant part of yolk protein. It makes up 15% of the total protein in the oocyte of *Manduca*. Other proteins are present in smaller amounts in addition to free amino acids.

Small amounts of carbohydrate are also present in the yolk. Some is associated with the vitellins; most of the rest is stored as glycogen. In the terminal oocyte of *Manduca*, free carbohydrates contribute about 2% of the dry weight.

Patterns of accumulation in the oocyte The accumulation of yolk in the oocytes occurs in the lower, proximal part of each ovariole, known as the vitellarium, and results in a very rapid increase in size of the oocyte. In *Drosophila* the oocyte volume increases about 100 000 times during the course of its development.

The process of yolk accumulation in insect oocytes is commonly called vitellogenesis. However, the same term is used to describe the synthesis of vitellogenin and it is also used in this sense in other animal groups. To avoid confusion, synthesis of the specific yolk protein will be referred to as "vitellogenin synthesis," while "yolk accumulation" will be used to describe the accumulation of yolk in the oocyte.

In those Lepidoptera, Ephemeroptera and Plecoptera that do not feed as adults, yolk accumulation is completed in the late larva or pupa. In most species, however, a period of maturation is required in the adult before the oocytes are ready for ovulation. Commonly this period is only a matter of days, but in cases of adult diapause it may be very prolonged. Hence it is necessary, in insects, to differentiate between becoming adult (that is, having the outward appearance of an adult with, in most species, wings, and being able to fly) and becoming sexually mature.

In many species yolk accumulation is largely restricted to the oocyte nearest the oviduct in each ovariole (known as the terminal oocyte). The succeeding oocytes remain relatively small until the first is discharged from the ovariole into the oviduct, a process known as ovulation. Hence there is an interval between successive ovulations from any single ovariole. However, in many Lepidoptera and Diptera, and in some other insects, yolk accumulation occurs simultaneously in a number of oocytes in each ovariole and fully developed oocytes are stored before oviposition.

Cyclical yolk accumulation is often dependent on nutrition. This is most obvious in blood-sucking insects like mosquitoes and triatomine bugs, which take discrete meals with relatively long intervals between them. These insects produce a batch of eggs after each meal and the number of eggs produced is proportional to the amount of blood ingested (Fig. 13.8).

Periodicity of yolk deposition also occurs in viviparous and ovoviviparous species, perhaps due to the physical constraints imposed by the developing embryos. The viviparous cockroach *Diploptera* retains its ootheca in a brood sac and mechanical

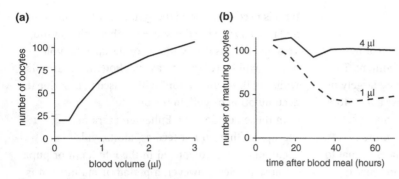

Figure 13.8 Affect of "meal" size on egg production and resorption (*Aedes* [Diptera]). Insects were given blood by enema. (a) Number of oocytes greater than 200 μm long in relation to the amount of blood placed in the midgut (Klowden, 1987). (b) Effect of amount of blood on resorption. Notice that insects given only 1 μl started to accumulate yolk in a similar number of oocytes to those given 4 μl but subsequently more than half the oocytes were resorbed. Oocytes were considered to be maturing if they were more than 100 μm long (after Lea *et al.*, 1978).

stimulation by the ootheca is largely responsible for the inhibition of yolk accumulation in the oocytes remaining in the ovarioles (see below).

Yolk proteins Most yolk protein is synthesized in the fat body independently of the ovaries. In addition to vitellogenins and lipophorin, some insects are known to produce smaller amounts of other yolk proteins in the fat body. The fat body of *Aedes aegypti* also produces proenzymes that are probably involved in yolk metabolism during embryogenesis. These enzymes are sequestered in the oocyte outside the yolk granules in which the vitellins are stored. Similar enzymes are produced by many insects.

Follicle cells also contribute proteins to the oocyte. In the higher Diptera they produce the same yolk proteins as the fat body, though in smaller amounts. Follicle-specific proteins are known to be produced by the follicle cells of several Lepidoptera. In the silk moth *Bombyx*, one of these proteins constitutes about 20% of soluble protein in the fully developed oocyte. In addition to these major components, the follicle cells may contribute other proteins in smaller amounts. Eggs of *Blattella*, and probably those of other insects, have high concentrations of the protein calmodulin in the cortical regions and outside the bounding membranes of yolk granules. Calmodulin is synthesized in the

follicle cells and may enter the oocyte through gap junctions. It probably binds calcium, which is involved in the uptake of vitellogenin and perhaps also in the ooplasmic transport of yolk vesicles.

Control of vitellogenin synthesis. Vitellogenin synthesis is initiated by juvenile hormone acting directly on the fat body in most insects, but in Diptera juvenile hormone acts indirectly. Here, it causes both the fat body and the resting-stage ovaries to acquire competence to respond to other signals.

The synthesis of juvenile hormone by the corpora allata of the female is regulated by signals from the brain, where relevant information from other tissues is integrated. Synthetic activity may be switched on or off. The signal to initiate juvenile hormone synthesis is generally related to nutrient or mated status. In locusts, mating and some food odors act via neurosecretory cells in the brain to initiate the synthesis and release of juvenile hormone (Fig. 13.9). Blood-feeding also leads to juvenile hormone synthesis in mosquitoes and in *Rhodnius*; a high-protein meal has the same effect in *Phormia* (Fig. 13.10). There is also some evidence of a direct component of regulation from the ovaries.

Juvenile hormone synthesis is regulated by allatotropins and allatostatins acting on the corpora allata directly via the neural connection to the brain

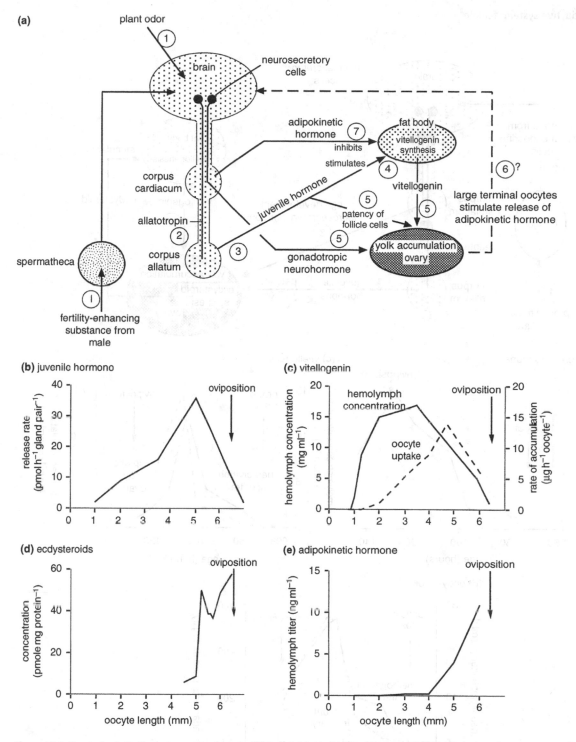

Figure 13.9 Control of vitellogenin synthesis and yolk accumulation in *Locusta*. (a) Diagram showing the sequence of events that follow insemination of the female and the transfer of fertility-enhancing substance by the male. Numbers show the sequence of events. Stage 6 is inferred (see text). (b) Release of juvenile hormone from the corpora allata during the first vitellogenic cycle (stage 3 in (a)) (data from Gadot and Applebaum, 1985). (c) Vitellogenin concentration in the hemolymph and rate of uptake by the terminal oocytes (stage 5 in (a)) (after Gellissen and Emmerich, 1978). (d) Accumulation of ecdysteroid (mainly ecdysone) in the posterior quarter of the terminal oocyte in the later stages of yolk accumulation. Ecdysone leads to a resumption of meiosis in the oocyte nucleus (data from Lanot *et al.*, 1987). (e) Hemolymph titer of adipokinetic hormone during yolk accumulation (stage 7 in (a)) (data from Moshitzky and Applebaum, 1990).

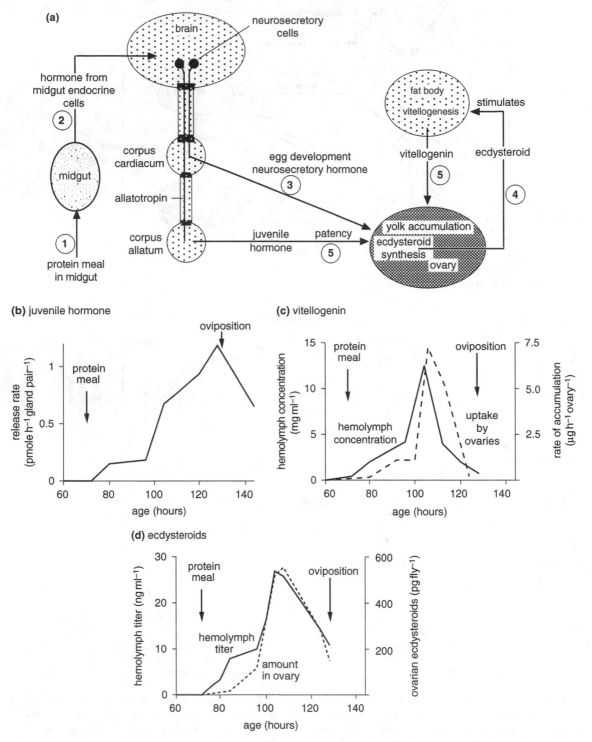

Figure 13.10 Control of vitellogenin synthesis and yolk accumulation in *Phormia*. (a) Diagram showing the sequence of events that follow a high-protein meal. Numbers show the sequence of events (based on Yin *et al.*, 1994). (b) Release of juvenile hormone from the corpora allata during the vitellogenic cycle (stage 5 in (a)) (after Zou *et al.*, 1989). (c) Vitellogenin concentration in the hemolymph and rate of uptake by the ovaries (stage 5 in (a)) (after Zou *et al.*, 1989). (d) Ecdysteroid: hemolymph titer and amount in the ovaries (after Yin *et al.*, 1990).

or via the hemolymph (Section 21.4). In the viviparous cockroach *Diploptera*, small terminal oocytes between 0.6 and 1.4 mm long enhance the rate of synthesis, but large, fully developed oocytes have an inhibitory effect which acts on the brain via the hemolymph. When an ootheca is present, it stimulates mechanoreceptors in the brood sac which inhibit juvenile hormone synthesis via the neural pathway.

Although there is a decline in juvenile hormone synthesis at the end of the first gonotrophic cycle in *Locusta*, vitellogenin synthesis in this insect is further reduced by adipokinetic hormone acting directly on the fat body (Fig. 13.9). The hemolymph titer of this hormone increases markedly when the terminal oocytes are large. This effect is produced by a hormone titer only about one-tenth of that required to initiate lipid mobilization in the fat body during flight (Section 6.2.4) so that its two functions do not conflict. Although vitellogenin synthesis by the fat body stops, the mRNA responsible for its synthesis is stored for the next cycle of synthesis.

In Diptera, synthesis of vitellogenin is regulated by ecdysone (Fig. 13.10). The prothoracic glands, which are the principal source of ecdysone in the immature stages, are no longer present in adults, and ecdysone is produced by the follicle cells. In the female mosquito, *Aedes*, ecdysone synthesis is itself regulated by a hormone, egg development neurosecretory hormone, produced by cells in the brain. Release of this hormone is triggered by a blood meal, but it stimulates the follicle cells to produce ecdysone only if they have already been rendered competent to respond by juvenile hormone. At the fat body, ecdysone is changed to 20-hydroxyecdysone, which is the active hormone. Ecdysone, or a precursor, is synthesized by the follicle cells in other insects such as the locust (Fig. 13.9d), but here its primary function is the regulation of molting during the embryonic period (Section 14.2).

In some Lepidoptera all the oocytes are fully developed at eclosion. This is the case, for example,

in the cecropia moth, *Hyalophora*. In this insect vitellogenin appears in the hemolymph when the juvenile hormone titer declines and the larva spins its cocoon. Vitellogenin concentration in the hemolymph remains high throughout the pupal period, finally declining in the pharate adult as the oocytes accumulate yolk. In *Plodia*, ovarian development early in the pupal period is correlated with high titers of ecdysone, but vitellogenin synthesis only proceeds when ecdysteroid titers are low.

Accumulation of proteins in the oocyte. The oocyte is surrounded by the follicle epithelium, but when the oocyte starts to accumulate vitellogenin, extensive intercellular spaces develop between the follicle cells, although the cells remain in contact with each other through gap junctions. The intercellular spaces permit direct access of the hemolymph to the surface of the oocyte and the follicle is said to be patent. Vitellogenin is taken up selectively from the hemolymph, which is made possible by specific receptors for vitellogenin on the plasma membrane of the oocyte. The receptors with their bound vitellogenin are taken into the oocyte by endocytosis. The protein is accumulated and stored in yolk bodies, while the receptors are recycled to the surface membrane.

The vitellogenins are changed following their accumulation in the oocyte, although in most insects the changes are small. For example, *Rhodnius* vitellin contains the same four subunits and has the same molecular weight as the vitellogenin, but has more phospholipids and triacylglycerol. Minor changes to the vitellogenins also occur in *Carausius*, but in *Blattella* the larger of two subunits comprising the vitellogenin is cleaved into two, and cleavage of the vitellogenin also occurs in the boll weevil, *Anthonomus*. Proteins from the follicle cells are also modified as they enter the oocyte.

The uptake of proteins from the hemolymph is regulated by juvenile hormone. This induces patency and also produces changes in the surface of the oocyte. It does not, however, determine the specificity of uptake, at least in *Drosophila*. The fact that, in many species, only the terminal oocyte in each ovariole accumulates yolk appears to occur because it produces a factor that inhibits yolk uptake by the younger follicles. This may act via a local intercellular interaction or via the hemolymph.

In addition to the selective uptake of vitellogenins, some non-specific uptake from the hemolymph also occurs. As a result of this, substances present in the hemolymph that are not specifically associated with embryonic development may be present in the yolk.

Yolk lipids Most of the lipid that accumulates in the oocyte is synthesized in the fat body; less than 1% is synthesized by the oocyte. In *Manduca* 90% of the lipid is carried to the oocyte by low-density lipophorins. At the oocyte membrane the lipophorin unloads its lipids and is recycled. Most of the remainder of the lipid comes from high-density lipophorin, which is taken up by the oocyte in *Manduca* so that it also contributes to the protein content of the yolk. In *Hyalophora*, however, apolipoprotein III is recycled and only a very high-density lipophorin is taken up. The uptake is receptor mediated. Much of the phospholipid that accumulates in the oocyte is also transported by lipophorin.

13.2.5 Vitelline envelope and chorionogenesis

The follicle cells start to produce the proteins of the vitelline envelope (Section 14.1) as yolk accumulation nears its end. In *Drosophila* it starts to form as a number of discrete plaques which subsequently coalesce, but in some other insects it is secreted uniformly over the surface of the oocyte from the beginning. It is possible that in some insects, the oocyte also contributes to formation of the envelope. In insects with polytrophic ovarioles, the vitelline envelope is only produced around the oocyte, not around the trophocytes. At about the same time the vitelline envelope is produced, the follicle cells also produce the ligands responsible for determining the extreme anterior and posterior ends of the embryo, and those that determine its dorsal–ventral axis (Section 14.2).

Synthesis of the vitelline envelope in mosquitoes is initiated and maintained by 20-hydroxyecdysone, which reaches a peak in the hemolymph at this time. In a number of insects the vitelline envelope becomes compacted after fertilization, as the egg is laid.

The chorion (Section 14.1) is produced wholly by the follicle cells. The various layers are laid down sequentially, usually by addition to the outer surface, but, in the lamellate chorion of Lepidoptera, additional material is added to layers already present by the insertion of new sheets of fibers into the lamellae. As a result, in *Lymantria*, the lamellae increase in thickness from about 0.04 µm when first laid down to 0.2 µm in the fully developed chorion. Subsequently, the fibers which form the lamellae thicken and ultimately fuse together so that the lamellae now appear homogeneous rather than fibrous. During this phase there is some compaction of the lamellae and the effect of this is to increase the density of the chorion, a process known as "densification." In many eggs, the outlines of the individual follicle cells producing the chorion are visible on the outside of the completed structure.

Proteins constitute over 90% of the endochorion and much of the exochorion when this is present. Many different proteins are involved: 19 have been recorded in the endochorion of *Drosophila* eggs, and 100 or more are present in the chorion of various Lepidoptera. In *Acheta* (Orthoptera), *Oncopeltus* (Hemiptera) and *Drosophila*, most of the proteins are large, with molecular weights in excess of 20 kDa. In the Lepidoptera, by contrast, most are in the range 10–20 kDa. These proteins include structural proteins and enzymes responsible for post-secretion changes in the chorion.

Some of the proteins are produced at different times by the follicle cells. As a result, successive layers of the chorion may be chemically different (Fig. 13.11b). Where marked structural differences

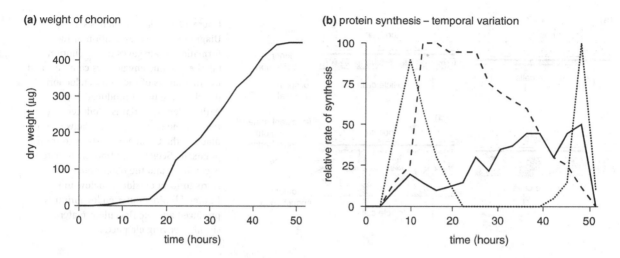

(a) weight of chorion

(b) protein synthesis – temporal variation

(c) protein synthesis – spatial variation

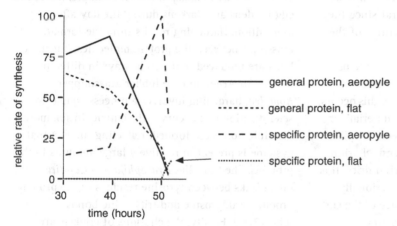

general protein, aeropyle

general protein, flat

specific protein, aeropyle

specific protein, flat

Figure 13.11 Chorionogenesis in a moth (*Antheraea*). (a) Increase in the weight of the chorion with time (data from Paul and Kafatos, 1975). (b) Temporal variation in the rates of synthesis of three chorionic proteins (indicated by full, dashed and dotted lines). These examples were selected to show different patterns of synthesis (after Mazur *et al.*, 1980). (c) Spatial variation in protein synthesis. The general protein is produced by all follicle cells. The specific protein is only produced in the final stages of chorionogenesis in the follicle cells producing the aeropyle region of the egg. Virtually none is produced by the other follicle cells (represented here by cells forming a "flat" area of chorion). The specific protein is only produced in small amounts relative to the general protein; in the figure, amounts are multiplied by ten compared with the general protein (after Mazur *et al.*, 1980).

occur in different parts of the chorion, specific proteins may be produced by different regions of the follicle epithelium. The specific proteins are additional to a population of proteins that provide the basic material of the chorion and which are produced by all the follicle cells. For example, the cells producing the aeropyle crown of the chorion of *Antheraea* synthesize eight proteins that are not made by any other follicle cells and a further 15 proteins that are also produced by cells forming some other specific areas of the chorion. These specific proteins are produced principally toward the

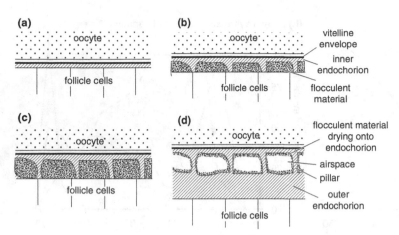

Figure 13.12 Chorionogenesis. (a)–(d) Diagrammatic representation of the formation of airspaces in the chorion. (a) The vitelline envelope is complete and the first layers of the inner endochorion (hatched) are being produced. (b) A flocculent material is produced over most of the surface. In gaps between this material the endochorion starts to form pillars. (c) Secretion of flocculent material is complete and the outer endochorion starts to form outside it (below in the figure). (d) More outer endochorion is produced and the flocculent material shrinks, leaving airspaces.

end of chorionogenesis (Fig. 13.11c). This may be true for area-specific proteins in general since they are often concerned with surface structures of the chorion.

The endochorion of many insect eggs contains extensive spaces where the outer endochorion is supported by pillars (see Fig. 14.3). Where this occurs, the follicle cells produce a "flocculent material" or "filler" which occupies the space between the pillars and supports the newly formed outer endochorion which roofs it (Fig. 13.12). As the chorion dries, this filler collapses, leaving an airspace. Functionally similar material is produced at the surface of the eggs of *Antheraea*, where it forms a mold for the production of crowns around the openings of aeropyles.

Aeropyles, micropyles and other pores in the chorion are formed by long cellular processes from one, two or three cells (Fig. 13.13). These processes contain bundles of microtubules and, if there is more than one, they may be held together by cell junctions or simply twist around each other, as in *Dacus*. Proteins of the chorion are laid down around the processes. Late in chorionogenesis the cells producing the processes degenerate, leaving pores through the chorion.

The chorion of many insect eggs hardens after it is produced, though sometimes not until the egg is laid.

This is the case in mosquitoes, for example, where the egg hardens and darkens during the day after oviposition. Hardening results from the formation of crosslinks between the protein molecules, but these links are produced in different ways in different insects. In *Aedes* and the lubber grasshopper, *Romalea*, hardening involves a process similar to the sclerotization that occurs in the cuticle. In silk moths, disulfide bonds are formed, reflecting the fact that cysteine is present in relatively large amounts in the proteins. The fruit flies *Drosophila* and *Ceratitis* form direct links between tyrosine residues in the proteins producing dityrosine and trityrosine bonds (Fig. 13.14). Finally, the chorions of crickets are hardened by disulfide linkages, but, in addition, crosslinks are formed by calcium.

Regulation of these processes that occur after secretion is not generally understood. Probably, enzymes that catalyze the reactions are incorporated into the chorion in inactive forms and subsequently activated. In *Drosophila*, for example, an inactive peroxidase in the chorion is activated by hydrogen peroxide produced at the plasma membrane of the follicle cells.

In grasshoppers, some beetles and the moth *Micropteryx* a coating is added to the outside of the egg as it passes down the oviduct. In grasshoppers it is called the extrachorion.

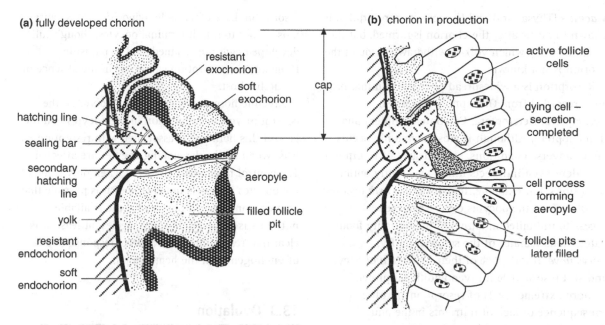

(a) fully developed chorion

(b) chorion in production

Figure 13.13 Chorionogenesis. Formation of pores in the chorion. Sections through the chorion of *Rhodnius* at the junction of the cap with the main shell. (a) The complete chorion; (b) formation of an aeropyle by a process from one of the cells. Notice also the differential activity of other cells at the junction and lines of weakness where the chorion breaks when the larva hatches (after Beament, 1946).

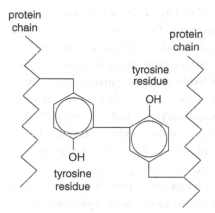

Figure 13.14 Hardening of the chorion by forming dityrosine bonds between protein chains, as in some fruit flies.

13.2.6 Resorption of oocytes

Oocytes in the ovarioles may be destroyed and their contents resorbed by the insect. Resorption,

or oosorption, may occur when an oocyte is at any stage of development, but it is most commonly observed in terminal oocytes during yolk accumulation. In extreme cases the terminal oocytes of all the ovarioles may be resorbed; in others, only some oocytes are destroyed while those in most of the ovarioles continue to develop normally.

During resorption the yolk spheres break down and protein and lipid yolk disappear from the oocyte. It is probable that vitellogenins (or vitellins) are released from the yolk spheres and returned to the hemolymph as a consequence of an increase in permeability of the oocyte membrane. The role of the follicle cells is not clear, but they become folded on each other as the oocyte shrinks, losing their connections with the oocyte. Finally, the whole follicle collapses to form a resorption body, which persists at the base of the ovariole. In grasshoppers the resorption body is frequently colored orange. In

Machilis (Thysanura) and Hymenoptera resorption is known to occur after the chorion is formed, but the mechanism by which yolk is withdrawn through the chorion is not known.

Resorption is a widespread phenomenon that has been recorded from Orthoptera, Blattodea, Dermaptera, Heteroptera, Coleoptera, Diptera and Hymenoptera. Although it most commonly occurs under adverse conditions, some degree of resorption is evident even in insects under apparently optimal conditions. For example, females of *Locusta* resorb about 25% of the oocytes even when they have access to unlimited amounts of high-quality food; autogenous females of the screwworm fly, *Chrysomya*, resorb 30% of their first batch of oocytes and still resorb 10% if given additional protein.

More extreme levels of resorption are often a consequence of lack of nutrients in the adult (Fig. 13.8b). In *Culicoides barbosai* (Diptera) the number of oocytes completing development for a second oviposition is proportional to the size of the blood meal taken by the insect – the remainder are resorbed; in *Locusta* the percentage of oocytes resorbed is inversely proportional to the quantity of food eaten; and low levels of protein in the food lead to oocyte resorption in *Cimex*. In all these cases, all the oocytes start to accumulate yolk, but then some, or even all, of them are resorbed (Fig. 13.8b). Resorption is increased if the insect is parasitized, and sometimes also in relation to season and age of the insects. All these effects are probably a consequence of reduction in, or competition for, the available nutrients in the adult. Food shortage in larval stages has not been recorded as a factor influencing the extent of resorption, except in Thysanura.

The inability to produce or lay fertile eggs also leads to resorption. Yolk accumulation begins in *Schistocerca* before the female mates, but in the absence of mating many oocytes are resorbed. In this species, mature males produce a maturation pheromone and even the presence of mature males, without mating taking place, reduces the amount of resorption. Lack of suitable oviposition sites causes mosquitoes to resorb terminal oocytes, though fully developed eggs are retained, and in parasitic Hymenoptera resorption can occur in the absence of a suitable host.

Juvenile hormone is generally involved in the control of yolk accumulation in the oocytes and the lack of this hormone is associated with resorption, at least when all the developing oocytes are affected. The control mechanism involved when only some of the oocytes regress is not clear, but it is possible that competition between oocytes for the available nutrients is involved. In some species, at least, it is clear that resorption is not a direct response to a lack of vitellogenins in the hemolymph.

13.3 Ovulation

The passage of the oocyte into the oviduct, a process known as ovulation, involves the oocyte's escape from the follicular epithelium and the breakdown of the epithelial plug at the entrance to the pedicel. In *Periplaneta* the elasticity of the tunica propria helps to force the oocyte into the oviduct, where it may be stored temporarily before oviposition. In species where the external ovariole sheath contains muscle fibers, these probably assist the movements of the oocyte. The elasticity of the tunica propria causes it to fold up after the oocyte is shed and this pulls the next oocyte down into the terminal position. The empty follicle epithelium of the first oocyte usually persists, but it becomes greatly folded and compressed and forms a new plug at the entrance to the pedicel. The compressed follicle epithelium is known as a corpus luteum and sometimes the corpora lutea of two or three successive ovulations may be present together, despite the fact that they break down progressively. In *Melophagus*, only a small relic of the follicle epithelium persists, as much of the debris from the follicle cells and trophocytes is passed out of the ovariole when this contracts after ovulation.

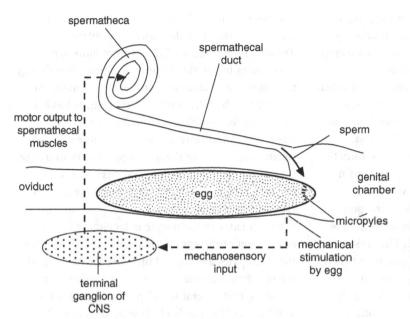

Figure 13.15 Control of fertilization in a grasshopper. As the egg enters the genital chamber it stimulates mechanoreceptors which trigger contraction of muscles around the spermatheca so that sperm are expressed onto the egg (based on Okelo, 1979).

Sometimes, as in Orthoptera, all the ovarioles ovulate simultaneously, but in other cases, as in viviparous Diptera, the ovarioles function alternately or in sequence. In Lepidoptera, which commonly lay large batches of eggs although possessing a total of only eight ovarioles, the oocytes may accumulate in the very long pedicels that join the ovarioles to the lateral oviducts until a large number is present. Similarly, in some Diptera and parasitic Hymenoptera, such as *Apanteles*, large numbers of eggs may be stored, in this case in the lateral oviducts, thus enabling the insect to lay a large number of eggs quickly when it finds a suitable oviposition site.

It is not clear how ovulation is controlled. Where it is followed immediately by oviposition, as in *Rhodnius*, both ovulation and oviposition may be regulated by the same factors (Section 13.5).

13.4 Fertilization of the egg

The egg is fertilized as it passes into the genital chamber when a few sperm are released from the spermatheca. The release of sperm in *Schistocerca* is in direct response to mechanosensory input produced by an egg entering the genital chamber. This input to the terminal abdominal ganglion results in contractions of muscles which compress the spermatheca and force sperm out (Fig. 13.15). A similar mechanism of sperm release occurs in the cricket, *Teleogryllus*. In this case, the egg, on entering the genital chamber, presses on the cuticle and stimulates subcuticular mechanoreceptors. The input from these receptors inhibits the activity of the oviducal muscles so that the egg stops moving, and, at the same time, activates the muscles of the spermathecal duct which squeeze the sperm, tail first, toward the egg. In other species, sperm may be forced out of the spermatheca by pulsed increases in hemolymph pressure. Sperm use can be a highly efficient process. In *Atta*, for example, founding queens use just two sperm per fertilized egg, rising to three sperm per fertilized egg in older queens. In this way queens can maintain their fertility without further mating for up to ten years, or even more.

In some insects, the female clearly has the capacity to release or withhold sperm according to environmental cues. Male Hymenoptera are haploid, developing from unfertilized eggs. Queen bees lay male eggs in large cells in the comb and female eggs in smaller cells. Females of some parasitic Hymenoptera tend to lay haploid (male) eggs in small hosts and diploid (female) eggs in larger hosts. For example, the ichneumonid wasp, *Coccygomimus*, lays its eggs in the pupa of the wax moth, *Galleria*. It determines whether or not to fertilize an egg from the size of the cocoon enclosing the host pupa, but it only differentiates between sizes if the cocoon exhibits some chemical characteristic of the pupa. The measurement of size made by the parasitoid is relative so that it tends to lay male eggs in the smaller members of the host population irrespective of their absolute size. Among the Aphelinidae, which parasitize homopterans, female eggs are laid in host nymphs, but male eggs are laid if the ovipositor enters a larval parasitoid already within the host. The parasitoid may be of the same species as the egg-laying female, or of a different species. The distinction between parasitized and unparasitized hosts is presumably made on the basis of chemicals encountered by the ovipositor.

How these insects measure size is not generally understood. Queen honey bees apparently use the forelegs to measure the sizes of the cells in which they are ovipositing. The parasitic wasp *Trichogramma minutum* lays its eggs in the eggs of other insects and regulates the number of eggs it lays (but not their sex) by determining the surface area of the host egg. It does this by measuring the curvature of the egg surface, probably determined from the angle between the antennal scape and the head, and the time taken to walk across the egg from one side to the other (the transit time). The mechanism by which the female permits or prevents the release of sperm is not known in any of these cases, although queen bees have a valve and pump on the spermathecal duct which presumably regulates release.

Females of some insect species have the capacity to selectively store subsets of sperm they receive at insemination. In gryllid crickets, for example, females preferentially store sperm from unrelated males, or from those they find more attractive. In *Tribolium* flour beetles females store more sperm from males better able to provide stimulation during post-copulatory courtship. In this way females can control which males fertilize their eggs. It has been suggested that multiple spermathecae may even allow females to store sperm from different males separately, affording them the opportunity to utilize sperm selectively after they have been stored. However, there is currently little evidence for this type of selective sperm use.

Sperm entry to the oocyte is facilitated by the orientation of the egg in the genital chamber, where the micropyles are aligned opposite the opening of the spermathecal duct. It is usually stated in the literature that several sperm penetrate each oocyte and that fertilization is effected by only one of the sperm while the rest degenerate. However, there is increasing evidence that monospermy is also common.

In a few insects fertilization occurs while the oocytes are still in the ovary. This is true of those Cimicoidea that practice hemocoelic insemination, and also occurs in *Aspidiotus*, a coccid, in which the sperm become attached to cells which proliferate in the common oviduct and then migrate to the pedicels.

13.5 Oviposition

13.5.1 Female genitalia: the ovipositor

The gonopore of the female insect is usually situated on or behind the eighth or ninth abdominal segment, but the Ephemeroptera and Dermaptera are exceptional in having the gonopore behind segment 7. In many orders there are no special structures associated with oviposition, although sometimes the terminal segments of the abdomen are long and telescopic, forming a type of ovipositor (see Fig. 11.1). Such structures are found in some

Lepidoptera, Coleoptera and Diptera. In *Musca* the telescopic section is formed from segments 6–9 and normally, when not in use, it is telescoped within segment 5. The sclerites of the ovipositor are reduced to rods in this species. In others, as in tephritids, the tip of the abdomen is hardened and forms a sharp point, which enables the insect to insert its eggs in small holes and crevices.

An ovipositor of a quite different form, derived from the appendages of abdominal segments 8 and 9, is present in Thysanura, some Odonata, Orthoptera, Homoptera, Heteroptera, Thysanoptera Terebrantia and Hymenoptera. As with male external genitalia, there has been considerable variation in the terminology used to describe female external genitalia. *Lepisma* can be viewed as having a basic form of ovipositor from which those of other insects have been derived. At the base of the ovipositor on each side are the coxae of segments 8 and 9. These are known as the first and second gonocoxae (also referred to as the first and second valvifers) (Fig. 13.16), and articulating with each of them is a slender process which curves posteriorly. These processes are called the first and second gonapophyses (or valvulae) and, together, they form the shaft of the ovipositor. In *Lepisma* (Fig. 13.16a) the second gonapophyses of the two sides are united, so that the shaft is made up of three elements fitting together to form a tube down which the eggs pass. At the base of the ovipositor there is a small sclerite, the gonangulum, which is attached to the base of the first gonapophysis and articulates with the second gonocoxa and the tergum of segment 9. The gonangulum probably represents a part of the coxa of segment 9. It is not differentiated in *Petrobius* (Archaeognatha).

In some Thysanura and in the Pterygota an additional process is present on the second gonocoxa. This is the gonoplac (or third valvula). It may or may not be a separate sclerite and may form a sheath around the gonapophyses. The gonoplacs are well developed in the Orthoptera, where they form the dorsal valves of the ovipositor with the second gonapophyses enclosed within the shaft as in tettigoniids (Fig. 13.16b) or reduced as in the gryllids. Throughout the Orthoptera the gonangulum is fused with the first gonocoxa.

The Hemiptera and Thysanoptera have the gonangulum fused with tergum 9, while the gonoplac may be present or absent. In Pentatomomorpha and Cimicomorpha the development of the ovipositor is related to oviposition habit. If the insect oviposits in plant or animal tissue the valves are sclerotized and lanceolate and the anterior strut of the gonangulum is heavily sclerotized. Species laying on leaf surfaces, however, have membranous and flap-like gonapophyses and the anterior strut of the gonangulum is membranous or absent.

In the Hymenoptera the first gonocoxae are usually absent (although they may be present in Chalcidoidea) and the second gonapophyses are united. The second gonapophyses slide on the first by a tongue-and-groove mechanism (Fig. 13.16c). In the Symphyta and parasitic groups, the ovipositor retains its original function, but in Aculeata it forms the sting. This does not involve any major modifications of the basic structure, but the eggs, instead of passing down the shaft of the ovipositor, are ejected from the opening of the genital chamber at its base. In *Apis* the first gonapophyses are known as the lancets and the fused second gonapophyses as the stylet. This forms an inverted trough, which is enlarged into a basal bulb into which the reservoir of the poison gland discharges.

Sensilla on the ovipositor. The terminal abdominal segments, and the ovipositor where one is present, have an array of mechanoreceptors. Although chemoreception may be an important factor in oviposition, the number of chemoreceptors on the ovipositor is usually very small. Insects receive most chemical information about their oviposition substrates from tarsal or antennal sensilla.

Trichoid mechanoreceptors are generally present and these may have directional sensitivity. In addition, small numbers of campaniform sensilla are also sometimes reported and may be a usual

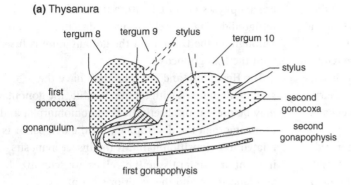

(a) Thysanura

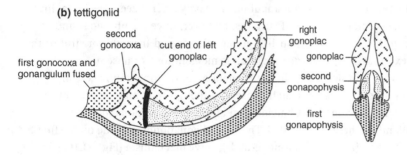

(b) tettigoniid

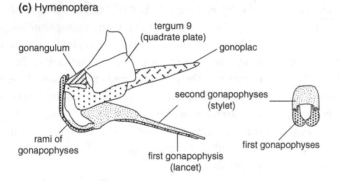

(c) Hymenoptera

Figure 13.16 Ovipositors derived from abdominal appendages. Homologous structures have the same shading in different diagrams. (a) Inner view of the genital segments of a thysanuran (*Lepisma*) (after Scudder, 1961). (b) Ovipositor of a tettigoniid. Left: lateral view with one gonoplac removed; right: transverse section (after Snodgrass, 1935). (c) Sting of a honey bee (*Apis*). Left: lateral view; right: transverse section through the shaft (after Snodgrass, 1956).

occurrence. Chemoreceptors have been recorded on the ovipositors of Orthoptera, Diptera, parasitic Hymenoptera and Lepidoptera. In most cases there are only a few contact chemoreceptors. The stemborer moth, *Chilo*, for example, has four, and the flour moth, *Ephestia*, ten. Multiporous, presumed olfactory, sensilla have been recorded around the ovipositor in *Musca* and other cyclorrhaphous flies, and the sunflower moth, *Homoeosoma*, has 59 on each side. By contrast, *Manduca* has no chemoreceptors associated with the ovipositor, nor

has the damselfly, *Sympecma*, even though it oviposits in plants.

13.5.2 Mechanisms of oviposition

In the majority of species that lack an appendicular ovipositor, eggs are simply deposited on a surface or, if the terminal segments of the abdomen are elongated or telescopic, may be inserted into crevices. In some cases, however, specialized structures are involved in the oviposition process.

(a) at rest

(b) ovipositing

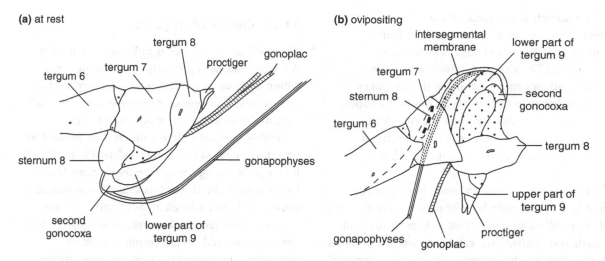

Figure 13.17 Oviposition. Changes in the positions of the basal parts of the ovipositor and the terminal abdominal segments during oviposition by an ichneumonid wasp (*Megarhyssa*) (after Snodgrass, 1935). (a) At rest; (b) during oviposition.

Many Asilidae, for instance, have spine-bearing plates called acanthophorites at the tip of the abdomen. These push the soil aside as the insect inserts the tip of its abdomen and oviposits and then, when the abdomen is withdrawn, the soil falls back and covers the eggs. The beetle *Ilybius* has two finely toothed blades forming an ovipositor. The points of these blades are pushed into the surface of a suitable plant and then worked upwards by a rapid, rhythmic, saw-like action so that a tongue of plant tissue is cut away at the sides. An egg is laid in the hole beneath the blades and is covered by the tongue of plant tissue when the blades are withdrawn.

Many species with an ovipositor derived from the appendages of segments 8 and 9 penetrate tissues using a sliding movement of the valves relative to each other, similar to the sting mechanism of *Apis*. In an ichneumon the tip of the abdomen is turned down at the start of oviposition so that the valves point ventrally instead of posteriorly (Fig. 13.17). The gonapophyses then work their way by rapid to-and-fro movements into the host tissue (or through the wood in which the host is boring in the case of *Rhyssa*). The gonoplacs do not enter the

wound, but become deflected outside it. In this way *Rhyssa* can bore through 3 cm of wood in 20 minutes.

In grasshoppers (Acrididae) the action of the valves is quite different, involving an opening and closing movement of the dorsal and ventral valves rather than a sliding movement (see Fig. 13.19a). These movements are produced by muscles inserted onto an apodeme at the base of the valves, and the whole structure is rhythmically protracted and retracted. The insect starts by raising the body on the first two pairs of legs and arching the tip of the abdomen downwards so that it presses more or less vertically on the ground. The opening movement of the valves scrapes particles of the substratum sideways and upwards and pressure is exerted down the abdomen so that the valves slowly dig a hole. As the hole deepens, the abdomen lengthens by the unfolding and stretching of the intersegmental membranes between segments 4 and 5, 5 and 6, and 6 and 7. The membranes are specialized to permit stretching, having a lamellate endocuticle under a thin epicuticle which is folded at right angles to the long axis of the body. As the abdomen lengthens, these folds become smoothed out and the endocuticle stretches. The intersegmental membranes of the male do not stretch to the same extent as in the mature female, nor do those of the immature female, indicating that some change occurs during maturation. The change is controlled by the corpora allata and it is probable that hydrogen bonding between the protein chains in the cuticle is reduced by an enzyme from the epidermis. The intersegmental muscles are also stretched and they are specialized to permit this (see Fig. 10.15).

The extension of the abdomen of grasshoppers may be very considerable. For instance, the abdomen of *Anacridium* stretches from 3–5 cm to 10 cm in length and *Schistocerca* can dig to a depth of 14 cm. During digging the ventral valves of the ovipositor lever the abdomen downwards while the upper valves push soil away and effect the excavation. The pull exerted by the ventral valves is transmitted to the abdomen and this results in the extension of the intersegmental membranes between segments 4 and 7. At intervals during digging the female partly withdraws her abdomen and the walls of the hole are smoothed and compacted by small movements of the ovipositor valves, together with twisting movements of the abdomen. Even in apparently suitable soil a female frequently abandons a hole and starts to dig again, but when a suitable hole has been constructed the process of oviposition proper begins. Just before an egg is laid, the female pumps air into her tracheal system by rapid movements of the head and then, with the thoracic spiracles closed, forces the air backwards so that the abdomen becomes turgid. It remains turgid until the egg is laid, then the head moves forwards again and the pressure is released. Eggs are passed out micropylar-end first and the abdomen is slowly withdrawn as more eggs are laid. When all the eggs have been laid the female fills the upper part of the hole with a frothy plug (see Fig. 13.21b) and, finally, after withdrawing her abdomen, she scrapes soil over the top of the hole with her hind tibiae. The whole process may take about two hours, of which egg-laying occupies some 20 minutes.

13.5.3 Control of oviposition

The readiness to oviposit is influenced by mating. A female *Bombyx*, for example, lays all her eggs within 24 hours of mating, whereas a virgin female retains most of her eggs for some days. An essentially similar change of behavior occurs in insects belonging to many different orders. It is induced by peptides or other substances transferred to the female by the male during sperm transfer (Section 12.4.2). Under normal conditions this substance presumably causes the female to seek an oviposition site and then, when she receives appropriate stimulation from her mechano- and chemoreceptors, to begin ovipositing. In *Bombyx* the eggs are normally laid close together in a single layer; if the trichoid mechanoreceptors on the anal papillae are damaged the eggs are deposited in uneven clumps. In the intact insect the information from the hairs is relayed to the brain (Fig. 13.18) and from the brain a command to motor neurons arising in the terminal abdominal ganglion leads to contractions of the oviduct so that eggs are deposited, and to movements of the abdomen which result in the eggs being spread regularly over a surface.

In grasshoppers the activity of the muscles moving the ovipositor valves is regulated by a central pattern generator in the terminal abdominal ganglion (Fig. 13.19). In the non-ovipositing insect the activity of the pattern generator is inhibited by neural activity from the head ganglia and the metathoracic ganglion. Sensory input from mechanoreceptors on the ovipositor valves probably helps to maintain and modulate the basic rhythm produced by the pattern generator. The ventral opener muscle is innervated by a motor neuron producing proctolin, and this may also be true of other muscles associated with the ovipositor valves. It is suggested that octopamine is also necessary for the normal functioning of the muscle and to maintain the activity of the central pattern generator.

In *Locusta* the movement of eggs down the oviducts appears to result from myogenic activity of

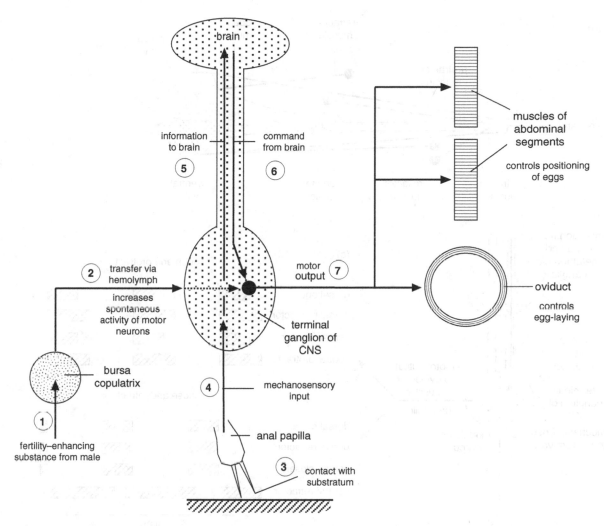

Figure 13.18 Control of oviposition in the silk moth (*Bombyx*). Numbers indicate the sequence of events (after Yamaoka and Hirao, 1971, 1977).

the oviducal muscles, and a myotropic peptide is present in the hemolymph during oviposition. These muscles are also controlled neurally, but neural stimulation has the effect of preventing the backwards movement of eggs. A second central pattern generator and octopamine from dorsal unpaired median (DUM) cells in the subterminal abdominal ganglion are also involved in integrating these activities, but their precise functions are not clear. A myotropic hormone is also present at

maximum titers during the period of oviposition of *Rhodnius*. It is presumed to have an important role in oviposition, as in *Locusta*.

13.5.4 Role of the accessory glands

The function of the female accessory glands is generally to fix eggs in position or protect them from desiccation and predators. Many insects attach their eggs on or close to the larval food source. Lice fix

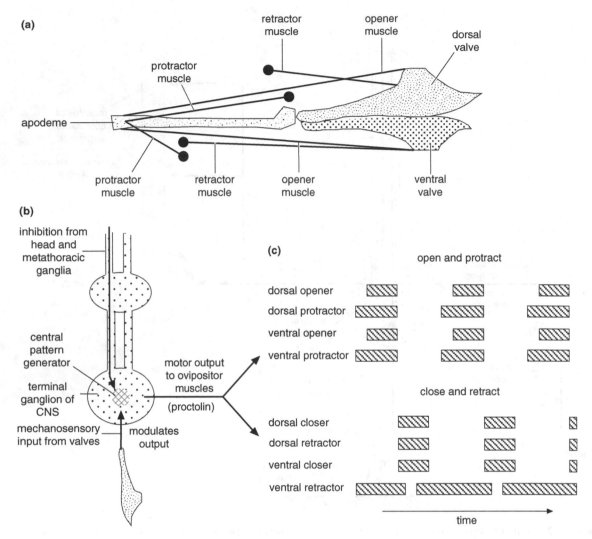

Figure 13.19 Control of oviposition in the locust (*Locusta*) (mainly after Thompson, 1986a,b). (a) Diagram of the ovipositor valves and the apodeme to which the dorsal valve hinges. Black spots indicate the origins of muscles on the body wall. Closer muscles and some accessory muscles are not shown. (b) Diagram of the terminal abdominal ganglion showing some features of regulation. (c) Pattern of activity in the principal muscles produced by the central pattern generator.

their eggs to hairs; many plant-feeding Heteroptera and Lepidoptera lay eggs on the host plant. The glue holding the eggs in position comes from the female accessory glands. Within the Lepidoptera, insects that drop their eggs from the air, such as the Hepialidae, lack accessory glands. However, the glands are well developed in species that cover their eggs with

secretion. In *Chrysopa* (Neuroptera) the eggs are laid on tall stalks formed from silk produced in the accessory glands.

The accessory glands of the parasitic wasp *Pimpla* (Hymenoptera) produce hyaluronic acid and a lipoprotein which coats the egg as it is laid into the host. It is suggested that this is a critical factor in preventing

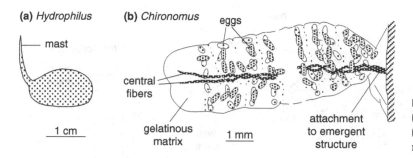

(a) *Hydrophilus*

mast

1 cm

(b) *Chironomus*

eggs

central fibers

gelatinous matrix

1 mm

attachment to emergent structure

Figure 13.20 Eggs of aquatic insects: (a) egg cocoon of *Hydrophilus* (Coleoptera) (after Miall, 1922); (b) egg rope of *Chironomus* (Diptera).

encapsulation by the host hemocytes and this may also occur in other parasitic insects (Section 5.4.3).

The accessory gland secretions of many aquatic insects form a gelatinous mass enclosing the eggs. For instance, *Chironomus dorsalis* produces a structure in which the eggs loop backwards and forwards around the circumference of the matrix while a pair of fibers, anchoring the mass to the surface, run through the center (Fig. 13.20b). Other species of chironomids and Trichoptera lay their eggs in masses or strands of gelatinous material.

Oothecae. Species from several insect groups lay their eggs in oothecae formed by secretions of the female accessory glands. Oothecae are produced by a majority of Blattodea, Mantodea and Acridoidea, and by the tortoise beetles, Cassidinae. These structures have precise and characteristic forms which permit respiration by the eggs and the escape of the newly hatched larvae while, at the same time, protecting them from desiccation and from predators. Among the Blattodea, *Blatta*, for instance, lays its eggs in two rows, each of eight eggs, inside a capsule which becomes tanned as it is formed (Fig. 13.21a). A crest along the top of the capsule contains cavities connected via small pores to the outside and via narrow tubes to the inside of the ootheca. Immediately below the opening of each tube, the chorion of an egg is produced into an irregular lobe containing airspaces which communicate with the airspace around the oocyte. Each egg is associated with an air tube in the ootheca.

Most Acridoidea lay their eggs in egg pods in the ground. A pod consists of a mass of eggs held together by a frothy secretion and is sometimes also enclosed by a layer of the same substance. The hole above the egg mass is plugged by more froth (Fig. 13.21b). The number of eggs in a pod reflects the number of ovarioles in the ovaries and may vary from 1 to over 300. The frothy plug permits respiration and provides an easy route to the soil surface for the hatching larvae.

Tortoise beetles also produce oothecae. The form and complexity of the ootheca varies from species to species, but *Basipta*, for example, attaches its ootheca to the stem of its food plant (Fig. 13.21c). The theca is formed of a large number of lamellae produced from an accessory gland secretion which is compressed into a plate-like form as it is extruded between the terminal sclerites of the abdomen. The lamellae form an open cup, occupied by about 30 chambers also formed by the lamellae. An egg is placed in each chamber as it is formed and around the outside of the cup the lamellae are plastered firmly together to form a hard "shell" with looser lamellae outside it.

Among aquatic insects, all the Hydrophilidae (Coleoptera) produce egg cocoons or oothecae. These cocoons are made from silk synthesized in modified ovarioles. The cocoon of *Hydrophilus* is molded to the shape of the abdomen with the aid of the forelegs; then the abdomen is withdrawn and the eggs are laid. Finally, the cocoon is sealed off and remains floating on the surface of the water. It is equipped with a silken "mast" about 2.5 cm high, which has a respiratory function (Fig. 13.20a).

In all these insects, the material from the glands, which is initially fluid, is modified as the chemicals

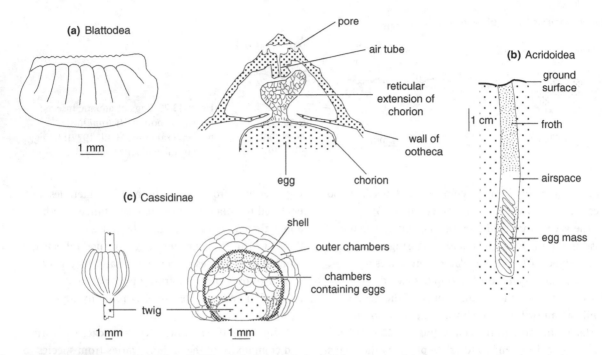

Figure 13.21 Oothecae. (a) Cockroach (*Blatta*). Left: lateral view of the ootheca; right: transverse section through the crest showing the passage by which air reaches the eggs (after Wigglesworth and Beament, 1950). (b) Grasshopper (*Acrida*) (after Chapman and Robertson, 1958). (c) Tortoise beetle (*Basipta*). Left: side view; right: view from above showing the chambers which are produced by secretions of the accessory glands. Each chamber contains one egg (after Muir and Sharp, 1904).

from different glands mix and come into contact with the outside air. The process is best understood in *Periplaneta*. The two accessory glands each consist of a mass of branched tubules lined with cuticle. The gland cells differ in different parts of the glands; in the left gland, which is larger than the right, three types are recognizable. These produce the protein from which the ootheca is formed, a β-glucoside of protocatechuic acid and an oxidase. The right-hand gland has two types of secretory cell. It secretes a β-glucosidase which liberates protocatechuic acid from its β-glucoside when the secretions of the two glands mix in the genital chamber. The protocatechuic acid is oxidized to a quinone by the oxidative enzymes and this tans the protein to produce a cuticle-like structure. A similar process occurs in the production of an ootheca by the mantis *Tenodera*.

Some insects cover the eggs with material that is not derived from the accessory glands. The heteropteran *Plataspis* forms a type of ootheca, laying its eggs in two rows and then covering them with hard elongate pellets of a secretion produced by specialized cells in the gut.

Interactions with male-derived proteins. Female accessory gland products may also interact with proteins derived from the male accessory glands or his sperm. In muscid flies the accessory gland secretion contains proteolytic enzymes and an esterase which have two essential roles in fertilization of the egg. They break down the acrosomal membrane of the sperm, and they lead to the digestion of a cap over the micropyle. Whether this also occurs in other insects is not known. In *Scatophaga* female accessory gland products are secreted during copulation, and an evolutionary history of multiple mating is associated with the coevolution of testes size in males and accessory gland size in females, suggesting that greater quantities of female secretions are required to process greater quantities of sperm and seminal fluid. In *Drosophila* post-mating responses to insemination, such as the storage of sperm, onset of oviposition and the loss of sexual receptivity, require the cleavage of seminal fluid proteins by female proteases that may be products of the female accessory glands.

Summary

- The internal anatomy of the female reproductive tract includes: the ovaries, where eggs are manufactured; the oviducts through which they pass; and a genital chamber referred to as the vagina and/or bursa copulatrix, into which males deposit their ejaculate. There are one or more spermathecae in which sperm are stored prior to their utilization, and frequently a pair of accessory glands.

- There are two basic types of ovary – panoistic and meroistic. In panoistic ovaries, ova develop directly from oocytes that have their origins in the germarium. In meroistic ovaries, nurse cells or trophocytes provide content to the developing oocyte. In telotrophic ovaries nurse cells remain in the germarium while in polytrophic ovaries they remain closely associated with the developing oocyte enclosed within the follicle.

- Accumulation of yolk in the oocytes occurs in the lower proximal region of each ovariole, and is under hormonal regulation.

- As yolk accumulation nears its end, follicle cells begin to produce proteins that will make up the vitilline envelope and the chorion.

- Oocytes may be resorbed when the adult insect experiences insufficient nutrition, or when they are unable to lay fertile eggs.

- The egg is fertilized as it passes into the genital chamber. Fertilization can be highly controlled, with an efficient use of stored sperm. In haplodiploid species females can regulate fertilization in order to adjust sex ratios, while in diploid species females can influence which sperm are stored and used at fertilization based on male mate quality.

- The external genitalia of some species form an ovipositor used to deposit eggs into the oviposition substrate. Chemoreceptors are often associated with the ovipositor and function in oviposition site selection.

- Reproductive accessory glands are responsible for the production of proteins that coat the egg surface. These proteins can serve to protect the egg, adhere it to the oviposition substrate or to encase the egg mass in a cocoon or oothecae. Accessory gland secretions can be involved in the capacitation of sperm, and undergo complex proteolytic interactions with seminal fluid products to facilitate post-mating changes in female physiology and behavior.

Recommended reading

Baer, B., Eubel, H., Taylor, N. L., O'Toole, N. and Millar, A. H. (2009). Insights into female sperm storage from the spermathecal fluid proteome of the honeybee *Apis mellifera*. *Genome Biology* **10**, r67.

Bell, W. J. and Bohm, M. K. (1975). Oosorption in insects. *Biological Reviews* **50**, 373–396.

Büning, J. (1994). *The Insect Ovary: Ultrastructure, Previtellogenic Growth and Evolution.* London: Chapman & Hall.

Hagedorn, H. H. (1985). The role of ecdysteroids in reproduction. In *Comprehensive Insect Physiology, Biochemistry and Pharmacology*, vol. 8, ed. G. A. Kerkut and L. I. Gilbert, pp. 205–262. Oxford: Pergamon Press.

Hinton, H. E. (1981). *Biology of Insect Eggs*, vol. 1. Oxford: Pergamon Press.

Koeppe, J. K., Fuchs, M., Chen. T. T., Hunt, L.-M., Kovalick, G. E. and Briers, T. (1985). The role of juvenile hormone in reproduction. In *Comprehensive Insect Physiology, Biochemistry and Pharmacology*, vol. 8, ed. G. A. Kerkut and L. I. Gilbert, pp. 165–203. Oxford: Pergamon Press.

Margaritis, K. (1985). Structure and physiology of the eggshell. In *Comprehensive Insect Physiology, Biochemistry and Pharmacology*, vol. 1, ed. G. A. Kerkut and L. I. Gilbert, pp. 153–230. Oxford: Pergamon Press.

Raikhel, A. S. and Dhadialla, T. S. (1992). Accumulation of yolk proteins in insect oocytes. *Annual Review of Entomology* **37**, 217–251.

Regier, J. C. and Kafatos, F. C. (1985). Molecular aspects of chorion formation. In *Comprehensive Insect Physiology, Biochemistry and Pharmacology*, vol. 1, ed. G. A. Kerkut and L. I. Gilbert, pp. 113–151. Oxford: Pergamon Press.

Scudder, G. G. E. (1961). The comparative morphology of the insect ovipositor. *Transactions of the Royal Entomological Society of London* **113**, 25–40.

Wheeler, D. (1996). The role of nourishment in oogenesis. *Annual Review of Entomology* **41**, 407–431.

Wolfner, M. F. (2009). Battle and ballet: molecular interactions between the sexes in *Drosophila*. *Journal of Heredity* **100**, 399–410.

References in figure captions

Beament, J. W. L. (1946). The waterproofing process in eggs of *Rhodnius prolixus* Stahl. *Proceedings of the Royal Society of London B* **133**, 407–418.

Büning, J. (1993). Germ cell cluster formation in insect ovaries. *International Journal of Insect Morphology and Embryology* **22**, 237–253.

Chapman, R. F. and Robertson, I. A. D. (1958). The egg pods of some tropical African grasshoppers. *Journal of the Entomological Society of Southern Africa* **21**, 85–112.

Gadot, M. and Applebaum, S. W. (1985). Rapid in vitro activation of corpora allata by extracted locust brain allatotropic factor. *Archives of Insect Biochemistry and Physiology* **2**, 117–129.

Gellissen, G. and Emmerich, H. (1978). Changes in the titer of vitellogenin and of diglyceride carrier lipoprotein in the blood of adult *Locusta migratoria. Insect Biochemistry* 8, 403–412.

Hopkins, C. R. and King, P. E. (1966). An electron-microscopical and histochemical study of the oocyte periphery in *Bombus terrestris* during vitellogenesis. *Journal of Cell Science* 1, 201–216.

Huebner, E. and Diehl-Jones, W. (1993). Nurse cell–oocyte interaction in the telotrophic ovary. *International Journal of Insect Morphology and Embryology* 22, 369–387.

Imms, A. D. (1957). *A General Textbook of Entomology*, 9th edn., revised by O. W. Richards and R. G. Davies. London: Methuen.

King, R. C. (1964). Studies on early stages of insect oogenesis. *Symposium of the Royal Entomological Society of London* 2, 13–25.

Klowden, M. J. (1987). Distension-mediated egg maturation in the mosquito, *Aedes aegypti. Journal of Insect Physiology* 33, 83–87.

Lanot, R., Thiebold, J., Lagueux, M., Goltzene, F. and Hoffmann, J. A. (1987). Involvement of ecdysone in the control of meiotic reinitiation in oocytes of *Locusta migratoria* (Insecta, Orthoptera). *Developmental Biology* 121, 174–181.

Lea, A. O., Briegel, H. and Lea, H. N. (1978). Arrest, resorption, or maturation of oocytes in *Aedes aegypti*: dependence on the quantity of blood and the interval between blood meals. *Physiological Entomology* 3, 309–316.

Mazur, G. D., Regier, J. C. and Kafatos, F. C. (1980). The silkmoth chorion: morphogenesis of surface structures and its relation to synthesis of specific proteins. *Developmental Biology* 76, 305–321.

Miall, L. C. (1922). *The Natural History of Aquatic Insects.* London: Macmillan.

Moshitzky, P. and Applebaum, S. W. (1990). The role of adipokinetic hormone in the control of vitellogenesis in locusts. *Insect Biochemistry* 20, 319–323.

Muir, F. and Sharp, D. (1904). On the egg-cases and early stages of some Cassididae. *Transactions of the Entomological Society of London* (1904), 1–23.

Okelo, O. (1979). Mechanisms of sperm release from the receptaculum seminis of *Schistocerca vaga* Scudder (Orthoptera: Acrididae). *International Journal of Invertebrate Reproduction* 1, 121–131.

Paul, M. and Kafatos, F. C. (1975). Specific protein synthesis in cellular differentiation: II. The program of protein synthetic changes during chorion formation by silkmoth follicles, and its implementation in organ culture. *Developmental Biology* 42, 141–159.

Scudder, G. G. E. (1961). The comparative morphology of the insect ovipositor. *Transactions of the Royal Entomological Society of London* 113, 25–40.

Snodgrass, R. E. (1935). *Principles of Insect Morphology.* New York, NY: McGraw-Hill.

Snodgrass, R. E. (1956). *Anatomy of the Honey Bee.* London: Constable.

Thompson, K. J. (1986a). Oviposition digging in the grasshopper: I. Functional anatomy and the motor program. *Journal of Experimental Biology* 122, 387–411.

Thompson, K. J. (1986b). Oviposition digging in the grasshopper: II. Descending neural control. *Journal of Experimental Biology* 122, 413–425.

Uvarov, B. P. (1966). *Grasshoppers and Locusts*, vol. 1. Cambridge: Cambridge University Press.

Wigglesworth, V. B. and Beament, J. W. S. (1950). The respiratory mechanisms of some insect eggs. *Quarterly Journal of Microscopical Science* 91, 429–452.

Yamaoka, K. and Hirao, T. (1971). Role of nerves from the last abdominal ganglion in oviposition behaviour of *Bombyx mori. Journal of Insect Physiology* **17**, 2327–2336.

Yamaoka, K. and Hirao, T. (1977). Stimulation of virginal oviposition by male factor and its effect on spontaneous nervous activity in *Bombyx mori. Journal of Insect Physiology* **23**, 57–63.

Yin, C.-M., Zou, B.-X., Li, M.-F. and Stoffolano, J. G. (1990). Ecdysteroid activity during oögenesis in the black blowfly, *Phormia regina* (Meigen). *Journal of Insect Physiology* **36**, 375–382.

Yin, C.-M., Zou, B.-X., Li, M.-F. and Stoffolano, J. G. (1994). Discovery of a midgut peptide hormone which activates the endocrine cascade leading to oögenesis in *Phormia regina* (Meigen). *Journal of Insect Physiology* **40**, 283–292.

Zou, B.-X., Yin, C.-M., Stoffolano, J. G. and Tobe, S. S. (1989). Juvenile hormone biosynthesis and release during oocyte development in *Phormia regina* Meigen. *Physiological Entomology* **14**, 233–239.

14 | The egg and embryology

REVISED AND UPDATED BY **MICHAEL R. STRAND**

INTRODUCTION

Embryogenesis is the process by which a single egg develops into a multicellular individual. Many of our most important discoveries in understanding the embryonic development of animals, including humans, derive from studies that were first conducted in insects. Some of these discoveries are also now being applied in the fields of medicine and agriculture. As discussed in Chapters 12 and 13, all future offspring produced by insects derive from specialized progenitor cells called germ cells, which migrate to the gonads, where they differentiate into gametes. The gametes produced by females are called oocytes (eggs), while the gametes produced by males are called spermatozoa (sperm). Most insects begin their embryonic development when genetic material from an egg and sperm fuse through the process of fertilization to form a zygote. The zygote then divides mitotically to produce all of the different cells that comprise the body of the nymph (exopterygote/ hemimetabolous species) or larva (endopterygote/holometabolous species), which will hatch from the egg. Embryogenesis proceeds through a similar series of steps in most insect species, but there are also a number of variations that in some cases are associated with unique life histories. In this chapter we first summarize key morphological and functional features of insect eggs (Section 14.1). Next we discuss the process of embryogenesis, including some of the molecular mechanisms that control axis formation and nutrient acquisition (Sections 14.2 and 14.3). We then consider sex determination (Section 14.4) and end the chapter by discussing parthenogenesis (Section 14.5), pedogenesis (Section 14.6) and other unique forms of embryonic development.

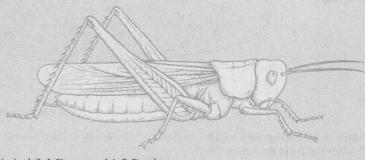

The Insects: Structure and Function (5th edition), ed. S. J. Simpson and A. E. Douglas.
Published by Cambridge University Press. © Cambridge University Press 2013.

14.1 The egg

Most insects produce large eggs relative to their own size. This is due to a majority of insects packaging their eggs with large amounts of yolk, which serves as the source of nutrients for growth and development of the embryo. In general, the eggs of Endopterygota contain less yolk and are smaller than those of Exopterygota. To some extent this may reflect differences associated with ovariole type (Section 13.2.1). For example, in two locust (Orthoptera) species, which have panoistic ovarioles, each egg weighs about 0.5% of female weight; among insects with telotrophic ovarioles, the egg of *Trialeurodes vaporarium* (Hemiptera) is over 1% of the female weight and that of *Callosobruchus maculata* (Coleoptera) 0.6%. By contrast, among insects with polytrophic ovarioles, comparable figures for *Apis mellifera* (Hymenoptera) and *Grammia geneura* (Lepidoptera) are 0.07% and 0.11%, respectively.

Several ecological and life history factors have also been suggested to affect the size and morphology of insect eggs. Among Lepidoptera from temperate regions, species overwintering in the egg stage have larger eggs than species that overwinter in some other stage, and species feeding on woody plants have bigger eggs than those feeding on herbaceous plants. Individual females of at least some butterflies lay smaller eggs as they grow older, and females of the corn borer moth, *Ostrinia*, lay smaller eggs if they do not receive adequate nutrition. Egg morphology of lycaenid butterflies also varies among populations in relation to habitat. Insects such as aphids package their eggs with little or no yolk because the mother provides nutrients necessary for embryonic development (Section 14.3). Many endoparasitic Hymenoptera also package eggs with reduced amounts or no yolk because they lay their eggs inside the body of another arthropod (the host), which provides nutrients for both embryonic and larval development.

Insect eggs have a wide variety of forms. Commonly, as in Orthoptera and many Hymenoptera, they are sausage-shaped (Fig. 14.1), but they may be conical, as seen for butterflies in the genus *Pieris*, rounded (other Lepidoptera) or barrel-shaped as in many Hemiptera. The eggs of Nepidae (Hemiptera) and some Diptera have respiratory horns (Fig. 14.3), while the eggs of many parasitic Hymenoptera have a projection called a pedicel, which accommodates some of the contents of the egg as it is being laid through the narrow opening of the ovipositor into a host.

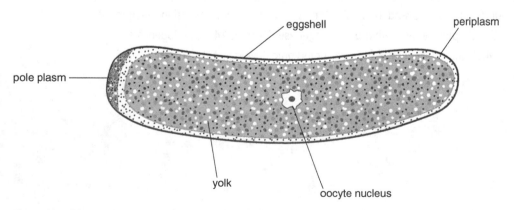

Figure 14.1 Diagrammatic sagittal section through the egg of *Bactrocera* (formerly *Dacus*) *tryoni* (Diptera) prior to fertilization (after Anderson, 1962).

14.1.1 Major structural features of insect eggs

Most insects produce eggs containing an oocyte nucleus, cytoplasm, posterior germ plasm and yolk, which is comprised of protein, lipids and carbohydrates (Section 13.2.4). Surrounding these components is the eggshell (Fig. 14.1). Key features of the other components include: (1) the cytoplasm, nucleus and pole plasm; and (2) the eggshell, which consists of several layers.

At the time of oviposition, the egg cytoplasm forms a bounding layer called the periplasm (Fig. 14.1), and an irregular reticulum within the yolk. The oocyte nucleus, surrounded by a halo of yolk-free cytoplasm, is usually located centrally within the egg, but in some species may be more anterior or posterior (Fig. 14.1). The posterior pole of dipteran, hymenopteran and lepidopteran eggs is frequently distinguished from the anterior pole by the presence of pole plasm (also called germ plasm), which differs from other cytoplasm in the egg because it contains specific mRNAs and proteins that determine the germ line (Section 14.2.3) (Fig. 14.1). Symbiotic microorganisms like *Wolbachia* (Section 14.5.1), which are transmitted vertically by insects, also often accumulate in the posterior pole of the egg.

The eggshell of insects consists of three distinct layers (vitelline membrane, wax layer and chorion), which are formed during the late stage of oogenesis (Section 13.2.5). The innermost layer of the eggshell is the vitelline membrane, which is approximately 300 nm thick in *Drosophila* eggs and 1–2 μm thick in many other insects. In parasitic Hymenoptera, such as *Habrobracon* and *Nasonia*, the vitelline membrane is only 150 nm, while in the damselfly, *Sympetrum*, it is 9 μm thick. The vitelline membrane may also vary in thickness in different parts of the egg, and is not necessarily uniform in structure over the entire surface of the oocyte. In *Drosophila* (Diptera), for example, there are spaces within the vitelline membrane where the chorion splits at the time of hatching, and in the beetle *Lytta* it is perforated

beneath the micropyles (Section 14.1.2). The vitelline envelope is the first layer of the eggshell to be made by the follicular epithelium and its deposition usually begins late in oogenesis when vitelline membrane proteins accumulate on the surface of the oocyte in small vesicles called vitelline bodies.

A wax layer is sometimes but not always present on the outside of the vitelline envelope. This has been observed in *Drosophila*, the hemipteran *Rhodnius*, grasshoppers and some Lepidoptera. It may be of general occurrence, at least among insects that lay their eggs in places where they are subject to desiccation. The wax layer is formed by accumulation of lipid-filled vesicles on the surface of the vitelline envelope, which create a water-impermeable layer.

The chorion forms the outermost layer of the eggshell, and is produced by the follicle cells while the egg is in the ovary. The primary constituents of the chorion are proteins that undergo a hardening process either before or shortly after oviposition. The hardening process is due to peroxidase-catalyzed protein crosslinking. In many mosquitoes, the chorion also hardens and darkens following oviposition due to the formation of melanin catalyzed by the enzyme phenoloxidase. The chorion of insect eggs varies in thickness from less than 1 μm in eggs of some endoparasitic Hymenoptera to over 50 μm in some orthopteran and lepidopteran eggs. Even closely related species may have chorions differing in thickness. In the silk moths, for example, the chorion of *Bombyx* is about 25 μm thick while that of *Hyalophora* is 55 μm thick. These differences are presumed to have ecological significance.

In some species, the chorion is subdivided into two anatomically distinct layers, called the endochorion and exochorion, but in other insects these distinct layers are absent. Even where both are present they are not necessarily homologous in different species. In the mosquito *Aedes aegypti* the endochorion is a homogeneous electron-dense layer, while the exochorion consists of a fibrilar network. In *Drosophila* proteins of the endochorion are grouped into octomers of four dimer pairs that are stabilized by disulfide or dityrosine bonds. Holes in the resulting crystal lattice are 2–4 nm across, which is large enough to permit the free movement of oxygen and carbon dioxide. Extensive airspaces are usually present in the chorion adjacent to the oocyte; in species that lay their eggs in moist environments these airspaces may extend through the entire thickness of the chorion. This is the case in the grouse locust, *Tetrix* (Fig. 14.2a). In the eggs of most species, however, openings through the chorion are restricted, and this limits water loss. For example, in *Musca* (Diptera), the inner meshwork in the chorion, which provides a continuous layer of air all around the developing embryo, connects with an outer meshwork by pores (aeropyles) through an otherwise solid middle layer (Fig. 14.2b). In many other species, the connections are even more restricted. For example, in *Calliphora* (Diptera), the outer meshwork and aeropyles connecting with the inner meshwork are absent over the greater part of the egg, and are present only between the hatching lines (Fig. 14.3). Eggs of *Ocypus* (Coleoptera) have an equatorial band of aeropyles connecting with the inner airspaces; *Rhodnius* eggs have a ring of aeropyles just below the

cap (Fig. 13.13); and those of the stick insect *Carausius* have a single small pore at which the reticular inner chorion is exposed at the egg surface. The respiratory horns present on the eggs of some Diptera and the Nepidae (Hemiptera) similarly serve the function of connecting the inner layer of air with the atmosphere outside, while at the same time restricting the area through which water loss can occur (Fig. 14.3). This is also true of the micropylar processes of many Hemiptera, which have a respiratory function (Fig. 14.4a).

The chorion outside the inner layer of airspaces is sometimes thick and, in Lepidoptera, is often lamellate. The lamellae are formed from sheets of protein fibers that are organized in the same orientation. The orientation of fibers in successive sheets is rotated so that they have a helicoidal arrangement through the thickness of the chorion. Rotation through 180° creates the appearance of lamellae, in exactly the same way as in chitinous cuticle (Section 16.2). The outer surface of the chorion is often sculptured, frequently with a basically hexagonal pattern reflecting the shapes of the follicle cells, which secrete it. Outside this, grasshoppers have an additional layer, the

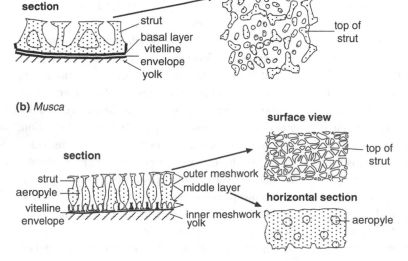

(a) *Tetrix*

surface view

section

strut
basal layer
vitelline
envelope
yolk

top of strut

(b) *Musca*

surface view

section

strut
aeropyle
vitelline
envelope

outer meshwork
middle layer
inner meshwork
yolk

top of strut

horizontal section

aeropyle

Figure 14.2 Airspaces in the chorion: (a) A chorion in which extensive airspaces extend through its thickness (*Tetrix* [Orthoptera]) (after Hartley, 1962) (b) A chorion in which the air-filled inner meshwork connects with the exterior through a limited number of aeropyles (*Musca* [Diptera]) (after Hinton, 1960).

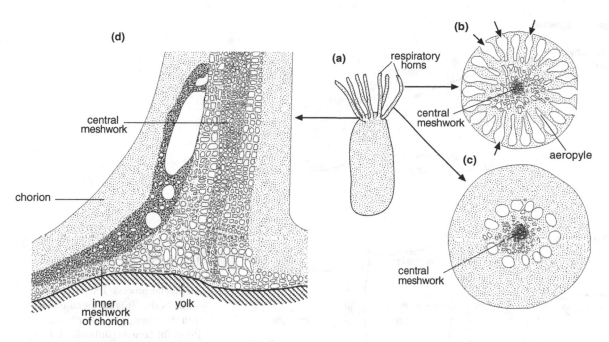

Figure 14.3 Respiratory horns on the egg of *Nepa* (Hemiptera) (after Hinton, 1961). (a) Whole egg. (b) Cross-section of a distal region of a horn. Arrows indicate the openings of the airspaces to the outside. (c) Cross-section of a proximal region of a horn. (d) Longitudinal section of the basal region of a horn where it connects to the chorion around the egg. Notice the continuity of the air-filled central meshwork of the horn with the inner meshwork that surrounds the egg.

extrachorion, which differs from the chorion in being produced by gland cells in the oviducts rather than by the follicle cells. The outer covering of the egg of the moth *Micropteryx* is also unusual in consisting of a forest of knobbed projections. The stalks carrying the knobs are about 50 μm high and 2–4 μm in diameter, and the knobs themselves are about 10 μm in diameter. These are produced in the hour following oviposition by exudation from the oocyte itself. The exudate forces the viscous outer layer of the shell, produced by the female accessory glands, outward to form projections.

14.1.2 Fertilization

Because the chorion is laid down during oogenesis, some provision is necessary to allow the entry of sperm when the egg is fertilized. This takes the form of structures called micropyles, which are

funnel-shaped pores, usually located near the anterior pole of the egg, that pass through the chorion and into which fit protuberances of the vitelline membrane. The pores are 1–2 μm in diameter, often with a wider funnel at the surface of the chorion. The vitelline envelope beneath the micropyles is also modified.

The number of micropyles varies: most dipteran eggs have only a single terminal micropyle, while Acrididae (grasshoppers) commonly have 30–40 arranged in a ring at the posterior end of the egg (Fig. 14.4b). The number of micropyles varies from 0 to 70 among species of Hemiptera, but is more or less constant within a species. When there is a large number of micropyles they are usually arranged in a circle around the anterior end of the egg. Among Hemiptera, most Cimicoidea have two micropyles, but different species of Coreidae have between 4 and 60, and pentatomids between 10 and 70. Eggs

(a) *Oncopeltus*

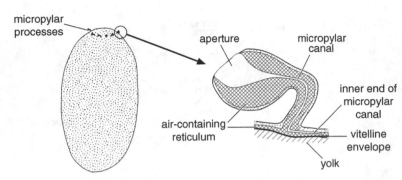

(b) *Locusta*

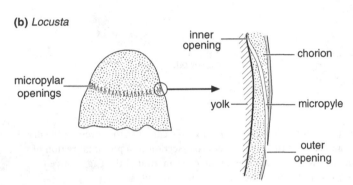

Figure 14.4 Micropyles. (a) *Oncopeltus* (Hemiptera). Left: whole egg; right: longitudinal section through a micropylar process. The process also connects the air layer round the egg with the atmosphere (after Southwood, 1956). (b) *Locusta* (Orthoptera). Left: posterior pole of egg; right: longitudinal section along the length of a micropylar canal (after Roonwal, 1954).

produced by older females of *Rhodnius* have a reduced number of micropyles compared with those laid by younger females. There are no micropyles in the eggs of species of Cimicoidea, which are fertilized in the ovary (Section 12.3). Sometimes the micropylar openings are associated with aeropyles and are raised above the surface of the egg. This occurs in Heteroptera (Hemiptera); in *Oncopeltus*, for example, each consists of a cup on a stem (Fig. 14.4a). The micropyle canal passes through the middle of the process and through the chorion and is surrounded by an open reticulum enclosing airspaces.

14.1.3 Respiration

Among the majority of terrestrial insects, gas exchange between the egg and environment occurs via the aeropyles and the chorionic airspaces of the eggshell. The layer of air in the inner chorion has direct access to the oocyte through pores and is also connected with the outside air via the aeropyles. Where only a single aeropyle is present, as in the beetle *Callosobruchus*, its size is critically important. Too small a pore will not permit the entry of an adequate supply of oxygen; the eggs of two strains of *Callosobruchus maculatus* with different metabolic rates have differently shaped aeropyles.

The eggs of some terrestrial insects laid in soil or detritus are subject to periodic flooding. Eggs of some species survive this because the chorion, with its hydrophobic characteristics, maintains a layer of air around the egg into which gas from the surrounding water may diffuse. Thus the chorion acts as a plastron (Section 17.6), but the effectiveness of a plastron depends on the area available for gas exchange – that is, on the extent of the air–water interface. In the eggs of Lepidoptera and many Hemiptera (including

Rhodnius), the air–water interface is too small to be of significance since it is limited to the openings of the aeropyles. Nevertheless, these eggs may survive some degree of flooding by virtue of the fact that they can tolerate great reductions in their metabolic rate. In other insects the plastron is larger and more effective. In the blowfly *Calliphora*, where the plastron occurs between the hatching lines, and in the fly *Musca*, where it extends over the whole egg, normal development continues if the egg is immersed in well-aerated water. The respiratory horns of many dipteran eggs also form an efficient plastron if the eggs are flooded. This is also true in the eggs of Nepidae (Hemiptera), which are essentially terrestrial in their respiration as the horns normally project above the surface of the water.

The surface tension of water contaminated with organic acids and other surface active substances is lower than that of clean water, and the ease with which a plastron is wetted, and hence ceases to function, is inversely proportional to the surface tension. Thus the plastron of insect eggs which are laid in organic materials subject to flooding needs to have a high resistance to wetting if it is to continue functioning under conditions of low surface tension. The plastron must also be able to withstand wetting by raindrops, which – momentarily – may exert a pressure approaching 50 kPa. Hence, although eggs in dung and similar situations can rarely be subject to flooding by more than a few centimeters of water, they possess a plastron capable of withstanding flooding by clean water to a much greater depth, and in some cases their resistance is greater than that exhibited by the plastron of some aquatic insects.

Eggs laid in water, such as those of dragonflies, obtain their oxygen from that dissolved in the water. The chorion of these eggs generally lacks any obvious system of spaces; oxygen passes through the solid material by diffusion. This is presumably because in the absence of any need to restrict water loss, the chorion is relatively porous. As noted above, the eggs of many endoparasitic Hymenoptera are very small and possess an extremely thin chorion. The eggs of these insects also respire by uptake of oxygen from the hemolymph of their hosts.

14.1.4 Water regulation

All insect eggs are subject to water loss, but the extent to which water loss occurs is affected by the choice of oviposition site and by the permeability of the chorion. Subsequently, during embryonic development and in the pharate first-stage larva, water loss may be further reduced by the formation of new structures within the egg. The eggs of some insects also take up water, which is often essential for continued embryonic development.

The rate of water loss from newly laid insect eggs is often very low, especially in species that do not take up water during development. In *Rhodnius*, for instance, water is lost at a rate of 2.26 μg cm^{-2} h^{-1} Pa^{-1} from the egg, but in species that lay their eggs in damp places, the rate of water loss may be much higher if the eggs are exposed to desiccating conditions. It is 45 μg cm^{-2} h^{-1} Pa^{-1} in exposed eggs of the beetle *Phyllopertha*, for example. The rate of loss may also change with time as the membranes covering the embryo alter (see below), and, in the beetle *Acanthoscelides* fertilization causes a change in the vitelline envelope that reduces water loss.

The chorion itself limits water loss to some extent, and differences in its thickness may be important in this respect. A thick chorion and a reduction in the number of respiratory horns is a characteristic of the eggs of heteropteran species that are laid in exposed situations subject to desiccation. Some tropical grasshoppers that survive the dry season in the egg stage have thick, tough chorions, and the inner chorion of the eggs of mosquitoes such as *Aedes*, which resist desiccation, is thicker and more melanized than that of non-resistant eggs, like those laid by mosquitoes in the genus *Culex*. However, it is generally considered that the principal factor in

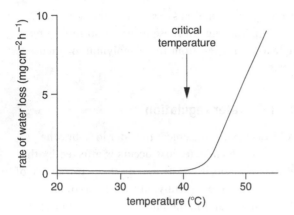

Figure 14.5 Water loss from the egg, showing the critical temperature (*Rhodnius* [Hemiptera]) (after Beament, 1946).

limiting water loss is the layer of wax on the inside of the chorion. Like the wax layer on the outside of the cuticle (see Chapter 16), it has a critical temperature above which water loss increases sharply (Fig. 14.5). The critical temperature for the eggs of *Rhodnius* is 42.5°C, and for the eggs of *Lucilia* (Diptera) and *Locustana* (Orthoptera) it is 38°C and 55–58°C, respectively.

A second layer of wax is subsequently laid down in the serosal cuticle (Section 14.2.10). Functionally, this secondary wax layer may supplement the primary wax layer on the inside of the chorion, or it may replace it if the latter becomes broken by the increase in size of the egg, which sometimes occurs. The rate of evaporation from the exposed egg of *Locustana* (Orthoptera) at 35°C and 60% relative humidity drops from 0.35–0.54 mg egg^{-1} d^{-1} at oviposition, when only the primary wax layer is present, to 0.11–0.36 mg egg^{-1} 24h^{-1} five days later, when the serosal cuticle and secondary wax layer are completed all over the egg except for the hydropylar region (see below). When this region is also sealed off, the evaporation rate falls to 0.03–0.04 mg egg^{-1} 24h^{-1}. In the mosquito *Anopheles gambiae* serosal cells (Section 14.2.7) secrete a cuticle that serves as another barrier to water loss.

14.1.5 Water absorption

The eggs of *Rhodnius*, and probably other Hemiptera and Lepidoptera that are laid in dry, exposed situations, develop without any uptake of water, but the eggs of other insect species absorb water from the environment in the course of development. Among terrestrial insects it is recorded in Orthoptera, some Hemiptera and some Coleoptera; in aquatic species it occurs in Odonata, Hemiptera, Coleoptera and some Diptera. Water absorption results in a considerable increase in egg volume and weight (Fig. 14.6).

Water uptake usually occurs over a limited period of development, which varies from species to species. For instance, among the grasshoppers it occurs before there has been any significant development in *Camnula*, during early embryonic development in *Locusta* and during later stages of embryogenesis in *Melanoplus differentialis*. No water is taken up by the eggs of these species immediately after oviposition, but then a period of rapid uptake occurs, followed by a period in which no marked change in water content is seen provided the egg is not subject to desiccation (Fig. 14.6). In the cricket *Allonemobius* water is taken up in an earlier stage of development at 30°C than at 20°C.

The mechanism of water uptake is not fully understood, but it is generally considered to be passive and to reflect changes in the permeability of the egg membranes, in the osmotic pressure of the yolk (which normally is greater than that of any surrounding water) and in the hydrostatic pressure within the egg.

The initial failure of grasshopper eggs to absorb water is due to the primary wax layer in the chorion. This forms a barrier to the osmotic uptake of water and is equally effective in preventing water loss. The period of water uptake begins at the time the serosa is completed and the serosal cuticle is first laid down (see below). The increase in volume resulting from water uptake produces cracking of the chorion and the primary wax layer must be disrupted. This leads to a great increase in permeability of the chorion and

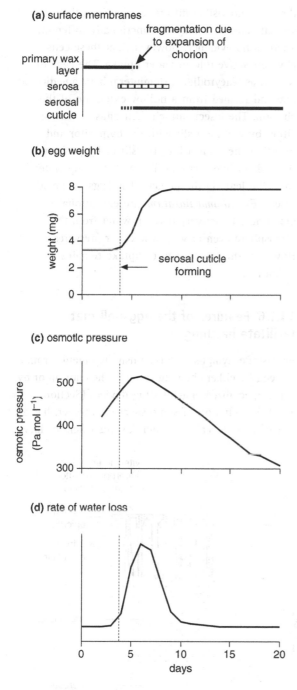

(a) surface membranes

primary wax layer

fragmentation due to expansion of chorion

serosa

serosal cuticle

(b) egg weight

weight (mg)

serosal cuticle forming

(c) osmotic pressure

osmotic pressure (Pa mol l⁻¹)

(d) rate of water loss

days

at the same time initiates mobilization of the yolk so that the osmotic pressure within the egg increases. These two factors also lead to a rapid uptake of water (Fig. 14.6). Following blastoderm formation in *Chortoicetes*, the osmotic pressure of the yolk falls, contributing to, but not wholly accounting for, the reduction in water intake. This reduction may also be a consequence of the completion of the serosal cuticle and the secondary wax layer or of the reduction in stress on the egg membranes so that they become less permeable. The eggs remain capable of taking up water during this period if they are subject at any stage to water loss.

The mechanism may be different in the cricket *Teleogryllus*. Here, the chorion remains permeable at all times, but water uptake is prevented initially, despite a high internal osmotic pressure, by the hydrostatic pressure within the egg. Subsequently, the structure of the inner chorion changes and its permeability increases, enabling the egg to take up water osmotically. The resulting increase in hydrostatic pressure may then be responsible for preventing further water uptake. In Hemiptera, it is suggested that water uptake by the eggs of *Notostira* does not begin until osmotically active substances are produced within the egg, while water uptake by the eggs of *Phyllopertha* stops following a modification of the chorion, which makes it waterproof.

Water uptake by the eggs of many species is limited to a specialized region of the shell called the hydropyle. It may involve a modification of the serosal cuticle, as it does in grasshoppers, or a modification of the chorion itself as in some Hemiptera and perhaps also some beetles. In some other insects, as in *Notostira* (Hemiptera) and *Gryllulus* (Orthoptera), water is absorbed over the whole of the egg and there is no hydropyle.

Figure 14.6 Water uptake by the egg of a grasshopper (*Chortoicetes*) (based on Lees, 1976). (a) Structural changes in the egg that affect water uptake and loss. (b) Change in egg weight due to water uptake. (c) Changes in the osmotic pressure of the yolk. (d) Changes in the rate at which water is lost if the egg is exposed to a desiccating environment, reflecting changes in permeability of the egg membranes.

In grasshoppers the hydropyle is at the posterior pole of the egg. The inner layer of the serosal epicuticle is deeply invaginated into the serosal endocuticle, and the outer epicuticle is a finely fibrous, proteinaceous material, which fills the folds (Fig. 14.7). The serosal endocuticle is thinner in this region than elsewhere. By contrast, in eggs of the water scorpion, *Nepa* (Hemiptera), the hydropyle comprises a fine meshwork within the chorion, which is isolated from the surrounding chorion. The region, which is normally in the water, provides a direct route through the chorion to the surface of the egg serosal cells. Whether the hydropyle is formed from the chorion or the serosal cuticle, the outer part is hydrophilic so that it readily holds water. In the grasshoppers, the outer serosal epicuticle performs this function and the thin, folded inner epicuticle is freely permeable, whereas it is impermeable over other parts of the egg. It becomes impermeable in species that have an egg diapause, and the end of diapause is associated with an increase in permeability.

In most cases the serosal cells underlying the hydropylar region are larger or more columnar than other cells in the serosal epithelium. Even where there are no visible differences between the serosal cells, they may be functionally different. There is no evidence, however, that these cells have an active role in water uptake. The eggs of whiteflies (Aleyrodidae, Hemiptera) have a pedicel at one end formed from a hollow extension of the chorion. The insects attach their eggs to leaves, either by cutting a slit with the ovipositor and inserting the pedicel into the slit or by pushing the pedicel into a stoma. The eggs take up water from the leaf via the pedicel. The eggs of the stick insect, *Extatosoma tiaratum*, are exceptional in being able to actively take up water from the atmosphere even at very low vapor pressures. However, the mechanism of uptake remains unknown.

14.1.6 Features of the eggshell that facilitate hatching

First-stage nymphs or larvae usually emerge from the egg by either chewing through the chorion or by using specialized egg-bursting devices (Section 15.1). The chorion of some species, however, has a line of weakness along which it splits when the larva

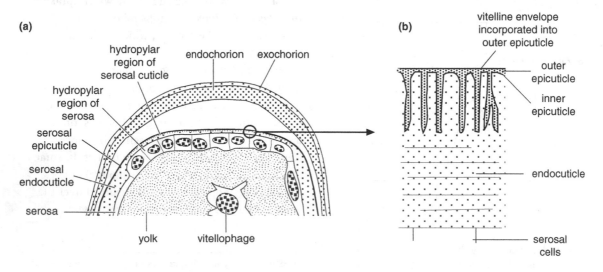

Figure 14.7 The hydropyle of a grasshopper. (a) Longitudinal section through the posterior pole of the egg (based on Roonwal, 1954). (b) Detail of the hydropylar serosal cuticle (based on Slifer and Sekhon, 1963).

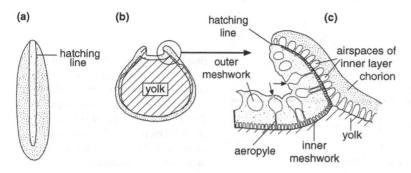

Figure 14.8 Hatching lines in the egg of *Calliphora* (Diptera) (after Anderson, 1960; Hinton, 1960). (a) Dorsal view of the whole egg. (b) Cross-section through the middle of the egg. (c) Detail of a section through one of the hatching lines where the air-filled spaces are back to back. Arrows indicate the openings of the airspaces to the outside.

exerts pressure from within. These hatching lines are usually visible from the outside, and sometimes define a cap or operculum.

A cap is present in the eggs of Phasmatodea, Embiidina (Embioptera), some Hemiptera and Phthiraptera. In *Rhodnius* (Hemiptera), the cap is joined to the rest of the chorion by a sealing bar, which is formed from a very thin layer of resistant endochorion (perhaps equivalent to the crystalline endochorion described above) and the thick amber-colored layer (Fig. 13.13). There are lines of weakness (hatching lines) in the amber layer above and below the sealing bar. Some pentatomid eggs also appear to have a cap, but this has the same structure as the rest of the chorion and is not joined to it by a sealing bar.

The eggs of muscoid Diptera have hatching lines in the form of two ridges running longitudinally along the length of the egg (Fig. 14.8). Along these lines, the inner layer of the chorion extends outward so that each ridge contains two inner layers which are back to back, creating a line along which the chorion is relatively weak. In *Drosophila melanogaster* there is a discontinuity in the endochorion where an operculum meets the body of the egg, and, along this line, the two parts are held together only by the thin innermost layer of the endochorion and a layer of exochorion. The vitelline envelope is also modified along this line. It is not certain how important these devices are as many species, even some *Drosophila* species, appear to have no weak lines in the chorion.

14.2 Embryogenesis

As noted in the introduction, female gametes are called oocytes. Before embryogenesis begins, the oocyte nucleus (primary oocyte) is initially diploid and must undergo meiosis in order to form a haploid cell. The process of meiosis occurs in two steps, the first and second meiotic divisions, which results in formation of four haploid nuclei surrounded by cytoplasm. Three of these haploid nuclei plus cytoplasm are called polar bodies, while the fourth is called the egg pronucleus. Prior to the onset of embryogenesis, the primary oocyte in most insect eggs initiates the first meiotic division while still in the ovary but is blocked at the metaphase stage of the cell cycle. Embryogenesis formally begins when the egg is activated, which refers to release of the first meiotic division from its blocked state and initiation of a cascade of events that results in formation of a nymph or larva. Several different stimuli have been implicated in the activation of insect eggs, including entry of the haploid sperm cell into the egg through the micropyle, and in some Hymenoptera compression of the egg in the ovipositor during the act of oviposition. Activation results in the oocyte rapidly completing meiosis. In most insect species, the polar bodies degenerate, whereas the egg pronucleus, surrounded by a halo of cytoplasm, approaches the sperm. Once in close proximity to one another, the egg pronucleus and sperm fuse, resulting in fertilization and formation of a diploid zygote. It is at this

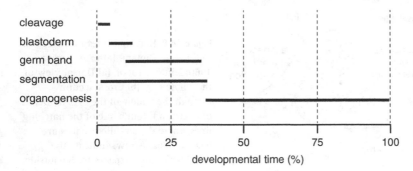

Figure 14.9 Timing of major events in the embryonic development of a long-germ-type insect (*Drosophila* [Diptera]) as a percentage of total development time from egg-laying (0%) to hatching (100%). Temporally, *Drosophila* eggs hatch by 24 hours after egg-laying under standard conditions (25°C).

point that embryogenesis proceeds through a series of steps referred to as cleavage, blastoderm and germ band formation, gastrulation, segmentation and organogenesis. In describing these steps, it is usual to refer to the timing of specific developmental events as a percentage of the total embryonic period from egg-laying to hatching (Fig. 14.9). This also makes it possible to compare in a relative sense when the different steps of embryogenesis are completed in species whose eggs develop at different absolute rates (Section 14.2.17).

14.2.1 Cleavage

Cleavage refers to the period after fertilization when the zygote undergoes a series of mitotic divisions to form an increasing number of smaller cells or nuclei. In most animals, early cleavage involves division of both cytoplasmic and nuclear material to form cells called blastomeres. This form of cleavage is also referred to as being total or complete, because mitosis results in true cells that are each enveloped by a plasma membrane. Most insects do not display total cleavage. Instead, cleavage involves only nuclear division (karyokinesis) without accompanied division of the cytoplasm (cytokinesis) or formation of a plasma membrane. This is known as syncytial cleavage. Each nucleus in the early insect embryo is thus surrounded by a halo of cytoplasm formed during syncytial cleavage that is called an energid (Fig. 14.10a).

The process of syncytial cleavage is best studied in *Drosophila*, where the first mitotic division occurs within ten minutes of zygote formation and results in formation of two energids. The next six nuclear divisions occur synchronously in the interior of the egg, but at the seventh division most energids begin migrating to the periplasm of the egg (Fig. 14.10b) where during cycle ten they distribute themselves as a monolayer at the egg surface. This monolayer is referred to as a syncytial blastoderm (Fig. 14.10c). Mitotic divisions are synchronous through cycle nine in *Drosophila*, whereas the next four mitotic division cycles (10–13) occur metachronously, which means mitosis is nearly synchronous but starts first in nuclei near the anterior and posterior poles of the egg and then spreads in a wave-like fashion toward the equator. Similar waves of mitoses have also been observed in *Apis* (Hymenoptera), *Calandra* (Coleoptera) and *Calliphora* (Diptera). Mitotic activity during the cleavage stage of embryogenesis is very high in eggs laid by insects from the phylogenetically advanced orders. For example, in *Drosophila* each complete mitotic cycle takes approximately ten minutes at 25°C, while in many Lepidoptera it takes approximately one hour. Generally, the time taken to complete a mitotic cycle is longer in the basal insect groups, such as Orthoptera, where it takes several hours.

The evolutionary origin of syncytial cleavage in insects is uncertain. Within the Hexapoda, one sister group of the insects, Collembola, exhibits total cleavage, whereas another, Diplura, exhibits

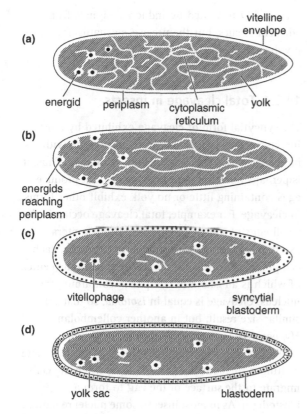

energid periplasm cytoplasmic yolk
reticulum

vitelline
envelope

(a)

(b)

energids
reaching
periplasm

(c)

vitellophage syncytial
blastoderm

(d)

yolk sac blastoderm

Figure 14.10 Early development in the egg leading up to the formation of the blastoderm.

syncytial cleavage. Other lineages of arthropods, such as the Chelicerata (horseshoe crabs, spiders and scorpions) also contain taxa that undergo total or syncytial cleavage, as do the Onychophora (velvet worms).

14.2.2 Formation of the blastoderm and vitellophages

In *Drosophila*, folds of the plasma membrane develop between adjacent nuclei in the periplasm during the fourteenth cycle of mitosis, which lasts much longer than the earlier nuclear divisions. The folds spread inward, and when they are about 40 μm below the nuclei, they spread laterally to form

approximately 6000 individual cells. This marks the end of the syncytial cleavage phase of embryogenesis; the resulting sheet of cells that forms at the periphery of the egg is called a cellular blastoderm, which completely surrounds the yolk (Fig. 14.10d). Following cellularization, the synthesis of proteins also shifts from translation of mRNAs that are pre-packaged into the egg during oogenesis to translation of mRNAs that are of zygotic origin.

Although most energids migrate to the surface of the egg to form the blastoderm, some energids remain in the yolk to form yolk cells called vitellophages (Fig. 14.10c). Thus, in the fruit fly *Bactrocera* (formerly *Dacus*) (Diptera), about 38 of 128 energids remain in the yolk to form the primary vitellophages and they subsequently increase in number to about 300. Commonly, the vitellophages begin to separate after the sixth or seventh mitotic cycles and develop an enlarged nucleus, which increases in size through endomitotic division of the chromosomes. In most orders some cells also migrate back from the blastoderm to form secondary vitellophages. In Blattodea, some Lepidoptera, and Nematocera, like mosquitoes and midges, all the vitellophages apparently have this secondary origin. There is also some evidence that vitellophages are derived from some of the pole cells (Section 14.2.13) and, in *Bactrocera*, some tertiary vitellophages are formed from the proliferating anterior midgut rudiment.

Vitellophages have a variety of functions, including the breakdown and engulfment of yolk into vacuoles. Later, when the yolk is enclosed in the midgut, vitellophages may form part of the midgut epithelium. Vitellophages are also involved in the formation of new cytoplasm. In the eggs of Orthoptera, Lepidoptera and Coleoptera, the yolk is sometimes temporarily divided by membranes into large masses, which contain one or more vitellophages. These yolk spherules are first formed close to the embryo and under the serosa, but ultimately extend throughout the yolk.

14.2.3 Establishment of the germ line

As previously noted (Section 14.1.1), pole plasm localizes to the posterior pole of most dipteran, hymenopteran and lepidopteran eggs. Energids that migrate to the posterior pole incorporate pole plasm when they cellularize. This specifies them to form primordial germ cells (also called pole cells), which are the exclusive progenitors of the germ line that gives rise to oocytes or sperm. In *Drosophila*, energids arrive at the posterior pole before reaching any other region of the egg periphery. These energids also cellularize earlier to form approximately 40 germ cells. Thereafter, germ cells are transcriptionally and translationally quiescent, rarely dividing or growing in size during the remainder of embryogenesis. Specification of germ cells early in embryogenesis via inheritance of pole plasm is called preformation. This appears to be the primary means by which the germ line is formed in the evolutionarily most advanced insect orders. Several of the components of pole plasm have been identified in *Drosophila* and other insect species. Some of these factors are essential for localizing pole plasm to the posterior of the egg, while others directly mediate formation of germ cells. Several components of pole plasm are also classified as maternal effect genes that regulate segmentation and formation of the anterior–posterior axis of the embryo (Section 14.2.8).

Among the more basal groups of hexapods, collembolans also appear to specify the germ line by preformation. In contrast, the basal orders of insects, e.g., Orthoptera, do not appear to specify the germ line by preformation. Instead, germ cells develop at about the time the mesoderm differentiates. In *Locusta*, germ cells first appear in the walls of the coelomic sacs in abdominal segments 2 and 5. In the case of Dermaptera, Psocoptera and some Hemiptera, the germ line differentiates during blastoderm formation. Determination of the germ line at a later stage of embryogenesis is called epigenesis. This process is presumably regulated by inductive signals from adjacent tissues, but the precise mechanisms involved remain unclear.

14.2.4 Total cleavage in insects

The syncytial form of cleavage exhibited by most insects appears to have arisen in association with the large amount of yolk packaged in their eggs. Thus, it is perhaps not surprising that insects that produce eggs containing little or no yolk exhibit other forms of cleavage. For example, total cleavage occurs in the small eggs of some endoparasitic Hymenoptera (Section 14.3.2) and Collembola. In the latter, each cell produced consists of a mass of yolk, in the center of which is an island of cytoplasm containing the nucleus. Cleavage is equal in *Isotoma*, so that cells of similar size result, but in another collembolan, *Hypogastrura*, cleavage is unequal, with the formation of micro- and macromeres. Total cleavage continues to about the 64-cell stage, when the cells migrate to the surface of the egg to form a blastoderm. As in most insects, some nuclei remain in the yolk forming the vitellophages and the original boundaries within the yolk disappear leaving a single central mass.

14.2.5 Germ anlage formation

The blastoderm of most insects is a single, uniform layer of cells that covers the entire surface of the egg and surrounds the yolk. Thereafter, a region of the blastoderm thickens to form more columnar cells that comprise what is called the germ anlage, which later develops into the germ band (Fig. 14.11). In advanced insects such as *Drosophila* (Diptera), the honey bee (Hymenoptera), *Bombyx mori* (Lepidoptera) and lacewings (Neuroptera), the germ anlage develops from nearly the entire surface of the blastoderm (Fig. 14.12a,b). This results in the germ anlage giving rise to the complete body plan (head, thoracic and abdominal segments) of the future

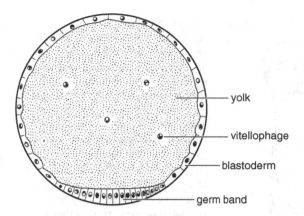

Figure 14.11 Diagrammatic cross-section of an egg showing the origin of the germ band as a ventral thickening of the blastoderm.

insect, with all body segments forming nearly simultaneously once formation of the germ band is complete (Section 14.2.6) (Fig. 14.13). Insects whose germ anlage develops from nearly the entire blastoderm are also referred to as producing long-germ-type embryos.

In contrast, most species in more basal orders such as the Isoptera and Orthoptera produce eggs in which the germ band arises from a relatively small proportion of the blastoderm located toward the posterior pole (Fig. 14.12c,d). In these short-germ-type embryos, the germ anlage patterns the head lobes (protocephalon) plus the most anterior segments of the future thorax and posterior terminus (Fig. 14.14a). However, the remaining abdominal segments of the body progressively form later as embryogenesis proceeds (Fig. 14.14b,c). Successive terminal addition of segments from a posteriorly localized presegmental zone appears to be the ancestral mode of segmentation in arthropods, whereas long-germ-type embryos, in which simultaneous formation of segments occur, are derived. Some species of Hemiptera and Coleoptera, such as *Tribolium*, also produce intermediate-germ-type embryos that fall between long- and short-germ-type embryos (Fig. 14.12e).

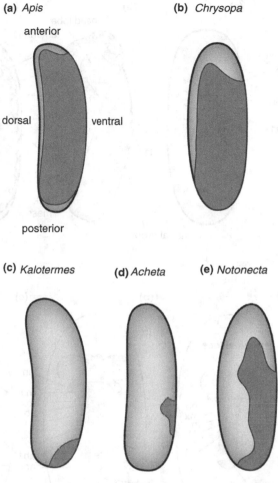

Figure 14.12 Diagram showing the relative size and position of the germ anlage (shaded) in eggs from different insects. Each egg is oriented with anterior at the top, posterior at the bottom, dorsal to the left, ventral to the right. (a) The honey bee (*Apis* [Hymenoptera]); (b) green lacewing (*Chrysopa* [Neuroptera]); (c) termite (*Kalotermes* [Isoptera]); (d) cricket (*Acheta* [Orthoptera]); (e) backswimmer (*Notonecta* [Hemiptera]). The honey bee and lacewing are examples of insects that produce long-germ-type embryos. The termite and cricket produce short-germ-type embryos. The backswimmer produces intermediate-germ-type embryos.

14.2.6 Gastrulation and segmentation

Gastrulation is the process by which the single-layered germ anlage becomes two-layered through invagination and proliferation of cells along

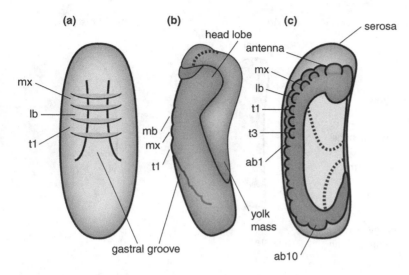

(a) **(b)** **(c)**

head lobe
antenna
serosa

mx
lb
t1
mb
mx
t1
t3
ab1
mx
lb
t1

gastral groove

yolk mass

ab10

Figure 14.13 Elongation and segmentation of the germ band in the honey bee (*Apis*). Note that the germ band of this long-germ-type embryo extends very little, with all segments forming near simultaneously from the germ anlage. (a) Ventral view at early gastrulation (anterior top, posterior bottom). (b) Lateral view at the end of gastrulation (anterior top, posterior bottom). (c) Lateral view after completion of segmentation. mb = mandibular segment; mx = maxillary segment; lb = labial segment; t1–t3 = thoracic segments; ab1–ab10 = abdominal segments (anterior top, posterior bottom) (based on Anderson, 1972).

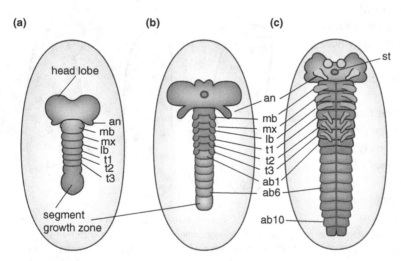

(a) **(b)** **(c)**

head lobe

an
mb
mx
lb
t1
t2
t3

segment growth zone

an
mb
mx
lb
t1
t2
t3
ab1
ab6
ab10

st

Figure 14.14 Elongation and segmentation of the germ band in the beetle *Tenebrio molitor* (Coleoptera). Note that in this short-germ-type embryo, the abdominal segments form from a segment-forming growth zone that elongates posteriorly. Each segment is delineated in antero-posterior succession. (a) Germ band at 25 hours post-oviposition. The head and thoracic segments are delineated at this time but no abdominal segments have formed. (b) At 36 hours, the germ band has further elongated with formation of the limb buds and delineation of six abdominal segments. (c) At 48 hours, the germ band is fully segmented. an = antennal bud; mb = mandibular segment; mx = maxillary segment; lb = labial segment; t1–t3 = thoracic segments; ab1–ab10 = abdominal segments.

the midline (Fig. 14.15). This results in formation of what is referred to as the germ band. The outermost cells in contact with the amniotic cavity become the ectoderm, while cells bordering the yolk become the mesoderm. However, considerable variation exists among insects in terms of the location and number of cells that invaginate along the midline to form the mesoderm. A third germ layer, endoderm, subsequently forms in association with two small rudiments of cells located at the anterior and

(a) *Clytra*

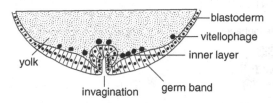

(b) *Apis*

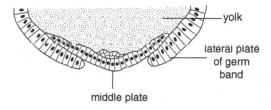

(c) *Locusta*

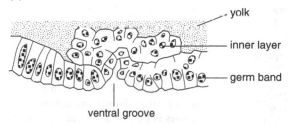

Figure 14.15 Different types of gastrulation.
(a) Invagination of the center of the germ band (*Clytra* [Coleoptera]). (b) Overgrowth from the sides (*Apis* [Hymenoptera]). (c) Proliferation of cells from the upper surface of the germ band (*Locusta* [Orthoptera]).

posterior edges of the germ anlage where the invaginations of the fore- and hindgut will later occur (Section 14.2.14). This layer gives rise to the midgut. Studies with *Drosophila*, the beetle *Tribolium* and selected other insects implicate *decapentaplegic* and other dorsal–ventral patterning genes in specification of the mesoderm while the receptor tyrosine kinase Torso along with terminal gap genes and GATA family transcription factors are implicated in specification of the endoderm (Sections 14.2.8, 14.2.9). With formation of a two-layered state, the

germ anlage enters the germ band stage. Externally, the germ band stage is characterized by the appearance of segments, which are repeated, metameric units of similar cells that are organized along the anterior–posterior axis.

In some beetles, as in *Clytra*, the invagination process is so marked that it is at first almost tubular, while in *Apis* a broad middle plate sinks in without rolling up and the ectoderm grows across to cover it. In Diptera the invagination of the posterior midgut rudiment deep into the yolk during the latter part of gastrulation bears a superficial resemblance to the process that occurs in other animals. Mesodermal invagination begins along the ventral surface, but extension of the mesoderm, and the ectoderm, which comes to cover it, pushes the invagination of the posterior midgut rudiment (endoderm) and the proctodeum anteriorly along the dorsal surface of the embryo (Fig. 14.16). Invagination of the proctodeum then carries the posterior midgut rudiment deep into the yolk (Fig. 14.16).

In the collembolan *Isotoma*, all the cells of the blastoderm divide tangentially to form outer and inner layers. Later, the cells of the inner layer migrate to the region of the germ band so that the extraembryonic region comes to consist of a single layer of cells.

14.2.7 Formation of embryonic membranes

The blastoderm cells that do not participate in formation of the germ band differentiate into an extraembryonic membrane called the serosa. In most species, the boundary between the future germ band and future serosa ruptures so that the serosa is able to migrate over the entire inner surface of the egg to fully envelope the germ band, yolk and vitellophages (Fig. 14.17). In addition, a second membrane called the amnion develops shortly after formation of the serosa from cells adjacent to the germ band (Fig. 14.17). Amnion formation begins when folds appear at the periphery of the germ band

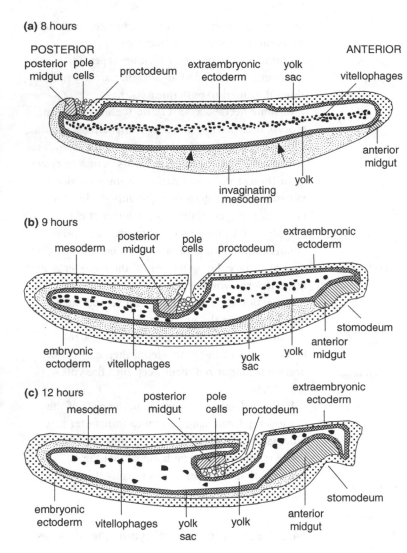

(a) 8 hours

POSTERIOR
posterior pole
midgut cells
proctodeum
extraembryonic
ectoderm
yolk
sac
ANTERIOR
vitellophages
anterior
midgut
invaginating
mesoderm
yolk

(b) 9 hours

posterior
mesoderm midgut
pole
cells proctodeum
extraembryonic
ectoderm
stomodeum
anterior
midgut
embryonic
ectoderm vitellophages
yolk yolk
sac

(c) 12 hours

posterior
mesoderm midgut
pole
cells proctodeum
extraembryonic
ectoderm
stomodeum
anterior
midgut
embryonic
ectoderm vitellophages
yolk yolk
sac

Figure 14.16 Early stages in the development of a dipteran with a long germ band as shown in sagittal section. Arrows in (a) indicate the movement of the mesoderm as it folds within the ectoderm (fruit fly, *Bactrocera* [formerly *Dacus*] [Diptera]). Each image is oriented with anterior at the right and posterior at the left (after Anderson, 1962).

(Fig. 14.17a). The amnion-forming cells then flatten and proliferate ventrally beneath the embryo until they meet and fuse at the ventral midline (Fig. 14.17b). The amnion and serosa may remain connected where the embryonic folds fuse or they may become completely separated, so that the embryo sinks into the yolk, and the yolk separates the amnion from the serosa (Fig. 14.17c). No further cell division occurs in the serosa once formed, but endomitosis may occur, which results in the nuclei becoming very large and polyploid. In *Gryllus*

(Orthoptera), serosal cells contain four times as much DNA as the nuclei in the germ band.

Several variations in serosa and amnion formation exist among insects. In *Drosophila*, for example, the serosa does not envelop the germ band but remains a cluster of cells on the side of the egg, while the amnion remains vestigial. Among thysanurans in the genus *Machilis*, the embryo does not become cut off in an amniotic cavity but rather invaginates within the yolk while remaining connected to the exterior. Adjacent to the embryo is

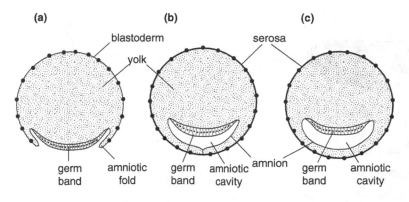

Figure 14.17 Development of the amniotic cavity as shown in cross-section. (a) Lateral folds beginning to grow over the germ band. (b) Lateral folds meet beneath the germ band. (c) Amnion and serosa separated; embryo immersed in yolk.

a zone of cells with small nuclei, while the rest of the egg is covered by cells with large nuclei (Fig. 14.18). From their superficial resemblance to the amnion and serosa in other insects, these zones are called the proamnion and proserosa. In Orthoptera, Phthiraptera and at least some Hemiptera, Coleoptera and Diptera, like mosquitoes, the serosa secretes a cuticle that enhances resistance to desiccation (Section 14.1.4). In the collembolan *Isotoma*, two successive serosal cuticles form on the outside of the serosa while some Coleoptera produce a non-cellular layer on the inside of the serosa.

14.2.8 Regulation of segmentation

The body of insects consists of multiple segments, which form during the segmentation stage of embryogenesis. A complex genetic regulatory cascade controls the segmental pattern that forms along the anterior–posterior axis of insect embryos. The segmentation gene cascade functions by progressively refining positional information, which ultimately results in specifying the cells that form each segment of the insect's body. Anterior–posterior axis formation is most fully understood in *Drosophila*, but comparative studies with other insects also provide insights into the evolution of the segmentation process in arthropods more generally. In *Drosophila*, the segmentation gene cascade begins when the oocyte is still in the ovary. Trophocytes

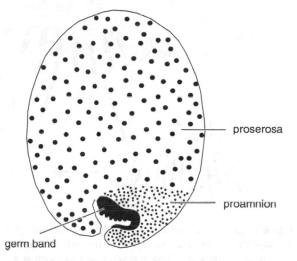

Figure 14.18 Early stage in the invagination of the embryo of *Machilis* (Thysanura), showing the differentiation of the extraembryonic membrane into a proamnion, which has small nuclei, and a proserosa, which has large nuclei (after Johannsen and Butt, 1941).

(nurse cells) express one mRNA called *bicoid*, which is transferred to the oocyte and accumulates at the anterior end, and a second mRNA called *nanos*, which accumulates at the posterior end together with other components that comprise pole plasm (Sections 14.1.1, 14.2.3). Soon after the egg is laid, both mRNAs are translated to produce Bicoid and Nanos proteins that diffuse away from the anterior and posterior ends respectively to form a gradient during the syncytial cleavage stage of embryogenesis

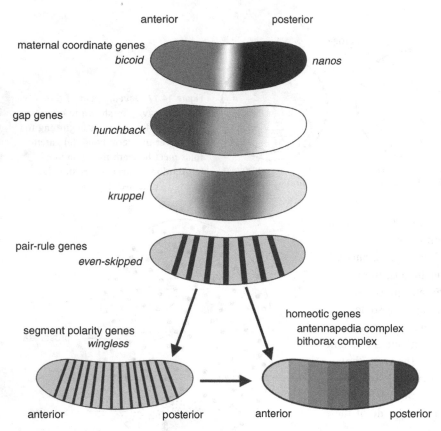

Figure 14.19 Generalized diagram of the segmentation gene cascade that establishes the anterior–posterior axis of the *Drosophila* embryo. The anterior–posterior axis is initially established by the maternal effect genes *bicoid* and *nanos*, which form protein gradients. These transcription factors cause gap genes such as *hunchback* and *krupple* to form gradients, which define broad regions of the embryo along the anterior–posterior axis. Gap genes regulate expression of several pair-rule genes including *even-skipped*, which specify the embryo into regions that are approximately two segments wide. The segment polarity genes then further subdivide the embryo into single-segment units along the anterior–posterior axis. The activity of pair-rule and segment polarity genes define where different homeotic genes associated with the antennapedia and bithorax complexes are activated, which define the identity of each segment.

(Fig. 14.19). These gradients of maternally derived proteins provide the coordinates that position the front and back of the embryo, which is why these factors are referred to as maternal coordinate genes.

The function of the maternal coordinate genes is to activate or suppress a second class of genes called gap genes (Fig. 14.19), because when their function is suppressed, the embryo exhibits large gaps in the segmental pattern relative to the segmental pattern that normally develops. An example of a gap gene is

hunchback, which is transcribed by zygotic nuclei in the syncytium in proportion to the quantity of Bicoid protein present. At the posterior end of the egg, however, *hunchback* RNA is disabled by Nanos protein. As a result of these activities, a gradient of Hunchback protein forms which is highest at the anterior pole of the egg and declines posteriorly (Fig. 14.19). In contrast, another gap gene, *kruppel*, is expressed primarily in "parasegments" 4–6 which are located approximately in the center of the embryo

(Fig. 14.19). The absence of *kruppel* causes the embryo to lack this region.

The products of the gap genes activate *even-skipped* and other genes that are collectively referred to as pair-rule genes. These genes are expressed in approximately every other segment, which results in formation of 14 parasegments that equate to the three gnathal segments of the future head, the three thoracic segments and the first eight abdominal segments (Fig. 14.14). Parasegments do not exactly correspond with the segments seen in the larva or the adult fly, but instead form units with what is called double-segment periodicity. Individuals with non-functional mutations in pair-rule genes produce larvae with only half of the segments normally present.

The proteins produced by the maternal coordinate, gap and pair-rule genes are transcription factors that function while the embryo is still in the syncytial stage and energids have not yet cellularized. Once cells form, interactions that further refine segmentation occur between cells. The first of these between-cell interactions is regulated by the segment polarity genes, which are activated by the pair-rule genes. Segment polarity genes are expressed in a segmentally reiterated pattern and establish the boundaries of each segment (Fig. 14.19).

The final class of genes in the segmentation cascade is the homeotic genes, whose expression is regulated by interactions between pair-rule and segment polarity genes. This regulation results in region-specific patterns of homeotic gene expression, which controls the characteristics of each segment of the head, thorax and abdomen (Fig. 14.19). There are two regions on chromosome 3 in *Drosophila* where the majority of the homeotic genes are located. The first region contains homeotic genes that collectively comprise the antennapedia complex, which controls formation of structures associated with head and thoracic segments, and the second chromosomal region contains the bithorax complex, which controls formation of structures associated with thoracic and abdominal segments (Fig. 14.19).

14.2.9 Regulation of dorsal–ventral axis formation

Insects also exhibit different structural features dorsally and ventrally, and thus must also specify formation of a dorsal–ventral axis. In *Drosophila* this begins while the oocyte is still in the ovary surrounded by follicle cells, and involves approximately 20 different gene products. However, these genes are completely different from those that control anterior–posterior axis formation and segmentation (Fig. 14.20). In brief, specification of the dorso-ventral axis begins when a signal from the oocyte nucleus causes follicle cells on the ventral side of the egg to produce a precursor protein called Pro-spaetzle, which is sequestered to the ventral side of the egg upon formation of the eggshell (Fig. 14.20a).

Early in embryogenesis, Pro-spaetzle is cleaved by a protease to form Spaetzle, which is the ligand for a receptor called Toll. After the egg is laid and formation of the blastoderm occurs, all blastoderm cells express the Toll receptor, but because Spaetzle is localized to only the ventral side of the egg, this factor only binds to the Toll receptor on ventrally located cells (Fig. 14.20b). Spaetzle binding activates the Toll receptor, which together with other proteins, like Tube and Pelle, causes the inhibitor κB protein Cactus to dissociate from a type of NF-κB transcription factor called Dorsal. Free Dorsal then translocates to the nucleus at high concentrations (Fig. 14.20b). On the lateral sides of the embryo, lower levels of Dorsal translocate to the nucleus due to lower level activation of the Toll receptor. In contrast, the complete absence of Spaetzle and activation of the Toll receptor results in Dorsal remaining bound to Cactus in the cytoplasm of dorsally located blastoderm cells (Fig. 14.20b).

In ventral cell nuclei, Dorsal binds to the promoters of several genes, which results in expression of *twist*, *snail* and *rhomboid* but represses expression of the genes *zerknullt* and *decapentaplegic* (Fig. 14.20c). With no Dorsal in the nucleus, cells on the dorsal side

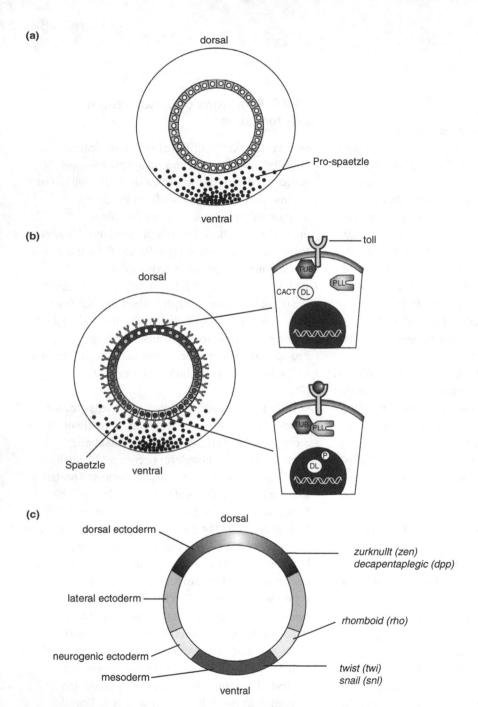

Figure 14.20 Generalized diagram illustrating establishment of the dorsal–ventral axis of the *Drosophila* embryo. (a) The precursor protein Pro-spaetzle is packaged into the egg while still in the ovary. Pro-spaetzle localizes to the ventral region of the egg. (b) In the early embryo Pro-spaetzle is processed to Spaetzle on the ventral side of the egg. The Toll receptor is expressed on the surface of all blastoderm cells but this receptor is only activated on the ventral side of the embryo by binding Spaetzle, which is only present ventrally. Spaetzle binding activates the Toll receptor, which together with the proteins Tube (Tub) and Pell (PLL) causes the inhibitor κB protein Cactus (Cact) to dissociate from the NF-kB-like transcription factor Dorsal (DL). Free Dorsal then translocates to the nucleus. In the absence of Toll activation, Dorsal remains bound to Cactus in the cytoplasm of blastoderm cells on the dorsal side of the embryo. (c) In ventral cells, Dorsal binds to the promoters of several genes, which results in expression of the genes *twist*, *snail* and *rhomboid*, but represses expression of the genes *zerknullt* and *decapentaplegic*. In the absence of Dorsal in the nucleus, dorsally located cells express *zerknullt* and *decapentaplegic*. Differential expression of these genes results in activation or repression of other genes that result in ventral blastoderm cells forming the mesoderm, laterally located cells forming neurogenic and lateral ectoderm, and dorsally located cells forming dorsal ectoderm.

express *zerknullt* and *decapentaplegic*. The result of these activities is that ventral cells form mesoderm, laterally located cells form neurogenic and lateral ectoderm, and dorsally located cells form dorsal ectoderm (Fig. 14.20c). Thus, the dorso-ventral polarity of the embryo forms in response to a gradient formed by the level of Dorsal protein that translocates to cell nuclei. This in turn specifies formation of the two major primordial tissues, ectoderm and mesoderm, that form most of the organs in the insect body.

14.2.10 Conservation of axis formation in other insects

While much has been learned about anterior–posterior and dorso-ventral axis formation in *Drosophila*, only certain components of these processes appear well conserved in insects generally. The structure and function of the segment polarity and homeotic genes, for example, are well conserved. Pair-rule gene orthologs also appear to be involved in segmentation of all arthropods, but it remains unclear whether gap genes do. In addition, while orthologs of the *nanos* gene have been identified in both basal and advanced lineages of insects, *bicoid* appears to exist only in cyclorrhaphous Diptera. Thus, while *nanos* is likely essential to specifying the anterior–posterior axis of most, if not all, insects, *bicoid* is not. Comparative studies of dorso-ventral axis formation are more limited, although processes first described in *Drosophila* appear to occur in other species of Diptera and Coleoptera.

14.2.11 Dorsal closure

The germ band is two-layered but essentially represents the ventral regions of the future insect. The lateral and dorsal parts of the embryo develop from narrow domains on either side of the germ band that at the end of the germ band stage begin extending laterally, eventually enveloping the yolk

when they meet at the dorsal midline. This process is referred to as dorsal closure.

As the ectoderm extends, the amnion and serosa shrink and become confined to an antero-dorsal region where the serosa finally invaginates into the yolk in the form of a tube. This is known as the dorsal organ; it is ultimately digested in the midgut (Fig. 14.21). In *Leptinotarsa* (Coleoptera) and other Chrysomelidae the amnion breaks and grows up inside the serosa (Fig. 14.21c). Later it is replaced by the ectoderm while the serosa remains intact around the outside. In Lepidoptera, Tenthredinidae (Hymenoptera) and Nematocera (Diptera), amnion and ectoderm grow dorsally together so that the ectoderm forms the dorsal closure from the beginning and the amnion forms a membrane all around the outside (Fig. 14.21b,c). The serosa is invaginated and destroyed in Nematocera, but it persists in Lepidoptera and Tenthredinidae so that a layer of yolk is present all around the embryo, held between the amnion and serosa. This provides the first meal for larval Lepidoptera when they hatch. Although there is a considerable increase in the area of the epidermis during dorsal closure, it appears to be produced by changes in the shapes of the epidermal cells, and not by the addition of more cells.

14.2.12 Organogenesis and structures formed by the ectoderm

Upon completion of gastrulation and segmentation of the germ band, the ectoderm and mesoderm differentiate during the remainder of embryogenesis into the organ systems and other structures present in the nymph or larva that hatches from the egg. These events are collectively referred to as organogenesis. Major structures formed by the ectoderm include the following.

Epidermis and the formation of appendages
The whole outer wall of the embryo forms from ectoderm. The majority of cells forming the outer wall are epidermal cells that together form

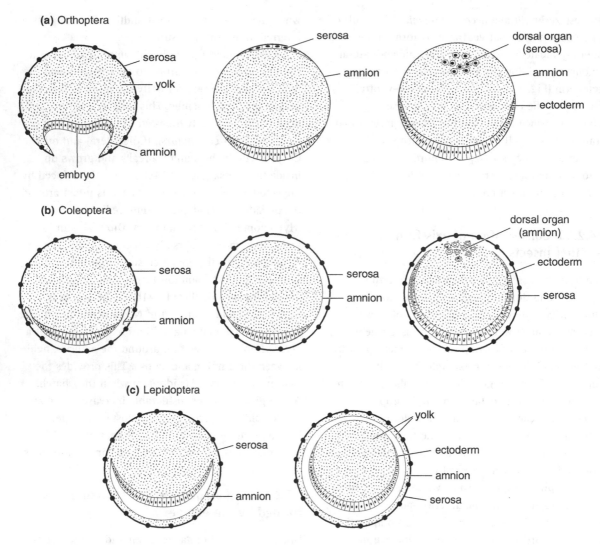

Figure 14.21 Dorsal closure and the fate of the embryonic membranes as shown in cross-section. The earliest stages, figures on the left of the diagram, can be derived from (b) or (c) in Fig. 14.17 (from Imms, 1957). (a) Amnion forms the provisional dorsal closure. Serosa contracts and is destroyed (*Oecanthus* [Orthoptera]). (b) Amnion forms the provisional dorsal closure. Serosa remains intact around the outside of the embryo (*Leptinotarsa* [Coleoptera]). (c) Ectoderm forms the dorsal closure, amnion and serosa remain intact, enclosing a layer of yolk outside the embryo (Lepidoptera).

the epidermis (Section 16.1). Later in embryogenesis, the epidermis produces the cuticle that covers the outside of the embryo and, in turn, the immature insect that hatches from the egg (Section 16.1). Other cell types present in the outer wall of the embryo include oenocytes (Section 16.1) that are isolated

from the epidermis of all the abdominal segments except possibly the last two.

All of the appendages of the immature insect form from outgrowth of the body wall. This growth occurs partly as a result of cell division and partly by changes in the shape of the appendages and rearrangement of the cells. The labrum develops in front of the stomodeum, and the antennal rudiments grow on either side on the protocephalon (Figs. 14.13, 14.14, 14.16). The protocorm becomes segmented and, in the lower insect orders, each segment extends laterally to form the rudiment of an appendage. The anterior appendages in grasshoppers (Orthoptera) are first apparent once about 30% of the total development time has elapsed (Section 14.2.16). Behind the protocephalon are the rudiments of the mandibles, maxillae and labium. The latter begins as a pair of limbs then later fuses along the midline to form a single labium. The appendages of the next three segments form the walking legs. These grow longer and become folded and grooved where later the cuticle will become segmented.

The abdominal appendages subsequently disappear except that in some insects the appendages of segments 8 and 9 contribute to the ovipositor and those on segment 11 form the cerci. In Orthoptera and some other orders, the appendages of the first abdominal segment also persist for a time. They are known as the pleuropodia and have a distal area in which the cells become very large. They then degenerate, becoming torn off when the insect hatches. The pleuropodia have a specialized function in *Hesperoctenes* (Hemiptera). The egg of this insect has no yolk or chorion as it develops within the female parent. Its pleuropodia grow and fuse together to form a membrane, which completely covers the embryo and makes contact with the wall of the oviduct to function as a pseudoplacenta in viviparous species (Section 14.3.1). The pleuropodia assume a variety of forms in Coleoptera, but in Dermaptera, Hymenoptera and Lepidoptera they are only small papillae, which soon disappear.

Among holometabolous insects in which the adult appendages develop from imaginal discs (see Chapter 15), the latter may already be apparent in the embryo. In *Drosophila* the discs invaginate beneath the surface of the embryonic epidermis at about 60% development. Initially the imaginal cells remain an integral part of the epidermis, connecting

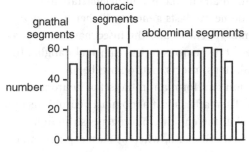

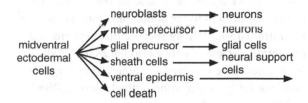

Figure 14.22 Development of the central nervous system in an embryonic grasshopper (*Schistocerca*). (a) Number of neuroblasts formed in the body segments (data from Doe and Goodman, 1985). (b) Fate of cells from the midventral ectoderm which forms the ganglia of the ventral nerve cord (based on Doe and Goodman, 1985).

with the surface by long, narrow extensions, which collectively form a small placode in the epidermis, but they subsequently become completely separated from the surface in the peripodial cavities (Section 15.3.2).

Nervous system The central nervous system arises from ectodermal cells (neurogenic ectoderm) on the ventral side of the embryo (Fig. 14.22). Individual cells divide tangentially, cutting off large cells that form neuronal stem cells called neuroblasts. This occurs at about 20% of development in *Drosophila*. The number of neuroblasts in each segment is constant across individuals of one insect, and it is similar among segments of one individual (Fig. 14.22) and even across species as widely different as a grasshopper (Orthoptera) and *Drosophila* (Diptera). Each neuroblast produces a characteristic lineage of neurons, which is controlled by several transcription factors that are sequentially expressed. In ametabolous and hemimetabolous species, the neuroblasts generate all their progeny during embryogenesis and the insect hatches with a complete set of neurons in its segmental ganglia. In holometabolous insects, neuroblasts begin proliferation and form an initial set of neurons for the larva. Thereafter, most abdominal neuroblasts are eliminated at the end of embryogenesis through programmed cell death, whereas most cephalic and thoracic neuroblasts become mitotically quiescent. These quiescent cells later produce the huge number of neurons present in the adult. Entry into quiescence is controlled spatially by specific Hox proteins and temporally by specific transcription factors. Exit from quiescence involves a combination of signals from the fat body and stimulation by insulin-like peptides.

In grasshoppers (Orthoptera), similar numbers of neurons are produced in all the segments, but differential mortality results in differing final numbers in the different ganglia. For example, 98 neurons are produced by the median neuroblast of the metathoracic segment of the grasshopper and 86 from that in the first abdominal segment. More of the latter die, however, which results in the ganglia finally having 93 and 59 midline neurons, respectively (Fig. 14.23). Cell death (apoptosis) corresponds with the times of embryonic molts, suggesting some hormonal regulation. By contrast, differences in the numbers of neurons in different ganglia of *Drosophila* result primarily from differences in their postembryonic production (see Fig. 15.23). The sheath cells surrounding the central nervous system are also ectodermal in origin, being derived from ventral cells in the same way as the neuroblasts.

As segmentation proceeds in the insect embryo, the ganglia become differentiated. Three paired groups of neuroblasts, corresponding with the protocerebrum (forebrain), deutocerebrum (midbrain) and tritocerebrum (hindbrain), develop in the protocephalon and, in addition, when the full

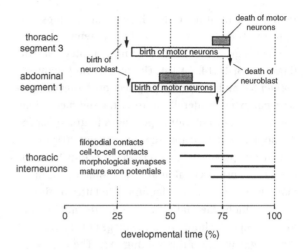

Figure 14.23 Development of neurons in the ventral ganglia of an embryonic grasshopper. Top: birth and death of neurons derived from the median neuroblasts of the metathorax and first abdominal segments (based on Thompson and Seigler, 1993). Below: maturation of median spiking local interneurons in the metathoracic ganglion (based on Leitch *et al.*, 1992).

complement of segments is present, 17 postoral ganglia may be recognizable: three in the gnathal segments (mandibular, labial and maxillary), three in the thorax and eleven in the abdomen. The first three fuse to form the subesophageal ganglion in all insects and some fusion of abdominal ganglia always occurs, but the extent to which fusion of other ganglia occurs varies with taxon.

Axons grow as processes known as growth cones from which transient extensions spread out, apparently sensing guidance cues in the surrounding environment. The first axons to grow along a particular pathway are called pioneer axons. They use cues on the surfaces of, or released by, other cells. These cues may attract or repel other cells. A variety of different cells may be involved, including other neurons, glial cells and specific epithelial tissues. Once the pathway is established, the growth cones of later axons follow the pioneer axons, and are able to distinguish between bundles of axons to follow the appropriate path. Synapses start to form in the central nervous system at about 70% of development and from this time normal action potentials occur in some interneurons, indicating that the nervous system is becoming functional (Fig. 14.23).

The optic lobes, which come to be associated with the protocerebrum, contain no neuroblasts, although similar-looking large cells are present. In Orthoptera, the optic lobes are formed by delamination from the ectodermal thickening which forms the eye, but, in Hymenoptera and Coleoptera, they develop from an ectodermal invagination arising outside the eye rudiment.

The cells of the stomatogastric system (the nervous system associated with the gut) are formed from the ectoderm of the stomodeum. The receptors of the peripheral nervous system are formed from epidermal cells. In *Drosophila* the cells giving rise to larval sensilla are already distinguishable by 23% development. Each of these cells divides twice to form cells that will become a neuron and three sheath cells. If the fully developed sensillum contains more than one neuron, additional neurons form by division of the initial neuron mother cell. In *Drosophila* this process is complete by about 40% development. Clusters of cells which will form sensilla in the grasshopper embryo are well developed by 55% development.

Tracheal system, gut and Malpighian tubules

Among the remaining tissues and organs formed by the ectoderm are the tracheal system, portions of the gut and Malpighian tubules. The tracheal system arises in each segment. Each metameric unit begins its development independently during germ band formation (Section 14.2.5) when groups of tracheal cells differentiate and are set aside from the neighboring epidermal cells. These tracheal cells then invaginate to form a sac-like structure that generates a luminal cavity, which is then expanded and remodeled through a process of branching. Growth of the tracheal system into a multi-branched network occurs through migration of tracheal cells toward non-tracheal cells that secrete a fibroblast growth factor-like protein (FGF) named Branchless in *Drosophila*. Initially, tracheal cells in the sac-like invagination that are close to sources of Branchless begin migrating while remaining attached to other tracheal cell neighbors. This results in bud-like extensions, which subsequently elongate to form branches composed of multicellular tubes that eventually form finer branches composed of individual cells arranged in a head-to-tail fashion. A second signaling system involving the protein Delta and its receptor Notch is also important in regulating the branching process. As branches from different segments connect with one another, specialized terminal tracheal cells form at the periphery of the tracheal system where gas is exchanged with surrounding tissues.

The foregut and hindgut are ectodermal in origin, being formed from the stomodeum and proctodeum, respectively. Their development is considered together with the midgut (Section 14.2.14). The Malpighian tubules arise as evaginations from the tip of the proctodeum. Early in their development, a large cell becomes apparent at the tip of each evagination. This cell does not divide, but regulates the continuous division of cells close to it so that the tubule becomes elongated. Development of Malpighian tubules begins at 36% of development in *Rhodnius*. In most insects, only two or three pairs of tubules develop in the embryo, but others may be produced during larval development.

Imaginal discs

In holometabolous species the formation of mouthparts, wings and other structures of the adult insect that are absent during the larval stage form during metamorphosis from imaginal discs (Section 15.3.2). Imaginal disc primordia comprising 30–40 cells form from ectoderm during organogenesis in blowflies and their close relatives. Each disc primordium is also specified into anterior-posterior and dorsal–ventral compartments during embryogenesis by a small number of genes encoding primarily transcription factors. The genes that pattern the axes of imaginal discs in *Drosophila* appear to be conserved in other holometabolous species. Unlike cyclorrhaphous Diptera, however, many holometabolous insects do not form imaginal disc primordia until the last larval stadium.

Embryonic cuticles As previously noted, the serosa, which envelops the yolk and blastoderm, often secretes a cuticle that helps protect the egg from desiccation. Molecular studies in the mosquito *Anopheles gambiae* indicate that the composition of the serosal cuticle is similar to the cuticle(s) that later forms around the embryo and larva. Soon after blastoderm formation, insects belonging to the hemimetabolous orders and some holometabolous species in the Neuroptera, Lepidoptera and Coleoptera produce a first cuticle that forms around the embryo. This cuticle has a thickness of approximately 6 μm in the grasshopper *Melanoplus*, and comprises an epicuticle and a chitinous endocuticle. This first embryonic cuticle becomes broken up and is no longer visible in the middle stage of embryogenesis, when a second embryonic cuticle sometimes forms. The period during which embryos of hemimetabolous species are enveloped by a first and potentially second cuticle is referred to as the pronymphal stage. In most hemimetabolous species, the second cuticle separates from the epidermis in the late stages of embryogenesis when the cuticle covering the first instar nymph is formed. The second embryonic cuticle is then shed when the insect hatches from the egg. Grasshoppers actually hatch as a pronymph while still enveloped by the second embryonic cuticle (Section 15.1). These pronymphs are incapable of walking but can effectively dig through soil. The embryonic cuticle then separates with formation of the nymphal cuticle and is shed when the insect undergoes ecdysis to the first nymphal stage. In contrast, most holometabolous insects do not produce embryonic cuticles, and the cuticle of the first larval stage starts to form at about 80% development as dorsal closure becomes complete.

A longstanding question in insect biology is how holometabolous insects, which exhibit three juvenile life-stages (larva, pupa, adult) and complete metamorphosis, evolved from a hemimetabolous ancestor that exhibits two life-stages (nymph, adult) and incomplete metamorphosis. One hypothesis is that the larval stage of holometabolous insects arose from the pronymphal stage of hemimetabolous species.

Cuticle production and molting in the embryo appears to be regulated by ecdysteroids, just as in postembryonic molts. In *Locusta* and *Bombyx* the three major ecdysteroids in the embryo are ecdysone, 20-hydroxyecdsone and 2-deoxyecdysone that are produced in the ovary of the female parent by the follicle cells and are packaged into eggs as inactive conjugates. Free ecdysone reaches peak levels when the serosal cuticle, the two embryonic cuticles and the first-stage larval cuticle are produced. It is not known if the prothoracic glands of the embryo (Section 15.4) produce ecdysteroids after they develop. Juvenile hormone is also present in the embryo, but its role in the embryonic molts is unclear.

14.2.13 Mesoderm structures and formation of the body cavity

The mesoderm is derived from the inner layer of the germ band, which forms two lateral strands running the length of the body and joined across the midline by a thin sheet of cells. In the basal orders the lateral strands become segmented and the somites separate off from each other, but in more advanced orders, e.g., Lepidoptera and Hymenoptera, the somites remain connected together. Among the cyclorrhaphous Diptera there is a tendency for the mesoderm to remain unsegmented. In *Bactrocera* (formerly *Dacus*), for example, the strands of mesoderm only become segmented as they differentiate into the definitive structures. The mesoderm in the protocephalon arises *in situ* in the lower orders, but moves forwards from a posterior position in the more advanced groups.

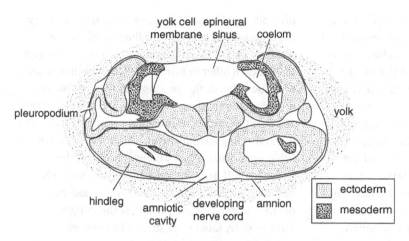

Figure 14.24 Transverse section through a grasshopper embryo showing coelomic cavities, pleuropodia and the yolk cell membrane.

Coelomic cavities develop as clefts in the somites in *Carausius* (Phasmatodea) and *Formica* (Hymenoptera), but by the mesoderm rolling up to enclose a cavity in *Locusta* (Orthoptera) and *Sialis* (Megaloptera) (Fig. 14.24). In the Heteroptera (Hemiptera), the coelomic sacs remain open to the epineural sinus, while in Diptera no coelomic cavities are formed. Where they are most fully developed a pair of coelomic cavities is present in each segment of the protocorm, while, in the protocephalon, pairs of cavities develop in association with the premandibular and antennal segments.

At the same time as the coelom forms, the primary body cavity develops as a space between the upper surface of the embryo and the yolk. This cavity is called the epineural sinus, and in Orthoptera and *Pediculus* (Phthiraptera) it is bounded dorsally by a special layer of cells that form the yolk cell membrane (Fig. 14.24). Soon thereafter the walls of the coelomic sacs break down as the mesoderm differentiates to form muscles and other tissues. As a result, the coelomic cavities and the epineural sinus become confluent. Thus, although the body cavity of insects is called a hemocoel, it is strictly a mixocoel in those insects developing coelomic cavities. Some of the coelomic sacs, particularly those associated with the antennae, may be quite large and make a significant contribution to the final cavity. When the midgut is formed, the mesoderm extends dorsally between the body wall and the gut so that the body cavity also extends until it completely surrounds the gut.

The outer walls of the coelomic sacs form the somatic muscles, the dorsal diaphragm, the pericardial cells and the subesophageal body. The latter is found in the Orthoptera, Plecoptera, Isoptera, Ischnocera (Phthiraptera), Coleoptera and Lepidoptera and consists of a number of large binucleate cells in the body cavity and closely associated with the inner end of the stomodeum. The cells become vacuolated and usually disappear at about the time of hatching, but in Isoptera they persist until the adult stage is reached. It is usually assumed that these cells are concerned with nitrogenous excretion, but they may be concerned with the breakdown of yolk.

Muscles, fat body and hemocytes Several tissues derive from mesoderm, including muscles and the fat body, which is the primary metabolic organ of insects (Chapter 6). The somatic muscles are formed from cells known as myoblasts. These fuse together to form syncytia, which are progressively enlarged by the incorporation of more myoblasts. The syncytia produce processes resembling axon growth cones that move over the inner surface of the epidermis until they reach their attachment points. The inner walls of the coelomic sacs form the visceral muscles.

The dorsal vessel (heart) is formed from special cells, the cardioblasts, originating from the upper angle of the coelomic sacs, while the aorta is produced by the approximation of the median walls of the antennal coelomic sacs on either side. Another group of cells derived from mesoderm are hemocytes (blood cells), which in many insects arise during two stages of development. The first population of hemocytes is produced during embryogenesis from mesoderm located in the head region of the embryo, while the second population is produced during the larval or nymphal stage in hematopoietic organs, which form bilaterally during embryogenesis from mesoderm located in proximity to the dorsal vessel.

Gonads The mesoderm that forms the gonads derives from several segments, but condenses to form a single structure on each side of the body in the future abdomen. The rudiment thickens ventrally and gives rise to solid strands of cells in which cavities appear to form the lateral ducts. The median ducts arise from ectodermal invaginations. In turn, the pole cells that form during early embryogenesis (Section 14.2.3) migrate to form the germ cells in the gonad. In the Nematocera (Diptera), all the pole cells migrate to form the germ cells in the gonads, but in Cyclorrhapha (Diptera), some are lost during migration, perhaps becoming vitellophages. Most pole cells migrate through the blastoderm before gastrulation, but others do so only during or after gastrulation and are carried forward with invagination of the proctodeum (see Fig. 14.16). Thereafter, the germ cells are enclosed by the mesoderm and increase in number before becoming separated into columns by ingrowth of mesoderm. These columns form the germaria of the ovarioles or the testis follicles.

14.2.14 Endoderm and formation of the alimentary canal

The foregut and hindgut arise early in development as ectodermal invaginations, the stomodeum and proctodeum (Fig. 14.25). These invaginations carry the anterior and posterior rudiments of endoderm (Section 14.2.6) into the embryo where they extend toward each other to form paired longitudinal strands of tissue beneath the yolk and above the visceral mesoderm. From these strands, midgut tissue spreads out over the surface of the yolk, eventually enclosing it and forming a complete gut. Thus, formation of the ectodermal parts of the alimentary canal, the stomodeum (future foregut) and proctodeum (future hindgut), precedes formation of the endodermal part of the alimentary canal, which is the midgut. Interestingly, functional studies in *Drosophila* suggest that yolk cells (vitellophages) are required for morphogenesis of the endoderm to form the midgut, and also express genes that suggest vitellophages might themselves derive from endoderm.

14.2.15 Metabolic changes

Oxygen uptake by the egg increases throughout development as the embryo increases in size, while the respiratory quotient, a value that measures basal metabolic rate from estimates of carbon dioxide production (Chapter 17), soon declines. This suggests that initially the small carbohydrate reserves of the egg are used and that lipid subsequently becomes the main resource for metabolic activity. It has been calculated that, in the egg of a grasshopper, 75% of oxygen uptake concerns the oxidation of fat. The oxidation of lipid is advantageous because it produces relatively large quantities of metabolic water compared with oxidation of carbohydrates, and there is no nitrogenous waste, which would be produced if the embryo used its protein reserves for energy metabolism.

14.2.16 Transcriptional and translational activity

With the complete sequencing of the *Drosophila* genome and the genomes of several other insect species (including the beetle *Tribolium*, several

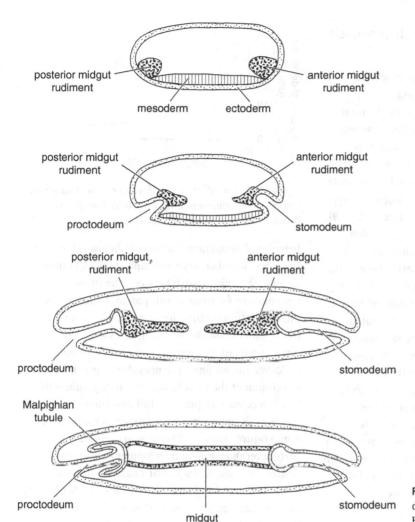

Figure 14.25 Diagrams illustrating the development of the midgut (after Henson, 1946).

mosquitoes, the honey bee, the silk moth *Bombyx mori*, the pea aphid and the parasitic wasp *Nasonia vitripennis*), detailed spatiotemporal information has been generated using DNA microarrays, high throughput sequencing and *in situ* hybridization methods that characterize the thousands of genes transcribed during embryogenesis. Together these genes represent the embryonic transcriptome. The rate of RNA synthesis in eggs is very high at first, but as new nuclei and DNA are produced, the rate declines. Subsequently, the rate of synthesis per

nucleus rises again so that, although the total amount of DNA in the embryo is stable, the total amount of RNA continues to slowly increase. Recent improvements in high throughput methods for protein analysis also provide insights into the identity of the thousands of proteins and peptides expressed during embryogenesis (the embryonic proteome). Many of these proteins are unique to the embryonic stage, while others are also expressed during later stages of development.

14.2.17 Temporal pattern of embryogenesis among insects

As noted at the beginning of our discussion of embryogenesis (Section 14.2), referring to developmental events as a percentage of the total embryonic period from egg-laying to hatching makes it possible to compare when different stages of embryogenesis are completed in insects whose eggs develop at very different absolute rates. For example, embryogenesis of *Drosophila*, whose long-germ-type eggs hatch 24 hours after oviposition (see Fig. 14.9), and *Locusta*, whose short-germ-type eggs hatch approximately 15 days after oviposition, are temporally very different. But as a percentage of the total embryonic period from egg-laying to hatching, the timing of major embryological events (cleavage, blastoderm and germ band formation, gastrulation, segmentation, organogenesis) is quite similar.

Overall, species that lay long-germ-type eggs (most holometabolous insects) generally exhibit shorter embryonic development times than species that lay short-germ-type eggs (Orthoptera and related orders). Thus, at 30°C, total embryonic development times for mosquitoes in the genus *Culex* (long-germ-type) takes about 30 hours, moths in the genus *Ostrinia* (Lepidoptera) (long-germ-type) take 82 hours and the hemipteran *Oncopeltus* (intermediate-germ-type) takes five days. In contrast, the grasshoppers *Schistocerca* and *Ornithacris* (short-germ-type) require 20 and 43 days, respectively, to complete embryogenesis under similar physical conditions.

The duration of embryonic development tends to decrease as temperature increases (Fig. 14.26). Development is not completed if the temperature exceeds a certain level, often in the range 35–40°C, nor if it falls below a certain level, which is about 14°C in *Oncopeltus* and 13°C in *Cimex* (both Hemiptera). Some development does occur at lower temperatures, however, and in *Oncopeltus* some morphogenesis occurs even at 5°C. In addition to these developmental thresholds, there is also a

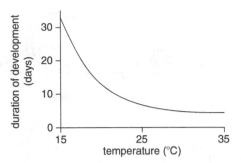

Figure 14.26 The effect of temperature on the duration of embryonic development (milkweed bug, *Oncopeltus* [Hemiptera]) (from Richards, 1957).

behavioral temperature threshold below which the fully developed embryo will not hatch. It is thus necessary to distinguish between the threshold temperature for some development, below which no differentiation occurs, the threshold for complete development and the hatching threshold. These distinctions are not always clear in the literature.

Above the minimum temperature for full development the total heat input (temperature by time) necessary to produce full development and hatching is approximately constant whatever the temperature. Thus, in *Shistocerca gregaria* (Orthoptera), complete development requires 225 degree days above 15°C. For example, at 30°C development takes about 15 days ([30 – 15°C] × 15 days = 225 degree days) and at 20°C about 45 days ([20 – 15°C]) × 45 days = 225 degree days). In *Schistocerca* this relationship also holds with fluctuating temperatures, including periods below the minimum for full development, but in *Oncopeltus* (Hemiptera) and some other insects this is not entirely the case as some development occurs below the minimum for complete development. Hence, periods of low temperature do have an influence on the total number of degree days necessary for development in some insects.

Humidity also influences egg development in some species. Eggs of *Musca* (Diptera) only develop at high humidities and even at 80% relative humidity only 15% of the eggs survive to hatching. In *Lucilia*

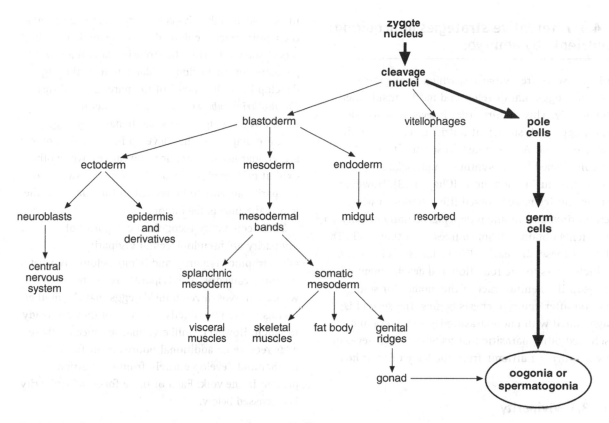

Figure 14.27 General summary of insect embryogenesis from fertilization to completion of organogenesis. The origins of different cell and tissue types, along with the major organs or cells they give rise to, are indicated.

(Diptera) there is a linear relationship between the time of development and saturation deficit. As previously noted (Section 14.1.4), the eggs of some insects absorb water, which is necessary for completion of embryogenesis. If there is sufficient moisture in the environment to prevent death through desiccation, but not enough for development to continue, the eggs may remain quiescent for some time. Under such circumstances the eggs of *Schistocerca* partially develop and then remain quiescent and viable for up to six weeks. Thereafter, completion of embryogenesis will proceed if more water becomes available.

In some species the embryonic period is greatly prolonged by an egg diapause. Many insect species overwinter in embryonic diapause (Section 15.6), and as an extreme example, the diapause eggs of the brown locust, *Locustana*, may survive for over three years. Embryonic diapause occurs at different stages of development in different species even within one family. For instance, in the family Acrididae (grasshoppers), it occurs just after blastoderm formation in *Austroicetes*, just before katatrepsis in *Melanoplus femur-rubrum*, but not until 80% development in several other species of *Melanoplus*.

In summary, although total embryonic development times vary temporally among species and with environmental conditions, the process of embryogenesis is very similar in the sense that it results in formation of the same specialized cells, tissues and organs that all insects share. The derivation of these shared tissues and organs is summarized in Fig. 14.27.

14.3 Alternative strategies of acquiring nutrients by embryos

Most insects are oviparous, which means they produce eggs that develop and hatch outside their own body. This requires that insects pre-provision their eggs with sufficient nutrients for embryonic development. As previously discussed, this is accomplished by the synthesis and packaging of yolk into eggs during oogenesis (Chapter 13). However, some species have evolved life histories in which embryonic development depends partially or fully on nutrients obtained from sources other than yolk. The first of these alternative life histories is viviparity, which refers to the retention and development of eggs in the genital tract of the mother for some period after embryogenesis begins. The second is associated with endoparasitic Hymenoptera and selected other parasitic insects whose eggs develop by acquiring nutrients from the body of their hosts.

14.3.1 Viviparity

Viviparity refers to insects that retain their eggs after fertilization and whose embryos receive nourishment directly from the parent. This nourishment can be either the only source of energy for the embryo or it may be in addition to nutrients that are present in yolk that was pre-packaged into the egg during oogenesis. Viviparous species commonly produce fewer offspring than related oviparous species and this may be associated with a reduction in the number of ovarioles. Thus, *Melophagus* and *Glossina* (Diptera) have only two ovarioles per ovary, whereas the related, oviparous, *Musca domestica* has 70 ovarioles per ovary. Similarly, among the viviparous Dermaptera, *Hemimerus* has 10–12 ovarioles per ovary, but only about half are functional, and *Arixenia* has only three ovarioles per ovary.

Sometimes the eggs are retained and development occurs in the ovariole, as in *Hemimerus*, aphids and some beetles in the family Chrysomelidae. In other insects, such as the viviparous Diptera, the opening to the genital tract is enlarged to accommodate birth of larval-stage offspring. In Strepsiptera and a few parthenogenetic Cecidomyiidae (Diptera) the eggs develop in the hemocoel of the parent. The thrips *Elaphothrips tuberculatus* (Thysanoptera) is facultatively viviparous: some females lay eggs (i.e., are oviparous) that develop into females; others are viviparous and produce only males, while others exhibit both modes of reproduction. The switch to viviparity appears to be related to the quality of the leaves the insects feed upon.

Three commonly recognized categories of viviparity are pseudoplacental viviparity, adenotrophic viviparity and hemocoelous viviparity. A fourth category, ovoviviparity, refers to species whose embryos develop inside eggs that the mother retains in her reproductive tract until they are ready to hatch. However, unlike viviparous species, these eggs receive no additional nourishment from the mother and develop entirely from the nutrients present in the yolk. Each of these forms of viviparity is discussed below.

Pseudoplacental viviparity Insects exhibiting pseudoplacental viviparity produce eggs containing little or no yolk. They are retained by the female and are presumed to receive nourishment via embryonic or maternal structures called pseudoplacentae. There is, however, no physiological evidence concerning the importance of these structures. Viviparous development continues up to the time of hatching, but the larvae are free-living.

In the earwig *Hemimerus* the fully developed oocyte has no chorion or yolk, but is retained in the ovariole during embryonic development. It is accompanied by a single nurse cell and enclosed by a follicular epithelium one cell thick. At the beginning of embryonic development, the follicle epithelium becomes two or three cells thick and at the two ends thickens still more to form the anterior and posterior maternal pseudoplacentae (Fig. 14.28a). As the embryo develops it comes to lie in a pseudoplacental cavity produced by the enlargement of the follicle, but it becomes connected with the follicle by cytoplasmic processes extending out from the cells of the amnion and, later, from those of the serosa (Fig. 14.28b). Furthermore, some of the embryonic cells form large trophocytes, which come into contact with the anterior maternal pseudoplacenta. The follicle epithelium and pseudoplacentae show signs of breaking down and this is taken to indicate that nutriment is being drawn from them.

Later in development, the serosa spreads all around the embryo and, with the amnion, enlarges anteriorly to form the fetal pseudoplacenta (Fig. 14.28c). By this time, dorsal closure is complete except anteriorly, where the body cavity is open to an extraembryonic cavity, the cephalic vesicle. It is presumed that nutriment passes from the pseudoplacenta to the fluid in the cephalic vesicle and then circulates into and around the embryo. The embryonic heart is probably functional at this time, thus aiding the circulation.

The parthenogenetic eggs (Section 14.5) of aphids also develop in the ovarioles. They are much smaller than eggs that are fertilized when these are produced by the same species of aphid, and they have very little yolk. The oocyte is only briefly surrounded by follicle cells, and it separates from them as it moves down the ovariole. As a result, the oocyte does not become surrounded by a vitelline envelope or a chorion since these are normally produced by the follicle cells. At first, nutrients are received via the nutrient cords as aphids have telotrophic ovarioles (Section 13.2.1), but later, when the nutritive cord is broken, the ovariole sheath appears to assume a trophic function. It thickens and shows obvious signs of metabolic activity. Extensive gaps develop in the sheath, which permit the direct access of hemolymph, and it is probable that the embryo obtains some nutrients from this source. When the embryo subsequently develops a serosa, this also appears to have a trophic function. The oocyte is invaded by bacterial endosymbionts such as *Buchnera aphidicola* before the ovariole sheath thickens,

while the embryo is at an early stage of development. They enter through a large pore in the blastoderm, and accumulate in cells, which become the mycetocytes where the symbionts reside during the nymphal and adult stage (Section 4.4). These cells are differentiated very early in development. Physiological studies and genomic analyses of aphid symbionts reveal that they provide essential amino acids and other factors necessary for growth.

Pseudoplacental viviparity also occurs among Psocoptera in the genus *Archipsocus*, where the serosa is an important trophic organ, and in a rare family of ectoparasitic Hemiptera, the Polyctenidae, where first the serosa and then the pleuropodia function as pseudoplacentae.

Adenotrophic viviparity In adenotrophic viviparity, fully developed eggs with chorions are produced and passed to the uterus, where they are retained. Embryonic development follows normally, but when the larva hatches it remains in the uterus and is nourished by special maternal glands. Birth occurs when the larva is fully developed and pupation follows within a short time, there being no free-living feeding phase. This type of viviparity only occurs in a few blood-sucking Diptera: Glossinidae (tsetse flies), Hippoboscidae, Streblidae and Nycteribiidae.

In the tsetse fly, *Glossina*, each ovary has only two ovarioles which function in sequence: first, one ovariole on the right produces a mature oocyte, then one on the left, then the second on the right and so on. Only one fully chorionated oocyte is produced at a time. It is fertilized in the uterus and embryonic development proceeds rapidly, lasting three or four days at 25°C. The larva hatches in the uterus, on the ventral wall of which is a small pad of glandular cells with a cushion of muscle beneath and other muscles running to the ventral body wall. This structure is known as the choriothete and it is responsible for removing the chorion and the cuticle of the first-stage larva. It undergoes cyclical development, degenerating during the later stages of larval development and starting to regenerate just

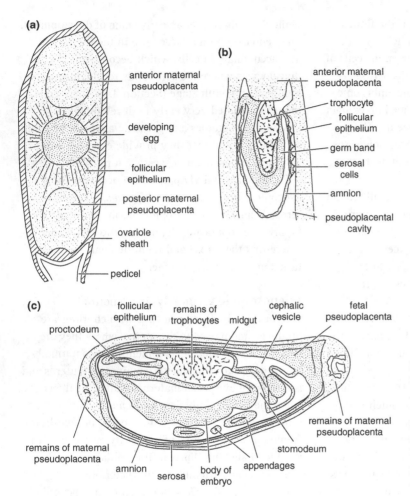

Figure 14.28 Pseudoplacental viviparity. Stages in the embryonic development of *Hemimerus* (Dermaptera) (from Hagan, 1951). (a) Early cleavage showing the maternal pseudoplacentae. (b) Fully developed germ band. (c) End of blastokinesis. The maternal pseudoplacentae have been replaced by an embryonic (fetal) pseudoplacenta.

before larviposition, so that it is fully developed by the time the next larva is ready to hatch. The choriothete adheres to the chorion and pulls the chorion off by muscular action when the chorion is split longitudinally by the larval egg burster. The chorion becomes folded up against the ventral wall of the uterus. The cuticle of the first-stage larva is pulled off in the same way. When the second-stage larva molts, its cuticle is not shed immediately but it is subsequently split by the growth of the third-stage larva. The female expels cast cuticles at parturition.

The larva feeds in the uterus on the secretion of specialized accessory glands, known as milk glands. The milk contains approximately equal quantities of lipids and proteins, plus amino acids. It also contains large numbers of bacterial symbionts (*Wigglesworthia glossinida*), which are transmitted to the developing offspring. Milk production is cyclical, beginning just before a larva hatches and continuing until the third larval stage is well developed (Fig. 14.29). The nutrients in the milk are derived from the blood meals taken by the parent fly. Most flies feed on the first day of pregnancy and thereafter at intervals of three or four days.

A relatively large proportion of the amino acids in the first blood feed are incorporated into lipids in the fat body and transferred to the milk gland during the

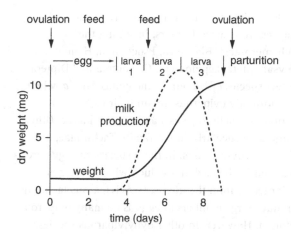

Figure 14.29 Adenotrophic viviparity. Growth of the tsetse fly, *Glossina*, through its embryonic and larval stages in the uterus of the adult female. The larva feeds on "milk" produced by the female accessory gland. Milk production is shown on an arbitrary scale. The female parent may take a blood meal on any day up to day six, but feeding after this is unusual (after Denlinger and Ma, 1974).

later stages of gestation. Amino acids from the second feed are probably taken up directly by the milk glands and there is little lipid synthesis. The adult fly does not usually feed in the later stages of pregnancy when the large larva occupies most of the space in its abdomen.

The larval respiratory system opens by a pair of posterior spiracles in the first two larval stages, but the system is highly specialized in the third, enabling the larva to respire while still in the uterus, but limiting water loss. The terminal segment of the abdomen bears two heavily sclerotized lobes (polypneustic lobes), each of which is crossed by three longitudinal bands of perforations leading into the tracheal system. Each of these perforations is guarded by a valve which permits air to be drawn into the system but does not allow it to be forced out. In addition to these openings in the polypneustic lobes, the spiracles of the second-stage larva on the insides of the lobes remain open because the second instar cuticle is not shed.

Indirectly acting dorso-ventral muscles produce a piston-like movement in a specialized part of the tracheal system in the polypneustic lobes and it is suggested that this movement sucks air in through the valved perforations and forces it forwards between the two linings of the tracheae. An exhalent current flows through the loose second-stage linings and out through the persistent spiracles of the second-stage larva. The respiratory muscles contract 15–25 times per minute. By this means the larva is able to draw air in through the genital opening of the parent, but this mechanism can only function while the second-stage cuticle persists, a period of four or five days at the beginning of the third larval stage. In the earlier larval stages, oxygen may be obtained at least partly by diffusion from the female tracheal system, which invests the uterus, while in the late third-stage larva the valves in the polypneustic lobes disappear and a two-way airflow through the perforations is possible.

The hindgut of the larval tsetse fly is occluded at its connection with the midgut and again at the anus, so that waste materials from the midgut are not voided and the hindgut forms a reservoir for nitrogenous waste. This arrangement prevents the larva from fouling the female's ducts. Following parturition, the larva moves to a suitable site for puparium formation; it does not feed during the short period for which it is active.

Maturation of the next oocyte occurs during pregnancy, and ovulation follows about 30 minutes after parturition, so maximum fecundity is achieved. As far as is known, the development of the Hippoboscidae and Nycteribiidae does not essentially differ from that of *Glossina*, but there is no evidence for complex air circulation similar to that which occurs in the larva of *Glossina*.

Hemocoelous viviparity Hemocoelous viviparity differs from the other forms of viviparity in that development occurs in the hemocoel of the parent female. This type of development occurs throughout the Strepsiptera and in some Cecidomyiidae (Diptera) that are pedogenic, which is defined as the ability to reproduce as a sexually mature larva (Section 14.6). Female Strepsiptera have two or three ovarial strands

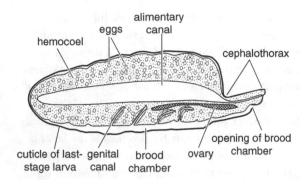

Figure 14.30 Hemocoelous viviparity. Sagittal section through a female strepsipteran showing eggs in the hemocoel (from Clausen, 1940).

on either side of the midgut, but there are no oviducts and mature oocytes are released into the hemocoel by the rupture of the ovarian walls. In *Stylops* the eggs contain very little yolk, but some is present in other strepsipteran genera such as *Acroschismus*. Sperm enter through the genital canals, which arise in the ventral midline of the female and open into the hemocoel (Fig. 14.30). Fertilization and development occur in the hemocoel with direct transfer of nutriment from the hemolymph to the embryo. The larvae hatch and find their way to the outside through the genital canals. In the case of pedogenic cecidomyiids (gall midges) in the genus *Miastor*, mature larvae produce eggs parthenogenetically (Section 14.5) that are similarly liberated into the hemocoel from simple sacs. The developing eggs are nourished via nurse cells, which arise independently of the oocyte, and later via the serosa, which becomes thickened and vacuolated. When the larvae hatch they feed on the tissues of the female and any unhatched eggs, finally escaping through a rupture in the wall of the parent.

Ovoviviparity

Many insect species are oviparous, retaining their eggs in the genital tract until just before the offspring are ready to hatch, at which time the eggs are laid. All the nourishment for the embryo, however, derives from yolk pre-packaged into the egg during oogenesis.

Ovoviviparity occurs sporadically in selected species from several orders, including the Ephemeroptera, Blattodea, Psocoptera, Hemiptera, Thysanoptera, Lepidoptera, Coleoptera and Diptera. A few species of Diptera in the genus *Musca*, which are normally oviparous, retain their eggs for the entirety of embryogenesis and deposit larvae. Other Diptera, particularly in the family Tachinidae, are always ovoviviparous. In these species the eggs are retained in the median oviduct, which becomes enlarged to form the uterus. Ovoviviparous tachinids produce large numbers of eggs as do many oviparous Diptera. However, in other ovoviviparous species, such as the flesh fly, *Sarcophaga*, small numbers of relatively large eggs are produced at each ovulation, and in *Musca larvipara* only one large egg is produced at a time. The increased size of the eggs permits the accumulation of more nutriment so that the embryo can develop beyond the normal hatching stage and larvae are born in a late stage of development. In *Hylemya strigosa* (Diptera), for instance, the larva passes through the first larval stage and molts to the second while still in the egg, casting the first-stage cuticle immediately after hatching. In *Termitoxenia* (Diptera), development in the uterine egg goes even further. A fully developed third-stage larva hatches from the egg immediately after it is laid. It pupates a few minutes afterwards and the free-living larva never feeds.

Fundamentally, all cockroaches (Blattodea) are oviparous, laying their eggs in an ootheca, which is extruded from the genital ducts. However, species vary in how long they carry oothecae, which results in life histories intermediate between oviparity and ovoviviparity. In some species, like *Periplaneta*, the ootheca may be carried, projecting from the genital opening, but it is finally dropped some time before the eggs hatch. Other species, such as *Blattella*, continue to carry the ootheca externally until the time of hatching. Some others extrude the ootheca, but then withdraw it into the body where it is held in a median brood sac extending beneath the rest of the reproductive system. In these species, the ootheca may be poorly developed and as the eggs increase in size they project beyond the ootheca. In most species the increase in egg size results only from the absorption of water, but in *Diploptera*, in which the eggs increase in length by five or six times during embryonic development, there is also an increase in dry weight, suggesting some nutrients may be obtained from the parent after ovulation.

14.3.2 Endoparasitoid eggs and polyembryony

Parasitoids are defined as insects that are free-living as adults but which are obligately parasitic during their immature stages. The largest diversity of parasitoids occurs in the Order Hymenoptera, but parasitoids also occur in several other groups including the Diptera, Coleoptera and Strepsiptera. Parasitoids exhibit two main developmental strategies: ectoparasitoids that lay their eggs externally on hosts and endoparasitoids that inject their eggs into the body of the host. In the Hymenoptera, ectoparasitism is the ancestral form of development, while endoparasitism has evolved independently at least eight times from different ectoparasitic ancestors.

Embryonic adaptations associated with an endoparasitic lifestyle Ectoparasitoids, like other terrestrial insects, lay eggs with a rigid chorion, which helps protect the embryo from desiccation, and an abundant yolk source that supplies the nutrients necessary to complete embryogenesis. Ectoparasitoid Hymenoptera also generally lay long-germ-type eggs that undergo syncytial cleavage and pattern the anterior–posterior axis in a manner similar to *Drosophila*. Endoparasitoids in contrast usually develop in the body cavity of another insect, where protection from desiccation and pre-packaging of a yolk source are not required. Unconstrained by the need for a pre-packaged source of nutrition or protection from desiccation, many endoparasitoids lay eggs with very thin

chorions and little or no yolk. For adults, the reduced energetic costs of laying reduced or yolkless eggs favor increases in fecundity. For offspring, yolkless eggs correlate with formation of specialized membranes derived from polar body or blastoderm cells that envelop the developing parasitoid embryo. Once formed, these membranes allow the embryo to rupture out of the chorion and increase greatly in size by absorbing nutrients from the host. Among three families of Hymenoptera, the Scelionidae, certain Braconidae and a few Aphelinidae, these membranes later differentiate into cells called teratocytes that are released into the host when the parasitoid larva hatches. The yolkless eggs laid by many endoparasitoids have also shifted away from syncytial cleavage and undergo complete cleavage (Section 14.2.4) as either long- or short-germ-type embryos.

Polyembryony Access to abundant nutritional resources combined with a shift from syncytial to total cleavage has likely been instrumental in the evolution of a second novelty called polyembryony. Polyembryony is defined as the formation of two or more embryos from a single egg. Sporadic polyembryony refers to the occasional formation of more than one embryo per egg. This occurs in a few groups of insects and several other animals, including humans, who occasionally give birth to identical twins. Obligate polyembrony is much rarer but occurs in hundreds of endoparasitoid species in genera from four families of Hymenoptera (Encyrtidae, Braconidae, Platygasteridae and Dryinidae) and a few taxa of Strepsiptera. The phylogenetic distance between these groups indicates that polyembryony has evolved independently multiple times, yet despite multiple origins, all polyembryonic wasps share the common biology of laying tiny, yolkless eggs into the egg or larval stage of insect hosts. After oviposition, each egg laid by a polyembryonic wasp develops into a single embryo. This embryo then clonally gives rise to additional embryos that together form an assemblage called a polygerm or polymorula. Broods produced by polyembryonic platygasterids and braconids range from as few as two offspring to approximately 100. In contrast, a few species of polyembryonic encyrtids produce thousands of offspring per egg. Polyembryonic encyrtids have evolved a second novelty: a caste system in which morphologically and functionally distinct larvae develop from the same egg.

Polyembryony is best known in the encyrtid *Copidosoma floridanum*, which parasitizes plusiine moths, e.g., *Trichoplusia ni* (Fig. 14.31). After oviposition the polar nucleus of the *C. floridanum* egg rapidly separates from the pronucleus, which migrates to the posterior pole of the egg along with cytoplasmic pole plasm (Section 14.2.3). The *C. floridanum* egg then initiates first cleavage that results in formation of two equal-sized daughter cells (blastomeres) and an anterior polar cell that contains the polar nucleus. Although aborted in most insects, polar nuclei remain viable and fuse in polyembryonic encyrtids to form a polar body cell. After first cleavage, all of the germ plasm associates with one of the blastomeres. Second cleavage is unequal, with the germ plasm-containing blastomere dividing into large and small daughter blastomeres, with the pole plasm always segregating to the smaller cell, which becomes determined as the primordial germ cell (Fig. 14.31a). The other blastomere, in contrast, divides into two equal-sized daughter cells (Fig. 14.31a). Subsequent cleavages are asynchronous, resulting in the formation of approximately 200 somatic cells and 4–8 germ cells. Concurrently, the nucleus of the polar cell divides without cytokinesis to form a syncytial compartment that migrates as an extraembryonic membrane over the dividing blastomeres. The embryo then emerges from the chorion and continues development in the host body cavity unconstrained by the eggshell. Enveloped by the extraembryonic membrane this embryo is referred to as the primary morula, which usually implants itself in the prothorax of the host embryo (Fig. 14.31b).

Upon hatching of the host larva, the *C. floridanum* embryo enters the proliferation phase of development (Fig. 14.31c). This phase begins when the primary morula divides itself into 2–5 embryonic masses called proliferative morulae, which together form a polymorula. Each proliferative morula consists of hundreds of round, seemingly non-differentiated cells surrounded by the extraembryonic membrane. Each proliferative morula in turn becomes further subdivided by invagination of the extraembryonic membrane, resulting ultimately in formation of more than 1000 proliferative morulae consisting of approximately 40 cells each by the time the host molts to the fourth instar. During this period, some secondary morulae initiate morphogenesis and develop into soldier larvae that are recognized by their elongate body form and larger mandibles (Fig. 14.31c). On average, about 4% of the total number of larvae produced per host are soldiers, which never molt and die from desiccation after their reproductive caste siblings consume the host.

Most embryos synchronously enter the third phase of development, morphogenesis, at the end of the host's fourth instar and develop into a second caste, called reproductive larvae. Morphogenesis is first recognizable by the rounded cells in proliferative morulae becoming fibroblastic and compacting. This is rapidly followed by formation of a primordium, followed by gastrulaton and segmentation. Almost all of the morulae that synchronously initiate morphogenesis in the host fourth instar develop into reproductive larvae that eclose in the host's fifth instar. These larvae morphologically have a rounded body form and small mandibles (Fig. 14.31d). Developmentally, reproductive larvae rapidly consume the host, pupate and subsequently emerge as adult wasps. In contrast, the soldier larvae die when the reproductive larvae finish consuming the host.

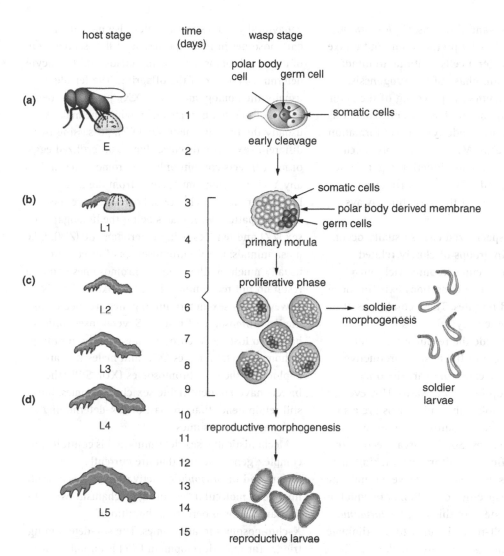

Figure 14.31 Development of the polyembryonic wasp *Copidosoma floridanum* (Hymenoptera; Encyrtidae). The left column shows the developmental stages of the host *Trichoplusia ni* (Lepidoptera; Noctuidae); the middle column shows development time in days; and the right column shows the developmental stages of *C. floridanum*. (a) The female wasp lays one or two eggs into the egg stage of the host. Soon after oviposition, the wasp egg undergoes two cleavage events that result in the formation of four blastomeres. One of these blastomeres inherits pole plasm and is specified as the first germ cell while the three other blastomeres become somatic cells. The polar body begins to envelope the four blastomeres. (b) The somatic and germ cells continue to divide. When the polar body fully envelops the embryonic blastoderm cells, the embryo ruptures out of the chorion to form a primary morula. The host egg hatches during this period to form a first instar larva (L1, days 3–4). (c) As the host grows and molts to form 2nd–4th instar larvae (L1–L4, days 5–10), the primary morula enters a proliferation stage of embryogenesis that results in the formation of an increasing number of secondary morulae. Most secondary morulae contain both somatic and germ cells but a few secondary morulae form that lack germ cells. These embryos initiate morphogenesis to form soldier larvae. (d) As the host molts to the 5th instar (L5), secondary morulae with germ cells initiate morphogenesis and develop into reproductive larvae. The reproductive larvae then eat the host (days 13–15), pupate and emerge as adult wasps. The soldier larvae die when their reproductive caste siblings finish consuming the host (adapted from Strand, 2009).

Similar to other advanced insects, *C. floridanum* establishes the germ line by preformation, but unlike more typical insects, germ cells continue to divide during the proliferation phase of embryogenesis. More remarkably, asymmetric parceling of the germ line during the proliferation phase of embryogenesis serves as the mechanism underlying caste formation in polyembryonic wasps. Most embryos produced during the proliferation phase inherit the germ line and develop into reproductive larvae. However, embryos that do not inherit the germ line always develop into soldiers.

The evolution of specialized castes usually occurs in species that live in groups of closely related individuals, and that occupy resource-rich but defensible resources. These conditions exist in social insects like ants and termites. These conditions also exist in polyembryonic encyrtid wasps that propagate clonally inside the resource-rich and defensible environment of the host. Reproductive larvae are so named because they are the only offspring that develop into adult wasps. However, hosts parasitized by polyembryonic wasps are also attacked by other species, resulting in intense competition for resources. Soldier larvae recognize and kill heterospecific competitors by attacking them with their mandibles. Field studies, however, indicate that risks from intraspecific competitors are equal or greater. Commensurate with this risk, *C. floridanum* soldiers exhibit a well-refined ability to discriminate kin from non-kin and aggressively attack and kill non-relatives. In both inter- and intraspecific competition, soldier larvae increase their own fitness by assuring the survival of their reproductive siblings.

14.4 Sex determination

Sex determination refers to the mechanisms underlying development of an egg into one sex or another. Most insects are sexually dimorphic, with males and females differing morphologically and behaviorally. Males and females also have distinct chromosomes in most species, such that segregation of the sex chromosomes during meiosis of the oocyte determines the sex of the offspring. The female is usually the homogametic sex (XX), possessing two copies of one sex chromosome, while the male is usually the heterogametic sex (XY), possessing two different sex chromosomes. Hence, unfertilized eggs of most insects contain only X-chromosomes and any Y-chromosome must come from the male. Lepidoptera and Trichoptera, however, have the opposite pattern with males being the homogametic sex (ZZ) and females being heterogametic (ZW). Like most animals, the Y-chromosomes of insects are usually much smaller than X-chromosomes and as a result do not recombine with X. Unlike mammals, however, few sex-determining genes are located on the Y-chromosome of insects. Several insect orders have also lost the Y-chromosome entirely, resulting in homogametic females (XX) and males that are haploid for the sex chromosomes (XO). Still other insects have no identifiable sex chromosomes, but still retain genes that are part of sex-determining pathways on autosomes.

Mechanistically, sex determination is controlled by complex gene cascades that are currently best understood in *Drosophila*. Activation of this cascade involves a molecular "counting mechanism" whereby the early embryo perceives the ratio of X-chromosomes to autosomes. The sex-determining trigger for male development (XY) is an imbalance between genes on the X-chromosome and one of the autosomes (X:A = 0.5), while female development is triggered by an equal balance between genes on the X-chromosomes and autosomes (X:A = 1) (Fig. 14.32). In females the molecular counting mechanism activates transcription of the gene *sexlethal*, whose corresponding protein regulates splicing of subsequent *sexlethal* mRNAs that are transcribed to produce a female-specific form (splice variant) of the Sexlethal protein. This results in expression of a female-specific form of the Transformer protein, which regulates other genes.

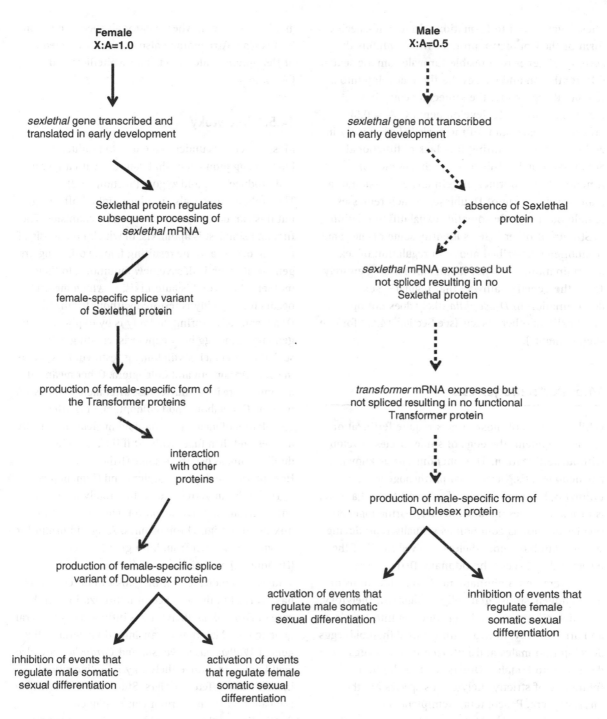

Figure 14.32 Heterogametic (XX:XY) sex determination pathway in *Drosophila* showing key genes and events. Females (XX) have an equal ratio of genes on the X-chromosome and autosomes (X:A = 1.0), while males (XY) have an imbalance between genes on the X-chromosome and autosomes (X:A = 0.5).

These events lead to formation of a female-specific form of the Doublesex protein, which inhibits the activity of genes responsible for male somatic sexual differentiation and causes the fly to develop into a female. Reciprocally, the molecular counting mechanism for male development (X: A = 0.5) prevents transcription of the *sexlethal* gene early in embryogenesis, resulting in a lack of functional Sexlethal protein. This leads to an absence of functional Transformer protein and expression of a male-specific form of Doublesex, which represses female but not male somatic sexual differentiation.

Studies of other insects identify some of the same components described above in regulation of sex determination, but other factors are missing entirely. Thus, the genetic pathway that regulates sex determination in *Drosophila* likely does not operate identically in other insects (see Section 14.5.2 for one such example).

14.5 Parthenogenesis

While the eggs of most insects require fertilization for development, the eggs of some species develop without fertilization. This phenomenon is known as parthenogenesis. Occasional parthenogenesis, following from the failure of a female to find a mate, is probably widespread. Obligately parthenogenetic species are not as common as sexually reproducing species, but have nonetheless evolved in all of the insect orders except the Odonata, Dermaptera, Neuroptera and Siphonaptera. Two major forms of parthenogenesis are generally recognized: thelytoky, in which all unfertilized eggs develop into females; and arrhenotoky, in which unfertilized (haploid) eggs develop into males and fertilized (diploid) eggs develop into females. Orders with the highest frequency of strictly thelytokous species are the Thysanoptera, Psocoptera, Hemiptera and Phasmatodea. Arrhenotoky, also known as haplodiploidy, is most common in the Hymenoptera

and Thysanoptera, where nearly all species develop in this way. Arrhenotoky also occurs in selected taxa of Hemiptera (scale insects and whiteflies) and Coleoptera.

14.5.1 Thelytoky

Most insect eggs undergo meiosis to produce a haploid egg pronucleus that can fuse with a sperm and produce a diploid zygote (Section 14.2). Thelytokous insects also produce diploid offspring, but this can occur by two different mechanisms. The first and simplest is apomixis, in which no meiosis of the egg occurs and the resulting female offspring are genetically identical, save new mutations, to their mother. The second is automixis, in which meiosis occurs but diploidy is restored without fertilization. One means of restoring diploidy is by duplicating the genome of the egg before meiosis, creating a 4N cell, such that after meiosis diploidy is restored. This occurs in some Orthoptera and Coleoptera. Other means of automixis are if two nuclei fuse after meiosis I, as seen in some Collembola and Lepidoptera, or if after completion of meiosis the haploid nucleus mitotically divides and then fuses with itself (i.e., gamete duplication), as occurs in some Orthoptera, Hemiptera, Lepidoptera, Diptera and Hymenoptera.

Thelytoky in vertebrates is invariably associated with interspecific hybridization, which results in mixing of foreign chromosomes. A hybrid origin for thelytoky is known from both grasshoppers (Orthoptera) and stick insects (Phasmatodea), but studies also indicate that thelytoky in other insects has evolved in the absence of hybridization. In the case of *Drosophila*, thelytoky is influenced by several genetic loci. Maternally transmitted bacteria in the genera *Wolbachia*, *Rickettsia* and *Cardinium* are also well known to induce thelytoky in a diversity of insects from different orders. Studies with *Wolbachia* indicate that the bacterium causes gamete duplication in the parasitoid wasp, *Trichogramma cacoeciae*. *Wolbachia* can also distort or alter the sex of offspring arthropods produce. In some species *Wolbachia* kills one sex (males) early in development, which constrains the mother to producing only daughters. The molecular mechanisms underlying how *Wolbachia* causes these effects are unknown.

14.5.2 Arrhenotoky

In arrhenotokous Hymenoptera, females determine whether or not to fertilize an egg by controlling the release of sperm from the spermatheca as eggs pass down the oviduct. This mechanism also provides a way the mother can control the sex ratio of her offspring. Many ecological and genetic factors have been implicated in sex ratio evolution, but at the physiological level, how females control sperm release is unclear.

Although arrhenotoky is distinguished by the development of haploid eggs into males and diploid eggs into females, chromosome number does not appear to be the causal mechanism. Complementary sex determination (CSD) is the mechanism for arrhenotoky in at least some Hymenoptera. One form of CSD involves a single locus with multiple alleles. Individuals that are heterozygous at this locus develop into females and individuals that are hemizygous (haploid) develop into males. However, inbreeding also leads to individuals that are homozygous (diploid) at the CSD locus and they also develop into males. This mode of sex determination has been shown to operate in more than 60 species, representing all of the major hymenopteran subgroups. The *csd* gene has been identified in the honey bee, which combined with a second recently identified factor, *feminizer*, compose a portion of a sex-determination pathway. Interestingly, *csd* appears to have evolved from *feminizer*, and *feminizer* is related to the sex-determining gene *transformer* in *Drosophila*, whose alternative splicing determines whether an individual develops into a male or female (Section 14.4). *Feminizer* is also alternatively spliced, with some evidence suggesting that the *feminizer* gene mediates the switch in

developmental pathways for female versus male development controlled by heterozygosity at the CSD locus. In contrast, sex determination in other groups of Hymenoptera, such as the Chalcidoidea, is not explained by single-locus CSD because homozygous diploid individuals still develop into females. Multi-locus CSD has been suggested as one possible explanation for sex determination in chalcidoids. Studies with *Nasonia vitripennis* suggest a maternal product and a parent-specific (imprinting) effect may also underlie sex determination in chalcidoids.

A rare form of development in insects related to arrhenotoky is hermaphrodism, where the same individual produces eggs and sperm and progeny develop by self-fertilization. The only known hermaphroditic insects are a few scale species in the genus *Icerya*. The coccid, *Icerya purchasi*, is of interest in that, apart from a few haploid males, the adult population consists entirely of hermaphrodites, which have diploid ovaries and haploid testes. No true females occur. When a hermaphrodite-forming larva hatches from the egg, all cells in its body are diploid, but after a time haploid nuclei appear in the gonad. They form a core from which the testis develops, surrounded by the ovary. Oocytes undergo a normal reduction division, but the spermatocytes, which are already haploid, do not. The hermaphrodites are normally self-fertilizing, but are sometimes fertilized by occasional males, which develop from unfertilized eggs. Cross-fertilization between hermaphrodites, however, does not occur.

Another unusual variant associated with some arrhenotokous wasps is the existence of supernumerary chromosomes that are transmitted paternally. In some strains of *Nasonia vitripennis* and *Trichogramma kaykai* a supernumerary chromosome called *paternal sex ratio* (*psr*) is transmitted by males and causes the paternal chromosomes to condense after fertilization. This results in survival of the maternal chromosomes and *psr* itself, but causes the egg to develop into a haploid male wasp.

14.5.3 Alternation of generations

A number of insects combine the advantages of parthenogenesis with those of sexual reproduction through an alternation of generations. This occurs, for instance, in gall wasps from the family Cynipidae (Hymenoptera), which are bivoltine (having two generations each year) and alternate parthenogenetic and arrhenotokous generations. *Neuroterus lenticularis*, for example, forms galls on the underside of oak leaves in which the species overwinters. Only females emerge in the spring, which lay unfertilized (haploid) eggs in catkins or young leaves. These eggs give rise to both haploid males and diploid females whose diploid condition is apparently restored during early embryogenesis. In this way the bisexual generation arises with male and female wasps emerging from catkin galls in early summer. After mating the females of this generation lay eggs that are fertilized and which produce the parthenogenetically reproducing females of the following spring generation.

As described in detail in Chapter 15, aphids have a more complex alternation of generations, with several parthenogenetic generations occurring during the summer (Fig. 15.37). An alternation of generations may also occur in Cecidomyiidae (Diptera).

14.6 Pedogenesis

As previously noted (Section 14.3), some insects sexually mature precociously and are able to reproduce while still larvae or pupae. This phenomenon is known as pedogenesis. Most pedogenic insects are also parthenogenetic and viviparous.

In the case of *Miastor* (Diptera, Cecidomyiidae) and *Micromalthus* (Coleoptera), larvae give birth to other larvae or, occasionally, lay eggs. Pedogenesis occurs in *Miastor* only under very good or poor nutritional conditions. Under intermediate nutritional conditions normal adults are produced. Young larvae are set free in the body cavity of the pedogenetic larva and they feed on the maternal tissues, eventually escaping through the body wall of the parent.

Micromalthus has five reproductive forms: adult males, adult females, male-producing larvae, female-producing larvae and larvae producing males and females. The species has a complex heteromorphosis (Fig. 14.33). The form emerging from the egg is a mobile first instar called a triungulin, which molts to an apodous (legless) larva that can develop in one of three ways. It can develop through a pupa to a normal adult female, it can molt to a larval form, which gives rise pedogenetically to a male, or it can molt to a pedogenetic larva, which produces triungulins. Male-producing larvae lay a

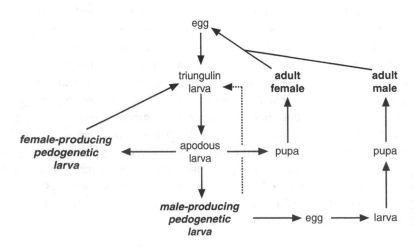

Figure 14.33 Pedogenesis. The life history of the beetle *Micromalthus*. Bold type: reproductive forms; italics: pedogenetically reproducing larvae (based on Pringle, 1938).

single egg containing a young embryo, but the egg adheres to the parent and when the larva hatches, it eats the parent larva. If, for some reason, the parent larva is not eaten, it subsequently produces a small brood of female larvae.

Among cecidomyiids in the genera *Tekomyia* and *Henria*, pupal forms give birth to larvae. The larvae of these insects are also of two types: one develops into a pupa and, ultimately, a normal adult, while the other develops into a hemipupa, which exhibits vestiges of wings and legs. A brood of larvae, commonly between 30 and 60, then escapes from the hemipupa by rupturing the cuticle. Pedogenesis is the normal method of reproduction in these insects and although normal adults are sometimes produced it is not

certain that they are capable of producing viable offspring.

Pedogenesis also occurs in aphids. Although the young are not born until the parent aphid has reached the adult stage, their development may begin before the parent is born. As a result, an aphid can bear both her daughters and, within them, her grand-daughters, resulting in the telescoping of generations. Development of the offspring continues throughout the larval life of the parent.

The hemipteran, *Hesperoctenes*, is an example of a pedogenetic form in which fertilization occurs. Some last-stage larvae are found with sperm in the hemocoel as a result of hemocoelic insemination. These sperm fertilize the eggs, which develop in the ovaries of the larva.

Summary

- Most insects are oviparous and lay eggs with a pre-packaged source of nutrients – yolk – that is surrounded by a rigid eggshell with features important for fertilization, respiration and water regulation.

- Embryogenesis begins with the formation of a zygote nucleus and ends with development and hatching of a nymph or larva. The extraordinary series of events that transpire during embryogenesis are regulated by a multitude of genes that form complex regulatory cascades and signaling pathways.

- Some insect groups have evolved other strategies for acquiring the nutrients necessary for embryonic development. These include viviparous species whose eggs obtain nutrients from the mother, and a large number of endoparasites whose eggs obtain nutrients from a host.

- Sex determination in most insects involves the inheritance of male- and female-specific sex chromosomes and is controlled by complex gene cascades. Insects have also evolved several forms of parthenogenesis.

- Taken together, we are far from fully understanding all of the processes involved in regulating embryonic development. However, we have a substantially better grasp of these complex events than existed 30 years ago. The events that take place during embryonic development also set the stage for the postembryonic physiological processes to be discussed in following chapters.

Recommended reading

Affolter, M. and Caussinus, E. (2008). Tracheal branching morphogenesis in *Drosophila*: new insights into cell behaviour and organ structure. *Development* **135**, 2055–2064.

Anderson, D. T. (1972a). The development of hemimetabolous insects. In *Developmental Systems: Insects*, vol. 1, ed. S. J. Counce and C. H. Waddington, pp. 95–163. New York, NY: Academic Press.

Anderson, D. T. (1972b). The development of holometabolous insects. In *Developmental Systems: Insects*, vol. 1, ed. S. J. Counce and C. H. Waddington, pp. 165–242. New York, NY: Academic Press.

Bate, M. (1993). The mesoderm and its derivatives. In *The Development of* Drosophila melanogaster, vol. 2, ed. M. Bate and A. M. Arias, pp. 1013–1090. New York, NY: Cold Spring Harbor Laboratory Press.

Bate, M. and Arias, A. M. (eds.) (1993). *The Development of* Drosophila melanogaster, 2 vols. New York, NY: Cold Spring Harbor Laboratory Press.

Beukeboom, L. W., Kamping, A. and Zande van de, L. (2007). Sex determination in the haplodiploid wasp *Nasonia vitripennis* (Hymenoptera: Chalcidoidea): a critical consideration of models and evidence. *Seminars in Cell & Developmental Biology* **18**, 371–378.

Blackman, R. L. (1987), Reproduction, cytogenetics and development. In *Aphids: Their Biology, Natural Enemies and Control*, vol. 2, ed. A. K. Minks and P. Harrewijn, pp. 163–195. Amsterdam: Elsevier.

Cavaliere, V., Bernardi, F., Romani, P., Duchi, S. and Gargiulo, G. (2008). Building up the *Drosophila* eggshell: first of all the eggshell genes must be transcribed. *Developmental Dynamics* **237**, 2061–2072.

Damen, W. G. M. (2007). Evolutionary conservation and divergence of the segmentation process in arthropods. *Developmental Dynamics* **236**, 1379–1391.

Donnell, D., Corley, L., Chen, G. and Strand, M. R. (2004). Inheritance of germ cells mediates caste formation in a polyembryonic wasp. *Proceedings of the National Academy of Sciences, USA* **101**, 10095–10100.

Extavour, C. G. and Akam, M. (2003). Mechanisms of germ cell specification across the metazoans: epigenesis and preformation. *Development* **130**, 5869–5884.

Foe, V. E., Odell, G. M. and Edgar, B. A. (1993). Mitosis and morphogenesis in the *Drosophila* embryo: point and counterpoint. In *The Development of* Drosophila melanogaster, vol. 1, ed. M. Bate and A. M. Arias, pp. 149–300. New York, NY: Cold Spring Harbor.

Goltsev, Y., Rezende, G. L., Vranizan, K., Lanzaro, G., Valle, D. and Levine, M. (2009). Developmental and evolutionary basis for drought tolerance of the *Anopheles gambiae* embryo. *Developmental Biology* **330**, 462–470.

Hadley, N. F. (1994). *Water Relations of Terrestrial Arthropods*. San Diego, CA: Academic Press.

Heimpel, G. E. and De Boer, J. G. (2008). Sex determination in the Hymenoptera. *Annual Review of Entomology*, **53**, 209–230.

Hoffmann, J. A. and Lagueux, M. (1985). Endocrine aspects of embryonic development in insects. In *Comprehensive Insect Physiology, Biochemistry and Pharmacology*, vol. 1, ed. G. A. Kerkut and L. I. Gilbert, pp. 435–460. Oxford: Pergamon Press.

Ji, S., Li, Y., Zhou, Z., Kumar, S. and Ye, J. (2009). A bag-of-words approach for *Drosophila* gene expression pattern annotation. *BMC Bioinformatics* 10, 119.

Lawrence, P. A. (1992). *The Making of a Fly: The Genetics of Animal Design*. Oxford: Blackwell.

Lecuyer, E., Yoshida, H., Parthasarathy, N., Aim, C., Babak, T., Cerovina, T., Hughes, T., Tomancak, P. and Krause, H. (2007). Global analysis of mRNA localization reveals a prominent role in organizing cellular architecture and function. *Cell* 131, 174–187.

Manning, G. and Krasnow, M. A. (1993). Development of the *Drosophila* tracheal system. In *The Development of* Drosophila melanogaster, vol. 1, ed. M. Bate and A. M. Arias, pp. 609–685. New York: Cold Spring Harbor Laboratory Press.

Margaritis, L. H. (1985). Structure and physiology of the eggshell. In *Comprehensive Insect Physiology, Biochemistry and Pharmacology*, vol. 1, ed. G. A. Kerkut and L. I. Gilbert, pp. 153–230. Oxford: Pergamon Press.

Marin, I. and Baker, B. S. (1998). The evolutionary dynamics of sex determination. *Science* 281, 1990–1994.

Moran, N. A., McCutcheon, J. P. and Nakabachi, A., (2008). Genomics and evolution of heritable bacterial symbionts. *Annual Review of Genetics* 42, 165–190.

Nagy, L. M. and Grbic, M. (2009). Embryogenesis. In *Encyclopedia of Insects*, 2nd edn., pp. 316–320. Amsterdam: Academic Press.

Normark, B. B. (2003). Evolution of alternative genetic systems in insects. *Annual Review of Entomology* 48, 397–423.

Patel, N. H. (1994). The evolution of arthropod segmentation: insights from comparisons of gene expression patterns. *Development*, 1994 supplement, 201–207.

Pearson, B. J. and Doe, C. Q. (2004). Specification of temporal identity in the developing nervous system. *Annual Review of Cell and Developmental Biology* 20, 619–647.

Pennacchio, F. and Strand, M. R. (2006). Evolution of developmental strategies in parasitic Hymenoptera. *Annual Review of Entomology* 51, 233–258.

Rehorn, K. P., Michelson, T. H. and Reuter, R. (1996). A molecular aspect of hematopoiesis and endoderm development common to vertebrates and *Drosophila*. *Development* 122, 4023–4031.

Sander, K. (1996). Variants of embryonic patterning mechanisms in insects: Hymenoptera and Diptera. *Seminars in Cell & Developmental Biology* 7, 573–582.

Sander, K., Gutzeit, H. O. and Jäckle, H. (1985). Insect embryogenesis: morphology physiology, genetical and molecular aspects. In *Comprehensive Insect Physiology, Biochemistry and Pharmacology*, vol. 1, ed. G. A. Kerkut and L. I. Gilbert, pp. 319–385. Oxford: Pergamon Press.

Skaer, H. (1993). The alimentary canal. In *The Development of* Drosophila melanogaster, vol. 2, ed. M. Bate and A. M. Arias, pp. 941–1012. New York, NY: Cold Spring Harbor Laboratory Press.

Sousa-Nunes, R., Yee, L. L. and Gould, A. P. (2011). Fat cells reactivate quiescent neuroblasts via TOR and glial insulin relays in *Drosophila*. *Nature* 471, 508–512.

St. Johnston, D. (1993). Pole plasm and the posterior group genes. In *The Development of* Drosophila melanogaster, vol. 1, ed. M. Bate and A. M. Arias, pp. 325–363. New York, NY: Cold Spring Harbor Laboratory Press.

St. Johnston, D. and Nusslein-Volhard, C. (1992). The origin of pattern and polarity in the *Drosophila* embryo. *Cell*, **68**, 201–219.

Strand, M. R. (2009). Polyembryony. In *Encyclopedia of Insects*, 2nd edn., pp. 821–825. Amsterdam: Academic Press.

Taraszka, J. A., Kurulugama, R., Sowell, R. A., Valentine, S. J., Koeniger, S. L., Arnold, R. J., Miller, D. F., Kaufman, T. C. and Clemmer, D. E. (2005). Mapping the proteome of *Drosophila melanogaster*: analysis of embryos and adult heads by LC-IMS-MS methods. *Journal of Proteome Research* **4**, 1223–1237.

Truman, J. W. and Riddiford, L. M. (2002). Endocrine insights into the evolution of metamorphosis in insects. *Annual Review of Entomology* **47**, 467–500.

Tsuji, T., Hasegawa, E. and Isshiki, T. (2008). Neuroblast entry into quiescence is regulated intrinsically by the combined action of spatial Hox proteins and temporal identity factors. *Development* **135**, 3859–3869.

Zrzavy, J. and Stys, P. (1995). Evolution and metamerism in Arthropoda: developmental and morphological perspectives. *Quarterly Review of Biology* **70**, 279–295.

References in figure captions

Anderson, D. S. (1960). The respiratory system of the egg-shell of *Calliphora erythrocephala*. *Journal of Insect Physiology* **5**, 120–128.

Anderson, D. T. (1962). The embryology of *Dacus tryoni* (Frogg.) [Diptera, Trypetidae (=Tephritidae)], the Queensland fruit fly. *Journal of Embryology and Experimental Morphology* **10**, 248–292.

Anderson, D. T. (1972). The development of holometabolous insects. In *Developmental Systems: Insects*, vol. **1**, ed. S. J. Counce and C. H. Waddington, pp. 165–242. New York, NY: Academic Press.

Beament, J. W. S. (1946). The waterproofing process in eggs of *Rhodnius prolixus* Ståhl. *Proceedings of the Royal Society of London B: Biological Sciences* **133**, 407–418.

Clausen, C. P. (1940). *Entomophagous Insects*. New York, NY: McGraw-Hill.

Denlinger, D. L. and Ma, W.-C. (1974). Dynamics of the pregnancy cycle in the tsetse *Glossina morsitans*. *Journal of Insect Physiology* **20**, 1015–1026.

Doe, C. Q. and Goodman, C. S. (1985). Early events in insect neurogenesis I: development and segmental differences in the pattern of neuronal precursor cells. *Developmental Biology* **111**, 193–205.

Hagan, H. R. (1951). *Embryology of the Viviparous Insects*. New York, NY: Ronald Press.

Hartley, J. C. (1962). The egg of *Tetrix* (Tetrigidae, Orthoptera), with a discussion on the probable significance of the anterior horn. *Quarterly Journal of Microscopical Science* **103**, 253–259.

Henson, H. (1946). The theoretical aspect of insect metamorphosis. *Biological Reviews of the Cambridge Philosophical Society* **21**, 1–14.

Hinton, H. E. (1960). Plastron respiration in the eggs of blowflies. *Journal of Insect Physiology* **4**, 176–183.

Hinton, H. E. (1961). The structure and formation of the egg-shell in the Nepidae (Hemiptera). *Journal of Insect Physiology* 7, 224–257.

Imms, A. D. (1957). *A General Textbook of Entomology*. 9th edn., revised by O. W. Richards and R. G. Davies. London: Methuen.

Johannsen, O. A. and Butt, F. H. (1941). *Embryology of Insects and Myriapods*. New York, NY: McGraw-Hill.

Lees, A. D. (1976). The role of pressure in controlling the entry of water into the developing eggs of the Australian plague locust *Chorthoicetes terminifera* (Walker). *Physiological Entomology* 1, 39–50.

Leitch, B., Laurent, G. and Shepherd, D. (1992). Embryonic development of synapses on spiking local interneurones in locust. *Journal of Comparative Neurology* 324, 213–236.

Pringle, J. A. (1938). A contribution to the knowledge of *Micromalthus debilis* LeC. (Coleoptera). *Transactions of the Royal Entomological Society of London* 87, 271–286.

Richards, A. G. (1957). Cumulative effects of optimum and suboptimum temperatures on insect development. In *Influence of Temperature on Biological Systems*, ed. F. H. Johnson, pp. 145–162. Washington, DC: American Physiological Society.

Roonwal, M. L. (1954). The egg-wall of the African migratory locust, *Locusta migratoria migratorioides* Reiche and Frm (Orthoptera, Acrididae). *Proceedings of the National Institute of Sciences, India* 20, 361–370.

Slifer, E. H. and Sekhon, S. S. (1963). The fine structure of the membranes which cover the egg of the grasshopper, *Melanoplus differentialis*, with special reference to the hydropyle. *Quarterly Journal of Microscopical Science* 104, 321–334.

Southwood, T. R. E. (1956). The structure of the eggs of the terrestrial Heteroptera and its relationship to the classification of the group. *Transactions of the Royal Entomological Society of London* 108, 163–221.

Strand, M. R. (2009). Polyembryony. In *Encyclopedia of Insects*, 2nd edn., pp. 821–825. Amsterdam: Academic Press.

Thompson, K. G. and Siegler, M. V. S. (1993). Development of segment specificity in identified lineages of the grasshopper CNS. *Journal of Neuroscience* 13, 3309–3318.

15 Postembryonic development

REVISED AND UPDATED BY **STUART REYNOLDS**

INTRODUCTION

Postembryonic development allows some of the most remarkable examples of the diversity and ecological success of insect. The distinction between juvenile and adult phenotypes can be extreme, allowing different life-stages to occupy very different ecological niches. Changes in the rate of development throughout the lifecourse can allow insects to wait out periods of inclement conditions by entering diapause, and some insects express alternative phenotypes under particular environmental conditions, allowing multiple phenotypes to be packed within the same genome. In Section 15.1 the process of hatching from the egg is considered, followed by aspects of larval development (Section 15.2), metamorphosis (Section 15.3) and control of postembryological development through to adulthood (Section 15.4). Section 15.5 then provides some examples of environmentally determined polyphenisms. The chapter ends with a final section on delayed development during diapause (Section 15.6).

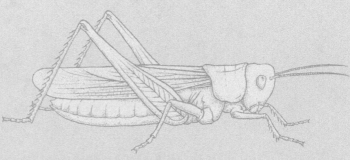

The Insects: Structure and Function (5th edition), ed. S. J. Simpson and A. E. Douglas.
Published by Cambridge University Press. © Cambridge University Press 2013.

15.1 Hatching

15.1.1 Mechanisms of hatching

Most insects force their way out of the egg by exerting pressure against the inside of the shell. The insect increases its volume by swallowing the extraembryonic fluid and in some cases by swallowing air which diffuses through the shell. Then, waves of muscular contraction pump hemolymph forwards so that the head and thoracic regions are pressed tightly against the inside of the shell. In grasshoppers, and perhaps in other insects, these muscular waves are interrupted periodically by a simultaneous contraction of the abdominal segments which causes a sudden increase in pressure in the anterior region. The dorsal membrane of the neck in grasshoppers has a pair of lobes, the cervical ampullae, which are inflated by the increase in hemolymph pressure (Fig. 15.1a). They serve to focus the pressure on a limited area of the shell. If the shell does not split, the ampullae are withdrawn and a further series of posterior to anterior waves of contraction follows, ending with another sudden abdominal contraction. One of these sudden contractions ultimately ruptures the shell.

The position of the rupture generally depends on where the insect puts pressure on the chorion. In grasshoppers, the chorion is split transversely above the ampullae; in the water beetle, *Agabus bipustulatus* (Coleoptera), the split is longitudinal, while in some species it is variable in position. The chorion of some species has a line of weakness along which it splits (Section 14.1). The egg of the blue blowfly *Calliphora vicina* (Diptera), for example, has a pair of longitudinal hatching lines running along its length (see Fig. 14.8), and in Heteroptera a hatching line runs around the egg where the cap joins the body of the chorion (see Fig. 13.13). In eggs of the mosquito *Aedes aegypti* (Diptera), there is a line of weakness in the serosal cuticle and a split in the chorion follows this passively, perhaps because the serosal cuticle and chorion are closely bound.

In species with a thick serosal cuticle, such as Acrididae and Heteroptera, hatching is aided by an enzyme that digests the serosal endocuticle before hatching begins. The enzyme is thought to be produced by the pleuropodia, and, because they are not covered by the embryonic cuticle, it is secreted directly into the extraembryonic space (Section 14.2).

Cuticular structures known as egg bursters aid hatching in a number of insects. These are usually on the head of the embryonic cuticle of Odonata, some Orthoptera, Heteroptera (Hemiptera), Neuroptera and Trichoptera, but are on the cuticle of the first-stage larva in Nematocera (Diptera), Carabidae (Coleoptera) and Siphonaptera. Their form varies, e.g., pentatomids (Hemiptera) have a T- or Y-shaped central tooth, while Cimicomorpha (Hemiptera) have a row of spines running along each side of the face from near the eye to the labrum (Fig. 15.1c). Some sucking lice (Anoplura) have a pair of spines, and several pairs of lancet-shaped blades arising from the embryonic cuticle over the head. Often, as in fleas, mosquitoes and tsetse flies, a cuticular tooth is in a membranous depression which can be erected by blood pressure (Fig. 15.1b). In *Agabus* the egg burster is a spine on either side of the head, but many Polyphaga (Coleoptera) have egg bursters on the thoracic or abdominal segments of the first-stage larva. For example, larvae of the pollen beetle, *Meligethes aeneus*, have a tooth on each side of the mesonotum and metanotum, while larval tenebrionid beetles have a small tooth on either side of the tergum of these segments and the first eight abdominal segments (Fig. 15.1d).

It is not clear how these various devices function and in some insects (e.g., *Agabus*, the egg of which has a soft chorion), they may no longer be functional. In other cases they appear to be pushed against the inside of the shell until it is pierced and then a slit is cut by appropriate movements of the head. The larva of the fruit fly *Bactrocera tryoni* (Diptera) uses its mouth hooks in a similar way, repeatedly protruding them until they tear the chorion. The blades in lice and the spines in the bed bug, *Cimex lectularius*

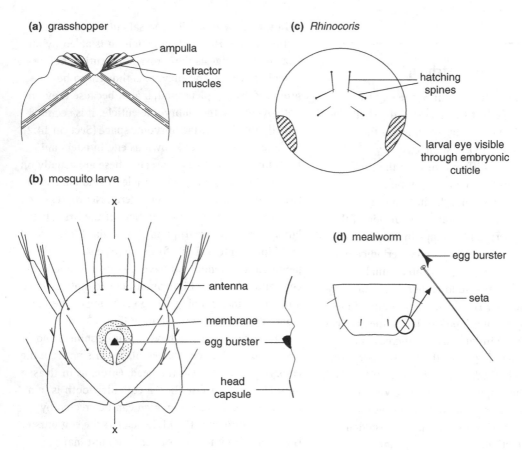

(a) grasshopper

ampulla
retractor muscles

(b) mosquito larva

X

antenna

membrane

egg burster

head capsule

X

(c) *Rhinocoris*

hatching spines

larval eye visible through embryonic cuticle

(d) mealworm

egg burster

seta

Figure 15.1 Hatching devices. (a) Cross-section through the neck membrane of a grasshopper showing the ampullae (after Bernays, 1972a,b). (b) The head of a first-instar larva of a mosquito. The position of the egg burster is shown from above and in a diagrammatic vertical section along the line XX (after Marshall, 1938). (c) Embryonic cuticle over the head of a sucking bug showing the hatching spines. The cuticle is shed as soon as the larva escapes from the egg (*Rhinocoris*, Heteroptera) (after Southwood, 1946). (d) Position of the egg burster spine at the base of a hair (seta) on the eighth abdominal segment of the first-instar mealworm (*Tenebrio*, Coleoptera) (after van Emden, 1946).

(Hemiptera), are used to tear the vitelline envelope, the chorion then being broken by pressure.

Larval Lepidoptera gnaw their way through the chorion and, after hatching, continue to eat the shell until only the base is left. In *Pieris brassicae* (Lepidoptera), where the eggs are laid in a cluster, newly hatched larvae may also eat the tops off adjacent unhatched eggs.

When the egg is enclosed in an ootheca, the larva escapes from this after leaving the egg. In cockroaches (Blattodea) the ootheca is split open

before hatching by the swelling of the eggs as they absorb water. The first-stage larvae of grasshoppers, still enclosed in the embryonic cuticle, wriggle through the frothy plug secreted above the eggs (see Fig. 13.21b). The head is thrust forwards through the froth by elongating the abdomen, the tip of which is pressed against the substrate to give a point of support, while the cervical ampullae are withdrawn. Then, when elongation is complete, the cervical ampullae are expanded to give a purchase while the abdomen is drawn up. By repeating this cycle the larva tunnels its way to the soil surface following the line of least resistance offered by the soft plug.

Special muscles which assist the hatching process are known to occur in some insects. They break down soon after hatching.

15.1.2 Intermediate molt

In those insects which possess an embryonic cuticle, this separates from the underlying epidermis some time before hatching, but it is not shed, so the insect hatches as a pharate first-stage larva (Section 15.2). The embryonic cuticle is shed during or immediately after hatching in a process commonly known as the intermediate molt. In bed bugs, *C. lectularius*, or lice, *Pediculus humanus*, as the larva emerges from the egg, it swallows air and, by further pumping, splits the embryonic cuticle over the head. The cuticle is shed as the larva continues to hatch because, in heteropteran Hemiptera, it is attached inside the chorion at two or three places.

The intermediate molt in acridids (Orthoptera) begins as the newly hatched larva reaches the surface of the soil. Ecdysis behavior is triggered by the lack of all-round contact with the substrate, which the larva has experienced up to that time. The larva swallows air, increasing its volume by about 25%, and, as it does so, waves of shortening and lengthening pass along the body of the larva causing it to move forwards within the intact embryonic cuticle. It maintains its forward position inside the embryonic cuticle by backwardly directed spines on the abdominal sternites. As a result of these movements the embryonic cuticle is pulled taut over the head and thorax, eventually splitting mid-dorsally. The first-stage larval cuticle expands so that the insect swells out of the embryonic cuticle, which is worked backwards. It is held at the tip of the abdomen by two spiny knobs, called brustia, at the bases of the cerci so that the hindlegs can be withdrawn and then it is finally kicked off by the hindlegs.

15.1.3 Hatching stimuli

The stimuli that promote hatching are largely unknown and, in many cases, insects appear to hatch whenever they reach the appropriate stage of development. Even in these instances, however, it is possible that some external stimulus influences hatching.

Suitable temperatures are necessary for any insect to hatch and there is a threshold temperature below which hatching does not occur. This temperature varies from species to species, but is about 8°C for *Cimex*, 13°C for the milkweed bug, *Oncopeltus fasciatus* (Hemiptera), and 20°C for the desert locust, *Schistocerca gregaria*. It is independent of the threshold temperature for full embryonic development, which may be either higher, as in *Cimex* (13°C), or lower, as in *Schistocerca* (about 15°C). The failure of fully formed larvae to hatch at low temperatures is related to inactivity. Newly hatched *Schistocerca* larvae, for instance, are not normally active below about 17°C and their activity remains sluggish below 24°C, and *Cimex* are not normally active below 11°C. Further, temperatures must be sufficiently high for the enzyme digesting the serosal cuticle to function efficiently (Section 15.1.1).

If temperature is suitable and the larva is fully formed, hatching may be induced by specific environmental stimuli. For example, larvae of the stem-boring moth, *Chilo partellus* (Lepidoptera), hatch just after dawn in response to the high light intensity (Fig. 15.2a). Among aquatic insects, the eggs of *A. aegyptus* mosquitoes hatch when immersed in deoxygenated water – the lower the oxygen tension the greater the percentage hatching, but responsiveness varies with age. The larvae are most sensitive soon after development is complete and will then hatch even in aerated water, but if they are not wetted for some time they will only hatch at very low oxygen tensions. Low oxygen tension is perceived by a sensory center in the head or thorax and maximum sensitivity coincides with a

(a) *Chilo*

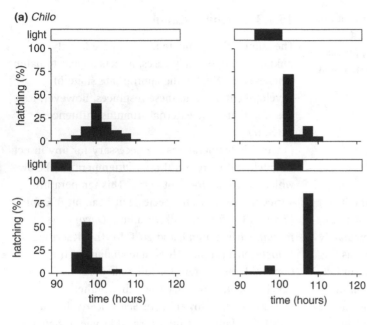

(b) *Schistocerca*

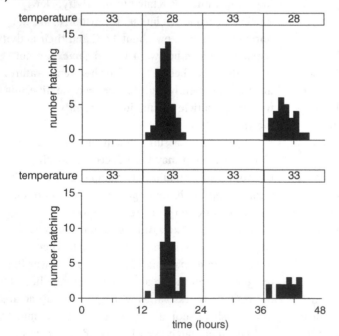

Figure 15.2 Hatching stimuli. (a) Direct response to light by unhatched caterpillars of the moth, *Chilo partellus*. Light (white) and dark (black) periods are shown by the bar above each graph. In continuous light (top left) the insects hatched over an extended period with a peak after about 100 hours of egg incubation. "Lights on" after about 94 hours of incubation stimulated hatching (bottom left), darkness after this delayed it (right-hand graphs). In all cases, eggs were maintained at a constant temperature of 30°C with 12-hour light:12-hour dark cycles up to the time of the experiment (based on Chapman *et al.*, 1983). (b) Entrainment to a temperature cycle leading to hatching by larvae of the desert locust, *Schistocerca gregaria*. Insects in the upper panel experienced cycling temperatures throughout life. Insects in the lower panel experienced cycling temperatures up to the beginning of the experiment (time 0 in figure), but then were maintained at 33°C. They hatched at the same time as insects experiencing cycling temperatures despite the fact that they were experiencing a constant temperature. The eggs were in constant darkness throughout the experiment (after Padgham, 1981).

period of maximum activity of the central nervous system, as indicated by the concentration of acetylcholine. Low oxygen tension has completely the opposite effect on the hatching of *Agabus* beetle larvae, which only occurs in oxygenated water. Some damselflies, *Lestes* spp. (Odonata), lay their eggs above the water. Fully formed larvae hatch when they are wetted.

In other species, hatching at a particular time of day results from the entrainment of activity to the cycle of changing light or temperature over the previous days. Unhatched larvae of the pink bollworm moth, *Pectinophora gossypiella* (Lepidoptera), for example, are entrained to the light cycle in the days before hatching so that they hatch at dawn. They hatch at the appropriate time of day even if they are in continuous light for the last day of development. Similarly, the hatching of larval *S. gregaria* locusts at dawn is not an immediate response to the changing temperature of the surrounding soil, but results from an entrainment of activity to the 24-hour cycle of temperature changes during the last days of embryonic development (Fig. 15.2b).

15.2 Larval development

Postembryonic development is divided into a series of stages, each separated from the next by a molt. The form that the insect assumes between molts is known as an instar; that follows hatching or the intermediate molt, being the first instar, which later molts to the second instar, and so on until at a final molt the adult or imago emerges. No further molts occur once the insect is adult except in Apterygota and non-insect hexapod groups. The periods between molts are also called stages, or stadia – for example, first larval stage, second larval stage and so on – although the term instar is commonly used to refer to this period as well as to the form of the insect during the period.

During larval development there is usually no marked change in body form, each successive stage being similar to the one preceding it, but the degree of change from last-stage larva to adult varies considerably and may be very marked. This change is called metamorphosis, and can be related to the loss of adaptive features peculiar to the larva, and the gain of features peculiar to the adult; the extent of change that occurs is a reflection of the degree of ecological separation of the larva from the adult.

In every case where the matter has been investigated in insects, metamorphosis is the consequence of a molt that is initiated in the absence of an adequate level of juvenile hormone (JH). As the name implies, the role of JH in larval stages is to prevent developmental progress from the larval stage toward adulthood, including toward the pupa where the insect's life history includes a pupal stage. Effectively, JH is a "morphostatic" hormone, and it prevents metamorphosis.

Insects can be grouped into three categories – ametabolous, hemimetabolous or holometabolous – according to the extent of the change at metamorphosis. Ametabolous insects have no metamorphosis, the adult form resulting from a progressive increase in size of the larval form. This is characteristic of the Apterygota and hexapods other than insects in which the larva hatches in a form essentially like the adult apart from its small size and lack of development of genitalia. At each molt the larva grows bigger and the genitalia develop progressively. Adults and larvae live in the same habitat.

In hemimetabolous or direct-developing insects (e.g., Orthoptera, Isoptera, Hemiptera), the larva hatches in a form that generally resembles the adult except for its small size and lack of wings and genitalia (Fig. 15.3), but, in addition, usually with some other features which are characteristic of the larva but which are not present in the adult. At the final molt these features are lost; quantitative analysis of growth changes shows a gradual transformation through the larval instars and a sharp discontinuity at the molt from larva to adult. This discontinuity applies not only to typical adult features such as the wings and genitalia, but to other features that are not regarded as typically adult. In the blood-sucking bug *Rhodnius prolixus* (Hemiptera), larval cuticle with its stellate folds and abundant plaques bearing sensilla is replaced by adult cuticle, which has transverse folds, a few sensilla and no plaques (Section 15.3.1).

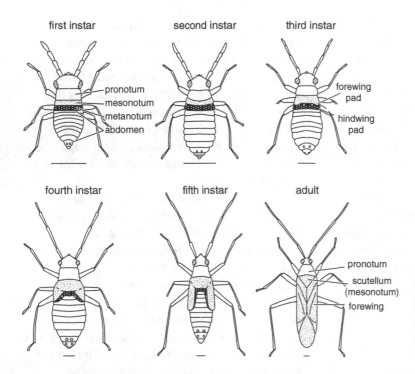

first instar

second instar

third instar

pronotum
mesonotum
metanotum
abdomen

forewing pad

hindwing pad

fourth instar

fifth instar

adult

pronotum

scutellum (mesonotum)

forewing

Figure 15.3 Successive stages in the postembryonic development of a hemimetabolous insect. The horizontal line under each stage represents 0.5 mm (*Cyllecoris*, Heteroptera) (after Southwood and Leston, 1959).

Plecoptera, Ephemeroptera and Odonata all have aquatic larvae and their specific larval adaptations are much more marked than in the previous groups. Hence these forms undergo a more conspicuous metamorphosis involving, among other things, loss of the larval gills (Fig. 15.4). The general body form, nevertheless, resembles that of the adult and these insects are also regarded as hemimetabolous.

By contrast, in holometabolous insects, the usually soft-bodied larvae are mostly quite unlike the adult and a pupal stage is present between the last larval stage and the adult (Fig. 15.5). The pupa is characteristic of holometabolous development, which occurs in all the Neuroptera, Trichoptera, Lepidoptera, Coleoptera, Hymenoptera, Diptera and Siphonaptera.

The morphological appearance of an insect, together with the way in which it interacts with its environment, constitute its phenotype; effectively the embryonic, larval, pupal and adult stages of a holometabolous insect are successively expressed alternative phenotypes. Many genes contribute to the specification of each of these phenotypes in any one insect, and the regulated expression of sets of genes associated with the larval, pupal and adult stages constitutes the essence of the holometabolous life history. A key finding illuminating the nature of metamorphosis is that the gene encoding the transcription factor *broad complex* (*brc*) is expressed during molting in several insects in response to ecdysteroid molting hormones, but only in the absence of JH. The gene product is required for the insect to develop a normal pupal phenotype at the larval–pupal molt, and then again for a normal adult phenotype at the time of the pupal–adult molt. Expression of *brc* when it would not normally be present (i.e., before the last larval stage) causes pupal genes to be expressed prematurely. *brc* thus appears to be a key determinant of metamorphic developmental progression.

Among the insect groups that typically have a hemimetabolous development, a few have life histories somewhat analogous to those of holometabolous insects. Hemiptera have a typical

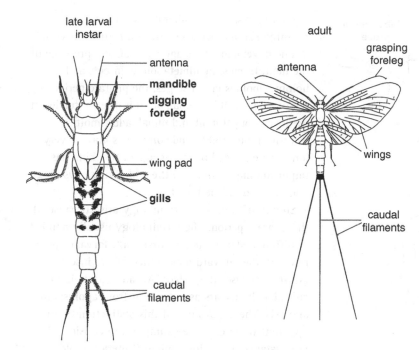

late larval
instar

antenna

mandible

**digging
foreleg**

wing pad

gills

caudal
filaments

adult

grasping
foreleg

antenna

wings

caudal
filaments

Figure 15.4 Late larval instar and adult of an aquatic hemimetabolous insect showing conspicuous adaptive features in the larva. These features are indicated in heavy type (*Ephemera*, Ephemeroptera) (Kimmins, 1950; after Macan, 1961).

hemimetabolous development, but in Thysanoptera, which are phylogenetically close to Hemiptera, the last two larval stages do not feed, and the final stage, which is often called a pupa, is sometimes enclosed in a cocoon. Both of these stages have external wing pads, but the earlier larval stages have none; instead, some development of the wings occurs internally. Within the Hemiptera, the last two larval stages of male scale insects (Coccoidea) also do not feed. These insects clearly have a life history that can be regarded as holometabolous even though they are phylogenetically far removed from the majority of holometabolous insects. Also within the Hemiptera, the final larval stages of whiteflies (Aleyrodidae) feed for a short period but then have an extended non-feeding period during which the pharate adult develops. This stage is commonly known as a pupa, and some texts refer to whitefly development as holometabolous, although there is really no resemblance to the holometabolous development in other groups.

Although the morphological changes that accompany the transition from the larval to the adult condition in hemimetabolous insects are much less extensive than in Holometabola, JH has a very similar morphostatic role in regulating development. The presence of JH maintains the immature larval condition, and its absence permits developmental progression to the adult stage.

To explain the origin of the holometabolous life history, it has been proposed that the pupa corresponds to the larval condition of Hemimetabola, while the holometabolous larva is the key evolutionary innovation, having its origin in the late-embryonic pronympal stage of direct-developing insects. It is suggested that this is due to the advancement of JH secretion into embryonic life in holometabolous insects, preventing an embryonic transition from pronymph to nymph that is normally experienced in direct-developing insects. This "pronymph hypothesis" is strongly supported by the finding that the metamorphic gene *brc* is expressed in the embryos of hemimetabolous insects in two restricted phases, first during early development, and then again during the pronymph–nymph transition. In accordance with this theory, *brc*

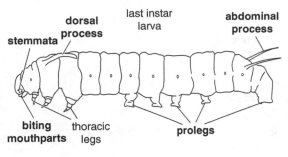

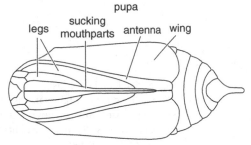

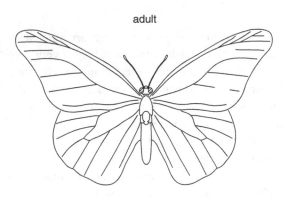

Figure 15.5 Successive stages in the postembryonic development of a holometabolous insect. Adaptive features of the larva are indicated in heavy type. Only one larval stage is shown (*Danaus*, Lepidoptera) (after Urqhart, 1960).

is expressed continuously in the ametabolous firebrat, *Thermobia domestica* (Apterygota, Thysanura).

The term "molt," as commonly used, includes two distinct processes: apolysis, the separation of the epidermis from the existing cuticle; and ecdysis, the casting of the old cuticle after the production of a new one. In the interval between apolysis and ecdysis

the insect is said to be pharate (from the Greek, meaning shroud) because during this time the cuticle of one developmental stage conceals the presence of the next. In most hemimetabolous insects and holometabolous larval stages, the pharate period is relatively short (Fig. 15.6a), but it is often extended at the larval–pupal molt and pupal–adult molt (eclosion) (Fig. 15.6b) and sometimes may be very long. An extended pharate stage of this kind has important implications for the underlying physiology and behavior of the insect.

Some difficulties in terminology arise because of the pharate period. The terminology usually applied to different stages, such as first-stage larva, or pupa, refers to the outward appearance of the insect between ecdyses (Fig. 15.6). An alternative is to consider the instars as extending from apolysis to apolysis. The importance of this distinction becomes apparent when extreme examples are considered. For instance, in cyclorrhaphan Diptera the old cuticle of the last larval stage forms the puparium and it is from this that the adult emerges. There is a normal pupal instar, but since it is always enclosed within the old larval cuticle it is in fact a pharate pupa (Figs. 15.6c, 15.16b). Similarly, some moths diapause as adults within the pupal cuticle, so that although this stage outwardly appears to be a pupa, it is actually a pharate adult. In practice, the pharate stage is not always readily recognizable without dissection, and the more usual terminology, with each stage extending from ecdysis to ecdysis, has been retained in this book, but with reference to the pharate condition where it is obviously important.

15.2.1 Types of larvae

Larvae of hemimetabolous insects essentially resemble adults; they are sometimes called "nymphs" to distinguish them from the more radically different larvae of holometabolous insects. The most conspicuous difference between hemimetabolous and holometabolous larvae is in the development of the

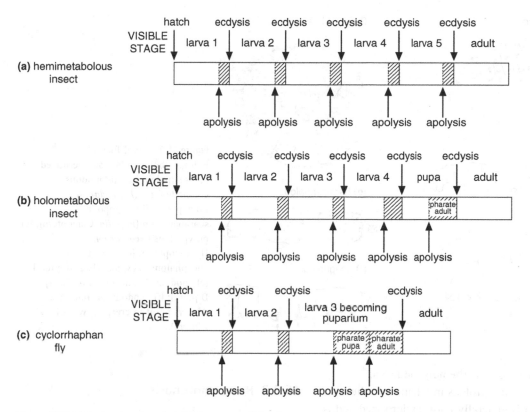

Figure 15.6 Diagram showing the visible stages of development and the times of apolysis and ecdysis, defining the pharate periods (hatched), in: (a) a hemimetabolous insect; (b) a holometabolous insect; (c) a cyclorrhaphan fly in which the pupa is always concealed within the puparium.

wings. In the former, the wings develop as external buds which become larger at each molt, finally enlarging to form the adult wings (Fig. 15.3). In the latter, however, the wings (and often other appendages) develop in invaginations of the epidermis beneath the larval cuticle and so are not visible externally (see Figs. 15.20, 15.21). The invaginations are finally everted so that the wings become visible externally when the larva molts to a pupa (Fig. 15.5). In this text, the term "larva" is used for the immature stages of both hemimetabolous and holometabolous insects to emphasize the basic similarities in physiological regulation of the different forms (see Section 15.4).

Many different larval forms occur among the holometabolous insects. The least modified with respect to the adult is the oligopod larva, which is hexapodous with a well-developed head capsule and mouthparts similar to the adult, but no compound eyes. Two forms of oligopod larvae are commonly recognized: a campodeiform larva, which is well sclerotized, dorso-ventrally flattened and is usually a long-legged predator with a prognathous head (Fig. 15.7a); and a scarabaeiform larva, which is fat with a poorly sclerotized thorax and abdomen, and which is usually short-legged and inactive, burrowing in wood or soil (Fig. 15.7b). Campodeiform larvae occur in the Neuroptera, Trichoptera, Strepsiptera and some Coleoptera, while scarabaeiform larvae are found in the Scarabaeoidea and some other Coleoptera.

(a) campodeiform

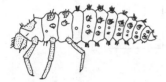

(b) scarabaeiform

(c) polypod

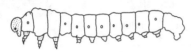

(d) eucephalous

(e) hemicephalous

(f) acephalous

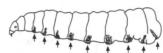

Figure 15.7 Larval forms of holometabolous insects. Sclerotized parts stippled. (a) Oligopodous, campodeiform (*Hippodamia*, Coleoptera); (b) oligopodous, scarabaeiform (*Popillia*, Coleoptera); (c) polypodous (*Neodiprion*, Hymenoptera); (d) apodous, eucephalous (*Vespula*, Hymenoptera); (e) apodous, hemicephalous (*Tanyptera*, Diptera); (f) apodous, acephalous. Arrows point to creeping welts (*Musca*, Diptera).

A second basic form is the polypod larva, which has abdominal prolegs in addition to the thoracic legs. It is generally poorly sclerotized and is a relatively inactive form living in close contact with its food (Fig. 15.7c). The larvae of Lepidoptera, Mecoptera and Tenthredinidae (sawflies; Hymenoptera) are of the polypod type.

The third basic form is the apodous larva, which has no legs and has a poorly sclerotized cuticle. Several different forms can be recognized according to the degree of sclerotization of the head capsule:

(1) Eucephalous – with a well-sclerotized head capsule (Fig. 15.7d); found in Nematocera (Diptera), Buprestidae and Cerambycidae (Coleoptera) and Aculeata (Hymenoptera).

(2) Hemicephalous – with a reduced head capsule which can be retracted within the thorax (Fig. 15.7e); found in Tipulidae and orthorrhaphous Brachycera (Diptera).

(3) Acephalous – without a head capsule (Fig. 15.7f); characteristic of the maggots of Cyclorrhapha (Diptera).

15.2.2 Heteromorphosis

In most insects, development proceeds through a series of essentially similar larval forms leading up to metamorphosis, but sometimes successive larval instars have quite different forms. Development which includes such marked differences is termed heteromorphosis. (Hypermetamorphosis is a term sometimes used for this type of development, but this implies the use of the term metamorphosis to refer to change of form throughout the life history rather than restricting it to the larva–adult transformation.) Heteromorphosis is common in predaceous and parasitic insects in which a change in habit occurs during the course of larval development. Two types of heteromorphosis occur, one in which the eggs are laid in the open and the first-stage larva searches for the host, and a second in which the eggs are laid in or on the host.

In the first type, the first-stage larva is an active campodeiform larva (Fig. 15.8a). In Strepsiptera this larva attaches itself to a host, often a bee or a sucking

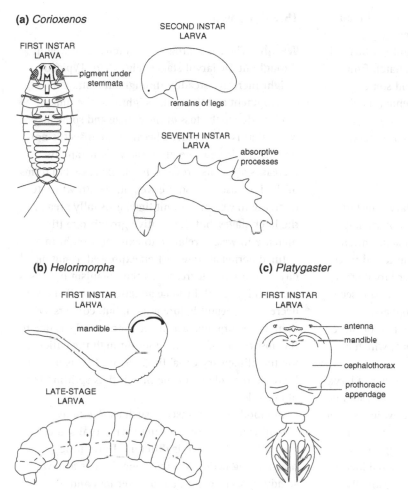

(a) *Corioxenos*

FIRST INSTAR LARVA

SECOND INSTAR LARVA

pigment under stemmata

remains of legs

SEVENTH INSTAR LARVA

absorptive processes

(b) *Helorimorpha*

FIRST INSTAR LARVA

mandible

LATE-STAGE LARVA

(c) *Platygaster*

FIRST INSTAR LARVA

antenna

mandible

cephalothorax

prothoracic appendage

Figure 15.8 Heteromorphosis. (a) Larval stages of *Corioxenos* (Strepsiptera) (after Clausen, 1940); (b) first and late larval stages of *Helorimorpha* (Hymenoptera) (after Snodgrass, 1954); (c) first-instar larva of *Platygaster* (Hymenoptera).

bug, when the latter visits a flower in which the larva is lurking. Subsequently, it becomes an internal parasite and loses all trace of legs, while developing a series of dorsal projections which increase its absorptive area. Later, in the sixth and seventh stages it develops a cephalothorax. A basically similar life history with an active first-stage larva followed by inactive parasitic stages occurs in Mantispidae (Neuroptera), Meloidae and some Staphylinidae (Coleoptera), Acroceridae, Bombyliidae and Nemestrinidae (Diptera), Perilampidae and Eucharidae (Hymenoptera) and Epipyropidae (Lepidoptera). The first-stage larvae of Meloidae and Strepsiptera are sometimes called triungulins

because, in some species, they have three pretarsal claws.

The second type of heteromorphosis occurs in some endoparasitic Diptera and Hymenoptera. Among the parasitic Hymenoptera exhibiting this type of development, the first-stage larva is known as a protopod larva. It has many different forms in different species and is often quite unlike a normal insect (Fig. 15.8b,c). The first-stage larva of *Helorimorpha* spp. (Hymenoptera, Braconidae), for instance, has a big head, a small unsegmented body and a tapering tail (Fig. 15.8b). The third-stage larva, on the other hand, is a fairly typical eucephalous hymenopteran larva. In the Platygastridae, the first-stage larva is even more specialized, with an anterior cephalothorax bearing rudimentary appendages, a segmented abdomen and various tail appendages (Fig. 15.8c). These larvae hatch from eggs which contain very little yolk and some authorities regard them as embryos which hatch precociously, but others believe them to be specialized forms adapted to their environment.

15.2.3 Numbers of instars

Primitive insects usually have more larval instars than advanced species. Thus Ephemeroptera may molt as many as 40 times, and cockroaches often have ten molts. In contrast, most hemipteroid insects and endopterygotes have five or fewer larval stages (Table 15.1). Even within a group of related insects there is variation. Among the grasshoppers, Acridoidea, the Pyrgomorphidae have five or more larval instars while the Gomphocerinae, which are also usually smaller, have four.

Despite these generalizations, the number of larval stages through which a species passes is not absolutely constant. In the Orthoptera, the female, which is often bigger than the male, commonly has an extra larval stage, and larvae hatching from small eggs grow slowly and have an additional stage. The red locust, *Nomadacris*, may have six, seven or occasionally eight larval instars, depending on its environment and that of the parents. In *Plusia interpunctella* and some other Lepidoptera, larvae reared in isolation may pass through five, six or seven instars, while nearly all of those reared in a crowd have only five. At least among the Lepidoptera, much larger numbers of molts may occur if the insect has a poor food supply, and the larva of the clothes moth, *Tineola bisselliella*, is recorded as molting as many as 40 times. The extra molts are associated with elevated JH levels consequent upon starvation. Not all insects have this flexibility to undertake extra molts, and in most Hemiptera and cyclorrhaphous Diptera and many Hymenoptera and Coleoptera, the number of instars is constant.

15.2.4 Growth

Weight There is a progressive increase in weight throughout the larval stages (Fig. 15.9). Typically, weight increases steadily throughout a stage of development and then falls slightly at the time of molting due to the loss of the cuticle and some loss of water that is not replaced because the insect is not feeding. Following the molt, the weight rapidly increases above its previous level. Expressed in terms of the increase in absolute weight, the growth rate (increase in weight per unit time) is usually greater in the later stages, but the relative growth rate (the increase in weight relative to existing weight in a defined period of time – often expressed as mg mg^{-1} day^{-1}) normally decreases as the organism increases in size (Fig. 15.9a). In some aquatic insects there is no decrease in weight before a molt, but, conversely, there is a sharp increase at the time of ecdysis due to the absorption of water, either through the cuticle or via the alimentary canal (Fig. 15.9c). This is used to increase the volume of the insect, thus splitting the old cuticle.

In blood-sucking insects, such as *Rhodnius*, which feeds only once during each larval stage, the pattern of growth is different. During the non-feeding period of each stage there is a slow, steady loss in weight due to water loss and respiration, but feeding is accompanied by a sharp increase in weight followed by a fairly rapid fall as water is eliminated. There is, of course, a net increase in weight from the beginning of one instar to the beginning of the next.

The final weight of the adult insect often varies according to the conditions under which the larva develops. Rapid development at high temperatures results in adults that are relatively light in weight. This may be due to restriction of the period available for feeding, but food utilization may also be more efficient at lower temperatures because metabolism proceeds at a lower rate. Crowding, possibly through its effect on the rate of development, also influences adult size, insects from crowded conditions being

Table 15.1 **Numbers of larval stages in different orders of insects. The commonly occurring numbers are given. Numbers outside this range may occur under some conditions**

Order	Common name	Number of larval stages
Archaeognatha		10–14
Thysanura	Bristletails	9–14
Ephemeroptera	Mayflies	20–40
Blattodea	Cockroaches	6–10
Mantodea	Mantids	5–9
Grylloblattodea	Rock crawlers	8
Orthoptera	Crickets	5–11
Phasmida	Stick insects	8–12
Isoptera	Termites	5–11
Dermaptera	Earwigs	4–6
Embioptera	Webspinners	4–7
Plecoptera	Stoneflies	22–23
Hemiptera	Bugs	3–5
Thysanoptera	Thrips	5–6
Psocoptera	Booklice	6
Phthiraptera	Lice	3–4
Neuroptera	Lacewings	3–5
Mecoptera	Scorpionflies	4
Siphonaptera	Fleas	3
Diptera	Flies	3–6
Trichoptera	Caddis flies	5–7
Lepidoptera	Butterflies	5–6
Coleoptera	Beetles	3–5
Hymenoptera	Bees	3–6

smaller than others reared in isolation. For example, in one experiment on *Locusta migratoria* (Orthoptera), adult females from larvae reared in isolation weighed 1.5 g, while others from larvae reared in crowds weighed only 1.2 g. Where isolation is correlated with the production of an extra larval instar, the difference in weight may be even more marked. Finally, adult weight may be influenced by

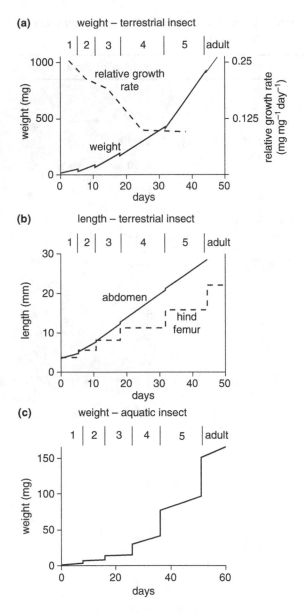

Figure 15.9 Larval growth patterns in hemimetabolous insects. Numbers at the top indicate successive larval stages and vertical lines the times of molting. (a) Change in weight and relative growth rate of a terrestrial insect (*Locusta*, Orthoptera) (data from Duarte, 1939). (b) Changes in linear dimensions of a terrestrial insect. The abdomen increases almost linearly by unfolding of intersegmental membranes, but the hind femur only increases in length at each molt (*Locusta*, Orthoptera) (after Clarke, 1957). (c) Change in weight of an aquatic insect (*Notonecta*, Heteroptera) (from Wigglesworth, 1965).

the food on which the larva has been nourished. This is particularly well illustrated in phytophagous insects, such as the grasshopper *Melanoplus sanguinipes* (Orthoptera), in which the weight of females varies from 140 mg to 320 mg, depending on the food available to the larvae. However, insects in which the timing of metamorphosis is determined by size (see below) are much less variable in size as adults.

Increase in size of the cuticle Fully sclerotized cuticle is stiff and inextensible, so growth in size (i.e., surface dimensions) of such hard body parts only occurs when an insect molts, the new, soft cuticle being expanded at the time of ecdysis. Consequently, these parts increase in size in a series of steps (Fig. 15.9b). Membranous regions can expand, however, by the pulling out of folds. Thus a structure with a wholly membranous cuticle, or one, such as the abdomen of *Locusta*, in which the membranes are extensible, may grow continuously. Other regions in which there is rather less membrane show an intermediate type of growth, with some surface extension occurring in the course of each stage, together with a marked increase at each molt. It should be noted, however, that the cuticle usually grows in thickness between molts, whether or not increase in surface size occurs; this occurs both by the addition of new layers to the inner (epidermal) surface of the cuticle, and by the incorporation of additional materials into existing layers (this is termed "intussusception").

In many hemimetabolous and ametabolous insects, the number of annuli in the antennal flagellum increases during postembryonic life. Thus, first-instar larvae of *Dociostaurus marrocanus* (Orthoptera) have antennae with 13 annuli, while in adults there are 25. The new annuli are added basally by division of the most proximal annulus of the flagellum, known as the meriston. In orthopteroid insects and Odonata some of the annuli adjacent to the meriston, and derived from it, may also divide. They are known as meristal annuli. At each molt in the cockroach

Periplaneta americana (Blattodae) the meriston divides to produce 4–14 new annuli and each meristal annulus may divide once.

Where the cerci are segmented, as in Blattodea and Mantodea, they increase in size at successive molts by the addition of segments basally.

It is often true that different parts of the body grow at different rates when compared with some standard such as total body length. If the rate of growth of the part, such as the head, is the same as that of the standard, growth is said to be isometric; if the rate is different, growth is said to be allometric. Relatively faster growth is said to be positive allometry, while slower growth is negative allometry. For example, in

the parasitic dermapteran *Hemimerus vicinus*, the meriston with the meristal annuli grows faster than the antenna as a whole, so that in the adult it contributes a greater proportion of the length than it does in the earlier larval instars (Fig. 15.10a). Conversely, the five apical segments grow more slowly than the whole antenna, so their final contribution is proportionately less than originally.

The straight-line relationship between two parts on a log-log plot as illustrated in Fig. 15.10a occurs if growth rates are consistent over time, but this is not always the case. The mesothorax of *Dysdercus fasciatus* (Hemiptera) increases in size at roughly the same rate as the body as a whole at the first molt, but

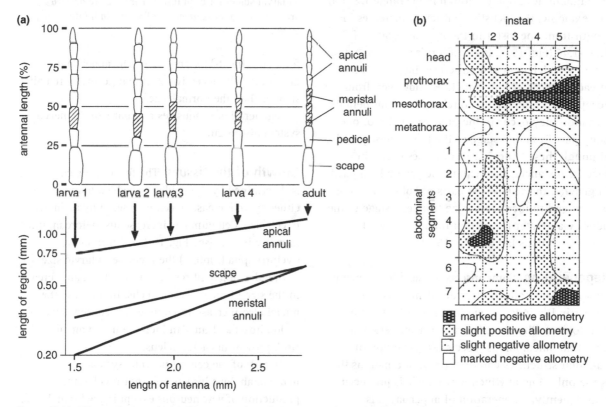

Figure 15.10 Allometric growth. (a) Changes in the relative proportions of different parts of the antenna of a hemimetabolous insect in different stages of development. Above: representation of the whole antenna with lengths of different parts shown as a percentage of the whole; below: changes in lengths of different parts plotted against the length of the whole antenna (log-log scale) (*Hemimerus*, Dermaptera) (after Davies, 1966). (b) Diagram showing the growth rates of different parts of the body relative to the growth of the body as a whole from one larval stage to the next in a hemimetabolous insect (*Dysdercus*, Heteroptera) (after Blackith *et al.*, 1963).

at later molts it exhibits positive allometry
(Fig. 15.10b); the seventh abdominal segment
increases at about the same relative rate as the whole
body in the early stages, but also exhibits marked
positive allometry at the final molt, as the genitalia
develop. In none of the segments is growth relative to
the body as a whole uniform throughout larval life.
Similar variations in relative growth rate occur in
Ectobius spp. (Blattodae) and are probably usual in
larvae of hemimetabolous insects.

Dyar's law suggests that a particular sclerotized
part of the body (such as the head capsule or a leg
segment) increases in linear dimensions from one
instar to the next by a ratio which is constant
throughout development (often in the range 1.2–1.4).
For example, the locust's hind femur increases in
length from one instar to the next by a ratio of
1.36–1.46 (Fig. 15.9). Przibam's rule suggests that the
ratio by which cuticular structures increase in length
at each ecdysis is constant (1.26). However, from the
account given above it is obvious that considerable
variation may occur in the increase in size of one
body part from one stage to the next. Consequently,
although these "rules" are sometimes useful in
specific instances, they cannot be generally applied.
In general, the sclerotized parts of holometabolous
larvae tend to increase in size from one stage to the
next by a greater proportion than do those of
hemimetabolous larvae (Fig. 15.11).

Regeneration Some insects are able to regenerate
an appendage following its accidental loss. This
occurs in Blattodae, some Orthoptera (but not in
Acridoidea), Phasmatodea, some Hemiptera and
larval holometabolous insects. Regeneration of
cuticular structures can only occur at a molt as this
is the only time at which new cuticle is produced.
Consequently, regeneration of appendages is
restricted to larval stages and does not occur in
adults. If the loss occurs early in a developmental
stage, before the production of molting hormone, the
appendage reforms at the next molt, but replacement
of a limb lost after this does not occur until the

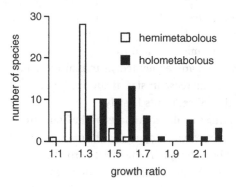

Figure 15.11 Average growth ratios from one larval
stage to the next for a range of hemimetabolous and
holometabolous insects. The growth ratio is the ratio of size
in one stage to the size in the previous stage averaged over
all larval stages for each insect. The data are derived
primarily from measurements of head width (after Cole,
1980).

following (next-but-one) molt. The appendage
grows at each successive molt, but generally remains
smaller than the normal size.

Regeneration of muscles and parts of the nervous
system also occurs.

Growth of the tissues The form of the cuticle
is determined by the epidermis, which may grow
either by an increase in cell number or by an increase
in cell size. Cell numbers increase just before molting
in many insects (see Fig. 16.15), but in larval
Cyclorrhapha (Diptera) the increase in larval size
during development results entirely from an increase
in the size of the epidermal cells. In this case the
nuclei also increase in volume as a consequence of
endomitosis and thus an increase in the amount of
DNA present in each nucleus.

Growth of the central nervous system of
hemimetabolous insects does not involve the
production of new neurons except in the brain. In the
terminal abdominal ganglion of the house cricket,
Acheta domesticus (Orthoptera), for example, there
are about 2100 neurons at all stages of development.
On the other hand, the number of glial cells in the
ganglion increases from about 3400 in the first-stage

larva to 20 000 in the adult and the volume of the ganglion is increased 40-fold. In most holometabolous insects there is extensive reconstruction of the nervous system at metamorphosis and undifferentiated neuroblasts persist through the larval period up to this time (see below).

Marked changes occur in the sensory system during larval development of hemimetabolous insects. Additional mechanoreceptors and chemoreceptors are added to those already present at each molt (Fig. 15.12). The numbers of ommatidia forming the compound eyes also increase. By contrast, the numbers of sensilla are constant through the larval life of holometabolous insects and compound eyes are only present in adults.

The musculature of most larval hemimetabolous insects closely resembles that of adults, and probably all the adult muscles, including those that become flight muscles, are represented at hatching. In addition, there may be muscles that are operative only during molting and which disappear after the final molt. The muscles grow by an increase in fiber size between molts and by the addition of new fibers at molts (see Fig. 10.13). The musculature of larval holometabolous insects is, by contrast, usually fundamentally different from that of the adults. During larval development, muscles increase in size, but there is no basic change in their arrangement.

As with the epidermis, an increase in the size of an internal organ may result from an increase in cell size or in cell number, or sometimes both. The fat body of larval *Aedes aegypti* (Diptera) grows by an increase in cell number, but most other tissues in this insect and in *Drosophila melanogaster* have a constant number of cells and grow by cell enlargement. This enlargement is accompanied by endomitosis. In the midgut both processes occur; the epithelial cells enlarge, but ultimately break down during secretion and each is replaced by two or more cells derived from the regenerative cells. In some insects the whole of the midgut epithelium is replaced at intervals by regenerative cells (see Chapter 3). In general, it

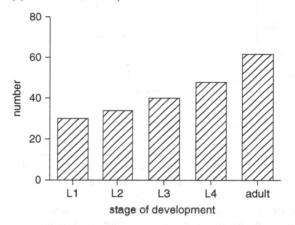

(a) contact chemoreceptors

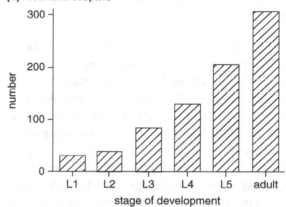

(b) mechanoreceptors

Figure 15.12 Changes in numbers of peripheral sensilla during the development of hemimetabolous insects. (a) Changes in the numbers of contact chemoreceptors on the tip of the maxillary palp of a grasshopper in successive stages of development (*Bootettix*, Orthoptera) (after Chapman and Fraser, 1989). (b) Changes in the numbers of mechanosensitive filiform hairs on the prosternum of a locust in successive stages of development (after Pflüger et al., 1994).

appears that tissues that are destroyed at metamorphosis grow by cell enlargement while those that persist in the adult grow by cell multiplication.

The development of the Malpighian tubules varies. In orthopteroid orders they increase in number throughout larval life. The primary tubules arise as diverticula from the proctodeum in the embryo

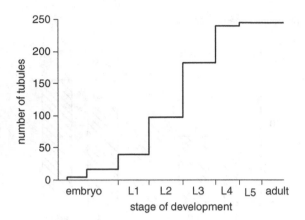

Figure 15.13 Changes in the number of Malpighian tubules during the development of a hemimetabolous insect (*Schistocerca*, Orthoptera) (based on data in Savage, 1956).

(see Fig. 14.24). There are four primary tubules in *Blatta orientalis* (Blattodae); some other insects have six. Secondary tubules develop later, largely postembryonically. The desert locust, *Schistocerca gregaria*, for example, has six primary tubules, but 12 more are added before the larva hatches and more develop in each larval stage up to the adult (Fig. 15.13). Secondary tubules appear as buds at the beginning of each larval stage, but after their initial development they increase in length without further cell division as a result of an increase in cell size. In holometabolous insects, the number of Malpighian tubules remains constant throughout larval life, although they do increase in length by increasing cell size and by cell rearrangement. New tubules only form at metamorphosis.

The increase in volume of internal structures, especially the fat body, is limited by the cuticle. In holometabolous larvae with soft, folded cuticle, considerable growth is possible. Extension of the abdomen by unfolding intersegmental membranes occurs in species with more rigid cuticles. In addition, in grasshoppers, and probably in some other insects, some growth of internal organs occurs at the expense of air sacs, which become increasingly compressed during each developmental stage.

The reverse happens during the pupal stage of holometabolous insects, when air sacs increase in size during the pupal–adult transformation.

Control of growth Larval growth is characterized by periodic molts and to some extent internal changes are correlated with the molting cycle. Molting is initiated by ecdysteroid molting hormone and at larval molts the effect of this hormone is modulated by JH, so larval genes are activated and hence larval characters produced. There is increasing evidence that growth is also regulated by insulin-like hormones.

While hormones exert an overall controlling influence, local factors control the form of particular areas of the body. For example, epidermal cells often show distinct polarity, secreting cuticle in a form giving an obvious anterior–posterior pattern. In the first-stage larva of *Schistocerca*, the cuticular plates associated with each epidermal cell on the sides of the abdominal sternites are produced into backwardly pointing spines; similarly, in *Oncopeltus*, a row of spines marks the posterior end of the area of cuticle secreted by each of the cells forming the abdominal sternites; and the scales of Lepidoptera grow out with a particular orientation. Experimental manipulation shows that the polarity of the cells within a body segment is produced by a gradient of a diffusible substance known as a morphogen. Similar gradients occur in the legs.

In addition to having a particular orientation, cuticular structures are dispersed in regular patterns characteristic of the species. For instance, the abdominal tergites of larval *Rhodnius* bear a number of evenly spaced sensilla. At each molt these increase in number, new sensilla being formed in the biggest gaps between the existing sensilla. This is consistent with the hypothesis that a determining substance present in the epidermis is absorbed by existing sensilla, but accumulates between them if they become widely spaced due to growth of the epidermis. If this concentration exceeds a certain threshold the development of a new sensillum is initiated. The development of sensilla on the adult cuticle of *Oncopeltus* can be accounted for in a similar way.

Where two or more integumental features are present in an integrated pattern they may be controlled by the same substance. In *Rhodnius*, for instance, it is suggested that a differentiating substance (morphogen) in high concentration produces the sensilla and that the same substance in low concentration initiates the development of dermal glands, which are thus arranged round each sensillum. Where the integumental features are not arranged in an integrated manner, as with the hairs and scales on the abdomen of *Ephestia kuehniella* (Lepidoptera), two determining substances might be involved.

In *Drosophila*, and almost certainly in the other insects, the boundaries of the parasegments are sources of signals that organize the patterning and orientation of associated epidermal cellular fields (this is discussed more fully in Chapter 14).

There is relatively little information on the control of growth of internal organs, but some show cyclical activity that coincides with the molt. The fat body cells of *Rhodnius*, for example, exhibit a marked increase in RNA concentration and mitochondrial number just before a molt, and the ventral abdominal intersegmental muscles become fully developed only at this time. In insects in which the Malpighian tubules increase in number, mitosis and development of new tubules are phased with respect to the molt.

15.3 Metamorphosis

The changes that occur in the transformation of the larva to the adult may be more or less extensive depending on the degree of difference between the larva and the adult. In hemimetabolous insects the changes are relatively slight; in holometabolous insects the changes are very marked, and a pupal stage is interpolated between the final larval stage and the adult.

15.3.1 Hemimetabolous insects

Epidermal mitosis and expansion only occurs at the time of a molt and, in hemimetabolous insects, progressive development of the wing buds occurs at each molt. Apart from being small, the larval wing buds differ from the adult wings in lacking a membranous basal region and accessory sclerites. These do not appear until the final molt.

In general, the wings arise in such a way that the lateral margins of the wing buds become the costal margins of the adult wings (Fig. 15.3), but in dragonflies (Odonata) the buds arise in an erect position, the margin nearer the midline ultimately becoming the costal margin (Fig. 15.14a). The wing buds of locusts originate as simple outgrowths of the terga, but at the antepenultimate molt they become twisted into the position found in the Odonata. This twisting results from the lower epidermis growing more rapidly than the upper. At the final molt the wings twist back so that the costal margins of the folded adult wings are ventro-lateral in position.

The genitalia develop progressively by modification of the terminal abdominal segments, but with a more marked alteration at the final molt (Fig. 15.14b).

Although accessory wing sclerites are not developed in the larva, the muscles that are attached to them in the adult are already attached to appropriate positions on the larval cuticle. For instance, in adult *Locusta migratoria* the promotor-extensor muscle of the mesothoracic wing is inserted into the first basalar sclerite, but the equivalent muscle of the metathoracic wing is inserted into both basalar sclerites. In the larva, although the sclerites are not developed, the muscle in the mesothorax has one point of attachment to the pleural cuticle, while that in the metathorax has two.

In grasshoppers and dragonflies all the flight muscles are present in the larva, although some lack striations and are presumably non-functional. These muscles increase in size throughout the larval period and further changes occur in the young adult. Many muscles disappear after the final molt. They are

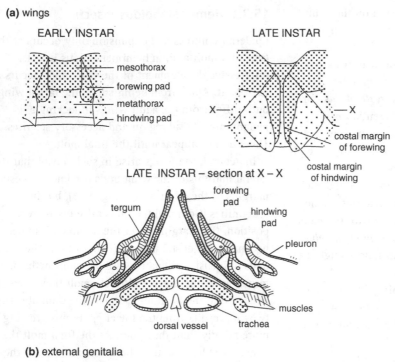

(a) wings

EARLY INSTAR

- mesothorax
- forewing pad
- metathorax
- hindwing pad

LATE INSTAR

X — X

- costal margin of forewing
- costal margin of hindwing

LATE INSTAR – section at X – X

- tergum
- forewing pad
- hindwing pad
- pleuron
- muscles
- dorsal vessel
- trachea

(b) external genitalia

INSTAR 1

- sternum 8
- ventral valve
- sternum 9
- dorsal valve
- paraproct
- cercus

0.5 mm

INSTAR 2

0.5 mm

INSTAR 4

- sternum 8
- ventral valve
- sternum 9
- inner valve
- dorsal valve
- paraproct
- cercus

0.5 mm

INSTAR 5

0.5 mm

ADULT

- sternum 7
- sternum 8 (subgenital plate)
- cercus
- ventral valve
- dorsal valve

0.5 mm

Figure 15.14 Development of adult external structures in hemimetabolous insects. (a) Wings: above: dorsal views of the pterothorax of young and older larvae of a dragonfly; below: transverse section through the dorsal metathorax of a dragonfly larva at the same stage as that shown above right. The section is in the plane marked X–X above right (after Comstock, 1918). (b) Genitalia: ventral view of the tip of the abdomen of a female grasshopper at different stages of development (*Eyprepocnemis*, Orthoptera) (after Jago, 1963).

necessary for the process of molting and their disappearance is accompanied by the death of the motor neurons supplying them. Other changes also occur in the nervous system. For example, changes in the arborization patterns of afferent sensory axons may occur within the central nervous system even though the peripheral sensilla remain functional. The effect of this is to alter the neural pathways of which the sensory neurons form a part, so that they can play a different role in adult behavior.

15.3.2 Holometabolous insects

The development of adult features in holometabolous insects varies with the degree of modification of larval features involved. In Neuroptera and Coleoptera, where the larvae have some resemblance to the adults, relatively little reconstruction occurs, but in Diptera the tissues are almost completely rebuilt following histolysis and phagocytosis of the larval tissues. This reconstruction occurs in the pupa.

Pupa All the features of the adult become recognizable in the pupa, which consequently has a greater resemblance to the adult than to the larva. At the larva–pupa molt, the wings and other features that up to now have been developing internally are everted and become visible externally, although not fully expanded to the adult form (Fig. 15.5). The appendages extend freely from the body in the pupae of some insects – a condition known as exarate – but in other species the appendages are glued down by a secretion produced at the larva–pupa molt. This is the obtect condition and obtect pupae are usually more heavily sclerotized than exarate pupae. A further differentiation can be made on the presence or

absence of articulated mandibles in the pupa. When articulated mandibles are present (the decticous condition), they have apodemes fitting closely within the adult mandibular apodemes (Fig. 15.15) and hence can be moved by the mandibular muscles of the pharate adult. The alternative condition, with immobile mandibles, is known as adecticous.

Decticous pupae are always exarate. They occur in Megaloptera, Neuroptera, Trichoptera and some Lepidoptera. Some adecticous pupae are also exarate as in Cyclorrhapha, Siphonaptera and most Coleoptera and Hymenoptera, but others are obtect. Most Lepidoptera, Nematocera and Orthorrhapha (Diptera), Staphylinidae, some Chrysomelidae (Coleoptera) and many Chalcidoidea (Hymenoptera) have obtect, adecticous pupae.

The last-stage larva is often quiescent for two or three days before the ecdysis to a pupa, and in some cases the insect is a pharate pupa for a part of this time. During this period, the insect is sometimes referred to as a prepupa, but this does not usually represent a distinct morphological stage separated from the last-stage larva by a molt. Such a separate morphological stage known as the prepupa does exist in Thysanoptera and male Coccidae (Hemiptera). In these insects the prepupa is a quiescent stage following the last-stage larva, and it is succeeded by a pupal stage.

Most insect pupae are immobile and hence vulnerable, and most insects pupate in a cell or cocoon, which affords them some protection. Many larval Lepidoptera (e.g., Sphingidae, Noctuidae) construct an underground cell in which to pupate, cementing particles of soil with a fluid secretion from the hindgut. The larva of the puss moth, *Cerura vinula* (Lepidoptera), constructs a chamber of wood fragments glued together to form a hard enclosing layer, and some beetle larvae pupate in cells in the wood in which they bore. Many larvae produce silk which may be used to hold other structures, such as leaves, together to form a chamber for the pupa, but in some other species a cocoon is produced wholly from silk (Fig. 15.16a). Silken cocoons are produced

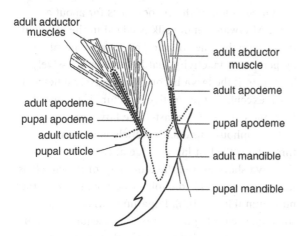

Figure 15.15 Mandible of a decticous pupa showing the pupal apodemes inside the adult apodemes. Adult cuticle shown as dotted lines (*Rhyacophila*, Trichoptera) (after Hinton, 1946).

(a) lepidopteran cocoon

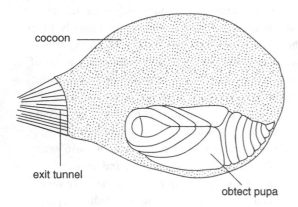

cocoon

exit tunnel

obtect pupa

(b) cyclorrhaphan puparium

exarate pupa

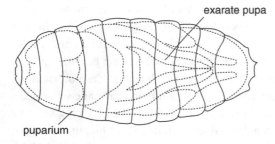

puparium

Figure 15.16 Protection of the pupa. (a) A lepidopteran pupa inside a silk cocoon (*Saturnia*). (b) A pupa (dotted) inside the puparium of a cyclorrhaphan dipteran. The puparium is formed from the cuticle of the last-stage larva.

by Bombycoidea among the Lepidoptera and by Siphonaptera, Trichoptera and some Hymenoptera.

Cocoon formation by the silk moth, *Antheraea pernyi* (Lepidoptera), takes about two days. It follows gut purging, in which the larva expels the gut contents by a series of waves of contraction passing along the abdomen from front to back. The larva subsequently enters an active wandering phase that ends when it finds a suitable site in which to pupate. The first phase of cocoon formation is the construction of a scaffold of silk threads between leaves of the food plant and the production of a stalk which attaches the cocoon to the leaf petiole. Subsequent behavior consists of a series of cycles in

which the insect weaves loops of silk using figure-of-eight movements of the head to construct one end of the cocoon and then turns through 180° to form the other end. A complete layer of silk has been produced after a period of about 14 hours, and the insect turns from one end of the cocoon to the other at much shorter intervals (5 minutes compared with 80 minutes at 23°C) and at the same time coats the inside of the cocoon with a liquid from the anus, containing crystals of calcium oxalate. This liquid accumulates in the hindgut after purging and the calcium oxalate is produced by the Malpighian tubules.

When the salivary glands produce silk, they also secrete compounds that will crosslink the silk proteins in the same way that proteins are linked in the cuticle (see Chapter 16). These tanning agents, which are phenolic compounds derived from the food, are produced as glucosides and so are inactive at the time of synthesis. The salivary gland also produces the enzymes to activate the tanning agents, but, as the silk dries very quickly, activation does not occur until it is rewetted by fluid from the caterpillar's hindgut. At this time the chains of silk protein are linked together and the wall of the cocoon becomes stiff and colored yellow-brown. This period of impregnation of the cocoon lasts for about an hour. Afterwards, more silk is added to the inside of the cocoon, but the spinning cycles are interrupted by periods of inactivity that become progressively longer until the larva becomes completely quiescent.

An exceptional protective structure is produced from the cuticle of the last-stage larva by cyclorrhaphous Diptera. Procuticle is laid down throughout the last larval stage at the end of which the larva shortens and the outer part of the cuticle is tanned to form a rigid ovoid structure known as the puparium (Fig. 15.16b). Puparium formation is sometimes called "pupariation." Subsequently, after apolysis, the pupal/adult head and appendages, previously concealed beneath the larval epidermis, are everted and the pupal cuticle is secreted. The pupa remains inside the puparium and the adult, when it emerges, escapes simultaneously from both the pupal and puparial cuticles.

A few insects form unprotected pupae. These are particularly well known in the Nymphalidae and Pieridae (Lepidoptera), where the pupae are suspended from a silk pad. These exposed pupae exhibit homochromy (Section 25.5.2), whereas protected pupae are normally brown or very pale in color.

The behavior of aquatic insects on pupation varies considerably. Some larvae, such as those of the aquatic Syrphidae and beetles of the Family Hydrophilidae, leave the water and pupate on land, but many others, particularly the aquatic Diptera, pupate in the water. Sometimes the pupae of aquatic Diptera are fastened to the substratum. For example, the pupae of Blepharoceridae have ventro-lateral pads on the abdomen with which they attach themselves to stones, while Simuliidae construct open cocoons attached to stones and rocks (Fig. 15.17a). The pupa projects from the open end of the cocoon, which is constructed more strongly in faster flowing water than it is in a weak current. Chironomidae pupate in their larval tubes or embedded in the mud, while *Acentropus* (Lepidoptera) form a silken cocoon with two chambers separated by a diaphragm. The pupa lies in the lower chamber, which is air-filled.

Other aquatic pupae obtain oxygen from the air, either directly or indirectly. The pupae of most Culicidae and Ceratopogonidae (Diptera) are free-living and active. They are buoyant so that, when undisturbed, they rise to the surface and respire via prothoracic respiratory horns (Fig. 15.17b). If disturbed, movements of the anal paddles drive them downwards. The pupae of some Culicidae and Ephydridae (Diptera) have their respiratory horns embedded in the tissues of aquatic plants, obtaining their oxygen via the aerenchyma.

The occurrence of a pupa is indicative of the broad differences which occur between larval and adult forms of holometabolous insects. It is a stage during which major internal reconstruction occurs and is of particular importance in permitting the development and attachment of adult muscles to the cuticle and the full development of the wings.

Most of the adult thoracic muscles are different from those of the larva. It has been suggested that muscles will only develop in an appropriate form and length if they have a mold (template) in which to do so. The pupa provides such a mold for the adult muscles. Further, most muscles are attached to the cuticle by filamentous projections from the epicuticle (see Chapter 10), which can only develop at a molt, in this case the pupa–adult molt.

(a) *Simulium* **(b)** mosquito

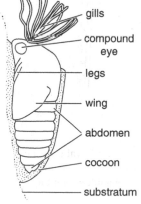
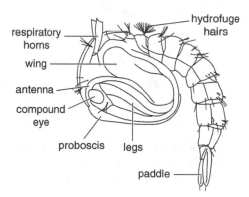

Figure 15.17 Aquatic pupae. (a) Fixed pupa of a simuliid (Diptera). (b) Mobile pupa of a mosquito (Diptera).

Development of the wings internally within the larva is restricted by lack of space; this problem becomes more acute as the insect approaches the adult condition and the flight muscles also increase in size. Thus, development can only be completed after the wings are everted and, for this reason, two molts are necessary in the transformation from larva to adult. At the first, from larva to pupa, the wings are everted and grow to some extent. Further growth occurs and the adult cuticle is laid down at the pupa–adult molt.

The importance of the pupa in wing development and associated changes is emphasized by the absence of a pupal stage in the life histories of female Strepsiptera and Coccidae, which are wingless and larviform. The males in these groups are winged and have a pupal stage (see Section 15.2).

Adult epidermal structures If an adult appendage does not differ markedly from that of the larva, it may be formed by a proliferation of the tissue within and at the base of the larval organ. This occurs in the legs of Lepidoptera, for example. Soon after the larva of the cabbage butterfly, *Pieris brassicae* (Lepidoptera), enters its final stage the epidermis of its thoracic legs becomes separated from the cuticle except at points of muscle attachment, and is now therefore free to thicken and fold. The first thickening, well supplied with tracheae, develops at the junction of the second and third leg joints (Fig. 15.18a) and from this differentiation center a wave of cell multiplication spreads out (Fig. 15.18b,c). As a result of the increase in area, the epidermis becomes folded and a particularly large fold develops basally. Later, when the epidermis expands to form the

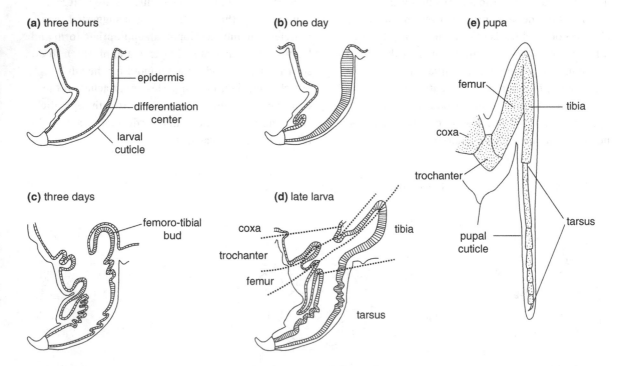

Figure 15.18 Development of the adult leg of a lepidopteran at different times after the molt to the last larval stage and in the pupa (*Pieris*) (based on Kim, 1959). (a)–(d) Changes in the epidermis within the larval leg. The presumptive areas of the adult leg are shown separated by dotted lines in (d). (e) Fully developed leg (shaded) of the pharate adult within the pupal cuticle.

pupal leg, this basal fold becomes divided by a longitudinal septum to form the femur and tibia. Epidermis from the more proximal parts of the larval leg forms the coxa and trochanter, and more distal tissue forms the tarsus. Further differentiation continues in the pupa to produce the adult leg (Fig. 15.18e).

Where the difference between larval and adult organs is more marked the adult tissues develop from epidermal thickenings called imaginal buds or discs. The founder cells comprising these discs divide to form clones. Since the production of adult organs is restricted to small groups of cells the remainder of the epidermis is free to undergo larval modifications. The discs may be regarded as islands of embryonic tissue that remain undifferentiated until they give rise to the adult structures. They do not produce cuticle in the larva and the cells continue to divide between molts (Fig. 15.19a), not just at the time of the molt, as occurs with larval epidermal cells.

The imaginal disc commonly becomes invaginated beneath the larval epidermis, forming a cavity known as the peripodial cavity (Fig. 15.20b). It is lined with epidermis (the peripodial membrane) and as the imaginal disc enlarges the appendage forms and pushes into the cavity (Fig. 15.20c), becoming folded as it increases in size. In Lepidoptera, the wings and antennae develop in peripodial cavities, but in Diptera all the main ectodermal features of the adult develop in this way (Fig. 15.19c).

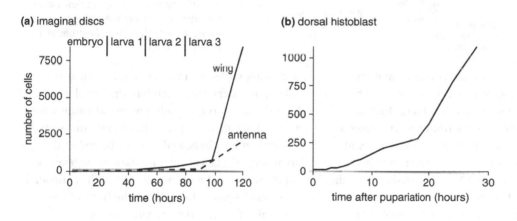

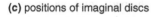

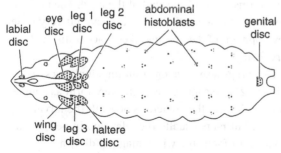

Figure 15.19 Development of adult epidermis in *Drosophila*. (a) Numbers of cells in antennal and wing imaginal discs during the embryonic and larval periods (based on Oberlander, 1985). (b) Number of cells in the anterior dorsal histoblasts during pupal development (after Madhavan and Madhavan, 1980). (c) The positions of imaginal discs and histoblast nests in the larva (after Nöthiger, 1972).

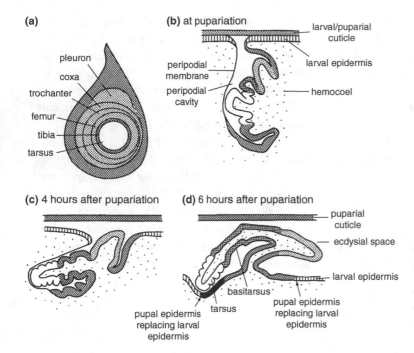

(a)

pleuron
coxa
trochanter
femur
tibia
tarsus

(b) at pupariation

larval/puparial cuticle

peripodial membrane

peripodial cavity

larval epidermis

hemocoel

(c) 4 hours after pupariation

pupal epidermis replacing larval epidermis

(d) 6 hours after pupariation

puparial cuticle

ecdysial space

larval epidermis

basitarsus

tarsus

pupal epidermis replacing larval epidermis

Figure 15.20 Development of an adult leg in *Drosophila*. (a) Fate map of a leg disc (after Schubiger, 1971). (b)–(d) Sections through the disc at different times after pupariation showing the elongation and eversion of the leg (from Fristrom and Fristrom, 1993).

At pupation the appendage is everted and the peripodial membrane comes to form part of the epidermis of the general body wall (Fig. 15.20c,d). Evagination probably results from a rearrangement of the cells of the disc which, at the same time, also tend to increase in surface area. This process is initiated by an increase in 20-hydroxyecdysone in the hemolymph; at earlier molts, when the titer of this hormone also increases, evagination is probably inhibited by JH. At some stage, the discs acquire the "competence" to undergo metamorphosis. Prior to this metamorphosis does not occur even in the appropriate humoral environment. The acquisition of competence appears to be associated with the number of divisions that the cells of the disc have undergone.

The details of development of the imaginal discs vary from one insect to another and from organ to organ. Where an appendage is present in the larva as well as the adult, the imaginal disc is closely associated with the larval structure. Thus, in *Pieris*, the adult antenna is first apparent in the first-stage larva as a thickening of the epidermis at the base of the larval antenna. The cells divide and, in the succeeding stages, an invagination is produced which pushes upwards, deep into the larval head. In the fifth-stage larva, the adult antennal tissue grows more quickly than the peripodial membrane, so that it is thrown into folds, and toward the end of the larval stage the larval antenna starts to degenerate and is invaded by imaginal cells. When the peripodial cavity (which opens by a slit on the front of the epidermis of the head) evaginates, the antenna is carried to the outside and the peripodial membrane now forms a part of the wall of the head. The maxilla develops in an essentially similar way, but very little development of the labium takes place until the fifth larval stage.

The wings also develop from imaginal discs (Fig. 15.21). In some Coleoptera they form as simple evaginations of the epidermis beneath the larval cuticle, but more usually they develop in peripodial cavities. In *Pieris* the wing imaginal discs are apparent in the embryo and invaginate in the second and third larval stages. During the fourth stage, the wing starts to develop in the peripodial cavity, finally becoming external at the larva–pupa molt.

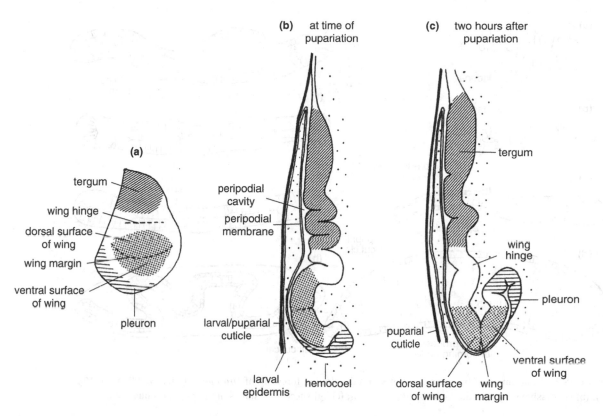

(b) at time of pupariation

peripodial cavity

peripodial membrane

larval/puparial cuticle

larval epidermis

hemocoel

(c) two hours after pupariation

tergum

wing hinge

pleuron

ventral surface of wing

puparial cuticle

dorsal surface of wing

wing margin

(a)

tergum

wing hinge

dorsal surface of wing

wing margin

ventral surface of wing

pleuron

Figure 15.21 Development of a forewing in *Drosophila*. (a) Fate map of a wing disc (after Bryant, 1975). (b) Section through the disc at the time of pupariation (from Fristrom and Fristrom, 1993). (c) Section through the disc two hours after pupariation (from Fristrom and Fristrom, 1993).

In *Drosophila*, on the other hand, invagination of the peripodial cavity is complete before the larva hatches, but the wing itself does not develop until the second larval stage, growing more extensively in the third stage.

During development, the cells of each imaginal disc become programmed for their ultimate phenotypic expression in the adult. They are said to become determined, and their subsequent development is fixed; they will produce the same structure even if they are moved experimentally to another position on the appendage. More detailed features are not determined until after the more basic characters. For example, the dorsal and ventral surfaces of the wing are determined before the details of the pattern, and cells become committed to a

segment of a leg before a specific structure on that segment. Some degree of determination of the adult tissues may already be apparent in the embryo, and determination of wing regions is advanced at the time of pupariation (Fig. 15.21). In *Drosophila*, cells along the dorso-ventral boundary of the wing have a critical organizational role.

The internal development of the wings is complex, involving great expansion and the formation of the veins. The development of a *Drosophila* wing is used here as an example. When the wing evaginates at pupation, the upper and lower surfaces have already come together, although they remain separated along certain lines called lacunae (Fig. 15.22a,b). Four lacunae run the length of the wing rudiment, the second dividing into two distally. A nerve and a

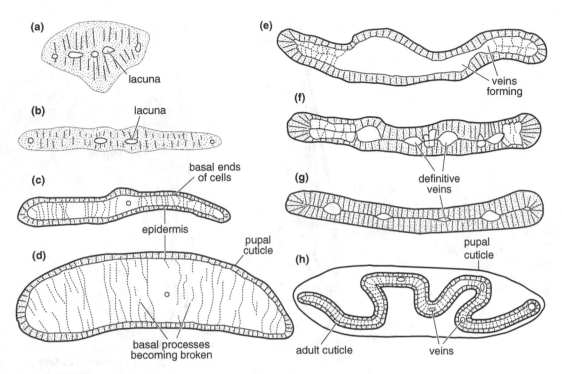

Figure 15.22 Diagrammatic transverse sections of a developing forewing of *Drosophila* (after Waddington, 1941): (a), (b) successive stages in the larva before pupation; (c)–(g) successive stages in the pupa; (h) pharate adult.

trachea enter the second lacuna, and at about this stage, some six hours after the formation of the puparium, the pupal cuticle is laid down. After this the upper and lower surfaces of the wing are forced apart by an increase in blood pressure (Fig. 15.22c,d). The cells at first become stretched across the gap as narrow threads connecting the two surfaces, but these connections are finally broken except at the wing margins. A less extensive inflation occurs in *Tenebrio molitor* (Coleoptera) and *Habrobracon juglandis* (Hymenoptera). Perhaps the inflation has the effect of expanding the newly formed pupal cuticle to the greatest possible extent so that the development of the adult wing can proceed.

Following inflation, the wing flattens again. The epidermal layers on the two sides first become apposed round the edges (Fig. 15.22e) and then the contraction spreads inward so that a flat double sheet of cells is produced (Fig. 15.22f). During this process

the definitive wing veins are formed along lines where the two epidermal layers remain separated (Fig. 15.22g). The veins are at first wide channels, but they become narrower as the wing continues to expand. Cell division proceeds actively, especially above the veins, so that here the cells become crowded and columnar, while elsewhere they are flattened. The fully developed wing finally secretes the adult cuticle (Fig. 15.22h).

Even in the imaginal discs some programmed cell death (apoptosis) occurs in the course of differentiation. This is especially obvious where normal adult structures are of reduced size. The female vaporer moth, *Orgyia antiqua* (Lepidoptera, Lymantriidae), for example, has only vestigial wings, but up to the pupal molt her wings develop normally as imaginal discs. Then, during apolysis, as the epidermis retracts from the pupal cuticle, extensive cell death occurs so that the wings become greatly reduced in size. Cell death on a much smaller scale occurs in the developing wings of the male, which is fully winged.

When the imaginal appendages are everted from the peripodial cavities during the pharate pupal period, the peripodial membrane contributes to the general epidermis of the adult body wall (Figs. 15.20, 15.21). The extent to which the larval epidermis is replaced varies. In Coleoptera there is no extensive replacement, but in Hymenoptera and Diptera the epidermis is completely renewed from imaginal cells. The epidermis of the head and thorax are formed by growth from the imaginal appendage discs, while the abdominal epithelium is formed from special groups of cells called histoblasts. In *Drosophila* the abdominal segments have pairs of dorsal, ventral and spiracular histoblasts (Fig. 15.19c). Unlike imaginal discs, histoblast cells secrete an overlying larval cuticle prior to metamorphosis. Each group of histoblasts initially comprises 5–15 cells which remain unchanged throughout larval development, but after pupariation the cells start to divide. They gradually replace the larval cells, but at no time is there any discontinuity in the epidermis. The larval cells are completely replaced and are sloughed off into the body cavity and phagocytosed.

Adult muscles The muscular system usually undergoes extensive modification at metamorphosis. The modifications fall into four categories:

(1) Larval muscles may pass unchanged into the adult.
(2) Existing larval muscles are reconstructed.
(3) Larval muscles may be destroyed and not replaced.
(4) New muscles, not represented in the larva, may be formed.

The adult muscles of some flies are already present in the larva as rudimentary non-functional fibers. The dorsal longitudinal muscles in *Simulium ornatum* (Diptera), for instance, are only about 4 μm in diameter in the first-stage larva. They grow throughout the larval period and their nuclei increase in number. During the pharate pupal period the muscle rudiment becomes divided to produce the definitive number of fibers and myofibrils appear for the first time. They continue to grow until some time after the final molt.

Reconstruction of larval muscles occurs in two ways. In the Neuroptera, Coleoptera and some Lepidoptera, larval muscles contain two sets of nuclei, the functional larval nuclei and other, small, nuclei which are scattered through the cytoplasm. At metamorphosis the small nuclei multiply and, with associated cytoplasm, form myocytes. These migrate into the body of the muscle and associate in strands to form new fibers. In Diptera, and some Lepidoptera and Hymenoptera, on the other hand, myoblasts originating outside the larval muscle are concerned in the production of adult muscle, adhering to the outside or penetrating the sarcolemma in order to form new fibers. The dorsal longitudinal flight muscles of adult *Drosophila* are formed by new myoblasts, but larval muscles form a scaffold that determines the appropriate development of the adult muscles.

Most larval muscles are completely histolyzed and disappear during the early pupal period, but the precise timing varies because some larval muscles have specific functions in pupal development or adult eclosion and are destroyed much later than others. For example, the larval leg muscles of *Manduca sexta* (Lepidoptera) start to degenerate when the larva enters its wandering phase and degeneration is complete by the time the larva pupates. Some abdominal muscles, however, do not degenerate until after pupation, and others persist until after adult ecdysis. In *Drosophila*, most muscles of the head and thorax start to break down before puparium formation and are fragmented before the larva pupates, but the dilator muscles of the pharynx remain unchanged until after pupation. They are apparently important in the evagination of the head region of the insect at pupation, and only degenerate after this has occurred. In addition, one pair of muscles persists in each abdominal segment for about half the pupal period. They may help to establish the segmentation of the pupal abdomen by telescoping each segment into the preceding one.

The first sign of muscle degeneration is liquefaction of the peripheral parts of the fiber. This is followed by separation of the fibrils and, in the flour moth, *Ephestia kuehniella* (Lepidoptera), phagocytes penetrate the sarcolemma and assist the destruction. The sarcolemma breaks down and the muscles separate from their attachments and fragment, the remains being consumed by phagocytes.

New muscles are always formed by free myoblasts that aggregate to form multicellular groups within which cells fuse to form the multinucleate muscle fibers. Where the new muscle is associated with an existing motor neuron that becomes respecified (see below), the myoblasts may migrate along the axon to their definitive positions and the neuron may influence the rate of myoblast proliferation. Some larval muscles in Diptera and Lepidoptera that persist through the pupal stage are specifically associated with eclosion; they break down after the adult insect has emerged from the pupal cuticle (see Fig. 10.14).

Adult nervous system The central nervous system of holometabolous insects is extensively restructured at metamorphosis. The brain becomes considerably enlarged, mainly due to the development of the antennal and optic lobes in relation to the much larger antennae of adult insects and the appearance of compound eyes. In most holometabolous insects, particularly those that are more specialized, fusion of some ventral ganglia commonly occurs at metamorphosis. This condensation is effected by a forward movement of the more posterior ganglia, resulting from the shortening of the interganglionic connectives. For instance, the larva of *Manduca* has, in addition to the head ganglia, three thoracic and seven separate abdominal ganglia. In the adult, the meso- and metathoracic ganglia are fused with the first two

abdominal ganglia to form a compound ganglion close behind the prothoracic ganglion. The next three abdominal ganglia remain separate, but the last three fuse together to form another compound ganglion (Fig. 15.23a). In the course of these changes the perineurium is histolyzed and the neural lamella digested, the former being redeveloped from remaining glial cells. Shortening of neural connectives in the wax moth, *Galleria mellonella* (Lepidoptera), can be studied in vitro through organ culture methods and has been shown to be triggered directly by ecdysteroid molting hormones. The higher Diptera are exceptional in having a more condensed central nervous system in the larva than in the adult.

Over 90% of the neurons in the nerve cord of adult *Drosophila* develop during the larval period from groups of neuroblasts that persist from the embryo. In the first-stage larva of *Drosophila*, there are 47 neuroblasts in each thoracic neuromere, but only six in each abdominal neuromere. *Manduca* has over 20 neuroblasts in each of the thoracic ganglia, but only four in the abdominal neuromeres behind the first. These cells start to divide at different times during larval development. The most rapid division occurs in the final larval stages (Fig. 15.23c). Some of these cells die just before and during pupal development, but, despite this, the neuroblasts contribute over 2500 new interneurons to each of the thoracic ganglia and about 50 to each abdominal ganglion in *Manduca*. Similar numbers are produced in *Drosophila*. All these new cells become interneurons; probably no new motor neurons are formed.

The motor neurons already present in the larval ganglion have two possible fates, depending on the muscles they innervate:

(1) If the larval muscle degenerates, but a new muscle develops in a similar position, the neuron may persist and innervate the new muscle. It is said to be respecified.
(2) If the larval muscle is destroyed and not replaced, the motor neuron dies. This may occur at various

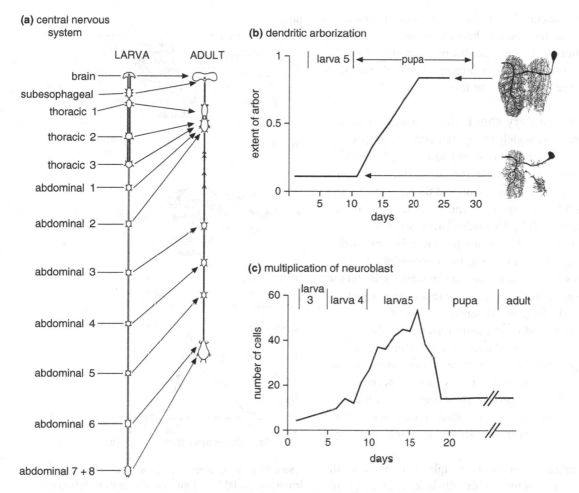

Figure 15.23 Metamorphosis of the central nervous system of *Manduca*. (a) Brain and ventral nerve cord of the larva and adult showing the fusion of ganglia. (b) Changes in the form of an abdominal motor neuron from larva to adult. The extent of the arbor is expressed on an arbitrary scale (from Truman and Reiss, 1988). (c) Changes in the number of neurons produced from a neuroblast nest in an abdominal ganglion. The decline in cell number at pupation is due to the deaths of many cells (after Booker and Truman, 1987).

times during metamorphosis. About 16% of all the motor neurons in the abdominal ganglia of *Manduca* die soon after pupation following the deaths of their target muscles.

Respecification of neurons occurs in relation to leg and flight muscles as well as to the newly formed abdominal intersegmental muscles. For example, the neuron innervating a femoral flexor muscle in the larva of *Manduca* is respecified to innervate the femoral extensor muscle in the adult leg. This involves regression and then regrowth of the dendritic arborization in the ganglion and the remaking of contact with the new target muscle by the axon.

The sensory system is almost entirely renewed at metamorphosis in holometabolous insects. The compound eyes and antennae of the adult are completely new structures, and this is true of the sensilla associated with the mouthparts, at least in a majority of insects. It is known, however, that a few mechanoreceptors persist during the pupal stage of *Manduca* and some stretch receptors and chordotonal organs are still present in the adult.

Adult alimentary canal The alimentary canal is extensively remodeled at metamorphosis in species that have different larval and adult diets. In larval Lepidoptera, for example, the gut is relatively simple and the midgut occupies most of the body cavity, but the adult has a large crop and rectal sac and only a small midgut (Fig. 15.24a,b). This change is associated with the change from leaf-feeding, with continuous access to food, to fluid-feeding, with the need to store nectar in the crop between feeds and the requirement to digest and absorb a diet principally composed of sugars. Reconstruction of the stomodeum and proctodeum results from the renewed activity of the larval cells, without any accompanying cell destruction in Coleoptera, but in Lepidoptera and Diptera new structures develop, at least partly, from proliferating centers known as imaginal rings at the inner ends of the foregut and hindgut. The larval cells are sloughed into the body cavity.

The midgut is probably completely renewed in all holometabolous insects, usually being reformed from the regenerative cells at the base of the epithelium. These cells proliferate and form a layer around the outside of the larval cells which thus come to lie in the lumen of the new alimentary canal. Sometimes this process occurs twice, once on the formation of the pupa and again when the adult tissues are forming. It is suggested that the special pupal midgut enables the insect to digest the sloughed remains of the larval midgut so that these can be assimilated and used in the reconstruction. In the silkworm *Bombyx mori* (Lepidoptera), a large number of anti-microbial peptides are expressed in the gut during the prepupal period. In *Manduca*, the newly formed epithelium contains large vacuoles containing the anti-microbial enzyme – lysozyme – which is

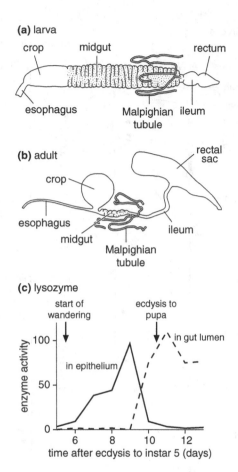

Figure 15.24 Metamorphosis of the alimentary canal in a lepidopteran. (a) Larval gut; (b) adult gut; (c) lysozyme activity in the midgut epithelium and lumen during the final larval and early pupal stages of *Manduca* (after Russell and Dunn, 1991).

discharged into the lumen when the epithelium is complete (Fig. 15.24c). It is presumed that these are adaptations to protect the insect from infection by gut bacteria, which might be released into the hemocoel during digestive tract reconstruction.

Malpighian tubules Sometimes the larval Malpighian tubules remain unchanged in the adult, or slight modifications may occur as in the Lepidoptera. Caterpillars have a cryptonephridial arrangement of the Malpighian tubules (Fig. 18.21), but at metamorphosis the parts associated with the rectum are histolyzed, while the more proximal parts form the adult tubules. In Coleoptera, the tubules are rebuilt from special cells in the larval tubules, while in Hymenoptera, the larval tubules break down completely and are replaced by new ones developing from the tip of the proctodeum.

Fat body The fate of the fat body at metamorphosis has been most fully studied in Lepidoptera and Diptera. In both groups, the cells of the fat body become disassociated at the larva–pupa molt through the activity of hemocytes. Subsequently, in *Drosophila* the cells persist independently but are progressively histolyzed. Some may persist in the adult's head, but the bulk of the adult fat body is formed from mesenchyme cells on the inside of the imaginal discs. In Lepidoptera, the cells of the peripheral fat body which synthesize storage proteins in the caterpillar are destroyed, but those of the perivisceral fat body that accumulate the proteins reassociate to form the adult fat body.

Other adult systems In general, the tracheal system shows little change other than the development of new branches to accommodate the particular needs of the adult, such as the supply to the flight muscles, and the elimination of some specifically larval elements. In cyclorrhaphous Diptera, however, an extensive reconstruction occurs. The pupal tracheal system consists of four main tracheal trunks which extend into the head and as far as the anterior segments of the abdomen from the thoracic spiracle. From these main trunks, tufts of fine, unbranched tracheae arise and each ends in several tracheoles which become tightly coiled about one-third of the way through the puparial period (Fig. 15.25). Up to this time, the developing adult flight muscles are still very short and their development is not dependent on an oxygen supply. Subsequently, however, they elongate rapidly and the tracheoles uncoil, extending with the muscle fibers whose metabolic processes at this time are at

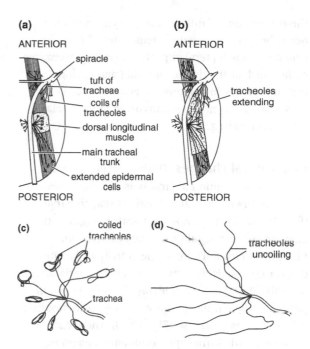

Figure 15.25 Metamorphosis of the tracheal system in the thorax of *Calliphora* (after Houlihan and Newton, 1979). (a), (b) Diagrams of the right half of the thorax of the pharate adult three and four days after pupariation showing the tracheal supply to the developing dorsal longitudinal muscle. At first, the muscle is very short, and connects with long extensions of the epidermal cells. As it elongates, the extensions become shorter. (c), (d) A tuft of tracheoles arising from the end of a trachea. The tracheoles are coiled on day 3 (c), but have uncoiled by day 4 (d).

least partially dependent on aerobic respiratory processes. Subsequently, the adult tracheal supply develops from tracheoblasts largely independently of the pupal supply, but it remains fluid-filled until about three hours before eclosion. The filling of the adult tracheae with air corresponds with a sharp increase in oxygen consumption, partly associated with increased metabolic activity in the tissues and partly with an increase in movements by the fly. Note that it is not obvious whether increased oxygen consumption is the cause or the consequence of tracheal air filling.

In many insects the circulatory system undergoes little change from larva to adult, although changes in the proportions of different hemocyte types may occur between life-stages. Hemocytes of *Heliothis armigera* (Lepidoptera) express the lysosomal enzyme cathepsin L at the time of molting and during metamorphosis in response to ecdysteroids. Presumably the enzyme is involved in molt-related tissue remodeling.

Biochemical changes During the pupal period, oxygen consumption at first falls and then rises again, following a characteristic U-shaped curve (Fig. 15.26). This is associated with changes in the enzyme systems regulating energy release. In *Calliphora vicina* (Diptera) the activity of lactate dehydrogenase is high two days after pupariation, but falls sharply about midway through the pupal period, while citrate synthase and glycerolphosphate dehydrogenase increase (Fig. 10.19). These changes indicate a switch from predominantly anaerobic catabolic processes to aerobic processes corresponding with the time at which the pupal tracheal system forms a well-developed air supply to the developing muscles. A comparable change in the relative emphasis of anaerobic and aerobic metabolic pathways occurs in *Samia cynthia* (Lepidoptera,

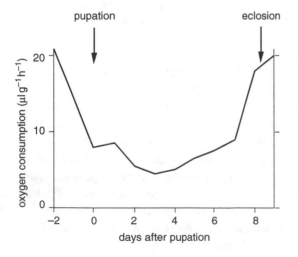

Figure 15.26 Reduced oxygen consumption during pupal development (*Galleria*, Lepidoptera) (after Sláma, 1982).

Saturniidae) after about two-thirds of the pupal period has elapsed.

The components of adult proteins are derived partly from the histolysis of larval tissues and partly from proteins built up and stored in the larval hemolymph for this specific purpose (Section 5.3.4).

The waste products of pupal metabolism accumulate in the rectum and are discharged as the meconium when the adult emerges. Uric acid accumulates throughout the pupal period, but especially during histolysis, while in Lepidoptera and Hymenoptera allantoic acid comprises an appreciable part of the nitrogenous waste of the pupa. In *Phormia regina* (Diptera), urea accumulates during the development of the adult, suggesting that it is the end product of nitrogen metabolism in this insect.

15.3.3 Eclosion

The escape of the adult insect from the cuticle of the pupa or, in hemimetabolous insects, of the last larval stage is known as eclosion. Where the pupa is enclosed in a cell or cocoon the adult also has to escape from this.

Escape from the cocoon or cell Sometimes the pharate adult is sufficiently mobile to make its escape from the cocoon or cell while still within the pupal cuticle. This is the case in species with decticous pupae which use the pupal mandibles, actuated by the adult muscles, to bite through the cocoon. Sometimes, as in Trichoptera, the adult mouthparts are non-functional and the sole function of the adult mandibular muscles is to work the pupal mandibles at eclosion; subsequently they degenerate. The pupa moves away from the cocoon before the adult emerges, aided by its freely movable appendages and backwardly directed spines on the pupal cuticle.

Other methods are employed in species with adecticous pupae. Among the Lepidoptera, the pupae of Monotrysia and primitive Ditrysia work their way forwards with the aid of backwardly directed spines on the abdomen, forcing their way through the wall of the cocoon with a ridge or tubercle known as a cocoon cutter on the head. The pupa does not escape completely from the cocoon, but is held with the anterior part sticking out by forwardly directed spines on the ninth and tenth abdominal segments. With the pupal cuticle fixed in this way the adult is able to pull against the substratum and so drag itself free more readily. Cocoon cutters are also present in Nematocera, although in this group they are usually multiple structures.

In many insects with adecticous pupae, the adult emerges from the pupa within the cocoon or cell, making its final escape later, often while those parts of its cuticle that will be inflated are still soft and unexpanded. This is true of many higher Ditrysia (Lepidoptera), whose escape is facilitated by the flimsiness of the cocoon or the presence of a valve at one end of the cocoon, through which the insect can force its way out while the ingress of other insects is prevented. The cocoon of *Saturnia pavonia* is of this type (Fig. 15.16a), while, in Megalopygidae, a trap door is present at one end.

Some Lepidoptera produce secretions which soften the material of the cocoon. *Cerura*, for instance, produces an oral secretion containing potassium hydroxide, which softens one end of its cell of agglutinated wood chips. This enables the adult insect to push its way out, protected by the remnants of the pupal cuticle. The silk moths *Bombyx mori* and *Antheraea pernyi* produce a proteinase which softens the silk wall of the cocoon sufficiently for the adult to push its way out. In *Antheraea* the proteinase is secreted onto the surface of the galeae two days before eclosion. It dries, forming a semicrystalline encrustation. At the time of eclosion a liquid is secreted from labial glands which open by a single median pore just below the mouth. This liquid dissolves the enzyme and wets the inside of the cocoon. It contains potassium and functions as a buffer, keeping the enzyme solution at about pH 8.5. The enzyme digests the sericin coating of the silk so that the fibroin threads are readily separated.

Cyclorrhaphous Diptera have an eversible sac called the ptilinum at the front of the head, which assists in their escape from the puparium (Fig. 15.27a). It can be expanded in the newly emerged fly by blood forced into the head from the thorax and abdomen and is then withdrawn again by muscles which force blood back to the thorax. Pressure of the ptilinum splits off the cap of the puparium and, if the puparium is buried in the soil, the ptilinum is also used by the fly to dig its way to the surface. Once the fly's cuticle has hardened, the ptilinum is no longer eversible and the muscles associated with it degenerate. Its position is indicated in the mature fly by the ptilinal suture.

The degree of hardening that these insects undergo before escaping from the cocoon varies. In some, most of the cuticle remains soft until after eclosion, but some parts, particularly those involved in locomotion, harden beforehand. Thus, in the blowfly *Calliphora vicina* the legs and apodemes harden, so do the bristles which protect the soft cuticle and such specialized parts as the halteres, antennae and genitalia. The remainder of the cuticle does not harden until after it is expanded when the insect is free. In Lepidoptera, however, most of the body does not expand greatly after eclosion and hardening of the cuticle is extensive before the insect emerges from the cocoon, although the wings remain unsclerotized and are inflated after emergence.

Other insects emerge from the pupa and harden fully before making their escape from the cocoon and they may have specialized features to assist this. Coleoptera and Hymenoptera use their mandibles to bite their way out. Some weevils of the Subfamily Otiorrhynchinae have an appendage, known as the false mandible, on the outside of the mandible (Fig. 15.27b), which is used in escaping from the cocoon and then, in most species, falls off. Among the Cynipidae (Hymenoptera), which do not feed as adults, the sole function of the adult mandibles is to allow the insect to escape from the host in which the larva pupated.

The cuticle of fleas also hardens before they escape from their cocoons and they may remain in the cocoon for some time after emergence. Their escape

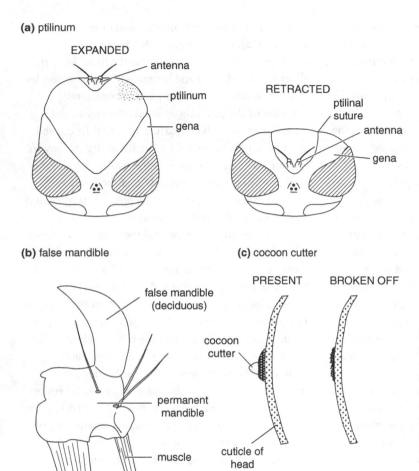

(a) ptilinum

EXPANDED

antenna

ptilinum

gena

RETRACTED

ptilinal
suture

antenna

gena

(b) false mandible

false mandible
(deciduous)

permanent
mandible

muscle

(c) cocoon cutter

PRESENT   BROKEN OFF

cocoon
cutter

cuticle of
head

Figure 15.27 Escape from pupal protective structures. (a) Ptilinum of a cyclorrhaphan fly. Dorsal views showing the ptilinum expanded (left) and retracted (right). (b) False mandible used by the beetle *Polydrosus* to escape from its cocoon. The false mandible is deciduous (after Hinton, 1946). (c) Cocoon cutter of the flea, *Trichopsylla*. The cutter is shed after escape from the cocoon (after Hinton, 1946).

is stimulated by mechanical disturbances and in many species is facilitated by a cocoon cutter on the frons. In *Trichopsylla* the cocoon cutter is deciduous (Fig. 15.27c). Finally, in the males of Strepsiptera the mandibles are used to cut through the cephalothorax of the last-stage larva in which pupation occurs. The larval cephalothorax is earlier extruded through the cuticle of the host so that the adult insect can easily escape.

Eclosion of aquatic insects Of the insects that pupate under water, some emerge under water and swim to the surface as adults; in others, the pupa rises to the surface before the adult emerges. In the Blepharoceridae (Diptera), the adult undergoes some

degree of hardening within the pupal cuticle, so that as soon as it emerges it rises to the surface and is able to fly. Black flies (Diptera, Simuliidae) and *Acentropus nivea* (Lepidoptera), which have aquatic larvae and pupae, also emerge beneath the surface, but come up in a bubble of air. In *Acentropus* this is derived from the air in the cocoon, while *Simulium* pumps air out into the gap between the pupal and adult cuticles. Thus the adult emerges into a bubble of air and is able to expand its wings before rising to the surface.

The pupae of mosquitoes (Diptera, Culicidae) are buoyant, while some other insects, such as *Chironomus* midges, whose pupae are normally submerged, increase their buoyancy just before emergence by forcing gas out of the spiracles into the space beneath the pupal cuticle, or by increasing the volume of the tracheal system. Aided by backwardly directed spines, these pupae then escape from their cocoons or larval tubes and rise to the surface. Many Trichoptera swim to the surface as pharate adults, the middle legs of the pupae of these species being fringed to facilitate swimming, and the insect may continue to swim at the surface until it finds a suitable object to crawl out on. In other Trichoptera, the pharate adults crawl up to the surface, while the last-stage larvae of Odonata and Plecoptera crawl out onto emergent vegetation so that the adult emerges above the water. Larval Ephemeroptera also come to the surface, but the form that emerges is a subimago, not the imago. The subimago resembles the imago, but its legs and caudal filaments are shorter and the wings are translucent instead of transparent and are fringed with hairs. The subimago flies off as soon as it emerges, but settles a short distance away and soon molts to the imago. This is the only known example of a fully winged insect undergoing a molt.

In mosquitoes the pharate adult rises to the surface and eclosion is associated with swallowing air. Before the pupal cuticle is split, the air is probably drawn from the space between the adult and pupal cuticles and is derived from air drawn into the tracheal system through the respiratory horns of the pupa. Once the adult is able to swallow air from outside, its volume increases still further and it can surge forward to free itself of the exuvia. A comparable process probably occurs in other insects with aquatic pupae.

Timing of eclosion It is important that eclosion is timed so that an insect's life history is synchronized with suitable environmental conditions and so that the meeting of the two sexes is facilitated. Temperature is a particularly important cue for such synchronization since, to a large extent, it governs the rate of development and the activity of the insect. In long-lived species, diapause may also be important

in synchronizing eclosion so that the sexes meet (see below).

Apart from these general seasonal effects, many insects emerge at particular times of day, often at night or in the early morning. This probably gives the insect some degree of protection against predators while it is at its most vulnerable in the period before it is able to fly. Thus, in Britain, when they are ready to molt, last-stage larvae of *Anax imperator* (Odonata) leave the water between 20.00 and 21.00 hours and by 23.00 hours most adults have emerged and are expanding their wings. The timing of emergence of *Sympetrum* spp. (Odonata) and most tropical dragonflies is similar, although some temperate species emerge during the day, perhaps because activity is limited by low temperature at night.

Moths are also known to emerge at particular times of day, the timing being controlled by the eclosion hormone acting in response to photoperiod or temperature cycles and differing in its time of release from species to species. For example, with a cycle of 17 hours light:7 hours dark, *Hyalophora cecropia* (Lepidoptera) escapes from the pupal cuticle in the first hours of the light period, while *Antheraea pernyi* does so toward the end of the light period. The emergence of *Manduca sexta* is also affected by light, but is regulated more effectively by an increase in temperature of 3°C or more. This is to be expected since the insect pupates beneath the soil surface.

It is probable that the time of emergence of many insects is determined by entrainment to light or temperature cycles. This is true of the moths described above and of *Drosophila melanogaster*. In other cases, emergence may be a direct response to environmental conditions, even if these act at some previous time. For example, synchrony within a brood of the mosquito *Aedes taeniorhynchus* is brought about by the tendency of the larvae to pupate at about sunset, but different broods emerge at different times depending on the temperature during the pupal period.

It is common for male insects to emerge as adults a little before the females, although the difference is not great. This probably reflects, in part, the smaller size of many male insects, but may also reflect adaptive evolution, reflecting the competitive advantage to be gained by males that locate females at the earliest possible time.

15.4 Control of postembryonic development

Most cuticular growth and changes in cuticular form only occur when an insect molts. These processes are governed, primarily, by two classes of hormone: molting hormones (ecdysteroids) and juvenile hormones (JH) (Chapter 21). Molting is induced and regulated by ecdysteroids. As their titer in the hemolymph rises, the epidermal cells exhibit a complex pattern of DNA and RNA synthesis. This is known as the preparatory phase; it ends with division of the epidermal cells (Fig. 15.28 shows some of the processes involved at the molt to adult in *Manduca*; similar changes occur in the epidermis and cuticle at

all molts). Except for cells in the imaginal discs of holometabolous insects, this is the only time when mitosis occurs in epidermal cells. At this time, with the ecdysteroid titer approaching its peak, apolysis occurs and the epidermal cells produce the new epicuticle. Then, the falling titer of ecdysteroid leads to the production of chitin and the proteins of the new procuticle. Ecdysis follows soon afterwards. (For a more complete account of the cuticular changes occurring during a molt, see Section 16.5.)

The type of cuticle produced, whether it is larval, pupal (in the case of holometabolous insects) or adult depends on whether or not JH is present during a critical period in each developmental stage. This critical period usually occurs when the ecdysteroid initiates the next molt. In hemimetabolous insects, if JH is present during the critical period, the insect retains its larval characteristics; if JH is absent, the insect becomes an adult (Fig. 15.29). It is apparent that, in the absence of JH, a program of gene expression is initiated which differs from that occurring when JH is present.

In holometabolous insects, similar critical periods occur (Fig. 15.30). Detailed accounts of the relevant

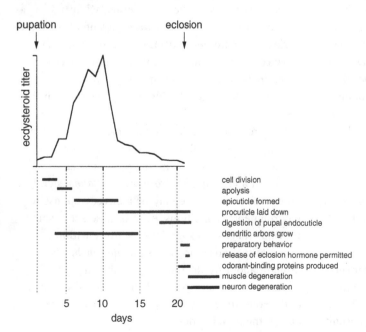

Figure 15.28 Changes in ecdysteroid titers in the hemolymph of *Manduca* in relation to changes in different tissues during the pupal period (data from various sources).

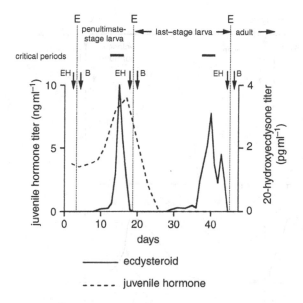

Figure 15.29 Changes in hormone titers regulating molting and metamorphosis in a hemimetabolous insect. At the molt from larva to larva, juvenile hormone is present during the critical period; at the molt from larva to adult, there is a critical period, but no juvenile hormone is present at this time. Experimental application of juvenile hormone during this sensitive period would lead to the production of another larval stage. Eclosion hormone (EH) and bursicon (B) are assumed to be produced for a brief period immediately before and after each ecdysis (E), respectively. Their production in this insect has not been demonstrated experimentally (Nauphoeta, Blattodea) (based on Lanzrein et al., 1985).

hormonal events are available for *Manduca sexta* and *Bombyx mori*, but the picture is similar in other species. In all larval stages except the last, the feeding period of the instar is terminated by a large surge of ecdysteroid that initiates molting. Because this occurs in the presence of high JH levels, a new larval cuticle is formed and premature development of the imaginal discs is prevented. In the last larval stage, however, JH levels fall before a small peak of ecdysteroid (called the commitment peak) occurs. Although the level of ecdysteroid in the commitment peak is in itself insufficient to initiate molting, it nevertheless causes the insect to become irreversibly

committed to metamorphosis because during the commitment process JH receptors are lost from most larval tissues. Subsequently (in *Manduca* it is 24 hours later), a second surge of ecdysteroid occurs, this time approximately ten times the size of the commitment peak. This is sufficient to cause the secretion of a new pupal cuticle. Remarkably, this second pre-molting ecdysteroid surge is accompanied by renewed JH secretion. The JH that is present at this stage fails to cause larval molting, however; because JH receptors are now absent from the epidermis it prevents the premature adult development of structures derived from imaginal discs. After the molt to the pupal stage, if environmental conditions are suitable, adult molting is initiated by further secretion of ecdysteroid, again in the absence of JH. The epidermal cells now switch to the production of adult cuticle, but this change can be prevented experimentally by applying JH during a critical period, when the insect responds by forming a second pupa, showing that JH receptors are again present in the epidermis.

Although the major changes are effected by these two hormones, many other factors are concerned in regulating molting. The sequence of events is: initiation of molting; switching on the production of ecdysteroids; controlling the process of ecdysis; and controlling sclerotization. These are described in the following sections and illustrated diagrammatically in Fig. 15.31.

15.4.1 Initiation of molting and metamorphosis

The factors responsible for initiating molting are only poorly understood. In general, it is probably true that the insect responds to reaching a certain size, but how it measures size is not known in most cases. In *Rhodnius* molting is induced by a full meal of blood, resulting in abdominal distension that stimulates receptors in the ventral body wall. Abdominal stretch is also important in the milkweed bug, *Oncopeltus*, although here the stretch is achieved progressively

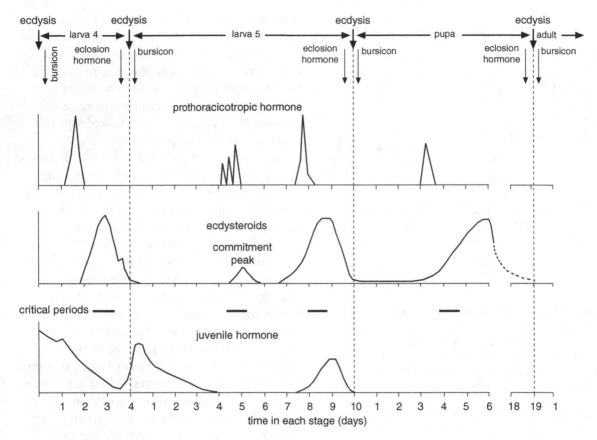

Figure 15.30 Changes in hormone titers regulating molting and metamorphosis in a holometabolous insect. At the molt from larva to larva, juvenile hormone is present during the critical period; at the molt from larva to pupa, no juvenile hormone is present at the first critical period. The second critical period of sensitivity to juvenile hormone in the fifth-stage larva regulates development of the imaginal discs. Eclosion hormone and bursicon are produced for a brief period before and after each ecdysis (based on data for *Manduca*, Lepidoptera).

over a prolonged period of feeding. In the larva of *Manduca*, however, the insect initiates molting when it reaches a certain body weight. It is not known how the caterpillar assesses its size, but in this case stretching of the body wall is not involved.

The size at which molting occurs is not absolute, but depends on the insect's size at the beginning of the stage. Some species molt if they are starved and they may become smaller. This is known to occur in beetles and caterpillars living in stored products such as grain or flour. Clearly some factor apart from size is involved in initiating molting in these insects.

Metamorphosis occurs when ecdysteroids are produced in the absence of juvenile hormone. In *Manduca* the switch to metamorphosis is related to head size at the beginning of the larval stage. If the head width is less than 5 mm at this time, JH production continues and the larva is committed to molting to another larva. If head width exceeds 5 mm, however, the larva is committed to become a pupa at the next molt (Fig. 15.32). Head size may not be the relevant parameter for this insect, but it seems likely that some correlated aspect of size is relevant. It is not known how the insect is able to assess its

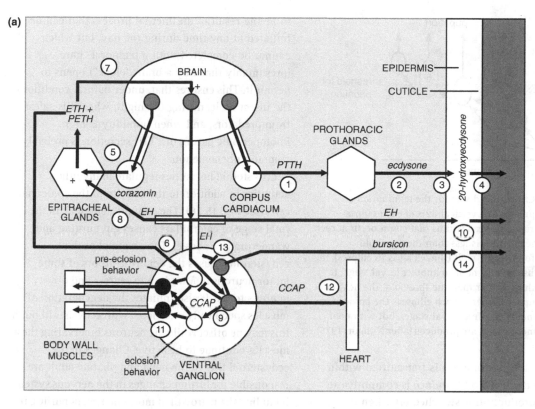

(a)

(b) ⬤ neurosecretory cell   ◯ interneuron   ● motor neuron

CONTROL OF APOLYSIS AND CUTICLE PRODUCTION

1 PTTH stimulates synthesis and release of ecdysone
2 Ecdysone in hemolymph
3 Ecdysone hydroxylated at tissues
4 20-hydroxyecdysone regulates genes producing cuticle

CONTROL OF ECDYSIS

5 Corazonin stimulates release of ETH and PETH
6 ETH and PETH act on ventral ganglia to switch on pre-eclosion behavior
7 ETH and PETH act on brain to cause release of EH
8 Positive feedback loop between EH and ETH/PETH results in massive central and peripheral release of EH

9 Central release of EH causes release of CCAP
10 EH in hemolymph causes cuticle plasticization
11 CCAP in ventral ganglia switches on eclosion behavior and switches off pre-eclosion behaviour

CONTROL OF EXPANSION AND SCLEROTIZATION

12 CCAP in hemolymph increases heartbeat
13 Bursicon release is part of eclosion behavior but subject to sensory control
14 Bursicon in hemolymph first plasticizes cuticle, then causes cuticle sclerotization

Figure 15.31 Hormones involved in regulation of events at a molt. (Juvenile hormone is not shown.) Names of hormones and regulatory peptides are italicized. Anatomical structures (i.e., sources and targets of hormones) are in capitals. CCAP, crustacean cardioactive peptide; EH, eclosion hormone; ETH, ecdysis-triggering hormone; PETH, pre-ecdysis-triggering hormone; PTTH, prothoracicotrophic hormone. "+" indicates the reciprocally excitatory actions of ETH/PETH and EH in the positive feedback loop involving these peptides. Sequential actions of the hormones are numbered as indicated in the key.

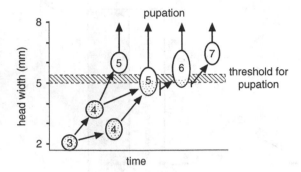

Figure 15.32 Critical head size for the termination of larval development. The vertical axis of each ellipse indicates the range of head widths that might occur at each stage. If the head width is less than the threshold immediately following a molt (shaded areas in ellipses), the insect will subsequently molt to another larval stage. If the head width is greater than the threshold, the next molt will be to a pupa. Number in each ellipse is the larval stage. *Manduca* normally has five larval stages, but with poor food quality more stages are produced (after Nijhout, 1975).

size, or how the information is transmitted within the body. If the larva of *Manduca* is committed to pupate, JH production is switched off when the insect reaches a weight of about 5 g.

15.4.2 Production of ecdysteroid hormones

The prothoracic gland is stimulated to produce ecdysone by the prothoracicotropic hormone (PTTH) (Chapter 21). Release of this hormone, usually via the corpora cardiaca (but from the corpora allata in *Manduca*), results from the presumed neural signal produced by size. Each period of ecdysteroid production is preceded by a peak of PTTH. During the feeding part of the final larval stage of *Manduca*, and perhaps also of other insects, the release of PTTH is inhibited by a high titer of JH in the hemolymph, although in earlier larval stages PTTH is released in the presence of JH.

In *Manduca* PTTH release only occurs during a restricted period during the day, corresponding to the expected "night." PTTH release is therefore said to be "gated" (Fig. 15.33), in the sense that it appears to be the result of an internal process that can be initiated at any time during the day, but which cannot be completed until a temporal "gate" (presumably timed by a brain "clock") opens to permit it. This ensures that, under natural conditions, the insect molts during the night, when it is safest from predators, and when humidity is high. Photoperiodic gating of PTTH secretion is probably a common phenomenon.

Ecdysteroid hormones regulate many other activities in addition to the activity of the epidermal cells (see Fig. 15.28). The commitment peak in the final stage of caterpillars causes gut purging and wandering (see above). The prepupal peak of ecdysteroids causes the dendritic arbors of some motor neurons in *Manduca* to regress and other neurons to die. It also initiates the degeneration of muscles specifically associated with ecdysis, although this may be offset while the neurons innervating these muscles continue to be active. Changes in the ecdysteroid titer in the pupa and pharate adult are responsible for further changes in the nervous system, including the regrowth of motor neuron dendrites, the development of olfactory glomeruli and of sensitivity in the adult olfactory system (Fig. 15.28).

The action of ecdysteroid hormones depends on the presence of ecdysteroid receptors in the target tissues. The occurrence of these varies temporally, and there are at least three isoforms. They may be present in many or all tissues, in which case the hormone provides a general cue, or they may be present only in specific cells of specific tissues, so that the hormone exerts tissue-specific regulatory effects in addition to the general one.

15.4.3 Control of ecdysis

Five neuropeptides are involved in the control of eclosion in *Manduca*, and the process is probably similar at other molts and in other insects. They are: corazonin, which is produced in neurosecretory neurons of the brain and nerve cord; the ecdysis-triggering hormones (ETH and PETH) from

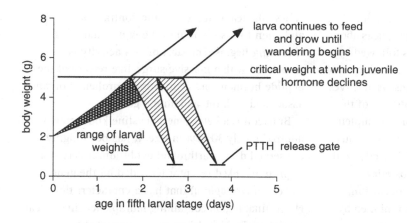

Figure 15.33 The critical body weight at which the molting program is initiated in the final larval stage of *Manduca*. The release of prothoracicotropic hormone (PTTH) is gated. Larvae developing more slowly miss the gate for PTTH release on day 3 and are delayed until day 4 (after Nijhout, 1981).

the epitracheal glands; eclosion hormone (EH) from neurosecretory cells in the brain; and crustacean cardioactive peptide (CCAP) from cells in the ventral nerve cord (see Chapter 21 for details of hormones and their origins). These neuropeptides interact to induce a specific set of coordinated behaviors that culminate in eclosion, and which are only performed at this time (Fig. 15.31).

The initiation of eclosion is dependent on signals indicating that the insect is physiologically in an appropriate state and that it is appropriately positioned for the process to be completed successfully (grasshoppers, for example, suspend themselves from their hindlegs before a molt). These signals are largely unknown, but the falling ecdysteroid titer is important in permitting release of the ecdysis-related neuropeptides as well as promoting the sensitivity to these hormones of their target tissues.

Corazonin appears to be the first of the ecdysis-related neuropeptides to be released into the hemolymph at the appropriate time. It acts directly on cells in the epitracheal glands to promote the release of ETH and PETH. These hormones in turn act on the ventral ganglia of the nerve cord to switch on pre-ecdysis behavior (see below). ETH and PETH also stimulate neurosecretory cells with soma in the brain to release EH from their central and peripheral neurohemal release sites. EH and ETH/PETH then participate in a positive feedback loop, each stimulating the release of the other, so their

concentrations rapidly build up to a peak about an hour before eclosion occurs, declining over the next few hours. This surge of EH and ETH/PETH means that ecdysis is initiated by an unambiguous timing signal. In emerging adult *Manduca*, the release of EH (and perhaps also ETH) is gated so that eclosion occurs at a specific time of day, but it is not gated at earlier molts. Gated production of EH probably accounts for the specific times of eclosion in many other insects mentioned in Section 15.3.3.

EH is released locally within the central nervous system as well as into the hemolymph, and the former leads to the release of CCAP in the ventral ganglia. CCAP switches off pre-eclosion behavior and switches on eclosion behavior. Acting via the hemolymph, it also causes an accelerated heartbeat, perhaps also facilitating the circulation of ecdysis-related hormones in the hemolymph.

The programs of motor neuron activity that produce pre-eclosion and eclosion behavior are built into the central nervous system. In some insects, fictive behaviors can be produced even in isolated nervous systems. Important components of this behavior are located in the frontal ganglion and stomatogastric system (air swallowing) and in the abdominal ganglia. In the abdomen, pre-eclosion behavior consists of a series of rotational movements. These are produced by the muscles of the two sides contracting in antiphase, while those of successive segments contract together (Fig. 15.34a). This behavior, which probably serves to free the adult abdomen from the overlying pupal cuticle, lasts about 30 minutes in *Hyalophora cecropia*. It is followed by a period of quiescence of about the same duration and then by eclosion behavior. The latter consists of waves of abdominal contraction and of "shrugging" of the wing bases that continues until eclosion is complete. These movements are produced by the synchronous activity of muscles on the two sides of the body, with activity spreading forwards from the posterior abdominal segments (Fig. 15.34b). After eclosion, this behavior is no longer useful and is inhibited by higher centers in the central nervous system.

In *Manduca* EH also causes plasticization of the cuticle of the wings and probably other parts of the cuticle that are inflated during ecdysis, as well as initiating neural switches that permit the later release of the tanning hormone bursicon. In saturniid silk moths, EH also directly causes the breakdown of certain muscles that degenerate after ecdysis, although in *Manduca* this is differently regulated by the declining titer of ecdysteroid prior to ecdysis.

15.4.4 Controlling expansion and sclerotization

After ecdysis in *Manduca*, pre-patterned wing-spreading behavior is triggered via the subesophageal ganglion. Tonic contraction of the abdomen forces blood into the wings and wing spreading begins. The start of this activity is associated with a massive surge-like release of the peptide hormone bursicon from neurohemal organs associated with the abdominal ganglia.

Bursicon is a heterdimeric cystine knot protein approximately 30 kDa in size. It has been highly conserved during arthropod evolution. It acts at a G-protein-linked receptor (encoded by the gene *rickets* in *Drosophila*) that has a characteristic N-terminal region containing multiple leucine-repeat motifs, and which drives production of the intracellular second messenger cyclic AMP. Bursicon has several actions. First, it produces a further increase in the plasticity of the cuticle of selected body parts (those that will be markedly expanded following ecdysis – notably the wings in adults). It also causes the programmed cell death (apoptosis) of certain cells, especially the epidermal cells within the wings. Finally, it initiates sclerotization (tanning) of the cuticle by promoting uptake of tanning precursors from the hemolymph.

In adult *Manduca*, premature release of bursicon is prevented by mechanical contact of the insect with its surroundings, such as occurs while the insect is still within its old cuticle. Similar inhibition occurs in insects, such as some flies, that pupate below the

(a) pre-eclosion

(b) eclosion

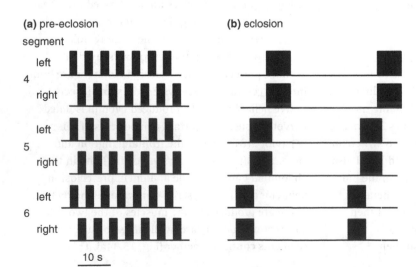

Figure 15.34 Patterns of motor activity in nerves to the abdominal muscles initiated by hormones at eclosion of an adult moth (*Manduca*) (see Fig. 15.31) (based on Truman, 1978). (a) Pre-eclosion behavior is switched on by ecdysis-triggering hormone. (b) Eclosion behavior is switched on by crustacean cardioactive peptide, the release of which is stimulated by eclosion hormone.

surface of the ground, and in first-stage grasshoppers when they hatch from their underground eggs. In both cases, bursicon release is inhibited while the insects are digging, but begins as soon as they are free on the surface. In other cases, such as the larval and pupal molts of *Manduca*, however, bursicon release occurs as an obligatory part of pre-patterned ecdysis behavior and cannot be delayed. In these cases, presumably, the environment in which ecdysis normally takes place is much more predictable.

In *Manduca*, *Drosophila* and a number of other insects, the bursicon-containing neurons in the abdominal ganglia, and their transverse nerve release sites, also contain CCAP, so that release of these two hormones into the hemolymph is coordinated. The released CCAP accelerates the pumping activity of the heart, directly facilitating wing expansion, and also assisting the distribution of bursicon to its target tissues.

15.5 Polyphenism

The term polymorphism has been used in a general sense to indicate variation in body form or color within a species. It is evident, however, that, while some instances of polymorphism are under strict genetic control, others are a product of the environment in which the insect develops. These instances are usually referred to as polyphenism. Thus, polyphenism refers to the occurrence of different phenotypes within a species where the development of the phenotype is governed by environmental conditions.

Polyphenism is a characteristic feature of social insects, where the different castes are determined by the needs of the colony, and of aphids, where different morphs occur seasonally. In addition, many insects exhibit seasonal variations in form that can be attributed to environmental conditions. Different color forms or differences in wing development are among the more common.

In all these cases, the environmental conditions are detected by the nervous system and modulate the amounts or timing of hormone secretion. JH or ecdysteroids are often involved, but, in other cases, neuropeptides from the brain provide the link between the nervous system and the tissues.

Castes of social insects Among social Hymenoptera, queens and workers are all female. While sex is determined genetically (females are diploid while males are haploid), caste is determined by conditions experienced during larval life. In the honey bee, *Apis mellifera* (Hymenoptera, Apidae), a newly hatched female larva has the potential to develop into either form. Its subsequent development is determined by the quality and quantity of the food supplied to it. If given a diet based primarily on secretions of the nurse bees' mandibular glands ("royal jelly"), it will become a queen. The key component of royal jelly that triggers queen development has been identified as a 57 kDa protein, royalactin. If a larva receives larger proportions of secretion from the hypopharyngeal glands of the workers, it will itself become a worker. Similarly, among the larvae of worker ants, the differentiation of minor or major workers, or of soldiers, is dependent on feeding, which is ultimately reflected in larval size.

These size differences are associated with different titers of JH, a hormone that controls insects' development. The switch from worker to queen in *Apis* is made in the late fourth and early fifth larval stages. Queen larvae develop a high titer of JH late in the fourth larval stage, whereas in worker larvae the level of JH is very low (Fig. 15.35). Among larval ants (Formicidae), major worker or soldier development is apparently the result of a high titer of JH early in the final larval stage. If the JH titer is above a threshold, the critical size at which pupation occurs is reset so that the larva continues to develop until it reaches a new threshold (Fig. 15.36a). The threshold level for soldier development is variable and soldiers already present in a colony can suppress the development of further soldiers, apparently by releasing a pheromone comparable with those (Section 27.2.7) that regulate

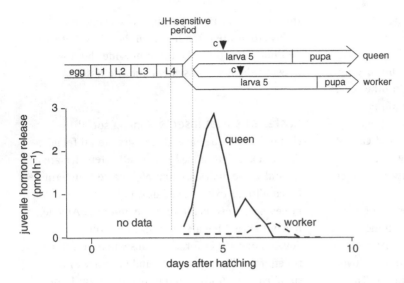

Figure 15.35 Polyphenism. Control of development of female honey bees. The development of larvae diverges in the final (fifth) stage as the production of juvenile hormone by the corpora allata becomes markedly higher in larvae destined to become queens. This corresponds with a JH-sensitive period late in the fourth stage when the switch from workers to queens occurs. The small peak in juvenile hormone production before pupation corresponds with a second JH-sensitive period, perhaps regulating development of the imaginal discs (c, time at which cells are capped by workers) (after Rachinsky and Hartfelder, 1990, and other sources).

colony structure in termites (Isoptera). It is believed that the ant pheromone raises the threshold for JH that must be reached for the switch to soldier production to occur. Not only does the high JH titer reset the critical size for pupation, it may also reprogram the development of imaginal tissues so that some parts of the body are allometically increased in size relative to the overall size of the insect (Fig. 15.36b).

Aphid morphs Aphids (Hemiptera) occur in a variety of different forms. Most generations consist entirely of parthenogenetic females, but a sexual generation usually occurs each year. Some, like *Aphis fabae* (Hemiptera, Aphididae), have separate winter and summer hosts. Movement between the two hosts demands the production of winged morphs (alatae or alates), whereas at other times the aphids are commonly wingless (apterae), although winged forms may also be produced on the secondary, summer host (Fig. 15.37).

The sexual generation is usually produced in response to the short days of autumn acting on the parent aphid. Males are produced by the loss of an X-chromosome during the single maturation division of the oocytes (female aphids have two

X-chromosomes, males only one [Section 14.4]). Under long photoperiods, the largest embryos in the ovarioles are determined for the production of more virginoparae (insects that will produce more parthenogenetic forms). Under short photoperiods, however, the embryos become determined as gynoparae (giving rise to sexual forms) or directly as oviparae (the sexual female). The production of virginoparae is influenced by light passing directly through the cuticle of the head, not via the compound eyes. Juvenile hormone appears to be involved in virginopara production, and low titers mimic short days and the production of sexual forms, but the mechanism is not well understood.

The production of winged (alate) forms is also under environmental control, and is induced by crowding, short photoperiods and low temperatures. The production of winged gynoparae in *Aphis fabae* requires exposure of both the parent female and the postnatal insect to short days. Initially, the embryo is affected indirectly by the light regime experienced by its parent, but well-developed embryos are themselves sensitive to photoperiod. It appears that JH acting early in development suppresses wing development, but how the environmental factors regulate JH titers is unclear.

(a)

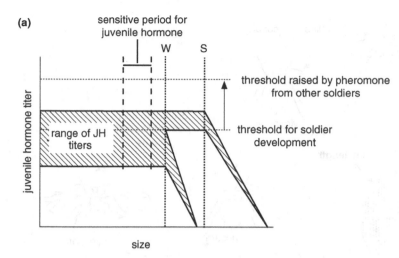

(b)

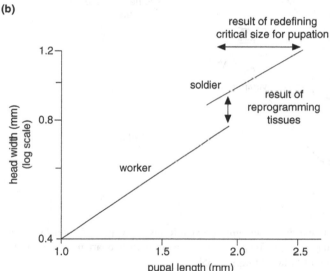

Figure 15.36 Polyphenism. Control of soldier development in the ant *Pheidole*. (a) Soldier development is determined by the hemolymph titer of juvenile hormone. Above a threshold titer, the critical size for completion of larval development (dotted vertical lines) is increased and the larva is destined to become a soldier. Soldiers already in the colony produce a pheromone which raises the threshold for juvenile hormone so that production of more soldiers is unlikely to occur. W, size threshold for pupation as worker; S, size threshold for pupation as soldier (after Wheeler and Nijhout, 1984). (b) Reprogramming of the critical size for completion of larval development shown in (a) results in the production of adults with greater head widths. At the same time adult tissues are reprogrammed to those of soldiers, resulting in a break in the allometric relationship between head width and body size (after Wheeler and Nijhout, 1983).

Brachyptery in other insects Short-winged forms are known to occur in many species of Orthoptera and Hemiptera. Long wings are often produced in response to crowding or poor food quality. Although wing development can be suppressed by experimentally elevating the JH titer in final-stage larvae, the interpretation of such experiments is not simple, since the time of application in relation to critical periods and the possibility of effects on the titers of other hormones must be taken into account.

Control of wing polymorphism may be more complex than a simple developmental switch operated by JH.

Color, form and behavior Many insects exhibit color variation in response to environmental changes. These may be regular occurrences, as in the seasonal forms of some butterflies, or more unpredictable, as in the phenomenon of homochromy (Section 25.5.2).

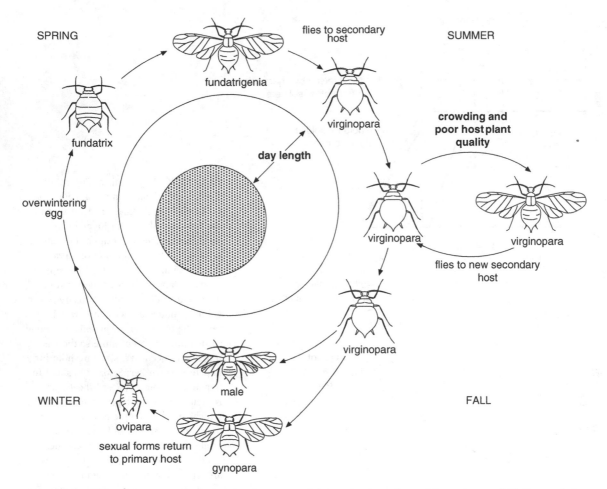

Figure 15.37 Polyphenism. Control of wing development and the production of sexual forms in an aphid. Seasonal changes in form result largely from responses to long or short day lengths. The production of winged virginopara in summer is a response to crowding or poor host plant quality (based on Hardie and Lees, 1985).

Crowding the larvae of some Lepidoptera causes them to become darker, or to have more dark patterning, than siblings reared in isolation. This is due to the deposition of greater quantities of melanin in the cuticle, which in caterpillars is controlled by pheromone biosynthesis activating neuropeptide (PBAN), so called because the same peptide controls pheromone biosynthesis in adults (Sections 21.1.3 and 27.5). Crowded larvae of *Spodoptera littoralis* have higher hemolymph titers of PBAN and the

inhibition of darkening by synthetic PBAN analogs strongly suggest that this peptide is the agent regulating the polymorphism in vivo.

Most (but not all) species of locusts (Orthoptera, Acrididae) also exhibit a color polymorphism associated with population density. Locusts reared in isolation are usually green with little black pigmentation; those reared in crowds are yellow or orange with extensive black pigmentation. The black patterning but not the yellow/orange coloration is regulated by the brain neuropeptide corazonin. All locusts exhibit marked changes in behavior with crowding, with gregarious insects being more active and attracted by others, whereas solitarious locusts are less active and avoid others. There are also morphological and anatomical differences; the pronotum of *Locusta* reared in isolation is produced into a prominent dorsal crest, while in crowded insects it is saddle-shaped. Locusts reared in crowds have larger brains and fewer antennal sensilla than their isolated siblings and levels of expression of olfaction-related genes (including *chemosensory protein* and *takeout*) in the insects' antennae are increased. The gregarized phenotype can be epigenetically transmitted by crowded females to their offspring. In the desert locust, *Schistocerca gregaria*, the transmission of behavioral characteristics is chemically mediated via the action of a compound (apparently an L-dopa analog) originating from the female accessory glands and acting upon developing eggs after ovulation and soon after laying. Color and size characteristics appear to be transmitted to hatchlings in a different manner from behavior, via effects on egg size, with crowded females laying larger eggs that yield larger, black hatchlings that are typical of the gregarious phase.

The key environmental signal regulating the immediate behavioral response to crowding is much better understood, and in *Schistocerca gregaria* has been shown to be mediated by a combination of visual and olfactory stimulation or by mechanical (touch) stimulation of the outer face of the hind femur alone, which occurs more frequently under crowded conditions. Behavioral changes occur within two hours of experimental stimulation of the femur, and include a higher level of spontaneous activity, as well as an increased tendency to approach other locusts. Crucially, it has been shown that the neurotransmitter serotonin is involved; gregarious locusts have transiently elevated levels of serotonin in their thoracic CNS, but not the brain, and serotonin antagonists inhibit gregarious behavior while exogenous serotonin increases it.

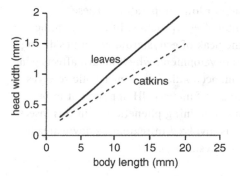

Figure 15.38 Polyphenism. The relationship between head width and body size in relation to feeding by caterpillars of *Nemoria arizonaria*. Insects feeding on the harder food (leaves) develop progressively bigger heads (after Greene, 1989).

An extreme example of polyphenism which involves differences in morphology as well as in color occurs in the caterpillar of *Nemoria arizonaria*. The spring generation of this species resembles the catkins of the oak trees on which it feeds. A second generation feeds on the same tree's leaves and are twig mimics rather than catkin mimics. They are greenish-gray in color, rather than yellow, and they lack many of the cuticular outgrowths which help to give insects in the spring generation their resemblance to catkins. The development of the two phenotypes is determined by the food eaten by the insects and, since the leaves of these oaks are very tough, the insects of the summer generation develop relatively larger heads (Fig. 15.38). It is possible that food quality affects form and color through a peptide comparable with that described above in *Spodoptera*.

Many adult butterflies also exhibit seasonal color polyphenisms according to day length or temperature. Production of the different morphs is often associated with the presence or absence of a pupal diapause (Section 15.6). A well-studied example is *Bicyclus anynana* (Lepidoptera: Nymphalidae). Different morphs differing in wing pattern (prominent ventral eyespots versus cryptic patterning) are produced by exposure of the larval stages to high or low temperatures. These environmental effects produce differences in the timing of the peak of ecdysteroid hormones during pupal–adult development, which in turn affects wing pigmentation, accumulation of metabolic reserves and reproductive function. JH appears not to be involved in determining phenetic fate in this insect (although it is involved in regulating downstream reproductive physiology).

15.6 Diapause

Diapause is a delay in development evolved in response to regularly recurring periods of adverse environmental conditions. It is not referable to immediately prevailing adverse environmental conditions, and thus differs from a delay produced by currently adverse conditions such as low temperature.

In temperate regions diapause facilitates winter survival; in the tropics it is commonly associated with surviving regularly occurring dry seasons. Diapause also contributes to the synchronization of adult emergence so that the chances of finding a mate are considerably improved. This is particularly necessary in long-lived species such as the dragonfly, *Anax*, in which larval development usually extends over two summers and during this time individual rates of development vary so that at any one time a wide variety of larval sizes is present.

15.6.1 Occurrence of diapause

Many insects living in regions where winter temperatures are too low for development enter a state of diapause at some stage of the life history, but the occurrence of diapause in insects from warmer regions depends on the severity of their environment and the conditions in their particular microhabitat. Many insects in the tropics survive without a diapause.

Diapause can occur in any stage of development from the early embryo to reproductive adult, but in the majority of diapausing species only one stage exhibits diapause (Fig. 15.39). Egg diapause is common in grasshoppers and Lepidoptera, occurring early in embryogenesis in some species, and much later in others. Many holometabolous insects diapause in the pupal stage. In eggs and pupae, morphogenesis comes to a standstill during diapause, reflected in the reduced oxygen consumption (Fig. 15.40). Diapausing larvae may stop feeding altogether, but some species remain active and feed periodically. They may molt, but such molts are not associated with growth, and are called stationary molts. Adult diapause is characterized by a lack of sexual development or behavior. Feeding may occur in some species.

Some insects have an obligate diapause, with every individual in every generation entering diapause. Insects with an obligate diapause usually have only one generation each year, a univoltine cycle. In other

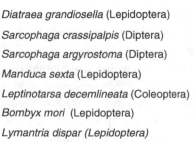

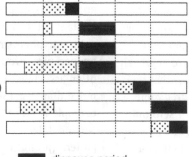

Figure 15.39 Diapause in different insects showing the separation of the sensitive stage (dotted) from the stage of diapause (black) (based on Nijhout, 1994).

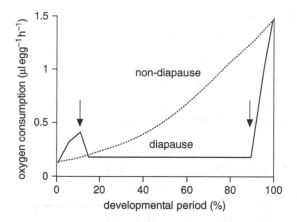

Figure 15.40 Oxygen consumption during diapause in the egg of *Melanoplus*. The dotted line shows the progressive increase in consumption that would occur if the egg developed without diapause. Note that the developmental period is expressed as a percentage of the total time for embryonic development. In the absence of diapause this would be about 30 days; with diapause it could be 200 days. Arrows mark the beginning and end of diapause (partly from Lees, 1955).

species, some generations may be completely free of diapause while in other generations some or all of the insects enter diapause. This is facultative diapause. It occurs in insects with two or more generations per year, a multivoltine cycle. Facultative diapause is well suited to regions with a long developmental season as it enables the insect to make the best use of the time available. The difference between obligate and facultative diapause is probably only one of degree. Insects with an obligate diapause apparently respond to such a wide range of environmental factors that they invariably undergo diapause, but in the laboratory it may be possible to avoid diapause by exposing them to conditions outside their normal range.

15.6.2 Initiation of diapause

The most reliable and consistent indicator of seasons is day length, or photoperiod, and this is the most important of the stimuli initiating

diapause. Other possible indicators are temperature, the state of the food and the age of the parent. The sensitive stage at which diapause is induced of necessity occurs before that at which diapause occurs (Fig. 15.39). The embryonic diapause of *Bombyx* is an extreme example. Here, diapause is primarily induced by exposure of the egg of the previous generation to diapause-inducing conditions.

The period of sensitivity to stimulation also varies in different species. Larval *Lacanobia oleracea* (Lepidoptera) are only sensitive to photoperiod for two days, while in *Bombyx mori*, although the well-developed embryo is the most sensitive stage, the first three larval stages are also sensitive, although to decreasing extents. Further, a number of photoperiodic cycles are necessary in order to produce an effect. The larva of *Dendrolimus pini* (Lepidoptera), for instance, must be subjected to about 20 short-day impulses to induce diapause, while exposure of the larva to 15 and 11 short-day impulses are required to induce diapause in pupal *Acronycta rumicis* (Lepidoptera) and *Pieris brassicae* (Lepidoptera), respectively. This number varies, however, under the influence of temperature and nutrition.

Photoperiod Outside the tropics, long days occur in summer and short days in winter, with increasing or decreasing day length in spring or autumn. The relatively short days of autumn herald the approach of winter and for many species they act as a stimulus initiating diapause. There is a critical day length around which small differences in photoperiod produce a complete change from non-diapause to diapause development (Fig. 15.41). As these insects develop without diapause under long-day conditions they are known as long-day insects.

Some insects react in the opposite way. *Bombyx* is an example of such a short-day insect. Here, exposure of the eggs of the bivoltine race to long days ensures that the eggs of the next generation will enter diapause (Fig. 15.42). Adults developing from

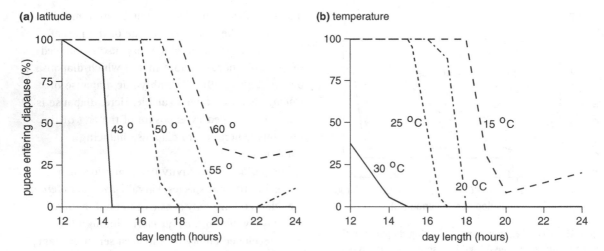

Figure 15.41 Diapause in the moth *Acronycta* (after Danilevskii, 1965). (a) The effects of photoperiod. Diapause is prevented by long days, but the critical day length, at which the switch from non-diapause to diapause development occurs, increases at higher latitudes. All insects were reared at 23°C. (b) The interaction of photoperiod with temperature. The critical photoperiod for diapause is reduced in insects reared at higher temperatures. All insects were from the same population, from 50 °N.

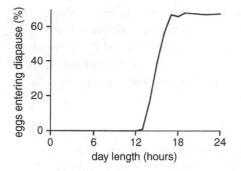

Figure 15.42 Diapause in the egg of the silk moth, *Bombyx*. Long days experienced by the embryo induce diapause in the eggs of the next generation (see Fig. 15.39) (after Danilevskii, 1965).

eggs exposed to a photoperiod of less than 14 hours lay non-diapausing eggs in the next generation. Thus the long-day conditions of spring acting on the eggs lead to the production of diapause eggs in the autumn, but these eggs, being subjected to short days, will ensure that the eggs of the next generation, laid the following spring, develop without diapause.

Photoperiod interacts with temperature, and in long-day insects the critical day length, below which diapause occurs, is often longer at lower temperatures. For example, the critical day length inducing pupal diapause in *Acronycta* is about 16 hours at 25°C, but almost 19 hours at 15°C (Fig. 15.41b). In *Pieris*, however, the critical day length is not influenced by temperature.

The response of many animals to photoperiod involves the perception of small changes in day length rather than the actual duration of the light period, but in insects this is not the case. Most insects respond to the absolute length of the photoperiod. A few cases in which the insect is believed to respond to dynamically changing day length are recorded, but even these can probably be accounted for in terms of reactions to the absolute day length.

Cells in the brain are directly sensitive to light in many insects, and, in these species, the compound eyes and ocelli do not mediate the photoperiodic response, and sometimes, as in the larva of *Pieris*, there is a relatively transparent area in the cuticle of the head above the sensitive area in the brain.

On the other hand, in the beetle *Pterostichus nigrita* (Coleoptera), in which the cuticle of the head is heavily pigmented, the compound eyes mediate the response.

Light intensity during the photoperiod is not important, provided that it exceeds a very low threshold value. This varies with the species, but commonly is about 170 lux or less. Hence daily fluctuations in light intensity due to clouds have no effect on photoperiod and the "effective day length" includes the periods of twilight. As a result of this high sensitivity, insects inside fruit and even the pupa of *Antheraea pernyi* inside its cocoon are affected by photoperiod. Some insects are so sensitive that they are stimulated by moonlight (about 5 lux) which might, therefore, contribute to the effective day length. This effect, however, might be offset by the relatively low temperatures occurring at night. This insensitivity to light during the photoperiod means that experimentally often all that is necessary to set the clock is a "skeleton" photoperiod in which pulses of light at the times of "light-on" and "-off" define the period that elapses between the beginning and end of the "day." Such experiments, together with the use of short pulses of light to interrupt the "night," reveal that the photoperiodic system is generally sensitive to the length of the dark period, rather than the light period. There are two types of photoperiodic system, one in which the nervous system is particularly sensitive to light at certain times of night, and another in which it is the phase relationship between two rhythms set respectively by dawn and dusk that is important. In most insects, only short wavelengths are concerned in the photoperiodic reaction.

Temperature Temperature also plays a part in the induction of diapause and, in general, in temperate regions high temperature suppresses and low temperature enhances any tendency to enter diapause. Thus, it will reinforce photoperiod. With changes in latitude, however, this reinforcement does not occur as at higher latitudes, although summer days are longer, the temperatures are lower than they are nearer the equator. Hence species with extensive geographical ranges are differently adapted to photoperiod in different parts of their range. For instance, in southern Russia, *Acronycta* enters diapause only when the day length falls below 15 hours, but with increasing latitude there is an increase in the critical day length so that specimens from St. Petersburg (60 °N) only avoid diapause when the photoperiod exceeds 18 hours (Fig. 15.41a). These differences are inherited characteristics of the populations.

Because of these differences, the life history of *Acronycta* varies in different parts of its range. In the south (latitude 43 °N), three partly overlapping generations occur and only in the last do the pupae undergo diapause. Farther north (latitude 50 °N), two generations occur, the whole of the second being determined for diapause, while around St. Petersburg (latitude 60 °N) only a very small number of individuals avoid diapause and the species is largely univoltine. Comparable differences in the life histories are known in other insects. *Bombyx* again differs from most other insects in that high temperature induces diapause and low temperature prevents it. In *Nasonia vitripennis* (Hymenoptera), temperature acts independently of photoperiod, and chilling the female causes her to lay eggs which will give rise to diapausing larvae.

Other factors There is evidence in a few cases that the amount or quality of the food can influence diapause. A shortage of prey results in reproductive diapause in the green lacewing, *Chrysopa carnea* (Neuroptera), and host plant condition, perhaps related to protein shortage, induces diapause in the larvae of a number of stem-boring moths.

Rearing conditions of the parents may sometimes be important. For example, solitary phase females of the brown locust, *Locustana pardalina* (Orthoptera), lay 100% diapausing eggs compared with only 42% by gregarious phase females. Further, old gregarious phase females lay more diapausing eggs than young females.

15.6.3 Preparations for diapause

Before insects become inactive in diapause there is usually a build-up of reserve food substances, particularly in the fat body, with a consequent reduction in the proportion of water in the body. Comparison of different forms determined for diapause or non-diapause shows that those destined for diapause build up bigger food reserves. In adult *Pyrrhocoris apterus* (Hemiptera), for example, diapause-associated storage proteins accumulate in the hemolymph, and, in *Bombyx*, eggs destined for diapause receive about 10% more lipid during vitellogenesis than non-diapause eggs, and the lipid contains a higher proportion of unsaturated fatty acids.

Pupal diapause in some moths and flies is associated with the presence of considerably greater quantities of hydrocarbons in the cuticular surface wax. In *Manduca*, wax continues to be secreted for about ten days after pupation in diapause pupae, whereas in non-diapause pupae wax production is complete in four or five days (Fig. 15.43). The puparium of diapausing *Sarcophaga crassipalpis* (Diptera) has about double the quantity of hydrocarbons compared with non-diapause insects,

most of it on the inside of the cuticle. This increase in wax probably reduces water loss.

In many insects from temperate regions, diapause is associated with an increase in cold tolerance, although the two phenomena are not necessarily closely linked. Diapause is frequently associated with the accumulation of polyhydric alcohols, such as sorbitol and glycerol, which can function as cryoprotectants. However, the amounts that accumulate initially may not be sufficient to affect the supercooling temperature of the insect. Rather, they appear to be formed as a result of the general depression of metabolic activity. Further increase in these cryoprotectants may occur later as part of a thermal acclimation process which is associated with, but is not a component of, diapause (see Chapter 19).

15.6.4 Diapause development

Except in the adult, a delay in morphogenesis is characteristic of diapause, but, although morphogenesis is at a standstill, physiological changes do occur. This physiological development is referred to as diapause development. As with other physiological processes, diapause development occurs most rapidly under certain environmental conditions of which temperature is often of overriding importance. The range of temperatures at which diapause development occurs, unlike that for morphogenesis, varies with the geographical distribution of the species and, in temperate regions, the optimum temperature for diapause development is commonly within the range 0–10°C, well below the temperature necessary for morphogenesis (Fig. 15.44). At higher or lower temperatures, diapause development proceeds more slowly and in the extremes stops altogether. On the other hand, in tropical species, where diapause is concerned with survival of the dry season, the temperature range for diapause development is often little, if any, lower than the range for morphogenesis.

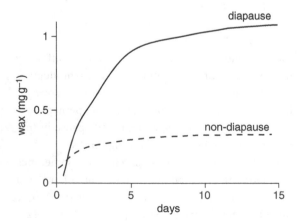

Figure 15.43 Preparation for diapause. The laying down of cuticular wax in the first few days of pupal life in diapause and non-diapause *Manduca* (after Bell *et al.*, 1975).

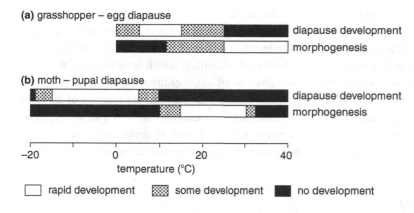

(a) grasshopper – egg diapause

diapause development
morphogenesis

(b) moth – pupal diapause

diapause development
morphogenesis

temperature (°C)

☐ rapid development   ▨ some development   ■ no development

Figure 15.44 Diapause development. Temperature ranges for diapause development and normal morphogenesis. (a) Eggs of the Australian grasshopper, *Austroicetes* (based on Andrewartha, 1952). (b) Pupae of the moth *Saturnia* (adapted from Lees, 1955).

Diapause is commonly associated with a relatively low water content of the tissues. This probably enhances the ability of the insect to survive periods of extremely low temperature, but the role of water in diapause development is not clear. In the eggs of many orthopteran species, morphogenesis is not resumed until water is available, but this water is only effective after a period of diapause development has been completed. Hence, although water is essential for the resumption of morphogenesis, as is true even of grasshopper eggs developing without diapause (Section 14.1), it is not concerned in diapause development. A similar restoration of the water balance accompanying reactivation after diapause is known to occur in other insects in other stages of development.

The duration of diapause development varies considerably with conditions and from species to species. Under optimal conditions the diapause development of *Gryllulus commodus* (Orthoptera, Gryllidae) is completed in 15 days; on the other hand, *Cephus cinctus* (Hymenoptera, Cephidae) requires a minimum of 90–100 days. Once diapause development is complete, morphogenesis is resumed provided the environmental conditions are suitable. If they are not, the insect remains in a state of quiescence until conditions become more favorable.

In the vast majority of insects, photoperiod is of no importance once diapause has been initiated, but in a few cases it directly affects the duration of diapause. For example, the larva of *Dendrolimus*

pini resumes its growth after a delay of only two weeks if the days during diapause are long, but remains dormant for twice as long in short days.

15.6.5 Control of diapause

Diapause in the larva and pupa, and probably also in late embryonic stages, results from a deficiency of ecdysone so that growth and molting do not occur. The release of ecdysteroids is normally triggered by prothoracicotropic hormone, and the lack of this hormone is ultimately responsible for the insect's failure to continue development. In the larvae of many species, high titers of JH may suppress PTTH secretion, but this is not always true. Sometimes JH seems to be necessary only at the very beginning of diapause, and in the wasp *Nasonia* it seems to play no part at all. Similarly in lepidopteran pupae, the suppression of PTTH secretion does not seem to be a consequence of high JH titers.

Diapause early in embryogenesis is produced by a diapause hormone in the parent female. In *Bombyx*, the hormone has been identified and is one of a number of peptides produced from a longer protein precursor (one of the others encoded by this gene is the pheromone biosynthesis activating neuropeptide, PBAN). Experiments in which embryos in dechorinonated eggs were cultured in vitro indicate that diapause hormone acts on the ovary, not on the egg or embryo. Only oocytes weighing about 500 µg enter diapause; smaller or larger oocytes are not affected. High levels of glycogen occur in diapause eggs, but it is not understood how this determines diapause. Diapause hormone results in an increase in the expression of the enzyme trehalase in the oocytes, leading to the conversion of hemolymph trehalose to glycogen in the yolk. Conversion of glycogen to sorbitol occurs in diapause eggs, but culture experiments show that although sorbitol can maintain diapause, it does not induce it. In addition, diapause hormone causes a change in pigmentation of the serosa due to the uptake of 3-hydroxykynurenine, which is subsequently converted to an ommochrome.

Adult diapause is linked to the absence of juvenile hormone, which is usually involved with vitellogenin synthesis or uptake, and, in males, with the development of reproductive behavior.

Summary

- Postembryonic development is divided into a series of stages, each separated from the next by a molt.

- In hemimetabolous insects the developmental changes that transform a larva into the adult are relatively slight; in holometabolous insects the changes are very marked and a pupal stage is interpolated between the final larval stage and the adult.

- Most cuticular growth and changes in cuticular form occur when an insect molts. The molt includes two distinct processes: apolysis, the separation of the epidermis from the existing cuticle, and ecdysis, the casting of the old cuticle after the production of a new one. The escape of the adult insect from the cuticle of the pupa or, in hemimetabolous insects, of the last larval stage is known as eclosion.

- Molting is governed, primarily, by two classes of hormone: molting hormones (ecdysteroids) and juvenile hormones (JH). The prothoracic gland is stimulated to produce ecdysone by prothoracicotropic hormone (PTTH). Five neuropeptides interact to induce a specific set of coordinated behaviors that culminate in ecdysis and eclosion. After ecdysis and eclosion, expansion of the cuticle is controlled by the peptide hormone bursicon.

- Polyphenism refers to the occurrence of different phenotypes within a species where the development of the phenotype is governed by environmental conditions.

- Diapause is a delay in development evolved in response to regularly recurring periods of adverse environmental conditions.

Recommended reading

HATCHING
Reynolds, S. E. (1980). Integration of behaviour and physiology during ecdysis. *Advances in Insect Physiology* 15, 474–593.

METAMORPHOSIS
Erezylmas, D. F., Rynerson, M. R., Truman, J. W. and Riddiford, L. M. (2009). The role of the pupal determinant *broad* during embryonic development of a direct-developing insect. *Development Genes and Evolution* 219, 535–544.

Truman, J. W. and Riddiford, L. M. (1999). The origins of insect metamorphosis. *Nature* 410, 447–452.

Truman, J. W. and Riddiford, L. M. (2002). Endocrine insights into the evolution of metamorphosis in insects. *Annual Review of Entomology* 47, 467–500.

GROWTH AND MOLTING
Hutchinson, J. M. C., McNamara, J. M., Houston, A. I. and Vollrath, F. (1997). Dyar's Rule and the Investment Principle: optimal moulting strategies if feeding rate is size-dependent and growth is discontinuous. *Philosophical Transactions of the Royal Society of London B* 352, 113–138.

BODY SIZE
Karl, I. and Fisher, K. (2008). Why get big in the cold? Towards a solution to a life-history puzzle. *Oecologia* 155, 215–225.

Mirth, C. K. and Riddiford, L. M. (2007). Size assessment and growth control: how adult size is determined in insects. *BioEssays* 29, 344–355.

Nijhout, H. F. (2003). The control of body size in insects. *Developmental Biology* 261, 1–9.

ECDYSIS BEHAVIOR
Ewer, J. and Reynolds, S. E. (2002). Neuropeptide control of molting in insects. In *Hormones, Brain and Behavior*, vol. 3, ed. D. Pfaff, A. Arnold, A. M. Etgen, S. Fahrbach and R. T. Rubin, pp. 1–92. San Diego, CA: Academic Press.

Žitňan, D., Kim, Y.-J., Žitňanová, I., Roller, L. and Adams, M. E. (2007). Complex steroid-peptide-receptor cascade controls insect ecdysis. *General and Comparative Endocrinology* 153, 88–96.

BURSICON AND SCLEROTIZATION
Honegger, H. W., Dewey, E. M. and Ewer, J. (2008). Bursicon, the tanning hormone of insects: recent advances following the discovery of its molecular identity. *Journal of Comparative Physiology A* 194, 989–1005.

HORMONAL CONTROL OF POSTEMBRYONIC DEVELOPMENT
Riddiford, L. M. (2008). Juvenile hormone action: a 2007 perspective. *Journal of Insect Physiology* 54, 895–901.

Spindler, K. D., Honl, C., Tremmel, C., Braun, S., Ruff, H. and Spindler-Barth, M. (2009). Ecdysteroid hormone action. *Cellular and Molecular Life Sciences* 66, 3837–3850.

POLYPHENISM

Hartfelder, K. and Emlen, D. J. (2005). Endocrine control of insect polyphenism. In *Comprehensive Molecular Insect Science*, vol. 3, ed. L. Gilbert, K. Iatrou and S. S. Gill, pp. 651–703. Amsterdam: Elsevier.

Tagu, D., Sabater-Muñoz, B. and Simon, J.-C. (2005). Deciphering reproductive polyphenism in aphids. *Invertebrate Reproduction and Development* 48, 71–80.

Zera, A. J. (2007). Endocrine analysis in evolutionary-developmental studies of insect polymorphism: hormone manipulation versus direct measurement of hormonal regulators. *Evolution and Development* 9, 499–513.

LOCUST PHASES

Anstey, M. L., Rogers, S. M., Ott, S. R., Burrows, M. and Simpson, S. J. (2009). Serotonin mediates behavioral gregarization underlying swarm formation in desert locusts. *Science* 323, 627–630.

Pener, M. P. and Simpson, S. J. (2009). Locust phase polyphenism: an update. *Advances in Insect Physiology* 36, 1–286.

Simpson, S. J. and Miller, G. A. (2007). Maternal effects on phase characteristics in the desert locust, *Schistocerca gregaria*: a review of current understanding. *Journal of Insect Physiology* 53, 869–876.

Tanaka, S. and Maeno, K. (2010). A review of maternal and embryonic control of phase-dependent progeny characteristics in the desert locust. *Journal of Insect Physiology* 56, 911–918.

DIAPAUSE

Saunders, D. S. (1982). *Insect Clocks*. Oxford: Pergamon Press.

Schiesari, L., Kyriacou, C. P. and Costa, R. (2011). The hormonal and circadian basis for insect photoperiodic timing. *FEBS Letters* 585, 1450–1460.

References in figure captions

Andrewartha, H. G. (1952). Diapause in relation to the ecology of insects. *Biological Reviews of the Cambridge Philosophical Society* 27, 50–107.

Bell, R. A., Nelson, D. R., Borg, T. K. and Cardwell, D. L. (1975). Wax secretion in non-diapausing and diapausing pupae of the tobacco hornworm, *Manduca sexta*. *Journal of Insect Physiology* 21, 1725–1729.

Bernays, E. A. (1972a). The muscles of newly hatched *Schistocerca gregaria* larvae and their possible functions in hatching, digging and ecdysial movements (Insecta: Acrididae). *Journal of Zoology, London* 166, 111–158.

Bernays, E. A. (1972b). The intermediate moult (first ecdysis) of *Schistocerca gregaria* (Forskål) (Insecta: Orthoptera). *Zeitschrift für Morphologie der Tiere* 71, 160–179.

Blackith, R. E., Davies, R. G. and Moy, E. A. (1963). A biometric analysis of development in *Dysdercus fasciatus* Sign. (Hemiptera: Pyrrhocoridae). *Growth* 27, 317–334.

Booker, R. and Truman, J. W. (1987). Postembryonic neurogenesis in the CNS of the tobacco hornworm, *Manduca sexta*: I. Neuroblast arrays and the fate of their progeny during metamorphosis. *Journal of Comparative Neurology* 255, 548–559.

Bryant, P. J. (1975). Pattern formation in the imaginal wing disc of *Drosophila melanogaster*: fate map, regeneration and duplication. *Journal of Experimental Zoology* 193, 49–78.

Chapman, R. F. and Fraser, J. (1989). The chemosensory system of a monophagous grasshopper, *Bootettix argentatus*. *International Journal of Insect Morphology and Embryology* 18, 111–118.

Chapman, R. F., Bernays, E. A., Woodhead, S., Padgham, D. E. and Simpson, S. J. (1983). Control of hatching time of eggs of *Chilo partellus* (Swinhoe) (Lepidoptera: Pyralidae). *Bulletin of Entomological Research* 73, 667–677.

Clarke, K. U. (1957). On the increase in linear size during growth in *Locusta migratoria* L. *Proceedings of the Royal Entomological Society of London A* 32, 35–39.

Clausen, C. P. (1940). *Entomophagous Insects*. New York, NY: McGraw-Hill.

Cole, B. J. (1980). Growth ratios in holometabolous and hemimetabolous insects. *Annals of the Entomological Society of America* 73, 489–491.

Comstock, J. H. (1918). *The Wings of Insects*. New York, NY: Comstock Publishing Company.

Danilevskii, A. S. (1965). *Photoperiodism and Seasonal Development of Insects*. Edinburgh: Oliver & Boyd.

Davies, R. G. (1966). The postembryonic development of *Hemimerus vicinus* Rehn & Rehn (Dermaptera: Hemimeridae). *Proceedings of the Royal Entomological Society of London A* 41, 67–77.

Duarte, A. J. (1939). Problems of growth of the African migratory locust. *Bulletin of Entomological Research* 29, 425–456.

Fristrom, D. and Fristrom, J. W. (1993). The metamorphic development of the adult epidermis. In *The Development of Drosophila melanogaster*, ed. M. Bate and A. M. Arias, pp. 843–897. New York, NY: Cold Spring Harbor Laboratory Press.

Greene, E. (1989). Diet-induced developmental polymorphism in a caterpillar. *Science* 243, 643–646.

Hardie, J. and Lees, A. D. (1985). Endocrine control of polymorphism and polyphenism. In *Comprehensive Insect Physiology, Biochemistry and Pharmacology*, vol. 8, ed. G. A. Kerkut and L. I. Gilbert, pp. 441–490. Oxford: Pergamon Press.

Hinton, H. E. (1946). A new classification of insect pupae. *Proceedings of the Zoological Society of London* 116, 282–328.

Houlihan, D. F. and Newton, J. R. L. (1979). The tracheal supply and muscle metabolism during muscle growth in the puparium of *Calliphora vomitoria*. *Journal of Insect Physiology* 25, 33–44.

Jago, N. D. (1963). Some observations on the life cycle of *Eyprepocnemis plorans meridionalis* Uvarov, 1921, with a key for the separation of nymphs at any instar. *Proceedings of the Royal Entomological Society of London A* 38, 113–124.

Kim, C.-W. (1959). The differentiation centre inducing the development from larval to adult leg in *Pieris brassicae* (Lepidoptera). *Journal of Embryology and Experimental Morphology* 7, 572–582.

Kimmins, D. E. (1950). Ephemeroptera. In *Handbooks for the Identification of British Insects*, vol. 1, part 9, *Ephemeroptera*. London: Royal Entomological Society of London.

Lanzrein, B., Gentinetta, V., Abbegglen, H., Baker, F. C., Miller, C. A. and Schooley, D. A. (1985). Titers of ecdysone, 20-hydroxyecdysone and juvenile hormone III throughout the life cycle of a hemimetabolous insect, the ovoviviparous cockroach *Nauphoeta cinerea*. *Experientia* **41**, 913–917.

Lees, A. D. (1955). *The Physiology of Diapause in Arthropods*. Cambridge: Cambridge University Press.

Macan, T. T. (1961). *A Key to the Nymphs of the British species of Ephemeroptera*. Ambleside: Freshwater Biological Association.

Madhavan, M. M. and Madhavan, K. (1980). Morphogenesis of the epidermis of adult abdomen of *Drosophila*. *Journal of Embryology and Experimental Morphology* **60**, 1–31.

Marshall, J. F. (1938). *The British Mosquitoes*. London: British Museum.

Nijhout, H. F. (1975). A threshold for metamorphosis in the tobacco hornworm, *Manduca sexta*. *Biological Bulletin* **149**, 214–225.

Nijhout, H. F. (1981). Physiological control of molting in insects. *American Zoologist* **21**, 631–640.

Nijhout, H. F. (1994). *Insect Hormones*. Princeton, NJ: Princeton University Press.

Nöthiger, R. (1972). The larval development of imaginal discs. In *The Biology of Imaginal Discs*, ed. H. Ursprung and R. Nöthiger, pp. 1–34. New York, NY: Springer-Verlag.

Oberlander, H. (1985). The imaginal discs. In *Comprehensive Insect Physiology, Biochemistry and Pharmacology*, vol. 2, ed. G. A. Kerkut and L. I. Gilbert, pp. 151–182. Oxford: Pergamon Press.

Padgham, D. E. (1981). Hatching rhythms in the desert locust, *Schistocerca gregaria*. *Physiological Entomology* **6**, 191–198.

Pflüger, H.-J., Hurdelbrink, S., Czjzek, A. and Burrows, M. (1994). Activity-dependent structural dynamics of insect sensory fibres. *Journal of Neuroscience* **14**, 6946–6955.

Rachinsky, A. and Hartfelder, K. (1990). Corpora allata activity, a prime regulating element for caste-specific juvenile hormone titre in honey bee larvae (*Apis mellifera carnica*). *Journal of Insect Physiology* **36**, 189–194.

Russell, V. W. and Dunn, P. E. (1991). Lysozyme in the midgut of *Manduca sexta* during metamorphosis. *Archives of Insect Biochemistry and Physiology* **17**, 67–80.

Savage, A. A. (1956). The development of the Malpighian tubules of *Schistocerca gregaria* (Orthoptera). *Quarterly Journal of Microscopical Science* **97**, 599–615.

Schubiger, G. (1971). Regeneration, duplication and transdetermination in fragments of the leg disc of *Drosophila melanogaster*. *Developmental Biology* **26**, 277–295.

Sláma, K. (1982). Inverse relationship between ecdysteroid titers and total body metabolism in insects. *Zeitschrift für Naturforschung* **37**c, 839–844.

Snodgrass, R. E. (1954). *Insect Metamorphosis*. Washington, DC: Smithsonian Institution.

Southwood, T. R. E. (1946). The structure of the eggs of the terrestrial Heteroptera and its relationship to the classification of the group. *Transactions of the Royal Entomological Society of London* **108**, 163–221.

Southwood, T. R. E. and Leston, D. (1959). *Land and Water Bugs of the British Isles*. London: Warne.

Truman, J. W. (1978). Hormonal release of stereotyped motor programmes from the isolated nervous system of the cecropia silkmoth. *Journal of Experimental Biology* **74**, 151–173.

Truman, J. W. and Reiss, S. E. (1988). Hormonal regulation of the shape of identified motorneurons in the moth *Manduca sexta*. *Journal of Neuroscience* **8**, 765–775.

Urquhart, F. A. (1960). *The Monarch Butterfly*. Toronto: University of Toronto Press.

van Emden, F. I. (1946). Egg-bursters in some more families of polyphagous beetles and some general remarks on egg-bursters. *Proceedings of the Royal Entomological Society of London* **21**, 89–97.

Waddington, C. H. (1941). The genetic control of wing development in *Drosophila*. *Journal of Genetics* **41**, 75–139.

Wheeler, D. E. and Nijhout, H. F. (1983). Soldier determination in *Pheidole bicarinata*: effect of methoprene on caste and size within castes. *Journal of Insect Physiology* **29**, 847–854.

Wheeler, D. E. and Nijhout, H. F. (1984). Soldier determination in *Pheidole bicarinata*: inhibition by adult soldiers. *Journal of Insect Physiology* **30**, 127–135.

Wigglesworth, V. B. (1965). *The Principles of Insect Physiology*. London: Methuen.

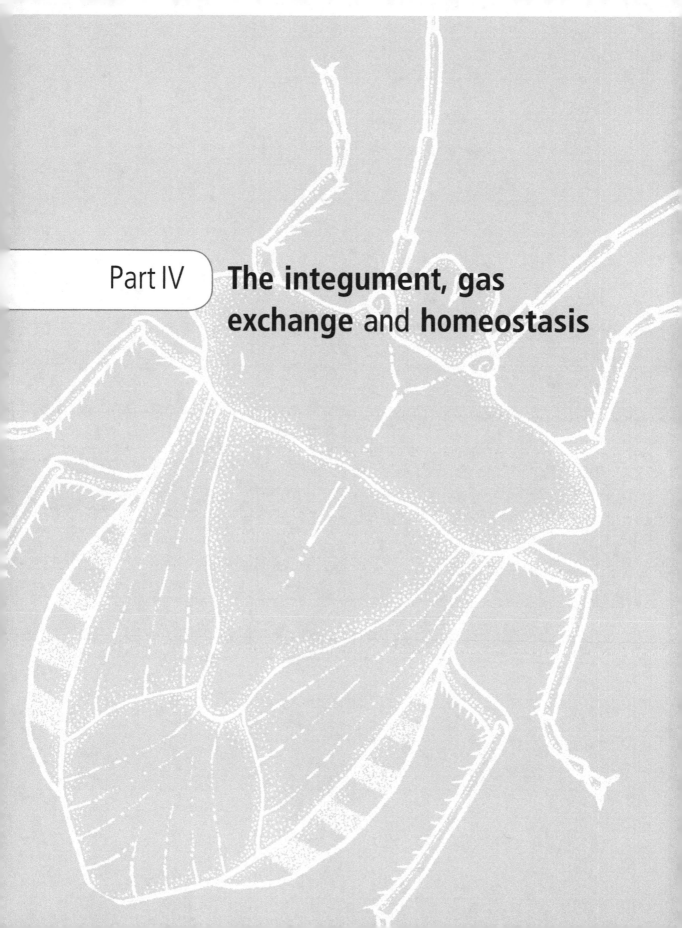

Part IV **The integument, gas exchange and homeostasis**

16 Integument

REVISED AND UPDATED BY **HANS MERZENDORFER**

INTRODUCTION

The enormous success of insects is partly based on the unique properties of their outermost layer, which is called the integument. It consists of a single layer of ectodermal cells (epidermis, hypodermis), which is covered by the cuticle, a chitinous apical extracellular matrix secreted by the epidermis. Epidermal cells are also involved in the formation of a basal extracellular matrix, the basement membrane, which effectively separates the integument from the hemocoel. The integument largely determines the outer shape of an insect and functions as exoskeleton, to which the muscles are attached. It forms a sensory interface with the environment and protects the insect from various harms, including mechanical damage, radiation, desiccation and invasion of pathogenic microorganisms. For this purpose almost all outer surfaces of the insect body are covered by the integument, including ectodermal invaginations like the oral cavity, the fore- and hindgut, the lower genital ducts and many glands.

In this chapter, insights into the cellular and molecular architecture of the integument are provided. The chapter is divided into seven sections. Section 16.1 describes organization and differentiation of the epidermis. Section 16.2 focuses on the architecture of the cuticle. In Sections 16.3 and 16.4 the chemical composition and its variation in different cuticle types is discussed. This is followed by Section 16.5 on molting and Section 16.6 on cuticle formation, including a section on sclerotization. In Section 16.7 the various functions of the integument are outlined.

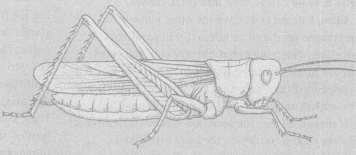

The Insects: Structure and Function (5th edition), ed. S. J. Simpson and A. E. Douglas.
Published by Cambridge University Press. © Cambridge University Press 2013.

16.1 Epidermis

16.1.1 Epidermal cells

The epidermis is the outer cell layer of the insect and is mostly single-layered. The density and shape of epidermal cells may change substantially during development. The apical plasma membranes of epidermal cells form short microvilli, which possess electron-dense plaques at their tips (Fig. 16.1a). These specialized membrane regions are known as plasma membrane plaques and believed to be the sites of secretion of chitin fibers. The plaques are connected to actin filaments that may help to orientate the chitin-synthesizing machinery. During and just after a molt, the epidermal cells also develop cytoplasmic processes on the outside, extending into the pore canals of the cuticle (see Fig. 16.3b), but these processes may be withdrawn as the cuticle matures.

The basal plasma membrane is commonly flat, but may have infoldings that increase the membrane surface facilitating transport of molecules (basal labyrinth or membrane reticular system). These infoldings are particularly apparent in gland cells (Fig. 16.1c,d).

All the epidermal cells are glandular in the sense that they secrete cuticle components and enzymes involved in cuticle modification and digestion at the time of molting. To accomplish these functions, they have an extensive rough endoplasmic reticulum and Golgi complexes, and they frequently contain membrane-bound pigment granules. Some epidermis cells have specialized glandular functions, and these cells may be categorized into three classes.

Class 1 gland cells have the outer plasma membrane produced as microvilli or parallel lamellae, which may abut directly onto the cuticle, but are often separated from it by a space in which it is presumed their secretion accumulates (Fig. 16.1c). The cuticle above the cells is usually unmodified and it is presumed that the secretion reaches the external surface of the cuticle via the pore canals and epicuticular filaments. Class 1 cells are often involved in pheromone production.

Class 2 gland cells are derived from epidermal cells, but have no direct contact with the cuticle, nor do they have a duct. They are only known from the sternal glands of termites.

Class 3 gland cells are also below the epidermis, but connect with the exterior by a duct (Fig. 16.1d). The apical plasma membrane of the gland cell possesses microvilli-like structures and forms a cavity. The duct projects into the cavity of the gland cell and it is perforated, forming a structure called the end apparatus. The duct is produced by a separate cell. Class 3 gland cells produce the cement on the outer surface of the cuticle (see below) and exhibit cycles of development that are synchronized with the molting cycle. In all dipterous larvae they form peristigmatic glands. These are composed of large cells with ducts opening to the outside, near the edges of the spiracles. Their secretion, which is produced continuously, is responsible for the hydrofuge properties of the cuticle surrounding the spiracles, which prevent the entry of water into the tracheal system. In addition, class 3 gland cells are often involved in the production of defensive secretions and pheromones (Chapter 27).

Epidermal cells and derived gland cells may combine to form gland units of various complexity, which are composed of a gland cell, reservoir and reaction compartments, and collecting and excretory ducts formed by canal cells. Sometimes gland units contain accessory cells, specialized secretory cells which add supplementary compounds to the secretion product. Insect glands can be built in various ways and serve numerous different functions. The list includes silk glands, defense glands, glandular hairs, wax glands, gland cells of adhesive organs (Section 8.1.3), milk glands of viviparous flies and various head glands. In addition to gland cells, a multitude of other specialized cells that reside within the integument are derived from epidermal cells. These cells include sensillum cells, duct cells, hairs, bristles and scales (formed by trichogen and tormogen cells) and oenocytes, which will be discussed next.

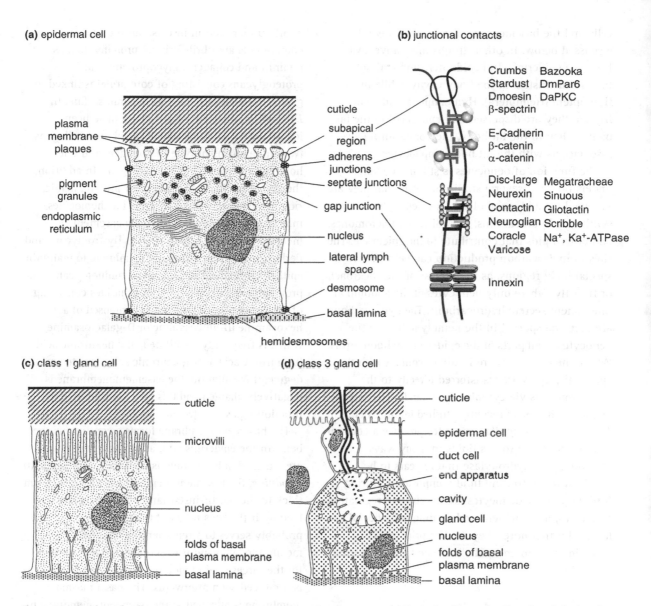

Figure 16.1 Epidermal cells, membrane organization and junctional contacts. (a) Typical epidermal cell as found during the intermolt period. (b) Membrane organization and junctional contacts; some proteins that have been identified in the respective region are indicated. (c) Class 1 gland cell lacking a duct to the exterior. (d) Class 3 glandular unit with a gland cell that is connected to the exterior via a duct formed by a duct cell.

16.1.2 Oenocytes

Oenocytes, named because of their pale amber color, are regarded as derived from epidermal cells, but despite their ectodermal origin they may be found associated with various tissues. They are often large, polyploid cells with an extensive tubular endoplasmic reticulum. In Ephemeroptera, Odonata and Heteroptera (Hemiptera), the oenocytes remain in the epidermis between the bases of the epidermal cells and the basement membrane, which will be discussed below. In other groups they move away from the epidermis. In Lepidoptera and Orthoptera they form clusters in the body cavity, while in Homoptera (Hemiptera), Hymenoptera and some Diptera they are dispersed and embedded in the fat body. When in the fat body, they often form close associations with the fat body trophocytes.

The function of oenocytes is still not completely understood, but several studies indicate that they are involved in lipid metabolism. The oenocytes synthesize hydrocarbons, including sex pheromones, and other lipids that contribute to the epicuticle. The sites of hydrocarbon production can be restricted to special body regions, as in the case of the cockroach or fruit fly, where only oenocytes of the abdominal integument secrete hydrocarbons. The hydrocarbons are then transported in the hemolymph from the oenocytes to all parts of the epidermis by lipophorins. Where the oenocytes are in direct contact with the epidermis, lipids are transferred directly to the epidermal cells via cytoplasmic strands. The role of oenocytes has been recently studied in more detail in *Drosophila*. *Drosophila* oenocytes produce a cocktail of enzymes that process lipids in many ways, including a ω–hydroxylase that appears to be involved in controlling lipid composition. During feeding periods, oenocytes are involved in regulation of growth, development and feeding behavior of the larvae. Furthermore, they are the principal cells depleting the stored lipids from the fat body during starvation.

16.1.3 The basement membrane

The epidermal cells stand on the basement membrane, which is not a biological membrane made of lipid mono- or bilayers, but a basal extracellular matrix. It consists frequently of two layers which are fused to each other, the basal lamina produced by epidermal (or other epithelial) cells and the reticular lamina synthesized by hemocytes and fat body cells. The chemistry of the basement membrane is still poorly understood in insects, but the primary components are fibril-forming proteins such as laminin and collagens, glycoproteins and proteoglycans consisting of core proteins linked to glycosaminoglycans such as heparan sulfate. In *Drosophila*, three conserved genes have been identified encoding basement membrane collagens, two alpha chains of type IV collagen, and one homolog of types XV/XVIII collagens. In addition, the glycoprotein SPARC (*s*ecreted *p*rotein, *a*cidic, *r*ich in *c*ysteine), which is secreted by hemocytes, interacts with the collagens of the basement membrane. Proteoglycans such as Dystroglycans and Perlecans act at the basement membrane to maintain epithelial organization. The glycosaminoglycan of the proteoglycans are heteropolysaccharides consisting of repeating disaccharide units composed of a hexosamine (D-glucosamine or D-galactosamine), which is frequently *N*-sulfated, and hexuronic acid (L-iduronic acid or D-glucuronic acid). Due to various degree of *N*-sulfation, the basement membrane is negatively charged and this may contribute to its functioning as a molecular sieve.

The basement membrane forms a continuous sheet beneath the epidermis and, at points where muscles are attached, it is continuous with the sarcolemma. In *Rhodnius*, the basement membrane is about 0.15 μm thick in the fourth-instar larva, but six days after feeding it thickens to about 0.5 μm. The thickening probably serves to strengthen the basement membrane, allowing it to serve as a stable platform for the forces used in molding the new epicuticle that is produced soon afterwards. The basal plasma membrane is attached to the basement membrane by hemidesmosomes, comprising adhesion molecules of the integrin family.

16.1.4 Membrane organization and junctional contacts

The epidermis is built of polarized cells exhibiting apical, subapical and basolateral membrane domains that differ in their protein composition. The apical domain in the embryonic epidermis includes membrane proteins such as Stranded-at-second, a putative cell surface receptor, or the Yellow protein, a dopachrome-converting enzyme secreted into the extracellular space, both of which frequently serve as markers for this membrane region. The subapical domain is specified by the presence of two protein complexes that are involved in establishing and/or maintaining epithelial polarity (Fig. 16.1b). In *Drosophila* one of these complexes is composed of Bazooka and DmPar6, which both contain PDZ domains mediating protein–protein interactions, and an atypical protein kinase, DaPKC. The other complex of the subapical region is formed by Crumbs, Dmoesin, β-spectrin and Stardust. Crumbs is a transmembrane protein that recruits the guanylate cyclase Stardust, the actin-binding proteins Dmosein and β-spectrin via its short cytoplasmic tail to the membrane. The protein complexes of the subapical region are required for the correct assembly of junctional complexes.

The formation of cell junctions is a crucial step during epidermis development as it ensures correct organization and function of the entire tissue. Epidermal cells are held together by several types of junctional contacts below the subapical region, which not only mediate adhesiveness but also prevent the passage of small molecules and facilitate cellular communication. Two distinct junctional complexes regulate cell–cell adhesion in the epidermis: more apical adherens junctions (zonula adherens) and more basal septate junctions (homologous to vertebrate tight junctions). These junctions are formed by numerous transmembrane, scaffolding and actin-binding proteins (Fig. 16.1b).

Adherens junctions are primarily formed by the transmembrane cell adhesion protein E-cadherin, with a large extracellular domain mediating homophilic interactions, and the cytoplasmic protein β-catenin, which binds to the cytoplasmic region of E-cadherin but also can bind the α-catenin that links the complex to actin filaments. In addition to this complex, the Nectin–Afadin complex is associated with adherens junctions, regulating epithelial organization.

Septate junctions are characterized by a ladder-like array of crossbridges that span the 20 nm gap between the basolateral membranes of the neighboring epidermal cells. They encircle the epithelial cells as a continuous belt, with the exception of tricellular contacts, where a septate junction's continuity is interrupted by channel-like structures of unknown function that span the epithelium. The structure of septate junctions appears to be more complex than that of adherens junctions. They are composed of transmembrane cell adhesion proteins (Neurexin, Contactin and Neuroglian), that form a tripartic extracellular complex that is not only observed in neural tissue but also in epidermal cells. Another transmembrane protein, the Na^+, K^+-ATPase, which is known to be involved in driving epithelial transport, localizes to septate junctions and is essential for their function as a paracellular diffusion barrier. The scaffolding proteins Coracle and Varicose may help to organize the junctional complex by binding the cytoplasmic domain of Neurexin and restricting its lateral mobility within the basolateral membrane. Several additional cytoplasmic proteins are required for septate junction formation, including Scribble, Gliotactin, Disc-large, Sinuous and Megatrachea. The latter two proteins have attained particular interest, because they are similar to vertebrate claudins that have a crucial role in paracellular electrolyte transport in different epithelia. However, it appears unlikely that they are real functional homologs. Another protein that associates with septate junction proteins and helps to organize the basolateral membrane is Yurt. Additional to its function in establishing and maintaining epithelial polarity, it has been shown to prevent passive diffusion of molecules larger than 10 kDa.

At greater distances from the cuticle, adjacent cells are no longer tightly bound to each other, and the spaces between them (lateral lymph spaces) are, to some extent, isolated from the hemolymph by desmosomes close to the basement membrane. Gap junctions between epidermal cells provide a pathway for the movement of low molecular weight substances (<1 kDa), such as ions (e.g., Ca^{2+}) and metabolites (e.g., inositol phosphates, cyclic nucleotides). They are made from different innexins (invertebrate connexin analogs) that, like vertebrate connexins, may be arranged as hexamers and form heteromeric hemichannels. Two of these hemichannels of neighboring cells eventually dock head-to head and form a channel connecting the neighboring cells (Fig. 16.1b). The gap junctions enhance coordination between cells. The most basal domains of the epidermis are specified by integrins that interact with components of the basement membrane. Genetic studies carried out in *Drosophila* revealed that the proteins found in different membrane domains are highly dynamic structures, whose formation and maintenance is interdependent.

Junctional complexes also play a crucial role in establishing polarization of epidermal cells during development. Although there may be variations between different insect species, this process follows a general pathway, which is initiated by the activity of genes already expressed in the early embryo. First signs for polarization can be detected in *Drosophila* embryos during cellularization, when transient junctions are established which contain junctional proteins such as E-cadherin, α/β-catenins and Disc-lost. Later, other protein complexes reinforce polarity by forming successive subdomains along the apical–basal axis.

16.1.5 Epidermis differentiation

Most of what we currently know about the membrane organization of insect epidermal cells derives from genetic studies performed with *Drosophila* embryos. Unlike many other systems, the *Drosophila* epidermis is a primary epithelium that is formed without passing through intermediate mesenchymal states. Therefore, *Drosophila* embryos have proved a highly suitable model to study epidermis formation. *Drosophila* development begins with multiple nuclear divisions yielding a syncytial organized early embryo (see Section 14.2; Fig. 16.2). During cellularization, which requires maternal products and the activities of several zygotic genes such as *nullo*, *slam* and *bottleneck*, membranes between the nuclei are formed, generating the first epithelial layer, which is called the blastoderm. The epidermis is formed from lateral parts of the blastoderm after enough cells have been generated by additional cell divisions. To enclose the entire body of the embryo, the epidermal cells move and stretch and finally fuse at the dorsal site (Section 14.2).

Determination of the body axis is initiated in the oocyte before fertilization by the asymmetrical distribution and activity of maternal gene products (mRNA and proteins, Section 14.2.8). Diffusion of these molecules generates morphogenic gradients that switch on the expression of primary segmentation genes (gap genes) in larger segments of the ectoderm. These segments get further subdivided into anterior and posterior compartments by the selective expression of "segment polarity genes" such as *wingless* and *hedgehog*, or the homeodomain genes *engrailed*. The products of the "gap genes" form short-range morphogenic gradients, which control primary pair-rule genes initializing the metameric segmentation process at the blastoderm stage. The activities of "segment polarity genes" lead to the formation of segmental primordia that correspond to future segments. It is the complex interplay of different morphogenic factors controlling gene expression in a concentration-dependent manner which finally establishes positional identity throughout each segment and controls further differentiation of the epidermis, such as the formation of trichomes, non-sensory hair-like cuticular extensions. The analysis of *Drosophila* trichome patterns helped to elucidate the underlying mechanism of genetic control. For instance, the dorsal cuticle of the thoracic segments exhibits four types of cuticular organization:

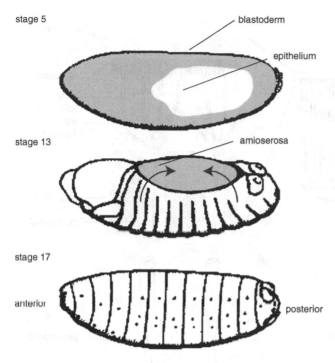

stage 5

blastoderm

epithelium

stage 13

amioserosa

stage 17

anterior

posterior

Figure 16.2 Origin and formation of the epidermis in *Drosophila* embryos. *Drosophila* epidermal cells differentiate from the lateral blastoderm. The region of the blastoderm that gives rise to epidermal cells from stage 5 embryos is highlighted in white. At stage 13 the epidermal cells from the left and right sites move and stretch toward the dorsal site to enclose the amnioserosa, a unique extraembryonic epithelium from higher flies. When the dorsal closure is completed (stage 17), the epidermis is the outermost layer and starts to produce cuticle material (modified after Payre, 2004).

a single row of large, pigmented trichomes; a narrow stripe of naked cuticle; up to three rows of massive trichomes; and 6–7 rows of thin trichomes. Formation of the thin trichomes requires the activities of Wingless and Line proteins. Formation of the other cuticular structures depends on Hedgehog, where the concentration of Hedgehog determines which structure is eventually formed. High Hedgehog levels yield large pigmented trichomes, intermediate levels result in naked cuticle or cuticle with massive trichomes. On the ventral side of the abdomen, each segment possesses 6–7 rows of pigmented denticles, which are larger pigmented cuticular extensions. Activation of the *wingless* pathway determines the naked fate, whereas activation of the EGF (epidermal growth factor) pathway promotes denticle formation. The activities of *wingless* and EGF pathways on epidermal differentiation finally regulate the expression of *shavenbaby*, a transcription factor, which is repressed in epidermal cells producing naked cuticles.

16.2 The cuticle

Insect cuticle is a natural biocomposite with unique physiochemical properties. It is secreted across the apical membrane of epidermal cells.

The cuticle is the outermost part of insects and covers the whole of the outside of the body, as well as lining ectodermal invaginations such as the foregut, hindgut and the tracheal system. It is composed of three distinct layers: epicuticle, exocuticle and endocuticle (Fig. 16.3a). Exo- and endocuticle are sometimes also referred to as procuticle. While the epicuticle is rather thin (1–4 μm), the thickness of the procuticle can exceed 200 μm. Epicuticle and procuticle differ in their composition. While the procuticle is a chitinous structure, the epicuticle does not contain significant amounts of chitin. In some insects, a fourth layer, the mesocuticle, is found between the exo- and endocuticle. The

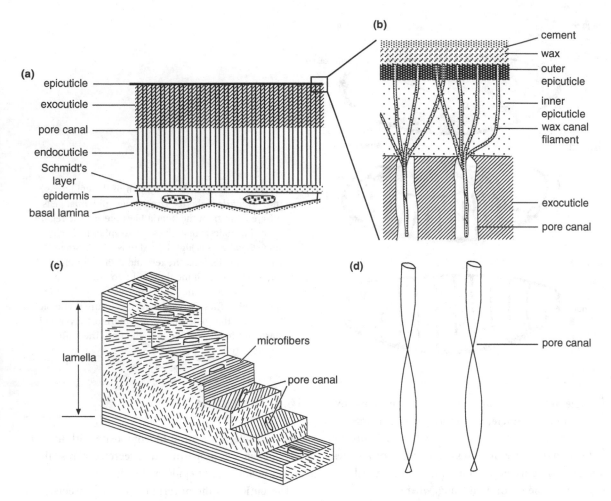

Figure 16.3 Basic structure of the integument. (a) Section through mature integument. (b) Section through the epicuticle at greater magnification. (c) Segment of cuticle with helicoidal arranged chitin microfibrils in which ribbon-shaped pore canals are embedded such that they lie parallel with the fibrils in each layer. (d) As a consequence of the helicoidal arrangement of microfibrils the pore canals adopt a twisted ribbon-shaped form (Neville *et al.*, 1969).

mesocuticle is not fully sclerotized and can be stained with acid fuchsin (a red dye). The cuticle varies in nature in different parts of the body and among insects with different life forms. The most obvious difference is between the rigid cuticle of the sclerites and the flexible cuticle of the membranes between them (Section 16.4).

16.2.1 Epicuticle

The outermost part of the integument is the epicuticle, a multilayered structure that is 1–4 μm thick. The thickest layer of the epicuticle is the inner epicuticle (0.5–2.0 μm) immediately outside the procuticle. External to this is a very thin outer epicuticle (*ca.*15 nm), and on this again is a wax layer of variable thickness. Some insects have an additional thin "cement" layer outside the wax (Fig. 16.3b). The epicuticle is primarily made from lipids, proteins and lipoproteins. The inner epicuticle is chemically complex and is known to consist primarily of tanned lipoproteins. During its production, phenolic substances and phenoloxidases are also present, which are presumably concerned with tanning the proteins. Phenoloxidases persist as extracellular enzymes in mature cuticle, producing further tanning if the epicuticle is damaged.

The outer epicuticle is a very thin trilaminar layer. It is a highly polymerized lipid, and probably also has a protein component. Polyphenols and phenoloxidase take part in its formation. It is the first-formed layer of new cuticle produced at each molt, protecting the new procuticle from the molting enzymes. The outer epicuticle is believed to be inextensible, setting a limit on any extension of the procuticle during growth or other activities. The material forming the outer epicuticle is often referred to as cuticulin, and sometimes as sclerotin.

The epicuticular wax layer contains many different compounds (Fig. 16.4). Hydrocarbons are universally present, and may make up over 90% of the wax, as in the cockroaches. Chain lengths range from around 12 to over 50 carbon atoms, and compounds with an odd number of carbon atoms in the chain are usually dominant. In larval Lepidoptera and Coleoptera, aliphatic alcohols are the most abundant compounds. In this case, compounds with even numbers of carbon atoms, in the range 12 to 34, are dominant.

In some species, the wax forms a bloom on the outside of the cuticle and, in a few, very large quantities are produced. This is the case with some Fulgoroidea and scale insects (Hemiptera), and the larva of *Calpodes* (Lepidoptera). Bees secrete large quantities of wax, which they use in the production of their larval cells. This wax is secreted by epidermal cells on the ventral surface of abdominal segments 4 to 7. These cells presumably contribute to the insect's epicuticular wax, but they become greatly enlarged when the bees have an adequate supply of honey, which provides the basic chemicals

from which the wax is produced. The secretion of the glands, which contains over 300 components, is pressed into flat scales between the overlapping sterna. The scales are removed by spines on the posterior basitarsi and are manipulated with the forelegs and mandibles to produce the cells of honeycomb. The wax is also important in waterproofing the cuticle and, in some insects, is the source of chemical signals important in intra- and, perhaps, interspecific signaling. It is synthesized by the oenocytes.

The cement is a very thin layer outside most of the wax, perhaps consisting of mucopolysaccharides closely associated with lipids. It may serve to protect the underlying wax, although it is sometimes present as an open meshwork. It is not produced by all insects. Honey bees, for example, appear to lack this layer. Cement is the product of type 3 gland cells in the epidermis.

16.2.2 Procuticle

The procuticle is a prime example for a natural biocomposite providing excellent mechanical properties. It is primarily made of proteins and chitin (Figs. 16.5, 16.6a, 16.7b). Mechanical rigidity and strength of insect cuticle is mainly achieved by fiber reinforcement of the procuticle using chitin as the fiber element. Chitin polymers crystallize in the form of microfibrils, which are embedded in a matrix of cuticular proteins forming a supramolecular chitoprotein complex. By this arrangement chitin stabilizes the complete structure in a way that resembles constructions of steel-reinforced concrete. In insect cuticles, chitin is always associated with proteins. Maybe it is even covalently linked to them as indicated by solid nuclear magnetic resonance spectroscopy. The formation of covalent bonds may require deacetylation of chitin to obtain primary amino groups that can be crosslinked by chinones. Although chitin is an important structural component of insect cuticles, their physicochemical properties are mainly determined by the cuticular

(a)

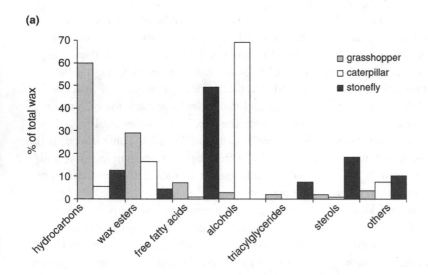

(b)

hydrocarbons	$CH_3-(CH_2)_x-CH_3$	(x=12–37)

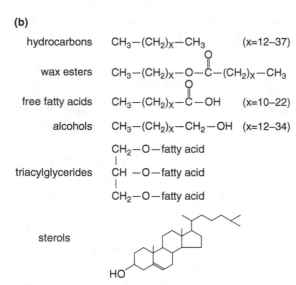

Figure 16.4 The main groups of compounds occurring in the epicuticular wax shown as percentage of total wax in different insect species. (a) In grasshoppers hydrocarbons, in caterpillars alcohols and in adult stoneflies free fatty acids predominate. (b) The structures of the main types of compounds.

proteins. Insect genomes contain numerous genes encoding cuticular proteins, and different types of cuticle contain different sets of cuticle proteins. Cuticles are initially formed as rather soft structures, but become hardened in a process that is called sclerotization, which means the crosslinking of cuticular proteins (Section 16.6.2). The degree of sclerotization is an additional factor that contributes to the mechanical properties of cuticles.

16.2.3 Pore canals and wax channels

Running through the cuticle at right angles to the surface are very fine pore canals, 1 μm or less in diameter (Fig. 16.3b). They extend from the epidermis to the inner epicuticle and, at least early in the development of the cuticle, contain cytoplasmic extensions of the epidermal cell at densities from 1.5×10^4 to 1.2×10^6 per square millimeter. In adult beetles and hemipteran bugs, the pore canals are more abundant in the outer procuticle and some 30 pore canals per cell in the outer region may join to a single canal lower down.

In *Rhodnius* (Hemiptera) the pore canals are circular in cross-section, but in many species they have a flattened, ribbon-like form, the plane of flattening being parallel with the microfibers in each layer of the cuticle. As the microfibers in successive layers change direction the canal also rotates, so that it has the form of a twisted ribbon (Fig. 16.3c,d). It is presumed that the cytoplasmic threads within the canals are subsequently withdrawn and the canals may be filled by chitin and protein, with the chitin microfibers oriented along the canal. Whether they are filled or not, the canals contain tubular filaments (called wax canal filaments, or epicuticular filaments) arising from the plasma membrane of the epidermal cell. At the epicuticle, these filaments diverge and extend to the surface of the outer epicuticle (Fig. 16.3b). The filaments range from 13 nm to 21 nm in external diameter in different insects, and have a lipid-filled lumen of about half this diameter. They possibly consist of tanned protein and are concerned with the transport of lipids from the epidermal cells to the surface of the cuticle. Lipids are present in the procuticle and, in *Rhodnius*, they impregnate the walls of the pore canals and are present in layers at intervals of 0.5–1.0 μm. These lipid layers become dispersed when the insect feeds and the cuticle stretches, but more layers are laid down with the new cuticle. Lipids, together with the sclerotized proteins, may play a significant role in cuticle hardening. However, experimental evidence for the latter assumption is still missing.

16.3 Chemical composition of the cuticle

16.3.1 Chitin

Insect cuticle contains chitin in the range of 20–50% of its dry weight. Pure chitin is a polysaccharide

made up of *N*-acetylglucosamine residues (Fig. 16.5a). In natural sources, however, chitin is found partially deacetylated, forming a heteropolysaccharide that also contains some glucosamine residues. Complete deacetylation yields chitosan, a homopolymer consisting exclusively of glucosamine. The sugar residues are linked by β(1,4) glycosidic bonds, and rotated with respect to each other by 180°; hence the repetitive unit is a dimer called chitobiose. Overall, a single polymer forms a helicoidal structure with six sugar residues per winding. However, chitin is found usually in the form of crystalline microfibrils of varying diameter and chain length that assemble in the extracellular space. Three allomorphic crystalline forms of chitin are found in nature: α-, β- and γ-chitin. They differ mainly in the orientation of the sugar chains, in the degree of hydration and in the size of the unit cell as it can be deduced from X-ray diffraction patterns. α-chitin is the predominant form in insect cuticles. It exhibits an antiparallel orientation of the sugar chains with respect to the reducing and non-reducing ends. This arrangement facilitates tight packaging of the sugar chains that are stabilized by intramolecular and intermolecular hydrogen bonds (Fig. 16.5b).

The chitin microfibrils of the insect cuticle are about 3.0 nm in diameter and about 0.3 μm long. Each microfibril consists of about 20 single chitin chains that are embedded in a protein matrix (Fig. 16.6a). They lie parallel to each other in the plane of the cuticle, forming a layer, but their orientation is often different in successive levels throughout the thickness of the cuticle. In most insects, the microfibrils in the outer parts of the procuticle, which subsequently becomes the exocuticle, rotate anticlockwise through a fixed angle in successive levels so that their arrangement is helicoidal and a series of thin lamellae is produced (termed Bouligand helicoids; Fig. 16.6b). This is called lamellate cuticle, which has a characteristic appearance in electron microscopy (Fig. 16.7a). The inner procuticle may also be lamellate throughout, or layers with helicoidally arranged microfibrils may

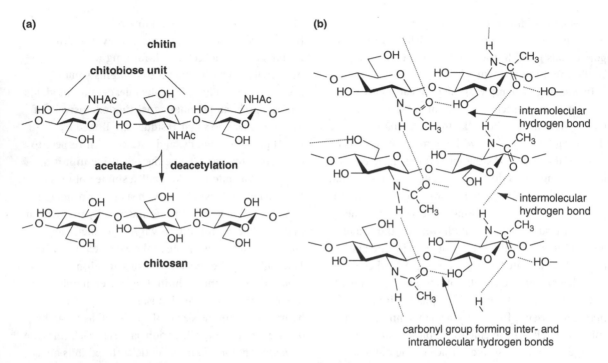

(a)

chitin

chitobiose unit

acetate ← deacetylation

chitosan

(b)

intramolecular hydrogen bond

intermolecular hydrogen bond

carbonyl group forming inter- and intramolecular hydrogen bonds

Figure 16.5 Chemical structures of chitin and chitosan and hydrogen bonding. (a) Chitin and chitosan are polymers of *N*-acetylglucosamine and glucosamine units, respectively. From a stereochemical view, the repetitive unit of the chitin polymer is not a single sugar moiety, but a disaccharide, which is called chitobiose. (b) Hydrogen bonding in α-chitin. Inter- and intramolecular hydrogen bonds contribute to the mechanical strength of the chitin microfibril (according to Kameda *et al.*, 2005). In β-chitin hydrogen bonds are frequently formed with intercalating water molecules.

alternate with layers in which the microfibrils are uniformly oriented. Wholly lamellate cuticle is found in Apterygota and in larval and pupal Lepidoptera, Diptera and Coleoptera. Where helicoidal and unidirectional layers alternate, all the unidirectional layers may have the same orientation, as in locusts and cockroaches, or they may have different orientations, as in beetles and bugs. In some such cases the intervening layers of helicoidal cuticle are very thin, so that the orientation appears to change suddenly from one layer to the next. This is called a pseudo-orthogonal or plywood-like arrangement (Fig. 16.6c). The alternate production of helicoidally arranged and unidirectional layers of microfibrils has a circadian periodicity in most species.

16.3.2 Chitin metabolism

Chitin synthesis requires the activities of different enzymes, but the key enzyme in this process is chitin synthase (EC 2.4.1.16), a membrane-integral family II glycosyltransferase. The chitin synthase transfers the sugar portion of UDP-*N*-acetylglucosamine to the non-reducing end of the growing chitin polymer. UDP-*N*-acetylglucosamine is synthesized in a sequence of enzymatic reactions following a variant of a biochemical pathway named after Louis Leloir, who discovered the importance of activated sugar polymers in glycosyl transferase reactions (Fig. 16.8). Due to the lack of structural data on chitin synthases, the precise catalytic mechanism is not known. From a stereochemical point of view, chitin synthesis follows

(a)

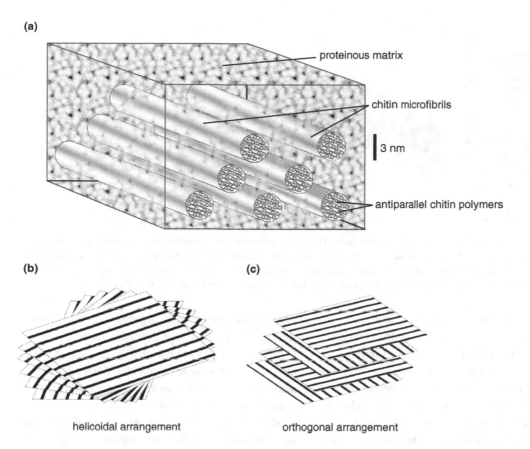

proteinous matrix

chitin microfibrils

3 nm

antiparallel chitin polymers

(b)

(c)

helicoidal arrangement

orthogonal arrangement

Figure 16.6 Chitin deposition in the cuticle. (a) Chitin microfibrils are formed of about 20 single sugar chains that are orientated in an antiparallel fashion. The microfibrils are embedded in a proteinous matrix of various cuticle proteins. (b) Helicoidal and (c) orthogonal arrangements of cuticle layers in insects.

an inverting reaction mechanism in which the nucleophilic attack by the acceptor hydroxyl group leads to an inversion of the anomeric carbon of the donor substrate. The reaction cycle likely involves an oxocarbenium ion-like transition state and requires a catalytic base.

Because the catalytic site of the chitin synthase faces the cytoplasm, the growing polymer has to be translocated across the membrane to reach the extracellular space, where it is deposited. It is likely that the extended transmembrane regions typically found in family II glycosyltransferases are involved in this transport process. As also reported for the related cellulose synthases, chitin synthases appear to form oligomeric complexes of single chitin synthase units. In all insects studied so far, only two genes encode chitin synthases (*ChsA* and *ChsB*). While *ChsA* synthesizes chitin for epidermal and tracheal cuticles, *ChsB* is required for chitin synthesis in the course of peritrophic matrix production in the midgut. Regulation of chitin synthesis is tightly coupled to development, not only in the cuticle but also in the midgut. *ChsA* expression is up-regulated at the beginning of all types of molts, where the old cuticle is shed and replaced by a newly synthesized cuticle. *ChsB*, however, is expressed in all feeding stages and down-regulated during molting and periods of starvation.

(a)

(b)

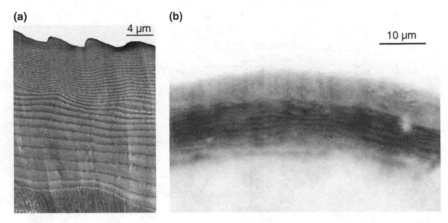

4 µm

10 µm

Figure 16.7 Larval cuticle of different body regions from *Tribolium castaneum*. (a) Transmission electron microscopy of ultra-thin sections stained with uranyl acetate. The lamellate procuticle shows Bouligand helicoids and pore canals (courtesy of S. Chaudhari and S. Muthukrishnan). (b) Cryosection of the larval integument stained with the fluorescent chitin-specific dye Calcofluor white (gray scales inverted; H. Merzendorfer). The procuticle is particularly rich in chitin.

During the molts the old cuticle is partially degraded before ecdysis. According to the classical view chitinases (EC 3.2.1.14) are secreted into the apolysial space along with other hydrolytic enzymes of the molting fluid. However, recent data from *Triolium* suggest that chitinases are not excluded from the newly synthesized cuticle, but that chitin is rather protected from chitinolytic degradation by the cuticle protein Knickkopf. All insect chitinases so far investigated are endo-splitting glycosidases, yielding oligosaccharides of different chain length that are finally degraded by β-*N*-acetylglucosaminidases (EC 3.2.1.52), which have essential functions during development in chitin turnover and/or *N*-glycan processing.

Insect endo-chitinases belong to family 18 of the glycosyl hydrolases, attacking preferentially the β-1,4-glycosidic linkage between the GlcNAc residues. Their structure is characterized by a $(\alpha/\beta)_8$ barrel. The reaction cycle appears to follow a substrate-assisted mechanism, in which the oxocarbonium ion intermediate is stabilized by anchimeric assistance of the sugar *N*-acetyl group after donation of the proton from the catalytic carboxylate. Family 18 chitinases produce chito-oligosaccharides, maintaining a β-anomeric configuration at the reducing ends of the products. Hence, they are retaining glycosidases.

Insect chitinases are encoded by multiple genes that have been grouped into eight classes based on similarities of their deduced amino acid sequence and domain architecture. They are built from 1–5 catalytic domains (GH18 domain, some of them can be catalytically inactive), 0–7 chitin-binding domains (peritrophin A domains), perA and serine/threonine-rich linker regions that are glycosylated. Most of the chitinases have a leader peptide and are therefore secreted into the extracellular space. Some of them, however, have a single transmembrane domain at the *N*-terminus. The expression of many different chitinases in a single insect suggests that individual chitinases have specialized functions. These functions have been carefully analyzed recently in *Tribolium castaneum* by RNAi-induced gene silencing. It turned out that some chitinases serve in the turnover of cuticular chitin, some in adult metamorphosis and others in degradation of the peritrophic matrix. In addition to the degradation of cuticle and peritrophic matrix, chitinases and chitinase-like proteins may have functions in regulating cell proliferation, tissue remodeling and immune responses.

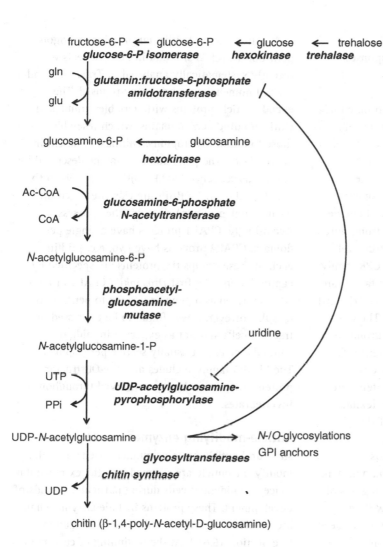

fructose-6-P ← glucose-6-P ← glucose ← trehalose
glucose-6-P isomerase   **hexokinase**   **trehalase**

gln
**glutamin:fructose-6-phosphate
amidotransferase**
glu

glucosamine-6-P ← glucosamine
hexokinase

Ac-CoA
**glucosamine-6-phosphate
N-acetyltransferase**
CoA

N-acetylglucosamine-6-P

**phosphoacetyl-
glucosamine-
mutase**

N-acetylglucosamine-1-P

uridine

UTP
**UDP-acetylglucosamine-
pyrophosphorylase**
PPi

UDP-N-acetylglucosamine → N-/O-glycosylations
glycosyltransferases GPI anchors
chitin synthase
UDP

chitin (β-1,4-poly-N-acetyl-D-glucosamine)

Figure 16.8 Chitin biosynthesis in insects. The pathway starts with trehalose, the main hemolymph sugar in most insects. Fructose-6-phosphate generated in the first reactions of glycolysis enters one route of the Leloir pathway, generating UDP-N-acetylglucosamine. The sugar moiety of UDP-N-acetylglucosamine is transferred by the chitin synthase to the non-reducing end of the chitin polymer. Feedback inhibition by UDP-N-acetylglucosamine and inhibition by uridine is indicated.

16.3.3 Proteins

Proteins are the major constituents of insect cuticles and determine their properties. Cuticle from any one part of an insect contains several different proteins which may differ from the proteins in the cuticle of other parts of the same insect. The most obvious difference between different types of cuticles is that between soft cuticles found in membranous regions of the exoskeleton and hard cuticles that have protective functions, such as the cuticle from coleopteran elytrae. These differences are apparent even before the cuticle is hardened, and hence the differing physical properties of different parts of the cuticle appear to be at least partly a consequence of the different proteins they contain, independent of the hardening process. Sometimes only few major proteins can be detected in cuticles of particular body regions. In other cases, 100 or even more are present in the cuticles. The same proteins are present in cuticle with similar properties from different stages of an insect. Thus, the proteins appear to characterize the type of cuticle rather than the stage of development. Insect cuticles contain structural proteins, cuticle organizing and anchoring proteins, cuticle-modifying enzymes and proteins related to pigmentation and immune responses (Table 16.1).

Structural cuticle proteins The structural cuticle proteins are encoded by several gene families that have been annotated and analyzed in different insect genomes, including those from *Apis*, *Nasonia* (Hymenoptera), *Anopheles*, *Drosophila*, (Diptera), *Bombyx* (Lepidoptera) and *Tribolium* (Coleoptera). The largest gene family encoding structural cuticle proteins is the CPR family. For instance, from more than 200 *Anopheles* genes that encode structural cuticle proteins, 156 genes belong to the CPR family. Cuticle proteins encoded by genes of this family are characterized by the RR consensus motif Gx(7)[DEN]Gx(6)[FY]xA[DGN]x(2,3)G[FY]x[AP]x(6). The CPR protein family can be divided into three groups that differ in the extended RR consensus sequence. The two major and well-established groups are the RR-1 and RR-2 cuticle proteins. RR-1 proteins are characteristically associated with soft or flexible cuticles found in membranous regions of the exoskeleton, while RR-2 cuticle proteins are frequently found in hard cuticles. The most prominent CPR protein is probably resilin, a protein that confers elastic properties to specific regions of cuticle. The unique properties of resilin will be discussed in more detail in Section 16.4.3. It has been suggested that CPR family proteins exhibit chitin-binding properties; however, whether this is a general property of CPR proteins has still to be clarified.

Other types of structural cuticle proteins have been reported that lack the RR consensus sequence but sometimes possess GGX or AAP(A/V) repeats. Non-CPR proteins include the highly hydrophobic apidermins, which were discovered in *Apis mellifera*, the *Drosophila* adult cuticle proteins Edg-91 (a glycine-rich *Drosophila* protein similar to chorion proteins and cytokeratine), the Tweedle proteins and the CPF, CPFL and CPLC family proteins.

Another group of cuticular proteins analogous to peritrophins (CPAPs) found in many insects was recently systematically analyzed in *Tribolium*, and a revised nomenclature has been proposed. They encode cuticle proteins with variable numbers of chitin-binding perA domains, which resemble those found in peritrophins, chitinases and chitin deacetylases. The perA domain may be described by consensus sequence Cx(11–20)Cx(5–6)Cx(9–19)Cx(10–17)Cx(4–14)C. CPAPs are classified according to the number of perA domains they possess. Accordingly, CPAP1 proteins have a single perA domain, CPAP2 proteins have two, etc. Within each of these groups the proteins are specified by capital letters. The first *Drosophila* CPAP3 protein was described as a gene analogous to peritrophins (*gasp*). However, it was found to be expressed in tracheal cells and not as expected in midgut epithelial cells that usually secrete peritrophins. The CPAP3 family includes also the Obstructor proteins, which are highly conserved throughout invertebrates.

Cuticle-modifying enzymes In addition to structural cuticle proteins, numerous enzymes that modify the cuticle are secreted into the extracellular space by epidermal cells during particular periods of development. These proteins include enzymes that are involved in sclerotization and melanization (see Section 16.6.2). At the beginning of each molt, hydrolytic enzymes, e.g., N-acetylglucosaminidases, chitinases and proteinases, are secreted into the exuvial space after apolysis, which is the separation of the cuticle from the epidermis during molt. If inactivated, they become activated in the molting fluid and facilitate shedding by digesting inner parts of the old cuticle (Sections 16.5.3 and 16.6).

One way to alter physicochemical properties of insect cuticles may be by changing the degree of deacetylation, which varies a great deal between different insect species and body parts. Alteration of the degree of deacetylation is achieved by chitin deacetylases (EC 3.5.1.41). Interestingly, chitosan has

Table 16.1 **Overview on different types of cuticle proteins**

Cuticle protein	Catalytic activity	Conserved domains	Gene copy numbers *Drosophila* (*D*), *Anopheles* (*A*), *Tribolium* (*T*), *Bombyx* (*B*)
Structural cuticle proteins			
CPR proteins (RR-1,2,3)	–	RR-1, RR-2 and RR-3 motifs	≥100 in *D*, *A*, *T*, *B*
CPF	–	44 aa CPF consensus motif	3 in *D*, 4 in *A*, 5 in *T*, 1 in *B*
CPFL	–	C-terminal CPFL consensus motif	3 in *T*, 7 in *A*, 4 in *B*
CPLC (-G, -W, -P, -A)	–	Low-complexity regions	27, 9, 4, 3 (respectively) in *A*
Tweedle	–	Tweedle consensus motif	27 in *D*, 12 in *A*, 4 in *B*, 3 in *T*
CPAP1–3	–	1–3 perA domains	9 CPAP1 in *T*, 7 CPAP3 in *T*
Glycine-rich	–		29 in *B*
Unclassified	–	Various	11 in *A*, 34 in *B*
Cuticle-modifying enzymes			
Hydrolases			
N-acetylglucosaminidase	glucosaminidase	Catalytic domain	4 in *T*, 2 in *B*
Chitinase groups I–V	glucosidase	Catalytic domain (various numbers, some maybe inactive) PerA domain (various numbers, some none) S/T-rich linker domain (various numbers, some none) Some possess a transmembrane domain	About 20 in *D*, *A*, *T*, *B*
Chitin deacetylases groups I–VIII	Aminohydrolase	Catalytic domain, some perA domain, some LDLa domain	5–9 in *D*, *A*, *T*, *B*
Serine peptidases Carboxypeptidases Metallopeptidases Cysteine peptidases	Peptidase	Catalytic domain, some perA domain	Numerous in *D*, *A*, *T*, *B*

Table 16.1 (*cont.*)

Cuticle protein	Catalytic activity	Conserved domains	Gene copy numbers *Drosophila* (*D*), *Anopheles* (*A*), *Tribolium* (*T*), *Bombyx* (*B*)
Proteinase inhibitors such as Serpins	–	Reactive center loop (variable sequence that determines specificity)	Numerous in *D, A, T, B*
Phenoloxidases			
Laccase	ortho- and para-diphenol:O_2 oxidoreductase	Copper oxidase domain, CxRxC motif	1–2 in *D, A, T, B*
Tyrosinase	ortho-diphenol:O_2 oxidoreductase	Copper oxidase domain	1–3 in *D, A, T, B*
Isomerases			
Dopachrome conversion enzyme	isomerase	–	1 in *A*, 2 in *D*
Quinone/quinone methide isomerases	Ortho-chinone isomerase Para-chinone methide Tautormerase	–	Several in *D, A, T, B*
Chitin-organizing and -anchoring proteins			
Knk	Oxidoreductase?	DM13, DOMON domain, GPI anchored	3 in *A, D, T*, 1 in *B*
Rtv	–	Transmembrane domain, putative chitin-binding motif	1 in *A, D, T, B*
ZP proteins such as Papillote, Dumpy, Miniature, Dusky	–	ZP domain, transmembrane domain (maybe cleaved off) Dumpy: numerous EGF repeats Papillote: half-ZP domain	14 in *D*, >10 in *A, T, B*
Pigmentation-related proteins			
Bilin/Biliverdin-binding proteins (insecticyanins) Cyanoproteins (hexamerins) Beta-carotene-binding proteins	– – –	Retinol-binding domain β-carotene-binding domain	Several in *D, A, T, B*

antibacterial properties that might also provide some protection against the invasion of bacteria.

Chitin deacetylases are secretory proteins that belong to a family of extracellular chitin-modifying enzymes catalyzing the N-deacetylation of chitin to form chitosan. Those insects whose genomes are sequenced possess 5–9 genes encoding chitin deacetylases. The variety of chitin deacetylases is increased by expression of alternately spliced mRNAs. Chitin deacetylase genes were classified into five ortholog groups based on phylogenetic analysis and the presence of additional motifs. Except for group V chitin deacetylases, they all contain a single chitin-binding domain (perA domain) in addition to the catalytic domain. *In situ* hybridization experiments performed in *Tribolium castaneum* revealed that the group I chitin deacetylases TcCDA1 and TcCDA2 are expressed in tracheal and epidermal cells. The RNAi-induced knockdown of *TcCDA1* and *TcCDA2* gene expression led to animals that were incapable of shedding their old cuticle at every type of molt and died because they were entrapped in their exuviae.

Cuticle-organizing and -anchoring proteins Cuticle-organizing proteins were discovered first by analyzing *Drosophila* mutants exhibiting defects in the formation of epidermal and tracheal cuticles. Due to the loss of cuticle integrity, mutant embryos stretch to several times the size of wild-type embryos after removing the vitelline membrane, a phenotype which is called blimp-like. A blimp-like phenotype is, for instance, obtained when loss-of-function mutations have occurred in the *kkv1* gene encoding chitin synthase 1. Blimp-like phenotypes were also reported for genes that were indirectly associated with chitin metabolism, such as *knickkopf* (*knk*) and *retroactive* (*rtv*). Mutations in these genes do not only disrupt the formation of the embryonic cuticle, but also interfere with tracheal development. The tracheal defects are caused by the failure to correctly form luminal chitin, which is required for tube size control in tracheal development. Luminal chitin has a filamentous texture in wild-type embryos. In *knk* and *rtv* mutants luminal chitin is still formed, but fails to form filamentous textures and appears amorphous. From this finding it was concluded that chitin assembly rather than chitin synthesis is affected by mutations in *knk* or *rtv*. The Knk protein is located at the plasma membranes of epidermal and tracheal cells, where it is attached to the extracellular site by a GPI anchor. The extracellular part of this protein exhibits a domain which may fold in a fibronectin type III-like fashion, a fold known to mediate adhesion to the extracellular matrix. In contrast to Knk, Rtv is a transmembrane protein that faces the extracellular site and may bind chitin by exposed aromatic amino acid residues. Their precise function in organizing the cuticle has still to be established. Very recent data suggest a chitin-protective role for Knk.

Adhesion between epidermal cells and the cuticle appears also to be the function of *zona pedicula* (ZP) proteins such as "Dumpy," "Piopio" and "Papillote." Loss-of-function mutations in one of their encoding genes cause defects in the innermost layer of the cuticle and the detachment from the epidermis. Two further *Drosophila* ZP proteins present in the apical membranes of wing epidermal cells, "Dusky" and "Miniature," have been suggested to mediate the interactions between the cytoskeleton and the forming cuticle.

16.3.4 Lipids

The lipid composition of insect cuticles is complex and varies extensively between different species and different developmental stages. The major lipids are hydrocarbons, which include saturated alkanes (with methyl branches) and unsaturated alkenes. To a minor degree also fatty acids, alcohols, glycerides, sterols, ketones, aldehydes and esters are found (Fig. 16.4). Cuticular lipids usually are found underneath the cement layer, but sometimes they are also on the cuticle surface. In the latter case they form either mosaics of cement lipids or wax blooms that function in camouflage, waterproofing and thermal protection.

In terrestrial insects, a prime function of cuticular lipids is the protection against water loss. However, insect cuticles get water-permeable above a critical temperature, which varies between different species. This change in permeability is caused by a phase transition of the cuticular lipids, which presumably involves the reorientation of lipid molecules, disrupting tight packing. The critical transmission temperature at which 50% of the hydrocarbons are reoriented or "melted" depends on the chain length, degree and type of branching and the ratio of saturated and unsaturated hydrocarbons. Generally, higher transmission temperatures are observed when cuticular lipids contain more saturated and unbranched longer-chain molecules. Conversely, lower transmission temperatures are observed when more unsaturated and branched short-chain molecules are present. It has to be noted that other factors also affect water loss, which explains why the melting temperatures do not correlate well with typical temperatures in the environment of some insects.

Hydrocarbon profiles may be highly specific for a particular insect species or population. Hence, they are useful features for taxonomic, systematic and biogeographical analysis. The hydrocarbons have been well characterized in many insect species, as they are easily extracted from the cuticles and identified by means of gas chromatography and mass spectrometry. Synthesis of most cuticular hydrocarbons occurs in oenocytes, not in epidermal cells. How precisely the carbon hydrates are transported from oenocytes to the epicuticle is still an open question, but studies performed with cockroaches suggest that the hydrocarbons, which are synthesized by abdominal oenocytes, are loaded onto lipophorins, special hemolymph carrier proteins that transport the hydrocarbons to the fat body and cuticle. Once they have passed the epidermis they likely enter pore canals and wax channels to reach the epicuticle. Lipid transport across the cuticle is obviously not unidirectional, as exogenous lipids can pass through the cuticle into the insect's body. According to one hypothesis, even pesticides may pass through the cuticle by pore canals and wax channels.

Many hydrocarbon compounds serve a function in communication between individuals, including those acting as sex pheromones. In social insects, the types and blends of epicuticular hydrocarbons can communicate information about age, sex, caste, rank, nest and relatedness, and trigger stereotyped patterns of behavior. Dead colony members are identified by their hydrocarbon composition, signaling workers to remove them from the nest, a behavior known as necrophoresis. In Argentine ants this behavior is induced by the loss of two epicuticular hydrocarbons, dolichodial and iridomyrmecin, which disappear from the hydrocarbon profile within one hour of death of the ant.

Among non-social insects, hydrocarbons are involved in mate recognition. For example, in male crickets saturated hydrocarbons trigger mating behavior and unsaturated hydrocarbons induce avoidance. Cuticle lipids also comprise anti-microbial substances that contribute to the immune defense against pathogenic microorganisms. For instance, water beetles impregnate their cuticle surface with anti-microbial substances such as benzoic acid, esters of p-hydroxybenzoic acid, phenylacetic acid, sesquiterpenes, α-hydroxy acids and tiglic acid.

16.3.5 Minerals

Most insect cuticles do not contain large amounts of minerals. However, some cuticles that are particularly exposed to mechanical stress are reinforced by incorporation of metals and minerals. This is, for instance, the case in the cutting edges of mandibles in the genera *Schistocerca* and *Locusta* (Orthoptera). These animals feed on various types of hard, silicium-rich grasses. Therefore, the mandibular cuticles in this region are significantly harder than

(a) mandibular cuticle

(b) zinc incorporation

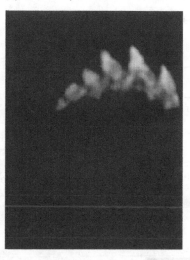

Figure 16.9 Cuticular hardening. Zinc reinforcement of the mandible of a caterpillar (Fontaine *et al.*, 1991). (a) Scanning electron microscopy of the inner surface of a mandible. (b) X-ray microanalysis of the same mandible showing the presence of zinc (white areas) at the cutting edge.

the cuticles from other body parts. Hardening appears to be at least partly due to the incorporation of larger amounts of zinc (Fig. 16.9). Zinc incorporation has also been reported for mandibular cuticles of leaf-cutter ants. In some other ants, termites and beetles, manganese is incorporated into cuticles as well as zinc. In termites these metals were only found in the edges of the mandibles. In cockroaches and earwigs, ferrous and siliceous inclusions have been reported. Calcium and magnesium, which are extensively incorporated as carbonates and phosphates into crustacean cuticles and mollusc shells, were detected in significant amounts in cuticles of muscid flies and tenebroid beetles.

16.4 Types of cuticles

Cuticles can have different mechanical properties in hard and rigid, membranous or elastic extendable cuticles. Stiffness can vary between 1 kPa in the extensible intersegmental membrane of the locust to 20 GPa measured in the tibial flexor apodeme of the locust. Many factors contribute to the differences in mechanical properties. Hard and stiff cuticles generally have a high protein content between 70–85% (dry weight) and a low chitin content of about 15–30% (dry weight). The water content is rather low at approximately 12%. Soft cuticles contain about 50% each of chitin and protein (dry weight) and exhibit a high water content of about 40–75%. The type of cuticle protein, the textures of microfibrils and the shape of the cuticle are additional factors that determine mechanical properties.

16.4.1 Rigid and hard cuticles

Rigidity of cuticles was for a long time mainly attributed to the process of sclerotization, which forms covalent crosslinks between proteins. Indeed, the stiffness of insect cuticle increases as a greater proportion of the cuticular protein becomes sclerotized (Fig. 16.10a). However, as the cuticle gets sclerotized it loses water (Fig. 16.10b). Interestingly, the sclerotized material can be softened by formic acid, which is known to break hydrogen bonds required to form stable β-sheets in cuticle proteins that contribute to stiffness. This observation indicates that hydrogen bonding is at least as important to cuticle stiffness as covalent bonding between cuticle proteins.

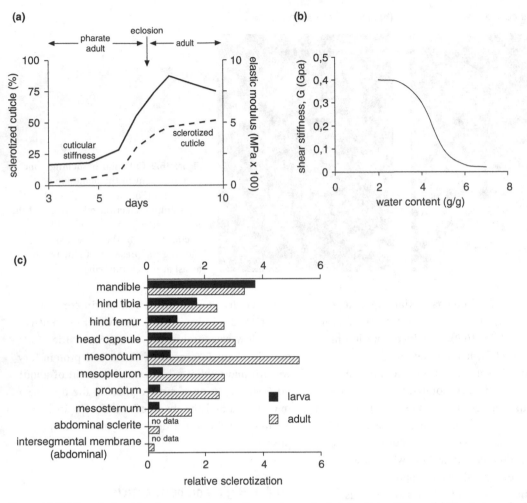

Figure 16.10 The relationships between sclerotization, water content and cuticle stiffness. (a) The degree of sclerotization of an abdominal tergite from the honey bee increases over the period of molt and is particularly high around the time of eclosion. Sclerotization is expressed as the inverse of the percentage of proteins that can be extracted (based on Richards, 1967). (b) Shear stiffness with change in water content of untanned cuticle from *Calliphora* sp. In the absence of tanning already a small reduction in water content leads to a significant increase in stiffness (based on Vincent and Wegst, 2004). (c) Degree of sclerotization of cuticles from different body parts of a mid-fifth-stage and ten-day-old adult locust. The amount of ketocatechols released upon hydrochloric acid hydrolysis was used as a measure for sclerotization (based on Andersen, 1974).

The extent of sclerotization varies between different developmental stages and in different parts of the cuticle. In larval *Schistocerca*, for example, the exocuticle becomes sclerotized within one day of emergence, and the endocuticle produced thereafter does not. In the adult, on the other hand, sclerotization continues for weeks so that exo- and endocuticle become sclerotized.

In the fully hardened larva of *Schistocerca*, the mandibles are much more heavily sclerotized than other parts of the cuticle; they are also heavily sclerotized in the adult, but to no greater degree than some other parts of the body. The dorsal mesothorax (mesonotum) is particularly heavily sclerotized in relation to its function in flight (Fig. 16.10c).

Additional rigidity of sclerotized cuticles is generated by its shape. Usually cuticles are not flat, but have an overall curved shape which contributes significantly to the overall stiffness. Furthermore, inflexions of the cuticle seen as grooves (sulci) on the outside and as ridges on the inside support the structure. They provide rigidity in the same way as a T-girder. Even greater rigidity is provided when inflexions meet internally to form an "internal" (endophragmal) skeleton. Deep, finger-like invaginations of the cuticle are called apodemes or apophyses. In the head, four apodemes join centrally to form the tentorium (Fig. 1.5), and in the thorax of winged insects, pleural and sternal apophyses join or are held together by muscles (Fig. 7.4). The tentorium provides rigidity to the head just above the level at which the mandibles are articulated. The junction of pleural and sternal apophyses gives lateral stiffness to the thorax, which is important for functioning of the flight muscles.

Hardness, as opposed to rigidity, increases relatively slowly after ecdysis and it may take several days for the cuticle to reach its final degree of hardness. During this time, the cuticle increases in thickness and this may contribute significantly to the increased hardness. It may also be that other changes occur in the cuticle following the initial rapid sclerotization. Parts of the cuticle that are especially hard may contain metals and minerals (Section 16.3.5).

16.4.2 Membranous cuticle

In flexible arthrodial membranes that join the sclerites, the procuticle remains largely unsclerotized (Fig. 16.11a). It differs from the cuticle of sclerites in containing different cuticle proteins. The extent of the membrane and the method of articulation of the two adjacent sclerites determine the degree of movement that can occur at the joint. Sometimes, as between abdominal segments, the

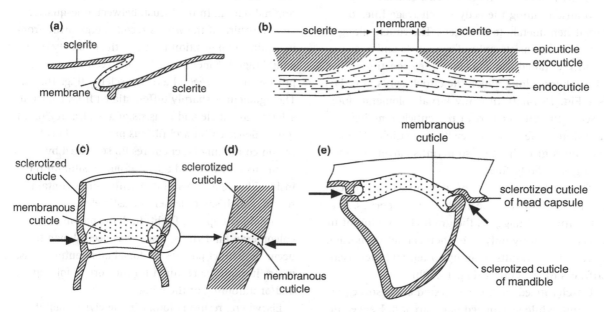

Figure 16.11 Flexibility of the cuticle. Arrows show points of articulation, at which the two sclerotized regions meet and are free to move relative to each other (partly after Snodgrass, 1935). (a) Intersegmental membrane; extensive membrane with no articulation between the sclerites. (b) Scheme of a membranous connection similar to that shown in (a). (c) An intrinsic articulation where the points of articulation are within the membrane, as in most leg joints. (d) Detailed view on one of the articulations shown in (c). (e) Extrinsic articulations where the sclerotized parts meet outside the membrane.

membrane is extensive and there is no specific point of contact between adjacent sclerites, so movement is unrestricted (Fig. 16.11a,b). More usually, the sclerites make contact with each other to form articulations and the joints are called monocondylic or dicondylic, depending on whether there are one or two points of contact (see Fig. 8.1). Monocondylic articulations, such as that of the antenna with the head, permit considerable freedom of movement (comparable to the human shoulder joint), whereas dicondylic articulations, which occur at many of the leg joints (as also in human knee or elbow joints), give more limited, but more precise movements, and a less complex array of muscles is necessary to provide fine control of movement. The articular surfaces may lie within the membrane (intrinsic), as in most leg joints (Fig. 16.11c,d), or they may lie outside it (extrinsic), as, for example, in the mandibular articulations (Fig. 16.11e).

Apart from membranous areas, exocuticle is also absent along the ecdysial cleavage lines of larval hemimetabolous insects. The cuticle along these lines consists only of epicuticle and undifferentiated procuticle, so they constitute lines of weakness along which the cuticle splits at ecdysis (see Fig. 16.14e,f). In many larval holometabolous insects, the greater part of the cuticle remains undifferentiated and somewhat extensible. This facilitates growth and also permits movement by changes in body form.

The water content in non-sclerotized cuticles can be altered to modify the mechanical properties. In the soft cuticle of maggots, the gradual reduction of the water content by only a few percent during normal development results in an up to ten-fold increased stiffness of the cuticle (Fig. 16.10b).

Unsclerotized cuticle is digested and resorbed at molting, while sclerotized parts are lost. Hence, an unsclerotized cuticle is more economic for a larva, as the bulk of material is conserved. In addition, many larval insects eat their exuviae so that they conserve as much as possible from the cuticle. Some insects are

known to remove material from unsclerotized cuticle during long periods without food, presumably using it for maintenance of other tissues.

16.4.3 Elastic and extensible cuticle

Elastic cuticles Some parts of the cuticle contain a colorless, rubber-like protein called resilin. The protein molecules are covalently linked by their tyrosine residues, and these linkages (di-/trityrosines) are produced continuously as the resilin is laid down, unlike sclerotization, in which the proteins become linked some time after they are produced. Resilin is extremely rich in glycine, which comprises 30–40% of the amino acids. Its amino acid sequence prevents other crosslinks forming; these would impair its rubber-like properties.

Like rubber, resilin can be stretched under tension and stores the energy involved so that when the tension is released it returns immediately to its original length. In the locust, between one-quarter and one-third of the wing's recoil energy away from its equilibrium position is due to the elasticity of the wing hinge ligament, which is between the pleural process and the second axillary sclerite (Fig. 16.12). The ligament is sharply differentiated from adjacent sclerotized cuticle and consists of a ventral region of tough, dense chitin and fibrous protein, a dorsal region containing layers of resilin separated by chitinous lamellae and a pad of pure resilin on the inside. The vein joints of damselfly wings contain resilin, which serves as serial elastic elements essential for flapping flight. In dermapteran and coleopteran hindwings, resilin has been shown to occur in distinct patches of dorsal and ventral veins, where it might be required for the correct folding and/or unfolding of the wing.

Elsewhere, resilin is found in the clypeo-labral spring, which keeps the labrum pressed against the mandibles, and in the food pump of reduviid bugs. In fleas, a pad of resilin above the hindleg stores energy for the jump. However, a recent study performed with

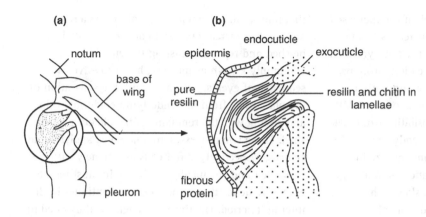

Figure 16.12 Elastic cuticle (modified after Andersen and Weis-Fogh, 1964). (a) Transverse section through the thoracic wall and wing base of a grasshopper, showing the position of the wing hinge. (b) The hinge enlarged showing the resilin pad.

jumping spittlebugs suggests that only a small part of the energy is stored by resilin in the pleural arch of the metathorax. Instead, bending of the arch's chitinous cuticle by muscle contractions stores much of the energy needed for jumping. The energy stored by resilin appears to be required for restoration of the original shape of the cuticle. In beetles, where there are no inspiratory muscles, the abdominal terga and sterna are held apart by ribs of cuticle that end, dorsally and ventrally, in pads of resilin. During expiration, these pads are compressed by the action of the dorso-ventral muscles; when these muscles relax, the pads resume their original shape, pushing the terga and sterna apart so that inspiration occurs.

Plasticization of cuticle Whereas resilin is an elastic protein, undifferentiated procuticle can sometimes be plasticized in order to facilitate stretching. This probably occurs in all insects at the time of ecdysis, where plasticization facilitates expansion of the new cuticle.

Plasticization at ecdysis facilitates the stretching of presumptive sclerites, before the cuticle is sclerotized. In the blowfly *Calliphora*, plasticization of the sclerites occurs when the newly eclosed fly swallows air and pumps rhythmically to expand the cuticle. Very little extension of the cuticle can be produced before this period, even though the procuticle is still undifferentiated. Once sclerotization begins, the cuticle is again inextensible. In *Manduca*, and

probably in other insects, these changes are regulated by two hormones, eclosion hormone, which is released into the hemolymph just before ecdysis and which also switches on ecdysis behavior, and bursicon, which is released just after eclosion and which subsequently initiates sclerotization.

Plasticization also occurs in female grasshoppers when ovipositing. Female locusts and grasshoppers lay their eggs in the ground, often extending the abdomen to more than twice its normal length by stretching some of the intersegmental membranes (see Section 13.5). The membranes are not extensible in males.

Plasticization is possibly best examined in the blood-sucking bug *Rhodnius*, whose larvae normally take only one very large blood meal in each developmental stage. This results in considerable distension of the abdomen, so that its surface area increases about four-fold and the cuticle becomes considerably thinner. This distension is facilitated by plasticization of the largely unsclerotized abdominal cuticle. Plasticization is apparent within two minutes of the start of feeding, and is initiated by the pumping activity associated with food intake. Information is passed via stretch receptors to the central nervous system from which neurosecretory axons extend to the body wall, where they branch extensively beneath the epidermis. They release serotonin so that a high local concentration occurs. This hormone signal triggers a pH change in the cuticle from 7 to below 6. As a result of the increased charge density, the water content increases from about 26% to 31%, presumably decreasing hydrogen bonding that stabilizes β-sheets in cuticle proteins. The increase in water content goes along with a significant decrease in stiffness from about 219 MPa to 10 MPa and an increase in extensibility from 10% to 100%, allowing expansion of the body volume for blood uptake. The increase in abdominal size in *Rhodnius* associated with plasticization is to a large extent reversible, and the abdomen slowly shrinks to approach its original size as the blood meal is digested and excess fluid is excreted. In vitro measurements of physicochemical properties of *Tenebrio* cuticle proteins from pre-ecdysial larval and pupal cuticles seem to be in line with the hypothesis outlined above, as small changes in the intracuticular pH (and ionic strength) can cause significant changes in the mechanical properties of unsclerotized cuticle by altering protein interactions.

16.5 Molting

The growth of insects is limited by the cuticle. Sclerotized cuticle is virtually inextensible, and membranous cuticle can stretch only to the extent that folds in the outer epicuticle allow. As a result, for any marked increase in size to occur, the cuticle must be shed and replaced. Casting the cuticle is commonly known as molting, but it involves a sequence of events beginning with the separation of the old cuticle from the underlying epidermal cells and ending with the remnants of the old cuticle being shed. These two processes are known as apolysis and ecdysis, respectively. After ecdysis, the new cuticle is expanded and chemically modified. In most insects, extensive deposition of new cuticle continues in the intermolt period. The timing of events is illustrated in Fig. 16.13 and changes in the integument in Fig. 16.14. The hormonal regulation of molting is discussed in Chapter 15.

16.5.1 Changes in the epidermis

In most insects with well-sclerotized cuticles, such as the larvae of most hemimetabolous insects, epidermal cell density is constant through most of the intermolt period. It increases as a result of mitosis at the onset of the molting cycle and then undergoes a sharp, relatively small decrease during cuticular expansion following ecdysis. There is a net increase in the total number of cells, so that although the

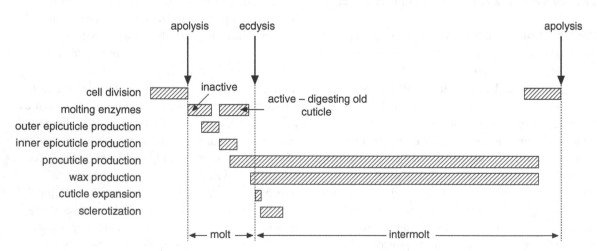

Figure 16.13 Generalized sequence of events involved in cuticle production. The timing of apolysis and the degree of overlap between procuticle production and cell division may vary significantly from species to species.

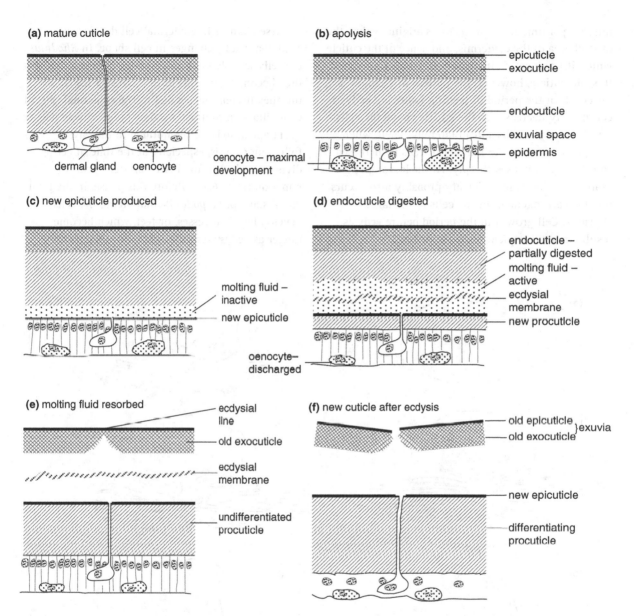

(a) mature cuticle

dermal gland   oenocyte

oenocyte – maximal development

(b) apolysis

epicuticle
exocuticle
endocuticle
exuvial space
epidermis

(c) new epicuticle produced

molting fluid – inactive
new epicuticle
oenocyte– discharged

(d) endocuticle digested

endocuticle – partially digested
molting fluid – active
ecdysial membrane
new procuticle

(e) molting fluid resorbed

ecdysial line
old exocuticle
ecdysial membrane
undifferentiated procuticle

(f) new cuticle after ecdysis

old epicuticle } exuvia
old exocuticle
new epicuticle
differentiating procuticle

Figure 16.14 Changes in the integument during molting. (a) Mature cuticle. (b) Apolysis occurs after cell divisions and morphological changes of epidermal cells. Microvilli of epidermal cells become disintegrated and the plasma membrane plaques are internalized. (c) Molting fluid is produced with partially inactive hydrolytic enzymes. New microvilli and plasma membrane plaques are formed. The new outer and inner epicuticles are deposited. (d) The inactivated enzymes in the molting fluid are activated, substrate becomes accessible and digestion of the old endocuticle starts. At the same time new procuticle is laid down. (e) All the old endocuticle is digested and partially resorbed. Only the epicuticle holds the cuticle together along the ecdysial line. New cuticle becomes thicker. (f) After ecdysis the new cuticle expands and the epidermal cells change their morphology. The procuticle is still undifferentiated but becomes increasingly sclerotized during the intermolt stage. The old cuticle may break up at the ecdysial line.

number per unit area returns to its original value, the overall area of the epidermis, and hence of the cuticle which it produces, is increased. In larval forms with flexible cuticle, however, the pattern of changes is different. In the period of feeding following ecdysis, cell density decreases as the cuticle above the cells unfolds and the cells spread laterally. Then, when mitosis occurs and the number of cells is increased, cell density increases sharply. At the larval–pupal molt, some increase in density probably also occurs due to rearrangement of the cells as the larva shortens. Cell growth in the period before ecdysis results in a subsequent decrease in density.

These changes in epidermal cell density are accompanied by changes in cell shape. In *Rhodnius* the cells are columnar immediately after molting, but then become squamous until the insect feeds; after this they become deeper again. The epidermal cells of caterpillars, in contrast, become progressively deeper (from apical to basal membrane) in the days following ecdysis, especially at the time of cell division and when the insect shortens prior to pupation (Fig. 16.15). Before this phase, in the final larval stage of *Calpodes* (Lepidoptera), the cells develop basal processes, or feet, which become longer as ecdysteroid titer rises, but then shorten

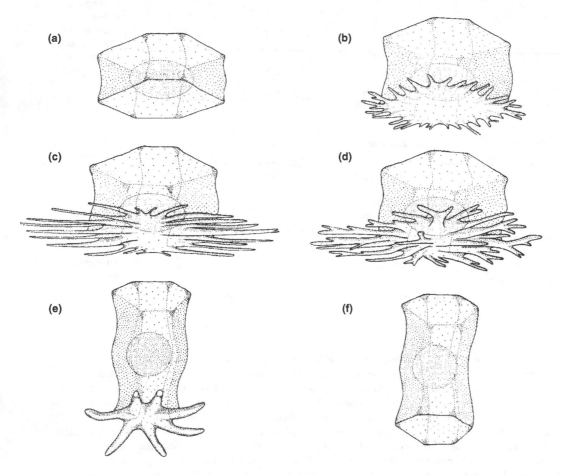

Figure 16.15 Changes of epidermal cell morphology during the final larval stage of *Calpodes* (Lepidoptera), according to Locke (1985).

rapidly as the larval segments also shorten prior to pupation. The feet are probably critically important in producing changes in the shape of an insect at a metamorphosis, and are probably formed in all insects.

16.5.2 Apolysis

Perhaps as a result of the changes in cell shape, tension is generated at the epidermal cell surface, resulting in its separation from the cuticle (Fig. 16.14b). In *Podura* (Collembola), however, the outer plasma membrane of the epidermal cells forms small outpushings that separate off as vesicles to form foam which lifts off the cuticle. The separation of the cuticle from the underlying epidermis is known as apolysis, and the space formed between the epidermis and cuticle is called the exuvial, or subcuticular space.

16.5.3 Digestion of the old endocuticle

Before the cuticle is shed, the endocuticle is digested by enzymes secreted by the epidermal cells (Fig. 16.14d). A mixture of enzymes is involved, including chitinases and different peptidases such as trypsin-like and chymotrypsin-like serine peptidases, carboxypeptidases, cysteine peptidases and metallopeptidases.

At least in some species, some of the enzymes are present in the cuticle even before apolysis. For example, in the larva of *Calpodes*, electron-dense droplets are secreted during the production of the last layers of procuticle. These exuvial droplets are probably inactive precursors of molting enzymes. In many insect orders chitinases are present in the integument. Probably most of the enzymes, however, are secreted into the exuvial space after apolysis. In *Podura* (Collembola) they are produced as granules, and in various Lepidoptera they are described as forming a gel.

These enzymes do not start to digest the endocuticle until the outer epicuticle of the new cuticle is formed. They may be inactive because they are in the form of a precursor (zymogen) that requires activation, their substrate is not available to them or their enzymatic activity is blocked by proteinase inhibitors present in the molting fluid. Proteolytic enzymes are probably secreted as precursors and require the secretion of a zymogen-activating enzyme before they become active. Chitinases, however, appear to have the potential to be active from the time of secretion, but depend on the prior action of proteolytic enzymes that free the chitin from its close association with protective proteins, such as Knickkopf. In *Manduca*, and probably in other insects, a chitinase attacks internal linkages in the chitin molecule, producing small oligosaccharides. These are then broken down by a β-N-acetylglucosaminidase. The two enzymes act synergistically (Section 16.3.2).

Activation is associated with the active transport of potassium into the exuvial space, accompanied by a bulk flow of water. The fluid produced is called molting fluid and its ionic composition may serve to buffer the enzyme systems against pH changes during the subsequent digestion of cuticle, as well as contributing to the solution of exuvial droplets. The enzymes digest all of the unsclerotized cuticle except the ecdysial membrane (see below), but have no effect on the exocuticle or on the muscle and nerve connections to the old cuticle. These connections persist so that the insect is still able to move and receive stimuli from the environment, but they are finally broken at ecdysis by the movements of the insect.

The products of cuticular digestion are absorbed through the mouth and anus, and possibly also directly through the integument, at least in *Manduca*. Up to 90% of the materials present in the cuticle may be conserved in this way. Surgical inactivation of the foregut revealed that during the molt between the fourth and fifth larval instars, foregut activity is modulated in order to ingest molting fluid prior to the onset of ecdysis.

As a result of the activity of molting fluid, the cuticle becomes very thin and weak along the ecdysial lines (Fig. 16.14). These vary in position, but in the locust there is an L-shaped line on the head (Fig. 1.3a) and a median dorsal line on the prothorax.

16.5.4 Ecdysis

When the molting fluid and the products of digestion of the molting fluid are resorbed, the old cuticle consists of little more than epicuticle and exocuticle and is quite separate from the new cuticle (Fig. 16.14e). Ecdysis usually follows as soon as digestion is complete (see Chapter 15).

A phase of preparatory activity loosens the old cuticle (pre-ecdysis behavior). It consists of a stereotyped pattern of relatively simple motor activities following each other in a more or less definite order. Seven motor programs are recognized in the preparatory phase of *Teleogryllus* (Orthoptera). Adult Lepidoptera make partial rotatory movements of the abdomen to loosen the pupal cuticle just before eclosion. The sequence of activities is controlled largely endogenously by the central nervous system, but feedback from peripheral sensilla can prolong activities. A stereotyped pattern of ecdysis movements is also observable in *Tribolium* (Coleoptera), which consists of three intervals. About ten hours before ecdysis is initiated, dorsal–ventral contractions begin. Then, ecdysis is promoted by vigorous bouts of reverse-bending. After the shedding of the old cuticle the beetles appear sluggish and are mostly at rest.

The insect splits the old cuticle by exerting pressure against it from within. An increase in blood volume, occurring before ecdysis in *Schistocerca* (Fig. 5.10) and probably in other insects, contributes to the insect's ability to do this. Then, in the preparatory phase of ecdysis, it usually swallows air or water, swelling the gut so that hemolymph pressure is increased. Blood is pumped into a particular part of the body, often the thorax, so that this expands and exerts pressure on the old cuticle, causing it to split along its lines of weakness. Special muscles may be concerned in these pumping movements.

When the old cuticle is split, the insect draws itself out, usually head and thorax first, followed by the abdomen and appendages. Many insects suspend themselves freely from a support so that emergence is aided by gravity. All the cuticular parts are shed, including the intima of fore- and hindgut, the endophragmal skeleton and the linings of the tracheae, except for some delicate parts which may break off. The old cuticle is referred to as the exuvia.

Immediately after emergence, the new cuticle is unexpanded and soft, so that it provides the insect with little support. The blood probably acts as a hydrostatic skeleton, since its volume is still high, and, in adult *Calliphora* (Diptera), it constitutes 30% of the body weight at this time. In the later stages of ecdysis and immediately afterwards the insect expands the new cuticle before it hardens. This again often involves swallowing air or water. In the blowfly, air is pumped into the gut by the pharyngeal muscles, producing a steady increase in blood pressure, while at the same time simultaneous contractions of the abdominal and ptilinal muscles produce transient increases in pressure. Expansion results partly from the opening out of deep folds in the new cuticle and partly from the pulling out of minor wrinkles in the epicuticle (Fig. 16.16). It is probably facilitated by a plasticization of the procuticle of the presumptive sclerites. The high pressure does not expand the membranous regions between the sclerites perhaps because the presumptive sclerites are held tightly together by accessory muscles.

When expansion is complete, blood volume is reduced so that it comprises only about 10% of the body weight. Some parts of the skeleton may be hardened before ecdysis. This pre-ecdysial hardening is usually restricted to small parts of the cuticle such as the claws, which are essential for the insect to hold

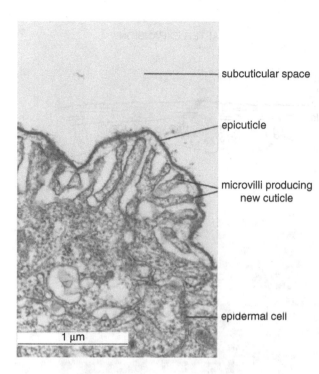

subcuticular space

epicuticle

microvilli producing
new cuticle

epidermal cell

1 μm

Figure 16.16 Formation of the epicuticle. Section through the distal regions of the epidermis during a molt at a stage approximately equivalent to Fig. 16.14c. The outer epicuticle is present as a complete layer. Above the epicuticle is the exuvial space, which is beneath the old cuticle missing in this preparation (courtesy of Dr. M. Locke).

on with, but it is more extensive in Cyclorrhapha and those Lepidoptera which have to escape from a pupal cell or cocoon.

16.6 Cuticle formation

16.6.1 Formation of the epi- and procuticle

Production of the new cuticle is generally believed to begin with the secretion of the outer epicuticle (cuticulin) as discontinuous patches at the tips of microvilli of the epidermal cells. The patches grow at their margins and eventually coalesce to form a continuous layer over the whole of the epidermis, which separates the old from the new, forming cuticle (Fig. 16.17).

As the outer epicuticle is produced, the ecdysial membrane forms from inner layers of the old endocuticle and becomes sclerotized, preventing its degradation. It has been suggested that formation of the ecdysial membrane is simply a consequence of

the presence of sclerotizing agents and phenoloxidases, and that it has no functional significance, but it may play some as yet unknown role. The inner epicuticle is secreted when the outer epicuticle is complete and the apical surfaces of the epidermal cells withdraw slightly. It is discharged in vesicles, which coalesce to form a discrete layer, and, in *Calpodes*, phenoloxidase is secreted at the same time. Polyphenols, however, are not present until about the time of ecdysis, so tanning of this layer does not take place until some time after its production. Production of procuticle begins after the inner epicuticle is laid down. The chitin microfibrils are produced at plaques on the surface membrane of the epidermal cells. At the same time, protein is laid down in the interstices between the fibrils. Some of this protein is synthesized in the epidermal cells, while some is taken up from the hemolymph, having been synthesized elsewhere. The zone of deposition of new cuticle is distinct and is sometimes called Schmidt's layer, but its precise nature is not known.

(a)

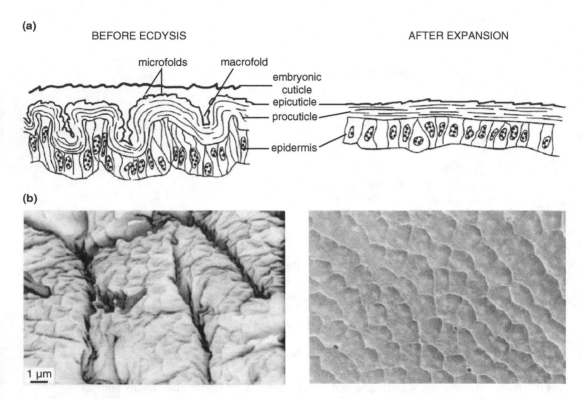

Figure 16.17 Expansion of the new cuticle in a first-stage larva of the locust, *Schistocerca gregaria*, immediately after hatching. Similar changes occur at subsequent molts. Left: before ecdysis of the old embryonic cuticle, the new larval cuticle exhibits macro- and microfolds; right: after expansion, about 30 minutes later, but before sclerotization is evident. The cuticle appears flat without any macrofolds, and the microfolds have almost disappeared. Outlines of the underlying epidermal cells producing each path of cuticle are clearly seen (after Bernays, 1972). (a) Diagrammatic sections through the integument. (b) Low-power scanning electron micrographs of the surface of the cuticle (courtesy of Dr. E. A. Bernays).

The first cuticle is produced by the embryo and then replaced during subsequent molting. Detailed insight into cuticle formation has been obtained from studies in *Drosophila* by carefully analyzing changes in the cuticle's ultrastructure during embryonic development. Cuticle differentiation of *Drosophila* embryos appears to occur in three phases. In the first phase, the different cuticular layers are established. The apical plasma membrane has protruded riffles, so-called undulae. The microfibrils of the procuticle are deposited perpendicular to these apical undulae, which appear to be necessary for microfibril orientation. In the *Drosophila* embryo the epicuticle and procuticle appear to be generated simultaneously. In the second phase the cuticular layers become thicker and in the third phase synthesis of cuticular material stops and the chitinous lamellae rotate to achieve their typical orientation. It has to be noted, however, that differentiation of the embryonic cuticle may vary significantly between different species.

16.6.2 Sclerotization of cuticle

Sclerotization of the cuticle is primarily a consequence of the formation of crosslinks between cuticle proteins so that they form a rigid matrix in which chitin microfibrils can be embedded. The process of crosslinking is also called hardening or tanning. The sclerotization process is frequently but not necessarily accompanied by a typical change in color to tan, brown or black. This color change may be due to the sclerotization process itself or to the formation of eumelanins, pheomelanins, pteridines or pigments that derive from carotenoids or from tryptophan (kynurenines, ommochromes).

Some hardening takes place before ecdysis, but the bulk occurs soon afterwards when the new cuticle is completely expanded. The proteins are laid down as the procuticle is formed, but the crosslinker molecules, the catechols dopa and dopamine

(Fig. 16.19), are not formed until immediately prior to sclerotization. Insects cannot synthesize the ring forming the skeleton of the catecholamines, and it is usually derived from the aromatic amino acids, tyrosine and phenylalanine.

Most insects store tyrosine over the intermolt period. Sometimes it is stored in the fat body, but often the bulk is in the hemolymph. As tyrosine is relatively insoluble it may be common for it to be conjugated with another compound so that its solubility is increased. In *Manduca* it is stored as a glucoside (Fig. 16.19), which, in addition to increasing its solubility, may also protect it from being metabolized in competing pathways.

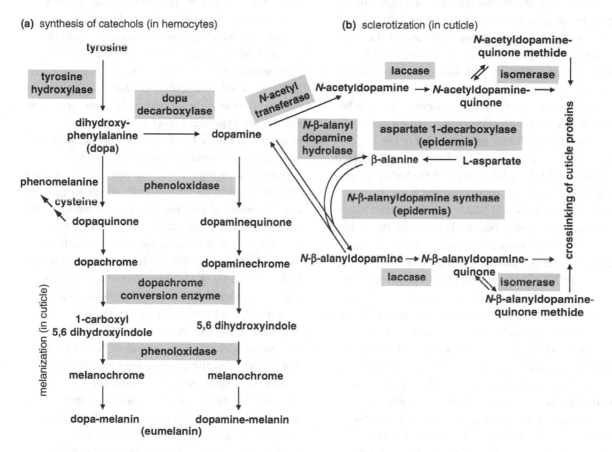

Figure 16.18 Melanization and sclerotization pathways in *T. castaneum*. The diagram illustrates the synthesis of catechols from tyrosine and their roles in melanization (a) and sclerotization (b). Formation of dopa-melanin is a minor reaction in *Tribolium* (modified according to Arakane *et al.*, 2009).

The initial steps of sclerotization seem to take place primarily in hemocytes (Fig. 16.18a). The first step is the conversion of tyrosine to dihydroxyphenylalanine (dopa) by the enzymatic activity of tyrosine hydroxylase. The next step is the decarboxylation of dopa to dopamine by dopa decarboxylase (EC 4.1.1.28). Dopa and dopamine serve also as precursors for melanin synthesis, while the agents causing sclerotization derive only from dopamine, which is enzymatically converted into N-acetyldopamine (NADA) and N-β-alanyldopamine (NBAD) by the enzymatic activities of N-acetyltransferase and N-β-alanyldopamine synthase, respectively (Fig. 16.18b).

NADA is probably the only tanning agent in the cuticles of grasshoppers. In *Manduca*, NADA predominates at larva–larva molts and at the pupa–adult molt, but at the larva–pupa molt, NBAD predominates (Fig. 16.19c). The latter has also been shown to be important in some flies. The kind of catecholamine used for sclerotization may also be crucial for the physical properties and the color of the cuticles. NADA is the main catechol in the unpigmented hard larval head capsule of *Manduca*, whereas NBAD is predominant in the hard pigmented cuticles of the larval mandible and pupa. In *Tribolium*, NADA and NBAD are major catechols in the rust-red colored cuticle. The finding that a *black* mutant of *Tribolium* has decreased concentrations of NBAD when compared with wild-type beetles is in line with NBAD's presumed role in pigmentation. However, the *black* mutation influenced the puncture resistance of the cuticle, suggesting that NBAD is also critical for the hardening of the beetle cuticle.

Immediately after ecdysis NADA and NBAD are transferred to the cuticle, where they are oxidized to their corresponding chinones by the catalytic activities of phenoloxidases (Fig. 16.18b). Two multicopper phenoloxidases, namely tyrosinase (EC 1.10.3.1) and laccase (EC 1.10.3.2), have been detected in insect cuticles so far. Laccase was hypothesized to play an important role in sclerotization by oxidizing the catechols. Consistent with its presumed role in cuticle sclerotization, one of two isoformic genes of the *Manduca* laccase (*MsLac2*) was found to be highly expressed in the epidermis of pharate pupae (i.e., pre-eclosed, still within the last larval integument), while the expression in the epidermis of late pupae was low. It is likely expressed in an active form rather than as a zymogenic form. Further experimental evidence for its presumed role in sclerotization was provided by the successful crosslinking of recombinant RR-1 cuticle proteins from *Manduca* using catechols that were oxidized in vitro by recombinant MsLac2. The biological significance of phenoloxidases with regard to sclerotization and pigmentation was also studied in the red flour beetle, *Tribolium*, using the RNA interference technique. While the knockdown of the mRNAs encoding one of two laccase isoforms (TcLac2) led to severe sclerotization and pigmentation defects at all molts, the two tyrosinase isoforms as well as a second laccase (TcLac1) isoform had no effect on cuticle tanning. The two tyrosinases and TcLac1 apparently serve other functions unrelated to cuticle tanning in this beetle, such as melanization, wound healing and immunity.

Quinones are highly reactive molecules and, once formed by phenoloxidases, their linkage to proteins is independent of enzyme activity. According to the current hypothesis of cuticle sclerotization they undergo nucleophilic addition reactions with the imidazole ring of histidyl residues (maybe also with lysine or less likely cysteine), which are found in higher numbers in RR-1 cuticle proteins of highly sclerotized, hard cuticles. Three modes of crosslinking have been suggested (Fig. 16.19a): (1) chinon sclerotization – covalent bonds form directly between the quinone ring and the cuticle protein; (2) chinon methide sclerotization – quinones are converted into unstable quinone methides by chinon isomerases – the linkage is between the quinone ring and the protein; (3) β-sclerotization – N-acetyldopamine chinone methide is converted to 1,2-dehydro-N-acetyldopamine chinon, leading to linkages

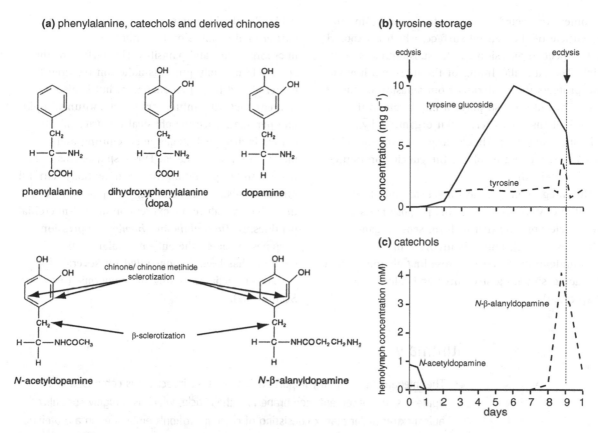

(a) phenylalanine, catechols and derived chinones

phenylalanine

dihydroxyphenylalanine (dopa)

dopamine

chinone/ chinone metihide sclerotization

β-sclerotization

N-acetyldopamine

N-β-alanyldopamine

(b) tyrosine storage

ecdysis ecdysis

tyrosine glucoside

tyrosine

(c) catechols

N-β-alanyldopamine

N-acetyldopamine

days

Figure 16.19 Precursors of sclerotizing agents and changes in their amounts in the hemolymph of *Manduca*. (a) Chemical structures of different tyrosine derived precursors of cuticle sclerotization. The arrows point to the sites where covalent bonds form between the protein and the chinones. (b) Tyrosine is stored largely as a glucoside, which is hydrolyzed before ecdysis. As it is rapidly converted into catechols, free tyrosine never reaches high levels (Ahmed *et al.*, 1983). (c) Catechols are only present in the hemolymph at the time of molting. *N*-acetyldopamine is most abundant at the molt to the final larval stage. At the larval–pupal molt only *N*-alanyldopamine is produced (Hopkins *et al.*, 1984).

involving the β-carbon of the quinone side-chain. β-sclerotization has been linked to the formation of colorless or transparent cuticles, while quinone tanning also causes the cuticle to darken.

16.7 Functions of the integument

The integument is one of the primary features responsible for the success of the insects. It plays an important part in supporting the insect, an essential requirement in terrestrial animals. The tubular, external skeleton of the legs provides great strength

and relative lightness compared with the internal skeleton of vertebrates. Further, the presence of hard, jointed appendages makes accurate movements possible with a minimum of muscle, and, by lifting the body off the ground, facilitates rapid movement. The rigidity and lightness of cuticle in the formation of wings has made flight possible.

The integument provides protection. Some insects, such as adult beetles, have hardened, heavily sclerotized cuticles which make them difficult for predators to catch or parasites to parasitize. Protection from the physical environment is also afforded. Again in beetles, the upper cuticle of the abdomen, protected by the elytra, is very thin but the cuticle of the ventral surface, which is exposed and subject to abrasion by the substratum, is very thick. The cuticular lining of the fore- and hindgut also protects the epidermis from abrasion by the food. The cuticle also plays a major role in the success of insects as terrestrial organisms by reducing water loss. This is largely a function of the wax layer, but the whole of the cuticle contributes (see Chapter 18).

The integument displays a sensory interface with the environment. For this purpose parts of the cuticle are modified to form sense organs (Chapters 23, 24), and it is involved in communication, either by providing pheromones or producing signaling structures and colors (see Chapter 24).

Finally, the integument is a highly effective barrier against invasion by pathogenic microorganisms and parasites. The surface of the epicuticle not only prevents adhesion required for bacterial and fungal colonization, but is also impregnated with antibacterial and antifungal lipids and peptides. Once the physical integrity of the cuticle is disrupted, hemolymph clotting and melanization are induced. This response closes the wound and traps and inactivates potentially harmful microorganisms. Furthermore, the epidermal cells immediately induce the expression of anti-microbial peptides. In *Drosophila* and *Bombyx*, expression of genes encoding the anti-microbial peptide Cecropin has been demonstrated for several surface epithelia, including the epidermis (see also Section 5.2).

Summary

- The integument has many important functions for insects. It is composed of the epidermis, the basement membrane and the cuticle, which is a highly specialized apical extracellular matrix consisting of chitin microfibrils embedded in a proteinous matrix.

- The cuticle is a multilayered biocomposite that can be divided into endo-, exo- and epicuticle. The epicuticle is covered with a wax layer that serves for waterproofing. In addition, the wax layer can be impregnated with hydrocarbons important for social communication.

- Most of the cuticle components are secreted by epidermal cells in a developmentally regulated manner, but some components are added by other cells such as oenocytes. Insect cuticles vary widely in their mechanical properties, mainly due to differences in the composition of cuticle proteins and degree of sclerotization. The latter process results in hardening of the cuticle due to the formation of crosslinks between cuticular proteins.

- The cuticle has only a limited capacity for expansion. To allow growth, it has to be shed regularly during molting, a process that involves the partial degradation of the old cuticle by chitinolytic and proteolytic enzymes and the biosynthesis of the new cuticle.

- Compounds that interfere with cuticle formation or degradation have turned out to be highly effective tools for the control of insect pests and vectors.

Recommended reading

Andersen, S. O. (2010). Insect cuticular sclerotization: a review. *Insect Biochemistry and Molecular Biology* **40**, 166–178.

Moussian, B. (2010). Recent advances in understanding mechanisms of insect cuticle differentiation. *Insect Biochemistry and Molecular Biology* **40**, 363–375.

Muthukrishnan, S., Merzendorfer, H., Arakane, Y. and Kramer, K. J. (2011). Chitin metabolism in insects. In *Insect Molecular Biology and Biochemistry*, ed. L. I. Gilbert, pp. 193–235. San Diego, CA: Academic Press.

Neville, A. C. (1975). *Biology of the Arthropod Cuticle*. Berlin: Springer-Verlag.

Willis, J. H. (2010). Structural cuticular proteins from arthropods: annotation, nomenclature, and sequence characteristics in the genomics era. *Insect Biochemistry and Molecular Biology* **40**, 189–204.

References in figure captions

Ahmed, R. F., Hopkins, T. L. and Kramer, K. J. (1983). Tyrosine and tyrosine glucoside titres in whole animals and tissues during development of the tobacco hornworm *Manduca sexta* (L.). *Insect Biochemistry* **13**, 369–374.

Andersen, S. O. (1974). Cuticular sclerotization in larval and adult locusts, *Schistocerca gregaria*. *Journal of Insect Physiology* **20**, 1537–1552.

Andersen, S. O. and Weis-Fogh, T. (1964). Resilin: a rubberlike protein in arthropod cuticle. *Advances in Insect Physiology* **2**, 1–66.

Arakane, Y., Lomakin, J., Beeman, R. W., *et al.* (2009). Molecular and functional analyses of amino acid decarboxylases involved in cuticle tanning in *Tribolium castaneum*. *Journal of Biological Chemistry* **284**, 16584–16594.

Bernays, E. A. (1972). Changes in the first instar cuticle of *Schistocerca gregaria* before and associated with hatching. *Journal of Insect Physiology* **20**, 281–290.

Fontaine, A. R., Olsen, N., Ring, R. A. and Singla, C. L. (1991). Cuticular metal hardening of mouthparts and claws of some forest insects of British Columbia. *Journal of the Entomological Society of British Columbia* **88**, 45–55.

Hopkins, T. L., Morgan, T. D. and Kramer, K. K. (1984). Catecholamines in haemolymph and cuticle during larval, pupal and adult development of *Manduca sexta* (L.). *Insect Biochemistry* **14**, 533–540.

Kameda, T., Miyazawa, M., Ono, H. and Yoshida, M. (2005). Hydrogen bonding structure and stability of alpha-chitin studied by 13C solid-state NMR. *Macromolecular Bioscience* **5**, 103–106.

Locke, M. (1985). The structure of epidermal feet during their development. *Tissue & Cell* **17**, 901–921.

Neville, A. C., Thomas, M. G. and Zelazny, B. (1969). Pore canal shape related to molecular architecture of arthropod cuticle. *Tissue & Cell* **1**, 183–200.

Payre, F. (2004). Genetic control of epidermis differentiation in *Drosophila*. *International Journal of Developmental Biology* **48**, 207–215.

Richards, A. G. (1967). Sclerotization and the localization of brown and black colors in insects. *Zoologische Jahrbücher: Abteilung für Anatomie und Ontogenie der Tiere* **84**, 25–62.

Snodgrass, R. E. (1935). *The Principles of Insect Morphology.* New York, NY: McGraw-Hill.

Vincent, J. F. V. and Wegst, U. G. K. (2004). Design and mechanical properties of insect cuticle. *Arthropod Structure & Development* **33**, 187–199.

17

Gaseous exchange

REVISED AND UPDATED BY **JON F. HARRISON
AND LUTZ T. WASSERTHAL**

INTRODUCTION

Gaseous exchange in insects occurs through a system of air-filled internal tubes, the tracheal system, the finer branches of which extend to all parts of the body and may become functionally intracellular in muscle fibers. Thus oxygen is carried in the gas phase directly to its sites of utilization. While the blood is not concerned with oxygen transport in most insects, some insects have now been shown to have hemocyanin, an oxygen-carrying pigment, in the blood. In terrestrial insects and some aquatic species, the tracheae open to the outside through segmental pores, the spiracles, which generally have some filter structures and a closing mechanism reducing water loss from the respiratory surfaces. Other aquatic species have no functional spiracles, and gaseous exchange with the water involves arrays of tracheae close beneath the surface of thin, permeable cuticle.

   This chapter is divided into ten sections. Section 17.1 describes the tracheal system, its structure, distribution and development. Section 17.2 deals with the number, structure and distribution of the spiracles. Section 17.3 follows with cutaneous gas exchange; Section 17.4 treats respiratory pigments; and Section 17.5 describes gaseous exchange in terrestrial insects, considering diffusion and ventilation in resting and flying insects and control of ventilation. Section 17.6 addresses the gaseous exchange in aquatic insects, with oxygen uptake from the air and by gills. Section 17.7 gives attention to insects subject to occasional submersion. Section 17.8 refers to the gas exchange in endoparasitic insects. Section 17.9 is concerned with other functions of the tracheal system, and Section 17.10 with gas exchange in insect eggs.

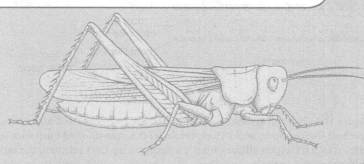

The Insects: Structure and Function (5th edition), ed. S. J. Simpson and A. E. Douglas.
Published by Cambridge University Press. © Cambridge University Press 2013.

17.1 Tracheal system

17.1.1 Tracheae

The tracheae are the larger tubes of the tracheal system, running inward from the spiracles and usually breaking up into finer branches, the smallest of which are about 2 μm in diameter. Larger tracheae are multicellular structures, while the smallest are formed by single cells. Tracheae are formed by invaginations of the ectoderm and so are lined by a cuticular intima that is continuous with the rest of the cuticle. The intima consists of outer epicuticle with a protein/chitin layer beneath it. A spiral thickening of the intima runs along each tube, called a taenidium (Fig. 17.1). In the taenidia the protein/chitin cuticle is sclerotized. The chitin microfibrils in

the taenidia run around the trachea, while between the taenidia they are parallel with the long axis of the trachea. A layer of resilin may be present beneath the epicuticle.

Some tracheae seem designed to remain expanded and to serve as conducting pipes, while others collapse easily, providing ventilation. The taenidia prevent collapse of the trachea if pressure within the tube is below air or hemocoel pressure (Section 5.2.1; Fig. 5.5). In the wing tracheae of some insects, the taenidia are themselves twisted or supercoiled, giving some elasticity to the wall of the trachea. The structural differences that account for collapsibility of some tracheae remain poorly known. Because the tracheal walls are relatively thick, it is thought that little oxygen passes through the walls; thus the tracheae serve primarily to transport oxygen from

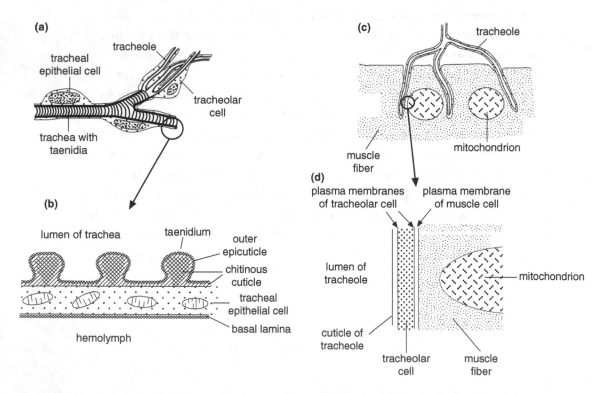

Figure 17.1 Tracheal system. (a) Diagram showing a trachea and tracheoles. (b) Longitudinal section of the tracheal wall; tracheal lumen above, hemolymph below. (c) Diagram showing tracheoles indenting a muscle fiber. (d) Enlargement of area circled in (c). Oxygen diffuses from the lumen of the tracheole (left) through the cuticle and the tracheolar cell to the mitochondrion in the muscle fiber.

the spiracles to tracheoles. However, because carbon dioxide is much more soluble in tissue, it is likely that a significant fraction of carbon dioxide transported does cross the tracheae. In some insects, the tracheae have an intimate contact with the ovaries by the cuticle between the taenidia, which bulges toward the oocyte. Wigglesworth called these "aeriferous tracheae" and suggested that they functionally replace tracheoles as the primary supply to the ovaries.

In some insects, the tracheae are expanded to form thin-walled air sacs (see Section 17.1.4) in which the taenidia are absent or poorly developed and often irregularly arranged. The air sacs collapse easily under pressure and serve as bellows for ventilation of the tracheal system; they also allow for tissue growth and increases in hemolymph volume. The distribution of air sacs is quite variable among insect species and with age. First-instar locusts virtually lack air sacs, while adults have many air sacs throughout the body. Similarly, caterpillars such as *Manduca sexta* lack air sacs while adult moths have many. *Drosophila* have air sacs in the head and thorax, but not in the abdomen. Many cockroaches and tenebrionid beetles apparently lack air sacs entirely.

17.1.2 Tracheoles

At various points along their length, especially distally, the tracheae give rise to finer tubes, the tracheoles. Tracheoles are blind-ended, air-filled extensions of terminal tracheal cells, and are the primary site of gas exchange. They are usually less than 2 μm in diameter, tapering to 0.3 μm and lack the longitudinal seam found on larger tracheae. Some tracheoles retain their cuticular lining at molting, which is not usually true of tracheae. They are formed in "tracheolar cells," which are derived from the epidermal cells lining the tracheae (Fig. 17.2). The intima of tracheoles is some 16–20 nm thick and may consist only of outer epicuticle. It is thrown into taenidial ridges, but, unlike the taenidia of tracheae, these ridges are not filled with chitin/protein matrix. The thin walls and high surface-to-volume ratio of the tracheoles enable their high diffusing capacity. Most tracheoles are extracellular, and in flight muscles they generally run parallel to the muscle fibers. In some flight muscles they penetrate deeply into the muscle via invaginations of the plasma membrane (Fig. 17.1c,d).

17.1.3 Tracheal development

The development of the tracheal system in *Drosophila* has been a focus of intense interest by developmental biologists, and these studies have revealed many fundamental new insights. In flies, the tracheal system arises from sacs of 80 embryonic cells at each segment. Remarkably, the entire larval tracheal system develops from growth and branching of these 80 cells. At metamorphosis, some adult tracheae are formed from imaginal discs, while others

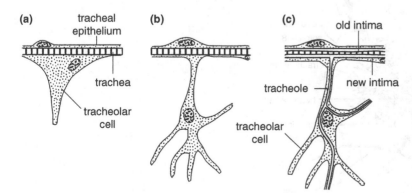

Figure 17.2 Development of a tracheole (after Keister, 1948). (a) Tracheolar cell developing from the tracheal epithelium. (b) Tracheolar cell with extensive cytoplasmic processes. (c) Tracheole within tracheolar cell and becoming connected to existing trachea at a molt.

(a)

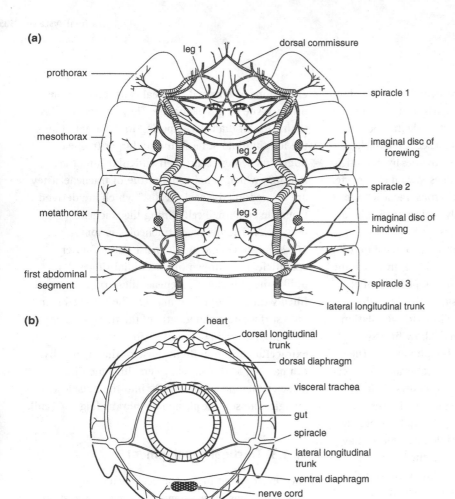

- prothorax
- leg 1
- dorsal commissure
- spiracle 1
- mesothorax
- leg 2
- imaginal disc of forewing
- spiracle 2
- metathorax
- leg 3
- imaginal disc of hindwing
- first abdominal segment
- spiracle 3
- lateral longitudinal trunk

(b)

- heart
- dorsal longitudinal trunk
- dorsal diaphragm
- visceral trachea
- gut
- spiracle
- lateral longitudinal trunk
- ventral diaphragm
- nerve cord
- ventral longitudinal trunk
- ventral commissure

(c)

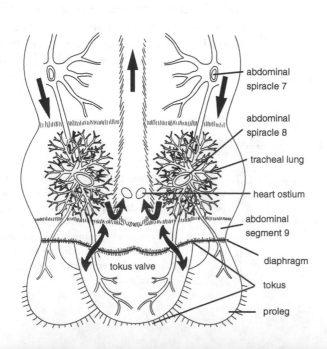

- abdominal spiracle 7
- abdominal spiracle 8
- tracheal lung
- heart ostium
- abdominal segment 9
- diaphragm
- tokus valve
- tokus
- proleg

arise from larval tracheal cells that re-enter the cell cycle and regain developmental potency. Each cluster of embryonic cells undergoes development and branching within that segment, leading to bilateral symmetry. There are four major morphological steps in tracheal development: sac formation, and then primary, secondary and tertiary branching. The first tube is formed by invagination of the embryonic cell cluster, which retains an opening connected to the embryonic surface that will form the spiracle. Primary branches form at bud sites, with multiple cells migrating out that branch and form tubes. Secondary branches are formed by single cells that form a tube as they extend. Terminal tertiary branching, forming the tracheoles, occurs as cytoplasmic extensions of the secondary tube. These terminal branches grow toward target tissues, and then form internal air-filled tubes that connect with the air-filled lumen of the secondary branch (Fig. 17.2).

Our knowledge of the genes and developmental program of the *Drosophila* tracheae is improving. Specific genes are associated with each of the three morphological steps. In addition, branching is controlled by feedback between target tissues which secrete fibroblast growth factor (FGF, encoded by the gene *branchless*), which binds to a fibroblast growth factor receptor (FGFR, encoded by the gene *breathless*). Local hypoxia is sensed by hypoxia inducible factor (HIF) within the tracheoles, which respond by increasing the number of FGFR. Enhanced activation of this signaling pathway leads to increased branching and growth of the terminal tubes. This oxygen regulation of terminal cell branching likely is responsible for the matching of tracheole supply to the metabolic needs of the tissues.

Formation of the tracheal lumen occurs under the control of specific regulatory genes such as *ribbon* and *synaptobrevin*. Both the lumen of the tracheae and its length change during development, and these two processes are under different types of physiological and genetic regulation. Lumen expansion occurs primarily by changes in the apical (luminal) membrane.

As noted above, different branches and regions of the tracheal system have very different properties. How are these generated? Specific developmental genes are activated early in embryonic development, which label cells according to location. For example, the gene *Dpp* labels cells as dorsal or ventral; *Spitz/EGF* and *Wingless/Wnt* label cells as central or peripheral; and *Hedgehog* determines anterior/posterior differences. These location signals induce expression of regional and branch-specific genes that help determine the characteristics of particular tracheae. Substrate interactions with other local cells are also important in determining the direction of migration and morphology of tracheae.

Creation of a continuous tracheal system from the segmental systems of the embryo requires fusion of multiple tracheae across segments. This is accomplished by fusion cells, developmentally marked by the gene *escargot*, at the ends of primary branches in each segment. These fusion cells contact and fuse with each other, and form a continuous lumen.

17.1.4 Distribution of the tracheae within the insect

In those Collembola that have tracheae and in the Archaeognatha, the tracheae from each spiracle form a tuft which remains separate from the tufts of other

Caption for Figure 17.3 Tracheal system (from Snodgrass, 1935). (a) Tracheation of the thorax and first abdominal segment of a caterpillar, dorsal view. (b) Diagrammatic cross-section of the abdomen of an orthopteran showing the principal tracheae and tracheal trunks. (c) Tracheal lung in the posterior larval abdomen of the skipper *Calpodes*. The tracheae of the eighth segment finely branch and terminate freely in the hemolymph near the posterior heart ostia and in the tokus, a separate compartment (after Locke, 1998).

spiracles. In the majority of insects, however, the tracheae from neighboring spiracles join to form longitudinal trunks running the length of the body (Fig. 17.3). Usually there is a lateral trunk on either side of the body connecting the spiracles. In addition, there are often longitudinal trunks along the gut, and on either side of the heart and nerve cord. Tracheae branch from the longitudinal trunks to the nearby tissues. Transverse commissures often connect both sides at each segment.

The arrangement of tracheae tends to follow a similar pattern in each body segment, especially in larval forms and in the abdomen of adults. This basic arrangement is modified, however, in the head, where there are no spiracles, in the abdomen beyond the most caudal spiracle and in the thorax. In grasshoppers, the head is supplied with air from spiracle 1 (often on the prothorax, see Section 17.2.2) through two main tracheal branches on each side, a dorsal branch to the antennae, eyes and brain and a ventral branch to the mouthparts and their muscles (Fig. 17.4).

In many caterpillars the tracheae arising from the eighth abdominal spiracles differ from the tracheae of the other segments. They are finer, more numerous and have an especially thin-walled intima with a reticulate basal lamina. Through the transparent integument of the skipper butterfly, *Calpodes ethlius*, hemocytes have been observed to adhere transiently to these tufts on their passage through the posterior hemocoel into the posterior heart ostia (Fig. 17.3c). The fine tracheal ramifications probably serve as lungs for aerating the hemocytes, cells without permanent supply by tracheoles. In adults the tracheal system of the eighth segment is lost. However, the thin walls of many air sacs offer similar conditions for the gas exchange with adhering hemocytes.

The thorax of many flying insects contains multiple air sacs (Fig. 17.4). In insects such as locusts, these are compressed by changes in thoracic volume caused by contractions of the flight muscles, producing ventilation. Transport of oxygen to the head entails special problems as the head contains many highly metabolic tissues (brain, mandibular

muscles), yet the head lacks spiracles. Many insects have large, collapsible, longitudinal tracheae that extend from the thorax to the head; these are regularly ventilated. Locusts have two large tracheae from the first prothoracic spiracle that are likely sites of inspiration of air to the brain (Fig. 17.4).

The abundance of tracheoles seems to match the metabolic demands of the tissues, though there is little quantitative data. Tracheoles are most abundant in flight muscles. However, they are also particularly abundant in the central nervous system, especially in the neuropile, perhaps suggesting that synapses have a high demand for oxygen. In the ganglia of the locust central nervous system, every point in the neuropile is within $10\,\mu m$ of a tracheole, but no tracheal branching occurs in the outer parts of the ganglion where the cell bodies are situated (Fig. 17.5).

Each flight muscle has a primary supply consisting of a large tracheal trunk or air sac running alongside or through the muscle (Fig. 17.6). If a trachea forms the primary supply it widens to an air sac beyond the muscle. From the primary supply, small, regularly spaced tracheae arise at right angles, running into the muscle. These form the secondary supply and they are often oval proximally, permitting some degree of collapse, and taper regularly to the distal end. Finer branches pass in turn from these tracheae into the muscles. In Odonata the terminal tracheolar branches run alongside and between the muscle fibers, but in close-packed and fibrillar flight muscle they sometimes indent the fiber membrane and the finest tracheolar branches can be closely associated with the mitochondria (Fig. 17.1c,d). Thus they are functionally within the muscle fiber, although anatomically they are still extracellular.

Major changes in tracheation occur at a molt and during the pupal period in endopterygote insects. This may involve changes in the functional spiracles as well as in tracheation. For example, the first-stage larva of *Sciara* (Diptera) has only a pair of posterior abdominal spiracles, the second-stage larvae has only a pair of thoracic spiracles and the fourth-stage larvae has eight pairs of spiracles.

(a)

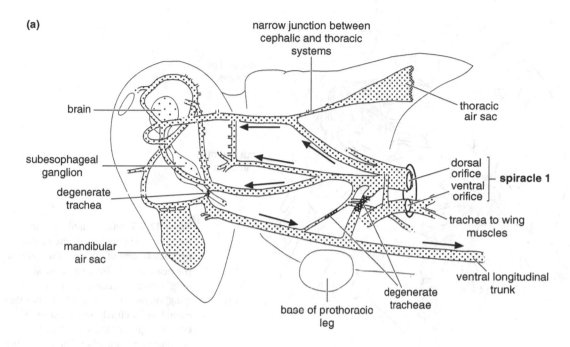

narrow junction between cephalic and thoracic systems

thoracic air sac

brain

subesophageal ganglion

dorsal orifice
ventral orifice ⎱ **spiracle 1**

degenerate trachea

trachea to wing muscles

mandibular air sac

base of prothoracic leg

degenerate tracheae

ventral longitudinal trunk

(b)

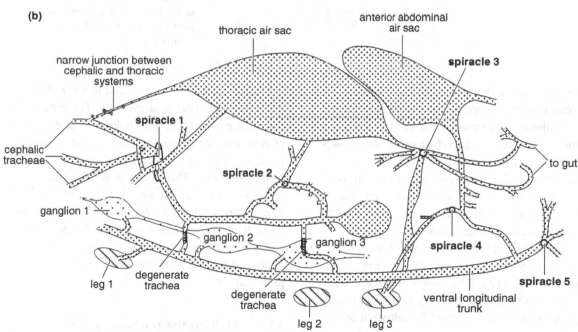

thoracic air sac

anterior abdominal air sac

narrow junction between cephalic and thoracic systems

spiracle 3

spiracle 1

cephalic tracheae

to gut

spiracle 2

ganglion 1

ganglion 2

ganglion 3

spiracle 4

leg 1

degenerate trachea

degenerate trachea

spiracle 5

ventral longitudinal trunk

leg 2

leg 3

Figure 17.4 Tracheal system of a locust (*Schistocerca*) (after Miller, 1960b). (a) Supply to the head from spiracle 1; arrows indicate the probable direction of airflow resulting from abdominal ventilatory movements. (b) Supply to the pterothorax; notice that the thoracic air sac is largely isolated from other parts of the tracheal system by narrow or degenerate tracheae.

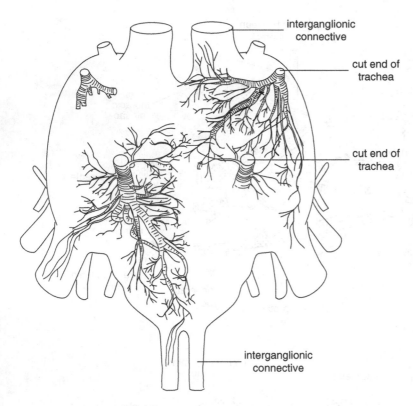

interganglionic connective

cut end of trachea

cut end of trachea

interganglionic connective

Figure 17.5 Tracheal supply to the metathoracic ganglion of the central nervous system of a locust. Two pairs of tracheae, arising from the ventral longitudinal tracheae, enter the ganglion on the ventral side. Notice that branching is largely restricted to the central region of the ganglion. The peripheral region, where the cell bodies of the neurons are present, does not have extensive branching (after Burrows, 1980).

Between molts, the relative tracheal volume decreases strongly because the tissues continue to grow without compensatory increases in the volume of the major tracheae. In *Schistocerca*, the air sacs also become compressed by growing tissues, decreasing absolute tracheal volume later in the instar. As a consequence, tissues become relatively less well supplied with oxygen with insect growth within each larval stage. Consistent with this hypothesis, late in the instar, metabolism and behavior of *Schistocerca* and *Manduca* are suppressed by very mild hypoxia, which is not the case early in the instar. Possibly sensing of tissue hypoxia is one of the triggers for initiation of the hormonal cascade that controls molting; *Manduca* and *Drosophila* molt at smaller sizes in hypoxia, and metabolic rates do not increase after the critical weight for molting, suggesting oxygen limitation (Fig. 17.7).

Changes in the distribution and number of tracheoles can occur between molts. For instance, in the event of damage to the epidermis, the epidermal cells produce cytoplasmic threads, which extend toward and eventually attach themselves to the nearest tracheole. These cytoplasmic threads, which may be 150 μm long, are contractile and drag the tracheole to the region of oxygen-deficient tissue (Fig. 17.8). During embryonic development and within an instar, hypoxia stimulates and hyperoxia suppresses tracheolar branching and proliferation.

17.1.5 Molting the tracheal system

Prior to the molt, the epithelial layer of the tracheae increases in size, with increases in cell size, and in some insects, cell division, leading to an increase in tracheal diameter. A new cuticle is formed under the

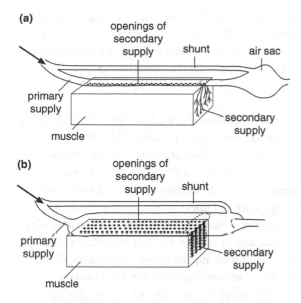

Figure 17.6 Tracheal supply to the flight muscles. Arrows indicate the inward flow of air (after Weis-Fogh, 1964). (a) Primary supply is a trachea ventilated by an air sac outside the muscle. (b) Primary supply is an air sac directly associated with the muscle.

old cuticle. Molting fluid fills the space between the new and old cuticle, and the old cuticle detaches (apolysis, Fig. 17.9). The old cuticle is pulled out of the tracheae through the spiracles with the rest of the exuvia, and the molting fluid is reabsorbed. To facilitate this, the longitudinal trunks break at nodes between segments. Where the number of functional spiracles is less than ten, the "non-functional" spiracles persist and facilitate the shedding of the tracheal intima so that this can occur even in apneustic insects. The "non-functional" spiracles may be visible as faint scars on the cuticle. From each scar, a strand of cuticle (formed from a collapsed trachea) connects with the longitudinal trachea and, at each molt, a tube of new cuticle is laid down around it. The old intima is withdrawn through the tube, which subsequently closes and forms the cuticular strand, connecting the new intima with the outer cuticle. Similar ecdysial tubes are formed next to functional spiracles when the structure of the spiracles is so complex that it does not permit the old intima to be drawn through it. This occurs, for

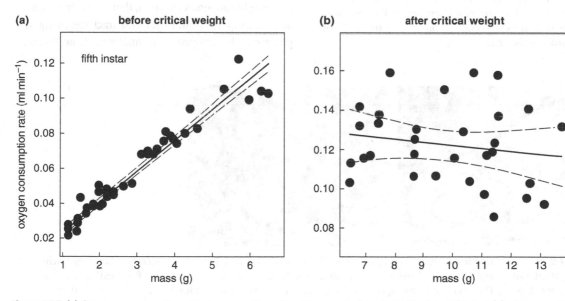

Figure 17.7 (a) Oxygen consumption rate increases linearly with mass before the critical weight, but (b) plateaus after the critical weight in *Manduca sexta* (after Callier and Nijhout, 2011).

example, in Elateridae and Scarabaeidae among the beetles and in some Diptera.

The cuticular lining of the tracheoles appears to be shed in some insects and some stages, but not in others. In *Sciara* (Diptera) the intima is shed at larva–larva molts, but not at the larva–pupa or pupa–adult molts, whereas in *Rhodnius* (Hemiptera) the linings of the tracheoles are never shed, and in grasshoppers even very small tracheoles shed their linings.

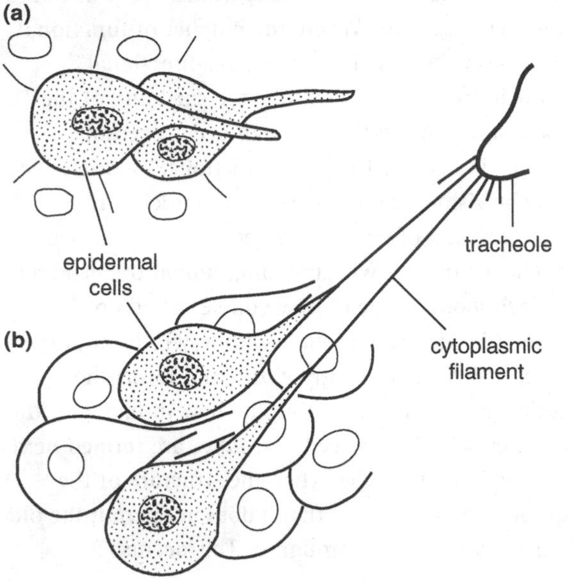

(a)

(b)

epidermal cells

tracheole

cytoplasmic filament

Figure 17.8 Change in the position of a tracheole (after Wigglesworth, 1959). (a) Two epidermal cells send out processes in the direction of a tracheole. (b) Cytoplasmic filaments from the epidermal cells attach to a tracheole and draw it toward the cells.

Pneumatization The tracheal system is liquid-filled in a newly hatched insect, and liquid fills the space between old and new cuticles before each ecdysis. The liquid is replaced by gas in a process known as pneumatization. In terrestrial insects, gas may enter via the spiracles, but pneumatization occurs in aquatic insects and in many terrestrial species without access to air. Gas usually appears first in a main tracheal trunk and then spreads rapidly through the system, which becomes completely gas-filled in 10–30 minutes. Recent research on this process in insect eggs has supported two mechanisms by which it may occur. First, there is evidence for an important role for cavitation. Evaporation from the eggs creates a negative pressure which draws air out of solution within the tracheae, creating a bubble inside the egg. Formation of the bubble coincides with the tanning of the tracheal cuticle, which apparently makes the inside of the tracheae hydrophobic. Second, there is evidence for active transport of fluid out of the tracheae, as liquid is progressively removed from smaller tracheae. It is not

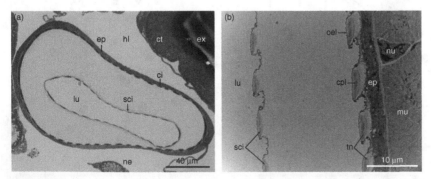

Figure 17.9 Images of molting tracheae in a grasshopper, *Locusta migratoria*. (a) Light micrograph of large leg tracheae, showing third-instar shed intima (sci) and its lumen (lu) within the larger fourth-instar trachea (ci = cuticular intima, ep = epithelial layer). (b) Electron micrograph of molting tracheae showing shed intima (sci) and the lumen of the third-instar tracheae (lu), next to the larger fourth-instar tracheal epithelia (tn = taenidia, cpl = cuticular protein layer, oel = outer epithelial layer, ep = epithelia, nu = nucleus, mu = muscle) (after Snelling *et al.*, 2011).

clear yet whether cavitation-type mechanisms can also function when tracheae are pneumatized after a molt.

In many insects, some liquid normally remains in the endings of the tracheoles. During periods of high energy consumption the liquid is withdrawn from the system and air is drawn further into the tracheoles. Possibly, fluid reabsorption is due to increases in the osmotic pressure of cells when metabolic rate is high.

17.2 Spiracles

The spiracles are the external openings of the tracheal system, found on the thorax or abdomen (never the head). They are lateral in position, and in the Insecta there is never more than one pair of spiracles on a segment, usually on the pleuron. Often, each spiracle is contained in a small, distinct sclerite, the peritreme.

17.2.1 Number and distribution

The largest number of spiracles found in insects is ten pairs, two thoracic and eight abdominal; this occurs in dragonflies, grasshoppers, cockroaches, fleas and some hymenopteran and dipteran larvae. Some endopterygote larvae have only nine pairs of spiracles, with only one thoracic pair. Mycetophilid (Diptera) larvae have eight pairs of spiracles; one mesothoracic and eight abdominal. Many larval Diptera, especially Cyclorrhapha, have only two pairs of spiracles, one mesothoracic and one posterior abdominal. Mosquito larvae have only one pair of spiracles; in some species these occur on the thorax and in others on the posterior abdominal segment. Many aquatic larval insects are apneustic, lacking functional spiracles.

Among other hexapod groups, some Diplura, such as *Japyx*, have 11 pairs of spiracles, including four pairs on the thorax, while the sminthurids (Collembola) have only a single pair of spiracles, between the head and prothorax, from which tracheae extend, without anastomoses, to all parts of the body. Most Collembola have no tracheae at all.

17.2.2 Structure

In its simplest form, found in some Apterygota, the spiracle is a direct opening from the outside into a trachea, but generally the visible opening leads into a cavity, the atrium, from which the tracheae arise. In this case the opening and the atrium are known collectively as the spiracle. Often the walls of the atrium are lined with hairs or filter lamellae (Fig. 17.10). In some Diptera and Coleoptera the spiracle is covered by a sieve plate containing large numbers of small pores. Hairs and sieve plates serve to prevent the entry of dust, microorganisms or, especially in aquatic insects, water into the tracheal system.

The spiracles of most terrestrial insects have a closing mechanism, which is important in the control of gas exchange and internal pressures. The closing mechanism may consist of one or two movable valves in the spiracular opening, or it may be internal, closing off the atrium from the trachea by means of a constriction. In most insects, closure results from the activity of one muscle; opening is produced by the elasticity of the cuticle associated with the spiracle, an elastic ligament or a second muscle. Even where there is an opener muscle, its activity is often associated with an elastic component so that, in the absence of any muscular activity, the spiracle opens.

The metathoracic spiracle of grasshoppers is an example of a "one-muscle" spiracle (Fig. 17.11). It lies in the membrane between the meso- and

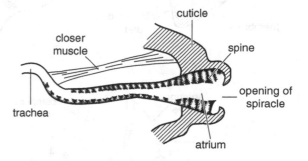

Figure 17.10 Section through a spiracle of a louse, *Haematopinus*, showing the atrium and branched spines in which dust accumulates (after Webb, 1948).

metathorax and has two movable semicircular valves, which are unsclerotized except for the hinge and a basal pad into which a muscle is inserted. This muscle, by pulling down on the valves, causes them to rotate and so to close. The spiracle normally opens because of the elasticity of the surrounding cuticle, but in flight it opens wider as a result of the slight separation of the mesepimeron and metepisternum. These two sclerites, which surround the spiracle, are normally held together by an elastic bridge, but, when the basalar and subalar muscles contract, the sclerites are pulled apart. This movement is transmitted to the spiracle largely through a ligament connecting the metepisternum to the anterior valve, making the spiracle open wide.

The mesothoracic spiracle of grasshoppers is a "two-muscle" spiracle (Fig. 17.12a). It is unusual in having two tracheae leading directly from the external opening. The dorsal opening is connected

with the head and prothorax and the ventral orifice is connected with the flight muscles and legs. A sclerotized rod runs along the free edge of the posterior valve, passing between the orifices and running around the ventral one. The closer muscle arises on a cuticular inflexion beneath the spiracle and is inserted into a process of the sclerotized rod, while the opener muscle, also from the cuticular inflexion, is inserted onto the posterior margin of the posterior valve. When the insect is at rest and the closer muscle relaxes, the spiracle opens to some 20–30% of its maximum as a result of the elasticity of the cuticle; the opener muscle plays no part. Contraction of the opener muscle occurs during slow, deep ventilatory movements and results in the spiracle opening fully.

Closure of abdominal spiracles usually involves a constriction method (Figs. 17.12b, 17.13). Commonly, the atrium is pinched between two sclerotized rods, or

(a)

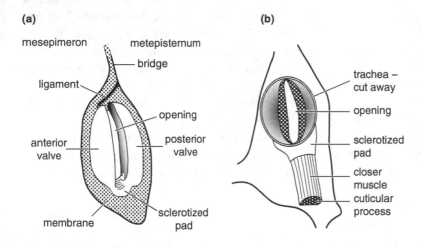

(b)

(c)

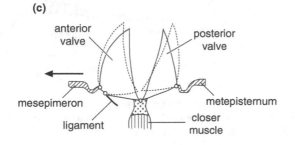

Figure 17.11 A one-muscle spiracle; second thoracic spiracle of a locust (*Schistocerca*) (after Miller, 1960a). (a) External view; (b) internal view; (c) diagrammatic section through the spiracle showing how movement of the mesepimeron (arrow) causes the valves to open wide (dotted).

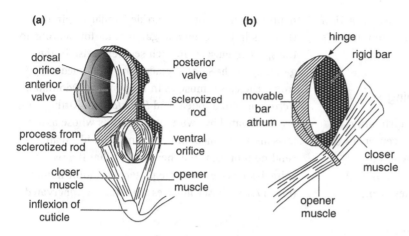

Figure 17.12 Two-muscle spiracles of a grasshopper. (a) First thoracic spiracle seen from inside (based on Snodgrass, 1935; Miller, 1960a). (b) Abdominal spiracle (after Snodgrass, 1935).

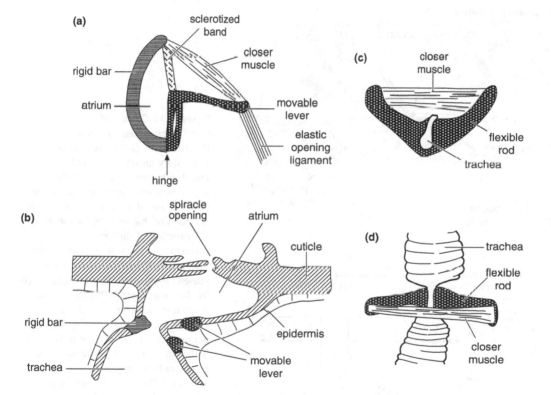

Figure 17.13 Closing mechanisms internal to the spiracle. (a),(b) Abdominal spiracle of a caterpillar seen from inside (a) and in horizontal section (b) (from Imms, 1957). (c),(d) Constricting mechanism on a trachea of a flea seen in horizontal section (c) and from above (d) (after Wigglesworth, 1965).

in the bend of one rod, as the result of the contraction of a muscle. In other instances, the atrium or trachea is bent so that the lumen is occluded.

17.2.3 Control of spiracle opening

In the insects that have been studied (primarily locusts, cockroaches and dragonflies), centers in the thoracic or first abdominal ganglia generate a respiratory rhythm that controls ventilation and coupled openings of the spiracles. These central neuronal centers often override similar respiratory rhythms in the segmental ganglia. Motor neurons to the spiracle muscles in each segment arise in the ganglion of the same segment or that immediately in front. The closer muscles in dragonflies (Odontata), *Periplaneta* (Blattodea) and *Schistocerca* (Orthoptera) are innervated by two motor neurons whose axons pass along the median nerve and then bifurcate, sending a branch to either side, so that the two spiracles receive the same pattern of motor impulses (Fig. 17.14). Closer muscles may also be innervated

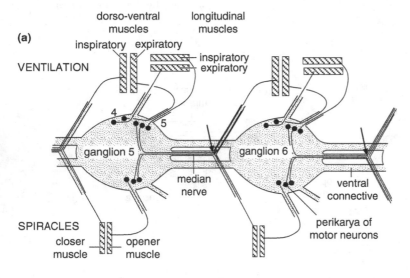

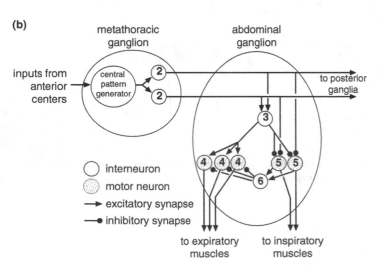

Figure 17.14 Control of spiracles and ventilatory movements in a locust (*Schistocerca*). (a) Innervation of ventilatory muscles (top) and spiracular muscles (bottom). The cell bodies of motor neurons controlling the muscles of one segment are not confined to the ganglion of that segment; some are in the ganglion of the next anterior segment. The axons in the median nerve divide (arrow) so that homologous muscles on the two sides are innervated by the same neurons. Notice that the ventilatory muscles and spiracular muscles are not controlled by the same neurons. Numbers adjacent to cell bodies in ganglion 5 relate to the cells shown in (b) (based on Lewis *et al.*, 1973). (b) Diagram showing the control of ventilatory muscles. The interneurons (2) are driven by a central pattern generator and extend to all the abdominal ganglia. Motor neurons are numbered to correspond with the same numbers in (a). See text for further explanation (based on Lewis *et al.*, 1973; Pearson, 1980).

by an inhibitory axon. The opener muscle, when it is present, is innervated from one or two neurons whose cell bodies are usually lateral in the ganglion and the spiracles of the two sides are independently innervated. In *Blaberus* and related cockroaches, a common inhibitory neuron innervates a number of abdominal spiracle opener muscles.

The closer muscle is caused to contract by the activity of its motor neurons, but the frequency of action potentials, determining the degree of contraction, may be altered by various factors acting on the central nervous system. Of particular importance are a high level of carbon dioxide and a low level of oxygen (hypoxia) in the tissues. Both of these conditions tend to arise while the spiracles are closed as carbon dioxide is produced and oxygen is utilized in respiration. Both conditions lead to a reduction in action potential frequency and so to spiracle opening. In two-muscle spiracles the action potential frequency to the opener muscle is increased by high carbon dioxide levels and hypoxia. Increases in the respiratory rhythm and spiracular opening can also be achieved by feed-forward regulation from the centers that control flight motor behavior.

Carbon dioxide acts directly on the closer muscle of one-muscle spiracles; with direct effects on muscle plus interference with neuromuscular transmission, the junction potential falls, muscle tension is reduced and the spiracle opens. The threshold of the peripheral response to carbon dioxide is set by the frequency of motor impulses from the central nervous system.

17.3 Cutaneous gas exchange

Some gaseous exchange takes place through the cuticle of most insects, but does not usually amount to more than a small percentage of the total gaseous movement. On the other hand, Protura and most Collembola have no tracheal system and must depend on cutaneous respiration together with transport by the hemolymph from the body surface to the tissues.

Cutaneous respiration is also important in many endoparasitic and aquatic insects, where it is coupled with an apneustic (no spiracles) tracheal system. Cutaneous respiration without an associated apneustic tracheal system can only suffice for very small insects with a high surface–volume ratio.

17.4 Respiratory pigments

Until recently it was thought that only a few aquatic insects specialized for extremely hypoxic environments had respiratory pigments. Larval midges of the genus *Chironomus*, and endoparasitic bot fly larvae in the genus *Gastrophilus* have hemoglobins that enable them to extract oxygen from extremely hypoxic media. Now it is clear that most insects have intracellular hemoglobin, and a wide variety of "lower" insects have hemocyanin in their blood. Our understanding of the function of most of these respiratory pigments remains limited.

Most insects that have been studied possess intracellular hemoglobin, primarily expressed in tracheal cells and fat body. Molecular analyses indicate that these proteins do bind oxygen, although their function remains controversial. In *Drosophila*, hypoxia exposure decreases expression of these hemoglobin genes, while hyperoxia increases expression, suggesting that these hemoglobins may function in protection against oxidative stress rather than as oxygen storage proteins.

Larvae of the genus *Chironomus* (Diptera) are the only insects yet known to have hemoglobin in the blood. *Chironomus* larvae live in burrows in the mud under stagnant water, which is commonly poor in oxygen. The hemoglobin of *Chironomus* has two heme groups, and is 50% saturated with oxygen at less than 0.1 kPa, compared to more than 3 kPa in vertebrates. A flow of water is directed through the burrow by dorso-ventral undulating movements of the body, and the current so produced provides food and oxygen. During such bouts of irrigation, the hemoglobin in the blood becomes fully saturated with oxygen, and delivers oxygen to the tissues. During the intervals between ventilatory bouts, the oxygen of the surrounding medium is quickly used up, and hemoglobin releases oxygen; with an oxygen store able to last about nine minutes.

The third-stage larva of *Gastrophilus* is an internal parasite in the stomach of the horse. Larval *Gastrophilus* larvae contain intracellular hemoglobin concentrated in large hemoglobin cells. As in *Chironomus*, the hemoglobin has a very high affinity for oxygen. Four pairs of tracheal trunks run from the posterior spiracles and give off short branches at intervals along their lengths. Each branch breaks up into numerous tracheoles, which are functionally, if not structurally, within a hemoglobin cell. Within the horse's stomach, the larva receives only an intermittent supply of air in gas bubbles with the food, and the hemoglobin of the hemoglobin cells enables the larva to take up more oxygen than is needed for its immediate requirements. This oxygen is used later when air bubbles are no longer available, providing a store of up to four minutes.

Diving beetles of the genus *Anisops* also have intracellular hemoglobin that aids in oxygen storage. When *Anisops* dives, it carries with it a small ventral air store, which is continuous with air under the wings. All the spiracles open into the air store, and the spiracles of abdominal segments 5–7 are very large and covered by sieve plates. From the atria of these spiracles several tracheae arise, branching repeatedly to form "trees," the terminal branches of which indent large cells, called hemoglobin (or tracheal) cells, which are filled with hemoglobin. The hemoglobin becomes oxygenated when the bug is at the surface and deoxygenated during a dive. The oxygen released enables the insect to remain submerged for longer than would otherwise be possible, while at the same time affecting its buoyancy. When the insect first dives, it is buoyant because of its ventral air store, but as the store is used up this buoyancy is reduced until the density of the insect is roughly the same as water and it is able to float in mid-water. This phase is maintained by the steady release of oxygen from the hemoglobin cells resulting from the reduction in partial pressure of oxygen in the store. After about five minutes, however, the hemoglobin is fully unloaded and the insect, now with a tendency to sink, swims up to the surface and renews its air store.

Hemocyanins are copper-containing respiratory pigments that occur in the hemolymph of many invertebrates. Among insects, hemocyanins have been identified in Collembola, Archaeognatha, Dermaptera, Orthoptera, Phasmatodea, Mantodea, Isoptera and Blattaria. The functional significance of this hemocyanin has been best studied in the stonefly, *Perla marginata*. Hemocyanin accounts for about 25% of all hemolymph proteins in nymphal and adult stoneflies. In *P. marginata*, the hemocyanin is 50% saturated with oxygen at about 1 kPa, suggesting it can be useful in both oxygen transport and storage. Phylogenetic analyses confirm that the insect hemocyanins are related to crustacean hemocyanins.

17.5 Gaseous exchange in terrestrial insects

Oxygen passes through the tracheal system from the spiracles to the tissues and ultimately must reach the mitochondria in order to play a part in oxidative processes. There are thus two distinct phases in the transport of gases, one through the tracheal system, known as air-tube transport, and one through the tissues in solution in the cytoplasm, known as tissue diffusion.

17.5.1 Diffusion

The rate of diffusion of a gas depends on a number of factors. It is inversely proportional to the square root of the molecular weight of the gas, so that in air, oxygen, with a molecular weight of 32, diffuses 1.2 times faster than carbon dioxide, with a molecular weight of 44. The rate of steady-state diffusion (J, with units such as moles s^{-1}) of a gas depends on the area available (A), the distance over which diffusion occurs (L), the diffusion constant for the molecule of interest in the particular media (D) and the concentration gradient for that molecule (ΔC):

$$J = A \times D \times \Delta C / L$$

The diffusion constant (D, $m^2\,s^{-1}$) varies with temperature (faster at higher temperature), molecular size (faster for smaller molecules) and differs dramatically between air and water (D is about 10 000 times larger in air than in water). The average distance traveled by a diffusing molecule increases with the square root of time; thus in tissues, diffusion works well over small distances (micrometers) but is generally inadequate at distances greater than a millimeter. Most tracheoles are within 50 μm of mitochondria (less than 20 μm in flight muscle), so diffusion works well. Many respiratory physiologists express gradients for oxygen and carbon dioxide in partial pressures, in this case the Krogh's constant, K (with units such as moles $s^{-1}\,kPa^{-1}\,m^{-1}$) is used instead of D. K is calculated by multiplying D by the solubility of the gas in the media; because solubility of gases decreases with temperature, K for tissues or water falls as temperature rises, unlike D. If the molecule of interest changes chemically in the media (e.g., oxygen binding to hemocyanin, or carbon dioxide forming bicarbonate), then K should be calculated by multiplying D by the capacitance of the media, which includes both physically dissolved and chemically bound molecules.

In some cases, $A \times D/L$ are combined into a single term called the diffusive conductance (G_{dif}), allowing an overall estimate of the diffusing capacity of the system:

$$J = G_{dif} \times \Delta C (or \Delta P)$$

The importance of diffusion in insect gas exchange remains a controversial topic. In the 1920s, Krogh calculated the partial pressure gradient for oxygen for a caterpillar by measuring oxygen consumption rate and the diameters and lengths of many of the tracheae. He concluded that diffusion should suffice for oxygen delivery, and for many years it was often stated that diffusion was the major mechanism of gas exchange in insects. However, we now know that many insects, and perhaps all large insects, utilize convective ventilation of the tracheal system. Nonetheless, the fact that most insects can recover from complete anoxia suggests that diffusion can provide the minimal oxygen necessary to activate motor behavior in insects.

It is often stated that increasing body size necessarily leads to problems with diffusion due to the increasing diffusion distance down longer tracheae. However, as long as the cross-sectional area increases in proportion to tracheal length, steady-state diffusion rates can be maintained. In the abdominal transverse tracheae to the gut of locusts, the diffusive conductance does not increase as much as metabolic rate with age/size, and thus transport of oxygen by diffusion requires an increasing gradient for oxygen. In locusts, convective ventilation becomes increasingly more important than diffusion with age/size; perhaps this is a general trend for insects as size increases.

Due to its greater solubility, the permeability constant of carbon dioxide in the tissues is 36 times greater than that for oxygen, so that despite its higher molecular weight, carbon dioxide travels more quickly than oxygen through the tissues for the same difference in partial pressure. Hence a system capable of bringing an adequate supply of oxygen to the tissues will also suffice to take the carbon dioxide away. As carbon dioxide is more soluble, it is present in much higher concentrations in the blood and tissues than oxygen, generally in the form of bicarbonate. Carbonic anhydrase is required to quickly convert bicarbonate to carbon dioxide; this enzyme is reported to occur in the tissues but not the blood.

17.5.2 Convection

We now know that many (likely most) insects, large and small, ventilate their tracheal system by convective movements of gases. Convection is the bulk flow of fluid, driven by a pressure gradient. Given that diffusion can suffice for gas exchange of many insects, why is convection so common? Theoretical studies have demonstrated that use of convection rather than diffusion can decrease water loss rates, even in small insects. In addition, convection is likely important for allowing cells in different parts of the body to experience similar oxygen and carbon dioxide levels. Ventilatory convection generally occurs by discrete pumping movements. For a completely convective system:

$$V_{O2} = f \times TV \times (\text{inspired } C_{O2} - \text{expired } C_{O2})$$

where V_{O2} is the rate of oxygen consumption (in units such as moles s^{-1}), f is the ventilatory frequency, TV is tidal volume (volume of air per pumping event) and C_{O2} is the concentration of oxygen in the inspired and expired air. For some insects (e.g., locusts exposed to hypoxia), it has been demonstrated that gas exchange through the spiracles is virtually completely convective. However, it is likely that for most insects gas exchange through the spiracles is a mixture of convection and diffusion.

Ventilatory convection can be accomplished by compression of the air sacs or tracheae. Some tracheae are circular in cross-section and resist any change in form because of their taenidia; these appear to be suitable to serve as conducting tubes for gas exchange. However, thanks to X-ray synchrotron imaging that allows visualization of the tracheal system of intact, living insects, we now know that many tracheae in a variety of insects compress substantially, apparently producing tidal ventilation. This phenomenon has been best studied in carabid beetles, in which longitudinal and secondary and tertiary tracheae of the head and thorax compress significantly (while abdominal tracheae do not). Combining measurement of such tracheal compression with high-resolution measurement of CO_2 emission has revealed that these tracheal compressions produce significant convection through the spiracles, but do not account for the majority of CO_2 emission (which may be diffusive). Because air sacs can be quite large, their compression and inflation can produce extensive convection.

Insects produce compression of the air sacs and tracheae in multiple ways. The best understood is abdominal pumping. Abdominal compression leads to a reduction of body volume and a rise in hemolymph pressure that causes expiration. Expansion of the air sacs and inspiration result from the reduction of pressure due to the muscular or elastic expansion of the abdomen. Changes in abdominal volume may be produced in various ways. In Heteroptera and Coleoptera, the tergum moves up and down (Fig. 17.15a); in Odonata, Orthoptera, Hymenoptera and Diptera, both tergum and sternum move (Fig. 17.15b), and this movement may be associated with telescoping movements of the abdominal segments (Fig. 17.15c); and in adult Lepidoptera, the movement is complex and involves movements of the pleural regions as well as terga and sterna.

Another mechanism for producing ventilation is by transfer of hemolymph back and forth between thorax and abdomen (Fig. 17.16). For example, in Lepidoptera and flies, septal structures such as connective tissue and fat body or air sacs at the front of the abdomen occlude the hemocoel. When the heart pumps forwards, it draws hemolymph from the abdomen and transfers it to the head and thorax. The effect of this is to cause expansion of air sacs in the abdomen and to compress those in the anterior compartment. When the heart reverses, it pumps blood into the abdomen and the changes in the tracheal system are reversed. During forward beating of the heart, air is sucked in through the abdominal spiracles and forced out through thoracic spiracles; when the heart reverses, the converse is true. The volume changes of the cephalic air sacs could be visualized in *Drosophila* and *Calliphora* using synchrotron X-ray videography (Fig. 17.16d,e). The hemolymph shift produces an alternating change in tracheal pressure and tracheal P_{O2}. At the transition from backward to forward beating, the abdomen performs a pumping stroke, which exerts a pressure pulse restricted on the abdominal tracheal system at the moment when most hemolymph is accumulated in the abdomen and the hydraulic effect of the

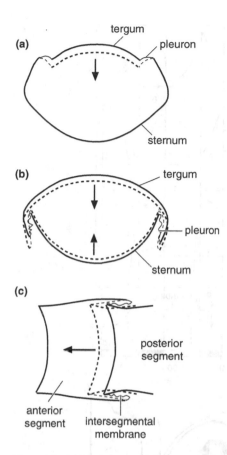

(a) tergum / pleuron / sternum

(b) tergum / pleuron / sternum

(c) posterior segment / anterior segment / intersegmental membrane

Figure 17.15 Ventilatory movements of the abdomen. Dashed lines represent the contracted position, compressing the abdominal air sacs and causing expiration. Arrows indicate directions of movement in the compression phase (from Snodgrass, 1935). (a) Dorso-ventral compression involving tergal depression, transverse section. (b) Dorso-ventral compression involving both tergum and sternum, transverse section. (c) Longitudinal compression by telescoping one segment within another anterior to it, horizontal section.

hemolymph is maximized, compressing air sacs and driving air through spiracles. In a resting blowfly at about 23°C, heartbeat reverses about twice each minute, in *Drosophila* about five times each minute.

Airflow may be tidal, moving in and out of the same spiracles, or unidirectional through the insect. Tidal airflow occurs in carabid beetles, *Periplaneta*

(Blattodea), in the wing tracheae of Lepidoptera and the elytral air sac of scarabaeid Coleoptera. In many insects, however, including other species of cockroach, opening and closing of certain spiracles is synchronized with the ventilatory pumping movements of the abdomen, so that air is sucked in through some spiracles and pumped out through others and a directed flow of air through the tracheal system is produced. This is a more efficient form of ventilation than tidal flow as much of the "dead" air, trapped in the inner parts of the system by tidal movements, is removed. In most insects the flow of air is from front to back and, in *Schistocerca* (Orthoptera), spiracle movements are always coupled with ventilatory movements. In many insects, however, the two activities may become uncoupled so that they are not always synchronized and the coupling may even be modified so as to produce a reversal of the airstream. In *Schistocerca*, spiracles 1, 2 and 4 are open during inspiration, but closed during expiration, when spiracle 10 opens (Fig. 17.17a). When the insect is more active, expiration takes place through spiracles 5–10. The spiracles for inspiration open immediately after the expiratory spiracles have closed and remain open for about 20% of the cycle while air is drawn in. Then, for a short time, all the spiracles are closed. The abdomen starts to contract while the spiracles are still closed, so the air in the tracheae is under pressure; this is known as the compression phase. Then the expiratory spiracles open and air is forced out. The expiratory spiracles are only open for some 5–10% of the cycle. During activity the frequency of ventilation and spiracular movements is increased. The periods for which the spiracles are open remain the same, but the intervals between openings are reduced and the compression phase eliminated (Fig. 17.17b). The net time for which the spiracles are open increases as a result.

A variety of mechanisms are known for producing ventilation of the head. *Schistocerca* sometimes exhibit protraction and retraction of the head on the prothorax (neck ventilation), and movement of the

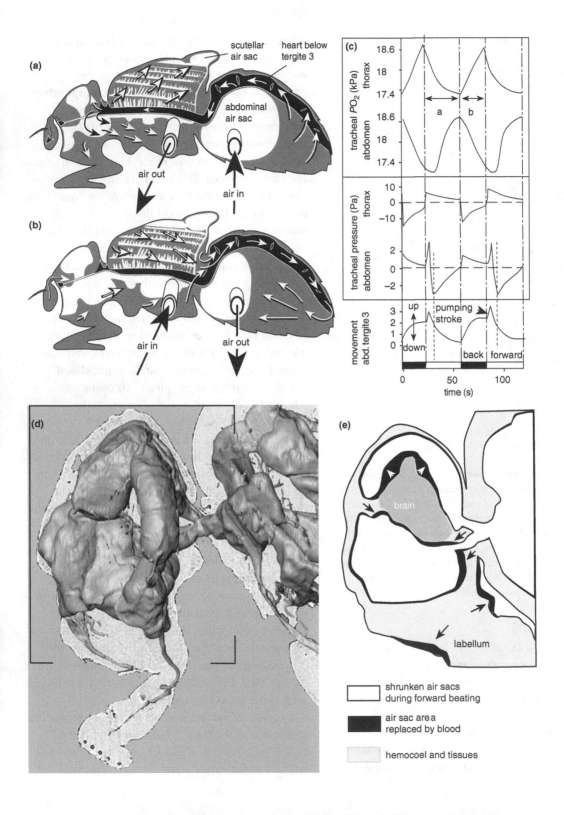

(a)

scutellar air sac
heart below tergite 3
abdominal air sac
air out
air in

(b)

air in
air out

(c)

tracheal PO_2 (kPa)

thorax
18.6
18
17.4

a | b

abdomen
18.6
18
17.4

tracheal pressure (Pa)

thorax
10
0
-10

abdomen
2
0
-2

movement abd. tergite 3
3
2
1
0

up
down

pumping stroke

back | forward

0   50   100
time (s)

(d)

(e)

brain

labellum

□ shrunken air sacs during forward beating

■ air sac area replaced by blood

▨ hemocoel and tissues

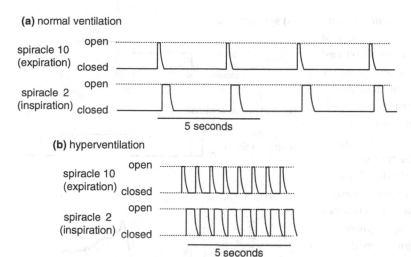

Figure 17.17 Spiracle activity during abdominal ventilation of a locust (after Miller, 1960a). (a) Normal ventilation; (b) hyperventilation.

prothorax on the mesothorax (prothoracic ventilation). *Drosophila* ventilate their head and thorax during flight by pumping of the proboscis under certain flight conditions. Other mechanisms for producing ventilation during flight are covered below.

17.5.3 Discontinuous gas exchange

Although ventilation is often continuous, there may be extended periods during which all the spiracles are closed. The movement of oxygen into the tracheae and carbon dioxide emission then occur in discrete bursts when the spiracles open; relatively little gas exchange occurs while they are closed. This

phenomenon is known as discontinuous gas exchange (DGE) or discontinuous ventilation. It is common in adult insects when they are inactive at moderate to cool temperatures, that is, when their metabolic rates are low. The pupae of many insects also exhibit DGE.

The periodicity of ventilatory bouts varies from insect to insect and with environmental conditions. In diapausing pupae of the moth *Hyalophora*, the interval between spiracle openings is as long as eight hours. In some resting grasshoppers at 15°C, carbon dioxide is released about once per hour, but at 30°C the period of spiracle closure is reduced to about five minutes. In the beetle *Psammodes*, each cycle lasts

Caption for Figure 17.16 Alternating gas exchange in anterior body and abdomen caused by reversals of the heartbeat in the blowfly (*Calliphora*). The tracheal system is represented diagrammatically (after Wasserthal, 1996). (a) Forward beating of the heart draws hemolymph from the abdomen and pushes it into the head and thorax. This expands abdominal air sacs and reduces the volume of those in the head and thorax so that air is inspired through abdominal spiracles and expired through thoracic spiracles. (b) Backwards beating of the heart draws hemolymph from the head and thorax and releases it into the abdomen. This compresses the abdominal air sacs and expands those in the head and thorax so that air is drawn in through thoracic spiracles and forced out through abdominal spiracles. (c) Effect of heartbeat reversals and coordinated abdominal movements on tracheal pressure and oxygen partial pressure alternating in the scutellar and abdominal air sac (after Wasserthal, 2012 and unpublished). (d) X-ray tomograph of the distended cephalic air sacs in *Drosophila melanogaster*. (e) Periodic volume changes of the cephalic air sacs by hemolymph shift in *Drosophila*. Diagram based on X-ray videos, lateral view of the head as in (d) (modified after Wasserthal et al., 2006). White: compressed air sac volume during accumulation of blood (black) during forward pulses of the heart. Arrows point at moving contours of shrinking air sacs and outwards-extending cuticle of the basal labellum by hemolymph increase during forward heartbeat (black).

about 15 minutes (Fig. 17.18), while in some ants and the honey bee it is less than five minutes. DGE is best understood in lepidopteran pupae. During the closed phase, the spiracles are closed and little gas exchange occurs (Fig. 17.19). During this time, much of the carbon dioxide produced dissolves and is converted to bicarbonate in the hemolymph and tissues. Thus, the utilization of oxygen without replacement by gaseous carbon dioxide causes a reduction in tracheal system pressure. Next is the flutter phase, in which the spiracles rapidly pulse open and closed, allowing bulk inflow of air and oxygen, with reduced carbon dioxide emission. Next is the open phase, when the spiracles open wide; oxygen enters the spiracles and carbon dioxide and water vapor are emitted, often by diffusion. In some insects, such as *Psammodes* (Coleoptera), ventilatory movements of the abdomen are made during this period so that gases are actively pumped in and out of the system until the spiracles close again (Fig. 17.18b).

In pupal Lepidoptera, the pattern of spiracular behavior can be explained by effects of oxygen and carbon dioxide on the spiracles. At the end of the open phase, tracheal oxygen and pressure levels are near atmospheric. During the closed phase, the partial pressure of oxygen and the barometric pressure of the tracheal system drop linearly with time; the elastic abdomen compresses in response to the pressure gradient (Fig. 17.19). When the partial pressure of oxygen reaches about 4 kPa, the spiracles begin to flutter, and tracheal pressures rise to near atmospheric. Carbon dioxide accumulates up to a threshold of about 5 kPa, leading to spiracular opening.

Many variations exist for DGE and its control. In some ants there is significant outward emission of CO_2 during the flutter phase. In grasshoppers oxygen has no effect on the duration of the phases of DGE, while hypoxia can either increase or decrease the duration of the flutter phase depending on species.

The functional significance of DGE in insects has been the subject of major controversy and investigation. There are at least three adaptive

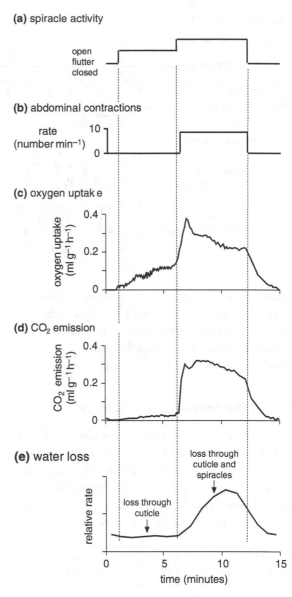

Figure 17.18 Discontinuous gas exchange in a beetle (*Psammodes*). Changes occurring in a single bout of ventilation (after Lighton, 1988). (a) Activity of the spiracles. During the flutter phase, they open slightly for brief periods and then close again. (b) Abdominal ventilation coincides with the open phase of the spiracles. (c) Oxygen uptake; some occurs during flutter, but the rate is greatly increased when the spiracles open fully. (d) Carbon dioxide emission is largely confined to the open phase of the spiracles; very little occurs during spiracular flutter. (e) Water loss increases when the spiracles open.

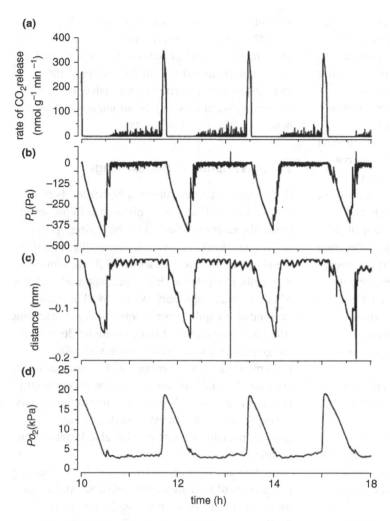

Figure 17.19 Discontinuous gas exchange cycle in *Atacus atlas* pupae. (a) Rate of CO_2 release vs. time; note the closed phase with zero CO_2 release flutter phase with low intermittent CO_2 release and then the open phase with a large burst of CO_2 emission. (b) Intra-tracheal pressure falls linearly with time during the closed phase, and then is near atmospheric during the flutter and open phases. (c) Abdomen length shortens during the closed phase, and remains at resting length during the flutter and open phases. (d) Intra-tracheal P_{O2} falls linearly with time during the closed phase, remains near 3–4 kPa during the flutter phase and rises to near atmospheric during the open phase (after Hetz and Bradley, 2005).

hypotheses for the occurrence of DGE, including the reduction of respiratory water loss (hygric hypothesis), facilitation of underground gas exchange (chthonic hypothesis) and reduction of oxidative damage. In addition, it has been proposed that the occurrence of DGE need not be adaptive, but could be simply a consequence of having neurosensory systems for the control of ventilation and spiracular opening by carbon dioxide and oxygen (or for matching oxygen supply to demand). The idea that DGE is adaptive because it reduces respiratory water loss has been prominent since the

papers of Buck and Schneidermann and colleagues in the 1950s. There are two basic mechanisms by which DGE can reduce respiratory water loss. As a consequence of DGE, the spiracles are closed for most of the time, greatly reducing respiratory water loss. Prolonged spiracular closure leads to greater gradients for oxygen diffusion, allowing more oxygen uptake relative to water loss during the open phase (for either diffusion or convection). In addition, during the flutter phase, convective inward airflow can allow oxygen uptake with reduced outward diffusion of water. This adaptive argument for DGE has been questioned on a number of grounds. First, the amount of water lost through the spiracles is usually very small relative to that lost via the cuticle. Second, in some comparative studies, the occurrence of DGE does not correlate well with expected exposure to desiccating conditions. Nonetheless, a variety of recent studies have supported the hypothesis that DGE reduces respiratory water loss, at least in some insects. In the caterpillar larvae *Erynmis propertius*, some individuals switch between continuous and discontinuous gas exchange, and use of DGE reduces water loss, without any change in metabolic rate. While some insects abandon DGE when acutely dehydrated (grasshoppers), acclimation to dry conditions reduces the duration of the open phase in cockroaches, and reduces water loss rate. Meta-analyses of 40 species in a phylogenetic context demonstrated that insect species from warmer habitats had longer DGE durations, especially when precipitation was low, consistent with the hygric hypothesis.

The hypothesis that DGE occurs to improve gas exchange underground is weakened by the fact that both hypoxia and hypercapnia tend to cause spiracular opening. The hypothesis that DGE functions to reduce oxidative damage is supported by the observation that some lepidopteran pupae regulate tracheal oxygen levels during the flutter phase at consistently low levels across a wide range of oxygen levels. Certainly, continuously open spiracles would expose insect tissues to levels of oxygen much higher than normally experienced by tissues of vertebrates, and higher tissue oxygen level is often correlated with increased mitochondrial reactive oxygen species production and oxidative damage.

There is also strong support for models that suggest that DGE arises non-adaptively from the properties of the control systems. For example, DGE is most common when the metabolic rates of insects are low (low rates of oxygen consumption and carbon dioxide production allow the spiracles to remain closed for longer periods of time), and DGE is even observed in semi-aquatic insects at high humidities. In summary, it seems likely that DGE is likely to arise from interactions between internal gas levels and the neural system for control of the spiracles in most insects, but that reduction in respiratory water loss may be an important adaptive benefit of DGC in some insects.

17.5.4 Variation in gas exchange

Higher metabolic rates demand higher levels of oxygen intake. This is most obvious in flight, when metabolic rates can rise 5–30-fold. Flying insects achieve maximal mass-specific metabolic rates that are about double those observed in flying birds and mammals, likely due at least in part to the advantages of the tracheal respiratory system. Walking and running also require more oxygen than just standing still. The faster an insect runs, the higher its rate of oxygen consumption, and running vertically upwards requires more energy, and so more oxygen, than running on a horizontal surface. It is generally believed that very little anaerobic respiration occurs during locomotion of most insects, although grasshoppers do generate considerable lactate when jumping.

Feeding and digestion increase metabolic rates due to the costs of anabolism, gut peristalsis and active transport. The increase in metabolic rate with feeding, often termed the specific dynamic action, averages 3.3-fold higher than resting values in insects. Metabolic rate increases more than ten-fold in *Rhodnius prolixus* after a huge blood meal. Dietary characteristics can also affect metabolic rates; locusts consuming food high in carbohydrate have higher metabolic rates, probably to eliminate excess caloric intake (known as adaptive thermogenesis, or facultative, diet-induced thermogenesis). During extreme food restriction (e.g., during seasonal dearths), insects exhibit torpor or diapause accompanied by decreases in metabolic rates of up to 98%. In diapausing pupae of the flesh fly, *Sarcophaga*, peaks of oxygen consumption occur at about four-day intervals when the temperature is constant at 25°C. Over a period of about 36 hours, oxygen consumption rises from a resting level of about $1.5 \, \mu l \, h^{-1}$ to 9–$10 \, \mu l \, h^{-1}$ and then falls to the original level. The peaks in oxygen consumption coincide with peaks of protein synthesis and release. It is reported that there are no parallel changes in carbon dioxide emission, suggesting CO_2 is accumulating as bicarbonate or being incorporated into larger compounds. Similar variation in oxygen consumption occurs in the diapausing pupa of the cabbage butterfly, *Pieris*, and it is probably a common phenomenon.

Another important source of variation in metabolic rates among insects is body size. Across species, larger insects have higher metabolic rates (Watts) but lower mass-specific metabolic rates (Watts g^{-1}). On log-log plots, metabolic rates (Watts) scale approximately with body mass$^{0.75}$. This relationship also seems to hold for social insect colonies within species. During ontogeny, the pattern of metabolic rate with body mass is more variable. While it is usually true that older/larger insects have higher metabolic rates, the scaling exponents can be quite variable, probably because development of insects is associated with a wide variety of morphological and physiological variation.

Temperature also strongly affects metabolic rates, with most inactive insects showing exponential increases in metabolic rates with rises in temperature across moderate ranges of temperature (often doubling or tripling with a 10°C rise in temperature). However, some insects such as honey bees in flight, honey bee swarms and bumble bee colonies sometimes exhibit increases in metabolic rate at lower temperatures, apparently for thermoregulatory purposes.

Gas exchange in flight The massive increase in oxygen consumption that occurs when an insect flies requires a greatly increased airflow through the tracheae to the flight muscles. The size of the spiracles is closely matched to oxygen need during flight. Experimentally sealing any thoracic spiracle causes a reduction in flight metabolic rate in *Drosophila*, and flight performance is tightly linked to the available spiracular area. During the flight of *Schistocerca*, abdominal ventilation increases in frequency and amplitude, but still only supplies about $50 \, ml \, air \, g^{-1} \, h^{-1}$, not sufficient to supply the needs of the flight muscles. However, the distortion of the thorax, and in particular the raising and lowering of the notal sclerites, produces large volume changes in the extramuscular air sacs of the pterothoracic tracheal system (Figs. 17.4b, 17.6), while changes in the diameters of shortening muscles may compress the intramuscular air sacs. This pterothoracic ventilation produces an airflow of about $350 \, ml \, air \, g^{-1} \, h^{-1}$, which is adequate for the needs of the flight muscles.

Thoracic pumping is also important in flight in Odonata and probably in Lepidoptera and Coleoptera. In Hymenoptera and Diptera, changes in the thoracic volume during flight are not very large and abdominal pumping is of greater importance in maintaining the air supply to the flight muscles. In a large cerambycid beetle, *Petrognatha*, there is evidence that a stream of air, resulting from the forward movement of the insect in flight, flows in through the forwardly directed second spiracles into large tracheae running directly to the third spiracles. This stream of air ventilates the primary supply to the flight muscles while the secondary supply is probably ventilated by muscular compression.

When *Schistocerca* starts to fly, the pattern of spiracle opening also alters (Fig. 17.20). At first, spiracles 1 and 4–10 close, but then they open and close rhythmically in synchrony with abdominal ventilation so that there is a flow of air via the head to the rest of the central nervous system. Increased abdominal ventilation may also improve the blood circulation and hence the fuel supply to the flight muscles. Spiracles 2 and 3 remain wide open throughout flight and although they show some incipient closures after a time these do not affect the airflow through the spiracles. These spiracles supply

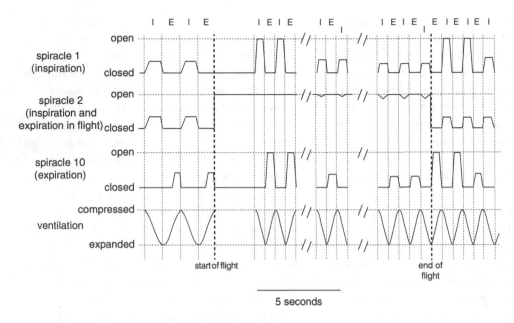

Figure 17.20 Spiracle activity and abdominal ventilation during flight of *Schistocerca*. I, inspiration; E, expiration (after Miller, 1960b).

the flight muscles, and, since the tracheal system of these muscles is largely isolated and the spiracles remain open all the time, there is a tidal flow of air in and out of them.

Hawk moths, with their powerful and long-lasting flight, have very high metabolic rates. The O_2 content of the flight muscles during steady flight exceeds even the resting level. This is facilitated by a unidirectional airstream with inspiration through the first spiracles and expiration through the second spiracles. The directed airflow is generated by the flight apparatus and abdominal up-and-down movements while inspiration is prevented through the posterior thoracic spiracles (Fig. 17.21). The different structure of the meso- and metathoracic spiracles reflects this unidirectional respiratory airstream: The inspiratory mesothoracic spiracle is protected by dense filter lamellae. The expiratory metathoracic spiracle is devoid of filter structures and has an external valve lip, which is directed outwards. During warm-up shivering of hawk moths, wingbeat amplitude and the deformation of

the pterothorax are small and the metathoracic spiracle does not become enclosed in the subalar cleft during incomplete downstroke. Ventilation pulses show the same mean pressure at the anterior spiracles and in the mesoscutellar air sacs. This tidal in- and outflow through both anterior and posterior thoracic spiracles occurs also in the steady flight of saturniid moths, with their low wingbeat frequency.

17.5.5 Control of ventilation

As in the control of spiracles, ventilatory movements are initiated by the accumulation of carbon dioxide and, to a lesser extent, the lack of oxygen in the tracheal system, acting directly on centers in the ganglia of the central nervous system. Each abdominal ganglion produces rhythmical sequences of activity controlling the movements. In *Schistocerca* and *Periplaneta* the metathoracic ganglion acts as a pacemaker and overrides the rhythms of the other ganglia.

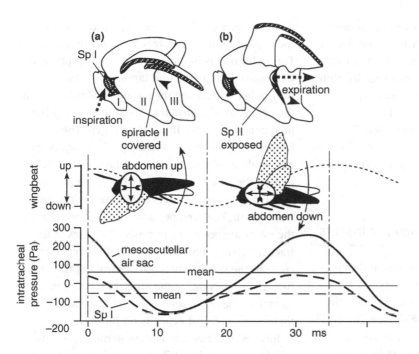

Figure 17.21 Unidirectional respiratory airstream in the flying hawk moth *Manduca* based on the pressure difference between anterior spiracles (Sp I) with a mean negative pressure and the mesoscutellar air sac with a mean positive pressure. Schematic lateral views of the thorax showing the deformation of the thorax (arrows) during (a) downstroke and (b) upstroke of the wings. This affects the adduction and retraction of the metathorax (arrowheads) enclosing or exposing the Sp II in the subalar cleft, and up-and-down moving abdomen. (a) Downstroke: volume increase of the metathoracic air sacs and decrease of intratracheal pressure. Inspiration through Sp I while Sp II are closed. (b) Upstroke: reduction of the metathoracic tracheal volume by flattening the thorax and increase of intratracheal pressure. Expiration through the open and exposed Sp II. I prothorax; II mesothorax; III metathorax (modified after Wasserthal, 2001).

In normal ventilation by *Schistocerca*, involving only dorso-ventral movements of abdominal sclerites, expiration is produced by tergosternal muscles innervated by motor neurons from the ganglion of the corresponding segment. Inspiration is produced by muscles inserted low on the tergum and innervated by axons in branches from the median ventral nerve. The perikarya of these motor neurons are in the ganglion of the preceding segment, so the alternation of inspiratory and expiratory movements is controlled from different ganglia (Fig. 17.14). Coordination is achieved by a coordinating interneuron, one in each ventral connective (cell 2 in Fig. 17.14b), which originates in the metathoracic ganglion and extends to the last abdominal ganglion. These interneurons are driven by a central pattern generator, probably in the metathoracic ganglion, whose activity is regulated by inputs from the head and other parts of the body. Action potentials in the interneuron have an excitatory effect on a local interneuron (cell 3) and probably have a weak direct inhibitory effect on the inspiratory motor neurons (cell 5 in Fig. 17.14b). The local interneuron excites the expiratory motor neurons (cell 4) and strongly inhibits the inspiratory motor neurons (cell 5). Inspiration occurs when the intersegmental interneurons are silent and the inspiratory motor neurons become spontaneously active, at the same time inhibiting the expiratory motor neurons via another local interneuron (cell 6).

The rate of ventilation is altered by reducing the interval between inspiratory bursts and is affected by sensory input from various sources. Centers sensitive to carbon dioxide are present in the head and thorax of *Schistocerca* and these modify the activity of the pacemaker, while the output is also modified by high temperature and nervous excitation generally. Proprioceptors may play some part in the maintenance of the frequency of ventilation. The coordination of the spiracles with the ventilatory movements is brought about by motor patterns derived from the ventilatory centers.

17.6 Gaseous exchange in aquatic insects

Aquatic insects obtain oxygen directly from the air or from that dissolved in the water. The former necessitates some semipermanent connection with the surface or frequent visits to the surface. Insects that obtain oxygen from water nearly always retain a tracheal system so that the oxygen must come out of solution into the gaseous phase. This is important because the rate of diffusion in the gas phase is very much greater than in solution in the hemolymph or tissues. Gaseous exchange with water takes place through thin-walled gills well supplied with tracheae, but in other cases a thin, permanent film of air is present on the outside of the body. The spiracles open into this film so that oxygen can readily pass from the water into the tracheae.

17.6.1 Aquatic insects obtaining oxygen from the air

Most aquatic forms obtaining air from above the water surface must make periodic visits to the surface, but a few have semipermanent connections with the air that enable them to remain submerged indefinitely. The larva of the hover fly *Eristalis* (Diptera) has a telescopic terminal siphon which can extend to a length of 6 cm or more in a larva only 1 cm long. By means of the siphon the larva can reach the water surface with its posterior spiracles, while the body remains on the bottom mud (Fig. 17.22).

Some other species obtain oxygen by thrusting their spiracles into the aerenchyma of aquatic plants. This habit occurs in larval *Donacia* (Coleoptera) and *Chrysogaster* (Diptera), and larvae and pupae of *Notiphila* (Diptera) and the mosquito *Mansonia*. With the exception of *Mansonia*, all of these species live in mud containing very little free oxygen. The functional spiracles are at the tip of a sharp-pointed post-abdominal siphon in larval forms (Fig. 17.23), and on the anterior thoracic horns of the pupae.

For most aquatic insects obtaining their oxygen from the air, however, periodic visits to the surface of the water are necessary to renew the gases in the tracheal system. Problems facing all insects that come to the surface are those of breaking the surface film and of preventing the entry of water into the spiracles when they submerge. The ease with which this is accomplished depends on the surface properties of the cuticle and, in particular, on its resistance to wetting. When a liquid rests on a solid or a solid dips into a liquid, the liquid–air interface meets the solid–air interface at a definite angle that is constant for the substances concerned. This angle, measured in the liquid, is known as the contact angle (Fig. 17.24). A high contact angle indicates that the surface of the solid is only wetted with difficulty; such surfaces are said to be hydrofuge. Under these

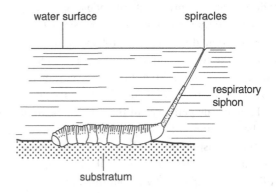

Figure 17.22 Semipermanent connection with the water surface. The larva (rat-tailed maggot) of the hover fly *Eristalis*, with its respiratory siphon partly extended (after Imms, 1947).

(a)

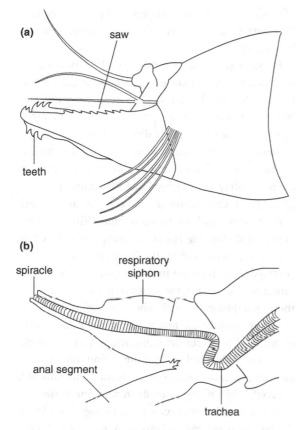

(b)

Figure 17.23 Semipermanent connection with the aerenchyma of a plant. The respiratory siphon of a mosquito larva (*Mansonia*) which connects with the aerenchyma of aquatic plants (after Keilin, 1944). (a) Lateral view showing saw which cuts into the plant tissue and recurved teeth which hold the siphon in place. (b) Longitudinal section showing a terminal spiracle and a trachea.

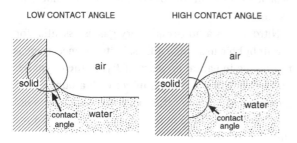

Figure 17.24 Diagrams showing low and high contact angles.

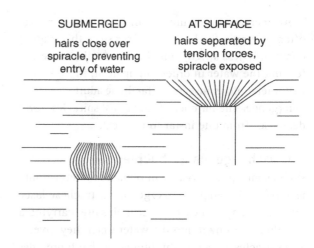

Figure 17.25 Diagrams showing the movements of hydrofuge hairs surrounding a spiracle when a Stratiomya larva (Stratiomyidae) is submerged and at the surface. The movement of the hairs is entirely passive, depending on physical forces acting between the hairs and the water (modified after Wigglesworth, 1965).

conditions, the cohesion of the liquid is greater than its adhesion to the solid, and so, when an insect whose surface properties are such that the contact angle is high comes to the air–water interface, the water falls away, leaving the body dry. The whole surface of the cuticle may possess hydrofuge properties, so it is not readily wetted at all, or these properties may be restricted to the region around the spiracles. In dipterous larvae, for example, perispiracular glands produce an oily secretion in the immediate neighborhood of the spiracle. Often, hydrofuge properties around the spiracle are associated with hairs, as in *Notonecta* (Hemiptera), or valves, as in mosquito larvae, which close when the insect dives, but open at the surface (Fig. 17.25), being spread out by surface tension.

In many of these insects, only the posterior spiracles are functional and they are often carried on the end of a siphon extending from the posterior body, as in larval Ephydridae and Culicidae (Diptera). In these insects, only the tip of the siphon breaks the surface film. In water scorpions (Hemiptera, Nepidae), the spiracles are at the base of an air-filled tube.

An increase in the number of functional spiracles often occurs in the last-stage larva since this stage is commonly less strictly aquatic than earlier stages, leaving the water in order to pupate or, in hemimetabolous insects, to facilitate adult emergence. The number of functional spiracles never decreases from one instar to the next.

Gas exchange via air bubbles Some insects, such as mosquito larvae, can remain submerged only as long as the supply of oxygen in the tracheae lasts, but others have an extratracheal air store, carrying a bubble of air down into the water when they dive. The spiracles open into this bubble, so that it provides a store of air additional to that contained in the tracheal system, enabling the insects to remain submerged for longer periods than would be possible without it.

The position of the store is characteristic for each species. In *Dytiscus* (Coleoptera), it is beneath the elytra, and experimental removal of the hindwings increases the space beneath the elytra and enables the insect to remain submerged for longer periods. In *Notonecta* (Hemiptera), air is held by long hydrofuge hairs on the ventral surface as well as in a store under the wings and in a thin film held by small bristles over the dorsal surface of the forewing. The related *Anisops* (Heteroptera) has ventral and subelytral stores supplemented by oxygen loosely associated with hemoglobin in large hemoglobin cells just inside the abdominal spiracles (Section 17.4).

An air store also gives the insect buoyancy, so that as soon as the insect stops swimming or releases its hold on the vegetation, it floats to the surface. The position of the store is such that the insect breaks the surface suitably oriented to renew the air. *Dytiscus*, for instance, comes to the surface tail first and renews the subelytral air from the posterior end of the elytra. In *Anisops*, the aquatic bug first must swim down against the upward force of the large air bubble. However, as oxygen is consumed, the bubble decreases in size. For a considerable period of time, the bubble is maintained at the size required for neutral buoyancy by release of oxygen from hemoglobin.

Physical gills. An air bubble provides an insect with more oxygen than that available at the time of the dive because it also acts as a temporary gill. When an insect dives, the gases in its air store are in equilibrium with the gases dissolved in the water, assuming that this is saturated with air. Normally, at the beginning of the dive, the bubble would contain approximately 21% oxygen and 79% nitrogen (Fig. 17.26). Carbon dioxide is very soluble, so there is never very much in the air bubble. Within a short period of diving, the P_{O2} in the bubble is reduced, as the oxygen is utilized by the insect, and there is a corresponding increase in the proportion and P_{N2} of nitrogen. As a result, the gases in the bubble and those in solution immediately surrounding it are no longer in equilibrium and movements of gases will occur, tending to restore equilibrium. Oxygen tends to pass into the bubble from the surrounding medium, because the oxygen tension in the bubble is reduced, while nitrogen tends to pass out of the bubble into solution, because the nitrogen tension in the bubble is increased. Thus more oxygen will be made available to the insect than was originally present in the bubble, which is, in fact, acting as a gill. This effect is enhanced by the fact that the oxygen passes into the bubble about three times more readily than nitrogen passes out into solution. As a result of this, the insect is able to remain submerged for longer periods than would be possible if it depended solely on the oxygen initially available in its store.

Nitrogen, as a non-respiratory gas, is essential for the air bubble to act as a gill as, in its absence, there is no change in partial pressure of P_{O2} as the oxygen is utilized. For this reason an insect with a bubble of pure oxygen in water saturated with oxygen does not survive for very long unless it is able to come to the surface.

With inactive insects at low temperatures, when the rate of utilization of oxygen is low, the

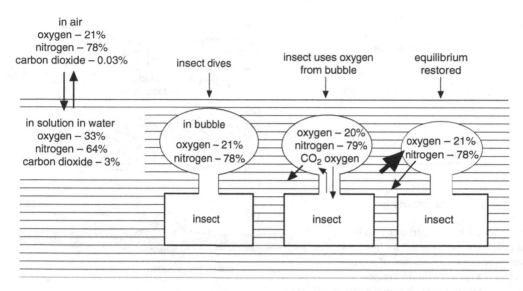

Figure 17.26 Diagram of an air bubble acting as a physical gill. As the insect dives, the gases in the bubble are in equilibrium with those dissolved in the water. As the insect uses oxygen, the equilibrium is perturbed. It is restored by the inward movement of oxygen and the outward movement of nitrogen. This is a continuous process; it is shown as two separate steps for clarity. Note that carbon dioxide produced by the insect is immediately dissolved in the water. Oxygen comes out of solution faster than nitrogen goes in. The bubble shrinks continuously as nitrogen goes into solution.

efficiency of the air store as a gill may be adequate for the insect to remain submerged for a long time. Thus *Hydrous* (Coleoptera) can remain submerged for some months during the winter, but in more active insects, with higher oxygen requirements, the bubble lasts for only a short time and the insect must come to the surface more frequently. If the temperature is above about 15°C, *Notonecta* (Hemiptera) uses oxygen so rapidly that the gill effect is of negligible importance and the insect soon surfaces to replenish its air store. At 10°C, however, it remains submerged for twice as long as would be possible with the initial oxygen supply, and below 5°C it can survive for a very long period without access to the surface.

The efficiency of the bubble as a gill depends on the oxygen content of the water adjacent to the gill. In water devoid of oxygen, or containing only a very little, the gas tends to pass out of the bubble into solution and will be lost to the insect. Even if the oxygen tension in the water exceeds that in the bubble, the amount entering the bubble will depend on the difference in tension. Hence, the higher the oxygen tension of the outside water, the more effective will be the bubble as a gill, and this must be true of any type of gill. Consequently, the frequency with which an insect visits the surface will vary inversely with the oxygen tension of the water (Fig. 17.27).

17.6.2 Insects obtaining oxygen from the water

In all insects living in water, some inward diffusion of oxygen from the water takes place through the cuticle, and in many larval forms gaseous exchange takes place solely in this way. Cutaneous diffusion depends on the permeability of the cuticle and a lower oxygen tension in the tissues as compared with the water. In many larval forms the cuticle is relatively permeable and in *Aphelocheirus* (Hemiptera: Heteroptera), for example, the cuticle of

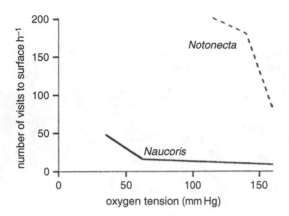

Figure 17.27 Because an air bubble carried by an insect acts as a physical gill, the time for which an insect can remain submerged increases as the amount of oxygen dissolved in the water (oxygen tension) increases. Consequently, the insect makes fewer visits to the surface at higher oxygen tensions. Data for *Naucoris* at 20°C; *Notonecta* at 17°C (data from de Ruiter *et al.*, 1952).

the last larval stage is about four times as permeable as that of the adult.

In very small larvae, such as the first-stage larvae of *Simulium* (Diptera) and *Chironomus* (Diptera), in which the tracheal system is filled with fluid, cutaneous diffusion into the hemolymph meets the whole oxygen requirement of the insect. In general, however, blood circulation is poor and the rate of diffusion through the blood slow, and diffusion into the blood would not suffice for most larger insects. Hence the majority of insects that obtain oxygen from water do have a tracheal system, although the spiracles are non-functional. This is called a closed tracheal system. As oxygen is used by the tissues of the insect, its partial pressure within the tracheae falls, creating a pressure gradient between the tracheoles just beneath the cuticle and the adjacent water. Under these conditions, oxygen from the water diffuses into the tracheal system, within which it can rapidly diffuse around the body to the tissues. An incompressible tracheal system is essential for this type of gas movement to occur as, otherwise, the tracheae would collapse under the pressure of water as oxygen was used.

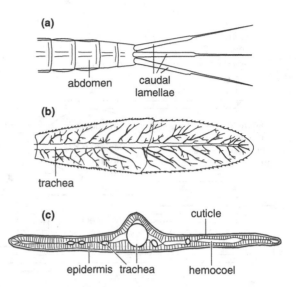

Figure 17.28 Tracheal gills (caudal lamellae) of a damselfly larva (Zygoptera). (a) Dorsal view of the posterior end of the abdomen. (b) One lamella showing the tracheae; only the major branches are shown (after Gardner, 1960). (c) Transverse section of a lamella (after Tillyard, 1917).

Tracheal gills In some insects, such as *Simulium* larvae (Diptera), there is a network of tracheoles just beneath the general body cuticle, but often there are leaf-like extensions of the body, forming gills. These are covered by very thin cuticle with a network of tracheoles immediately beneath and are known as tracheal gills. In most larval Zygoptera there are three caudal gills (Fig. 17.28); larval Trichoptera have filamentous abdominal gills, while, in larval Plecoptera, the position of the gills varies from species to species.

Larval Anisoptera (Odonata) have gills in the anterior part of the rectum, known as the branchial chamber (Fig. 17.29). Water is drawn in and out of the chamber by the activity of muscles largely unconnected with the gut. The hemocoel in the posterior part of the abdomen is isolated from the rest of the body by a muscular diaphragm across the fifth abdominal segment. Because of this, contraction of dorso-ventral muscles in the posterior abdominal segments exerts pressure on the branchial chamber.

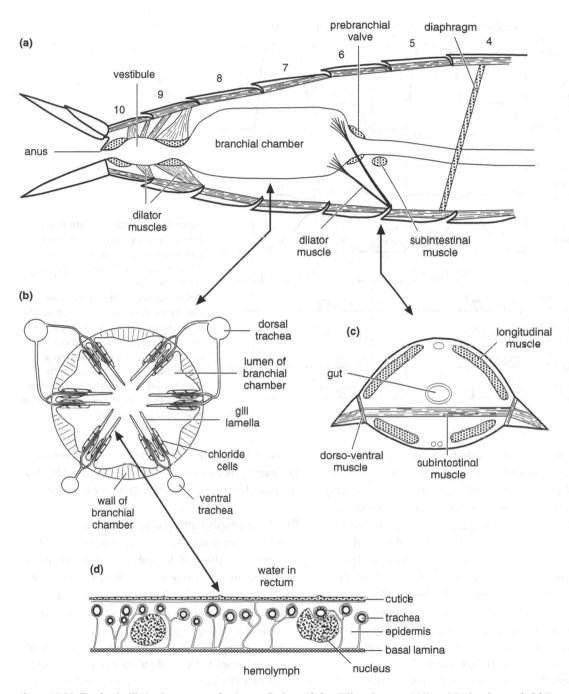

Figure 17.29 Tracheal gills in the rectum of a dragonfly larva (after Tillyard, 1917; Mill and Pickard, 1972). (a) Longitudinal section through the abdomen; numbers indicate abdominal segments. (b) Transverse section through the branchial chamber. (c) Transverse section through abdominal segment 6 showing the subintestinal muscle. (d) Detail of gill epithelium showing the tracheae close beneath the cuticle (after Schmitz and Komnick, 1976).

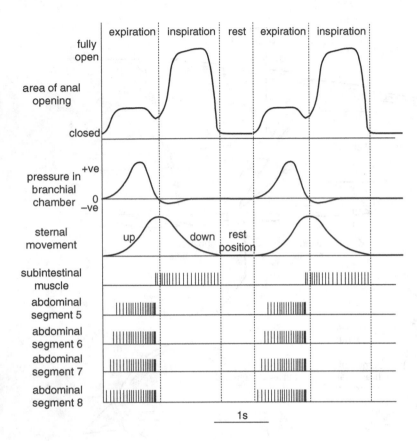

Figure 17.30 Rectal ventilation in a dragonfly larva. Lower traces show action potential frequency in the subintestinal muscle and in the dorso-ventral muscles of successive abdominal segments. As the dorso-ventral muscles contract they raise the sterna and compress the branchial chamber and water is forced out. When the subintestinal muscle contracts, the sterna are lowered, pressure in the chamber becomes negative and water flows in through the widely open anus (after Mill and Pickard, 1972).

Water is forced out through the partially open anus and drawn in again when the volume of the abdomen is restored, partly by the elasticity of the cuticle and partly by contraction of muscles in the diaphragm and of a transverse subintestinal muscle. During inspiration the anal valves are wide open (Fig. 17.30). A muscular prebranchial valve presumably prevents water from moving forwards into the midgut. Some 85% of the water in the rectum is renewed at each cycle of compression and relaxation, and the frequency of pumping in *Aeschna* varies from about 25 to 50 cycles per minute. At the higher rates the interval between inspiration and expiration is reduced.

In general, although a good deal of gaseous exchange probably takes place through tracheal gills (in *Agrion* [Odonata] larva 32–45% of the oxygen absorbed normally takes this route) some insects are able to survive without gills under normal oxygen tensions. Where the oxygen tension of the water is low, however, gills are of importance as they considerably increase the area available for gaseous exchange.

Plastron respiration Some insects have specialized structures holding a permanent thin film of air on the outside of the body in such a way that an extensive air–water interface is always present for gaseous exchange. This film of gas is called a plastron and the tracheae open into it so that oxygen can pass directly to the tissues.

The volume of the plastron is constant and usually small as it does not provide a store of air, but acts solely as a gill. The constant volume is maintained by various hydrofuge devices spaced very close together so that water does not penetrate between them except under considerable pressure. Excess external pressures that may occur during the normal life of the insect are resisted. Such pressures may develop through the utilization of oxygen from the plastron so that the internal pressure is reduced, or through the insect being in deep water and therefore subjected to high hydrostatic pressure.

In adult insects the plastron is held by a very close hair pile in which the hairs resist wetting because of their hydrofuge properties and their orientation. The most efficient resistance to wetting would be achieved by a system of hairs lying parallel with the surface of the body. *Aphelocheirus* (Hemiptera) approaches this condition in possessing hairs that are bent over at the tip (Fig. 17.31a); in other insects, such as *Elmis* (Coleoptera), the hairs are sloping (Fig. 17.31c). The ability of sloping hairs to withstand collapse depends on their being slightly thickened at the base and also on their close packing. As the air in the plastron is compressed, the closely packed hairs become pressed together so that their overall resistance to further compression increases.

In the adult *Aphelocheirus*, the plastron covers the ventral and part of the dorsal surface of the body. The hairs which hold the air are 5–6 µm high and about 0.2 µm in diameter (Fig. 17.31a). They are packed very close together, at about 2 500 000 mm^{-2}, and are able to withstand a pressure of about 400 kN m^{-2} before they collapse. Hence, this is an extremely stable plastron that will only be displaced by water at excessive depths. The spiracles open into a series of radiating canals in the cuticle and these connect with the plastron via small pores. These canals are lined with hairs so that the entry of water into the tracheal system is prevented. The basal rate of oxygen utilization at 20°C is about 6 µl h^{-1} and this is readily provided by the plastron, so that, except in poorly oxygenated water, the insects need never come to the surface.

The plastrons of other insects are generally less efficient than that of *Aphelocheirus* as they have a less dense hair pile from which the air is more readily displaced. However, a number of insects with a hair density of 3×10^4 to 1.5×10^5 mm^{-2} have plastrons that are usually permanent and adequate for the insect's needs. In these cases, the plastron is often supplemented by a less permanent macroplastron, consisting of a thicker layer of air outside the plastron and held by longer hairs than the plastron, as in *Hydrophilus* (Coleoptera) (Fig. 17.31b), or by the erection of the plastron hairs, as in *Elmis* (Fig. 17.31c). The macroplastron is an air store and acts as a physical gill. It becomes reduced in size and finally eliminated when the insect is submerged, leaving only the plastron. The hairs holding the macroplastron are relatively long and flexible, so as the gas bubble is reduced they tend to clump, leaving patches of exposed cuticle liable to wetting. To avoid this, these insects groom their hair pile and, in *Elmis*, there are brushes on the legs for this purpose. These brushes are also used to capture air bubbles and add them to the macroplastron. Insects that only have a plastron do not make grooming movements of this type.

17.6.3 Ventilation in aquatic insects

Close to a solid surface, water movement becomes laminar with little mixing between successive layers. The region of laminar flow is called a boundary layer and, in general, is greater in depth in still or slowly moving water than in fast-flowing, turbulent water. This has important consequences for the availability of oxygen at any respiratory interface. In turbulent water, oxygen is moved by convection and tends to be uniformly distributed. Movement across the boundary layer, however, depends on diffusion. As a result, water adjacent to an air bubble or a gill will tend to become depleted of oxygen because the rate of diffusion into the layer adjacent to the gill surface may not match the rate at which oxygen is taken into the gill. As movement into the gill is also a passive, diffusive process, it is obviously important to maintain the concentration of oxygen in the boundary layer at a high level. This is achieved by the insect making ventilatory movements that renew the layer of water adjacent to the gill.

In many insects this is achieved by movements of the gills themselves. In *Ephemera* (Ephemeroptera) a

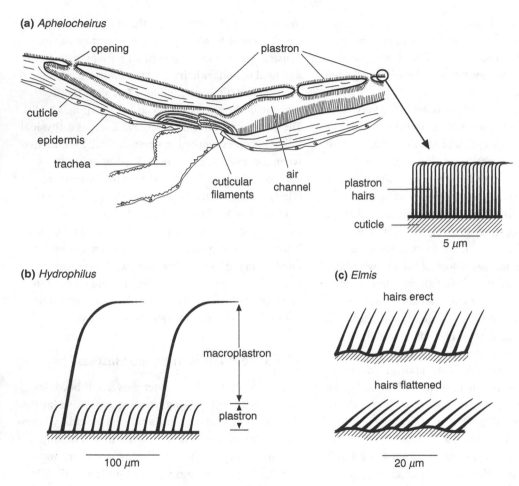

Figure 17.31 Plastron. (a) Plastron of *Aphelocheirus* (Hemiptera). Section through a spiracular rosette showing the junction of a trachea with a system of channels in the cuticle connecting with the plastron. Below right, detail of the plastron (after Thorpe and Crisp, 1947). (b) Macroplastron of *Hydrophilus* (Coleoptera) (after Thorpe and Crisp, 1949). (c) Macroplastron of *Elmis* (Coleoptera) showing the hairs erect, forming a macroplastron (above) and with the hairs compressed so that only the true plastron remains (after Thorpe and Crisp, 1949).

good deal of gaseous exchange takes place through the gills, but in *Cloeon* (Ephemeroptera) they function almost entirely as paddles, pushing water over the real respiratory surface on the abdomen. The larva of *Corydalis* (Neuroptera) ventilates the gills using rhythmic movements of the tubercles on which the gills are mounted. Anisopteran (Odonata) larvae achieve the same result by pumping water in and out of the rectum. Zygopteran larvae also pump water in and out of the rectum, and, in most species, this is

associated with ionic regulation (see Chapter 18). In *Calopteryx* (Odonata: Zygoptera), however, pumping activity is enhanced in water deficient in oxygen. The flow of water from the rectum probably stirs the water around the caudal gills, but it is possible that some gaseous exchange also occurs in the rectum.

Some insects using an air bubble as a gill or that have a plastron also ventilate. For example, the bug *Naucoris* (Hemiptera: Heteroptera) holds onto vegetation and makes swimming movements with its back legs, directing a current over its air bubble; *Phytobius relatus* (Coleoptera) ventilates its plastron with its middle legs.

By changing the frequency with which these movements are performed, the insect is able to maintain its oxygen uptake at a constant level even in water containing small amounts of dissolved oxygen. For example, in *Ephemera*, the gills beat actively in water containing little oxygen and more slowly when the oxygen tension is high (Fig. 17.32). If the oxygen content of the water is very low, however, oxygen consumption falls. Similar changes occur in the rates of ventilatory movements made by a variety of other aquatic insects.

Control of these ventilatory movements appears to depend on similar phenomena to those occurring in terrestrial insects. The ventilatory movements of *Corydalis* are produced by hypoxia, while those of *Naucoris* are probably stimulated by a high concentration of carbon dioxide in the tracheal system. At least in the former, there is evidence for central pattern generators in the ganglia of abdominal segments 2 and 3, with that in segment 3 being dominant. This drives the interneurons that activate or inhibit the motor neurons.

For insects living in fast-flowing water, active ventilatory movements are not necessary, but their body forms and orientation in the current may be important in maintaining the oxygen supply. The cocoons of black flies (Diptera: Simuliidae) are oriented so that their posterior, pointed ends face upstream while the spiracular gills, extending from the open end of the cocoon, are pointed downstream or upwards. The cocoons lie within the boundary layer of water flowing over the stones to which they are attached, and the flow around the cocoon produces vortices around the gills, ensuring a steady supply of oxygen (Fig. 17.33a).

Vortices create regions of low pressure and, if the water is already saturated with air, gas will readily come out of solution, facilitating oxygen uptake by gills. In the extreme, the reduction in pressure may result in air coming out of solution as bubbles. In the pupae of Blephariceridae (Diptera), bubbles of gas appear on the gill surfaces, and this phenomenon of "out-gassing" appears to maintain an air bubble around the adult beetle *Potamodytes* (Fig. 17.33b). It is possible that the facilitation of gas exchange by reduction in water pressure is a widespread phenomenon in aquatic insects living in flowing water.

17.7 Insects subject to occasional submersion

It is relatively common for terrestrial insects to fall into water, but in general this does not occur sufficiently regularly for special respiratory adaptations to have evolved. Insects living close to the edge of water are, however, subject to more frequent alternations of submersion and emergence and so tend to be adapted for respiration in air or water. This is true for insects living intertidally, where submersion occurs regularly, and for those living at the edges of streams, where submersion is much less regular.

Some intertidal beetles, such as *Bledius* and *Dichirotrichus*, effectively avoid submergence by living in burrows and crevices where sufficient air is trapped to enable them to survive aerobically for

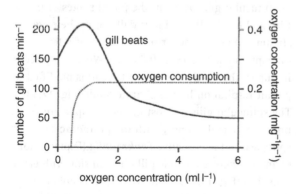

Figure 17.32 Ventilation. The relationship between gill beat and oxygen uptake in relation to the oxygen content of the water in a mayfly larva (*Ephemera*) (after Eriksen, 1963).

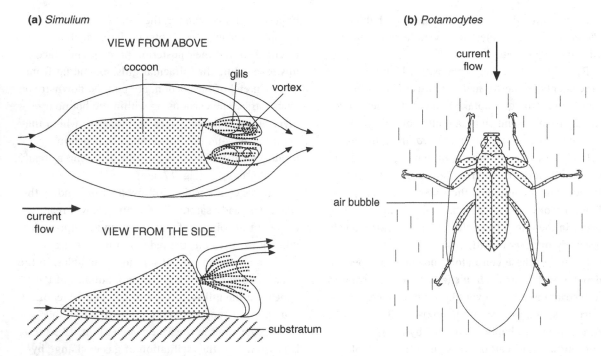

(a) *Simulium*

VIEW FROM ABOVE

cocoon

gills

vortex

current
flow

VIEW FROM THE SIDE

substratum

(b) *Potamodytes*

current
flow

air bubble

Figure 17.33 Insects in fast-flowing water are oriented with respect to the current. The orientation produces vortices in the flow. These result in regions of reduced pressure so that gases tend to come out of solution. (a) Pupal cocoon of a blackfly (*Simulium*) from above and the side. The orientation of the cocoon in the water current results in the production of vortices around the gills (after Eymann, 1991). (b) The beetle, *Potamodytes*, holds onto the substratum facing into the current. The flow round the legs and body creates regions of low pressure and gases come out of solution to form an air bubble round the posterior part of the insect (after Stride, 1955).

several hours. If they do become submerged they become quiescent as a result of anoxia. The time for recovery when returned to air is then proportional to the period of anoxia. Lice living on aquatic mammals live in the layer of air trapped in the fur, so that these, too, probably encounter no particular respiratory problems.

17.7.1 Spiracular gills

A spiracular gill is an extension of the cuticle surrounding a spiracle and bearing a plastron connected to the tracheal system by aeropyles. In water, the plastron provides a large gas–water interface for diffusion, while in air the interstices of the gill provide a direct route for the entry of oxygen, and water loss is limited because the gill opens into

the atrium of the spiracle. Thus, in air, water loss through the spiracles is scarcely greater than in terrestrial insects.

Spiracular gills occur in the pupal stages of many flies and beetles living intertidally or at the edges of streams. They also occur in the larvae of a few Coleoptera and in *Canace* (Diptera). Where they occur in pupae, the basal structure of the gill is modified to permit respiration by the pharate adult (Fig. 17.34). The spiracular gills of most dipteran pupae are prothoracic and connect with the prothoracic spiracles. In the cranefly, *Taphrophila* (Diptera), for example, there is a single gill on each side with eight branches (Fig. 17.34a). The whole gill is about 1.5 mm long. Blackfly pupae (Diptera: Simuliidae) also have two prothoracic gills. In this case each has at least two main branches and in some species there

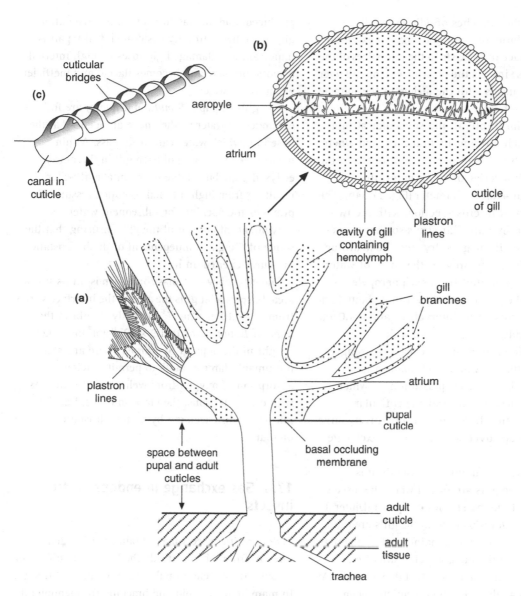

Figure 17.34 Spiracular gills of a pharate adult cranefly (*Taphrophila*) (after Hinton, 1957). (a) Diagram showing basic structure and connection of gill to the tracheal system; the two left-hand branches represent the gill as seen from the outside; the remainder have the upper layer of cuticle removed to show the extent of the atrium. (b) Transverse section through a gill branch. (c) Detail of a plastron line.

is abundant secondary branching. In the pupae of Psephenidae (Coleoptera), however, the gills are associated with the abdominal spiracles and in *Psephenoides volatilis* (Coleoptera) they are on abdominal segments 2–7. Each gill has 4–10 long, slender branches. Abdominal spiracular gills also

occur in the larvae of beetles of the genera *Torridincola*, *Sphaerius* and *Hydroscapha*.

The form of the plastron differs in different insects. In *Eutanyderus* (Diptera) and many other species it is held by hydrofuge cuticular struts running at right angles to the surface of the gill. These struts branch at the apices and the branches of adjacent struts anastomose to form an open network. An extensive air–water interface thus exists in the interstices of the network. In these insects the plastron extends all over the gills and connects with the atrium of the spiracle at the base of each gill.

In *Taphrophila* (Diptera), the spiracular atrium extends into each gill and its branches, and where the atrium meets the wall of the gill it opens to the outside through a series of small pores, the aeropyles, about 4 μm in diameter (Fig. 17.34b). The atrium is flattened in cross-section, with the two walls connected by cuticular struts so that it does not collapse even if the gills dry, and the hydrofuge properties of the struts prevent the entry of water into the atrium. Running from each aeropyle, on the outside of the gill, is a shallow canal about 4 μm wide, which is crossed at intervals of about 1.0 μm by cuticular bridges about 0.5 μm wide (Fig. 17.34c). The lining of the canals is strongly hydrofuge, so that in water each holds a long cylinder of air, known as a plastron line, which is not easily displaced because of the cuticular bridges. The plastron lines provide a relatively large air–water interface over which gaseous exchange occurs.

For the plastron to function efficiently it is important that the gills are turgid when the insects are submerged. In many species of tipulid (Diptera) and in *Psephenoides* (Coleoptera) the gills are rigid, but in other species turgor is maintained by the high osmotic content of the gills, which results from the hemolymph and epidermal cells that they contain. At the pupa–adult apolysis the pupal cuticle becomes separated from that of the adult, but the gills become cut off from the rest of the tissues by a basal occluding membrane so they are still lined by an epithelium and contain hemolymph (Fig. 17.34a). The epithelium within the gills disintegrates and the cells form loose, irregular clusters in the middle of the gill. This isolated tissue is important because it is able to repair damage to the surface of the gill, forming cuticular plugs in any holes that are produced, and this facility is retained even after the gills have been strongly desiccated. Gill repair is important as a damaged gill loses its high internal osmotic pressure and becomes flaccid and inefficient when in the water.

The gills of pupal *Simulium* (Diptera) are fully expanded in water as they have an opening at the base into which water can freely pass. A thin membrane at the base of each gill in the newly ecdysed pupa bursts due to the intake of water resulting from high internal osmotic pressure. This provides the opening for subsequent water movement into or out of the gill, ensuring that the shape of the gill is independent of the hydrostatic pressures exerted on it.

The efficiency of a plastron depends on its surface area. In spiracular gills the area of the interface varies from $1.5 \times 10^4 \, \mu m^2 \, mg^{-1}$ net body weight in the pupa of *Eutanyderus* to $1.9 \times 10^{-6} \, \mu m^2 \, mg^{-1}$ net weight in the pupa of *Simulium*. These areas are presumably large enough to permit efficient absorption of oxygen from well-aerated water. As with other plastrons, the plastron of a spiracular gill resists displacement by high hydrostatic pressures.

17.8 Gas exchange in endoparasitic insects

Endoparasitic insects may obtain their oxygen directly from the air outside the host or by diffusion through the cuticle from the surrounding host tissues. In many ichneumonid and braconid (Hymenoptera) larvae, the tracheal system of the first instar is liquid-filled and, even when it becomes gas-filled, the spiracles remain closed until the last instar. Thus, these insects and the young larvae of most parasitic Diptera depend entirely on cutaneous diffusion. In braconid larvae the hindgut is everted through the anus to form a caudal vesicle. This is variously developed in different species, but in some, such as *Apanteles*, it is relatively thin-walled and closely associated with the heart (Fig. 17.35) so that oxygen passing in is quickly carried around the body. In these insects the vesicles are responsible for about one-third of the total gaseous exchange.

When the tracheal system becomes air-filled, networks of tracheoles may develop immediately beneath the cuticle, thus facilitating the diffusion of gases away from the surface. In *Cryptochaetum iceryae* (Diptera), a parasite of scale insects, there are two caudal filaments, which in the third-instar larva are ten times as long as the body and are packed with tracheae. These filaments often get entangled with the host tracheae and so provide an easy path for oxygen transfer.

Other insects, and particularly older, actively growing larvae with greater oxygen requirements, connect with the outside air by penetrating the host's

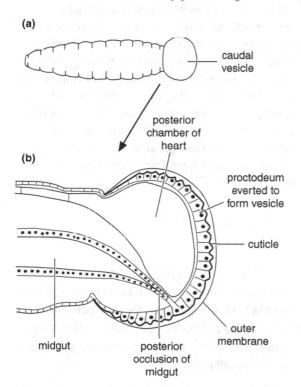

(a)

caudal vesicle

posterior chamber of heart

(b)

proctodeum everted to form vesicle

cuticle

outer membrane

midgut

posterior occlusion of midgut

Figure 17.35 Caudal vesicle of a parasitic hymenopteran larva (*Apanteles*) (from Wigglesworth, 1965). (a) Whole larva showing the caudal vesicle. (b) Longitudinal section of caudal vesicle.

body wall or respiratory system. The majority of these insects use the posterior spiracles to obtain their oxygen. Chalcidid (Hymenoptera) larvae are connected to the outside from the first instar onwards by the hollow egg pedicel, which projects through the host's body wall. The posterior spiracles of the larva open into the funnel-shaped inner end of the pedicel and so make contact with the outside air. Many tachinid (Diptera) larvae that are parasitic in other insects tap the host's tracheal supply or pierce its body wall from within using their posterior spiracles. The host epidermis is stimulated to grow and spreads around the larva, almost completely enclosing it, and secreting a thin, cuticular membrane over its surface. The larva of *Melinda* (Diptera), parasitic in snails, respires by sticking its posterior spiracles out through the respiratory opening of the snail.

Parasites of vertebrates also often use atmospheric air. The larva of *Cordylobia* (Diptera) bores into the skin and produces a local swelling, but it always retains an opening to the outside into which the posterior spiracles are thrust. Similarly in the larva of the warble fly, *Hypoderma*, the warble opens to the outside, but here the larva bores its way out to the surface from within the host's tissues.

17.9 Other functions of the tracheal system

Apart from gaseous exchange, the tracheal system has a number of other functions. The whole system, and the air sacs in particular, lower the insect's specific gravity. In aquatic insects it also gives some degree of buoyancy and in the larvae of the phantom midge, *Chaoborus* (Diptera), the tracheae form hydrostatic organs enabling the buoyancy to be adjusted.

The tracheal system in some insects has a major role in hemolymph circulation (Section 5.2).

Air sacs, being collapsible, allow for the growth of organs within the body without any marked changes in body form. Thus, at the beginning of the final-stage larva, the tracheal system of *Locusta* (Orthoptera) occupies 42% of the body volume, but by the end of the stage it only occupies 3.8% due to the growth of other organs causing compression of the air sacs. The air sacs also permit changes in gut volume as a result of feeding. In *Locusta*, the increase in crop volume following a meal is accompanied by a corresponding decrease in the volume of the thoracic air sacs.

In nocturnal moths tracheae form a reflecting tapetum beneath the eye (Section 22.1.2), and tympanal membranes are usually backed by an air sac which, being open to the outside air, allows the tympanum to vibrate freely with a minimum of damping (Section 23.2.7).

Expansion of the tracheal system may also assist when an insect inflates itself after a molt. Thus, in dragonflies, spiracle closure, preventing the escape of gas from the tracheae, accompanies each muscular effort of the abdomen during expansion of the wings.

The air sacs of some insects insulate the thorax, and therefore the flight muscles, from the abdomen. This makes it possible for thorax and abdomen to have significantly different temperatures and gives the insects the capacity to regulate thoracic temperature (Section 19.2).

An important general function of tracheae and tracheolar cells is in acting as connective tissue, binding other organs together.

Finally, the tracheal system may be involved in defense. In the cockroach *Diploptera*, quinones, which probably have a defensive function, can be forcibly expelled from the second abdominal spiracle; and in *Gromphadorrhina* (Blattodea) sounds are produced by forcing air out through the spiracles.

17.10 Gas exchange in insect eggs

Gas exchange in insect eggs occurs by diffusion through wax and protein layers. The eggs of many terrestrial insects are subject to occasional flooding and those of many species have a chorionic plastron. Physiological studies suggest that the egg chorion is a high-resistance structure, designed to minimize water loss while allowing adequate oxygen delivery. Older and warmer *Manduca sexta* eggs have lower internal oxygen levels. The oldest, warmest eggs are nearly anoxic internally, and survival and development in these eggs is increased by hyperoxia, indicating oxygen-limitation of development. It seems likely that similar trade-offs between reduction of water loss and uptake of oxygen occur in the eggs of many terrestrial insects, especially those from warm, dry habitats.

Summary

- Gas exchange in insects is facilitated by a tracheal system, an air-filled cuticle-lined invagination. While most insects appear to have intracellular hemoglobin, and some groups of insects have hemocyanin in the hemolymph, the dominant pathway of gas exchange is via the air-filled tracheae. The tracheal system provides a lightweight, high-capacity oxygen-delivery system that enables the highest rates of gas exchange in animals, and recovery from anoxia by diffusion.

- Most insects have spiracles with closing mechanisms that are actively regulated by the nervous system; these respond to oxygen and carbon dioxide levels, and often are coordinated to allow unidirectional flow.

- Most insects utilize convection to move air through the larger parts of the tracheal system. Usually, convection is accomplished by skeletal muscle-driven body volume changes or hemolymph transfers between compartments; these raise local pressures, compressing tracheae and air sacs. Inspiration may be due to passive elasticity or active increases in body volume.

- Tracheal system branching and growth is controlled by interactions between growth factors secreted by the tissues and receptors on the tracheal cells. Local hypoxia stimulates increases in growth factor receptors on the tracheae, increasing diameters, branching and proliferation.

- Aquatic insects commonly modify their tracheal system to enable oxygen uptake from the water, such as via a thin cuticle (gills) or structures that hold air against a spiracle (plastrons or bubbles). Gas exchange across gills is enhanced by pumping mechanisms that move water across the gill.

Recommended reading

Burmester, T. and Hankeln, T. (2007). The respiratory proteins of insects. *Journal of Insect Physiology* 53, 285–294.

Ghabrial, A., Luschnig, S., Metzstein, M. M. and Krasnow, M. A. (2003). Branching morphogenesis of the *Drosophila* tracheal system. *Annual Review of Cell and Developmental Biology* 19, 623–647.

Harrison, J., Frazier, M. R., Henry, J. R., Kaiser, A., Klok, C. J. and Rascon, B. (2006). Responses of terrestrial insects to hypoxia or hyperoxia. *Respiratory Physiology & Neurobiology* 154, 4–17.

Kestler, P. (1985). Respiration and respiratory water loss. In *Environmental Physiology and Biochemistry of Insects*, ed. K. H. Hoffmann, pp. 137–183. Berlin: Springer.

Miller, P. L. (1981). Ventilation in active and in inactive insects. In *Locomotion and Energetics in Arthropods*, ed. C. F. Herreid and C. R. Fourtner, pp. 367–390. New York, NY: Plenum Press.

Resh, V. H., Buchwalter, D. B., Lamberti, G. A. and Eriksen., C. H. (2008). Aquatic insect respiration. In *An Introduction to the Aquatic Insects of North America*, ed. R. W. Merritt, K. L. Cummins and M. B. Berg, pp. 39–54. Dubuque, IA: Kendall Hunt Pub Co.

Socha, J. J., Forster, T. D. and Greenlee, K. J. (2010). Issues of convection in insect respiration: insights from synchotron X-ray imaging and beyond. *Respiratory Physiology & Neurobiology* 173, S65–S73.

Wasserthal, L. T. (1996). Interaction of circulation and tracheal ventilation in holometabolous insects. *Advances in Insect Physiology* 26, 297–351.

References in figure captions

Burrows, M. (1980). The tracheal supply to the central nervous system of the locust. *Proceedings of the Royal Society of London B* **207**, 63–78.

Callier, V. and Nijhout, H. F. (2011). Control of body size by oxygen supply reveals size-dependent and size-independent mechanisms of molting and metamorphosis. *Proceedings of the National Academy of Sciences USA* **108**, 14664–14669.

de Ruiter, L., Wolvekamp, H. P., Tooren, A. J. and Vlasblom, A. (1952). Experiments on the efficiency of the "physical gill" (*Hydrous piceus* L., *Naucoris cimicoides* L., and *Notonecta glauca* L.). *Acta Physiologica et Pharmacologica Neerlandica* 2, 180–213.

Eriksen, C. H. (1963). Respiratory regulation in *Ephemera simulans* (Walker) and *Hexagenia limbata* (Serville) (Ephemeroptera). *Journal of Experimental Biology* **40**, 455–468.

Eymann, M. (1991). Flow patterns around cocoons and pupae of black flies in the genus *Simulium* (Diptera: Simuliidae). *Hydrobiologia* **215**, 223–229.

Gardner, A. E. (1960). A key to the larvae of the British Odonata. In *Dragonflies*, ed. P. S. Corbet, C. Longfield and N. W. Moore, pp. 191–225. London: Collins.

Hetz, S. K. and Bradley, T. J. (2005). Insects breathe discontinuously to avoid oxygen toxicity. *Nature* **433**, 516–519.

Hinton, H. E. (1957). The structure and function of the spiracular gill of the fly *Taphrophila vitripennis*. *Proceedings of the Royal Society of London B* **147**, 90–120.

Imms, A. D. (1947). *Insect Natural History*. London: Collins.

Imms, A. D. (1957). *General Textbook of Entomology*, revised O. W. Richards and R. G. Davies. London: Methuen.

Keilin, D. (1944). Respiratory systems and respiratory adaptations of larvae and pupae of Diptera. *Parasitology* **36**, 1–66.

Keister, M. L. (1948). The morphogenesis of the tracheal system of *Sciara*. *Journal of Morphology* **83**, 373–424.

Lewis, G. W., Miller, P. L. and Mills, P. S. (1973). Neuro-muscular mechanisms of abdominal pumping in the locust. *Journal of Experimental Biology* **59**, 149–168.

Lighton, J. R. B. (1988). Simultaneous measurement of oxygen uptake and carbon dioxide emission during discontinuous ventilation in the tok-tok beetle, *Psammodes striatus*. *Journal of Insect Physiology* **34**, 361–367.

Locke, M. (1998). Caterpillars have evolved lungs for hemocyte gas exchange. *Journal of Insect Physiology* **44**, 1–20.

Mill, P. J. and Pickard, R. S. (1972). Anal valve movement and normal ventilation in aeshnid dragonfly larvae. *Journal of Experimental Biology* **56**, 537–543.

Miller, P. L. (1960a). Respiration in the desert locust: II. The control of the spiracles. *Journal of Experimental Biology* **37**, 237–263.

Miller, P. L. (1960b). Respiration in the desert locust: III. Ventilation and the spiracles during flight. *Journal of Experimental Biology* **37**, 264–278.

Pearson, K. G. (1980). Burst generation coordinating interneurons in the ventilatory system of the locust. *Journal of Comparative Physiology* **137**, 308–313.

Schmitz, M. and Komnick, H. (1976). Rectal Chloridepithelien und osmoregulatorische Salzaufnahme durch den Enddarm von Zygopteren und Anisopteren Libellenlarven. *Journal of Insect Physiology* 22, 875–883.

Snelling, E. P., Seymour, R. S. and Runciman, S. (2011). Moulting of insect tracheae captured by light and electron-microscopy in the metathoracic femur of a third instar locust *Locusta migratoria*. *Journal of Insect Physiology* 57, 1312–1316.

Snodgrass, R. E. (1935). *Principles of Insect Morphlogy*. New York, NY: McGraw-Hill.

Stride, G. O. (1955). On the respiration of an aquatic African beetle, *Potamodytes tuberosus* Hinton. *Annals of the Entomological Society of America* 48, 344–351.

Thorpe, W. H. and Crisp, D. J. (1947). Studies on plastron respiration: I. The biology of *Aphelocheirus* (Hemiptera, Aphelocheiridae [Naucoridae]) and the mechanism of plastron respiration. *Journal of Experimental Biology* 24, 227–269.

Thorpe, W. H. and Crisp, D. J. (1949). Studies on plastron respiration: IV. Plastron respiration in the Coleoptera. *Journal of Experimental Biology* 26, 219–260.

Tillyard, R. J. (1917). *The Biology of Dragonflies*. Cambridge: Cambridge University Press.

Wasserthal, L. T. (1996). Interaction of circulation and tracheal ventilation in holometabolous insects. *Advances in Insect Physiology* 26, 297–351.

Wasserthal, L. T. (2001). Flight-motor-driven respiratory air flow in the hawkmoth *Manduca sexta*. *Journal of Experimental Biology* 204, 2209–2220.

Wasserthal, L. T. (2012). Influence of periodic heartbeat reversal and abdominal movements on hemocoelic and tracheal pressure in resting blowflies *Calliphora vicina*. *Journal of Experimental Biology* 215, 362–373.

Wasserthal, L. T., Cloetens, P. and Fink, R. (2006). Synchrotron x-ray-videography and -tomography combined with physiological measurements for analysis of circulation and respiration dynamics in insects (*Drosophila* and *Calliphora*). *Deutsche Tagung für Forschung mit Synchrotronstrahlung, Neutronen und Ionenstrahlung an Großgeräten*, University of Hamburg, October 4–6, F-V55.

Webb, J. E. (1940). The origin of the atrial spines in the spiracles of sucking lice of the genus *Haematopinus* leach. *Proceedings of the Zoological Society of London* 118, 582–587.

Weis-Fogh, T. (1964). Functional design of the tracheal system of flying insects as compared with the avian lung. *Journal of Experimental Biology* 41, 207–227.

Wigglesworth, V. B. (1959). The role of the epidermal cells in the migration of tracheoles in *Rhodnius prolixus*. *Journal of Experimental Biology* 36, 632–640.

Wigglesworth, V. B. (1965). *The Principles of Insect Physiology*. London: Methuen.

18 | Excretion and salt and water regulation

REVISED AND UPDATED BY **JULIAN A.T. DOW**

INTRODUCTION

Cells operate most efficiently within a narrow range of conditions. It is therefore important that the environment within the cell and in the animal in general should be kept as near optimal as possible, a process known as homeostasis. This involves the maintenance of a constant level of salts and water and osmotic pressure in the hemolymph ("osmoregulation"), the elimination of toxic nitrogenous wastes derived from protein and purine metabolism and the elimination of other toxic compounds which may be absorbed from the environment. The excretory system is primarily responsible for homeostasis, and for the excretion of toxic compounds – often following metabolic modification to chemicals more readily excreted or which can be safely stored. The success of insects in exploiting a wide range of habitats can be attributed – at least in part – to the robustness of their osmoregulatory systems. Although osmoregulation and excretion are distinct functions, they overlap both in the tissues and the ions used, and so they are considered together in this chapter. Section 18.1 covers the structure of the excretory system; Sections 18.2–18.4 cover the production of primary urine by the Malpighian tubules, its modification in the lower tubules and hindgut and its endocrine control. In Section 18.5, the handling of nitrogenous waste is discussed, followed by other roles of the excretory system in detoxification (Section 18.6) and non-excretory roles (Section 18.7). Excretory functions of cells outside the gut (such as nephrocytes) are covered in Section 18.8, and water regulation at an organismal level in Section 18.9.

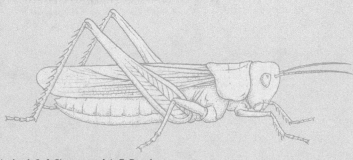

The Insects: Structure and Function (5th edition), ed. S. J. Simpson and A. E. Douglas.
Published by Cambridge University Press. © Cambridge University Press 2013.

18.1 Excretory system

Excretion involves the production of urine that removes potentially toxic materials from the body. This process is carried out in two stages: the relatively unselective removal of substances from the hemolymph, forming the primary urine, and the selective modification of this primary urine by the reabsorption of useful compounds or the addition of others that may be in excess in the body. This "excretory cycle" is the primary task of the hindgut, comprising the Malpighian tubules, ileum and rectum (Fig. 18.1). In insects, the primary urine is produced by the Malpighian tubules. Its selective modification usually occurs in the rectum, but may also take place in the Malpighian tubules or in the ileum. In aquatic insects, epidermal cells outside the gut may also be involved in maintaining the insect's ionic balance.

18.1.1 Malpighian tubules

Malpighian tubules are long, thin, blindly ending tubes arising from the gut near the junction of midgut and hindgut (Fig. 18.1) and lying freely in the body cavity. (They are named after their discoverer, Marcelo Malpighi, a medical pioneer of the seventeenth century.) Malpighian tubules are derived from the ectodermal hindgut, although they are not lined with cuticle. They are found in nearly all insects, and in Myriapods, Arachnids and tardigrades; and are therefore an ancient invertebrate specialization. Malpighian tubules are absent (probably secondarily) from Collembola and aphids, and represented only by papillae in Diplura, Protura and Strepsiptera.

When present, Malpighian tubules vary in number from two in coccids to about 250 in the desert locust, *Schistocerca gregaria*. The number may increase during postembryonic development (see Fig. 15.13). They can range in length from 2 mm to 100 mm and from 30 μm to 100 μm in diameter. In most insects, they open independently into the gut, but in some species they first come together in groups at ampullae. Because of the large number of tubules and the infolding of the basal plasma membrane (see below) their total outer surface area is large, facilitating rapid transport. In adult *Periplaneta*, with more than 150 tubules, the total outer surface area of the tubules (not taking account of the infolding of the basal plasma membrane) is over 500 mm^2; the infoldings increase the surface area at least ten times to more than 5000 mm^2.

The wall of the tubule is one cell thick, with one or a few cells encircling the lumen (Fig. 18.2). The primary urine is produced by the cells. This cell type bears close-packed microvilli on the lumen side,

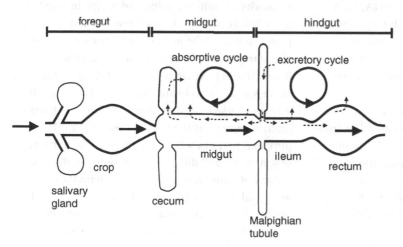

Figure 18.1 Overview of the insect alimentary canal, showing the absorptive and excretory cycles of water movement (Berridge, 1970). Heavy arrows show the path of food through the alimentary canal, and dotted lines show cycles of fluid movement. Heavier lines denote the cuticle lining of the foregut and hindgut.

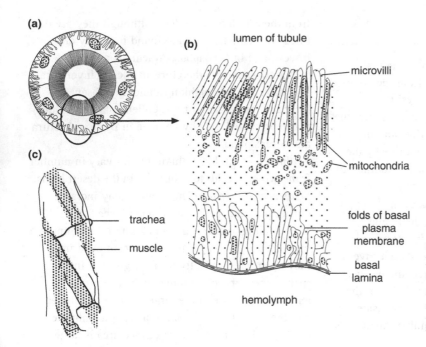

(a)

(b)

lumen of tubule

microvilli

mitochondria

folds of basal plasma membrane

basal lamina

hemolymph

(c)

trachea

muscle

Figure 18.2 Structure of a typical Malpighian tubule. (a) Cross-section of a tubule. (b) Detail of part of one cell. (c) End of a Malpighian tubule of *Apis* showing the spiral muscle strands and the tracheal supply (after Wigglesworth, 1965).

while the basal plasma membrane is deeply infolded. Laterally, the cells are held together by septate junctions (analogs of vertebrate tight junctions) near their apices. Mitochondria are abundant within and immediately beneath the microvilli and in association with the basal plasma membrane (Fig. 18.2b). The principal cells are frequently binucleate (as in *Rhodnius* [Hemiptera]) or polytene (containing multiple copies of nuclear DNA, with aligned chromatids, as in Diptera). Polyteny may allow cells to grow bigger without dividing, so maximizing their surface area:junctional perimeter ratio, and limiting the passage of water or solutes between cells.

In many insects, the tubules can contain multiple regions or cell types; or some tubules can differ from others in the same animal. Most Diptera, for example, have small stellate cells scattered among the principal tubule cells. Stellate cells have small microvilli and few mitochondria. In *Drosophila*, genetic mapping reveals six cell types and six regions within the four Malpighian tubules, of which the right-hand pair of tubules always ramify anteriorly, and the left-hand pair always posteriorly, in the body cavity. These domains and cell types have distinct, specialized functions. The initial segments (more prominent in the anterior tubules) are packed with calcium-rich spherites, presumably for storage excretion (Section 18.5.2); the main segment generates the primary urine and the lower segment reabsorbs the primary urine and is rich in alkaline phosphatase (though the reason why is not clear). The principal cells of the main segment are specialized for active transport (discussed below), whereas the interdigitating stellate cells control chloride flux and express aquaporins (water channels). The Malpighian tubules of *Rhodnius* also have different cell types in discrete regions of the tubules. The cells in the proximal part of each tubule (nearest the point of origin at the gut) differ from those in the distal part in having more widely separated microvilli which are variable in length; less complex infoldings of the basal membrane; and mitochondria concentrated in the basal region. Mixtures of cell types are also known to occur in tubules of *Periplaneta* (Blattodea), *Carausius* (Phasmatodea) and *Tenebrio* (Coleoptera).

The domestic cricket, *Acheta*, has Malpighian tubules with three distinct regions: proximal, comprising about 5% of the length of the tubule; middle, 75% of the length; and distal, 20%. Each region is made up of one cell type. The cells of the middle section, forming the bulk of the tubule, resemble the principal cells described above. Those of the distal part have close-packed microvilli which almost fill the lumen of the tubule, but the basal plasma membrane is not as extensively folded as in the cells of the middle section and most mitochondria are concentrated just below and within the microvilli. The distal and middle parts of the tubules are both secretory, but their rates of secretion are controlled differently. The short proximal section of tubule is probably resorptive. The water boatman, *Cenocorixa* (Hemiptera), has four discrete regions in its Malpighian tubules; and the tubules of larval Lepidoptera have conspicuous regional differences.

The Malpighian tubules of *Rhodnius* and Lepidoptera and Diptera in general have no muscles other than a series of circular and longitudinal muscles proximally, while those of Coleoptera and Neuroptera have a continuous muscular sheath. In orthopteroids, Odonata and some Hymenoptera, strands of muscle spiral round the tubule (Fig. 18.2). These muscles produce writhing movements of the tubules in the hemocoel. These movements help to maintain the concentration gradients across the tubule wall by continually bringing tubules into contact with fresh hemolymph. They, perhaps, also assist fluid movement within the tubules.

In some larval Coleoptera and Lepidoptera the distal parts of the Malpighian tubules are closely associated with the rectum, forming a convoluted layer over its surface. This is known as a cryptonephridial arrangement, and is associated with the ability to produce strongly hyperosmotic excreta (Section 18.9.1).

18.1.2 Hindgut of terrestrial insects

Among terrestrial insects, modification of the urine often occurs primarily in the rectum, which is structurally modified for this function (Fig. 18.3). In particular, the major osmotic work is performed in the rectum (see Table 18.1). The structure of the ileum is not well known, but in *Schistocerca* the apical and basal plasma membranes are extensively folded with large numbers of associated mitochondria, and much fluid reabsorption occurs here. In this species, the rectum also retains the structure typical of terrestrial insects and contributes to modification of the urine (see Fig. 18.10).

In fluid-feeding insects, where water conservation is of limited or no importance, the cells of the ileum resemble those just described, but the rectum does not possess the usual arrangement of enlarged cells forming rectal pads.

18.1.3 Sites of ion exchange in aquatic insects

Freshwater insects tend to lose ions to the environment; insects in saltwater tend to gain them. Thus, insects from either environment must compensate for these changes in the ionic composition of the hemolymph in addition to offsetting food-induced changes. They do this by modifying the urine as it passes along the hindgut and by moving ions directly out of or into the surrounding medium.

In those groups where the hindgut has been examined in detail, either the ileum or the rectum appears to be associated with ionic uptake or secretion. In addition, all aquatic insects, whether they occur in fresh- or saltwater, possess structures which are known or inferred to be involved in ion exchange with the medium. They are present in the epidermis and may occur as isolated cells or complexes of a few cells, or as more extensive areas. Even cells in the gut may be involved in ion exchange with the medium if the insect drinks or

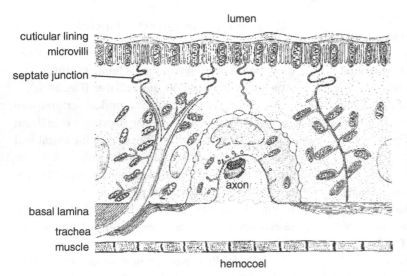

Figure 18.3 Typical rectal structure. Cockroach rectum, showing specializations for active reabsorption. The rectum is lined by a semipermeable cuticle, secreted by the epithelium. The apical microvilli are densely packed with mitochondria, to energize active ion uptake from the lumen. Ions are then pumped into lateral spaces between the cells, where local osmotic gradients favor the movement of water from lumen to basolateral space. The reabsorbed fluid then issues into the hemocoel (after Oschman and Wall, 1969).

takes in water via the anus. In some cases it is impossible to determine the source of fluid in the gut, so structures concerned with modification of the primary urine and those concerned with ion exchange with the medium are considered together.

Cells that absorb ions from, or secrete them into, the surrounding medium are known as chloride cells, although they transport other ions in addition to chloride. They are characterized by a deeply folded plasma membrane, either at the apical or basal surface, and mitochondria are abundant, sometimes being closely associated with the folded plasma membrane (Fig. 18.4). The cuticle above the cell is often perforated. In the mayfly, *Coloburiscoides*, for example, it consists only of epicuticle and is less than 0.5 μm thick. A thin outer layer of the epicuticle is perforated by openings less than 9.5 nm in diameter; the pores in a thicker, inner layer are larger. Single chloride cells are scattered over the body surface of larval mayflies (Ephemeroptera) and stoneflies (Plecoptera) and of both larvae and adults of aquatic bugs (Hemiptera: Heteroptera).

In some other aquatic insects the chloride cells are grouped together in discrete regions. Larval Trichoptera have groups of chloride cells forming chloride epithelia on the dorsal surface of several abdominal segments. In mosquito larvae, the epithelia form discrete external structures, the anal papillae (Fig. 18.5). The cuticle covering these papillae is always very thin, 3.0 μm or less, and the tips of the short microvilli of the cells beneath are attached to it by hemidesmosomes. Folds of the basal plasma membrane extend halfway across the cell. There are many mitochondria, indicating that the cells are very active, many of them associated with the bases of the microvilli. The cells in the papilla appear to form a syncytium, both in mosquito and chironomid larvae. The overall size of these papillae varies inversely with the ionic concentration of the surrounding water, at least in some species (Fig. 18.5). Their ultrastructure also changes; the cells have fewer microvilli and mitochondria in insects living in water with higher concentrations of salts.

Chloride cells do not seem to be a uniquely insect phenomenon; similar chloride cells (or "mitochondrion-rich cells" – MRCs) are found in fish gills, show highly similar morphology, and are also capable of absorbing ions from freshwater, and secreting them into saltwater. Again, the structural signature is of transport-active cells with deep apical crypts linked to the outside by a narrow channel, and

Table 18.1 **Comparison of osmotic pressures and ionic concentrations in hemolymph, Malpighian tubules and rectum**[a]

Species	Compartment	Osmotic pressure[b]	Na	K	Cl
Terrestrial insects					
Carausius	Hemolymph	171	11	18	87
	Tubule	171	5	145	65
	Rectum	390	8	327	–
Schistocerca	Hemolymph	214	103	12	107
	Tubule	226	47	165	88
	Rectum	433	1	22	5
Rhodnius	Hemolymph	206	174	7	155
	Tubule	228	114	104	180
	Rectum	358	161	191	–
Freshwater insect					
Aedes aegypti (larva)	Hemolymph	138	87	3	–
	Tubule	130	24	88	–
	Rectum	12	4	25	–

Notes:
[a] Notice that the osmotic pressure of the tubule fluid (primary urine) is the same as that of the hemolymph although concentrations of particular ions are different. In terrestrial insects the rectal fluid becomes hyperosmotic due to reabsorption of water whereas in freshwater insects it becomes hyposmotic due to reabsorption of salts (partly after Stobbart and Shaw, 1974; concentrations are in mEq l^{-1}).
[b] Expressed as the equivalent solution of NaCl.

situated within an epithelium that is otherwise highly impermeable.

In addition to externally situated chloride cells, at least some of these insects also have epithelia concerned with ionic regulation lining parts of the hindgut. For example, in the bug *Cenocorixa*, the cells of the ileum have basal and apical folds with associated mitochondria, but the rectum is unmodified. On the other hand, the rectum of freshwater mosquito larvae is lined by a continuous layer of cells in which both apical and basal membranes are deeply infolded and there are many mitochondria. The epithelium of the anterior part of the rectum also has this structure in saltwater mosquitoes, but the cells of the posterior rectum are deeper and have more extensive apical folding. Most mitochondria are associated with these folds. In both fresh- and saltwater insects, the rectum usually lacks the complex lateral scalariform junctions between cells that are such a characteristic feature of the rectum in terrestrial insects and which are associated with water uptake.

Dragonfly larvae (Anisoptera) have no external chloride cells, but comparable epithelia are associated with their rectal gills (see Fig. 17.29), and line most of the rectum in damselfly larvae (Zygoptera). In larvae of dytiscid beetles, the

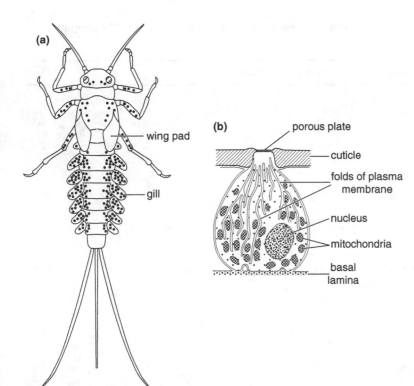

(a)

— wing pad

— gill

(b)

porous plate

— cuticle

folds of plasma membrane

nucleus

mitochondria

basal lamina

Figure 18.4 Chloride cells in a mayfly larva. (a) Dorsal view of larva. Dots show positions of the chloride cells. Their size is greatly exaggerated. (b) Diagrammatic section through a chloride cell. In some chloride cells, the basal plasma membrane, rather than the apical membrane, is infolded (after Komnick, 1977).

whole of the epithelium lining the ileum is modified, but the cells of the rectum do not have folded membranes.

18.2 Urine production

A fluid must be produced in the Malpighian tubules to carry excreted substances to the hindgut. This fluid is called the "primary urine" (sometimes also tubule fluid) to differentiate it from the modified excreta that leaves the insect via the anus, having been modified on its passage through the hindgut.

Terrestrial insects that feed on solid foods probably rarely excrete wholly liquid urine; it is usually mixed with undigested food in the rectum so the feces may be more or less fluid depending on the amount of urine it contains. Fluid-feeders and aquatic insects, however, normally excrete liquid urine.

18.2.1 Formation of the primary urine

In all insects, the movement of water into the Malpighian tubules from the hemolymph depends on the active transfer of cations into the lumen of the tubule (Fig. 18.6). Potassium is usually the predominant cation (Table 18.2), but in insects feeding on vertebrate blood which is high in sodium, sodium has a major role. A V-type H^+-ATPase on the apical plasma membrane, activated by the mitochondria in the microvilli, pumps hydrogen ions into the lumen of the tubule. The hydrogen is then exchanged for potassium or sodium from the cell, probably by a CPA2 class of exchanger. Chloride ions follow, moving down the electrochemical gradient, and water flows through the cells (transcellularly) down the osmotic gradient created by the accumulation of ions between the microvilli. Because the wall of the tubule is highly permeable to water,

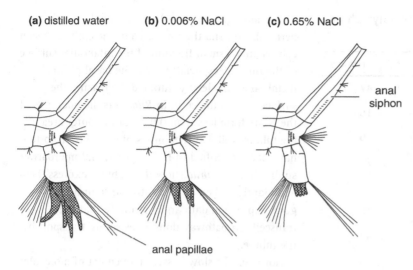

(a) distilled water **(b)** 0.006% NaCl **(c)** 0.65% NaCl

anal siphon

anal papillae

Figure 18.5 Chloride epithelium. Posterior end of a mosquito larva showing the anal papillae (shaded), which contain the chloride cells. Their size is reduced in larvae reared in water with higher salt concentrations (after Wigglesworth, 1972).

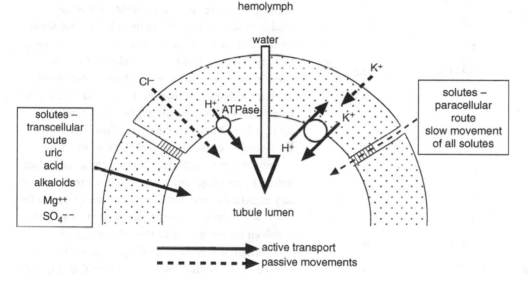

hemolymph

water

Cl^-

K^+

H^+ ATPase

K^+

H^+

solutes – transcellular route uric acid alkaloids Mg^{++} SO_4^{--}

solutes – paracellular route slow movement of all solutes

tubule lumen

→ active transport
- - → passive movements

Figure 18.6 Overview of production of primary urine. Representation of the processes involved in the movement of water into the tubule and associated transcellular and paracellular movements of solutes (based on Phillips, 1981; Maddrell and O'Donnell, 1992).

this requires a difference of only a few mosmol l^{-1} between the lumen and the hemolymph.

18.2.2 Movement of solutes

Once the fluid is produced within the tubule, solutes in the hemolymph also move in (Fig. 18.7). This involves passive diffusion down the concentration gradient as well as, in some cases, active pumping. Passive diffusion is believed to occur both between the cells (the paracellular route) and through the cells (the transcellular route). Movement between the cells is slow for two reasons: the cells are held together by septate junctions (analogs of the

Table 18.2 **Composition (mM) of hemolymph and primary urine of *Schistocerca*[a]**

Constituent	Hemolymph	Primary
Na$^+$	103	47
K$^+$	12	165
Mg^{++}	12	20
Ca^{++}	9	7
cr	107	88
Phosphate	6	12
Glucose	2.5	4.6
Alanine	1.0	1.0
Aspartate	0.1–0.9	0.5
Asparagine	1.0	0
Arginine	1.5	0
Glutamate	0.1–1.0	0.8
Glutamine	4	0.5
Glycine	14	4.0
Histidine	1.4	0
Isoleucine	0.4	0
Leucine	0.4	0
Lysine	1.0	0
Methionine	0.4	0
Phenylalanine	0.7	0
Proline	13	38
Serine	2–4	1.0
Threonine	0.5	0
Tyrosine	1.0	0
Valine	0.6	0

Note:
[a] Notice that potassium and proline are in much higher concentrations in the urine because they are actively secreted. Most other substances are in lower concentration in the urine; they diffuse into the tubule (after Phillips *et al.*, 1986).

vertebrate tight junctions) which greatly reduce the permeability; and the total area of the clefts between cells is only a small fraction of the total outer surface of the tubules, especially since the basal plasma membranes are deeply infolded, increasing the surface area of the cells. In *Rhodnius* the openings of the clefts form less than 0.05% of the outer area of the tubule wall. Large families of solute transporters have been identified in recent years, and microarray studies in *Drosophila* show that tubules express them abundantly. Taken together, the structural and gene-expression data argue strongly that the transcellular pathway dominates solute transport by the tubule.

However, the slow passive movement of molecules through the intercellular gaps is also important because it ensures that any foreign molecules of moderate size that enter the hemolymph are slowly removed. This provides the insect with some capacity to eliminate novel, toxic molecules, for which specific excretory mechanisms have not evolved. Such a mechanism, however, could also result in the loss of molecules that are important to the insect, such as hormones and nutrients. However, lipid-soluble compounds, including some of the hormones, are transported through the hemolymph bound to proteins, and so are too large to be lost at any significant rate. Smaller molecules of use to the insect are resorbed at some site downstream of that involved in production of the primary urine.

Although it is hard to distinguish experimentally between paracellular and transcellular flux, Fig. 18.7 clearly shows that most molecules sit on one of two curves, consistent with either a low-permeability paracellular flux or a high-permeability transcellular flux.

18.2.3 Excretion of ions

Inorganic ions other than potassium and sodium commonly diffuse into the urine transcellularly. In adult blowflies (*Calliphora*), the movement of most anions occurs in inverse proportion to their hydrated

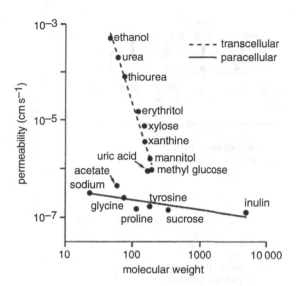

Figure 18.7 Movement of solutes across the tubule. Passive movement of solutes into the unstimulated Malpighian tubules of *Rhodnius*. Charged substances, such as sodium, glycine and tyrosine, as well as larger molecules, move between the cells (paracellularly). (Note that sodium is actively transported through the cells when the Malpighian tubules are stimulated by diuretic hormone.) Uncharged substances, such as urea and mannitol, move much more quickly, although they have similar molecular weights to glycine and tyrosine. They probably diffuse through the cells (transcellularly) (based on O'Donnell *et al.*, 1983).

size. In some insects, however, some of the ions are actively transported. This is true of phosphate in *Carausius* (Phasmatodea) and sulfate in *Schistocerca* (Orthoptera).

Chloride generally follows the active transport of sodium or potassium in order to balance the positive charge on these ions. It is thus not necessary to invoke active transport of chloride: a high, selective permeability is all that is required. Both paracellular and transcellular flux of chloride has been suggested, and both undoubtedly occur; different species may make different relative use of the two pathways. In *Drosophila*, the diuretic neuropeptide leucokinin stimulates chloride flux, and its receptors have been shown to be exclusively located on the stellate cells, implicating them as the route of chloride flux; in the

mosquito *Aedes aegypti*, electrophysiological data implicate the paracellular pathway.

18.3 Modification of the primary urine

18.3.1 Terrestrial insects

The production of primary urine depends on the active movement of potassium or sodium into the Malpighian tubules followed by a passive movement of anions, primarily chloride, to restore electrical equilibrium. The insect cannot normally sustain such a high loss of ions from the hemolymph, and they are recovered by reabsorption from the fluid. In *Schistocerca* (Orthoptera), and perhaps in many other terrestrial insects eating solid food, this occurs in the ileum and the rectum and depends on the active reabsorption of chloride and sodium (Fig. 18.8). Potassium follows along the electrical potential gradient generated by the movement of chloride (see Fig. 3.23). In *Carausius* (Phasmatodea), about 95% of the sodium and 80% of the potassium in the primary urine may be reabsorbed. On the other hand, the rectum of adult *Pieris* (Lepidoptera) shows none of the structures associated with reabsorption which, in this insect, occurs in the ileum.

In some other insects, most or all of the reabsorption occurs in the Malpighian tubules themselves. Absorption of potassium, chloride and water occurs in the lower, proximal regions of the tubules of *Rhodnius*, although the structure of the rectum suggests that some further modification of the urine occurs there.

Useful organic compounds are also recovered from the primary urine. Glucose is actively reabsorbed from the Malpighian tubules of *Locusta* so the primary urine is modified, even as it is being produced. The uptake of amino acids in the rectum may also be active, and this is certainly the case with proline in *Schistocerca*.

The cuticle lining the rectum limits the size of molecules which can be absorbed as, in the locust for

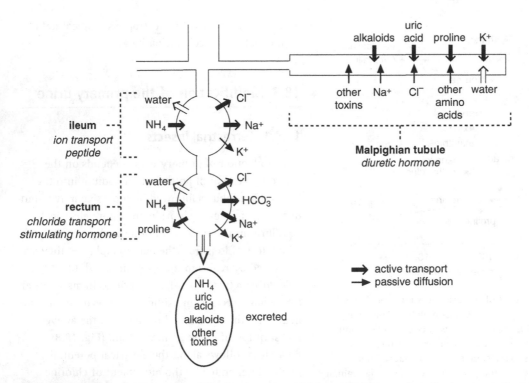

Figure 18.8 Urine production and modification in a terrestrial insect (*Schistocerca*). Active transport of potassium into the Malpighian tubule (through V-ATPase and exchanger) leads to the osmotic movement of water, and most other solutes follow passively. Many of the solutes are recovered as the urine moves toward the hindgut, but ammonia is actively secreted into it. The hormones regulating the processes are shown in italics (partly based on Phillips and Audsley, 1995).

example, it is impermeable to molecules with a radius greater than 0.6 nm. As a consequence, glucose passes readily through it, but trehalose does so only at a low rate, and larger molecules are unable to do so. The effect of this is to protect the rectal cells from toxic molecules whose concentrations may increase in the rectum and to ensure that these substances are excreted.

18.3.2 Freshwater insects

Freshwater insects tend to lose salts to the environment because most species have a highly permeable cuticle. Potassium, sodium and chloride are reabsorbed in the rectum, but water is not. As a result, the rectal fluid is hypotonic to the hemolymph (Table 18.2).

Salts are gained from the food, but also from the environment by the chloride cells. In the larvae of *Aedes*, *Culex* and *Chironomus* (Diptera), potassium, sodium, chloride and phosphate are actively taken up (Fig. 18.9a). Larvae of *Aedes aegypti* are able to maintain their hemolymph osmotic pressure in a medium containing only 6 μmol l^{-1} of sodium, indicating that they are able to take up salt from extremely dilute solutions. Salts are also taken up by the rectal gills of larval Anisoptera (Odonata) and the rectal chloride epithelium of Zygoptera (Odonata). A deficiency of inorganic ions in the environment causes larval Zygoptera to pump water in and out of the rectum, as anisopteran larvae do for respiratory purposes. The effect of this is to supply water, potentially containing ions, to the chloride epithelium in the rectum. Not all freshwater insects

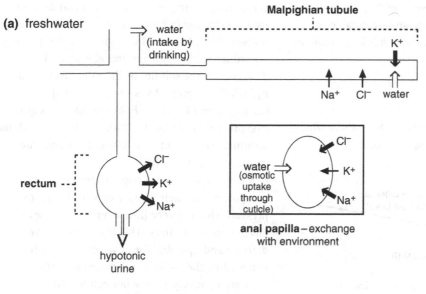

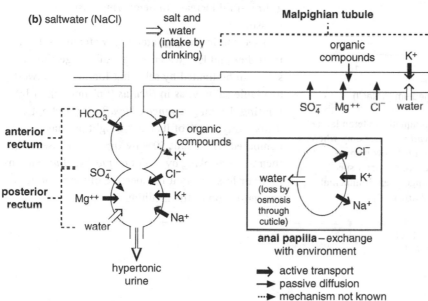

Figure 18.9 Urine production and modification in aquatic insects (mosquito larvae). Nitrogenous excretory products not shown. (a) A freshwater insect, such as *Aedes aegypti*. The larva gains water by drinking and by absorption through the permeable cuticle of the anal papillae; excess water is removed as urine. (b) A saltwater insect, such as *Aedes campestris*. The gain of water due to drinking is greater than osmotic loss through the cuticle. Further water is lost in the urine. Note that although water is moved into the posterior rectum, the fluid produced there is hypertonic to the hemolymph and the medium (partly based on Bradley and Phillips, 1977).

have the capacity to take up ions from the environment. Larval *Sialis* (Megaloptera), for example, are unable to take up chloride.

Some freshwater insects are able to offset changes in ionic concentrations in the hemolymph with compensating changes in the non-electrolyte fraction. Probably, amino acids are produced from hemolymph proteins in sufficient quantity to maintain osmotic pressure.

Freshwater insects are able to regulate the composition and osmotic pressure of the hemolymph over the range of conditions to which they are normally subjected, but in hypertonic media regulation breaks down and the hemolymph rapidly becomes isotonic with the medium (Fig. 18.9, *Aedes aegypti*). These insects are, apparently, unable to produce a fluid in the rectum which is hypertonic to the hemolymph.

18.3.3 Saltwater insects

A number of aquatic insects live in habitats in which the salinity varies widely (Figs. 18.10, 18.11). *Aedes*

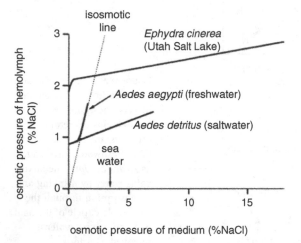

Figure 18.10 Osmotic regulation in aquatic dipteran larvae. The freshwater species *Aedes aegypti* regulates only when the osmotic pressure of the medium is below its own hemolymph osmotic pressure. The two saltwater species regulate moderately well over a range of environmental osmotic pressures (Shaw and Stobbart, 1963).

detritus, for example, occurs in salt marshes, and the fly *Coelopa frigida* breeds in seaweed washed up on the sea shore. In both situations the salinity varies according to the degree of inundation and desiccation. The salinity of the salt pans in which *Ephydrella* (Diptera) lives varies seasonally from 0.3 to 1.3 mol l^{-1} NaCl. *Ephydra cinerea* is quite exceptional, living in the Utah Salt Lake, which has a salinity equivalent to a 20% sodium chloride solution (3 mol l^{-1} NaCl).

Most of these insects regulate the ionic composition of their hemolymph so that its osmotic pressure changes very little over wide ranges of environmental salinity. This is the case in *Aedes detritus* and *Ephydra*, for example (Fig. 18.10). Others, like the caddis fly, *Limnephilus affinus*, regulate relatively poorly but can tolerate a three-fold increase in hemolymph osmotic pressure.

Insects living in saltwater gain water and salts with their diet and lose water osmotically. Ingestion of salts can be limited by selecting food of the lowest available salinity, as in *Bledius* (Coleoptera), or by limiting drinking. *Limnephilus affinis* (Trichoptera) drinks only 3–7% of its body weight per day compared with about 50% by freshwater insects, but the rate of drinking by *Aedes taeniorhynchus* remains more or less constant irrespective of the osmotic concentration of the medium.

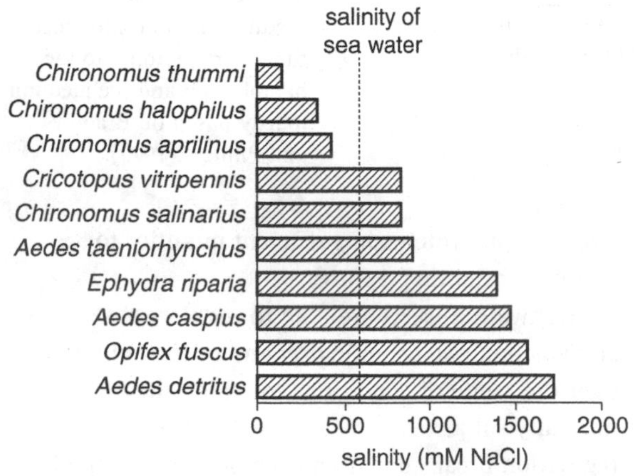

Figure 18.11 Tolerance to salinity among aquatic dipteran larvae. Each bar shows the range of salinities that each species can tolerate. *Chironomus thummi* is a freshwater species included for comparison (Foster and Treherne, 1976).

Saltwater insects get rid of excess ions by excreting urine which is hypertonic to the hemolymph. In saltwater mosquitoes, some reabsorption occurs in the anterior rectum, but sodium, potassium, magnesium and chloride ions are secreted into the posterior rectum to create a hypertonic fluid (Fig. 18.9b). The osmolarity of the urine produced in the posterior rectum is proportional to that of the medium. If the insect is in a hypotonic medium, the posterior rectum is inactive and the insect produces hypotonic urine. In this way the hemolymph osmotic pressure is regulated. At least in *Aedes campestris*, the insect adapts to the presence of different salts in the medium by excreting the appropriate ions. Thus, a larva living in water with a high concentration of sodium chloride actively secretes sodium and chloride ions into the posterior rectum; if bicarbonate is the dominant anion in the medium, it is probably secreted into the rectum; if it is sulfate, this is secreted into the Malpighian tubules.

18.4 Control of diuresis

The rate of urine production can vary enormously to match the changing needs of the insect; the excretory system is thus under endocrine control. Many species reduce urine production and increase blood volume just before a molt and then increase urine production and rapidly reduce blood volume after cuticle expansion. Diuresis is most obvious at eclosion of adult holometabolous insects. *Pieris* (Lepidoptera), for example, reduces its blood volume to less than 33% in the three hours after eclosion; and the blood volume of *Sarcophaga* (Diptera) is reduced to 25% in 36 hours. Insects that eat wet food also produce relatively large amounts of urine after feeding. This is most marked in blood-feeding insects. In the course of a blood meal, *Rhodnius* may ingest more than ten times its own weight; it eliminates most of the water in the plasma, about half of the ingested volume, in the next hour.

These changes in the rates of fluid loss depend on changes in the rate of primary urine production that also necessitate changes in reabsorption to ensure that an increase in fluid loss does not result in excessive loss of ions or other useful compounds. An increase in the rate of primary urine production by the Malpighian tubules does not necessarily result in an increase in water loss as active recovery may occur in the lower part of the system. Individual *Schistocerca* deprived of food for one day, and so subject to some degree of water shortage, recover about 95% of the fluid from the primary urine, and, in both *Schistocerca* and the desert beetle, *Onymacris*, the fluid from the Malpighian tubules is directed forwards into the midgut and resorbed. High rates of urine flow may be maintained in these instances to remove potentially toxic materials from the hemolymph.

Production of the primary urine and subsequent reabsorption are regulated independently. Diuretic hormones stimulate increased fluid production by the Malpighian tubules and other hormones regulate reabsorption.

The trigger for release of the diuretic hormone in *Rhodnius* is abdominal stretching produced by the meal; abdominal distension resulting from feeding may also be the trigger in some other insects. In *Rhodnius*, stretching results in the release of serotonin from the abdominal nerves in the hemocoel at the same time as diuretic hormone is released. The two substances act synergistically to regulate primary urine production by the Malpighian tubules.

With the advent of genomics, microarrays and mass-spectroscopic identification of neuropeptides, our understanding of neuroendocrine control of excretion has advanced enormously. It is clear that – just as in mammals – the control of osmoregulation and excretion is one of the most subtle, multifactorial processes in the whole body, and the excretory system integrates signals from many nervous and neuroendocrine factors (Table 18.3). Understanding is particularly well developed for *Drosophila* (Fig. 18.12). In the main segment of the tubule, the

Table 18.3 **Summary of major factors acting on the Malpighian tubule**

Name	Structure	Discovered	Effects
Kinin	DPAFNSWG-NH$_2$	Myotropic effect on *Leucophaea* hindgut (Holman *et al.*, 1986)	Diuretic, act through calcium to activate chloride shunt conductance, in all species studied to date
DH$_{31}$	GLDLGLSRGFSGSQAAKHLMG LAAANYAGGP-NH$_2$	Diuretic factor in *Diploptera punctata* (Furuya *et al.*, 2000)	Diuretic, acting through cAMP to stimulate cation transport, in all species to date. Also natriuretic in *Anopheles*
DH$_{44}$	RMPSLSIDLPMSVLRQKLSLEKERK VHALRAAANRNFLNDI-NH$_2$	Diuretic factor in *Manduca sexta* (Kataoka *et al.*, 1989)	Diuretic, acting through cAMP to stimulate cation transport, in all species to date
CAP$_{2b}$	*pyro*QLYAFPRV-NH$_2$	Diuretic factor from *Manduca*, assayed in *Drosophila* (Davies *et al.*, 1995)	Diuretic in Diptera, acting through calcium to stimulate cation transport. Antidiuretic in *Rhodnius*
Tyramine	Biogenic amine OH ... NH$_2$	Diuretic factor in *Drosophila* (Blumenthal, 2003)	Acts (probably) through calcium to activate chloride shunt conductance
5HT	Biogenic amine H$_2$N ... OH ... HN	Diuretic factor in *Rhodnius prolixus* (Maddrell *et al.*, 1971)	No effect in *Drosophila*
Tenmo-ADFa	VVNTPGHAVSYHVY-COOH	Antidiuretic factor in beetle, *Tenebrio molitor* (Eigenheer *et al.*, 2002)	Acts through cGMP
Tenmo-ADFb	YDDGSYKPHIYGF-COOH	Antidiuretic factor in beetle, *Tenebrio molitor* (Eigenheer *et al.*, 2003)	Acts through cGMP. Diuretic in *Acheta*

Source: Based on Coast (2007).

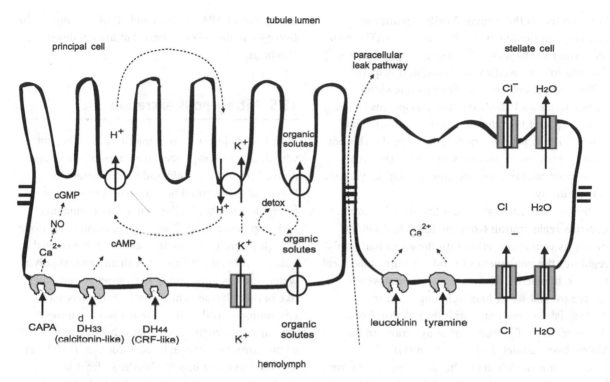

Figure 18.12 Summary of transport and signaling in the main segment of the *Drosophila* Malpighian tubule. The principal cells are packed with mitochondria, and energize active transport. The apical V-ATPase actively transports H^+ into the tubule lumen and this is exchanged for K^+ and some Na^+ through (probably separate) exchangers. K^+ enters the cell from the blood through inward rectifier K^+ channels, and also through cotransports; the net effect is an electrogenic movement of cations from hemolymph to lumen, making the lumen electrically positive. Transport in the principal cell is stimulated by CAPA, DH_{31} and DH_{44} neuropeptides, which between them elevate the second messengers cAMP, cGMP, calcium and nitric oxide, probably all acting mainly to stimulate the V-ATPase. The principal cells are also packed with broad-specificity organic solute transporters and detoxification/metabolism enzymes, and so probably provide the major route for elimination of unwanted organic solutes. The stellate cells control the transcellular chloride and water fluxes, probably through channels. The neuropeptide leucokinin and the biogenic amine tyramine both act through intracellular calcium to increase chloride conductance. The paracellular pathway provides an alternative route for flux of ions such as chloride, and of solutes that are not actively transported by the principal cells.

principal cell is acted on by DH_{31} – a neuropeptide which very distantly resembles the vertebrate hormone calcitonin – and by DH_{44}, another neuropeptide which resembles vertebrate cortocotropin releasing factor. Both act through G-protein-coupled receptors to activate adenylate cyclase, and so make the second messenger cyclic AMP, which stimulates fluid secretion. Interestingly, although both act through cAMP, their actions may

not be identical; in the mosquito *Anopheles*, DH_{31} is natriuretic (preferentially causes the excretion of sodium). Another class of peptide, the CAPA neuropeptides (originally identified as CAP_{2b} in *Manduca*), act to generate an intracellular calcium signal, which stimulates nitric oxide synthase to produce nitric oxide, and thence to activate a soluble guanylate cyclase to produce cyclic GMP. In *Drosophila*, this also acts to stimulate fluid production. As the principal cells contain the cation-pumping apparatus (the apical V-ATPase and its partner exchanger), all these neuropeptides are thought to act specifically on cation transport.

The stellate cell controls the chloride shunt conductance and also expresses aquaporins (water channels), and is under separate neurohormonal control. Kinin peptides rapidly increase the chloride conductance (in about one second) via the second messenger calcium, and the biogenic amine tyramine acts similarly.

Rhodnius is unusual in that the diuretic hormone controls reabsorption from the lower Malpighian tubule as well as secretion into the upper part. It also regulates the movement of fluid from the blood meal into the hemocoel. There are also factors which reduce overall water loss; it is important to distinguish between an antidiuretic action (reducing the formation of primary urine by action on the Malpighian tubule), and the stimulation of reabsorption (which helps the insect conserve water, but is not the same as antidiuresis). Reabsorption is regulated in most insects by one or more separate hormones. In *Schistocerca*, separate factors control reabsorption in the ileum and the rectum and this may also be true in other insects. An ion transport peptide from the corpora cardiaca stimulates chloride absorption in the ileum, although it is not so effective on the rectum, suggesting the existence of a further chloride transport stimulating hormone. ITP is a large peptide of 72 amino acids, with structural similarity to arthropod hyperglycemic, vitellogenesis-inhibiting and molt-inhibiting hormones.

True antidiuretic factors have been identified in the beetle *Tenebrio molitor*. These act through cyclic GMP to reduce fluid production by the tubule. However, Tenmo-ADFb and cGMP are diuretic in the cricket, *Acheta domesticus*. It is worth noting that although cyclic AMP and calcium both act to stimulate fluid production, cyclic GMP can be either a diuretic or antidiuretic second messenger according to species. For example, both *Manduca* CAP$_{2b}$ and endogenous CAPA peptides and cGMP are diuretic in *Drosophila* and other Diptera, but are antidiuretic in *Rhodnius*.

18.5 Nitrogenous excretion

Most insects (exceptions being those that feed on very dilute plant sap) need to excrete excess nitrogen derived from amino acid and purine metabolism. Potentially, this could be excreted as any of several molecules: uric acid, allantoin, urea or ammonia. Although it is a normal metabolite, ammonia is toxic at high levels. Consequently, ammonia is excreted in quantity primarily by insects with an ample supply of water, such as those living in freshwater and others, like blowfly larvae, which live in extremely moist environments (Table 18.4, *Lucilia* larva). However, ammonia also comprises a relatively large proportion of the nitrogen excreted by *Schistocerca*, *Periplaneta* and the larva of a bruchid beetle. At least in *Schistocerca*, it is excreted as salts, probably combined with organic anions. These salts are very insoluble, so ammonia is removed from solution. In *Periplaneta* it is likely that bacteria in the hindgut produce at least some of the ammonia. These findings suggest that the production of ammonia by terrestrial insects may be more widespread than has been supposed.

For most terrestrial insects, water conservation is essential and the loss by excretion must be minimized. Hence it has generally been argued that terrestrial insects must produce a less toxic substance than ammonia so that less water is required for its safe elimination. The substance produced by a majority of these insects is uric acid, which often appears to comprise over 80% of the total nitrogenous material excreted by these insects (Table 18.4). Urea, although present in the urine of a number of insects, is usually only a minor component; indeed, *Drosophila* lacks an enzyme to convert allantoic acid to urea. It is, however, the principal component of the nitrogenous excrement of

Table 18.4 **The distribution of nitrogen in the excreta of insects expressed as a percentage of the total nitrogen in the excreta**[a]

Insect	Order	Uric acid	Allantoin/allantoic acid	Urea	Ammonia
Terrestrial insects – solid food					
Schistocerca	Orthoptera	55	–	+	40
Periplaneta	Blattodea	–	–	–	Up to 90
Melolontha (A)	Coleoptera	100	–	–	–
Attagenus (A)	Coleoptera	25	–	35	20
Pieris (L)	Lepidoptera	95	4/1	–	–
Terrestrial insects – liquid food					
Rhodnius	Hemiptera	90	–	+	–
Dysdercus	Hemiptera	–	61/–	12	–
Pieris (A)	Lepidoptera	20	10/70	–	–
Lucilia (L)	Diptera	–	10/–	–	90
Glossina (L)	Diptera	100	–	+	–
Aedes (A)	Diptera	43	–	13	18
Freshwater insects					
Aeschna (L)	Odonata	8	–	–	74
Sialis (L)	Megaloptera	–	–	–	90

Notes:
[a] Where the total does not equal 100, other components, not shown in the table, make up the balance.
(A), adult; (L), larva; –, not present; +, present in small amounts.

a carpet beetle, *Attagenus*, and the larva of a bruchid seed beetle, *Caryedes*.

Urate is generated through the purine degradation pathway. Urate, or its precursor xanthine, is probably actively pumped into the Malpighian tubules of all insects that normally excrete uric acid. In the tsetse fly, *Glossina*, the activity of the uric acid pump is proportional to the quantity of urate in the hemolymph. Subsequently, uric acid may be precipitated in the tubule by the action of V-ATPases; in *Drosophila* mutants for V-ATPase subunits, uric acid crystals are not observed in the tubule lumen. Urate is not always the end-point of purine

metabolism: it may be further metabolized to allantoin and allantoic acid. Allantoin is produced in small amounts by many insects, but is sometimes present in larger amounts. It is the form in which most nitrogenous waste is excreted by most species of Heteroptera that have been studied (*Rhodnius* is an exception), and by the stick insect *Carausius* and a beetle, *Chrysobothris*. The larvae of *Lucilia* accumulate uric acid in their tissues, but excrete it after conversion to allantoin. Allantoic acid occurs in the excreta of larval and adult Lepidoptera. It usually constitutes less than 1% of the nitrogen excreted by these insects, but may contain as much as 25% of the nitrogen in the meconium (waste products from the pupal stage). It is also produced by larval Hymenoptera and Diptera. In adult *Calliphora*, uric acid is further converted to allantoin by the Malpighian tubules, which also actively pump it into the lumen.

Uricase (or urate oxidase), the enzyme which performs the first step in the conversion from urate to allantoin, is known to be completely tubule-specific in *Drosophila melanogaster*, suggesting that while many tissues may make xanthine, later stages in purine metabolism are performed uniquely by the tubule. The proportions of these three related compounds, uric acid, allantoin and allantoic acid, may vary with the food and stage of development. For example, larvae of the moth *Lasiocampa* produce mainly allantoic acid when fed on a grass, but predominantly uric acid when fed on birch. Larvae of the tortoiseshell butterfly, *Aglais*, produce mainly uric acid; this also predominates in the pupa; but here, appreciable amounts of allantoin and allantoic acid are also formed. In the adult, allantoic acid predominates. Uricase expression can vary even within a genus; it is expressed only in third instars and adults in *Drosophila melanogaster*, only in the adult of *D. pseudoobscura* and only in the third instar larva of *D. virilis*. The adaptive significance of this difference is not yet clear.

18.5.1 Using genomics to discover metabolic pathways

It is now possible to work out the possible pathways for nitrogenous excretion by comparing the *Drosophila* genome against known metabolic pathways, to see which enzymes are present and which are absent (Fig. 18.13). *Drosophila* has several ways of making hypoxanthine, an early compound in the purine metabolic pathway, and enzymes are abundantly expressed that can convert this to xanthine, urate, allantoin and allantoic acid. However, the genome does not encode either allantoicase or urease, which would convert allantoic acid to urea, and urea to ammonia; so such products

are unlikely to come from purine metabolism in flies. Arginase, however, is abundantly expressed, particularly in the fat body; this allows the conversion of arginine to urea. Similarly, aromatic amino acids can be catabolized (via cystathionine) to ammonia. Hence, the *Drosophila* genome encodes genes that would allow all the recognizable end-metabolites of nitrogenous metabolism to be generated.

Urate is generated from xanthine by the enzyme xanthine dehydrogenase/oxidase. In *Drosophila* this enzyme is encoded by the eye color gene, *rosy*. It is expressed widely, but mainly in the fat body and particularly the Malpighian tubules. It is likely, therefore, that most cells excrete xanthine as their waste purine, and that this is taken up by tubules and metabolized to urate within the tubules. Urate is then transported across the tubules, and acidification of the tubule lumen by the apical V-ATPase precipitates uric acid crystals; mutation of any of the genes encoding the tubule V-ATPase abolishes the formation of uric acid crystals. In *rosy* mutants, no uric acid is formed, and xanthine accumulates to such high levels in the tubule that it forms orange concretions that block the tubule lumen, causing it to swell and deform. Interestingly, these symptoms exactly recapitulate those of the corresponding human disease, xanthinuria type I.

Amino acids are often present in the feces of insects, but usually in small amounts that probably result largely from incomplete absorption. Under some circumstances, however, large quantities of amino acids are present. Thus, in the tsetse fly *Glossina*, most of the arginine and histidine from the blood of the host are excreted unchanged after absorption. These are substances with high nitrogen contents which would require a considerable expenditure of energy if they were to be metabolized along the normal pathways. *Schistocerca* actively secretes proline into the Malpighian tubules. It is subsequently reabsorbed in the rectum where it is the principal substrate for oxidative metabolism.

There is also evidence that other insects void some amino acids in large quantities when they have an

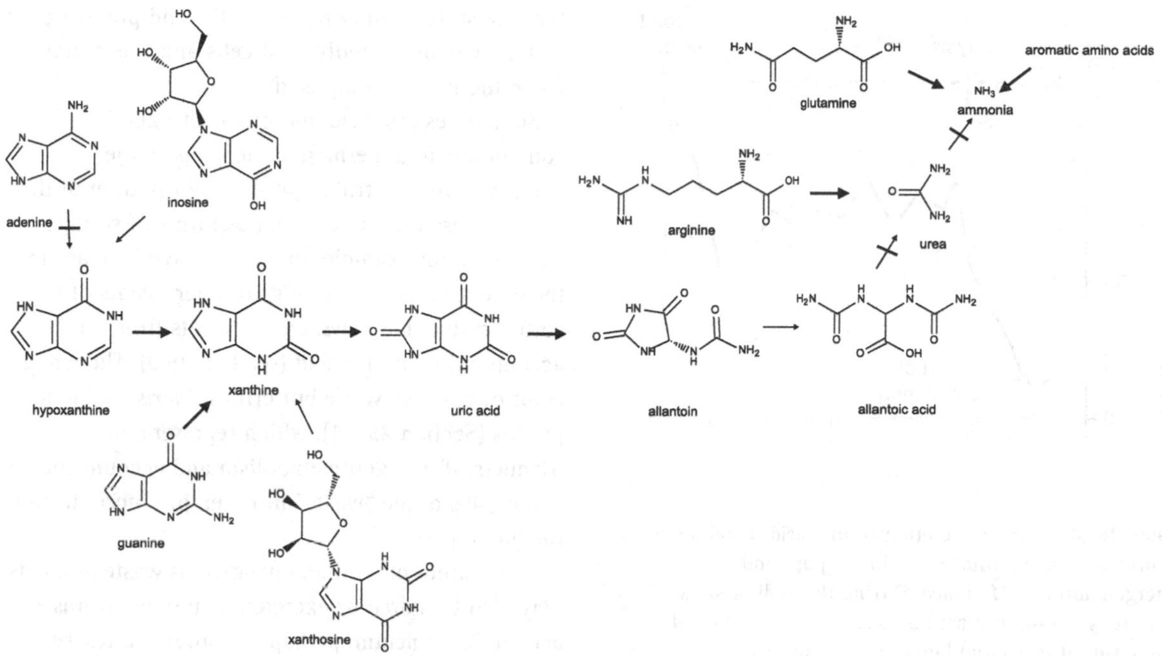

Figure 18.13 Potential pathways for nitrogen excretion. The major metabolic pathways are shown, and heavy arrows indicate the corresponding enzyme is particularly strongly expressed in *Drosophila* (FlyAtlas.org). Where an arrow is blocked, this means that the *Drosophila* genome does not encode such an enzyme, so the step is unlikely to occur in *Drosophila* or related insects.

abundance of nitrogen in the diet, but in none of these cases is it certain that these are a result of specific elimination from the hemolymph rather than differential absorption from the midgut. Ommochromes, derived from tryptophan, are certainly excreted if high levels of tryptophan occur in the hemolymph. For example, fecal pellets of locusts are colored red by ommochromes during molting or starvation. At these times, tryptophan is probably liberated from proteins metabolized during structural rearrangement or used in energy production. Ommochromes also color the meconium of adult Lepidoptera.

18.5.2 Storage excretion

Waste materials may be retained in the body in a harmless form instead of being passed out with the urine. This is known as storage or deposit excretion. Uric acid, because it is very insoluble, can be stored as a solid and is found in the fat body of many insects at some stages of development (Fig. 18.14). These deposits are not always permanent. For example, in the cockroach *Periplaneta*, uric acid accumulates in the fat body when there is ample nitrogen in the diet, but this store is depleted if the insect feeds on a diet deficient in nitrogen. Similarly, there is evidence that the shield bug, *Parastrachia japonensis*, uses uric acid as a reserve of nitrogen during diapause. The use of uric acid as a nitrogen store is probably limited to insects with symbiotic microorganisms which can synthesize amino acids from uric acid.

Urate is a powerful antioxidant, and may provide local protection for the Malpighian tubule, which is packed with mitochondria that are likely to

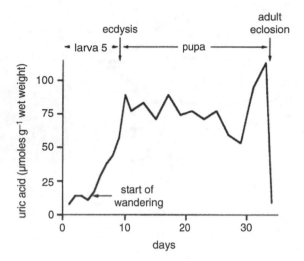

Figure 18.14 Storage excretion of uric acid. Total amounts of uric acid in the final-stage larva, pupa and newly emerged adult of *Manduca*. During the feeding stage of the larva (days 1–4), uric acid is excreted. At the start of wandering it is accumulated in the body, primarily in the fat body. Immediately following eclosion, most is excreted in the meconium (based on Levenbook *et al.*, 1971; Buckner and Caldwell, 1980).

generate high levels of potentially damaging reactive oxygen species.

Some storage of nitrogenous end products must presumably occur in all insects during embryonic development and, in holometabolous insects, in the pupal stage when normal excretion is not possible. During embryonic development of the desert locust *Schistocerca*, the uric acid content of an egg increases progressively from about 0.5 μg egg^{-1} to 25 μg egg^{-1}. At first it is present in the yolk, but when the fat body develops it contains much of the uric acid. Uric acid also builds up in the fat body of larval *Manduca* when it stops excreting it before pupation (Fig. 18.14). Then, in the days before eclosion, when morphogenesis is almost complete, the accumulation of uric acid increases sharply. Most of it is excreted with the meconium immediately following eclosion. Uric acid accumulates in epidermal cells of caterpillars and of *Rhodnius* during molting. In these cases it is possible

that the stores simply represent the end product of metabolism of the individual cells and it is removed when the molt is completed.

Sometimes uric acid and other nitrogenous compounds form permanent deposits in the epidermis and contribute to the color pattern of the insect. These can be regarded as forms of storage excretion. For example, the progressive increase in the extent of white markings in later instars of the cotton stainer bug, *Dysdercus*, results from an accumulation of uric acid (see Fig. 25.9). The white color of cabbage white butterflies, *Pieris*, is due to pterins (Section 25.3.1), which represent end products of nitrogen metabolism and account for about 14% of the "waste" nitrogen produced during the pupal period.

Compounds other than nitrogenous waste products may also be stored or excreted. A number of insects accumulate calcium, perhaps to provide a reserve. In *Drosophila*, all regions of the tubule, especially the distal segment of the anterior tubule, can excrete calcium; 85% is sequestered. Calcium is stored as intracellular or luminal concretions, or spherites; in *Drosophila* these may be derived from peroxisomes, and can pack the lumen of the initial (distal) segment of the anterior Malpighian tubules. Vertebrate blood contains a significant amount of calcium, which *Rhodnius* stores in the cells of the upper Malpighian tubules. The accumulation increases over successive stages of development. Xylem also contains relatively high concentrations of calcium and magnesium, and xylem-feeding cicadas and cercopids (Homoptera) accumulate them in part of the tubular region of the midgut. Sometimes calcium is stored in specialized regions of the Malpighian tubules. For example, larvae of the fly *Ephydra* living in water containing high levels of calcium salts store calcium in enlarged Malpighian tubules; in the stick insect *Carausius* the terminal regions of the inferior tubules are expanded to store it. The common theme is that the distal (or initial) segments of insect tubules appear specialized for calcium storage excretion.

Other species also use calcium temporarily stored in the Malpighian tubules for specific functions. For example, in the face fly, *Musca autumnalis*, and other Cyclorrhapha it contributes to hardening of the puparium; in the stick insect *Carausius* it is later taken into the hemolymph and deposited in the chorion of the eggs. In other cases, as in *Acheta* (Orthoptera) and *Calpodes* (Lepidoptera), the store probably provides a metabolic reserve.

Heavy metals accumulate in midgut cells of lepidopteran caterpillars and cercopids (Hemiptera). Cadmium is sequestered by the posterior midgut cells of larvae of the midge *Chironomus*. Metallothioneins bind heavy metals; in *Drosophila* metallothionein genes are abundantly expressed in midgut and tubules, and are further up-regulated on exposure to copper, cadmium or zinc. Copper accumulation is particularly pronounced in a region of the midgut that contains goblet-shaped, acid-secreting cuprophilic cells. The pathway for response to heavy metals seems similar to other organisms; exposure increases expression of the metal-responsive transcription factor MTF1, which then increases expression of metallothioneins and other genes. Flies mutant for MTF1 survive well under laboratory conditions, but are sensitive to heavy metals in the diet; a similar result is seen when the four *Drosophila* metallothionein genes are knocked out in combination.

18.6 Detoxification

Insects can be exposed to toxins in their diet, or through the application of xenobiotics, such as insecticides. In a majority of insects, toxic chemicals that enter the tissues are excreted, often after being metabolized. The changes frequently result in the compound becoming less toxic, but the converse is sometimes true and toxicity is enhanced. Some compounds, notably those that are water soluble, are metabolized to components that are subsequently incorporated into the insect's primary metabolic

pathways. This may be true of some non-protein amino acids, for example. Most lipophilic substances, however, are first converted to a water-soluble product and then excreted or sequestered. These steps are similar to the three phases of drug metabolism in humans. Phase I (or non-synthetic modification) typically involves oxidation, reduction or hydrolysis by drug-metabolizing enzymes (DMEs) such as cytochrome P450s, alcohol dehydrogenase, esterase or monoamine oxidase. If the resulting molecule is sufficiently polar, it can be directly excreted; if not, it can be modified by phase II (conjugating) DMEs. These enzymes are typically transferases, such as methyltransferase, glutathione-S-transferase, sulfotransferase, N-acetyltransferase or UDP-glucuronosyltransferase. The actual transport (excretion) of the modified molecule is sometimes termed phase III, and is mediated by broad-specificity organic solute transporters, such as the multi-drug resistance (MDR) and multi-drug resistance-associated protein (MRP) families of ABC transporters.

Many different enzyme systems are known to be involved in these reactions and some systems are almost certainly ubiquitous. The best known is the system of polysubstrate monooxygenases (also called mixed-function oxidases). The terminal component of this system is cytochrome P450, so called because it absorbs light maximally at around 450 nm when complexed with carbon monoxide. Cytochrome P450 combines with the substrate (which may be a toxin) and with molecular oxygen, catalyzing the oxidation of the substrate. This system is important because the cytochrome will combine with many different lipophilic substrates and, in addition, exists in different forms, called isoenzymes, which vary in their substrate specificity. The system is, therefore, able to metabolize a wide range of substances.

The activity of the polysubstrate monooxygenase system toward a particular substrate is usually low or may not be detectable at all if the insect has not previously encountered the substrate. Exposure of the system to an appropriate substrate, however, induces an increase in cytochrome P450, leading to

Table 18.5 **Induction of cytochrome P450 by a single compound results in increased metabolism of many others**

Substrate (type of compound)	Without toxin in food	Amount	Percentage increase
Indole 3-carbinol[a,b]	0.46	1.80	391
α-pinene (monoterpene)	3.18	16.56	521
Farnesol (sesquiterpene)	3.70	34.96	945
Phytol (diterpene)	3.70	29.97	810
β-carotene (carotenoid)	0	1.97	>1000
Caffeine (alkaloid)	0.47	1.22	256
Sinigrin (glucosinolate)	0.30	0.70	233
Digitoxin (cardiac glycoside)	0	0.70	>500

Notes:
[a] Larvae of the fall armyworm, *Spodoptera frugiperda*, were reared on an artificial diet without or with a potentially toxic compound, indole 3-carbinol. The activity of the monooxygenase system in the midguts of these insects was measured against various potential toxins. Activity is measured as mmol NADPH oxidized min^{-1} mg^{-1} protein (Yu, 1987).
[b] Compound on which the insects feed.

more rapid metabolism of the substrate on subsequent encounters. Because of the broad specificity of the system, it becomes active against a range of substrates, not just that which caused the induction (Table 18.5). Induction is transient, and persists only as long as an inducing agent is present. Some of the other enzyme systems involved in metabolizing toxins are also inducible.

These processes may occur in a variety of tissues as there is no organ comparable with the liver, which is the focus for comparable reactions in vertebrates. Activity of the appropriate enzymes often occurs in the midgut, fat body and Malpighian tubules, with the highest level of activity often in the midgut. Where does detoxification occur? In mammals, most tissues have some detoxification capability, but the liver is the main organ. The FlyAtlas.org microarray resource allows the tissue distribution of phase I and phase II DMEs to be mapped unambiguously (Fig. 18.15). Most of the very large P450 and glutathione-S-transferase (GST) gene families are expressed in one or a few tissues, most commonly the Malpighian tubules, midgut, hindgut or fat body.

Conspicuously, the tubule is highly enriched in P450s that have been implicated in xenobiotic resistance, confirming its importance as a metabolic, as well as an excretory, organ.

Different species differ widely in their ability to metabolize toxic compounds. Among plant-feeding insects, this variation contributes to host plant specificity. The caterpillar of *Manduca sexta*, for example, habitually feeds on alkaloid-containing plants, including tobacco, and it is able to do this because it detoxifies the alkaloids. Nicotine from tobacco is taken into the midgut cells, where most of it is metabolized to cotinine-*N*-oxide by a microsomal oxidase (Fig. 18.16). The cotinine-*N*-oxide enters the hemolymph and passes into the urine in the Malpighian tubules, which have a non-specific alkaloid pump. Any nicotine that enters the hemolymph is also rapidly oxidized. Conversely, the Malpighian tubules of *Drosophila* excrete the cardiac glycoside ouabain (an inhibitor of the Na^+/K^+ ATPase) unchanged, by active transport mediated by a member of the organic anion transport peptide (OATP) family, thus making the tubule resistant to this transport inhibitor.

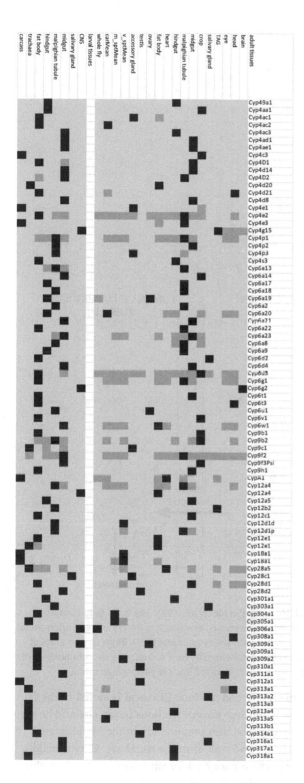

Although the end products of these metabolic processes are commonly excreted, they are sometimes sequestered. An alternative approach is sequestration of a compound within the insect (storage excretion). Sequestration may also occur without any prior metabolism of toxic compounds and can provide protection against predation. In some species the compound is stored in the cuticle, perhaps minimizing the risk to the insect itself, but other species store the defensive substances in glands or in the hemolymph. There is little understanding of how these insects avoid autotoxicity, but in the milkweed bug, *Oncopeltus*, and the monarch butterfly, *Danaus*, which sequester cardiac glycosides, the sodium/potassium pumps responsible for maintaining the ionic environment of the nervous system are relatively insensitive to ouabain (a cardiac glycoside). The pumps of insects which do not sequester cardiac glycosides are much more sensitive.

18.7 Non-excretory functions of the Malpighian tubules

18.7.1 Specialized secretions and light production

In a few insect species the Malpighian tubules are modified for silk production. The tubules of larval lacewings, *Chrysopa* (Neuroptera), are thickened

Figure 18.15 Tissue specificity of the cytochrome P450 family in 18 adult and 9 larval tissues of *Drosophila*, based on the FlyAtlas.org online resource (Chintapalli *et al.*, 2007). For each gene (rows) abundant expression is coded dark gray, and the tissue with the maximal expression for the gene in each life-stage is coded black. This "heatmap" shows (1) most P450 genes are abundantly expressed in only a few tissues – relatively few genes show dark gray rows; (2) the midgut and Malpighian tubule are major sites of P450 expression in the adult; (3) epithelia (midgut, tubule, hindgut) and fat body are the major sites in larvae – as evidenced by the columns enriched with black squares.

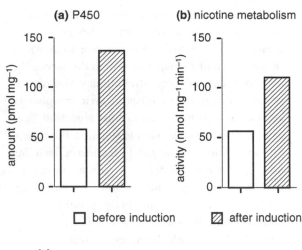

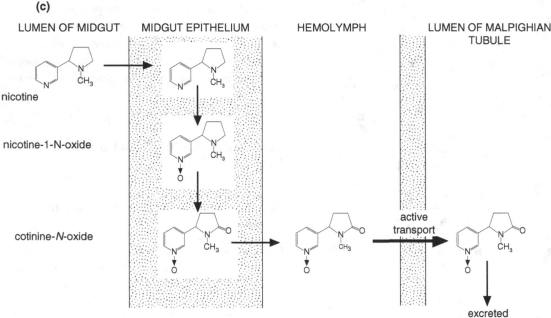

Figure 18.16 Metabolism of a toxic compound (nicotine) by larval *Manduca* (Snyder *et al.*, 1993, 1994). (a) Induction of cytochrome P450 by feeding on an artificial diet containing 0.75% nicotine. Expressed as amount of P450 per mg of microsomal protein in the midgut. (b) Rate of metabolism of nicotine in relation to induction of cytochrome P450. Expressed as nmol of product formed per mg of microsomal protein in the midgut. (c) Nicotine is metabolized by the P450 enzyme to cotinine-*N*-oxide in the midgut cells. Cotinine-*N*-oxide enters the hemolymph and is actively transported into the Malpighian tubules.

distally and the nuclei of the cells become branched after the second stadium. These thickened regions produce the silk of the pupal cocoon. During larval life, the tubules produce a proteinaceous substance that acts as an adhesive on the anal sucker during locomotion and may, at the same time, be an excretory end product. Uric acid is stored in the fat body of these insects. Antlions (myrmeleontid larvae, Neuroptera) also produce silk in the Malpighian tubules and store it in the rectal sac. Chrysomelid beetles produce a sticky substance in the Malpighian tubules for covering the eggs.

Among spittle bugs (Cercopidae, Homoptera) the proximal region of the larval tubules is enlarged, consisting of large cells without microvilli. These cells produce the material which, when mixed with air, forms the spittle within which the larvae live. Some other cercopids build rigid tubes, that of *Chaetophyes compacta* being conical and attached to the stem of the host plant. The proximal part of each Malpighian tubule in this insect is divided into two zones, one that produces the fibrils forming the basis of the tube, and another, more distal zone, that produces spittle. Fibrils pass out through the anus and are laid down by characteristic semicircular movements of the tip of the abdomen accompanied by radial pushes from the inside that push the tube into its polygonal form. Other organic materials and calcium and magnesium from the Malpighian tubules are deposited on the meshwork of fibers to form the hardened tube.

In the larva of the fly *Arachnocampa luminosa* the enlarged distal ends of the Malpighian tubules form luminous organs (Section 25.7).

18.7.2 Immunity

Traditionally, the fat body was considered to be the cardinal immune tissue in an insect, but it is clear that most of the barrier epithelia possess at least some innate immune capability, termed "epithelial immunity." It is perhaps not surprising that the tubules are key immune organs, since they ramify throughout the hemocoel, and so are uniquely positioned to monitor the whole organism's health. This provides some defense against pathogens. In *Drosophila*, excised tubules exposed to bacteria are able to up-regulate expression and secretion of the anti-microbial peptide Diptericin, and so kill bacteria completely independently of other tissues, including the fat body. Furthermore, overexpression of the gene for nitric oxide synthase in just the tubules protected the whole fly against infection, implying that nitric oxide is an important immune signal. Consistent with this, nitric oxide synthase is up-regulated in tubules of mosquitoes after a malaria-infected blood meal. The response in tubules is similar to that described for the fat body: Peptidoglycans of Gram-negative bacteria are detected through the PGRP receptors on the tubule, signaling through Imd, Kenny and Relish, and eliciting up-regulated expression of anti-microbial peptides, notably Diptericin.

18.8 Nephrocytes

Nephrocytes are cells in the hemocoel that take up foreign chemicals of relatively high molecular weight, with which the Malpighian tubules may be incapable of dealing. They take up dyes and colloidal particles – but not bacteria or larger foreign objects, which are targets of macrophages, a class of hemocyte. Some nephrocytes are usually present on the surface of the heart (where they are known as pericardial cells); others lie on the pericardial septum or the alary muscles. In larval Odonata they are scattered throughout the fat body, and in the louse *Pediculus* they additionally form a group on either side of the esophagus. Nephrocytes of caterpillars form a sheet, one cell thick, in the hemocoel, on either side of the body. The sheet is closely associated with the fat body and extends through the thorax and abdomen. In larval Cyclorrhapha (Diptera) they form a conspicuous chain, sometimes called a garland, running between the salivary glands (Fig. 18.17). In addition, larval *Calliphora* have 12–14 large pericardial cells, 140–200 μm in diameter, on each side of the posterior part of the heart, and hundreds of smaller pericardial cells, 25–60 μm in diameter, around the anterior heart.

Nephrocytes are characterized by deep, labyrinthine infoldings of their plasma membrane, with the external surface of the folds held together by desmosome-like structures, so that the outer layer of the cell is more or less smooth (Fig. 18.17). The cells are surrounded and held together by a basal lamina. Coated vesicles arising from the invaginations become associated with an intracellular system of tubular elements. Different types of vesicles, including lysosomes, lie within this zone, while a perinuclear zone contains abundant rough endoplasmic reticulum and Golgi. Nephrocytes usually have one or two nuclei.

The basal lamina and membrane junctions at the outer ends of the invaginations limit the size of molecules that can enter the labyrinth, but it is known that the nephrocytes of *Rhodnius* can absorb hemoglobin. Molecules are taken into the cell by pinocytosis and, following degradation, may be returned to the hemolymph. In addition, the nephrocytes release lysozyme, which is important in resistance to disease.

The basal infoldings of *Drosophila* nephrocytes are guarded by proteins (nephrin and podocin) that are characteristic of the "slit diaphragm" of vertebrate podocytes, the cells which regulate the filtration of blood in Bowman's capsule of the kidney. This has led to the intriguing suggestion that insects may represent an intermediate stage in the evolution of the vertebrate glomerular kidney; tubules, podocytes and a simple vascular system coexist in insects, but have yet to come together in a rudimentary glomerulus. However, it must be recalled that insects are not "primitive" or "ancestral": their excretory systems are extremely potent and are key determinants of insect success, so a glomerular kidney is simply not necessary in insects.

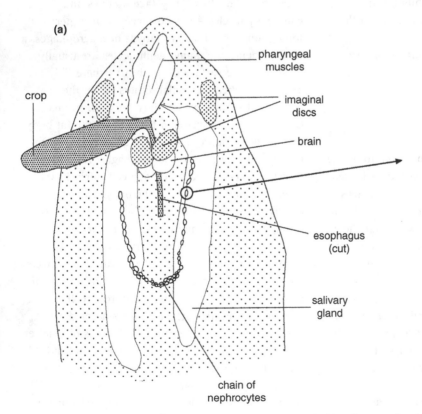

Figure 18.17 Nephrocytes. (a) Dissection of the anterior part of a blowfly larva showing the chain of nephrocytes known as a garland. (b) Nephrocyte from the garland of *Drosophila* larva. α-vacuoles are formed from the amalgamation of pinocytotic

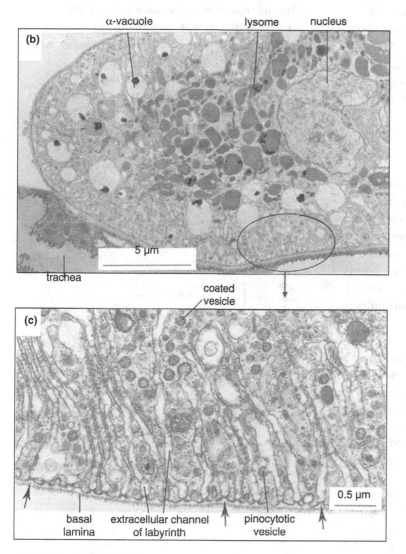

α-vacuole   lysome   nucleus

(b)

5 μm

trachea

coated
vesicle

(c)

basal   extracellular channel   pinocytotic
lamina   of labyrinth   vesicle

0.5 μm

Caption for Figure 18.17 (*cont.*) vesicles formed in the labyrinth (after Koenig and Ikeda, 1990). (c) Cortical region of a nephrocyte of *Drosophila* larva showing the labyrinth. Arrows show the desmosome-like connections between adjacent projections of the cell (Koenig and Ikeda, 1990).

18.9 Water regulation

The water content of different insects commonly ranges from about 60% to 80% of the body weight, although values as low as 40% are recorded. As these values include the cuticle, which has relatively low water content, the water content of the living tissues is higher than these figures suggest. The degree of tolerance to water loss varies from species to species, but, ultimately, loss of water leads to death. For example, among the beetles, *Carabus* dies if it loses more than about 20% of its body water; the tenebrionid *Phrynocolus* can tolerate up to 50% loss; and the rice weevil, *Sitophilus*, over 80%.

The problems faced by insects in regulating water content vary according to their habitat and so terrestrial, freshwater and saltwater insects will be considered separately.

18.9.1 Water loss

Terrestrial insects lose water by evaporation from the general body surface and the respiratory surfaces, as well as in the urine. If they are to survive, these losses must be kept to a minimum and must be offset by water gained from other sources. Most water is gained from the food, but many insects drink if they are deprived of food and water is available. Metabolic water always contributes to the available water.

Water loss through the cuticle Cuticular permeability varies greatly between insects, depending on their habitat and lifestyle. Thus the desert cicada, *Diceroprocta*, that feeds on xylem and so has ample water available to it, loses 100 μg cm^{-2} mm Hg^{-1} h^{-1} through the cuticle (at 30°C in dry air), while in the beetle *Eleodes* from the same habitat the rate of cuticular loss is 17 μg cm^{-2} mm Hg^{-1} h^{-1}, and in *Onymacris*, a beetle from the very arid Namib desert, it is 0.75 μg cm^{-2} mm Hg^{-1} h^{-1}. Among grasshoppers the rate of water loss through the cuticle varies from 15 μg cm^{-2} mm Hg^{-1} h^{-1} in a desert species (*Trimerotropis*) to 67 μg cm^{-2} mm Hg^{-1} h^{-1} in an alpine species (*Aeropedellus*). Very little water moves through the puparial cuticle of the tsetse fly, *Glossina* (0.3 μg cm^{-2} mm Hg^{-1} h^{-1}).

The rate of evaporation of water from insect cuticle varies with the temperature and humidity of the adjacent air (Fig. 18.18). In many species, evaporation through the cuticle shows very little increase with temperature up to a certain point but, above this, the rate of water loss increases sharply although there is probably not usually a

distinct transition temperature at which the change occurs, as was once commonly believed. The increase in the rate of water loss usually occurs at temperatures well above the normal environmental temperatures that the insect is likely to meet. It may be due to melting of the wax on the surface of the cuticle. The change produced by high temperature is permanent. Water loss from the cuticle is limited by high humidity of the adjacent air.

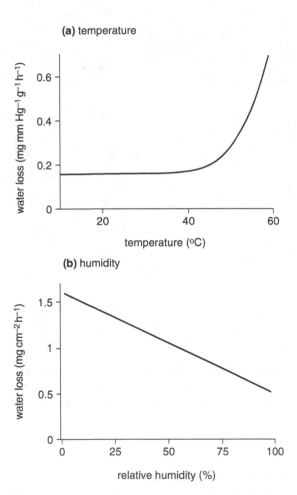

Figure 18.18 Water loss through the cuticle. (a) Effects of temperature on water loss from *Locusta* (after Loveridge, 1968). (b) Effects of relative humidity on water loss from *Periplaneta* (after Appel *et al.*, 1986).

Water loss through the cuticle is restricted by the epicuticular lipids of the epicuticle and primarily by the outer layer of wax. The precise relationship between permeability and the wax is not understood. Earlier studies suggested that a monolayer of lipid molecules was primarily responsible for waterproofing the cuticle, but there is no direct evidence for this. On the contrary, insects with more impermeable cuticles generally have greater quantities of wax per unit area than species with more permeable cuticles, suggesting that the thickness of the wax is important. This is supported by evidence of changes within a species in relation to the need to resist desiccation. For example, diapause pupae of *Manduca* have greater quantities of cuticular wax than non-diapause pupae (see Fig. 15.43) and there is three times as much wax on the surface of the desert beetle *Eleodes* in summer, when the beetle is exposed to high temperatures, than in winter. If winter beetles are kept at high temperatures, however, they produce more wax.

The specific composition of the wax is probably also important. Lipids with longer carbon chains melt at higher temperatures than those with shorter chain lengths. In *Drosophila*, insects exposed to higher temperature during the pupal period have relatively more long-chain alkadienes and methyl-branched alkanes than insects from pupae kept at lower temperatures. They also lose water less rapidly (Fig. 18.19). Individuals of the grasshopper *Melanoplus sanguinipes* reared at higher temperatures have a higher proportion of *n*-alkanes to methyl-branched alkanes of similar chain length than siblings reared at lower temperatures; *n*-alkanes melt at higher temperatures. In this case, rearing temperature has no effect on the chain lengths.

The ducts of dermal glands and pore canals appear to be the main channels via which water moves through the cuticle and the rate of water loss is positively correlated with the number of dermal glands associated with the cuticle. Loss of water by this route may be hormonally regulated, perhaps by occluding and opening the channels.

Water loss from the respiratory surfaces Respiratory surfaces, being permeable, are a potential source of water loss. The loss from this source is reduced by the invagination of the respiratory surfaces as tracheae and is further limited by the spiracles. When the spiracles are open, the rate of water loss is increased (Fig. 18.20, and see Fig. 17.17). Consequently, conditions which result in more prolonged or more frequent spiracle opening result in increased water loss. High metabolic rates

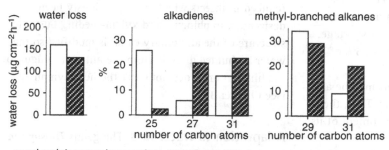

pupal maintenance temperature ☐17°C ▨24°C

Figure 18.19 Effects of epicuticular wax on water loss through the cuticle of adult *Drosophila*. Larvae were reared at 17°C and pupae maintained at either 17°C or 24°C. Reduced water loss in the insects from the higher temperature is associated with higher proportions of longer-chain (number of carbon atoms) components of the cuticular surface wax. Amounts of cuticular components are expressed as percentages of the total epicuticular wax. Only the major constituents are shown (data from Toolson, 1982).

due to increased temperature or activity necessitate longer overall periods of spiracle opening to permit the exchange of respiratory gases; they also lead to increased water loss (Fig. 18.20). Spiracle opening becomes especially important during flight when some spiracles may be continuously open. In *Schistocerca* the rate of water loss at 30°C and 60% relative humidity increases from 0.9 mg g^{-1} h^{-1} when the insect is at rest to 8.0 mg g^{-1} h^{-1} with the insect in tethered flight. Loss through the spiracles accounts for a large part of this increase, although not all of it. Discontinuous ventilation has been considered an adaptation to limit water loss, although not all recent studies support this thesis (Section 17.5.3); an alternative model is that discontinuous ventilation serves to match oxygen supply to demand.

Water loss in excretion Nitrogenous excretion necessitates the production of a fluid urine and, although some reabsorption of water may occur as the urine moves toward the anus, there is always some water loss. The insect can control its water loss by regulating the relative rates of urine production and reabsorption. This is demonstrated by differences in the rates of fluid excretion occurring at different times during an insect's development. A locust feeding on wet food produces feces with a water content of up to 90%, but this is reduced to 25% when the insect feeds on dry food in a dry habitat. When experiencing some degree of water shortage, the insect probably reabsorbs as much as 95% of the water in the primary urine.

Insects with cryptonephridial systems may be able to reabsorb water even more effectively. Thus, the larva of *Tenebrio* can reduce the water content of its feces to 15%.

Fluid-feeding insects need to eliminate excess fluid rapidly. In discontinuous feeders, such as *Rhodnius*, this may be achieved physiologically (Section 18.4). Even here, the insect can adjust the loss to its current needs. For example, the percentage of a blood meal eliminated by *Glossina* just after feeding is

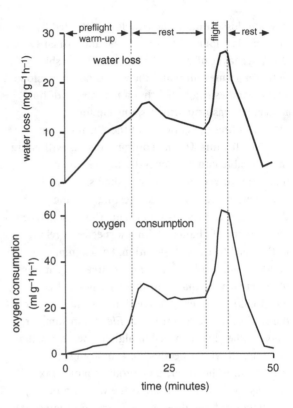

Figure 18.20 Evaporational water loss. Changes in water loss during preflight warm-up and flight of a carpenter bee, *Xylocopa*. The increases are mainly due to increased loss through the spiracles as oxygen consumption increases when the insect increases its metabolic rate (after Nicolson and Louw, 1982).

reduced in insects which have previously been desiccated. In phloem- and xylem-feeding bugs, the structure of the alimentary canal is modified so that water is shunted directly from the anterior midgut to the hindgut and does not enter the hemolymph (see Chapter 3).

Adaptation to water loss The genus *Drosophila* has species which range from cool temperate to desert cactophiles, and so is a good example for studies of adaptation to desiccation. Some of the adaptation is behavioral; *Drosophila* exposed to higher temperatures rest for longer in the middle of the day. Subtle physiological adaptations are also possible. A laboratory population resistant to adult desiccation was selected over many generations. Before desiccation, the resistant adults had slightly lower hemolymph osmotic pressure. Both wild-type and resistant flies osmoregulated tightly, but the resistant flies had greater reserves, and so could survive longer. Although the selection pressure was applied to adults, the key trait to evolve was a longer larval development period, allowing larvae to grow bigger and adults to emerge with higher water reserves.

18.9.2 Gain of water

To offset the inevitable loss by transpiration and excretion, water must be obtained from other sources. Most insects normally obtain sufficient water with their food. Locusts and some other insects adjust their food intake in relation to its water content and their own water balance; there is evidence that, if the moisture content of the food is very low, some insects consume more food than needed in order to extract the water from it. For example, larvae of the flour moth, *Ephestia*, and the carpet beetle, *Dermestes*, eat more food at low humidities, but it is clear that the bulk of their water is obtained as a result of the metabolism of this food rather than directly from its original water content (see Fig. 18.23). Some other insects, however, appear to maintain a relatively constant nutrient intake irrespective of its water content. Caterpillars, for example, eat more artificial diet when it is diluted with water, regulating their water content by changes in urine production.

Drinking Many insects drink water if they are dehydrated in the laboratory and it is assumed that they also do so under natural conditions. For example, locusts without food for 24 hours, or with access only to dry food, drink if they encounter free water, but turn away from it if they have had access to moist food. Cockroaches, termites, beetles, flies,

Lepidoptera and Hymenoptera are also known to drink.

Some tenebrionid beetles living in the Namib desert obtain water by drinking water that condenses from fogs. They may simply take condensate from any surface, but some species exhibit behavioral adaptations for collecting condensate. *Onymacris unguicularis* is normally diurnal, but during nocturnal fogs the insects adopt a head-down posture on the tops of sand dunes. Water condenses on the body and trickles down to the mouth, so that insects may increase their weight by as much as 12% overnight. Another species, *Lepidochora discoidalis*, constructs ridges of sand to trap moisture that it subsequently extracts.

The physiological control of drinking varies in different insect species. In adult *Phormia* (Diptera) and the locust *Locusta*, drinking is initiated by reduced blood volume, presumably measured by stretch receptors. However, in the fly *Lucilia*, the amount of water imbibed is correlated with the concentration of chloride ions in the hemolymph. At least in *Locusta*, the termination of drinking is regulated independently of hemolymph volume, and is related to the reduction in hemolymph osmotic pressure following water intake.

Uptake through the cuticle All terrestrial insects passively absorb some water from the air as a result of water molecules striking the surface of the cuticle. The rate of uptake increases with humidity (reflecting the number of water molecules in the air) and the permeability of the cuticle. In most cases, the rate of uptake by this route does not balance water loss, but, in some species under some circumstances, there may be a net gain of water.

Some insects and related hexapods have special structures concerned with the absorption of water. The larva of *Epistrophe* (Diptera) can evert an anal papilla into a drop of water and absorb it. In Collembola the ventral tube and in *Campodea* (Diplura) the eversible vesicles on the abdomen have this function.

A number of insect species actively take up water vapor from the air if they are short of water and the air humidity exceeds a critical equilibrium level (Table 18.6). Most of these insects are larval forms living in environments where water is in short supply, such as deserts, or environments created by vertebrates, but the phenomenon is also exhibited by a wide range of Psocoptera – including winged adults – that live in more moist environments. For most species studied, the critical equilibrium humidity is above 70% relative humidity, but Thysanura and several species of biting lice can take up water even when humidity is less than 50%.

Water uptake in these insects either occurs in the rectum or through a modified structure on the mouthparts. In tenebrionid beetle larvae, uptake involves the close apposition of the Malpighian tubules to the rectum, a cryptonephridial arrangement, while in Thysanura and the larval rat flea, although the rectum is modified and is involved in water uptake, it is not closely associated with the Malpighian tubules. In both these cases, air enters the rectum via the open anus. Water uptake in the desert cockroach, *Arenivaga*, and in the bark lice and biting lice, occurs via modifications of the hypopharynx which is extended so that the structures are exposed between the other mouthparts during periods of water uptake.

Rectal uptake is most fully understood in tenebrionid beetles. Here the distal ends of the Malpighian tubules are in close contact with the rectum, enclosed by layers of flattened cells that form the perinephric sheath (Fig. 18.21). It is assumed that the sheath is impermeable, except perhaps at specific points (see below), but anteriorly it is not tightly bound to the rectal wall and hemolymph probably moves through the gap between the rectum and sheath into the perinephric space immediately surrounding the tubules. At intervals, the Malpighian tubules are swollen into flattened chambers, called boursouflures. These are so numerous in larval *Onymacris* that they cover the whole surface of the rectum; in larval *Tenebrio* they are less extensive. Each boursouflure has a cell, called a leptophragma cell, which is closely apposed to a thin region of the perinephric sheath. At these points the sheath may be permeable.

Water uptake from the rectum results from the development of a very high osmotic pressure in the Malpighian tubules in the perinephric space. Potassium, sodium and hydrogen ions are all actively

Table 18.6 **Uptake of water from water vapor. Critical equivalent humidities above which some insects can absorb water**

Species	Order	Habitat	Equivalent humidity (%)	Temperature (°C)
Thermobia	Thysanura	Domestic	47	30
Arenivaga	Blattodea	Desert	80	30
Elipsocus	Psocoptera	Bark	76	20
Stenopsocus	Psocoptera	Leaves	76	20
Columbicola	Phthiraptera	Feathers	43	30
Xenopsylla (larva)	Siphonaptera	Rat nest	65	25
Tenebrio (larva)	Coleoptera	Stored products	88	21
Onymacris (larva)	Coleoptera	Desert	82	27

(a)

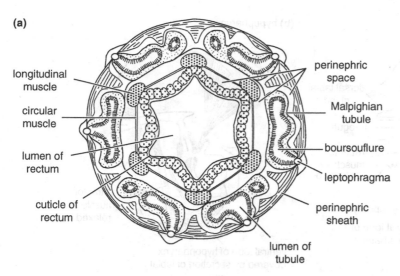

(b)

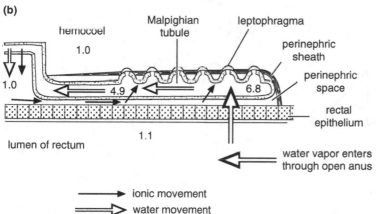

→ ionic movement
⇒ water movement

Figure 18.21 Absorption of water vapor from the rectum of a beetle larva (*Tenebrio*). (a) Transverse section of the cryptonephridial complex (after O'Donnell and Machin, 1991). (b) Diagrammatic representation of ionic and water movements. It is assumed that ions are pumped into the Malpighian tubules from the hemolymph, but not via the leptophragmata, whose function is unknown. Numbers show the osmotic pressure in osmol kg^{-1}. Note the very high levels in the Malpighian tubule (Machin, 1983; O'Donnell and Machin, 1991).

transported into the tubules; chloride ions follow passively. It is believed that the solution may become supersaturated with potassium chloride. The osmotic pressure is sufficiently high that water molecules are drawn from the air in the rectum into the perinephric space and then into the tubules (Fig. 18.21b). Water and ions are presumably reabsorbed into the general body cavity in more proximal parts of the tubules.

In caterpillars, which also have a cryptonephridial system, although they are not known to absorb water vapor, the Malpighian tubules pass beneath the muscle layers of the rectum and then double back on themselves to form a more convoluted outer layer. In these insects the cryptonephridial system

may be primarily important in ionic regulation, although it may also help to maintain the high blood volume needed for its function as a hydrostatic skeleton.

In the bark lice (Psocoptera), the hypopharynx bears a pair of bladders which are normally concealed by the labrum and labium. During periods of water uptake, however, these bladders are exposed (Fig. 18.22) and are covered with a hygroscopic fluid believed to be secreted by labial glands. At humidities above the critical equilibrium humidity, water vapor condenses onto the fluid covering the bladders and the condensate is pumped into the gut by a cibarial pump. In the desert cockroach,

(a) hypopharynx concealed

(b) hypopharynx exposed

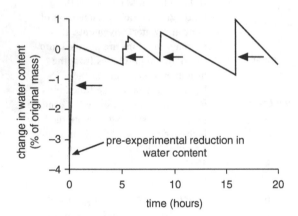

pump muscle

cibarial pump

cibarial cavity

duct through hypopharynx

labrum

labium

dorsal labial gland

foregut

retractor muscle – contracted

hypopharynx

ventral lobe of hypopharynx

retractor muscle – relaxed

ventral lobe of hypopharynx – covered by secretion of labial gland

(c) water content

pre-experimental reduction in water content

Figure 18.22 Absorption of water vapor via the hypopharynx. (a),(b) Diagrammatic sagittal sections through the head of a bark louse showing the hypopharynx concealed (a) and exposed (b). In the latter position the distal lobes of the hypopharynx are covered by a hygroscopic secretion (stippling in (b)) into which water vapor condenses. The condensed water is drawn through a duct in the hypopharynx into the gut by the cibarial pump. Arrows show the direction of water movement (after Rudolph and Knülle, 1982). (c) Regulation of body water in a biting louse. Immediately before the experiment, the louse was exposed to drying conditions which reduced its water content by 3.5%. At time 0 it was exposed to 91% relative humidity at 30°C. The critical equilibrium relative humidity for this species is 52%. The insect absorbed water when the hypopharyngeal bladders were exposed (arrows, comparable with (b)), but slowly lost water when they were concealed (comparable with (a)). Notice that the insect was able to regulate within ±1% of its normal water content (Rudolph, 1983).

Arenivaga, the cuticle of the hypopharyngeal bladders is covered by a dense mat of hydrophilic hairs. Water vapor condenses onto the hairs, which swell as they take up water. Water is released from the cuticle by a solution secreted onto its surface

from a gland on the epipharyngeal face of the labrum.

The activities associated with water uptake – such as opening the anus; exposing the hypopharyngeal bladders; muscular pumping activity – only occur at humidities above the critical equilibrium humidity if the insect's water level is suboptimal. Thus the insect responds both to ambient humidity, presumably measured by humidity receptors (Section 19.8), and its own water content, perhaps measured volumetrically and involving stretch receptors. As a consequence, it is able to maintain its water content within narrow limits (Fig. 18.22c).

Metabolic water Water is an end product of oxidative metabolism and the water so produced contributes to the water content of the insect; some species are dependent on this water for survival. The amount of metabolic water produced depends on the amount and nature of the food utilized. The complete combustion of fat leads to the production of a greater weight of water than the weight of fat from which it is derived (100 g of palmitic acid produces 112 g of water; 100 g of glycogen only 56 g of water). However, metabolism is tied to energy demands of the insect and the release of one kilocalorie leads to the production of less metabolic water if fat is oxidized to meet these demands than if carbohydrate is the substrate. Thus the insect does not gain more water from the combustion of fat unless it also expends more energy. This may be of significance during flight, but perhaps not in other circumstances.

The larvae of *Tribolium* (Coleoptera) and *Ephestia* (Lepidoptera) normally obtain much of their water from the oxidation of food, especially at low humidities. Metabolic water contributes over 90% of the body water in a pupa of *Ephestia* when the larva is reared in dry air on dry food; the food alone does not contain sufficient water to account for most of the water in the pupa (Fig. 18.23). In order to produce this water, these insects eat and metabolize greater quantities of food at lower humidities. Metabolic water is also of particular importance to starved insects and it enables them to survive for short

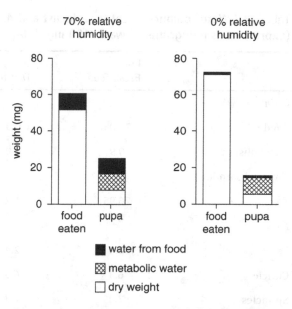

Figure 18.23 Metabolic water provides about 50% of the water in a pupa of the flour moth, *Ephestia*, developing at 70% relative humidity. In dry air (0% relative humidity), the food contains very little water. The dry weight of the pupa is less than that of a pupa developing at 70% relative humidity despite the fact that the larva eats more. Nearly all the water in the pupa is metabolic water (Fraenkel and Blewett, 1944).

periods where they would otherwise die from desiccation.

18.9.3 Water balance

Water balance is the net result of the various gains and losses experienced by the insect. A net gain is necessary for normal growth; a net loss, if sustained, will eventually lead to death. The balance will vary with the quality of food, the environmental conditions and the physiological capabilities of the insect.

Water is normally obtained with the food or by drinking, and metabolic water is insignificant, but, when external water is not available, metabolic water may become the major source. This may occasionally be true for insects, like the locust, which are deprived

Table 18.7 **Water balance sheet for a locust and *Arenivaga* under different conditions (expressed as milligrams of water 100 mg^{-1} day^{-1})**

	Locust Fresh food	Dry food	*Arenivaga* Dry air	88% relative humidity
Gain				
Food	76.05	0.3	0.22	0.44
Metabolism	0.9	0.9	0.87	0.87
Vapor absorption	0	0	0	2.14
Total	76.95	1.2	1.09	3.45
Loss				
Feces	32.4	2.4	0.19	0.19
Cuticle	6.3	6.3	>5.43	>0.65
Spiracles	7.35	2.4		
Total	46.05	11.1	5.62	0.84
Net change	+30.9	−9.9	−4.53	+2.61

Source: Edney (1977).

of normal food (Table 18.7). In this case, however, the amount of metabolic water would not be sufficient to enable the insect to survive or develop normally. By contrast, the larva of *Ephestia* appears to metabolize excess food under dry conditions and so produces enough metabolic water for survival (Fig. 18.23). Metabolic water is also particularly important for insects that engage in long-distance flights. During flight, the rate of water loss is greatly increased because of the increased airflow through the spiracles and over the surface of the body, while at the same time, the insect is unable to replace this water from external sources. Nevertheless, the high rate of metabolism produces relatively large quantities of metabolic water and this enables the insect to sustain an appropriate water balance. For example, an aphid in flight for six hours uses 15% of its dry weight. Almost all of this will be lipid which produces more than its own weight of metabolic

water. As a result, body water only declines by about 7% (Fig. 18.24); and the wet weight of the insect, expressed as a proportion of total weight, rises slightly. A consequence of this is that water loss is unlikely to be a constraint on the insect's flight; this also appears to be true for other species of insect.

In a well-hydrated insect, excretion is the major source of water loss. However, in insects that are short of water because of inadequate food, or because their habitat is very dry, transpirational losses through the cuticle and spiracles become dominant (Table 18.7).

The rate of water loss through the spiracles of resting insects is generally low relative to that occurring through the cuticle. Some studies show that over 90% of the water loss occurs through the cuticle (Fig. 18.25). However, there are exceptions. Among tenebrionid beetles, more

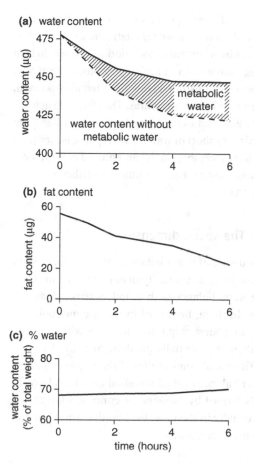

(a) water content

(b) fat content

(c) % water

Figure 18.24 Metabolic water in the flight of an aphid. It is assumed that all dry weight loss is due to the combustion of fat and this provides metabolic water (Cockbain, 1961). (a) The water content falls during flight. The initial rapid fall is due to defecation. Metabolic water (hatched area) contributes significantly to the total water content after the second hour of flight, when the water content is almost constant. (b) Decline in fat content. (c) The percentage water content of the insect rises slightly, despite the considerable loss of water.

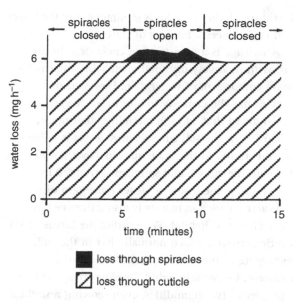

Figure 18.25 Water loss from the cuticle and spiracles of a resting grasshopper (*Romalea*). The insect was ventilating discontinuously with the spiracles open for relatively long periods at extended intervals. At the time of opening and closing, the spiracles fluttered slightly, producing the slow rise and fall in water loss. *Romalea* lives in relatively moist environments and its cuticle may be more permeable than that of insects from drier habitats (based on Hadley and Quinlan, 1993).

than 40%, and up to 70%, of water loss may occur through the spiracles. To some extent, these values vary with cuticular permeability, as, if the cuticle is very impermeable, a given respiratory loss must represent a greater proportion of the total water loss. The proportion of water lost through respiration also increases as the temperature rises, but still, for many insects, most water is lost through the cuticle.

Table 18.7 gives examples of the overall balance in two insects under different conditions. For the locust feeding on normal wet food, water intake greatly exceeds water loss. Balance is achieved by adjusting the rate of water loss, although, given a choice of wet and dry foods, the insect will also adjust its water intake. Metabolic water produces only an insignificant amount. However, with dry food and reduced food intake, the same amount of metabolic water provides the greater part of the water gained. Under these conditions, dry feces are produced and loss through the spiracles is minimized, but the insect still experiences a net loss of water.

Arenivaga is a desert-living cockroach. In dry air, with relatively dry food, metabolic water is the main component of water gain. However, this species has the capacity to absorb water vapor from moist air so that, at high humidities that might occur below the soil surface, most water is gained via this route. At the same time, loss by transpiration through the cuticle and spiracles is reduced in moist air, so the insect achieves a net gain of water.

Water balance in insects is also influenced by their choice of habitat. Wireworms, the larvae of the beetle *Agriotes*, which normally live in the soil, aggregate in the wettest parts of a humidity gradient. Conversely, adult *Tenebrio* always choose the drier of two humidities, even showing a slight preference for 5% over 10% relative humidities. Other species, like *Schistocerca*, show a preference for intermediate humidities, in this case 60–70%.

Several species have been shown to change their responses to humidity according to their state of hydration, and this is probably a common phenomenon. For example, fully hydrated *Periplaneta* tend to turn toward a dry (40% relative humidity)

airstream, but after a period of desiccation they move more readily toward a wetter airstream. Coupled with differences in the extent of locomotion, this would lead to aggregation in drier or moister regions, respectively, as has been demonstrated with the Oriental cockroach, *Blatta* and the beetle *Sitophilus*. The effect of such behavior will tend to minimize further water loss from insects already short of water. The response is mediated by humidity receptors on the antennae. The structure and functioning of these sensilla is described in Section 19.8.

18.9.4 The social dimension

This discussion of water balance has so far focused on individual insects. Social Hymenoptera need to achieve water balance both as individuals and as colonies. In flight, bees need to balance metabolic water gain against respiratory loss, but with a diet of nectar there is potentially an abundance of dietary water. The initial composition of the nectar can affect the water balance of the individual and the colony, as honey is formed by evaporative concentration of nectar, usually in the hive. Evaporative water loss can also be used for cooling.

Summary

- The basic excretory system is composed of the Malpighian tubules and hindgut, which generate urine and selectively reabsorb water or nutrients, respectively. Some insects lack tubules, or have other specializations (e.g., anal papillae).

- Urine production is driven by a plasma membrane V-ATPase, combined with exchangers and channels. It is under sophisticated neuroendocrine control.

- The excretory system also plays key roles in detoxification and metabolism; and in innate immunity.

- Nephrocytes and macrophage-like hemocytes provide further protection against toxic compounds or foreign material in the hemocoel.

- Although organismal water balance depends critically on the excretory system, there are a range of physiological and behavioral strategies that can contribute to optimal osmoregulation.

Recommended reading

Beyenbach, K. W., Skaer, H. and Dow, J. A. T. (2010). The developmental, molecular, and transport biology of Malpighian tubules. *Annual Review of Entomology* 55, 351–374.

Coast, G. (2007). The endocrine control of salt balance in insects. *General and Comparative Endocrinology* 152, 332–338.

Dow, J. A. T. (2009). Insights into the Malpighian tubule from functional genomics. *Journal of Experimental Biology* 212, 435–445.

Maddrell, S. H. P. (1981). The functional design of the insect excretory system. *Journal of Experimental Biology* 90, 1–15.

O'Donnell, M. J. (2009). Too much of a good thing: how insects cope with excess ions or toxins in the diet. *Journal of Experimental Biology* 212, 363–372.

References in figure captions and tables

Appel, A. G., Reierson, D. A. and Rust, M. K. (1986). Cuticular water loss in the smokybrown cockroach, *Periplaneta fuliginosa*. *Journal of Insect Physiology* 32, 623–628.

Berridge, M. J. (1970). A structural analysis of intestinal absorption. *Symposia of the Royal Entomological Society of London* 5, 135–150.

Blumenthal, E. M. (2003). Regulation of chloride permeability by endogenously produced tyramine in the *Drosophila* Malpighian tubule. *American Journal of Physiology: Cell Physiology* 284, C718–C728.

Bradley, T. J. and Phillips, J. E. (1977). The location and mechanism of hyperosmotic fluid secretion in the rectum of the saline-water mosquito larva *Aedes taeniorhynchus*. *Journal of Experimental Biology* 66, 111–126.

Buckner, J. S. and Caldwell, J. M. (1980). Uric acid levels during last-larval instar of *Manduca sexta*, an abrupt transition from excretion to storage in fat body. *Journal of Insect Physiology* 26, 27–32.

Chintapalli, V. R., Wang, J. and Dow, J. A. T. (2007). Using FlyAtlas to identify better *Drosophila* models of human disease. *Nature Genetics* 39, 715–720.

Coast, G. (2007). The endocrine control of salt balance in insects. *General and Comparative Endocrinology* 152, 332–338.

Cockbain, A. J. (1961). Water relationships of *Aphis fabae* Scop. during tethered flight. *Journal of Experimental Biology* 38, 175–180.

Davies, S. A., Huesmann, G. R., Maddrell, S. H. P., *et al.* (1995). CAP2b, a cardioacceleratory peptide, is present in *Drosophila* and stimulates tubule fluid secretion via cGMP. *American Journal of Physiology* 269, R1321–R1326.

Edney, E. B. (1977). *Water Balance in Land Arthropods*. Berlin: Springer-Verlag.

Eigenheer, R. A., Nicolson, S. W., Schegg, K. M., Hull, J. J. and Schooley, D. A. (2002). Identification of a potent antidiuretic factor acting on beetle Malpighian tubules. *Proceedings of the National Academy of Science USA* **99**, 84–89.

Eigenheer, R. A., Wiehart, U. M., Nicolson, S. W., *et al.* (2003). Isolation, identification and localization of a second beetle antidiuretic peptide. *Peptides* **24**, 27–34.

Foster, W. A. and Treherne, J. E. (1976). Insects of marine saltmarshes: problems and adaptations. In *Marine Insects*, ed. L. Cheng, pp. 5–42. Amsterdam: North-Holland Publishing Co.

Fraenkel, G. and Blewett, M. (1944). The utilisation of metabolic water in insects. *Bulletin of Entomological Research* **35**, 127–139.

Furuya, K., Milchak, R.J., Schegg, K.M., *et al.* (2000). Cockroach diuretic hormones: characterization of a calcitonin-like peptide in insects. *Proceedings of the National Academy of Science USA* **97**, 6469–6474.

Hadley, N. F. and Quinlan, M. C. (1993). Discontinuous carbon dioxide release in the Eastern lubber grasshopper *Romalea guttata* and its effect on respiratory transpiration. *Journal of Experimental Biology* **177**, 169–180.

Holman, G. M., Cook, B. J. and Nachman, R. J. (1986). Isolation, primary structure and synthesis of two neuropeptides from *Leucophaea maderae*: members of a new family of cephalomyotropins. *Comparative Biochemistry and Physiology Part C: Comparative Pharmacology* **84**, 205–211.

Kataoka, H., Troetschler, R. G., Li, J. P., Kramer, S. J., Carney, R. L. and Schooley, D. A. (1989). Isolation and identification of a diuretic hormone from the tobacco hornworm, *Manduca sexta*. *Proceedings of the National Academy of Science USA* **86**, 2976–2980.

Koenig, J. H. and Ikeda, K. (1990). Transformational process of the endosomal compartment in nephrocytes of *Drosophila melanogaster*. *Cell and Tissue Research* **262**, 233–244.

Komnick, H. (1977). Chloride cells and chloride epithelia of aquatic insects. *International Review of Cytology* **49**, 285–329.

Levenbook, L., Hutchins, R. F. N. and Bauer, A. C. (1971). Uric acid and basic amino acids during metamorphosis of the tobacco hornworm, *Manduca sexta*, with special reference to the meconium. *Journal of Insect Physiology* **17**, 1321–1331.

Loveridge, J. P. (1968). The control of water loss in *Locusta migratoria migratorioides* R. & F: I. Cuticular water loss. *Journal of Experimental Biology* **49**, 1–13.

Machin, J. (1983). Water vapor absorption in insects. *American Journal of Physiology* **244**, R187–R192.

Maddrell, S. H. P. and O'Donnell, M. J. (1992). Insect Malpighian tubules: V-ATPase action in ion and fluid transport. *Journal of Experimental Biology* **172**, 417–429.

Maddrell, S. H. P., Pilcher, D. E. M. and Gardiner, B. O. C. (1971). Pharmacology of the Malpighian tubules of *Rhodnius* and *Carausius*: the structure–activity relationship of tryptamine analogues and the role of cyclic AMP. *Journal of Experimental Biology* **54**, 779–804.

Nicolson, S. W. and Louw, G. N. (1982). Simultaneous measurement of evaporative water loss, oxygen consumption, and thoracic temperature during flight in a carpenter bee. *Journal of Experimental Zoology* **222**, 287–296.

O'Donnell, M. J. and Machin, J. (1991). Ion activities and electrochemical gradients in the mealworm rectal complex. *Journal of Experimental Biology* **155**, 375–402.

O'Donnell, M. J., Maddrell, S. H. P. and Gardiner, B. O. C. (1983). Transport of uric acid by the Malpighian tubules of *Rhodnius prolixus* and other insects. *Journal of Experimental Biology* 103, 169–184.

Oschman, J. L. and Wall, B. J. (1969). The structure of the rectal pads of *Periplaneta americana* L. with regard to fluid transport. *Journal of Morphology* 127, 475–509.

Phillips, J. (1981). Comparative physiology of insect renal function. *American Journal of Physiology* 241, R241–R257.

Phillips, J. E. and Audsley, N. (1995). Neuropeptide control of ion and fluid transport across locust hindgut. *American Zoologist* 35, 503–514.

Phillips, J. E., Hanrahan, J., Chamberlin, M. and Thomson, B. (1986). Mechanisms and control of reabsorption in insect hindgut. *Advances in Insect Physiology* 19, 329–422.

Rudolph, D. (1983). The water-vapor uptake system of the Phthiraptera. *Journal of Insect Physiology* 29, 15–25.

Rudolph, D. and Knülle, W. (1982). Novel uptake systems for atmospheric water vapor among insects. *Journal of Experimental Zoology* 222, 321–333.

Shaw, J. and Stobbart, R. H. (1963). Osmotic and ionic regulation in insects. *Advances in Insect Physiology* 1, 315–399.

Snyder, N. J., Hsu, E.-L. and Feyereisen, R. (1993). Induction of cytochrome P-450 activities by nicotine in the tobacco hornworm, *Manduca sexta. Journal of Chemical Ecology* 19, 2903–2916.

Snyder, N. J., Walding, J. K. and Feyereisen, R. (1994). Metabolic fate of the allelochemical nicotine in the tobacco hornworm, *Manduca sexta. Insect Biochemistry and Molecular Biology* 24, 837–846.

Stobbart, R. H. and Shaw, J. (1974). Salt and water balance: excretion. In *The Physiology of Insecta*, vol. 5., ed. M. Rockstein, pp. 361–446. London: Academic Press.

Toolson, E. C. (1982). Effects of rearing temperature on cuticle permeability and epicuticular lipid composition in *Drosophila pseudoobscura. Journal of Experimental Zoology* 222, 249–253.

Wigglesworth, V. B. (1965). *The Principles of Insect Physiology*, 6th edn. London: Methuen.

Wigglesworth, V. B. (1972). *The Principles of Insect Physiology*, 7th edn. London: Chapman & Hall.

Yu, S. J. (1987). Microsomal oxidation of allelochemicals in generalist (*Spodoptera frugiperda*) and semispecialist (*Anticarsia gemmatalis*) insect. *Journal of Chemical Ecology* 13, 423–436

19 | Thermal relations

REVISED AND UPDATED BY **JOHN S. TERBLANCHE**

INTRODUCTION

The responses of insects to temperature are of increasing interest to a wide range of
research fields. This is at least partly a consequence of the need to accurately forecast the
effects of climate change on the abundance and distribution of insect pests of agriculture
and vectors of human and animal disease, and also the need to predict the impacts of
climate change on biodiversity and ecosystem function. Insect systems function optimally
within a limited range of temperatures. For many insects, enzyme activity, tissue function
and the behavior of the whole insect is optimal at a relatively high temperature, often in the
range 30–40°C (see Figs. 3.15, 10.20). This chapter considers the factors that determine an
insect's body temperature (Section 19.1), how body temperature is regulated (Section 19.2)
and how insect performance varies as a function of temperature (Section 19.3). We next
examine how behavior and survival are affected by temperature extremes (Sections
19.4–19.7), the mechanisms and processes affecting performance and survival at
whole-animal, tissue, cell and nervous system levels (Sections 19.8–19.9), and finally,
some of the large-scale patterns identified in insect thermal biology (Section 19.10).
Two key terms that appear in the chapter are defined as follows:
Ectothermal body temperature depends on heat acquired from the environment;
endothermal body temperature depends on heat produced by the animal's own
metabolism.

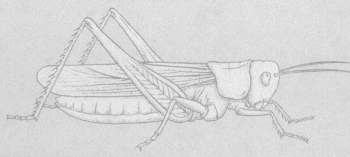

The Insects: Structure and Function (5th edition), ed. S. J. Simpson and A. E. Douglas.
Published by Cambridge University Press. © Cambridge University Press 2013.

19.1 Body temperature

19.1.1 Heat gain

The body temperature of an insect is always a reflection of ambient conditions coupled with any heat that may be produced by metabolic activity. Because the mechanical efficiency of muscles is very low, any muscular activity produces heat. However, in insects, because of the small size of the muscles and the high rate of heat loss from the organism, the effects of muscular activity on body temperature are usually insignificant. The flight muscles, however, are relatively large and oscillate at high frequencies when generating the power needed for flight. Consequently, their activity produces a significant amount of heat and the thoracic temperatures even of quite small insects are elevated above ambient temperature during flight. The extent to which thoracic temperature is elevated depends on body size because smaller insects produce less heat and lose it more rapidly (Fig. 19.1a), but the relationship between size and temperature is not a simple one as other factors also affect rates of heat loss. For example, insects with high wing loading, in general, have higher wingbeat frequencies (see Fig. 9.22). This results in more heat production and higher body temperatures (Fig. 19.1b).

The flight muscles may be used to generate heat even when the insect is not flying. Because power output for flight for most species is only achieved at relatively high muscle temperatures, usually above 30°C, flight at lower ambient temperatures is preceded by a period of warming up. Warm-up behavior is especially common in night-flying insects and those day-flying insects, such as bumble bees, that fly early in the year in temperate climates where the ambient temperature, even during the day, is often too low for flight to occur without warm-up. Similar behavior occurs in some beetles, dragonflies and occasionally in locusts. Such endothermic warm-up occurs even in small insects, including a bee, *Lasioglossum*, weighing

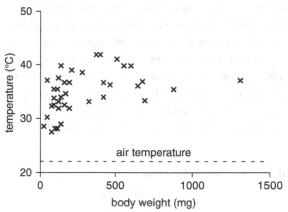

(a) body weight in bees

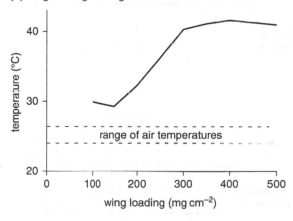

(b) wing loading in dung beetles

Figure 19.1 Elevated thoracic temperatures during flight due to heat production by the flight muscles. (a) Effects of size. Smaller insects have lower thoracic temperatures. Data for bees (Apoidea) flying at an air temperature of 22°C. Each point relates to a different species (after Stone and Willmer, 1989). (b) Effects of wing loading. Higher wing loading results in higher thoracic temperatures. Data for five species of dung beetle flying at air temperatures of 24–26°C (based on Bartholomew and Heinrich, 1978).

only 10 mg, and a hover fly, *Syrphus*, weighing 20 mg. However, not all insects exhibit such behavior even though they may be theoretically capable of doing so. For example, horse flies appear to rely wholly on external sources of heat to produce the body temperatures necessary for flight and most Lepidoptera do not exhibit preflight warm-up.

During warm-up, the depressor and levator muscles of the wings contract together. During flight, however, these muscles contract in antiphase. This difference results from a change in the pattern of firing of the motor neurons. In Lepidoptera, with synchronous flight muscles, the motor neurons fire in phase instead of in antiphase (Fig. 19.2a). Diptera and Hymenoptera, however, have asynchronous flight muscles in which muscle contractions are not determined by the rate of firing of the motor neurons, but depend on stretch activation of one set of muscles by its antagonists (Section 10.5.2). During warm-up, however, the muscles function as synchronous muscles. At least in the bees, *Apis* and *Bombus*, the rates of stimulation to the two sets of muscles are not exactly the same, so they tend to contract irregularly. However, the uncoordinated twitches

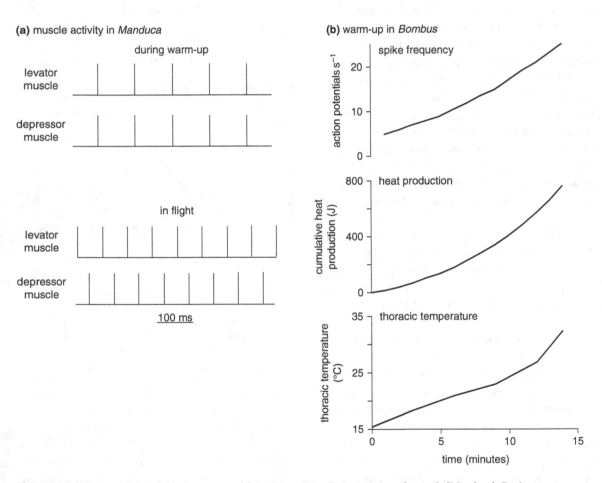

Figure 19.2 Heat production during warm-up. (a) Activity of the flight muscles of a moth (*Manduca*). During warm-up levator and depressor muscles contract at the same time; during flight they are in antiphase. (b) Warm-up in a bumble bee (*Bombus*). Time 0 is the start of warm-up behavior. Muscle activity (shown as spike frequency) increases over time as the thorax warms up. Notice that the total amount of heat produced, about 800 joules, is sufficient to raise the thoracic temperature by about 200°C; it does not do so because more heat is lost as the difference between thoracic and air temperature increases. Air temperature in this experiment was 11°C. The thorax was already hotter than this at the beginning of the experiment (after Heinrich and Kammer, 1973).

that result are too slow to produce stretch activation. At higher rates of stimulation the muscles go into tetanus. As the thorax warms up, the motor neurons fire more rapidly (Fig. 19.2b). This results in a progressive increase in the rate of heat production, but the rate at which the thorax heats up does not match the rate of heat production because heat is lost more quickly as the difference between thoracic and air temperatures increases.

Because the antagonistic muscles contract almost in phase with each other, wing movements are very small and this warm-up behavior is described as "shivering." In Hymenoptera and Diptera, the wing articulations appear to become uncoupled so that no wing movements occur at all.

In many insects that exhibit warm-up behavior, the thorax is insulated from the abdomen by air sacs or, in Hymenoptera, by the waist (Fig. 19.3a). As a result, heat is not readily transferred to the abdomen except in the hemolymph. The heart loops between the dorsal longitudinal flight muscles and opens in the head. During its passage through the muscles the hemolymph heats up so that hot hemolymph is pumped into the head and forced back into the abdomen. At the junction between thorax and abdomen (or propodeum and gaster in Hymenoptera), this backward-flowing hemolymph surrounds the aorta in which cooler hemolymph from the abdomen is flowing forwards. This system functions as a heat exchanger, and heat from the thoracic hemolymph warms the abdominal hemolymph (Fig. 19.3b). In *Apis* the heart forms a spiral where it passes through the waist. This increases the time and surface area for temperature exchange and so increases the efficiency of the exchanger. Heat is conserved in the thorax by its insulation and the heat exchanger, and the temperature of the abdomen (gaster in Hymenoptera) is usually much lower than that of the thorax (Fig. 19.3c).

Some insects also use the flight muscles to warm up for activities other than flight. The katydid, *Neoconocephalus*, begins singing soon after dark and will sing at air temperatures as low as 17°C. The sound is produced by the activity of the wing muscles (Section 26.3). At low temperatures shivering raises the thoracic temperature to 30°C or above before singing starts. During shivering the muscles contract in synchrony, but during singing they act in antiphase. Some cicadas that sing at night also warm up, presumably using the flight muscles to do so. In this case, the muscles producing the heat are separate from the timbal muscles that produce the sound (Section 26.3).

Scarab beetles that make dung balls are able to raise their body temperatures when walking or rolling the dung. However, these activities sometimes occur without an increase in body temperature so the muscles used in walking do not produce the temperature increase. Again, it is likely that heat is produced by oscillations of the flight muscles.

In the absence of solar radiation, the body temperature of a resting insect is close to ambient, but solar radiation will elevate body temperature above ambient. The amount by which body temperature exceeds air temperature, sometimes called the temperature excess, is directly proportional to the intensity of radiation. Small insects heat up more rapidly than larger ones because of their smaller mass, but attain a lower final body temperature because they lose heat more rapidly, having a greater surface–volume relationship (Figs. 19.4, 19.5).

The extent to which body temperature is raised depends on the reflectance properties of the insect's cuticle. Dark insects, or dark parts of insects, reflect relatively little of the radiation that falls on them, while light colors reflect more. In relation to this, alpine and arctic insects are often black, and many insects of temperate regions exhibit seasonal variation in the extent of their black markings. Butterflies in the genus *Colias* illustrate these points. Species living at higher elevations have more dark scales on the underside of the hindwings and absorb more radiation than those at lower elevations, and *Colias eurytheme*, which occurs at lower elevations,

(a)

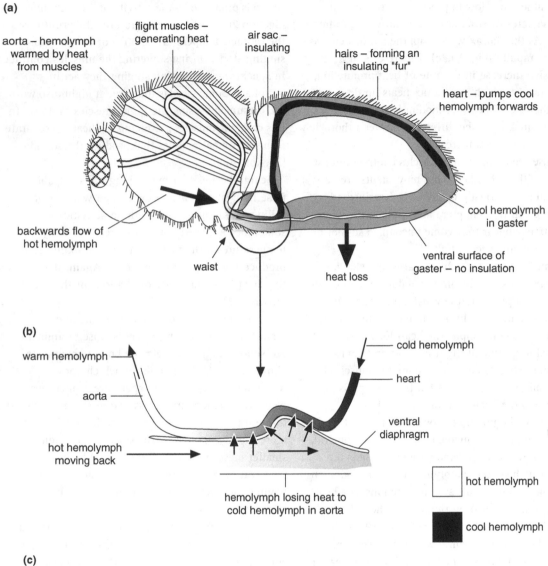

flight muscles – generating heat

aorta – hemolymph warmed by heat from muscles

air sac – insulating

hairs – forming an insulating "fur"

heart – pumps cool hemolymph forwards

cool hemolymph in gaster

ventral surface of gaster – no insulation

heat loss

waist

backwards flow of hot hemolymph

(b)

warm hemolymph

aorta

hot hemolymph moving back

cold hemolymph

heart

ventral diaphragm

hemolymph losing heat to cold hemolymph in aorta

hot hemolymph

cool hemolymph

(c)

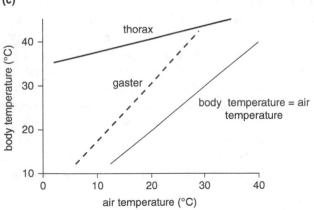

thorax

gaster

body temperature = air temperature

Figure 19.3 Thermoregulation in a bumble bee (*Bombus*). (a) Sagittal section of a bee showing the features involved in regulating thoracic temperature (after Heinrich, 1976). (b) Detail of the heat exchange system in the waist region. As hot hemolymph flows back under the ventral diaphragm it loses heat to the cool, forwardly flowing hemolymph in the aorta (after Heinrich, 1976). (c) Temperature of the thorax and gaster of bumble bees in flight. As air temperature rises, the gaster gets markedly hotter. The insect cannot fly without overheating when thoracic temperature exceeds 45°C (after Heinrich, 1975).

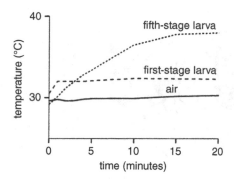

Figure 19.4 Effect of solar radiation on body temperature of a locust (*Schistocerca*). At time 0, the insects were placed in the sun, having previously been in shade. The larger, fifth-stage larva reaches a higher temperature but heats up more slowly than the smaller, first-stage larva (after Stower and Griffiths, 1966).

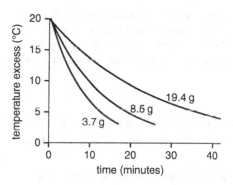

Figure 19.5 The effects of size on heat loss from adult beetles (*Heliocopris dilloni*). Three beetles of different weights were heated until their metathoraces were 20°C above ambient (the temperature excess) and then allowed to cool. The beetles were all of the same species so their body forms were similar. Differences in the rate of cooling result from the larger surface:volume ratio of the smaller insects (data from Bartholomew and Heinrich, 1978).

is darker in the spring and fall than during the summer. The change is triggered by different day lengths experienced by the later-stage larvae. If the day length is longer than 14 hours, as it is in summer, the adults have relatively few dark scales; when the day length experienced by the larvae falls to 12 hours, as in spring and fall, the adults have more dark scales (Fig. 19.6).

Developmental temperature has a direct effect on coloration in many insects. It is common for insects reared at low temperatures to be very dark, while siblings reared at high temperatures are pale, sometimes almost white. These differences, which are well known in grasshoppers and caterpillars, will, in turn, affect body temperature.

Insects sometimes gain heat by conduction from the substratum. This sometimes occurs in locusts, which lower their bodies so that they are in contact with the ground, and has also been recorded in flies. A novel example of heat gain by insects has been found in the scarab beetle, *Cyclocephala colasi*, pollinating thermogenic flowers, *Philodendron solimoesense*, in tropical French Guiana. At nightfall the flowers produce heat, which aids scent dispersal and, consequently, insect attraction. However, the flowers remain 5–10°C warmer than ambient temperatures throughout the entire night. The beetle *C. colasi* typically spends the night in the floral chamber, during which time pollen becomes attached, before the beetles disperse to another flower the following day. The relatively higher temperatures inside the flower allow the insects to remain active (e.g., for mating) at a lower metabolic cost than would be the case if *C. colasi* had to produce the heat through metabolism. This energy saving would be approximately 2–4.8-fold, depending on the time of night, and suggest that this behavior could provide a major benefit to the insect's energy budget.

19.1.2 Heat loss

Evaporation cools insects because the latent heat of vaporization is drawn from the body. In stationary insects in still air this is the most important source of heat loss. Hence factors affecting evaporation will also affect heat loss.

The rate of evaporation from a body is limited by the humidity of the immediate environment. If environmental humidity is high, little evaporation occurs, and little heat is lost by this route. In dry air,

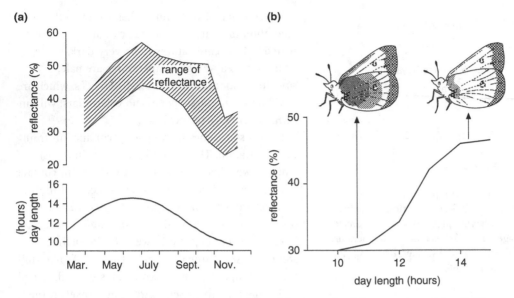

Figure 19.6 Seasonal variation in black pigmentation of the butterfly, *Colias eurytheme* (after Hoffmann, 1973). (a) Above: seasonal variation in the amount of reflectance at 650 nm from the underside of the hindwing. Reflectance is inversely proportional to the abundance of black scales and so to the degree of heating. Below: seasonal changes in day length. (b) Effects of experimental day lengths during the last larval stages on the dark patterning and average reflectance from the hindwing.

on the other hand, evaporation is much faster and an insect's body temperature in dry air may be 3–4°C below ambient. Evaporation may be reduced by local accumulations of water vapor around the body and this effect may be enhanced by hairs and scales holding a layer of still air adjacent to the body. Restricting evaporation from the tracheal system by closure of the spiracles also tends to maintain body temperature. Air movement tends to remove local accumulations of water vapor and so to increase evaporation. Thus, at higher wind speeds, body temperature is reduced.

Evaporation is generally slight at low air temperatures, so that, even in the absence of radiation, body temperature slightly exceeds air temperature because of the heat produced by the insect's metabolism. At higher temperatures, evaporation is increased and body temperature falls below ambient. For example, in a resting grasshopper at a constant relative humidity of 60%, the

temperature excess at 10°C is 0.6°C, at 20°C it is 0.4°C, while at 30°C body temperature is 0.2°C below ambient.

Convection of heat away from the body is also important if an insect's body temperature is above ambient and especially in moving air. Heat is lost more rapidly by smaller insects with a larger surface-volume relationship (Fig. 19.5). Convection is increased at higher wind speeds and is a major source of heat loss in a flying insect. In a flying locust, for example, 60–68% of heat loss results from convection compared with less than 10% by evaporation and 10% by long-wave radiation.

Conduction is probably unimportant as a means of heat loss except in the transfer of heat between the body and the layer of air immediately adjacent to it.

Heat loss from the body is reduced by insulation at the surface. Many species have the thorax clothed with seta which hold an insulating layer of air around the body. In species of *Colias* living at high altitudes the hairs are longer than in species from low altitudes, creating a layer of seta which, in *C. meadii*, living above 3000 m, is almost 1.5 mm thick. The effectiveness of the insulation depends on the density of seta, but, in general, the temperature excess of flying insects is increased 50–100% by their insulation, amounting to as much as 9°C in a large hawk moth (e.g., *Manduca*). Insulation is achieved in dragonflies by a layer of air sacs beneath the thoracic cuticle.

19.2 Thermoregulation

Insects are relatively small animals; the vast majority weigh less than 500 mg. As a result, their surface-volume relationships are large compared with other, larger animals, and, in consequence, most insects are unable to maintain a constant body temperature solely through metabolic heat production. Many species can, however, regulate their temperature so as to approximate the optimal one under some conditions. This may be done behaviorally, or, in some cases, physiologically.

19.2.1 Behavioral regulation

Extreme temperatures are typically avoided. At temperatures over 44°C, approaching the upper lethal temperature, larvae of the desert locust, *Schistocerca*, become highly active. Similarly, movement into an area of low temperature promotes a brief burst of activity. This activity is undirected, but may tend to take the insect out of the immediately unfavorable area so that it is neither killed by extreme heat nor trapped at temperatures too low for its metabolism to continue efficiently.

Within the normal range of temperature in which they are active, insects have a preferred range in which, given the choice, they tend to remain for relatively long periods. This preferred temperature range is toward the upper end of the normal range of temperatures experienced by the insect; in

Schistocerca (Orthoptera), for example, it extends from 35°C to 45°C, with a peak at 40–41°C (Fig. 19.7). The tendency to remain still in this preferred range may be regarded as a mechanism tending to keep the insects within a range of temperatures which is optimal for performance. Interestingly, *Locusta migratoria* prefer 38°C, a temperature which maximizes rate of growth, rather than lower temperatures (32°C) where efficiency of energy use is highest. This suggests that locusts make physiological trade-offs (e.g., between quality of nutrition and growth rate) to maintain preferred body temperature. Therefore, the relative costs and benefits of behavioral thermoregulation should be carefully considered within the context of each species in its particular habitat, including the impact of predators or other biotic and abiotic constraints.

In the field, many insects are known to regulate body temperature by moving into areas of sunlight when the air temperature is low, or moving to the

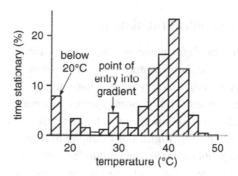

Figure 19.7 Temperature preference of second-stage larvae of the desert locust, *Schistocerca*, as shown by the amount of time for which the insects were stationary at different points in a temperature gradient. The insects had previously experienced a temperature of 35°C. Between 32°C and 43°C the insects often had the abdomen in contact with the floor of the apparatus. Above 42°C, if they remained stationary, the abdomen was usually lifted clear of the floor. Above 45°C they were usually continuously active, often hopping. This high level of activity resulted in their moving back to a lower temperature. If they remained below 20°C for more than a few minutes, the insects became sluggish (after Chapman, 1965).

shade when ambient temperatures are high. They also vary posture and orientation to the sun, exposing large areas of the body if it is cool, or as little as possible when body temperature is high. Using these devices, insects are able to extend their active periods and maximize behaviors such as foraging or mating. Examples of such behavior are recorded in insects from many orders: grasshoppers, some caterpillars, butterflies, flies and other groups.

Because insects have an asymmetric thermal performance curve (see Section 19.3), overheating by a few degrees is more deleterious than cooling an equivalent amount below the optimum temperature. Insects may be able to compensate for variation in environmental temperatures to some degree by behavioral adjustments, but the few studies which have directly tackled this question suggest behavioral thermoregulation is insufficient for complete compensation and instead, a suite of mechanisms, including physiological and biochemical alterations, is necessary to offset potential fitness costs.

19.2.2 Physiological regulation

During flight, the flight muscles raise the thoracic temperature above ambient and, for many insects, the only way to avoid overheating at high air temperature is to stop flying. Some species, however, have the ability to regulate thoracic temperature physiologically during flight. For example, in flight, *Manduca* keeps its thoracic temperature between 38°C and 42°C over a range of ambient temperatures from 12°C to 36°C. Some other moths, bees, beetles and dragonflies are known to have a similar capacity (Fig. 19.3c).

When insects are warming up they are able to alter heat production by varying the activity of the flight muscles (Section 19.1.1). In flight, however, the flight muscles are dedicated to producing the aerodynamic power necessary for flight, and their output can be varied only within narrow limits. For this reason, flying insects are generally thought to be unable to modulate heat production by altering wingbeat

frequency, but exceptions are known. The large dragonfly *Anax* reduces its wingbeat frequency from about 35 Hz at an air temperature of 20°C to 25 Hz at 35°C, at the same time reducing its flight speed. Dragonflies also make more intermittent flights and spend more time gliding at higher temperatures. In honey bees, reduction of wingbeat frequency at high temperatures is also known to contribute to regulation of thoracic temperature.

In general, however, flying insects control their body temperature by regulating heat loss. Moths and bumble bees do this by varying the heat transfer to the abdomen, which acts as a radiator of heat as it is less well insulated than the thorax. As the thorax warms up, the heart beats more rapidly and with greater amplitude. The effect of this is to increase the rate of circulation to the abdomen and reduce the efficiency of the heat exchanger. In bumble bees, the efficiency of the heat exchanger may be greatly reduced at high ambient temperatures by the hemolymph being pushed back into the gaster in discrete slugs. Consequently, the difference in temperature between the gaster and the air is greater at high temperatures (Fig. 19.3c). In honey bees, however, this is not true. The temperature of the gaster is always very similar to ambient, probably because the efficiency of the heat exchanger is so high that little or no thoracic heat reaches the gaster.

Only evaporative cooling can reduce body temperature below ambient, and most insects do not have sufficient water to use this method. Xylem-feeding insects, however, have an abundant water supply and a few cicada species are known to exhibit an increased rate of water loss at about 38°C, which can reduce body temperature to as much as 5°C below ambient (Fig. 19.8). The water is lost by an active process through pores almost 10 μm in diameter on the dorsal surface of the thorax and abdomen. This process is only effective when the water content of the insect is high and humidity is low. Honey bees foraging for nectar also use evaporative cooling. When the head reaches a temperature of about 44 °C, a flying worker

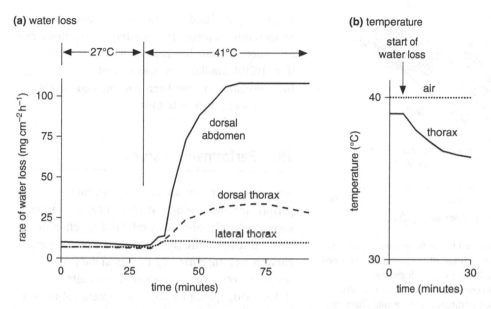

Figure 19.8 Cooling by evaporative water loss in a desert cicada (*Diceroprocta*). (a) Rates of water loss from different parts of the insect. At moderate temperatures water loss is similar from all parts. When the insect is transferred to high temperature only a slight increase occurs in loss from the lateral thorax, but much greater losses occur from the dorsal thorax and abdomen. Water loss at high temperature occurs primarily through large pores, which are present at higher density on the dorsal abdomen than the dorsal thorax; there are none on the lateral thorax (after Hadley *et al.*, 1989). (b) The cooling effect of water loss from a feeding cicada (after Hadley *et al.*, 1991).

regurgitates a drop of nectar from the honey stomach and holds it on the mouthparts. Evaporation of the drop cools the head and as the head and thorax are in broad continuity, the thorax is also cooled (Fig. 19.9). At high ambient temperatures, honey bees switch from pollen gathering to nectar collecting, presumably because pollen gatherers do not have an adequate supply of nectar for cooling.

It is possible that some other insects use evaporative cooling very selectively. Grasshoppers have patches of thin cuticle (known as Slifer's patches) on the dorsal sides of their thoracic and abdominal segments. The patches are most extensive on the middle segments of the abdomen. The rate of evaporation through these patches is high and it has been suggested that this produces local cooling, perhaps of the gonads, when temperatures are very high. A similar function has been ascribed to patches of thin cuticle on the sternites of some Pentatomidae (Hemiptera).

19.2.3 Regulation of colony temperature by social insects

Social insects regulate the temperature of their nests so that their larvae develop under relatively constant conditions. Ants, for example, move their larvae to the most favorable situations in the nest. On warm days in summer, the older larvae are brought near to the surface, while in winter they may be 25 cm or more below the surface and so avoid the lowest temperatures. On hot days, *Formica* blocks its nest entrance, stopping the entry of warm air.

Regulation of colony temperature is also well known in several honey bee species (e.g., *Apis mellifera* and *A. cerana*). At high temperatures, workers stand at the entrance of the hive fanning with their wings and creating a draught through the nest. This is sufficiently effective to keep the temperature of the brood down to 36°C when the hive is heated to 40°C. Water may also be carried in to

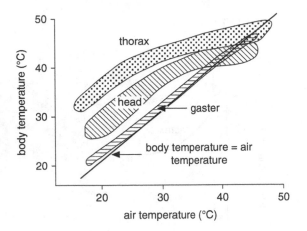

Figure 19.9 Thermoregulation in the honey bee (*Apis*). Each shaded area represents the range of temperatures recorded from bees in free flight. Notice that at higher air temperatures, head and thoracic temperatures rise less sharply than at lower temperatures. This results from the cooling effects of evaporative water loss from nectar droplets on the mouthparts. The temperature of the gaster is always close to air temperature (body temperature = air temperature). It plays no part in cooling. (Compare Fig. 19.3c) (after Heinrich, 1980a,b).

help cool the hive by evaporation and, at excessively high temperatures, the bees leave the combs and cluster outside so that further heating due to their metabolism is avoided. On the other hand, in winter when there is little or no brood, the bees cluster together on and between a small number of combs. This behavior is seen when air temperature drops below 15°C, and their metabolic heat maintains the inside of the cluster at 20–25°C. By packing closer when the temperature is very low and spreading out when it is higher, the bees are able to regulate this temperature. In addition, individual bees at low temperature increase their metabolic rate, and so increase heat output, by contracting their flight muscles without moving the wings.

Queen bumble bees also warm up using the flight muscles and rest with the underside of the gaster, which is not insulated, closely pressed against brood cells. Heat is transferred to the gaster and from it to the brood. As in *Apis*, the lower the temperature, the

more energy is used to generate heat, and at temperatures approaching freezing the metabolic rate of a queen bumble bee is as high as during flight (Fig. 19.10). Similar processes of nest thermoregulation have been documented in other social insects, such as termites.

19.3 Performance curves

Because most insects are ectothermic, their performance (e.g., wingbeat frequency, locomotion speed, metabolic or growth rate) is a function of ambient temperature (Fig. 19.11). The performance curve is well supported by empirical data from several insect species for a variety of traits (e.g., *Acheta* [Orthoptera] *Drosophila* [Diptera]; *Aphidius* [Hymenoptera]; *Pieris* [Lepidoptera]). Two characteristics of the performance curve should be noted. First, this is an asymmetric curve, meaning that for a 3°C increase in temperature above the optimum there will be a greater decline in performance, and consequently higher risk of overheating, compared with a 3°C decrease below the optimum. Second, above and below the optimum temperature, performance declines to extinction. These end points, or thermal limits of performance, can be regarded as the functional limits to insect activity, but they are not necessarily immediately lethal, particularly at low temperatures. The extremes of the performance curve are also sometimes referred to as critical thermal limits (or critical thermal maximum or minimum) and the difference between the critical thermal maximum and critical thermal minimum is the thermal range across which insects can operate. As insects are cooled (or heated) from optimum temperatures, a range of typical behavioral responses occur, including reduced performance and eventual cessation of activity, chill stupor or coma (or heat stupor and coma) and, ultimately, death. Remarkably, even small differences in genotype of key muscle enzymes, such as phosphoglucose isomerase, can make large differences to performance

(a) oxygen consumption

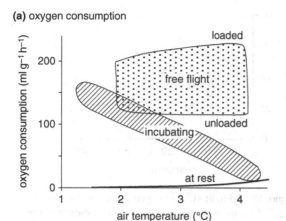

(b) brood temperature

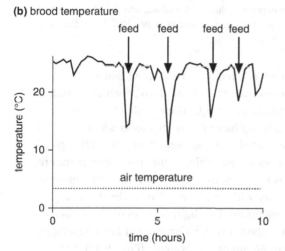

Figure 19.10 Thermoregulation in a social insect, a bumble bee (*Bombus*). (a) Oxygen consumption of queen bees performing different activities. The shaded areas show the ranges of a series of measurements. At very low temperatures, the queen generates heat by shivering, and her metabolic rate, as indicated by oxygen consumption, is as high as in flight. At higher temperatures, muscle contractions are presumably at a lower frequency, so she produces less heat and uses less oxygen. In this way she is able to regulate the amounts of heat transferred to the brood, and so keeps the brood at a relatively constant temperature despite wide fluctuations in air temperature (after Kammer and Heinrich, 1974; Heinrich, 1975). (b) Temperature of a clump of bumble bee cells (brood) is maintained close to 25°C, although the air temperature is only about 4°C. Whenever the queen leaves the cells in order to feed, the temperature drops sharply (after Heinrich, 1974).

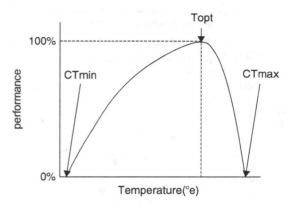

Figure 19.11 A simplified insect thermal performance curve. Insect performance (e.g., traits such as metabolic rate, locomotion speed, jumping distance) decreases from the temperature where performance is highest relative to other temperatures (i.e., optimal temperature [Topt]) toward the lower limit or critical thermal minimum (CTmin). Increasing temperature toward the upper limit, or critical thermal maximum (CTmax) results in a more rapid loss of performance relative to decreasing temperatures below Topt.

of insects at different temperatures, suggesting fundamentally different thermal reaction norms. Such variation in thermal reaction norms has significant implications for the evolution of insect performance under natural conditions.

Importantly, insects can experience trade-offs between various aspects of the performance curve. For example, performing well at low temperatures results in poorer performance at high temperatures, and vice versa (Fig. 19.12). In a release–recapture experiment on *Drosophila melanogaster* it was demonstrated that cold-acclimated flies had a significantly higher probability of recapture on cold days relative to warm-acclimated flies. While the cold-acclimated flies tended to find bait stations under cold conditions, they were not simultaneously able to perform well under warm conditions. Another type of performance trade-off that can occur over evolutionary time scales is that the optimum performance temperature could be permanently changed through natural selection. Thus, evolving a high optimum temperature can lead to a narrowing of

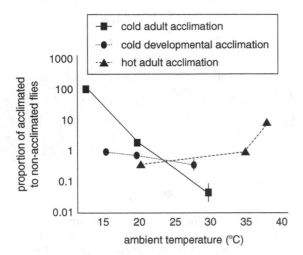

Figure 19.12 Effect of ambient temperature on the proportion of recaptured cold-acclimated to non-acclimated *Drosophila melanogaster* females under varying environmental temperatures (redrawn from Kristensen *et al.*, 2008, with kind permission from the National Academy of Sciences USA).

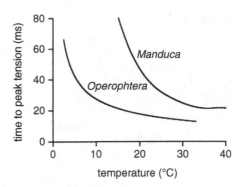

Figure 19.13 Flight at low temperatures. The time for the flight muscles to reach peak tension is shorter in the winter moth, *Operophtera*, which flies at low thoracic temperatures, than in *Manduca*, which only flies when the thoracic temperature is above 30°C (after Marden, 1995).

the thermal range for key behaviors (i.e., "generalist" and "specialist" thermal behavior are traded off against each other). In insects a general pattern is emerging which shows they are more capable of adjusting their low temperature tolerance independently of high temperature tolerance and suggests responses at either thermal extreme are decoupled.

19.4 Behavior and survival at low temperatures

Although most insects are active at moderately high temperatures, some exhibit feeding, mating or activity at temperatures close to freezing and many temperate species survive low temperatures during the winter.

19.4.1 Flight at low temperature

A number of moths fly habitually when the air temperature is close to 0°C. Some of these, such as

Eupsilia, operate by raising the thoracic temperature to 30°C or above, even at an ambient temperature of 0°C. These moths have a thick insulating hair pile on the thorax and a well-developed countercurrent heat exchange system for maintaining the thoracic temperature. Other species, like the winter moth, *Operophtera*, can fly without having to raise the thoracic temperature. The flight muscles of these insects have shorter twitch durations at low temperatures than *Manduca*, for example (Fig. 19.13), but, nevertheless, even at 5°C a cycle of contraction and relaxation may take 100 ms. Thus wingbeat frequency cannot exceed about 10 Hz and at lower temperatures must be even less. These moths are able to fly by virtue of their low wing loading. For *Operophtera* this is about 3.2 mg cm^{-2}, ten times lower than most other moths of similar size. Thus a wingbeat frequency as low as 4 Hz produces enough power to keep the insects air-borne.

Many small Diptera also fly at low air temperatures. Their small size probably precludes any significant warming, so their flight muscles must be operating at low temperature, but relatively little is known of their muscle physiology.

19.4.2 Survival at low temperatures (cold hardiness)

Survival at low temperatures is affected by the duration and severity of exposure. Low temperatures above freezing are not immediately lethal for most insects, although the insects may ultimately die if conditions persist. For example, *Locusta* larvae move very sluggishly and do not feed if the temperature is below 20°C, but remain alive for many days even at temperatures close to 0°C. These insects recover and develop normally if returned to a high temperature after a short interval. Similarly, last-stage larvae of the wax moth, *Galleria*, grow and pupate at 25°C and above, but at 20°C most larvae die before pupation, although some feeding does occur and a few larvae live as long as 160 days. Below 20°C all the larvae die, living for progressively shorter periods at lower temperatures.

Insects overwintering in temperate and subarctic regions are exposed to temperatures below freezing for some part of the year and at certain stages of development. Insect responses to subzero temperatures are usually classified as either freeze-tolerant or freeze-susceptible, meaning that they can, or cannot, survive freezing of the tissues. The temperature at which body fluids freeze is usually well below 0°C, partly because the freezing point is lowered by antifreeze solutes, but also because water may cool to temperatures well below its freezing point before ice crystals begin to form. This is known as supercooling and the temperature to which an insect supercools, the supercooling point (SCP), is influenced by many factors controlling the formation of ice crystals.

There are several ways in which to classify insect cold hardiness. Here, the main method of categorization is essentially based on the relationship between proportion mortality and SCP temperatures. Freeze-intolerant insects can be categorized into freeze-avoiding, highly chill-tolerant, moderately chill-tolerant, chill-susceptible or opportunistic species. Freeze-avoiding species experience no mortality above their SCP (e.g., *Epiblema*). Highly chill-tolerant species experience some mortality above their SCP, although they are capable of surviving fairly long-duration exposures to low temperature. Moderately chill-tolerant species have a relatively low SCP, but survival is low at temperatures above the SCP. Chill-susceptible species show relatively high mortality even after only short exposures, despite low SCP. Opportunistic species experience high mortality around 0°C and typically use behavioral means to avoid low temperatures. Freeze-tolerant species are classified into partial, moderate and strong freeze tolerance, and finally, freeze-tolerant with low supercooling points. An example of strong freeze tolerance is *Eurosta solidaginis*, which has a SCP of −10°C but can survive below −50°C.

A range of genes have been implicated in cold tolerance and recovery from chilling stress, although there is considerable variation among species and even populations within species (Tables 19.1 and 19.2). In addition, several studies have pursued chromosomal loci or candidate genes underlying chill tolerance. For example, quantitative trait loci mapping can identify genome regions in which adaptive change has taken place, including genetic variation which has resulted in changes in gene expression and protein structure.

The SCP for most freeze-tolerant insects is in the range −5°C to −10°C, although some species supercool to much lower temperatures. Some examples of freeze-tolerant insects are given in Table 19.3. Once frozen, these insects may be able to withstand very low temperatures. An extreme example is the caterpillar of *Gynaephora groenlandica* from Greenland. This insect, which freezes at −7°C, can remain frozen for nine months and survive temperatures as low as −70°C.

Intracellular ice formation destroys cells, and if many cells are affected the insect dies when it thaws. Any mechanism which reduces the rate of cooling at the surface of the cells is important in avoiding intracellular freezing. Thus, the regulated freezing

Table 19.1 **Genes involved in low-temperature tolerance of insects**

Organism	Gene	Function	Reference
Diptera			
Delia antiqua	Δ9-acyl-CoA desaturase	Phospholipid biochemistry	Kayukawa *et al.*, 2007
Drosophila melanogaster	Hsp83	Stress protein	Goto, 2000; Qin *et al.*, 2005
	Hsp26	Stress protein	
	Hsp23	Stress protein	
	CG10912	Membrane protein	
	CG9568	Membrane protein	
	CG13510	Membrane protein	
	Enoyl-CoA hydratase	Fatty acid oxidation	
	Smp-30	Cytosolic Ca^{2+} maintenance	
	Hsr-omega	RNA transcripts related to heat shock	Collinge *et al.*, 2008
	Proline	Energy metabolism	Misener *et al.*, 2001
Culicoides variipennis sonorensis	Hsp23	Stress protein	Nunamaker *et al.*, 1996
	Hsp40	Stress protein	
	Hsp43	Stress protein	
	Hsp48	Stress protein	
	Hsp60	Stress protein	
	Hsp70	Stress protein	
	Hsp92	Stress protein	
Epiblema scudderiana	EsMlp	Muscle protein	Bilgen *et al.*, 2001
Eurosta solidaginis	HIF-1α	Hypoxia/stress control	Morin *et al.*, 2005
Sarcophaga crassipalpis	Small Hsps	Stress protein	Rinehart *et al.*, 2007
	Hsp23	Stress protein	
	Hsp70	Stress protein	
	Hsp60	Stress protein	
	Hsp90	Stress protein	

Table 19.1 (*cont.*)

Organism	Gene	Function	Reference
Liriomyza sativa	Ls-Hsp19.5	Stress protein	Huang *et al.*, 2009
	Ls-Hsp20.8	Stress protein	
	Ls-Hsp21.7	Stress protein	
Hymenoptera			
Aphidius colemani	Pupalcuticular protein homolog	Cuticle component	Colinet *et al.*, 2007
	Gasp precursor	Chitin metabolic process	
	Bellwether ATP synthase	Energy production and conversion	

of a large amount of the extracellular body fluids before intracellular freezing can take place might be considered beneficial for survival of extremely low temperatures.

Ice-nucleating agents in the hemolymph or cells induce freezing in a controlled way. Ice-nucleating proteins and a lipoprotein are known to occur in the hemolymph of some freeze-tolerant adult beetles and larval Diptera. Larvae of the goldenrod gall fly, *Eurosta*, lack ice nucleators in the hemolymph, but a similar function is performed by crystals of calcium phosphate in the Malpighian tubules. Crystals of uric acid may perform an ice-nucleating function in other insects. As these compounds limit the effects of nucleation occurring around food in the gut, they may be especially important in species, such as the beetle *Phyllodecta*, that are subject to regular nightly freezing, enabling them to feed during the day and retain food in the gut at night. The effects of nucleating agents are not influenced by cryoprotectants, which may also be present even though in freeze-susceptible insects these same compounds act as supercooling agents.

Glycerol and sorbitol are the most widely occurring cryoprotectants in insects, although other polyhydric alcohols and some sugars (e.g., trehalose) serve similar functions. These compounds, by adding to the pool of solute molecules, affect the osmotic pressure of the hemolymph and help to regulate cell volume during extracellular ice formation, and also help to stabilize proteins. They are derived from glycogen stored in the fat body and, in many insects, begin to accumulate at the onset of an overwintering period (Fig. 19.14). Exposure to temperatures approaching 0°C is often the trigger for the production of cryoprotectants and ice-nucleating compounds (Fig. 19.14).

Freezing of the tissues quickly kills freeze-susceptible insects, but many freeze-susceptible species have very low SCPs, often below −20°C. The SCP often varies seasonally, being lowest at the coldest times of year and varying inversely with the accumulation of polyhydric alcohols (Fig. 19.14). Differences also occur between different stages of a species, and eggs generally have lower SCPs than active stages (Table 19.4). The overwintering stages of insects from temperate regions that are not exposed annually to very low temperatures nevertheless have SCPs as low as

Table 19.2 **Genes/proteins involved in recovery from low-temperature tolerance of insects**

Organism	Gene	Function	Reference
Diptera			
Drosophila melanogaster	Frost	Secreted protein	Goto, 2000; Sinclair *et al*, 2007
	Hsp70	Stress protein	
Aphidius colemani	Hsp70/Hsp90	Stress protein	Colinet *et al.*, 2007
	ATP synthase	Energy production and conversion	
	Fumarase	Energy production: TCA cycle	
	Phosphoglycerate kinase	Energy production: glycolytic cycle	
	Aldolase	Energy production: glycolytic cycle	
	Arginine kinase	Energy production	
	Guanine nucleotide-binding protein	Regulation of signal transduction	
	Mitochondrial malate dehydrogenase	Energy production: TCA cycle	
	Proteasome	Proteolytic complex	
	Cofilin/actin-depolymerizing factor	Actin regulatory protein	
	Hsp70	Stress protein	
	Glyceraldehyde-3-phosphate	Energy production: glycolytic cycle	
	Aconitase	Energy production: TCA cycle	
Belgica antarctica	Small Hsps	Stress protein	Rinehart *et al.*, 2006a
	Hsp70		
	Hsp90		
Culex pipiens	Hsp70	Stress protein	Rinehart *et al.*, 2006b

some arctic species. For example, the SCP of eggs of the aphid *Myzus persicae* in Britain is below −35°C, similar to that of arctic aphid species.

It is important to recognize that while the SCP is a lower lethal limit, high mortality may occur for reasons other than freezing at temperatures much

Table 19.3 **Supercooling points and cryoprotectants of some freeze-tolerant insects occuring in the Northern Hemisphere at high latitudes. The insects overwinter in the stage shown**

Species	Stage	Super-cooling point (°C)	Lowest survival temperature (°C)	Cryo-protectant
Blattodea				
Cryptocercus	Adult	−6	?	Ribitol
Lepidoptera				
Gynaephora	Larva	−7	−70	Glycerol
Papilio	Pupa	−20	−30	Glycerol
Coleoptera				
Dendroides	Larva	−10	?	Glycerol, sorbitol
Pterostichus	Adult	−10	−80	Glycerol
Diptera				
Eurosta	Larva	−10	−50	Glycerol, sorbitol

higher than this. For example, in the aphid *Myzus*, the SCP of first-stage larvae is −26°C, but half the insects are killed if exposed to −8°C for only one minute. They do not die immediately, but do so progressively over the next three days. On the other hand, there are some examples, such as the larva of the goldenrod gall moth, *Epiblema*, where freezing at the SCP is the principal cause of death.

Previous exposure to moderately low temperatures often enhances an insect's ability to survive low temperatures. If *Myzus* is reared at 5°C instead of 20°C, half the insects survive exposure to −20°C instead of −8°C. Even a brief acclimation period at low temperature can protect an insect from later cold shock (that is, the injury produced by a brief exposure to low, non-freezing temperatures). For example, most pharate adults of the flesh fly, *Sarcophaga*, reared at 25°C are killed by a two-hour exposure to −10°C, but if they are first exposed to 0°C for two hours they survive the lower temperature. Previous exposure to a moderately high temperature may have the same effect on low-temperature survival and it is possible that heat shock proteins contribute to the change in sensitivity.

Freeze-susceptible insects lack ice-nucleating agents in the hemolymph and, in many species, the likelihood of ice formation is reduced by emptying the gut before the cold period as food in the gut may form nuclei for ice formation.

The reduction in SCP is commonly associated with a reduction in water content, which increases the osmotic pressure of the hemolymph. Further reduction of the SCP is produced by polyhydric alcohols, especially glycerol. These are present at much higher concentrations than in freeze-tolerant species (Fig. 19.14), sometimes accounting for 20% of the wet weight of the insect. In cold-tolerant insects, temperatures close to 0°C inhibit the conversion of glycogen phosphorylase to its inactive form so that most of the enzyme is in the active form. As a result, glycogen is phosphorylated to produce glucose-1-phosphate. At the same time, low temperature reduces the activity of glycogen synthase, just as it reduces the activity of most other enzymes, and glycogen is not resynthesized. Instead, the glucose-1-phosphate is converted in a series of steps to the sugar alcohols (Fig. 19.15). At high temperature, the enzyme balance is reversed and the alcohols are reconverted to glycogen.

In addition, some insects surviving winter conditions by supercooling are known to have compounds that stabilize the SCP. These are peptides or glycopeptides that perhaps adsorb to the surface of newly formed ice crystals and prevent new water molecules from reaching the crystals. In

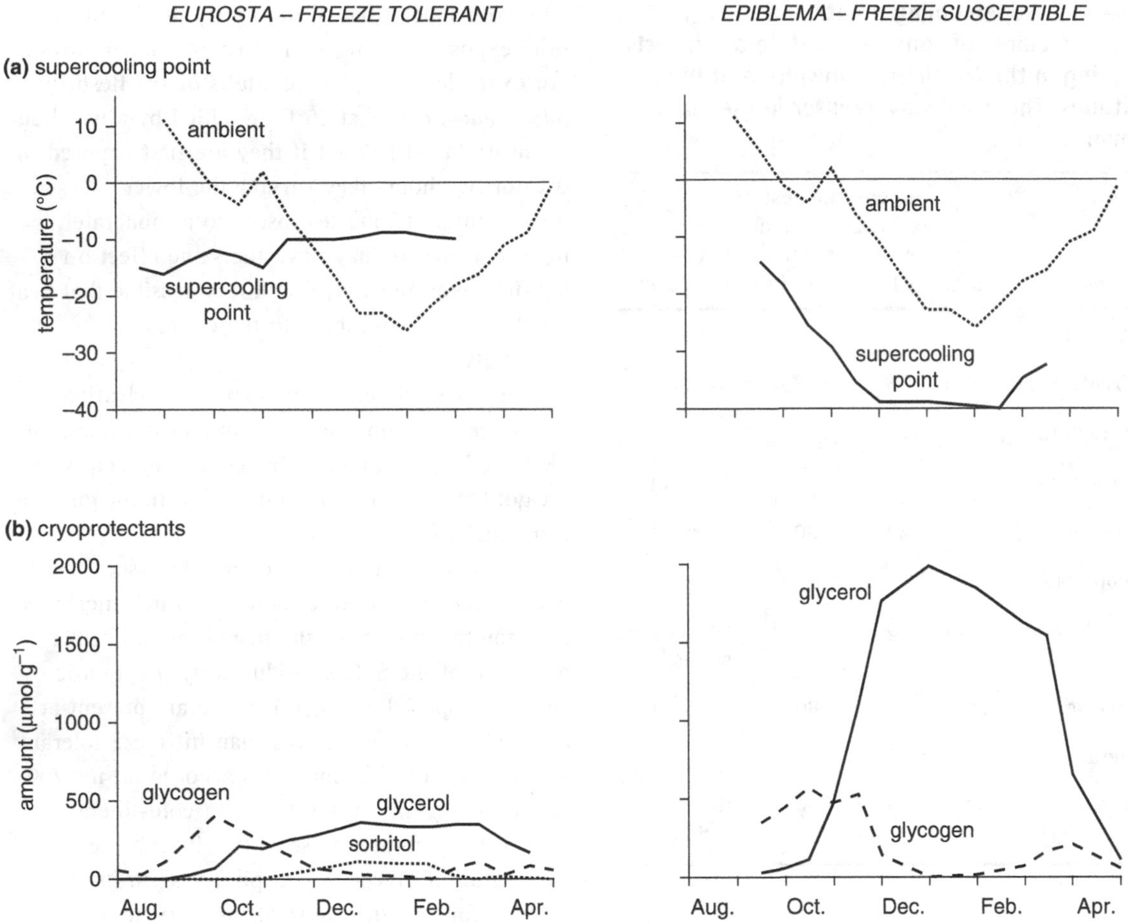

Figure 19.14 Supercooling points and cryoprotectants of two insects that use different strategies to survive low temperatures during the winter. *Eurosta* is a gall fly that tolerates freezing. *Epiblema* is a gall moth that is freeze susceptible, but has a low supercooling point. Both species overwinter in the last larval stage in galls on the host plant, and both may occur on the same plant. Data in the figure relate primarily to insects overwintering near Ottawa, Canada. (a) Variation in supercooling points over the winter. Ambient temperatures are the minima occurring over a two-week period in one year. Notice that minimum temperatures are below the supercooling point of *Eurosta* for several months; the supercooling point of *Epiblema* is always well below the environmental minima (based on Morrissey and Baust, 1976; Rickards *et al.*, 1987). (b) Seasonal changes in glycogen and the cryoprotectants, glycerol and sorbitol, that are derived from it. Notice that glycerol levels in the freeze-susceptible insects reach levels five times higher than in the freeze-tolerant insect. *Epiblema* produces very little sorbitol. Amounts are expressed in relation to the wet weights of the insects (based on Storey and Storey, 1986; Rickards *et al.*, 1987).

this way they contain the spread of ice. These compounds are called thermal hysteresis factors or antifreeze proteins and are present in many insect species. Their production is induced by low temperatures and short photoperiods. In larval *Tenebrio* these environmental features produce a brief elevation in hemolymph titer of juvenile hormone, which may have a regulatory role in the production of the antifreeze proteins.

Table 19.4 **Supercooling points and cryoprotectants of some freeze-sensitive insects occuring in northern latitudes**[a]

Species	Stage	Supercooling point (°C)	Cryo-protectant
Hemiptera			
Rhopalosiphum	Egg	−40	?
Myzus	Larva	−26	?
Lepidoptera			
Malacosoma	Egg	−41	Glycerol
Laspeyresia	Larva	−37	Glycerol
Pieris	Pupa	−26	Sorbitol
Vanessa	Adult	−21	?
Coleoptera			
Dendroctonus	Larva	−34	Glyccrol
Coccinella	Adult	−24	None?
Diptera			
Hylemya	Larva	−25	?
Sarcophaga	Pupa	−23	Glycerol
Anopheles	Adult	−17	?
Hymenoptera			
Neodiprion	Egg	−32	Glycerol
Cephus	Larva	−27	None?
Megachile	Pupa	−42	Glycerol
Camponotus	Adult	−28	Glycerol

Notes:
[a] The insects overwinter in the stage shown. Values for supercooling points are the lowest values observed in mid-winter where data are available.
? Not known.

19.5 Activity and survival at high temperatures

At the upper end of the temperature range, above the preferred or optimum temperature, insects typically show a sharp rise in locomotor activity. At still higher temperatures this is followed by an inability to move, a phase known as heat stupor, and then by irreversible heat injury and death. The temperature at which death occurs depends on the species, the duration of exposure and interaction with other factors, in particular, humidity. The rate of heating and ability of the insect to mount physiological responses to reduce heat injury are also likely to play an important role in high-temperature tolerance. Such rapid physiological responses to high temperature that result in protection from heat injury include heat hardening. Hardening may be defined as the ability of a thermal treatment to reduce thermal injury or death on a subsequent exposure to thermal extremes, and usually take place within minutes to hours of transfer to new conditions. However, heat hardening only improves high-temperature survival by a relatively small amount compared with, for example, low-temperature hardening responses where effects are typically more pronounced.

Insects are cooled by evaporation, so that for short periods they can withstand higher air temperatures if the air is dry. *Periplaneta*, for example, dies at 38°C at high humidities, but can survive a short exposure of up to 48°C if the air is dry. Few insects can tolerate the high rate of water loss that cooling necessitates for very long, but some desert cicadas and honey bees with access to ample water supplies do habitually reduce their body temperatures by evaporative cooling (see Section 19.1.2). Over longer periods at high temperatures, however, low humidity can result in negative effects on survival due to desiccation. Thus, *Blatta* can survive for 24 hours at 37–39°C if the air is moist, but die as a result of similar exposure in dry air. The effect of ambient humidity is likely to be more pronounced for small rather than large insects since the former are more susceptible to dehydration stress. Humidity does not affect the lethal temperature of small insects such as lice as the volume of water available for evaporation is small while the surface taking up heat is relatively large.

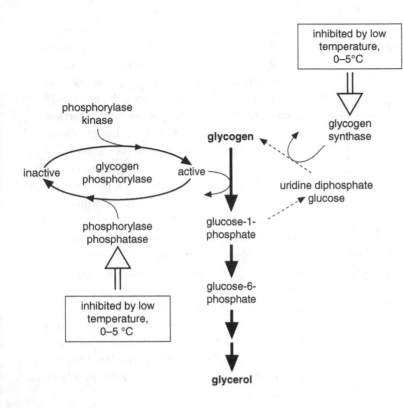

Figure 19.15 Synthesis of glycerol from glycogen. At low temperatures, the conversion of glycogen phosphorylase from the active to the inactive form is inhibited; the production of the active form continues. As a result, the proportion of this enzyme in the active form is greatly increased so glycogen is metabolized to glucose-6-phosphate. Other enzymes (not shown) then lead to the synthesis of glycerol or sorbitol. Reconversion of glucose-1-phosphate to glycogen does not occur because the activity of glycogen synthase is inhibited by low temperature.

For many insects the lethal temperature for short-term exposure is within the range 40–50°C, but for insects from particular habitats, lethal temperatures may be very different. Thus *Grylloblatta* (Grylloblattodea), living at high altitudes in the Rocky Mountains where ambient temperatures remain cool (in the range of 5–15°C), dies when its temperature gets as high as 20°C; *Thermobia* (Thysanura), the fire brat, dies at 51°C; and for chironomid larvae living in hot springs at 49–51°C the lethal temperature must be even higher.

Some modification of the upper lethal temperature occurs, depending on the previous experience of the insect. Thus *Drosophila* reared at 15°C and maintained at 15°C as adults survive for about 50 minutes in dry air at 33.5°C, but if they are maintained at 25°C beforehand they survive for about 130 minutes. If the larvae are also reared at the higher temperature the period of exposure that the insects can survive is still further increased to 140 minutes in adults maintained at 15°C and to 180 minutes in adults maintained at 25°C. Thus, two types of acclimation can be recognized: long-lasting acclimation due to conditions during development, and short-term physiological acclimation depending on the more immediate conditions. The latter is thought to be more easily reversible. The effect of physiological acclimation to high temperatures might also be affected by moisture availability during acclimation.

At least a part of the short-term acclimation to high temperatures is a result of the production of heat shock proteins. These have been demonstrated in several insect species, as well as in many other organisms. *Drosophila* and *Locusta* each produce six heat shock proteins, and the larva of the gypsy moth, *Lymantria dispar*, has seven. They belong to three different families of proteins with different molecular weights. Probably the most well investigated are heat shock proteins Hsp70, Hsp90 and Hsp30. Some expression of these proteins occurs at optimal rearing temperatures, but expression is enhanced within minutes of a sharp rise in temperature. At high temperatures, proteins become denatured and clump together as insoluble aggregates. Heat shock proteins may protect other proteins from denaturing or they may bind to the surface of an aggregate and promote its dissolution, at the same time causing the proteins to refold. The same proteins also give some protection against low temperatures for short periods and are also induced by toxic chemicals or other environmental stresses, such as dehydration. Their production should, thus, probably be regarded as a general response to stress. Heat shock proteins are rapidly produced and yet high rates of expression are only sustained for periods of a few hours after the shock. Remarkably, even though elevated heat shock protein production may cease soon after transfer to new temperature conditions, improved high-temperature tolerance may persist for much longer periods (up to days). Ongoing heat shock protein expression is detrimental, and development, fecundity and survival may be negatively affected.

Workers of the ant *Cataglyphis bombycina* are active foragers in the Sahara desert at temperatures above those tolerated by other small animals. They forage for a very brief period in the middle of the day when the average ground surface temperature is above 46°C; other species stop foraging when the surface temperature is above 45°C. These ants have long legs that allow them to lift their bodies off the surface and they also frequently move up stalks of dry vegetation where the temperature is slightly lower than at the ground surface. In addition, they apparently have a high constitutive level of heat shock proteins so that they are already protected for the brief periods when they are exposed.

Death at high temperatures may result from various factors. Proteins may be denatured or the balance of metabolic processes may be disturbed so that toxic products accumulate. Loss of membrane function, pH changes and damage to DNA may all result from high temperature exposure and lead to irreparable damage. Blowfly larvae kept at high temperatures accumulate organic and inorganic phosphates and adenyl pyrophosphate in the hemolymph, suggesting loss of cellular ion homeostasis. Nuclear magnetic resonance metabolomic profiling of heat stress and recovery in *Drosophila* has shown that hardening by prior heat stress allows faster metabolite homeostasis. In some cases food reserves may be exhausted; *Pediculus* (Phthiraptera), for instance, survives better at high temperatures if it has recently fed. Sometimes, particularly over long periods, death at high temperatures may result from desiccation.

Candidate genes underlying high temperature tolerance of insects have begun to be examined. A variety of approaches, including heat shock factor (*hsf*) or *hsp* mutants, which either are unable to produce heat shock proteins or produce them constantly, and quantitative trait loci (QTL) mapping have been used to target traits of insect heat tolerance. Studies suggest that while one or a few key genes (e.g., *hsps*) can underlie marked variation in heat stress resistance, there are likely several physiological levels and biochemical adaptations involved in whole-organism thermal tolerance.

19.6 Acclimation

As mentioned briefly in the previous section, responses to temperature are not static, but vary according to the previous experience of the insect.

Such modification is known as acclimatization in the field, or acclimation in the laboratory, and has been discussed briefly above in relation to survival at extreme temperatures. Hardening can also be considered a form of acclimation. However, acclimation also occurs within the normal temperature range of an insect and affects both physiology and behavior. For example, the oxygen consumption of *Melasoma* (Coleoptera) adults increases with temperature, but the level of consumption depends on the temperature to which the insects were previously exposed and, for any given temperature, oxygen consumption is higher in insects acclimated to lower temperatures (Fig. 19.16). Note that at a given temperature (e.g., 25°C) and for an individual of similar mass, metabolic rate of cold-acclimated individuals is higher than that of warm-acclimated individuals. The temperature at which maximum oxygen consumption occurs is also lower in the insects preconditioned at the lower temperature. Two adaptive explanations have been suggested: first, that individuals living at low temperatures have upregulated metabolic rate to enable activity in cooler conditions; or, second, that individuals living at warmer temperatures downregulate metabolic rate to limit respiratory water loss. Similarly, the spontaneous output from the central nervous system is related to preconditioning temperature as is the level of activity of the whole insect. Thus, adults of *Ptinus* (Coleoptera) previously maintained at 15°C are less active at all temperatures than others previously maintained at 28°C. Acclimation is a continuous process and can occur remarkably quickly in *Drosophila*, where it has been examined under natural variable thermal conditions.

19.6.1 Rapid thermal responses

Acclimation responses of insects occur over a range of time scales (e.g., minutes, hours, days, weeks, months). While it has been known for some time that insects are capable of adjusting their thermal tolerance rapidly, just how well they do so, what

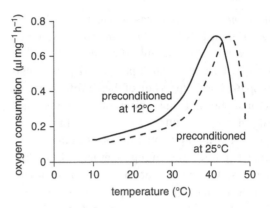

Figure 19.16 Effects of acclimation (preconditioning) at different temperatures on oxygen consumption by adult *Melasoma* (after Wigglesworth, 1972).

mechanisms are employed and how widespread these responses are, are only just beginning to be fully appreciated. These responses can also occur remarkably quickly. One such mechanism is rapid cold hardening, and can be defined as a rapid improvement in survival at a low lethal temperature after brief pre-treatment (1–2 hours) to a sub-lethal low-temperature shock. In some species as much as 80–100% improvement in survival has been reported after a period of less than two hours (e.g., *Drosophila melanogaster*). Similarly, at warm temperatures, brief exposure to high sub-lethal temperatures result in greater survival at a lethal high temperature after a few hours. For example, this has been demonstrated in the wasp *Trichogramma carverae* and several fly species. Such mechanisms are likely employed by insects to quickly alter their physiological tolerances over the course of a day, probably to increase mating and feeding time, and may be critical to the survival of sudden, unpredictable cold snaps in the wild. These rapid thermal responses usually last for a few hours but are typically reversible. Rapid cold hardening may also be accompanied by small but consistent alterations in membrane phospholipids, which likely aid in maintaining cell function at low temperatures.

However, not all species examined to date have inducible thermal tolerance, especially over short time scales. For example, little evidence for rapid cold hardening has been found in the tsetse fly *Glossina pallidipes*, despite the fact that it shows longer-term acclimation responses. A range of other insect species do not show rapid adjustments to temperature stress, including some Coleoptera and Lepidoptera. Furthermore, of the terrestrial arthropod species that do show hardening responses, considerable variation in the magnitude of the protection afforded exists, both within and among species and depends also on the trait examined.

19.6.2 Is acclimation beneficial?

For many organisms it is considered that a prior experience of a novel environment can give rise to a performance advantage, with fitness benefits, when exposed to that environment on a subsequent occasion. Formally, this is known as the beneficial acclimation hypothesis (BAH), and can be defined as acclimation to a particular environment giving a performance advantage in that environment over another organism that has not had the opportunity to acclimate to that particular environment. While the BAH is intuitively appealing, it has come under close scrutiny. In insects, many tests of the BAH have not found support, partly because at least one of the alternative hypotheses could not be rejected. By contrast, several recent studies have either demonstrated or claimed support for the BAH in insects. However, support for environment-specific advantages (e.g., colder or hotter is better) has been documented for some insect species including *Drosophila*, and different acclimation hypotheses may even be supported among life-stages within the same species (e.g., in the kelp fly, *Paractora dreuxi*), possibly as the ability to behaviorally thermoregulate develops.

Either way, it is clear that insects have a range of potential responses to acclimation conditions and that support for the BAH is not as widespread as previously thought. Clearly, further consideration is required to assess the potential acclimation hypotheses and to link these with adaptive phenotypic plasticity in an evolutionary context.

19.7 Cryptobiosis

Cryptobiosis is the term used to describe the state of an organism when it shows no visible signs of life and metabolic activity is brought reversibly to a standstill. The only insect in which this is known to occur is the larva of *Polypedilum* (Diptera), a chironomid living in pools on unshaded rocks in Nigeria. In the dry season these pools dry up and the surface temperatures of the rock probably reach 70°C. Active larvae of *Polypedilum* die after an exposure of one hour at 43°C, but if they are dehydrated so that their water content is less than 8% of its original value they can survive extreme temperatures for long periods. Some recovery occurs even after exposure for one minute at 102°C or several days in liquid air at −190°C. At room temperatures larvae can withstand total dehydration for three years and some showed a temporary recovery after ten years. Evidence for cryptobiosis also occurs in some larvae of *Sciara* (Diptera) and ceratopogonid larvae. Some other insects may possess specific tissues which exhibit the phenomenon, although the insect as a whole may not. For example, blood cells in the gills of *Taphrophila* (Diptera) pupae (see Section 17.7.1) and *Sialis* larvae can be desiccated for long periods, but when rehydrated show some vital activities such as clotting.

19.8 Temperature and humidity receptors

Sense cells responsive to temperature and humidity are usually present together in a single sensillum. All insects that have been studied have such receptors on the antennae. They consist of a short peg without any pores and an immovable socket. Projecting into the peg are the distal dendrites of two neurons

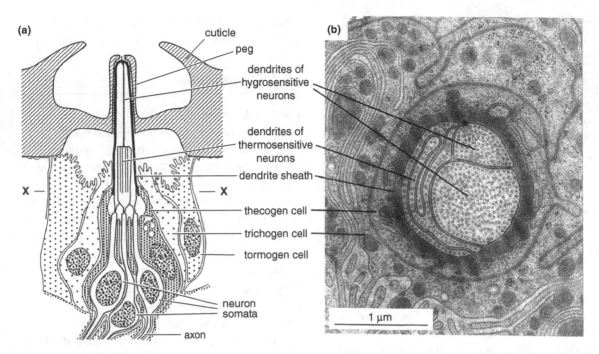

Figure 19.17 Thermohygroreceptor. (a) Diagrammatic longitudinal section through a coeloconic sensillum. (b) Electron micrograph of a transverse section of a thermohygroreceptor at the level of X–X in (a). Notice the dendrites of two hygrosensitive neurons and folds of the dendrite of a thermosensitive neuron completely occupying the space in the dendrite sheath (*Bombyx*) (after Steinbrecht *et al.*, 1989).

(Fig. 19.17). The dendrites usually occupy the whole of the space within the dendrite sheath that forms the lining of the peg and are attached to it. Microtubules within the dendrite are joined together, forming a structure with some resemblance to the tubular body of mechanosensitive dendrites (Section 23.1.1). A third dendrite ends below the base of the peg. In many insects it is produced into a series of lamellae or it may be flattened and in the form of a whorl. Proximally, the dendrites of both types have a ciliary region. On the basis of their responses, the cells with dendrites extending into the peg are believed to be hygrosensitive, while the cell which does not enter the peg is probably thermosensitive.

The sensilla may be exposed at the surface of the antenna or concealed in a pit, forming one type of coeloconic sensillum (see Fig. 24.3). When exposed at the surface, they are associated with trichoid

mechanoreceptors which form a protective umbrella above them, preventing direct contact with any outside object. The number of sensilla on each antenna is small. There is often a group on the terminal annulus and one or two on each of the proximal annuli. Worker honey bees have about 50 sensilla on each antenna, and *Periplaneta* has about 100 sensilla. Presumably, since temperature and humidity are ubiquitous properties of the air surrounding the antenna, a large number of receptors is not necessary for the insect to obtain accurate information.

In addition to these sensilla, some multiporous olfactory sensilla on the antennae of *Periplaneta* and *Locusta* contain a cell that is thermosensitive. It is structurally indistinguishable from the chemosensitive neurons. About 100 multiporous sensilla with a thermosensitive neuron are also present on each maxillary palp of *Periplaneta*.

The transduction mechanisms in hygrosensitive neurons are not understood. One possibility is that changes in ambient humidity affect the water content of the cuticle associated with the peg, causing a small deformation that is perceived by the dendrites connected to the inner wall of the peg; they effectively function like mechanoreceptors. However, the presumed hygrosensitive dendrites in the silk moth *Bombyx* shorten when the insect is dry adapted, suggesting that there may be some direct effect of humidity on the cells.

One of the hygroreceptors responds to a rise in humidity with an increased firing rate, while the other responds to a fall in humidity (Fig. 19.18); they are known as moist and dry receptors, respectively. The greater the change in humidity, the higher the firing rate of the cell responding. These receptors also fire continuously at constant humidities. In *Periplaneta*, for example, the dry receptor has a higher tonic firing rate at low humidities. It is also temperature sensitive, with a higher firing rate at higher temperatures.

The thermoreceptor responds to a fall in temperature; the greater the fall, the higher the firing rate (Fig. 19.19). It is called a cold receptor, and such cells may respond to temperature changes of less than 1°C. In the larva of *Speophyes*, a cave-dwelling beetle, the firing rate of the cold receptor cell follows slow changes of less than 1°C occurring over five minutes (Fig. 19.19c).

The temperature-sensitive cells in cockroach olfactory sensilla differ in their responses from other thermoreceptors that have been described. They have a maximum rate of firing in the range 18–27°C, with lower rates at lower and higher temperatures. They thus have the potential to provide the insect with an absolute measurement of air temperature.

It is unclear how this information is integrated within the nervous system. As some of the cells are both thermo- and hygrosensitive there is potential for ambiguity. Nevertheless, it is clear from their behavior that insects are able to detect changes in both temperature and humidity. Bed bugs, *Cimex*, are sensitive to changes of less than 1°C, and honey bees can be trained to select one of two temperatures differing by only 2°C, and may, perhaps, also detect absolute levels of temperature. This is suggested by their ability to move to, and remain in, areas of preferred temperature. In *Drosophila*, however, the cyclic adenosine monophosphate-dependent protein kinase A (cAMP-PKA) signaling pathway in mushroom bodies of the CNS plays an important role in thermoregulatory behavior.

The beetle *Melanophila* responds to infrared radiation, which is perceived by sensilla in thoracic pits. Blood-sucking bugs may also have this capacity.

19.8.1 Cellular mechanisms of temperature sensation

The exact cellular mechanisms of temperature sensation have long remained an open question. However, demonstrations that ion channels of the transient receptor potential (TRP) family can be activated by variation in ambient temperature suggest an important role in temperature sensing. These TRP channels have a range of temperature sensitivity depending on the exact subfamily of ion channel, and appear to be present across a range of vertebrate taxa and in the few invertebrate taxa which have been examined thus far. In insects, however, two main families of TRPs appear to play a role in thermosensation, namely the melastatin and vanilloid receptors. In *Drosophila* larvae, RNA interference of TRPA-1 reduced behavioral avoidance of high temperatures, and thus suggests an important role for these TRPs in insect thermoregulation. A separate subfamily of TRPs is required for cool temperature avoidance (TRPC subfamily proteins). Thus, distinct TRP channels are required for warm and cool avoidance. Although particular TRP channels occur across a wide range of taxa, their function may vary significantly for a similar ion channel. For example, in *Drosophila*, a mammalian cold-activated ion channel (ANKTM-1) responded to warming rather than cooling.

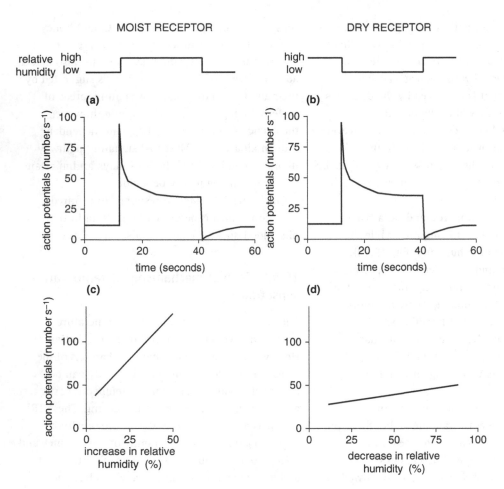

Figure 19.18 Responses of hygrosensitive neurons to rapid changes in humidity. (a) Response of moist receptor to a rapid change from dry (0%) to saturated (100%) air and back again. The firing rate (action potentials s^{-1}) increases sharply immediately following the change and then maintains a higher sustained rate of firing at the higher humidity (cockroach, *Periplaneta*) (based on Yokohari and Tateda, 1976). (b) Response of dry receptor to a rapid change from saturated (100%) to dry (0%) air and back again. The firing rate (action potentials s^{-1}) increases sharply immediately following the change and then maintains a higher sustained rate of firing at the lower humidity (cockroach, *Periplaneta*) (based on Yokohari and Tateda, 1976). (c) Response of moist receptor following rapid increases in humidity (stick insect, *Carausius*). The number of action potentials is proportional to the magnitude of the increase in relative humidity (based on Tichy, 1979). (d) Response of dry receptor following rapid decreases in humidity. Note that the x-axis shows the percentage decrease in relative humidity. Larger decreases produce slightly higher firing rates (stick insect, *Carausius*) (based on Tichy, 1979).

19.9 Temperature-related changes in the nervous system

The rate of firing of peripheral mechanosensory neurons increases with temperature in both *Periplaneta* and *Schistocerca* and this is,

perhaps, a general phenomenon. The extent of the change differs in sensilla on different parts of the body. For example, the number of action potentials produced by a standard stimulation of the wind-sensitive hairs on the head of *Schistocerca* more than doubles over the range 25–40°C, while

(a) response to falling temperature

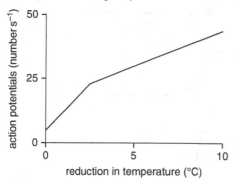

(b) response to different
constant temperatures

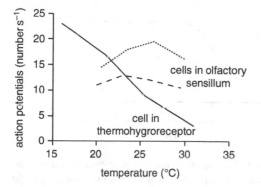

(c) response to cycling temperature

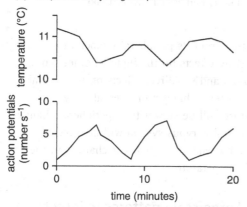

changes in the input from sensilla elsewhere on the body are less marked (Fig. 19.20). Hairs on the hind tarsi exhibit a broad maximum of response over the range 30–39°C, firing less actively at lower and higher temperatures. This corresponds approximately to the preferred temperature range (Fig. 19.7) and these sensilla possibly provide the insect with information by which it is able to determine this range when in a temperature gradient.

Changes also occur in the activities of motor neurons and interneurons. For example, the conduction velocity of the motor axon innervating the fast extensor tibiae muscle of *Schistocerca* increases with temperature, while the synaptic delay between this neuron and the flexor muscle motor neurons decreases.

The firing rate of an interneuron receiving input from the wind-sensitive hairs also increases with temperature, and its output is a resultant of its own temperature and that of the sensory neurons on the outside of the head. Within the central nervous system of *Periplaneta*, unidentified units, possible interneurons, fall into four categories with respect to their responses to temperature (Fig. 19.21). In one, the output is directly proportional to temperature; others show a similar response with the addition of a transient increase if the temperature falls, or decrease if the temperature rises; in a third type the output is inversely proportional to temperature; a fourth class is not affected by temperature.

The response of the insect as a whole is affected by temperature in a variety of ways; by

range of constant temperatures. A cell in a thermohygroreceptor of *Periplaneta* is more active at lower temperatures, but cold receptors in olfactory sensilla exhibit only slight differences with temperature, tending to have maxima at intermediate temperatures (based on Nishikawa *et al.*, 1992). (c) Response to cycling temperature of a cold receptor on the antenna of a beetle larva, *Speophyes*. The response follows changes in temperature, increasing as the temperature falls, and decreasing as it rises. Notice that the temperature changes are slow, only about 0.5°C over five-minute periods (after Altner and Loftus, 1985).

Figure 19.19 Responses of cold receptor neurons. (a) Response to continuously falling temperature of a cell in a thermohygroreceptor of the cockroach, *Periplaneta*. Cold cells in olfactory sensilla respond in essentially the same way (based on Nishikawa *et al.*, 1992). (b) Response to a

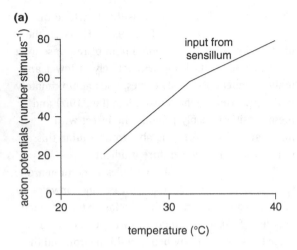

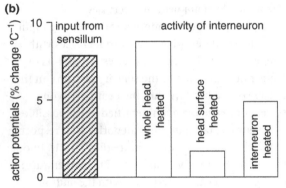

Figure 19.20 Effect of temperature on mechanosensory input from the wind-sensitive hairs on the head and the activity of a giant interneuron in the brain of a grasshopper, *Schistocerca americana* (after Miles, 1992). (a) Sensory input (action potentials per stimulus) from a single hair deflected by 29 μm at different temperatures. (b) Changes in responses, expressed as percentage change in the frequency of action potentials for each degree (°C) change in temperature. Input from the wind-sensitive hairs increased sharply when they were heated, but the central response, as indicated by the activity of the interneuron, was dependent on the temperatures of both the sensory cells and of the interneuron. If both experienced a rise in temperature (whole head heated), the interneuron responded sharply to a change in temperature; if only the temperature of the mechanoreceptors was affected (head surface heated), the interneuron showed little change in activity; if the temperature of the interneuron increased, but that of the sensory neurons did not (interneuron heated) its activity increased, but to a lesser extent than if the whole head was heated.

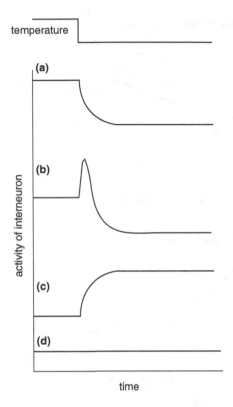

Figure 19.21 Diagrammatic representations of the response elicited by a sharp fall in temperature on the activity of four different types of unidentified neurons in the cockroach nerve cord (after Kerkut and Taylor, 1958).

information from its peripheral temperature receptors, by effects on the input of other types of receptors and via direct effects on the central nervous system. Changes in internal body temperature will be slower than peripheral changes, so the central nervous system will be affected primarily by relatively persistent changes in the external temperature.

19.10 Large-scale patterns in insect thermal biology

Interest in large-scale biogeographic patterns has led to the conceptual reunification of ecology and physiology, a discipline known as

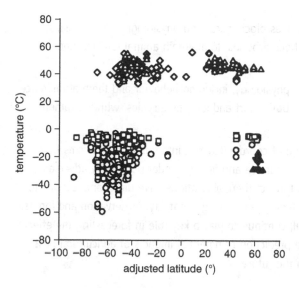

Figure 19.22 Large-scale patterns of thermal tolerance in insects are related to latitudinal variation in climate (redrawn from Addo-Bediako *et al.*, 2000 with permission from the Royal Society of London). Open circles: supercooling points in freezing-intolerant species; open squares: supercooling points in freezing-tolerant species; open triangles: CTmax; open diamonds: upper lethal temperatures; filled triangles represent Collembola and Acari). Note that latitude has been adjusted for altitude in this figure.

macrophysiology. Such large-scale investigations have resulted in a number of significant insights into thermal biology and climate relationships of terrestrial arthropods. For example, it is clear that thermal tolerance is related to climate in a given geographic location and that, generally, high-temperature tolerance varies less than low-temperature tolerance across the planet (Fig. 19.22). A similar approach has revealed that ectotherms at different latitudes faced with increasing ambient temperatures are likely to be affected in different ways owing to variation in thermal safety margins. For example, equatorial species are more likely to face extinction from global warming than temperate or polar organisms, since current air temperatures are closer to these tropical species' thermal maxima. Recently it was found that across European diving beetle species, variation in geographic range sizes could be explained on the basis of thermal tolerance. Finally, large-scale patterns in freeze-tolerance strategy have also been investigated, showing that freeze tolerance has evolved for different reasons in the Northern and Southern Hemispheres. This reflects the more variable temperatures in the Southern Hemisphere and the long, hard winters of the Northern Hemisphere. In conclusion, macrophysiological studies are likely to play an important role in discovering novel biogeographic patterns in insect physiology in the future.

Summary

- Insects are capable of sensing, integrating and responding to temperature in their environments. Many insects are capable of modifying the environmental temperatures they experience using behavioral or physiological adjustments.

- Insect performance is a function of body temperature. This function is non-linear and asymmetric. Overheating poses a greater survival threat than cooling an equivalent amount.

- Insects can survive low temperatures by either allowing or preventing freezing of body fluids. There is considerable diversity within these two general strategies of cold hardiness among insect species.

- Insects survive high temperatures by using behavioral strategies, having high levels of heat resistance, rapidly developing heat resistance or some combination of these.

Improving heat tolerance involves biochemical and physiological mechanisms, such as production of heat shock proteins, to maintain animal function under extreme conditions.

- Thermal history affects insect physiology, including behavior and thermal tolerance. These responses can occur at both short and long time scales, within or among generations.

- Insects clearly possess a range of coordinated and integrated mechanisms that have evolved to allow them to survive and flourish under potentially adverse thermal conditions. Studies of insect thermal relations have direct applications to numerous research fields, including pest management, cryopreservation and forensic entomology. Such studies will continue to play a key role in forecasting the effects of climate change and in the prediction of potential impacts of agricultural pest species or disease vectors in the future.

Recommended reading

Angilletta, M. (2009). *Thermal Adaptation: A Theoretical and Empirical Synthesis.* London: Oxford University Press.

Chown, S. L. and Nicolson, S. W. (2004). *Insect Physiological Ecology: Mechanisms and Patterns.* London: Oxford University Press.

Chown, S. L. and Terblanche, J. S. (2007). Physiological diversity in insects: ecological and evolutionary contexts. *Advances in Insect Physiology* 33, 50–151.

Denlinger, D. L. and Lee, R. E. (2010). *Low Temperature Biology of Insects.* Cambridge: Cambridge University Press.

Heinrich, B. (1993). *The Hot-blooded Insects.* Cambridge, MA: Harvard University Press.

References in figure captions and tables

Addo-Bediako, A., Chown, S. L. and Gaston, K. J. (2000). Thermal tolerance, climatic variability and latitude. *Proceedings of the Royal Society of London B* 267, 739–745.

Altner, H. and Loftus, R. (1985). Ultrastructure and function of insect thermo- and hygroreceptors. *Annual Review of Entomology* 30, 273–295.

Bartholomew, G. A. and Heinrich, B. (1978). Endothermy in African dung beetles during flight, ball making, and ball rolling. *Journal of Experimental Biology* 73, 65–83.

Bilgen, T., English, T. E., McMullend, D. C. and Storey, K. B. (2001). *EsMlp,* a muscle-LIM protein gene, is upregulated during cold exposure in the freeze-avoiding larvae of *Epiblema scudderiana. Cryobiology* 3, 11–20.

Chapman, R. F. (1965). The behaviour of nymphs of *Schistocerca gregaria* (Forskål) in a temperature gradient, with special reference to temperature preference. *Behaviour* 24, 283–317.

Colinet, H., Hance, T., Vernon, P., Bouchereau, A. and Renault, D. (2007). Does fluctuating thermal regime trigger free amino acid production in the parasitic wasp *Aphidius colemani* (Hymenoptera: Aphidiinae)? *Comparative Biochemistry and Physiology A* 147, 484–492.

Collinge, J. E., Anderson, A. R., Weeks, A. R., Johnson, T. K. and McKechnie, S. W. (2008). Latitudinal and cold-tolerance variation associate with DNA repeat-number variation in the *hsr-omega* RNA gene of *Drosophila melanogaster*. *Heredity* 101, 260–270.

Goto, S. G. (2000). Expression of *Drosophila* homologue of senescence marker protein-30 during cold acclimation. *Journal of Insect Physiology* 46, 1111–1120.

Hadley, N. F., Toolson, E. C. and Quinlan, M. C. (1989). Regional differences in cuticular permeability in the desert cicada *Diceroprocta apache*: implications for evaporative cooling. *Journal of Experimental Biology* 141, 219–230.

Hadley, N. F., Quinlan, M. C. and Kennedy, M. L. (1991). Evaporative cooling in the desert cicada: thermal efficiency and water/metabolic costs. *Journal of Experimental Biology* 159, 269–283.

Heinrich, B. (1974). Thermoregulation in bumblebees: I. Brood incubation by *Bombus vosnesenskii* queens. *Journal of Comparative Physiology* 88, 129–140.

Heinrich, B. (1975). Thermoregulation in bumblebees: II. Energetics of warm-up and free flight. *Journal of Comparative Physiology* 96, 155–166.

Heinrich, B. (1976). Heat exchange in relation to blood flow between thorax and abdomen in bumblebees. *Journal of Experimental Biology* 64, 561–585.

Heinrich, B. (1980a). Mechanisms of body-temperature regulation in honeybee, *Apis mellifera*: I. Regulation of head temperature. *Journal of Experimental Biology* 85, 61–72.

Heinrich, B. (1980b). Mechanisms of body-temperature regulation in honeybee, *Apis mellifera*: II. Regulation of thoracic temperature at high air temperature. *Journal of Experimental Biology* 85, 73–87.

Heinrich, B. and Kammer, A. E. (1973). Activation of the fibrillar muscles in the bumblebee during warm-up, stabilization of thoracic temperature and flight. *Journal of Experimental Biology* 58, 677–688.

Hoffmann, R. J. (1973). Environmental control of seasonal variation in the butterfly *Colias eurytheme*: I. Adaptive aspects of a photoperiodic response. *Evolution* 27, 387–397.

Huang, L.-H., Wang, C.-Z. and Kang, L. (2009). Cloning and expression of five heat shock protein genes in relation to cold hardening and development in the leafminer, *Liriomyza sativa*. *Journal of Insect Physiology* 55, 279–285.

Kammer, A. E. and Heinrich, B. (1974). Metabolic rates related to muscle activity in bumblebees. *Journal of Experimental Biology* 61, 219–227.

Kayukawa, T., Chen, B., Hoshizaki, S. and Ishikawa, Y. (2007). Upregulation of a desaturase is associated with enhancement of cold hardiness in the onion maggot *Delia antiqua*. *Insect Biochemistry and Molecular Biology* 37, 1160–1167.

Kerkut, G. A. and Taylor, B. J. R. (1958). The effect of temperature changes on the activity of poikilotherms. *Behaviour* 13, 259–279.

Kristensen, T. N., Hoffmann, A. A., Overgaards, J., Sorensen, J. G., Hallas, R. and Loeschcke, V. (2008). Costs and benefits of cold acclimation in field-released *Drosophila*. *Proceedings of the National Academy of Sciences USA* **105**, 216–221.

Marden, J. H. (1995). Evolutionary adaptations of contractile performance in muscle of ectothermic winter-flying moths. *Journal of Experimental Biology* **198**, 2087–2094.

Miles, C. I. (1992). Temperature compensation in the nervous system of the grasshopper. *Physiological Entomology* **17**, 169–175.

Misener, S. R., Chen, C.-P. and Walker, V. K. (2001). Cold tolerance and proline metabolic gene expression in *Drosophila melanogaster*. *Journal of Insect Physiology* **47**, 393–400.

Morin, P., McMullen, D. C. and Storey, K. B. (2005). HIF-1α involvement in low temperature and anoxia survival by a freeze tolerant insect. *Molecular and Cellular Biochemistry* **280**, 99–106.

Morrissey, R. E. and Baust, J. G. (1976). The ontogeny of cold tolerance in the gall fly, *Eurosta solidagensis*. *Journal of Insect Physiology* **22**, 431–437.

Nishikawa, M., Yokohari, F. and Ishibashi, T. (1992). Response characteristics of two types of cold receptors on the antennae of the cockroach, *Periplaneta americana* L. *Journal of Comparative Physiology A* **171**, 299–307.

Nunamaker, R. A., Dean, V. C., Murphy, K. E. and Lockwood, J. A. (1996). Stress proteins elicited by cold shock in the midge *Culicoides variipennis sonorensis* Wirth and Jones. *Comparative Biochemistry and Physiology B* **113**, 73–77.

Qin, W., Neal, S. J., Robertson, R. M., Westwood, J. T. and Walker, V. K. (2005). Cold hardening and transcriptional change in *Drosophila melanogaster*. *Insect Molecular Biology* **14**, 607–613.

Rickards, J., Kelleher, M. J. and Storey, K. B. (1987). Strategies of freeze avoidance in larvae of the goldenrod gall moth, *Epiblema scudderiana*: winter profiles of a natural population. *Journal of Insect Physiology* **33**, 443–450.

Rinehart, J. P., Hayward, S. A. L., Elnitsky, M. A., Sandro, L. H., Lee, R. E. and Denlinger, D. L. (2006a). Continuous up-regulation of heat shock proteins in larvae, but not adults, of a polar insect. *Proceedings of the National Academy of Sciences USA* **103**, 14223–14227.

Rinehart, J. P., Robich, R. M. and Denlinger, D. L. (2006b). Enhanced cold and desiccation tolerance in diapausing adults of *Culex pipiens* and a role for HSP70 in response to cold shock but not as a component of the diapause program. *Journal of Medical Entomology* **43**, 713–722.

Rinehart, J. P., Li, A., Yocum, G. D., Robich, R. M., Hayward, S. A. L. and Denlinger, D. L. (2007). Up-regulation of heat shock proteins is essential for cold survival during insect diapause. *Proceedings of the National Academy of Sciences USA*, **104**, 11130–11137.

Sinclair, B. J., Gibbs, A. G. and Roberts, S. P. (2007). Gene transcription during exposure to, and recovery from, cold and desiccation stress in *Drosophila melanogaster*. *Insect Molecular Biology* **16**, 435–443.

Steinbrecht, R. A., Lee, J.-K., Altner, H. and Zimmermann, B. (1989). Volume and surface of receptor and auxiliary cells in hygro-/thermoreceptive sensilla of moths (*Bombyx mori, Antheraea pernyi* and *A. polyphemus*). *Cell and Tissue Research*, **255**, 59–67.

Stone, G. N. and Willmer, P. G. (1989). Warm-up rates and body temperatures in bees: the importance of body size, thermal regime and phylogeny. *Journal of Experimental Biology* 147, 303–328.

Storey, J. M. and Storey, K. B. (1986). Winter survival of the gall fly larva, *Eurosta solidaginis*: profiles of fuel reserves and cryoprotectants in a natural population. *Journal of Insect Physiology* 32, 549–556.

Stower, W. J. and Griffiths, J. F. (1966). The body temperature of the desert locust (*Schistocerca gregaria*). *Entomologia Experimentalis et Applicata* 9, 127–178.

Tichy, H. (1979). Hygro- and thermoreceptive triad in antennal sensillum in the stick insect, *Carausius morosus*. *Journal of Comparative Physiology* 132, 149–152.

Wigglesworth, V. B. (1972). *The Principles of Insect Physiology*. London: Methuen.

Yokohari, F. and Tateda, H. (1976). Moist and dry hygroreceptors for relative humidity of the cockroach, *Periplaneta americana*. *Journal of Comparative Physiology* 106, 137–152.

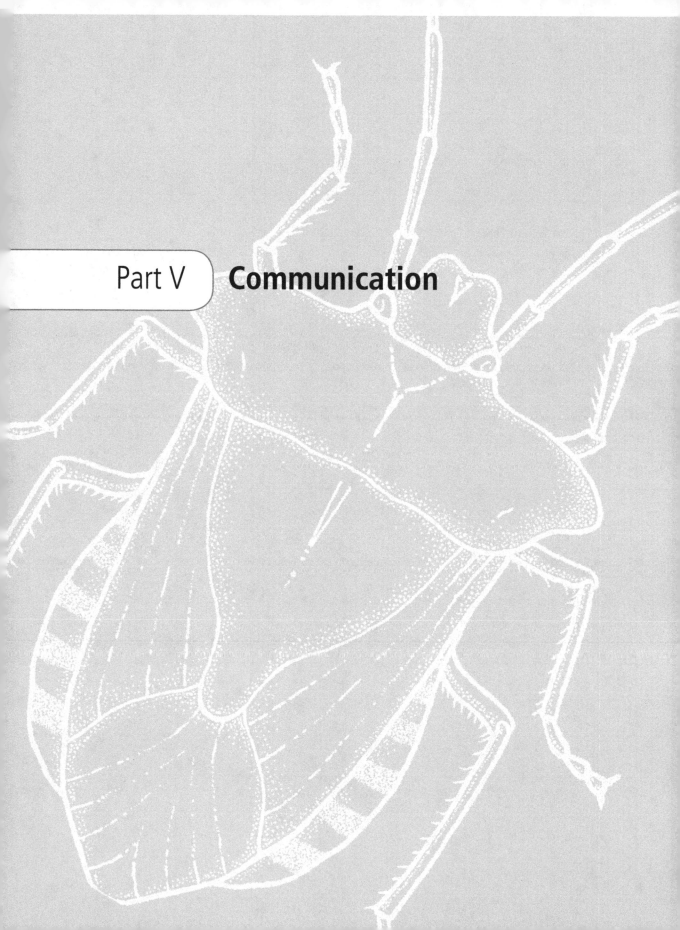

Part V **Communication**

20 Nervous system

REVISED AND UPDATED BY **STEPHEN ROGERS**

INTRODUCTION

The central nervous system is ultimately responsible for producing behavior. It synthesizes inputs from arrays of sensory neurons that individually can only encode or represent tiny parts of the total environment and produces sophisticated representations of the outside world and the internal state of the insect. It is responsible for integrating these many different sensory inputs and deciding upon and organizing appropriate behavioral responses. It does this by coordinating the activity of the approximately 300 skeletal muscles that articulate the body in a precise temporal and spatial manner, all while monitoring the consequences of its own activity during ongoing behavior. The sensory systems are considered in Chapters 22–24, and the stomodeal system, regulating gut activity, is described in Section 3.1.7. In this chapter the basic cellular components of the nervous system, both nerve cells and the supporting glial cells, are described in Section 20.1. The means by which neurons carry signals and communicate with each other is covered in Section 20.2. In Section 20.3 the anatomy of the ventral nerve cord along with the structure and function of some of the different kinds of neurons found within are considered. Section 20.4 deals with the large-scale anatomy of the brain, together with some discussion of how different regions are specialized for different functions. Finally, in Section 20.5 a few examples of how networks of neurons work together to integrate and analyze sensory information, coordinate complex movements and learn new information are described.

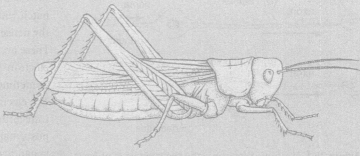

The Insects: Structure and Function (5th edition), ed. S. J. Simpson and A. E. Douglas.
Published by Cambridge University Press. © Cambridge University Press 2013.

20.1 Basic components

20.1.1 Neuron

The basic element in the nervous system is the nerve cell, or neuron (Fig. 20.1). This consists of a cell body containing the nucleus from which one or more long cytoplasmic projections extend to make contact with other neurons or with effector organs, principally muscles. The cell body is called the soma or perikaryon. It contains abundant mitochondria, Golgi complexes and rough endoplasmic reticulum. Information is conducted from one cell to another along the processes of the neuron. Neurons from most animals typically have distinct regions called dendrites, specialized for the reception of incoming signals, and a long process called an axon, which conducts signals over long distances and which terminates in regions for the passing on of signals to other neurons or effectors. In a textbook neuron the dendrites arise from the soma and the axon is a single large fiber. Nearly all insect central nervous system neurons are, however, monopolar, having only a single projection from the soma (Fig. 20.1a). This

projection, or neurite, subsequently branches to form axonal and dendritic regions, but it can be very difficult to differentiate between input and output regions by anatomy alone. The peripheral sense cells are bipolar, with a short, and usually unbranched, distal dendrite receiving stimuli from the environment and a proximal axon extending to the central ganglia (Fig. 20.1b). Some multipolar cells, with a number of branches, (Fig. 20.1c) occur in the ganglia of the nervous system and are also associated with stretch receptors. In the central nervous system it is common for both dendritic and axonal regions to branch extensively; they are said to arborize. Golgi complexes and rough endoplasmic reticulum are absent from the dendrites and axon. No protein synthesis occurs in these parts so all organelles and proteins within them are manufactured in the soma. Numerous microtubules, about 20 nm in diameter, run along the branches, possibly providing pathways for transport of material from the soma.

The points at which neurons receive information from, or convey it to, other cells are known as synapses. At most synapses, where transmission from one cell to another involves a chemical messenger, the plasma membranes of the two cells lie parallel with each other and are close together, separated by a gap of 20–25 nm, known as the synaptic cleft. These chemical synapses are characterized, when viewed in section with the electron microscope, by an area of electron-dense material adjacent to the membrane of the fiber transmitting information, the presynaptic fiber. The dense material, which possibly represents proteins involved in the release of chemical transmitter, forms a patch 150–500 nm across. In the cytoplasm adjacent to the dense material are synaptic vesicles (Fig. 20.2a). These are often small, 50–80 nm in diameter, and electron-lucent, but may be larger and electron-dense. Sometimes both types are present at a single synapse. The electron-lucent vesicles contain the chemical transmitter that conveys information to the following postsynaptic cell. They release their contents into the synaptic cleft by exocytosis. The electron-dense granules contain neuropeptides with a modulatory

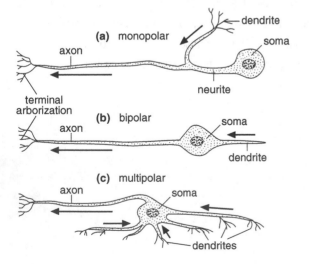

Figure 20.1 Neurons. Diagrammatic representation of basic types of neuron. Arrow shows direction of conduction.

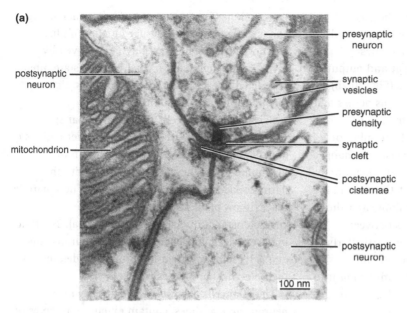

(a)

presynaptic neuron

postsynaptic neuron

synaptic vesicles

presynaptic density

mitochondrion

synaptic cleft

postsynaptic cisternae

postsynaptic neuron

100 nm

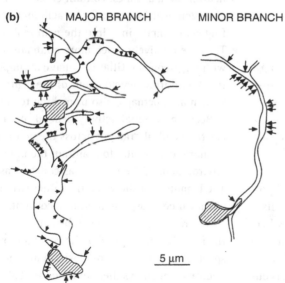

(b)   MAJOR BRANCH   MINOR BRANCH

5 µm

Figure 20.2 Synapses. (a) Synapse in the optic lobe of *Drosophila*. The presynaptic cell is a retinula cell; the two postsynaptic cells are monopolar interneurons (see Fig. 20.21) (after Meinertzhagen and O'Neil, 1991). (b) Synapses on small sections of branches of a non-spiking neuron. Inward-pointing arrows indicate positions of input synapses from other neurons; outward-pointing arrowheads show positions of output synapses. Notice that the major branch has a mixture of outward and inward synapses; the minor branch has only input synapses. Other branches (not shown) have only output synapses. The total number of synapses on this cell would be many thousands. Hatched areas show points at which branches extend out of the plane of view (after Watson and Burrows, 1988).

function (Section 20.2.3). The postsynaptic cell has electron-dense material adjacent to the membrane opposite the presynaptic specialization; sometimes, as in Fig. 20.2a, there are areas of disc-like vesicles called cisternae.

A distinction needs to be made between physical and functional synapses. Cells may make many synaptic contacts with each other, but physiologically they may represent a single synapse, evoking a single postsynaptic potential (see below) in the connecting neuron. A single retinula cell in the eye of a fly has about 200 physical output synapses, while some interneurons of the locust may have as many as one million physical output synapses.

Sometimes, a distinction can be made between regions containing only output synapses (axonal) and others containing solely input branches (dendritic), but in many neurons input and output synapses occur on the same branches (Fig. 20.2b). Sometimes a single output synapse spans across and therefore communicates with more than one postsynaptic cell, as in Fig. 20.2a. The number of synapses in a cell is not necessarily constant; changes may occur during postembryonic development and also with the experience of the insect. Changes in the number of physical synaptic connections alter the strength of physiological connections between neurons and are important in learning, particularly in the formation of long-term memories. A single neuron typically has input synapses connecting it to more than one presynaptic neuron and generates output to more than one postsynaptic cell.

Some neurons are in direct electrical contact with each other, instead of communicating chemically. At these electrical synapses the membranes of the two cells are separated by a gap of only 3.5 nm, and current can flow directly across the gap. This type of connection between cells is called a gap junction.

20.1.2 Glial cells

Each neuron is, except for its finest branches, almost wholly invested by folds of one or more glial cells. Synaptic contacts between neurons occur only where glial folds are absent, although the absence of a glial fold does not necessarily indicate a synapse. Glial cells are much more numerous in the central nervous system than neurons. Some are closely associated with the somata, others are at the surface of the neuropile with processes extending inward to invest the axons and dendrites. In the case of small axons, several may be enclosed within one glial fold (Fig. 20.3a, axon B), but larger axons are usually individually enveloped. The enveloping sheath may consist of a single fold or the fold may coil around several times so that there are several layers forming

the envelope (Fig. 20.3a, axon A). In general, larger axons are invested by a larger number of glial windings. The same effect may be achieved by several overlapping cells. Glial processes are often connected to each other by desmosomes, tight junctions and gap junctions.

Between the glial cells are extracellular spaces. These are relatively extensive in the outer regions of ganglia, but much more restricted within the neuropile, where they are continuous with the narrow spaces between the glial folds. These narrow spaces show periodic lacunae, which are characteristic of insect nerves (Fig. 20.3a). The fluid in the extracellular spaces bathes the neurons directly and is therefore of great importance in determining their electrical properties (see below).

The glial cells pass nutrient materials to the neurons and, at times, contain extensive reserves of glycogen. Nutrient transfer is facilitated by finger-like inpushings into the neuronal somata. They are relatively much more abundant in neurons with large somata. Glial cells have an important function during development of the central nervous system and, perhaps, also in making repairs.

Because neuronal activity depends on an electrochemical gradient across the neuronal membrane, it is vital to maintain the immediate environment of the neurons as constant as possible. The hemolymph shows marked fluctuations in chemical composition and ionic concentration. It is therefore necessary to shield the nervous system from the hemolymph to provide the constant extracellular space of correct ionic composition around the neurons that allows them to function. This constancy is provided by a layer of specialized glial cells, known as the perineurium, around the outside of the nervous system (Fig. 20.3a), which are mesodermal in origin in contrast to the ectodermal origin of other glial cells.

The cells of the perineurium are held together by tight junctions (in adult insects) and septate desmosomes. On the inner side, the cells have relatively extensive processes which penetrate

(a)

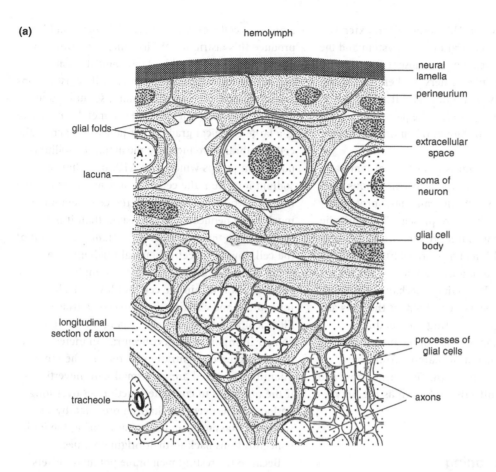

(b)

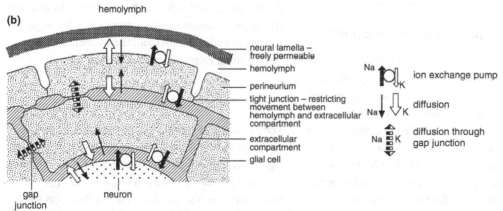

Figure 20.3 Glia. (a) Diagrammatic section of part of an abdominal ganglion showing glial cells (light shading) that play different roles. Axons A and B are referred to in the text (after Smith and Treherne, 1963). (b) Representation of the role of glial cells in maintaining a constant ionic composition of the fluid bathing the neurons in a ganglion. The extracellular compartment is shown hatched to emphasize its isolation from the hemocoel. The pumps maintain a high concentration of sodium and low potassium in this compartment despite the diffusive movements of the ions. Broad open arrows indicate greater permeability to potassium than to sodium ions (after Treherne and Schofield, 1981).

between other glial cells. The perineurium extends over the whole of the central nervous system and the larger peripheral nerves, but it is absent from the finer branches of peripheral nerves where the axons are wrapped by individual glial cells. It forms the "blood–brain barrier," allowing the passage only of specific substances into the neural environment (see below).

Outside the perineurium is a thick basal lamina known as the neural lamella. It is an amorphous layer of neutral mucopolysaccharide and mucoprotein in which collagen-like fibrils are present. The fibrils lie parallel with the surface, but are otherwise randomly oriented. The neural lamella is probably secreted by cells of the perineurium, although other cells may also contribute to it. It provides mechanical support for the central nervous system, holding the cells and axons together while permitting the flexibility necessitated by the movements of the insect. It offers no resistance to diffusion of material from the hemolymph into the nerve cord. The perineurium and neural lamella are collectively known as the nerve sheath.

20.2 Basic functioning

20.2.1 Electrical properties of the neuron

A difference in electrical potential is maintained across the plasma membrane of all animal cells; neurons are no exception, with the inside being negatively charged with respect to the outside. This resting membrane potential is often about -70 mV. Neurons, muscle cells and some glandular cells stand apart from most cell types in that they have the property of *excitability* – that is, the electrical potential across the cell membrane can change rapidly, and this is the basis of most neuronal signaling.

The resting membrane potential arises from the differential distribution of positively and negatively charged ions inside and outside the cell (Table 20.1), creating an electrochemical gradient

across the cell membrane. Several factors combine to produce this distribution: First, there is a Donnan equilibrium established by large, non-diffusible organic anions within the neuron, while most cations are small ions, such as potassium$^+$, sodium$^+$ and calcium^{++}, which can cross the cell membrane. There is a large chemical gradient for potassium across the cell membrane produced by the action of sodium–potassium pumps which uses ATP to exchange sodium ions from the cell for potassium ions outside it in a 3:2 ratio. Not only that, the cell membrane is much more permeable to potassium than it is to sodium, with the result that it constantly leaks out of the cell down its electrochemical gradient. The net result is that potassium ions are at a high concentration within the cell, but low outside, while for sodium and chloride the converse is true. Electrical activity by neurons changes their ionic composition; the sodium concentration increases and the potassium concentration decreases. The ionic quantities involved are very small, but, nevertheless, if the neuron is to continue functioning over long periods the recovery mechanism provided by the sodium–potassium pumps is vital to bring the ionic concentrations back to their original values.

Because the resting membrane potential differs from zero, the cell is said to be polarized. A change in the potential toward zero, a decrease in its negative charge, is called depolarization; an increase in negativity is called hyperpolarization.

20.2.2 Signal transmission

Transmission of signals via the nervous system involves three different processes. First, the incoming signal, which may be visual, mechanical or chemical, for example, is converted to electrical energy (transduction). The transduction mechanisms used by sensory cells are considered in Chapters 22–24. The change in membrane potential produced by the incoming signal is called the receptor or generator potential in the case of sensory neurons, the postsynaptic potential at synapses within the

Table 20.1 **Ionic concentrations (mmol kg^{-1} tissue) within the nerve sheath and in hemolymph of some insects**

Species	Region	Na	K	Ca	Mg
Periplaneta	Hemolymph	156	8	4	5
	Nerve cord	76	132	–	–
Romalea	Hemolymph	56	18	–	–
	Nerve cord	69	89	–	–
Carausius	Hemolymph	15	18	7	53
	Nerve cord	64	313	30	22

Source: After Treherne (1974).

central nervous system or the end plate potential in muscles. These potentials spread *passively* through the neuron, but they show an exponential decay in amplitude as current leaks back across the neural membrane. Two terms, the space constant and the time constant, define how far the current will passively spread and how long it takes to decay. This decay in the incoming signal is the reason why the second process is required where the electrical signal is actively conducted along the axon as action potentials in a self-sustaining wave of depolarization that covers large distances without diminishing. There are some neurons, however, which do not produce action potentials and rely entirely on the passive spread of the postsynaptic potential. Finally, the electrical signal is usually converted to a chemical signal for transmission to the following cell at a synapse.

Action potentials Unlike the receptor or postsynaptic potential, which typically varies considerably in amplitude depending on the strength of the incoming signal (see below), the action potential, or spike, is of near constant amplitude. The number of action potentials produced is proportional to the size of the postsynaptic potential (or receptor potential in the case of sensory neurons). As all the action potentials produced by one neuron have the same amplitude, information about the magnitude of any signal is conveyed by their number and frequency. When a generator or postsynaptic potential is converted into action potentials, an analog signal coded by voltage amplitude is converted into a digital signal coded by spike rate and number.

The current produced by the receptor or postsynaptic potential produces a small depolarization of the membrane in the axon as it spreads passively from its origin. This small depolarization is sufficient to cause voltage-sensitive sodium channels to open, and this results in a rapid positive swing in the charge on the inside of the membrane, amounting to 80–100 mV in the cockroach. This is the rising phase of the action potential (Fig. 20.4a). The period of permeability to sodium is short-lived because the sodium channels close automatically and become inactivated. Potassium channels, called delayed rectifiers, are also activated by the change in voltage, but do so more slowly than sodium channels, starting to open as the sodium channels close. As a result, potassium flows out of the fiber and down its electrochemical gradient, which again becomes more negatively charged on the inside. This is the falling phase of the action potential. The total duration of the action potential is very brief, only 2–3 ms.

(a)

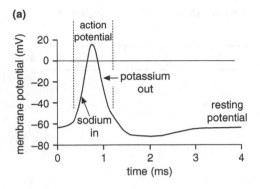

(b)

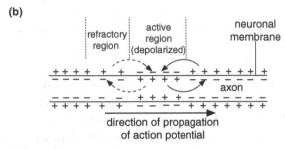

(c)

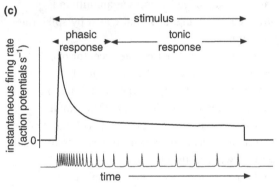

Figure 20.4 Action potential. (a) Changes in the membrane potential associated with the production of an action potential. (b) Diagram showing the electrical charge across the axonal membrane and propagation of the action potential. At the extreme ends, the membrane is polarized; the inside is negatively charged with respect to the outside. The action potential results from depolarization, the inside of the membrane becoming less negatively charged and even becoming positive. This depolarization spreads passively along the membrane (shown by arrows). In the region through which it has recently passed (the refractory region), the outward flow of potassium offsets the build-up of positive charge. Consequently, the membrane repolarizes in this region, but becomes depolarized in the previously

A current flows in a local circuit away from the point of depolarization inside the axon, and toward it on the outside (Fig. 20.4b). Where this current reaches an area of resting membrane, it produces a slight depolarization, leading to the further changes in permeability, first to sodium and then to potassium, that produce an action potential. This is a continuous process so that the action potential is propagated along the fiber. Although current flows out from the action potential in both directions, no further change occurs in the membrane potential of the area through which the action potential has just passed because the sodium channels remain inactivated and the continued outflow of potassium negates any tendency for depolarization.

The frequency at which action potentials can follow each other is limited by the refractory condition of the axon immediately following the passage of an action potential. The period for which no further action potentials can be produced is known as the absolute refractory period; it lasts for 2–3 ms (Fig. 20.4b). After this, an action potential can be produced, but only by a very strong stimulus (a large receptor potential) and, as the axon recovers, progressively weaker stimuli are able to produce action potentials until the level of excitation returns to normal. This relative refractory period lasts for 10–15 ms.

The rate at which action potentials are produced by a neuron decreases with time after arrival of the stimulus even if that stimulus remains at constant strength (Fig. 20.4c). At first, action potentials are often produced in rapid sequence, but the intervals between them get longer within a few milliseconds. This type of response is called phasic. In many neurons it is followed by a more sustained response during which the rate of action potential production

resting membrane (to the right of center in the diagram). (c) Below: a train of action potentials from a peripheral sensory neuron showing the phasic and tonic periods; above: the same information shown as the instantaneous firing rate.

remains relatively constant, or declines only slowly. This is called a tonic response. There is a continuum of responsiveness between different neurons. Some are highly phasic and stop firing entirely after a brief period, others show little variation in the face of a constant level of stimulation, but most neurons fall somewhere midway between these two extremes. The rate of action potential production is often called the firing rate. This is often expressed as the total number of action potentials produced in one second (or some other appropriate unit of time). However, because the rate of firing may vary considerably during the phasic period of a response, it is often expressed as the instantaneous firing rate, which is based on the intervals between action potentials. For example, a cell producing two action potentials separated by 10 ms is said to have an instantaneous firing rate of 100 s^{-1} at that moment (the number of 10-ms intervals in one second).

The production of an action potential in a neuron normally results from stimulation of receptors, but many neurons also fire spontaneously in the absence of any such input. This spontaneous activity is frequently achieved by the activity of separate classes of ion channel than those used to produce action potentials. These additional channels prevent full repolarization of the membrane occurring. This maintains the system in a highly active state so that it is more easily excited. Stimuli which would be subthreshold in an unexcited fiber may produce a response in a fiber already discharging spontaneously. The firing rate of a spontaneously active neuron may be increased, decreased or remain unaffected by the input of other neurons, in particular from neuromodulators (Section 20.2.3).

The velocity of conduction in a nerve fiber is proportional to the square root of its diameter. Giant fibers, with diameters of 8–50 μm, have conduction speeds of 3–7 m s^{-1}; fibers of about 5 μm diameter conduct at speeds of 1.5–2.3 m s^{-1}. The velocity of conduction also varies with temperature. Not all neurons produce action potentials. In some cells the passive spread of the receptor or postsynaptic site reaches throughout the cell and is sufficient to produce changes in membrane potential at presynaptic sites that lead to the release of a transmitter substance. The retinula cells of the eye are of this type, as are some interneurons in the central nervous system. The latter are called non-spiking interneurons (Section 20.3.2). Non-spiking communication is limited by the rapid decay of passive signals with time and distance from their origin. This means that non-spiking neurons tend to be relatively small, or at least they have a short distance between their input and output synapses. Non-spiking neurons, however, do allow output synapses to respond directly to the strength of input signal without having to integrate a train of action potentials in order to gauge stimulus strength, which takes longer to perform.

Events at the synapse *Chemical synapses.*

The arrival of an action potential at a presynaptic terminal of a chemical synapse depolarizes it and results in the opening of voltage-sensitive calcium channels so that calcium ions move into the neuron, which initiates the complex series of events that eventually leads to synaptic transmission. In summary, calcium increases the probability that a synaptic vesicle, a bubble of membrane in the cytoplasm containing transmitter substance, will fuse with the neuronal membrane and release the transmitter into the synaptic cleft. The greater the number or frequency of arrival of action potentials, the more calcium enters the cell and the more synaptic vesicles release their contents.

Each vesicle contains a similar amount of transmitter so that the release of multiple vesicles increases the total amount of transmitter released in a stepwise manner. The basic quantity of transmitter per vesicle is called a quantum. Because the release of transmitter is essentially a probabilistic event, it occurs occasionally even in the absence of any electrical activity in the presynaptic axon.

The transmitter affects the permeability of the postsynaptic membrane either directly, causing ligand-gated ion channels to open (ion tropic signaling), or indirectly, binding to membrane receptors and activating second messenger systems (metabotropic signaling). In the former case the postsynaptic change may be a depolarization or a hyperpolarization of the membrane, depending on the nature of the transmitter substance and the receptors present on the postsynaptic membrane. Depolarization is produced by opening cation channels in the membrane. As sodium and calcium ions flow in faster than potassium ions move out, the magnitude of the negative charge on the inside of the membrane is reduced. Hyperpolarization is produced by an inward movement of chloride ions, increasing the negative charge. A depolarization is excitatory; it increases the probability that action potentials will be produced. A hyperpolarization is inhibitory; it decreases the probability that action potentials will be produced. The changes are known as excitatory postsynaptic potentials (EPSPs) and inhibitory postsynaptic potentials (IPSPs), respectively. Metabotropic signaling across a synapse may not cause any immediate change in potential across the neural membrane but can have profound effects on the excitability of neurons and hence how they respond to ionotropic synaptic transmission.

The magnitude of the postsynaptic potential is variable (graded), and is proportional to the amount of transmitter substance crossing the synapse from the presynaptic terminal. Action potentials are only produced if the change in membrane potential at the action potential initiation zone exceeds a certain magnitude. The action potential initiation zone is normally near the soma in sensory neurons, but its placement is highly variable in interneurons and motor neurons, to the extent that some large intersegmental neurons in insects are known to have two or more separate spike initiation zones. Action potential initiation does not depend on the activity of a single synapse, but is the sum of all the postsynaptic events occurring in the neuron; the PSPs are said to summate. Summation may be either spatial or temporal (Fig. 20.5).

Electrical synapses. An electrical synapse is formed by a gap junction between two neurons. At an electrical synapse there is a direct flow of current from the pre- to the postsynaptic cell. Since there is no chemical cascade leading to transmitter release and diffusion of a transmitter across a synaptic cleft, transmission is extremely rapid. Some electrical synapses in some animals are known to conduct in either direction, but it is not known if this is also true in insects.

20.2.3 Chemical messengers of neurons

A variety of chemical messengers are produced and released by insect neurons. These substances can be grouped into five classes, based on their chemical structure: acetylcholine, biogenic amines, amino acids, nitric oxide and peptides (Fig. 20.6; Table 20.2). Some 100 different neuropeptides have been identified in insect nervous systems, although in many cases their functions are not known. Neuropeptides are synthesized in the cell soma and transported to release points along the axon; most other chemical messengers of neurons are synthesized in the terminal regions of the axon. The chemical messengers have three types of function, depending on the spatial scale over which they operate.

Neurotransmitters are only released into the synaptic cleft and have a transient effect on the electrical potential of the postsynaptic membrane. The effect is transient due to enzymatic degradation of the transmitter molecule or its re-uptake into the presynaptic terminal.

Neuromodulators are released in the vicinity of the synapse, modifying synaptic transmission. Their effects are relatively slow and long-lasting.

Neurohormones are released into the hemolymph from neurohemal release areas and function as hormones.

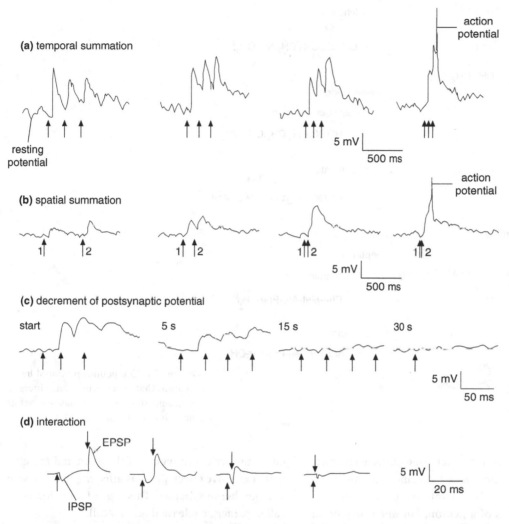

Figure 20.5 Postsynaptic changes in membrane potential. Each trace shows changes in the membrane potential of the postsynaptic interneuron following stimulation by one or more presynaptic neurons. Arrows indicate the arrival of an action potential in the presynaptic axon. Notice the different time scales. (a) Temporal summation in an interneuron. Stimulation via a single axon results in a progressive increase in the postsynaptic potential as the intervals between stimuli are reduced. At very short intervals, an action potential is produced (extreme right) (input from locust tibial mechanoreceptor) (after Newland and Burrows, 1994). (b) Spatial summation in an interneuron. Stimulation via two separate axons (labeled 1 and 2) produces an action potential when the stimuli are close together (extreme right) (input from locust tibial mechanoreceptors) (after Newland and Burrows, 1994). (c) Depression due to previous activity. Continuous stimulation over a 30-second period results in a progressive decline in the postsynaptic potential. Figure shows the changes in membrane potential of an interneuron at various times after the start of stimulation (after Newland and Burrows, 1994). (d) Interaction of excitatory and inhibitory effects. Summation between an excitatory postsynaptic potential (EPSP) and an inhibitory postsynaptic potential (IPSP) is dependent on the interval between the presynaptic events shown by arrows (cockroach giant interneuron) (after Callec, 1985).

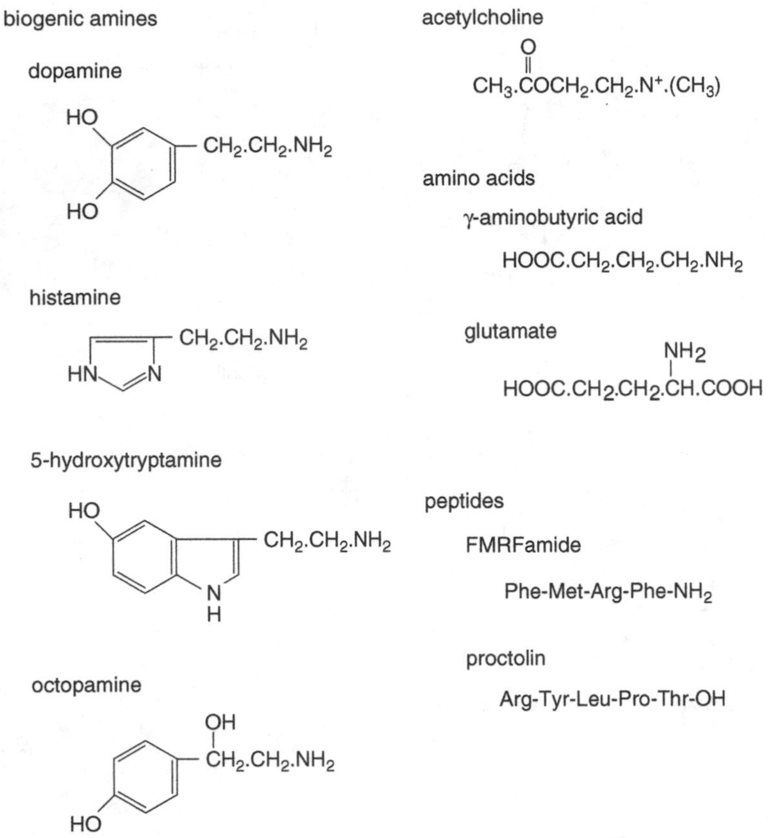

biogenic amines

dopamine

HO— (benzene ring) —CH₂.CH₂.NH₂
HO—

histamine

(imidazole ring) —CH₂.CH₂.NH₂
HN—N

5-hydroxytryptamine

HO— (indole ring) —CH₂.CH₂.NH₂
N
H

octopamine

OH
(benzene ring)—CH.CH₂.NH₂
HO—

acetylcholine

$$CH_3.COCH_2.CH_2.N^+.(CH_3)$$
(with O double-bonded above the first C)

amino acids

γ-aminobutyric acid

$$HOOC.CH_2.CH_2.CH_2.NH_2$$

glutamate

$$HOOC.CH_2.CH_2.CH.COOH$$
(with NH₂ on the CH)

peptides

FMRFamide

Phe-Met-Arg-Phe-NH₂

proctolin

Arg-Tyr-Leu-Pro-Thr-OH

Figure 20.6 Compounds produced by neurons that act as neurotransmitters or neuromodulators. Some may also act as neurohormones (see Fig. 21.3).

It is common for more than one of these chemical messengers to be produced by a single neuron, sometimes at spatially separate release sites on different branches of a neuron, but when they occur together at the same synapse they are said to be co-localized. As their release from the neuron is regulated by its electrical activity they may be co-released, although the release of neuromodulators may only occur with higher levels of electrical activity. While the neurotransmitter is released only into the focal area of the synaptic cleft, neuromodulator release occurs over a broader area. It may be limited to the presynaptic membrane immediately surrounding the synapse. This is called parasynaptic release (Fig. 20.7b) and the effects of the neuromodulator occur only within a limited area. In other cases, the neuromodulator is released over a larger unspecialized region of the axon and has a more extensive effect, perhaps affecting transmission at a number of synapses. This type of secretion is called paracrine release (Fig. 20.7c,d).

Neurotransmitters Acetylcholine is probably the most widespread excitatory transmitter in the insect nervous system (Table 20.2). It is the major transmitter of olfactory and mechanosensory neurons and of many interneurons. Serotonin appears to be a co-transmitter in some chordotonal and multipolar sensory neurons, while histamine is the transmitter in retinula cells of the compound eyes and ocelli. The salivary glands of cockroaches receive innervations from two types of neuron containing either serotonin or dopamine. The type of saliva produced depends on the neurotransmitter

Table 20.2 **Substances secreted by neurons in the central nervous system and their principal functions**

Chemical	Neurotransmitter	Neuromodulator	Neurohormone
Acetylcholine (excitatory in CNS)	+++		
Biogenic amines			
Dopamine	+	+	
Histamine	+		
Serotonin (5-hydroxytryptamine)		++	+
Octopamine	+	+++	+
Amino acids			
γ-aminobutyric acid	+++		
(GABA)	(inhibitory)		
Glutamate (excitatory at muscles)	+++		
Peptides			
Adipokinetic hormone			+++
Allatotropin		+	+
Allatostatin		+	+
Cardioactive peptide			+++
Diuretic peptide			+++
FMRFamide		+	
Leucokinin		+	
Pheromone biosynthesis (PBAN)		+	++
Proctolin	+	+++	+
Prothoracicotropic hormone			+++

Note:
+++, major function; +, occasional function.

released – serotonin leads to the release of protein-rich saliva, while dopamine leads to the release of protein-free saliva. Octopamine functions as a neurotransmitter in the light organs of fireflies.

Glutamate is the principal excitatory transmitter at insect nerve–muscle junctions, but also occurs in some interneurons. γ-aminobutyric acid (GABA) is the principal inhibitory neurotransmitter both in the central nervous system and at neuromuscular junctions. Other compounds among those listed in Table 20.2 may also sometimes function as neurotransmitters, but the distinction between neurotransmitter and neuromodulator is not always clear.

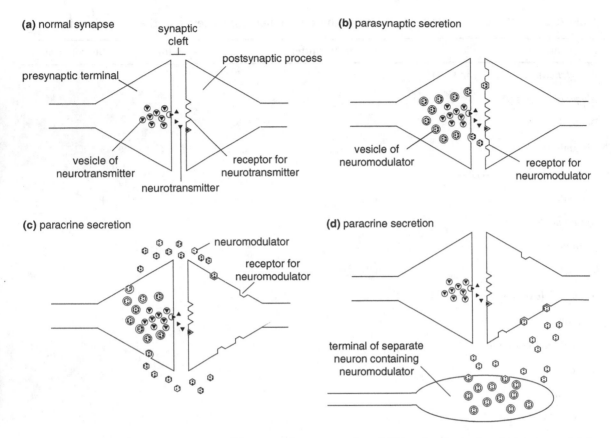

(a) normal synapse

synaptic cleft

postsynaptic process

presynaptic terminal

vesicle of neurotransmitter

receptor for neurotransmitter

neurotransmitter

(b) parasynaptic secretion

vesicle of neuromodulator

receptor for neuromodulator

(c) paracrine secretion

neuromodulator

receptor for neuromodulator

(d) paracrine secretion

terminal of separate neuron containing neuromodulator

Figure 20.7 Neuromodulation at a synapse. Diagrammatic representation of different pathways by which neuromodulators may affect the postsynaptic element. Transmitter compound is shown as black triangles; neuromodulator as shaded hexagons. (a) A synapse with no neuromodulators. (b) Parasynaptic secretion. The neuromodulator co-occurs with the transmitter and acts close to the synapse. (c) Paracrine secretion. The neuromodulator co-occurs with the transmitter but is released around the area of the synapse. It may affect other neurons in the vicinity. (d) Paracrine secretion. The neuromodulator is produced by a separate neuron and affects the general environment around the synapse.

Neuromodulators The synaptic interaction of one neuron with another neuron, or with a muscle, mediated by a neurotransmitter can be altered (modulated) by chemicals known as neuromodulators. Neuromodulators can act presynaptically, altering the tendency for vesicles of transmitter substance to be released, or postsynaptically, altering the response of the postsynaptic membrane to a given quantity of neurotransmitter by making it more or less excitable. They produce these effects by modifying the ionic permeability of the pre- or postsynaptic cell by binding to specific membrane receptors, affecting the cell's ion channels via second messenger systems.

Biogenic amines and a large number of different neuropeptides are widely distributed in the central nervous system, suggesting that they are of considerable importance. In some cases they are known to act as neuromodulators, but their function is only inferred in a majority of instances. Given that many biogenic amines and peptides function by modifying how neurons function, it is unsurprising that many of them have been found to be implicated in learning or other experience-driven modifications of behavior.

(a) interneuron

in flight

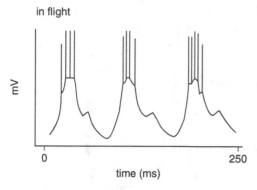

quiescent

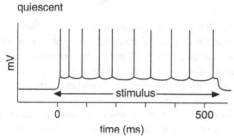

quiescent + octopamine

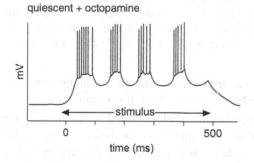

(b) stretch receptor

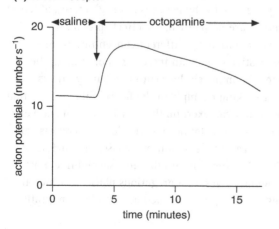

The overall effects of neuromodulation may be more profound than changes at single synapses might suggest. The stomatogastric ganglion of Crustacea, which consists of a small network of just 30 neurons, is a model system for analyzing the effects of neuromodulation, which can deliver several radically different output patterns depending on the relative amounts of up to 15 different neuromodulators released into the system.

Octopamine. Octopamine has a widespread modulatory effect on the activity of skeletal muscles (see Section 10.3.2), but it also modulates the activity of some interneurons and sensory neurons. One example of its effect on interneurons concerns the generation of the pattern of activity of locust flight muscles. During flight the interneurons organizing the wing-beat pattern produce action potentials in bursts (Fig. 20.8a), but a similar bursting pattern cannot be generated experimentally when the insect is at rest unless octopamine is applied to the preparation. It is known that the concentration of octopamine in the metathoracic ganglion increases during the early part of flight, and it is likely that this increase is responsible for a change in the properties of the cell membrane of these interneurons.

Octopamine can also affect muscle properties themselves. In locusts, the dorsal and ventral unpaired median cells in the thoracic and abdominal ganglia and some interneurons in the brain are octopaminergic. Octopamine, released by efferent

Figure 20.8 Neuromodulatory effects of octopamine in the locust. (a) Effect on an interneuron associated with the flight muscles. When the insect is in flight the action potentials in the interneuron occur in bursts, but stimulation of the interneuron in a quiescent insect produces a train of action potentials. A bursting pattern similar to that occurring in flight can be produced experimentally by applying octopamine to the quiescent preparation (after Orchard *et al.* 1993). (b) Output from a wing stretch receptor with the wing oscillating at 18 Hz (similar to the normal wingbeat frequency). When octopamine was added the output from the stretch receptor increased for a time (after Ramirez and Orchard, 1990).

dorsal unpaired neurons that innervate the flight muscles, stimulates glycolysis immediately before and during takeoff when rapidly available energy is required. During sustained flight, when locusts switch to lipid metabolism, the octopaminergic neurons cease activity, with a corresponding decrease in glycolysis.

Octopamine may act via the hemolymph as a neurohormone modulating the activity of sensory neurons. This occurs with the neuron of the locust forewing stretch receptor. Its activity is enhanced by octopamine (Fig. 20.8b), which is released into the hemolymph at the beginning of flight.

Octopamine has a role in regulating aggressive behaviors during social encounters in insects such as crickets, although the mechanism by which it acts is still unclear. In another context, octopamine-containing neurons are involved in learning, specifically in linking novel stimuli to food or other rewarding outcomes, which has been shown in bees, flies and crickets. Another biogenic amine, dopamine, has been implicated in learning to link novel stimuli with aversive or "punishment" outcomes, often in the same sensory systems as octopamine mediates rewards.

Serotonin. Relatively small numbers of serotonergic neurons, mostly interneurons, are present in the central nervous system. Adult *Calliphora*, for example, have only 154 such neurons, some occurring as homologous repeats in each neuromere. Despite the relatively small number, most regions of neuropile are invaded by the branches of serotonergic neurons. In locusts, serotonergic axons run to the gut, the reproductive organs and the salivary glands. Comparable innervation does not occur in *Calliphora*, but in this insect serotonin is released as a neurohormone to act on the same organs.

An example of neuromodulation by serotonin is known from studies on the olfactory lobe of *Manduca*, where a single serotonergic neuron innervates all the glomeruli (see Fig. 20.23c). A low concentration of serotonin reduces the activity of

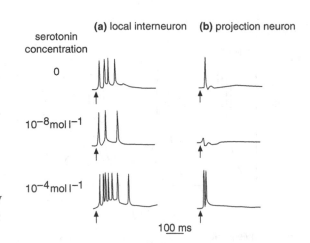

Figure 20.9 Neuromodulatory effects of serotonin on interneurons in the olfactory lobe of *Manduca*. Notice that a low concentration (10^{-8} mol l^{-1}) of serotonin reduced the number of action potentials produced by both neurons when the antennal nerve was stimulated, but a higher concentration (10^{-4} mol l^{-1}) had the opposite effect. Arrows show the time at which the antennal nerve was stimulated (after Kloppenburg and Hildebrand, 1995). A diagrammatic representation of the two types of interneuron is shown in Fig. 20.23b. (a) Local interneuron restricted to the antennal lobe. (b) Projection neuron connecting a glomerulus to other parts of the brain.

olfactory-lobe neurons, while a higher concentration enhances it (Fig. 20.9). It is known that experimentally administered serotonin alters the olfactory responses of the honey bee, *Apis*, and it may have similar effects on *Manduca* and other insects. Although the concentrations of serotonin experienced by the neurons in the glomeruli of living insects are not known, the effects produced experimentally by different concentrations are suggestive of how an insect's response might be altered completely by neuromodulatory effects.

A striking example of the far-reaching effects that serotonin can exert on the nervous system and hence on behavior is found in desert locusts. Locusts can transform from a solitarious phase, in which they shun the company of other locusts and live mostly sedentary lives, to a gregarious phase, in which they aggregate together, sometimes into huge migrating swarms. Phase change is a long process that eventually encompasses many changes in morphology and physiology, but the start of the process of swarm formation is a realignment of behavior away from being repelled by other locusts toward being mutually attracted. Serotonin is both sufficient and necessary to cause this initial change in behavior.

Nitric oxide. The gaseous neurotransmitter nitric oxide differs from other neurochemicals in several key respects. It is synthesized (from the amino acid arginine) immediately before release and diffuses rapidly, readily crossing cell membranes and therefore bypassing conventional synaptic machinery. It is a short-lived molecule, which serves to limit how far it can spread from its release site. Neurons containing the enzyme nitric oxide synthase are found throughout the central nervous system, with a particularly high density of fibers in the olfactory system, mushroom bodies and central complex in all insect species studied. By contrast, the amount of nitric oxide synthase found in the visual system appears to differ considerably between species. In the ventral nerve cord, the mechanosensory neuropiles which receive inputs from sensory structures on the legs are richly supplied with fibers from nitric oxide-producing neurons. Being a comparatively recently discovered neurotransmitter, the functions of nitric oxide in the nervous system are still only partially known. Nitric oxide promotes the habituation, that is, the progressive weakening, of the proboscis extension reflex of bees on repeated applications of odors. It has also been implicated in modifying diverse neuronal circuits, such as increasing the digging rhythm of ovipositing locusts, and in the periphery of specifically altering the sensitivity of the taste receptors to salts.

Neuropeptides. Although neuropeptides occur in cells of the central nervous system, their functions are not generally understood except for those that are neurohormones. One exception is proctolin, which functions as a neuromodulator at some visceral and skeletal muscle junctions (Section 10.3.2). About 500 proctolinergic neurons, both interneurons and motor neurons, are found in various parts of the central nervous system of *Calliphora*. The advent of full genome sequencing has made finding neuropeptides somewhat easier. The *Drosophila* genome contains at least 35 neuropeptide precursor genes, some of which can generate several final neuropeptide transmitters. Individual neuropeptides are often confined to relatively small numbers of interneurons, but collectively a great number of neurons express one or more neuropeptides, often in conjunction with other transmitter types. The accessory medulla in the visual system and the central complex (Section 20.4.1) are particularly richly supplied with neuropeptide-containing neurons, as are some regions of the protocerebrum that contain the cell bodies of neurosecretory cells.

Pigment-dispersing factor is a neuropeptide that is a principle transmitter of the small number of clock neurons responsible for setting the circadian rhythm of physiological and behavioral activity of insects. Intrinsic molecular feedback loops regulate the activity of clock genes contained within these neurons, which can be modulated by inputs from the visual system that respond to ambient light levels. Pigment-dispersing factor is in part responsible for harmonizing the activity of the different populations of clock neurons in the brain, as well as signaling an unknown number of other neuronal targets.

Neurohormones Neurohormones are produced by neurons in various ganglia of the central nervous system. Their distribution and functions are considered in Chapter 21. The cells producing neurohormones are commonly called neurosecretory cells. With advancing knowledge it has become clear that many neurons produce comparable secretions which do not necessarily function as hormones, but rather as neuromodulators (see above). In this text, the terms neurosecretion and neurosecretory cell will be used only in cases where the secretion is known to have a hormonal function, being transported in the hemolymph.

20.3 Anatomy of the nervous system

20.3.1 Ganglia

The ganglia of the central nervous system contain the somata of all the interneurons and motor neurons, as well as their axonal and dendritic processes. The axons of motor neurons leave through peripheral nerves while incoming sensory axons terminate and arborize within the ganglion. Within the ganglia, the somata are grouped peripherally, forming the cortex. The core of each ganglion is called the neuropile and is occupied by the arborizations of all the neurons. The neuropile is the site of synaptic integration. Within this mass of fibers, some with a common orientation are grouped together to form distinct tracts running in an anterior–posterior direction, with commissures connecting the left and right halves of the ganglion. In general, though, the fibers lack a common orientation. Axons and dendrites are complexly interwoven and it is here that synapses occur. No synapses, apart from neuromuscular junctions, occur outside the central nervous system.

Primitively, a ganglion is presumed to have been present ventrally in each segment, each responsible for processing local sensory information from receptors and organizing the motor output of the segmental limb and body wall musculature. This arrangement is partially visible in many embryonic insects, but in postembryonic stages some segmental ganglia fuse with a presegmental ganglion to form the brain, and some degree of fusion of postcephalic ganglia also occurs. Where such fused compound ganglia occur, each segmental element is referred to as a neuromere. The central nervous system thus consists of the brain, which is pre-oral, and is positioned dorsally in the head above the foregut, followed by a series of segmental ganglia lying close above the ventral body wall. Adjacent ganglia are joined by a pair of interganglionic connectives that contain only axons and glia; there are no somata, and no synapses.

The first ganglion in the ventral chain is the subesophageal. This is a compound ganglion, lying ventrally in the head, formed by the fusion of the neuromeres of the mandibular, maxillary and labial segments. In some insects, particularly Hymenoptera and Diptera, the subesophageal ganglion is itself fused to the brain, leaving only a small hole through which the esophagus passes. The ganglion sends nerves to the mandibles, maxillae and labium, and an additional one or two pairs to the neck and salivary glands.

There are typically three thoracic ganglia, but in some insects two or all three of these are fused together (Fig. 20.10). The metathoracic ganglion is fused with at least the first abdominal neuromere in nearly all pterygota. Each thoracic ganglion has some five or six nerves on each side, which branch to innervate the muscles and sensilla of the thorax and its appendages. The arrangement of nerves varies considerably, but usually the last nerve of one segment forms a common nerve with the first nerve of the next.

The largest number of abdominal ganglia occurring in larval or adult insects is eight, as in Thysanura, male *Pulex* (Siphonaptera) and many larval forms, but the last ganglion is always compound, being derived

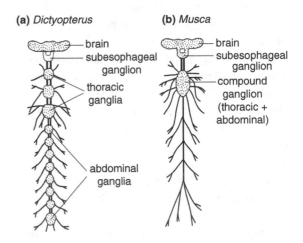

(a) *Dictyopterus*   **(b)** *Musca*

Figure 20.10 Numbers of ventral ganglia (from Horridge, 1965). (a) Most ganglia remain separate (*Dictyopterus*, Coleoptera). (b) All the ventral ganglia except the subesophageal are fused into a single compound ganglionic mass (*Musca*, Diptera).

from the ganglia of the last four abdominal segments. The abdominal ganglia are smaller than those of the thorax and, in general, fewer peripheral nerves arise from each of them than from the thoracic ganglia. In addition, the branching of the nerves is less diverse and variable, reflecting the relative simplicity, lack of wings and legs, and similar segmental organization of the abdominal musculature.

Sometimes, most or all of the ventral ganglia are fused to form a single compound ganglion, as in the blood-sucking bug *Rhodnius*, and in cyclorrhaphan flies (Fig. 20.10b). Adult holometabolous insects often have more ganglia fused than the larvae (see Fig. 15.23a), but the cyclorrhaphous Diptera are an exception. The degree of fusion can be highly variable between even quite closely related taxa. Moreover, the degree of ganglion fusion can even differ between male and female and social cast of *Melipona* stingless bees.

In most cases the muscles of a segment are innervated by neurons from the ganglion of the same segment, but some innervation by axons arising in neighboring ganglia also occurs (see Fig. 10.8). Some afferent fibers may also be intersegmental.

20.3.2 Neurons of the central nervous system

Motor neurons Each segmental ganglion contains the somata of motor neurons concerned with control of muscles, primarily in the same body segment. The number of motor neurons in each ganglion is relatively small, corresponding to the relatively small number of muscle units that insects possess and the small motor pool of typically 2–11 neurons innervating each muscle. Nevertheless, there are about 500 in the mesothoracic ganglion of *Periplaneta*.

The somata of motor neurons are relatively large; in the locust metathoracic ganglion they range from about 20 μm to 90 μm in diameter. They are roughly constant in position and homologous neurons on the two sides of a ganglion are placed more or less symmetrically. The neurite of each motor neuron increases in diameter on entering the neuropile, within which it gives rise to many branches (Fig. 20.11). The shape of this dendritic tree is characteristic for each motor neuron, and its complexity is an indication of the complexity of its synaptic connections with other neurons. The axon branches from the neurite and runs into a peripheral nerve. Insects differ from vertebrates in that they contain both excitatory motor neurons releasing glutamate and inhibitory motor neurons that release GABA. Inhibitory motor neurons decrease the amount of force that can be produced by a muscle and increase its rate of relaxation after contraction. They may act on muscle fibers directly but may also presynaptically affect the activity of excitatory motor neurons. Inhibitory motor neurons all innervate several muscles, often including antagonistic pairs, and their axons are highly

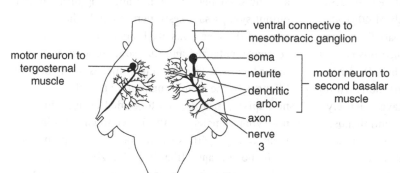

motor neuron to tergosternal muscle

ventral connective to mesothoracic ganglion

soma
neurite
dendritic arbor
axon
nerve 3

motor neuron to second basalar muscle

Figure 20.11 Two motor neurons in the metathoracic ganglion of a locust showing the main dendritic branches and axons extending into lateral nerve 3. The somata are in the outer part of the ganglion while the dendrites are in the central neuropile. Left: motor neuron to a tergosternal muscle; right: motor neuron to the second basalar muscle (after Burrows, 1977).

branched. A locust's leg, for example, is innervated by 70 excitatory motor neurons but only three inhibitory motor neurons. The excitatory motor neurons themselves can be subdivided into slow, fast or intermediate classes depending on the speed of muscle contraction they evoke. In insects the strength and rapidity of muscle contraction depends more on the individual motor neuron activated than on the aggregate activity of a large motor pool of functionally homogeneous neurons as in vertebrates.

Interneurons Direct synaptic connections are known to occur between afferent (sensory) fibers and motor neurons, but these monosynaptic pathways are rare because this arrangement allows for little variability of response, although this response is rapid. In general, interneurons are interpolated between sensory and motor neurons, and most of the neurons in the central nervous system are interneurons.

Anatomically, interneurons can be divided into local and intersegmental interneurons. Local interneurons are restricted to a single ganglion; they may be spiking or non-spiking. It is estimated that there are approximately 1500 intraganglionic interneurons in the mesothoracic ganglion of *Periplaneta*, compared with 200 interganglionic interneurons. Non-spiking interneurons form a significant proportion of these local interneurons.

Spiking local interneurons. Three populations of spiking local interneuron are known from the thoracic ganglia of locusts, though more doubtless remain to be discovered. Each consists of 40–100 neurons whose somata all occur in spatially discrete clusters in the ventral cortex. These neurons cannot be individually identified, although there is some variation in the location of the arborizations within each population. All these neurons have a densely branched ventral arborization in the ventral-most region of the neuropile that receives the terminal arborizations of exteroceptive sensory neurons. There is an extremely short dorsally orientated axon that gives rise to a more sparsely branched dorsal

arborization in regions of neuropile where sensory-motor integration occurs. One group of spiking local interneurons, which has somata near the midline of the ganglion, has inhibitory outputs; another with somata in an antero-medial position has excitatory effects on postsynaptic neurons.

Non-spiking local interneurons. Non-spiking local interneurons in the ventral ganglia often arborize extensively in the ipsilateral side of the ganglion (that is, on the same side as the soma) (Fig. 20.12a). Passive changes in membrane potential are adequate to produce synaptic transmission; spike production does not occur. Postsynaptic potentials are produced by these neurons by the release of a chemical synaptic transmitter, just as in spiking neurons, and in some cases a tonic postsynaptic potential is produced. Non-spiking interneurons in the thoracic ganglia of locusts receive a barrage of both excitatory and inhibitory inputs, whose frequency and intensity increases when the locust moves its leg, hyperpolarizing the membrane potential by as much as 20 mV and depolarizing it by up to 10 mV. An experimental change in the membrane potential of as little as 2 mV may have an effect on a postsynaptic neuron. An individual non-spiking neuron in a thoracic network makes extensive connections to several motor neurons and other interneurons and so can have a profound effect on the flow of information through neuronal networks. Notwithstanding this, the lack of active signal propagation within a single non-spiking interneuron and the sometimes close intermingling of input and output synapses makes it very possible that at least some of these neurons are functionally compartmentalized and that some processing may occur locally within discrete branches of the neuron. Changes in the membrane potential of the non-spiking neuron lead to graded changes in postsynaptic potential and hence to changes in membrane properties and the rate of spike production in the postsynaptic fiber (Fig. 20.12b).

Interganglionic or intersegmental interneurons. Intersegmental interneurons transmit information

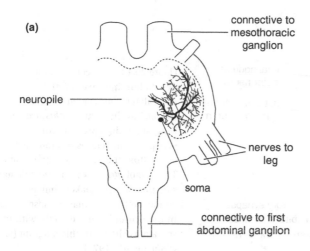

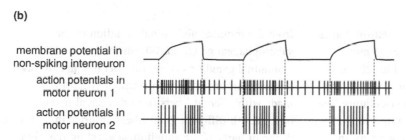

Figure 20.12 Non-spiking neurons. (a) A non-spiking neuron from the metathoracic ganglion of a locust (after Siegler and Burrows, 1979). (b) A non-spiking neuron affecting the activity of two motor neurons (labeled 1 and 2). An increase in the membrane potential (depolarization) causes neuron 1 to fire more actively and neuron 2 to start firing (after Pearson, 1977).

along the nerve cord. Fibers that carry information from anterior to posterior ganglia are called descending fibers; those transmitting information to the brain or anterior ganglia are called ascending fibers. These neurons are responsible for coordinating the activities of different ganglia and so bring about the coordinated activity of the whole insect. The axons of these neurons may have arbors, indicating that they make synaptic connections, in several ganglia or may extend through some ganglia without branching, arborizing only in the ganglion where they end.

One of the most thoroughly studied intersegmental interneurons is the descending contralateral movement detector (DCMD) of the locust. There is one DCMD in each neural connective, running from the brain to the metathoracic ganglion. The soma lies posteriorly in the protocerebrum. The DCMD itself is a relay for another interneuron, the lobula giant

movement detector (LGMD), whose processes are entirely confined to the brain. The LGMD has six major dendrites within the lobula, part of the optic lobe which diverge so that their fine branches occupy a sphere about 200 μm in diameter. The LGMD is most strongly driven by looming visual stimuli – that is, objects on apparent collision course with the locust – with the firing rate increasing near exponentially as the object approaches and its image expands across more and more of the retina. The DCMD synapses with the LGMD with 100% fidelity; each and every LGMD spike is conveyed to this postsynaptic interneuron whose axon crosses to the contralateral side of the brain and then down the ventral connective. The DCMD gives rise to a single branch in the prothoracic ganglion, and to further branches in both the mesothoracic and metathoracic ganglia. These branches form less extensive arborizations than the dendritic fields of the motor

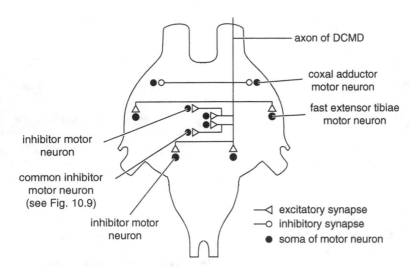

Figure 20.13 Connections of a descending interneuron, the contralateral movement detector (DCMD) of the locust (*Schistocerca*). Diagrammatic representation of synaptic connections to motor neurons controlling the muscles of the hindleg. The homologous axon on the left-hand side (not shown) makes similar connections. Note that synapses occur in the neuropile, not directly with the somata as shown in this diagram (after O'Shea *et al.*, 1974).

neurons with which they connect. The neuron makes excitatory output synapses to a number of motor neurons, some of which are shown in Fig. 20.13. These include neurons of the extensor tibiae muscle and the common inhibitor of the leg musculature (see Fig. 10.9). It also provides inhibitory input to the coxal adductor muscle motor neuron. Activity in DCMD, such as occurs during the approach of a colliding object, is sufficient to induce stalling and diving maneuvers during flight and may have a role in initiating other escape and predator avoidance behaviors.

The ventral nerve cords of cockroaches, various Orthoptera, dragonfly larvae, *Drosophila* (Diptera) and possibly all insects contain a number of axons which are much bigger than the majority. They are called giant fibers. In the cockroach, there are 6–8 giant fibers, 20–60 μm in diameter, in each connective as well as 10–12 large axons between 5 μm and 20 μm in diameter (the axons of most interneurons are less than 5 μm in diameter). In the cockroach, the somata of all the giant fibers are in the terminal abdominal ganglion. In contrast, adult *Drosophila* have only two giant interneurons and they have somata in the brain.

Giant fibers run for considerable lengths of the nerve cord. In *Periplaneta* the largest fibers extend from the terminal abdominal ganglion to the subesophageal ganglion. Within the terminal abdominal ganglion each giant fiber has an extensive dendritic arborization within which synapses are made with afferent fibers from the cercal nerve (Fig. 20.14). Different giant fibers arborize in different parts of the ganglion, suggesting that they receive input from different groups of sensilla.

The giant fibers allow for the rapid transmission of information over long distances because of their high conduction rates due to their large size and lack of synapses. As such, they are often implicated in escape behaviors.

Homologies Because the early patterning of neurogenesis is similar in different segments and even in different insect species, it is sometimes possible to recognize homologous neurons in different ganglia and across taxa, despite the differences that develop postembryonically as a consequence of functional differentiation. For example, specific interneurons in the meso-, metathoracic and first abdominal neuromeres of the locust that have somewhat similar arrangements of their axons and arborizations (Fig. 20.15) develop from precisely homologous cells during early development. The DCMD can be easily recognized in

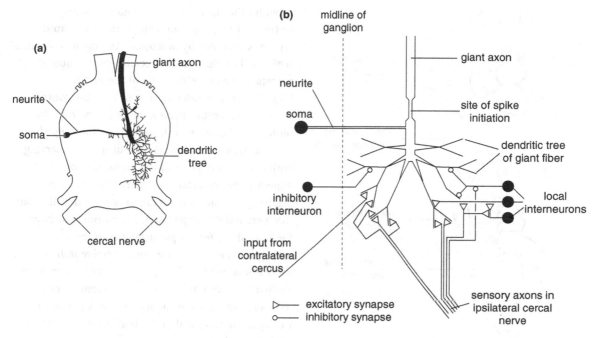

Figure 20.14 Integration at synapses. A giant fiber in the terminal abdominal ganglion of *Periplaneta* (after Callec, 1985). (a) Diagram showing the dendritic arbor and contralateral position of the soma. (b) The neuron with the giant axon receives direct input onto its dendritic branches from sensory neurons in both the ipsilateral and contralateral cerci as well as from local interneurons. Inputs are both excitatory and inhibitory.

both *Schistocerca* and *Locusta* despite not being closely related. In other cases, where the embryonic origins are not known, possible homologies are suggested by the similarity of position and anatomy of neurons, sometimes together with their immunocytochemical characteristics.

20.3.3 Peripheral nerves

The peripheral nerves radiating from the ganglia are made up of large numbers of axons. Within these nerves, the axons remain independent of each other; they do not branch or form synapses. Axons conveying information to the ganglia from the peripheral sensory neurons are said to be afferent or sensory fibers; those that distribute information from the central nervous system are called efferent or motor fibers. Most nerves contain both afferent and efferent axons, but the antennal and cercal nerves contain only sensory axons.

Most nerves are left and right pairs, but there is a small median nerve which runs from each ganglion. In the abdomen it extends from one ganglion to the next, but in the thorax it extends only part of the way before branching transversely to the spiracles and some ventilatory muscles. These nerves contain small numbers of sensory and motor axons. Each motor axon divides where the median nerve branches, so that the homologous muscles on either side of the body are innervated by the same neuron (see Fig. 17.21).

20.4 Brain

The brain is the principal association center of the body, receiving sensory input from the sense organs of the head and, via ascending interneurons, from the more posterior ganglia. Motor neurons to the antennal muscles are also present in the brain, but

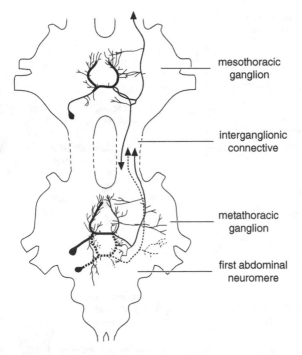

mesothoracic
ganglion

interganglionic
connective

metathoracic
ganglion

first abdominal
neuromere

Figure 20.15 Homologous interneurons. The three interneurons shown are derived from homologous neuroblasts in the meso- and metathoracic and first abdominal ganglion in the embryo. The major branches of the neuron in the first abdominal neuromere are shown as broken lines for clarity. Arrows indicate that the axon continues in the direction shown (*Locusta*) (after Pearson *et al.*, 1985).

the majority of nerve cell bodies found here belong to interneurons and the bulk of its mass is composed of their fibers. Many of the interneurons are concerned with the integration of activities; others extend down the nerve cord to the more posterior ganglia, transmitting information that controls the insect's behavior from the brain.

Three regions are recognized in the brain: the protocerebrum, deutocerebrum and tritocerebrum (Fig. 20.16).

20.4.1 Protocerebrum

The protocerebrum is bilobed and is continuous laterally with the optic lobes. In hypognathous insects it occupies a dorsal position in the head and,

as with other ganglia, the somata are largely restricted to a peripheral zone while the central region is occupied by neuropile. Antero-dorsally, in a region called the pars intercerebralis, a group of somata occurs on either side of the midline (Fig. 20.17). The anterior cells of the pars contribute fibers to the ocellar nerves, while fibers from the more lateral cells enter the protocerebral bridge (pons cerebralis), a median mass of neuropile connecting with many other parts of the brain. Neurons that appear to be important in regulating sleep and ensuring its continuation are located within the pars intercerebralis; octopaminergic neurons that have the opposite effect of promoting alertness and wakefulness project to the pars intercerebralis from a more dorsal region of the protocerebrum. Also within the pars intercerebralis are neurosecretory cells, the axons of which decussate (cross over each other to the opposite side) within the brain and extend to the corpora cardiaca.

Mushroom bodies At the sides of the pars intercerebralis are the mushroom bodies, or corpora pedunculata (Fig. 20.17). Each consists of a cap of neuropile, the calyx, from which a stalk (peduncle) runs ventrally before dividing into two or sometimes three lobes, known as the α, β and γ lobes. The terminology can be somewhat confusing since in some insects the two major lobes are designated the ventral and medial lobes, according to their orientation. The calyx may be single or double, which appears to have occurred early in the evolutionary history of the insects. Insects as distantly related as *Apis* and *Periplaneta* have double calyces, which comprise three concentric rings of neuropile known as the lip, collar and basal ring (Fig. 20.18b). Double calyces are thought to present a larger surface area for incoming neurons to make synaptic connections with and are a feature of insects with large olfactory systems. Insects that do not have a sense of smell, such as cicadas, or have a reduced sense of smell, such as dragonflies, do not have calyces on their mushroom bodies. Some insects have

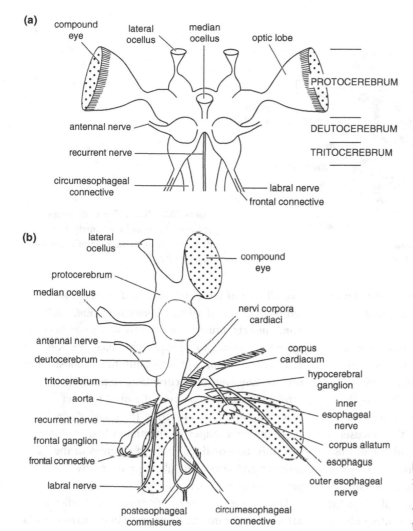

Figure 20.16 Brain. Diagrammatic representations (*Locusta*) (after Albrecht, 1953). (a) Anterior view; (b) lateral view, which also shows the major ganglia of the stomodeal nervous system.

a small accessory calyx that is located adjacent to the pedunculus. This receives gustatory inputs and outputs to a distinctive loblet or layer that wraps around the other lobes.

The mushroom bodies are given their form by a large number of interneurons, called Kenyon cells, which have their somata above the calyx. A typical Kenyon cell has dendrites in the calyx and an axon running down the stalk that divides ventrally to form the α and β lobes (Fig. 20.18a). Morphologically, several different types of Kenyon cells are

recognizable, reflecting functional differentiation. For example, some classes of Kenyon cells do not branch in the calyces, or have a distinctive "claw-like" arborization in this region and send an axon down into the γ lobe. Different populations of Kenyon cells also differ in gene expression and this may reflect differences in signaling. Not all classes of Kenyon cell are present in all insect taxa. The relative size of the mushroom bodies is related to the complexity of behavior shown by the insects. They are small in Collembola, Heteroptera, Diptera and

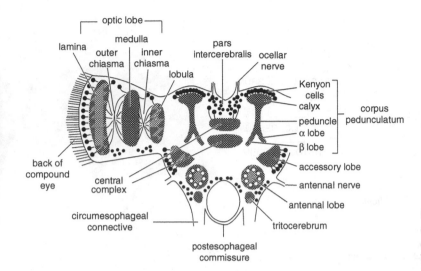

optic lobe

lamina
medulla
outer chiasma | inner chiasma
lobula
pars intercerebralis
ocellar nerve
Kenyon cells
calyx
peduncle
α lobe
β lobe
corpus pedunculatum
back of compound eye
central complex
accessory lobe
antennal nerve
antennal lobe
circumesophageal connective
tritocerebrum
postesophageal commissure

Figure 20.17 Brain. The main regions of organized neuropile (shaded). The distribution of somata is indicated by black dots.

Odonata, of medium size in Coleoptera, Orthoptera, Blattodea, Lepidoptera and sawflies, and most highly developed in social insects (Fig. 20.19a). Plant-feeding Coleoptera species that have a broad dietary range have larger mushroom bodies than related species with more restricted diets, and gregarious-phase desert locusts, which have both broader diets and more complex social interactions, likewise have larger mushroom bodies than locusts in the solitarious phase. In termites the volume of the peduncle and α and β lobes is large and the calyx represents only about one-third of the total volume of the mushroom bodies. In bees and social wasps the calyx is large and very complex, representing about 70% of the total volume. Worker honey bees have about 170 000 Kenyon cells in each mushroom body, representing about 40% of all the neurons in the brain; the fly *Calliphora* has about 21 000 Kenyon cells on each side, about 12% of the total neuron population.

Changes occur in the mushroom bodies with age and experience. Over the first week of adult life, the number of fibers in the mushroom bodies of *Drosophila* increases by more than 20% (Fig. 20.19b) and is then maintained at a plateau until the flies are very old. These changes are reflected in an increase in volume of the calyx. Changes also occur in the

overall size of the mushroom bodies over the first ten days of life of worker honey bees (Fig. 20.19c). Some insects, principally crickets and some beetles, have been shown to be able to produce new Kenyon cells over the course of their adult life. In most other insects, however, the neuroblasts responsible for producing the Kenyon cells die on the insect reaching adulthood or during the pupal stage. In these insects changes in mushroom body volumes can only be explained through changes in the number and density of fibers produced by pre-existing neurons.

The increase in size of the mushroom bodies is affected by the insects' experience in both *Drosophila* and *Apis*. The number of fibers in the peduncle increases less in flies kept in isolation compared with those in a crowd, while among *Apis* workers, the lip and collar of the calyx increase more in foragers compared with nurse bees of the same age (Fig. 20.19b,c).

The calyces of the mushroom bodies receive input from projection neurons carrying information from the antennal lobes (Fig. 20.18b). Within the calyx, each of these neurons communicates with a large number of Kenyon cells. Social Hymenoptera appear to be exceptional in that the calyces also receive a large input from the visual system. In *Apis*

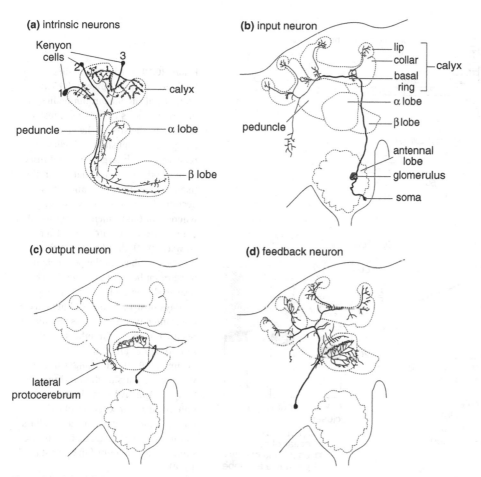

(a) intrinsic neurons

Kenyon cells

calyx

peduncle

α lobe

β lobe

(b) input neuron

lip
collar — calyx
basal ring
α lobe
peduncle
β lobe
antennal lobe
glomerulus
soma

(c) output neuron

lateral protocerebrum

(d) feedback neuron

Figure 20.18 Mushroom body. Examples of neurons associated with the mushroom bodies. (a) Intrinsic neurons (Kenyon cells: labeled 1, 2 and 3) showing different types of dendritic arborization in the calyx. Notice that the axon of neuron 1 extends only into the β lobe; for neuron 2, only the proximal part is shown; the axon of neuron 3 extends to both the α and β lobes (*Apis*) (based on Mobbs, 1985). (b) Input neuron from the antennal lobe. The dendritic arbor is restricted to a single glomerulus, but this is not true of all the input neurons from the antennal lobe to the mushroom bodies. (Note that this would be called an *output projection* neuron from the antennal lobe, but an *input* neuron to the mushroom body.) The axon has branches to both parts of the calyx (*Manduca*) (based on Homberg, 1994). (c) Output neuron with a dendritic arbor in the α lobe and an axon running to the lateral protocerebrum (*Manduca*) (based on Homberg, 1994). (d) Feedback neuron connecting different parts of the mushroom body (*Manduca*) (based on Homberg, 1994).

connections from the olfactory system are made in the lip of the calyx, while inputs from the visual system connect in the collar. Another group of GABA-containing neurons, known as feedback neurons, arborize in both the lobes and the calyx (Fig. 20.18d). The calyces are also innervated by terminals of octopamine- and dopamine-containing neurons. The pedunculus and lobes are also sites of major multimodal synaptic input. The terminals of the Kenyon cells converge onto a number of large output interneurons in the lobes, the nature of which are still largely unknown. These interneurons run primarily to other parts of the protocerebrum, including the central body (Fig. 20.18c).

The mushroom bodies are involved in olfactory learning, and visual learning in some insects. There is

(a)

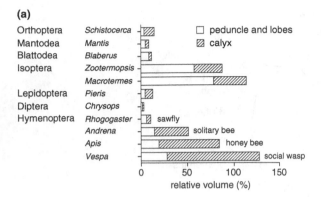

(b) *Drosophila*

(c) *Apis*

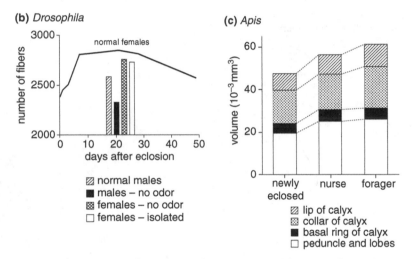

Figure 20.19 Mushroom bodies. Variation in development. (a) The volumes of the mushroom bodies in different species of insect expressed relative to the size of the central complex. The mushroom bodies are relatively much larger in social insects than in others. Notice that in termites (Isoptera) the peduncle and lobes account for most of the volume, whereas in the Hymenoptera the calyx contributes the greater part (after Howse, 1974). (b) Changes in the number of fibers in the peduncle of a mushroom body of *Drosophila*. The insects shown by histograms were all 21 days post-eclosion. Maintenance in environments offering less stimulation resulted in smaller numbers of fibers in both males and females (after Technau, 1984). (c) Changes in the volumes of parts of the mushroom body of *Apis*. All parts increase in volume as the insects get older. Foragers have bigger lips and collars in the calyces than nurse bees of the same age and parents (after Durst *et al.*, 1994).

some evidence that they may also have a role in some decision-making processes. *Drosophila* mutants with compromised mushroom body function or missing components show mostly normal behavior but are unable to learn certain kinds of associations. They also display more stereotyped behaviors and some difficulties in responding to new conditions.

Central complex The central complex is a series of four interconnected neuropiles which span the midline of the protocerebrum. It is the only unpaired region of organized neuropile. From dorsal to ventral these are the protocerebral bridge, the fan-shaped body (also known as the upper division of the central body), the ellipsoid body (also known as the lower

division of the central body) and a pair of noduli at the base (Fig. 20.20a). The neuropiles of the central complex show an obviously structured organization, with an array of 16 columns, eight in each half of the brain, represented in the protocerebral bridge and throughout the central body. The central body also shows a highly laminar organization in the orthogonal plain, with distinct layers of input neurons. There are large populations of columnar neurons, each consisting of 16 neural elements which interconnect the columns of different layers in a precise and systematic manner, each bringing different sets of columns into contact with each other, including information from each hemisphere of the brain (Fig. 20.20b,c). The principal outputs of

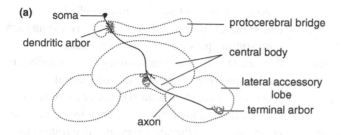

(a) soma — protocerebral bridge
dendritic arbor
central body
lateral accessory lobe
terminal arbor
axon

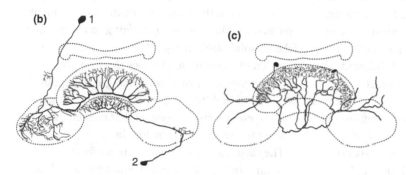

(b) 1
(c)
2

Figure 20.20 Examples of neurons of the central complex (*Schistocerca*) (after Homberg, 1991, 1994). (a) A neuron linking the protocerebral bridge to the lower part of the central body. (b) Two neurons connecting the lateral accessory lobes with parts of the central body. Neuron 1 contributes to the system of columns in the upper central body (see text). (c) Two symmetrical neurons with extensive arborization in the upper central body.

the central complex are to a region called the lateral accessory lobe, which in turn is thought to have connections to major populations of descending interneurons. Some of the probable input fibers are serotonergic and have extensive arborizations throughout the central complex. Most of the cells of the central body contain neuropeptides, probably used as neurotransmitters, and many of the transverse neurons innervating the lower part of the central body are GABAergic.

The possible functions of the central complex are slowly becoming clearer. Despite containing neurons known to respond to mechanosensory, visual and chemical stimuli and considerable evidence that interference with its function disrupts normal motor patterns of behavior, the path of information in and out of the central complex is still far from fully established. The most fully characterized role of the central complex, with anatomical pathways and physiological data, is in the processing of visual information from the dorsal rim region of the eye, which is sensitive to polarized light. The plane of polarization varies systematically across the sky

depending on the position of the sun, but can still be detected even on a cloudy day. This information is used to make a sky compass, which is used by some insects to navigate. In locusts a class of interneurons conveys polarized light information from a small visual neuropile called the posterior optic tubercle to the protocerebral bridge. These neurons show a systematic pattern of branching with different members of the population having major branches in one ipsilateral and one contralateral column, such that in any one neuron these major arborizations are always eight columns apart in the protocerebral bridge. These neurons likewise show a systematic difference in the plane of polarized light that most strongly excites them, giving rise to a topographic map across the central complex. Polarized light-sensitive columnar neurons convey information from the protocerebral bridge to the central body and then out into the lateral accessory lobes. Polarized light-sensitive neurons constitute at most 20% of the whole cell population of the central complex. There is increasing evidence that the central complex is involved in monitoring and organizing overall motor patterns used in locomotion. Recordings of cockroaches from multiple neurons in the central complex using multi-electrode arrays showed that their activity was correlated with the stepping rate of walking, and that in many cases the change in activity of the neurons preceded the change in behavior. Furthermore, electrical stimulation of these neurons through the electrodes could alter walking patterns. Other work, including several studies using *Drosophila* mutants, has suggested a general role for the central complex in sensory to motor integration, in particular in relation to the position of the body and different body regions with reference to outside space.

Optic lobes The optic lobes are lateral extensions of the protocerebrum to the compound eyes. Each consists of three successive neuropiles (Fig. 20.17), known as the lamina, the medulla and the lobula complex which, in Lepidoptera, Trichoptera and Diptera, is subdivided into the lobula and the lobula plate. Within these neuropiles the arrangements of arborizations of different sets of neurons produce a layered appearance. There is a strong degree of retinotopy maintained in all three principal optic neuropiles, i.e., the pattern of neural signals is a precise representation of any image falling on the eye, although this gets looser as processing progresses from the lamina to the lobula. Between the successive neuropiles the fibers cross over horizontally, forming the outer and inner optic chiasmata so that the neural map of the visual image is reversed along the antero-posterior axis and then re-reversed.

In the lamina of most insects, the axons of retinula cells from one ommatidium remain together and are associated with neurons originating in the lamina and the medulla to form a cartridge. In insects with an open rhabdom (Section 22.1.2), however, each cartridge contains the axons from retinula cells with the same field of view rather than from the same ommatidium. In either case, the number of cartridges in the lamina is the same as the number of ommatidia.

Axons from most of the retinula cells in the eye end in the cartridges, although one or two from each ommatidium pass through to the medulla (Fig. 20.21a). In flies, the six retinula cells ending in the lamina have the same range of spectral sensitivity, with maximum sensitivity to green and another peak in the ultraviolet. They form part of a color-insensitive (because they all arise from cells with the same visual pigment) pathway which converges onto the giant interneurons in the lobula (see below). The two cells extending to the medulla (shown solid black in Fig. 20.21a) have different wavelength sensitivities. In flies there are four different pigments, each with a single absorption peak found in different ommatidia. These form a coarser, color-sensitive system that converges on the smaller interneurons of the lobula.

The cartridges of one insect are uniform in structure. In the fly, 19 cells contribute to each one, and each group of neurons is wrapped by a glial cell which effectively isolates the cells within from extraneous inputs (Fig. 20.21a). The retinula cell axons that terminate in the lamina synapse with monopolar interneurons whose axons run to the medulla, and with other neurons that have cell bodies in the medulla (centrifugal fibers). The monopolar cells are of two types: some receive input only from retinula axons in a single cartridge, and they are known as small-field cells; others receive input from several cartridges, and are known as wide-field cells.

Over 60 neurons contribute to each column in the medulla. The axons of the ultraviolet-sensitive retinula cells, as well as axons of lamina neurons, terminate within these columns, each penetrating to different layers in the medulla. This suggests that visual information is beginning to be split into several parallel pathways that may be specialized for different functions, such as shape detection, ultraviolet, color and motion processing. Many of the neurons in the medulla are very small, which has made it extremely difficult to analyze their physiological function.

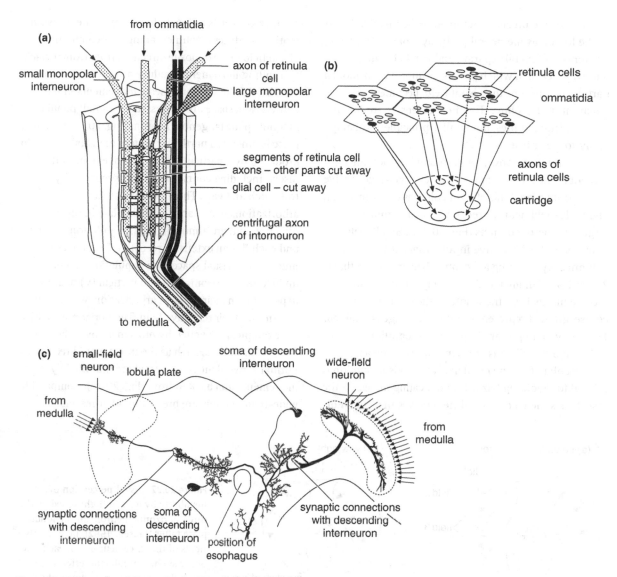

Figure 20.21 Neurons of the optic lobe. (a) Diagram of an optic cartridge in the lamina of a fly. The side facing out of the page has been cut away to show the arrangement of fibers within the cartridge (after Laughlin, 1975). (b) Relationship between a cartridge in the lamina and the overlying ommatidia. In flies, which have an open rhabdom (see Fig. 22.3), the axons of retinula cells with the same visual field (shown here in black) in adjacent ommatidia come together in one cartridge. In other insects each cartridge receives sensory axons only from the ommatidium immediately above it. (c) Representation of wide-field and narrow-field interneurons with arborizations in the lobula plate of a fly and their connections with descending interneurons (based on Gronenberg and Strausfeld, 1992).

Some of the precision retained in the medulla is lost in the lobula as the neural pathways converge. The existence of several separate and parallel neural pathways, each of which processes information about a particular aspect of the visual field, becomes very apparent in the lobula. The most extensive studies are on cyclorraphan flies, but similar principles probably apply to other insects.

As mentioned above, two types of interneuron are present in the lobula: small-field neurons which receive inputs from a small number of columns in the medulla; and wide-field neurons receiving inputs from a very large number of columns (Fig. 20.21c). Small-field neurons in the fly receive inputs from about 12–150 columns, representing a receptive field in the eye that is 20° or less in diameter. Collectively, these neurons receive inputs from the whole of the retina so that a coarse-grained representation of the image is retained. They may be important in pattern recognition.

For highly agile, fast-moving insects such as cyclorraphan flies the rapid analysis of visual motion is critical for avoiding hazards and locating mates. As the insect flies the whole visual field moves in a coherent manner known as a flow field. For example, forward motion produces a countervailing movement of the visual field in the opposite direction, from front to back, and rolling movements produce a counter rotation in the visual field. The large-scale movements of the entire visual space are processed by a small population of lobula plate tangential neurons (Fig. 20.21c). Each of these is tuned to a particular pattern of visual flow field produced by a particular kind of movement of the insect, including course deviations produced by roll, pitch and yaw. These neurons all have arborizations that span the lobula plate in one direction, but are almost flat in the orthogonal plane and each has an extensive receptive field covering much of the visual space. The population can be divided into horizontal (H) and vertical (V) neurons, depending on their plane of arborization, which in turn determines their receptive field. The horizontal cells, for example, are excited by movement over the eye from front to back, such as the insect would receive in normal forward movement, and are inhibited by movement from back to front (Fig. 20.22). Comparable wide-field neurons are present in other insects.

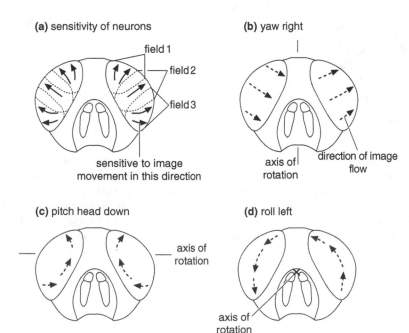

(a) sensitivity of neurons

field 1
field 2
field 3

sensitive to image movement in this direction

(b) yaw right

axis of rotation
direction of image flow

(c) pitch head down

axis of rotation

(d) roll left

axis of rotation
axis of rotation

Figure 20.22 Visual perception of instability by a fly in flight based on the three horizontal neurons in the lobula plate (after Hausen, 1982). (a) Receptive fields of the three neurons on each side. Some slight contralateral effects also occur, but these are not shown. (b) Yaw to the fly's right (to the left in the diagram) produces a movement of images to the left across the eyes. The horizontal neurons of the left optic lobe would be excited while those of the right would be inhibited. (c) Pitching head down produces an upward movement of the image; the upper horizontal neurons of both sides (field 1 in (a)) would be excited and the lower ones inhibited. (d) Rolling does not stimulate the horizontal neurons.

Both small- and wide-field interneurons have axons projecting to the deutocerebrum. Here they synapse with descending interneurons, regulating the motor systems of the thorax that are involved in walking and flight (Fig. 20.21c; see also Fig. 9.18).

20.4.2 Deutocerebrum

The deutocerebrum contains the antennal (olfactory) lobes and the antennal mechanosensory and motor center (Figs. 20.16, 20.17). The latter is a relatively poorly defined region of neuropile containing the terminal arborizations of mechanosensory neurons from the scape and pedicel, and perhaps also from the flagellum of the antenna. It also contains dendritic arborizations of the motor neurons controlling the antennal muscles.

The antennal lobes are regions of neuropile, one in relation to each antenna, within which are discrete balls of dense synaptic neuropile called glomeruli, the number of which varies greatly between species (Fig. 20.23a). *Periplaneta* has 125 and *Manduca* about 64, while *Aedes* has fewer than ten glomeruli. Axons from olfactory sensilla on the antenna terminate in the glomeruli, and each axon only goes to one glomerulus in most insects. Locusts are somewhat exceptional in that they have about 1000 poorly defined glomeruli, but each olfactory afferent innervates several of these. Within a species, individual glomeruli appear to be constant in form and position. In the males of many insects that find mates by detecting pheromones released by females, there are two or three larger glomeruli grouped together to form a macroglomerular complex to which the axons of sex pheromone-specific sensory neurons converge. The axons from olfactory receptors on the maxillary and labial pulps also terminate in a group of glomeruli contiguous to those that receive inputs from the antennae.

A massive condensation of the number of neurons conveying olfactory information from the antennae to higher olfactory centers occurs in the antennal lobe, which gives the glomeruli their characteristic shape. Olfactory afferents expressing the same olfactory receptor proteins all project to the same glomerulus, where they synapse onto projection neurons that convey the information to other olfactory neuropiles, principally the calyces of the mushroom bodies (Section 20.4.1) and a region called the lateral horn on the dorso-lateral edge of the protocerebrum (Figs. 20.18b, 20.23b). For example, in the honey bee, 60 000 olfactory sensory neurons synapse onto 800 projection neurons dispersed among 160 glomeruli in each antennal lobe. Most projection neurons have uniglomerular arborizations and a large population of them use acetylcholine as a transmitter. Some other projection neurons are multiglomerular and many of these use the inhibitory transmitter GABA.

There is a second population of interneurons, the local neurons whose projections are entirely confined to the antennal lobe. These neurons differ considerably in morphology, patterns of connectivity, physiology and neurotransmitters. Local interneurons interconnect different glomeruli and may innervate several or even all them (Fig. 20.23b). Bees have approximately 4000 local interneurons in each antennal lobe. Many local interneurons are inhibitory, using either GABA or histamine as a transmitter; others are excitatory and contain acetylcholine. It is common for the local interneurons to inhibit activity in the projection neurons. The extent to which odor discrimination is dependent on labeled lines versus across-fiber patterning is not known. A labeled line responds only to a specific compound or combination of compounds; such a system conveys information totally unambiguously. Pheromonal processing by the macroglomerular complex and associated interneurons appears anatomically and physiologically to be a distinct labeled line system. Across-fiber patterning requires the assessment of the overall input from an array of neurons with different ranges of sensitivity. The relatively large numbers of glomeruli and output interneurons

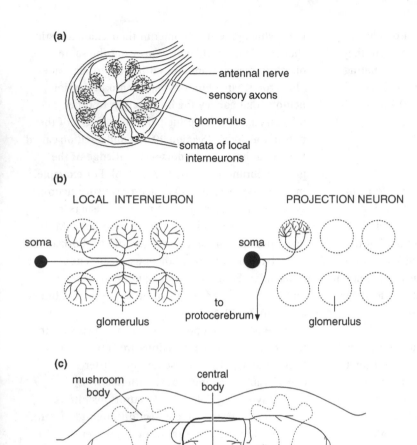

(a)

antennal nerve
sensory axons
glomerulus
somata of local interneurons

(b)

LOCAL INTERNEURON

soma

glomerulus

to protocerebrum

PROJECTION NEURON

soma

glomerulus

(c)

mushroom body

central body

soma

antennal lobe

axon terminals in each glomerulus

Figure 20.23 Antennal lobe. (a) Diagram of an antennal lobe showing glomeruli. (b) Representation of a local interneuron and an output (projection) neuron (after Matsumoto and Hildebrand, 1981). (c) A serotonergic neuron with terminals in all the glomeruli and extensive branching through other parts of the brain. Notice the soma in the contralateral olfactory lobe. A similar neuron innervates the left-hand lobe (after Kent *et al.*, 1987).

provide the insect with the potential to distinguish a very large number of odors using across-fiber patterning.

The physiological characteristics of the projection neurons vary. Some respond only when the antennal receptors are stimulated by a specific odor that is a regular feature in the life of the species, like a pheromone component. A few may respond when the antenna is stimulated by specific mixtures of compounds that are of particular importance to the species, such as a pheromone blend. Other projection neurons, however, respond when the antenna is exposed to any one of a wide range of odors and it is probably these neurons that provide the insect with general information about odors in the environment. Studies using fluorescent dyes that change in the presence of calcium have allowed the activity of large parts of the antennal lobes to be sampled simultaneously. These have revealed that different odors will activate a different and consistent set of glomeruli. The coarse activity patterns of projection neurons are often quite complex, with alternating periods of bursts of spikes and periods of inhibition after exposure to an odor. Superimposed on this large-scale temporal variation is an underlying regular 20–30 Hz oscillation in the membrane potential that occurs synchronously across the entire population of projection neurons. This underlying oscillation, meditated by inputs from wide-field local interneurons, serves to regulate the timing of spikes across the projection neuron population, such that they tend to occur simultaneously in all projection neurons receiving olfactory input. This complex ascending pattern of projection neuron firing is transformed into a sparse code of neuron firing in the mushroom body, where odors are represented by just one or two spikes each across a large array of Kenyon cells.

Other interneurons provide input to the antennal lobes from other brain regions. Several populations of neuromodulatory neurons, containing serotonin, dopamine, octopamine and/or peptides arborize in part or all of the antennal lobes, the pattern of which differs considerably between insect species. A group of ventral unpaired neurons in the subesophageal ganglion which releases either dopamine or octopamine is thought to be important in associating odors with appetitive or aversive taste qualities. In addition to this neuromodulatory input there are also feedback neurons that project back from the mushroom bodies into the antennal lobes.

20.4.3 Tritocerebrum

The tritocerebrum is a small part of the brain consisting of a pair of lobes beneath the deutocerebrum. From it, the circumesophageal connectives pass to the subesophageal ganglion, and the tritocerebral lobes of either side are connected by a commissure passing behind the esophagus (Fig. 20.17). Anteriorly, nerves containing sensory and motor elements connect with the frontal ganglion and the labrum.

20.5 Controlling behavior

The nervous system does not act as a simple relay between receptors and effectors. It makes internal neuronal representations of features in both the external and internal environment, depending on sensory inputs across large populations of sensory receptors. It also coordinates and integrates the activities of different parts of the body so that appropriate behavioral responses occur, and constantly updates its internal state in the light of sensory feedback induced by those behavioral responses. Integration occurs at many levels within the central nervous system, as has already been alluded to above. This ranges from the spread of signals within individual neurons to the connectivity between networks of cells in neuronal circuits and the modifications of signals that occur at synapses. Integration also involves the modulation of signals at synapses such that the strength of connections between neurons can be actively varied and information is transformed as it flows through the nervous system.

20.5.1 Integration at the synapse

Changes in synaptic transmission may occur as a result of changes in either the pre- or postsynaptic neuron, or both. They may be produced either as a result of activity of the neurons themselves, by the activity of neuromodulators or through inhibitory synapses from other neurons that occur near the presynaptic terminals. Neuromodulators are considered in Section 20.2.3.

Synapses from different neurons often occur in extremely close proximity, such that there appears to be an interleaving of pre- and postsynaptic membranes. Therefore it is common for the activity of a presynaptic neuron at a synapse to be modified by synaptic inputs from other neurons close by, which may be either ionotropic or metabotropic in nature. The net result is a change in the number of transmitter vesicles releasing their contents at the synapse and thus alteration in the effect on the postsynaptic neuron.

An important class of this kind of synaptic modification is known as presynaptic inhibition, which has been extensively studied during sensory feedback following leg movements in insects. The femoro-tibial chordotonal organ of the locust monitors movement of the joint during walking. There are about 90 sensory axons from the chordotonal organ and the effect of each one is regulated by the activity of the whole ensemble via inhibitory inputs from interneurons, which control the gain of the signal reaching the central nervous system. The unusual feature of these inhibitory inputs is that by altering the permeability of the axon to chloride ions they actually depolarize the membrane but lower its resistance. Any further depolarization of the membrane such as that produced by an advancing spike traveling down the axon reverses the flow of chloride ions so that it counteracts the depolarizing effect that would normally be produced by the voltage-gated depolarization of sodium ions that normally propagates the spike down the axon. This reduces the amplitude of the spike or even abolishes its further transmission. Since the amount of transmitter released is a function of how long a spike depolarizes the synaptic terminal for, reduced spike amplitudes lead to less transmitter being released.

Postsynaptic integration results from temporal and spatial summation. As contact between any two neurons involves multiple synapses, spatial summation is a normal feature of neuronal activity. It may also involve the inputs from more than one neuron and the inputs may be excitatory, inhibitory or both. Thus, spatial summation may result in an increase (Fig. 20.5b) or a decrease (Fig. 20.5d) in the magnitude of the postsynaptic potential, either increasing or decreasing the likelihood that an action potential will be produced by the postsynaptic neuron. Since the space constant of a passive membrane is dependent upon its diameter, the shape of the dendritic arbor has a profound effect on how spatial integration occurs. Inputs occurring on large-diameter branches near an action potential initiation zone have a much stronger weighting than those occurring on fine branches far away.

Temporal summation results from the arrival of successive action potentials at a synapse (Fig. 20.5a). The postsynaptic potential produced by the arrival of a single action potential at an excitatory synapse may not be big enough to initiate a new action potential and will decay as dictated by the time constant of the membrane. If, however, a second action potential arrives at the synapse and causes a further depolarization of the postsynaptic fiber before the first potential has completely decayed, the total potential resulting from the two successive depolarizations is greater than would have been produced by either event by itself. It may exceed the threshold and initiate a postsynaptic action potential.

20.5.2 Integration by interneurons

Some measure of integration is achieved by individual interneurons as a result of the diversity of inputs that they receive from elsewhere, and the diversity of effector units which they supply. For instance, Fig. 20.14 shows some of the known inputs to one of the giant fibers in the terminal abdominal ganglion of the cockroach. These include the input from mechanoreceptors, some of which exhibit a purely phasic input while others are tonic or phaso-tonic. Some cercal afferents synapse onto inhibitory interneurons, and there is input from both ipsilateral and contralateral cerci. Any one or all of these input routes may be in action at the same time, and all this information is integrated across the dendritic arbor, leading to the production of spikes at the initiation site located at the narrow point on the axon. The number and frequency of spikes varies with the overall input. The insect's behavior, moreover, does not depend on the activity of a single interneuron, but on that of a whole population of interneurons.

Fig. 20.13 shows, for comparison, the motor connections of the DCMD interneuron in the metathoracic ganglion of *Schistocerca*. These include excitatory synapses with the motor neurons of the extensor tibia muscle and inhibitory neurons of the flexor muscle. Contact is made with both ipsilateral and contralateral neurons.

20.5.3 Neural mapping

For the insect to make appropriate responses to external stimuli it is important that incoming information is represented in a spatially coherent manner within the central nervous system. In the auditory system of crickets, which has a broad range of frequency sensitivity, the afferents have a tonotopic organization, with the fibers arrayed in ascending order according to their best frequency of response. All olfactory sensory neurons expressing the same odorant receptor protein project to the same glomerulus in the antennal lobe. Both the visual and mechanosensory systems preserve the external spatial organization of incoming sensory signals within the central nervous system. In the eye this is achieved by the system of cartridges and columns in the lamina and medulla forming a retinotopic map and feeding into the wide-field or narrow-field interneurons of the lobula (Section 20.4.1).

In the mechanosensory system, preservation of the spatial pattern is achieved by the spatial distribution of sensory arborizations within the ganglia. Hairs on the femur, tibia and tarsus of the locust *Schistocerca* arborize in progressively more posterior and lateral regions of the ventral neuropile in the relevant ganglion (Fig. 20.24a). Dorsal, ventral, medial and lateral positions of the hairs on the legs also tend to be retained within the neuropile. Thus

the sensory neuropile contains a three-dimensional representation of the positions of the hairs on the legs. This is called a somatotopic map. Moreover, some of the spiking local interneurons with which the sensory axons connect retain the pattern from a single leg (Fig. 20.24b,c), but others incorporate information from different leg regions, losing some of the spatial content of the signal so the map becomes more coarse-grained. Some intersegmental interneurons also retain some positional information in the location of the arborizations of their input regions.

The effect of neuro-anatomical organization on neuronal physiology can be clearly seen in the cockroach escape response. The trichoid sensilla on the cerci of the cockroach are differentially sensitive to deflection in different directions. Their axons connect with four giant interneurons, two on each side, in a manner which retains the major components of directionality. Thus the interneurons are most responsive to deflection of the hairs backwards or forwards, to the left or to the right. This information is conveyed by the interneurons to the thorax, where the output of motor neurons, producing a turn to the left or right, depends on the balance of activity between the interneurons on the two sides.

As most insects have large numbers of contact chemoreceptors on the tarsi and legs as well as scattered over other parts of the body surface, it is to be expected that positional information is also necessary for the gustatory system, in addition to quality coding. In general, gustatory sensilla also contain a mechanosensitive neuron (Section 24.2). In Diptera projections of the neurons from gustatory sensilla appear to be clearly separable into two anatomical classes. One neuron clearly has a thicker axon than the others and projects to a ventro-medial region of neuropile, which is also the destination of neurons from purely mechanosensory sensilla. The other finer neurons project to an adjacent dorso-lateral region where they have partially overlapping arborizations with each other.

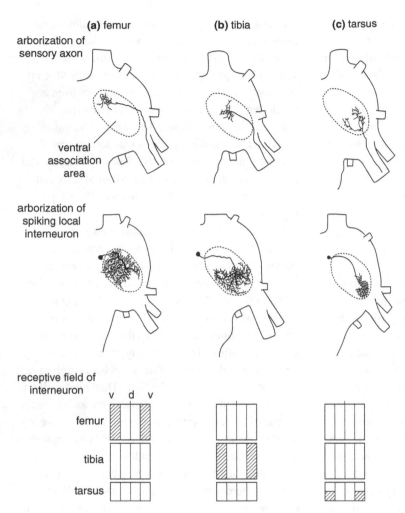

Figure 20.24 Somatotopic mapping of mechanosensory input from the hindleg of a locust. All the mechanosensory axons from the leg arborize in a ventral region of the neuropile known as the ventral association area (indicated by the dotted line). The diagrams at the bottom represent the leg as if slit along the ventral side and flattened so that the dorsal midline is in the center, ventral midline at the outer edges. Cross-hatching shows the area in which mechanical stimulation results in excitation of the interneuron shown in the central panel (after Burrows and Newland, 1993). (a) Input from the femur. Top: the arborization of a mechanosensitive neuron at the base of the femur. Center: the arborization of a spiking local interneuron which is excited by stimulation of the ventral area of the femur, but not other parts of the leg (shown at bottom). (b) Input from the tibia. Top: the arborization of a mechanosensitive neuron at the base of the tibia. Center: the arborization of a spiking local interneuron which is excited by stimulation of the ventral area of the tibia, but not other parts of the leg (shown at bottom). (c) Input from the tarsus. Top: the arborization of a mechanosensitive neuron on the basal tarsomere. Center: the arborization of a spiking local interneuron which is excited by stimulation of the ventral area of the distal tarsus, but not other parts of the leg (shown at bottom).

It seems that in Diptera there is a separation between gustatory and mechanosensory processing, but gustatory information from each leg or the mouthparts is processed in separate segmentally organized regions of the central nervous system. In Orthoptera and perhaps Lepidoptera there is no anatomical separation between mechanosensory and gustatory afferents from gustatory sensilla. All the neurons from one sensillum project to the same somatopically defined region of neuropile. Furthermore, recordings from the same spiking local interneurons in locusts that receive synaptic inputs from mechanosensory sensilla have revealed that these neurons also receive gustatory inputs. As stimulation of the gustatory neurons depends on contact with the substrate, the gustatory inputs summate with the mechanosensory inputs and alter the amplitude and duration of response produced by purely tactile stimuli. These interneurons therefore process a combined mechano- and chemosensory signal which encodes the location of the stimulus. Gustatory inputs into these local networks controlling leg movement seem to code for aversion since they promote the removal of the leg from the source of stimulus. It must be inferred that some separation of taste quality, at least with respect to being acceptable or unacceptable, is maintained in the central nervous system, but there is no physiological data on this.

20.5.4 Pattern generation

Many behaviors consist of regular, repeated activities that occur in a rhythmic fashion, such as flight, walking, ventilation and, at least in grasshoppers, oviposition. The basic patterning of motor neuron activity in these instances is produced by arrays of interneurons within the central nervous system that form a central pattern generator (CPG). These networks can produce an oscillating or rhythmic output even in the absence of sensory input when the central nervous system is experimentally isolated from all peripheral receptors (see Fig. 8.19). In some CPG networks, one or several of the constituent neurons have intrinsic

oscillatory properties so that they spontaneously produce regular bursts of spikes which drive activity in the rest of the network. In other instances the rhythm emerges purely from the pattern of connections within the network and the excitatory and inhibitory connections between the constituent neurons.

The locust flight generator was one of the earliest model systems in which the principal features of the CPGs were established. Fig. 20.25 shows a highly simplified model of how the basic oscillation of wing levator and depressor muscles may be generated in the locust central nervous system. So far some 46 mesothoracic and 39 metathoracic identified morphotypes of interneuron have been characterized as having some participation in flight generation, and the existence of others have been inferred from

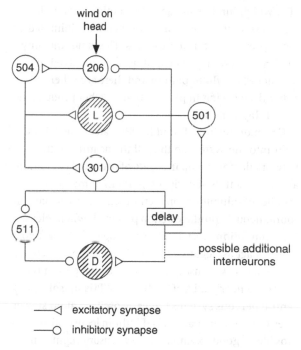

Figure 20.25 Pattern generator. Diagrammatic representation of the interactions between a group of interneurons producing alternations in the contractions of the locust flight muscles. Interneurons represented by numbered circles; motor neurons hatched. L, motor neuron to levator muscle; D, motor neuron to depressor muscle (based on Robertson and Pearson, 1985).

analyses of the network properties. In the absence of tarsal contact with the ground, wind acting via the aerodynamic mechanoreceptor hairs on the head stimulates interneuron 206, which in turn excites interneuron 504. This excites the levator muscle motor neurons (L), so that the wings are raised, and, at the same time, excites interneuron 301. This interneuron has a critical function; it disinhibits (removes inhibition from) the depressor muscle motor neurons (D) by inhibiting another inhibitory interneuron and also excites them, but with a time delay caused by the signal needing to pass through other interneurons first, with each synapse in the network adding a substantial time delay. Thus, it first puts the depressor motor neurons in a state of readiness to respond (since they are no longer inhibited) and then induces the response. Also acting via the time delay, it excites interneuron 501, which inhibits the activity of other cells in the system so that the levator motor neuron is inhibited as the depressor motor neuron fires. This is not the only rhythm-generating system of neurons involved in producing the flight pattern and there are other networks working in parallel that together produce the actual flight rhythm.

Common features found in CPGs are: time delays built into the system so that all the neurons in the network do not end up in a state of perpetual excitation, as well as setting a basic temporal rhythm; a large number of inhibitory connections and neurons that will spontaneously produce action potentials when released from inhibition (post-inhibitory rebound). These two features act as negative feedback regulators of activity in the network. Sensory inputs are a vital part of the system in nearly all CPG networks. CPGs in isolated central nervous systems typically oscillate at far slower rates than in intact animals. Flight in locusts also provides a good example of how sensory input can control the rate at which CPGs operate. Each wing has a stretch receptor associated with it that monitors the position of the wing. These sensory neurons make direct excitatory connections with the depressor motor neurons and indirect connections with the elevator neurons via inhibitory interneurons (among other

interneuronal connections). This forms a simple negative feedback loop regulating the range of movement of the wing; the more the wing is elevated the greater the excitation to the depressors and inhibition of further activity in the elevators. More than this, the activity of the stretch receptors and other sensory receptors on the wings dramatically increases the rate of oscillation between depressor and elevator activity by promoting the next phase of movement. In the intact insect, the basic pattern is also modified by sensory input that enables the insect to adjust its movements to differences in the environment, such as negotiating obstacles and compensating for differences in wind speed.

Neuromodulatory neurons also play an important role in CPG networks. Not only can they alter the rate at which activity occurs, by altering the strengths of synapses in networks, they can dramatically reconfigure the way in which the entire network works. CPGs should not be viewed as fixed anatomical and physiological constructs within the central nervous system: they can be created and removed from the nervous system by the actions of neuromodulators, peptides and other plasticity-inducing mechanisms. Flight, which consists of moving just four non-jointed wings, is comparatively simple compared to the organization needed to control the legs during locomotion. Each leg consists of multiple joints, each under control of antagonistic muscle sets whose activity needs to be rhythmically altered in phase with those controlling other joints. On top of that, the activities of the different legs need to be coordinated together to produce a gait. Add in the number of different rates and types of movement the legs can produce in insects and the problem in understanding how multiple-nested CPG networks coordinate with each other to produce multiple coherent behaviors is formidable.

20.5.5 Neural basis of learning

Many, if not all, insects are able to modify their behavior depending upon previous experience – that is, they learn, to a greater or lesser degree. Some insects such as honey bees show remarkable learning abilities, remembering, for example, the location of flowers visited so they can return to them, and associating particular colors and patterns with nectar rewards. They, and other insects, can also learn to associate certain stimuli with aversive outcomes, such as electric shocks or – more relevant in the natural world – noxious tastes.

The mechanisms underlying learning involve the modification of neuronal properties, particularly the strength of synaptic connections. As such, almost any pair of neurons could in principle undergo some form of plasticity concomitant with learning. Some degree of learning can occur in isolated segmental ganglia. Headless cockroaches and locusts can learn to keep a leg flexed to avoid electrical shocks. Flexion involves several muscles of which one, the coxal adductor muscle, has been most fully investigated. A reduction in the activity of the motor neuron to the adductor muscle leads to a reduction in muscle tension so that the leg drops. As a result, the leg receives a shock. This leads to increased activity in the neuron and the leg is raised (Fig. 20.26). Maintenance of the high level of activity in the motor neuron keeps the leg raised so that further shocks are avoided.

A distinction can be made between short-term memory, which involves modification of existing synaptic proteins to alter their function, and long-term memory, which requires new gene transcription and protein synthesis. In practice, short-term memories may last many days and new gene expression can occur within one or two hours of exposure to a learning regime. Modification of existing structures frequently involves the phosphorylation of receptors and ion channels, hence changing their dynamics. An incoming depolarization, produced by a spike arriving in the presynaptic terminal, may have its amplitude or duration changed by modifying the sensitivity or kinetics of voltage-gated ion channels, for example. At the postsynaptic membrane neurotransmitter

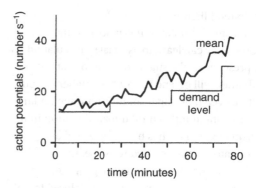

Figure 20.26 Learning in an isolated ganglion. Each time the frequency of action potentials arriving at the coxal adductor muscle fell below an arbitrary "demand level," the muscle was given an electric shock. After a number of shocks, the frequency of action potentials rose. By increasing the demand level, the firing rate of the motor neurons activating the muscle was increased progressively. This is the equivalent of the ganglion "learning" to keep the leg raised in order to avoid shocks (after Hoyle, 1965).

receptors may also have their kinetics altered and modifications to ion channels may affect the amplitude of evoked postsynaptic potentials. These modifications are often brought about by the actions of protein kinases, which are activated by cAMP or cGMP synthesized in response to the action of metabotropic neurotransmitter or modulator receptors on bound adenyl or guanyl cyclase enzymes. Altered gene transcription may alter the number of presynaptic vesicles or postsynaptic receptors present, change the number of ion channels present or even the type of channel produced. In the longer term the entire synapse may be remodeled, with the contact area between neurons expanding or contracting and even entirely new synaptic terminals being made.

If learning is to occur, then at some level neurons conveying information about the qualities that need to be associated have to make contact with each other in the central nervous system, either directly or via intermediary neurons. Many of the neurons containing neuromodulators such as octopamine, dopamine and serotonin have extremely extensive arborizations throughout the central nervous system and so can connect disparate parts of the neuropile to each other.

Bees will extend their proboscis to drink a sugar solution if a small droplet of it is placed on the mouthparts. They can learn to associate particular odors with the presence of the sugar solution so that eventually they will extend the proboscis when the odor is presented, even if no sugar solution is present. A large octopamine-containing ventral unpaired neuron in the subesophageal ganglion has been implicated as a critical mediator of this learned response. Its arborizations span the subesophageal ganglion and olfactory system in the brain, so it is well placed to link the gustatory and olfactory systems with the motor output producing the proboscis extension. Using an intracellular electrode to activate this neuron and make it spike is sufficient to induce a learned response, even in the absence of an actual odor. Different neurons containing dopamine are thought to mediate aversive learning in a similar behavioral context.

Certain brain neuropiles such as the mushroom bodies and the central complex bring together processes from large numbers of neurons conveying different sensory signals and arrange them into an intercalated lattice so that there are a large number of possible anatomical contacts between neuron pairs. These structures therefore lend themselves to the formation of associative memories. The mushroom bodies have been shown experimentally to be involved in associative learning in several insects, particularly bees and *Drosophila*. The use of the many mutants that show deficits in learning and memory in *Drosophila* and the controlled recovery of expression of the gene product in different brain regions have allowed dramatic progress in understanding the mechanistic pathways and anatomical location of memories. There is some evidence that short- and long-term aversive memories are encoded in different lobes of the mushroom bodies. Anatomical changes in the mushroom bodies are correlated with differences in experience (Fig. 20.19b,c) and even a single flight lasting about ten minutes is sufficient to produce changes in the dendritic spines of Kenyon cells through which synaptic connections are made with other interneurons. Learned associations between

visual patterns and shapes with aversive stimuli have been localized to the fan-shaped body of the central complex.

Habituation to a repetitive stimulus is a common phenomenon and is an example of a simpler kind of non-associative learning. For example, a honey bee extends its proboscis if a drop of sucrose solution is applied to one of its antennae, but with frequent repetition of the stimulus the response wanes and finally disappears completely. As stimulation of the opposite antenna causes the response to reappear (dishabituation), the change in the nervous system that produces habituation cannot be occurring in the interneurons or motor neurons immediately responsible for proboscis extension, suggesting that changes may be occurring within the antennal lobe.

Habituation by grasshoppers to certain auditory and visual stimuli is correlated with changes in specific interneurons which convey information to other parts of the nervous system and so affect the behavioral response. For example, the locust's DCMD interneuron habituates if the insect is repeatedly presented with the same visual stimulus to the same part of the eye, and fewer action potentials are produced by each successive stimulation (Fig. 20.27).

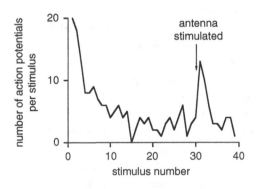

Figure 20.27 Habituation in the central nervous system. Responses of the descending contralateral movement detector (DCMD) interneuron to repeated visual stimulation by a moving black disc. Stimuli were presented at eight-second intervals. After stimulus number 30 an antenna was stimulated electrically. This produced a short-lived dishabituation (after Rowell, 1974).

However, stimulating another part of the eye or changing the state of arousal of the insect by providing other stimuli results in a recovery to the original rate of firing.

20.5.6 Rhythms of behavior

Insects, along with most other organisms, exhibit rhythms of behavioral or physiological activity with an approximately 24-hour cycle. These are called circadian rhythms. These rhythms are driven by neurons, known as pacemakers, found in the protocerebrum and optic lobes. The neuronal and molecular bases of circadian rhythms were first established in *Drosophila*. In *Drosophila* the circadian clock comprises approximately 150 neurons. Their pacemaker properties arise from the action of a group of proteins whose transcription and translation is regulated by mutual feedback loops, so that their production varies in a rhythmic manner. Two transcriptional activators Clock (CLK) and Cycle (CYC), which on binding together initiate the transcription of *period* (*per*) and *timeless* (*tim*) are central to the mechanism of the clock. After translation the PER and TIM proteins interact with each other and can come together to form a dimer. PER and TIM interact with a number of phosphorylating proteins that regulate their activity and location within the cell. Over time the PER–TIM dimer proteins move from the cytoplasm into the nucleus. Here they accumulate and increasingly repress the CLK–CYC dimer by modifying its structure and affinity for DNA. This in turn leads to a feedback loop where PER/TIM in the nucleus leads to the curtailment of their own expression, with the repressor activity reaching a peak at the end of the night. A second, less well understood feedback loop centers on the cyclical expression of *clk* itself by two other transcriptional regulators that also undergo similar regular changes in expression level. The circadian rhythm is rarely exactly 24 hours long when animals are kept under constant light or constant dark

conditions and the clock needs to be entrained to actual day lengths by changes in light levels and/or temperature associated with the onset of day and night. The TIM protein is destroyed by light, which is mediated by a blue-light-sensitive pigment cryptochrome activating a ligase enzyme known as JETLAG. Consequently, very little TIM is present during the day (Fig. 20.28a), and this therefore frees CLK/CYC from repression, leading to increased transcription of *per* and *tim*.

If the onset of darkness is experimentally delayed, the accumulation of proteins is also delayed and the clock is set back (Fig. 20.28b); if the photophase begins earlier than normal, the proteins are destroyed and, in the absence of RNA, no more is produced and the clock is set forward (Fig. 20.28c). Although the same genes are present in the moth *Antheraea*, the PER protein does not enter the nucleus and the diurnal rhythm of activity is clearly produced in a different way. Analyses of the molecular machinery of circadian clocks in other insects have revealed many differences with that first established in *Drosophila*, with most having a greater number of clock components and substantial variation in the timing of accumulation of the various gene products.

In the cockroach *Periplaneta*, the environmental light cycle is perceived via the compound eyes, but in the adult aphid *Megoura* and pupae of the silk moth *Antheraea*, light acts directly on cells in the brain.

The circadian clock neurons direct periods of activity over the daily period. In *Drosophila* there are peaks of activity at dawn and just before dusk (or just before lights on and lights off in a laboratory setting). Different groups of neurons appear to be responsible for inducing these different periods of activity. A group of neurons with cell bodies in the ventro-lateral cortex of the protocerebrum, near the optic lobes, is critical for the morning period of activity (called morning or M cells). All these neurons contain the neuropeptide

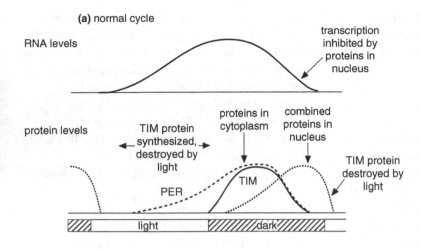

(a) normal cycle

RNA levels

transcription inhibited by proteins in nucleus

protein levels

TIM protein synthesized, destroyed by light

proteins in cytoplasm

combined proteins in nucleus

PER

TIM

TIM protein destroyed by light

light | dark

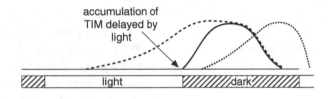

(b) delayed onset of dark

accumulation of TIM delayed by light

light | dark

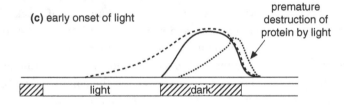

(c) early onset of light

premature destruction of protein by light

light | dark

Figure 20.28 Circadian rhythm of *Drosophila*. (a) In the normal cycle TIM protein is destroyed by light. The entry of PER protein into the nucleus is dependent on an association with TIM protein. Within the nucleus, the two proteins inhibit transcription of their RNAs. (b) If the onset of darkness is delayed, TIM protein is destroyed, but since its RNA is present it starts to accumulate at the onset of darkness. (c) If the light period starts early, TIM protein is destroyed. No more is synthesized immediately because no RNA is present.

pigment-dispersing factor. Another group of neurons, which do not contain pigment-dispersing factor and have more widely dispersed cell bodies, govern the evening period of activity (called evening or E cells). These two populations of cells are coupled and between them coordinate daily periods of activity. A subset of the evening cells with dorsal cell bodies has an important role in entraining intrinsic circadian rhythms to the actual onset of day and night as detected by changing light conditions and thereby regulating the entire clock neuron network.

Summary

- The central nervous system is composed of neurons, which are cells specialized for communication, and glial cells that provide a supporting and nutritional role. The entire central nervous system is surrounded by a sheath which provides a blood–brain barrier.

- Neurons communicate with each other and with effectors at specialized structures called synapses. Most synapses use a chemical transmitter which travels from the pre- to postsynaptic membrane. Signals within neurons are conveyed from input synapses in dendritic regions to output regions on axonal terminals by graded changes in membrane potential and by action potentials which can actively propagate signals over long distances without decaying.

- The ventral nerve cord is composed of a chain of segmental ganglia which show a greater or lesser degree of fusion in different insect species. Each ganglion contains populations of local and intersegmental interneurons, as well as motor neurons responsible for controlling activity in limbs and other muscles associated with the segment.

- The brain is composed of protocerebral, deutocerebral and tritocerebral segements. It receives and processes sensory inputs from the major sense organs of the head – the eyes and antennae. Other specialized neuropiles such as the mushroom bodies and central complex have important roles in higher-level integration, organizing executive behavioral output and in learning and memory.

- Behavior arises from complex interactions between networks of neurons. Behavior can be modified by experience and is manifested in the nervous system by changes in neuronal function and in particular the strength of synaptic connections.

Recommended reading

STRUCTURE OF NEURONS AND GLIA

Carlson, S. D. and Saint Marie, R. L. (1990). Structure and function of insect glia. *Annual Review of Entomology* 35, 597–621.

Edwards, J. S. and Tolbert, L. P. (1998). Insect Neuroglia. In *Microscopic Anatomy of Invertebrates, vol. 11B, Insecta*, ed. F. W. Harrison and M. Locke, pp. 449–466. New York, NY: Wiley-Liss.

Ito, K., Urban, J. and Technau, G. M. (1995). Distribution, classification and development of *Drosophila* glial cells in the late embryonic and larval ventral nerve cord. *Roux's Archives in Developmental Biology* **204**, 284–307.

Lane, N. J. (1985). Structure of components of the nervous system. In *Comprehensive Insect Physiology, Biochemistry and Pharmacology*, vol. 5, ed. G. A. Kerkut and L. I. Gilbert, pp. 1–47. Oxford: Pergamon Press.

Strausfeld, N. J. and Meinertzhagen, I. A. (1998). The insect neuron: types, morphologies, fine structure, and relationship to the architectonics of the insect nervous system. In *Microscopic Anatomy of Invertebrates, vol. 11B, Insecta*, ed. F. W. Harrison and M. Locke, pp. 487–538. New York, NY: Wiley-Liss.

BLOOD–BRAIN BARRIER

Carlson, S. D., Juang, J.-L., Hilgers, S. L. and Garment, M. B. (2000). Blood barriers of the insect. *Annual Review of Entomology* **45**, 151–174.

SIGNAL TRANSMISSION

Wicher, D., Walther, C. and Wicher, C. (2001). Non-synaptic ion channels in insects: basic properties of currents and their modulation in neurons and skeletal muscles. *Progress in Neurobiology* **64**, 431–525.

CHEMICAL SIGNALS

Golding, D. W. (1994). A pattern confirmed and refined: synaptic, nonsynaptic and parasynaptic exocytosis. *BioEssays* **16**, 503–508.

Holman, G. M., Nachman, R. J. and Wright, M. S. (1990). Insect neuropeptides. *Annual Review of Entomology* **35**, 201–217.

Homberg, U. (1994). Flight-correlated activity changes in neurons of the lateral accessory lobes in the brain of the locust *Schistocerca gregaria*. *Journal of Comparative Physiology A* **175**, 597–610.

Nässel, D. R. (1988). Serotonin and serotonin-immunoreactive neurons in the nervous system of insects. *Progress in Neurobiology* **30**, 1–85.

Orchard, I., Belanger, J. H. and Lange, A. B. (1989). Proctolin: a review with emphasis on insects. *Journal of Neurobiology* **20**, 470–496.

GANGLIA

Niven, J. E., Graham, C. M. and Burrows, M. (2008). Diversity and evolution of the insect ventral nerve cord. *Annual Review of Entomology* **53**, 253–271.

Weevers, R. de G. (1985). The insect ganglia. In *Comprehensive Insect Physiology, Biochemistry and Pharmacology*, vol. 5, ed. G. A. Kerkut and L. I. Gilbert, pp. 213–297. Oxford: Pergamon Press.

NON-SPIKING INTERNEURONS

Burrows, M. (1996). *The Neurobiology of an Insect Brain*. Oxford: Oxford University Press.

NEURAL HOMOLOGIES

Kutsch, W. and Breidbach, O. (1994). Homologous structures in the nervous systems of arthropods. *Advances in Insect Physiology* **24**, 1–113.

MUSHROOM BODIES

Fahrbach, S. E. (2006). Structure of the mushroom bodies of the insect brain. *Annual Review of Entomology* 51, 209–232.

Heisenberg, M. (2003). Mushroom body memoir: from maps to models. *Nature Reviews Neuroscience* 4, 266–275.

OPTIC LOBES

Borst, A., Haag, J. and Reiff, D. F. (2010). Fly motion vision. *Annual Reviews of Neuroscience* 33, 49–70.

DEUTOCEREBRUM

Galizia, C. G. and Rössler, W. (2010). Parallel olfactory systems in insects: anatomy and function. *Annual Review of Entomology* 55, 399–420.

LEARNING

Dukas, R. (2008). Evolutionary biology of insect learning. *Annual Review of Entomology* 53, 145–160.

Keene, A. C. and Waddell, S. (2007). *Drosophila* olfactory memory: single genes to complex neural circuits. *Nature Reviews Neuroscience* 8, 341–354.

RHYTHMS OF BEHAVIOR

Sandrelli, F., Costa, R., Kyriacou, C. P. and Rosato, E. (2008). Comparative analysis of circadian clock genes in insects. *Insect Molecular Biology* 17, 447–463.

References in figure captions and tables

Albrecht, F. O. (1953). *The Anatomy of the Migratory Locust*. London: Athlone Press.

Burrows, M. (1977). Flight mechanisms in the locust. In *Identified Neurons and Behaviour of Arthropods*, ed. G. Hoyle. New York, NY: Plenum Press.

Burrows, M. and Newland, P. L. (1993). Correlation between the receptive fields of locust interneurons, their dendritic morphology, and the central projections of mechanosensory neurons. *Journal of Comparative Neurology* 329, 412–426.

Callec, J. J. (1985). Synaptic transmission in the central nervous system. In *Comprehensive Insect Physiology, Biochemistry and Pharmacology*, vol. 5, ed. G. A. Kerkut and L. I. Gilbert, pp. 139–179. Oxford: Pergamon Press.

Durst, C., Eichmüller, S. and Menzel, R. (1994). Development and experience lead to increased volume of subcompartments of the honeybee mushroom body. *Behavioral and Neural Biology* 62, 259–263.

Gronenberg, W. and Strausfeld, N. J. (1992). Premotor descending neurons responding selectively to local visual stimuli in flies. *Journal of Comparative Neurology* 316, 87–103.

Hausen, K. (1982). Motion sensitive interneurons in the optomotor system of the fly: II. The horizontal cells – receptive field organization and response characteristics. *Biological Cybernetics* **46**, 67–79.

Homberg, U. (1991). Neuroarchitecture of the central complex in the brain of the locust *Schistocerca gregaria* and *S. americana* as revealed by serotonin immunocytochemistry. *Journal of Comparative Neurology* **303**, 245–254.

Homberg, U. (1994). *Distribution of Neurotransmitters in the Insect Brain*. Stuttgart: Gustav Fischer Verlag.

Horridge, G. A. (1965). The Arthropoda: general anatomy. In *Structure and Function of the Nervous System of Invertebrates*, ed. T. H. Bullock and G. A. Horridge, pp. 801–964. San Francisco, CA: Freeman.

Howse, P. E. (1974). Design and function in the insect brain. In *Experimental Analysis of Insect Behaviour*, ed. L. Barton Browne, pp. 180–194. Berlin: Springer-Verlag.

Hoyle, G. (1965). Neurophysiological studies on "learning" in headless insects. In *The Physiology of the Insect Nervous System*, ed. J. E. Treherne and J. W. L. Beament, pp. 203–232. London: Academic Press.

Kent, K. S., Hoskins, S. G. and Hildebrand, J. G. (1987). A novel serotonin-immunoreactive neuron in the antennal lobe of the sphinx moth *Manduca sexta* persists throughout postembryonic life. *Journal of Neurobiology* **18**, 451–465.

Kloppenburg, P. and Hildebrand, J. G. (1995). Neuromodulation by 5-hydroxytryptamine in the antennal lobe of the sphinx moth *Manduca sexta*. *Journal of Experimental Biology* **198**, 603–611.

Laughlin, S. B. (1975). The function of the lamina ganglionaris. In *The Compound Eye and Vision of Insects*, ed. G. A. Horridge, pp. 341–358. Oxford: Clarendon Press.

Matsumoto, S. G. and Hildebrand, J. G. (1981). Olfactory mechanisms in the moth *Manduca sexta*: response characteristics and morphology of central neurons in the antennal lobes. *Proceedings of the Royal Society of London B* **213**, 249–277.

Meinertzhagen, I. A. and O'Neil, S. D. (1991). Synaptic organization of columnar elements in the lamina of the wild type in *Drosophila melanogaster*. *Journal of Comparative Neurology* **305**, 232–263.

Mobbs, P. G. (1985). Brain structure. In *Comprehensive Insect Physiology, Biochemistry and Pharmacology*, vol. **5**, ed. G. A. Kerkut and L. I. Gilbert, pp. 299–370. Oxford: Pergamon Press.

Newland, P. L. and Burrows, M. (1994). Processing of mechanosensory information from gustatory receptors on a hind leg of the locust. *Journal of Comparative Physiology A* **174**, 399–410.

Orchard, I., Ramirez, J.-M. and Lange, A. B. (1993). A multifunctional role for octopamine in locust flight. *Annual Review of Entomology* **38**, 227–249.

O'Shea, M., Rowell, C. H. F. and Williams, J. L. D. (1974). The anatomy of a locust visual interneurone; the descending contralateral movement detector. *Journal of Experimental Biology* **60**, 1–12.

Pearson, K. G. (1977). Interneurons in the ventral nerve cord of insects. In *Identified Neurons and Behaviour of Arthropods*, ed. G. Hoyle, pp. 329–337. New York, NY: Plenum Press.

Pearson, K. G., Boyan, G. S., Bastiani, M. and Goodman, C. S. (1985). Heterogeneous properties of segmentally homologous interneurons in the ventral nerve cord of locusts. *Journal of Comparative Neurology*, 233, 133–145.

Ramirez, J.-M. and Orchard, I. (1990). Octopaminergic modulation of the forewing stretch receptor in the locust *Locusta migratoria*. *Journal of Experimental Biology* 149, 255–279.

Robertson, R. M. and Pearson, K. G. (1985). Neural circuits in the flight system of the locust. *Journal of Neurophysiology* 53, 110–128.

Rowell, C. H. F. (1974). Boredom and attention in a cell in the locust visual system. In *Experimental Analysis of Insect Behavior*, ed. L. Barton Browne, pp. 87–113. Berlin: Springer-Verlag.

Siegler, M. V. S. and Burrows, M. (1979). The morphology of local non-spiking interneurones in the metathoracic ganglion of the locust. *Journal of Comparative Neurology* 183, 121–148.

Smith, D. S. and Treherne, J. E. (1963). Functional aspects of the organisation of the insect nervous system. *Advances in Insect Physiology* 1, 401–484.

Technau, G. M. (1984). Fiber number in the mushroom bodies of adult *Drosophila melanogaster* depends on age, sex and experience. *Journal of Neurogenetics* 1, 113–126.

Treherne, J. E. (1974). The environment and function of insect nerve cells. In *Insect Neurobiology*, ed. J. E. Treherne. Amsterdam: North-Holland Publishing Co.

Treherne, J. E. and Schofield, P. K. (1981). Mechanisms of ionic homeostasis in the central nervous system of an insect. *Journal of Experimental Biology* 95, 61–73.

Watson, A. H. D. and Burrows, M. (1988). Distribution and morphology of synapses in nonspiking local interneurones in the thoracic nervous system. *Journal of Experimental Biology* 272, 605–616.

21 | Endocrine system

REVISED AND UPDATED BY **STUART REYNOLDS**

INTRODUCTION

Hormones are chemical signals that are produced by the insect and circulate
in the blood to regulate long-term physiological, developmental and behavioral
activities. These signals complement those of the nervous system, which provides
short-term coordination. The activities of the two systems are closely linked and sometimes
not clearly distinguishable. General aspects of hormones are discussed in this chapter,
including their chemical structures (Section 21.1); endocrine organs that secrete hormones
(Section 21.2); the means by which hormones are transported in the hemolymph
(Section 21.3); regulation of hormone titers (Section 21.4); and the mode of action of
hormones on their target tissues (Section 21.5). Specific actions of hormones regulating
particular functions are considered in other chapters; notable examples include molting
and metamorphosis in Section 15.4, yolk synthesis in Section 13.2.4, embryonic cuticles in
Section 14.2.10, diuresis in Section 18.3.3, mobilization of fuel for flight in Section 9.6.2,
polyphenism in Section 15.5 and diapause in Section 15.6.

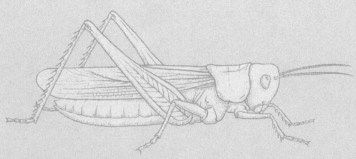

The Insects: Structure and Function (5th edition), ed. S. J. Simpson and A. E. Douglas.
Published by Cambridge University Press. © Cambridge University Press 2013.

21.1 Chemical structure of hormones

Apart from molting hormones (polyhydroxylated steroids) and juvenile hormones (sesquiterpenes), most known insect hormones are polypeptides. Some biogenic amines are also known to function as hormones (see Section 20.2.3).

21.1.1 Molting hormones

Molting hormones are ecdysteroids (Fig. 21.1a) which, in immature insects, are produced by the prothoracic glands. In most adult insects the prothoracic glands degenerate, but gonadal and other tissues may produce ecdysteroids. In most insects, as in *Drosophila melanogaster* (Diptera), the main ecdysteroid secreted is ecdysone. As noted in Chapter 15, ecdysteroids act on the epidermis to promote molting. Ecdysone is generally considered to be a prohormone, being converted in the fat body or epidermis in most insects to the active hormone 20-hydroxyecdysone, by a cytochrome P450 enzyme CYP314A1, encoded by the gene *shade*. It remains possible, however, that ecdysone itself may have actions distinct from those of 20-hydroxyecdysone.

Insects cannot synthesize steroids. Consequently, sterols, usually cholesterol or a closely related structure, are essential dietary constituents (see Chapter 4). The rate of sterol uptake into ecdysteroidogenic tissues may regulate ecdysteroid titers to some extent. Conversion of cholesterol to the secreted ecdysteroid is accomplished by a series of steps that involve several P450 enzymes of the *Drosophila* Halloween family, as shown in Fig. 21.1b. Steroidogenesis involves two sequential migrations from the cytoplasm into the mitochondria and then out again, and thus depends on the uptake and transport of these rather hydrophobic compounds within the cell. In *D. melanogaster*, mutations in the *npc1* and *npc2* genes, which encode proteins known to be important for sterol trafficking, cause defective ecdysteroid biosynthesis.

Phytophagous insects lack cholesterol in their diet and must make do with other phytosterols. As a result, in some insects ecdysteroidogenesis begins with a different precursor and the prothoracic glands secrete ecdysteroids other than ecdysone. For example, in *Manduca sexta* (Lepidoptera), the main secreted ecdysteroid is 3-dehydroecdysone, which is then converted to ecdysone by a 3-β reductase enzyme in the hemolymph, while in the honey bee, *Apis mellifera* (Hymenoptera), and in some Heteroptera (Hemiptera), the principal 20-hydroxylated ecdysteroid is makisterone A, which replaces 20-hydroxyecdysone.

21.1.2 Juvenile hormone

Juvenile hormone (JH) is an acyclic sesquiterpene (Fig. 21.2a) produced by the corpora allata using the mevalonate pathway (Fig. 21.2b). Several slightly different forms, known as JH0, I, II and III containing 19, 18, 17 and 16 carbon atoms, respectively, have been isolated. JHIII is the form occurring in most insects; JHI and II are known primarily from Lepidoptera, and JH0 only from lepidopteran eggs. In the wandering stage larva of *M. sexta* and in pharate adult males of *Hyalophora cecropia* (Lepidoptera), the corpora allata produce juvenile hormone acid (epoxyfarnesoic acid) that is subsequently converted to juvenile hormone in the imaginal discs and accessory sex glands, respectively.

In addition to JHIII, the corpora allata of cyclorrhaphous Diptera secrete two related compounds, a bisepoxide of JHIII (JHB3) and methylfarnesoate, and it is possible that similar compounds are also produced by other insects. These compounds have JH-like functions in bioassays but the extent to which these compounds function in the same manner as other JHs is not well understood.

21.1.3 Peptide hormones

Insects use a great diversity of protein or peptide hormones, the extent of which is shown in Table 21.1. Peptides are distinct from proteins only in that

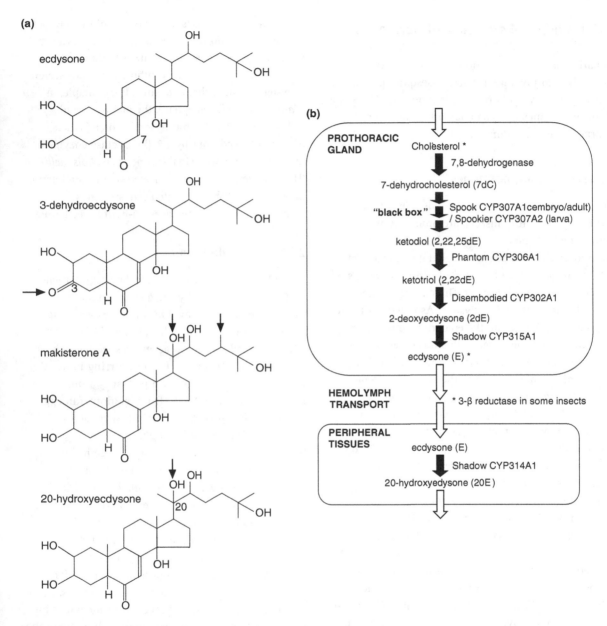

Figure 21.1 (a) Ecdysteroids. Structures of ecdysone and related ecdysteroids produced by insects. Arrows indicate differences from ecdysone. Numbers relate to positions of carbon atoms. (b) Ecdysteroid biosynthesis and the involvement of cytochrome P450 (CYP) enzymes. The pathway shown is the one that occurs in the prothoracic gland equivalent part of the *Drosophila melanogaster* ring gland, and the names of the enzymes are those of the corresponding genes. The reactions occurring in the "black box" during transformation of 7dC to 2,22,25dE have not been fully elucidated but involve multiple steps; the very similar P450 enzymes, Spook and Spookier, are involved in this stage of steroidogenesis, the former acting only in embryos and adults, the latter only in larvae. After Rewitz *et al.* (2006). Note that in other insects the steroid at the start of the pathway may be a phytosteroid rather than cholesterol, and this may cause the final ecdysteroid product to differ. Thus in *Manduca sexta*, 3-dehydroecdysone replaces ecdysone as the secretory product of the prothoracic gland, and this is then converted to ecdysone by a 3-β reductase enzyme in the hemolymph.

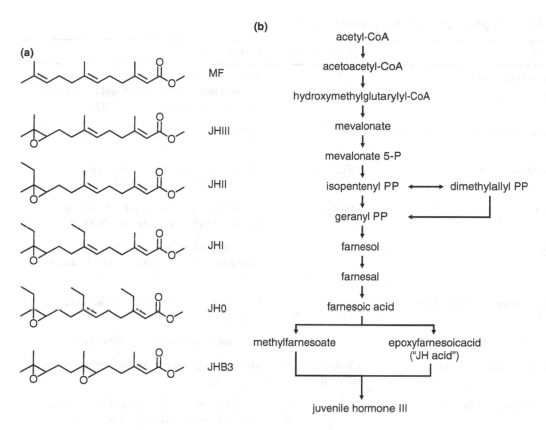

Figure 21.2 (a) Juvenile hormones and related molecules. MF, methylfarnesoate; JHB3, juvenile hormone III bisepoxide. See text for details of occurrence and function. (b) Juvenile hormone biosynthesis. The diagram shows the mevalonate pathway leading to juvenile hormone. For details, see Bellés *et al.* (2005). Two different routes are available from farnesoic acid to juvenile hormones. In Orthoptera, Dictyoptera, Coleoptera and Diptera, farnesoic acid is first esterified to yield methylfarnesoate and then an epoxide group is added to yield JHIII. In lepidopteran insects, epoxidation takes place first, and epoxyfarnesoic acid is then esterified. Where methylation is required to form JHII, etc., it takes place after this. In some adult moths, epoxyfarnesoic acid is secreted and stored outside the corpora allata.

their amino acid sequences are relatively shorter. The primary structures (amino acid sequences) of some selected examples are shown in Fig. 21.3. The fantastic variety of bioactive peptides in insects presents endless opportunity for confusion over names, and calls for an agreed system of nomenclature that would ideally include useful information about the nature and biological role of the peptide in question. A system has been proposed (Table 21.1) that allocates each peptide to one of 32 structurally distinct families and also specifies the identity of the insect species from which the peptide was first isolated; the system uses a five-letter code to indicate genus and species names, and intraspecific isoforms are identified by numerals (see examples in Fig. 21.3). There are also existing widely understood common names for many insect peptides, however, which have also been used in this book.

Table 21.1 **Insect peptide and protein hormone families**

Peptide family	Abbreviation	Function/activity
1 Adipokinetic hormone	AKH	Mobilize stored lipids and/or carbohydrates; inhibit protein synthesis; modulate immune reactions; stimulate visceral muscle
2 Adipokinetic hormone/corazonin-related peptide	ACP	Unknown
3 FGLa-related allatostatin	FGLa/AST	Inhibit JH biosynthesis; inhibit visceral muscle activity; regulate food intake
4 PISCF-related allatostatin	PISCF/AST	Inhibit JH biosynthesis; inhibit visceral muscle activity; regulate food intake
5 Allatotropin	AT	Promote JH biosynthesis; inhibit ion transport in gut; stimulate visceral muscle activity
6 Arginine-vasopressin-like peptide	AVLP	Diuretic (but acts indirectly on Malpighian tubules)
7 Bursicon (a and b subunits)		Promote cuticle tanning; cuticle plasticization
8 Calcitonin-like diuretic hormone	CT-like DH (CT/DH)	Diuretic (acts directly on Malpighian tubules)
9 Cardioacceleratory peptide 2b	CAP2b	Cardioaccelatory; diuretic and antidiuretic (acts directly on Malpighian tubules)
10 Corazonin	CRZ	Cardioaccelatory; initiate ecdysis; phase transition in locusts
11 Corticotropin releasing factor (CRF)-related diuretic hormone	CRF-related DH (CRF/DH)	Diuretic (acts directly on Malpighian tubules)
12 Crustacean cardioactive peptide	CCAP	Specific CNS neuromodulatory role in ecdysis behavior; stimulate visceral muscle activity; stimulate AKH release; diuretic
13 Ecdysis-triggering hormone and pre-ecdysis-triggering hormone	ETH and PETH	Initiate pre-ecdysis and ecdysis behavior; promote release of EH
14 Eclosion hormone	EH	Promote ETH release; cuticle plasticization
15 FMRFamide-like peptide	FaLP	Stimulate visceral muscle activity; inhibit ecdysteroidogenesis
16 Insulin-related peptide	IRP	Regulate cellular processes involved in metabolism, growth and reproduction; stimulate ecdysteroidogenesis
17 Crustacean hyperglycaemic hormone-(chh)-related ion transport peptide	CHH-related ITP (CHH/ITP)	Stimulate ion and water transport in ileum

Table 21.1 (*cont.*)

Peptide family	Abbreviation	Function/activity
18 Kinins	K	Stimulate visceral muscle activity; regulate meal size
19 Myoinhibitory peptide	MIP	Inhibit JH biosynthesis; inhibit visceral muscle activity; inhibit ecdysteroidogenesis
20 Myosuppressin	MS	Inhibit visceral muscle activity; diuretic (acts directly on Malpighian tubules); regulate food intake; inhibit ecdysteroidogenesis
21 Neuroparsin	NP	Uncertain; may be involved in female reproductive cycles
22 Neuropeptide F	NPF	Stimulate feeding
23 Orcokinin	OK	Light entrainment of circadian clock
24 Pigment dispersing factor	PDF	Output of circadian clock
25 Proctolin		Modulate visceral and skeletal muscle activity
26 Prothoracicotropic hormone	PTTH	Stimulate ecdysteroidogenes
27 Pyrokinin/diapause hormone/pheromone biosynthesis activating neuropeptide (= melanization and reddish coloration hormone)	PK, PK-related DH (PK/DH), PBAN (PK/ PBAN)	Stimulate visceral muscle activity; promote diapause; promote pheromone biosynthesis; initiate cuticle melanization
28 Sex peptide	SP	Male accessory gland secretions; modulate female sexual behavior
29 Short neuropeptide F	sNPF	Control food intake and body size; inhibit JH biosynthesis
30 SIFamide	SIFa	Modulate sexual behavior
31 SulfaKinin	SK	Stimulate visceral muscle activity
32 Tachykinin-related peptide	TRP	Stimulate visceral muscle activity; stimulate feeding

Note:
The peptides are classified according to the scheme proposed by Coast and Schooley (2011). The attribution of "Function/ activity" in the third column of the table implies neither that the indicated activity is necessarily a normal physiological function, nor that this activity is universal in all insects in which one or more members of the peptide family occur. As discussed in the text, peptides vary in their ability to exert these functions in different insects, and even at different times in the life of the same insect.

Bioactive peptides are produced by individual cells, mainly in the central nervous system and the midgut epithelium (Section 21.2.1). Some are not strictly hormones, being used in more local signaling. A wide range of functions is well established for many peptides, but the in vivo roles

(a) Proctolin (present in many insects)

R-Y-L-P-T

(b) *Manduca sexta* eclosion hormone

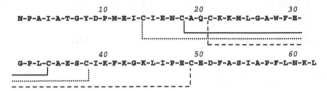

```
              10                    20                    30
N-P-A-I-A-T-G-Y-D-P-M-E-I-C-I-E-N-C-A-Q-C-K-K-M-L-G-A-W-F-E-
                                 |         |
                                 |         |..................
                                 |         '- - - - - - - - - - - - -

              40                    50                    60
G-P-L-C-A-E-S-C-I-K-F-K-G-K-L-I-P-E-C-E-D-F-A-S-I-A-P-F-L-N-K-L
        |                       |
        |.......................|
        '- - - - - - - - - - - - -'
```

(c) Adipokinetic hormones from various insects

Locusta migratoria	(Locmi-AKH-1)	pQ-L-N-F-T-P-N-W-G-T-NH₂
Locusta migratoria	(Locmi-AKH-2)	pQ-L-N-F-S-A-G-W-NH₂
Manduca sexta	(Manse-AKH)	pQ-L-T-F-T-S-S-W-G-NH₂
Carausius morosus	(Carmo-AKH)	pQ-L-T-F-T-P-N-W-G-T-NH₂
Gryllus bimaculatus	(Grybi-AKH)	pQ-L-N-F-S-A-G-W-NH₂

(d) *Locusta migratoria* sulfakinins

```
                          SO₄
                           |
(Locmi-SK)       pG-L-A-S-D-D-Y-G-H-M-R-F-NH₂
```

(e) *Diploptera punctata* FGLa-related allatostatin family peptides

Dippu-FGLa/AST-1	L-Y-D-F-G-L-NH₂
Dippu-FGLa/AST-2	A-Y-S-Y-V-S-E-Y-K-R-L-P-V-Y-N-F-G-L-NH₂
Dippu-FGLa/AST-3	S-K-M-Y-G-F-G-L-NH₂
Dippu-FGLa/AST-4	D-G-R-M-Y-S-F-G-L-NH₂
Dippu-FGLa/AST-5	D-R-L-Y-S-F-G-L-NH₂
Dippu-FGLa/AST-6	A-R-P-Y-S-F-G-L-NH₂
Dippu-FGLa/AST-7	I-A-P-S-G-A-Q-R-L-Y-G-F-G-L-NH₂
Dippu-FGLa/AST 8	G-G-S-L-Y-S-F-G-L-NH₂
Dippu-FGLa/AST-9	G-D-G-R-L-Y-A-F-G-L-NH₂
Dippu-FGLa/AST-10	P-V-N-S-G-R-S-S-G-S-R-F-N-F-G-L-NH₂
Dippu-FGLa/AST-11	Y-P-Q-E-H-R-F-S-F-G-L-NH₂
Dippu-FGLa/AST-12	P-F-N-F-G-L-NH₂
Dippu-FGLa/AST-13	I-P-M-Y-D-F-G-I-NH₂

(f) *Gryllus bimaculatus* MIP family peptides

Grybi-MIP-1	W-Q-D-L-N-G-G-W-G

(g) *Manduca sexta* PISCF-related allatostatin family peptides

Manse-PISCF/AST	pQ-V-R-F-R-Q-C-Y-F-N-P-I-S-C-F

(h) *Drosophila melanogaster* bursicon (Bu) and partner of bursicon n(pB)

```
Bu QPDSSVAATDNDITHLGDDCQVTPVIHVL-QYPGCVPKPIPSFACVGRCASYIQVSGSKI
pB LRYSQGTGDENCETLKSEIHLIKEEFDELGRMQRTCNADVIVNKCEGLCNSQVQPSVITP

Bu WQMERSCMCCQESGEREAAVSLFCPKVKPGERKFKKVLTKAPLECMCRPCTSIEESGIIP
pB TGFLKECYCCRESFLKEKVITLTHCYDPDGTR-----LTSPEMGSMDIRLR-------EP

Bu QEIAGYSDEGPLNNHFRRIALQ
pB TECKCFKCG-----DFTR----
```

(i) PTTH from *B. mori*, *M. sexta* and *D. melangoaster*

```
Bommo NIQVEN--QAIPDPPCTCKYKKEIEDLGENSVPRFIETRNCNKTQQ---PTCRPPYICKE
Manse NIKVEEYNQAIPDPPCSCEYKKGFINLGENVFPSNIETINCSTNQQ---QSCPPPYICKE
Drome --------NDVHSAGCDCKVTNELVDLGGLHFPRFLMNAVCESGAGRDLAKCSHGSNCRP

Bommo SLYSITILKRRETKSQESLEIPNELKYRWVAESHPVSVACLCTRDYQLRYNNN
Manse SIYEIKILRKRKSMAEKSLARPTDLEIGWVAESLPISVGCICTRDYVI-----
Drome LEYKVKVLAQT-SQSDHPYSWMNK-DQPWQFKTVTVTAGCFCTK--------
```

of many others remain obscure even though they may have been shown, experimentally, to produce certain physiological effects when injected or tested in vitro.

Insect peptides vary greatly in size (i.e., the number of amino acid residues they contain). Some have as few as five residues (e.g., proctolin, which occurs with the same sequence in many insects), but others are larger (e.g., *M. sexta* eclosion hormone, 62 residues) (Fig. 21.3). These two peptides are simple amino acid chains, but some others are post-translationally modified (Fig. 21.3). Probably the majority of insect peptide hormones are C-terminally amidated, many (e.g., adipokinetic hormones) have a cyclized N-terminal pyroglutamate residue (also called pyrrolidone carboxylic acid), while a few have other post-transcriptional modifications (e.g., sulfakinins have a sulfated tyrosine residue in the middle of the heptapeptide sequence).

Other peptide hormones incorporate more than one polypeptide chain in their molecular structure. For example, *Drosophila* prothoracicotrophic hormone (PTTH) is a homodimer of two identical subunits linked by cysteine crossbridges between the polypeptide subunits. By contrast, *Drosophila* bursicon is a heterodimeric protein comprising a 141 amino acid residue subunit linked to a 121 residue subunit. The two polypeptides are encoded on different genes (*bursicon* and *partner of bursicon*) that show sufficient sequence similarity to suggest they have evolved from the same ancestral gene (Fig. 21.3).

Peptide hormones with similar functions in different insect species often have similar sequences, indicating that they are members of a peptide family (i.e., they are products of evolutionarily related genes). Although regions of the sequence may differ, there are conserved regions (motifs) in which the amino acid sequences are identical or closely similar. The multiple allatostatins of the cockroach *Diploptera punctata* are a good example. This insect can potentially generate 13 distinct peptide sequences of the FGLa-related allatostatin family (Dippu-FGLa-AST 1–13), ranging in length from 6 to 18 amino acids. All but one have an identical C-terminal pentapeptide sequence that is important for biological activity (Fig. 21.3). The remaining

Caption for Figure 21.3 Some examples of insect peptide hormones with their amino acid sequences. (a) Proctolin – this peptide is present with the same sequence in many insects; (b) eclosion hormone from *M. sexta* – lines connecting three pairs of cysteine residues as intramolecular disulfide bridges are shown. The lines are drawn differently only for clarity; (c) adipokinetic hormone (AKH) family peptides from a number of different insects (many more examples with slightly different sequences are known); (d) sulfakinin from *L. migratoria*, showing post-translational modification of the peptide's structure; (e) *D. punctata* FGLa-related allatostatin family peptides showing the large range of similar peptides with (apparently) the same function that may exist in a single insect species; (f) an MIP family peptide from *G. bimaculatus*, which also has allatostatic function; (g) a *M. sexta* PISCF-related allatostatin family peptide that also has an allatostic function; (h) *D. melanogaster* bursicon (Bu) and partner of bursicon (pB) – these two peptides associate to form the tanning hormone heterodimer. The amino acid sequences of the peptides are aligned in order to show their similarity (identical residues are shaded); some gaps have been introduced to facilitate the alignment; (i) PTTH from *B. mori*, *M. sexta* and *D. melanogaster* – the amino acid sequences of PTTH from the two lepidopteran species are more closely related than the two bursicon subunits. The extent of similarity between the lepidopteran and the dipteran proteins is quite low – nevertheless, certain key amino acid residues are conserved (identical residues in all three hormone molecules are shaded), reflecting the retention of the structural framework of the molecule; some gaps have been introduced to facilitate the alignment. In (c), (d), (e), (f) and (g) the five-letter code proposed by Coast and Schooley is used to indicate the insect species in which the peptide was first described. The abbreviated name of the peptide indicates the insect peptide family to which it belongs (see Fig. 21.3). Amino acid sequences are shown using the one-letter code: A, alanine; C, cysteine; D, aspartic acid; E, glutamic acid; F, phenylalanine; G, glycine; H, histidine; I, isoleucine; K, lysine; L, leucine; M, methionine; N, asparagine; P, proline; Q, glutamine; R, arginine; S, serine; T, threonine; V, valine; W, tryptophan.

peptide changes only one of these conserved residues and substitutes a structurally very similar amino acid. Peptides with this same consensus sequence (i.e., members of the same family) are found in other orthopteroid species.

In other cases, however, peptides that have functionally similar roles in their "own" species have completely different structures – i.e., they belong to a different peptide family. Thus, some other orthopterans – e.g., the cricket *Gryllus bimaculatus* (Orthoptera, Gryllidae) – as well as insects of other families, not only have allatostatins of the FGLa/AST family (*G. bimaculatus* produces 13 of these from a single precursor) but also have other allatostic peptides with dissimilar sequences that are produced from a different precursor and are members of a different peptide family. These were first discovered because of their ability to inhibit visceral muscle activity, and are thus called myoinhibitory or MIP family peptides. In *G. bimaculatus* there are six genes encoding MIP family peptides, but because some genes are the same, there are only four different MIP family peptide sequences (Fig. 21.3). The allatostatins of Lepidoptera are different again (the figure shows an example from *M. sexta*). These belong to the PISCF-related family of allatostatins and are structurally completely different from either of the other two allatostatin peptide families.

The endocrine activity of peptides may take one of two forms: they may act directly on effector organs or they may act on other endocrine organs, which, in turn, are stimulated to produce hormones. Adipokinetic hormone is an example of a peptide acting directly on an effector organ. Prothoracicotropic hormone, allatotropins and allatostatins are examples of peptides regulating the synthesis of other hormones.

A given family of peptide hormones can serve a wide variety of functions. The adipokinetic hormones (AKHs) illustrate this point (Fig. 21.3). AKH family peptides mobilize lipids in a wide range of insects, but also increase hemolymph trehalose levels,

increase heart rate, inhibit protein synthesis and a number of other functions. Not all of these functions are usually found in the same insect, but multiple physiological effects are common. The various members of the peptide family are found to have different potencies in the various insect species in which they are tested. Generally (but not always) a given peptide will be most effective in the species of origin, presumably because its molecular structure best matches the specificity of the receptor from that species. But the same peptide can have different physiological effects in different insects, e.g., AKH family peptides cause lipid mobilization in locusts and moths, but elevate hemolymph carbohydrate levels in the stick insect *Carausius morosus* (Phasmatodea); or effects may differ in different developmental stages, e.g., AKH family peptides mobilize carbohydrates in *M. sexta* larvae, but increase lipid levels in adults of the same species. In most cases we do not know if this multiplicity of physiological actions is due to multiple receptor types or differential coupling of the same receptor type to different cellular machinery (see Section 21.5.3).

When two hormones with similar functions are present within a species, they may have similar molecular structures, or be different. For example, as noted above, all but one of the FGLa-related allatostatins produced by *D. punctata* have the same amino acid sequence at the C-terminal end of the molecule, although other parts of their sequences differ. The PTTHs of different insects, on the other hand, differ considerably, to the extent that little similarity can be seen in the primary amino acid sequences of insects from different Families (Fig. 21.3). It appears nevertheless that certain important amino acid residues (especially cysteine residues) that determine the three-dimensional structure of the molecule have been conserved, and all PTTHs characterized to date appear to be members of the growth factor superfamily that includes mammalian nerve growth factor; they possess a

C-terminal cystine-knot motif typical of members of the Trunk family of proteins, other members of which are embryonic developmental signaling ligands.

Peptide hormones are derived from protein precursors, which are cleaved and modified post-transcriptionally to produce the active peptide (Fig. 21.4). Sometimes only a single copy of a single active peptide is formed from one precursor protein. This is the case with eclosion hormone, for example. On the other hand, several different peptides may be produced from a single precursor protein; these may have similar or dissimilar functions. The 13

different FGLa-related allatostatins of *D. punctata*, all of which are derived by proteolytic cleavage from the same precursor protein, appear to have similar physiological functions. In the case of the *D. melanogaster* gene encoding FMRFamide-like (FaLP) family peptides, not only are several different peptides produced in this way, but the precursor also includes multiple copies of a single peptide, DPKQDFMRFamide (which, like the other peptides produced from this precursor, stimulates visceral muscle activity). It is not clear in these two cases why so many copies of the DNA sequence encoding

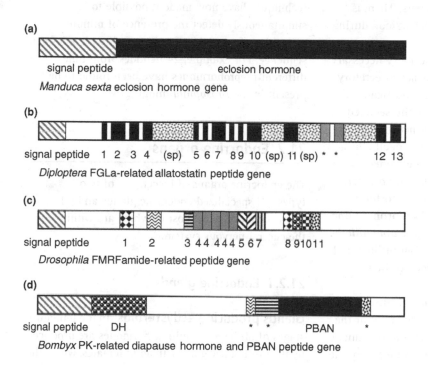

(a)

signal peptide   eclosion hormone

Manduca sexta eclosion hormone gene

(b)

signal peptide   1 2 3 4   (sp)   5 6 7   8 9 10 (sp) 11 (sp) * *   12 13

Diploptera FGLa-related allatostatin peptide gene

(c)

signal peptide   1   2   3 44 444 5 6 7   8 9 10 11

Drosophila FMRFamide-related peptide gene

(d)

signal peptide   DH   *   *   PBAN   *

Bombyx PK-related diapause hormone and PBAN peptide gene

Figure 21.4 Precursor proteins of selected insect peptide hormones. The diagram shows the domain structures of the precursor proteins for several peptides. (a) *Manduca sexta* eclosion hormone gene: the precursor contains only a single bioactive peptide; (b) *Diploptera* FGLa-related allatostatin peptide gene: many different pepides with the same function are produced from the precursor; (sp), acidic spacer region; 1–13, FGLa-related allatostatin family peptides (their sequences are shown in Fig. 21.3); asterisk, predicted peptides of unknown function, not members of the FGLa-related allatostatin family); (c) *Drosophila* FMRFamide-related peptide gene: the precursor contains multiple copies of one peptide; 1–11, FMRFamide-related peptides, there are five copies of peptide 4 (the sequence of this peptide is given in Fig. 21.3); (d) *Bombyx* PK-related DH (diapause hormone) and PBAN (pheromone biosynthesis activating neuropeptide) peptide gene: this precursor contains two different polypeptides with different functions. (Note: the sizes [number of amino acid residues] of the different proteins are not proportional to their lengths in the diagram.)

similar or even identical peptides are required; it may be because large amounts of the peptide are needed. But a single precursor protein may also encode different peptides with different functions. This is the case in the silkworm *Bombyx mori* (Lepidoptera), and also other insects, where a single precursor protein is the source of both diapause hormone and pheromone biosynthesis activating neuropeptide (PBAN). As mentioned above, sometimes more than one gene product may be needed for a single protein hormone, as is the case for bursicon. This gives the opportunity for the hormone to change its function by pairing with different combinations of subunits. There is some evidence that this occurs with bursicon during development.

The expression of the cellular machinery necessary for the specialized phenotype of the neurosecretory cell – post-transcriptional cleavage, modification, packaging and regulated secretion of the secreted peptide(s) – appears in at least some neurosecretory cell types to be under the control of a specific basic helix–loop–helix transcription factor encoded by the gene *dimmed*. In the fruit fly *D. melanogaster*, the cleavage of neuropeptide precursors to form the active peptide is accomplished by a prohormone convertase enzyme encoded by the gene *amontillado*. Flies deficient in either Dimmed or Amontillado fail to process a number of neuropeptides to their final form.

The great diversity of peptides in each insect species, combined with the huge biodiversity of the Insecta, makes it effectively impossible to document all the peptides used by insects. The scope of peptide signaling in a single insect species, however, is well illustrated by genomic analysis of *D. melanogaster*. It has been estimated that the *Drosophila* genome encodes at least 42 neuropeptide precursors, which can be post-translationally processed to form 75 predicted regulatory peptides. Most of these genes are also present in most other fully sequenced insect genomes, indicating that peptide signaling systems have been strongly conserved over probably hundreds of millions of years of insect evolution.

Not all of the predicted *Drosophila* peptides are known to have functional roles in vivo, and it is also possible that different peptides within the same insect species may actually perform redundant functions. Nevertheless, this gives some idea of the diversity of the insect peptidome. That the estimate is probably reasonably complete is indicated by the fact that no fewer than 64 separate putative peptide receptors have been identified in the same insect's genome – in the same order as the predicted number of peptides. Not all of the predicted peptides have been detected biochemically. Advances in analytical techniques have now made it possible to simultaneously detect the presence of minute quantities of numerous peptides each with a defined sequence. For example, 24 peptides from nine different preprohormones have been found to be present in the *Drosophila* midgut.

21.2 Endocrine organs

The endocrine organs of insects are of two types: (1) specialized endocrine glands and (2) neurosecretory cells, most of which are within the central nervous system.

21.2.1 Endocrine glands

Glands producing ecdysteroids In the immature stages of all insects, molting hormones are produced by the prothoracic glands. In adult females, where the same hormones are produced to regulate embryonic development, the follicle cells in the ovary are the principal source. Ecdysteroids are also produced in the testes of some insects, where they regulate development of the male tract. There is evidence that in some insects ecdysteroids are also produced elsewhere in the abdomen, possibly by oenocytes. The prothoracic or thoracic glands are a pair of usually diffuse glands at the back of the head or in the thorax in most insects (Fig. 21.5), but in the base

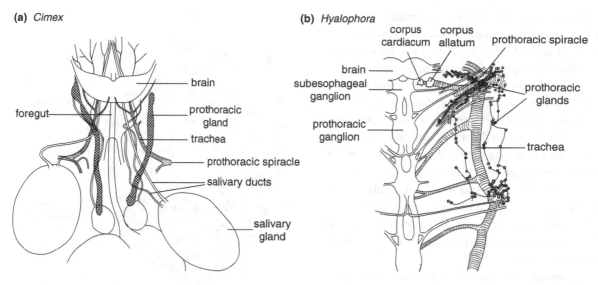

Figure 21.5 Prothoracic glands. General arrangement: (a) in larval bed bug, *Cimex lectularius* (after Wells, 1954); (b) in pupal silk moth, *Hyalophora cecropia* (after Herman and Gilbert, 1966).

of the labium in Thysanura. Each gland has a rich tracheal supply and often a nerve supply. In most insects, the nervous connection is with the subesophageal ganglion, but in some there is also a connection to the prothoracic ganglion, and, in cockroaches, there is a connection to the brain. The prothoracic glands of Heteroptera have no nerve supply.

The glandular cells making up the prothoracic gland in most hemimetabolous insects and in Coleoptera are small, up to 25 μm across; they are present in large numbers and undergo cycles of mitosis. In larval *Tenebrio molitor* (Coleoptera), there are up to 1000 cells. However, in most other holometabolous insects and in Heteroptera (Hemiptera), the cells are large, and are present in relatively small numbers, ranging from about 30 in the larva of *Diatraea grandiosella* (Lepidoptera) to almost 250 in *Hyalophora cecropia*. The cells in these species do not divide, but undergo endomitotic division so that they become polyploid, increasing greatly in size through larval life. The glands show cycles of development associated with secretion. Active prothoracic gland cells have highly developed

rough endoplasmic reticulum, associated with intense protein synthetic activity; the glands probably secrete other products in addition to ecdysteroids, and in vitro experiments show that protein factors are released with autocrine stimulatory effects.

The prothoracic glands degenerate in the adults of nearly all insects, but persist in Apterygota, which continue to molt throughout life, and in the adults of solitarious locusts. As is the case for other tissues that degenerate during adult development, the apoptotic breakdown of the prothoracic glands at this time is triggered by the presence immediately before the final molt of ecdysteroid in the absence of JH.

Corpora allata The corpora allata are glandular bodies, usually one on either side of the esophagus (see Fig. 20.16), although they may be fused to form a single median organ, as in higher Diptera. Each is connected with the corpus cardiacum on the same side by a nerve carrying fibers from neurosecretory cells of the brain. In addition, a fine nerve connects each corpus allatum with the subesophageal ganglion. This is a major nerve in Ephemeroptera, where the nerve from the corpus cardiacum is absent. In Thysanura, the corpora allata are in the bases of the maxillae and, in addition to a fine nerve direct from the subesophageal ganglion, they are innervated by branches from the mandibular and maxillary roots of the subesophageal ganglion.

In Thysanura and Phasmatodea, the corpora allata are hollow balls of cells, with gland cells forming the walls. Elsewhere they are solid organs of glandular secretory cells, often with lacunae between the cells. In hemimetabolous insects the glands contain large numbers of small glandular cells. By contrast, in holometabolous species the gland cells are large and few in number; there are only about 20 in *Drosophila* and some beetles.

The corpora allata produce JH, whose principal functions are in the regulation of metamorphosis (Section 15.4) and, in the adult female of some species, yolk synthesis and deposition in the oocytes (Section 13.2.4).

Corpora cardiaca

The corpora cardiaca are a pair of organs often closely associated with the aorta, and forming part of its wall (see Fig. 20.16). In higher groups such as Lepidoptera, Coleoptera and some Diptera they become separated from the aorta. In adult *Manduca* the corpora cardiaca become fused with the corpora allata. Corpora cardiaca are not known to be present in Collembola. Each organ contains the endings of axons from neurosecretory cells in the brain and other axons passing through to the corpora allata. In addition, they contain intrinsic secretory cells with long cytoplasmic projections extending toward the periphery of the organ. These cells are probably derived from neurosecretory cells, rather than being glandular cells with a separate origin. Their projections probably facilitate the release of secretions into the hemolymph.

The corpora cardiaca store and release hormones from neurosecretory cells in the brain, to which they are connected by one or two pairs of nerves. They serve as neurohemal organs for several different hormones. In addition, the intrinsic secretory

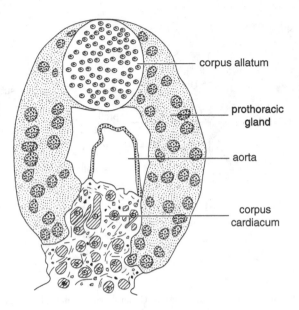

Figure 21.6 Ring gland of cyclorraphous Diptera. From early pupa of the hover fly, *Eristalis* sp. In this species the ring gland is fused with the hypocerebral ganglion, but this is not always the case (after Cazal, 1948).

cells produce adipokinetic hormone and some other peptides whose functions are often unknown. Sometimes storage and secretory cells are intermingled, but in the desert locust *Schistocerca gregaria* (Orthoptera) and some Heteroptera one part of the corpus cardiacum is concerned with secretion and another part with storage.

Ring gland of cyclorrhaphous Diptera

In the larvae of cyclorrhaphous Diptera the ring gland surrounds the aorta just above the brain (Fig. 21.6). It is formed from the corpora allata, corpora cardiaca and prothoracic glands all fused together, although the component elements can still be identified. The ring gland is connected to the brain by a pair of nerves and it also has a connection with the recurrent nerve.

The larvae of Nematocera have completely separate endocrine glands, but larval Orthorrhapha approach the cyclorrhaphan condition, except that the corpus allatum tends to be separated from the rest as a single median lobe.

Endocrine cells of the midgut All insects have cells in the midgut epithelium, which by virtue of their ultrastructure and immunocytological properties are believed to be endocrine cells, although in no case has their endocrine function been rigorously proved. These are isolated cells scattered among the principal midgut cells. They stand on the basal lamina of the midgut and are, therefore, in direct contact with the hemolymph. The basal plasma membrane is flat, not deeply infolded like that of the principal midgut cells, and they release their contents into the hemolymph by exocytosis. The cells are of two morphological types: those that extend to the lumen of the midgut by a slender process, and those that apparently do not. These are called open and closed cells, respectively. Where the open cells reach the lumen, they have microvilli. It is probable that the endocrine cells which extend to the gut lumen respond directly to substances in the lumen. Immunostaining indicates that these cells produce biologically active peptides.

The adult female mosquito *Aedes aegypti* (Diptera) has about 500 endocrine cells scattered through the midgut and a ring of cells in the pyloric region (Fig. 21.7). By contrast, at least 30 000 endocrine cells are present in the midgut epithelium of the large cockroach *Periplaneta americana*. The endocrine cells in the midgut of *A. aegypti* are immunoreactive to four different antisera, each cell containing only one peptide. These are perhaps the only peptides present in the midgut cells of *Aedes*, but other insects are known to have as many as ten different peptides in the midgut endocrine cells.

Although the functions of these peptides are not known, they are presumed to have a hormonal role relating to digestion and absorption, probably regulating the synthesis of digestive enzymes. Some are similar to peptide hormones secreted by endocrine cells in the vertebrate gut. For example, half of the 500 endocrine cells in the midgut of *A. aegypti* are stained by an antiserum to the peptide sequence phenylalanine–methionine–arginine–phenylalanine–amide (FMRFamide). A large family

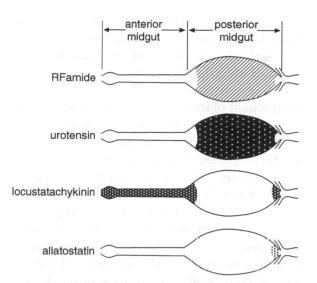

Figure 21.7 Endocrine cells in the midgut of a female mosquito, *Aedes aegypti*. Diagrams show the distribution of endocrine cells whose contents are immunoreactive to antibodies against four different peptides. The functions of these substances in the mosquito are not known (after Veenstra *et al.*, 1995).

of neuropeptides having this N-terminal sequence is thus recognized. All the cells stained in this way are situated in the part of the midgut where the blood meal is digested (equivalent to the RFamide-staining region in Fig. 21.7); the cells appear to release their contents during the six hours after a meal. It is possible that these cells are involved in the regulation of digestive enzyme synthesis and release.

Other midgut endocrine cells contain diuretic peptides, with known experimental stimulatory activity on ion and water transport by Malpighian tubules. It seems doubtful, however, that these cells are the sole regulators of the water balance of the whole insect, and the peptides they secrete may instead have more local effects on gut pH and water content. Still other cells may have different functions, perhaps regulating gut wall motility. For example, a prominent category of midgut endocrine cells in many insects contains peptides of the tachykinin family, which are active in visceral muscle bioassays. Finally, some gut peptides inhibit feeding when injected into lepidopteran larvae, and it is possible that the midgut endocrine system may participate in the regulation of food intake by signaling when the gut is "full."

Epitracheal glands Epitracheal glands are present only in insects from the higher orders (some Coleoptera and Hymenoptera, and all Lepidoptera and Diptera). They consist of 16–18 large neurosecretory ("Inka") cells attached to a trachea near each spiracle. These cells are the source of two hormones, ecdysis-triggering hormone (ETH) and pre-ecdysis-triggering hormone (PETH), which are specifically associated with ecdysis and used at no other time in the insect's life (see Section 15.4.3). But isolated Inka cells can be detected in most insects by immunostaining with an ETH antibody. They are numerous, small secretory cells that are scattered throughout the tracheal system. The size and synthetic activity of these neuroendocrine cells varies with the molting cycle. In *Manduca* one of the epitracheal gland cells increases in size prior to ecdysis, reaching a diameter up to 250 μm, regressing again afterwards.

21.2.2 Neurosecretory cells

Neurosecretory cells normally occur in the ganglia of the central nervous system. They generally resemble monopolar nerve cells, but are characterized by showing cytological evidence of secretion and, usually, by discharging their products into the hemolymph. There is no sharp distinction between neurosecretory cells, as defined here, and neurons with a neuromodulatory function. They may secrete similar compounds (those secreting peptides are termed "peptidergic"), but the effects of neuromodulatory cells are generally restricted to the nervous system (see Section 20.2.3).

Each neurosecretory cell has a dendritic arbor in the neuropile of the ganglion in which the soma occurs, through which the cell receives neuronal input governing its activity. The axon extends through the central nervous system, but at some point, characteristic for a particular neurosecretory cell, it penetrates the blood–brain barrier, so that the cell's secretion is released into the hemocoel and is then transported to the target cells in the hemolymph. At its terminal outside the blood–brain barrier, the axon divides into fine branches which end in swellings. Within these swellings are synapse-like structures (called synaptoids) with closely associated clusters of secretory vesicles. The secretion is presumed to be released into the hemolymph at the synaptoids.

The areas at which secretions of neurosecretory cells are released into the hemolymph are called neurohemal areas, or, if a well-defined structure is formed, neurohemal organs. The corpora cardiaca are the neurohemal organs at which many of the hormones produced in the brain enter the hemolymph (see below).

The somata of neurosecretory cells occur in all the ganglia of the central nervous system. Examples of a few of the cells in the ganglia of the caterpillar of *Manduca* for which the nature of their secretion is known are given in Fig. 21.8.

In the brain there are often two main groups of neurosecretory cells on each side. One group is in the pars intercerebralis, near the midline. The number of apparent neurosecretory cells in this group is very different in different species. *Schistocerca*, for example, has about 500, whereas *Aphis fabae* (Homoptera: Hemiptera) has only four or five. The axons from these cells pass backwards through the brain and some or all of them cross over to the opposite side (decussate), emerging from the brain as a nerve which runs back to the corpus cardiacum. Most of the fibers end here, but a few pass through the corpus cardiacum to the corpus allatum and, in the locust, to the foregut and ingluvial ganglion. In most Apterygota, these median neurosecretory cells are contained in separate capsules of connective tissue, known as the lateral frontal organs, on the dorsal side of the brain, but in some Machilidae (Archaeognatha) the cells are intercerebral as in

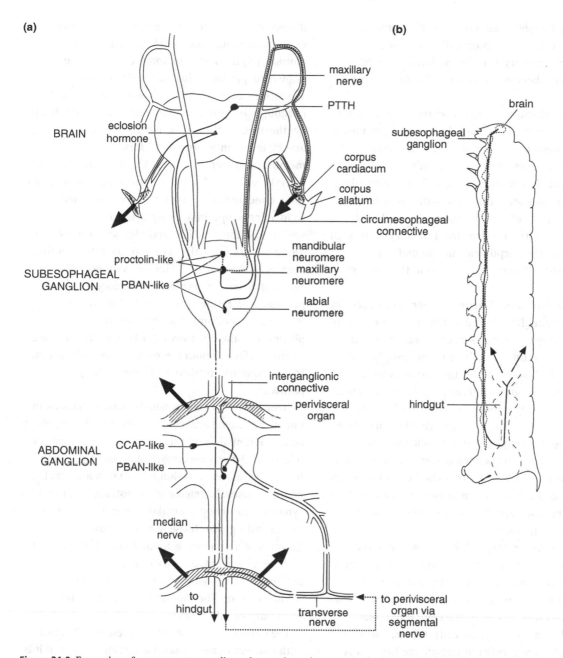

Figure 21.8 Examples of neurosecretory cells and neurohemal organs in the larva of the lepidopteran, *Manduca sexta*.
(a) Connections of some neurosecretory cells with their neurohemal organs. Bold arrows indicate release of peptides into
the hemolymph. Only a few of the cells are shown. Note that in most insects the corpora cardiaca are the neurohemal organs
for the prothoracicotropic hormone (PTTH); the Lepidoptera are unusual in using the corpora allata for this purpose.
It is important to note the suffix "like" on some labels indicates that the peptides in the labeled cell are similar to, but
not necessarily the same as the named peptide. Abbreviations: CCAP, crustacean cardioactive peptide; PBAN, pheromone
biosynthesis activating neuropeptide (based largely on data of Dr. N. T. Davis). (b) Eclosion hormone is released
from neurohemal areas on the hindgut. The position of the soma in the brain is shown in (a) (after Truman and
Copenhaver, 1989).

Pterygota. *Petrobius maritimus* (Archaeognatha), however, occupies an intermediate position with some neurosecretory cells in the lateral frontal organs and others in an adjacent frontal zone of the brain.

The second group of neurosecretory cells in the brain is variable in position. It is sometimes medial to the corpora pedunculata, and sometimes between the latter and the optic lobes. In some Diptera and Hymenoptera, the cells corresponding with this group are associated with the other neurosecretory cells in the pars intercerebralis. A second axon tract passes from these cells through the brain to the corpus cardiacum and, in *S. gregaria*, some fibers also extend to the corpus allatum.

Variable numbers of neurosecretory cells occur in the ventral ganglia. Their products are released into the hemolymph at various neurohemal sites. For example, the axons of cells in the subesophageal ganglion of *M. sexta* run to the corpora cardiaca via the brain or the lateral nerves (Fig. 21.8). The axons of some cells in the abdominal ganglia leave the ganglion in the median interganglionic connective and diverge laterally, forming swollen structures called perivisceral or perisympathetic organs; others may reach the same structures via the lateral nerves. In some cases the neurohemal areas occur along the lateral nerves without the development of discrete neurohemal organs.

Some neurosecretory cell bodies occur outside the central nervous system. In *Carausius morosus* there are 22 neurosecretory cells with somata on the peripheral nerves in each abdominal segment. Unlike the neurosecretory cells in the central nervous system, they are multipolar cells. Others, called cardiac cells, are present alongside the heart. It is uncertain if these cells are connected to the central nervous system.

The secretions of neurosecretory cells can be determined by diverse techniques, including immunostaining and microanalysis of individually dissected neurosecretory cells using bioassays or even mass spectrometry. These cells usually contain peptides, but in some cases they are known to produce biogenic amines, and sometimes both are present in the same cell (although in different granules). The products are synthesized in the soma, where they are associated with a protein to form membrane-bound granules. The granules can be directly visualized using electron microscopy, but their concentrated contents can be revealed in light microscope images by staining with specific antibodies. The granules are transported along the axon to the terminals at rates varying from 5 to 21 mm h^{-1}. Granule contents are released from the terminals by exocytosis, a process that results from the fusion of the membrane enclosing the granule with the plasma membrane, causing a temporary increase in the surface membrane of the axon. Membrane is subsequently retrieved via the endocytic pathway.

Neurosecretory cells resemble normal neurons in having an excitable plasma membrane. They conduct action potentials, and this electrical activity leads to release of the neurosecretory material at the axon terminals, just as an action potential in a normal neuron causes the release of neurotransmitter at a synapse. The action potentials are generally much longer lasting than those in other neurons, however, with durations up to 20 ms. The secretory activity of a neurosecretory cell is regulated by neurons making synaptic connections with its dendritic arbor. The inputs to the cell may be excitatory or inhibitory.

The contents of some neurosecretory cells react with antisera of more than one neuropeptide and it is likely that these cells secrete cocktails of peptides, each producing a different, though probably related, effect. Some neurosecretory cells are known to co-release both peptides and biogenic amines (e.g., proctolin and octopamine).

21.3 Transport of hormones

The steroid hormones of vertebrates are relatively insoluble in water and are transported through the plasma and tissue fluids bound to carrier proteins. Little is known about how they are carried in insect hemolymph. Although ecdysteroids have been shown to bind to hemolymph proteins in *Locusta migratoria* (Orthoptera), the significance of this observation is uncertain. Ecdysteroids are much more highly hydroxylated, and are therefore more hydrophilic than vertebrate steroids; carriage by specific binding proteins may not be necessary.

Although JH is slightly soluble in water, in the hemolymph most is bound to a protein. Calculations suggest that, in the locust, less than 0.1% of the JH is not associated with protein, and this is probably usual for other insects. This is important because free JH tends to bind non-specifically to surfaces. Its association with protein both solubilizes it and prevents non-specific binding.

In many insects the binding protein is a lipophorin; although this may have a relatively low affinity for JH, the large quantities of the protein present in hemolymph may provide an efficient means of circulation. At least in Orthoptera and Lepidoptera, however, specific high-affinity JH-binding proteins are present that are quite different from lipophorins. That in Orthoptera is a high-density lipoprotein, while that of Lepidoptera is a low molecular weight protein which is not associated with significant amounts of lipid or carbohydrate. Orthopteran binding protein has a higher affinity for JHIII, the normal JH of orthopterans, than it has for JHI or II, while the binding protein of Lepidoptera has the opposite properties.

The high-affinity binding protein also gives JH some degree of protection from enzymic degradation. In *M. sexta* it protects the hormone from non-specific esterases in the hemolymph, but not from the specific JH esterase.

Peptides enter the hemolymph in neurohemal areas or organs formed by the terminals of neurosecretory cells (Section 21.2.2). Once in the hemolymph, being water soluble, they require no special transport mechanisms to enable them to reach their targets.

21.4 Regulation of hormone titer

Hormone titers in the hemolymph vary in specific temporal patterns necessary for the proper regulation of the activities they govern (see Figs. 13.9, 13.10, 15.29, 15.30). These variations in titer are produced by changes in their synthesis, release, degradation and excretion.

21.4.1 Molting hormones

Ecdysteroids are made but not stored in the prothoracic gland; their appearance in the hemolymph reflects their synthesis in and immediate release from the gland. Ecdysteroid synthesis is principally triggered by a pre-molt surge of PTTH, but is also subject to multiple hormonal and neural controls. The competence of the prothoracic glands to respond to PTTH and other regulating factors varies through the molting cycle (Fig. 21.9). The complexity of regulation of ecdysteroidogenesis allows the potentially dangerous activity of molting to be accurately timed to coincide with favorable internal and external conditions.

In *Drosophila*, PTTH actions on prothoracic gland cells are mediated (Fig. 21.10) by binding to specific cell surface receptors that are members of the Torso family of receptor tyrosine kinases; binding leads to activation of the intracellular mitogen-activated protein kinase (MAPK) signaling pathway, which includes the proteins Ras and Raf. An important downstream step is the phosphorylation of extracellular-signal-regulated protein kinase (ERK) within prothoracic gland cells. Genetic deletion of

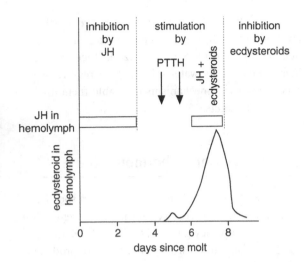

Figure 21.9 Prothoracic glands: regulation of ecdysteroid biosynthesis in the final-stage larva of *M. sexta* (based on Smith, 1995).

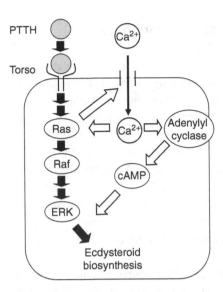

Figure 21.10 Mode of action of prothoracicotrophic hormone (PTTH). The diagram shows the action of PTTH on cells of the *Drosophila* ring gland. PTTH is a polypeptide of the Trunk family of proteins and binds to a membrane-bound receptor protein, Torso. The receptor is a tyrosine kinase that phosphorylates an unknown target. Solid arrows show signaling through the canonical MAPK pathway, including the proteins Ras, Raf and ERK. Multiple steps are required, and lead to the activation of ecdysteroid biosynthesis through increased levels of Halloween family enzymes (see Fig. 21.1b). Open arrows show that PTTH also acts through Torso to cause an influx of Ca^{2+} ions through the plasma membrane, which in turn activates membrane-bound adenylyl cyclase to increase synthesis of intracellular cAMP. Both calcium and cAMP modulate MAPK signaling by mechanisms unknown in this case, but which have been extensively studied in similar mammalian cellular systems (based on Smith, 1995; Rewitz *et al.*, 2009).

either Torso or ERK in *Drosophila* has the same phenotypic effect as deletion of PTTH, confirming the dominating importance of this pathway in PTTH action. The actual increase in ecdysteroid synthesis results from the coordinated up-regulation of several Halloween family P450 enzymes (Spook, Phantom, Disembodied and Shadow) known to be involved in ecdysteroid biosynthesis (Fig. 21.1b). Regulation of both transcription and translation appears to be involved.

PTTH has additional actions on prothoracic gland cells, however. In *M. sexta* PTTH causes an influx of calcium ions into the prothoracic gland cells, which in turn causes an increase in the level of cyclic adenosine monophosphate (cAMP) and therefore cAMP-dependent phosphorylation of (probably several) protein kinases. The importance of this pathway and its relationship to MAPK signaling is uncertain; calcium and cAMP affect Ras-mediated signaling in other systems, and may here affect the dynamics of the steroidogenic response. PTTH also promotes the production of β-tubulin, which may be important in the transport of ecdysteroid precursors within the cells.

Other hormones also influence steroidogenesis in the prothoracic glands. Insulin-like peptides have been shown to stimulate ecdysteroid synthesis in some but not all insects studied. More than 30 genes encoding insulin-like Bombyxin peptides are present in the *Bombyx mori* genome, but their in vivo roles in regulating molting hormone titer are as yet unclear. They probably exert their steroidogenic effects through the phosphatidylinositol 3-kinase (PI3-K) pathway. It is known that nutritional status is important in permitting the synthesis of ecdysteroids by the prothoracic glands; nutrients may interact with the insulin-signaling pathway, probably through the TOR (Target of Rapamycin) pathway. Several brain–gut peptides derived from the Orcokinin precursor sequence are also highly effective in stimulating ecdysteroid synthesis in *Bombyx* prothoracic glands. These peptides are present in nerves innervating the prothoracic gland, providing yet another route of molting control.

Peptides inhibiting ecdysteroid synthesis have also been described. In *Bombyx* there are at least three types: prothoracicostatic peptide (PTSP – a member of the myoinhibitory peptide [MIP] family) and myosuppressin (a member of the FLRFamide family) appear to act hormonally, while another group of peptides (FRMFamides) are present in nerves innervating the prothoracic gland. It has been shown in some insects that prothoracic glands themselves may secrete growth factor-like peptides that influence ecdysteroid secretion. In adult gonadal tissues, ecdysteroid synthesis is regulated by peptides such as testis ecdysiotropin of the gypsy moth *Lymantria dispar* (Lepidoptera) and the ovary ecdysteroidogenic hormone of mosquitoes, which are distinct from PTTH.

As if this were not already complex enough, ecdysteroids also exert both positive and negative feedbacks on their own production by the prothoracic gland. Additionally, JH has both positive and negative effects on ecdysteroid production, depending on the stage of development. These regulatory interactions affect baseline activity and also cause the competence of the prothoracic gland to respond to PTTH to vary through the developmental cycle (Fig. 21.9). During the feeding period of final-instar larvae of the lepidopteran *M. sexta*, juvenile hormone inhibits ecdysone synthesis. Subsequently, the ecdysteroids produced by the glands have a positive feedback effect which is enhanced when JH is also present. Finally, however,

when the glands are highly active, the secreted ecdysteroids have a negative feedback and contribute to the subsequent rapid decline in hemolymph titer. The rate at which the prohormone ecdysone is converted to the active 20-hydroxyecdysone also varies during the molting cycle, and this is influenced by the 20-hydroxyecdysone titer.

Production of 20-hydroxyecdysone is offset by its degradation and excretion, as well as by its conversion to conjugated forms which are inactive. As a result, the period for which the active hormone remains in the hemolymph is limited. For example, in the third-stage larva of *Calliphora vicina* (Diptera), ecdysteroid half-life (the time within which half is converted to other compounds or excreted) is only about three hours, while in the pupa it is more than one day. The midgut is a major site of ecdysteroid metabolism. The Malpighian tubules excrete both ecdysone and 20-hydroxyecdysone, as well as various metabolites.

Ecdysteroid degradation is functionally important, because normal development during molting and metamorphosis is dependent on the removal of this hormone (see Chapter 15); if the molting hormone titer does not fall from its pre-ecdysial peak level, then events late in the molt, including activation of molting fluid enzymes and ecdysis, are prevented. The molt-disrupting effects of non-steroidal ecdysteroid agonists, typified by the insecticide tebufenozide, are due at least in part to their long persistence in the insect because they are not metabolized. The rate of ecdysteroid breakdown is itself regulated by the presence of active molting hormone; paradoxically, tebufenozide actually depresses the titer of endogenous ecdysteroids because it stimulates the synthesis of ecdysteroid-metabolizing enzymes.

Conjugates of ecdysteroids are often phosphates or glucosides. Conjugation serves to inactivate the hormone, and also enables it to be sequestered. The resulting stores of ecdysteroids can be released to boost the titer of active hormone. In eggs, most of the ecdysteroid that is present is in conjugated form, except for brief periods when active hormone is present (see Fig. 14.9).

21.4.2 Juvenile hormone

Juvenile hormone is released from the corpora allata as it is produced; it is not stored. Its hemolymph titer, which varies through the course of development, is consequently a product of the rate of synthesis and the rate at which it is degraded or excreted. Estimates of its half-life in various insects are usually less than two hours, so sustained high titers must reflect high rates of synthesis. Fig. 21.11a illustrates the close correspondence between synthesis and hemolymph titer in an adult cockroach.

The structure of the corpora allata varies in relation to JH synthesis. At times of maximum JH production, the number of cells increases, cell organelles proliferate and the enzymes necessary for the synthesis increase in abundance. All these factors decrease when the rate of synthesis is low, and it is possible that their regulation involves different mechanisms, but this is not known. The overall rate of synthesis is regulated by peptides (Fig. 21.12); allatotropins enhance synthesis, allatostatins decrease it. These peptides are produced by neurosecretory cells in the brain, and allatostatin receptors are known to be present in the corpora allata of *Diploptera* (Blattodea), but the mode of action of these peptides is uncertain. They may act in part as neuromodulators as the axon terminals are present in the corpora allata. They may also act directly on the glandular cells of the corpora allata as neurohormones, however, as they are also present in hemolymph. Inhibition of JH production appears to act at an early stage in JH biosynthesis, probably before the production of mevalonate (see Fig. 21.2b).

These peptides are not the only regulators of corpus allatum activity. The sensitivity of the glands to the peptides varies, probably in relation to the numbers of receptors. It is not known how this is regulated, but independent neural input from the brain may be involved. There is also evidence that

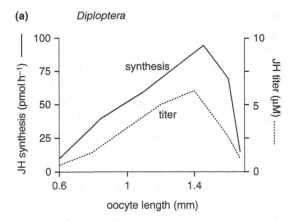

(a) *Diploptera*

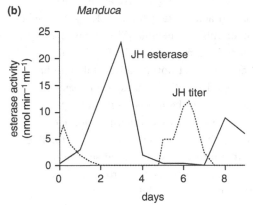

(b) *Manduca*

Figure 21.11 Juvenile hormone: regulation of hemolymph titer. (a) Synthetic activity of the corpora allata and hemolymph titer of juvenile hormone during an ovarian cycle in the cockroach *Diploptera punctata* (after Tobe and Stay, 1985). (b) JH esterase activity and hemolymph titer of juvenile hormone throughout the final larval stage of *M. sexta* (after Jesudason *et al.*, 1990).

octopamine affects JH synthesis, enhancing synthesis in *Locusta* and *Apis*, but inhibiting it in adult *Diploptera*.

JH in the hemolymph has a feedback effect on its own production, at least in adult *Diploptera* (Fig. 21.12); low titers enhance synthesis, and high titers inhibit it. Factors produced by the ovary have similar effects; an ovary with small oocytes produces a stimulating factor while one with large oocytes makes an inhibitory factor. The inhibitory

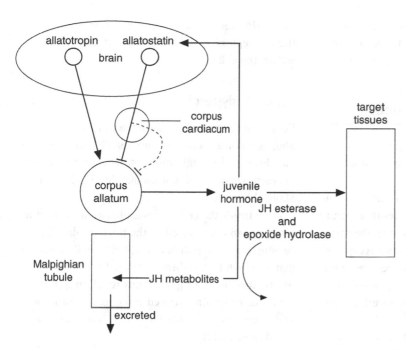

Figure 21.12 Juvenile hormone: regulation of hemolymph titer involves the balance between synthesis in the corpora allata and degradation and excretion by the Malpighian tubules.

effects are mediated by enhanced release of allatostatins; whether allatotropins are also involved is not known.

Once in the hemolymph, the short half-life of JH is due to its enzymic degradation. At least in Lepidoptera, its breakdown is catalyzed by substrate-specific enzymes: an esterase and an epoxide hydrolase. The activity of these enzymes varies through development, and peaks of activity coincide with the disappearance of JH from the hemolymph (Fig. 21.11b). In larval *Manduca* JH esterase is usually dominant, but in the mosquito *Culex* the epoxide hydrolase predominates.

21.4.3 Neuropeptides

The release of peptide neurohormones is determined by electrical activity in the neurosecretory cells that make them. This is presumably controlled synaptically or through neuromodulation by other neurons in the central nervous system. But circulating hormones may also affect neuropeptide release from neurosecretory cells directly.

For example, ecdysis-triggering hormone (ETH) promotes the release of eclosion hormone (EH), while EH promotes the release of ETH; the result is a positive feedback cycle causing massive release of both hormones (see Section 15.4.3).

Insect neuropeptide signals are terminated by enzymatic breakdown of the signaling peptides. This has not been systematically investigated for all insect neurohormones, but it is likely that multiple enzymes are involved. It is possible that different peptides are degraded by different enzymes, but more than one enzyme may be involved in degrading a single peptide. A variety of membrane-bound peptidase enzymes has been described that includes both endopeptidases, which cleave between amino acid residues within the peptide sequence, and exopeptidases (both aminopeptidases and carboxypeptidases), which cleave only residues at the ends of the peptide chain. Many of these enzymes are metalloenzymes that require zinc, but some are members of the serine proteinase family of enzymes. Most of those that have been investigated in detail are membrane-bound enzymes located in neural tissue or in Malpighian tubules, but in a few cases peptidase activity has been shown to be present in hemolymph plasma.

21.5 Mode of action of hormones

When a hormone reaches a target cell that has appropriate receptors, intracellular biochemical changes are initiated that underlie the cell's physiological responses. Not all cells respond in the same way, the specificity of response being a feature of the receptor proteins that are present in the target organs. Only those cells with appropriate receptors will be affected by a hormone and since a receptor may exist in different forms and be expressed at different times, hormones have different effects on different tissues and at different stages of development, even at different times of day.

Depending on the nature of the hormone, these receptors are in the cell membrane or within the cell.

21.5.1 Ecdysteroids

Ecdysteroids are usually assumed to be lipophilic and, as a consequence, to pass readily through cell membranes. Although attempts have been made to characterize their cellular uptake, little is known about this process.

Once inside the cell, ecdysteroids bind to specific receptor proteins located in the nucleus, directly causing the activation of a relatively small number of genes, which then indirectly cause the activation or inactivation of other genes, thus regulating the synthesis of molting-related proteins in a process that is understood at the molecular level in some detail (Fig. 21.13).

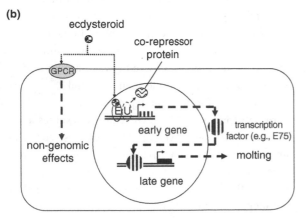

Figure 21.13 Mode of action of ecdysteroids. (a) Domain structures of the three *D. melanogaster* isoforms of Ecdysone Receptor (EcR) and Ultraspiracle (Usp); the mature receptor is a heterodimer of EcR and Usp (after Hu *et al.*, 2003). (b) Ecdysteroids act through Ecr/Usp to promote transcription of "early" and "late" molting-related genes in temporal succession. Initially the EcR–Usp complex is bound together with a co-repressor protein to Ecdysone response elements (ERE) in the genome. Binding of ecdysteroid to the complex causes the co-repressor protein to dissociate and this leads to increased transcription of target genes. Some actions of ecdysteroids appear to be mediated differently, without effects on gene transcription, via a G-protein coupled receptor (GPCR).

The ecdysteroid receptor is a heterodimer of two similar but distinct proteins, each containing a DNA-binding region with two characteristic metalloprotein motifs called zinc fingers, as is characteristic of nuclear receptor DNA-binding transcription factors. The two subunits are Ecdysone Receptor (EcR) and a protein homologous to the vertebrate retinoid X receptor (RXR) called Ultraspiracle (Usp). The presence of ecdysteroid is recognized by the "E" region of EcR (Fig. 21.13a), but EcR is unable to bind to ecdysteroid in the absence of Usp, which allosterically modulates the structure of this region of the EcR protein to allow ecdysteroids to interact with it. Although it has a typical ligand-binding "E" region, Usp itself does not interact with ecdysteroid directly. This has led to suggestions that Usp may bind to another ligand and thus modulate ecdysteroid action (see Section 21.5.2).

The heterodimeric ecdysteroid receptor recognizes and binds to specific regulatory DNA sequences in the genome (ecdysone response elements – ERE) that have an imperfectly palindromic consensus nucleotide sequence (5'-AGGTCANTGACCT-3'), where N is any nucleotide. These ERE sites are located close to molting-related genes. In the absence of molting hormone, EcR associates with other proteins, e.g., the SMRT-related Ecdysone Receptor-interacting Factor (SMRTER), resulting in the repression of target gene expression; but in the presence of ecdysteroid, the receptor undergoes a change in molecular shape that results in dissociation of the co-repressor and strong stimulation of transcription. This is due to two separate transcriptional activation function (AF) regions in the EcR molecule; AF1 is located in the A/B domain, while AF2 is in the E domain. Although AF1 is not essential to EcR function, this region of the receptor protein alone strongly promotes transcription in reporter assays where only a target gene and its associated ERE are present. AF2 also strongly promotes transcription, and its experimental deletion from EcR results in a "dominant negative" receptor.

In this way, ecdysteroids regulate the rate of transcription of a relatively small number of "early response genes" (e.g., E74, E75 and broad complex), which themselves encode transcription factors (Fig. 21.13b). The Early proteins now act to regulate the expression of a larger number of "late" genes that are directly involved in the events of the molt. The expression of many of these molting-related genes is stimulated, but some are repressed, being released from this repression when the ecdysteroid titer falls again. This process of direct and indirect gene regulation leads to the complex time- and tissue-specific pattern of gene regulation that is required to coordinate molting and metamorphosis. For example, in the pharate pupal epidermis of Manduca, the elevated ecdysone levels that initiate molting turn on transcription of broad complex, the protein product of which is required to trigger the de novo synthesis of DOPA decarboxylase and to activate the enzyme which catalyzes the synthesis of a phenol oxidase from its proenzyme, making available the essential enzymes for cuticular tanning.

Early studies of the pattern of response to ecdysteroid hormones in dipteran insects like Chironomus and Drosophila used the "puffing pattern" of polytene chromosomes as an indicator of transcriptional activity, and this was very important in the realization that developmental processes involve multi-level controls of differential gene activation. Modern molecular techniques such as "gene chips" allow transcriptional control to be studied in much greater qualitative and quantitative detail. We now know that large numbers of mRNAs either rise or fall in level in response to the pre-molt ecdysteroid surge.

The tissue specificity and timing of this response is due at least in part to the presence in cells of other transcription factors that permit only a subset of the ecdysteroid-responsive genes in the genome to respond to the hormone. Moreover, the EcR protein is produced in a number of different variant isoforms (there are three in Drosophila: A, B1 and B2) due to differential post-transcriptional processing ("splicing") of the EcR mRNA (Fig. 21.13a). These splice variants are differentially expressed in different tissues and at different stages of development. Since the isoforms differ in the A/B region of the protein, they differ in their ability to promote target gene transcription, and indeed EcR-A has no AF1 function at all. Different EcR variants vary in their ability to promote the expression of different target genes. Although the pattern of expression of the EcR isoforms clearly has the potential to be important in development, no general picture has yet emerged as to their separate roles.

A small number of actions (including the apoptosis of *Bombyx* anterior silk glands) have been ascribed to ecdysteroids acting via a completely different and much faster-acting signaling pathway (Fig. 21.13b) that involves cell surface G-protein-coupled receptors (GPCRs). The mechanism and significance of these actions is not yet certain.

21.5.2 Juvenile hormone

The morphostatic (or "status quo") action of JH in immature insects is to modify the transcriptional responses to ecdysteroid molting hormones (Section 15.4), and it is usually said that JH has no effect by itself. It is evident, though, that like ecdysteroids, JH must enter the cell to exert effects of this kind. JHs are very hydrophobic and are presumed to be able to enter target cells simply by diffusing through the plasma membrane.

The nature of the intracellular receptor to which JH must bind once it is inside the cell was unknown for many years; attempts to identify the receptor by biochemical techniques involving binding of radiolabeled hormone were either unsuccessful or misleading. A more successful approach has been to use genetics. A *Drosophila melanogaster* gene, *methoprene-tolerant* (*met*) was identified, which when mutated confers resistance to the JH-mimic insecticide methoprene, an insecticidal chemical that induces lethal malformations by disrupting metamorphosis. This gene encodes a transcriptional

regulator (Met) belonging to the bHLH-PAS family, which binds radiolabeled JHIII with nanomolar affinity and regulates transcription of known target genes in a JH-dependent way, both of which are attributes required of any functional JH receptor.

Although reducing adult female fertility, *met* loss of function mutations are not lethal in *Drosophila*, strongly implying that in this insect Met mediates only some actions of JH, or that its function is shared by at least one other JH receptor protein. A second *Drosophila* gene encoding a probable JH receptor was later identified as *germ cell expressed* (*gce*), which encodes another bHLH-PAS protein very closely related to Met. It is now known that many insects have only one gene of this type, closer to *gce* than to *met*, suggesting that a Gce-like protein may mediate typically morphostatic JH actions in insects other than *Drosophila*. It was initially suggested that Met and Gce might interact to form a heterodimer that acted as a juvenile hormone receptor, but it now appears that a third bHLH-PAS protein, FISC, is a more likely candidate as the heterodimeric partner for Met (Fig. 21.14).

Once JH has bound to its receptor, it must influence the expression of genes affecting development. For the morphostatic actions of JH, these genes are also responsive to ecdysteroids. It is not known whether these two hormones act together on the same "Early" genes, or whether their effects converge on genes that are downstream in both regulatory pathways. One gene that appears to be required early in JH action is the transcription factor *krüppel*, which is up-regulated by JH in larvae and pupae of both *Drosophila* and *Tribolium* and also during reproductive maturation of adult female *Aedes aegypti* mosquitoes. A JH response element (JHRE) has been identified that is required for JH-dependent gene expression in mosquitoes; tellingly, the Met–FISC complex binds to this DNA sequence in vitro. We may conclude at present that this is good evidence that at least some actions of JH are mediated by a heterodimeric nuclear receptor,

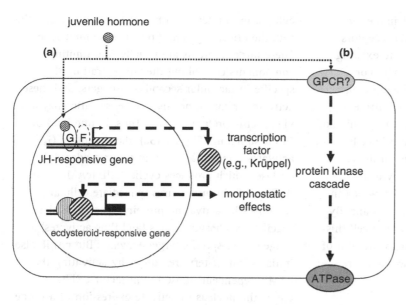

Figure 21.14 Mode of action of juvenile hormones. F, FISC; G, Gce; GPCR, G-protein coupled receptor; JH, juvenile hormone. (a) Morphostatic actions of JH whereby the hormone promotes the expression of specific transcription factors (e.g., *krüppel*), which then modulate the downstream actions of ecdysteroid-responsive transcription factors. Other possibilities for JH morphostatic actions (e.g., co-regulation of Early ecdysteroid response genes through JH binding to Ultraspiracle) are not shown. Note that some actions of JH in female reproductive physiology do not require ecdysteroids, so that in these cases downstream interaction with ecdysteroid-responsive transcription factors is not required (see Riddiford, 2008; Li *et al.*, 2011). (b) Non-nuclear actions of JH as in *Rhodnius* female reproduction, where JH causes shrinkage of ovarian follicle cells through activation of a cell surface ATPase (the involvement of a GPCR is conjectural) (see Davey, 2007).

probably including Gce (or a Gce-like protein) and FISC (Fig. 21.14).

A further molecular-level puzzle, however, concerns the possible involvement in JH action of Usp, the heterodimeric partner of EcR in the ecdysteroid receptor. *Drosophila* Usp binds specifically to labeled methylfarnesoate (a JH-like product of the ring gland in this insect), implying that this protein may also be involved in JH action. An obvious hypothesis is that JH modulates responses to ecdysteroids by acting directly on the ecdysteroid receptor. Consistent with this, the phenotypic effects of *Usp* mutations are similar to those of genetically deleting synthesis of all farnesoid products of the ring gland. But it is uncertain that methylfarnesoate plays a physiologically relevant JH-like role in *Drosophila*, and it is hard to reconcile this work with that on Met

and Gce without supposing that both schemes are too simple. It is possible that the JH receptor and Usp may act together in at least some JH actions.

In adult insects, however, where it affects sexual maturation and behavior, in at least some cases JH acts independently of ecdysteroids. Here, JH may function as a primer or as a regulator of the target tissue. As a primer, it acts directly on the nucleus so that the cell becomes responsive to a subsequent regulatory signal, which may be JH itself or another hormone. For example, fat body cells do not respond to JH (or ecdysteroids in the Diptera) by synthesizing vitellogenin until they have been previously exposed for some time to JH. The process of priming may include the development of specific receptors in the cell membrane and the initiation of transcription.

As a regulator, JH may bind to a nuclear receptor to promote the synthesis of proteins (for example, vitellogenin), essentially as described above for morphostatic actions, or it may act via receptors located on the cell membrane to activate existing enzymes via a second messenger cascade. For example, in adult female *Rhodnius* (Hemiptera), JH not only causes synthesis of the yolk protein precursor vitellogenin, but also causes the opening of intercellular gaps between ovarian follicle cells ("patency"), thus allowing hemolymph proteins including vitellogenin to reach the oocyte. JH stimulates a protein kinase in these cells, which in turn activates an ATPase in the cell membrane. By pumping sodium and potassium out of the cell this produces an osmotic gradient, drawing water out of the cells so that they shrink, and spaces appear between them.

21.5.3 Peptide hormones and biogenic amines

Peptides and amines are mostly lipophobic, and will not pass through cell membranes. Via specific receptor proteins for these hormones, which are present in the cell membranes, they activate cascades of second messenger synthesis and/or protein phosphorylation, which leads to the activation of intracellular enzymes that are already present but not ordinarily active. Many of these membrane receptors have been identified and studied. A large number are members of the G-protein coupled receptor (GPCR) superfamily (transmembrane proteins with seven membrane-spanning domains), but there is a smaller number of single transmembrane domain-containing receptors. As the name implies, GPCRs associate with special heterotrimeric GTP-binding proteins (G-proteins). When a hormone binds to the extracellular part of its own GPCR, a conformational change occurs, resulting in the release from the G-protein of bound GDP, which is exchanged for GTP. The G-protein's α subunit, together with its bound GTP, now dissociates from the β and γ subunits and diffuses within the cytoplasm to interact with target proteins within the cell, the effect depending on the α subunit type. These target proteins are usually rate-limiting components controlling the concentration of specific intracellular second messengers, and these activate, or sometimes inactivate, specific enzymes via specific protein kinases. Thus, in one kind of GPCR signaling (Fig. 21.15a), the G-protein G_s activates a membrane-bound enzyme adenylyl cyclase, which produces cyclic AMP (cAMP) from ATP. cAMP in turn causes immediate cellular responses by activating protein kinase A, which specifically phosphorylates (and thus activates) a range of cell-specific target enzymes. But cAMP also initiates longer-term responses by activating the cAMP-dependent transcription factor CREB, which enters the nucleus to activate expression of a range of target genes. An alternative cell GPCR-linked signaling pathway (Fig. 21.15b) uses the receptor associated G-protein G_q. This activates phospholipase C, which then cleaves membrane-associated phospholipids to produce the intracellular second messengers inositol trisphosphate (IP_3) and diacylglycerol (DAG); IP_3 promotes the release into the cytoplasm of Ca^{2+} ions from intracellular stores. The resulting elevated calcium levels act via the calcium-binding protein calmodulin and protein kinase C (activated by DAG) to activate another set of cellular targets.

Knowledge of the intracellular mechanism(s) of action of insect peptide hormones generally lags behind determination of their structures and the identities of their receptors. Although there is a good deal of experimental information concerning the possible roles of cellular signaling pathways in insect hormone action (e.g., reporting the actions of pharmaceutical agents that interfere with the pathways), it is necessary to be cautious in drawing conclusions, because the same hormone may exert its effects through different pathways in different insects or at different stages of development, and simultaneous multiple cellular actions may in fact be required.

(a)

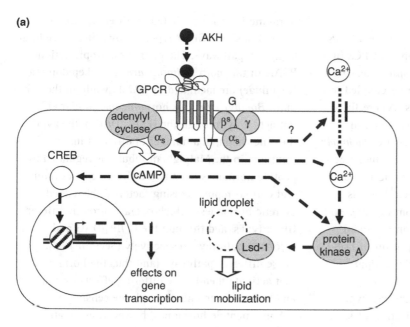

(b)

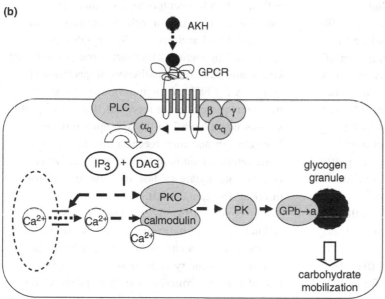

Figure 21.15 Mode of action of adipokinetic hormone. (a) Lipid-mobilizing action of adipokinetic hormone (AKH) in fat body cells of (e.g.) adult *Manduca sexta*. (b) Carbohydrate-mobilizing action of AKH in fat body cells of (e.g.) *Periplaneta americana*. α, β and γ are G-protein subunits; cAMP, cyclic AMP; CREB, cAMP response element-binding protein; DAG, diacylglycerol; GPa and GPb, active and inactive forms of glycogen phosphorylase; GPCR, G-protein coupled receptor protein; G_s, GTP-binding protein of class "s" (stimulatory); IP_3, inositol trisphosphate; Lsd-1, lipid storage droplet 1 protein; PK, phosphorylase kinase; PKC, protein kinase C; PLC, phospholipase C (see Gade and Auerswald, 2003; Arrese and Soulages, 2010).

Adipokinetic hormone is probably the best-studied example. Although the actions of this hormone are mediated by a single type of GPCR (it is related to the mammalian gonadotropin-releasing hormone receptor), the receptor can be coupled to multiple cellular signaling pathways. Where this hormone causes mobilization of lipids (e.g., in adult *M. sexta*) or proline (e.g., the beetle *Pachnoda sinuata* [Coleoptera, Cetoniidae]), the hormone acts on fat body cells via G_s to cause elevation of both cAMP levels, and also intracellular Ca^{2+} levels through enhanced calcium entry from outside. Both actions are necessary for mobilization of stored lipids, which depends on phosphorylation of Lsd-1, a protein that is associated with cellular lipid droplets and regulates lipase activity by an unknown mechanism. But in other insects, where adipokinetic hormone causes mobilization of stored carbohydrates (e.g., the cockroaches *Blaberus discoidalis* and *P. americana*), the hormone's effects are not mediated by cAMP, but by DAG and IP_3, which promote the release into the cytoplasm of Ca^{2+} ions from intracellular stores (this contrasts with the hormone's lipid-mobilizing action, where calcium movements are IP_3-independent). During the hyperglycemic response, the adipokinetic hormone-stimulated increase in intracellular calcium concentration causes activation of protein kinase C, which phosphorylates phosphorylase kinase, enabling it to convert glycogen phosphorylase *b* to *a*. This last change results in the liberation of glucose from stored glycogen. It is uncertain whether the two different types of response to adipokinetic hormone in different insects are due to the presence in their fat body cells of only G_s or G_q, or whether both types of signaling cascade are present, but certain types of downstream targets are absent.

The cellular actions of other insect hormones have mostly not been studied in this level of detail, but in the case of pheromone biosynthesis activating neuropeptide (PBAN), a rather similar picture emerges, in which the actions of a single peptide

hormone type are mediated in different insects through a single GPCR type, but by different cellular signaling pathways. The pheromonotropic actions of PBAN in the moth *Heliothis armigera* (Lepidoptera, Noctuidae) are mediated by cAMP, while in the silk moth *Bombyx mori* this hormone has similar effects but acts only through the $G_q/IP_3/DAG$ pathway. On the other hand, in a number of different insects, fluid transport in the Malpighian tubules is regulated by peptide hormones from two different classes (kinins and corticotropin releasing factor (CRF)-related diuretic hormones), which operate through different GPCR types, and through the $G_q/IP_3/DAG$ and $G_s/cAMP$ pathways, respectively (Section 18.4). These different pathways (and thus the hormones that activate them) have differing effects on ion and water movements through the cells.

Some peptide hormones, however, associate with non-GPCR receptors and act through completely different intracellular pathways. For example, PTTH and insulin-like peptides as mentioned in Section 21.4.1 act through the MAP kinase and PI3 kinase pathways, respectively. The receptors for both of these hormones are single transmembrane domain receptors that act as protein kinases when bound to the appropriate hormone. The eclosion hormone receptor also has a single transmembrane domain which, when activated, acts as a guanylate cyclase to cause an increase in cGMP (Section 15.4.3).

It is important to realize, however, that cellular signaling pathways are rarely completely independent. This can make it very difficult to identify the primary cellular signaling route, and indeed the very concept of such a "primary" route may not be very meaningful. For example, signaling via cAMP or cGMP frequently results in changes in intracellular calcium levels, and this in turn influences the level of IP_3, feeding back onto both cAMP synthesis and calcium release. A good example of this kind of cross talk is the steroidogenic action of PTTH (see Section 21.4.1). Inter-pathway interactions of this kind are often time-dependent, and may result in cyclic, even oscillatory, changes in the state of the cell in response to a single hormone; multiple hormones make response dynamics even more complex. In this way the various different activities of the cell can be regulated by different hormones according to the physiological and developmental requirements of the whole insect.

Summary

- Numerous hormonal substances control the physiological, developmental and behavioral activities of insects. These belong to three major chemical classes: steroids (molting hormones), sequiterpenes (juvenile hormones) and polypeptides.

- Hormones are secreted by specialized glandular tissues in various body regions and by modified neurons (neurosecretory cells).

- Regulation of hormone titers involves changes in synthesis, release, activation, degradation and excretion.

- Hormones act upon target tissues via cell membrane and nuclear receptors, eliciting cellular responses and changes in gene expression patterns.

Recommended reading

GENERAL ASPECTS OF INSECT HORMONES

De Loof, A. (2008). Ecdysteroids, juvenile hormone and insect neuropeptides: recent successes and remaining major challenges. *General and Comparative Endocrinology* 155, 3–13.

Nijhout, H. F. (1994). *Insect Hormones*. Princeton, NJ: Princeton University Press.

CHEMICAL NATURE AND BIOSYNTHESIS OF HORMONES

Bellés, X., Martín, D. and Piulachs, M.-D. (2005). The mevalonate pathway and the synthesis of juvenile hormone in insects. *Annual Review of Entomology* 50, 181–199.

Bendena, W.G. (2010). Neuropeptide physiology in insects. In *Neuropeptide Systems as Targets for Parasite and Pest Control*, ed. T. G. Geary and A. G. Maule, pp. 166–191. Austin, TX: Landes Bioscience.

Brown, M. R., Sieglaff, D. H. and Rees, H. H. (2009). Gonadal ecdysteroidogenesis in Arthropoda: occurrence and regulation. *Annual Review of Entomology* 54 105–125.

Coast, G. M. and Schooley, D. S. (2011). Toward a consensus nomenclature for insect neuropeptides and peptide hormones. *Peptides* 32, 620–631.

Defelipe, L. A., Dolghih, E., Roitberg A. E., *et al.* (2011). Juvenile hormone synthesis: "esterify then epoxidize" or "epoxidize then esterify"? Insights from the structural characterization of juvenile hormone acid methyltransferase. *Insect Biochemistry and Molecular Biology* 41, 228–235.

Jones, G., Jones, D., Li, X., *et al.* (2010). Activities of natural methyl farnesoids on pupariation and metamorphosis of *Drosophila melanogaster. Journal of Insect Physiology* **56**, 1456–1464.

Nässel, D. R. and Winther, A. M. E. (2010). *Drosophila* neuropeptides in regulation of physiology and behavior. *Progress in Neurobiology* **92**, 42–104.

Rewitz, K. F., Rybczynski, R., Warren, J. T. and Gilbert, L. I. (2006). The Halloween genes code for cytochrome P450 enzymes mediating synthesis of the insect moulting hormone. *Biochemical Society Transactions* **34**, 1256–1260.

Roller, L., Yamanaka, N., Watanabe, K., *et al.* (2008). The unique evolution of neuropeptide genes in the silkworm *Bombyx mori. Insect Biochemistry and Molecular Biology* **38**, 1147–1157.

ENDOCRINE ORGANS

Beaulaton, J. (1990). Anatomy, histology, ultrastructure, and functions of the prothoracic (or ecdysial) glands in insects. In *Morphogenetic Hormones of Arthropods, vol. 1, part 2, Embryonic and Postembryonic Sources*, ed. A. P. Gupta, pp. 343–435. New Brunswick, NJ: Rutgers University Press.

Cassier, P. (1990). Morphology, histology and ultrastructure of JH-producing glands in insects. In *Morphogenetic Hormones of Arthropods, vol. 1, part 2, Embryonic and Postembryonic Sources*, ed. A. P. Gupta, pp. 83–194. New Brunswick, NJ: Rutgers University Press.

Meola, S., Sittertz-Bhatkar, H., Langley, P., Kasumba, I. and Aksoy, S. (2003). Abdominal pericardial sinus: a neurohemal site in the tsetse and other cyclorraphan flies. *Journal of Medical Entomology* **40**, 755–765.

Orchard, I. and Loughton, B. G. (1985). Neurosecretion. In *Comprehensive Insect Physiology, Biochemistry and Pharmacology*, vol. 7, ed. G. A. Kerkut and L. I. Gilbert, pp. 61–107. Oxford: Pergamon Press.

Park, D. and Taghert, P. H. (2009). Peptidergic neurosecretory cells in insects: organization and control by the bHLH protein DIMMED. *General and Comparative Endocrinology* **162**, 2–7.

Veenstra, J. A., Agricola, H.-J. and Sellami, A. (2008). Regulatory peptides in fruit fly midgut. *Cell and Tissue Research* **334**, 499–516.

Žitňan, D., Kingan, T. G., Hermesman, J. L. and Adams, M. E. (1996). Identification of ecdysis-triggering hormone from an epitracheal endocrine system. *Science* **271**, 88–91.

TRANSPORT OF HORMONES

Kolodziejczyk, R., Bujacz, G., Jakob, M., Ozyhar, A., Jaskolski, M. and Kochman, M. (2008). Insect juvenile hormone binding protein shows ancestral fold present in human lipid-binding proteins. *Journal of Molecular Biology* **377**, 870–881.

REGULATION OF HORMONE TITER

Audsley, N., Matthews, H. J., Price, N. R. and Weaver, R. J. (2008). Allatoregulatory peptides in Lepidoptera, structures, distribution and functions. *Journal of Insect Physiology* **54**, 969–980.

Gilbert, L. I., Rybczynski, R. and Warren, J. T. (2002). Control and biochemical nature of the ecdysteroidogenic pathway. *Annual Review of Entomology* **47**, 883–916.

Huang, X., Warren, J. T. and Gilbert, L. I. (2008). New players in the regulation of ecdysone biosynthesis. *Journal of Genetics and Genomics* **35**, 1–10.

Kaneko, Y., Kinjoh, T., Kiuchi, M. and Hiruma, K. (2011) Stage-specific regulation of juvenile hormone biosynthesis by ecdysteroid in *Bombyx mori*. *Molecular and Cellular Endocrinology* **335**, 204–210.

Marchal, E., Vandersmissen, H. P., Badisco, L., *et al.* (2010). Control of ecdysteroidogenesis in prothoracic glands of insects: a review. *Peptides* **31**, 506–519.

Nijhout, H. F. and Reed, M. C. (2008). A mathematical model for the regulation of juvenile hormone titers. *Journal of Insect Physiology* **54**, 255–264.

Rewitz, K. F., Yamanaka, N., Gilbert, L. I. and O'Connor, M. B. (2009). The insect neuropeptide PTTH activates receptor tyrosine kinase Torso to initiate metamorphosis. *Science* **323**, 1403–1405.

Ueda, H., Shinoda, T. and Hiruma, K. (2009). Spatial expression of the mevalonate enzymes involved in juvenile hormone biosynthesis in the corpora allata in *Bombyx mori*. *Journal of Insect Physiology* **55**, 798–804.

BREAKDOWN AND METABOLISM OF HORMONES

Isaac, R. E., Bland, N. D. and Shirras, A. D. (2009). Neuropeptidases and the metabolic inactivation of insect neuropeptides. *General and Comparative Endocrinology* **162**, 8–17.

Kamita, S. G. and Hammock, B. D. (2010). Juvenile hormone esterase: biochemistry and structure. *Journal of Pesticide Science* **35**, 265–274.

Sonobe, H. and Ito, Y. (2009). Phosphoconjugation and dephosphorylation reactions of steroid hormone in insects. *Molecular and Cellular Endocrinology* **307**, 25–35.

Williams, D. R., Fisher, M. J., Smagghe, G. and Rees, H. H. (2002). Species specificity of changes in ecdysteroid metabolism in response to ecdysteroid agonists. *Pesticide Biochemistry & Physiology* **72**, 91–99.

RECEPTORS AND MODE OF ACTION OF HORMONES

Fan, Y., Sun, P., Wang, Y., *et al.* (2010). The G protein-coupled receptors in the silkworm, *Bombyx mori*. *Insect Biochemistry and Molecular Biology* **40**, 581–591.

Li, M., Mead, E. A. and Zhu, J. (2011). Heterodimer of two bHLH-PAS proteins mediates juvenile hormone-induced gene expression. *Proceedings of the National Academy of Sciences USA* **108**, 638–643.

Nakagawa, Y. and Henrich, V. C. (2009). Arthropod nuclear receptors and their role in molting. *FEBS Journal* **276**, 6128–6157.

Riddiford, L. M. (2008). Juvenile hormone action: a 2007 perspective. *Journal of Insect Physiology* **54**, 895–901.

Spindler, K.-D., Hönl, C., Tremmel, C., Braun, S., Ruff, H. and Spindler-Barth, M. (2009). Ecdysteroid hormone action. *Cellular & Molecular Life Sciences* **66**, 3837–3850.

Van Hiel, M. B., Van Loy, T., Poels, J., *et al.* (2010). Neuropeptide receptors as possible targets for development of insect pest control agents. In *Neuropeptide Systems as Targets for Parasite and Pest Control*, ed. T. G. Geary and A. G. Maule, pp. 211–226. Austin, TX: Landes Bioscience.

References in figure captions and table

Arrese, E. L. and Soulages, J. L. (2010). Insect fat body: energy, metabolism, and regulation. *Annual Review of Entomology* **55**, 207–225.

Bellés, X., Martín, D. and Piulachs, M.-D. (2005). The mevalonate pathway and the synthesis of juvenile hormone in insects. *Annual Review of Entomology* **50**, 181–199.

Cazal, P. (1948). Les glandes endocrines rétro-cérébral des insectes (étude morphologique). *Bulletin Biologique de la France et de la Belgique*, supplement **32**, 1–227.

Coast, G. M. and Schooley, D. S. (2011). Toward a consensus nomenclature for insect neuropeptides and peptide hormones. *Peptides* **32**, 620–631.

Davey, K. G. (2007). From insect ovaries to sheep red blood cells: a tale of two hormones. *Journal of Insect Physiology* **53**, 1–10.

Gade, G. and Auerswald, L. (2003). Mode of action of neuropeptides from the adipokinetic hormone family. *General and Comparative Endocrinology* **132**, 10–20.

Herman, W. S. and Gilbert, L. I. (1966). The neuroendocrine system of *Hyalophora cecropia*. *General and Comparative Endocrinology* **7**, 275–291.

Hu, X., Cherbas, L. and Cherbas, P. (2003). Transcription activation by the ecdysone receptor (EcR/USP): identification of activation functions. *Molecular Endocrinology* **17**, 716–731.

Jesudason, P., Venkatesh, K. and Roe, R. M. (1990). Haemolymph juvenile hormone esterase during the life cycle of the tobacco hornworm, *Manduca sexta* (L.). *Insect Biochemistry and Molecular Biology* **20**, 593–604.

Li, M., Mead, E. A. and Zhu, J. (2011). Heterodimer of two bHLH-PAS proteins mediates juvenile hormone-induced gene expression. *Proceedings of the National Academy of Sciences USA* **108**, 638–643.

Rewitz, K. F., Rybczynski, R., Warren, J. T. and Gilbert, L. I. (2006). The Halloween genes code for cytochrome P450 enzymes mediating synthesis of the insect moulting hormone. *Biochemical Society Transactions* **34**, 1256–1260.

Rewitz, K. F., Yamanaka, N., Gilbert, L. I. and O'Connor, M. B. (2009). The insect neuropeptide PTTH activates receptor tyrosine kinase Torso to initiate metamorphosis. *Science* **323**, 1403–1405.

Riddiford, L. M. (2008). Juvenile hormone action: a 2007 perspective. *Journal of Insect Physiology* **54**, 895–901.

Smith, W. A. (1995). Regulation and consequences of cellular changes in the prothoracic glands of *Manduca sexta* during the last larval instar: a review. *Archives of Insect Biochemistry and Physiology* **30**, 271–293.

Tobe, S. S. and Stay, B. (1985). Structure and regulation of the corpus allatum. *Advances in Insect Physiology* **18**, 305–432.

Truman, J. W. and Copenhaver, P. F. (1989). The larval eclosion hormone neurons in *Manduca sexta*: identification of the brain-proctodeal neurosecretory system. *Journal of Experimental Biology* **147**, 457–470.

Veenstra, J. A., Lau, G. W., Agricola, H.-J. and Petzel, D. H. (1995). Immunohistological localization of regulatory peptides in the midgut of the female mosquito *Aedes aegypti*. *Histochemistry and Cell Biology* **104**, 337–347.

Wells, M. J. (1954). The thoracic glands of Hemiptera Heteroptera. *Quarterly Journal of Microscopical Science* **95**, 231–244.

22 | Vision

REVISED AND UPDATED BY **MICHAEL F. LAND
AND LARS CHITTKA**

INTRODUCTION

Light is perceived by insects through a number of different receptors. Most adult insects and larval hemimetabolous insects normally have a pair of compound eyes, whose structure (Section 22.1) and function in form and motion vision (Section 22.2) are described below in turn. Section 22.3 covers the molecular and physiological function of photoreceptors and mechanism of regulating light sensitivity before explaining the processes of color vision and polarization vision. Adult insects also typically have three single-lens eyes, called ocelli, whose optics and function are described in Section 22.4. Larval holometabolous insects have one or more single-lens eyes, known as stemmata, on the sides of the head (Section 22.5). Some insects also possess epidermal light receptors, and, in some cases, light is known to have a direct effect on cells in the brain (Section 22.6). Magnetic sensitivity aids orientation in at least some insects, and has known interactions with light sensitivity (Section 22.7).

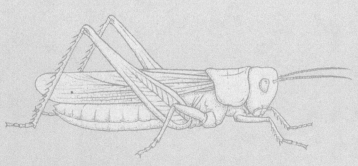

The Insects: Structure and Function (5th edition), ed. S. J. Simpson and A. E. Douglas.
Published by Cambridge University Press. © Cambridge University Press 2013.

22.1 Compound eyes

22.1.1 Occurrence

Compound eyes are so called because they are constructed from many similar units called ommatidia. They are present in most adult pterygote insects and the larvae of hemimetabolous insects, but are strongly reduced or absent in wingless parasitic groups, such as the Phthiraptera and Siphonaptera, and in female coccids (Hemiptera). This is also true of cave-dwelling species. Among termites (Isoptera), compound eyes are greatly reduced or absent from stages that are habitually subterranean, and, although present in winged reproductives, the sensory components of the eyes degenerate during the permanently subterranean reproductive life. Among Apterygota, compound eyes are lacking in some Thysanura, but Lepismatidae have 12 ommatidia on each side. Well-developed compound eyes are present in Archaeognatha. In the non-insect orders of Hexapoda, Collembola have up to eight widely spaced ommatidia, while Protura and Diplura have no compound eyes.

Each compound eye may be composed of several thousand ommatidia. There are up to 30 000 in the eyes of dragonflies, 10 000 in drone honey bees, 5500 in worker honey bees and 800 in *Drosophila*. At the other extreme, workers of the ant *Ponera punctatissima* have only a single ommatidium on each side of the head. Usually the eyes are separate on the two sides of the head, but in some insects, such as Anisoptera (Odonata) and male Tabanidae and Syrphidae (Diptera), the eyes are contiguous along the dorsal midline, this being known as the holoptic condition.

22.1.2 Ommatidial structure

Each ommatidium consists of an optical, light-gathering part and a sensory part, which transforms light into electrical energy. The sensory receptor cells of most diurnal insects end close to the lens, and, because of the method of image formation, these are called apposition eyes (Fig. 22.1). Most night-flying insects, however, have eyes with a clear zone between the lenses and the sensory components; they are called superposition eyes, and produce brighter images than apposition eyes (Fig. 22.2).

The cuticle covering the eye is transparent and colorless and usually forms a biconvex corneal lens. In surface view, the lenses are usually closely packed together, forming an array of hexagonal facets. Each corneal lens is produced by two epidermal cells, the corneagen cells, which later become withdrawn to the sides of the ommatidium and form the primary pigment cells. Beneath the cornea are four cells, the Semper cells, which, in many insects, produce a second lens, the crystalline cone. This is usually a hard, clear, intracellular structure bordered laterally by the primary pigment cells.

The sensory elements are elongate photoreceptor neurons. Generally there are eight receptor cells in each ommatidium, but some species have seven, and others nine. Each receptor cell extends basally as an axon, which passes out through the basal lamina backing the eye and into the lamina of the optic lobe (see Fig. 20.17). The margin of each receptor cell nearest the ommatidial axis is differentiated into close-packed microvilli extending toward the central axis of the ommatidium at right angles to the long axis of the photoreceptor cell. The microvilli of each receptor lie parallel with each other and are often aligned with those of the receptor cell opposite, but are set at an angle to those of adjacent receptor cells (Fig. 22.1d). The microvilli of each receptor cell collectively form a rhabdomere. The visual pigment (rhodopsin) is located within the microvillar membrane. In many insects such as bees and flies (but not butterflies), most receptor cells have a twist along their lengths. Thus, the orientation of the microvilli of each rhabdomere changes regularly through the depth of the eye, eliminating polarization sensitivity where it isn't needed (see below).

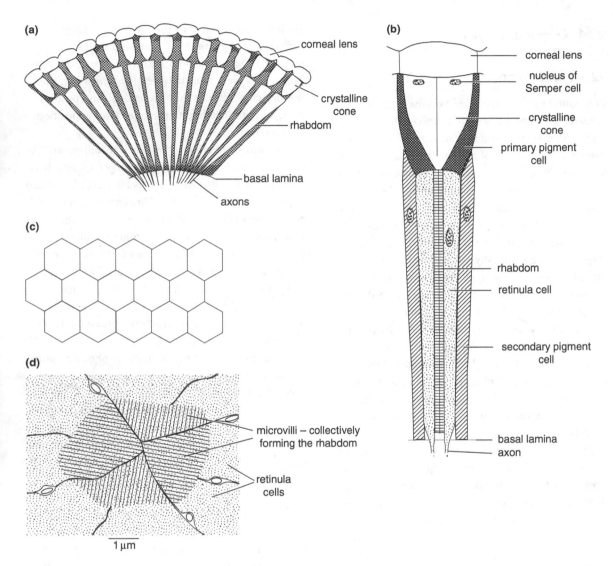

(a)

corneal lens

crystalline cone

rhabdom

basal lamina

axons

(b)

corneal lens

nucleus of Semper cell

crystalline cone

primary pigment cell

rhabdom

retinula cell

secondary pigment cell

basal lamina

axon

(c)

(d)

microvilli – collectively forming the rhabdom

retinula cells

1 μm

Figure 22.1 Apposition eye. (a) Diagrammatic section through an apposition eye showing the rhabdoms extending to the crystalline cones. (b) Ommatidium. (c) Surface view of part of an eye showing the outer surfaces of some corneal lenses (facets). (d) Cross-section through a fused rhabdom (*Apis*) (after Goldsmith, 1962).

In most insects the rhabdomeres abut on each other along the axis of the ommatidium, forming a "fused" rhabdom (although the cells are not actually fused), but Diptera, Dermaptera, some Heteroptera (Hemiptera) and some Coleoptera have widely separated rhabdomeres forming an "open" rhabdom (Fig. 22.3). Because a fused rhabdom acts as a light guide, all the photoreceptor cells within one ommatidium have the same field of view. In species with open rhabdoms, each receptor cell within an ommatidium has a separate visual field, shared by individual cells in each of the adjacent ommatidia (see Fig. 20.21b).

The rhabdom of apposition eyes usually extends the full length of the photoreceptor cells between the crystalline cone and the basal lamina. It is 150 μm

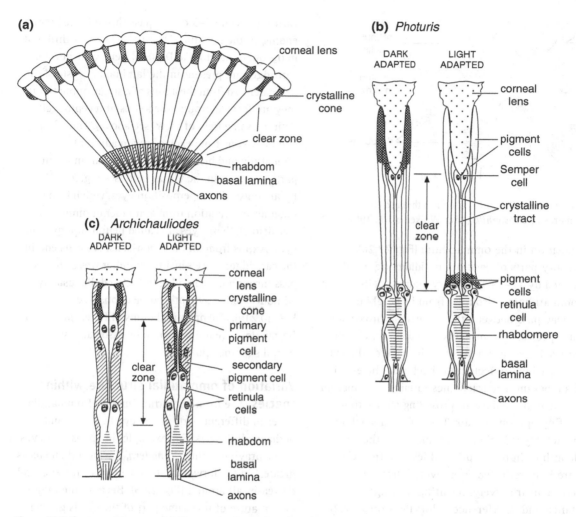

Figure 22.2 Superposition eye. (a) Diagrammatic section through a superposition eye showing the clear zone between the rhabdoms and the lens systems. (b) Exocone eye in which the clear zone is crossed by a tract formed from the Semper cells. Note that there is no crystalline cone in exocone eyes. Left: dark-adapted; right: light-adapted (*Photuris*, Coleoptera). (c) Eucone eye in which the clear zone is bridged by photoreceptor cells. Left: dark-adapted; right: light-adapted (*Archichauliodes*, Megaloptera) (after Walcott, 1975).

long in the ant *Camponotus* and, in *Drosophila*, with an open rhabdom, each rhabdomere is 80 µm long. It is usually shorter in superposition eyes, and even in apposition eyes one of the rhabdomeres may be very short (see Fig. 22.4a, cell 9).

There is much variation in the way that the clear zone in superposition eyes is bridged. In many Lepidoptera and Coleoptera, the receptor cells extend to the crystalline cone as a broad column, but the

rhabdom is restricted to the basal region (Fig. 22.2c), but, in Carabidae and Dytiscidae, one of the receptor cells also has a short distal rhabdomere just below the cone. In other Lepidoptera (the Bombycoidea and Hesperioidea), the receptor cells of each ommatidium form a thin strand, which may be only 5 µm across, to the lens. Beetles with exocone eyes (see below) have a similar structure, but it is formed by the Semper cells and the receptor cells are restricted to a

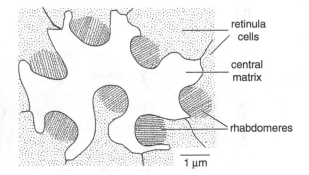

retinula
cells

central
matrix

rhabdomeres

1 µm

Figure 22.3 Open rhabdom. A section through the rhabdomeres of *Drosophila* (after Wolken *et al.*, 1957).

basal position in the ommatidium (Fig. 22.2b). The sensory parts of each ommatidium are usually surrounded by 12–18 secondary pigment cells so that each ommatidium is isolated from its neighbors.

Tracheae pass between the ommatidia proximally in some species, and, in many Lepidoptera, closely packed tracheae form a layer at the back of the eye. This layer, which reflects light back into the eye, is called a tapetum. The tracheoles forming the tapetum are flattened and enlarged, providing alternating layers of cytoplasm and air. These form a stack in which each layer is about one-quarter of the wavelength of light in optical thickness. In such a structure light reflected from every interface interferes constructively, resulting in a high reflectance and interference colors (Section 25.2.2). In butterflies the reflected light from the back of the eye doubles the effective length of each rhabdom as the light passes through it twice. In moths, with superposition eyes, the basal region of each ommatidium is surrounded by tracheae. Here, their function is probably to contain the light within the rhabdom, increasing sensitivity and also resolution via isolating the ommatidia from each other (sheathing).

In insects with fused rhabdoms, the axons passing back from each ommatidium are kept together in the lamina of the optic lobe, each ommatidium being represented by a separate cartridge (Section 20.4). In general, 6–7 retinal axons terminate in the lamina cartridge, while 1–3 others pass through the lamina, ending in the medulla. The arrangement is different in the open rhabdom of Diptera. Here, well-developed cartridges are present in the lamina, but instead of being derived from the axons of a single ommatidium, they are formed around the axons of receptor cells with the same field of view (see above). Thus they contain axons from each of seven adjacent ommatidia and bring together information about a particular area in the visual field (see Fig. 20.21b). By analogy with superposition eyes, which bring together information from a number of ommatidia (see below), these eyes are called neural superposition eyes. Axons from all the green-sensitive cells end in the cartridge; those sensitive to other wavelengths pass through the cartridge to the medulla, usually without making synaptic connections (see Fig. 20.21a). The open rhabdoms of Heteroptera and Dermaptera are not known to be associated with neural superposition.

Variation of ommatidial structure within species

The form and arrangement of ommatidia differ in different parts of the eye in many, and perhaps all, insects. For example, in apposition eyes of the praying mantis, *Tenodera*, the facet diameter is greatest in the forwardly directed part of the eye, and decreases all around (Fig. 22.5). Because the radius of curvature of this same part of the eye is greater (so the surface is flatter) than elsewhere, the angle between the optical axes of adjacent ommatidia (the interommatidial angle) is less than elsewhere and the rhabdoms are longer, but thinner. This area of the eye (the acute zone) is functionally equivalent to the vertebrate fovea (see below) and similar regions are known to be present in the eyes of other insect species with a particular need for good resolution. In dragonflies, for example, there is a wedge of enlarged facets across the fronto-dorsal region of the two eyes, which provides a strip of high resolution for detecting prey insects against the sky. In water surface-living insects, such as the pond skater *Gerris* and empid flies that scavenge over the surface,

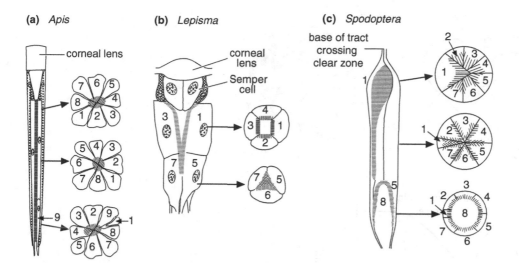

(a) *Apis*   **(b)** *Lepisma*   **(c)** *Spodoptera*

Figure 22.4 Different rhabdoms showing a longitudinal section of an ommatidium with transverse sections at the positions shown. Numbering of photoreceptor cells is for clarity; the numbers have no other significance. (a) Twisted rhabdom. The receptor cells twist through 180° from top to bottom of the rhabdom. Each receptor cell is given the same number in cross-sections at different levels in the eye. Twisting is not shown in the longitudinal section, but note the short ninth cell which only contributes to the rhabdom proximally (*Apis*). (b) Tiered rhabdom. Four photoreceptor cells contribute to the more distal rhabdom, three others to the proximal rhabdom (*Lepisma*, Thysanura) (after Paulus, 1975). (c) Tiered rhabdom in a superposition eye. Only the proximal part of the eye is shown. Cell 8 only contributes to the rhabdom proximally; distally, cell 1 contributes a major proportion (*Spodoptera*, Lepidoptera) (after Langer *et al.*, 1979).

the high-resolution strip is around the equator of the eye, imaging the horizon.

In males of many species, the eye is differentiated into a dorsal region with relatively large facets and small interommatidial angles and a ventral region with much smaller facets and bigger interommatidial angles. In drone honey bees and hover flies the change in facet size is gradual, but in others, such as male bibionids (Diptera), it is abrupt. The division is complete in the male of *Cloeon* (Ephemeroptera), where each eye is in two parts quite separate from each other. Not only are the ommatidia in these two parts different in size, they are also different in structure. Those of the dorsal part are relatively large and of the superposition type, while in the lateral part the ommatidia are smaller and of the apposition type. The eyes are also divided into two in the aquatic beetle *Gyrinus*, where the dorsal eye is above the surface film when the insect is swimming and the ventral eye is below the surface. The significance of these differences is discussed below.

Many species have a band of ommatidia along the eye's dorsal rim which differ from those in the rest of the eye. Here, the photoreceptor cells are not twisted and the microvilli of different receptor cells are at right angles to each other (Section 22.3.4). This area of the eye is polarization-sensitive (Section 22.3.4). Differences may also occur in the nature of the screening pigments in different parts of the eye. These differences relate to the wavelengths absorbed by metarhodopsin and the regeneration of rhodopsin (Section 22.3.1).

Interspecific variation in ommatidial structure The origin and form of the crystalline cone vary in different insects. Most species have eucone eyes in which the structure is intracellular in the Semper cells. It is usually conical, but in some groups, notably in Collembola and Thysanura, it is

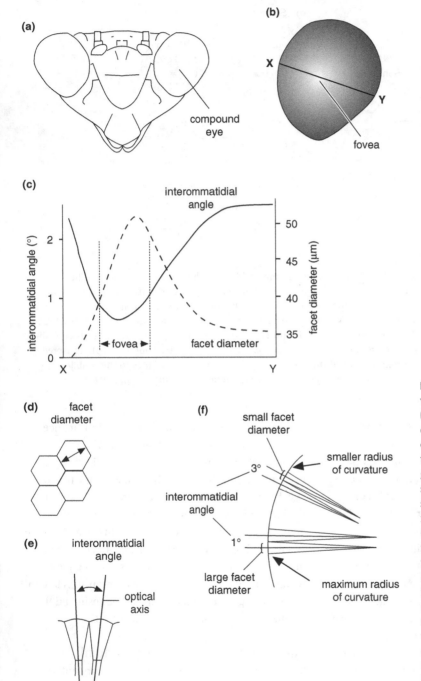

Figure 22.5 Variation in ommatidia within an eye (*Tenodera*, Mantodea) (after Rossel, 1979). (a) Frontal view of the head showing the compound eyes. (b) Compound eye. White area is the fovea in which facet diameter is greatest and interommatidial angle smallest. Increasing density of shading shows decreasing facet size and increasing interommatidial angles. (c) Variation in facet size and interommatidial angle along the transect X–Y shown in (b). (d) Facet diameter. (e) The interommatidial angle is the angle between the optical axes of adjacent ommatidia. (f) Diagrammatic section through the eye showing variation in the curvature of the surface of the eye and the associated differences in facet diameter and interommatidial angle.

more or less spherical. In a few beetles, some Odonata and most Diptera, the Semper cells secrete an extracellular cone which is liquid-filled or gelatinous rather than crystalline. Ommatidia with this type of lens are called pseudocone ommatidia. The Semper cells do not produce a separate lens in some species, but their cytoplasm is clear and they occupy the position of the cone. These acone eyes are present in various families of Coleoptera – for example, Coccinellidae, Staphylinidae and Tenebrionidae – in some Diptera and in Heteroptera (Hemiptera). In Elateridae and Lampyridae (Coleoptera), the Semper cells do not contribute to the lens. Instead, the corneal lens forms a long cone-shaped projection on the inside (Fig. 22.2b). This is known as the exocone condition.

Although it is common for all the receptor cells to be of similar length and to form a rhabdomere all along the inner margin, this is not always the case. The ommatidia of *Apis* have nine receptor cells. Eight of them have more or less similar rhabdomeres distally, but proximally two cells (cells 1 and 5 in Fig. 22.4a) do not contribute to the rhabdom while the rhabdomere of the short, ninth cell is present. Many other insects have photoreceptor cells of different lengths so that they have a tiered arrangement, as in *Lepisma* (Thysanura) (Fig. 22.4b), or as in *Spodoptera* (Lepidoptera) (Fig. 22.4c). Many other arrangements are known.

22.2 Form and motion vision

22.2.1 Image formation

Image formation depends on the optical properties of the corneal lens and the crystalline cone. Refraction of light occurs at any interface with a difference in refractive index on the two sides. In most apposition eyes, the outer surface of the corneal lens is the principal or only refracting surface, although in butterflies further refraction occurs in the crystalline cone.

In an apposition eye, each ommatidium is separated from adjacent ommatidia by screening pigment, so each functions as an independent unit. Each lens produces a small inverted image of the object in its field of view, which is in focus at the tip of the rhabdom (Fig. 22.6a). Because the rhabdoms of apposition eyes are fused, they function as light guides, within which image detail is lost, and all the photoreceptor cells from one ommatidium share the same small field of view. The light reaching the rhabdom in each ommatidium has an overall intensity which varies from one ommatidium to the next, depending on the amount of light reflected by objects in the field of view, and so collectively the rhabdoms transmit an erect mosaic image made up of the adjacent contributions from all the ommatidia.

By bringing the light to a focus at the tip of the rhabdom, the insect maximizes the amount of light entering the rhabdom. In flies, with neural superposition eyes, the amount of light available from each point in space enters through seven ommatidia (see Fig. 20.21b), so the signal contains more photons than light entering through a single lens. Consequently, flies have greater sensitivity at low light intensities than insects with fused rhabdoms, without this compromising the eye's resolution.

Dark-adapted superposition eyes function in a very different way from apposition eyes. The screening pigments are withdrawn so that light leaving one lens system is not confined within a single ommatidium, but can reach the rhabdoms of neighboring units (Fig. 22.6b). This enables the eye to function at low light intensities. Whereas in apposition eyes stray light is absorbed by the screening pigment, in superposition eyes it is optically redirected and utilized for image formation.

The superposition eye forms a single upright image. This requires that light is refracted not just on entering each lens, but also within it, so that it follows a curved pathway. The lenses of superposition eyes each possess a gradient of refractive index from the lens axis to its edge (Fig. 22.7). This gradient causes light rays to undergo greater refraction as they travel toward the lens axis, and less refraction as they travel away

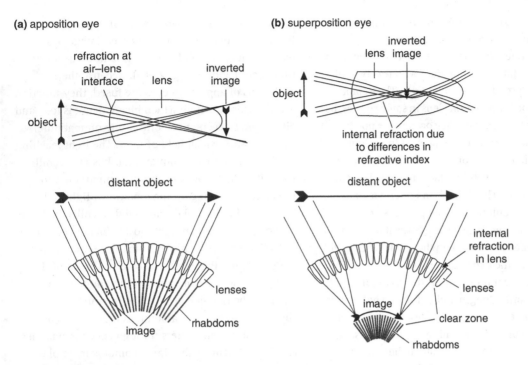

(a) apposition eye

refraction at
air–lens
interface

lens

inverted
image

object

object

distant object

lenses

rhabdoms

image

(b) superposition eye

inverted
lens image

object

internal refraction due
to differences in
refractive index

distant object

internal
refraction
in lens

lenses

image

clear zone

rhabdoms

Figure 22.6 Image formation. (a) In an apposition eye each lens forms an inverted image at the tip of the rhabdom (above), but because the rhabdomeres function as a single unit the impression of this image is not retained. Consequently, the output from each rhabdom is a response to the overall intensity of light that reaches it. Information concerning the object is thus represented as a series of spots differing in intensity (below; suggested by the dotted arrow). (b) In a superposition eye light rays are refracted internally within the lens (above). They are unfocused as they exit the lens, but collectively form a single upright image at the tips of the rhabdoms (below).

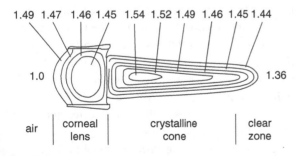

1.49 1.47 1.46 1.45 1.54 1.52 1.49 1.46 1.45 1.44

1.0

1.36

air

corneal
lens

crystalline
cone

clear
zone

Figure 22.7 Lens cylinder from a superposition eye. The refractive index changes within the corneal lens and crystalline cone so that refraction occurs within the lenses as well as at the air–corneal lens and crystalline cone–clear zone interfaces (after Horridge, 1975).

from the axis, thus making the light path curved. The effect of this is to bend light continuously as it passes down the lens, and the

whole structure behaves as though it were a two-lens telescope, accepting parallel light, focusing it internally and then emitting it again as a redirected parallel beam (Fig. 22.6b). Sigmund Exner, who first described superposition optics, referred to such a structure as a lens cylinder. The net result is that parallel light passing through many elements converges in the clear zone onto the same rhabdom.

The eyes of butterflies are interesting in that they operate as apposition eyes with optically isolated ommatidia, but their optical system is essentially of the superposition type, with a lens-cylinder type of lens, and light emerging parallel into the rhabdom, rather than focused as in a typical apposition eye. Such a system is known as afocal apposition.

22.2.2 Resolution

Resolution refers to the degree of fineness with which an eye forms an image of an object. In compound eyes, resolution is determined by the interommatidial angle ($\Delta\phi$), and by the angle over which each ommatidium accepts light ($\Delta\rho$; Fig. 22.8a). For

acceptance angle ($\Delta\rho$)

(a) ray optics   (b) wave optics

d/f λ/D

Airy disc

(c) light guide   (d) wave guide

(first two modes)

total reflection   partial reflection

LP_{01}
LP_{11}

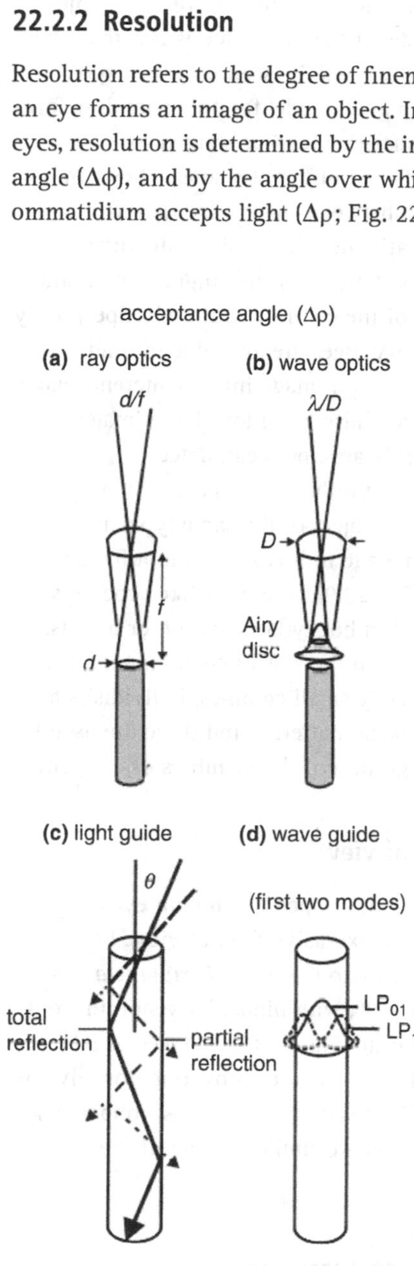

Figure 22.8 Ommatidial optics. (a) Ray and (b) wave optical components of the acceptance angle of an ommatidium ($\Delta\rho$). By ray optics the acceptance angle is the angle subtended by the rhabdom (diameter d) at the nodal point of the lens (focal length f). In wave optics a point source produces a diffraction pattern (known as the Airy disc) whose angular diameter at half-maximum intensity is

diurnal insects these two angles are approximately matched.

The interommatidial angle is the angular separation of the visual axes of adjacent ommatidia. In apposition eyes where each rhabdom functions as a unit, the fineness of the image will be greater the smaller the interommatidial angle. The interommatidial angle is often between 1° and 3°, but is greater than 5° in many beetles, and as little as 0.24° in parts of the eye of the dragonfly *Anax* (Odonata). In many insects it varies in different parts of the eye (see above). Resolution generally decreases when the eye becomes dark-adapted, because the ommatidial acceptance angle expands to admit more light (Section 22.3.2).

There is nothing to be gained, in terms of resolution, by having interommatidial angles smaller than the acceptance angle ($\Delta\rho$) of individual ommatidia. This angle is determined partly by the angular width of the rhabdom (d/f in Fig. 22.8a), and partly by the spread of light in the image (Fig. 22.8b).

given in radians by the wavelength of light (λ) divided by the lens diameter (D). The interaction between the Airy disc and the wave guide modes in the rhabdom (Fig. 22.8d) is complex (see Stavenga, 2006). For most purposes ray optics (a) gives a good approximation to the true acceptance angle, although ultimately it is diffraction (b) that limits the available resolution. (c), (d) Rhabdoms can behave as light guides and wave guides. Light guides (c) will trap all rays up to the critical angle (θ) which for a rhabdom is about 12°. For rays making larger angles with the rhabdom axis reflection is only partial and so their light is progressively lost from the rhabdom. In a wave guide (d) light travels down the structure in the form of transverse interference patterns known as modes. The number of modes present depends on the diameter, wavelength and refractive index difference. Thus for a fly rhabdomere with a diameter of 0.9 µm the first two modes are present (known as LP_{01} and LP_{11} as depicted) at $\lambda = 400$ nm, but at $\lambda = 600$ nm only a single mode (LP_{01}) is present. Wider rhabdoms sustain more complex modes. Note that a proportion of the modal energy lies outside the rhabdom, and this is available for capture by screening pigment (Fig. 22.13). Because this proportion is greater for the higher-order modes they are lost first, which has the effect of narrowing the acceptance angle.

This spread is a consequence of the wave nature of light, and it occurs because different parts of the wave converging on the focus interfere with each other. The result is a diffraction pattern known as the Airy disc, after its discoverer. The angular width of this pattern (in radians; 1 radian is 57.3°) is given by λ/D, where λ is the wavelength of light (about 0.5 μm) and D is the diameter of the lens. In compound eyes D is small, typically around 25 μm, which means that the finest resolvable image point is about 0.5/25 radians, or just over 1°. For comparison, humans, with a 2.5 mm pupil, can resolve better than 1 minute of arc, 100 times finer than a bee. The only way to improve the resolution of a compound eye is to increase the diameters of the facets (D), and this can only be achieved for small regions of the eye if the size of the eye as a whole is not to become impossibly large. High resolution is needed in predatory insects such as dragonflies (Odonata: Anisoptera) and robber flies (Diptera: Asilidae) that hunt insects on the wing, and male insects of various orders that capture their mates on the wing. They have all developed limited "acute zones" of enlarged facets. An overall increase in resolution is not an option for a compound eye because this would require both larger facets and larger numbers of facets; the size of the eye increases as the square of resolution, so that a compound eye with 1 minute resolution would have a diameter of 12 meters!

22.2.3 Form perception

The eye's ability to detect the form of an object depends on its resolving power (Section 22.2.2). As, in diurnal insects, the interommatidial angle, and hence the angular resolution, is often 1° or less in some parts of the eye. This sets the limits of resolution. In predatory insects, such as mantids (Fig. 22.4), the acute zone provides a region of high-quality form perception in the front of the eye where prey is detected before the strike. In other insects, ommatidia in the dorsal part of the eye have small angles of acceptance, giving better form perception than other parts of the eye. If an object subtends an angle at the eye which is less than the ommatidial angle, it will be seen only as a spot.

Bees provide a good model for understanding form perception because they can be trained to discriminate between different shapes. They can not only distinguish between shapes and patterns that differ in fine spatial detail, but also categorize vertical visual patterns by stripe angular orientation (independently of the color or width of stripes) or by plane of symmetry. Bees are also able to bind separate features of an image into a coherent image. In an example reminiscent of top-down image recognition in primates, bees can detect a camouflaged object only after exposure to a non-camouflaged object of the same type, and subsequently manage to break the camouflage of novel objects (Fig. 22.9); attention-like processes have been found in honey bees and other insects, such as fruit flies. In the wasp *Polistes dominulus*, which has relatively small colonies, individuals have highly distinct facial patterns, and these are used for individual recognition of the members of a colony.

22.2.4 Field of view

Insects with well-developed compound eyes generally have an extensive field of view. For example, in the horizontal plane *Periplaneta* has vision through 360°, with binocular vision in front and behind the head (Fig. 22.10). In the vertical plane the visual fields of the two eyes overlap dorsally, but not ventrally. The visual fields of grasshoppers and many other insects are similar to that of the cockroach.

22.2.5 Distance perception

Insects are able to judge distance with considerable accuracy. This is most obvious in insects such as grasshoppers, which jump to a perch, or in visual predators like mantids, but is also true for any insect landing at the end of a flight. Two possible

Figure 22.9 Pattern recognition by honey bees involves cognitive processes. Bees are typically trained to associate a visual pattern with a reward, and in a subsequent unrewarded test, must discriminate the trained from a different pattern. If bees are tested for discrimination of two camouflaged targets (top panels), they fail even after extensive training. However, the task can be introduced using non-camouflaged objects (bottom panels), guiding bees so that they "know what to look for." If bees so entrained are then tested with the camouflaged objects in the top row, they will discriminate them with high accuracy (after Zhang and Srinivasan, 1994).

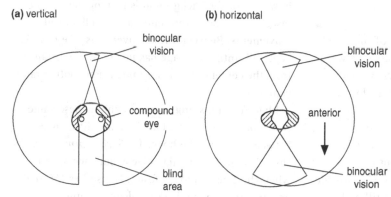

Figure 22.10 Visual field of *Periplaneta* (after Butler, 1973). (a) In the vertical plane, the visual fields of the two eyes overlap above the head, giving binocular vision. (b) In the horizontal plane, the visual fields of the two eyes overlap in front of and behind the head, giving binocular vision in both directions.

mechanisms exist that enable insects to estimate distance: a stereoscopic mechanism and motion parallax.

Some large insects with high spatial resolution have binocular vision in front of the head and so have the potential to assess distance stereoscopically – for example, praying mantids. This depends on the angle which the object subtends at the two eyes (Fig. 22.11). Errors can arise in the estimation of distance due to the size of the interommatidial angle, as this is important in determining resolution. Fig. 22.11 shows the error of estimation which might arise if the interommatidial angle was 2°; larger interommatidial angles will result in greater errors.

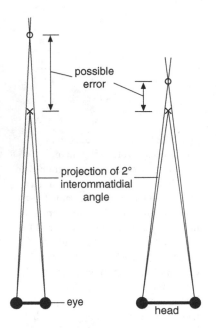

Figure 22.11 Distance perception. Diagrams illustrate how a wider head with greater separation of the eyes improves the estimation of distance. If the interommatidial angle is 2°, an object might lie anywhere between the cross and the open circle. Doubling the width of the head greatly reduces the possible error.

Errors will also be larger if the distance of the insect from its prey is long relative to the distance between the eyes, and in many carnivorous insects which hunt visually, such as mantids and Zygoptera (Odonata), the eyes are wide apart.

Insects, such as grasshoppers, that jump to perches and need to judge distances accurately, make peering movements while looking at their proposed perch. Peering movements are side-to-side swayings of the body, keeping the feet still and with the head vertical, but moving through an arc extending 10° or more on either side of the body axis. In such cases distance is estimated by motion parallax, the extent of movement over the retina as the head is swayed from side to side; big movements indicate that the object is close to the insect, while small movements show that it is at a greater distance. Grasshoppers are able to judge

distances even when blinded in one eye, which is consistent with the idea that they are using motion parallax.

22.2.6 Visual tracking

Visual tracking refers to an animal's ability to keep a moving target within a specific area of the retina, often when the animal itself is also moving. It occurs, for example, when a predator such as a mantis or dragonfly catches its prey, or when a male fly pursues a female. To do this, the insect must move its head or body to minimize the angle subtended by the object relative to some reference point on the retina. This point is usually in the center of an acute zone where the ommatidia have bigger facets and smaller ommatidial angles than elsewhere in the eye, giving better resolution of the object.

When watching slowly moving prey against a homogeneous background, a mantis moves its head smoothly to keep the image in the high-acuity regions (Fig. 22.12). If the prey moves rapidly, however, or the background is heterogeneous, the insect makes rapid intermittent (saccadic) head movements. Between these movements, the head is kept still until the image has moved some distance from the center of the acute zone, when another rapid change occurs.

Male flies pursuing females exhibit these same two types of behavior, but there are differences between species. The hover fly *Syritta pipiens* tracks smoothly at low angular speeds but saccadically when the female moves faster. Houseflies (*Musca*) and the dolichopodid, *Poecilobothrus*, appear to reorient their bodies continuously in flight, but are probably making head saccades at the same time. This saccadic behavior was also demonstrated in blowflies (*Calliphora*). Small-target motion detectors responding selectively to target motion against the motion of the background have been identified in the lobula of *Eristalis tenax* hover flies

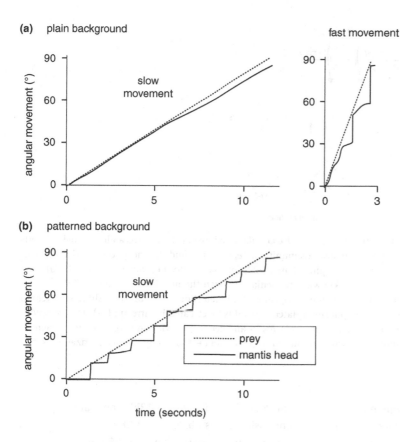

(a) plain background

fast movement

(b) patterned background

time (seconds)

Figure 22.12 Visual tracking of prey by a mantis. The dotted line shows the steady movement of a potential prey item; the solid line shows the orientation of the mantid's head (after Rossel, 1980). (a) When the prey moves slowly against a plain background, the mantis moves its head steadily, keeping the prey within the foveal areas of the eyes (left). When the prey moves quickly (right), the mantis makes saccadic movements of the head. (b) With the prey against a patterned background, the mantis makes saccadic movements whatever the speed of the prey. In a saccade, the head remains still (horizontal lines) until the image of the prey has passed outside the fovea; it is then moved rapidly (vertical lines) so that the image of the prey is now in its original position. In this example, each movement of the head is through approximately 10°.

22.2.7 Visual flow fields

During forward locomotion, objects in the environment appear to move backwards with respect to the organism. This is especially important in flight, and insects are unable to make controlled flights without visual input, requiring image movement across the eye from front to back. Like all other behaviors associated with vision, response to the visual flow field is possible because the neural circuitry within the optic lobe permits the extraction of the appropriate information from image movement over the retina. In flies this is mediated by the lobula plate tangential cells, which have large receptive fields and are specialized for detecting particular directions of motion. Specific interneurons within the central nervous system relay appropriate information to the circuits controlling locomotion.

22.3 Receptor physiology, color and polarization vision

22.3.1 Transduction

The conversion of light to electrical energy involves the visual pigment. Known as rhodopsin, this is a G-protein coupled receptor molecule consisting of seven helices spanning the membrane of the microvilli that make up each rhabdomere. There are between 500 and 2000 rhodopsin molecules on each of the 40 000 microvilli in a *Drosophila* rhabdomere. Bound inside the rhodopsin molecule is a small chromophore; in vertebrates this is 11-*cis* retinal (the aldehyde of vitamin A), but in Diptera and some other insects it is 11-*cis*-3-hydroxy retinal (Fig. 22.13). It is the absorption of a photon by this chromophore that sets in train a cascade of events

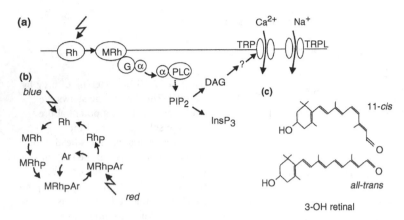

Figure 22.13 Signal transduction. (a) The transduction cascade in *Drosophila*. Rhodopsin (Rh) absorbs a photon and converts to metarhodopsin (MRh). This activates the G-protein whose α subunit is released and bonds to phospholipase C (PLC). This complex enzymatically splits phosphatidyl inositol 4,5-bisphosphate (PIP2) into two molecules, diacylglycerol (DAG) and inositol 1,4,5-trisphosphate (InsP₃). DAG acts via an unknown intermediary to open the membrane channels TRP to Ca²⁺ ions and TRPL to Na⁺ and Ca²⁺ ions. (b) The rhodopsin/metarhodopsin cycle. When rhodopsin absorbs a short-wavelength photon it converts to metarhodopsin, which becomes phosphorylated (P) and is inactivated by arrestin (Ar). This complex can be reactivated by a long-wavelength photon, and after losing arrestin and dephosphorylation is reconverted to active rhodopsin. (c) The structure of the chromophore 3-hydroxy retinal in its 11-*cis* and – following photoisomerization – its *all-trans* form. Simplified from Hardie (2006).

leading first to the conversion of rhodopsin to metarhodopsin, and ultimately to the depolarization of the microvillar membrane (Fig. 22.13a,b). In dipteran photoreceptors the situation is complicated by the presence of a second ultraviolet-absorbing chromophore (3-hydroxy retinol). This is a sensitizing pigment that is capable of transferring the energy of the absorbed ultraviolet photons to the primary photopigment, resulting in enhanced overall sensitivity.

The first step in the phototransduction cascade is a change of the 11-*cis* form of the chromophore to the *all-trans* form (Fig. 22.13c), resulting in the conversion of rhodopsin to active metarhodopsin, which can then activate associated G-proteins. Potentially, metarhodopsin can continue to activate G-proteins, but it is inactivated by binding with the small protein arrestin, after activating 5–10 G-protein molecules.

Metarhodopsin is, itself, photosensitive, absorbing light at a different wavelength from the rhodopsin

that gave rise to it (Fig. 22.13b). The main rhodopsin in flies absorbs at 480 nm and the corresponding metarhodopsin at 560 nm. Photoactivation reconverts the metarhodopsin to inactive rhodopsin which, after release from arrestin, is available to transduce light again. Thus light of a different wavelength from that to which the photopigment is sensitive contributes to the efficiency of the response by regenerating rhodopsin. Under constant light conditions the amounts of rhodopsin and metarhodopsin are in equilibrium. This is very different from vertebrates, where rhodopsin has to be regenerated via a complex cycle involving the pigment epithelium.

Following the activation of the G-protein the transduction cascade becomes complex, but in outline it proceeds as in Fig. 22.13a. Activated metarhodopsin causes the α subunit of the G-protein to detach and bind to phospholipase C (PLC). This combined molecule then acts as an enzyme that converts phosphatidyl inositol 4,5-bisphosphate

(PIP$_2$) into two second messenger molecules, diacylglycerol (DAG) and inositol 1,4,5-trisphosphate (InsP$_3$). Current evidence in *Drosophila* favors DAG as the main agent responsible, indirectly, for causing the cation channels in the membrane (the so-called TRP and TRPL channels) to open, resulting in a membrane depolarization. It seems that the components of this cascade do not simply diffuse freely in the microvillar cytoplasm, but many, including PLC and the TRP channels, are anchored onto a protein scaffold known as the INAD complex, which is in turn attached to the actin cytoskeleton of the microvillus.

The Ca^{2+} that enters the microvillus through the TRP and TRPL channels has two opposite effects on the electrical response. Initially it rapidly sensitizes the remaining channels in the microvillus (about 25 per microvillus in flies), which open and thus amplify the response. This further Ca^{2+} influx then causes a complete inactivation of the channels of the microvillus, and a refractory period lasting about 100 ms. For a single photon capture, the net effect of these two process is to produce a single "quantum bump," a rapid electrical depolarization of 1–2 mV, with a current of about 10 pA and lasting about 30 ms. The responses to larger photon numbers sum across the receptor as a whole to give peak responses as high as 70 mV. The range over which a single photoreceptor can operate before reaching saturation is 3–4 decades of light intensity.

The probability that a photon will encounter a visual pigment molecule is increased by the length of the rhabdomere or rhabdom. The proportion of white light absorbed is given approximately by $kL/(2.3 + kL)$ where L is the length and k the absorption coefficient which, for rhabdomeric receptors, is about 0.01 μm^{-1}. Thus a *Drosophila* rhabdomere 80 μm long absorbs about 26% of the light reaching it. In the much longer rhabdoms of insects such as dragonflies nearly all the incident light is absorbed. A tapetum, when present, reflects light back through the rhabdom and so further increases the chances that the photons and receptor molecules will interact.

Rhabdoms have a slightly higher refractive index than the surrounding fluid, and so act as light guides, retaining by total internal reflection most of the light that enters them within about 12° of their long axis (Fig. 22.8c). However, because of their narrow diameter – only a few times greater than the wavelength of light – rhabdoms also behave as wave guides. Within wave guides the light waves interfere, producing a series of stable interference patterns, known as modes, in which the light is not uniformly distributed (Fig. 22.8d). Importantly, a proportion of the modal light travels outside the rhabdom (the boundary wave), and this light is available for capture either by adjacent photoreceptors, or by screening pigment (see Section 22.3.2). This proportion increases as the rhabdoms get narrower, and this limits the useful width of a rhabdom to about 1 μm.

Action potentials are not typically produced by photoreceptor cells and depolarization is transmitted along their axons by passive conduction. In the lamina, most of the receptor cells synapse with large monopolar cells, and with other interneurons (see Fig. 20.21c). The neurotransmitter at these synapses is histamine and the signal is inverted. That is, a depolarization of the receptor cell produces a hyperpolarization of the interneuron. Any new signal is amplified at these synapses while the effects of the general level of illumination are reduced. The mechanisms producing these effects are not fully understood, but involve presynaptic interactions.

22.3.2 Adaptation

The natural change from darkness to full sunlight involves a change in light intensity of some 10 log units: a white surface in bright sunlight reflects about 4×10^{20} photons m^{-2} sr^{-1} s^{-1} compared with about 10^{10} photons m^{-2} sr^{-1} s^{-1} on an overcast night with no moon, and 10^{14} photons m^{-2} sr^{-1} s^{-1} on a moonlit night. However, the response of the photoreceptor cells is fully saturated by an increase

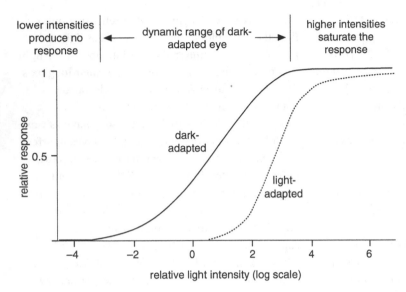

lower intensities produce no response | dynamic range of dark-adapted eye | higher intensities saturate the response

relative response

relative light intensity (log scale)

dark-adapted

light-adapted

Figure 22.14 Response of visual cells. Different light intensities can only be distinguished within the dynamic range. In the light-adapted eye, the response is shifted to the right (higher intensities). Notice that within the dynamic range of the light-adapted eye a small change in intensity results in a much bigger change in the response level than in the dark-adapted eye. In other words, contrasts are more readily differentiated by the light-adapted eye. Relative response is the receptor potential expressed as a proportion of the maximum response in dark- or light-adapted eyes (based on Matic and Laughlin, 1981).

in intensity of about four log units (Fig. 22.14). Various devices are used to regulate the amount of light reaching the receptors so that the insect can operate over a wider range of intensities. The eye is said to adapt to the light conditions.

This type of adaptation occurs at two different levels in the eye: (1) the amount of light reaching the photoreceptors is regulated; and (2) the receptor sensitivity can be changed. In addition, changes occur at the synapses between receptor cells and interneurons in the lamina. Although the latter does not influence the immediate response to the amount of light, it does influence the information reaching the central nervous system.

Regulation of light reaching the receptors The amount of light reaching the rhabdom is regulated by movement of pigment in the screening cells, sometimes associated with anatomical changes in the ommatidium. Pigment movements are most clearly seen in superposition eyes. In the dark-adapted eye, pigment is withdrawn to the distal ends of the pigment cells and light leaving the lenses can move between ommatidia in the clear zone (Figs. 22.2; 22.6b). At high light intensities, however, the pigment moves so that light can now reach the

rhabdom only via a narrow tract across the clear zone and the effective aperture of the ommatidium through which light reaches the rhabdom is very small. The eye is now functioning as an apposition eye with the ommatidia now acting as separate units. In superposition eyes where the photoreceptor cells form a broad column to the lens system, these cells undergo extensive changes in shape (Fig. 22.2c). Dark adaptation results in their extension and in compression of the crystalline cone so that the screening pigment is restricted to the most peripheral parts of the eye. In the light, the receptor cells are short and the primary pigment cells extend below the lens.

Comparable pigment movements occur in apposition eyes of some insects. For example, when the ant *Camponotus* is in the light, the primary pigment cells extend proximally (Fig. 22.15). At the same time, the proximal end of the crystalline cone is compressed to form a narrow crystalline tract surrounded by the pigment cells. Light must pass through the narrow opening to reach the rhabdom. At the same time, the rhabdom shortens, further increasing the likelihood that only light entering the ommatidium directly along its axis will reach the rhabdom.

(a) dark-adapted   **(b)** light-adapted

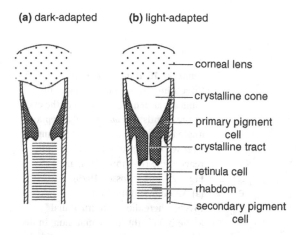

corneal lens

crystalline cone

primary pigment
cell

crystalline tract

retinula cell

rhabdom

secondary pigment
cell

Figure 22.15 Adaptation. Changes in the apposition eye of an ant, *Camponotus*. Only the most distal part of an ommatidium is shown (after Menzi, 1987): (a) dark-adapted; (b) light-adapted.

In the acone eyes of many Diptera and Heteroptera the photoreceptor cells extend in the dark, carrying the distal end of the rhabdom to a more peripheral position and at the same time causing the Semper cells to become shorter and broader. This may involve a movement of 15 μm or even more by the rhabdom. At the same time, the primary pigment cells are displaced laterally so that the aperture of the optic pathway is increased.

The changes associated with dark adaptation begin within minutes of an insect's entry into darkness, but take longer to complete. In the ant *Camponotus*, the first change is already apparent within 15 minutes of an individual moving from light to dark, but completion takes about two hours. The changes are more rapid in some other insects. In addition to a direct response to environmental conditions, pigment movements are commonly entrained to the light cycle, occurring even in the absence of any environmental change. Thus the eye of the codling moth normally starts to become light-adapted about 30 minutes before sunrise and dark-adapted just before sunset, the process taking about an hour to complete. A consequence of these changes is that the rhabdom receives light from a wide acceptance angle

in dark-adapted eyes (Fig. 22.16), resulting in a reduction in image resolution. The range of light intensities over which these pigment movements occur varies according to the insect's normal habits. For example, pigment movements in the eye of the cave-dwelling tenebrionid beetle *Zophobas*, occur at light intensities about five orders of magnitude lower than they do in the diurnal *Tenebrio*. Pigment movements are regulated individually within each ommatidium. In the moth *Deilephila* a structure near the tip of the crystalline cone absorbs ultraviolet light, the principal wavelength producing pigment movements, but how it effects movement is not known.

Regulation of receptor sensitivity Regulation of sensitivity in many apposition eyes is achieved by structural changes in the photoreceptor cells that vary the amount of light available for capture by rhodopsin, and thus act as a pupil. During light adaptation, granules of an absorbing pigment are present within the receptor cells (in addition to that in the screening pigment cells) close to the inner ends of the microvilli of the rhabdoms or rhabdomeres (Fig. 22.17). In the light these pigment granules migrate toward the surface of the rhabdom, and absorb the modal light traveling along the outside (see Fig. 22.8d), thereby bleeding light progressively out of the interior of the rhabdom and so reducing the amount available for photoreception. These movements are rapid, taking only seconds to complete in dipterans and butterflies. In the dark, large vesicles develop in the endoplasmic reticulum so that a clear space, known as the palisade, is formed around the rhabdom, effectively isolating it from the absorbing pigment.

In addition to this pupil mechanism, there are numerous points in the transduction cascade at which sensitivity reduction can occur. One is the rate at which arrestin inactivates metarhodopsin, thereby varying the number of G-protein activations following photon capture. Another is the blocking effect of physiological concentrations of Mg^{2+} on the

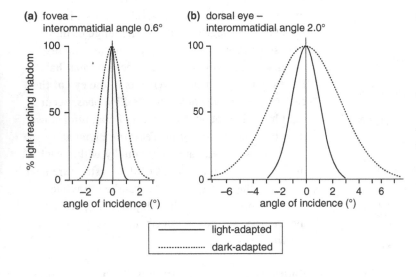

(a) fovea –
interommatidial angle 0.6°

(b) dorsal eye –
interommatidial angle 2.0°

% light reaching rhabdom

angle of incidence (°)

———— light-adapted

·············· dark-adapted

Figure 22.16 Dark adaptation.
Effects on the acceptance angles of
ommatidia in different parts of the eye
of the mantis, *Tenodera*. The acceptance
angle is usually measured as the angle
over which 50% of the incident light
reaches the rhabdom. Compare
Fig. 22.4 (after Rossel, 1979).
(a) An ommatidium in the
fovea, where the interommatidial
angle is 0.6°. (b) An ommatidium in the
dorsal eye, where the interommatidial
angle is 2.0°.

(a) light-adapted

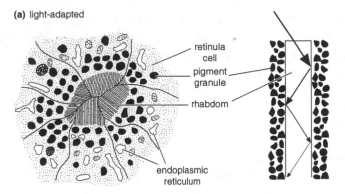

retinula
cell

pigment
granule

rhabdom

endoplasmic
reticulum

(b) dark-adapted

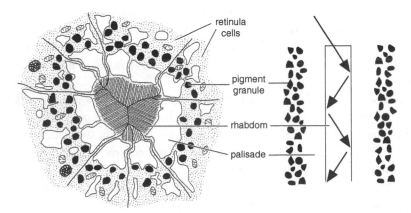

retinula
cells

pigment
granule

rhabdom

palisade

Figure 22.17 Adaptation. Changes in the
photoreceptor cells in an apposition
eye. Each diagram represents (left) a
cross-section of the rhabdom and
associated receptor cells, and (right) a
longitudinal section showing the extent
to which light is internally reflected
within the rhabdom (partly after Snyder
and Horridge, 1972): (a) light-adapted;
(b) dark-adapted.

TRP channels; this intensifies as the cell depolarizes, thus decreasing the channel conductance as a function of light intensity.

Adaptation also occurs at the synapses between the photoreceptor cells and the interneurons in the lamina. The mechanisms are not understood, but probably include modulation of the quantity of transmitter released by the receptor cells. As a result of this adaptation, the input to the lamina is believed to be more or less constant despite differences in overall light intensity. Consequently, changes in stimulation of comparable magnitude are registered by similar activity in the medulla, irrespective of the background level of stimulation.

22.3.3 Spectral sensitivity and color vision

Photoreceptor cells can have wavelengths of maximum sensitivity (λ_{max}) from ~330 nm to 640 nm. When plotted over a linear wavelength scale, photoreceptors typically have roughly Gaussian sensitivity profiles around the peak, with sensitivity falling off smoothly over several dozen nanometers to both sides from the value of maximum sensitivity. In longer wavelength receptors there is a smaller β sensitivity peak at short wavelengths (Fig. 22.18). λ_{max} depends fundamentally on the amino acid sequence of the opsin protein, where specific substitutions at sites interacting with the

chromophore will induce spectral tuning. Other important factors in spectral tuning and the shape of the sensitivity function are the choice of the chromophore (retinal or 3-hydroxy retinal), accessory pigments, tapeta, as well as various filtering mechanisms that limit the spectrum of light available to any one receptor type.

The ancestral set of photoreceptors in the pterygote insects appears to have been an ultraviolet, a blue and a green receptor. This set is retained by many extant insects. These receptor types are distributed over various types of ommatidia, all of which have six green receptors in many species. In bumble bees, honey bees and trichromatic Lepidoptera (*Vanessa, Manduca*), type I ommatidia contain a blue and an ultraviolet receptor, type II ommatidia have two ultraviolet receptors and type III ommatidia possess two blue receptors. A modification occurs in flies, which have two types of ommatidia (named "pale" and "yellow"). Both of these contain six receptor cells (R1–6) with unusually broad spectral sensitivities covering the range from 300 nm to 600 nm. This broad sensitivity is mediated by a combination of a blue-sensitive opsin with an ultraviolet-absorbing sensitizing pigment. In addition, "pale" ommatidia contain an ultraviolet (R7p) and a blue receptor cell (R8p), while "yellow" ommatidia contain an ultraviolet (R7y) and a green receptor (R8y). R7 is located directly on top of R8, and where the

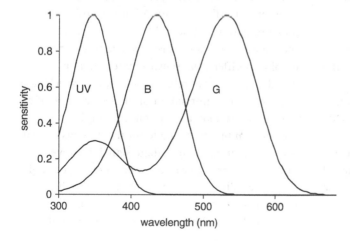

Figure 22.18 Photoreceptor spectral sensitivity. Absorption of light of different wavelengths by three commonly occurring photoreceptor types in insect eyes with peak absorption in the ultraviolet (UV), blue (B) and green (G) ranges of the spectrum, shown here for the bumble bee *Bombus terrestris* (after Skorupski *et al.*, 2007). The sensitivity function of each receptor is normalized to a maximum of unity.

rhabdomeres are tiered in such a way (Fig. 22.4b,c) that the distal rhabdomeres function as filters for those situated more proximally. Specialized eye regions aside, the various types of ommatidia are often randomly distributed across the eye, although not necessarily in equal proportions.

In many insects, however, the occurrence of different visual pigments varies in different parts of the eye apparently in relation to the eye's different functions. For example, in the drone honey bee, ommatidia in the dorsal part of the eye have only ultraviolet- and blue-sensitive pigments, whereas in the ventral part of the eye, green-sensitive pigment is also present. This distinction appears to occur in many flying insects, perhaps because dorsal ommatidia are skyward-pointing and the ventral ones face the vegetation. Sexual differences may also occur, as in the butterfly *Lycaena rubidus*, where the dorsal ommatidia in females have a visual pigment absorbing maximally at 568 nm; this pigment is absent from the dorsal ommatidia of males.

Many Lepidoptera, Odonata and possibly also some Hymenoptera and Coleoptera possess more than three color receptor types – for example, cells with peak sensitivity in the red or in the violet. Such higher diversity of spectral receptor types can be achieved via increases of opsin types (mediated via gene duplication and subsequent spectral tuning) or by systematic usage of filtering mechanisms that control the light reaching the photopigments, or a combination of both strategies. The butterfly *Pieris rapae*, for example, has four opsins, including the typical ultraviolet- and green-sensitive ones, but the blue-sensitive opsin is duplicated so that peak sensitivities are at $\lambda_{max} = 360$, 425, 453 and 563 nm. However, there exist also red receptors ($\lambda_{max} = 620$ nm) and deep red receptors ($\lambda_{max} = 640$ nm), but these express the same visual pigment as the green receptors ($\lambda_{max} = 563$ nm). Diversity here is achieved by pairing cells containing the same opsin with three different filtering pigments arranged around the rhabdom. Conversely, the swallowtail butterfly,

Papilio xuthus, has five different opsins where spectral fine-tuning is again achieved by peri-rhabdomal filtering. Here, red sensitivity is mediated by an opsin with peak sensitivity of $\lambda_{max} = 575$ nm, i.e., at longer wavelengths than the human "red" receptor ($\lambda_{max} = 565$ nm). Typically, one color receptor type expresses only one opsin, but this species also has receptor cells with a very broad-band spectral sensitivity, and containing two opsins with $\lambda_{max} = 520$ and 575 nm. In addition, as in flies, *Papilio* photoreceptors are arranged in tiers, meaning that the light reaching more proximal photoreceptors is first filtered by the distal ones.

In a few insects the lenses are colored so that the rhabdoms of the associated ommatidia can receive only those wavelengths transmitted by the lenses. Male dolichopodid flies, for example, have alternating vertical rows of red and yellow facets. As a result, the same visual pigment in different ommatidia will be differentially activated, providing a basis for color vision even if only one type of visual pigment is present. Likewise, differently colored tapeta paired with identical photoreceptors could generate different spectral sensitivities.

Color vision is the ability to see images in which objects have color attributes. This requires image-forming eyes and the ability to recognize targets based on their spectral identity and independently of intensity. A mechanistic requirement is the existence of at least two photoreceptor cell types with distinct spectral sensitivities. This is because a single receptor type, despite its differential sensitivity to different wavelengths, cannot reliably disentangle wavelength from intensity. Information on the wavelength of a photon is lost upon absorption, and all the information that is available for the nervous system is a change in receptor voltage signal. However, an intermediate change can be induced by a weak light at peak sensitivity, or a strong light where the receptor is less sensitive. Thus, the signals from two receptors (at least) must be related to one another in order to extract reliable information about spectral identity, and this is done by so-called color opponent neurons, which have been found, for example, in the optic lobes of flies and bees. In addition, there must be the neural machinery for binding color attributes with other image features.

However, the existence of these mechanisms is not in itself sufficient evidence for color vision. For example, two distinct spectral receptor types in an animal might be used for color vision, but they could also be used to each drive a separate behavior routine, such as phototaxis, mate search or oviposition. In many motion-related behavioral contexts (such as edge detection, assessment of speed, motion parallax), insects behave as if color blind – in bees such behaviors are mediated only by green receptor input, whereas in flies the broad-band R1–6 receptors provide the input to such behaviors. Thus behavioral tests are necessary to prove that receptor responses are integrated in a color vision system. It is essential that an animal can respond to targets based on their spectral composition, and independently of intensity. Many insects, including Hymenoptera, Lepidoptera and Diptera can be trained to associate colors with certain outcomes, even some nocturnal insects. The conjunction of color with shape (as required by a strict definition of color vision) requires that insects can disentangle shape from color and moreover to categorize distinctly shaped objects by color; this has been demonstrated in bees and wasps.

Many insects have distinct color preferences when they first search for food; however, these can typically be modified by learning. Interestingly, herbivorous insects often prefer not only green, but more strongly yellow when searching for suitable forage. This bias can be understood when one considers that a spectrally opponent neural mechanism with antagonistic inputs from green receptors and more short-wave receptors drives the behavioral response: typically, yellow targets provide a stronger signal in such a mechanism than the relatively broad-band reflectance of green foliage.

Many species of bees prefer colors in the violet to blue range, an evolutionary response to the fact that flowers with such colors often contain high nectar rewards.

22.3.4 Discrimination of the plane of vibration (polarization sensitivity)

Light waves vibrate in planes at right angles to the direction in which they are traveling. These planes of vibration may be equally distributed through 360° about the direction of travel, or a higher proportion of the vibrations may occur in a particular plane. Such light is said to be polarized, and if all the vibrations are in one plane the light is plane-polarized.

Light coming from a blue sky is polarized. The degree of polarization and the plane of maximum polarization from different parts of the sky vary with the position of the sun. It is, consequently, possible to determine the position of the sun, even when it is obscured, from the composition of polarized light from a patch of blue sky. Certain insects are able to make use of this information in navigation. It is particularly important in the homing of social Hymenoptera, and is best known in *Apis* and the ant *Cataglyphis*. In other insects, including Odonata and some Diptera, the ability to perceive polarized light probably enables the insects to maintain a constant and steady orientation.

Detection of the plane of polarization is possible because photopigment molecules are preferentially oriented along the microvilli of the rhabdomere and maximum absorption occurs when light is vibrating in the same plane as the dipole axis of the pigment molecule. If the rhabdom is twisted, as in most ommatidia, there can be no preferred plane of absorption; polarization sensitivity depends on a uniform orientation of the visual pigment molecules within a rhabdomere. In ants and bees, and other insects responding to the plane of polarization, straight rhabdomeres are present only in a small group of ommatidia along the dorsal rim of each eye. These polarization-sensitive ommatidia constitute only 6.6% of the total ommatidia in *Cataglyphis*, and 2.5% in *Apis*.

When light is polarized all wavelengths are affected, but the polarization receptors of ants and bees are responsive to ultraviolet light. In *Cataglyphis*, six of the eight photoreceptor cells in the dorsal rim ommatidia are ultraviolet-sensitive; four have a common orientation of their microvilli, those of the other two are at right angles (Fig. 22.19a). Consequently, when one set of cells is responding maximally because its microvilli are parallel with the plane of polarization, the other set is responding minimally. Each ommatidium in the dorsal rim responds maximally to polarization in one plane, and the population of receptors exhibits a range of different orientations (Fig. 22.19b). In the medulla of the optic lobe, the photoreceptor cell axons synapse with interneurons responding maximally to polarization in a particular plane (Fig. 22.19c). It is believed that, by scanning the sky, the insect is able to match the inputs of its polarization receptors to the pattern in the sky. When reorienting, it matches the sky pattern to the remembered neural input. A neural map that could achieve this kind of comparison has been located in the central body of the locust brain.

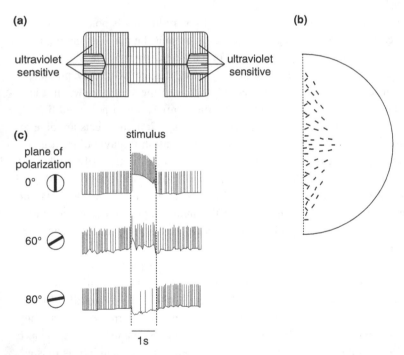

Figure 22.19 Detection of the plane of polarization. (a) Arrangement of microvilli in the rhabdom in the polarization-sensitive area. Six of the eight photoreceptor cells are ultraviolet-sensitive, with orthogonally arranged microvilli (*Cataglyphis*) (after Wehner, 1982). (b) Orientation of the rhabdoms in the polarization-sensitive area along the dorsal margin of the right eye of *Cataglyphis* (after Wehner, 1989). (c) Response of an interneuron in the medulla to stimulation of the eye with light polarized in different planes. Vertically polarized light (0°) results in depolarization of the interneuron and an increase in its firing rate. Near horizontally (80°) polarized light causes a hyperpolarization and inhibition of the cell. These effects are thought to be produced by interaction of the input from receptor cells with different microvillar orientation, as in (a) (*Gryllus*) (after Labhart, 1988).

22.4 Dorsal ocelli

Dorsal ocelli are found in adult insects and the larvae of hemimetabolous insects. Typically there are three ocelli forming an inverted triangle antero-dorsally on the head (Fig. 22.20a), although in Diptera and Hymenoptera they occupy a more dorsal position on the vertex. The median ocellus shows evidence of a paired origin, as the root of the ocellar nerve is double and the ocellus itself is bilobed in Odonata and *Bombus* (Hymenoptera). Frequently, one or all of the ocelli are lost and they are often absent in wingless forms.

A typical ocellus has a single thickened cuticular lens (Fig. 22.20b), although in some species, such as *Schistocerca* (Orthoptera) and *Lucilia* (Diptera), the cuticle is transparent, but not thickened, and the space beneath it is occupied by transparent cells. Each ocellus contains a large number of photoreceptor cells packed closely together without any regular arrangement; in the locust ocellus there are 800–1000. A rhabdomere is formed on at least one side of each receptor cell, and the rhabdomeres of between two and seven cells combine to form rhabdoms. The rhabdomeres usually occupy much of the cell boundary and, in the case of *Rhodnius* (Hemiptera), are present all around the cells, forming a hexagonal meshwork similar to that in the stemmata of *Cicindela* (Coleoptera) (Section 22.5). The structure of the rhabdomeres in the dorsal ocelli is the same as that in the compound eye. Pigment cells sometimes invest the whole ocellus, but in some species, e.g., cockroaches, they are lacking. A reflecting tapetum, probably consisting of urate crystals in a layer of cells, may be present at the back of the receptor cells.

Each photoreceptor cell gives rise, proximally, to an axon which passes through the basal lamina of the ocellus and terminates in a synaptic plexus immediately behind the ocellus. Two anatomical classes of ocellar interneurons originate here. Some have giant axons up to 20 μm in diameter, often called large (L) fibers, others are of small diameter (S fibers). About ten large fibers and up to 80 small ones are associated with each ocellus. In most insects studied the large interneurons end in the brain, but in bees and flies some extend to the thoracic ganglia

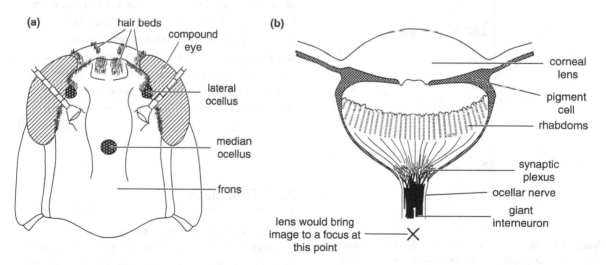

Figure 22.20 Ocellus. (a) Frontal view of the head of a grasshopper, showing the positions of the ocelli. (b) Diagrammatic longitudinal section through an ocellus of a grasshopper. The diagram shows the light-adapted condition with the pigment of the pigment cells restricting the entry of light. Notice that the image is focused below the rhabdoms (after Wilson, 1978).

(Fig. 22.21). Where these descending interneurons are absent, the pathway to the thoracic motor centers is completed by second-order descending interneurons. The small interneurons connect with several other centers in the brain, including the optic lobes, mushroom bodies and the central body.

The receptor cell axons synapse repeatedly and reciprocally with each other and with the

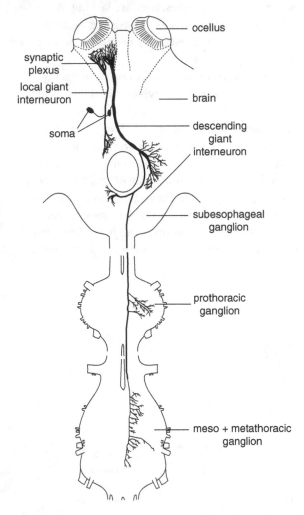

Figure 22.21 Giant interneurons from the ocelli. A diagram of the anterior ganglia of the central nervous system of a honey bee showing examples of giant interneurons, a local neuron that does not extend beyond the brain and a descending neuron that runs from the ocellus to the pterothoracic ganglion (based on Mobbs, 1985).

interneurons, which also synapse with each other. Some of the synapses between interneurons and receptor cells are input synapses to the receptor cells, indicating that the interneurons may modulate the activity of the receptor cells as well as receiving information from them.

Illumination produces a sustained depolarization of the photoreceptor cell which is proportional to light intensity. No action potentials are produced in the receptor cells, and graded receptor potentials are transmitted along the axons to the synapses. As at the first synapses behind the compound eye, the signal is amplified and the sign is reversed. Also as in the compound eye, the input signals arising from contrasts in illumination are of similar amplitude even though the background level of illumination is different. The giant interneurons transmit information to the brain, either electrotonically or by spiking.

Because the image produced by the lens is not in focus on the retina in some species, the function of ocelli was long thought to be only light detection. However, it is now known that in dragonflies and nocturnal wasps, the median ocelli are focused, and, at least in dragonflies, adapted for accurate detection of the horizon. Here, the median ocellus can even detect the direction of moving gratings and is involved in early-stage motion processing and subserving pitch control. Even in species where the lens is underfocused there is still potential for form vision. In the locust the ocelli are involved in detecting roll, their sensitivity to rapid changes in light intensity being well suited for the perception of changes in the position of the horizon.

22.5 Stemmata

Stemmata are the only visual organs of larval holometabolous insects. They are sometimes called lateral ocelli, but this term is better avoided as it leads to confusion with the dorsal ocelli from which they are functionally and often structurally distinct. Extraocular photoreceptor organs in the optic lobes of adult insects are also sometimes called stemmata, and indeed it appears that they are developmentally based on larval precursors (Section 22.6.2). Larval stemmata occur laterally on the head and vary in number from one on each side in tenthredinid larvae (Hymenoptera: Symphyta) to (most typically) six on each side, for example in larval Lepidoptera (Fig. 22.22). The larvae of fleas and most hymenopterans, other than Symphyta, have no stemmata. In larval Cyclorrhapha (Diptera) they are represented only by internal receptors. Some stemmata are simple visual organs, while others are complex camera-type eyes.

Stemmata are of two types, those with a single rhabdom, and those with many rhabdoms. The former occur in Mecoptera, most Neuroptera, Lepidoptera and Trichoptera. They are also present in Diptera, but in some species several stemmata are fused together to form a compound structure with a branching rhabdom. In Coleoptera the stemmata of many species have a single rhabdom, but some species, such as those of larval Adephaga (Coleoptera) have multiple rhabdoms. Stemmata with multiple rhabdoms also occur in larval Symphyta (Hymenoptera).

In caterpillars each stemma has a cuticular lens beneath which is a crystalline lens (Fig. 22.22b,c). Each lens system has seven photoreceptor cells associated with it. Commonly, three form a distal rhabdom and four form a proximal rhabdom. A thin cellular envelope lies around the outside of the sense cells and is, in turn, shrouded by the extremely enlarged corneagen cells. All the distal cells contain a visual pigment with maximal absorbance in the green part of the spectrum, while some proximal cells contain a blue- or ultraviolet-sensitive pigment. The rhabdomeres within the stemmata of caterpillars have different visual fields, and the acceptance angles of the distal rhabdomeres are close to 10° so they have low spatial resolution. The proximal rhabdomeres have much smaller acceptance angles of less than 2°. This, together with the fact that the focal plane of the lens is at the level of the proximal cells, gives them better spatial resolution. The visual fields of adjacent stemmata do not overlap so the caterpillar perceives an object as a very coarse mosaic, which is improved by side-to-side movements of the head, enabling it to examine a larger field. It is known that caterpillars can

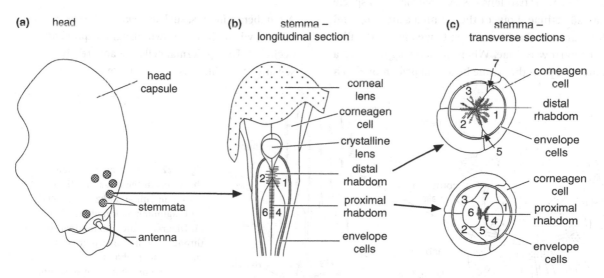

Figure 22.22 Stemmata with a single rhabdom in a caterpillar (mainly after Ichikawa and Tateda, 1980). (a) Side of the head showing the positions of the stemmata. (b) Longitudinal section of one stemma. (c) Transverse sections through the distal and proximal rhabdom.

differentiate shapes and orient toward boundaries between black and white areas.

The situation is different in larval symphytans (Hymenoptera), which have only a single stemma. These have large numbers of rhabdoms, each formed by eight photoreceptor cells and each group of cells is isolated by pigment from its neighbors. The lens produces an image on the tips of the rhabdoms which, in *Perga*, are oriented at about 5° from each other (equivalent to the interommatidial angle in compound eyes). Consequently, this type of eye is capable of moderately good form perception.

The larvae of the tiger beetle, *Cicindela*, have six stemmata, like caterpillars, but with a large number of photoreceptor cells in each stemma, as in the Hymenoptera. The largest of the stemmata has about 5000 receptor cells, each of which forms a rhabdomere on all sides so that the rhabdoms are in the form of a lattice (Fig. 22.23). It is possible that spatial resolution in these eyes is limited because of optical pooling and perhaps electrical coupling. In the larvae of the visual-oriented, predatory sunburst diving beetles *Thermonectus marmoratus*, several of the 12 stemmata have multiple retinae so that, together with two lensless eye-patches, this species has 28 retinae. Four of the stemmata are long and tubular, with horizontally extended but vertically very narrow retinae. When the larva approaches a potential prey the whole head and body move up and down in the sagittal plane so that the four horizonal retinae scan vertically across the target before a strike is made. The remaining stemmata probably act as movement detectors, allowing the animal to orient the tubular stemmata toward potential prey.

The optic lobes of larval insects consist of a lamina and medulla comparable with those associated with compound eyes of adults and, at least in caterpillars, each stemma connects with its individual cartridge in the lamina. In all these types of stemmata, the photoreceptor cells contain screening pigment granules in addition to the visual pigment. Movement of the granules – away from the rhabdomeres in the dark and toward them in the light – provides sensitivity adjustment. Caterpillars have three visual pigments and the neural capacity to distinguish colors. The larvae of several holometabolous species have been shown, experimentally, to respond to the plane of polarization of incident light. In neither case is the behavioral importance of these abilities understood.

22.6 Other visual receptors

22.6.1 Dermal light sense

A number of insects, such as *Tenebrio* larvae, respond to light when all the known visual receptors are occluded. The epidermal cells are apparently sensitive to light. This is also suggested by the

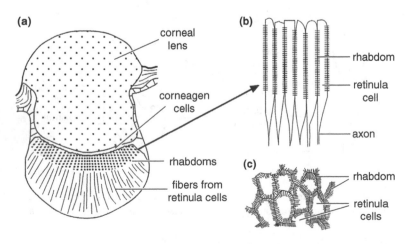

Figure 22.23 Stemma with multiple rhabdoms in the larva of a tiger beetle (based on Toh and Mizutani, 1994). (a) Diagrammatic longitudinal section. (b) Diagrammatic longitudinal section through the photoreceptor cells, showing the rhabdoms. (c) Diagrammatic transverse section through the receptor cells showing the rhabdomeres forming a continuous lattice.

pigment movements which occur in isolated epidermal cells of some insects, i.e., cells which do not receive neural input from the central nervous system (Section 25.5.1), and by the production of daily growth layers in the cuticle (Section 16.2).

Several families of butterflies are known to have photoreceptors on the genitalia of both sexes. In *Papilio* there are two receptors on each side, each consisting of a single neuron lying on a nerve a short distance below the epidermis. The cuticle above the neuron is transparent. The cells are called phaosomes and they probably monitor the positions of the genitalia during copulation.

22.6.2 Sensitivity of the brain

In several insect species, light affects neural activity directly by acting on the brain, not via the compound eyes or ocelli. This commonly occurs in the entrainment of diurnal rhythms. In some species day length – regulating diapause or polyphenism – is registered directly by the brain. Typically, such rhythmicity is mediated by cerebral cell clusters containing rhabdomeric structures and sometimes screening pigments, sometimes called adult stemmata. One such structure, the Hofbauer-Buchner eyelet in *Drosophila*, has been identified as an essential circadian photoreceptor. Similar structures have been found in several species of beetles, bees and hawk moths, where they contain two classes of opsin, sensitive to ultraviolet and blue. Other extraretinal cells in the optic lobes express green-sensitive opsin. Together, these various spectral inputs potentially facilitate the detection of changes in ambient light to control various photoperiodic rhythms.

22.7 Magnetic sensitivity and photoreception

A number of insect species have been shown to respond to changes in magnetic field, and it is possible that they use the Earth's magnetic field in navigation. Two mechanisms have been proposed to account for this response. Some insects are known to contain particles of magnetite, an iron oxide, which might be affected by the magnetic field. In worker honey bees, the magnetite is contained in innervated trophocytes in the abdomen. In many species, however, geomagnetic orientation has been demonstrated to depend on the spectral composition of ambient light, suggesting a role of photoreceptor proteins. The blue-light photoreceptor cryptochrome protein (CRY), also involved in setting the circadian clock in insects, is required for magnetoreception in *Drosophila*. This protein is expressed in various parts of the body, but also in the eyes, which, at least in birds, seem to mediate sensitivity to the Earth's magnetic field by generating a neural map of this field. In insects, CRY is also expressed in the eyes, but the precise location and physiology of magnetoreception, as well as its complex interaction with light sensitivity, remains to be determined.

Summary

- Insects have compound eyes, which in part mediate capacities similar to those of lens eyes in vertebrates, but sometimes with wholly different mechanisms. However, some of the visual abilities of insects are wholly different to those of humans – for example, the sensitivity to polarized and ultraviolet light, and the capacity (found in some insects) for 360° vision.

- Compound eyes typically have relatively poor spatial resolution when compared with vertebrate eyes, which can be compensated in part by relatively higher temporal resolution and rapid scanning of the visual scene.

- Many components of compound eyes, such as the resolution of the ommatidial array and color vision, show dramatic variation between species and are often exquisitely tuned to the environment in which their bearers operate.

- Insects also have single-lens eyes, the dorsal ocelli, which typically occur in triplicate in adults. Their function is in motion processing and horizon detection, but also the potential for form vision, albeit at relatively poor spatial resolution. Many insect larvae have stemmata (sometimes called larval ocelli) which enable form detection in at least some species.

- In addition, many insects have extraocular photoreceptors in a variety of locations, with varied functions related to magnetoreception, circadian rhythm control and sexual behavior.

Recommended reading

Borst, A. (2009). *Drosophila*'s view on insect vision. *Current Biology* 19, R36–R47.

Briscoe, A. D. and Chittka, L. (2001). The evolution of colour vision in insects. *Annual Review of Entomology* 46, 471–510.

Horváth, G. and Varjú, D. (2003). *Polarized Light in Animal Vision: Polarization Patterns in Nature*. Berlin: Springer-Verlag.

Land, M. F. and Nilsson, D.-E. (2012). *Animal Eyes*. 2nd edn. Oxford: Oxford University Press.

References in figure captions

Butler, R. (1973). The anatomy of the compound eye of *Periplaneta americana* L: I. General features. *Journal of Comparative Physiology* 83, 263–278.

Goldsmith, T. H. (1962). Fine structure of the retinulae in the compound eye of the honey-bee. *Journal of Cell Biology* 14, 489–494.

Hardie, R. C. (2006). Phototransduction in invertebrate photoreceptors. In *Invertebrate Vision*, ed. E. Warrant and D.-E. Nilsson, pp. 43–82. Cambridge: Cambridge University Press.

Horridge, G. A. (1975). Optical mechanisms of clear-zone eyes. In *The Compound Eye and Vision of Insects*, ed. G. A. Horridge, pp. 255–298. Oxford: Clarendon Press.

Ichikawa, T. and Tateda, H. (1980). Cellular patterns and spectral sensitivity of larval ocelli in the swallowtail butterfly, *Papilio*. *Journal of Comparative Physiology* 139, 41–47.

Labhart, T. (1988). Polarization-opponent interneurons in the insect visual system. *Nature* 331, 435–437.

Langer, H., Hamann, B. and Meinecke, C. C. (1979). Tetrachromatic visual system in the moth, *Spodoptera exempta* (Insecta: Noctuidae). *Journal of Comparative Physiology* 129, 235–239.

Matic, T. and Laughlin, S. B. (1981). Changes in the intensity-response function of an insect's photoreceptors due to light adaptation. *Journal of Comparative Physiology* 145, 169–177.

Menzi, U. (1987). Visual adaptation in nocturnal and diurnal ants. *Journal of Comparative Physiology A* 160, 11–21.

Mobbs, P. G. (1985). Brain structure. In *Comprehensive Insect Physiology, Biochemistry and Pharmacology*, vol. 5, ed. G. A. Kerkut and L. I. Gilbert, pp. 299–370. Oxford: Pergamon Press.

Paulus, H. F. (1975). The compound eye of apterygote insects. In *The Compound Eye and Vision of Insects*, ed. G. A. Horridge, pp. 3–19. Oxford: Clarendon Press.

Rossel, S. (1979). Regional differences in photoreceptor performance in the eye of the praying mantis. *Journal of Comparative Physiology A* 131, 95–112.

Rossel, S. (1980). Foveal fixation and tracking in the eye of the praying mantis. *Journal of Comparative Physiology A* 139, 307–331.

Skorupski, P., Doring, T. F. and Chittka, L. (2007). Photoreceptor spectral sensitivity in island and mainland populations of the bumblebee, *Bombus terrestris*. *Journal of Comparative Physiology A* 193, 485–494.

Snyder, A. W. and Horridge, G. A. (1972). The optical function of changes in the medium surrounding the cockroach rhabdom. *Journal of Comparative Physiology A* 81, 1–8.

Stavenga, D. G. (2006). Invertebrate photoreceptor optics. In *Invertebrate Vision*, ed. E. Warrant and D.-E. Nilsson, pp. 1–42. Cambridge: Cambridge University Press.

Toh, Y. and Mizutani, A. (1994). Structure of the visual system of the larva of the tiger beetle (*Cicindela chinensis*). *Cell and Tissue Research* 278, 125–134.

Walcott, B. (1975). Anatomical changes during light adaptation in insect compound eyes. In *The Compound Eye and Vision of Insects*, ed. G. A. Horridge, pp. 20–33. Oxford: Clarendon Press.

Wehner, R. (1982). Himmelsnavigation bei Insekten. Neurophysiologie und Vehalten. *Neujahrsblatt, Naturforschende Gesellschaft in Zürich* 184, 1–132.

Wehner, R. (1989). The hymenopteran skylight compass: matched filtering and parallel coding. *Journal of Experimental Biology* 146, 63–85.

Wilson, M. (1978). The functional organisation of locust ocelli. *Journal of Comparative Physiology* 124, 297–316.

Wolken, J. J., Capenos, J. and Turano, A. (1957). Photoreceptor structures: III. *Drosophila melanogaster*. *Journal of Biophysical and Biochemical Cytology* 3, 441–447.

Zhang, S. W. and Srinivasan, M. V. (1994). Prior experience enhances pattern discrimination in insect vision. *Nature* 368, 330–333.

23 | Mechanoreception

REVISED AND UPDATED BY **TOM MATHESON**

INTRODUCTION

Mechanoreceptors detect mechanical distortions that can arise from touching an object or from the impact of vibrations borne through the air, water or the substratum. They underpin the senses of touch and hearing, and monitor distortions of the body or limbs which arise from the stance or movements of the insect or from the force of gravity. Some mechanoreceptors are exteroceptors (which respond to external stimuli), some are interoceptors (responding to internally generated stimuli) and others are proprioceptors (which respond to body position and movement). Surprisingly, in some insects, mechanoreceptors even permit the detection of infrared radiation. Mechanoreceptors can have exquisite sensitivity and precise tuning to specific stimuli, permitting rapid and appropriate behavioral responses. Groups of similar mechanoreceptors often act together as "populations," each member being tuned differently, so that their summed neuronal output can retain the benefit of high sensitivity while permitting the representation of a very wide range of possible inputs.

Three broad structural categories of mechanoreceptor are present in insects: cuticular structures with bipolar neurons, discussed in Section 23.1; subcuticular structures with bipolar neurons, known as chordotonal organs, described in Section 23.2; and internal multipolar neurons which function as stretch or tension receptors, detailed in Section 23.3. The descriptions of this chapter are broadly divided into discussions of structure and distribution, mechanisms of action and function. Roles of mechanoreceptors in the generation of motor patterns are detailed in Chapter 9, and their involvement in sound communication is considered in Chapter 26.

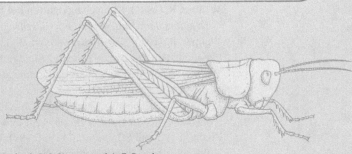

The Insects: Structure and Function (5th edition), ed. S. J. Simpson and A. E. Douglas.
Published by Cambridge University Press. © Cambridge University Press 2013.

23.1 Cuticular mechanoreceptors

Cuticular mechanoreceptors fall mainly into two classes: hair-like projections from the cuticle with a basal socket, and dome-like campaniform sensilla. Both types have similar arrangements of neurons and sheath cells.

23.1.1 Structure

Cuticular components *Hair-like structures.* Most of the larger hairs on an insect's body are mechanoreceptors. There are no pores (apart from the molting pore, see below) in the cuticle of hairs that function solely as mechanoreceptors, and they are, therefore, known as aporous sensilla. Many contact chemoreceptors also have a mechanoreceptor function and, although they have a terminal pore, it is not relevant to their mechanoreceptive function (Section 24.1). The wall of the hair consists of exocuticle with an outer layer of epicuticle.

Most commonly, the hairs taper from base to tip, and they are often known as trichoid sensilla (this term also applies to chemoreceptors with a similar form, see Chapter 24). Shorter, peg-like multimodal chemo- and mechanosensory hairs with a single pore at the tip are referred to as basiconic, though usage of some of these terms in the literature is often confused. Mechanosensory hairs on the cerci of crickets and cockroaches, for example, are very long relative to their diameter and do not taper; they are called trichobothria. Some insects possess club-shaped hairs that can signal body orientation relative to the gravitational field. The position-sensitive hairs on the cerci of some cockroaches consist of a sphere of cuticle on a short stalk. Other forms occur in relation to specific functions.

Whatever the external form, the hair-like structure arises from a socket permitting movement of the hair, and neural stimulation results from the displacement of the hair in its socket. To give this flexibility and hold the hair in its normal position, the socket usually has three components (Fig. 23.1). Externally,

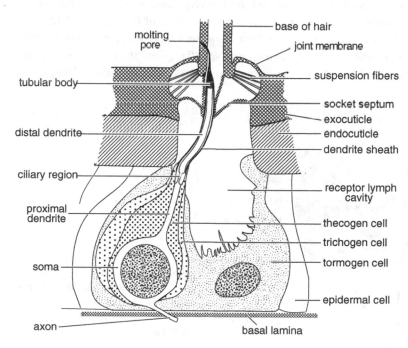

Figure 23.1 Mechanoreceptor hair sensillum. Diagram showing the principal features of a mechanosensitive sensillum.

a thin cuticular joint membrane is continuous with the general body cuticle and that of the hair. Beneath this, bridging a gap between the body cuticle and the hair, are rings of suspension fibers. Finally, on the inside, is the socket septum. This is also fibrous, but appears more delicate than the suspension fibers. It runs across the outer part of the receptor lymph cavity to the dendrite sheath.

These suspensory structures restore the hair to its original position following imposed movement. They often also restrict the extent to which the hair can be moved in different directions, so it is common for hair sensilla to have directional sensitivity. Directional sensitivity is also conferred by the position at which the dendrite is inserted into the hair base (see Fig. 23.5b).

In addition to scattered mechanoreceptor hairs, all insects have groups of small hairs, known as hair plates, at some joints in the cuticle. These act as proprioceptive organs when stimulated by relative movements of adjacent limb segments. Some hair-like structures – on the wings for example – are neither innervated nor articulated in a socket and presumably serve non-mechanoreceptive functions. They may have straight or, more commonly, twisted shafts.

Campaniform sensilla. Campaniform sensilla are areas of thin cuticle, domed and usually oval in shape, with a long diameter usually in the range 5–30 μm. The dome of thin cuticle consists of an outer homogeneous layer with the appearance of exocuticle, and an inner lamellar or fibrous layer (Fig. 23.2). Sometimes these two layers are separated by a layer of transparent (in the electron microscope) cuticle referred to as spongy cuticle. The dome is connected to the surrounding cuticle by a joint membrane. The dendrite sheath, enclosing a single dendrite, is inserted into the center of the dome, and a molting pore may be visible on the outside. Details of the structure vary according to the position of the sensillum even within a species. There are nine different types in adult *Calliphora* (Diptera).

Campaniform sensilla are situated in areas of the cuticle that are subject to stress. On the appendages, where they occur most commonly, they are usually close to the joints. Fig. 8.7a shows their distribution on one leg of a cockroach. In some orthopteroid insects there is a single campaniform sensillum at the base of each of the tibial spines. Campaniform sensilla also occur on the mouthparts, on the basal segments of the antennae, on veins close to the wing base and on the ovipositor, while large numbers are present on the halftere of Diptera (Fig. 9.12). An adult *Calliphora* has a total of approximately 1200 campaniform sensilla: about 36 on each leg, 140 on each wing base and 340 on each haltere. They often

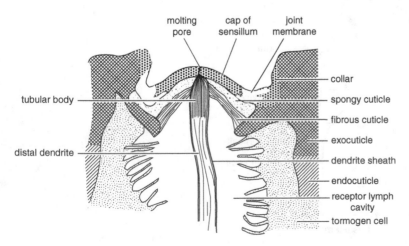

Figure 23.2 Campaniform sensillum. Diagram showing the essential cuticular features. The arrangement of cells beneath the cuticle would be similar to that in Fig. 23.1.

occur in groups, all those within a group having the same orientation and, possibly, functioning as a unit. During postembryonic development of the cockroach *Periplaneta americana*, both the number and the size of campaniform sensilla on the legs increase. Receptors present in the earliest instars remain through successive molts, with their caps progressively increasing in size. They remain innervated by the same sensory neuron throughout life. In addition, new caps (and, presumably, sensory neurons) appear in each group on the legs. For the tibial group, the total number increases from two in the first instar to 10–12 in the adult.

Infrared receptors. Some fire-seeking beetles (e.g., the buprestid beetle genus *Melanophila*) and flat bugs (e.g., *Aradus*) have highly modified cuticular mechanoreceptors that are exceptionally sensitive to infrared radiation such as that emitted by forest fires. In *Melanophila* there are about 80 dome-shaped sensilla in each of a pair of metathoracic pit organs found just behind the coxae of the mesothoracic leg. In *Aradus* there are about 12 sensilla per side interspersed among hair mechanoreceptors on membranous extensions of the prothorax called propleurae. Each infrared sensillum is dome-shaped with a small apical recess, and is innervated by a single bipolar sensory neuron.

Dendrite sheath. In all cuticular mechanoreceptors the distal part of the dendrite is enclosed within a sheath, known as the cuticular or dendrite sheath, which inserts at the hair base or in the center of a campaniform sensillum.

When an insect molts, the dendrite sheath extends across the space between old and new cuticles (Fig. 23.3), breaking off at ecdysis. The point of breakage is marked on the outside of the sensillum by a molting pore (Figs. 23.1, 23.2). During the molt the sheath presumably protects the dendrites from enzymes in the molting fluid, making it possible for the sensillum to continue functioning after apolysis until the time of ecdysis. In caterpillars of *Barathra* (Lepidoptera), for example, the sound-sensitive hairs are non-functional for only about 30 minutes before ecdysis, and the new hair is functional within a few minutes of the old cuticle being shed (Fig. 23.3).

Cellular components The cells of a sensillum are derived from epidermal cells and lie within the epidermal layer. One of the derivatives becomes the neuron, and two or three others are sheath cells.

Neuron. In a majority of cuticular mechanosensilla there is only one neuron. Its soma is large, with a large nucleus, and distally a single unbranched dendrite extends to the cuticle. It is divided into three sections. The proximal dendrite contains mitochondria and other organelles. Some distance beneath the cuticle it narrows abruptly to the ciliary region (Fig. 23.1), which contains a ring of nine doublets of microtubules but usually lacks the central tubules typically present in motile cilia. The tubules arise from a basal body which also extends roots into the proximal region of the dendrite. After a short ciliary region the dendrite widens again. In this distal region the neurotubules increase in number, but no other organelles are present. At the extreme tip of the dendrite is a dense mass of up to 1000 microtubules known as a tubular body. The cytoplasm around the tubules of the tubular body is electron-dense (when seen in section in the electron microscope) and there often appear to be cross-connections between the tubules as well as connections with the cell membrane.

Sheath cells. The sheath cells wrap around the neuron. When three sheath cells are present, the innermost is called the thecogen or neurilemma cell. It closely surrounds the soma and proximal dendrite and encloses an extracellular cavity around the ciliary region. Distally it is attached to the dendrite sheath, which it produces during development of the sensillum. It often contains a complex labyrinth of extracellular spaces within its folds. Sometimes there is apparently no separate neurilemma cell and its position and roles are taken over by the trichogen cell.

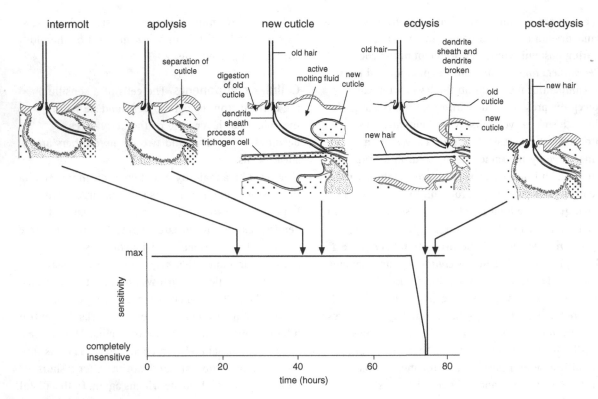

Figure 23.3 Sensitivity of a hair sensillum in relation to molting. Diagrams in the upper panel illustrate the structural changes occurring through one larval stage from one ecdysis (time 0) to the next (compare Fig. 16.17). Lower panel shows the sensitivity of the hair to stimulation at different times. Sensitivity is completely lost only for about 30 minutes at ecdysis (partly based on Gnatzy and Tautz, 1977).

Outside the thecogen cell is the trichogen cell. A finger-like projection from this cell secretes the cuticle of the hair (Fig. 23.3, center diagram). The projection then withdraws, leaving the hair as a hollow structure. In some adult insects, the trichogen cell dies after it has secreted cuticle so the sensillum only has two remaining sheath cells.

The outermost sheath cell is the tormogen cell. It secretes the cuticle of the socket and, when it withdraws, comes to surround a cavity, known as the receptor lymph cavity, through which the dendrite sheath passes and which is continuous with the cavity of the hair. The inner margin of the tormogen cell, where it forms the receptor lymph cavity, is often elaborated into irregular microvilli.

All these cells are held together distally by maculae adherens desmosomes and septate desmosomes, and they are connected by gap junctions. The thecogen cell is also connected to the neuron by desmosomes, but there are no gap junctions. The desmosomes isolate the receptor lymph cavity from the hemocoel.

23.1.2 Functioning

Three processes are involved in the functioning of cuticular mechanoreceptors. The mechanical distortions of the cuticle produced by bending a hair or compressing a campaniform sensillum must be transmitted to the dendrite, in a process known as coupling. The mechanical energy is then transformed, or transduced, into an electrical signal, the generator or receptor potential. Finally, the receptor potential is encoded into action potentials for transmission along the sensory axon. The central nervous system does not exert direct efferent (outward) control over receptor sensitivity, but most sensory neurons are subject to neuromodulatory effects. The transmission of trains of sensory action potentials to postsynaptic neurons in the central nervous system is subject to presynaptic inhibition so that sensory signals can be gated out or tuned to a prevailing behavior.

Coupling Coupling in cuticular mechanoreceptors probably involves distortion of the tip of the dendrite sheath containing the tubular body, and hence of the tubular body itself. In hair sensilla, movement of the base of the hair produces this distortion, although the precise mechanism differs between hair types. The hair acts as a lever and amplifies force exerted near its tip. Because of this, hair sensilla can respond to very low forces. The position at which the tubular body is inserted also influences the effect of bending and, along with other asymmetries of the hair and socket, probably contributes to the directional sensitivity commonly found in hairs.

In campaniform sensilla, coupling is effected by distortion of the cuticular dome resulting from compression. Oval-shaped sensilla are sensitive to compression along their short axis. In some circular sensilla, directional sensitivity is conferred by having a fan-shaped dendrite. Mechanical properties of the cuticle, and the coupling between a dome and its surrounding cuticle, determine how forces act to distort the dome. The precise location of the dome (e.g., on a wing vein) may thus have an important effect on the responsiveness of the receptor to natural stimuli. In nature, campaniform sensilla respond both to increasing strains in their preferred direction and to decreasing strains in the opposite direction. The dozen or so campaniform sensilla within one group on a cockroach leg differ not only in orientation but also in size, with larger receptors appearing in later molts. It is not yet known if the different sizes are

correlated with different thresholds or response properties, although this seems likely.

In the modified cuticular mechanoreceptors that detect infrared radiation in some insects, deformation of the sensory dendrite is achieved by thermal expansion of a fluid within a tiny, rigid cuticular sphere.

Transduction Transduction involves the deformation of the dendrite and production of a depolarizing receptor potential. Although not yet demonstrated in insects, it seems certain that the plasma membrane of the neuron contains mechanically sensitive (or activated) ion channels (MACs); that is, channels that are opened as a result of distorting the membrane so that an inward movement of ions occurs. In insects, potassium is likely to carry most of the receptor current.

It also seems certain that the tubular body plays an essential role in this process, and a distortion of as little as 5 nm is sufficient to produce a receptor potential. However, some mechanoreceptor dendrites do not have a fully developed tubular body, and a receptor potential is sometimes produced even when the microtubules are destroyed experimentally.

Action potential encoding Once the depolarizing receptor potential exceeds a threshold voltage, action potentials are generated in the sensory neuron. Their pattern of firing is determined by many factors, including the time course of the receptor potential, the nature of the ion channels activated and their possible inactivation, and circulating neuromodulatory substances.

Overall response to stimulation Hair sensilla differ in their responses to bending. In some, a response is only produced during or immediately following the movement; if the hair remains bent, no further response occurs (Fig. 23.4a). This is known as a phasic response, and the process underlying it as sensory adaptation. Other sensilla have a tonic response in which action potentials are produced as

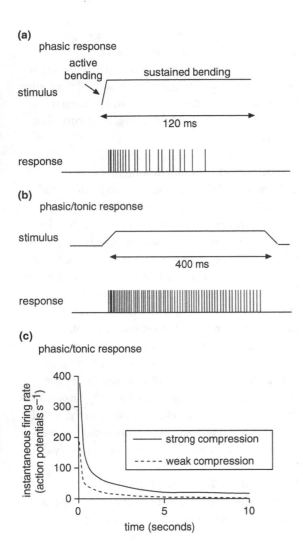

Figure 23.4 Sensory input from cuticular mechanoreceptors. (a) Phasic response of a hair sensillum on the trochanter of a cockroach. The hair was bent and remained bent throughout the recording. Lower trace shows the action potentials in the axon. Notice that the sensillum stops firing even though the hair remained bent (based on Spencer, 1974). (b) Phasic/tonic response of a campaniform sensillum on the proximal tibia of a cockroach. Action potentials were produced as long as compression of the sensillum due to bending the leg was sustained (based on Zill and Moran, 1981). (c) Phasic/tonic response of the same sensillum as in (b), showing the instantaneous firing rate. With stronger compression, more action potentials were produced and the neuron was still active after ten seconds of continuous stimulation (based on Zill and Moran, 1981).

long as the hair remains bent. In many cases the response contains both components, and it is then described as being phaso-tonic (Fig. 23.4b,c). Hairs with a tactile function often exhibit only a phasic response, while those with a proprioceptive function and campaniform sensilla usually exhibit phaso-tonic responses. The nature of the response is affected by all three stages of sensory reception but, in at least some cases, most of the sensory adaptation occurs during action potential encoding. As a result of many interacting biomechanical and electrical processes, individual hair receptors display characteristic angular and velocity thresholds, directional sensitivities and rates of adaptation. Moreover, these properties can change during postembryonic development. For example, wind-sensitive hairs on the cercus of a juvenile cricket respond best to wind acceleration, but over successive molts become progressively more sensitive to wind velocity. These changes are underpinned in part by the increasing length of each hair at each molt, and are also reflected in progressively modified output synaptic connections made by the sensory neurons within the central nervous system.

23.1.3 Functioning in the living insect

In all insects, some hair sensilla are concerned with tactile sensation of the environment (exteroception), whereas others respond to the animal's own movements and thus act as proprioceptors.

Exteroception Hair sensilla responding to external stimuli are usually phasic. They are concerned with the sense of touch, with sensing movement of the surrounding air or water, or even with orientation in relation to gravity.

Tactile hairs. Hairs responding to tactile stimulation are widely distributed over the insect's body, may be present in large numbers and may vary considerably in length. On the middle leg of *Schistocerca* (Orthoptera), for example, there are 1500–2000 mechanosensitive hairs. These hairs usually taper from the base and are generally curved and set at an angle to the surface of the cuticle. On the wing of *Schistocerca* there are tactile hairs of at least two different length classes distributed in specific patterns, as well as shorter multimodal basiconic hairs. Contact of any part of the body with an external object is detected by the exteroceptive hairs, and the signals conveyed to a highly ordered spatial array of sensory axonal arborizations within the central nervous system (see Fig. 20.24). This central somatotopic map of the body surface underlies the insect's ability to make an appropriate response.

Air movement detectors. Some insects have hairs that are specialized for the detection of air movements, sometimes including sounds.

Grasshoppers and some Lepidoptera have groups of small trichoid mechanoreceptors on specific regions at the front and top of the head (see Fig. 22.20a) or on the prothorax. These hairs are stimulated by the flow of air over the head and body when the insect flies. They are responsible for the maintenance of wingbeat (Section 9.7.2), and are sometimes called aerodynamic sense organs. Since they are directionally sensitive, they also contribute to the control of yaw (Section 9.7.2).

Among orthopteroid insects, hairs sensitive to air movement are present on the cerci and, in some cases, these are also responsive to sound. They are usually long (up to 1.5 mm on the cerci of the cricket *Acheta domestica*) and slender, and have flexible mounting in their sockets so that they vibrate freely rather than bending along their shaft. The filiform hairs of the cricket cercus have preferred directions of stimulation, and their sensory neurons project into the terminal abdominal ganglion in a highly ordered way to form a complex three-dimensional map of wind stimulus direction. Here they make synaptic outputs onto a small number of local and ascending interneurons (see Fig. 20.14). There is very low inter-animal variability in the placement and characteristics of the cercal hairs. Air moving at only $4\,\text{cm s}^{-1}$ is sufficient to stimulate the cercal hairs of

the locust. Stimulation of the cercal hairs can lead to a range of behavioral responses in addition to escape.

Whereas tympanal organs respond to changes in air pressure associated with a sound (see below), hair sensilla respond to movements of the air created by the sound source. Because the energy associated with air movement falls off rapidly with distance from the source, these receptors usually function only when the insect is close to the source, in the so-called near field. This is often no more than a few centimeters. Hairs responding to sounds in the range $32–1000\,\text{Hz}$ are present in some caterpillars (Fig. 23.5).

Orientation with respect to gravity. In most insects, orientation with respect to gravity involves proprioceptors (see below), but the desert cockroach, *Arenivaga*, and a few other insects, including the true crickets (Gryllidae), have modified hair sensilla which contribute specifically to this function. The hairs, which are suspended beneath the cercus, are swollen at the tips. Most of each hair's mass is presumably in this swelling so they hang vertically to the extent that their sockets permit. Consequently, any change in the insect's orientation with respect to gravity can be perceived by their movements. First-instar juvenile crickets, *Gryllus bimaculatus*, possess only one club-shaped hair on each cercus. Stimulation of just one of these gravity-receptive hairs is sufficient to elicit a compensatory head roll, but in adults, which have many such hairs, stimulation of any single one is ineffective in eliciting the head roll behavior.

Pressure receivers. Most aquatic insects are buoyant because of the air they carry beneath the water surface when they submerge, but *Aphelocheirus* (Hemiptera: Heteroptera), living on the bottom of streams and using a plastron for gas exchange, is not buoyant. The plastron only functions efficiently in water with a high oxygen content, such as normally occurs in relatively shallow water, so some mechanism of depth perception is an advantage to *Aphelocheirus*.

On the ventral surface of the second abdominal segment of *Aphelocheirus* is a shallow depression

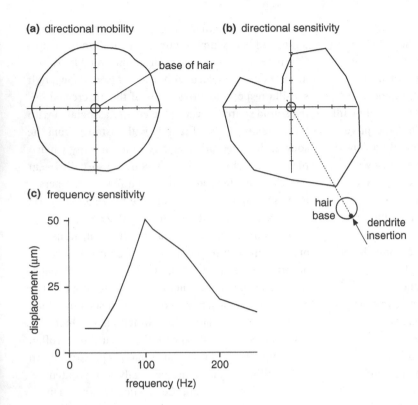

(a) directional mobility

base of hair

(b) directional sensitivity

hair base

dendrite insertion

(c) frequency sensitivity

displacement (µm)

frequency (Hz)

Figure 23.5 Sound reception by a hair sensillum of the caterpillar of *Barathra* (after Tautz, 1977, 1978). (a) The hair is equally mobile in all directions. The circle shows the extent to which it can be deflected in each direction. (b) The sensory neuron associated with the hair shows marked directional sensitivity. It produces most action potentials when the hair is bent toward the side on which the dendrite is inserted, and the fewest when bent in the opposite direction. Sensitivity is shown by the distance from the intersection of vertical and horizontal axes. (c) The tip of the hair exhibits maximum displacement, that is, the hair is most sensitive, at a sound frequency close to 100 Hz. At lower and higher frequencies, displacement (sensitivity) sharply declines.

containing hydrofuge hairs (see Fig. 17.31). These hairs are inclined at an angle of about 30° to the surface of the cuticle and mechanosensory hairs are dispersed among them. The volume of air trapped by these groups of hairs changes with pressure as the insect moves up and down in the water column, bending the sensilla. The insect responds to an increase in pressure by swimming up, but there is no behavioral response to a decrease in pressure.

Mushroom-shaped pressure receptors associated with external openings of the tracheal system in *Nepa* (Hemiptera) detect body tilt.

Proprioceptors *Trichoid sensilla.* Groups or rows of small hairs often occur at joints between leg segments (see Fig. 8.7a), on the basal antennal segments, on the cervical sclerites, at the wing base and elsewhere. The hairs are positioned so that movements of one part of the cuticle with respect to

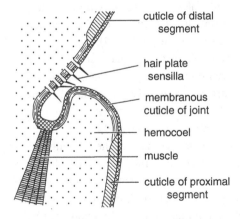

cuticle of distal segment

hair plate sensilla

membranous cuticle of joint

hemocoel

muscle

cuticle of proximal segment

Figure 23.6 Diagram showing how the sensilla of a hair plate are stimulated by coming into contact with adjacent cuticle.

another cause the hairs to bend (Fig. 23.6). Individual hairs may have preferred directions of response. Often the hairs in a hair plate are graded in size, which may reflect different response properties, and arranged in rows so that they are recruited in a graded way during a progressive movement of the joint. In the locust *Schistocerca*, long hairs of the trochanteral hair plate are in continuous contact with the overlying joint membrane and are thought to signal joint position, whereas shorter hairs are only stimulated when the leg is almost fully levated. Within the segmental ganglion, hair plate sensory neurons branch in patterns and locations that more closely resemble those of internal proprioceptors (e.g., chordotonal organs) than those of exteroceptive tactile hairs.

Hair plates on the first cervical sclerite and on the front of the prothorax provide the insect with information about the position of its head relative to the thorax. This aids the stability of the locust in flight when the head is oriented by a dorsal light reaction and the thorax is aligned with the head (Section 9.7).

A primary function of leg hair plates is to signal joint position, particularly the extremes of joint angle when movements must stop or reverse – i.e., at the start and end of the stance and swing phases of each step cycle. They are thus crucial in maintenance of posture and locomotion. Hair plates on the legs and antennae and between body segments may also be important in orientation with respect to gravity because the extent to which their hairs are stimulated can depend on the insect's orientation. Several different hair plates probably contribute to the response at any one time, although one group may be dominant. In the stick insect *Carausius*, hair plates on the trochantin which are stimulated by the coxa are of principal importance in influencing geotactic behavior. In some ants, although hair plates are involved in graviception, their removal does not impair the animal's ability to measure the slope of the terrain over which it is navigating.

In some Orthoptera, modified mechanosensory hairs on the legs or abdominal tergites can act as stridulatory pegs. In grasshoppers their afferent signals can modify the ongoing stridulatory leg movements and presumably affect the sound production.

Campaniform sensilla. In the insect skeleton all stresses can be expressed as shearing stresses in the plane of the surface. Such stresses produce changes in the shape of the campaniform sensilla, resulting in their stimulation. They thus act as load sensors, and can respond to externally applied forces (loads) or to forces that result from the animal's own muscle activity. Sensilla with different orientations are sensitive to bending in different directions, and individual sensilla within a group can have characteristic thresholds and response properties, so the group as a whole could provide detailed information about the magnitude and direction of any force.

Campaniform sensilla occur singly or in groups and many are oriented at right angles to the principal lines of stress (see Fig. 8.7a). For example, campaniform sensilla on the leg are so oriented that they are stimulated when the foot is on the ground and the leg bears the insect's weight. Those on the wings are aligned along veins, and those on the antennae are often arranged in rings around various segments, particularly the pedicel.

Campaniform sensilla on the tibiae are particularly sensitive to body load, and their very rapid responses precede any major changes in body position. Increased firing in proximal tibial campaniform sensilla during increased loading appears to contribute to increased firing of the trochanteral depressor muscle, which would act to compensate against the increase in body load. This is thus a positive feedback effect, although in other cases activation of campaniform sensilla can lead to negative feedback effects. Trochanteral and femoral campaniform sensilla affect the patterns of rhythmic motor output generated by central pattern-generating networks that underlie locomotion – for example by resetting the rhythm, or influencing the timing of the transitions between stance and swing phases of stepping. Load signals are integrated centrally with other sensory signals (e.g., joint position), and may be gated so that their effects are more or less prominent at different phases of an ongoing motor rhythm.

Campaniform sensilla on the wing veins of a fly, *Calliphora vomitoria*, respond similarly to both up and down movements of the trailing edge of the wing. They probably generate only one action potential per wingbeat, and are likely to be involved in the control of flight stability (see Sections 9.2.2 and 9.7). Campaniform sensilla on the haltares of a fly provide very rapid sensory signals to flight motor neurons through both monosynaptic electrical synapses and mono- or polysynaptic chemical pathways. They thus have a powerful influence on wing stroke kinematics during flight maneuvers.

Campaniform sensilla on the antennae are likely to be involved in the control of flight steering and maintenance, and in both passive and active tactile sensing. The antennal flagellum of the honey bee *Apis* contains many more campaniform sensilla than reported for other species, particularly at the tip. It is suggested that this may be related to bees' complex use of their antennae in comb building.

23.2 Chordotonal organs

23.2.1 Structure

Chordotonal, or scolopophorous, organs are subcuticular receptors that act most commonly as joint proprioceptors or as hearing organs. They consist of single units or, more commonly, groups of similar units, called scolopidia, which are attached to the cuticle at one or both ends. In many insects, the scolopidia are clustered into distinct groups, called scoloparia, that are identifiable morphologically and functionally (Fig. 23.7). Each scolopidium consists of three cells: a sensory neuron, an enveloping or scolopale cell and an attachment (cap) cell (Fig. 23.8). The dendrite narrows distally to a ciliary process containing a peripheral ring of nine double filaments

and with roots extending proximally within the dendrite. In some scolopidia, and perhaps in all, the doublets are connected with the cell membrane near their origins at the basal body by a structure known as the ciliary necklace due to its appearance. The distal part of the dendrite and the ciliary extension are surrounded by the scolopale cell, which produces a barrel-shaped sleeve of scolopale rods containing actin. These rods in turn insert distally into a dense extracellular scolopale cap that is produced by the attachment cell. In some scolopidia the tip of the cilium lies in a hollow of the cap. In others, the dendrite dilates distal to the cilium. The ciliary region of the dendrite lies in an extracellular space within the folds of the scolopale cell. Some scolopidia contain more than one neuron (with a maximum of five), in which case the additional dendrites all insert into the single scolopale cell. The attachment cell connects the sensillum to the epidermis in non-connective chordotonal organs, or to a connective tissue strand in connective chordotonal organs.

23.2.2 Functioning

The coupling mechanism by which external movement is transmitted to the dendrite probably differs in different chordotonal organs, but where the organ bridges the junction between two cuticular plates, movements of one relative to the other distort the scolopidium. Where the scolopidium is only connected with the cuticle at one end – as in tympanal organs – other factors are responsible for the distortion (see below). Whatever the origin of the distortion, the effect is perhaps to displace the scolopale cap, producing sliding of the tubule doublets which leads to bending at the base of the cilium. The actin of the scolopale may be involved in restoring the position of the cap and dendrite after bending. There are numerous hypotheses concerning putative mechanisms by which forces applied to the attachment cells might be transmitted to the dendritic membrane – many invoking active movements of different components of the scolopale or cilia – but

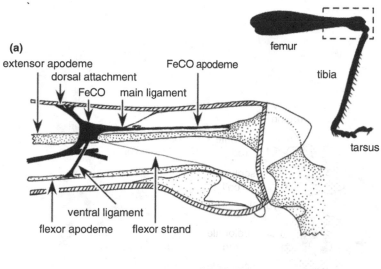

Figure 23.7 The femoro-tibial chordotonal organ (FeCO) of the locust hindleg is located distally in the femur ((a), inset), where it is anchored to the dorsal femoral cuticle. (a) A main ligament with fine supporting strands connects the body of the organ to the cuticular FeCO apodeme, which is an extension of the dorsal tibia. A ventral ligament and the slender flexor strand connect the organ to the flexor muscle apodeme (after Matheson and Field, 1995). (b) The bipolar sensory neurons form a proximal scoloparium consisting of generally larger cells and a distal scoloparium of smaller cells. All have their scolopidia oriented toward the main ligament, which is formed by all of the attachment cells of the organ. Flexion movements of the tibia pull on the main ligament and relax the flexor strand and ventral ligament, while the converse is true of tibial extensions (after Matheson and Field, 1990). (c) During tibial extension it is apparent that the attachment cells form a graded series of connections with the stiff apodeme, so some become slack before others. The unloading strands, attached distally to the hypodermis, take up tension during tibial extension to prevent the whole FeCO structure being pulled proximally by its intrinsic elasticity (after Shelton *et al.*, 1992).

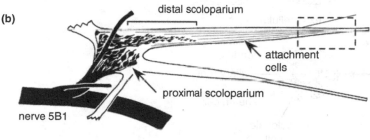

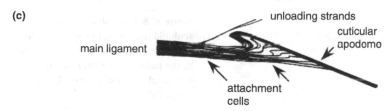

there are few direct and conclusive experimental findings. *Drosophila* mutants that have disrupted cilia in the scolopidia of Johnston's organ have markedly reduced auditory responses. Mutations in *nompA* are associated with disconnection of the cap from the cilia and lead to mechanosensory dysfunction, as do mutations that cause defects in the axoneme.

Transduction is probably dependent on mechanically activated ion channels (MACs), although this has not been demonstrated. Cultured chordotonal neurons contain MACs but it is not known whether these occur in the regions of membrane responsible for mechanotransduction or whether they have some other function.

Intracellular recordings made from different parts of the receptor neurons and attachment cells of auditory chordotonal sensilla demonstrate that the first electrical responses following transduction are a train of subthreshold depolarizations of the dendritic membrane. These events resemble quantum bumps seen in photoreceptors, occur randomly in the absence of stimulation and sum to form a noisy receptor potential. The receptor potential elicits small

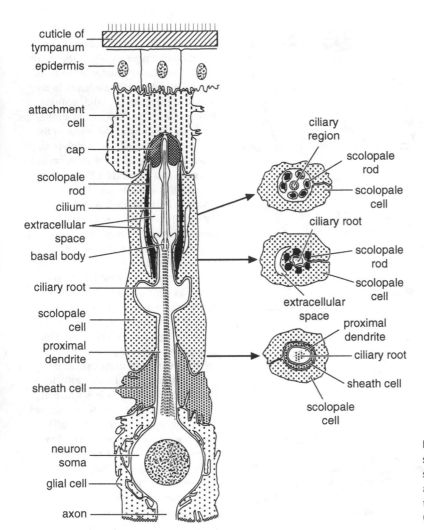

Figure 23.8 Basic structure of a scolopidium. Diagrams on the right show the arrangement in cross-section at the points indicated. From the tympanal organ of *Locusta* (after Gray, 1960).

amplitude, passively propagating spikes that originate in the distal dendritic membrane. These propagate electrotonically to the basal part of the inner dendritic segment, where they give rise to conventional "basal dendritic spikes" that in turn depolarize the spike-initiating zone on the axonal membrane to produce axonal spikes (action potentials) that propagate toward the central nervous system. Both the dendritic unitary potentials and the later processes leading to spike generation show time-dependent processes such as adaptation which govern the phasic or tonic nature of the final response to a stimulus.

23.2.3 Distribution and functions in the living insect

Chordotonal organs occur throughout an insect's body, with functions including joint proprioception, substrate vibration detection, hearing and, in the antennae, wind and gravity sensation. In larval *Drosophila* there are 90 such organs, each containing between one and five scolopidia, arranged in a segmental pattern and suspended between points on the body wall. *Rhodnius* (Hemiptera) has a single scolopidium on either side of the midline in four abdominal segments. The scolopale and associated dendrite are attached to the basal lamina of the epidermis. They record pressure from within the abdomen such as that resulting from feeding. An unusual situation is found in the genital chamber of the cricket *Teleogryllus commodus*, where a number of individual scolopidia insert directly into the cuticle rather than forming a distinct organ. They are stimulated when the wall is distended by the arrival of an oocyte in the genital chamber. Their activation leads to muscular changes that delay the backwards movement of the oocyte and move sperm down the spermathecal duct so that fertilization is effected.

In the thorax of many insects there are large chordotonal organs containing about 20 scolopidia that record the movements of the head on the thorax. Others in the wing bases of some insects record some of the forces that the wings exert on the body. In *Apis* there are three such organs, each with 15–30 scolopidia, at the base of the radial and subcostal veins and in the lumen of the radial vein. There is a complex system of chordotonal organs in the ventral thorax that is associated with the coxal leg joint and thoracic sternites. Praying mantids have a single "cyclopean ear" on the ventral midline that is sensitive to bat echolocation calls, and some parasitic flies have prosternal tympanal organs innervated by 140 scolopidia, which are sensitive to their orthopteran host's stridulations. Some lepidopterans have chordotonal organs associated with thoracic tympanal membranes that act as ears. A chordotonal organ associated with the tegula on the wing base of locusts is involved with regulation of wingbeat, and a chordotonal organ in the dipteran haltere might similarly contribute to the control of flight, but this has not been studied.

Four or more chordotonal organs are usually present in each leg (see Fig. 8.7a). The first is attached within the femur and is inserted distally into the knee joint (Section 23.2.4). In *Machilis* (Archaeognatha) this organ contains seven scolopidia; in grasshoppers there are about 300. The second, proximally in the tibia, is the subgenual organ (Section 23.2.5) and the third, which in *Apis* contains about 60 scolopidia,

arises in the connective tissue of the tibia and is inserted into the tibio-tarsal articulation. Finally, a small organ with only about three scolopidia extends from the tarsus to the pretarsus. In bushcrickets (Tettigoniidae) the subgenual is the most proximal of three chordotonal organs in the tibia (see Fig. 23.12b). The most distal is associated with the tympanal membranes in the forelegs (Section 23.2.7), but it is also present, although containing fewer scolopidia, in the middle and hindlegs, which have no tympanal organs.

Another chordotonal organ that occurs almost universally in insects is Johnston's organ, which lies in the antennal pedicel (Section 23.2.6). Some insects have tympanal organs (Section 23.2.7).

23.2.4 Femoral chordotonal organs

Insects of all orders appear to possess a femoro-tibial chordotonal organ (FeCO), which may be composed of more than one scoloparium. These have been particularly well studied. In Orthoptera, the organ is anchored proximally to the femoral hypodermis, and is attached by a ligament and apodeme to the head of the tibia near the insertion of the extensor tibiae muscle. There may be other attachments to the tibia or muscle apodemes that act to maintain the shape of the organ or bias its responses to pull on the main ligament (see Fig. 23.7). Simpler arrangements are found in Diptera, Hemiptera and Lepidoptera, where the attachment cells may insert directly onto the extensor muscle apodeme or muscle fibers. In the forelegs and middle legs of Orthoptera the organ is found proximally, with a long apodeme stretching the length of the femur, whereas in the hindleg of the locust it is located very distally, with a much shorter apodeme. The ligament, composed of attachment cells, is a complex visco-elastic structure that provides a mechanical basis for sequential transmission of force from movements of the tibia to different scolopidia in the organ (Fig. 23.7c). This contributes to a range-fractionation of the sensory responses.

The FeCO functions in leg postural reflexes involving muscles of the femoro-tibial joint as well as other joints of the same leg. Flexion of the tibia pulls on the main apodeme and ligament, stretching the FeCO and activating some of the sensory neurons. Tibial extension conversely relaxes the organ and activates a different subset of the receptors. Responses are phasic, phaso-tonic or, in some cases, purely tonic with no response during imposed movements. These responses signal acceleration, velocity or position of the tibia relative to the femur. Yet other sensory neurons respond to imposed vibrations.

In quiescent Orthoptera, stretch and relaxation of the main ligament leads to resistance reflexes in the tibial extensor and flexor muscles that contribute to postural stability. The stick insect FeCO underpins the behavior of catalepsy (death feigning). The strength of the reflexes is dependent on joint position, and varies with the internal state of the animal. In active animals, the sign of the reflexes can change so that they contribute to, rather than hinder, active movements. Surgical shortening of the FeCO apodeme leads to significant and predictable errors in aimed leg movements, which gradually diminish over a week following the manipulation, suggesting that mechanical or neuronal compensatory mechanisms come into play.

Stimulation of the FeCO generates reflexes in many other muscles of the same leg, including those at the tibio-tarsal and coxo-trochanteral joints. It also influences coordinated movement of adjacent ipsilateral and contralateral legs in stick insects.

23.2.5 Subgenual organs

The subgenual organ is a chordotonal organ in the proximal part of the tibia, usually containing 10–40 scolopidia, but up to 400 in some parasitoid wasps. It is not associated with a joint. Processes from the attachment cells at the distal ends of the scolopidia are packed together as an attachment body, which is fixed to the cuticle at one point, while the proximal

ends of the scolopidia are supported by a trachea (Fig. 23.9). The organ is often in two parts, one more proximal, sometimes called the true subgenual organ, and the other slightly more distal. Both are present in Odonata, Blattodea and Orthoptera. In Homoptera, Heteroptera, Neuroptera and Lepidoptera only the distal organ is present. Among Hymenoptera, the scolopidia enclose an extracellular space filled with acid mucopolysaccharide, giving the structure an irregular spherical shape. In Ensifera (Orthoptera) the subgenual organ is accompanied by an intermediate organ and the crista acustica to form the so-called complex tibial organ. Subgenual organs have not yet been reported in Coleoptera or Diptera.

Subgenual organs respond to vibrations of the substratum. They are extremely sensitive and, in the cockroach *Periplaneta*, for instance, respond to a displacement of only 0.2 nm at a frequency of 1.5 kHz. In this species vibrations up to 8 kHz are detected. At low frequencies – up to about 50 Hz – the neural response is synchronous with the stimulus, but at higher frequencies it is asynchronous. These sensilla also respond to air-borne sound of high intensities.

In the honey bee it appears that vibrations along the axis of the leg accelerate the hemolymph, which in turn exerts force on the subgenual organ.

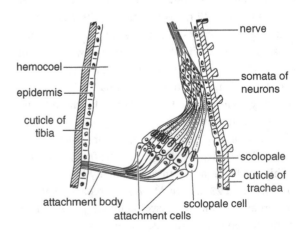

Figure 23.9 Subgenual organ of an ant (from Horridge, 1965).

It consequently oscillates backwards and forwards along the tibia, but with a lag due to inertia of the hemolymph and the organ itself. Because the organ's movements lag behind those of the stiff tibial cuticle, forces are exerted on the flexible attachment points, and these forces could lead to excitation of the scolopidia. The subgenual organ can thus be thought of as acting like a leaky piston (or diaphragm) in a fluid-filled tube. The compressibility of large air sacs associated with all subgenual organs is likely to be important in permitting movements of the relatively incompressible hemolymph within the stiff tube formed by the tibia.

In most insects, the detection of substrate-borne vibrations is perhaps concerned primarily with predator avoidance, but in green lacewings, *Chrysoperla* (Neuroptera) and many Auchenorrhyncha (Hemiptera) which communicate via the substrate, the subgenual organs are probably also important in reception of intraspecific signals (Section 26.2.1). Parasitoid wasps (Ichneumonidae) appear to use vibrations to detect their wood-boring prey hidden within trees. Some have highly modified legs and subgenual organs that might contribute to this behavior, but direct evidence is still lacking.

23.2.6 Johnston's organ

Johnston's organ is a chordotonal organ in the antennal pedicel with its distal insertion in the articulation between the pedicel and flagellum. It occurs in all adult insects and, in a simplified form, in many larvae. It consists of a single mass or several groups of scolopidia which respond to movements of the flagellum with respect to the pedicel. It is highly elaborated in some Diptera, where there may be 20 000 sensory neurons accommodated in a large swelling of the pedicel.

In the blowfly *Calliphora*, most of the sensilla comprising the organ give phasic responses, so that a single to-and-fro movement of the flagellum produces an "on" and an "off" response, the pattern of which changes with stimulus intensity. Some of the sensilla respond to movement in any direction, others only if they are moved in a particular direction.

Movement of the flagellum relative to the pedicel may have a number of causes, so Johnston's organ can serve a variety of functions in any one insect. In grasshoppers, bees, Lepidoptera, some flies and probably other insects, it acts as an air-speed indicator. In *Calliphora*, for example, wind blowing on the face causes the arista to act as a lever, rotating the third antennal segment on the second. Even in a steady airflow the antenna trembles, and this is sufficient to stimulate some sensory neurons of Johnston's organ even though they are stimulated primarily by changes in the degree of rotation of the third antennal segment, rather than by a steady deflection. With different angles of rotation, more or different scolopidia are stimulated, so Johnston's organ can give a measure of the degree of static deflection of the third antennal segment as well as changes in its position.

In some insects, Johnston's organ functions as a particle movement detector, perceiving near-field sound. This occurs during courtship of *Drosophila melanogaster*, when the female responds to the male's wing vibration, and during orientation dances of honey bees when workers respond to the sounds produced by an incoming dancing worker (Section 26.2.1). In *Drosophila* the sensitivity of Johnston's organ to specific frequencies of sound vibration is enhanced by active cellular processes analogous to the active contractile properties of vertebrate auditory hair cells. This amplification of sensitivity results from active vibrations of the antennal flagellum, possibly generated by molecular motors underlying sensory transduction in the membranes of individual mechanoreceptor neurons, or by active contractions of the cilia or other components of the scolopidium.

Males of some Diptera detect females by their flight tone using Johnston's organ. In males of Culicidae (mosquitoes) and Chironomidae (midges), the pedicel is enlarged to house the organ. In

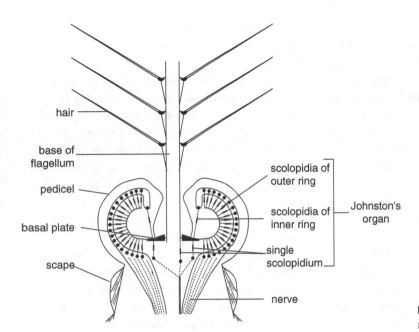

Figure 23.10 Johnston's organ of a male mosquito (from Autrum, 1963).

Culicidae the base of the antennal flagellum forms a plate from which processes extend for the insertion of the scolopidia (Fig. 23.10). The latter are arranged in two rings around the axis of the antenna and, in addition, there are three single scolopidia which extend from the scape to the flagellum. The males of these insects have plumose antennae with many fine, long hairs arising from each annular joint. These hairs are caused to vibrate by sound waves and their combined action produces a movement of the flagellum. The amplitude of flagellar movement is greatest near its own natural frequency, which approximately corresponds to the flight tone of the mature female. Stimulation at this frequency leads to the seizing and clasping response in mating. Males of the mosquito *Aedes aegypti* are most readily induced to mate by sound frequencies between about ~400 HZ and ~650 Hz, but the limits to which they respond become wider as the insect gets older, and are wider in unmated than in mated males.

Gyrinid beetles are able to perceive ripples on the water surface, apparently due to displacements of the antennal flagellum, so that they are able to avoid collisions with other insects and, by "echolocating"

using their own ripples, also avoid the sides of their container. In order to do this, the insect must also be able to detect the direction from which the ripples arrive.

Johnston's organ also contributes input relevant to an insect's orientation with respect to gravity, and, in the water boatman, *Notonecta* (Hemiptera), it is concerned with orientation in the water. In this species an air bubble extends between the head and the antennae so that when the insect is correctly oriented on its back the antennae are deflected away from the head. If, however, the insect is the wrong way up the antennae are drawn toward the head and Johnston's organ registers the displacement.

The multifunctional nature of Johnston's organ has been shown in *Drosophila* to be underpinned by five distinct subgroups of sensory neurons. The groups have different patterns of gene expression and have different central projections within the brain. Sound-sensitive neurons respond preferentially to antennal vibrations and express the putative mechanotransducer channel NompC. Other neurons respond preferentially to static displacements, do not express NompC and contribute to gravity and wind sensation. Those that respond to the smallest static deflections might underlie gravity responses, while others that respond to large-amplitude deflections are wind-speed indicators. Subsets of the wind-sensitive sensory neurons respond to different directions of wind flow and, because they project to different regions of neuropile in the brain, form a simple map of wind direction.

23.2.7 Tympanal organs

Structure and occurrence of tympanal organs Tympanal organs are chordotonal organs specialized for sound reception. Each consists of an area of thin cuticle, the tympanum (or tympanic membrane), which is generally backed by an air sac so that it is free to vibrate. Attached to the inside of the tympanum or adjacent to it is a chordotonal organ that contains from one scolopidium, in *Plea* (Hemiptera), to over 2000 in some Cicadidae. Tympanal organs occur on the neck membrane of some scarab beetles, on the prothoracic legs of Grylloidea and Tettigonioidea, on the prothorax of at least one parasitic fly, on the mesothorax of some aquatic Hemiptera, such as *Corixa* (Hemiptera) and *Plea*, on the metathorax in Noctuoidea and Mantodea and on the abdomen in Acrididae, Cicadidae, Pyralidoidea and Geometroidea (Lepidoptera) and Cicindellidae (Coleoptera). In *Chrysopa* (Neuroptera) the tympanum is on the ventral side of the radial vein of the forewing. Hearing in the primitively atympanate bladder grasshopper, *Bullacris membracioides* (Orthoptera), is mediated by six pairs of serially repeated abdominal ears derived from pleural chordotonal organs. The posterior organs, containing as few as 11 scolopidia, are simpler than the anterior ones which contain up to 2000, providing a revealing glimpse into the transitions in form and function leading from proprioceptive to auditory organs. Despite the lack of a tympanum, the organs are highly responsive to

behaviorally relevant air-borne sounds, and elicit appropriate reactions. The anterior organs are more sensitive than the posterior ones, which may be due to their overall shorter length.

Acrididae (grasshoppers) have an ovoid tympanum in a recess on either side of the first abdominal segment (Fig. 23.11). It is about 2.5×1.5 mm in size in adult *Schistocerca*. Most of its cuticle is weakly sclerotized mesocuticle, but small islands of well-sclerotized cuticle serve for the attachment of different groups of scolopidia. The cuticle of the air sac which backs the tympanum is only 0.2 µm thick and perhaps consists only of epicuticle. The chordotonal organ, known as Muller's organ, is complex, containing, in *Schistocerca*, about 80 neurons with their cell bodies aggregated into an "end organ" sometimes referred to as a "ganglion." The dendrites of the sensory units connect with the tympanum in four separate groups with different orientations (shown by arrows in Fig. 23.11b). The whole chordotonal organ and the auditory nerve which runs from it to the metathoracic ganglion are enclosed in folds of the air sac that backs the tympanum. Air sacs are contiguous right across the body between the two tympani.

The tympanal organs of Grylloidea (crickets) and Tettigonioidea (bushcrickets) are similar to each other in basic structure, being situated in the base of the fore tibia, which is slightly dilated and typically has a tympanum on either side, anterior and posterior, with the leg at right angles to the body (Fig. 23.12a). Often the posterior tympanum is bigger than the anterior one and sometimes, as in *Gryllotalpa* (Orthoptera), only the former is present.

In both crickets and bushcrickets (Orthoptera: Ensifera), the cavity of the leg between the two tympani is almost entirely occupied by a trachea divided into two by a rigid membrane; the blood space of the leg being restricted to canals anteriorly and posteriorly (Fig. 23.12b). In bushcrickets (Tettigoniidae), the trachea runs proximally through the femur without branching and widens into a vestibule opening at the prothoracic spiracle. Its

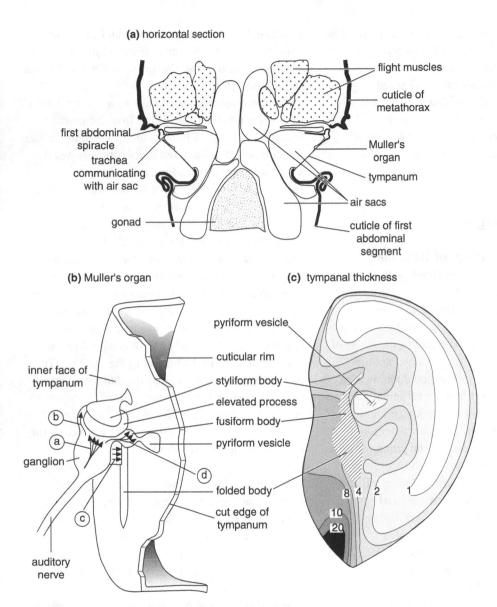

(a) horizontal section

flight muscles

cuticle of metathorax

first abdominal spiracle

trachea communicating with air sac

gonad

Muller's organ

tympanum

air sacs

cuticle of first abdominal segment

(b) Muller's organ

(c) tympanal thickness

pyriform vesicle

cuticular rim

inner face of tympanum

styliform body

elevated process

fusiform body

pyriform vesicle

ganglion

folded body

cut edge of tympanum

auditory nerve

Figure 23.11 Tympanal organ of a grasshopper. (a) Diagrammatic horizontal section through the metathorax and first abdominal segments showing the positions of the tympanal membranes. Notice that the air sacs form a continuum across the body (*Oedipoda*) (from Schwabe, 1906). (b) Diagram showing the attachment of the scolopidia to the inside of the tympanum. The cuticle of the air sac which normally covers Muller's organ and the tympanum has been removed. The orientations of scolopidia in different parts of the chordotonal organ are indicated by arrows. Letters show the positions of cells, some of whose sensitivities are shown in Fig. 23.16a. The styliform body, elevated process and folded body are cuticular structures continuous with the tympanum (*Locusta*) (after Gray, 1960). (c) Thickness of the tympanum. Lines mark contours of thickness and numbers show the thickness in microns. Hatched areas are regions of well-sclerotized cuticle (*Schistocerca*) (after Stephen and Bennet-Clark, 1982).

(a) transverse section of tibia

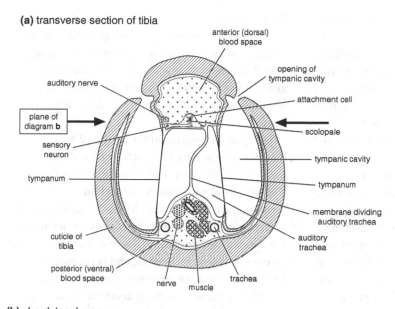

anterior (dorsal) blood space

auditory nerve

opening of tympanic cavity

attachment cell

plane of diagram **b**

scolopale

sensory neuron

tympanic cavity

tympanum

tympanum

membrane dividing auditory trachea

cuticle of tibia

auditory trachea

posterior (ventral) blood space

trachea

nerve   muscle

(b) chordotonal organs

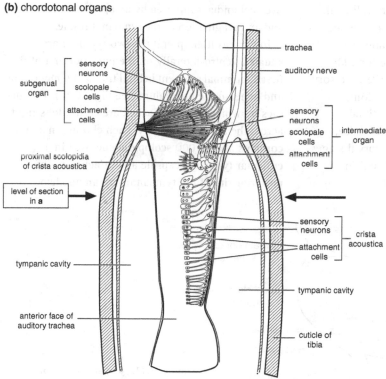

trachea

sensory neurons

auditory nerve

subgenual organ

scolopale cells

attachment cells

sensory neurons

scolopale cells   intermediate organ

proximal scolopidia of crista acoustica

attachment cells

level of section in **a**

sensory neurons   crista acoustica

tympanic cavity

attachment cells

tympanic cavity

anterior face of auditory trachea

cuticle of tibia

Figure 23.12 Tibial chordotonal complex of a bushcricket. (a) Transverse section through the tibia at the level shown in (b) (*Decticus*) (from Schwabe, 1906). (b) The complex of chordotonal organs exposed when the dorsal (anterior) cuticle of the tibia is removed (*Decticus*) (from Schwabe, 1906).

cross-sectional area changes from tibia to spiracle in a way that approximates the shape of an exponential horn so that sound entering the spiracle is amplified. This trachea is isolated from other tracheae and the spiracle has no closing mechanism. In crickets the trachea does not expand in this way, but the tracheae of the two sides join across the midline (see Fig. 23.14), although the air spaces are separated by a delicate septum.

The chordotonal organ associated with the tympanal organs is in the anterior hemocoelic space of the tibia (Fig. 23.12b). It is called the crista acoustica, and, in *Decticus*, it contains 33 scolopidia which lie parallel with each other in a vertical row, the sensilla becoming progressively smaller toward the distal end (see Fig. 23.16). In *Teleogryllus* they are within a tent-like membrane which extends along the tympanal organ between the wall of the trachea and the dorsal wall of the tibia. The neuron somata lie on the wall of the trachea and the dendrites project into the cavity of the tent. More proximal scolopidia are attached to large accessory cells within the tent, but, more distally, long attachment cells extend directly to the tibial epidermis. In these insects, as in the Acrididae, tympanal organs are present in all the larval stages, but they probably only become functional in the later larvae and adults. The complex tibial organ of

the atympanate ensiferan *Ametrus tibialis* consists of a subgenual organ, intermediate organ and a crista acustica homolog with 23 sensory neurons. The tibial organ as a whole responds to vibrations and only very poorly to sound, but the specific responsiveness or function of the crista acustica homolog is unknown.

The tympanal organs of the Noctuoidea (Lepidoptera) occupy the posterior part of the metathorax (Fig. 23.13) and the tympanum faces into a space between the thorax and abdomen roofed over by the alula of the hindwing. Medial to the tympanum, and resembling it structurally, is a second membrane, but it has no sense organ. This second membrane is the counter tympanic membrane, and is probably an accessory resonating structure. The sense organ attached to the back of the tympanum contains only two scolopidia, supported by an apodemal ligament and an invagination of the tympanal frame.

In cicadas (Hemiptera) the two tympani are situated ventro-laterally on the posterior end of the first abdominal segment behind the folded membrane and beneath the operculum (see Fig. 26.10). The air sacs by which they are backed are continuous right across the abdominal cavity. Each chordotonal organ contains about 1000 scolopidia enclosed in a cuticular tympanic capsule and attached to the posterior rim of the tympanum by an apodeme.

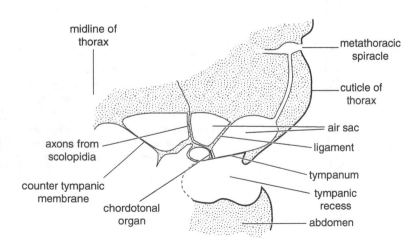

Figure labels: midline of thorax; metathoracic spiracle; cuticle of thorax; air sac; ligament; axons from scolopidia; tympanum; counter tympanic membrane; tympanic recess; chordotonal organ; abdomen

Figure 23.13 Tympanal organ of a noctuid moth. Diagrammatic horizontal section through the right-hand side of the metathorax. The tympanum faces toward the abdomen (after Roeder and Treat, 1957).

In the lacewing *Chrysopa*, the tympanum is on the ventral side of the radial vein of the forewing. This vein has a bulbous swelling near the base that is thick-walled and rigid above, but has a thin tympanum below. Unlike most other tympanal organs, the tympanum is in direct contact with the hemolymph; it is not backed by an air sac. There are about 28 scolopidia associated with the tympanum, but it is possible that as few as six of them forming a distal group are acoustically responsive.

Functioning of the tympanal organs Sound impinging on the tympanic membrane causes it to vibrate, the amplitude of the vibrations varying with the intensity of sound and the structure of the membrane. In *Locusta*, different parts of the membrane vibrate to some extent independently of each other.

Tympanal organs may be pressure receivers, in which sound impinges only on one side of the tympanum, or pressure-gradient (pressure-difference) receivers, in which sound reaches both sides of the membrane. In the latter, vibration of the tympanum depends on differences in sound pressure on its two sides resulting from differences in phase of the vibrations. The sound reaching the inner face of the tympanum is not the same as that reaching the outside because some frequencies may be differentially absorbed in transit through the system, which acts as a filter, or because the intensities of some frequencies are amplified more than others.

Sound can reach the inner side of a tympanum via the tracheal system or through the tissues, but, because the intensity and phase of the sound reaching the inside of the tympanum depends on frequency, each tympanum can function either as a pressure receiver or as a pressure-difference receiver. The change from one to the other occurs at about 10 kHz in locusts. At frequencies below this, sound reaches the tympanum from both sides and it functions as a pressure-difference receiver. At frequencies above 10 kHz, most sound reaches the

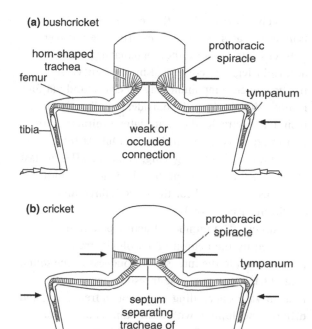

Figure 23.14 Tympanal organs functioning as pressure-difference receivers. Diagrammatic cross-section through the prothorax and the prothoracic legs. Arrows indicate how sound arriving at different points of the body may reach the tympanal organ of the right-hand leg. The relative magnitudes of these effects vary with the sound frequency (see text). (a) In a bushcricket, the tympanal organs of the two sides operate in isolation from each other. The trachea from the prothoracic spiracle forms a horn in which some frequencies are amplified. Sound reaches the tympanum directly from the outside and via the trachea of the same side. (b) In a cricket the two sides are part of a continuous system through the transverse trachea. Sound reaches the tympanum directly, and indirectly through the spiracles of both sides, and possibly via the contralateral tympanum.

tympanum from the outside because sound entering the air sacs is filtered out and so the tympanum functions as a pressure receiver. In bushcrickets, high-frequency sound (above 20 kHz) entering the prothoracic spiracle (Fig. 23.14a) is amplified in the tracheal horn so that it dominates the vibrations induced in the tympanic membrane, which acts as a pressure receiver. At low frequencies, however, relatively little amplification occurs and the tympanum acts as a pressure-difference receiver.

In crickets, because the tympanal organs of left and right legs are connected by the transverse trachea, each tympanum may receive sound via four routes: directly from the outside, or indirectly via the ipsilateral spiracle, the contralateral spiracle or the contralateral tympanum (Fig. 23.14b). At the frequency of the calling song, about 4.7 kHz, the last is relatively unimportant, but the input via both spiracles is essential for the insect's directional sensitivity (see below).

Loudness. The number of action potentials produced by the neuron of a scolopidium is proportional to the intensity or loudness of the sound (Fig. 23.15). The response threshold for a single neuron varies according to the sound frequency, and different scolopidia within the same chordotonal organ may have different thresholds. Between threshold and the point of saturation is the dynamic range in which the magnitude of the response is a measure of sound intensity. The dynamic range of a neuron is often around 30 dB, but by possessing cells with different thresholds the dynamic range for the

insect can be greatly extended. This is clearly evident in noctuid tympanal organs in which there are only two scolopidia. These have similar frequency responses but their thresholds differ by about 20 dB (Fig. 23.16c). The more sensitive cell enables the insect to hear bat sounds from greater distances, while stimulation of the less sensitive cell provides immediate information on the close proximity of the bat. Scolopidia of auditory organs may be stimulated by the insect's own normal activities. In *Locusta* abdominal ventilation and flight may both modulate responses to sound, in some cases completely inhibiting the response.

Frequency. The tympanal organs of different groups of insects respond to different frequency ranges. Acrididae respond to sounds with frequencies from 100 Hz to 50 kHz, tettigoniids from 1 kHz to 100 kHz, gryllids from 200 Hz to 15 kHz and cicadas from 100 Hz to 15 kHz. Within these ranges, sensitivity is generally greatest over a more limited range, corresponding with the sound frequencies to which the insects are adapted, commonly the sound produced by conspecifics. However, in many insects the tympanal organs are sensitive to much higher frequencies and respond to the sounds emitted by bats which, during their search phase, often cover a broad band of frequencies from about 25 kHz to over 100 kHz. Noctuids, for example, respond to frequencies from 1 kHz to 140 kHz, with maximal sensitivity in the range 20–40 kHz (Fig. 23.16c).

Frequency discrimination by tympanal organs results from differences in the mechanical parts of the system that convey vibrations to the sensory cells. Different parts of the tympanum of grasshoppers tend to vibrate maximally at different frequencies. At the point of attachment of the (a) cells in Fig. 23.11b, the membrane oscillates maximally when stimulated by sound in the range 5–10 kHz. Sound in this range is a major component of the sounds of many grasshoppers. At the point of attachment of the (d) cells in Fig. 23.11b, the membrane exhibits a maximum response in the same

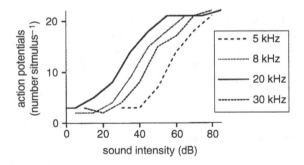

Figure 23.15 Sensitivity. Responses of one scolopidium in the crista acoustica of a bushcricket to sound pulses of similar duration at different frequencies. At higher intensities the number of action potentials produced increases until the response becomes saturated. The dynamic range extends over about 50 dB at all frequencies, but the scolopidium is most sensitive (lowest sound intensity to produce a response) at 20 kHz (after Oldfield, 1982).

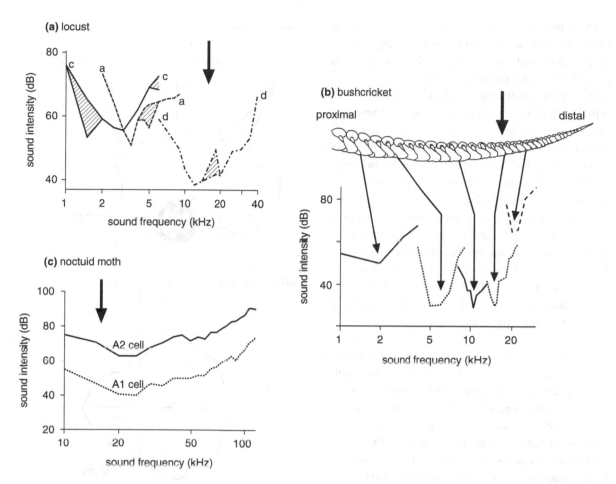

Figure 23.16 Frequency thresholds of individual scolopidia in tympanal organs. Note that the *lowest* threshold equates with the *highest* sensitivity. Bold arrows at the top indicate the approximate upper limit of human hearing (about 15 kHz). (a) A locust. Letters, a, c and d refer to the groups of scolopidia with the same lettering in Fig. 23.11b. Hatched areas show the range of sensitivities of different cells in the same group (after Michelsen and Nocke, 1974). (b) A bushcricket. Above: the crista acoustica. Below: responses of scolopidia at different distances along the crista acoustica. The more distal the scolopidium, the higher the frequencies to which it responds. Notice that cells responding to the highest and lowest frequencies are very insensitive, responding only at high sound intensities (after Oldfield, 1982). (c) A noctuid moth. The two scolopidia in the chordotonal organ respond to a wide range of frequencies, but cell A1 is much more sensitive than cell A2 (after Waters and Jones, 1996).

range, but has a second maximum between 15 kHz and 20 kHz. Consequently, different scolopidia of Muller's organ tend to be stimulated by sounds of different frequencies (Fig. 23.16a) and sensory signals recorded in the tympanal nerve – representing the sum of membrane oscillation and the sensitivity of the sensory neurons – are maximal

at 5–10 kHz, with a second maximum between 25 kHz and 35 kHz.

In bushcrickets the intermediate organ may be responsible for the perception of low-frequency sound, while, in the crista acoustica, proximal scolopidia are most sensitive to low frequencies and more distal cells to higher frequencies (Fig. 23.16b).

Pulse rate. Sound communication among insects often depends on the production and reception of discrete pulses of sound (Chapter 26). The ability to distinguish sound pulses following each other in rapid succession depends on the characteristics of the tympanum as well as those of the sensory neurons. The chordotonal organs of the tympanal organs of cicadas and noctuids can separate sound pulses up to about 100 Hz, but above this the response becomes continuous. In bushcrickets each pulse of sound caused by the impact of one tooth of the file of a singing conspecific elicits highly synchronous firing of afferents from the crista acoustica in a receiver. This synchronized firing, also observed in the tympanal nerve of locusts, is maintained in the auditory neuropile of the prothoracic ganglion. When animals are exposed to artificial songs, the degree of synchrony in the population response increases as the temporal pattern of the sounds approaches the normal species-specific song pattern.

Directional sensitivity. Because insects are small they are unable to determine the direction from which a sound comes by using differences in arrival times on the two sides of the body. Instead, directional sensitivity results from the ear's properties as a pressure-difference receiver, or, at higher frequencies, as a result of sound diffraction.

The response of a pressure-difference receiver may be greatly affected by the direction from which the sound reaches the body (Fig. 23.17). As a result, even a single ear may be capable of determining direction, and this effect is reinforced by the integration within the central nervous system of information from ears on opposite sides of the body. In bushcrickets the tympanic chambers resonate at the frequency of the insect's calling song, and sound at this frequency is amplified if it enters the chambers directly through the outer openings (Fig. 23.17a), but other frequencies are not amplified. This gives the insects great directional sensitivity to sound of the appropriate wavelengths. In crickets the septum across the transverse trachea in the prothorax (Fig. 23.14b) produces a large phase shift in sound

(a) bushcricket

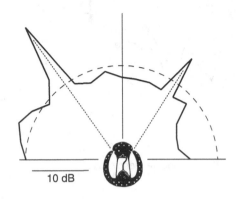

10 dB

(b) grasshopper

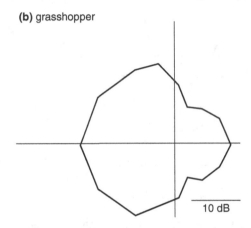

10 dB

Figure 23.17 Directional sensitivity of tympanal organs. Sensitivity is measured as decibels of sound pressure at the tympanum. The greater the distance from the point at which the vertical and horizontal axes cross, the greater the sensitivity. (a) Bushcricket, left foreleg. A cross-section through the proximal tibia containing the tympanal organ is shown at the center. The smooth arc shows the intensity of the signal (15 dB). Notice that when the sound comes from directly opposite the slit-like openings to the tympanic cavities (see Fig. 23.12) the sound is amplified. This is only true at sound frequencies close to 7.4 Hz, the principal frequency of the calling song (after Stephen and Bailey, 1982). (b) Grasshopper, left tympanum. The insect was exposed to sound of equal intensity from a series of positions all around its body. The tympanum was most strongly stimulated when the source was on the ipsilateral side (left of midline) (from Bailey, 1991).

reaching a tympanum from the spiracles. The effect is greatest at the frequency of the calling song. However, the difference in phase from the sound reaching the tympanum directly from outside varies with the direction of the sound source. As a result, the amplitude of vibration of the tympanum varies as the insect swings its leg forwards and back during walking. This enables the insect to determine the position of the sound source.

Sound diffraction is the reflection of sound from the body or nearby objects. Low-frequency sounds have wavelengths much longer than an insect's body and so are relatively undisturbed, but high-frequency sounds may be shadowed significantly. The result is that the tympani on the two sides of the body receive signals differing in intensity, providing the basis for directional sensitivity. Tettigoniids, which produce high-frequency songs (Section 26.1.2), and moths that respond to bat sounds depend on diffraction for their directional sensitivity. Noctuids are able to locate a source of high-frequency sound when they are flying and so they have to localize sounds in a vertical as well as a horizontal direction. The wings tend to screen the tympani, so there is a marked difference in the responses of the two organs to asymmetrical stimulation when the wings are raised, but when the wings are in the lower half of their beat there is little difference between the two sides. When the wings are down the insect is more sensitive to sounds coming from below it than to sounds from above. Thus, during flight, the auditory input from low-intensity sounds of high frequency will vary cyclically and in different ways according to the position of the sound source.

Functions of tympanal organs

Orthoptera and cicadas produce sounds for attracting mates and, in some species, for aggregation (Section 26.2.1). Recognition of these sounds is dependent primarily on sound patterns rather than on variation in frequency, although the hearing systems of different species respond maximally to the frequencies of the sounds emitted (see above). Directional sensitivity is obviously a critical element in these behaviors.

The accuracy with which the insects orient to sound direction is not very great. If the sound source is within about 10° of its body axis, the grasshopper *Chorthippus* makes turns of up to 60° (sometimes more) to either side. If the angle of incidence is greater, the insect turns toward the correct side, but the turning angle is anywhere from 60° to 120°, irrespective of the angle of incidence. This often results in the insect making a zigzag course toward the source because successive turns tend to put the sound source on the opposite side of the body.

Female crickets (*Gryllus bimaculatus*) walk toward singing males and in this case the steering is achieved on a sound-pulse-by-sound-pulse basis. Each pulse detected by the ears on the prothoracic legs elicits a tiny turning bias in the stepping pattern of the front legs, so the overall walking direction is governed by the sum of many small course corrections. The sensitivity of the response is modulated over a longer time scale of seconds by song pattern recognition centers in the brain so that the female orients best to conspecific sound but can continue to respond even if the signal is degraded by its passage through the noisy environment. The auditory signals underlying the turning movements are passed to the leg motor neurons by indirect neuronal pathways, perhaps even via the brain.

Sound is also important in predator avoidance, and insects in several groups have tympanal organs sensitive to high-frequency sounds. This sensitivity permits the avoidance of echolocating bats, which emit such high-frequency sounds. Noctuid moths can detect bats about 30 m away, and at such distances, where the intensity of stimulation is low, tend to turn away from the source of sound (see above). At high sound intensities, such as would occur when a bat was within about 5 m, a moth may close its wings and drop to the ground, or power dive, or follow an erratic, weaving course.

Most of the scolopidia in the tympanal organs of cicadas are maximally sensitive in the range 1–5 kHz, well below the major frequencies in the calling songs of many species. It is possible that they are important in predator recognition.

Some parasitic Diptera use their auditory sense to locate their singing hosts. This has been extensively studied in the case of *Ormia*, a parasitoid of crickets, and it also occurs in *Emblemasoma*, a species which parasitizes cicadas.

23.3 Stretch and tension receptors

Stretch and tension receptors differ from other insect mechanoreceptors in consisting of a multipolar neuron with free nerve endings, while all the others contain one or more bipolar neurons. They have a variety of forms. Sometimes they are an integral part of an oriented structure such as a muscle fiber or a strand of connective tissue, but in other cases they have no specific orientation or associated structures. The term "stretch receptor" is sometimes used rather loosely to include all multipolar receptors, but can be misleading as some respond to isometric tension rather than degree of stretch.

23.3.1 Unspecialized receptors

Multipolar cells without an associated specialized structure are often associated with epithelia. The cell body, ensheathed in glial cells, is in the hemocoel, but the basal lamina around the dendrites becomes fused with that of the epithelium. In the multipolar cells of the bursa copulatrix of *Pieris* (Lepidoptera), for example, each dendrite swells at intervals; the swellings, with their basal lamina, apparently contributing to the attachment. In this example the finer dendritic branches, free of glial covering, end in the basal lamina of the epithelial cell (Fig. 23.18), but in fly larvae, the naked dendrites of stretch receptors on the epidermis end within invaginations of the epidermal cells.

In soft-bodied dipterous larvae and perhaps in all insects with largely unsclerotized cuticle, multipolar cells are attached to the inside of the epidermis. The number of these neurons in cyclorrhaphan fly larvae is constant within a

segment, varying from 24 in the prothorax to 30 in each of the abdominal segments. They collectively form a subepidermal nerve net, although they do not anastomose to form a true net. These receptors monitor changes in body shape.

Insects with a hard external cuticle are usually considered to have no subepidermal nerve net, but there are numerous multipolar neurons in the abdomen of the house cricket, *Acheta*. Blood-sucking insects have small numbers of multipolar neurons beneath the abdominal epidermis. In adult tsetse fly, *Glossina* (Diptera), there are three pairs of such cells associated with the ventral body wall. They presumably monitor abdominal distension when the insect feeds. In locusts, five multipolar cells present beneath the arthrodial membrane at the femoro-tibial joints respond to movement and position of the tibia with respect to the femur, and another nine or more are present elsewhere in the leg, although their functions are unknown.

Some insects are known to have stretch receptors associated with the fore- and hindguts, and these perhaps are universally present, monitoring the movement of food along the gut, regulating meal size and probably fecal production. A pair of multipolar stretch receptors is also present on the bursa copulatrix of the cabbage butterfly, *Pieris*. These receptors monitor distension of the bursa, which is normally caused by the placement of the spermatophore. This distension causes the female to refuse further matings.

Multipolar stretch receptors found on or within the nerves of many insects respond to stretching of the nerve itself.

A very unusual and highly branched multipolar receptor is found associated with two chordotonal organs in each infrared organ of *Merimna atrata* (Coleoptera). The several hundred or so terminal dendrites form a "terminal dendritic mass" under a region of specialized cuticle on the ventro-lateral second and third abdominal sternites. The cell responds to changes in temperature and is thus considered to be an infrared detector – but the coupling and transduction mechanisms underlying this

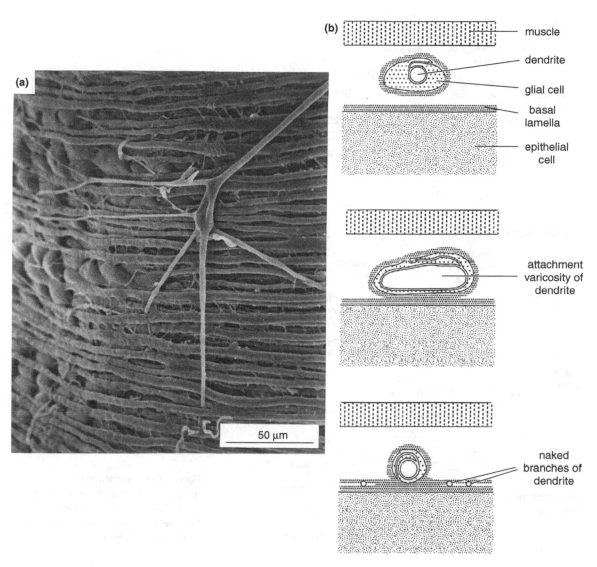

Figure 23.18 Unspecialized stretch receptor on the outside of the bursa copulatrix of *Pieris* (after Sugawara, 1981).
(a) The epithelium of the bursa is covered by a layer of muscles. The soma of the multipolar neuron is free in the hemolymph, although clothed by glial cells (not distinguishable in the photograph). Dendritic endings of the stretch receptor pass between the muscle strands and end on the surface of the epithelium. (b) Diagrammatic cross-sections of a dendrite beneath the muscle layer. Top: it approaches the epithelium clothed by a glial cell. Middle: it swells to form a varicosity. Its basal lamina is confluent with that of the epithelial cell beneath. Bottom: fine branches of the dendrite extend into the basal lamina of the epithelial cell. They are no longer covered by a glial cell.

(a) strand receptor

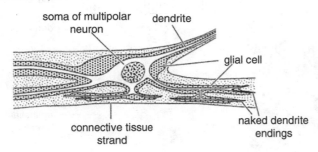

(b) muscle receptor

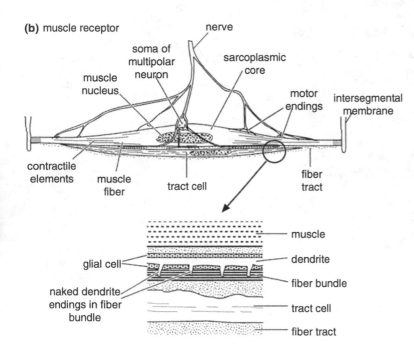

Figure 23.19 Specialized stretch receptors (based on Osborne, 1970). (a) Strand receptor; (b) muscle receptor. Inset below shows detail of dendrite endings in fiber bundle.

responsiveness are not known. There are putative serial homologs of the multipolar neurons in other abdominal segments, but these are relatively unspecialized and thought not to be infrared receptors.

23.3.2 Receptors with accessory structures

Where multipolar receptors are not immediately associated with an epithelium or a nerve, they are usually supported by a strand of connective tissue

or a modified muscle fiber, and are known as strand receptors or muscle receptor organs, respectively. A third type includes tension receptors that are associated with unspecialized muscle fibers.

A strand receptor consists of a strand of connective tissue, extending from one structure to another, with a multipolar neuron on the outside or embedded within it (Fig. 23.19a). The neuron is sheathed in glial cells except for the tips of the dendrite, which are embedded in the connective tissue. Some strand receptor neurons of orthopteroid insects – for example, one at the femur–tibia joint of the hindleg of *Schistocerca* – very unusually have their somata in the central nervous system.

Multipolar neurons of muscle receptor organs are associated with a modified muscle fiber which may be a component of a functional muscle, or may be a separate structure (Fig. 23.19b). Attached to the muscle is a tube of connective tissue known as the fiber tract, which is secreted by the tract cell. The neuron gives rise to 2–4 main dendrites, which run along the length of the fiber tract. From these main dendrites, side branches, which are not clothed in glial cells at the tips, are embedded in dense bundles of fibers. Muscle receptor organs have been described in Orthoptera, Neuroptera, Coleoptera, Trichoptera and Lepidoptera.

Multipolar neurons that act as muscle tension receptors are also associated with muscle fibers, but their dendrites generally insert close to the relatively unspecialized fibers' point of attachment to the cuticle. The location of the dendrites means that they are well placed to signal the tension exerted by the fibers against their attachment rather than degree of stretch. Such receptors are found in the flexor tibiae muscles of the prothoracic and mesothoracic legs, and at the base of the accessory flexor tibiae muscle of the metathoracic leg of *Schistocerca*. In all but one case, there is just one multipolar neuron associated with any given muscle; however, the dorsal ovipositor muscle of *Locusta* has around 200 multipolar and unipolar sensory neurons scattered among its muscle fibers. The responses of the individual receptors have not been recorded, but overall they signal muscle tension and act to inhibit the oviposition rhythm.

Occurrence of strand receptors, muscle receptor organs and tension receptors

Stretch receptors of these types are probably present in the abdomen of all insects. Dorsally, a pair extends from the tergum to the intersegmental membrane or from one intersegmental membrane to the next. Orthopteroid insects and beetles also have a vertical stretch receptor on each side of each segment, and some species also have ventral stretch receptors. Comparable dorsal receptors are present in the thorax of some insects.

Elsewhere in the body, strand or muscle receptors are associated with the mouthparts, legs and wings. The wing stretch receptors of locusts play a key role in the control of wing movements (Section 9.3.6).

23.3.3 Functioning of stretch receptors

Stretch receptors have none of the obvious structural features, either within the dendrites, such as the tubular body, or outside them, such as the scolopale, that occur in other mechanoreceptors. In all cases, however, a number of naked dendritic endings are embedded in connective tissue or in the basal lamina of the tissue with which the receptor is associated. It is inferred that transduction results from stretching the membrane of those endings.

These receptors exhibit tonic activity under sustained tension that is sometimes proportional to

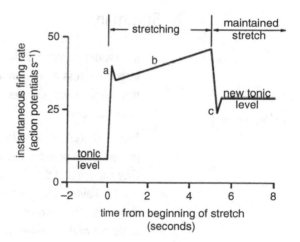

Figure 23.20 Instantaneous action potential frequency from a muscle receptor during stretching. The neuron fires tonically at a frequency dependent on the degree to which the receptor is already extended. As stretching begins and accelerates the response rises sharply (a). Continued stretching at a constant velocity is superimposed on an increasing tonic level as the receptor lengthens (b). When active stretching stops, the deceleration produces a sharp reduction in the firing rate (c), which then stabilizes at the new tonic level (after Weevers, 1966).

the degree of stretching (Fig. 23.20), but in other cases more closely related to the force. The flexor tibiae muscle receptors of locust legs signal active tension generated by the associated muscle, but respond only poorly to imposed stretch. The metathoracic receptor fires steadily at around ten spikes per second at resting muscle tensions, but at rates of up to 175 spikes per second during voluntary contractions of the accessory flexor muscle. Slow imposed stretch of the muscle causes only modest increases of firing up to 15 spikes per second, although more rapid stretches which generate transiently high forces cause the firing to briefly increase to 40 spikes per second. The slow contraction dynamics and visco-elastic properties of the muscle fibers thus act as a low-pass filter. These flexor tension receptors seem to enhance tibial flexion in postural reflexes.

The role of the muscle in muscle receptor organs is not clear. In caterpillars the activity of the motor neuron to the stretch receptor muscle is inhibited when the stretch receptor is stretched, but exhibits a transient decrease during stretching and a transient increase in activity during relaxation. It is suggested that this might protect the receptor from excessive stretch and take up the slack if stretch is suddenly reduced.

The normal role of these receptors, at least in some cases, is not simply to monitor changes in position; they may also promote negative feedback effects which tend to restore the system to its initial state. This is most clearly demonstrated in the dorsal muscle stretch receptors of caterpillars, where stretching a muscle receptor organ of one segment affects the activity of at least 32 different motor units. These are on the contralateral side of the same segment and the ipsilateral side of adjacent segments.

Summary

- Mechanoreceptors are classified as exteroceptors, interoceptors or proprioceptors. Insects possess a great many mechanoreceptors of many different types which, through their various specializations, underpin the senses of touch, hearing and proprioception.

- Mechanoreceptive sensilla are comprised of a cuticular structure (which is not always present), one or more sensory neurons and associated sheath cells with the cavities they enclose and the structures they produce. In most cases, the sensory neurons are bipolar, having a single dendrite, but some are multipolar with numerous branching dendrites. Chordotonal organ sensilla are characterized by a structure called a scolopale and are referred to as scolopidia.

- Mechanoreceptive sensilla can occur singly, but are often grouped together to form functional units.

- Activation of exteroceptors can lead to avoidance, orientation or escape reactions; activation of proprioceptors leads to postural adjustments and modifies the production of motor outputs for locomotion and other behaviors; and the activation of interoceptors regulates the passage of food through the gut, eggs through the oviducts and other internal processes.

Recommended reading

Field, L. H. and Matheson, T. (1998). Chordotonal organs of insects. *Advances in Insect Physiology* **27**, 1–228.

French, A. S. (1992). Mechanotransduction. *Annual Review of Physiology* **54**, 135–152.

Hoy, R. R. and Robert, D. (1996). Tympanal hearing in insects. *Annual Review of Entomology* **41**, 433–450.

Staudacher, E. M., Gebhardt, M. and Dürr, V. (2005). Antennal movements and mechanoreception: neurobiology of active tactile sensors. *Advances in Insect Physiology* **32**, 49–205.

Zill, S. N., Schmitz, J. and Büschges, A. (2004). Load sensing and control of posture and locomotion. *Arthropod Structure & Development* **33**, 273–286.

References in figure captions

Autrum, H. (1963). Anatomy and physiology of sound receptors in invertebrates. In *Acoustic Behaviour of Animals*, ed. R.-G. Busnel, pp. 412–433. Amsterdam: Elsevier.

Bailey, W. J. (1991). *Acoustic Behaviour of Insects: An Evolutionary Perspective.* London: Chapman & Hall.

Gnatzy, W. and Tautz, J. (1977). Sensitivity of an insect mechanoreceptor during moulting. *Physiological Entomology* **2**, 279–288.

Gray, E. G. (1960). The fine structure of the insect ear. *Philosophical Transactions of the Royal Society of London B* **243**, 75–94.

Horridge, G. A. (1965). The Arthropoda: receptors other than eyes. In *Structure and Function in the Nervous Systems of Invertebrates*, ed. T. H. Bullock and G. A. Horridge, pp. 1005–1062. San Francisco, CA: Freeman.

Matheson, T. and Field, L. H. (1990). Innervation of the metathoracic femoral chordotonal organ of *Locusta migratoria*. *Cell and Tissue Research* **259**, 551–560.

Matheson, T. and Field, L. H. (1995). An elaborate tension receptor system highlights sensory complexity in the hind leg of the locust. *Journal of Experimental Biology* **198**, 1673–1689.

Michelsen, A. and Nocke, H. (1974). Biophysical aspects of sound communication in insects. *Advances in Insect Physiology* **10**, 247–296.

Oldfield, B. P. (1982). Tonotopic organisation of auditory receptors in Tettigoniidae (Orthoptera: Ensifera). *Journal of Comparative Physiology* **147**, 461–469.

Osborne, M. P. (1970). Structure and function of neuromuscular junctions and stretch receptors. *Symposium of the Royal Entomological Society of London* **5**, 77–100.

Roeder, K. D. and Treat, A. E. (1957). Ultrasonic reception by the tympanic organ of noctuid moths. *Journal of Experimental Zoology* **134**, 127–157.

Schwabe, J. (1906). Beitrage zur Morphologie und Histologie der tympanalen Sinnesapparate der Orthopteren. *Zoologica, Stuttgart,* 50, 1–154.

Shelton, P. M. J., Stephen, R. O., Scott, J. J. A. and Tindall, A. R. (1992). The apodeme complex of the femoral chordotonal organ in the metathoracic leg of the locust *Schistocerca gregaria. Journal of Experimental Biology* 163, 345–358.

Spencer, H. J. (1974). Analysis of the electrophysiological response of the trochanteral hair receptors of the cockroach. *Journal of Experimental Biology* 60, 223–240.

Stephen, R. O. and Bailey, W. J. (1982). Bioacoustics of the ear of the bushcricket *Hemisaga* (Saginae). *Journal of the Acoustical Society of America* 72, 13–25.

Stephen, R. O. and Bennet-Clark, H. C. (1982). The anatomical and mechanical basis of stimulation and frequency analysis in the locust ear. *Journal of Experimental Biology* 99, 279–314.

Sugawara, T. (1981). Fine structure of the stretch receptor in the bursa copulatrix of the butterfly, *Pieris rapae crucivora. Cell and Tissue Research* 217, 23–36.

Tautz, J. (1977). Reception of medium vibrations by thoracic hairs of caterpillars of *Barathra brassicae* L. (Lepidoptera, Noctuidae): I. Mechanical properties of receptor hairs. *Journal of Comparative Physiology* 118, 13–31.

Tautz, J. (1978). Reception of medium vibrations by thoracic hairs of caterpillars of *Barathra brassicae* L. (Lepidoptera, Noctuidae): II. Response characteristics of the sensory cell. *Journal of Comparative Physiology* 125, 67–77.

Waters, D. A. and Jones, G. (1996). The peripheral auditory characteristics of noctuid moths: responses to the search-phase echolocation calls of bats. *Journal of Experimental Biology* 199, 847–856.

Weevers, R. de G. (1966). The physiology of a lepidopteran muscle receptor: I. The sensory response to stretching. *Journal of Experimental Biology* 44, 177–194.

Zill, S. N. and Moran, D. T. (1981). The exoskeleton and insect proprioception: I. Responses of tibial campaniform sensilla to external and muscle-generated forces in the American cockroach, *Periplaneta americana. Journal of Experimental Biology* 91, 1–24.

24 | Chemoreception

REVISED AND UPDATED BY **BRONWEN W. CRIBB**
AND DAVID J. MERRITT

INTRODUCTION

Stimulation by chemicals involves the senses of smell (olfaction) and taste (gustation). Olfaction implies the ability to detect compounds in the gaseous state. Insects have a range of receptors sensitive to odors. Insects have taste receptors on many parts of the body, often using them for purposes unrelated to feeding, and they have the ability to detect chemicals on dry surfaces as well as in solution. For these reasons it is usual to refer to "contact chemoreception" in insects rather than "taste" as used for mammals. The functional distinction between olfaction and contact chemoreception is usually clear, although olfactory receptors can respond to substances in solution and contact chemoreceptors can respond to high concentrations of some odors. The molecular receptors that provide the "lock and key" mechanism allowing detection of both forms of chemicals also show some overlap between the two chemosensory modalities. Processing within the central nervous system is, however, quite different for olfaction and contact chemoreception.

This chapter is divided into six sections. Section 24.1 describes the external structure of the chemosensory sense organs (chemosensilla). Section 24.2 deals with the cellular components within these structures. The distribution and numbers of chemosensilla are addressed in Section 24.3. Section 24.4 investigates the function of chemosensilla. The way function fits with insect behavior is examined in Section 24.5. The final section, Section 24.6, deals with neural projection of the chemosensory organs to the central nervous system.

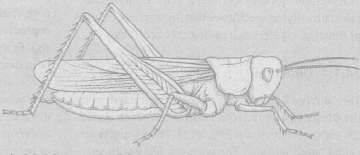

The Insects: Structure and Function (5th edition), ed. S. J. Simpson and A. E. Douglas.
Published by Cambridge University Press. © Cambridge University Press 2013.

24.1 External structure of chemosensory sense organs

In insects, as in other arthropods, sensing of the external environment requires specializations of the general cuticle and underlying epidermis. The external exoskeleton must be breached to allow chemoreception. Olfaction and contact chemoreception are mediated by sense organs that have a characteristic external cuticular component (Figs. 24.1–24.4) within which neural processes lie in close proximity to fine pores in the cuticle. A sense organ is a developmental unit composed of neural cells and support cells that secrete the cuticular component of the sense organ. The external structure is referred to as a sensillum (plural: sensilla). The word stems from the Latin *sensus*, meaning "sense." Sensilla have a range of appearances, but the hair-like form is the most common and is termed a trichoid sensillum (or sensillum trichodeum; plural: sensilla trichodea). Other shapes have different names: Stouter hairs such as spines are called chaetoid sensilla (or sensilla chaetica); pegs are referred to as basiconic sensilla; pegs sunken in pits are termed coeloconic sensilla; those in deep flasks are called ampullaceal sensilla; and if the surface structure of the sensillum is plate-like, it is termed a placoid sensillum.

Overall shape and length have been shown to be predictors of a sensillum's specific function. For example, blunt-tipped hairs are candidate contact chemoreceptors, whereas sharp-tipped hairs are more likely to be touch receptors (see Chapter 23). The benefit of a morphological description is that it simplifies the job of identifying specific sensilla for later investigation of function via electrophysiological recordings.

24.1.1 Olfaction

The cuticle of olfactory receptors is characterized by the presence of numerous small pores which permit the entry of chemicals (Fig. 24.1). These sensilla are called "multiporous." The pores are filled with an epicuticular lipid that coats the entire sensillum.

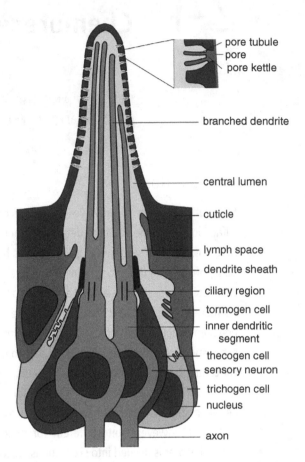

Figure 24.1 Olfactory sensillum. Schematic longitudinal section through a sensillum of the single-wall type. The pore region is enlarged to the right.

Pores can occur in rows at the base of grooves as seen in the lactic acid-receptive grooved pegs on the antennae of the mosquito *Aedes aegypti* (Diptera). Pores range from 10 nm to 25 nm in diameter, and pore density and overall distribution varies depending on the type of sensillum. The functional significance of variation in pore density has not yet been determined. Each pore, in many cases, opens into a cavity called the pore kettle, from which fine pore tubules, 10–20 nm in diameter, pass for a short distance into the lumen of the sensillum (Fig. 24.1). Different types of sensillum have different numbers of pore tubules per pore (0–50). These tubules are the remnants of wax canals, remaining after the trichogen cell has withdrawn from the cuticle. They sometimes appear to extend inwards to make contact with the dendrite, but this is not generally true. The role of pore tubules in sensing odors is still unclear.

Olfactory hairs and pegs usually do not possess the specialized socket region seen in dedicated mechanoreceptors and some contact chemoreceptors. The wall of the sensillum surrounds a central lumen within the hair shaft. This lumen contains the dendrites of the receptor neurons and is filled with fluid, called receptor lymph. Wall structure varies among olfactory sensilla. There are two forms, termed single-wall and double-wall. Trichoid (hair) and some basiconic (peg) sensilla are single-walled (Fig. 24.1). Wall thickness can be 0.5 μm or greater in long trichoid types, but in pegs may be less that 0.3 μm. A double-walled structure is seen in pegs with externally grooved walls (Fig. 24.2). Here, a transverse section reveals an outer and an inner cuticular wall, joined where pore channels traverse through both walls. The whole structure has the appearance of a wagon wheel in cross-section. A sheath (see Fig. 24.1) surrounds the dendrites within the receptor lymph cavity, but usually ends at the hair base, the dendrites emerging from it and extending up into the hair shaft.

Another class of sensilla, the humidity receptors, have an external appearance somewhat similar to olfactory pegs, but with no external pores. Because they usually respond to changes in temperature as well as humidity, they are dealt with in Chapter 19, but see Section 24.4.2 for a discussion of transduction mechanism. Examples of hygroreceptive sensilla are the coelocapitular sensillum, a mushroom-shaped protrusion set in a narrow cylindrical pit on the honey bee antennae; the sensillum capitulum on the antennae of the cockroach *Periplaneta americana* (Blattodea); and the styloconic sensilla on the tips of the antennal branches of the silk moth *Bombyx mori* (Lepidoptera).

24.1.2 Contact chemoreception

A characteristic of the contact chemoreceptor or gustatory sense organs in insects is the presence of a single pore at the tip of a hair-shaped (trichoid) sensillum, or more rarely, a cone-shaped basiconic sensillum (Figs. 24.3, 24.4). All contact chemosensory sensilla are hollow inside. In those of grasshoppers, the dendritic sheath lies in the center of the hair surrounded by the receptor lymph (Fig. 24.3a). In flies, however, the lumen of the hair is divided into a smaller lateral compartment containing the dendrites and a larger one occupying most of the space within the hair (Fig. 24.3b). The dendritic sheath in this latter type is continuous with the walls of the lateral compartment. The pore at the tip measures about 0.2 μm at its widest diameter and opens into the tube containing the dendrites. The pore can be circular, oval or slit-shaped. In some instances, a porous plug of cuticular fibrils fills the opening above the ends of the dendrites.

Hair-like sensilla usually have sockets similar to those of mechanoreceptor hairs, giving them some mobility (Fig. 24.4). Cone-shaped sensilla are found primarily on the inner mouthparts and internally on the cuticle-lined pharynx. They lack a socket, the cone being continuous with the surrounding cuticle. Presence of a socket indicates that, along with a gustatory function, the sensillum will be likely to contain a mechanoreceptive neuron at the base (Fig. 24.3; also see Chapter 23).

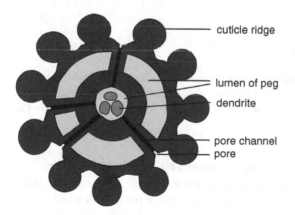

Figure 24.2 Olfactory sensillum. Schematic transverse section through a sensillum of the double-wall type, showing the wagon-wheel arrangement of pore channels.

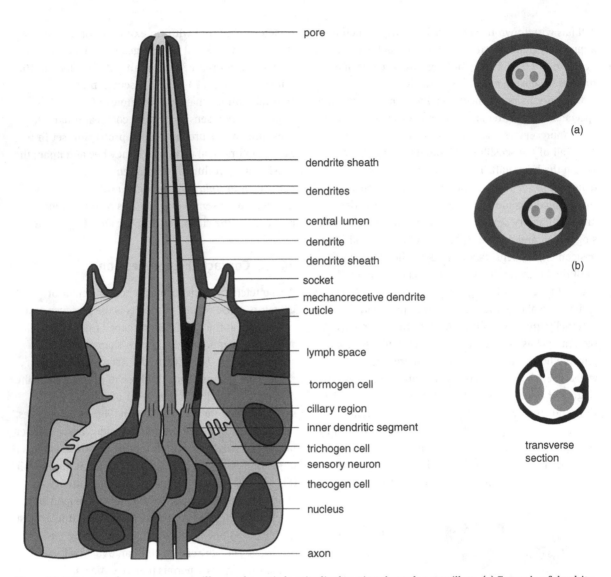

Figure 24.3 Contact chemosensory sensillum: schematic longitudinal section through a sensillum. (a) Example of dendrite sheath lying centrally in a transverse section of the sensillum shaft. (b) Example of dendrite sheath lying to one side in a transverse section of the sensillum shaft. Component labeled transverse section shows how axons lie in a sheath below the shaft as shown in the diagram.

24.2 Cellular components

The cells of contact chemosensory and olfactory sensilla are derived from sense organ precursor cells, as are other cuticular sensilla. A cluster of accessory or support cells is associated with each sensillum.

The support cells and neuron(s) arise from a single precursor cell in a fixed lineage. Their role is to secrete the cuticular components of the sensillum and to maintain electrophysiological function (Figs. 24.1, 24.3). The sensory neurons are bipolar cells whose dendrite innervates the cuticular apparatus of the sensillum.

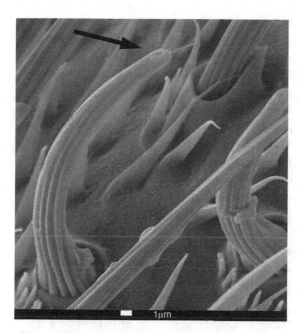

Figure 24.4 Sensilla on the leg of a water strider imaged using a scanning electron micrograph. Note the grooved peg with a pore at the tip (arrow) and the sensillum shaft nearby sitting in a deep socket. The very small surrounding hairs (microtrichia) are not likely to be innervated by any nerve cells.

Innermost, and surrounding the nerve cell bodies, is the thecogen cell. At its apex it produces the dendritic sheath. The trichogen cell surrounds the thecogen cell. There may be one or two such cells. Outermost is a tormogen cell. Trichogen and tormogen cells of olfactory sensilla in the silk moth *Antherea* (Lepidoptera) are known to produce molecules that bind with moth pheromone molecules. Accessory (support) cells are also thought to produce an enzyme that breaks down odor molecules. The role of these molecules and enzymes is discussed in Section 24.4.

The number of nerve cells associated with each sensillum varies. Contact chemoreceptors usually have 2–4 gustatory receptor cells. In addition, a mechanoreceptive neuron may be present, making five cells in total, as seen in the labial trichoid sensilla of the blowfly *Phormia regina* (Diptera).

The prominent taste hairs present on the labellum of adult *Drosophila* (Diptera) contain one chemosensory neuron activated by sucrose and other sugars, a second activated by salts, a third activated by pure water and a fourth neuron that responds to aversive compounds such as caffeine, which is named the "bitter" neuron. The same basic sense organs are present in fly larvae. More than four gustatory neurons are not common in flies but do occur; for example, eight neurons are associated with one type of gustatory sensillum on the mouthparts of adult *Drosophila*.

In olfactory sensilla, the numbers of nerve cells are more variable. Two or three sensory neurons are associated with the thick-walled pheromone-receptive sensilla of moths. Coeloconic sensilla often have three, and larger numbers occur; for example, 22 sensory neurons are associated with one type of sensillum trichodium in the biting midge *Forcipomyia (Lasiohelea) townsvillensis* (Diptera). In this case the sensory neurons are found in discrete bundles below the hair shaft, each bundle with its own set of accessory cells, suggesting they are formed from the fusing of multiple sensilla. Fusion of sensilla into complex olfactory organs occurs in Homoptera, such as aphids, and some beetles. Larval Diptera also have compound olfactory sensilla.

The structure of a typical chemosensory nerve cell (neuron) is illustrated in Figs. 24.1 and 24.3. A restricted "neck region" called the ciliary region lies distal to the neuronal cell body and separates the sensory dendrite into inner and outer segments. Olfactory sensilla typically have branching dendrites in the region of the outer dendritic segment. The dendrites of contact chemoreceptor neurons do not branch. An axon projects from the cell body of each sensory neuron to the central nervous system. Contact chemoreceptor neurons send projections to the ganglion of the segment on which the sensillum occurs. Olfactory neurons, commonly present on the antennae and maxillary palps, send projections to olfactory glomeruli in the antennal lobes of the brain.

24.3 Distribution and numbers of sensory sensilla

Olfactory sensilla are chiefly clustered on the antennae, maxillary and labial palps, but also occur on other body regions such as the genitalia of some insects (e.g., the sheep blowfly *Lucilia cuprina* [Diptera]). Contact chemosensory sensilla are distributed across the entire insect body, but cluster most densely in specific regions. They are grouped on the mouthparts, where they occur on the labrum, maxillae and labium (Chapter 2). Sensory hairs and pegs occur both outside and inside the proboscis of moths and butterflies. They are also present on the internal wall of the cibarium of insects as pegs, to allow sampling of food and fluids that flow into the pre-oral cavity. No chemoreceptors are present on the mandibles. A high density of contact chemoreceptors is found on the legs and some occur on the antennae, and some insects use antennal tapping to sample surface chemicals. Contact chemoreceptors are found on the ovipositor of some insects, especially those with long, extensible ovipositors such as parasitic wasps. Gustatory sensilla are also documented on the wings of some insects, including *Drosophila*, where they lie on the anterior wing margin.

Sensillum types and numbers can vary between juvenile and adult, and between male and female insects, reflecting their differing chemosensory needs. Females lay eggs (or live larvae) onto or into suitable substrates and rely on sensory information to choose these sites. Finding mates by males often involves chemoreception. Queen bees and ants have different chemosensory needs from workers and from reproductive males. In the carpenter ant, *Camponotus japonicus* (Hymenoptera), the structure, numbers of sensory neurons and distribution patterns of external sensilla have been mapped for the antennal flagellum of all types of colony members. Total numbers of sensilla per flagellum vary from 6000 in males and 7500 in workers, to 9000 in unmated queens.

Differences based on sex are most obvious in the silk moths. Males have pectinate (feathered) antennae, whereas the female has an unbranched flagellum. The branches of the male antenna hold many thousands of trichoid sensilla that are sensitive to female sex pheromone. In contrast the female has few pheromone-sensitive sensilla. For example, there are 17 000 pheromone-receptive neurons per antenna for the male silk moth *Bombyx mori*, and it has been calculated that the male moth responds to concentrations in air of 3000 molecules ml^{-1} – a vanishingly low concentration.

24.4 How the chemosensory sensillum functions

A series of processes is involved in producing the neuronal response in a chemosensory sensillum. The first step is the capture and uptake of the stimulating chemical and its transport to the dendritic membrane. At the dendrite membrane the chemical couples with molecular receptors, leading to transduction, which is the conversion of chemical to electrical energy. The initial steps that take place around the dendrite are collectively referred to as the perireceptor events. The suggestion has been made that it is these peripheral events, rather than intracellular signaling, that may manage the kinetics of the response. While the precise processes involved remain uncertain, the moth olfactory (pheromone) detection system has been modeled. Less is known about the process in contact chemoreception.

24.4.1 Perireceptor events

Fig. 24.5 shows the current model of perireceptor events for an olfactory sensillum, based mainly on data from silk moths. Most odor molecules are to some extent lipophilic, so they will dissolve in the epicuticular lipid forming the outer coating of a sensillum, but they do not readily dissolve in the

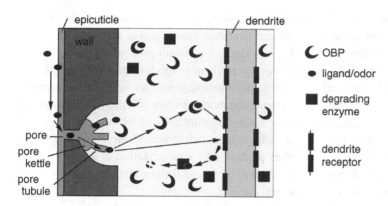

Figure 24.5 Schematic demonstrating perireceptor events in an olfactory sensillum.

insect's body fluids. A consequence of the former is that the whole outer surface of the sensillum is involved in the capture of odor molecules; it is not necessary for the molecules to strike the pores directly. A consequence is that odor molecules must be solubilized in order to reach the dendritic membrane. Transport of molecules is achieved by combining the odor molecule with specific water-soluble proteins called pheromone-binding proteins (PBPs) or general odor-binding proteins (OBPs). The former usually occur only in male moths, where they bind specifically with components of the female sex pheromone. The general OBPs are present in the olfactory sensilla of both sexes and are less specific. Binding is also thought to protect the odor molecules from degradation by enzymes, which serve to inactivate the odor molecules to stop them from continuously triggering the system after they have done their job. Studies using living antennae show two rates of pheromone molecule degradation within pheromone-sensitive sensilla. The first is a very fast decay of about 3 ms and the second is a slow decay, with a half-life of 4.5 minutes. This slower form of degradation is explained by invoking the protection of pheromone molecules by binding to the PBPs.

It is suspected that some OBPs change their conformation after interacting with the odor molecule, and it is this conformational change that activates the receptor sites on the dendrites. However, it should be noted that there are examples (e.g.,

Drosophila olfactory receptors) where OBPs appear unnecessary for stimulus transduction.

OBPs are present in such extremely high concentrations (10 mM) in the receptor lymph that diffusion of odorants or OBP–odorant complexes would be expected to be retarded. Indeed, it is difficult to explain how unbound or even bound pheromone and odorant molecules could travel from a pore to a dendrite membrane fast enough to explain the rapid responses recorded. Alternative hypotheses involve handing the odorant from one OBP to the next, or possibly forming some kind of hydrophobic channel that would allow the odor molecule to travel freely to the receptor. While possible for some, the structure of most OBPs excludes this latter option. Current theories support the prospect of multiple functions: solubilizers and carriers of the lipophilic odorants; semi-selective filters in odorant discrimination; receptor activators in consort with ligands; and/or deactivators of odorants after stimulation.

Currently, 44 OBP genes have been identified for the silkworm genome, 51 estimated for *Drosophila melanogaster* and 57 for *Anopheles gambiae* (Diptera). It is also apparent that PBPs and OBPs are not the only proteins present in the receptor lymph space. A second family of soluble polypeptides has been identified in sensilla lymph of insects. Known as chemosensory proteins (CSPs), they are smaller than OBPs and bear no sequence similarity. Eighteen CSPs have been identified in the silkworm genome.

By comparison to olfactory sensilla, the perireceptor events associated with gustatory sensilla are less well understood. In insects such as flies that regurgitate saliva onto the surface and others that feed on liquid food, gustatory sensilla on the external mouthparts will be wetted during feeding and are therefore in contact with dissolved food chemicals. In other cases, sensilla on the mouthparts, tarsi and ovipositor have to sense chemicals in solid form on dry substrates such as leaf surfaces. The tips of the dendrites within single-pore sensilla are covered by a viscous fluid containing mucopolysaccharides that sometimes exudes from the terminal pore. This "plug" may be important in the uptake of solids through the terminal pore. In this respect, the pore of the contact chemoreceptors works in much the same way as pores in olfactory sensilla.

The dendrites of the gustatory sensilla lie within a full-length sheath (Fig. 24.3). It might be expected to find the gustatory version of the OBPs within this sheath, not outside it. The gustatory peg sensilla found in *Drosophila melanogaster* meet this expectation, but in other cases labeling studies have found that binding proteins generally occur in the lymph outside the sheath, not within it. In some ways this is not surprising since it is the accessory cells which produce the proteins and they do not have access to the internal region of the dendrite sheath. The fluid within the dendrite sheath is, instead, maintained by the nerve cells. It is indisputable that gustatory binding proteins exist, but more investigation is needed regarding their distribution and function.

24.4.2 Transduction process and receptor molecules

Transduction occurs at the dendritic membrane. The odor molecule, either with or without its chaperoning protein, binds with a relatively specific receptor molecule found in the dendritic membrane. The coupling of stimulating chemical and receptor molecule opens ion channels and typically produces a depolarization of the cell membrane, called a receptor potential. This potential spreads electrotonically (i.e., is not propagated by voltage-gated channels) to the soma (cell body), leading to the production of action potentials (carried via depolarization-sensitive voltage-gated channels) at the spike-initiating zone of the soma if the potential is of sufficient magnitude. However, some odors produce a hyperpolarization of the cell membrane, inhibiting the sending of action potentials.

Only a small number of types of receptor molecules is present on the dendritic membranes of most sensory neurons, whether olfactory or gustatory. The impinging chemical (plus chaperoning protein), called the ligand, is the key to the molecular receptor lock. However, there is not a one-to-one relationship since many environmental chemicals will interact with multiple receptors and, conversely, receptors will often interact with multiple ligands. The ligand may be a food-related chemical or odor, an environmental toxin, a pheromone, carbon dioxide or even water. In contact chemoreceptors, generally speaking, each of the neurons is specific for one category of tastants (e.g., sugar or bitter, etc.), and can respond to many compounds within this taste category. There are notable exceptions, however. For example, the water-sensitive cell in the blowfly *Protophormia terraenovae* (Diptera) also responds to fructose at a high concentration. For odors there are no equivalents to the taste categories. Instead, perception is regulated by a much larger suite of specific receptors.

The first sets of molecular receptors discovered were termed either olfactory or gustatory (abbreviated in the literature to Ors and Grs, respectively). There are more olfactory than gustatory receptors, and numbers vary depending on the species. The current range is 62–341 Ors and 13–88 Grs. Delineation of receptor molecules into either gustatory or olfactory is somewhat unfortunate. Overlap has been found since the initial discovery of apparently discrete receptor molecule

Table 24.1 Receptors and ligands in the insect olfactory and gustatory organs of *Drosophila melanogaster*

Sensory system	Organ	Sensilla	Receptors[1,2,3,4]	Ligands	Proposed functions	Schematic membrane domains
Olfactory	Antennae	Trichoid (hairs)	Ors (Or67d and Or65a) in heteromeric complex with Or83b + LUSH and SNMP[5]	Cis-vaccenyl acetate	Deterrent for courting males and mated females, aggregation pheromone	Extracellular Intracellular
	Antennae	Trichoid (hairs)	Ors in heteromeric complex with Or83b	Cuticle components	Gender and conspecific detection	As above
	Antennae	Basiconic (pegs)	Ors in heteromeric complex with Or83b	Many volatile compounds, food odors	Odor recognition, discrimination, attraction/repulsion	As above
	Antennae	Basiconic (pegs)	Gr21a and Gr63a (co-expressed)	CO_2 atmosphere	Avoidance behavior	As above
	Antennae	Coeloconic (pegs)	Or35a in heteromeric complex with Or83b	Many volatile compounds, food odors	Unknown	As above
	Antennae	Coeloconic (pegs)	IRs[2]	Volatile amines, carboxylic acids, a few food odors	Unknown	
	Antennae	Coeloconic (pegs)	Nanchung and Waterwitch (part of the TRP channel family)[3]	Humidity	Desiccation avoidance	

Table 24.1 (*cont.*)

Sensory system	Organ	Sensilla	Receptors[1,2,3,4]	Ligands	Proposed functions	Schematic membrane domains
	Maxillary palps	Basiconic (pegs)	Ors in heteromeric complex with Or83b	Many volatile compounds, food odors	Taste enhancement	
Gustatory	Mouthparts (labial palps), pharynx, legs	Trichoid (hairs); some pegs	Gr5a, Gr61a, Gr64a; Gr64f and Gr61a both co-expressed with Gr5a	Sugars; Gr5a is required for response to one subset of sugars and Gr64a for response to a complementary subset	"Sweet taste," feeding regulation, feeding stimulant	
	Mouthparts		Gr93a co-expressed with Gr66a	Caffeine	"Bitter taste," avoiding harmful, noxious and toxic compounds	As above
	Labial palps, forelegs		Gr32a	Male inhibitory pheromone, compounds in seminal fluid	Necessary for inhibiting male-to-male courtship	As above
	Labella, legs		Gr33a	Co-receptor, non-volatile repulsive chemicals including caffeine and inhibitory male pheromone	Avoidance	As above
	Forelegs		Gr68a	Long-chain hydrocarbon	Female pheromone receptor, male-specific, required for the tapping step during courtship	As above

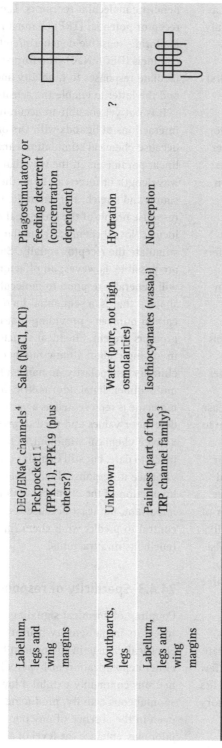

Labellum, legs and wing margins	DEG/ENaC channels[4] Pickpocket11 (PPK11), PPK19 (plus others?)	Salts (NaCl, KCl)	Phagostimulatory or feeding deterrent (concentration dependent)
Mouthparts, legs	Unknown	Water (pure, not high osmolarities)	Hydration
Labellum, legs and wing margins	Painless (part of the TRP channel family)[3]	Isothiocyanates (wasabi)	Nociception

Notes:

[1] Ors and Grs are olfactory and gustatory receptors predicted to contain seven transmembrane domains.

[2] IRs are ionotropic receptors, and related to ionotropic glutamate receptors and are predicted to contain three transmembrane domains.

[3] DEG/ENaC are amiloride-sensitive degenerin/epithelial Na^+ channels and contain two transmembrane domains.

[4] TRPs are part of the transient receptor potential channel family and contain six transmembrane domains.

[5] LUSH is a member of the odorant-binding protein family and SNMP is a sensory neuron membrane protein.

Source: Modified from Su *et al.* (2009) and Yarmolinsky *et al.* (2009); additional data from Montell (2009), Benton (2008) and Benton *et al.* (2009).

types. For example *Gr21a* and *Gr63a* are defined as members of the *Drosophila* gustatory receptor family but are expressed in antennal sensilla, where they respond to the olfactory stimulant, CO_2.

Receptor molecules for the general olfactory (Ors) and gustatory receptors (Grs) contain seven transmembrane domains, similar to the seven transmembrane domain G-protein-coupled receptor superfamily occurring in the vast majority of other organisms. However, experiments with *Drosophila* imply that the membrane topology is upside down when compared to those other organisms: The N-terminus is placed within the dendrite and the C-terminus outside. Also, these receptors act as ligand-gated (ionotropic) ion channels, which differs from olfactory and some gustatory molecular receptors in mammals, which act indirectly on ion channels via second messenger cascades (metabotropic). Having a more direct interaction between ligand and membrane-bound ion channels allows considerably faster responses than seen in mammalian receptors. Olfactory receptor molecules do not act independently (Table 24.1); they form heteromultimer complexes. The same may be the case for gustatory receptor molecules, which are known to be co-expressed in some cases.

There is much to be discovered about molecular receptors. An understanding of their function and their ligands lags behind their identification. Table 24.1 shows that, as well as Ors and Grs, a number of other types of receptors are present in insect sensilla. Recently, another class of ionotropic receptors (IRs) have been shown to confer odor responsiveness in *Drosophila*. These are predicted to contain only three transmembrane domains and a pore loop and are related to, but not the same as, the ionotropic glutamate receptors (iGluRs) that mediate communication at synapses throughout invertebrate (and vertebrate) nervous systems. Just as two or a few Ors or Grs are expressed in each sensory neuron, IRs are also expressed in small numbers in other sensory neurons, raising the possibility that they form functional units. There are two other forms of

dendritic molecular receptors. These are the transient receptor potential (TRP) channel family and the amiloride-sensitive degenerin/epithelial Na^+ channels (DEG/ENaC). The former family appears to mediate responses to humidity and isothiocyanates, and the latter to enable the sensing of salts.

It is not yet possible to accurately predict the interactions of ligands with Ors or Grs. This is because chemical stimulation cannot be classified by linear parameters in the way that frequency, wavelength or force can be in the sensing of light, sound and stretch. For example, Fig. 24.6 shows the response profile of the green-leaf odor receptor of the locust. Not all chemically similar compounds stimulate the receptor equally. Some generalizations are possible, however: an olfactory sensory neuron will generally respond to molecules with the same shape – that is, a sequential increase or decrease in carbon number – providing that the same functional group is present. Chemical structure is complex, involving carbon atomic number, functional group, chirality and polarity, to name a few parameters. The multi-dimensional odor metric, in which each molecule is represented as a vector of 1664 molecular descriptor values and used to predict the response to a given chemical stimulant, is the best approach taken to date, but still explains only about 50% of the variance in response. Being able to predict the interaction of the chemical with the receptor is important, not least because it can be used in pest control to predict what chemicals are likely to act as repellents or attractants.

24.4.3 Specificity of response

Coupling of chemical signaling molecules with receptors in the sensory neurons produces an electrical change in the neuron that can result in either depolarization or hyperpolarization. Olfactory neurons commonly exhibit a low level of spontaneous activity, producing action potentials even in the absence of any odor. Consequently, an odor may enhance the level of activity, have no effect

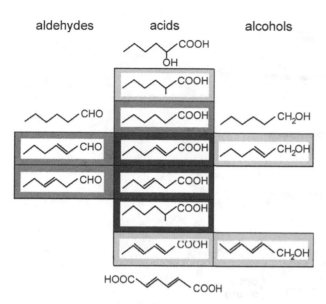

Figure 24.6 The response spectrum of a sensory neuron in a coeloconic sensillum on the antennae of *Locusta* that is sensitive to hexanoic acid and related compounds known as green-leaf odors. The degree of shading indicates intensity of response to the chemical (darkest = greatest sensitivity). Compounds not in boxes did not stimulate (after Kafka, 1971). (With kind permission from Springer Science+Business Media: *Gustation and Olfaction, Specificity of Odor-molecule Interaction in Single Cells*, 1971, p. 65, Kafka, W.A., Fig. 3.)

or reduce activity. All these effects may occur in one neuron in response to different compounds. Fig. 24.7a shows trains of spikes (action potentials) produced by stimulation with a range of odors on a background of spontaneous spike activity. Unlike olfactory neurons, contact chemoreceptor neurons generally fire only when stimulated, but if they do exhibit spontaneous activity, the firing rate is very low.

The response to a compound that increases the rate of firing is often phasic/tonic (i.e., showing an initial high response upon onset of the stimulus, followed by a substantial drop to a persisting, attenuated response with continued stimulation) and it may continue for a brief period after removal of the stimulus. In cases where the response is entirely phasic, the input from a chemosensory neuron may fall to zero very quickly, sometimes in less than one second. The flow of sensory information can be maintained by breaking off and remaking contact with the substrate. In the intervals between contacts the neurons disadapt and thus maintain a high level of input for one second or more. This behavior is seen in many insects during drumming movements with the legs, palps or antennae, in which they touch a surface repeatedly for very brief periods. Butterflies use tarsal drumming to determine acceptability of plant material for oviposition and parasitoid wasps explore suitability of host eggs through antennal drumming. When a grasshopper drums on a leaf with its palps (palpates) the contact chemoreceptors on the tips of the palps touch the surface about ten times a second, each contact lasting less than 20 ms.

Discontinuous stimulation also occurs in sensing some odors. Air-borne odors do not form a continuous gradient because of air turbulence; instead they occur as bursts separated by "clean" air. The ability to distinguish successive contacts with such odor bursts is important in odor-modulated flight and the ability to do so depends on the characteristics of the sensory neurons.

Males of the moth *Helicoverpa* (Lepidoptera) have large numbers of olfactory neurons on the antennae which respond to the principal component of the female sex attractant pheromone. Some exhibit a strong phasic response, which rapidly falls to zero. These neurons give discrete responses to 20-ms pulses of pheromone at 1.5 Hz; at 6 Hz the response is continuous, but distinct peaks still correspond with the arrival of the stimulus. Only at 12 Hz do the

(a)

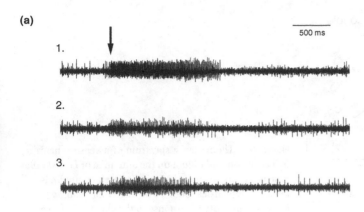

(b)

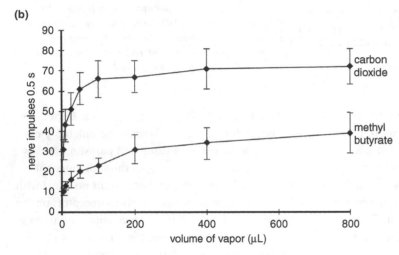

(c)

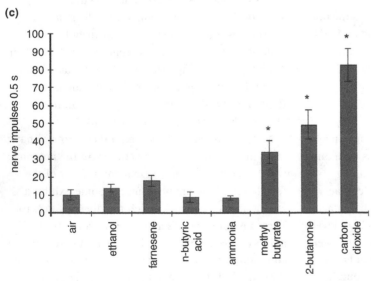

Figure 24.7 Response profiles for carbon dioxide receptors in antennal sensilla of the Queensland fruit fly, *Bactrocera tryoni*. (a) Electrophysiological recordings in response to stimulation with: (1) 400 μl of carbon dioxide saturated air; (2) 400 μl of methyl butyrate saturated air; (3) 400 μl of 2-butanone saturated air. Arrow shows onset of odor delivery. (b) Response profiles to carbon dioxide (top line) and methyl butyrate (bottom line) across increasing concentrations of the odors. (c) Differences in activation of the carbon dioxide receptor when exposed to a range of odors presented as 400 μl. Relative responses of carbon dioxide sensory receptors on the antenna of females. The asterisks indicate those chemicals that were significantly different from stimulation with clean air. Only methyl butyrate, 2-butanone and carbon dioxide could be detected by the sensory cell (after Hull and Cribb, 2001).

responses merge into a continuum from which the insect probably cannot distinguish the pulsed nature of the stimulus from continuous stimulation. Thus, the peripheral neurons have the capacity to allow the insect to recognize a stimulus that is pulsed at quite high frequency.

The dynamic range of odor and taste neurons usually extends over two or three orders of magnitude in concentration. At low concentrations, below threshold, the neuron does not respond. The magnitude of the neuronal response increases with stimulus concentration until the receptor responds maximally, seen for an odor-sensitive neuron in Fig. 24.7b. Here it can be seen that sensilla show different response profiles to different chemicals, and that sensilla sensitive to general odors respond to only a few chemicals (Fig. 24.7c). Threshold concentrations vary considerably. Pheromone-specific neurons often have much lower thresholds than neurons responding to other types of odor. In taste cells, those responding to major nutrient compounds, such as sugars and amino acids, are usually sensitive only to moderately high concentrations, but cells responding to chemicals with specific ecological relevance apart from nutrition tend to be much more sensitive. Even within a single insect, neurons responding to the same compound may have different thresholds. For example, in *Phormia* the neurons in the tarsal hairs that respond to sucrose have thresholds at concentrations one or two orders of magnitude lower than those in labellar sensilla (Fig. 24.8). Such differences provide the insect with a great dynamic range of response.

Generally, insects possess sensory neurons responsive to deterrent class chemicals (see Table 24.1). Some chemicals are deterrent at any concentration, whereas others become deterrent only at high concentrations. Some species have contact chemoreceptor neurons that respond only to specific chemicals that are key components of the food, or to contact pheromones. For example, many larval and adult brassica-feeding insects have a cell that

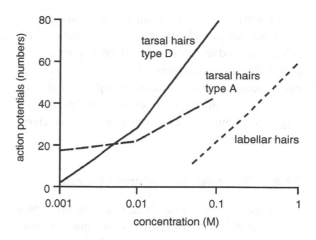

Figure 24.8 Differences in threshold of sucrose-sensitive neurons in different sensilla of the blowfly (data from McCutchan, 1969; Omand and Dethier, 1969).

responds to glucosinolates. Glucosinolates are characteristic compounds of Brassicaceae. Both sexes of adult fruit flies *Rhagoletis* (Diptera), in which the female marks fruit with a pheromone after oviposition, have neurons in tarsal receptors that are specific for the compound. These different cell types are often called "salt," "sugar," "amino acid" or "deterrent" receptors according to the class of compound to which they respond or the type of behavioral response induced when they are stimulated, but these response ranges are not absolute. A "sugar" receptor may also sometimes be stimulated by an inorganic salt, for example. Even within a class of cells, the specificity varies from sensillum to sensillum. For example, the "sugar" cell in the 14 contact chemoreceptive sensilla on the fifth tarsomere of *Helicoverpa armigera* adult moths have different response profiles, with some responding to a range of sugars as well as amino acids (e.g., sensillum F5a: five different sugars and seven amino acids), and others more narrowly tuned to specific sugars (e.g., sensillum F5d: glucose and myo-inositol).

Characterizing the response profiles of sensory neurons and the relevance to insect behavior can be difficult. A notable example of this is the sensory perception of the repellent DEET

(N,N-diethyl-3-methylbenzamide) by mosquitoes. DEET was patented by the US Army in 1946 and is still considered to be the most effective substance for personal protection against blood-feeding insects ever discovered, but instead of acting as a masking agent for human odors attractive to insects as was originally thought, it now seems that it is perceived as a repellent.

24.4.4 Response to mixtures of chemicals

The response to a normal food, usually containing a variety of different chemicals, is not simply the sum of the inputs occurring when the various components are tested singly. Also, biologically significant odors usually occur among a background of other environmental odors, and important odor signals may themselves be mixtures of compounds. Interactions occur between the chemicals in mixtures, and in some cases between the peripheral sensory neurons that detect them. The signals transmitted to the central nervous system are a reflection of these interactions.

Sometimes the effect of multiple chemicals is synergistic, as in a pheromone receptor of *Spodoptera litura* (Lepidoptera), where stimulation with both *cis*-9,*trans* 11,tetradecadenyl acetate and *cis*-9,*trans*-12,tetradecadenyl acetate produces a greater response than with *cis*-9,*trans* 11, tetradecadenyl acetate alone, but in other cases there are negative interactions. For example, the response of cells in some placoid sensilla of *Apis* (Hymenoptera) to the flower odor, geraniol, is inhibited by another flower odor, linalool. Sequential presentation of chemicals can also affect neuronal response: *Drosophila melanogaster* shows an avoidance response when exposed to carbon dioxide alone, but pre-exposure to food-associated odors inhibits this response. The food-associated odors are proposed to interact directly with the carbon dioxide receptors. In other cases interactions occur between neurons in the same sensillum. The physiological basis for such interactions is not fully understood,

but sometimes a compound that stimulates one cell is known to hyperpolarize another in the same sensillum. This suppresses the response to a stimulating compound in a mixture.

Two or more olfactory receptor neurons with identical sensitivity have never been found in the same sensillum, and it is rare that greatly overlapping sensitivities occur. Receptor neurons with different sensitivities are grouped in sensilla. In an attempt to understand the interactions among multiple sensory neurons, the system has been modeled and showed that the more dissimilar the neurons are, the greater the amplitude of the transmembrane receptor potential of the most sensitive neuron. The threshold is also slightly lower. In other words, sensory neurons are most sensitive when they work alone and tend to interfere with one another if they are similar in response profile. By clustering sensory neurons that respond to different ligands together within a sensillum, the neurons can remain sensitive without interfering with one another. Another advantage of clustering sensory neurons is illustrated by the case of some moth pheromones, which are composed of blends of chemicals where ratios of certain components are critical to recognition. In the crambid, tortricid and yponomeutid moth species receptive neurons are paired in the same sensilla, present in the same ratio as in the pheromone blend. Since overall numbers of sensilla can change over generations, such pairing would ensure the ratio of receptive neurons remains constant.

There may be a benefit to having multiple neurons per sensillum in terms of inhibition as well. As discussed in Section 24.2, bitter, sugar and water-sensitive neurons occur together in the same sensilla. When a bitter chemical (quinine), a sugar and water are presented to *Drosophila* adults as a mixture, the bitter compound inhibits the response of the sugar and water neurons. The conclusion is that bitter compounds interact directly with the transduction pathways of the sugar and water neurons. Such a system would allow an insect to avoid food materials on the basis of the deterrent chemical present, whether or not phagostimulatory sugars were present. This effect is not restricted to alkaloids or to deterrent compounds; interactions involving many different compounds occur. For example, in *Stomoxys* (Diptera), the activity of neurons responding to inorganic salts in the labellar hairs is inhibited by adenine nucleotides for which they appear to have purinoceptors. The nucleotides are phagostimulants for these insects; and sucrose and inositol both inhibit the response to sinigrin in caterpillars of *Mamestra* and *Trichoplusia* (Lepidoptera).

24.4.5 Temporal changes in neuronal responsiveness

The responsiveness of the sensillum can be modified as a result of physiological condition, such as mating status, nutritional state, age and time of day. In blood-feeding insects, feeding status has been shown to affect sensitivity of the peripheral receptor system. In mosquitoes, satiation affects the receptivity of the lactic acid receptor. The electrophysiological activity of the lactic acid-excited neurons is depressed following a blood meal, meaning that mosquitoes will be less likely to respond to a nearby host. A strong inhibition of host-seeking in *Anopheles gambiae* has also been demonstrated, associated with an inhibition of olfactory sensitivity to the odor of incubated sweat and indole, following blood-feeding. Host-seeking returns 72 hours later.

Peripheral modulation of olfaction has been investigated in detail for the Egyptian leaf worm *Spodoptera littoralis*. Two- to four-day-old moths show greater antennal sensitivity than newly emerged moths. Mating reduces antennal sensitivity by 19%. Feeding status has no effect on sensilla sensitivity. The reduction caused by mating may be mediated via activation of biogenic amine receptors coupled to olfactory receptors. Hormones and peptides may also play a role. Juvenile hormone has been shown to increase antennal response to sex

pheromone in reproductively inactive male moths of the species *Caloptilia fraxinella* (Lepidoptera). Responsiveness of the peripheral sensory system of the Egyptian leaf worm can also be affected by season. Olfactory responsiveness can be related to time of day, as is the case for *Drosophila* and cockroaches (*Leucophaea maderae* [Blattodea]). In the latter case, circadian rhythms are involved, although it is not yet clear how the change in sensitivity affects behavioral outcomes.

Feeding usually begins with the stimulation of gustatory receptors. The presence of sugars, amino acids or specific feeding stimulants will initiate and maintain feeding, whereas deterrent substances will inhibit feeding. Insects can regulate their nutrient intake by utilizing nutritionally modulated peripheral chemosensory neurons. Insects can select food sources which balance nutritional requirements via changes in sensitivity of the peripheral sense organs involved in monitoring sugars and proteins. When the grasshopper *Locusta migratoria* (Orthoptera) and the caterpillar *Spodoptera littoralis* (Lepidoptera) were restricted to either a protein diet or a carbohydrate diet, insects which had been exposed to a carbohydrate-restricted diet chose a diet high in carbohydrate, while protein-restricted insects chose a diet high in protein, and response (measured as the number of action potentials) of the contact chemoreceptors changed (Fig. 24.9). Mouthpart chemoreceptors showed an increase in responsiveness to stimulation with amino acids and sugars as titers of both nutrient groups dropped in the hemolymph.

Not only is responsiveness of sensilla variable, but the number of sensilla that develop on the mouthparts and antennae can also be affected by diet and odor. Numbers of sensilla on the maxillary palps and antennae of locusts vary across stadia with diet and odor experience during development. The number of sensilla that develop in a given sensory field appears to be linked to the variety of chemical stimuli experienced, as well as the chemical complexity of the environment.

(a) temporal variation

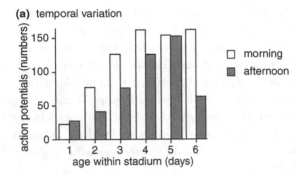

(b) nutritional variation

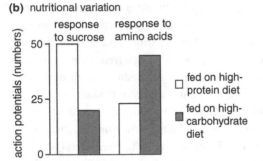

Figure 24.9 Variation in response profiles for contact chemoreceptors. (a) Variation in response of contact chemoreceptors in *Spodoptera* over six days and within each day (after Schoonhoven *et al.*, 1991). (b) Response of chemoreceptors in the maxillary palp sensilla of *Locusta* after access to a specific diet for four hours. The insects were previously fed on foods either high in protein or high in carbohydrate (data from Simpson *et al.*, 1991).

24.5 Integrating function and behavior

The olfactory and contact chemosensory systems are integrated in the behavior of insects. Olfaction is a sense that can be used at a distance as well as locally. Contact chemoreceptors are used in the last stages of assessing oviposition or feeding sites, and in regulating feeding.

Insects are capable of detecting a variety of odorants in their environment. Grasshoppers have a sensory response to plants of many different kinds, not only to their host plants, and antennal receptors of adult Lepidoptera respond to a wide range of plant-derived compounds. Many of these odors may

produce no overt behavior; others may be critically important in locating a mate, a food source or an oviposition site from a distance. They may also affect behavior close to the source, and responses to odors may be learned.

Because of air turbulence, continuous odor gradients cannot persist under natural conditions except for very short distances. Consequently, although odor may provide the signal that initiates upwind flight, the insect depends on a combination of mechanical and visual stimuli to achieve upwind movement. This type of behavior is known to occur in response to sex-attractant pheromones, such as those produced by many female Lepidoptera. It may also occur in relation to food or oviposition sites, especially in night-flying insects where a direct visual response may only be possible close to the source.

Although single compounds can sometimes induce these behaviors, it is generally true that specificity of attraction, where it occurs, results from particular combinations of compounds. For example, although a single compound can attract males to a pheromone-producing female, a mixture of two or three specific compounds provides the insect with an unambiguous signal (see Chapter 27). Similarly, herbivorous insects use blends. The black bean aphid, *Aphis fabae* (Homoptera), responds negatively to ten odors from its host when delivered as single odors, but a blend elicits a positive response (acting as an arrestant or attractant). Sheep blowflies are attracted to injured sheep by the sulfur-containing breakdown products of wool. Dimethyldisulfide and ethanediol, together with hydrogen sulfide, are major components of this odor, but individually neither dimethyldisulfide nor ethanediol has any effect on the fly's behavior.

Carbon dioxide is an attractant for many species that suck vertebrate blood and there are carbon dioxide receptors on the antennae of tsetse flies and other blood-sucking species. The carbon dioxide acts, together with acetone, octenol and other compounds commonly found in animal breath, to attract tsetse flies. Interestingly, many Lepidoptera have large numbers of carbon dioxide receptors in a pit on the labial palps. In *Heliothis* (Lepidoptera), these are sensitive to very small changes in the ambient level of carbon dioxide. This ability may be involved in assessing suitability of plants for oviposition.

There are relatively few examples of odors affecting behavior at close range, even though such behavior may be widespread. Flower scents on the bodies of bees returning from foraging help other workers to recognize the food source. Male monarch butterflies produce an aphrodisiac pheromone which enhances the female's readiness to mate. Females of the black swallowtail butterfly, *Papilio polyxenes* (Lepidoptera) are stimulated to land on artificial leaves by the odor of carrot leaves, the normal host plant of the larvae, but this behavior is disrupted if the odor of cabbage is also present.

Examples of insects being repelled by odors are also known. For example, nepetalactone, the chemical giving a characteristic odor to catnip, repels a number of phytophagous Homoptera and Coleoptera; linalool is a repellent for the aphid *Cavariella* (Homoptera).

Contact chemoreceptors are of primary importance in relation to feeding. In *Phormia*, for example, stimulation of the tarsal chemoreceptors with sugar leads to proboscis extension. This brings chemoreceptors on the labellum into contact with the food. Their stimulation causes the fly to spread its labellar lobes so the interpseudotracheal sensilla (see Fig. 2.6c) contact the food and sucking begins. A final monitoring occurs by the cibarial receptors. Continuous sensory input is necessary for continued feeding. Essentially similar sequences occur in other insects and, for many, sugars are major phagostimulants.

Sometimes, a specific "sign" chemical is necessary to induce feeding. This is often true for insects with a relatively specific host range. Feeding is inhibited in many plant-feeding insects by various plant secondary compounds, especially those in non-host plants. These components stimulate deterrent cells or inhibit the activity of nutrient-specific cells.

Contact chemoreception is also important as an oviposition cue in some insects. For example, the diamondback moth, *Plutella xylostella* (Lepidoptera), is induced to oviposit on cruciferous plants by non-volatile indole glucosinolates. Tarsal chemoreceptors are important in regulating such behavior in Lepidoptera.

Insects use contact chemoreception in the perception of some pheromones. Trimethylheptatriacontane, a component of the surface wax of the female, acts as a sexual recognition pheromone for the tsetse fly, *Glossina morsitans* (Diptera). Host-marking pheromones are also perceived by tarsal receptors.

24.6 Projections to the central nervous system

Nerve projections (axons) from olfactory and contact chemoreceptors innervate the central nervous system (CNS) with different endpoints.

24.6.1 The olfactory system

Axons from the olfactory system terminate in the paired olfactory lobes, which are part of the deutocerebrum (see Chapter 20) in the CNS. Each lobe is made up of spherical clusters of neural connections, called glomeruli. Generally speaking, receptor neurons with the same response profile converge and terminate in the same glomeruli. In this way the dispersed collection points for information on chemicals, namely the separate sensory neurons, gather together into the same processing regions in the brain. The structure of glomeruli varies between species and sex. Glomeruli can form functional complexes (clusters), as in the honey bee, or the macroglomerular complex of male moths, which are used to sense sex pheromones. There is also some degree of somatotopic organization of sensory afferents. In other words, within a glomerulus there appears to be some mapping of spatial information. For example, the proximal segment on the antenna of the cockroach *Periplaneta americana* tends to end up in more proximal regions of individual glomeruli. But there is variation among insect groups.

The olfactory system must handle mixtures of odor that change in ratio and concentration across time and space. For example, neighboring flowers may differ subtly in the ratios and concentrations of odors produced. The insect must be able to maintain stimulus identity in the face of such differences and changes. An example is found with the Sacred Datura plant (*Datura wrightii*) and the pollinating moth (*Manduca sexta* [Lepidoptera]). Of the 60 compounds present in the flower odor, only nine are involved in mediating flower-foraging behaviors, and only when present as a mixture. When the concentrations of these nine were varied, the brain retained the same pattern of firing across time. Such a temporal pattern may be used in the brains of moths to even out environmental changes in the source odor, ensuring the insect can still recognize its host.

Even though there is hard wiring of axons, chemosensory processing of odors in the central nervous system is flexible and can be modified. It has been shown for *Drosophila* that the antennal lobe network includes inhibitory and excitatory neurons that transform olfactory information. During olfactory learning, the release of octopamine or dopamine associated with reinforcing rewarding or punitive stimuli can modify synapses within the olfactory pathway, altering the activity of processing regions – for example, output neurons from the paired mushroom bodies. Early olfactory experience has been shown to modify neural activity in the antennal lobes of social insects as adults. Insects can learn to recognize odors associated with food, and innate preferences can change with experience, as seen with the noctuid moth *Helicoverpa armigera* when it is foraging on flowers.

24.6.2 The contact chemosensory system

Less is known about the contact chemosensory projections to the CNS than those of the olfactory system. The gustatory system is not challenged as greatly by temporal and spatial shifts in chemical composition as the olfactory system. Food selection depends mainly on the balance of phagostimulatory and deterrent inputs, although in the case of oligophagous and monophagous species, there may be a role for a specific host-related chemical.

The contact chemosensory projections mostly connect into the local ganglion or neuromere, although there are examples of projections terminating in the brain. In *Drosophila* the gustatory receptor neuron axons from the labellum, pharynx and some leg chemoreceptors travel to the subesophageal ganglion and the tritocerebrum. Here, axons coding bitter and sweet taste modalities terminate in different areas and there appears to be some spatial distribution of sensilla mapped into the system. Gustatory receptor neurons on the antennae and proboscis of the moth, *Heliothis virescens* (Lepidoptera), send axonal projections to the subesophageal ganglion and the tritocerebrum. Again, spatial information is mapped onto the distribution in the brain, with projections of the antennal gustatory receptor terminating posterior-laterally to those from the proboscis. In the locust there is local processing. The gustatory and mechanosensory signals from the hindleg terminate in the same metathoracic ganglia, indicating that they may be processed together. Such local circuits do not need complex gustatory coding and may be commonly involved in shaping feeding preferences on the basis of averseness or phagostimulatory suitability.

There is still uncertainty about how chemical identity is coded in phagostimulatory and deterrent neurons. One theory is that "labeled" lines are used, tuned to specific chemicals that allow for what is termed "across-fiber patterning," whereby the lines carrying the same chemical identities come together for later processing. Alternatively, more abstract qualities may be processed, such as sweet, salty, sour and bitter. But it has not been discounted that there is some other more direct measure involved.

Summary

- The first stage in the hierarchy allowing insects to discriminate one chemical from another involves molecules entering multiporous olfactory sensilla and single-pore contact chemosensory ("taste") sensilla.

- Chemicals then interact with molecular receptors on the dendritic membranes of the multiple chemosensory nerve cells housed within these cuticular structures, leading to opening of ion channels and generation of receptor potentials. Depolarizing receptor potentials trigger generation of action potentials, which are transmitted along chemosensory axons to the central nervous system.

- Sensory neurons show a degree of electrical interaction with one another within a sensillum, hence only those with differing sensitivity occur in the same sensillum.

- Olfactory neurons terminate in the glomeruli of the antennal lobes, whereas gustatory neuronal projections often terminate at local ganglia.

- Feedback mechanisms can affect sensitivity of the chemosensory system. Physiological condition, time of day, season and previous experience all affect the function of the peripheral chemoreception system. Changes in sensitivity to chemicals enable insects to select foods containing nutrients which are underrepresented or missing in the diet.

Recommended reading

STRUCTURE OF SENSILLA

Zacharuk, R. Y. (1985). Antennae and sensilla. In *Comprehensive Insect Physiology, Biochemistry and Pharmacology*, vol. 6, ed. G. A. Kerkut and L. I. Gilbert, pp. 1–69. Oxford: Pergamon Press.

CONTACT CHEMORECEPTION

Rogers, S. M. and Newland, P. L. (2003). The neurobiology of taste in insects. *Advances in Insect Physiology* 31, 141–204.

OLFACTORY PHYSIOLOGY

Su, C.-Y., Menuz, K. and Carlson, J. R. (2009). Olfactory perception: receptors, cells, and circuits. *Cell* 139, 45–59.

ODOR DETECTION CODES

de Bruyne, M. and Baker, T. C. (2008). Odor detection in insects: volatile codes. *Journal of Chemical Ecology* 34, 882–897.

CHEMORECEPTION PHYSIOLOGY

Yarmolinsky, D. A., Zuker, C. S. and Ryba, N. J. P. (2009). Common sense about taste: from mammals to insects. *Cell* 139, 234–244.

References in figure captions and table

Benton, R. (2008). Chemical sensing in *Drosophila. Current Opinion in Neurobiology* 18, 357–363.

Benton, R., Vannice, K. S., Gomez-Diaz, C. and Vosshall, L. B. (2009). Variant ionotropic glutamate receptors as chemosensory receptors in *Drosophila. Cell* 136, 149–162.

Hull, C. D. and Cribb, B. W. (2001). Olfaction in the Queensland fruit fly, *Bactrocera tryoni*: II. Response spectra and temporal encoding characteristics of the carbon dioxide receptors. *Journal of Chemical Ecology* 27, 889–906.

Kafka, W. A. (1971). Specificity of odour-molecule interaction in single cells. In *Gustation and Olfaction*, ed. G. Ohloff and A. F. Thomas, pp. 61–72. London: Academic Press.

McCutchan, M. C. (1969). Responses of tarsal chemoreceptive hairs of the blowfly, *Phormia regina. Journal of Insect Physiology*, 15, 2059–2068.

Montell, C. (2009). A taste of the *Drosophila* gustatory receptors. *Current Opinion in Neurobiology* 19, 345–353.

Omand, E. and Dethier, V. G. (1969). An electrophysiological analysis of the action of carbohydrates in the sugar receptor of the blowfly. *Proceedings of the National Academy of Sciences USA* 62, 136–143.

Schoonhoven, L. M., Simmonds, M. S. J. and Blaney, W. M. (1991). Changes in the responsiveness of the maxillary styloconic sensilla of *Spodoptera littoralis* to inositol and sinigrin correlate with feeding behaviour during the final larval stadium. *Journal of Insect Physiology* 37, 261–268.

Simpson, S. J., James, S., Simmonds, M. S. J. and Blaney, W. M. (1991). Variation in chemosensitivity and the control of dietary selection behaviour in the locust. *Appetite* 17, 141–154.

Su, C.-Y., Menuz, K. and Carlson, J. R. (2009). Olfactory perception: receptors, cells, and circuits. *Cell* 139, 45–59.

Yarmolinsky, D. A., Zuker, C. S. and Ryba, N. J. P. (2009). Common sense about taste: from mammals to insects. *Cell* 139, 234–244.

25 Visual signals: color and light production

REVISED AND UPDATED BY **PETER VUKUSIC AND LARS CHITTKA**

INTRODUCTION

Insects generate a spectacular variety of visual signals, from multicolored wing patterns of butterflies, through metallic-shiny beetles to highly contrasting warning coloration of stinging insects and their defenseless mimics. Section 25.1 explains what colors are and the subsequent sections describe how insect colors result from a variety of physical structures (Section 25.2) and pigments (Section 25.3). Often, several pigments are present together, and the observed color depends on the relative abundance and positions of the pigments, as well as control signals generating color patterns during development (Section 25.4). The position of color-producing molecules relative to other structures is also important, and this may change, resulting in changes in coloration (Section 25.5). The many biological functions of color in insect signaling are covered in Section 25.6. Table 25.1 lists the sources of color in some insect groups. A small selection of insects also exhibits fluorescence or luminescence (Section 25.7).

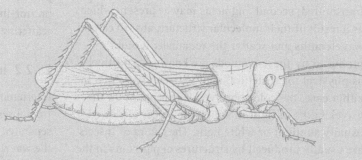

The Insects: Structure and Function (5th edition), ed. S. J. Simpson and A. E. Douglas.
Published by Cambridge University Press. © Cambridge University Press 2013.

25.1 The nature of color

Color is not an inherent property of objects; it is a perceptual attribute that depends on illumination, the spectral reflectance of an object and its surroundings, as well as the spectral receptor types and further neural processing in the animal in question. Thus the same object might appear differently colored to different viewing organisms. A red poppy, for example, is red to human observers, but appears as a UV-reflecting object to a bee pollinator, which does not have a red receptor and, like all insects studied to date, sees UV-A light between 300 nm and 400 nm. For reasons of simplicity, the color terminology in this chapter specifies what a human observer will perceive under daylight conditions. Information about UV is provided separately where available.

Color is generated from white light incident upon an insect when some of the incident wavelengths are eliminated, usually by absorption in its pigmentation, and the remainder are scattered. These scattered wavelengths of the reflected or transmitted components determine the color observed (Fig. 25.1). If all wavelengths are reflected equally then the reflecting surface appears white; if all are absorbed the appearance is black.

Differential reflection of light to produce color occurs in one of two ways. First, the physical nature of the surface may be such that only certain wavelengths are reflected, the remainder being transmitted; second, pigments may be present which, as a result of their molecular structure, absorb certain wavelengths and scatter the remainder. Colors produced by these methods are known, respectively, as physical (or structural) and pigmentary colors. In either case, the color observed is dependent on the wavelengths of light in which the object is viewed, usually sunlight or white light. The colors of insects are usually produced by structures or pigments in the cuticle or epidermis, but, if the cuticle is transparent, the hemolymph, fat body or even gut contents may contribute to the insect's color.

25.2 Structural colors

Various forms of surface and epidermal structures can be responsible for the production of whites and iridescent colors. They also assist in creating the appearance of many black surfaces. These colors may be produced by random scattering or by coherent scattering, such as by interference.

25.2.1 Scattering

Light may be scattered (i.e., reflected in all directions) by granules on, or irregularities in, a surface. If these scattering centers are randomly spatially distributed in or on the surface then virtually all incident light is reflected diffusely and the surface appears as a matt white. In some Lepidoptera, such as Pieridae, this results from the presence of densely packed arrays of pterin granules across each white scale's surface, the effect of which is coupled with that of the structure of the scale itself. In some Coleoptera, such as *Cyphochilus* sp., a bright, diffuse white appearance is generated by scattering from interconnected cuticular filaments that are randomly arranged within each white-colored scale. Pearly whites, such as those of *Morpho sulkowski* (Lepidoptera), are produced by the addition of a degree of specular (mirror-like) reflection in addition to the diffuse scattering.

25.2.2 Interference in multilayers

Interference colors result from the reflection of light from a series of neighboring interfaces that are separated by distances comparable with a quarter of the wavelength of light. As a result of wave superposition from these reflections, some

Table 25.1 **Some of the principal causes of colors in different insects**

Color	Taxon	Cause of color
Black*	Homoptera, Aphididae	Aphins
	Coleoptera	Melanin
	Diptera	Melanin
	Lepidoptera (larvae)	Melanin
Red	Odonata	Ommochromes
	Hemiptera, Heteroptera	Pterins and carotenoids
	Hemiptera, Coccoidea	Anthraquinones
	Coleoptera, Coccinellidae	Carotenoids
	Diptera, Chironomidae (larvae)	Porphyrin
	Lepidoptera	Ommochromes
Brown*	Lepidoptera	Ommochromes
	Many orders – eye colors	Ommochromes + pterins
Orange	Hemiptera, Heteroptera	Pterins
Yellow	Orthoptera, Acrididae	Carotenoids, flavonoids
	Hymenoptera	Pterins
	Lepidoptera, Papilionidae	Papiliochromes
Brassy yellow	Lepidoptera	Interference color
Bronze	Coleoptera, Scarabaeidae	Interference color
Gold	Coleoptera, Cassidinae	Interference color
	Lepidoptera, Danaidae (pupae)	Interference color
Green	Orthoptera	Insectoverdin (carotenoid + bilin)
	Lepidoptera (caterpillars)	Insectoverdin
	Lepidoptera, Zygaenidae	Interference color
	Diptera, Chironomidae (adults)	Bilin
Blue	Odonata	Interference color and ommochromes
	Lepidoptera	Interference color
	Orthoptera, Acrididae	Carotenoids
Ultraviolet	Lepidoptera, Pieridae	Interference color

Table 25.1 (*cont.*)

Color	Taxon	Cause of color
White	Hemiptera, Heteroptera	Uric acid
	Lepidoptera, Pieridae	Scattering + pterins
	Lepidoptera, Satyridae	Flavonoids
Silver	Lepidoptera, Danaidae (pupae)	Interference color
Iridescence	Coleoptera	Diffraction
	Coleoptera, Scarabaeidae	Interference colors

Note:
* Darkening may result from sclerotization.

wavelengths are reflected or transmitted in phase and are therefore reinforced, while others are out of phase and are cancelled out. The net result is that only certain wavelengths are reflected or transmitted and the surface appears colored.

Interference colors are common in adult Lepidoptera where the layers producing interference are formed by modifications of the scales. Each of the blue scales of the *Morpho rhetenor* butterfly, for instance, consists of a flat basal plate carrying a large number of near-parallel vertically aligned ridges that run parallel with the length of the scale (Fig. 25.2). Within each ridge are series of horizontal layers, separated by air spaces. Collectively, the horizontal layers in each adjacent ridge form a series of reflecting surfaces, which are spaced such that a blue color is produced by interference. Various optical appearances are produced by a great variety of scale modifications in many different Lepidoptera. For instance, strong ultraviolet reflectance at a peak wavelength of approximately 345 nm is generated in the butterfly *Colias eurytheme* by horizontal layered structures in its scale ridges that are 30–40 % thinner that those typical of iridescent *Morpho* butterflies.

Interference colors in other insects are produced by reflection at the interfaces of layers in the cuticle which differ in refractive index. The refractive indices of the alternating layers in the pupa of the danaid butterfly, *Euploea*, are 1.58 and 1.37. In jewel beetles (Buprestidae) and tiger beetles (Cicindellidae), these layers are in the exocuticle, but in tortoise beetles (Cassidinae) and some butterfly pupae they are in the endocuticle.

In some scarab beetles the reflecting surfaces are layers of chitin/protein microfibrils with a common orientation in the transparent exocuticle. Because their arrangement changes progressively in successive layers, creating helical orientation (Section 16.2), any given orientation recurs at intervals and, if the intervals are within the range of wavelengths of light, interference colors are produced. Due to this helical fibril orientation, the reflected light is circularly polarized (CP). The clockwise or anticlockwise orientation of fibrils through the exocuticle determines the right-handed or left-handed nature of the circular polarization. In the majority of cases, left-handed CP light is reflected from scarabs that exhibit this helical structure. However, in very few cases, *Plusiotis resplendens* for instance, both left- and right-handed CP light are reflected concurrently due to the presence of a multi-component helical structure.

(a)

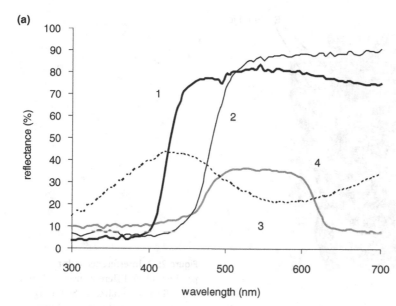

(b)

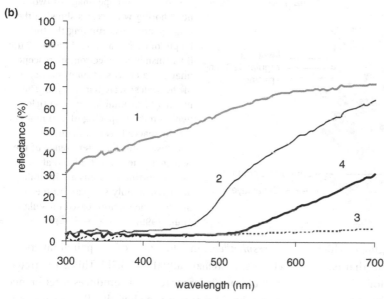

Figure 25.1 Representative reflectance spectra of insect body surfaces. Colors are produced by the absorbance of certain wavelengths, for example, by pigments in the insect cuticle; wavelengths that are not absorbed may reach a viewer's eye via reflectance or transmission. (a) Reflectance spectra from butterfly male forewings: (1) the white (and UV-absorbing) wing of the European cabbage white, *Pieris rapae*; (2) the yellow brimstone, *Gonepteryx rhamni* (Europe); (3) the purple (and UV-reflecting) spots on the wings of the Malayan jungleglory, *Thaumantis odana*; and (4) green parts of the wings of the Southeast Asian swallowtail butterfly, *Papilio lorquinianus*. (b) Reflectance of European bumble bee workers: (1) the white (and UV-reflecting) tip of the abdomen of the large earth bumble bee, *Bombus terrestris dalmatinus*; (2) the yellow band on the thorax of *B. t. dalmatinus*; (3) black parts of the abdomen of *B. t. dalmatinus*; (4) the brownish-red tip of the abdomen of the red-tailed bumble bee, *B. lapidarius*.

Whatever the nature or position of the reflecting surfaces which form the layer interfaces, the wavelength of the reflected light depends on their spacing. Viewing the surface from a more oblique angle is equivalent to increasing the optical path length between successive surfaces so that the color changes in a definite sequence as the angle of viewing is altered. This change in color with the angle of viewing is called iridescence and is a characteristic of interference colors.

A constant spacing between the reflecting surfaces produces relatively pure colors (Fig. 25.3a), as in the scales of blue *Morpho* butterflies (Fig. 25.2), or the moths *Urania leilus* and *Chrysiridia rhipheus*. In the majority of insect multilayer examples, spectral purity is enhanced when the reflecting layers are

(a)

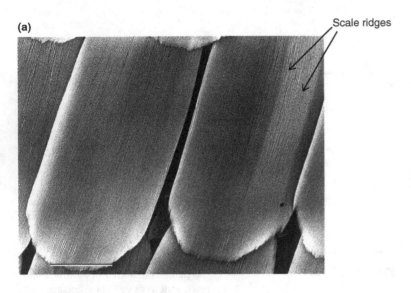

Scale ridges

(b)

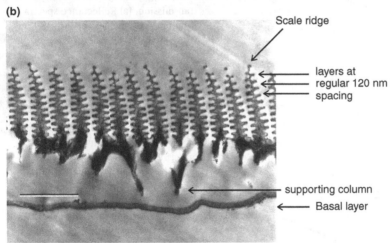

Scale ridge

layers at regular 120 nm spacing

supporting column

Basal layer

Figure 25.2 Interference color production in different insects. *Morpho* butterfly wing scales. (a) Scanning electron microscope image of two neighboring wing scales showing the ridging structures running the full length of each scale; scale bar = 50 μm. (b) Transmission electron microscope image of a cross-section of part of one of the scales; scale bar = 1 μm. The parallel horizontal layers of cuticle (with a regular spacing of 120 nm) and the air spaces between them are responsible for the interference of the reflected blue color. Notice that the supporting columns are irregularly arranged so only some are in the plane of the section (based on Vukusic *et al.*, 1999).

backed by a layer of melanin-based pigment which offers broad-band absorption to the light that is transmitted through the multilayer. Color conspicuousness is often enhanced by the presence of a strongly black border surrounding the brightly colored area, particularly so in some butterfly and weevil color patterning.

Broad-band colors such as gold, silver and bronze, by contrast, are produced if the spacing changes systematically with increasing distance from the surface of the cuticle. For example, in the endocuticle of the pupa of *Euploea*, and the exocuticle of

P. resplendens, the thickness of the paired layers changes systematically (Fig. 25.3b). This is referred to as a *chirped* layer structure. It produces interference over a broad range of wavelengths (Fig. 25.3b) and, in the cases of these insects, creates a gold mirror appearance to human eyes. Silver interference colors can be produced in the same way.

The peak reflected wavelength produced by interference, and also to some extent the brightness of the reflected colors, also depends on the refractive indices of the materials forming the layers. The horizontal layers in the ridges on blue *Morpho* wing

(a)

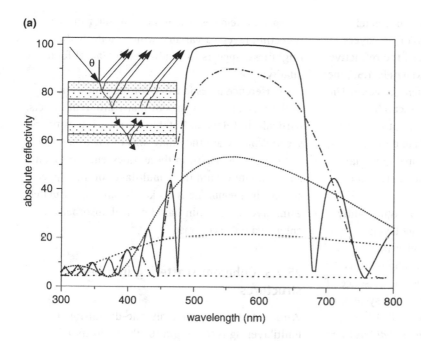

(b)

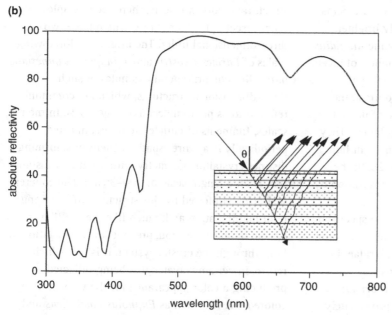

Figure 25.3 Theoretical modeling showing the spectral distribution of reflected intensities at multiple cuticular layers. (a) Reflected intensities from 1, 2, 3, 5 and 10 layers of cuticle, respectively, resulting in the increasing reflectances shown. In each case, air is the spacer layer and the normal incidence case is presented (dimensions used for modeling: cuticle layers 90 nm, and air layers 140 nm). Inset: a schematic diagram representing multiple reflections within a multilayer system leading to constructive interference. This multilayer comprises a single set of layer dimensions and this leads to a relatively narrow band of reflected wavelengths. (b) Reflection from a chirped multilayer (comprising ten cuticle layers of consecutively increasing thickness, spaced by air) producing the appearance of metallic gold. Inset: schematic diagram of multiple reflections in a chirped multilayer, showing the increasing thickness of cuticle layers through the system.

scales are uniformly separated by approximately 120 nm. Together with the thickness of the layers themselves, approximately 80 nm, and the refractive index of the cuticle material, approximately 1.56, the system produces a bright blue interference color. The paired layers in the exocuticle of the scarab beetle *Heterorrhina* are of similar thickness, but the interference color produced is green due to the different refractive indices of the paired cuticular layers compared with the air–cuticle interfaces in *Morpho*.

The brightness of the reflected color significantly increases with the number of reflecting layers. Generally, the insects that exhibit higher layer numbers are brighter in appearance, although this additional brightness may be offset by the presence of absorbing pigment within the layers in the system. In *Morpho rhetenor*, for instance, there are 8–12 horizontal layers per ridge and in *Morpho didius* there are 6–8 layers per ridge; quantitative measurements indicate that *M. rhetenor* is significantly brighter, reflecting 80% of incident blue light, while *M. didius* reflects approximately 40%. Similar numbers of layers are present in the cuticles of *Chrysochroa raja* and other iridescent buprestid beetles. By contrast, the metallic gold or silver cuticles of pupae of danaid butterflies can have over 200 reflecting layers. They too reflect approximately 80% of the incident light over a broad band of wavelengths (Fig. 25.3); the reflectivity is not higher due to optical absorption by the layers' constituent materials.

Apart from colors of the human visible spectrum, ultraviolet is also produced as an interference color. This occurs in some butterflies of the Pieridae. In *Eurema*, for example, certain scales possess analogous scale structures to those of *Morpho*; however, their reflecting layers are approximately 55 nm thick and are separated by air spaces of approximately 80 nm. This system reflects light with a peak wavelength of approximately 365 nm. The brightness of a similar ultraviolet-reflecting system in male *Colias eurytheme* butterflies has been shown to be an indication of their fitness to conspecific females. The number of their intra-ridge reflecting layers, a variable which controls this brightness, appears to be affected by larval food quality.

Interference is responsible for the iridescence of the membranous wings of many different insects, particularly Odonata. Notable for their intensity and spectral purity are the hindwings of the dragonfly *Neurobasis chinensis*. In these iridescent wings, an approximately 1 μm thick multilayer on the upper side of the membrane is backed by an approximately 2 μm layer of melanin-based optical absorber for enhanced color contrast.

25.2.3 Coherent scattering in other structures

While interference in mainly one-dimensional multilayering is the origin for the majority of structural colors in insects, there are examples of other insect optical structures which have two- or three-dimensional order. The brightly colored wing scales of *Parides sesostris* and *Teinopalpus imperialis* butterflies comprise good examples of such three-dimensional structures, which are commonly referred to as photonic crystals (Fig. 25.4). In their scales, filaments of cuticle are interconnected periodically in all three spatial dimensions, forming a distinct crystalline geometry with a lattice constant that is on a length scale of 200–30 nm. The coherent scattering conferred by this structure offers strong light manipulation and, much like the multilayer system in one dimension, prevents the propagation of light through the crystal system. It is this which results in certain wavelengths being reflected, producing a color appearance for the viewer. Many colored weevils, such as *Eupholus magnificus* and *E. schoeneri*, also exhibit this form of structure in their elytral and body scales. Commonly, in the lepidopteran and coleopteran scales that exhibit this three-dimensional photonic crystal structure, small grains of individual photonic crystal regions within each scale are oriented in different directions so that

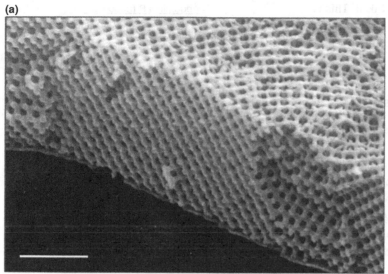

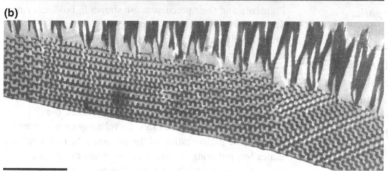

Figure 25.4 The three-dimensional photonic crystal present in the wing scales of the butterfly *Parides sesostris*. (a) Scanning electron microscope image showing the filaments of periodically interconnected cuticle with a fixed lattice constant enabling the scattering of a narrow band of color; scale bar = 1 μm. (b) Transmission electron microscope image of the scale region shown in (a); scale bar = 1.5 μm. This image reveals the highly geometric pattering that is associated with the two-dimensional projections of the actual three-dimensional crystal structures captured in various orientations from micron-sized domain to neighboring intra-scale domain within the scale. This variation in the angular orientation of each neighboring domain's photonic crystal leads to an overall color averaging effect.

they reflect specific different colors. As an ensemble, their overall effect is to create additive color mixing and render the color as angle-independent rather than truly iridescent.

25.3 Pigmentary colors

Pigments appear colored because they absorb certain wavelengths of light, the unabsorbed remaining light being scattered by various nanostructures. The energy of absorbed light is dissipated as heat. Which particular wavelengths are absorbed depends on the molecular structure of the compound. Particularly important in the production of color are the number and arrangement of double bonds, C=C, C=O, C=N and N=N. Particular functional groups are also important. The $-NH_2$ and $-Cl$ radicals, for example, shift the absorptive region of a compound so that it tends to absorb longer wavelengths. The color-producing molecule, known as a chromophore, is often conjugated with a protein molecule, forming a chromoprotein. Insects are able to synthesize most of their pigments, but not flavonoids or carotenoids which are, consequently, acquired through diet. The sources of some other pigments, found only in a few insects, are unknown. The black or brown of hardened cuticle often results from sclerotization (Chapter 16). However, cuticular hardening and darkening are not necessarily tightly linked. This is illustrated by albino strains of the locust *Schistocerca*, which have a hard but colorless cuticle.

25.3.1 Pigments that are synthesized

Melanin Dark cuticle is often the result of the presence of the pigment melanin, a nitrogen-containing compound which has been demonstrated in the cuticles of Blattodea, Diptera, adult and some larval Lepidoptera, and in Coleoptera. It is typically present as granules in the exocuticle. Insect melanin is a polymer of indole derivatives of tyrosine. Like sclerotization, melanization involves the production of DOPA and dopaquinone (Fig. 25.5). In *Drosophila*, various genes involved in pigment synthesis (*ebony*, *tan*) and patterning (e.g., *bric-a-brac*) have been identified. The synthesis of black melanin synthesis from DOPA requires the *yellow* gene in *Drosophila*, which is in turn regulated by the Engrailed and Abdominal-B proteins. Both bind to *cis*-regulatory sequences of the *yellow* gene; the former represses expression, while the latter activates it.

There are numerous pleiotropic interactions between color appearance and other traits, because the biochemical pathways for synthesis of cuticular pigmentation and other traits are interlinked, as demonstrated by studies of pigmentation mutants in various species of *Drosophila*. The genes *yellow*, *ebony* and *tan* all affect behavior in addition to pigmentation; they are not only expressed in epidermal cells but also in neurons, where their products are involved in synthesis of neurotransmitters; *yellow* has been implicated in sex-specific behavior. Melanin is also involved in wound healing, immune responses and the prevention of desiccation. These pleiotropies might place important constraints on evolutionary change in expression of

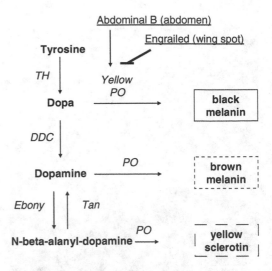

Figure 25.5 Melanin synthesis pathways in *Drosophila*. Pigments and their precursors are shown in bold, enzymes are shown in italics (DDC, dopadecarboxylase; PO, phenyloxidases; TH, tyrosine hydroxylase). In different cells different components of the pathway might be expressed, resulting in different pigmentation of the cuticle. Two proteins controlling expression of the *yellow* gene by binding to its *cis*-regulatory sequences are shown on top (underlined): in *D. melanogaster* males, the Abdominal-B protein activates the *yellow* gene, resulting in sex-specific pigmentation of the abdomen. In *D. biarmipes*, males but not females have a conspicuous black spot on the wing, and Engrailed protein represses the *yellow* gene and thus the generation of wing spots. Note that DOPA is also a key intermediate in cuticular sclerotization and, because of this, melanization and sclerotization may sometimes be linked together (redrawn from Wittkopp and Beldade, 2009).

these genes because a response to selective pressures on any one of the traits involved, e.g., cuticular pigmentation, might affect multiple other traits that all contribute to the fitness of the insect.

Pterins Pterins are nitrogen-containing compounds, all having the same basic chemical structure, a pyrazine ring and a pyrimidine ring, but differing in the radicals attached to this nucleus. They are synthesized from the purine guanosine triphosphate, and this synthesis involves the *rosy* and *purple* genes, which control eye color in *Drosophila*. Not all pterins appear colored. Some pterins are important metabolically as cofactors of enzymes concerned with growth and differentiation and may act as controlling agents in these processes. They often occur together with pigments of another group, the ommochromes, because they are cofactors of the enzymes involved in ommochrome synthesis. The vitamin folic acid also contains a pterin.

Pterin pigments can be white (leucopterin and isoxanthopterin, which absorb only in the ultraviolet), yellow (e.g., xanthopterin or dihydroxanthopterin, which absorb in the blue, but not necessarily as strongly in the UV-A) and orange to red (e.g., erythropterin, which absorbs blue, but not as much in the UV-A). They are important pigments in lepidopteran scales, where they are concentrated in pigment granules located on the crossribs. Leucopterin and xanthopterin are common in the wings of Pieridae, supplementing the structural white (Section 25.2.1). Males and females of the Japanese butterfly, *Pieris rapae crucivora*, both appear white to human observers, but only males' wings are adorned with beads, resulting in a sexual dichroism when viewed by conspecifics, such that only the males strongly absorb ultraviolet. In the orange sulfur butterfly, *Colias eurytheme*, pterin pigments interact with iridescent structural coloration, effectively amplifying the ultraviolet signal in males but not females. The yellow of the brimstone butterfly, *Gonepteryx*, is due to chrysopterin, the brighter color of the male resulting from a higher pigment concentration than is present in the female, while the red of the orange-tip butterfly, *Anthocharis*, is due to erythropterin. The yellows of Hymenoptera are produced by granules of pterins in the epidermis.

The pterins are also abundant in compound eyes, occurring with ommochromes in the screening pigment cells of the ommatidia. In this situation they are sometimes components of granules, but are often in solution. Several different pterins are present, not all of them colored. Their functions in the eyes are not fully understood, although some pterins are engaged in circadian rhythm regulation. They accumulate with age in the eyes of higher Diptera, and, because they are products of purine degradation, they may provide a means of storage excretion. The progressive accumulation provides a means of aging these insects. However, no comparable accumulation occurs in the eyes of the moth *Pectinophora*, although changes do occur in the pterins.

Ommochromes Ommochromes are yellow (xanthommatin), red (ommatin) and brown pigments usually occurring in granules coupled with proteins. Ommochromes (and their precursors such as kynurenin) typically absorb ultraviolet and might thus have a photoprotective function. The granules also contain accumulations of calcium. Xanthommatin is the most widely distributed ommochrome and is usually present wherever ommochromes are found.

Ommochromes, principally xanthommatin, are widely distributed in insects as screening pigments in the accessory cells of the eyes, usually associated with pterins. They are also present in the photoreceptor cells. It is thought that this is their original function in insects, and that its function in integument coloration in several insect taxa is derived. Yellow, red and brown body colors are produced by ommochromes in the epidermis. Xanthommatin turns from yellow to red upon reduction. The pink of immature adult *Schistocerca* is due to a mixture of ommochromes. Red Odonata, and probably also the reds and browns of nymphalid butterflies, are due to ommochromes, while in blue Odonata a dark brown ommochrome provides the background for the production of structural blue believed to stem from coherent

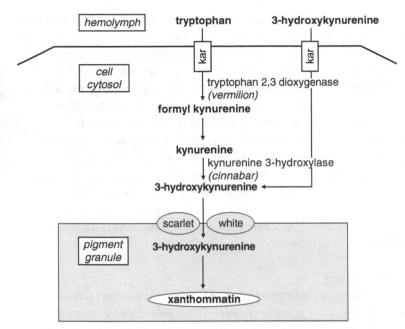

Figure 25.6 Ommochromes synthesis in the eye of *Drosophila melanogaster* (and similarly in butterfly wing patterns). Tryptophan is apparently moved into the cell by the amino acid transporter karmoisin (kar). The *vermilion* gene encodes for tryptophan 2,3 dioxygenase, an enzyme that catalyzes tryptophan's conversion to formyl kynurenine. Hydrolysis of formyl kynurenine results in kynurenine, whose transformation into 3-hydroxykynurenine in turns requires the *cinnabar* gene (and its product, kynurenine 3-hydroxylase). 3-hydroxykynurenine is also generated elsewhere in the body and taken up directly from the hemolymph. White and scarlet are transporter molecules that form a dimer localized in the surface of the pigment granules; they have been implicated in transporting ommatin precursors into the granule. Oxidative condensation of 3-hydroxykynurenine generates the ommochromes. Synthesis of other ommochromes involves the same pathway, differing in the final steps (after figure 1 in Reed and Nagy, 2005).

scattering from quasi-ordered arrays of particles in the endoplasmatic reticulum of pigment cells underlying the cuticle. Epidermal ommochromes sometimes directly underlie cuticular melanin, and in these cases they do not contribute to the insect's color.

The ommochromes are a group of pigments derived from the amino acid tryptophan via kynurenine and 3-hydroxykynurenine (both of which can also function as yellow pigments themselves). The biochemical pathway leading to their production involves the two enzyme genes *vermilion* and *cinnabar* (coding for tryptophan 2,3 dioxygenase and kynurenine 3-monooxygenase, respectively) and the ommochrome precursor

transporter gene *white*. Oxidative condensation of 3-hydroxykynurenine gives rise to the ommochromes (Fig. 25.6). In larval *Drosophila*, kynurenine production takes place primarily in the fat body, but its conversion to 3-hydroxykynurenine occurs in the Malpighian tubules, where it is stored. At metamorphosis, the 3-hydroxykynurenine is transported to the eyes, where the ommochromes are formed. In the adult fly, the whole process can take place in the cells of the eye, although this is normally supplemented by kynurenine and 3-hydroxykynurenine synthesized elsewhere. The scale-forming cells on butterfly wings also have the capacity to synthesize ommochromes from tryptophan or from

3-hydroxykynurenine. Transport of ommochromes in the hemolymph is achieved by specialized binding proteins.

Ommochrome production is the only way in which insects can remove tryptophan, which is toxic at high concentrations such as may occur at times of high protein turnover. A transitory increase in tryptophan occurs at metamorphosis in holometabolous insects, often followed by the production of ommochromes. In Lepidoptera ommochromes are accumulated in the meconium, the accumulated waste products of the pupal period which are voided immediately following eclosion. They are responsible for its characteristic red/brown coloration. Accumulation of ommochromes in the integument causes the larva of the puss moth, *Cerura*, to turn red just before pupation. Some of the ommochrome produced at the time of pupation contributes to the screening pigment in the eyes of the adult. Red fecal pellets containing ommochromes are produced by locusts during molting or starvation. At these times, excess tryptophan is likely to be liberated from proteins that are broken down during structural rearrangement or that are used for energy production.

Tetrapyrroles There are two major classes of tetrapyrroles: the porphyrins, in which the pyrroles form a ring, and the bilins, which have a linear arrangement of the pyrroles. The bilins are usually associated with proteins to form blue chromoproteins. Biliverdin occurs in many hemimetabolous insects, but is also found in Neuroptera and some Lepidoptera, although the latter usually contain other types of bilins. Associated with a yellow carotenoid, these pigments are responsible for the greens of many insects. Sometimes the pigments themselves are green. In *Chironomus* (Diptera), bilins derived from the hemoglobin of the larva accumulate in the fat of the adult and impart a green color to the newly emerged fly. In *Rhodnius*, the pericardial cells become green due to the accumulation of bilins derived from ingested hemoglobin.

A porphyrin having an atom of iron in its center is called a heme molecule and this forms the basis of two important classes of compounds, the cytochromes and the hemoglobins. In each case the heme molecule is linked to a protein. All insects are able to synthesize cytochromes, which are essential in respiration, the different cytochromes differing in the forms of their heme groupings. Normally they are only present in small amounts and so produce no color, but where they are present in high concentrations, as in flight muscle, they produce a reddish-brown color.

Some insects living in conditions subject to low oxygen tensions contain hemoglobin in the hemolymph, and these are colored red by the pigment showing through the integument. In *Chironomus* (Diptera) larvae the hemoglobin is in solution in the hemolymph, while in the larvae of *Gasterophilus* (Diptera) it is in hemoglobin cells. Many other insects contain hemoglobin in tracheal tissue and the fat body. Hemoglobin serves a respiratory function, but perhaps also serves as a protein store and enables the aquatic hemipteran *Anisops* to regulate its buoyancy (Section 17.9).

Papiliochromes Papiliochromes are yellow and reddish-brown pigments known only from the swallowtail butterflies, Papilionidae. Papiliochrome II is pale yellow and is formed from one molecule of kynurenine, derived from tryptophan, and one molecule of β-alanine.

Quinone pigments The quinone pigments of insects fall into two categories: anthraquinones (violet, blue and green) and aphins (red or purple). Both occur as pigments only among Homoptera (Hemiptera), the former in coccids (Coccoidea) and the latter in aphids only.

Anthraquinones are formed from the condensation of three benzene rings. In the coccids they give the tissues a red, or sometimes yellow, coloration. The best known is cochineal from *Dactylopius cacti*. The purified pigment is called carminic acid. It is present in globules in the eggs and fat body of the female, constituting up to 50% of the body weight. The larva contains relatively little pigment.

Aphins are quinone pigments formed from the condensation of three, in the monomeric forms, or seven benzene rings, in dimeric forms. They are found in the hemolymph of aphids, sometimes in high concentration, and impart a purple or black color to the whole insect. Neriaphin is a monomeric form. Quinone tanning is known to produce dark cuticle.

25.3.2 Pigments obtained from the food

Carotenoids Carotenoids are a major group of pigments that are lipid-soluble and contain no nitrogen. They are built up from two diterpenoid units joined tail to tail. In nearly all insect carotenoids, the central chain contains 22 carbon atoms with nine double bonds, and each of the end groups contains nine carbon atoms. There are two major groups of carotenoids: the carotenes, and their oxidized derivatives, the xanthophylls.

Yellow, orange and red are commonly produced by carotenoids, the color depending largely on whether or not the terminal groups are closed rings and on their degree of unsaturation. If the carotenoid is bound to a protein, the color may be altered, sometimes even resulting in a blue pigment. Insects cannot synthesize carotenoids and consequently must obtain them from the diet. Uptake from the food is, at least to some extent, selective. Orthopteroids preferentially absorb carotenes, while lepidopterans favor xanthophylls. Some post-ingestive modification of the carotenoids may occur.

Carotenoids can occur in many different tissues and in all stages of development. A number of structurally different carotenoids may be present in one insect. The possible metabolic functions of carotenoids in insects are not well understood. In other organisms they protect cells from damage due to photo-oxidation by light, but their importance to insects in this regard is not known. They are also the source of retinal in the insect's visual pigments. Apart from producing the reds and yellows of many insects, in combination with a blue pigment, often a bilin, they produce green. Green produced by a carotenoid with a bilin is sometimes known as insectoverdin. In some insects, the ability to sequester carotenoids might have functions beyond pigmentation; carotenoids also confer advantages with respect to photo-oxidative stress induced by ultraviolet radiation.

Flavonoids Flavonoids are heterocyclic compounds commonly found in plants. In insects they are mainly found among the butterflies and are common in Papilionidae, Satyridae and Lycaenidae as cream or yellow pigments. At least in some species, flavonoids are present in all developmental stages, including the egg, indicating that the flavonoids are stored in internal tissues of the adult as well as in the scales. Although some flavonoids function as deterrents to herbivores, it is clear this is not efficient for all insects. Where they occur in insects, flavonoids are acquired exclusively from herbivory, so the flavonoids present in their bodies reflect what is present in the host plants. Some post-ingestive modification of structure does occur so that the flavonoids in the insect are not exactly the same as those in the host. Perhaps these changes are produced by the insects themselves, but it is possible that their gut flora is responsible.

25.4 Color patterns

In general, insects do not have a uniform coloring, but have specific and often finely detailed patterns. The genetics and evolutionary developmental biology have been especially well studied in *Drosophila* and in the Lepidoptera. The development of black spots on fruit fly wings as well as the widespread lepidopteran eyespot patterns (with concentric rings of different colors) is controlled by a

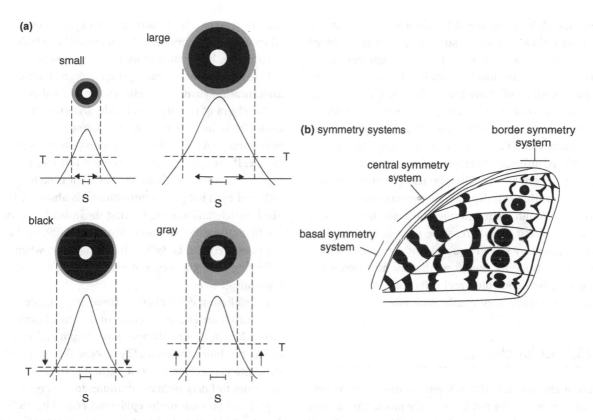

Figure 25.7 Color of eyespot patterning in the wings of Lepidoptera. (a) Control of variation in eyespot morphology. A signal (S), i.e., a "morphogen" diffuses outward (as indicated by horizontal arrows in the top two panels) from a central group of cells (the "focus" or organizing center), forming a concentration gradient as indicated by the quasi-Gaussian curves. It is thought that eyespot size is controlled by the strength of morphogen signal from the central cell group (top panels). In the surrounding epidermal cells, each scale-forming cell has the genetic capacity to form any of the pigments. In general, each scale cell only produces one pigment, depending on its response threshold (T) to one or more morphogens. Suppose that black pigment is synthesized if the response threshold to the morphogen is exceeded: a high threshold will result in a smaller area of black, whereas a lower threshold will generate a larger black area (bottom two panels; after Beldade and Brakefield, 2002). (b). A string of organizing centers on the wing's border generates a series of circular eyespots. Whether similar organizing centers generate the more proximal bands is not known (after Nijhout, 1994).

series of organizing centers from which morphogens diffuse outwards, creating a concentration gradient. Many of the wing-patterning genes first identified in *Drosophila*, such as *apterous*, *wingless*, *Distal-less*, *engrailed* and *cubitus interruptus*, are also involved in lepidopteran eyespot formation, where they are deployed in specific regions of the wings, often at specific times during development. In butterflies the wing pattern is produced by thousands of small overlapping scales. Morphogens govern the development of pigment in each scale-forming cell and different concentrations induce different pigments. Studies with tissue transplants have shown that the size of the eyespot is determined primarily by properties of tissue at the center, while the specific colors in each ring depend on response thresholds with which the host tissue responds to the chemical signal diffusing from the center. As diffusion extends uniformly outwards in all directions around an organization center, a symmetrical pattern is produced. It is also possible that a band of color rather than a series of separate rings can be produced by organization centers that are close together so that the patterns they produce grade into each other. On the forewings of many butterflies (Papilionoidea) there are rows of organizing centers that produce series of spots across the wing (Fig. 25.7). In general, it appears that each scale contains only a single pigment, although examples of scales with two pigments are known. The pigment produced by each scale-forming cell depends on synthesis within the cell. Although some intermediates may be manufactured elsewhere, scale color is not determined by selective uptake of presynthesized pigment. It seems, therefore, that the morphogens moving between epidermal cells regulate the synthetic machinery within each cell.

25.5 Color change

Color changes are of two kinds: short-term reversible changes which do not involve the production of new pigment, and long-term changes which result from the formation of new pigments and are not usually reversible. Short-term reversible changes are called physiological changes. Color changes involving the metabolism of pigments are called morphological color changes.

25.5.1 Physiological color change

Physiological color change is unusual in insects. It may occur where colors are produced physically as a result of changes in the spacing between reflecting layers, or where color is produced by pigments as a result of pigment movements.

Tortoise beetles change color when they are disturbed. Normally, the beetles are brassy yellow or green, but can change to brown-orange or red within minutes, presumably a form of aposematic coloration. In the Panamanian species *Charidotella egregia*, the gold coloration of the undisturbed state is generated by a chirped multilayer reflector, where porous patches in each layer contain humidity. A reduction in the state of hydration makes the cuticle translucent, revealing the red pigmentation underneath. Rehydration restores the original color.

The elytra of the Hercules beetle, *Dynastes*, are normally yellow due to a layer of spongy, yellow-colored cuticle beneath a transparent layer of cuticle. If the spongy cuticle becomes filled with liquid, as it does at high humidities, light is no longer reflected by it but passes through and is absorbed by black cuticle underneath, making the insect black. In the field, these changes probably occur daily, so that the insect tends to be yellow in the daytime, when it is feeding among leaves, and dark at night, making it less conspicuous.

Physiological color changes involving pigment movements are known to occur in the stick insect *Carausius*, in the grasshopper *Kosciuscola* and in a number of blue damselflies (Zygoptera, Odonata). All these insects become black at night due to the movement of dark pigment granules to a more superficial position in the epidermal cells (Fig. 25.8). *Kosciuscola* is blue during the day as a result of coherent scattering from small granules, less than 0.2 μm in diameter, believed to be composed mainly of white leucopterin and uric acid. At night, the blue is masked by the dispersal of larger pigment granules among the small reflecting granules which also disperse through the cell. Similar masking of structurally scattered blues occurs in a number of damselflies; in brown individuals of the stick insect *Carausius*, ommochrome granules occupy a superficial position at night, causing the insect to become darker, while during the day they occupy a proximal position in the epidermal cells, making the insects paler. Green specimens of *Carausius* do not change color because they lack ommochromes. The granules move along the paths of microtubules which may be responsible for the movements. Similar microtubules are present in the epidermal cells of *Kosciuscola*.

The color change in *Kosciuscola* and damselflies is temperature dependent. The insects are always black

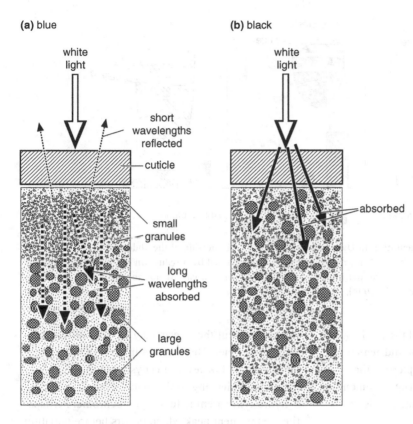

Figure 25.8 Physiological color change in *Kosciuscola*. The two diagrams have the same numbers of large and small granules (based on Filshie *et al.*, 1975). (a) At moderately high temperatures, the larger granules are restricted to the proximal parts of the epidermal cells. Short wavelengths of light are scattered by the small granules in the more distal parts of the epidermal cell, longer wavelengths travel further into the cell and are absorbed. The insects appear blue. (b) At lower temperatures, the pigments are generally dispersed through the cell. All the light is absorbed by the larger granules and the insects appear black.

below a temperature which is characteristic of the species, but usually about 15°C. At higher temperatures they tend to become blue, and, in some species, the change to blue is enhanced by light. Here, the epidermal cells are to some extent independent effectors, responding directly to stimulation. This is true of the change from black to blue in the damselfly *Austrolestes*, but the reverse change is controlled by a secretion released from the terminal abdominal ganglion. The significance of these changes is unknown, but they may be thermoregulatory. Dark insects absorb more radiation than pale ones so they may warm up more rapidly in the mornings and become active earlier than would be the case if they remained pale.

25.5.2 Morphological color change

Changes in the amounts of pigments can occur in response to a variety of external and internal factors.

Ontogenetic changes Many insects change color in the course of development. For example, the eggs of the plant-sucking bug *Dysdercus nigrofasciatus* (Hemiptera) are white when laid, becoming yellow as the embryo develops. The first-stage nymph is a uniform yellow color when it hatches, but becomes orange and then red. In the second instar, white bands appear ventrally on some of the abdominal segments (Fig. 25.9). These become more extensive and white bands are also present dorsally in the later stages. In the final larval instar, the red becomes less

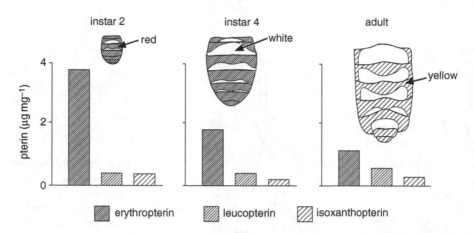

Figure 25.9 Ontogenetic changes in pigmentation in *Dysdercus nigrofasciatus*. Underside of the abdomen of second- and fourth-instar larvae and adult male. The change from red to yellow is accompanied by a reduction in the relative amount of erythropterin present (expressed as µg of pterin per mg body weight). The white areas are due to uric acid in the epidermal cells (data from Melber and Schmidt, 1994).

intense, especially in the female, and the adults are yellow with white stripes. The yellow and red colors of this insect are produced by three pterins, the proportions of which change through development to produce the different colors. The white bands are formed from uric acid.

It is also common for caterpillars to exhibit a regular change in color during development. The larva of the puss moth, *Cerura*, turns from green to red just before pupation as a result of the production of ommochrome in the epidermis. The early larva of the swallowtail butterfly, *Papilio demodocus*, is brown with a white band at the center, whereas the late larva is green with purple markings and a white lateral stripe.

In adult hemimetabolous insects, color change is often associated with aging and maturation. Adult male *Schistocerca* change from pink to yellow as they mature. The pink is produced by ommochromes in the epidermis which decrease in amount as the insect gets older, and the yellow is due to β-carotene, which increases with age. Color changes related to sexual maturation also occur in some Odonata.

These changes are controlled by changes in hormone levels associated with molting and sexual maturation, but the effects may be very different in different species. The changes in larval *Cerura* are initiated by a low level of ecdysteroid in the hemolymph, leading to the metabolism of 3-hydroxykynurenine to an ommochrome. Perhaps the commitment peak which occurs before pupation is normally responsible for the change (Section 15.4.2). Juvenile hormone leads to the accumulation of xanthommatin in *Bombyx* caterpillars, but prevents it in *Manduca*. Juvenile hormone, concerned in sexual maturation in adult *Schistocerca*, also regulates the accumulation of carotenoids which make the insects yellow as they become sexually mature. The hormone bursicon, which controls sclerotization, probably also regulates darkening occurring immediately after ecdysis.

Homochromy The colors of several insect species change to match the predominant color of the background. This phenomenon is called homochromy. The changes may involve the basic color of the insect or may involve a general darkening. For example, larvae of the grasshopper *Gastrimargus* tend to assume a color more or less matching the background when reared throughout

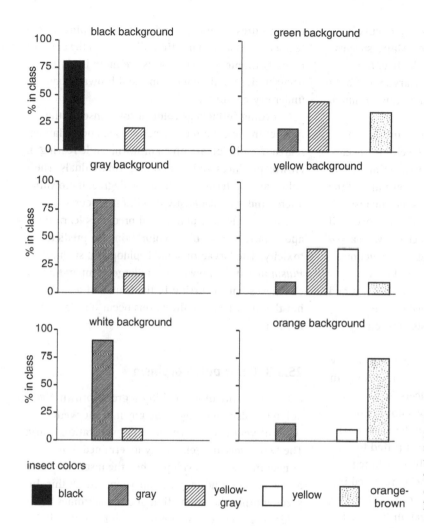

insect colors

Figure 25.10 Homochromy in the grasshopper *Gastrimargus*. Each graph shows the ground color of insects in the final larval stage after rearing from the first stage in containers with different colored backgrounds (after Rowell, 1970).

their lives on that background (Fig. 25.10). Individuals reared on black become black, those reared on white are pale gray. On a green background, however, most of these grasshoppers develop a yellowish coloration. The differences are produced by different amounts of black pigment, possibly melanin, in the cuticle and a dark ommochrome in the epidermis, together with yellow and orange pigments in the epidermis. The darkening of the stick insect *Carausius* that occurs on a dark background is also due to the accumulation of two ommochromes, xanthommatin and ommin. In this insect, changes

occur at any time during the course of a stadium, but in most other examples the change is first seen at a molt. Ommochrome deposition begins after about five days on the new background, and the changes in other insects generally require a similar period of exposure. Caterpillars, too, may exhibit some degree of homochromy. Larvae of the hawk moth *Laothoe populi* may be almost white or yellow-green, depending on the plants on which they are reared, and the pupae of some butterflies that are not concealed in a cocoon or in the soil may be green or dark or pale brown, according to their surroundings.

An extreme example of homochromy occurs in African grasshoppers after bush fires. Many species become black within a few days of the fire, the change occurring in adults as well as larvae, but only in bright sunlight. In diffuse light the change is much less marked.

All these changes depend on visual stimulation, although the details may vary from species to species. In *Carausius* darkening results from weak stimulation, or the absence of stimulation of the lower parts of the compound eye, such as would occur when the insect is on a dark surface. In *Gastrimargus* darkening occurs if a high incident light intensity is associated with a low reflectance background, and changes in the yellow and orange pigments are dependent on the wavelength of light. Homochromy in caterpillars and lepidopteran pupae is also determined by the wavelength of light and the contrasts in intensity reaching the different stemmata.

Visual inputs affect the activity of neurosecretory cells in the central nervous system. Homochromy in pupae of the peacock butterfly, *Inachis io*, is produced by changes in the relative amounts of melanin and lutein (a carotenoid) in the cuticle. The accumulation of both pigments is controlled by a single neuropeptide, which is widely distributed in the central nervous system before being released into the hemolymph. A low hormone titer immediately before pupation stimulates melanization, a higher titer results in increased incorporation of lutein into the pupal cuticle. In the grasshopper *Oedipoda miniata*, the dark-color-inducing neurohormone (DCIN) has been identified as (His[7])-corazonin.

The stink bug *Nezara viridula* (Hemiptera) changes from green to brown during diapause when overwintering under leaf litter or bark, and changes back to green again in spring, presumably to aid camouflage; these changes are induced in part by day length; temperature can also have an effect in some species.

Other factors affecting color Temperature is important in pigment development and there is a general tendency for insects reared at very high temperatures to be pale, and those developing at low temperatures to be dark. Humidity affects the color of many Orthoptera. Green forms are more likely to occur under humid conditions, and brown forms under dry conditions.

Crowding influences color in some insects, the most extreme examples being locusts. Locust larvae reared in isolation are green or fawn, while rearing in crowds produces yellow and black individuals. The colors and patterns change as the degree of crowding alters; under high-density conditions, locusts consume toxic host plants, and predators learn the aposematic yellow–black coloration as a predictor of toxicity. The larvae of some Lepidoptera, such as *Plusia* and the armyworms, undergo comparable changes, some occurring in the course of a stadium, but the most marked alterations occur only at molting.

25.5.3 Color polymorphism

Insects of many orders exhibit a green–brown polymorphism, tending to be green at the wetter times of year and brown when the vegetation is dry. The two forms are genetically determined, but homochromy that develops when the insect moves to a new background may be superimposed on this. In green morphs the production of ommochromes in the epidermis is largely or completely inhibited; if it occurred, the green would be obscured.

In some insects there are marked differences in color between successive generations, correlated with seasonal changes in the environment. Such seasonal polyphenism and its physiological regulation is discussed in Section 15.5. Extreme cases of color polymorphism occur in some butterflies in relation to mimicry (Section 25.6.1).

25.6 Significance of color

Insect pigments are often the end products of metabolic processes and may have evolved originally as forms of storage excretion. Pterins, for example, may be derived from purines, such as uric acid. Similarly, melanin production might be a method of disposing of toxic phenols ingested or arising from metabolism and it may be significant that melanin in the cuticle is often produced above metabolically active tissue such as muscle. Tryptophan in high concentrations reduces the rate of development of *Drosophila* and *Oryzaephilus* and it is noteworthy that ommochrome production follows the appearance of unusually high levels of tryptophan in the tissues. However, it is clear that storage excretion is not the sole, or even primary, function of pigments in most insects. For example, most insects excrete most of the end product of the biologically active hydrogenated pterins and do not store them. *Pieris*, on the other hand, synthesizes much larger amounts than would result from normal metabolism and accumulates them in the wings, where they contribute to the color.

In most cases, the colors of present-day insects are ecologically important. Color is most frequently involved as a defense against vertebrate predators and is also important in intraspecific recognition. It also has important consequences for body temperature, and color changes may contribute to thermoregulation (Chapter 19).

25.6.1 Predator avoidance

In many species, color and color patterns have evolved as part of a strategy to avoid predation. The color patterns may function in a variety of ways in different insects: for concealment (crypsis), to startle a predator (deimatic behavior), to deflect attack from the most vulnerable parts of the body or to advertise harmfulness or distastefulness (aposematism).

Crypsis Color often helps to conceal insects from predators. Many insects are known to select backgrounds on which they are least conspicuous. Often, homochromy is associated with some appropriate body form and behavior as in stick insects and many mantids and grasshoppers which

may be leaf-like or twig-shaped according to the backgrounds on which they normally rest. Protection may also be afforded by obliterative shading. Objects are made conspicuous by the different light intensities which they reflect as a result of their form. A solid object usually looks lighter on the upper side and darker beneath because of the effect of shadows, but by appropriate coloring this effect can be eliminated. If an object is shaded in such a way that when viewed in normal lighting conditions all parts of the body reflect the same amount of light, it loses its solid appearance. Such countershading is well known in caterpillars, where the surface toward the light is most heavily pigmented and the side normally in shadow has least pigment. To be successful in concealing the insect, this type of pigmentation must be combined with appropriate behavior patterns; if the larva were to rest with the heavily pigmented side away from the light it would become more, not less, conspicuous. Color may also afford protection if the arrangement of colors breaks up the body form. Such disruptive coloration is typically most efficient when some of the color components match the background and others contrast strongly with it. The color pattern should introduce high-contrast internal boundaries on the body surface, which surpass the salience of the outline. Disruptive coloration occurs in some moths which rest on tree trunks.

Deimatic behavior Some insect species have colored wings, or other parts of the body, which are normally concealed, but are suddenly displayed when a potential predator approaches. This behavior, sometimes associated with the production of a sound, has been shown, in a few cases, to startle the predator. It is known as deimatic behavior. In many insects exhibiting this behavior, the hindwing is deep red or black. It is normally concealed beneath the forewing, but is revealed by a sudden partial opening of the wings. This occurs in a number of stick insects and mantids, and in moths that normally rest on vegetation. Moths in the family Arctiidae have bright red or yellow abdomens, often with black markings. When disturbed, the abdomen is displayed by opening the wings. The deimatic display of some mantids involves the front legs, which have conspicuous marks on the inside. *Galepsus*, for example, displays the insides of the fore femora and coxae, which are orange, at the same time exposing a dark mark in the ventral surface of the prothorax.

Some Lepidoptera have a pattern of scales on the hindwing forming an eyespot which is displayed when the insect is threatened. One of the best-known examples is the peacock butterfly, *Inachis io*. It has one eyespot on the upper surface of each wing. These eyespots are primarily black, yellow and blue, surrounded by dark red. The butterfly rests with its wings held up over its back, the upper surfaces of the forewings juxtaposed so that the eyespots are concealed. If the insect is disturbed by visual or tactile stimuli it lowers the wings so that the eyespots on the forewings are displayed and then protracts the forewings to expose the hindwing eyespots. At the same time the insect makes a hissing sound with a timbal on the forewing (Section 26.3.3). The forewings are then retracted and partly raised and the sequence of movements repeated, sometimes for several minutes. While displaying, the body is tilted so that the wings are fully exposed to the source of stimulation and at the same time the insect turns so as to put the stimulus behind it. This eyespot display causes flight behavior in at least some birds. Some mantids also have a large eyespot on the hindwing. It remains controversial whether these eyespots obtain their startling function by resembling eyes, or whether their highly contrasting patterns present a more general aposematic pattern (see below) or induce neophobia.

Deflection marks Small eyespots, often present on the underside of the hindwings of butterflies, appear to deflect the attention of birds away from the head

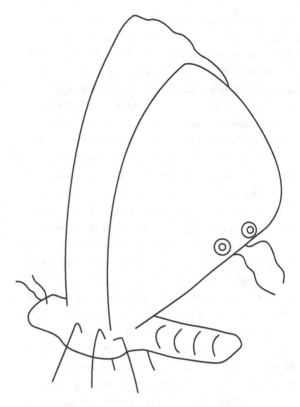

Figure 25.11 Deflection marks on the hindwing of a lycaenid butterfly. Two eyespots paired with false "antennae" generate the impression of a head, deflecting predator attacks away from body parts where a bite might be fatal (figure 3 from Stevens, 2005).

of the insect. There is no sharp distinction between eyespots used for deimatic behavior and those concerned with deflection. In general, deflecting spots are probably smaller than those used in intimidation, but it is possible that some may serve either function, depending on the nature and experience of the predator. Some butterfly species (in the families Lycaenidae, Riodinidae and Nymphalidae) have appendages on the hindwings that look like antennae or legs, paired with eyespot patterns, creating the illusion of the insect's head (Fig. 25.11). Observations from predator-induced wing damage indicate that these "false heads" are indeed efficient in deflecting vertebrate attacks to

Table 25.2 **Aposematic insects; examples from different orders**

Order and species	Stage	Color	Basis of unpalatability
Orthoptera			
Zonocerus variegatus	Adult	B/Y and R markings	Pyrrolizidine alkaloids
Hemiptera			
Aphis nerii	All	Bright Y	Cardiac glycosides
Oncopeltus fasciatus	Adult	Y/B spots	Cardiac glycosides
Coleoptera			
Coccinella septemfasciata	Adult	R/B spots	Alkaloids
Tetraopes oregonensis	Adult	R	Cardiac glycosides
Hymenoptera			
Vespula vulgaris	Adult	Y/B stripes	Sting
Lepidoptera			
Tyria jacobaeae	Larva	B/Y stripes	Pyrrolizidine alkaloids
Tyria jacobaeae	Adult	B/R marks	Pyrrolizidine alkaloids
Battus philenor	Larva	R/B spots	Aristolochic acids
Danaus plexippus	Larva	W/B and Y stripes	Cardiac glycosides
Zygaena filipendula	Adult	B/R spots	Cyanogenic glycosides

Note:
B, black; Y, yellow; R, red; W, white.

less vulnerable body parts, giving the butterfly a chance to escape.

Aposematic coloration Many insects are distasteful by virtue of chemicals they produce themselves, or that are sequestered from their food; other insects sting (Section 27.9). Such insects are commonly brightly colored, and are usually red or yellow combined with black. Such coloration is a signal to predators that the potential prey is distasteful and should be avoided. It is called aposematic coloration. In ladybird beetles, for example, the degree of carotenoid-based redness in the elytra of individuals correlates with alkaloid content. For such coloration to be effective, the predator must exhibit an innate or a learned avoidance response. Aposematic species occur in many orders of insects (Table 25.2). Development of aposematic coloration is dependent upon population density in some species of locust and grasshopper.

Mimicry Predators learn to avoid distasteful insects with distinctive colors. If the color patterns of some species are similar to each other, learning to avoid one species because it is distasteful also produces an avoidance of the other. Resemblance of one species to another is called mimicry. Mimicry takes two forms, Müllerian and Batesian. Species exhibiting Müllerian mimicry are all distasteful. Here, the advantage to the insects is that predation on any one species is reduced. For example, many social wasp species – such as *Vespa* and *Vespula*, which all have a sting – have the same basic black and yellow pattern; if a predator learns to avoid one species, it is likely to avoid others with a similar appearance. Sometimes the mimics have different lifestyles. Cotton stainer bugs of the genus *Dysdercus* usually have similar red and black coloration and all the species are distasteful. The predaceous reduviid bug *Phonoctonus* lives with *Dysdercus* and preys on them. It, too, is unpalatable to predators, and is a color mimic of *Dysdercus* so that avoidance of one leads to avoidance of both species. Müllerian mimicry is also common among Lepidoptera, with the genus *Heliconius* having been especially well studied. The adaptive polymorphism in *H. erato* is orchestrated by a precise temporal and spatial expression pattern of the *cinnabar* and *vermilion* genes, coding for enzymes in ommochrome production.

In Batesian mimicry, only one of a pair of species is distasteful, the other is not. They are called the "model" and "mimic," respectively. Here, the palatable species gains some advantage from a resemblance to a distasteful species. If edible mimics become too common relative to the model, a predator might learn to associate a particular pattern with palatability rather than distastefulness. This limits the numbers or distribution of a mimetic form, but the limit may be circumvented by the mimic becoming polymorphic, with each of its morphs resembling a different distasteful species. The best-known example of such polymorphism is that of the female *Papilio dardanus* (Lepidoptera), which has a large number of mimetic forms mimicking a series of quite different-looking butterflies. It appears that alleles of a single gene explain all the natural variants of this species.

Because the unpalatability of an individual may be affected by the nature of the food it eats, mimicry may vary temporally and spatially. For example, the viceroy butterfly (*Limenitis archippus*) and queen butterfly (*Danaus gilippus*) of North America are both distasteful and normally exhibit Müllerian mimicry. However, sometimes larvae of the queen butterfly feed on plants that are so low in the cardenolides sequestered by the insect that the resulting adults are not distasteful. It may be supposed that the palatable queen butterflies now depend for protection on their resemblance to other members of the species that have sequestered cardenolides, and on unpalatable viceroy butterflies. A special case of Batesian (deceptive) mimicry occurs where some insects mimic the visual appearance of their own predators, such as jumping spiders.

25.6.2 Intraspecific recognition

Color is important in intraspecific recognition in many diurnally active insects, and its role is most fully understood in damselflies and dragonflies (Odonata), and in butterflies. It often has two principal functions in male behavior: the recognition of females and the recognition of conspecific males; females also select males by coloration.

In some dragonflies, male color is important in defending a territory against other males. For example, the dorsal side of the abdomen of male *Plathemis lydia* is blue. If a second male enters a territory, the resident male faces the intruder and displays the color by raising the abdomen toward the vertical. This has an inhibiting effect on the intruding male. In the presence of a female, the abdomen is depressed so that the blue is not visible as the male approaches her. The females of many species change color as they become sexually mature and this is associated with a change in male behavior toward them. Among the damselflies, the females of some species are dimorphic, with one of the morphs resembling the male, and there are distinct differences in the behavior of males toward these two morphs.

In butterflies the general color of a female may be more important than the details of pattern in attracting males, although size and movement are also important. For example, males of the African butterfly, *Hypolimnas misippus*, are attracted by the female's red-brown wing color; the black and white markings sometimes present on the forewings are unimportant, although white on the hindwing may have an inhibitory effect. Males of *Colias eurytheme* are attracted by the yellow underside of the female hindwing, which is exposed when the female is at rest. In this species the male reflects ultraviolet from the upper side of its wings. This signal is used by the female in interspecific mate recognition, and it also inhibits attraction by other males, reducing the likelihood of further male intervention when a male is copulating. Such female mate recognition can have an important role in reproductive isolation between species. Strength of pigmentation or structurally based ultraviolet iridescence can depend on an individual's ability to acquire resources, and therefore be indicative of phenotypic or genotypic mate quality. In some social wasps, facial patterns differ between members of the nest, and colony members can remember the individual appearance of a nest mate.

25.7 Light production

Intrinsic luminescence, i.e., light produced by the insects, is known to occur in various Coleoptera, primarily in the families Lampyridae (fireflies), Elateridae (click beetles) and Phengodidae (railroad worms), as well as in a few Collembola, such as *Onychiurus armatus*, in the homopteran *Fulgora lanternaria*, some larval Diptera of the families Platyuridae and Bolitophilidae.

The light-producing organs occur in various parts of the body. The collembolan *Onychiurus* emits a general glow from the whole body. In most beetles the light organs are relatively compact, and are often on the ventral surface of the abdomen. In male

Photuris (Coleoptera) there is a pair of light organs in the ventral region of each of the sixth and seventh abdominal segments. In the female the organs are smaller and often only occur in one segment. The larvae have a pair of small light organs in segment eight, but these are lost at metamorphosis when the adult structures form. Larvae and females of railroad worms (Phengodidae, Coleoptera) have 11 pairs of dorso-lateral light organs on the thorax and abdomen and another on the head. In *Fulgora* (Hemiptera) the light organ is in the head. The light organs are generally derived from the fat body, but in the glow worm fly *Arachnocampa* (Diptera) they are formed from the enlarged distal ends of the Malpighian tubules.

25.7.1 Structure of light-producing organs in Coleoptera

Each light organ of an adult firefly consists of a number of large cells, the photocytes, which lie just beneath the epidermis and are backed by several layers of cells called the dorsal layer cells (Fig. 25.12). The cuticle overlying the light organ is transparent. The photocytes form a series of cylinders at right angles to the cuticle, with tracheae and nerves running through the core of each cylinder. Each trachea gives off branches at right angles and as these enter the region of the photocytes they break up into tracheoles, which run between the photocytes parallel with the cuticle. The tracheoles are spaced 10–15 µm apart, and as the photocytes are only about 10 µm thick the diffusion path for oxygen is short. The origin of the tracheoles is enclosed within a large tracheal end cell, the inner membrane of which is complexly folded where it bounds the tracheolar cell (Section 17.1.2). In some species the end cells are only poorly developed. The neurons entering the photocyte cylinder end as spatulate terminal processes between the plasma membranes of the end cell and the tracheolar cell within which the tracheoles arise. In adult *Pteroptyx* (Coleoptera)

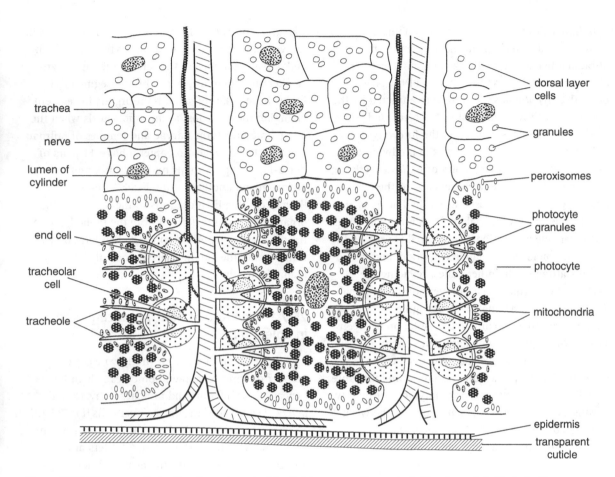

trachea

nerve

lumen of cylinder

end cell

tracheolar cell

tracheole

dorsal layer cells

granules

peroxisomes

photocyte granules

photocyte

mitochondria

epidermis

transparent cuticle

Figure 25.12 Diagrammatic section through part of the light organ on the ventral side of an adult *Photuris*. The tracheoles pass between the photocytes, but do not penetrate into the cells (based on Smith, 1963).

and some other genera, nerve endings occur on the photocytes as well as on the tracheal end cells.

The photocytes are packed with photocyte granules, each containing a cavity connecting with the outside cytoplasm via a neck. It is presumed that the reactants involved in light production are housed in these granules. Smaller granules also occur dorsally and ventrally. Mitochondria are sparsely distributed except where the cell adjoins the end cells and tracheoles. The dorsal layer cells also contain granules, generally understood to comprise urate

crystals, which form a reflecting layer at the back of the light-producing region which directs the luminescence more efficiently in the outward direction. It is estimated that the two lanterns of *Photinus* together contain about 15 000 photocytes, forming some 6000 cylinders, each with 80–100 end cells. The lanterns of larval fireflies contain the same elements, but their organization is simpler. The tracheal system is diffuse and there are no tracheal end cells. Nerve endings occur on the photocytes, and are not separated from them by the tracheal end cells as in adult *Photuris*.

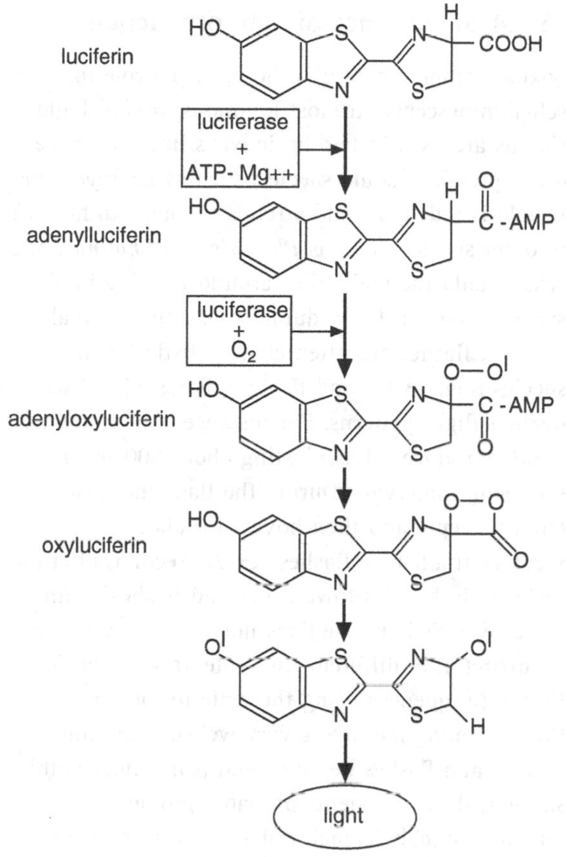

Figure 25.13 Basic reactions involved in light production from luciferin.

25.7.2 Mechanism of light production

Light is produced in organelles called peroxisomes, noted as centers for enzymatic oxidation reactions. A two-stage reaction occurs within these peroxisomes. First, adenylation of substrate luciferin (which is dependent on the presence of magnesium-ATP) occurs under the catalytic action of luciferase. The subsequent oxygenation of luciferyl adenylate by molecular oxygen results in the emission of light and the production of oxyluciferin (Fig. 25.13).

The reaction is very efficient, some 98% of the energy involved being released as light. Furthermore, in fireflies the metabolic cost of the light-flash production is low. It increases metabolic rates by approximately 40%, compared to 60% for walking activity.

In many insects the light produced by the bioluminescence organs is yellow-green in color, extending over a relatively narrow band of wavelengths, 520–650 nm in *Photinus* and *Lampyris* (Coleoptera). In larval and adult female railroad worms (Coleoptera) such as *Phrixothrix*, the light organs on the thorax and abdomen produce green to orange light, depending on the species, in the range 530–590 nm. That on the head produces red light with peak emission at 620 nm, but extending from about 580 nm to over 700 nm. The light produced by *Arachnocampa* (Diptera) is blue-green, that of *Fulgora* (Hemiptera) white. In fireflies, despite noticeable spectral differences in light production, the luciferin substrate is identical and their DNA similarity is 70–99%. The observed differences in spectral output are caused by single amino acid substitutions; namely, with variations in the luciferase active site that controls the conformation of the bound substrate.

25.7.3 Control of light production

The light organs of adult *Photuris* (Coleoptera) are innervated by three and four dorsal unpaired median (DUM) cells in the last two abdominal ganglia, respectively, which release octopamine. Like other DUM cells (Section 10.3.2), the axons from these cells divide to send symmetrical branches to the lanterns on each side. In most adult fireflies the axons terminate on the tracheal end cells, but in larvae, where there are no end cells, they innervate the photocytes directly. The lantern end cells, nerve endings and tracheolar cells form the "end organ" complexes. These, together with the enclosed tracheal channels, play a dominant role in the control of light production since they are only observed in species exhibiting flashing lanterns. No flashing lantern is observed in *Photuris* larva, but there is a slowly rising and falling light production capacity. Peroxisomes and mitochondria are also present in the larva but are unsegregated and no specializations are observed in the tracheal cell compartments.

In adult *Photuris*, light production appears to be regulated by the availability of oxygen. As the DUM neurons terminate on the tracheal end cells it is presumed that neural activity causes a change in these cells, facilitating the flow of oxygen to the photocytes. Flash duration in adult fireflies is generally very brief, a few hundred milliseconds, with flashes following each other at regular intervals. This implies that the oxygen supply to the photocytes is closely regulated. It is believed that hydrogen peroxide plays a role in this regulation: the peroxisome oxidases use oxygen arriving at open mitochondria to create hydrogen peroxide that builds up explosively due to the shutdown of the catalase. This completes the oxidation reaction and triggers the flash. The precise temporal control of firefly flashing is understood to be regulated by nitric oxide (NO). NO synthase is localized near synaptic terminals within the firefly lantern and measurements indicate that externally added NO stimulates bioluminescence production while the addition of NO scavengers inhibits light production. Furthermore, NO is known to control respiration by photocyte mitochondria reversibly. The proposed mechanism comprises neural stimulation resulting in the transient release of NO that diffuses into the periphery of adjacent photocytes. This inhibits mitochondrial respiration and permits oxygen to diffuse into the photocyte, which holds the bioluminescence reactants.

Species of fireflies of the genus *Pteroptyx* (Coleoptera), notably in Southeast Asia, form male groups in which the individual insects flash in synchrony. The males of these species, in isolation, can flash regularly with almost constant intervals between flashes. If an individual detects a flash within a critical period of having produced its own flash, it immediately flashes again. As the individuals produce flashes at regular intervals, this resetting rapidly results in synchronous flashing by all the insects.

25.7.4 Significance of light production

Sexual communication is the principal role for self-luminescence in most luminous insects. Light signals are used in two basic ways in Lampyridae (Coleoptera). In some species, such as *Lampyris*, the female is sedentary and attracts the male to herself; in other species, such as *Photuris* and *Photinus*, one sex, usually the male, flies around flashing in a specific manner. Flash duration and the interval between flashes are often characteristic for the species (Fig. 25.14) and flashing is associated with distinct flight patterns. For instance, male *Photinus pyralis* produce a flash lasting about 500 ms at six-second intervals. During the flash the male climbs steeply and then hovers for about two seconds. If a female flashes 1.5–2.5 seconds after the male flash, he flies toward her and flashes again three seconds later. He does not respond to flashes occurring after different time intervals. Repeated flashing sequences bring the male to the female. Precise timing requires a very well-defined time marker and flashes begin or end (sometimes both) sharply; these transients probably provide the temporal signals. Females of *Pteroptyx* are attracted by the flashes of the male group.

After mating, the females of some *Photuris* species change their behavior so their patterning of flashing comes to resemble that of females of other species. In this way they lure males of those species on which they then feed. The luminescence of *Arachnocampa* (Diptera) larvae also serves as a lure, attracting small insects into networks of glutinous silk threads on which they become trapped. The larvae then eat them.

In railroad worms (Coleoptera) the lateral light-producing organs may be suddenly illuminated if the insect is attacked, and they possibly have a defensive function. It is suggested that the red head light provides these insects with illumination, presumably when they are searching for the millipedes on which they prey. The relatively long wavelengths emitted by this organ may not be visible

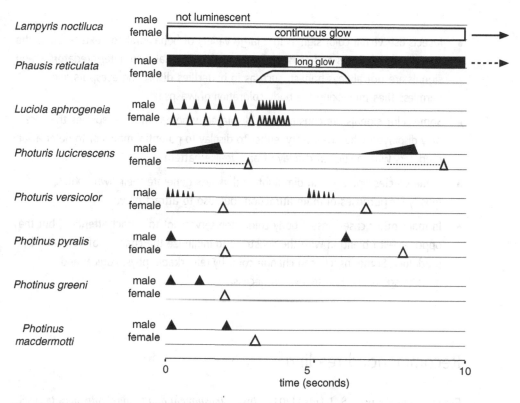

Figure 25.14 Flash patterns of different fireflies (Lampyridae). The height of each symbol represents the intensity of the flash; the shape shows intensity rising to a maximum and then declining. The dotted line in female *Photuris lucicrescens* shows the interval during which the female flash may occur (based on Case, 1984).

to most other insect predators, but the eyes of the railroad worms probably can detect them.

The sensitivity of the dark-adapted compound eyes of fireflies varies diurnally, as it does in at least some other insects. It increases rapidly by about four orders of magnitude at a time approximately corresponding with the time at which the insects normally flash.

Summary

- There is a spectacular variety of insect color patterns, such as in the butterflies, dragonflies and beetles. The ways in which butterflies, for example, generate iridescence by physical structures show nature's nanotechnology at its best. The way in which pigment patterns are generated by the interaction of morphogens and response thresholds in flies and butterflies has become a model system in evolutionary developmental biology.

- Insects use visual color signals in a large variety of behavioral contexts, such as the recognition of mates and conspecific competitors, but also to deter predators – and signals are not always honest, such as in butterflies displaying eyespots and harmless flies mimicking the body coloration of wasps.

- Some color signals are constitutively displayed, whereas in other species they are only displayed when necessary, either to display to potential mates or to deter avian predators with a sudden display of an eye-like pattern.

- Some species operating in dim light or darkness generate their own light, mostly for purposes of mate attraction, but also to attract prey.

- In many other cases, insect body coloration serves not to attract attention, but the opposite – to blend in with the substrate to minimize the chances of detection by predators. Some insects can change color by remarkable physiological and morphological mechanisms when necessary.

Recommended reading

Casas, J. and Simpson S. J. (eds.) (2011). *Insect Integument and Colour, Advances in Insect Physiology*, vol. **38**. London: Academic Press.

Ruxton, G. D., Sherratt, T. N. and Speed, M. P. (2004). *Avoiding Attack*. Oxford: Oxford University Press.

Vukusic, P. and Sambles, J. R. (2003). Photonic structures in biology. *Nature* **424**, 852–855.

Wittkopp, P. J. and Beldade, P. (2009). Development and evolution of insect pigmentation: genetic mechanisms and the potential consequences of pleiotropy. *Seminars in Cell & Developmental Biology* **20**, 65–71.

References in figure captions

Beldade, P. and Brakefield, P. M. (2002). The genetics and evo-devo of butterfly wing patterns. *Nature Reviews Genetics* **3**, 442–452.

Case, J. F. (1984). Vision in mating behavior of fireflies. *Symposium of the Royal Entomological Society of London* **12**, 195–222.

Filshie, B. K., Day, M. F. and Mercer, E. H. (1975). Colour and colour change in the grasshopper, *Kosciuscola tristis*. *Journal of Insect Physiology* **231**, 1763–1770.

Melber, C. and Schmidt, G. H. (1994). Quantitative variations in the pteridines during postembryonic development of *Dysdercus* species (Heteroptera: Pyrrhocoridae). *Comparative Biochemistry and Physiology B* **108**, 79–94.

Nijhout, H. F. (1994). Symmetry systems and compartments in lepidopteran wings: the evolution of a patterning mechanism. *Development* (1994 Supplement), 225–233.

Reed, R. D. and Nagy, L. M. (2005). Evolutionary redeployment of a biosynthetic module: expression of eye pigment genes vermilion, cinnabar, and white in butterfly wing development. *Evolution and Development* **7**, 301–311.

Rowell, C. H. F. (1970). Environmental control of coloration in an acridid, *Gastrimargus africanus* (Saussure). *Anti-Locust Bulletin* **47**, 1–48.

Smith, D. S. (1963). The organization and innervation of the luminescent organ in a firefly, *Photuris pennsylvanica* (Coleoptera). *Journal of Cell Biology* **16**, 323–359.

Stevens, M. (2005). The role of eyespots as anti-predator mechanisms, principally demonstrated in the Lepidoptera. *Biological Reviews* **80**, 573–588.

Vukusic, P., Sambles, J. R., Lawrence, C. R. and Wootton, R. J. (1999). Quantified interference and diffraction in single *Morpho* butterfly scales. *Proceedings of the Royal Society of London B* **266**, 1403–1411.

Wittkopp, P. J. and Beldade, P. (2009). Development and evolution of insect pigmentation: genetic mechanisms and the potential consequences of pleiotropy. *Seminars in Cell & Developmental Biology* **20**, 65–71.

26 Mechanical communication: producing sound and substrate vibrations

REVISED AND UPDATED BY **RALF HEINRICH**

INTRODUCTION

Insects can produce sounds that spread through air or water, and vibrations transmitted through the substrate on which they are resting. Sounds and vibrations may result from the insect's normal activities, such as the sounds produced by wing vibration in flight or by the mandibles of an insect chewing. In addition, numerous species produce specific acoustic and vibrational signals for intra- and interspecific communication that are generated independently of other activities by specialized structures and behaviors. Acoustic and vibrational communication signals can mediate information about the localization, species, size, physiological state and genetic quality of the sender and serve important functions for reproduction, competition and coordination of activities in social insects.

This chapter is divided into five sections. Section 26.1 introduces the nature and transmission of acoustic and vibrational signals. Section 26.2 describes the behavioral significance of acoustic and vibrational signals and Section 26.3 explains the mechanisms that insects use to generate these signals. An overview of acoustic and vibrational patterns and their information content is provided in Section 26.4. Section 26.5 describes the neural regulation of sound production in a few well-investigated insect species. To make use of acoustic and vibrational communication signals, insects have evolved various types of hearing organs and central nervous auditory processing strategies, which are considered in Chapter 23.

The Insects: Structure and Function (5th edition), ed. S. J. Simpson and A. E. Douglas.
Published by Cambridge University Press. © Cambridge University Press 2013.

26.1 Nature and transmission of acoustic and vibrational signals

Insects produce communication signals by stridulation, percussion, vibration, click mechanisms and air expulsion. In most species neuromuscular activity leads to mechanical vibration of some exoskeletal structure. These vibrations are transduced (either directly or indirectly) as cycles of compression and rarefaction to the surrounding medium or a contacted solid substrate. Mechanisms producing substrate vibration may also generate air-borne sounds and either one or both components of the signal may be detected and evaluated by the receiver.

26.1.1 Vibrational signals

Substrate-borne communication has been described in a variety of insects such as bees, termites, ants, lacewings, stoneflies and beetles. Substrate vibration may be produced by percussion, stridulation, timbal mechanisms or by contractions of the flight muscles. Except in the case of percussion, the vibrations are usually transmitted to the substrate via the insect's legs, although queen bees contact the combs with the ventral surface of their body.

The transmission of vibrational signals is critically determined by the mechanical properties of the substrate. In soil, vibrations are transmitted for only a few centimeters due to heavy damping. In contrast, insect-derived vibrational signals may be transmitted for more than 2 m through plants to be perceived by sensory organs (usually located in legs or antennae) and to elicit a behavioral response. Vibrations are propagated as bending waves at velocities ranging from less than $0.5 \, \text{m s}^{-1}$ to over $75 \, \text{m s}^{-1}$, depending on the characteristics of the plant material (stems, veins, leaves) and on the wavelength of the vibrations. Higher-frequency signals propagate faster than lower-frequency signals. Most insects produce vibrations in plants with dominant frequencies from below 200 Hz to 2000 Hz. Higher dominant

frequencies or higher harmonic frequencies seem to be irrelevant as communication signals since they are not detected by the typical vibration-sensitive sensory organs. Amplitudes of vibrational signals vary, but may be as high as $4 \, \mu\text{m}$ peak to peak, with accelerations measuring as much as $8 \, \text{m s}^{-2}$. If signal frequency is well tuned to the transmission properties of a plant, low attenuations of less than $0.05 \, \text{dB cm}^{-1}$ may enable long-distance communication in the range of meters.

26.1.2 Acoustic signals

Acoustic signals that are transmitted through air or water include both pressure waves and particle movements. In the near field – a distance of less than one wavelength away from the source (wavelengths are 3.43 m for a 100 Hz tone and 34.3 cm for a 1 kHz tone) – most of the energy of the signal involves particle displacement. In water, the near field is about four times greater than in air for a given wavelength. In the far field, pressure waves carry most of the energy and represent the more important component of the acoustic signal. Distinction between near and far field characteristics is important. Very low-intensity sounds (e.g., the courtship songs of fruit flies, acoustic signals of ants) with insufficient energy to stimulate a pressure receiver may have sufficient energy in the near field to stimulate particle displacement receptors. As a rule of thumb, sound pressure levels of air-borne acoustic signals are attenuated by 6 dB with each doubling of the distance from their source in the far field, while sound intensity remains almost constant throughout near fields. Sound propagation in the far field which limits communication distances is influenced by air temperature gradients, wind, humidity and vegetation, with different frequencies being differently affected by these parameters.

Insects using air-borne signals for long-distance communication employ specialized structures and behaviors to generate acoustic signals of high intensity, specific spectral components and temporal patterns. Anatomical structures whose vibrations are transmitted to the air are known as sound radiators. The sound radiator is often a part of the wing (e.g., in crickets and bushcrickets) or a timbal backed by an air sac (e.g., in cicadas) that is suspended from the surrounding cuticle in a way that facilitates its free vibration. A sound radiator functions most efficiently in producing sounds with wavelengths similar to or less than its own dimensions. Consequently, because of their small size, insects are constrained toward producing high-intensity sounds with short wavelengths (and high frequencies), which limits communication distances. In practice, air-borne sounds produced by different insects cover a very wide range of frequencies and are often ultrasonic to humans (Fig. 26.1).

Sound is produced on both sides of a vibrating membrane. If the membrane is small relative to the wavelength of the sound produced, sound waves from the two sides tend to interfere destructively because they are in antiphase. Such interference can be reduced or eliminated by a baffle separating the air on the two sides of the membrane. In many insects using part of the wing to produce sound, the non-vibrating parts of the wing surrounding the vibrating membrane may be used as a baffle. If the sounds from the two sides of the vibrating membrane are brought into phase with each other, constructive interference occurs, increasing the intensity of the sound produced. This is achieved by the cricket *Anurogryllus* (see below). Some species employ environmental objects to extend the size of the baffle (e.g., tree crickets of the genus *Oecanthus* sing while being positioned in a hole in a leaf that tightly encloses their tegmina) or use resonant properties of specially shaped burrows (mole crickets, Section 26.3.2) to amplify the intensity and tune the frequency spectrum of the generated sound signals.

26.2 Significance of acoustic and vibrational signals

The acoustic and vibrational signals produced by insects can be classified according to whether they are intraspecific, used to communicate with other members of the same species, or interspecific, giving signals to other species.

26.2.1 Signals having intraspecific significance

Signals having intraspecific significance are usually organized sounds and vibrations with a regular pulse repetition frequency. Temporal patterns of communication signals and their spectral composition are species-specific – and in some cases also gender-specific – and may contribute to reproductive isolation between species. Sometimes an individual insect produces different sounds or elaborate song patterns in different contexts (Fig. 26.2a). Sounds and vibrations may be used to attract reproductive partners from a distance, to establish copulatory readiness during courtship, to defend territories or other valuable resources and to recruit conspecifics to a feeding site or to defend against intruders. The communication signals

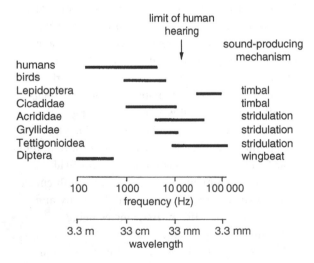

Figure 26.1 Range of air-borne sounds produced by some insect groups, birds and humans. Not all species within a group produce the full range of frequencies.

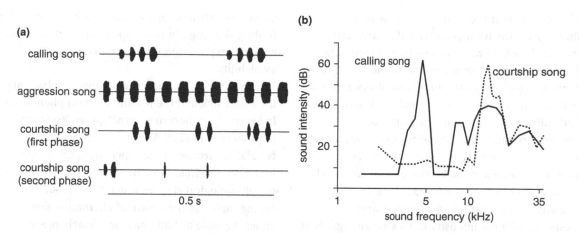

Figure 26.2 Acoustic signals of the cricket *Gryllus campestris*. (a) Patterns of sounds produced by tegminal stridulation by a male *Gryllus campestris* in different situations (after Huber, 1962). (b) Frequency spectra of calling and courtship songs. The sound intensity of the calling song is greatest at approximately 4 kHz, with additional less intense harmonics at higher frequencies. The courtship song has its highest intensity at about 16 kHz (after Nocke, 1972).

may contain information about the sender's physiological state (e.g., some grasshopper, bushcricket and land bug females signal copulatory readiness) and genetic (e.g., in *Drosophila montana*) or other qualities (e.g., about size in the house cricket, *Acheta domesticus*). In many species, only the males stridulate, but in some species female response songs can initiate male phonotaxis and queens, workers and even larvae of some social hymenopteran species employ acoustic and/or vibrational signals in social competition, recruitment for foraging and defense and initiation of care.

Reproductive isolation Acoustic and vibrational signals probably contribute to sexual isolation in many insects, but their critical role in the prevention of interbreeding has been experimentally determined in only a few species. Some grasshoppers of the genus *Chorthippus* are morphologically very similar. In the laboratory, different species interbreed and produce fertile offspring when females are stimulated with courtship songs of their own species and sound production of males from a different species is prevented. Field studies on the central European sympatric species *Ch. biguttulus*, *Ch. mollis* and

Ch. brunneus suggested that interbreeding is largely prevented and the prominent differences in temporal song patterns play a major role in this pre-mating sexual isolation. The same is probably true in *Drosophila*, where females are most effectively stimulated to accept male copulation attempts by backplayed courtship songs of their own species; heterospecific songs may even elicit female rejection (e.g., in *D. biauraria*). Some populations of the brown planthopper of rice, *Nilaparvata lugens*, exhibit little interbreeding, even in the laboratory, and it appears that differences in the patterns of substrate-borne vibrations produced by the different populations are primary isolating mechanisms.

These three examples demonstrate that air-borne sounds, in both the far field (*Chorthippus*) and near field (*Drosophila*), and substrate-borne vibrations can contribute to sexual isolation of sympatric species.

Attraction from a distance Attraction to a source of sound or vibration is called phono- or vibrotaxis, respectively. Whether insects recognize a particular acoustic or vibratory signal depends on the sensitivity of their sensory organs and their capability to extract behaviorally relevant information from the sum of all acoustic and vibratory information perceived at a particular location. Effective ranges of signals to stimulate phonotaxis depend on the nature of the signal itself (intensity, frequency content, temporal structure), the medium that propagates the signal (air, water, type of solid substrate), environmental conditions (temperature gradients, humidity, wind, etc.) and the degradation of signals as affected by vegetation, ground structure and background noise. Maximal distances that initiate phonotaxis have been estimated by comparing an insect's hearing thresholds with the intensity of its acoustic signals or determined in laboratory experiments in more or less ideal sound fields. Only few studies have measured effective communication ranges in natural habitats. As one example for attraction of mating partners, female grasshoppers of the species *Chorthippus biguttulus* produce signals with peak intensities of 65–69 dB SPL (measured at a 15 cm distance). These signals elicit male approach at distances of up to 2 m in the laboratory, up to 1.3 m in short and sparse vegetation and of less than 1 m in tall and dense vegetation. The longest acoustic communication distances have been reported in the Orthoptera and the cicadas. At night, females of the bladder grasshopper, *Bullacris membracioides*, have been estimated to recognize the male calling signal at distances of up to 450 m, mole crickets of the genus *Gryllotalpa* generate sounds that are audible to humans over several hundred meters and some cicadas are audible to the human ear for more than 1 km.

Substrate vibrations are also used to attract members of the opposite sex. This occurs in Delphacidae and Cicadellidae among the Homoptera, and in Plecoptera and Chrysopidae (Neuroptera). In many of these species, the females also produce vibrations, returning the calls of the males. Communication distance is usually restricted to a particular plant that the insects feed on (host plant) or that females may be attracted to by male pheromones. In addition to mate finding, substrate

vibrations may also regulate cooperative foraging for feeding sites (e.g., in treehoppers) in order to assure effective exploitation during limited periods of availability.

Frequently, as in crickets and most cicadas, only the males sing and the females perform phonotaxis. In several gomphocerine grasshoppers, however, both sexes stridulate. After longer periods without copulation, female grasshoppers (e.g., *Chorthippus brunneus*, *Ch. biguttulus*, *Gomphocerus rufus*) enter the physiological state in which they respond to calling males. In the course of alternating duets, either the male or both male and female orient and stepwise move toward each other until visual contact is established. In the bushcricket *Ancistrura nigrovittata*, females have to respond to a particular element of the male calling song with a precisely timed short-latency response song in order to initiate male phonotaxis.

Sometimes the sounds produced by male insects attract other males as well as females. Males may aggregate at preferred sites for feeding or oviposition and sing together (chorusing) during particular daily periods. Chorusing occurs in a few species of bushcrickets, grasshoppers and cicadas and enables direct comparison of several males during mate choice of the attracted females. It has been suggested that chorusing may increase the effective range of male calling signals, enhances female sexual responsiveness and reduces the risk of predation.

In unsynchronized choruses, the insects only sing when they hear a conspecific singing and stop when there is no feedback and individual songs are not precisely timed relative to each other. Consequently, sound production in unsynchronized choruses consists of irregularly alternating periods of silence and acoustic signaling. Some other species exhibit synchronized chorusing in which the calls of individuals are precisely timed with respect to others. In these species, a follower male starts its song at a precise interval after the start of another insect's song, and it adjusts the timing by altering the duration of its chirps and silent intervals. Females select individual males for reproduction on the basis of sound intensity, position of a male within the area covered by the chorus or whether a male leads or follows in the presentation of song sequences.

In some other Orthoptera such as the grasshopper *Ligurotettix planum*, chorusing involves the alternation of the songs of individual insects. Perfect alternation of songs is established with other nearby individuals, while more distant individuals have less impact on the rhythm of sound production.

Courtship When the two sexes are close together, many insect species exhibit courtship behavior before mounting and copulation. Courtship may have several functions, including establishment of sexual arousal that initiates (males) and permits (females) subsequent copulation, species recognition or providing the opportunity for sexual selection on the basis of traits involved in courtship behaviors.

In crickets and grasshoppers the calling song that is used to attract a mate changes to the courtship song as soon as the sexes establish visual contact. The courtship song is often of lower intensity (i.e., quieter) than the calling song and, in many crickets, the principal component of the courtship song is also at a higher frequency than that of the calling song (Fig. 26.2b). Compared with their calling song patterns, various grasshoppers, including several species of gomphocerine grasshoppers (e.g., *Chorthippus albomarginatus, Ch. oschei, Omocestus viridulus*), produce complex courtship songs containing several patterns that may progressively be included during the progress of courtship and seem to be associated with increasing sexual arousal (Fig. 26.3).

In insects that do not produce calling songs for acoustic mate attraction, courtship songs can serve as major cues for species recognition. This has been demonstrated for various species of *Drosophila*, where homospecific songs most effectively promote copulations, while heterospecific songs are either less stimulating on females or even elicit female rejection.

Signaling physiological state or genetic quality Calling and courtship songs are signals of reproductive readiness. These song types are usually generated by sexually mature males, but their production may be suppressed during post-copulatory periods. Female grasshoppers only stridulate in the physiological state of copulatory readiness, which establishes after extended periods without mating. Stridulating females are highly attractive to grasshopper males, since they signal high reproductive motivation. Following copulation male grasshoppers, crickets and bushcrickets may enter a sexually refractory period during which the production of calling songs is suppressed. Notably, bushcrickets that transfer large spermatophores to females remain silent until a new spermatophore of sufficient size has been generated.

Whether insect calling or courtship songs transmit information about male genetic quality that

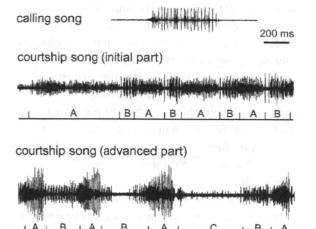

Figure 26.3 Calling and courtship song patterns of the grasshopper *Ch. albomarginatus*. Calling songs consist of repetitions of short (<1 second) sequences separated by pauses of 1.5–3 seconds. Courtship songs, which may last from several minutes to more than one hour of continuous stridulation, consist of three elements (A, B, C) generated in alternation. Appearance and frequency of C elements is associated with increasing arousal during the progress of courtship.

influences a female's selection of its reproductive partner has only been determined in a few species. Female *Drosophila melanogaster* prefer males that produce short song pulses with high carrier frequency, which have been demonstrated to correlate positively with the survival rate of the progeny of the respective males from egg to adulthood. Female house crickets (*Acheta domesticus*) preferentially approach males that include high pulse numbers in chirps of their calling songs. This song parameter is connected to male body size, a heritable trait that increases the attractiveness of the offspring. A similar situation may exist in the bushcricket *Phaneroptera nana*, in which females preferentially answered the longer chirps produced by larger males.

Territorial behavior and competition

Agonistic stridulation often occurs in relation to reproductive behavior. It has the effect of spacing males in the environment. The calling song, while attracting females, may, at the same time, serve as an aggressive signal to other males (e.g., in noctuid moths of the genus *Hecatesia* and in the grasshopper *Ligurotettix coquiletti*), but many species have a separate aggression song. Long-range acoustic signals enable males to claim relatively large territories without showing physical presence in all its areas. The territories may include burrows or high-quality food sites that may serve as additional cues for females to select a singing male for reproduction.

Aggressive stridulation has been studied in various territorial cricket species such as *Gryllus campestris*, *G. bimaculatus* and *Oecanthus nigricornis*. If a male intruder enters the territory, both resident and intruder generate aggression songs. The following agonistic interactions include threat displays, physical combat and occasional sequences of aggression song. Encounters are decided by the retreat of the loser, who finally leaves the disputed territory, while the winner usually continues to present aggression song.

Some insect larvae also use acoustic and vibrational signals to defend their nest sites or feeding grounds against conspecifics. Larvae that live within silken nests, such as some caddis flies and caterpillars of the moth *Drepana arcuata*, compete for these valuable retreats with series of aggressive sounds.

Communication in social insects

Acoustic and vibrational signals are used to mediate coordinated group actions (foraging, defense) and complex social interactions (social hierarchy, attendance).

Both worker and queen honey bees (*Apis mellifera*) use sounds and vibrations for signaling within a colony. Foragers returning to their hives can communicate the nature, direction and distance of food sources to their nest mates. This dance involves signals of various modes, including sounds and vibrations. Dancing bees emit low-frequency signals from their thoraces that attract dance-followers from a distance. Special characteristics of the combs that propagate the vibratory signals create a phase reversal of wall oscillations between neighboring combs at a particular distance from the dancing forager. The forager produces series of sound pulses through side-to-side movements of the gaster that other workers, which closely follow the dancer, perceive with their antennae. The period of sound pulse production and the number of sound pulses generated are thought to represent the distance to the food source. Honey bee queens generate "piping" or "quaking" sounds by flight muscle-driven vibration of the thorax. Piping may occur prior to the emergence of young queens from their cells, may be produced by immature queens that compete against other immature queens or may accompany the initiation of swarming.

Other bee species are also known to use vibratory signals for communication inside their hives, especially to recruit foragers to profitable feeding sites. Even larvae may produce rhythmic signals (e.g., by rubbing their mandibles against the cell wall of the comb in which they grow) to signal hunger and stimulate attendance of hive workers.

Several ants produce audible acoustic signals through stridulation that serve functions in recruitment to food sites, recruitment to defend against intruders and reinforcement of social status. Since ants are insensitive to air-borne sound, substrate-borne vibrations seem to be the crucial components of the signals. Stridulation-derived body vibrations of leaf-cutter ants, *Atta*, are transmitted to the leaf via the head. These substrate vibrations attract other workers to the site to ensure cooperative feeding. Queens and workers of the species *Myrmica schencki* generate distinctive sounds by scraping the first abdominal segment against a file located on the first segment of the gaster. Both signals attract other workers, but only the queen-derived signals elicit attendance behavior in workers.

Upon disturbance, termite workers and soldiers of *Zootermopsis* species generate percussion sounds by drumming their heads against the substratum. Temporal features of these vibrational alarm signals are detected by other members of the colony that subsequently retract to remoter parts of the nest.

Besides hymenopterans, other species such as the sap-feeding treehopper *Calloconopha pinguis* coordinate their foraging activities and recruit conspecifics by vibrational announcement of profitable feeding sites.

26.2.2 Signals having interspecific significance

Sounds having interspecific significance are usually unorganized, having no regular pulse repetition frequency and covering a broad spectrum of frequencies with sharp transients. They may be produced by both males and females and sometimes also by larvae. Sounds of this type are concerned with defense against predators (startle sounds, warning sounds, recruitment calls), with territoriality or with disguise.

Startle sounds function to alarm a potential predator, causing it to hesitate momentarily and increasing the chance for escape. Peacock butterflies, *Inachis io*, produce intense ultrasonic clicks that intimidate bats, but the same signal has little effect on bird predators. Moths have also been demonstrated to respond to echolocating bats with ultrasonic clicks that confuse prey localization and may cause the bats to veer away. Sounds associated with predator deterrence have also been described for click beetles, some mantids and grasshoppers.

Warning sounds that signal distastefulness or harmfulness of a prey to the potential predator usually accompany visual displays or regurgitation. Depending on the predator species, visual or acoustic displays or other behaviors may be the more important components of these multimodal signals.

Insect acoustic signals may also be used as aggressive mimicry. The Australian predatory tettigoniid *Chlorobalius leucoviridis* (Orthoptera) imitates the species-specific wing-flick responses of female cicadas to male calling songs. The bushcricket recognizes general features of different cicada species' calling songs and places its acoustic signal into the correct temporal window to initiate phonotaxis of male cicadas, which finally end as prey instead of finding a female for reproduction.

Insects that live in mutualistic association with tending ants may generate distinct substrate vibrations in response to predator threat to attract ants for their defense. This has been demonstrated for treehoppers *Publilia concava* and the larvae of Riodinidae and Lycaenidae (Lepidoptera) that develop in association with ants, while species not tended by ants do not produce alarm calls.

There is also evidence that sound mimicry occurs. Certain syrphid flies that look like bees or wasps also produce sounds by similar mechanisms to their models. However, since frequency content and temporal patterns do not specifically match with the sound of the mimicked species, the behavioral relevance of these acoustic signals remained questionable. Sound mimicry has been confirmed in butterfly larvae (*Maculinea rebeli*) that infiltrate ant colonies of the species *Myrmica schencki*. The caterpillars generate sounds that mimic those produced by ant queens to demonstrate social superiority to workers and stimulate their benevolent behavior. Instead of being attacked like other intruders, butterfly larvae and pupae that sound like ant queens are regarded as royalty and receive the same care by workers.

26.3 Mechanisms producing sounds and vibrations

Insects have evolved several mechanisms to produce acoustic and vibrational communication signals. Because of their small body size, high-frequency sounds can be generated most efficiently by specialized structures that produce high-frequency oscillations of sound-radiating structures from relatively slow muscle contractions.

26.3.1 Percussion

Percussion refers to vibrations produced by the impact of part of the body against the substrate or by clapping two parts of the body against each other. A single tap on a solid surface produces a complex wave pattern whose spectral composition and propagation largely varies according to the nature of the solid. Though air-borne sounds may also be generated, substrate vibrations are probably more important than sounds for most insects using percussion for communication.

Many insects produce vibrational signals by striking unmodified parts of their bodies against objects in their environment. Some termites and beetles use their heads for this purpose. As an example, soldiers of *Zootermopsis* produce repeated series of taps by rocking their heads up and down and banging their mandibles on the ground and/or their dorsal heads against the roof of their galleries. Workers and larvae produce a lower-intensity sound by hitting their heads on the gallery roof in similar

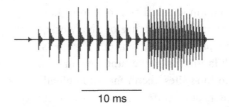

10 ms

Figure 26.4 Drumming signals produced by male *Isoperla* (Plecoptera). The figure shows air-borne sounds recorded close to the substratum in an experiment. The pattern resembles that occurring in the substratum. The insects respond to substrate vibrations (after Szczytko and Stewart, 1979).

vertical oscillating movements. Both sexes of death watch beetles, *Xestobium*, produce tapping sounds by banging their frons against the wood in which they are burrowing to produce a series of taps that last almost one second.

Booklice (Psocoptera) and some stoneflies (Plecoptera) of both sexes generate substrate vibrations through abdominal movements (Fig. 26.4). Some species have evolved a small cuticular knob on the ventral surface of the abdomen, which they rhythmically drum against the ground. In contrast, some stoneflies (*Siphonoperla*) and lacewings (*Chrysoperla*) do not directly touch the substrate with their abdomen, but rather transmit abdominal oscillations to the ground via their legs.

Some orthopteran species drum with their hindlegs. Males and females of some oedipodine grasshoppers drum both hindlegs on the ground to generate signals with sex-specific frequencies. In contrast, males of the bushcricket *Meconema* only use one leg for the production of leaf vibrations with various temporal patterns.

A few insects are known to produce air-borne sounds by striking one part of the body against another. The males of whistling moths, *Hecatesia*, produce sound by banging together specialized areas of the forewings known as castanets. Each castanet consists of a series of cuticular knobs on the costal area of the forewing separated from the rest of the forewing by a crescent of corrugated cuticle. Some noctuid moths generate sounds when their wing tips touch above the body during flight. Though this signal is perceived by other moths, it may be generated accidentally and not used as a communication signal.

Oedipodine grasshoppers often produce clicking or crackling sounds in flight, known as crepitation. These sounds may either result from impacts of the wings against the legs or from changes in tension in parts of the wing membrane when the wings are in certain positions. Grasshoppers of the species *Stenobothrus rubicundus* include crepitation sounds in their calling songs and final phases of courtship stridulation that have been demonstrated to critically determine the success of subsequent male mating attempts.

26.3.2 Stridulation

Stridulation refers to the production of vibrations by moving a cuticular ridge (called the scraper or plectrum) on one part of the body over a toothed ridge (known as the file or strigil) on another. Repeated contacts of the scraper against the teeth of the file cause part of the body to vibrate. These vibrations of the insect's body may give rise to substrate vibrations (e.g., after transmission through the legs) or, if the vibrating membrane is of appropriate dimensions, will produce air-borne sounds for long-range communication. This mechanism of producing sounds and vibrations is common for insects of many different orders, but is particularly associated with Orthoptera, Heteroptera and Coleoptera. Stridulatory organs have evolved through modifications of various exoskeletal structures and appendages of insect bodies, including wings, legs, antennae, mouthparts and segmental joints.

Stridulation in Orthoptera The Orthoptera employ two main methods of stridulation: tegminal stridulation in crickets (Grylloidea) and bushcrickets (Tettigonioidea) and femoro-tegminal stridulation in

grasshoppers (Acridoidea). Other mechanisms of sound production by abdomino-alary stridulation are present in mantids.

Crickets. In male Grylloidea, each tegmen has a file on the underside of the second cubital vein near its base, while the ridge forming the scraper is on the edge of the opposite tegmen (Fig. 26.5a,b). In field crickets the right tegmen always overlaps the left so that only the right file and left scraper contribute to sound production. In contrast, mole crickets can vary the relative positions of the wings. For sound production, the tegmina are raised at an angle of 15–40° to the body and rhythmically open and close so that the scraper rasps along the file. Each wing closure produces one syllable of sound by vibrations of both tegmina.

The principal sound radiator on the tegmen is the harp, an area of thick sclerotized wing membrane with no particularly sclerotized cross-veins (Fig. 26.5a). The harp is enclosed by the first and second cubital veins but separated from the rigid subcostal, radial and medial veins by a band of flexible cuticle so that it is free to vibrate as a plate. The mirror, a more distal area of thin cuticle completely transparent in *Gryllus*, appears not to be specifically involved in sound production, although a comparable area in tettigoniid bushcrickets is very important (see below). The resonant frequency of the harp is approximately the same as the impact frequency when the scraper rasps along the file, and this contributes to the pure tones of most cricket songs (Fig. 26.5c), although the impact frequency may slightly vary during a single wing stroke and from stroke to stroke. The air space between the body and the raised tegmina also forms a resonator (Fig. 26.5b) which both filters and amplifies the generated sound. By adjusting the volume of the space between stridulating wings and dorsal surface of the body, crickets may be able to modulate the frequency of the emitted sound. Female crickets generally lack a stridulatory apparatus.

Some gryllids use environmental features to improve sound radiation. Species of oecanthine

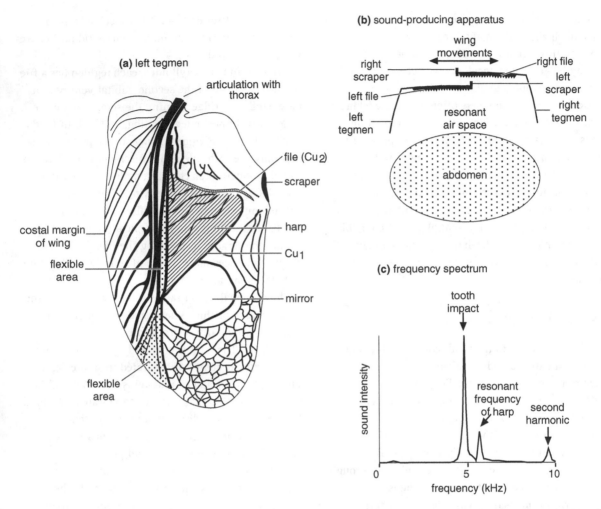

Figure 26.5 Stridulation in crickets. (a) Left tegmen of *Gryllus campestris* seen from above. The file is on the underside of Cu_2 and so would not be seen in this view (after Bennet-Clark, 1989). (b) Diagrammatic cross-section through an anterior abdominal segment of a stridulating cricket. The sizes of the files and scrapers are exaggerated. The size of the air space can be varied so that it resonates at the frequency of the harp (not shown in this diagram). (c) Frequency spectrum of the sound produced by *Gryllus bimaculatus* (after Stephen and Hartley, 1995).

crickets use leaves as baffles. They may either choose suitable singing positions between two leaves, which then function as baffles, or, as in *Oecanthus burmeisteri*, create acoustic baffles by chewing a hole into a leaf. When singing, this tree cricket sits with its head projecting through the hole and the dorsal edges of its tegmina tightly pressed against the opposite leaf surface. The sound-producing membrane on the wings is only about 3 mm in diameter, but the leaves, ranging in size from 70 × 80 mm to 170 × 300 mm, serve as efficient baffles that enable the relatively small insect to generate rather low-frequency signals (2 kHz with wavelengths of approximately 170 mm) with high intensities (Fig. 26.6a).

Mole crickets, including *Gryllotalpa* and *Scapteriscus* species, amplify their songs by singing

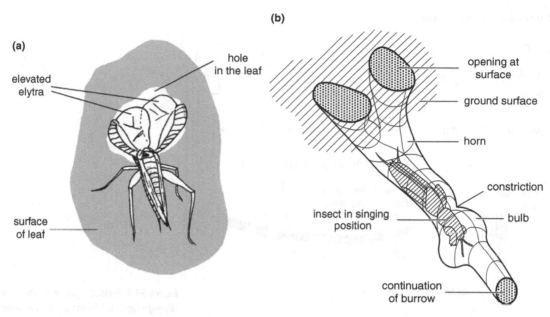

(b)

opening at
surface

ground surface

horn

constriction

bulb

continuation
of burrow

insect in singing
position

(a)

hole
in the leaf

elevated
elytra

surface
of leaf

Figure 26.6 Amplification of radiated sound intensity by environmental objects. (a) A male tree cricket *Oecanthus burmeisteri* singing on a leaf in front of a self-made hole. (b) Diagram of a singing burrow of *Gryllotalpa vineac* (after Bennet-Clark, 1970).

at particular positions in burrows shaped like exponential horns with two openings (Fig. 26.6b). The dimensions of the horn are adjusted to the body size. When the forewings are raised into the singing position they form a diaphragm across the diameter of the burrow. The sound produced by the cricket has a fundamental frequency of 3–5 kHz, and is both amplified and purified by the resonant properties of the burrow. As a result, the sound propagates over great distances of up to 600 m and is more sharply tuned to a particular frequency than the sound of a mole cricket singing outside its burrow.

Bushcrickets. The stridulatory apparatus of Tettigonioidea is similar to that of gryllids, but the left tegmen always overlaps the right during stridulation. The file on the right and the scraper on the left tegmen may still be present, though not functional, in species with short tegmina and reduced hindwings (e.g., *Ephippiger*, Fig. 26.7), but they are absent in most fully winged forms. The file of *Platycleis* contains an average of 77 teeth; similar numbers of teeth (~65–100) are present in other species, though the most lateral portions of a file may not contribute to sound production. Close to the plectrum on the right tegmen lies the mirror, an area of thin, clear cuticle supported by a surrounding frame of thickened veins (Fig. 26.7a). As the scraper is repeatedly stopped and released while passing over each tooth of the file, the frame is first distorted and then returns to its original form, causing the thin cuticle of the mirror to vibrate at its natural frequency. In most bushcrickets the natural frequency of the mirror and its frame is higher than the tooth impact frequency and each impact produces a short burst of sound waves (an impulse, Fig. 26.7a), which is rapidly damped before the next tooth impact occurs. The sound produced includes a much wider range of frequencies than that of crickets (Fig. 26.7b), perhaps due to vibrations produced by different parts of the wing. The range of frequencies often extends well into the human ultrasonic range and may be entirely inaudible to

(a) stridulatory mechanism

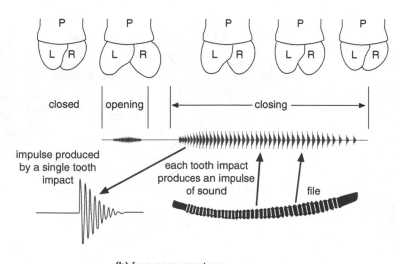

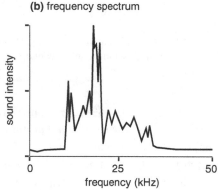

(b) frequency spectrum

Figure 26.7 Stridulation in bushcrickets (*Ephippiger*). (a) Stridulatory mechanism. Top: opening and closing of the forewings (tegmina). In this species the wings are very short and not used for flying. P, pronotum; L, left tegmen; R, right tegmen. Bottom: the file showing how each tooth impact produces a sound impulse. Detail of a single impulse to the left (based on Pasquinelly and Busnel, 1955). (b) Frequency spectrum of the sound produced (based on Dumortier, 1963b).

humans. The songs of *Conocephalus* species, for example, contain frequencies between 20 kHz and more than 100 kHz.

Most bushcrickets produce sound exclusively by wing closure; however, as shown for *Ephippiger* (Fig. 26.7b), additional syllables may arise on wing opening in certain species. The stridulatory wing movements are driven by the flight muscles, opening by the direct flight muscles and closure by the indirect muscles. In some species, as in *Neoconocephalus robustus*, the frequency of wing movements is as high as 200 Hz, much higher than the normal wingbeat frequency of these insects in flight.

Females of tettigoniid species usually lack a stridulatory apparatus, while females of most phaneropterinae and ephippigerinae express well-developed files and scrapers and use them in the production of acoustic signals. The morphological apparatus for sound production has evolved independently in males and females. In contrast to males, female bushcrickets stridulate by rubbing a prominent vein on the underside of the left wing over a field of modified spines on the upper surface of the right wing. The importance of female stridulation has been determined in various species such as *Leptophyes punctatissima*, *Poecilimon ornatus* and *Ancistrura nigrovittata*, in which females need to answer the male song within a particular time window to elicit male phonotaxis.

In addition to tegminal stridulation, some bushcrickets stridulate by rubbing ribs on the anal veins against a series of ridges on some abdominal tergites. In *Panteophylus* this abdomino-alary mechanism produces a broad-band acoustic signal that functions as a disturbance signal.

Short-horned grasshoppers. Many Acridoidea produce sounds by rubbing the hind femora against the tegmina. In Acridinae, a ridge on the inside of the hind femur rasps against an irregular intercalary vein, while in Gomphocerinae a row of pegs on the femur is rubbed against ridged veins on the tegmen (Fig. 26.8). A single tooth impact produces a highly damped vibration of the tegmen that lacks a specialized sound-radiating area. The sound produced by one leg movement consequently has a principal frequency at the tooth impact frequency. Sounds may be produced on both the up- and downstrokes. Intact grasshoppers simultaneously stridulate with both hindlegs. In most species studied,

both legs may produce slightly different song patterns and move slightly out of phase in order to mask silent periods at the top and the bottom of the stroke that may reduce attractiveness of the songs to females (e.g., in *Chorthippus biguttulus*).

The stridulatory leg movements are produced by coxal promoter and remotor muscles. These are bifunctional muscles. In flight, where they contribute to wing depression, both sets of muscles contract simultaneously, but during leg movement they alternate.

A stridulatory apparatus is often present in both male and female acridids. While male gomphocerine grasshoppers use pegs on the inner sides of their femora to generate sounds, females instead rub the thickened frame at the base of the bristles against a vein on the tegmen. Apart from this, the general mechanism of femoro-tegminal sound production is identical. Females usually produce a similar basic pattern as do males, but lack the diverse repertoires of elaborate song patterns that males may perform during courtship. In addition to femoro-tegminal sound production,

(a) stridulatory mechanism

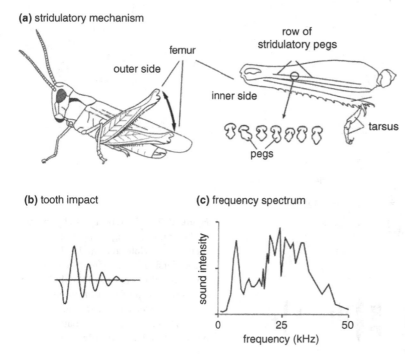

(b) tooth impact **(c)** frequency spectrum

Figure 26.8 Stridulation in grasshoppers. (a) Left: rhythmic movements of the hindlegs strike the stridulatory pegs against a prominent vein on the forewings. Right: the inner side of a hind femur showing the position of the row of stridulatory pegs (the file) with detail of a few pegs (*Stenobothrus*) (after Roscow, 1963). (b) Sound produced by a single tooth impact. (c) Frequency spectrum of the sound produced by *Chorthippus biguttulus* (based on Dumortier, 1963b).

different stridulatory mechanisms occur in other Acridoidea.

Stridulation in other insects *Hemiptera.*

Stridulation is common among the Heteroptera and different species may employ different parts of the body for sound production. In Pentatomomorpha, the most common mechanisms involve a file on the ventral surface rubbed by a scraper on the leg, or a file on the wing rubbed against a scraper on the dorsal surface of the body. For example, both sexes of *Kleidocerys resedae* have a vein-like ridge on the underside of the hindwing (Fig. 26.9a). This ridge bears transverse striations about 1.7 μm apart and is

rubbed on a scraper projecting from the lateral edge of the metapostnotum. Whereas the Cimicomorpha generally do not stridulate, nearly all Reduvioidea have a file between the front legs which is rasped by the tip of the rostrum (Fig. 26.9b). This apparatus is present in males, females and larvae.

Many aquatic Heteroptera stridulate under water using a variety of parts of the body. *Ranatra*, for example, has a file on the fore femora with a scraper on the coxa; *Buenoa* has a similar arrangement in addition to a file on the fore tibia which it scrapes with the rostrum. In these aquatic bugs, the air bubble normally carried beneath the surface for respiration acts as a resonator. In corixids, the bubble

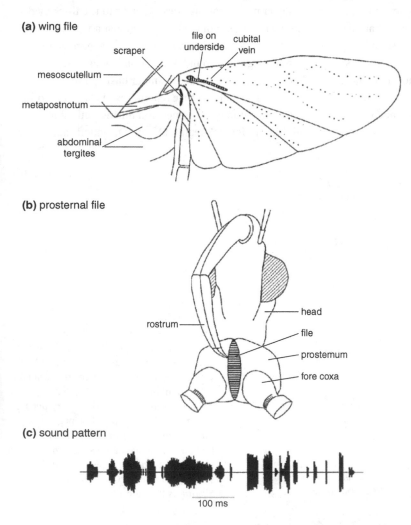

(a) wing file

scraper · file on underside · cubital vein · mesoscutellum · metapostnotum · abdominal tergites

(b) prosternal file

rostrum · head · file · prostemum · fore coxa

(c) sound pattern

100 ms

Figure 26.9 Stridulation in Heteroptera. (a) File on hindwing, scraper on metathorax. Note that the file is on the underside of the wing (*Kleidocerys*) (modified after Leston, 1957). (b) File on prosternum, tip of rostrum forms scraper (*Coranus*) (after Dumortier, 1963a). (c) Irregular sound pattern produced by *Coranus* (after Haskell, 1961).

is caused to vibrate by the movements of the head as it is rubbed by the femoral file. Because the bubble gets smaller the longer the insects remains submerged, its resonant frequency increases. Thus, the sound frequency produced by these insects varies in relation to the diving duration-related size of the air bubble.

Among Homoptera, frictional methods of stridulation occur in the aphid, *Toxoptera*, and most Psyllidae.

Coleoptera. Stridulation occurs in many beetles, especially among Carabidae, Scarabaeidae, Tenebrionidae and Curculionidae, and in the aquatic Dytiscidae and Hydrophilidae. Different species use different parts of the body to produce sounds, but most commonly the elytra are involved. In *Oxycheila*, for instance, there is a striated ridge along the edge of the elytron which is rubbed by a ridged area on the hind femur.

Larval Lucanidae, Passalidae and Geotrupidae stridulate by rubbing a series of ridges on the coxa of the middle legs with a scraper on the trochanter of the hindleg. In larval passalids the hindleg is greatly reduced to exclusively function as a scraper and is no longer used in locomotion (Fig. 26.10).

Lepidoptera. All developmental stages – larvae, pupae and imagos – may generate acoustic or vibrational signals through stridulation in respective species. Stridulation of caterpillars seems to be associated with ant mutualism. Lycaenid and riodinid butterflies stridulate with a file at the borders of abdominal segments. In *Arhoplala*, for example, a file on segment 6 is rubbed by segment 5.

Pupae of various groups, notably the Hesperiidae, Papilionidae, Lymantriidae and Saturniidae, exhibit coarse transverse ridges on the anterior edges of certain abdominal segments which are rubbed against fine tubercles on the posterior edges of the preceding segments. The pupa of *Gangara thyrsis*, a hesperiid, has a pair of transverse ridges on either side of the ventral midline of the fifth abdominal segment. The long proboscis extends between and beyond these ridges and is itself transversely striated,

so that when the abdomen contracts the proboscis rubs against the ridges and produces a hiss.

Stridulation appears to be rare in adult Lepidoptera. Few examples known to date include wing-to-leg stridulation in the noctuid *Rileyana fovea*, genital stridulation in *Syntonarcha minoralis* and *Psiligramma* sp. and foreleg coxa-to-femur stridulation in males of the genus *Urania*.

Other groups. Relatively isolated instances of frictional stridulation are widespread in other groups of insects. A few examples are given here.

The larva of *Epiophleha* (Odonata) has lateral ridged areas on abdominal segments 3 to 7. These are rubbed by the ridged inner side of the hind femur to produce a sound. Similarly, larval Hydropsychidae (Trichoptera) have ridges on the side of the head and a scraper on the front femur.

Among ants, stridulation occurs in Ponerinae, Dorylinae and primitive Myrmicinae. Usually, signal production results from movement of a scraper on the petiole against striations at the base of the gaster (Fig. 26.11). Finally, in the Tephritidae (Diptera) stridulation is probably widespread. For example, in the male of *Dacus tryoni* the cubito-anal area of each wing vibrates dorso-ventrally across two rows of 20–24 bristles on the third abdominal segment, thus producing a noise.

26.3.3 Timbal mechanisms

A timbal is an area of thin cuticle surrounded by a rigid frame. Vibrations are produced when the timbal buckles, meaning that it gets alternately distorted and relaxed by a muscle attached to its inner surface. Timbals are found in Hemiptera, especially in the Homoptera, and in some adult Lepidoptera, primarily in Arctiidae.

Timbals in Hemiptera The timbal mechanism has been most extensively studied in male Cicadidae (Homoptera). The timbal, an irregular area of thin cuticle, is located dorso-laterally on each side of the first abdominal segment. It consists of a membrane of

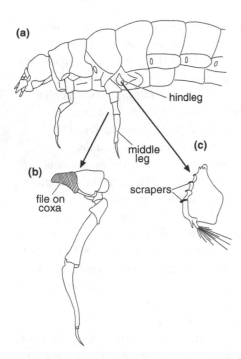

Figure 26.10 Stridulation in a beetle larva (*Passalus*) (after Haskell, 1961). (a) Outline drawing of the anterior end of the larva showing the position of the stridulatory apparatus. (b) File on posterior face of the coxa of the middle leg. (c) Scrapers on reduced hindleg.

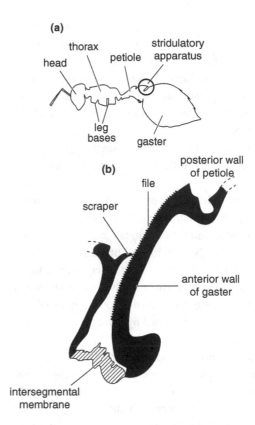

Figure 26.11 Stridulation in ants (*Myrmica*) (after Dumortier, 1963a). (a) Outline drawing showing the position of the stridulatory apparatus. (b) Diagrammatic sagittal section through the cuticle of the stridulatory apparatus. The size of the file teeth is exaggerated for clarity.

resilin supported by a sclerotized rim. Inserted in the membrane is a series of dorso-ventral sclerotized ribs and, posteriorly, a more extensively sclerotized area called the timbal plate. Dorsally, the membrane consists of a thickened pad of resilin (Fig. 26.12). The timbal is usually protected by a forward extension of the abdominal cuticle forming the timbal cover. Internally, a cuticular compression strut runs from the ventral body wall to the posterior edge of the supporting rim. A timbal muscle originates ventrally, runs parallel with the compression strut and connects to an apodeme attached to the timbal plate. Contraction of the timbal muscle causes the timbal to buckle inwards. Additionally, a tensor muscle connects the anterior rim of the timbal with a knob on the metathorax. Contraction of the tensor muscle pulls the rim of the timbal so that the

curvature of the latter is increased. Tension of the tensor muscle modulates the resistance of the timbal to buckling upon timbal muscle contraction.

The timbal is backed by an air sac which surrounds the muscle and communicates with the outside via the metathoracic spiracle. The presence of the air sac leaves the timbal free to vibrate with a minimum of damping. Projecting back from the thorax on the ventral surface lies the operculum, which encloses a cavity containing the tympanum (the hearing organ) and an area of thin, corrugated cuticle, the folded membrane, which separates the air sacs from the cavity beneath the operculum.

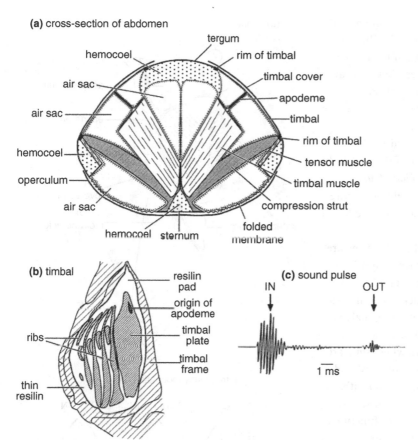

(a) cross-section of abdomen

tergum

hemocoel

rim of timbal

air sac

timbal cover

air sac

apodeme

hemocoel

timbal

rim of timbal

operculum

tensor muscle

timbal muscle

air sac

compression strut

hemocoel   sternum

folded membrane

(b) timbal

resilin pad

origin of apodeme

timbal plate

ribs

timbal frame

thin resilin

(c) sound pulse

IN   OUT

1 ms

Figure 26.12 Timbal organs of cicadas. (a) Diagrammatic transverse section through the first abdominal segment of a cicada showing the arrangement of the timbals, air sacs and timbal muscles (based on Pringle, 1954). (b) Detail of the tympanum showing the sclerotized elements (timbal plate and ribs, shaded) set in an unsclerotized membrane containing resilin (male *Cyclochila*) (after Young and Bennet-Clark, 1995). (c) Sound produced by the timbal buckling in and out (after Young and Bennet-Clark, 1995).

Sound is produced through in- and outward buckling of the timbal membrane. Depending on the arrangement of the vertical ribs on the timbal and on the tension exerted by the tensor muscle, the buckling may occur in a single movement or in a series of graded movements as the ribs give way. Each component of the buckling may produce a pulse of sound (Fig. 26.12b,c). When the muscle relaxes, the timbal moves back to its original position by the dorsal resilin pad. This outward movement of the timbal also produces sound, which may either represent a particular component of the sound pattern or, because of its low intensity, does not seem to contribute to the biological signal.

Buckling of the timbal causes it to vibrate at its resonant frequency and this vibration is transferred to the air sacs backing the timbal. If their resonant frequency is close to that of the timbal, the sound is greatly amplified. In the bladder cicada, *Cystosoma*, the air sacs are fused and fill the whole abdomen, except for a small dorsal space that contains the viscera. Some cicadas can vary the volume of the air sacs, thereby modulating their tuning and the intensity of acoustic signals. In some cicadas, such as *Cyclochila*, the timbals themselves are unimportant as direct sound radiators. Instead, sound radiates from the tympani, which also function as hearing organs. In other species both timbals and tympani, as well as other parts of the abdomen wall, act as radiators.

Repeated oscillations of the timbal muscle produce a sustained sound. In most species, including *Psaltoda* where they oscillate at 225 Hz, the timbal muscles are driven neurogenically (Section 10.3.2), but in *Platypleura*, with an oscillation frequency of 390 Hz, the timbal muscles are myogenic fibrillar muscles. In some species the muscles of the two sides contract in synchrony, while in others they contract alternately, thus doubling the pulse repetition frequency. Some species are known to vary the degree of synchrony between the two sides.

The frequency of the sound produced is determined by the natural frequency of the timbal. In *Cystosoma* this is 850 Hz, while in *Platypleura* it is about 4500 Hz. The sound may be a relatively pure tone with few harmonics, but in other species a broad range of frequencies is produced.

Among other Homoptera, timbal mechanisms are present in the males of all the Auchenorrhyncha examined and in both sexes of many families. In these insects the timbals are not backed by air sacs and so damping of the vibrations of the timbals is very high. Timbals are also present in some pentatomids, cydnids and reduviids, where they are on the dorso-lateral surfaces of the fused first and second abdominal terga, with an air sac beneath. In these insects timbals usually occur in both sexes, although often only those of the male are functional. The vibrations are transmitted to the substrate, presumably through the legs.

Timbals of Lepidoptera

Some Arctiidae have a timbal on each side of the metathorax (Fig. 26.13). It is covered by scales posteriorly, but anteriorly has a band of parallel horizontal striations, comparable with the vertical ribs on the timbals of cicadas. These vary in number in different species, from 15 to as many as 60.

Buckling of the timbal is not produced by a muscle directly attached to it, as it is in cicadas, but by the coxo-basalar muscle attached above and below it. When this muscle contracts, the timbal buckles inwards, starting dorsally and proceeding down the striated band, each rib being stressed to the point of buckling and then suddenly giving way and producing a pulse of sound. When the muscles relax, the timbal springs out due to elasticity and a

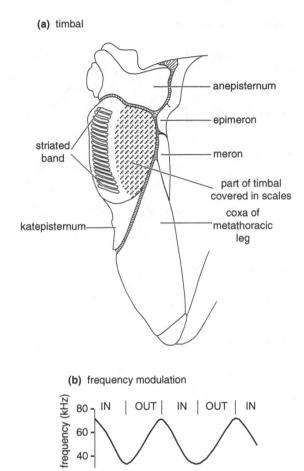

(a) timbal

anepisternum

epimeron

meron

striated band

part of timbal covered in scales

coxa of metathoracic leg

katepisternum

(b) frequency modulation

10 ms

Figure 26.13 Timbal organ of an arctiid moth (*Melese*) (after Blest *et al.*, 1963). (a) Diagram of the left side of the metathorax showing the position of the timbal relative to the surrounding sclerites. (b) Modulation of the principal frequency component of the sound produced. Each cycle is the result of one in–out movement of the timbal.

further series of pulses is produced. In *Melese* the sounds produced on the "in" movement show a progressive fall in frequency, those on the "out" movement a progressive rise (Fig. 26.13b). Most of the sound produced lies within a range of frequencies from 30–90 kHz, but the overall spectrum extends from 11 kHz to 160 kHz.

Timbals are also present in some Pyralidae. In the wax moth, *Galleria*, the tegula at the base of the forewing is modified to form a timbal. No muscles are directly associated with it, but it is actuated by twisting the wing base, presumably by the direct flight muscles. In males of another pyralid, *Symmoracma*, sound is produced by a timbal on the underside of the terminal abdominal segment.

Some butterflies, of which *Inachis io* (the peacock butterfly) is the best-known example, produce sounds as they open their wings due to the buckling of a small area of the forewing between the costal and subcostal veins.

26.3.4 Signals produced by the flight muscles

In addition to the stridulatory movements of some Orthoptera that involve the flight muscles (Section 26.3.2), oscillation of the flight muscles produces thoracic and wing vibrations which may be used in communication by a number of insect species.

Vibration of the wings in flight produces a sound with a fundamental frequency the same as the frequency of the wingbeat. However, other components may be added to this fundamental frequency as a result of the varied structure of different parts of the wing and the vibration of the thorax, so the overall sound produced is complex, consisting primarily of harmonics of the wingbeat frequency. The flight noise of a locust, *Schistocerca*, with a wingbeat frequency of about 25 Hz, is a complex sound with frequencies extending from 60 Hz to 6400 Hz, although mainly falling between 3200 Hz and 5000 Hz. Pulses of sound are produced at the rate of 17–20 per second, corresponding to the wingbeat frequency.

Insects such as Lepidoptera with very low wingbeat frequencies, of the order of 20 Hz, produce a flight tone that is inaudible to humans, but insects with higher wingbeat frequencies produce clearly audible sounds and these may sometimes provide relevant signals to the insects. The flight tone of *Apis* is about 250 Hz, and that of culicine mosquitoes ranges from about 200 Hz to over 500 Hz. In general, smaller species have a higher wingbeat frequency and flight tone than larger species. Hard-bodied insects usually produce a higher intensity of sound than soft-bodied insects.

Another moth, *Hecatesia exultans*, produces sound by vibrating its wings while at rest, and only small wing movements are necessary to produce the impacts. This species produces a much purer tone – with a peak frequency of approximately 30 kHz – than *H. thyridion*, probably because the space between the wings forms a resonator.

Many *Drosophila* species and some trypetid flies produce sounds when they are not flying by vibrating the partially opened wings. During courtship male *D. melanogaster* produce sound with one wing extended sideways at 90° to the body; the sound frequency is between 140 and 280 Hz (Fig. 26.14). *Drosophila erecta*, which sings with its wings only partially open, at 20–40° from the body axis, produces a high-frequency sound, about 450 Hz. These sounds are of very low intensity and are perceived as particle movements by the antennae of the female. Courtship songs consist of sequences with continuous tone (sinus song), or as a series of sound pulses (pulse song). In contrast, aggressive songs produced by both wings during male competition of *D. melanogaster* only contain series of pulses with rather irregular intervals.

Honey bees use their flight muscles to generate vibrations important in social communication (Fig. 26.15a). A worker bee returning from foraging performs a waggle dance during which air-borne vibrations are produced by the wings, which are held horizontally over her back. These vibrations have a frequency of 250–300 Hz, close to the wingbeat frequency, and are produced in 30-second pulses. They are detected as near-field sounds (particle displacement) by the antennae of following workers who, at the same time, may produce stop signals (previously known as begging calls). These, too, are produced by vibration of the flight muscles, but are transmitted through the insects' legs to the comb of the hive, which transmits the vibrations.

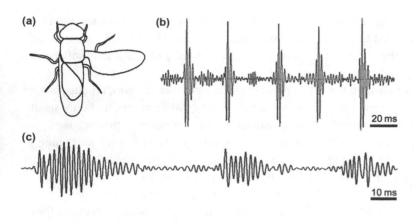

Figure 26.14 Sound-producing wing vibration by male *Drosophila melanogaster.* (a) Diagram showing the position of the wing during sound production. (b) Pulse song. (c) Sine song. The sounds are of very low intensity and are perceived in the near field.

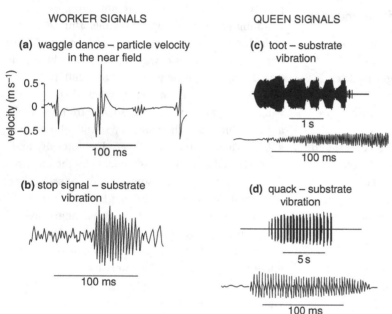

Figure 26.15 Vibrations produced by wing muscles of worker and queen honey bees (*Apis*). (a) Particle displacement in the near field produced by workers performing the waggle dance. The variation in amplitude of alternate pulses results from the side-to-side movement of the gaster (after Michelsen *et al.*, 1987). (b) Substrate vibration produced as a stop signal ("begging") by a bee following a dancer (after Michelsen *et al.*, 1986b). (c) Tooting. A substrate-borne vibration produced by a young queen. Detail of the vibrations below (after Michelsen *et al.*, 1986b). (d) Quacking. A substrate-borne vibration produced by a young queen. Detail of the vibrations below (after Michelsen *et al.*, 1986a).

Young virgin future queen honey bees produce piping sounds by thoracic vibration. The thorax is pressed to the comb so that vibrations are transferred directly to the comb. Queens that have escaped from the cells produce a "toot"; those still in the cells produce a "quack," which is a rapid train of short periods of vibration (Fig. 26.15c,d).

26.3.5 Air expulsion

Expulsion of air is a relatively rare mechanism of sound production in insects.

The death's head hawk moth, *Acherontia*, repeatedly sucks in and expels air through the mouth by dilating and contracting its pharynx. Through vibrations of the epipharynx the intake of air produces a series of sound pulses with maximum intensities at frequencies of 7–8 kHz (Fig. 26.16). Subsequent contraction of the pharynx with the epipharynx held erect expels the air, producing a whistle.

The hissing cockroach, *Gromphadorhina*, produces acoustic signals in male–male competition and courtship. Males generate sounds by expelling air

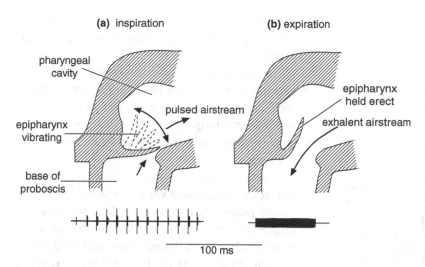

(a) inspiration   **(b)** expiration

pharyngeal cavity

epipharynx held erect

pulsed airstream

exhalent airstream

epipharynx vibrating

base of proboscis

100 ms

Figure 26.16 Sound produced by a vibrating column of air. Diagrammatic sagittal sections through the head of the death's head hawk moth, *Acherontia atropos*, showing the method of sound production (after Dumortier, 1963a). (a) On inspiration the epipharynx vibrates, producing a pulsed airstream and intermittent sounds. (b) On exhalation the airstream is continuous, producing a continuous whistle.

through the enlarged fourth abdominal spiracle. The trachea leading from the spiracle tapers to a narrow connection with the longitudinal tracheal trunk. Air is forced through this connection when the expiratory muscles contract and all other spiracles are closed, producing a hissing noise. The frequency spectrum of the sound has a maximum at about 10 kHz but extends up to about 40 kHz.

26.4 Patterns of acoustic and vibrational signals

26.4.1 Organized patterns

The acoustic and vibrational signals of insects are often modulated to produce discrete patterns characteristic of the species, and some species also vary their signal in relation to particular functions. Since similar organs for sound production and similar dimensions of body sizes within closely related species determine similar frequency spectra of acoustic communication signals, a large variety of temporal patterns has evolved that contribute to the separation of species. In addition, various species are able to produce songs that differ in intensity, frequency content and temporal patterning as

specific signals in different behavioral situations, such as attraction of mating partners from a distance, courtship or agonistic interactions (Figs. 26.2, 26.3).

Stridulating orthopteran species may produce different acoustic signals by varying the tooth impact frequency, the number of file teeth that impact the scraper on each complete wing or leg movement and the frequency of the movements. Fig. 26.17 shows a commonly used classification to describe the hierarchies of sounds produced by orthopterans. The sound of a single file tooth striking the scraper is called an impulse; the sound produced by a single up or down movement of a leg or by opening or closing a wing is called a syllable; and the sound produced by one whole sequence of leg movements is a chirp. Repeating continuous series of chirps, usually of a typical duration interrupted by silent periods, are called first-order sequences and several of these may be combined to create higher-order sequences. A typical calling song of the grasshopper *Chorthippus biguttulus* consists of groups of three first-order sequences of 2–5 seconds duration separated by pauses of 1–2 seconds. Each of these sequences includes up to 70 chirps that typically consist of six syllables, each resulting from three up and down movements of the hindlegs. Sound patterns corresponding to one syllable are

(a) impulse – one tooth impact

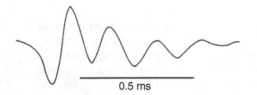

0.5 ms

(b) syllable – one down or up movement of leg
(opening or closing of wings)

tooth
impacts

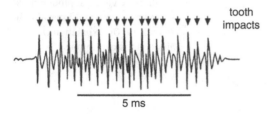

5 ms

(c) chirp – one complete cycle of leg (wing)
movements

down   up   down   up   down   up

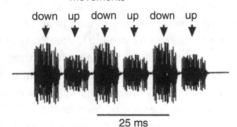

25 ms

(d) first-order sequence – continuous
series of chirps

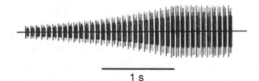

1 s

(e) second-order sequence – unit consisting of
several sequences

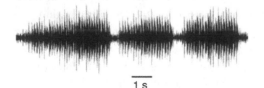

1 s

produced by striking approximately 20 from a total of approximately 100 stridulatory pegs on each hindleg against a wing.

The bushcricket *Platycleis intermedia* has, on average, 77 teeth on the file and produces a chirp of three syllables, one syllable on opening the wings and two on subsequent closures. These pulses employ, respectively, only 31%, 41% and 67% of the file teeth.

In tettigoniids the frequency of muscle oscillations producing wing closure varies considerably, even in closely related species. For example, while the wing muscles of *Neoconocephalus robustus* oscillate at about 200 Hz during singing, those of *N. ensiger* oscillate at only 10–15 Hz. These differences result in associated differences in the frequencies of syllable production.

Because the rate of muscular contraction is temperature dependent, the movements involved in stridulation are also affected by temperature. Thus the frequency with which sound impulses or syllables are produced by some crickets and bushcrickets varies linearly with temperature (Fig. 26.18).

Although the mechanism of sound production is similar among different species of phylogenetically related groups such as the grasshoppers, the crickets and the cicadas, temporal patterns of their

Figure 26.17 Terminology used to describe orthopteran sounds (grasshopper, *Chorthippus biguttulus*) (after Elsner, 1974). Similar terminology is applied to sounds produced by wing movements in crickets and bushcrickets.
(a) Impulse: sound produced by the impact of a single file tooth against a scraper. (b) Syllable: sound produced by a single up or down movement of the leg (or opening or closing the wings in crickets and bushcrickets). (c) Chirp: sound produced by a complete cycle of leg movements. (Fig. 26.16 shows the leg movements in a cycle, but note that the species is different.) (d) First-order sequence: sound produced by a series of chirps following in rapid succession with only brief, less than 100 ms, intervals between them. (e) Second-order sequence: typical calling song represented by a series of three first-order sequences with short interruptions.

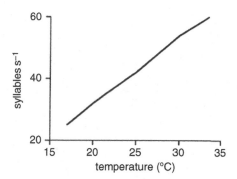

Figure 26.18 The effects of temperature on sound production. The frequency of syllable production (wing movements) produced by the cricket *Oecanthus* (after Walker, 1962).

acoustic signals vary considerably between species (Figs. 26.19, 26.20a).

Among cicadas, repetition frequency of rhythmical timbal deformations is variable between species due to differences in the frequencies of muscle oscillation. To provide some examples, each timbal muscle oscillates with frequencies of approximately 40 Hz in *Cystosoma*, 75 Hz in *Abricta* and 225 Hz in *Psaltoda*. The timbal muscles of the two sides may function in or out of phase; in the latter case the repetition rate of timbal-derived signals is doubled. An individual cicada may vary the repetition rate of timbal deformations by changing the tension of the timbal through altered activity of the tensor muscle. Rhythmic distortions of the timbal frame by the tensor muscle change the elastic properties of the timbal and may modulate the amplitude of the generated sounds. Sound intensity may also be modulated by varying the space between the abdomen and the operculum. This alters the resonant frequency of the air sacs, changes their tuning to the timbals and thus modulates the intensity of radiated sound.

Organized patterns of vibrations are also produced by some insects that communicate via the substrate. This occurs in Plecoptera and Neuroptera, in many Homoptera and in honey bees. These insects generate vibrations using different mechanisms –

abdominal oscillations in Plecoptera and Neuroptera, timbals in Homoptera (Fig. 26.20b) and the flight muscles in *Apis*. In most cases the vibrations are transmitted to the substrate via the legs. Since the frequency content of vibrational signals is largely determined by the transmission properties of the substrate, species-specific features are usually represented in the rate and rhythmicity of oscillations. Each pulse of the signal is produced by a single impact or bout of muscle activity. Modulation of muscle force may change the intensity and transmission range of vibrational signals.

The same applies to substrate-borne signals that are generated by percussion. Fig. 26.4 shows the pattern produced when the male stonefly *Isoperla* strikes the substrate with its abdomen. The female responds with a different pattern and other species also have distinct sequences.

26.4.2 Unorganized signals

Many insect species produce air-borne or substrate-borne signals that are not organized in precise temporal patterns (Fig. 26.9c). In these cases the vibrations themselves, in context, are sufficient to convey information to conspecifics or other species. Such sounds may be produced by any of the mechanisms described above.

26.5 Neural regulation of sound production

Neural regulation of sound production involves two steps that are usually mediated by different parts of the insect's central nervous system. Sound production has to be selected in appropriate situations from a repertoire of behavioral choices. This function is mediated by the brain, which selects appropriate sound patterns on the basis of sensory information about the environment and the internal physiological condition of the insect. In a second step, sound patterns have to be generated by

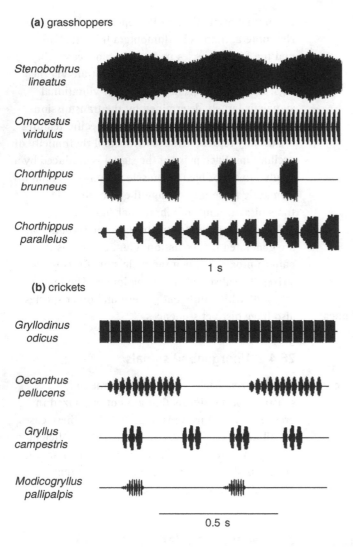

(a) grasshoppers

Stenobothrus lineatus

Omocestus viridulus

Chorthippus brunneus

Chorthippus parallelus

1 s

(b) crickets

Gryllodinus odicus

Oecanthus pellucens

Gryllus campestris

Modicogryllus pallipalpis

0.5 s

Figure 26.19 Patterns of sound produced by stridulatory mechanisms in Orthoptera. (a) Femoro-tegminal stridulation in grasshoppers. All the species are members of the Gomphocerinae and employ a similar method of sound production (after Haskell, 1957). (b) Tegminal stridulation in crickets. All the species employ a basically similar mechanism (after Popov *et al.*, 1974).

neuromuscular activity driving the sound-producing organs of the insect. This function is usually mediated by neural assemblies in ventral nerve cord ganglia of the segment that contains the sound-producing apparatus.

The patterns of muscular activity resulting in sound production are generated by pattern generators in the central nervous system. Pattern generators are networks of interneurons that are sufficient to generate rhythmic neuromuscular excitation that activates the sound-producing organs to generate the basic pattern of a species' acoustic signal. Only some interneurons and their contribution to pattern generation have been identified in a few species, but none of the networks has been analyzed in sufficient detail to fully understand its mechanism. Insects that employ bilateral pairs of sound-producing neurons contain hemiganglionic pattern generators with two identical sets of interneurons in each hemisphere that are interconnected to establish coordinated function. In crickets, where the sound is produced by movements of the forewings, motor neurons in the mesothoracic ganglion that drive the

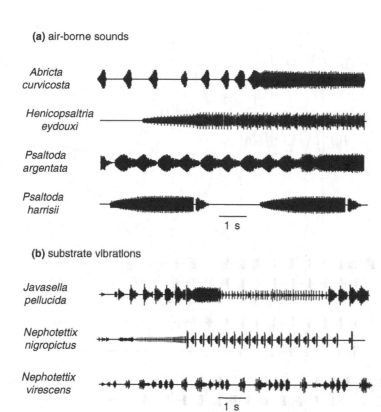

(a) air-borne sounds

Abricta curvicosta

Henicopsaltria eydouxi

Psaltoda argentata

Psaltoda harrisii

1 s

(b) substrate vibrations

Javasella pellucida

Nephotettix nigropictus

Nephotettix virescens

1 s

female

Nilaparvata lugens

male

1 s

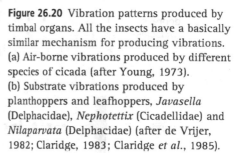

Figure 26.20 Vibration patterns produced by timbal organs. All the insects have a basically similar mechanism for producing vibrations. (a) Air-borne vibrations produced by different species of cicada (after Young, 1973). (b) Substrate vibrations produced by planthoppers and leafhoppers, *Javasella* (Delphacidae), *Nephotettix* (Cicadellidae) and *Nilaparvata* (Delphacidae) (after de Vrijer, 1982; Claridge, 1983; Claridge *et al.*, 1985).

forewing muscles are activated by central pattern-generating interneurons in the metathoracic ganglion. In grasshoppers, where stridulation involves coordinated movements of the hindlegs, the pattern generator is in the metathoracic ganglion. *Drosophila* fruit flies, which generate sounds by wing vibration, contain pattern-generating networks in the mesothoracic neuromere of the thoraco-abdominal ganglion complex.

The grasshopper *Chorthippus mollis* provides an example of the muscular activity underlying a distinctive song pattern (Fig. 26.21). A chirp starts with a pronounced, high-amplitude syllable of sound by moving one leg sharply downwards. Subsequently, both hindlegs perform oscillating up and down movements with the femur, which gradually increase and then decrease in amplitude. These oscillations are superimposed on a slow raising and lowering of the general position of the femur. As a result, different sections of the file strike the scraper and produce sound syllables that vary in intensity. Toward the end of the chirp the rapidity

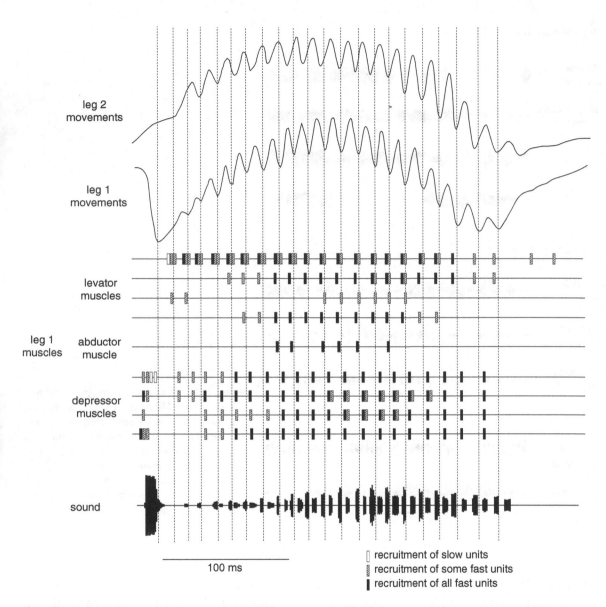

leg 2
movements

leg 1
movements

leg 1
muscles

levator
muscles

abductor
muscle

depressor
muscles

sound

100 ms

⬜ recruitment of slow units
▨ recruitment of some fast units
▮ recruitment of all fast units

Figure 26.21 Muscular activity during stridulation by a grasshopper (*Chorthippus mollis*). Top: the activity of the two hindlegs during a single chirp which consists of a sequence of up and down movements of each leg superimposed on a slower upward and downward movement. The effect of the slow upward movement is to bring progressively more proximal teeth of the femoral file into contact with the scraper. In this species the two legs are slightly out of phase. Center: the pattern of activity in muscles moving leg 1. Each horizontal line represents a separate muscle. Each muscle is innervated by fast and slow axons whose activity in recruiting the muscle units is shown by rectangles. Note that the greatest number of muscles and muscle units are activated by the fast axons during the middle of the chirp when the amplitude of leg movements is greatest (above) and the amplitude of sound greatest (below). The muscles are activated shortly before leg movement is observed. Bottom: sounds produced by the movements of leg 1. The vertical axis shows the amplitude of sound. Syllables of sound are produced on both the up- and downstrokes, but sound intensity is greatest on the downstroke (after Elsner, 1975).

with which each leg movement is made becomes reduced and the syllables occur at greater intervals. Except for the initial pronounced downstroke, both hindlegs perform similar sequences of movements that are slightly out of phase. The number of syllables in each chirp increases with the progress of the song sequence.

However, each of the hindleg muscles is made up of several units innervated by fast and slow axons. As a result, the force exerted by a muscle can be varied, and this affects sound production. When a grasshopper produces a chirp, the number of muscle units involved in the oscillating movements of the femur, and their activation via fast axons, first increases and then decreases (Fig. 26.21). These changes, together with slight changes in the timing, produce the alterations in up and down movements during the chirp. Variation in the activity of other muscles, the coxal adductor and abductor muscles, produces changes in sound intensity by pulling the coxa toward or away from the body and varying the pressure of the femur against the tegmen.

The degree of sensory feedback involved in the maintenance of proper pattern generation for the production of acoustic signals varies between insect species. The impact of acoustic feedback from self-generated sounds on the produced acoustic patterns is small or absent in most species studied. Some species, like the cricket *Gryllus bimaculatus*, even suppress the responsiveness of their auditory system during sound production to prevent adaptation and maintain sensitivity to other acoustic stimuli. However, proprioceptive information from mechanoreceptive hairs and campaniform sensilla on the wings and filiform sensilla on the cerci is necessary to maintain regular pulse patterns and intensity modulation of acoustic signals in crickets. In the cicada *Cystosoma* the frequency of oscillation of the timbal, maintained by a central pattern generator in the metathoracic ganglion, is modulated by feedback from a chordotonal organ inserted on the rim of the timbal.

Singing is initiated by descending interneurons from the brain, known as command neurons (Fig. 26.22). In grasshoppers and crickets, different types of command neurons invariantly initiate different song patterns. Thus, the command neurons transmit the activating signal from the brain to the thoracic pattern generators, but brain neuropiles located presynaptically to the command neurons mediate the decision about when and which pattern to sing. Initiation of courtship song production in male *Drosophila melanogaster* critically depends on groups of neurons in the posterior lateral protocerebrum that integrate female-derived olfactory and visual and other sensory inputs which promote or suppress courtship singing. In addition, mushroom bodies and the central complex influence the persistence of song production, regularity of sound patterns and the selection of sound signals with respect to the behavioral context (e.g., courting a female in contrast to fighting a male).

In grasshoppers, the central complex selects and coordinates song production on the basis of pre-processed acoustic and visual information. Stridulation can be stimulated by injections of small volumes of cholinergic agonists, proctolin, dopamine, GABA antagonists and inhibitors of nitric oxide formation into the central complex, suggesting that these transmitters activate (acetylcholine, proctolin, dopamine) or suppress (GABA, nitric oxide) sound production in respective behavioral situations. All of these signaling pathways converge on output neurons of the central complex whose activity is relayed to the descending stridulatory command neurons.

Selection and coordination of type, intensity, and timing of sound signals is mediated by the central complex, a highly structured brain neuropil known to integrate multimodal pre-processed sensory information by a large number of chemical messengers.

Especially drugs that effectively activate muscarinic acetylcholine receptors can stimulate prolonged sound production involving several different pattern elements in the correct temporal sequence. Sensory stimuli associated with behavioral

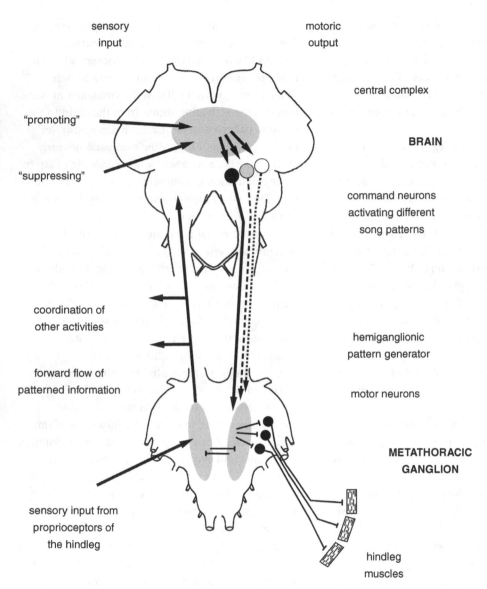

sensory input

motoric output

central complex

BRAIN

"promoting"

"suppressing"

command neurons
activating different
song patterns

coordination of
other activities

hemiganglionic
pattern generator

forward flow of
patterned information

motor neurons

**METATHORACIC
GANGLION**

sensory input from
proprioceptors of
the hindleg

hindleg
muscles

Figure 26.22 Neural control of sound production in a grasshopper. Stridulatory movements of the hindlegs arise from neuromuscular excitation patterns of hemiganglionic metathoracic pattern generators. Pattern-generating circuits responsible for a particular pattern from a species' repertoire are activated by a particular type of command interneuron that connects the decision-making neuropiles in the brain with the thoracic pattern generator. The decision about type of song pattern, timing and intensity of stridulation is made by the central complex on the basis of converging sensory information of various modalities. Stridulation involves particular activities of various other parts of the grasshopper body (e.g., fore- and middle legs, respiration) getting adjusted to the rhythmic stridulatory activity via proprioceptive sensory input and efference copies of motor commands (as demonstrated in crickets).

situations that favor or exclude sound production (e.g., hearing the response song of a conspecific female or being restrained during a physiological experiment) mediate the release of excitatory or inhibitory transmitters (acetylcholine or nitric oxide) in the central complex that promote or suppress the behavior. Inactivation of chloride channel-mediated inhibition causes the simultaneous production of different song patterns, suggesting that the central complex selects one song type that fits to the behavioral context while at the same time suppressing other song patterns and other behaviors that interfere with sound production. Output neurons of the central complex connect, either directly or indirectly via additional interneurons, to the command neurons that activate the thoracic pattern generators for sound production.

Besides the central pattern generators, sound production requires synchronized contributions by other parts of the insect body. Insects may need to establish and maintain particular postures, stabilize their position on the substrate and adjust their respiratory activity; all activities that interfere with sound production have to be inhibited.

Both external stimuli and internal factors are important in regulating song production. Males only sing when they are sexually mature and their calling activity may be interrupted after copulation until a new spermatophore has been produced. Female grasshoppers only sing when they are in a sexually responsive state and, like other aspects of sexual maturation in these insects, singing is regulated by juvenile hormone.

Summary

- Many insects produce air-borne sounds and substrate vibrations in order to transmit information to conspecifics, other insects or even non-insect species.

- Insect acoustic and vibrational signals contribute to reproductive behaviors by attracting potential mates from a distance, advertising individual quality, signaling physiological state and establishing copulatory readiness. Among other purposes they also serve important functions in territorial behaviors, defense against predators and coordination of social insects' behaviors.

- Insect acoustic signals differ in frequency composition, intensity and temporal patterns. Most sound signals are highly species-specific, especially between sympatric species, and many insects produce different signals in different behavioral contexts (e.g., mate attraction, courtship and agonistic interactions).

- Signals may be generated by unspecialized parts of the body (e.g., vibration of wings, drumming of head or appendages against the substrate) or by diverse and highly specialized sound-producing organs that insects have evolved on different body parts to enhance sound radiation (e.g., timbal organs of cicadas and moths, stridulatory files on wings and legs of orthopteran species).

- The sound-producing organs are activated by rhythmic neuromuscular output from central pattern generators in thoracic or abdominal ganglia. Activity of these pattern generators is controlled by the brain. The central complex selects timing and song type on the basis of sensory information about the environment and the internal physiological state and transmits this decision through descending command neurons to the thoracic or abdominal pattern generators.

Recommended reading

INSECT VIBRATIONAL SIGNALS AND THEIR PROPAGATION

Casas, J., Magal, C. and Sueur, J. (2007). Dispersive and non-dispersive waves through plants: implications for arthropod vibratory communication. *Proceedings of the Royal Society of London B* **274**, 1087–1092.

Hill, P. S. M. (2001). Vibration and animal communication: a review. *American Zoologist* **41**, 1135–1142.

Virant-Doberlet, M. and Cokl, A. (2004). Vibrational communication in insects. *Neotropical Entomology* **33**, 121–134.

INSECT ACOUSTIC SIGNALS

Bailey, W. J. (1991). *Acoustic Behaviour of Insects: An Evolutionary Perspective*. London: Chapman & Hall.

Bennet-Clark, H. C. (1995). Insect sound production: transduction mechanisms and impedance matching. *Symposia of the Society for Experimental Biology* **49**, 199–218.

Ewing, A. W. (1989). *Arthropod Bioacoustics: Neurobiology and Behaviour*. Ithaca, NY: Comstock Publishing Associates.

Greenfield, M. D. (2002). *Signalers and Receivers: Mechanisms and Evolution of Arthropod Communication*. Oxford: Oxford University Press.

COMMUNICATION IN SOCIAL INSECTS

Barth, F. G., Hrncir, M. and Jarau, S. (2008). Signals and cues in the recruitment behaviour of stingless bees (Meliponini). *Journal of Comparative Physiology A* **194**, 313–327.

Cocroft, R. B. (2005). Vibrational communication facilitates cooperative foraging in a phloem-feeding insect. *Proceedings of the Royal Society of London B* **272**, 1023–1029.

Hickling, R. and Brown, R. L. (2000). Analysis of acoustic communication by ants. *Journal of the Acoustic Society of America* **108**, 1920–1929.

Kirchner, W. H., Broeker, I. and Tautz, J. (1994). Vibrational alarm communication in the damp-wood termite *Zootermopsis nevadensis*. *Physiological Entomology* **19**, 187–190.

Michelsen, A., Kirchner, W. H. and Lindauer, M. (1986). Sound and vibrational signals in the dance language of the honeybee, *Apis mellifera*. *Behavioural Ecology and Sociobiology* **18**, 207–212.

Tautz, J., Casas, J. and Sandeman, D. (2001). Phase reversal of vibratory signals in honeycomb may assist dancing honeybees to attract their audience. *Journal of Experimental Biology* **204**, 3737–3746.

NEURAL REGULATION OF SOUND PRODUCTION

Gerhardt, H. C. and Huber, F. (2002). *Acoustic Communication in Insects and Anurans: Common Problems and Diverse Solutions*. Chicago, IL: University of Chicago Press.

Hedwig, B. (2001). Singing and hearing: neural mechanisms of acoustic communication in orthopterans. *Zoology* **103**, 140–149.

References in figure captions

Bennet-Clark, H. C. (1970). The mechanism and efficiency of sound production in mole crickets. *Journal of Experimental Biology* 52, 619–652.

Bennet-Clark, H. C. (1989). Songs and the physics of sound production. In *Cricket Behaviour and Neurobiology*, ed. F. Huber, T. E. Moore and W. Loher, pp. 227–261. Ithaca, NY: Comstock Publishing Associates.

Blest, A. D., Collett, T. S. and Pye, J. D. (1963). The generation of ultrasonic signals by New World arctiid moth. *Proceedings of the Royal Society of London B* 158, 196–207.

Claridge, M. F. (1983). Acoustic signals and species problems in the Auchenorrhyncha. In *Proceedings of the First International Workshop on Leafhoppers and Planthoppers of Economic Importance*, ed. W. J. Knight, N. C. Pant, T. S. Robertson and M. R. Wilson, pp. 111–120. London: Commonwealth Institute of Entomology.

Claridge, M. F., den Hollander, J. and Morgan, J. C. (1985). Variation in courtship signals and hybridization between geographically definable populations of the rice brown planthopper, *Nilaparvata lugens* (Stål). *Biological Journal of the Linnean Society* 24, 35–49.

de Vrijer, P. W. F. (1982). Reproductive isolation in the genus *Javasella* Fenn. *Acta Entomologica Fennica* 38, 50–51.

Dumortier, B. (1963a). Morphology of sound emission apparatus in arthropods. In *Acoustic Behaviour of Animals*, ed. R.-G. Busnel, pp. 277–345. Amsterdam: Elsevier.

Dumortier, B. (1963b). The physical characteristics of sound emissions in arthropods. In *Acoustic Behaviour of Animals*, ed. R.-G. Busnel, pp. 346–373. Amsterdam: Elsevier.

Elsner, N. (1974). Neuroethology of sound production in gomphocerine grasshoppers (Orthoptera: Acrididae): I. Song patterns and stridulatory movements. *Journal of Comparative Physiology* 88, 67–102.

Elsner, N. (1975). Neuroethology of sound production in gomphocerine grasshoppers (Orthoptera: Acrididae): II. Neuromuscular activity underlying stridulation. *Journal of Comparative Physiology* 97, 291–322.

Haskell, P. T. (1957). Stridulation and associated behaviour in certain Orthoptera: I. Analysis of the stridulation of, and behaviour between males. *Animal Behaviour* 5, 139–148.

Haskell, P. T. (1961). *Insect Sounds*. London: Witherby.

Huber, F. (1962). Central nervous control of sound production in crickets and some speculations on its evolution. *Evolution* 16, 429–442.

Leston, D. (1957). The stridulatory mechanisms in terrestrial species of Heteroptera. *Proceedings of the Zoological Society of London* 128, 381–400.

Michelsen, A., Kirchner, W. H., Andersen, B. B. and Lindauer, M. (1986a). The tooting and quacking vibration signals of honeybee queens: a quantitative analysis. *Journal of Comparative Physiology* 158, 605–611.

Michelsen, A., Kirchner, W. H. and Lindauer, M. (1986b). Sound and vibrational signals in the dance language of the honey bee, *Apis mellifera*. *Behavioral Ecology and Sociobiology* 18, 207–212.

Michelsen, A., Towne, W. F., Kirchner, W. H. and Kryger, P. (1987). The acoustic near field of a dancing honeybee. *Journal of Comparative Physiology A* 161, 633–643.

Nocke, H. (1972). Physiological aspects of sound communication in crickets (*Gryllus campestris* L.). *Journal of Comparative Physiology* 80, 141–162.

Pasquinelly, F. and Busnel, M.-C. (1955). Études preliminaries sur les mécanismes de la production des sons par les Orthoptères. In *Colloques sur l'acoustique des Orthoptères*, ed. R.-G. Busnel, pp. 145–153. Paris: Institut National De La Recherche Agronomique.

Popov, A. V., Shuvalov, V. F., Svetlogorskaya, I. D. and Markovich, A. M. (1974). Acoustic behavior and auditory system in insects. In *Mechanoreception*, ed. J. Schwartzkopff, pp. 281–306. Opladen: Westdeutscher Verlag.

Pringle, J. W. S. (1954). A physiological analysis of cicada song. *Journal of Experimental Biology* 31, 525–560.

Roscow, J. M. (1963). The structure, development and variation of the stridulatory file of *Stenobothrus lineatus* (Panzer) (Orthoptera: Acrididae). *Proceedings of the Royal Entomological Society of London A* 38, 194–199.

Stephen, R. O. and Hartley, J. C. (1995). Sound production in crickets. *Journal of Experimental Biology* 198, 2139–2152.

Szczytko, S. W. and Stewart, K. W. (1979). Drumming behaviour of four western Nearctic *Isoperla* (Plecoptera) species. *Annals of the Entomological Society of America* 72, 781–786.

Walker, T. J. (1962). Factors responsible for intraspecific variation in the calling songs of crickets. *Evolution* 16, 407–428.

Young, D. (1973). Sound production in cicadas. *Australian Natural History* 1973, 375–80.

Young, D. and Bennet-Clark, H. C. (1995). The role of the timbal in cicada sound production. *Journal of Experimental Biology* 198, 1001–1019.

27 Chemical communication: pheromones and allelochemicals

REVISED AND UPDATED BY **JEREMY N. MCNEIL AND JOCELYN G. MILLAR**

INTRODUCTION

Insects are prodigious users of chemical signals and cues, which play diverse
and fundamental roles in the transfer of information both within and between species.
Indeed, it is likely that no other group of animals makes such sophisticated use of chemical
signaling in their biology. This chapter begins by defining the different classes of signals
(Section 27.1), before describing the nature of intraspecific chemical signals (pheromones)
(Section 27.2), the information content of such pheromones (Section 27.3), their
biosynthesis (Section 27.4) and the mechanisms regulating their production (Section 27.5),
as well as their sensory perception by conspecifics (Section 27.6). In the next section
(Section 27.7) interspecific signals are discussed (allelochemicals), followed by their
mechanisms of production and release (Section 27.8). Section 27.9 concerns defensive
compounds, and the chapter ends with chemical mimicry (Section 27.10).

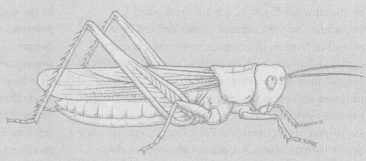

27.1 Defining chemical signals

Chemical signals and cues have been collectively called semiochemicals, derived from the Greek word "semeon" for signal. However, it has been suggested that the term "infochemical" may be more appropriate, based on the argument that nomenclature should be based on a "cost–benefit analysis" rather than the actual source of the signal. While there has not been complete acceptance of either term and both are used in the current literature, we will use infochemical when referring to "a chemical substance, which in a natural context, is implicated in the transfer of information during an interaction between two individuals that results in a behavioral and/or physiological response in one or both." In this chapter the term "signal" is applied to an infochemical produced by an emitter which has been shaped by evolution to transmit a specific message to the intended receiver. An example of this would be the release of a sex pheromone for the specific purpose of attracting a mate. The term "cue" is used to describe an infochemical that conveys information to a receiver, but was not shaped by natural selection for this purpose – that is, it is exploited by receivers, often to the detriment of the emitter. For example, the sex pheromone emitted by an insect to attract a mate may also be exploited as a kairomone by a predator or parasitoid.

Compounds that produce an immediate behavioral change following reception are referred to as "releaser" infochemicals. An example of this would be the upwind flight of a male moth when he encounters the pheromone plume emitted by a receptive female. In contrast, when the chemical signal stimulates a physiological change, the effects are not immediate. Instead, a cascade of events is initiated and must be completed before the change becomes evident, and hence these signals are known as "primer" infochemicals. For example, mature male desert locusts produce a primer pheromone that accelerates the sexual maturation of conspecifics, a process that obviously takes some time. In a different context, the honey bee queen mandibular pheromone, among numerous other functions, acts as a primer pheromone by inhibiting the development of ovaries in worker bees.

Infochemicals that mediate interactions between members of the same species are collectively referred to as pheromones, whereas those involved in interactions between different species are known as allelochemicals. Within each of the two categories there are a number of subdivisions, classified according to the actual roles that the infochemicals play. Some of the various types of pheromones are defined in Table 27.1, with more detailed descriptions and examples in the following text. Similarly, different types of allelochemicals are defined and described later on in the chapter. It is important to note that many infochemicals fall into more than one category. For example, volatile sex or aggregation pheromones may also be exploited as kairomones by predators and parasitoids to find their prey or hosts, respectively.

27.2 Pheromones used in intraspecific communication

27.2.1 Sex pheromones

Volatile sex pheromones serve as long-distance signals, emitted by receptive individuals to attract suitable mates. They can be effective over distances of hundreds of meters and possibly even kilometers. In the majority of cases studied, volatile sex pheromones are produced by females to attract conspecific males. However, the situation is sometimes reversed, as seen in some species of weevils and cerambycid beetles, wax moths and phytophagous stink bugs (Hemiptera), where it is the males that produce sex pheromones rather than females. There are also a few cases where both sexes produce pheromones to attract the opposite sex, but during different temporal windows. For example, in

Table 27.1 **Classes of pheromones that mediate interactions between conspecific individuals**

Sex pheromones	Released by one or both sexes; generally benefit both emitter and receiver, but not always, because pheromones also may be used to assess the quality of the individual as a mate.
Anti-aphrodisiac pheromones	Used by a male to mark a mated female, to render her unattractive to other males; benefits male by helping to ensure his paternity, benefits female by reducing harassment by courting males.
Epideictic or marking pheromones	Deposited on the surface or inside a resource, such as a fruit or parasitized host egg, to indicate that it has already been exploited; benefits the marker by reducing competition for its progeny.
Trail pheromones	Used by social and semi-social species to indicate the location of exploitable resources with respect to the position of the colony; benefits the colony by increasing the efficiency of foraging.
Alarm pheromones	Emitted in response to danger, such as an attack by a predator; benefits receivers as they modify their behavior to defend the colony, or to reduce the probability of being captured.
Aggregation pheromones	Results in aggregations of conspecifics; potentially beneficial to all individuals because it decreases the risk of predation or parasitism, improves the exploitation of a resource and/or increases the probability of locating a mate.
Anti-aggregation pheromones	Emitted when a resource is in danger of being over-exploited; benefits both emitters and receivers because it reduces intraspecific competition.
Pheromones of social insects	Numerous pheromones produced to coordinate activities within the colony. For example, the cuticular lipids contain signals that serve in recognition of nest mates, caste, reproductive status and even whether an individual is alive or dead. Queen pheromones serve many roles, including the suppression of worker reproduction and the organization of workers to feed and groom the queen.

the arctiid moth, *Estigmene acrea*, mating early in the scotophase occurs when females are attracted to the pheromones released by lekking males from scent organs known as coremata, whereas later in the night, mating occurs when males are attracted to calling females.

Volatile pheromones that are intended to carry a message over long distances are released by different mechanisms and structures than short-range or contact pheromones. The simplest method used to release volatile pheromones is passive evaporation of the pheromone from an exposed gland surface, but even here specific behavioral and morphological adaptations make the process more efficient. Thus,

pheromone-producing female moths call from exposed positions in the upper part of the canopy and also adopt a specific calling posture in which the pheromone gland is maximally exposed to air currents. For example, in pyralid moths, this involves curling the abdomen up above the body to allow maximum exposure of the everted gland. In many cases calling females also vibrate their wings, which may also facilitate pheromone emission. Male cerambycid beetles in the subfamily Cerambycinae release volatile pheromones from gland pores on the metathorax, and to help dissipate the pheromone they adopt a stereotyped "pushup" stance with the front part of the body raised, to maximize passage of wind currents over the pheromone release area so that the scent is rapidly dissipated. Other insects, such as male stink bugs, release pheromone onto the abdominal surface, which is covered with fine hairs to provide increased surface area to maximize evaporation rates. Calling female arctiid moths (e.g., *Utetheisa ornatrix, Pyrrharctia isabella*) actively pulse the pheromone gland, and the spiny morphology of the pheromone-releasing structures helps to form a fine aerosol spray of pheromone droplets that vaporize as they move downwind.

The use of volatile pheromones for long-distance attraction of mates has been extensively studied in the Lepidoptera. Female pheromone glands are of ectodermal origin, and are found in close association with the cuticle. As seen in many moth species, they may be concealed in sites such as the folds of the intersegmental membranes of abdominal segments (Fig. 27.1), but the location varies with species. For example, the pheromone glands of the female bagworm, *Tryridopteryx ephemeraefirmis*, are found on the thorax rather than the abdominal intersegmental membranes. In female aphids and mealybugs (Hemiptera), sex pheromones are produced in glands on the tibia of the metathoracic legs, whereas the sex pheromone of female houseflies is produced by oenocyte cells associated with the abdominal cuticle, and is subsequently transported to the cuticular surface, the site of release, by transport proteins called lipophorins. Most pheromone glands are ductless and lack distinct reservoirs, so the pheromone products are released as they are biosynthesized. However, there are many variations with respect to the glands that produce insect sex pheromones. For example, the pheromone of males of the pentatomid bug *Nezara viridula* is produced in a series of single cells whose ducts open onto the ventral surface of the abdomen. Furthermore, in each cell there is a lumen with secretions, suggesting that in this species at least, the pheromone may be stored.

Sex pheromones are usually emitted at a specific time of day or night, one of the mechanisms used to limit responses by males of congeneric species. In the few cases that have been studied in detail, there is evidence that calling behavior is governed by an endogenous circadian rhythm that may persist for several days under continuous darkness (Fig. 27.2). A subsequent study on *Pseudaletia unpuncta* showed that pheromone synthesis was also governed by an

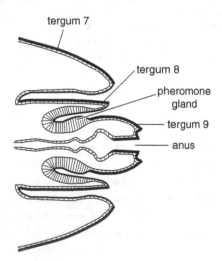

Figure 27.1 Horizontal section of the tip of the abdomen of a female moth (*Plodia*) in which the glands producing the sex-attractant pheromone are in a lateral position (from Wigglesworth, 1972).

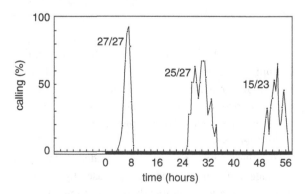

Figure 27.2 Evidence of an endogenous circadian rhythm in the calling behavior of the armyworm *Pseudaletia unipuncta* when entrained under 16L:8D and then held under constant darkness. The proportion of females calling is indicated for each calling period (from Turgeon and McNeil, 1982).

endogenous circadian rhythm. However, despite these endogenous circadian rhythms, the time when armyworm females call may be modified by prevailing abiotic conditions. The effect of temperature on the calling of virgin oblique banded leafroller females, *Choristoneura rosaceana*, is seen in Fig. 27.3: Females call at night at a mean ambient temperature of 20°C, but initiate calling in the latter part of the photophase at a mean ambient temperature of 15°C. This is considered an adaptation to ensure that pheromone is emitted at a time when males are still able to fly to receptive females. There

was also a significant age effect observed in this study, with older females initiating calling earlier that younger ones. This age-related trend has been reported in a number of other lepidopteran species and may represent an adaptation by older females to reduce competition with younger females, because the latter generally emit more pheromone.

Relatively non-volatile sex pheromones also play a significant role in mating over short distances. These compounds are typically a subset of the cuticular lipids that protect insects from desiccation. They are probably biosynthesized in oenocyte cells along with

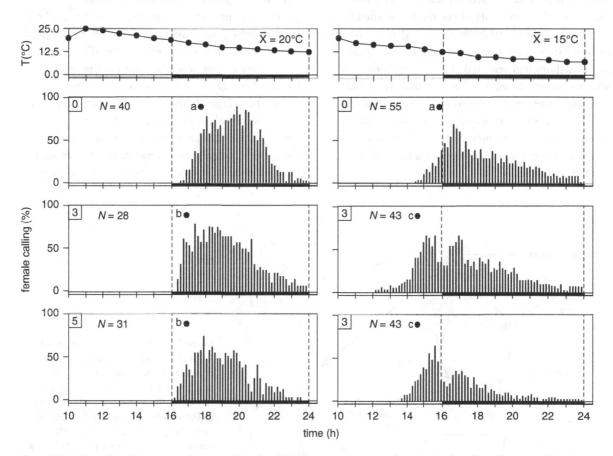

Figure 27.3 The calling behaviors of 0-, 3- and 5-day-old *Choristoneura rosaceana* virgin females when reared under simulated summer (mean daily temperature 20°C) and fall (mean daily temperaure 15°C) temperature conditions at a photoperiod of 16L:8D. The dark circle in each graph indicates the mean onset time of calling for the population (from Delisle, 1992).

the other cuticular hydrocarbons, and spread over the body surface by grooming. Thus, when the antennae of a male insect contacts any part of a female, he can immediately recognize her as a female of his species.

Female *Drosophila* fruit flies stimulate conspecific male courtship behavior with a pheromone composed of cuticular hydrocarbons. These pheromones are only detected by the antennae, tarsi or mouthparts when the male is in close proximity or in direct contact with the female. Once the male initiates courtship, a receptive female partially extrudes her ovipositor and produces a droplet of liquid (a presumed pheromone source) that further stimulates the male. Video analysis of courtship has shed light on an apparent contradiction with regard to the role of ovipositor extrusion in *Drosophila* mating, as some authors had reported that it stimulated males, whereas others categorized it as female rejection behavior. The detailed video analysis showed that partial ovipositor extension is associated with the production of the droplet that stimulated males, whereas full extension resulted in a marked decrease in male courtship.

The entire male courtship behavior may involve more than one pheromone produced by a receptive female. Males of the aphid parasitoids *Aphidius ervi* and *A. nigripes* (Hymenoptera) initiate upwind flight toward virgin females from several meters away, but once in close proximity they exhibit an intricate courtship ritual that includes wing fanning. The first step involves a long-distance pheromone, whereas the second is initiated by a short-range pheromone on the female's cuticle that only acts over a distance of a few centimeters.

In general, once a pheromone-emitting female has attracted a potential mate, short-distance pheromones produced by the male play a crucial role in mate recognition and mate choice. For example, many male moths have scent brushes (Fig. 27.4), which they expose when in close proximity to a female. A study on mating of the European corn borer, *Ostrinia nublialis*, showed that ablation of the male scent brushes, or removing the antennae of calling virgin females so that they could not perceive the male-produced compounds, markedly reduced the incidence of mating. Recently, the male corn

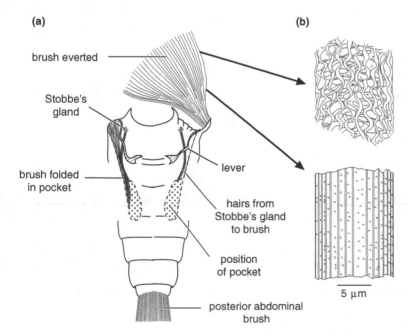

(a)

brush everted

Stobbe's gland

brush folded in pocket

lever

hairs from Stobbe's gland to brush

position of pocket

posterior abdominal brush

(b)

5 μm

Figure 27.4 Pheromone brushes of a male moth (*Phlogophora*). Pheromone from Stobbe's gland is transferred to the brushes through tubular hairs (after Birch, 1970). (a) Ventral surface of the abdomen showing one brush folded in its cuticular pocket (left), and the other partially everted (right). (b) Details of a hair from the brush. Above: the distal region forms an open lattice from which the pheromone readily evaporates. Below: the proximal part of the hair has a solid wall.

borer pheromone has been identified and assays demonstrated that the mating success of males without scent brushes could be restored if they were treated with the synthetic blend, providing experimental evidence that this chemical signal enhanced male mating success. It was also shown that there were significant differences between the male scents of the two pheromone strains of the European corn borer, as well as the closely related Asian corn borer, supporting the hypothesis that these male pheromones also have a role in reproductive isolation. Work on *Helicoverpa zea*, *Heliothis virescens* and *H. subflexa* (Lepidoptera) provided further evidence that male pheromones are not only important in female acceptance of conspecifics, but also in preventing heterospecific mating. It has been proposed that male pheromones might also have a role in male–male competition, by causing a male flying upwind to a calling female to veer off upon detecting the pheromone of a conspecific male. However, the data from the limited number of species that have been studied are contradictory, so the potential role of hair pencil pheromones in male–male interactions remains an open question.

27.2.2 Anti-aphrodisiac pheromones

Males of some insect species do, however, produce anti-aphrodisiac pheromones which are transferred to females during mating and reduce the attractiveness of mated females to other courting males. Anti-aphrodisiac pheromones have been reported in *Drosophila* fruit flies and in the plant bug, *Lygus hesperus*, but the most detailed studies have been carried out in the Lepidoptera. For example, *Pieris napi* males transfer methyl salicylate (MeS) in their ejaculate, which is released by the mated female when she is courted, and which rapidly deters males. The presence of the anti-aphrodisiac pheromone benefits both sexes; the mated female suffers less harassment by males and delaying remating reduces sperm competition, assuring males a higher

probability of paternity. The anti-aphrodisiac pheromone becomes depleted as the female repeatedly exhibits rejection behavior, so over time its effectiveness declines and she will remate, thereby increasing her lifetime reproductive output. Males of other *Pieris* species produce chemically different anti-aphrodisiac pheromones; *P. rapae* uses MeS and indole, whereas *P. brassicae* produces benzyl cyanide. In a good example of a pheromone being exploited as a kairomone, the egg parasitoids *Trichogramma brassicae* and *T. evanescens* use benzyl cyanide to locate ovipositing female moths. Anti-aphrodisiac pheromones have also been documented in other butterflies, such as *Heliconius melpomene*, where males mark females with β-ocimene. The source of these pheromones has not been determined, but is probably associated with the male accessory glands.

27.2.3 Aggregation pheromones

Aggregation pheromones may be produced by adults of one or both sexes and mediate the behavior of both sexes. In such situations the line between the roles of aggregation pheromones and sex pheromones becomes blurred because extensive mating often occurs in aggregations. Immature insects, such as first-instar nymphs of stink bugs, also produce aggregation pheromones, but the ready formation of heterospecific aggregations suggests that the signals are not species-specific.

Aggregation pheromones are used in a variety of contexts where individuals benefit by forming groups. These advantages may include the ability of the group to modify the habitat, or provide more effective defense. One of the most well-known uses of aggregation pheromones is by bark beetles, where the aggregation pheromones elicit mass attack on a host tree, so that the host's defenses are overwhelmed and the beetles can successfully colonize the tree. Mass attack is crucial, because a healthy tree can easily defend itself from one or a few of the tiny bark beetles, but its resin defenses are overwhelmed when it is simultaneously attacked by thousands of beetles. The system becomes even more complex because after a host tree is mass-attacked and fully colonized, the resident beetles begin producing anti-aggregation pheromones from the midgut which deter new arrivals from landing, and instead divert them to nearby trees. Thus, the anti-aggregation pheromones serve to prevent overcrowding so that the developing brood will not face starvation due to the limited nutritional resources of the host being completely consumed by a surfeit of larvae. Similarly, the anti-aggregation pheromones benefit the late arrivals by diverting them away from the already colonized tree, in which their larvae would have little chance of survival when matched against the much more developed larvae of the first arrivals.

The various steps, and the chemical signals controlling each step, are shown diagrammatically in Fig. 27.5.

In a somewhat different context, grain beetles (Laemophloeidae) that infest stored grains and similar commodities produce aggregation pheromones consisting of large cyclic esters known as macrolides. A critical mass of individuals is needed to ensure the growth of molds and fungi which produce nutrients and heat essential for survival of the beetles, particularly in extreme environments. The source of these pheromones is not known.

A number of insect species, such as ladybird beetles, stink bugs and leaf-footed bugs (*Leptoglossus* spp.), form overwintering aggregations, and there is evidence that their formation and maintenance

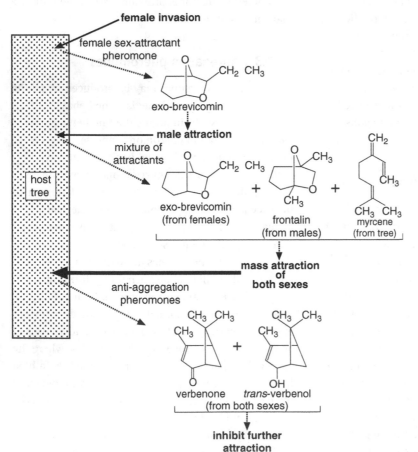

Figure 27.5 The role of aggregation and anti-aggregation pheromones in the invasion of a tree by the bark beetle *Ips paraconfusus*.

involves chemical signals. For example, the seven-spot ladybird beetle *Coccinella septempunctata* forms overwintering aggregations at the same sites year after year, implying that these sites are marked. Interestingly, aggregations may contain several species of ladybird beetles, suggesting that the aggregation signals are similar or even the same among related species.

Males of plant-feeding stink bugs produce volatile pheromones that attract females, males and even nymphs. According to the strict definition, these compounds would be considered aggregation pheromones because both sexes are attracted. However, the main role of the pheromones appears to be attraction of mates, and so the attraction of other males may be the result of these individuals exploiting the signal of the calling male. In this way, the sneaky males can not only intercept females that are attracted, but also reduce their chances of being attacked by the parasitoids and predators that exploit the pheromone as a kairomone to locate their stink bug prey.

A variety of aggregation pheromones play a critical role in the lifecycle of one of the world's most devasting insect pests, the desert locust, *Schistocerca gregaria*. These are produced by different developmental stages and not only influence densities, but also rates of sexual maturation and oviposition behavior. The combined effects of these and other pheromones is to synchronize the development and behaviors of the locusts that results in the formation of large swarms comprising individuals of the same age and physiological status.

27.2.4 Marking pheromones

Marking pheromones are used in a number of different contexts. One of the most common of these is the use of marking pheromones by females of both herbivorous and entomophagous species to mark hosts in which they have laid an egg. Deterring further oviposition serves to reduce intraspecific larval competition. Marking pheromones have been reported in a number of different orders, including the Coleoptera, Diptera, Hymenoptera and Lepidoptera, but in many instances the site of production remains unknown. In one of the few cases where the production site is known, females of the white spruce conefly, *Strobilomyia neanthracina*, produce a marking pheromone in the head and/or thorax, and mark the cone surface with their mouthparts following oviposition.

The marking pheromones of tephritid fruit flies have been extensively studied, with much of the original work being carried out with the apple maggot fly, *Rhagoletis pomonella*. Having laid an egg, the female drags her ovipositor over the fruit surface, leaving a chemical signal that is synthesized by cells in the midgut, which indicates that the site has been exploited. When exploiting hawthorn fruit, the original host of the apple maggot fly, it is important for the female to reduce the probability that she or other conspecific females will oviposit at the same site because a typical fruit contains sufficient resources for the optimal development of only one larva. However, the proportion of the total surface area of the host that is marked may depend on the size of the fruit, as well as the age and hunger level of the fly. Conspecific females encountering the pheromone spend less time foraging on marked fruit and generally leave in search of unmarked fruits. However, the process is dynamic because the probability that a female will avoid a marked fruit decreases as a function of the time since she last laid an egg. Furthermore, the pheromone is water soluble and so its persistence varies with the weather although, for both the apple maggot fly and another dipteran, the alfalfa blotch leafminer, *Agromyza frontella*, a 24–48 hour period between the deposition of the first and second egg is sufficient to provide a significant advantage to the first individual.

Many hymenopteran parasitoids mark hosts that they have parasitized to reduce the probability that their progeny will suffer intraspecific competition. However, unlike the phytophagous insects that mark the plant surface, parasitoids may mark the host externally or internally using chemicals from the Dufour's or poison glands, the lateral oviducts or the ovaries. The placement of the mark varies depending on the life-stage of the host being attacked and on the strategy used by the female parasitoids when inspecting potential hosts in order to ensure maximum efficacy.

Pollinating Hymenoptera, including the honey bee, *Apis mellifera*, as well as several bumble bee and sweat bee species, mark flowers with a repellent signal after nectar feeding. This reduces the time foragers spend at a site with a depleted source of nectar, but as the signal is relatively short-lived it will have dissipated by the time the plant has replenished the nectar source. Conversely, there is evidence that pollinators like the bumble bee, *Bombus terrestris*, can mark flowers with high-quality resources with an attractant found in the tarsi. Various hypotheses have been put forward to explain these observations, including the idea that there are two different signals, or that the same chemical signal is used but with the response varying depending on the concentration. Recently it has been suggested that there is one signal, but the response exhibited may involve associative learning that is dependent on the forager's experience with respect to the level of reward obtained. As with some other pheromones, the marking pheromone of one pollinator species may be used as a kairomone by sympatric species that exploit the same nectar sources. Whether this means that they use identical signals or share at least some pheromone components has not been determined.

27.2.5 Trail pheromones

A variety of insect species use trail pheromones, with their use by ants probably being most well known. Foraging strategies involving trails have been described for nearly 100 ant species and, as one would imagine, there are many different pheromone blends, produced from a variety of sources on the abdomen, including the poison, Dufour's, Pavan's and post-pygidial glands, as well as the hindgut and glands on the legs (Fig. 27.6). A simple scenario is that a foraging ant, upon locating a suitable food resource, deposits a chemical signal by dragging the abdomen on the substrate while returning to the nest (Fig. 27.7). The signal subsequently guides recruited nest mates to the resource. The recruited ants deposit more pheromone as they in turn return to the nest so that the trail is reinforced, creating a positive feedback loop that induces additional foragers to visit the resource. Then, as the resource is depleted, returning ants cease marking and over time the trial dissipates. Thus, a delicate balance is struck in the chemical and physical properties of the marking pheromone, which must persist for long enough for recruitment to occur, but not for so long that workers continue to be recruited after the resource is consumed. A variation of the trail pheromone is seen in species with small numbers of workers, with recruitment to a resource occurring by a process called tandem running (Fig. 27.8). Here, each ant returning from the resource recruits another worker, and leads the recruit to the resource by staying in physical contact.

Recent research on the pharaoh ant, *Monomorium pharaonis*, has shown that the functioning of trail pheromones may be rather more complex than the scenario described above. In this monomorphic species there are two forager castes ("pathfinders" and "non-pathfinders") that differ in behavior, as well as in the long- and short-lived trail pheromones that they produce. There is considerable spatial and temporal variation in the composition of the pheromones and it has been proposed that these maximize the establishment and utilization of trails by foragers.

Termites also produce trail pheromones to aid in the efficient exploitation of resources. Unlike ants, in which trail pheromones are produced from a number of glandular sources, termite trail pheromones are produced only from the sternal glands. Identified structures include (3*Z*,6*Z*,8*E*)-dodecatrienol and its analogs (3*Z*,6*Z*)-dodecadienol and (*Z*)-3-dodecenol,

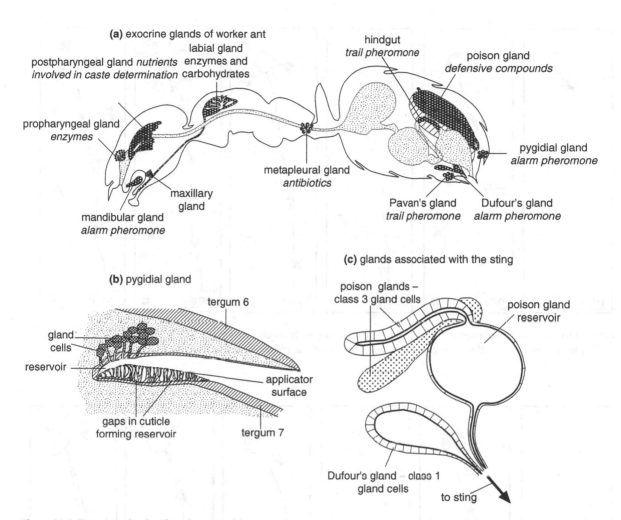

Figure 27.6 Exocrine glands of worker ants. (a) Longitudinal section of a worker ant showing the principal glands with dark shading. All the glands shown may not be found in one species. Examples of the functions of secretions from the glands are shown in italics, but note that the secretion from a particular gland may serve different functions in different species (based on Hölldobler and Wilson, 1990). (b) Pygidial gland of a worker ant, *Pachycondyla* (based on Hölldobler and Traniello, 1980). (c) Poison gland and Dufour's gland of a worker ant (*Myrmica*). The organization of the gland is simpler than that depicted in (a) (based on Billen, 1986).

the diterpenoid neocembrene, (5*E*)-2,6,10-trimethyl-5,9-undecadien-1-ol, and 4,6-dimethyldodecanal. Unlike the ants, termite trail pheromone structures are highly conserved, with many species apparently using the same major component(s) for their trail pheromones.

The gregarious caterpillars of certain lepidopteran species use trail pheromones to exploit food

resources. This cooperative foraging has been intensively studied in the eastern tent caterpillar, *Malacosoma americanum*, which form large silken tents in the trees on which they feed. Larvae exiting the tent to forage, especially the earlier instars, do so collectively, but when following the trail do not maintain physical contact with individuals in front of or behind them on the trail. As they move from the

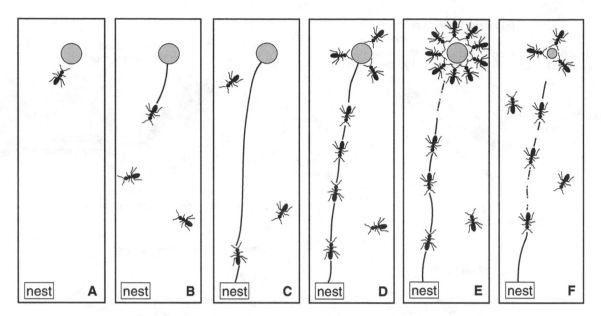

Figure 27.7 A schematic representation of the events leading to the establishment and subsequent dissipation of an ant trail. The thickness of the lines corresponds to the intensity of the chemical signal (from Billen, 2006).

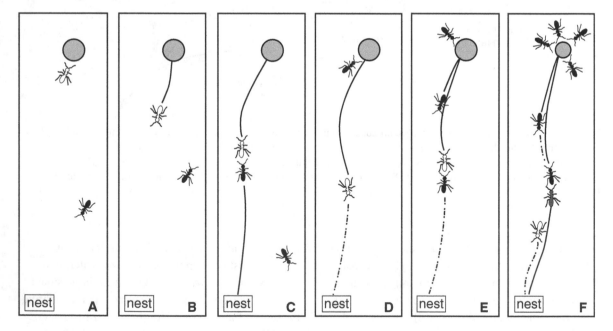

Figure 27.8 A schematic representation of the events leading to the establishment of an ant trail through tandem running. Scouts are represented as white ants and the recruited workers as black ants. The thickness of the lines corresponds to the intensity of the chemical signal (from Billen, 2006).

tent the larvae deposit a pheromone by repeatedly touching the substrate with the tip of the abdomen, producing what are referred to as exploratory trails. If the larvae do not locate a suitable food source or do not feed to satiation they do not deposit additional pheromone on the return trip. However, if a good foraging source is found, the caterpillars overmark the existing trail upon their return, and may also mark the tent itself. As the result of overmarking, these trails have higher concentrations of the pheromone and are known as recruitment trails, because hungry larvae preferentially follow them rather than following exploratory trails. Furthermore, larvae are clearly capable of judging the age of trails, selecting more recently laid trails over older ones.

The trail pheromone of the eastern tent caterpillar is a blend of the steroids 5β-cholestan-3-one and 5β-cholestan-3,24-dione, with the monoketone eliciting the strongest trail-following behavior. The diketone may have a role in species specificity because 5β-cholestan-3-one has also been identified as the trail pheromone of the sympatric forest tent caterpillar *M. disstria*, and it even stimulates trail following by caterpillars of *Eriogaster lanestris*, a European species in the same family.

Processionary behavior has been reported in a number of lepidopteran species. This is a variation on the trail-following behavior seen in species like the eastern tent caterpillar. However, as seen with the larvae of *Hylesia lineata*, they not only produce a trail pheromone from the abdominal tip but maintain direct contact with the individuals in front and behind them. Thus, both chemical and tactile signals play a role in the maintenance of the procession.

27.2.6 Alarm pheromones

Alarm pheromones are released in response to danger and result in behaviors that mitigate the potential impact of the threat. These behaviors can include increased levels of alertness, adoption of aggressive postures, recruitment to the source of the pheromone or, conversely, movement away from the signal. Alarm pheromones are common in eusocial species like bees, wasps and termites. They are typically small molecular weight and very volatile chemicals, such as hexanal or isopentyl acetate, which diffuse rapidly, and they are not typically species-specific. As an example, when a soldier (nasute) termite encounters danger it sprays the intruder with a viscous liquid produced in the frontal gland. This secretion not only contains defensive compounds, but also an alarm pheromone that serves to recruit other soldiers to the site. In termites, responses to alarm pheromones may differ between castes, but all the behaviors expressed collectively serve to protect the colony. In many bees, wasps and ants the worker caste produces the alarm pheromone and the site of production varies with species. However, in all cases the release of the alarm pheromone will recruit others to the site. As seen in the wasp *Polybia occidentalis*, the response produced by the alarm pheromone present in the venom may be affected by additional factors, such as the color of the perceived intruder and whether it is stationary or in motion.

Alarm pheromones are also produced by non-social insects that aggregate or feed in groups, such as aphids, thrips and some true bugs. When an aphid detects danger it releases an alarm pheromone, which in many species consists of the sesquiterpene (E)-β-farnesene, from its cornicles. The behavioral responses elicited by the pheromone can include retreating, waggling the abdomen in the air or even dropping off the plant. The particular behaviors expressed vary as a function of the concentration of the alarm pheromone, the age and/or morph of the individual, whether it is actively feeding or not and even the geographic origin of the species.

27.2.7 Pheromones of social insects

Social insects such as ants, termites and bees frequently live in colonies comprising thousands or even millions of individuals. The organization and regulation of the complex activities within each colony are controlled by chemical signals, sometimes augmented with tactile or acoustic signals. Within each nest or colony, the caste of each individual and the division of labor is modulated by chemical signals, with each individual primed to carry out specific tasks so that the colony functions smoothly as a single superorganism. However, tasks and behaviors are plastic, and will change to meet the changing needs of the colony.

The wide variety of different activities that must be regulated requires a large number of different types of signals (Fig. 27.9), whose chemistry is correlated with the context in which the signals are used (Fig. 27.10). Individuals assess nest mates by antennation to update themselves on the conditions within, and the needs of, the colony. In contrast, in situations where immediate action is required, such as when a colony is under attack, volatile alarm pheromones that diffuse rapidly are released, eliciting recruitment to the source of the pheromone and stimulating aggressive defensive behaviors. Considerable progress has been made in identifying chemicals that control specific behaviors, but our understanding of how all these individual signals and behaviors are integrated to give a single smoothly functioning colony is still limited. Below, the roles of some of these pheromones not discussed in previous sections, and their chemical identities, are examined in more detail.

Nest mate recognition Social insects share three fundamental characteristics: (1) the young are reared by conspecifics; (2) there are castes, with each caste being responsible for specific functions; and (3) generations overlap, so that non-reproductive workers assist their mother in rearing their siblings. The ability to distinguish nest mates (i.e., siblings) readily from non-nest mates is of primary importance, to ensure that rearing efforts are not wasted on more distantly related individuals, as well as to recognize heterospecific intruders. In large colonies, recognition of individuals is both unlikely and unnecessary: what matters is to identify an individual as being a nest mate, and secondarily, to identify its caste and/or task. Recognition requires that each individual must have a label, which is composed of a blend of chemicals distributed over the cuticle. The basic chemical blend, referred to as the colony odor, arises from a genetically determined cuticular chemical profile which is modified by factors such as diet and the odor of the material used to build the nest. Thus, each colony has a unique but dynamic chemical profile, which is under constant modification by environmental variables. Frequent contact between nest mates results in all individuals within the colony being able to continuously track and update their chemical profiles as the colony odor changes. The basic colony odor is further modified by subsets of cuticular chemicals that identify the caste of individuals. The chemical profiles of individuals are detected and sampled by mutual antennation or contacts with the mouthparts, and if the profiles do not match the colony odor, aggressive behavior is triggered. Interestingly, nest mate recognition signals are not present on newly emerged callow workers, so there is the potential for callows to be attacked. This is avoided by a combination of factors, which appear to include pheromonal substances which both appease worker aggression and render callows attractive to workers so that they can rapidly pick up colony odor from contact with the workers.

Division of labor: the role of primer pheromones Careful regulation of the number of reproductive individuals and workers within a social insect colony is essential for maximizing the overall fitness of the colony. This is accomplished with primer pheromones, which have proved to be much more difficult to study than releaser pheromones because of their delayed and often subtle actions, and because many primer pheromones are produced by relatively scarce and long-lived reproductives. It is often difficult to obtain enough of a primer pheromone to fully identify it, and subsequent

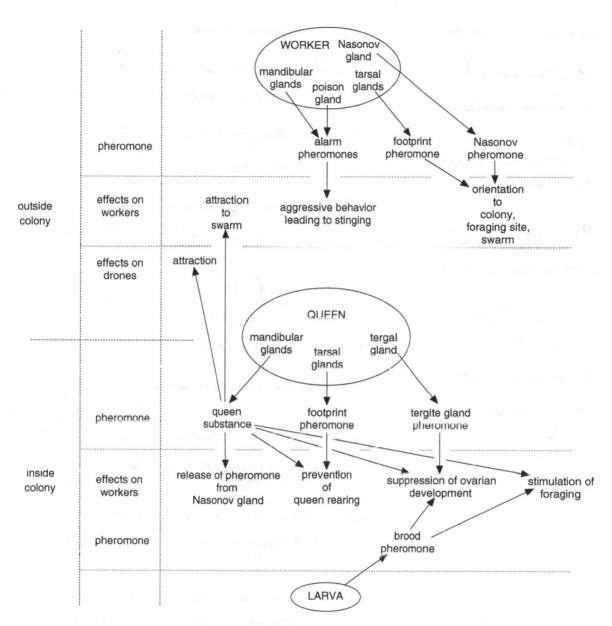

Figure 27.9 Roles of different pheromones used by the honey bee, *Apis mellifera*.

bioassays may last for days or even weeks per assay. Thus, the effects of primer pheromones have been demonstrated in numerous social insect species, but few have been identified.

The roles and identities of primer pheromones have been most well characterized in the honey bee, in which the queen provides the chemical signals that regulate colony homeostatis, growth and reproduction. Key to this is the multifunctional queen mandibular pheromone (QMP) that has both primer and releaser effects (Fig. 27.10). Among other things, QMP suppresses the development of ovaries in

queen mandibular pheromone

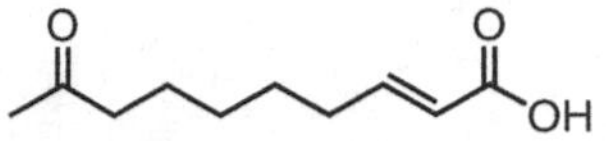

(2*E*)-9-oxodec-2-enoic acid

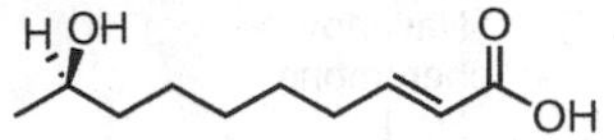

(2*E*,9*R*)-9-hydroxydec-2-enoic acid

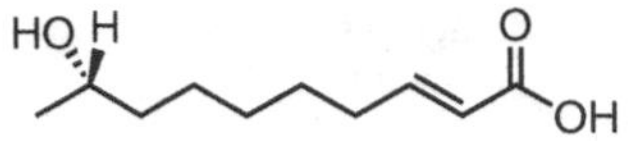

(2*E*,9*S*)-9-hydroxydec-2-enoic acid

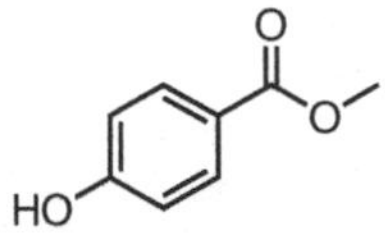

methyl *p*-hydroxybenzoate

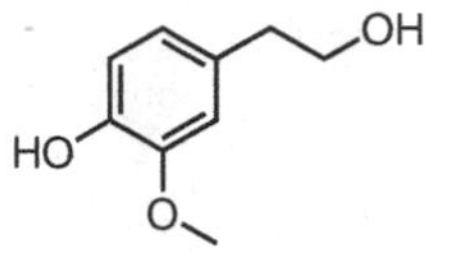

4-hydroxy-3-methoxyphenylethanol

queen retinue pheromone

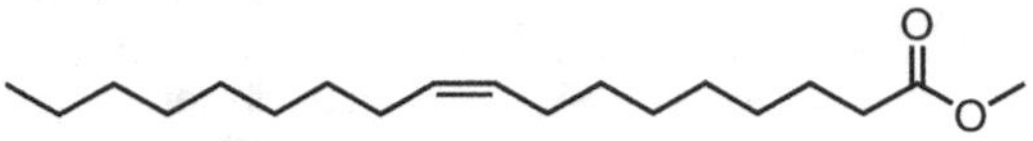

methyl (9*Z*)-octadec-9-enoate

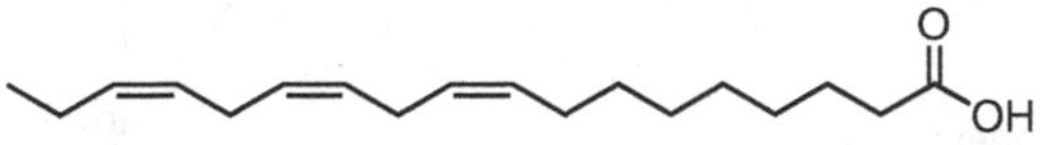

(9*Z*,12*Z*,15*Z*)-octadeca-9,12,15-trienoic acid

hexadecan-1-ol

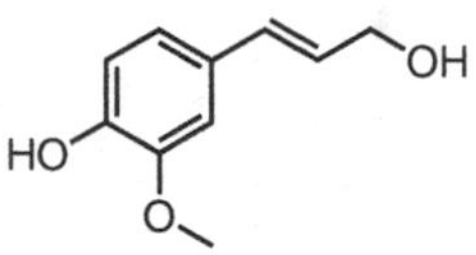

coniferyl alcohol

some components of brood pheromone

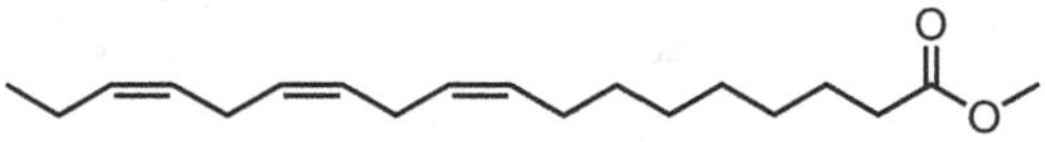

methyl (9*Z*,12*Z*,15*Z*)-octadeca-9,12,15-trienoate
(= methyl linolenate)

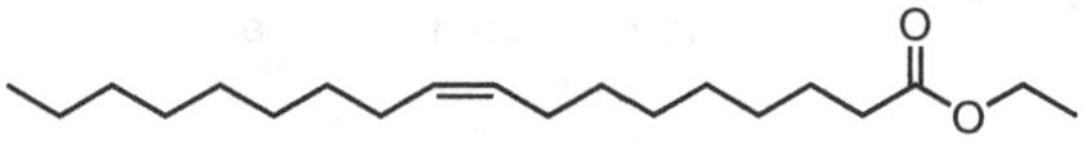

ethyl (9*Z*)-octadec-9-enoate
(= ethyl oleate)

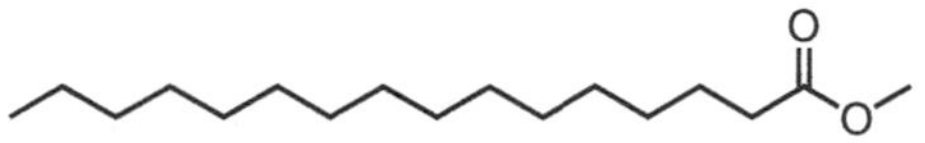

methyl hexadecanoate
(= methyl palmitate)

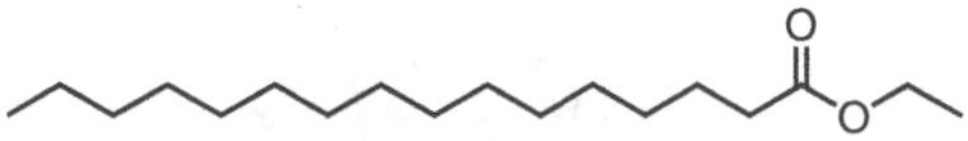

ethyl hexadecanoate
(= ethyl palmitate)

Figure 27.10 Some of the different types of honey bee pheromone components.

workers, attracts young workers to groom and feed the queen, inhibits the construction of the specialized cells used for development of new queens and drones and regulates the switch of workers from nurse bees that work in the hive to foragers that collect resources outside the hive. QMP and other queen pheromones are dispersed throughout the colony by continuous contacts between the retinue of young workers tending the queen and other members of the hive. The first component of QMP, (2E)-9-oxodec-2-enoic acid, was identified more than 50 years ago, but it was clear from its relatively weak biological activity that there must be additional components. Some 30 years later, both enantiomers of (2E)-9-hydroxydec-2-enoic acid (9-HAD), methyl p-hydroxybenzoate (HOB) and 4-hydroxy-3-methoxyphenylethanol (HVA) were identified in QMP (Fig. 27.10), subsequently followed by another four components (methyl [Z]-9-octadecenoate [methyl oleate], [E]-3-[4-hydroxy-3-methoxyphenyl]-prop-2-en-1-ol [coniferyl alcohol], hexadecanol and [9Z, 12Z, 15Z]-octadecatrienoic acid [linolenic acid]). The latter compounds are not active by themselves, but synergize the effects of the other components. Additional compounds from the Dufour's gland signal fertility, reinforcing the message of the QMP to suppress ovary development in workers. If the queen dies, the resulting rapid decrease in QMP and other queen infochemicals results in worker ovarian development, but worker eggs will not produce functional queens. Workers must feed royal jelly to existing larvae in order to produce a replacement queen before the queenless colony collapses.

Recent work has indicated that this complex signaling system has further nuances because queens from which the mandibular glands have been removed, and thus lacking ODA, were still able to suppress ovarian development in workers. In addition, other functions controlled by QMP were not disrupted. This suggests that there is redundancy in the complex blends of signals produced by queens,

with compensatory mechanisms when one or more components are lacking.

The developing brood also produces primer pheromones that affect worker development and physiology. These include ethyl palmitate and methyl linolenate, which inhibit ovarian development in workers, and methyl palmitate and/or ethyl oleate, which stimulate protein biosynthesis in the hypopharyngeal glands of workers, the secretions of which are fed to the brood (Fig. 27.10). The system is self-regulating because low levels of brood pheromone (and so small quantities of brood that do not require large numbers of nurses) accelerate the development of nurse bees into foragers, whereas high levels of brood pheromone (signaling a large brood that requires many tenders) has the opposite effect. There is an additional regulatory loop: High levels of ethyl oleate produced by foragers (signaling a sufficiency of foragers) inhibit the continued development of nurses into foragers, whereas low levels of this compound enhance the development of foragers.

Similar patterns of colony organization and development are found in other social insects, although the details may differ and few other primer pheromones have been identified to date. As one example, primer pheromones produced by queens of the termite *Reticulitermes speratus* inhibit the development of additional reproductives. However, when the queen dies, secondary reproductives known as neotenics are produced from either developing nymphs or adult workers. The neotenics produce a primer pheromone blend composed of the simple compounds butyl butyrate and 2-methylbutan-1-ol, which inhibits the development of additional reproductives. These two compounds are also produced by eggs, for which they serve the dual function of attracting workers as a releaser pheromone, and inhibiting worker development into neotenics as a primer pheromone.

Stingless bees in the genus *Melipona* appear to use a slightly different mechanism for regulating development of queens. Similar to honey bees, developing female larvae of *M. beecheii* have the potential to become either workers or queens, but those destined to become queens are not reared in special cells. Instead, their developmental outcome is determined by the primer pheromone geraniol, which is fed to them by nurse bees. The mechanism by which nurses assess whether additional queens are needed remains to be determined.

Social insects use pheromones for a wide variety of other purposes. For example, whereas it is clear that sex pheromones must be involved in reproductive behaviors, surprisingly few sex pheromones have been identified from social insects, in part because unmated queens are only present for very short periods, and they form a small percentage of the overall population. Identification can be further hindered by the reproductive biology because in many species mating only occurs during nuptial flights, which considerably complicates the design and execution of bioassays. The identification of QMP in honey bees was greatly facilitated by the fact that this species is domesticated and easily manipulated, so that large numbers of reproductives could be obtained for experiments. For other very important hymenopterans such as fire ants, the queen's sex pheromone has not yet been identified. To date, the sex pheromones of only three ant species have been identified, for the hairy wood ant, *Formica lugubris* (a blend of undecane, tridecane and (*Z*)-4-tridecene), and for two species of slave-making ant, *Polyergus breviceps* and *P. rufescens*, both of which use a blend of methyl 2-hydroxy-6-methylbenzoate and (3*R*)-3-ethyl-4-methylpentanol. The fact that the latter two are North American and European species, respectively, which have been separated for millions of years, suggests that sex pheromone structures may be highly conserved in the ants. In certain termite species, trail pheromone components have an additional role as the sex pheromone when released

at substantially higher rates, whereas other species may have distinct sex pheromones.

One of the most interesting chemical signals in social insects is that used to identify dead nest mates so they can be removed from the colony. The warm and humid nests of social insects are ideal for propagation of pathogens, so rapid removal of dead and potentially infectious carcasses, a behavior known as necrophoresis, is critically important to the health of the colony. It was originally thought that dead individuals might be recognized by compounds associated with decomposition products that accumulated after death, but recent studies suggest otherwise. With Argentine ants, *Linepithima humile*, it has been shown that chemicals produced by live ants (dolichodial and iridomyrmecin) inhibit the necrophoretic response by nest mates. After death, these signals associated with life rapidly dissipate (within one hour), so the individual is recognized as dead and is carried off to the refuse pile. It is likely that similar mechanisms operate with other social insect species.

27.3 Information content of pheromonal signals

Pheromones are used to convey many types of information, such as the location of the emitting individual, its sex, species and physiological state. To be most effective, signals should be unambiguous, and only elicit responses from the intended receivers. However, closely related congeneric species share some, or even all, of their pheromone components, and so there is considerable potential for mistakes. Thus, insects have evolved several mechanisms to minimize misinterpretation of their pheromonal signals. For example, closely related sympatric species may have different seasonal or daily activity patterns that allow them to use similar pheromone blends without risk of interference from congeners. However, when closely related sympatric species are not separated temporally or geographically, subtle changes in the chemistry of their pheromones have evolved to ensure the uniqueness of communication channels and the fidelity of the signal. First, a species may use unusual or even unique pheromone components, as suggested by existing data on scales and mealybugs. Of the more than 20 pheromone components identified from these insects, each compound is unique to the species that produces it, even though the compounds are clearly derived from similar precursors. This strategy avoids any possible chance of cross-attraction and, as a consequence, may be in part responsible for the fact that these insects are usually insensitive to analogs or stereoisomers of their pheromones. This is in sharp contrast to many other insects, where the presence of even traces of incorrect stereoisomers or other analogs can strongly inhibit the normal response to the pheromone (see below).

Second, a species may use a pheromone that consists of a specific blend of components, which is the most common method of generating unique pheromone messages. Furthermore, the increased complexity of the signal also allows for increased information content, which can be encoded in a number of different ways. For example, congeneric species may use different combinations or different subsets of a group of pheromone components that are common to many members of the genus to create a unique message. Alternatively, two species may use the same set of pheromone compounds, but in different ratios.

It should be emphasized that insects are masters at using minor modifications of a basic pheromone structure to generate different compounds and unique signals. For example, insects may take advantage of stereochemical features of signal molecules, using only one of the two possible enantiomers of a molecule (enantiomers are mirror images of each other, and are usually perceived by biological receptors as different compounds), or using only the *cis*- or *trans*-stereoisomer of a

molecule with a double bond (also referred to as *Z*- or *E*-isomers, respectively). There are even examples where insects use an unequal mixture of both enantiomers or both *E*- and *Z*-isomers in their pheromone blends, with both stereoisomers being necessary in order to elicit a proper response. In at least one species (the olive fruit fly, *Bactrocera oleae*), females produce a racemic mixture, where one enantiomer appears to elicit a response from males, and the other affects the behavior of conspecific females. Thus, by manipulating structural features such as chirality, double-bond position and geometry and minor modifications of functional groups, insects can create a large number of different signal molecules from relatively small and simple molecular skeletons. Consequently, when these are combined in different subsets and ratios, an essentially unlimited number of unique signals can be created.

A third method of ensuring signal specificity is the inclusion of a minor component in the pheromone blend that has a strongly inhibitory effect on sympatric congeners that use similar or even the same major pheromone components. The use of these behavioral antagonists is highly developed in families such as the Lepidoptera and Coleoptera. These effects have usually been discovered during screening trials or in lure optimization. For example, (*Z*)-11-hexadecenal, which is present in trace amounts in the pheromone of the Mediterranean corn borer, *Sesamia nonagrioides*, strongly inhibits attraction of the sympatric European corn borer, *Ostrinia nubilalis*, which shares a major component with the former species. In a related example, the Japanese beetle, *Popillia japonica*, uses the (R)-enantiomer of a chiral lactone, japonilure, as its pheromone, and even 1% of the (S)-enantiomer strongly inhibits the response. The reason for this is that a sympatric congener, the Osaka beetle, uses the (S)-enantiomer as its pheromone. All of the various permutations of pheromone stereochemistry as they relate to biological activity have been nicely reviewed by Kenji Mori (see the recommended reading section).

27.4 Biosynthesis of pheromones

In the most literal sense, all insect pheromones are derived from the diet, but there is a wide diversity in how they are produced. In some cases, close precursors are assimilated or sequestered from the food and then converted to the active pheromone chemicals by minor changes to the structure. Conversely, other species synthesize pheromones *de novo* from the most basic building blocks, such as acetate, malonate or mevalonic acid. As might be expected, the overall production of a pheromone will depend upon the physical and chemical properties required of the compound, the context in which it is used and the quantity in which it is released. Thus, pheromones vary from being highly volatile to completely non-volatile. Furthermore, as noted earlier, many pheromones consist of blends rather than single components, so in the interests of biosynthetic parsimony, the blend components are often closely related in structure and are made by minor variations of one main biosynthetic route.

Biosynthetic pathways of pheromones have now been studied in considerable detail in the Lepidoptera, Coleoptera, Diptera and Hymenoptera, using two complementary approaches. First, pheromone-producing tissues are carefully analyzed to look for any likely biosynthetic precursors of the active pheromone components. Second, possible biosynthetic intermediates are prepared in an isotopically labeled form (e.g., with one or more deuterium or radioactive tritium atoms in place of hydrogen(s), or with ^{13}C or radioactive ^{14}C in place of the normal ^{12}C), and are applied to the gland or injected into the hemolymph. After an incubation period to allow the labeled compounds to be taken up by the tissues, the pheromone glands are analyzed. If the pheromone is now isotopically labeled, this provides strong evidence that the applied compound is indeed a precursor in the biosynthetic pathway.

27.4.1 Lepidoptera

Bombykol, or (10E,12Z)-hexadecadienol, was the first insect pheromone identified (in 1959), and since then pheromone structures for approximately 2000 lepidopteran species have been identified, or at least suspected because of strong attraction in field screening trials. The majority of lepidopteran sex-attractant pheromones fall into two general classes. The first, designated as Type I pheromones, consist of 10–18 carbon chain aldehydes, alcohols or acetates, with 0–4 double bonds along the chain. Type I pheromones are synthesized in a gland located between the eighth and ninth abdominal segments of female moths. The biosynthesis of Type I pheromones starts with acetate being used to construct hexadecanoic (16:COOH) and octadecanoic (18:COOH) acids. These then undergo stereo- and regiospecific dehydrogenation, catalyzed by a Δ11-desaturase enzyme, to produce (Z)-11-hexadecenoate (Z11–16:COOH) and (Z)-11-octadecenoate (Z11–18:COOH) (Fig. 27.11). These acids, along with 16:COOH, then undergo chain-shortening steps, removing two carbons at a time from the carboxyl end, to give, for example, Z9–14:COOH and Z7–12:COOH from Z11–16:COOH. Once the proper chain length has been achieved, the carboxyl group is reduced to the aldehyde, or to the alcohol, which may then be acetylated. Alternatively, the carboxyl group may be reduced to the alcohol, and then oxidized again to the aldehyde at the moment of release. Thus, a combination of two initial substrates (16:COOH and 18:COOH), acted on by a single desaturase, followed by one or several chain-shortening steps, and then final modification of the terminal functional group, can produce a considerable number of pheromone components from a limited number of substrates and steps. Furthermore, the specificity of the enzymes and the

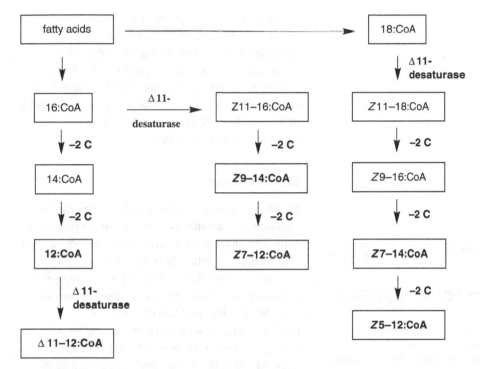

Figure 27.11 Biosynthesis of Type I pheromone components of the cabbage looper, *Trichoplusia ni*, from hexadecanoic (16:COOH) and octadecanoic (18:COOH) acids. The "CoA" suffix indicates that the acid is attached to coenzyme A as it is processed by the relevant enzymes. After reduction to the corresponding alcohols, and acetylation, the acetate esters of the compounds shown in bold can be found in the pheromone gland, and make up the pheromone blend.

substrates that they will accept provides a high degree of predictability, so that given a particular pheromone structure, we can make very good guesses as to how it was made.

Besides Δ11-desaturases, a number of other desaturases are known, including Δ5-, Δ9-, Δ10-, Δ13- and Δ14-desaturases, so that double bonds can be placed in other positions along the chain. Conjugated dienes can be formed by two separate desaturation steps, such as occurs in the biosynthesis of *E*9,*E*11–14:OAc, which is made by Δ11-desaturation of 16:COOH, chain shortening to *E*9–14:COOH, then a second Δ11-desaturation to give *E*9,*E*11–14:COOH, which is then reduced and acetylated. Alternatively, other conjugated dienes appear to be produced by allylic oxidation adjacent to the double bond of a monoene (e.g., *E*9–12:COOH), followed by 1,4-elimination of the elements of water with concurrent isomerization of the double bond to give *E*8,*E*10–12:OH, after reduction of the functional group. The acetylenic triple bond found in a few

lepidopteran pheromones is formed by further desaturation of a double bond.

Type II lepidopteran pheromones consist of somewhat longer-chain hydrocarbons from C17 to C25, with 2–5 double bonds with the *Z*-configuration in the 3, 6, 9, 12 or 15 positions, or derivatives thereof, such as mono- or diepoxides. There are some minor variations on these themes, such as unusual functional groups (e.g., a nitro group or an ester other than an acetate, a triple bond instead of a double bond or a methyl branch somewhere along the chain), but by and large most of the known lepidopteran pheromones can be accommodated by these two groupings.

The polyunsaturated hydrocarbons and their derivatives that constitute Type II lepidopteran pheromones are also derived from fatty acids, but there are a number of important differences between the biosyntheses of Type I and Type II pheromones (Fig. 27.12). First, the starting materials for Type II pheromones are linoleic (9*Z*,12*Z*–18:COOH) and

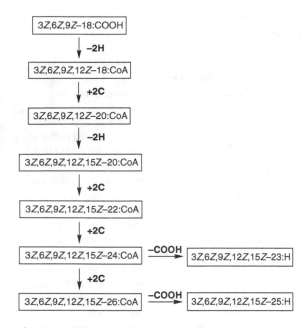

Figure 27.12 Proposed biosynthesis of polyunsaturated hydrocarbon, Type II lepidopteran pheromones. The hydrocarbons are made in oenocyte cells, then transported through the hemolymph to the pheromone gland by lipophorin carrier proteins. In the pheromone gland, the hydrocarbons may be epoxidized once or twice before being released.

linolenic (9Z,12Z,15Z 18:COOH) acids, with either two or three of the double bonds already in place, respectively. These precursors are then chain extended by sequential addition of acetate units, followed by decarboxylation to give an odd-numbered hydrocarbon. For those species that produce even-numbered chain lengths, the evidence suggests that these are produced by α-oxidation and decarboxylation to produce an odd-numbered acid, which is then either reduced to an aldehyde and decarbonylated or decarboxylated to produce the final even-numbered chain-length hydrocarbon.

Second, the polyunsaturated hydrocarbons constituting Type II pheromones are synthesized in oenocyte cells on the inner surface of the abdominal cuticle rather than in the pheromone gland, and the oenocytes also synthesize cuticular hydrocarbons. The polyunsaturated hydrocarbons are transported

through the hemolymph by lipophorin carrier proteins, and selectively released into the pheromone gland situated between the eighth and ninth abdominal segments, as in Type I insects. The hydrocarbons are then released from the gland, or epoxidized once or in some cases twice by epoxidases in the gland before being released, often in a blend with the parent hydrocarbons.

27.4.2 Coleoptera

The Order Coleoptera is large and diverse, and the pheromone structures used by its various species are correspondingly diverse. Most are likely derived from modifications of either fatty acid or terpenoid biosynthetic pathways. For example, the large-ring lactones produced as aggregation pheromones by grain beetles (Laemophloeidae) and the five-membered ring lactones with a long pendant chain produced by some scarab beetles in the Subfamily Rutellinae bear obvious similarities to fatty acids. Similarly, some of the pheromones used by bark beetles (Scolytidae) are obviously of terpenoid origin. However, there are also some more unusual types of pheromones. For example, some chafers (Scarabaeidae, Subfamily Melolonthinae) use methyl esters of the amino acids valine and isoleucine as female-produced sex pheromones. Sap beetles (Nitidulidae) produce multiple-branched and highly conjugated aggregation pheromones and more highly functionalized derivatives thereof, such as stegobinone, the pheromone of the drugstore beetle *Stegobium paniceum*. The basic skeletons of these pheromones are assembled *de novo* from acetate, priopionate and butyrate, with the latter two compounds providing the methyl and ethyl branches, respectively (Fig. 27.13). Most of the steps involve the typical sequence of fatty acid anabolism of chain elongation by two carbons by attack of an enolate on a second acid bound to coenzyme-A, followed by reduction of the carbonyl and dehydration. The cycle is then repeated as many times as necessary, building up the chain two carbons at a time. The

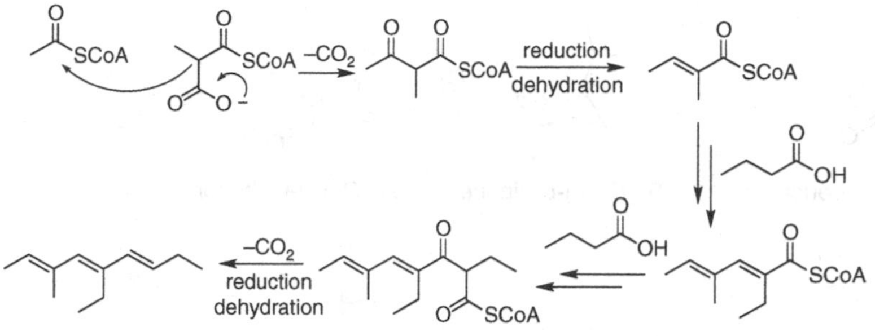

Figure 27.13 Biosynthesis of the polyunsaturated, methyl-branched hydrocarbon pheromones produced by nitidulid beetles. Ethyl branches arise from incorporation of butanoic acid building blocks.

decarboxylation in the final steps prevents any further chain extension, resulting in the final product.

Bark beetles (Scolytidae) produce a wide array of pheromones that are used in numerous different contexts (aggregation, sex and anti-aggregation pheromones are known). Consequently, the pheromones include a wide variety of chemical classes, including terpenoids, fatty acid derivatives, cyclic ethers, ketals and even compounds derived from amino acids. It was originally thought that most of these compounds were produced primarily by autoxidation of host tree compounds (e.g., oxidation of α-pinene to verbenone, Fig. 27.14a) or by symbiotic organisms in the gut. Alternatively, pheromone components could be made by relatively minor modifications of host volatiles, such as selective oxidation of myrcene to one or the other enantiomer of ipsdienol, in some cases followed by reduction to ipsenol (Fig. 27.14b), or the oxidation of α-pinene to cis- or trans-verbenol (Fig. 27.14c). However, more recent work has shown that a substantial proportion of the terpenoid pheromone components are actually synthesized de novo, via the typical mevalonic acid pathway that is common to numerous animals and plants (Fig. 27.14d). Many of the specific enzymes involved in de novo biosynthesis have been characterized, and the mechanisms of up- and down-regulation of the gene products involved in biosynthesis are reasonably well understood.

One of the more interesting types of pheromone compounds produced by bark beetles are the bicyclic ketals, and some related cyclic ethers such as pityol

and vittatol. Detailed biosynthetic studies have shown that these are produced from unsaturated ketones via ketoepoxide intermediates (Fig. 27.15). Of particular note is the extraordinary finding that the enzymes involved in the biosynthesis of one of these compounds, exo-brevicomin, are able to make the same enantiomer of this compound from either of the two enantiomers of the ketoepoxide precursor. For the (6R,7S)-enantiomer, the ketone oxygen attacks epoxide carbon 6 from the back (Fig. 27.15), inverting its configuration, whereas the configuration of carbon 7 remains unchanged. In contrast, with the (6S,7R)-enantiomer, the ketone oxygen now attacks carbon 7 from the back, so its configuration is inverted while carbon 6 remains unchanged. The net result is that either enantiomer of the ketoepoxide precursor produces only (+)-exo-brevicomin.

27.4.3 Diptera

Diptera use short-range or contact sex pheromones consisting of long-chain hydrocarbons and related compounds that are part of the cuticular wax layer. The pheromone components, which are synthesized by the same general mechanisms that are used to make long-chain and methyl-branched cuticular hydrocarbons, are important in maintaining water balance. For example, the housefly pheromone (Z)-9-tricosene and homologs are made by successive additions of malonyl CoA (derived from the amino acid valine) to acetyl CoA by a fatty acid synthase to give the 18-carbon octadecanoyl CoA (Fig. 27.16).

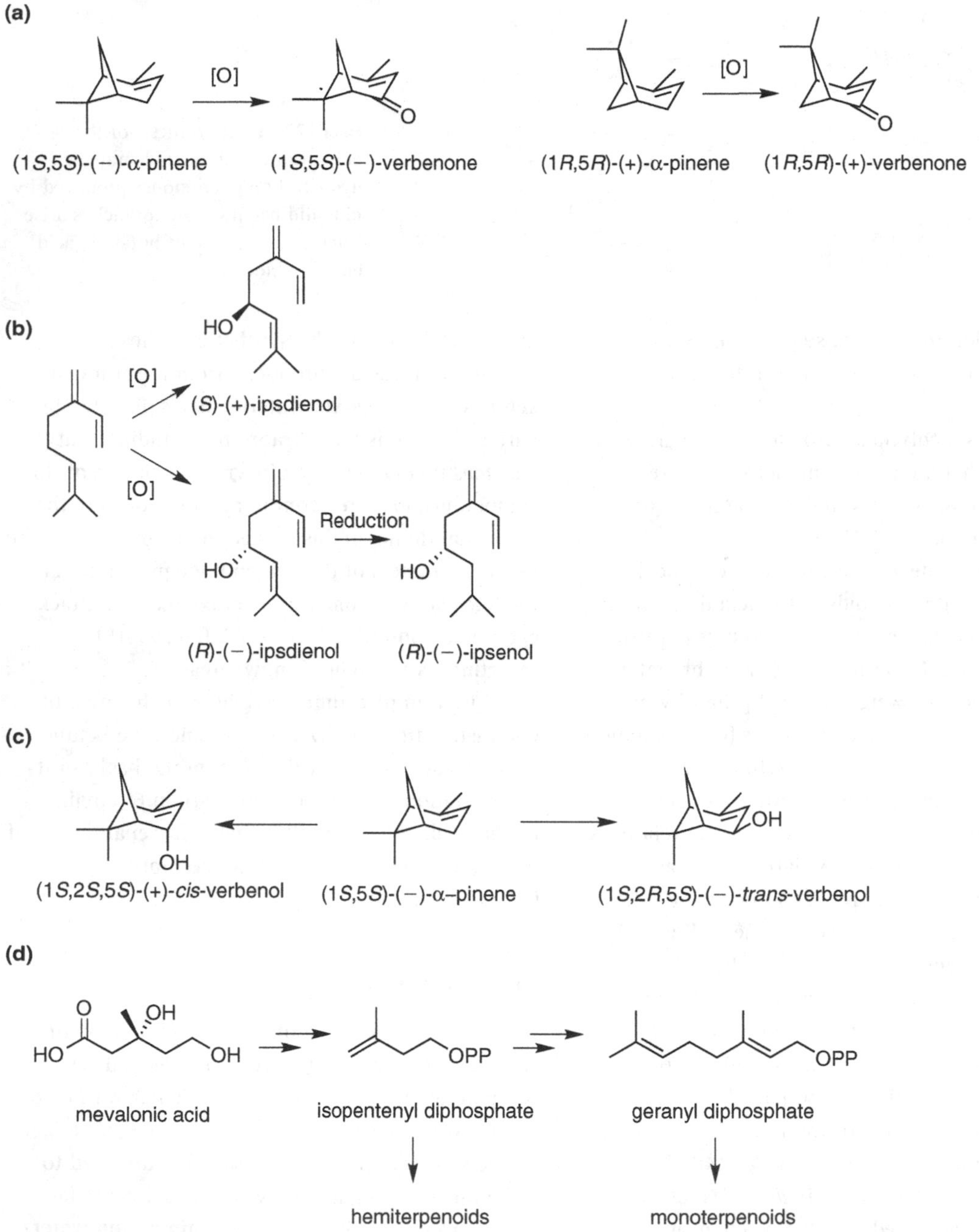

Figure 27.14 Sources and biosynthesis of some bark beetle pheromones. (a) Generation of verbenone enantiomers by air oxidation of α-pinene enantiomers. (b) Generation of ipsdienol by oxidation of myrcene. (c) Generation of *cis*- and *trans*-verbenol by oxidation of α-pinene. (d) *De novo* biosynthesis of terpenoid pheromones from mevalonic acid.

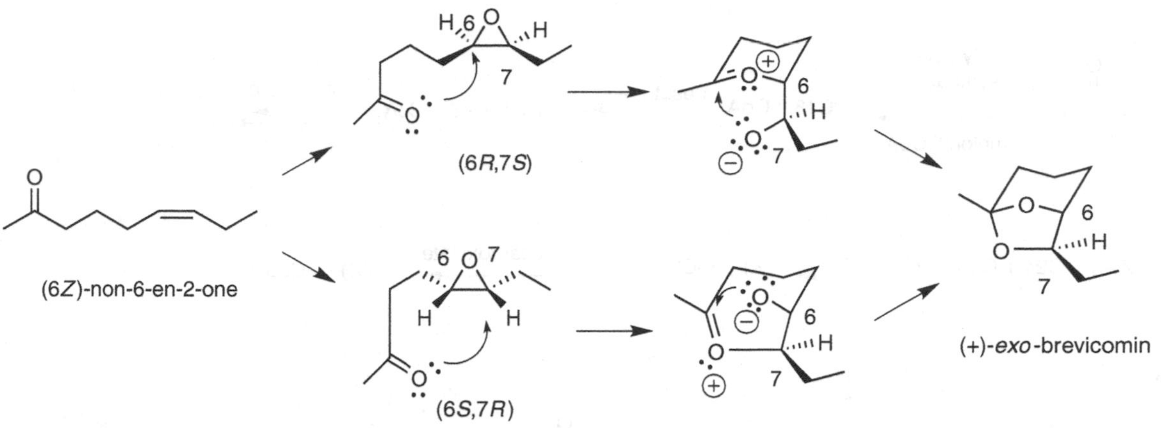

Figure 27.15 Biosynthesis of the bark beetle pheromone (+)-*exo*-brevicomin.

The (*Z*)-double bond is then specifically introduced at the 9 position, followed by additional two-carbon chain extensions as required, reduction of the fully chain-extended fatty acyl CoA to an aldehyde, oxidation to the acid and decarboxylation, giving a monounsaturated hydrocarbon with an odd number of carbons as the final product. As needed, this can then be further modified by cytochrome P450-mediated oxidation to the corresponding *cis*-9,10-epoxytricosane or the ketone (*Z*)-14-tricosen-10-one, both of which are found on the fly cuticle. Methyl-branched saturated hydrocarbons are also found as components of fly pheromones, with the methyl branches being introduced by reaction of the growing chain with methylmalonyl CoA instead of malonyl CoA.

27.4.4 Social insects

Social insects use numerous volatile and non-volatile pheromones for regulation of all aspects of the lives of individuals and colonies. At one end of the scale, non-volatile cuticular hydrocarbons provide signals for recognition of nest mates and, at a finer scale, for recognition of sex and caste. Although the biosynthesis of these compounds has not been studied directly in social insects, their close similarity

to the cuticular hydrocarbon pheromones of the Diptera suggests that they are made by similar pathways.

As described above, the blend of compounds known as the honey bee queen mandibular pheromone (QMP) has a number of effects in different contexts. Three of the major components of the QMP are (2*E*,9*R*)-9-hydroxydec-2-enoic acid and its enantiomer (2*E*,9*S*)-9-hydroxydec-2-enoic acid, and the corresponding ketone, (2*E*)-9-oxodec-2-enoic acid (Fig. 27.17). These three compounds are produced almost exclusively by the queen, with female worker bees instead producing analogs such as (2*E*)-10-hydroxydec-2-enoic acid and (2*E*)-dec-2-ene-1,10-dioic acid, which are incorporated into food for the developing brood. The biosynthesis of these compounds in both workers and queens starts from the same compound, octadecanoic acid, which is first oxidized at the terminal position by workers, and at the penultimate position by queens (Fig. 27.17), followed by four sequential rounds of β-oxidation and chain shortening to ten-carbon chains, then introduction of the double bond and, finally, oxidation of the hydroxy group to the ketoacid or diacid in queens and workers, respectively. Thus, tight control of the biosynthesis results in each caste producing the correct chemicals

(a)

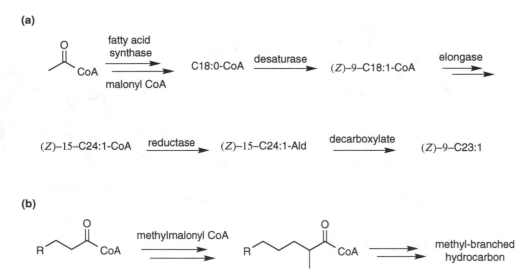

(b)

Figure 27.16 Biosynthesis of housefly pheromone components. (a) Biosynthesis of (Z)-9-tricosene. (b) Insertion of methyl branches from methylmalonyl CoA into branched-chain hydrocarbon pheromone components.

required for maintaining the social structure and function of the hive.

27.5 Regulation of pheromone production

In a number of moth species, pheromone biosynthesis is under neuroendocrine control, modulated by pheromone biosynthesis activating neuropeptide (PBAN), a member of the PBAN/ pyrokinin family of peptides that are implicated in processes as diverse as pheromone synthesis and diapause regulation. PBAN is synthesized in and released from the subesophageal ganglion (SOG), but the manner in which the neuropeptide works varies depending on the species. In the heliothine moths, PBAN is released into the hemolymph and acts directly on receptors located on the surface of the pheromone gland. However, in the gypsy moth, *Lymantria disparlur*, PBAN moves from the SOG to the pheromone gland via the ventral nervous system.

Conversely, in the redbanded leafroller, *Argyrotaeni velutinina*, PBAN is released into the hemolymph but acts on the corpus bursa, leading to the release of another peptide that stimulates the pheromone gland to synthesize the sex pheromone. While PBAN has not been explicitly implicated in the synthesis of pheromones by male moths, when both sexes use very similar compounds it is possible that PBAN is involved and plays a role in mate choice (Fig. 27.18).

In species of moths that migrate seasonally, the timing of pheromone production and ovarian development are tightly linked, and both are modulated by juvenile hormone (JH) titers. Under ecological conditions suitable for reproduction (i.e., summer), increasing JH titers result in ovarian development and stimulate the SOG to release PBAN, leading to production of the sex pheromone. Under unsuitable ecological conditions (short day length and decreasing temperatures typical of fall), JH titers remain low so the onset of reproduction is delayed and sexually immature moths emigrate in search of more suitable habitats. JH is also involved in

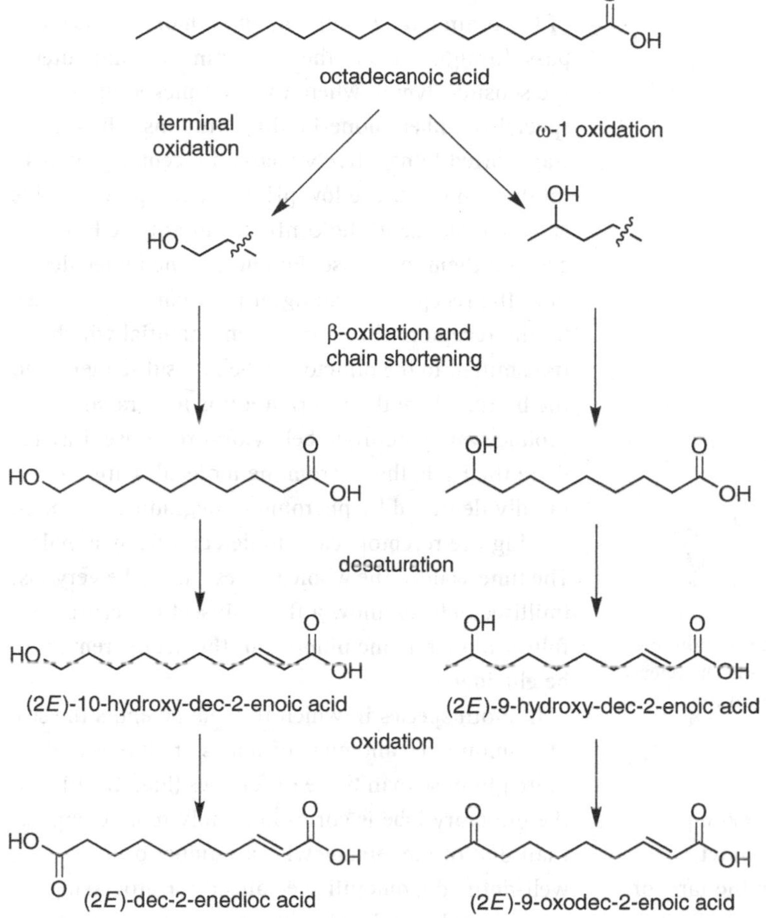

Figure 27.17 Biosynthesis of honey bee pheromone components by workers (left-hand path) and queens (right-hand path).

pheromone production by certain bark beetles, directly modulating the pheromone biosynthetic pathway, with no PBANs involved.

Following mating, most female insects exhibit a permanent (monoandrous species) or temporary (polyandrous) refractory period during which no pheromone is produced. A number of different factors transferred by the male at the time of mating, including sex peptides, JH and ecdysone, inhibit pheromone biosynthesis through effects on humoral and/or neural pathways. In the case of polyandrous species, the duration of pheromonostasis varies depending on male quality, with high-quality males inducing a longer refractory period.

27.6 Perception of pheromones and other infochemicals

An insect perceives a constantly changing array of infochemicals of importance to various aspects of its biology, such as finding a mate or a suitable food source, or avoiding enemies. Thus, insects have a multitude of specialized receptors to collect and integrate the information available from chemical signals and cues. Infochemicals are detected by receptors found on many parts of an insect's body, and the locations of specific types of receptors are closely associated with the role of the particular

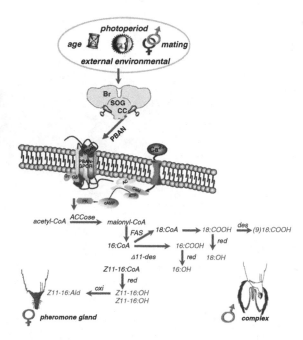

Figure 27.18 The biosynthetic pathway of sex pheromone biosynthesis in *Heliocoverpa armigera* following the release of PBAN from the subesophageal ganglion into the hemolymph (Jurenka and Rafaeli, 2011).

neurons and the axons lead back to the olfactory lobe of the brain (Fig. 27.19). The pheromone molecules pass through pores in the sensillum wall and enter the sensillar lymph where they are picked up by specialized pheromone-binding proteins (PBPs) and transported to negatively charged receptor proteins on the dendrites. The low pH of the receptor proteins causes a change in the configuration of the PBPs, causing them to release the pheromone molecules onto the receptors. Binding of pheromone molecules by the receptors elicits an action potential which is transmitted to the olfactory lobe and subsequently to the brain, where the information is integrated to produce an appropriate behavioral response. Having done their job, the pheromone molecules are then rapidly degraded by pheromone-degrading enzymes, leaving the receptor ready to detect a fresh signal. The time scale of the whole process has to be very fast (milliseconds) to allow a flying insect to accurately follow a pheromone plume, but the details remain to be elucidated.

In moth species in which the female emits the sex pheromone, the antennae of males are larger and more plumose than those of females (Fig. 27.20), and the olfactory lobe is correspondingly more complex than that of the female, with a number of well-defined glomeruli specialized for processing pheromonal signals. The axons of receptors that detect the different components in the pheromone blend terminate in glomeruli within the olfactory lobe, so stimulation of the different receptors must occur simultaneously in order for the signal to be correctly interpreted.

Reception and processing of other types of olfactory signals and cues follow a similar process, with odor molecules entering sensilla on the antennae through pores, followed by transport of the molecules through the sensillar lymph by odorant-binding proteins (OBPs), and perception by receptor proteins tuned to specific infochemicals (see also Chapter 24).

Insects also require receptors for detection of non-volatile infochemicals. The basic characteristics

signals and cues that they perceive. For example, marking pheromones are detected by contact chemoreceptors located on sites such as the tarsi or antennae, whereas cues associated with egg-laying sites are sampled by receptors on the ovipositor. Receptors associated with the detection of volatile chemical signals/cues are typically located on the antennae or maxillary palps, whereas gustatory receptors for recognition of suitable food resources are associated with the mouthparts (see Chapter 24).

The detection of sex pheromones has been extensively studied, and detection of female-produced sex pheromones by male moths will be used to describe the process whereby these species-specific signals are detected and integrated. The peripheral receptors of pheromones are located in the sensilla trichodea on the male antenna. The sensilla contain the dendrites of the olfactory

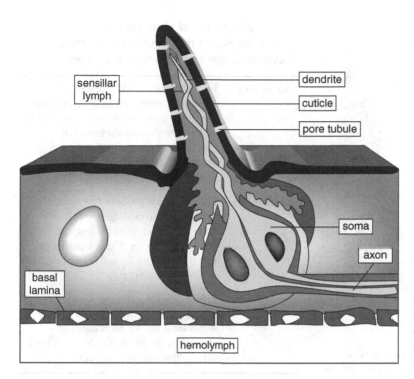

Figure 27.19 A diagrammatic representation of a pheromone-detecting sensillum trichodeum on the antenna of a male moth (from Leal, 2005).

of insect gustatory or contact chemoreceptors are highly conserved, with the receptors housed in sensilla with a single large pore at the tip (Chapter 24). The sensilla are classified into two morphological types: hairs or pegs. In addition to being associated with the mouthparts, they can also be located on the tarsi, wing margins, antennae and, in females, on the ovipositor. When the insect is feeding, the sensilla are often bathed in liquid from the food, providing a medium for the chemical stimuli to reach and enter the gustatory pore. During feeding the insect perceives both stimulant and deterrent molecules from a particular substrate, and it is the balance between these conflicting stimuli that determines the behavioral outcome, i.e., to feed or not to feed. When sampling dry surfaces, non-volatile infochemicals may be picked up by liquid exuded from the pore at the tip of a sensillum, or possibly by dissolution in and diffusion through the cuticular lipids surrounding the pore.

27.7 Information transfer between species: allelochemicals

In addition to the pheromones produced for sending specific messages to individuals of the same species, numerous other chemicals are used by insects to transmit or receive information between individuals of different species, or to disrupt the behaviors of other species. These compounds that mediate interspecific interactions are called allelochemicals, and they can be broken down into subclasses based on their function and the context in which they are used (Table 27.2). Thus, allomones benefit the producing organism, and generally have negative effects on the receiving organism. For example, defensive allomones vary in their effects from a simple warning to serious harm and even death when defensive compounds are wiped, squirted, sprayed on or injected into the receiving organism. In contrast, kairomones benefit the receiving organism to the

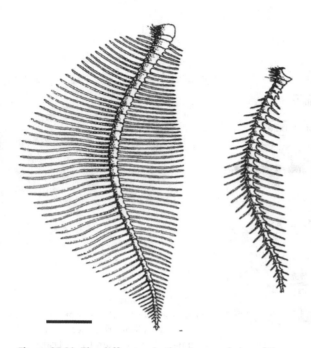

Figure 27.20 The difference in the shape and size of the antennae in the male and female giant silk moth of *Anthera polyphemus*. The scale bar is 2 mm (modified from Figure 3 in Keil, 1999).

Table 27.2 **Classes of chemical cues (allelochemicals) that mediate interactions between organisms of different species**

Allomones	Infochemicals, such as defensive chemicals, produced by species A that upon reception result in a behavioral and/or physiological change in species B that benefits the emitter, species A, and is usually detrimental to species B.
Kairomones	Infochemicals produced by species A that upon reception result in a behavioral and/or physiological change in species B that benefits the receiver, species B, and is detrimental to the emitter, species A.
Synomones	Infochemicals produced by species A that upon reception result in a behavioral and/or physiological change in species B that benefits both the emitter and the receiver.

detriment of the producer; several examples of pheromones being exploited as kairomones have already been mentioned. The kairomone emitter often has no option but to produce these chemicals, such as the release of carbon dioxide during respiration by terrestrial vertebrates, which mosquitoes exploit to locate their hosts for a blood meal. The host animals have to breathe, and so the exhaled CO_2 provides a reliable cue for host location. Synomones constitute a third class of allelochemicals, where both the producer and the receiver benefit.

As mentioned earlier, there is a lot of overlap in function with many semiochemicals, with a particular chemical having different roles in different contexts. For example, a compound that is released in small quantities with other compounds may serve as a pheromone, but when released in large quantities, it may serve as an alarm pheromone or a defensive secretion. Alternatively, a compound released as a pheromone to attract mates can be exploited by parasitoids or predators to locate their prey. The use of chemical signals can also go one step further – for example, when a predator mimics the pheromonal signals of its prey. Examples of various types of chemical mimicry will be discussed in more detail below.

Probably the most ubiquitous and well-known example of the use of synomones is in the attraction of pollinators to flowers by means of floral odors. Because the pollinator is usually rewarded with nectar or pollen, both participants benefit. Some relationships are very general, with a particular flower attracting numerous different types of pollinators, whereas other relationships are highly specific, with the floral odor being shaped by evolution to attract only one or a small group of pollinators.

Another excellent example of the use of synomones is provided by the coevolved relationships between some Central American acacia trees and the ants that live on them. The trees produce sweet secretions from nectaries and protein from specialized organs called Beltian bodies. In combination, the plant exudates provide most of the ants' nutritional needs. In return, the ants protect their host trees from invertebrate and vertebrate herbivores. In response to movements of the foliage, as would occur during feeding by a herbivore, the ants emerge, releasing a repellent as well as physically attacking the herbivore. Similar relationships exist between African ant and acacia species; in a further refinement, unpollinated flowers of the acacia produce an ant repellent so that pollinators are not attacked by the ant guards. After the flowers are pollinated, the repellent dissipates, meaning the ants can guard the developing seeds.

27.8 Producing, storing and releasing allomones

By their very nature as defensive compounds, allomones are typically rather nasty chemicals, so allomone-producing organisms have special adaptations to prevent self-intoxication. Allomones are produced by a wide variety of glands whose positions and size are determined by the method and context in which the compounds are used, as well as the amounts that are used. Thus, for defensive compounds that are released in relatively large amounts, such as sticky substances and glues that gum up attackers, or nasty-tasting compounds or irritants, the glands and their reservoirs can be large, with their contents comprising a significant fraction of an insect's body mass. In contrast, potent venoms or toxins that are injected by bites or stings are produced by smaller glands associated with the head or abdomen, respectively. Many glands producing defense substances consist of invaginations of the

epidermis, and in some cases the glands can be everted by hemolymph pressure or inflated with air, or even burst to rapidly release their contents for greatest effect. In non-eversible glands, the gland cells may line a reservoir, or the secretory region and reservoir may be separate but joined by a duct (Fig. 27.21). When released, the defensive compounds ooze onto the surface of the cuticle, where they are spread around during grooming behavior. Release of volatile components of the secretions is often accelerated by the presence of hairs or an uneven or sculpted cuticle to increase the surface area, thus enhancing evaporation.

Instead of being stored in specific glands, some allomones are stored in the hemolymph or deposited on the cuticle. In the former case, the compounds may be released by reflexive bleeding from the joints (for example, coccinellid and meloid beetles) or even the tarsi (spittle bugs) when the producer is attacked. Alternatively, the attacker may only be exposed to them when the integument is punctured and the hemolymph is released. In the latter case, the compounds are usually bitter or otherwise repugnant, so the producer is released as soon as an attacker experiences the nasty taste. Even if an individual with such cuticular or hemolymphal defenses is killed, the attacker rapidly learns to avoid such unpleasant or toxic prey.

Chemically defended insects have two interrelated problems: how to obtain or produce their defensive compounds, and how to avoid being poisoned by their own defenses. The first problem is dealt with in several ways. Some insects are able to synthesize their chemical defenses *de novo*, varying from very simple molecules like formic acid (derived from the amino acids serine and glycine) to complex chemicals such as alkaloids, and biochemicals such as enzymes and proteins (Fig. 27.22). Symbiotic microorganisms have been implicated in the biosynthesis of some compounds, such as the highly toxic compound pederin, found in *Paederus* spp. rove beetles.

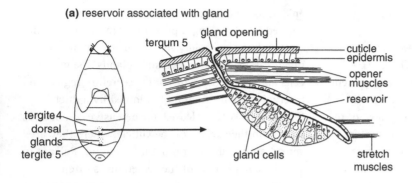

(a) reservoir associated with gland

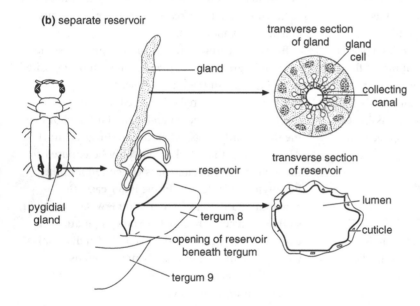

(b) separate reservoir

Figure 27.21 Non-eversible exocrine glands. (a) Gland cells directly associated with reservoir. Abdominal glands of larval Heteroptera, *Pyrrhocoris* (after Staddon, 1979). (b) Gland cells separate from reservoir. Coleopteran pygidial gland, *Cicindela* (based on Forsyth, 1970).

However, many insects acquire their defensive compounds or precursors to those compounds from their host plants. The majority of these insects have become specialists on toxic plants, and have evolved resistance or tolerance to the toxins of their hosts, as well as developing mechanisms for sequestering the toxins for their own use. For example, toxins may be absorbed through the gut membrane and transported through the hemolymph to storage sites or deposited on the cuticle. Many plant toxins are relatively polar compounds such as glycosides, or free acids or bases (Fig. 27.22), which will not diffuse easily through the relatively hydrophobic gut membrane. This problem has been circumvented by the development of specific carrier mechanisms to transport the target compounds through the gut membrane. Larvae of arctiid moths (e.g., *Creatonotos transiens* and *Tyria jacobaeae*) have developed an alternative solution to sequestration of pyrrolizidine alkaloids (Fig. 27.22). These alkaloids are present as the polar N-oxide forms in the host plant, but in the larval gut they are reduced to the much more lipophilic tertiary amines, which diffuse through the gut membrane into the hemolymph. There, they are reoxidized to the N-oxides by an enzyme specific for this purpose, then transported to their final destinations in the tissues. Insects appear to be able to regulate the sequestration of adequate but not excessive amounts of defensive

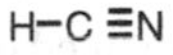
hydrogen cyanide

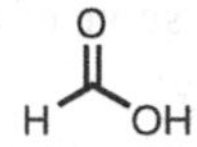
formic acid

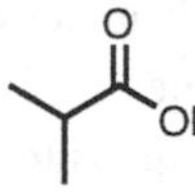
isobutyric acid

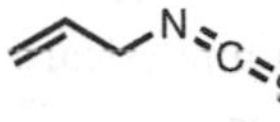
allyl isothiocyanate

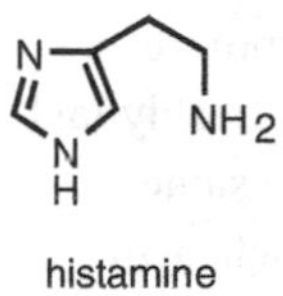
histamine

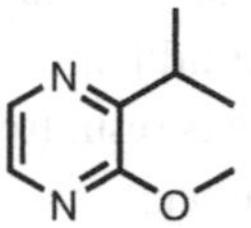
2-isopropyl methoxypyrazine

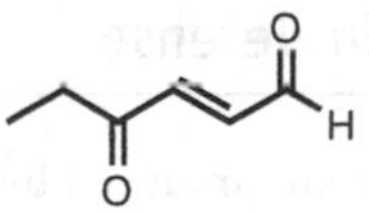
(*E*)-4-oxo-2-hexenal

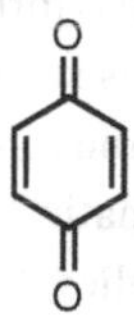
1,4-benzoquinone

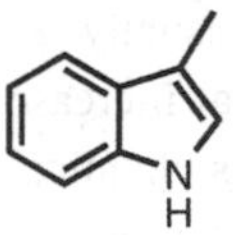
3-methylindole

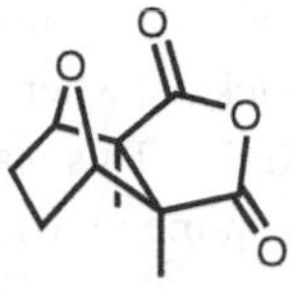
cantharidin

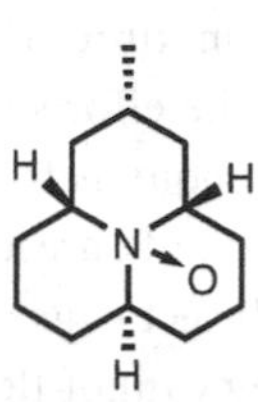
coccinellene

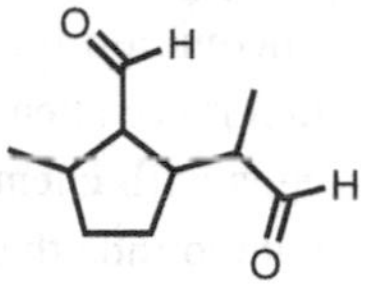
iridodial

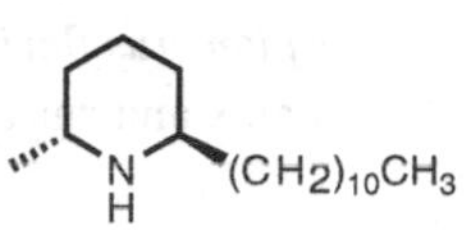
solenopsin A

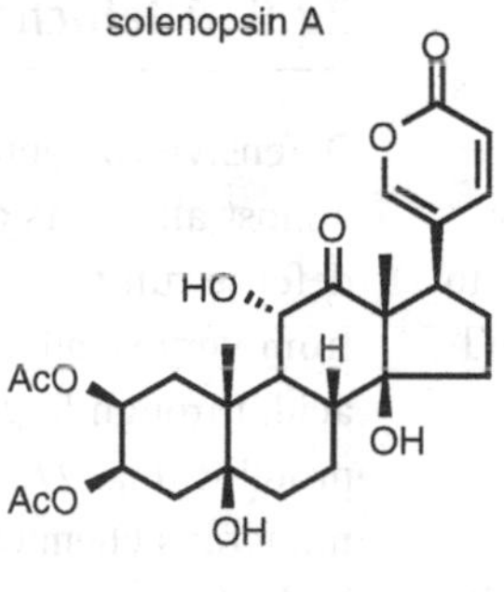
lucibufagin C

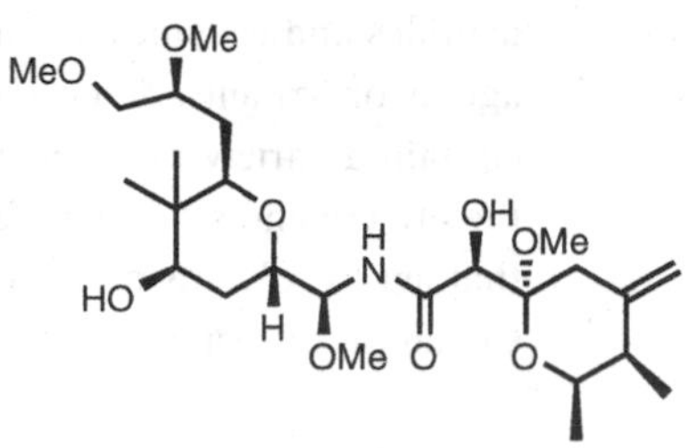
pederin

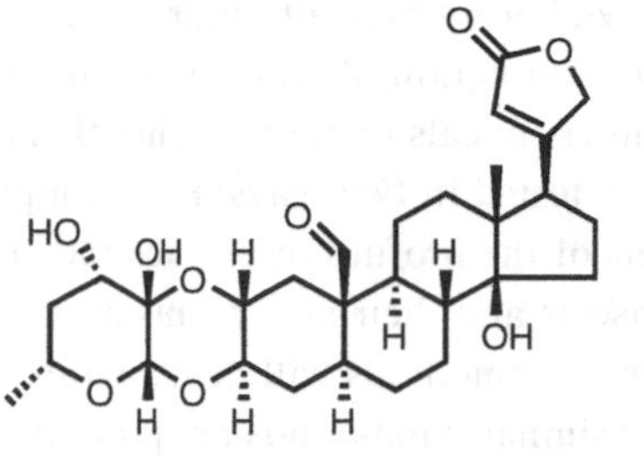
calotropin

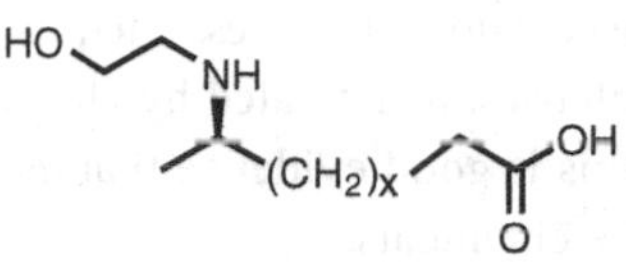
2-hydroxyethylamino carboxylic acid building blocks for polyazamacrolides x = 5, 6, or 7

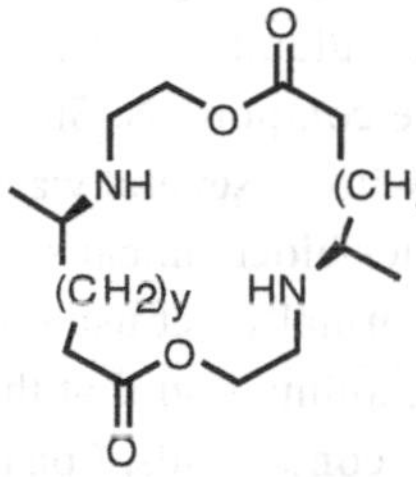
a polyazamacrolide assembled from two 2-hydroxyamino alkanoic acid subunits

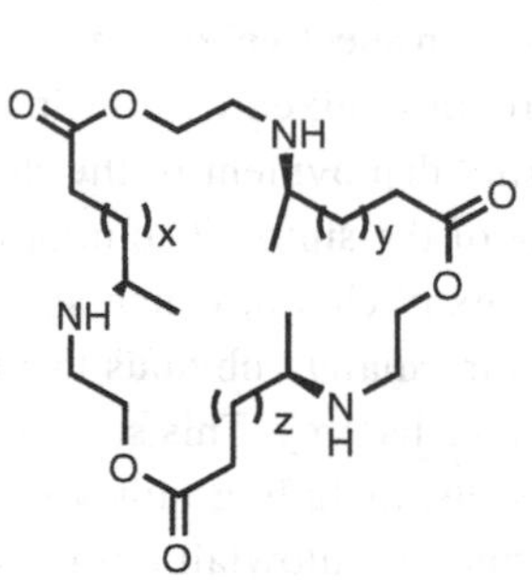
a polyazamacrolide assembled from three 2-hydroxyamino alkanoic acid subunits

Figure 27.22 Examples of defensive compounds produced by various types of insects, from very simple molecules like hydrogen cyanide and formic acid to very complex structures such as the polyazamacrolides produced by some beetle chemicals by altering rates of uptake or excretion as required. Determination of whether an insect is sequestering chemicals or synthesizing them *de novo* can be easily tested in two ways. First, simple comparisons of the profiles of extractable chemicals from the insects and their host plants by various chromatographic methods will demonstrate whether the same or similar compounds are present in both. Alternatively, insects can be reared on a different host plant or artificial diet that does not contain the chemicals suspected of being sequestered. If the resulting insects do not possess their normal chemical defenses, as revealed by chemical tests or bioassays, this is good evidence that the host is the source of the chemicals.

Having made or sequestered toxic or corrosive chemicals, insects then have the problem of how to prevent intoxication or incapacitation by their own defenses. For corrosive chemicals such as formic acid, the problem is solved by having storage reservoirs, ducts and glands lined with cuticle that is impervious to these compounds. Storage of toxic chemicals is dealt with in several ways. For example, the enzymes or other biochemical systems on which the toxins would normally act usually have mutated in the sequestering animals so that they are insensitive to these compounds. Compounds such as hydroquinones, cyanogenic glycosides and glucosinolates can be stored in different glands or compartments than the enzymes that convert them to their active forms (quinones, hydrogen cyanide and isothiocyanates, respectively). Consequently, the precursors are only mixed with their activating enzymes during deployment of the chemical defenses. One of the simplest solutions is to deposit toxins on the exoskeleton, which also makes the compounds immediately obvious to a predator as it attacks and bites its prey. This strategy is used by a variety of insects, including milkweed bugs (*Oncopeltus* spp.), swallowtail butterflies and zygaenid and some pyralid moths. For chemical defenses that consist of glues or sticky substances that entangle attackers, the glues are stored in liquid form in specialized glands and squirted or wiped onto attackers, where they rapidly become more viscous or solidify. This may occur from reaction of some of the components with oxygen in the air, or by evaporation of the more volatile components that solubilize the larger and less volatile components in the mixture. The efficacy of the defenses is increased by the efforts of the attacker to remove the compounds by grooming, which has the opposite effect of spreading them further.

Insect eggs are theoretically defenseless because they cannot flee from danger or mount a physical defense, nor can they sequester defensive compounds. To overcome this deficiency, it is not uncommon for female insects to incorporate defensive chemicals into eggs, or coat freshly laid eggs with chemicals. These may be the same compounds that the females use for their own defense (e.g., beetles in several families, arctiid moths and fireflies), or in the case of the reduviid bug *Apiomerus flaviventris*, the female collects resin from plants and smears it onto the newly laid eggs.

27.9 Allelochemicals used in defense

Defensive compounds of some type are produced by almost all orders of insects. The compounds used in defense run the gamut in terms of molecular size, from very small and volatile compounds like formic acid, through large non-volatile compounds such as proteins (Fig. 27.22). They include compounds from numerous chemical classes, including aliphatic hydrocarbons, terpenoids, alkaloids, phenolics, quinones and aromatics, and biochemicals such as peptides and enzymes. In addition to the actual toxic agents or irritants, defensive secretions usually contain a variety of other substances that increase the effectiveness of the defenses, such as chemicals that act as solvents or carriers, or enhance the spreading of secretions over surfaces. The mechanisms of action are diverse, varying from behavioral effects such as repellence or deterrence caused by irritating or bad-tasting compounds, through to a variety of highly toxic effects caused by poisons or enzymes. Many insects with effective chemical defenses are also brightly colored to make themselves obvious to potential predators (known as aposematicism), advertising their unpalatability and helping predators to quickly learn to avoid them. Alternatively, or sometimes in addition, aposematic insects may advertise their unpalatability or toxicity by strong odors. Given that many of these compounds are irritating or toxic in addition to smelling bad, it is likely that these compounds have a dual role as both warning signals and as actual deterrents, comprising a multileveled defense. For example, some coccinellid beetles and true bugs release foul-smelling methoxypyrazines when threatened. A wide variety of insects, including cockroaches, earwigs, grasshoppers and beetles release pungent benzoquinones. Lacewings exude 3-methylindole (the aptly named skatole, from its powerful fecal odor) and stick insects release indole. Still other insects, such as the caterpillars of swallowtail butterflies, produce volatile and noxious-smelling short-chain fatty acids such as isobutyric acid.

Below, a few specific examples are described to illustrate the diversity of chemical defenses used by insects, but these represent only a tiny fraction of the known defensive compounds and mechanisms. The interested reader is referred to a number of excellent books and reviews (see the recommended reading section at the end of this chapter).

Ants have evolved a wide array of chemical defenses and morphological and behavioral adaptations to render those defenses most effective. For example, the tiny but aptly named fire ants, *Solenopsis invicta* (Subfamily Myrmicine), clamp onto their victims with their mandibles so that they can more effectively drive home their sting. The venom, produced in a gland located in the abdomen,

contains piperidine alkaloids that cause instant pain in vertebrates. However, when competing for resources with other species, a droplet of venom is secreted and rapid vibration of the abdomen tip disperses the venom as a vapor. The venom constituents also form part of the queen's pheromone (see above). In contrast, ants in the Subfamily Formicinae have lost their stingers, and the main constituent of many of their venoms is formic acid, which their abdomen tips are adapted for spraying. They also attempt to bite because contact of the concentrated formic acid with open wounds will render the defense even more effective. Another group of formicine species in the genus *Camponotus*, the so-called exploding ants, have developed a truly extraordinary defensive adaptation whereby workers have hypertrophied mandibular glands that extend the full length of their bodies. When threatened, workers rapidly contract their abdomens, rupturing the gland and spraying its gluey contents, a mixture of alkanes, terpenoids and phenolics, in all directions. This type of self-destruction for the benefit of the colony is only adaptive because of the high degree of kinship of the colony members. It must also be mentioned that the chemical defenses of some tropical ant species have been co-opted by some of their specialist predators. Tropical frogs in the Family Dendrobatidae secrete venoms used for making poisoned arrows and darts. The frogs apparently do not synthesize these poisons *de novo*, but sequester them from their insect prey.

Other Hymenoptera, such as honey bees, also display altruistic, suicidal behavior in defense of the colony. Honey bee venom is produced in paired glands in the abdomen and stored in a poison sac. The venom consists of a cocktail of enzymes optimized through evolution to produce immediate pain, inflammation and tissue breakdown to enhance the spread and effectiveness of the venom. When a worker bee attacks, it drives home its barbed stinger and then pulls away, leaving the detached sting apparatus behind. The detached apparatus continues to pump, driving the stinger deeper into the flesh of its victim while simultaneously injecting venom. The detached sting also releases alarm pheromone, alerting other bees and inducing further attack. Loss of its abdominal tip is eventually lethal to the bee, which sacrifices itself for the greater good of the colony. Interestingly, the venom of the queen differs from that of workers, being better adapted to killing insects (i.e., competing queens) than deterring vertebrate attackers. Numerous other bees, wasps and hornets also have very effective stings, although in many species the act of stinging is not lethal, so the insect can sting a number of times.

Parasitic wasps make use of defensive compounds to increase their oviposition opportunities, as seen in the cynipid wasp *Alloxysta brevis*, a hyperparasite of aphids. The aphid hosts containing the primary parasitoid exploited by *A. brevis* are often protected by ants. If a foraging wasp is attacked by an ant, it releases a repellent from its mandibular glands to keep the ant at bay long enough for it to oviposit into a suitably parasitized aphid.

Among the more primitive Hymenoptera, sawfly larvae sequester compounds such as iridoid glycosides, glucosinolates and alkaloids from their host plants. *Neodiprion* species that feed on conifers sequester the resin oils from the needles in a pair of esophageal pouches. When threatened, the larvae regurgitate the oil to ward off attackers. A similar strategy is used by larvae of Australian spitfire sawflies in the genus *Perga*. They sequester oils from their eucalyptus host plants in a single large pouch in the esophagus and regurgitate it when attacked. At the end of larval development the oil is used to anoint the cocoon as it is spun, to provide protection during the vulnerable pupal stage.

Within the Order Hemiptera, the aptly named stink bugs (Pentatomidae), as well as related families such as the Coreidae, are specialists in chemical defense. Pentatomid nymphs have dorsal abdominal glands that release a mixture of volatile irritants including short-chain esters, alcohols, aldehydes and 4-oxo-2-alkenals. In the first instars these compounds also have a secondary role as aggregation pheromones, inducing the nymphs to huddle together for mutual defense and to decrease the probability of dessication. In adults the defensive glands are relocated to the metathorax during the final molt because dorsal abdominal glands would be hidden under the wings. The ducts open between the legs and the cuticle close to the duct openings is highly sculpted, increasing the surface area and thus accelerating the evaporation of the secretions.

The Coleoptera are masters of chemical defense, in part because their escape in response to danger is delayed by the time required to open the elytra protecting the wings and deploy the wings for flight. The defenses of a number of families, including the Carabidae, Silphidae, Staphylinidae, Gyrinidae, Tenebrionidae and Dytiscidae, often consist of secretions discharged from glands in the abdomen, via openings at the abdomen tip or via the rectum. The discharges are aimed by directing the abdomen tip toward the perceived threat. For example, carabid beetles spray concentrated formic acid toward attackers, and can do so several times in succession. One group of carabids, the bombardier beetles, have taken this type of defense one step further. The abdomen of the insect contains what is essentially a reaction chamber. When molested, the beetle releases a mixture of hydroquinones and hydrogen peroxide into the chamber, where catalase and peroxidase enzymes instantaneously convert the hydroquinones to quinones accompanied by the production of oxygen and sufficient heat to vaporize water in the secretion. The heated mixture of highly irritating quinones is explosively ejected in a rapidly pulsed spray, with a high degree of accuracy, toward an attacker.

The staphylinids, or rove beetles, also have abdominal defensive glands, but instead of squirting material, the secretion from the glands is mixed with rectal fluid, and the highly flexible abdomen is used to dab the mixture onto attackers. The secretion contains mixtures of compounds such as short-chain alcohols and esters, along with oxygenated terpenoids such as iridodial and dihydronepetalactone.

Another family of beetles, the blister beetles (Meloidae), produce the anhydride compound cantharidin, which is released by reflexive bleeding from the joints. This compound, the notorious aphrodisiac "Spanish fly," is a powerful irritant and highly toxic, and people have died from ingesting potions made from it. Despite the toxicity of this compound, a number of insects from several orders are attracted to cantharidin (so-called "cantharidiphiles") and some species actually use the compound as a pheromone. Males of *Neopyrochroa flabellate* (a fire-colored beetle, Pyrochroidae) do not produce cantharidin, but use it extensively, acquiring it through feeding on unknown sources. Males release some of the ingested cantharidin into a specialized cephalic gland where it serves a major role in female acceptance of courting males. During mating, the male transfers additional cantharidin to the female in the ejaculate. This nuptial gift not only protects her, but is also incorporated into her eggs, protecting them from predators. A similar strategy of acquiring one's chemical defenses from another species is practiced by the firefly *Photuris versicolor* (Lampyridae). The adult females are predatory, and they mimic the flashes of sympatric female fireflies in the genus *Photinus* to attract heterospecific males. The hapless males are devoured, and in so doing, the *Photuris* female gains a meal and acquires her prey's chemical defenses, which consist of distasteful steroids called lucibafagins.

Larvae, pupae and adults of leaf beetles (Chrysomelidae) can have powerful chemical defenses. Adults of many species are brightly colored and diurnally active, and clearly aposematic. Their defenses can include compounds sequestered from their host plants, ones that have been modified from host plant compounds, chemicals synthesized *de novo* or some combination thereof. The chemistry of these compounds is extraordinarily diverse, and includes glycosides of steroids, triterpenoids and oxazolinones, amino acid derivatives and

pyrrolizidine alkaloids, as well as simpler structures such as the monoterpenoids chrysomelidial and plagiolactone or simple aromatics such as salicylaldehyde (Fig. 27.22). The compounds are usually exuded from glands, or may be present systemically throughout the insect (Fig. 27.23).

No discussion of the chemical defenses of beetles would be complete without including the coccinellids, many of which are voracious predators of aphids, mealybugs and other soft-bodied insects. Larvae, pupae and adults are often protected with alkaloids such as coccinelline. In the adults, the compounds, which are present in the hemolymph, are released by reflexive bleeding from the leg joints, whereas larvae and pupae deploy chemical defenses in glandular hairs or spines. From a biosynthesis perspective one of the most remarkable chemical

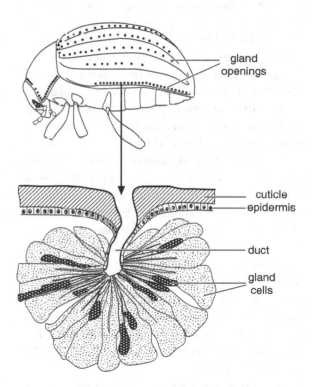

Figure 27.23 Non-eversible exocrine gland with no reservoir. Coleopteran pronotal and elytral glands, *Leptinotarsa* (after Deroe and Pasteels, 1977).

defenses identified in insects is that used by the phytophagous ladybird beetle *Epilachna borealis*. In a process known to synthetic organic chemists as combinatorial chemistry, the insect combines three hydroxyethylamino acid building blocks in different patterns and different numbers to make a family of several hundred different large ring structures called polyazamacrolides (Fig. 27.22). Variations of this biosynthetic strategy have been found in related species.

Chemical defenses are common in all life-stages of the Lepidoptera, including the eggs, which are often protected by toxins deposited on them by the adult female during oviposition. Caterpillars are soft-bodied and their typical defensive strategies consist of elaborate camouflage, the formation of protective structures such as rolled up leaves or the deployment of powerful chemical defenses. The latter can take the form of the stinging spines or hairs that are found in caterpillars in at least 13 lepidopteran families. The spines are hollow, with a gland at the bottom, and breakage of the spines on contact releases the venoms. The venoms contain histamine and other biogenic amines that cause pain and inflammation, as well as proteins that have various unpleasant effects on vertebrate predators. In an extreme example, larvae of the South American saturniid moth *Lonomia oblique* and some congeners produce anti-clotting agents, and hundreds of people have died as a result of hemorrhaging following contact with the caterpillars. Other caterpillars, such as those of swallowtail butterflies, have eversible glands called osmeteria that are coated with strong-smelling repellents. Still others, such as notodontid species in the genus *Schizura*, spray formic acid from glands associated with the head, whereas caterpillars of the saturniid *Attacus atlas* spray a cocktail of biogenic amines and aromatic compounds from projections on the thoracic and last abdominal segments.

Many Lepidoptera are specialist feeders on plants that contain compounds that are toxic to most other organisms. Thus, the female moths or butterflies use the characteristic toxins of their host plants as recognition and oviposition cues, while the developing larvae sequester the plant toxins as they feed. These compounds are subsequently retained into the pupal and adult stages. In the adult stage, the toxins can have multiple roles; for example, in arctiid moths such as *Utetheisa ornatrix* the male attracts and advertises his quality to females with a pheromone released from a pair of specialized eversible glands called coremata located on the terminal abdominal segments. The pheromone, hydroxydanaidal, is produced from pyrrolizidine alkaloids from the larval host plant, and the male transfers pyrrolizidine alkaloids to the female as a nuptial gift during mating. The female uses these compounds for her own protection, as well as incorporating them into the eggs to provide them with protection from predation. In addition to the compounds carried through from the larval stage, adults of danaine and ithomiine butterflies, and arctiine and ctenuchiine moths, can acquire further stores of pyrrolizidine alkaloids from ingestion of nectar or from plant surfaces, a process known as pharmacophagy because the chemicals acquired have no nutritional value and serve only for defense.

One of the most well-studied examples of the sequestration of host plant toxins by a specialist insect is that of the monarch butterfly, *Danaus plexippus*, probably the most well-known butterfly in North America. The larvae of this insect are specialist herbivores of milkweed plants (*Asclepias* spp.), which contain poisonous cardiac glycosides. These are sequestered by the larvae and afford defenses during the larval, pupal and adult stages. Gravid female adults then use the cardiac glycosides as oviposition cues, to start the cycle anew.

All life-stages of a number of zygaenid moths and nymphalid butterflies protect themselves through the production of hydrogen cyanide from the enzymatic breakdown of cyanogenic glycosides that are either sequestered from the host plant, or which may be produced directly by the insects from common amino acids. Cyanogenic glycosides or cyanide are also used in defense by some nymphalid butterflies, leaf beetles in the genus *Paropsis*, the carabid beetle *Megacephala virginica* and rhopalid bugs in the genera *Leptocoris* and *Jadera*.

It must be mentioned that chemical defenses are by no means restricted to terrestrial insects; aquatic insects also have chemical arsenals. For example, whirligig beetles (Gyrinidae) produce a deterrent, and probably toxic, norterpenoid called gyrinidal, while dytiscid beetles produce steroidal compounds related to hormones of potential predators. Giant water bugs (Belostomatidae) produce similar defensive compounds, in addition to the formidable defense represented by injection of toxins and enzymes with their proboscis. This venomous cocktail is normally used to rapidly subdue their typical prey (insects, tadpoles and small fish), but it also serves a useful purpose for defense.

In addition to direct defense of an individual, chemical defenses can also be used as more general repellents, deterrents or antibiotics. As an example of the former, green lacewings lay their eggs on stalks above a leaf surface. They coat the stalk with a mixture of carboxylic acids, esters and aldehydes, which repels ants and other predators that might otherwise eat the eggs. An analogous strategy is used by vespid wasps, which coat the pedicel of their nests with a carboxylic acid secretion from the van der Vecht's gland, to deter foraging ants.

Antibiotic substances are produced to prevent infection by bacteria or fungi, particularly in the warm and humid conditions found in the nests of social insects, which would otherwise be ideal breeding grounds for these organisms. The antibiotics used can be very simple, with broad-spectrum efficacy such as formic acid, phenolics or quinones, or more complex structures targeting a specific group of pathogens. One such case represents an extraordinary example of coevolution. Tropical ants in the tribe Attini harvest leaves and use the leaf fragments as a substrate to grow gardens in their nests, using specialized fungi in the Family Lepiotaceae. It is an obligate mutualism; the ants feed on the fungus, while the fungus gains protection and a means of dispersal. However, the fungus gardens are susceptible to infection by a parasitic fungus in the genus *Escovopsis*, and to keep this fungus in check the ants culture microbial symbionts on specialized body structures. A recent study identified the symbiont from three leaf-cutter ant species as a *Streptomyces* species that produces candicidin macrolides. These compounds inhibit the growth of the parasitic *Escovopsis*, but not the ant's fungus gardens. This amazing coevolution of ants, fungi, parasites and bacteria is found throughout the attine group of ants that comprises more than 200 described species.

27.10 Mimicry

There are some spectacular cases of chemical mimicry by both plants and insects. A number of orchid species practice deceptive pollination, whereby they attract males of their bee or wasp pollinators by emitting an odor that mimics the female insect's sex pheromone, while providing no nectar reward. In some species the attractive scent is a complex floral bouquet while in others the attraction may be due to a single compound. In a variation on this theme, flowers of other plant species emit odors that mimic a food source or an oviposition site to attract pollinators, again without providing any reward. For example, the Solomon's lily attracts several *Drosophila* species with a bouquet that mimics rotting fruit, whereas the dead-horse arum attracts female calliphorid flies by mimicking the odor of rotting meat.

Several different types of aggressive chemical mimicry are also known to be used by insects. For example, aggregations of newly emerged meloid beetle larvae (triungulins) emit odors that attract males of solitary bee species by mimicking the female bee's sex pheromone. The entire aggregation transfers to the male bee in a fraction of a second when he attempts to mate with it, and the triungulins subsequently transfer to the female bee when the male finds a receptive conspecific. The female then carries the beetle larvae back to her nest, where they feed on the nest provisions and the bee's eggs. This is a spectacular example of coevolution because both species live in a very harsh desert environment, and the triungulins are entirely dependent on being able to summon a male bee to begin their journey to the bee's nest. Triungulins that are not picked up and carried off will die within a few days.

A variation of luring males with sex pheromones has been reported in bolas spiders. The droplet at the extremity of the sticky silk bolas produced by a mature female contains components of the sex pheromones of female moths, and males are captured when they respond to what they perceive as a receptive conspecific female. Even more remarkable, the spider is able to change the attractant blend over the course of a few hours, allowing the spider to attract males of different moth species that have different activity periods during the scotophase. The male bolas spiders are much smaller and do not use a bolas to capture prey, but there is evidence that they, and the immatures of both sexes, use chemical mimicry to attract small fly species as prey.

In an entirely different kind of chemical mimicry, there have been changes in the facultative or obligatory mutualistic associations many insects have with ants, whereby the ants act as bodyguards in exchange for resources. Mutualistic relationships are frequently established between ants and scales or mealybugs, with the ants protecting the latter from natural enemies in exchange for nutritious honeydew excretions. The ants can be so effective at guarding their charges that it becomes necessary to control ants in vineyards and other crops to prevent serious outbreaks of scales and mealybugs.

Similar mutualisms are evident with aphids, where the presence of guarding ants reduces the incidence of parasitism or predation, with the ants being rewarded with honeydew. The strength of these mutualistic relationships spans a continuum, with some aphid species being strongly myrmecophilous (ants always present) through to others that are only weakly or not myrmecophilous. Myrmecophilous species have a greater tendency to remain on the plant when exposed to aphid alarm pheromone compared to weakly mymercophilous ones. Similarly, species considered to be myrmecophilous respond differently to alarm pheromone when ants are present or absent. With ants present, they exhibit a high frequency of abdominal waggling, a behavior associated with the release of alarm pheromone, whereas in the absence of ant guards the aphids are more likely to retreat or drop off the plant.

Many species of lycaenid butterflies also have mutualistic relationships with ants. For example, *Polyommatus icarus* larvae are tended by *Lasius flavus* ants, and the larvae are able to manipulate the number of attending ants by altering the production of droplets (the food reward collected by attending ants) from a dorsal nectar organ located on the seventh abdominal segment, and by the emission of chemical signals from tentacles located on the eighth abdominal segment. As seen in Fig. 27.24, the rate of tentacle eversion decreases as the number of attending ants increases, while the production of droplets initially increases and then declines. Furthermore, following a simulated attack, the rate of both tentacle eversion and droplet production are higher when there are a low rather than a high number of attending ants (Fig. 27.25). Similarly, the number of ants recruited following a simulated attack increased when the larvae could evert their tentacles, whereas no increase was

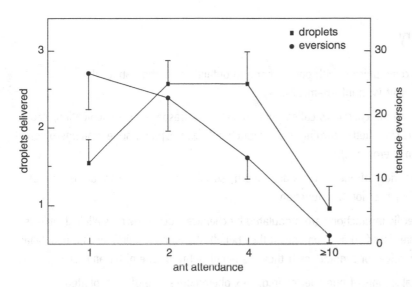

Figure 27.24 Changes in the release of droplets and evagination of tentacles by *Polyommatus icarus* larvae during a 15-minute assay as a function of the number of attending *Lasius flavus* ants (Axen *et al.*, 1996).

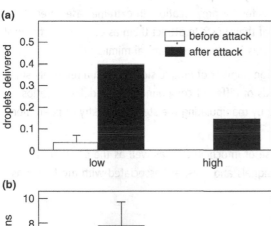

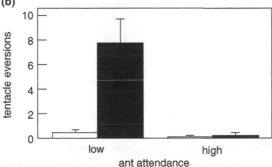

Figure 27.25 The response of *Polyommatus icarus* larvae to a simulated attack as a function of the number of attending *Lasius flavus* ants (Axen *et al.*, 1996).

observed when tentacle eversion was inhibited. Based on the behaviors exhibited by ants when the caterpillars extrude their tentacles, it has been postulated that the odor released is similar to the ant alarm pheromone. In this case, the chemical signal would fit the definition of a synomone, because both emitter and receiver benefit.

In some cases mutualistic interactions have become exploitative. The lycaenid larvae are actually raised by ant workers (cuckoo species) or have become true predators, feeding on the ant brood. In both scenarios, the lycaenid larvae use chemical mimicry to gain access to the ant colony by producing chemical signals that induce ant foragers to carry them into the nest, and once inside the caterpillars use chemical camouflage to avoid being attacked by their ant hosts. There is also evidence that the larvae are capable of synthesizing the colony odor chemicals of the host ant to some extent, in addition to acquiring them passively from attending workers.

Insects from other orders have also evolved into specialist parasites that infiltrate the nests of social insects, where they feed on their hosts. As with the lycaenids, these parasites, or inquilines as they are known, escape detection by mimicking the hydrocarbon recognition signals of their hosts. At least some of these inquilines, such as the staphylinid beetle *Trichopsenius frosti*, are capable of biosynthesizing their hosts' cuticular hydrocarbon profiles *de novo* from acetate, but the mechanism by which the beetle detects and mimics the blend of its hosts is not known.

Summary

- Insects communicate with each other and obtain information about their environment with infochemicals.

- Insects use infochemicals called pheromones for intraspecific communication, and these are classified according to their function (e.g., sex pheromone, trail pheromone or alarm pheromone).

- A wide variety of pheromones are used by social insects to mediate all aspects of colony organization and function.

- Interspecific interactions are modulated by chemical cues known as allelochemicals. These are classified as allomones if they benefit the emitter, kairomones if beneficial to the receiver or synomones if they are beneficial to both emitter and receiver.

- Chemical signals of one species (e.g., sex pheromones) are often exploited as kairomones by predators or parasites for prey location. In extreme cases predators actually mimic the pheromones of their hosts to attract them as prey, or as chemical camouflage to avoid detection. This is known as chemical mimicry.

- Insects are able to produce a large number of unique signals from a relatively small pool of chemicals by using blends of different compounds, using different ratios of the same set of compounds and by manipulating the stereochemistry of pheromone components.

- The sites of synthesis and release of infochemicals, as well as the position of receptors used to detect these signals and cues, are associated with the functions served by the infochemicals.

Recommended reading

PHEROMONE CHEMISTRY

Francke, W. and Schulz, S. (2010). Pheromones of terrestrial invertebrates. In *Comprehensive Natural Products II: Chemistry and Biology*, vol. 4, ed. L. Mander and H. W. Lui, pp. 154–223. Oxford: Elsevier.

EFFECTS OF PHEROMONE STEREOCHEMISTRY

Mori, K. (2007). Significance of chirality in pheromone science. *Bioorganic & Medicinal Chemistry* 15, 7505–7523.

CONTACT PHEROMONES

Blomquist, G. J. and Bagneres, A.-G. (eds). (2010). *Insect Hydrocarbons: Biology, Biochemistry, and Chemical Ecology.* Cambridge: Cambridge University Press.

CHEMICAL DEFENSES

Eisner, T., Eisner, M. and Siegler, M. (2005). *Secret Weapons: Defenses of Insects, Spiders, Scorpions, and other Many-legged Creatures.* Cambridge, MA: Harvard University Press.

PHEROMONE BIOSYNTHESIS

Jurenka, R. A. (2004). Insect pheromone biosynthesis. *Topics in Current Chemistry* **239**, 97–132.

Tillman, J. A., Seybold, S. J., Jurenka, R. A. and Blomquist, G. J. (1999). Insect pheromones: an overview of biosynthesis and endocrine regulation. *Insect Biochemistry and Molecular Biology* **29**, 481–514.

CHEMICAL MIMICRY

Vereecken, N. J. and McNeil, J. N. (2010). Cheaters and liars: chemical mimicry at its finest. *Canadian Journal of Zoology* **88**, 725–752.

PERCEPTION OF INFOCHEMICALS

Hansson, B. S. and Stensmyr, M. C. (2011). Evolution of insect olfaction . *Neuron* **72**, 698–711.

Leal, W. S. (2005). Pheromone reception. *Topics in Current Chemistry* **240**, 1–36.

References in figure captions

Axen, A. H., Leimar, O. and Hoffman, V. (1996). Signalling in a mutualistic interaction. *Animal Behavior* **52**, 321–333.

Billen, J. P. J. (1986). Morphology and ultrastructure of Dufour's and venom gland in the ant, *Myrmica rubra* (L.) (Hymenoptera: Formicidae). *International Journal of Insect Morphology and Embryology* **15**, 13–25.

Billen, J. P. J. (2006). Signal variety and communication in social insects. *Proceedings of the Netherlands Entomological Society* **17**, 9–25.

Birch, M. C. (1970). Structure and function of the pheromone-producing brush-organs in males of *Phlogophora meticulosa* (L.) (Lepidoptera: Noctuidae). *Transactions of the Royal Entomological Society of London* **122**, 277–292.

Delisle, J. (1992). Age related changes in the calling behaviour and the attractiveness of obliquebanded leafroller virgin females, *Choristoneura rosaceana*, under different constant and fluctuating temperature conditions. *Entomologia Experimentalis et Applicata* **63**, 55–62.

Deroe, C. and Pasteels, J. M. (1977). Defensive mechanisms against predation in the Colorado beetle (*Leptinotarsa decemlineata* Say). *Archives de Biologie, Bruxelles* **88**, 289–304.

Forsyth, D. J. (1970). The structure of the defence glands of the Cicindellidae, Amphizoidae, and Hygrobiidae (Insecta: Coleoptera). *Journal of Zoology* **160**, 51–69.

Hölldobler, B. and Traniello, J. F. A. (1980). The pygidial gland and chemical recruitment communication in *Pachycondyla* (=*Termitopone*) *laevigata*. *Journal of Chemical Ecology* **6**, 883–893.

Hölldobler, B. and Wilson, E. O. (1990). *The Ants*. Cambridge, MA: Harvard University Press.

Jurenka, R. and Rafaeli, A. (2011). Regulatory role of PBAN in sex pheromone biosynthesis of heliothine moths. *Frontiers in Endocrinology* **2**, 1–8.

Keil, T. A. (1999). The morphology and development of the peripheral olfactory organs. In *Insect Olfaction*, ed. B. S. Hansson, pp. 6–44. Berlin: Springer.

Leal, W. S. (2005). Pheromone reception. *Topics in Current Chemistry* **240**, 1–36.

Staddon, B. W. (1979). The scent glands of Heteroptera. *Advances in Insect Physiology* **14**, 351–418.

Turgeon, J. J. and McNeil, J. N. (1982). Calling behaviour of the armyworm, *Pseudaletia unipuncta*. *Entomologia Experimentalis et Applicata* **31**, 402–408.

Wigglesworth, V. B. (1972). *The Principles of Insect Physiology*. London: Methuen.

INDEX

Index entries in bold refer to figure captions.

Printed in the United States
By Bookmasters